GeoPlanet: Earth and Planetary Sciences

More information about this series at http://www.springer.com/series/8821

Indrajit Chowdhury · Shambhu P. Dasgupta

Earthquake Analysis and Design of Industrial Structures and Infra-structures

Springer

Indrajit Chowdhury
Civil Engineering Department
Future Institute of Technology
Kolkata, West Bengal, India

Shambhu P. Dasgupta
Civil Engineering Department
IIT Kharagpur
Kharagpur, West Bengal, India

The GeoPlanet: Earth and Planetary Sciences Book Series is in part a continuation of Monographic Volumes of Publications of the Institute of Geophysics, Polish Academy of Sciences, the journal published since 1962 (http://pub.igf.edu.pl/index.php).

ISSN 2190-5193 ISSN 2190-5207 (electronic)
GeoPlanet: Earth and Planetary Sciences
ISBN 978-3-030-08110-2 ISBN 978-3-319-90832-8 (eBook)
https://doi.org/10.1007/978-3-319-90832-8

Softcover re-print of the Hardcover 1st edition 2019

This Springer imprint is published by the registered company Springer Nature Switzerland AG
The registered company address is: Gewerbestrasse 11, 6330 Cham, Switzerland

Series Editors

Managing Editor

Anna Dziembowska
Institute of Geophysics, Polish Academy of Sciences

Advisory Board

Preface

If you are standing in a bookshop and have picked up this book for a casual rummaging, or surfing the network and staring at the book cover and wondering "*another book on earthquake*!!!".

No, we would not certainly hold that against you as having a preconceived bias.

For, it is indeed a fact that if you go to typical search engines like Amazon.com or Abebook.com and type "*Earthquake Engineering*," you would certainly be given access to at least thousands of books written on the topic and available in the market.

So why this book? When, significant amount of literature is already available in the market?

What value addition we hope to achieve by presenting this book?

Notwithstanding the authors' experience spanning more than three decades in international industry and academic research dealing with different types of structures and foundation, the bottom line needs to be seriously reflected upon.

Despite all these books available in the market, does not yet stop structures and foundations getting destroyed with every major occurrence of earthquake having magnitude greater than 6.5, with obvious outcome being human lives getting tragically extinguished untimely, or people becoming paupers overnight when dwellings he/she had built slogging for years together getting irrevocably destroyed.

People do not get surprised at all to read in newspapers or see in television catastrophic damages to infrastructures like bridges, retaining structures, landslides, or failures of dams when earthquake intensities of 7.0 or more occur even today.

All these instances only go on to show that despite the number of available books in the market, our knowledge on earthquake resistant design is yet not full proof. Otherwise, so many damages and loss of human lives would not have happened with monotonic regularity.

There is a saying among engineering consultants around the world

Earthquake does not kill, but ill conceived structures do.

Reconnaissance/forensic investigations on structures damaged after a number of earthquakes around the world have churned out different reasons for its failure.

From simple reasons like ill-planned buildings, poor construction, etc., to reasons as complex as *inertial* or *kinematical interaction* with soil and foundation, resulting in its damage.

However, some of the recent researches have also come up with some very surprising results too. It shows that many of the established theories that have been in practice for different structural systems for last 50 years or more are much too simplified and thus inadequate, to the extent *some of them significantly underrate the response and are still in practice till date.*

A classic case would be earth retaining structures under seismic load. The theory of dynamic earth pressure on gravity retaining wall by Mononobe and Okabe developed ninety three years ago is still in practice and considered sacrosanct. Though the characteristics of the retaining walls itself has changed significantly (due to advent of reinforced concrete) to thin cantilever and counterfort walls and does not make them behave like a rigid gravity wall any more.

There are no theories around to exactly predict the dynamic response of such flexible walls under seismic force, and we surely want to share this story with you in this book. It is our story yes, but certainly established through publications in peer-reviewed journals and having applied it to real-world situations that has survived a major shock at 6.8 in recent past.

In many cases, engineers still suffer from the misnomer that structures built below ground can be given reprieve of designing against seismic force. This concept is totally misconstrued and can seriously affect the performance of a substructure like metro tunnels box culvert. This problem has been taken up in detail, and solution to the same is put forward in this book.

How do you solve an earth dam problem under earthquake force with clay cores thus making it a heterogeneous section? Only solution would then be a FEM analysis which can be difficult to execute sometimes for various reasons like, lack of adequate software, paucity of trained staff in FEM analysis, to lack of detailed geotechnical data. We can certainly make you privy to technology where none of those constraints as cited above would stand in your way for a detailed analysis of such problems using standard utility software like Mathcad or even just a spreadsheet!!

How do you estimate dynamic pressures on rectangular bunker walls or circular silos under seismic force?

Rummage literature, there are no solutions to this problem, but this book has one (see Chap. 13).

In many such chapters in this book, we have illustrated such novel technologies vetted through publications in peer-reviewed journals or being implemented in real-life project with consensus agreement with client's consultant who agreed to our argument regarding inadequacy in recommendations as furnished in codes.

While presenting the book, three levels of readers have been kept in mind.

1. Graduate students taking an advanced coursework in earthquake engineering for whom Chaps. 1, 2, 3, and 6 would cover most of the course including some selective chapters like 7, 8, 9.
2. Hard core consulting engineers who carry out detailed engineering of different industrial structures and infrastructures and would like to give a more realistic solution to a client stretching beyond stipulations as furnished in country codes.
3. Doctoral researchers in earthquake who would like to carry out further research in linear and nonlinear analysis of systems under earthquake force and would like to dabble in analysis of coupled systems constituting different structures.

The book is primarily meant for being *self-taught* and thus is addressed more to the students rather than an instructor (who anyway we believe would find it equally stimulating).

While writing this book, it is assumed that the reader has a reasonably good background in advanced calculus, differential equations, and matrix algebra. Some background in fourth-order partial differential (plate) equations would certainly help (especially while studying Chaps. 4, 13 and 14).

We have not made any demarcation in the book in terms of structural or geotechnical engineering, for we are of the firm opinion that an earthquake engineer should have a complete reading of the problem and it is high time when treating *the superstructure and foundation separately*—this deplorable design culture is abolished for good.

The book remains a fruit of years of labor and grueling research (outside office work) for thirty years or more. The work is certainly a testimony to the excitement, challenge, joy as well as pain we endured together in this stupendous journey, trying to unravel the mystery of this fearsome fury of Mother Nature and has been an unforgettable experience indeed.

People who have graciously lent their support in this arduous journey are gratefully acknowledged.

Professor Chandrakant Desai (USA), Subhomoy Bhattacharya (UK), D. E. Beskos (Greece), Leroy Emkin (USA), Hemanta Hazarika (Japan), Farhang Ostadan (Bechtel USA), Ignacio Arango (Bechtel USA), D. K. Paul, R. N. Dubey, M. L. Sharma, Dipankar Choudhury, Nirmalya Bandopadhay (India): Thank you all for your constructive criticism on some of our research work at some point of its development.

Like to thank Mrs. Tapati Dasgupta for her endurance and cooperation in our research days and while we are near our goal of completion.

Our heartfelt thanks, to two people from the Institute of Geophysics of Polish Academy of Sciences, who went out of their way to see the book take shape. Prof. Paweł Rowiński, the Editor-in-Chief for his unstinting support from start to finish of this work, and Ms. Anna Dziembowska, the Managing Editor, for her amazing eagle eyed editing, that significantly improved the final presentation.

Last but not the least, the next-generation Satarupa Dasgupta and Rohan Chowdhury deserve special mention because of their stoic admiration for the wondrous book world that their fathers have unfolded before them.

Kolkata, India — Indrajit Chowdhury
Kharagpur, India — Shambhu P. Dasgupta

Contents

1 Global Seismicity and History of Earthquake Occurrence in India ... 1
1.1 Introduction ... 1
1.1.1 Global Occurrence of Seismicity ... 3
1.2 Seismic Activity in the Indian Subcontinent ... 6
1.2.1 The Kutch Earthquake of 1819 ... 6
1.2.2 The Assam Earthquake 1897 ... 6
1.2.3 Kangra Earthquake 1905 ... 7
1.2.4 Bihar Nepal Earthquake 1934 ... 7
1.2.5 The Quetta Earthquake 1935 ... 7
1.2.6 Assam–Tibet Earthquake 1950 ... 8
1.2.7 The Koyna Earthquake 1967 ... 8
1.2.8 The Bihar Nepal Earthquake 1988 ... 8
1.2.9 Uttarkashi Earthquake 1991 ... 8
1.2.10 Killari (Latur) Earthquake 1993 ... 9
1.2.11 Bhuj Earthquake 2001 ... 9
1.2.12 Indo-Nepal Earthquake 2015 ... 9
1.3 Development of Earthquake-Resistant Technology Around the World and India ... 10
1.4 Evolution of Views on Earthquakes from Mythology to Modern Science ... 14
1.4.1 Molten Magma Theory ... 15
1.4.2 Elastic Rebound Theory ... 17
1.4.3 Theory of Plate Tectonics ... 18
1.5 Intensity and Magnitude of an Earthquake ... 20
1.5.1 Magnitude of Earthquake (M_L, M_s and M_w) ... 26
1.6 Relationship Between the Length of Fault Rupture and Earthquake Magnitude ... 29
1.7 Relationship Between Duration of an Earthquake and Its Magnitude ... 30

1.8 Expected Peak Ground Acceleration of a Site 34
1.9 Attenuation of Earthquake with Distance 35
1.10 Summary and Conclusion 37
References 39

2 Basic Mechanics of Earthquake-Resistant Design 41
2.1 Introduction 41
2.2 Response Spectrum Method 42
2.2.1 The Codal Interpretation of Response Spectrum Method 47
2.2.2 Response Spectrum Method as Per Eurocode 8 56
2.2.3 Response Spectrum Method as Per UBC 97 64
2.3 Dynamic Theory of Structures Related to Earthquake Engineering 74
2.3.1 Free Vibration of a Body with a Single Degree of Freedom 74
2.3.2 Free Vibration of a Body Having Single Degree of Freedom with Damping 78
2.3.3 Characteristics and Response of a Body Under Transient or Impulse Loading 82
2.3.4 Response of Single-Degree of Freedom System Under Earthquake-Type Loading 83
2.3.5 Interpretation of Duhamel Integral with Respect to Time History 85
2.3.6 Earthquake Response of Structures Having Multi-degrees of Freedom 91
2.3.7 Decoupling of Matrix that Makes Life Easy 100
2.3.8 Decoupling of Equations with Damping Incorporated 103
2.3.9 Response of Multi-degree Freedom System Under Earthquake-Type Force 104
2.3.10 Some Other Interpretations of the Mass Participation Factor κ 108
2.3.11 Modal Combination of Forces 108
2.4 Various Computation Techniques Adapted for Eigensolutions 121
2.4.1 Method of Visual Inspection 121
2.4.2 Newton–Raphson Method 123
2.4.3 Matrix Deflation Method 125
2.4.4 Stodola–Vianello Method or Sweeping Matrix Technique 134
2.4.5 Eigenvalue Computation Based on Vector Transformation 138

2.5 Modal Analysis of Structures Assuming Continuum Model . . . 153
2.5.1 Vibration of a Flexural Beam 153
2.5.2 Derivation of Stiffness and Mass Matrix by Eigenvector Basis 162
2.5.3 Response of Flexural Beam to Earthquake-Type Load 165
2.5.4 Stiffness and Mass Matrix for Transversely Vibrating Beam Elastically Restrained at Free End with Lumped Mass 166
2.5.5 Vibration of a Shear Beam 166
2.5.6 Derivation of Stiffness and Mass Matrix of a Shear Beam by Eigenvector Basis 169
2.5.7 Response of Shear Beam Under Earthquake-Type Load 171
2.6 Time History Analysis a Different Approach 171
2.6.1 Central Difference Technique (CDT) 174
2.6.2 Wilson-Theta Method 184
2.6.3 Newmark Beta Method 197
References 207

3 Soil Dynamics and Earthquake Engineering 209
3.1 Introduction 209
3.2 Soil Dynamics and Earthquake Engineering 210
3.3 Characteristics of Wave Propagating Through an Elastic Medium 211
3.3.1 Propagation of Waves in Three Dimensions in an Infinite Elastic Medium 212
3.3.2 Compressive or P Waves Propagating Through Soil Medium 214
3.3.3 Shear or S Waves Propagating Through Soil Medium 216
3.3.4 Propagation of Rayleigh Waves Through Soil Medium 217
3.3.5 Propagation of Love Waves Through Soil Medium 224
3.4 Propagation of Waves in Polar Coordinates 227
3.5 Mathematical Background of Actual Ground Response, and Design Response Spectra 229
3.5.1 Plotting the Responses in Log Scale 237
3.5.2 Computation of the Integral Having Random Acceleration Response $\ddot{u}_g(\tau)$ 238
3.6 Amplification of Response Due to Soil Overlying Bedrock . . . 242
3.6.1 The SHAKE Method 242
3.6.2 Idriss and Seed's (1968) Model 247

3.6.3 Chowdhury and Dasgupta's (2007, 2013) Model 249
3.6.4 Equivalent Linear and Nonlinear Constitutive Models 260
3.7 Liquefaction of Soil Under Earthquake 261
3.7.1 Liquefaction Analysis of Cohesionless Sandy Soil 263
3.7.2 Liquefaction Based on CPT Value 278
3.7.3 Settlement of Soil Due to Liquefaction 287
3.8 Do Clays Liquefy Under Earthquake? 291
3.9 Effect of Liquefaction on Bearing Capacity of Soil 291
3.9.1 Local Punching Failure of Foundation Due to Soil Liquefaction 292
3.9.2 General Shear Failure of Foundation Due to Soil Liquefaction 294
3.10 Ground Subsidence Due to an Earthquake 296
3.11 Concluding Remarks 298
References 298

4 Earthquake Analysis of Earth Retaining Structures 301
4.1 Introduction 301
4.2 Mononobe and Okabe's (M–O) Method (1926) for Analysis of a Gravity-Type Retaining Wall 301
4.2.1 Some Comments on M–O Method and Its Validity 306
4.3 Seed and Whitman's Method 308
4.4 Steedman and Zeng's (1990) Method 309
4.5 Dynamic Analysis of Cantilever Retaining Wall 311
4.5.1 Dynamic Response of Dry Cohesionless Backfill ($C = 0$) 312
4.5.2 Effect of Vertical Acceleration S_v 315
4.5.3 Soil Inclined at an Angle i with the Vertical 316
4.5.4 Dynamic Response of Wall with c-ϕ Backfill 317
4.5.5 Dynamic Response of Wall with Pure Intact Clay ($\phi = 0$) as Backfill 321
4.5.6 Dynamic Response of Wall with c-ϕ Backfill and Overburden Surcharge q 322
4.5.7 Dynamic Response of Wall with c-ϕ Backfill Partially Submerged Below Water 325
4.5.8 Special Case of c-ϕ Soil When It Loses Contact for Some Portion at Top 327
4.5.9 Damping of Soil and that of the Wall 328
4.6 Dynamic Analysis of Counterfort Retaining Walls 332
4.6.1 Dynamic Response of Wall Having Dry Cohesionless Backfill ($C = 0$) 333
4.6.2 Dynamic Bending Moment and Shear Force 338

4.6.3 Soil Inclined at an Angle i with the Horizontal 341
4.6.4 Dynamic Response of Wall with c-ϕ Backfill 342
4.6.5 Dynamic Response of Wall with c-ϕ Backfill and Surcharge q 345
4.6.6 Dynamic Response of Wall with c-ϕ Backfill Partially Submerged Below Water 348
4.6.7 Validation of the Plate Theory 350
4.7 Dynamic Response of Substructures Under Earthquake Force 356
4.8 Ostadan's Method 366
References 368

5 Dynamic Analysis of Earth Dams Under Earthquakes 369
5.1 Introduction 369
5.2 Mononobe's (1936) Method for Analysis of Earth Dam 369
5.3 Gazetas's (1982) Method for Analysis of Earth Dam 374
5.4 Makdisi and Seed's (1977) Method for Analysis of an Earth Dam 375
5.5 Chowdhury and Dasgputa's (2010) Model for Dynamic Analysis of a Homogeneous and Clay-Cored Dam 377
5.5.1 Effect of Inner Core on Dynamic Response of the Dam 384
5.5.2 Estimation of Dynamic Amplitude and Stress 388
5.5.3 Earthquake Stability and Estimation of Lateral Seismic Coefficient (α_H) 390
5.6 Stability Analysis of a Dam 393
5.6.1 Effect of Vertical Seismic Coefficient α_V on FOS 395
5.6.2 Stability of a Dam with Impermeable Core 396
5.6.3 Effect of Hydrodynamic Pressure on Stability 399
References 403

6 Seismic Design of Shallow and Deep Foundations 405
6.1 Introduction 405
6.2 Basis of Increment in Bearing Capacity Under Seismic Force 405
6.3 Terzaghi's (1943) Basis of Ultimate Bearing Capacity of a Shallow Foundation Under Static Loading 407
6.3.1 Determination of Passive Pressure Due to Surcharge Load Q 408
6.3.2 Determination of Passive Force P_{pc} for Cohesion c 411
6.3.3 Determination of Passive Pressure Due to Weight of Soil $P_{p\gamma}$ 412

6.4 The Mechanics of Equilibrium of a Surface Footing Under Earthquake Force 415
6.4.1 Model for Dynamic Bearing Capacity Factors Under Seismic Load 417
6.4.2 Some Comments on the Above Method 424
6.5 A Proposed Model to Compute Bearing Capacity of Foundation Under Seismic Force 425
6.5.1 Determination of Passive Force P_{pq} Under Surcharge Load 427
6.5.2 Determination of Passive Force P_{pc} for Cohesion C 429
6.5.3 Determination of Passive Force $P_{p\gamma}$ for Weight Density γ_S 432
6.5.4 Dynamic Ultimate Bearing Capacity of Soil 432
6.5.5 Dynamic Bearing Capacity Factors as Per Meyerhof's Expression 433
6.5.6 Design Bearing Capacity of Foundation Under Seismic Load 434
6.5.7 Computation of Free Field Time Period of Soil Having G Varying with Depth 435
6.5.8 Selection of Appropriate Value of Damping Ratio 436
6.6 What Do We Do When Structures Are Supported on Pile Foundations? 438
6.6.1 Estimation of Pile Capacity Under Vertical Load 440
6.6.2 Conventional Methods for Estimation of Lateral Load Capacity of Piles 449
6.6.3 The Sub-grade Modulus Approach 466
6.6.4 Matlock and Reese' (1960) Method for Lateral Load Analysis of Long Slender Piles 472
6.6.5 Numerical Methods for Lateral Load Analysis of Piles 476
6.7 Dynamic Response of Pile Foundations Under Earthquake Force 493
6.7.1 Prakash and Chandrashekharan's (1980) Model for Dynamic Analysis of Piles 494
6.7.2 Chowdhury and Dasgupta's (2012) Model for Seismic Response of Pile 498
References 526

7 Dynamic Soil–Structure Interaction in Earthquake Analysis 531
7.1 Introduction 531
7.2 A Short History on Development of Dynamic Soil–Structure Interaction (DSSI) 533
7.3 The Difference in Approach Between Soil and Structural Dynamics 536

7.4 Why Are We Reluctant to Apply DSSI as an Analytical Tool? . . . 537
7.4.1 How Do We Know Whether DSSI Effect Is Significant or not? . . . 538
7.5 Coupling of Soil and Structure . . . 541
7.5.1 The Correct Modeling of Foundation and Structure . . . 543
7.6 Unraveling the Popular Myth . . . 548
7.7 Extension of Veletsos's Equation to Multi-degree of Freedom . . . 555
7.7.1 Why Does Some Commercially Available Software Show Attenuation of Response Considering DSSI? . . . 556
7.7.2 Extending the Proposed Model for a Space Frame . . . 568
7.8 Alternate Model for DSSI Based on Lagrange's Equation . . . 569
7.9 DSSI Model for Foundations that Are Massive and Embedded in Ground . . . 585
7.9.1 Model 1—Shallow Foundation . . . 586
7.9.2 Model 2—Foundation with Large Depth Supporting Heavy/High Structure . . . 597
7.10 Kinematical Interaction with Foundation and Structure . . . 605
7.11 Different Mechanical Analogs of Foundation and Soil . . . 628
References . . . 637

8 Geotechnical Considerations for Seismic Analysis . . . 641
8.1 Introduction . . . 641
8.2 Geotechnical Parameters that Influence Earthquake-Resistant Design Including DSSI . . . 641
8.3 Field Tests . . . 644
8.3.1 Block Vibration Test . . . 645
8.3.2 Seismic Cross-Hole Test . . . 649
8.3.3 Cyclic Plate Load Test . . . 650
8.4 Techniques to Compute Dynamic Shear Modulus When No Dynamic Soil Test Has Been Carried Out . . . 652
8.4.1 Theoretical Corelation from Other Soil Parameters . . . 653
8.5 Procedure to Estimate Large Strain in Soil Under Earthquake Force . . . 663
8.5.1 Estimation of Soil Strain for Earthquake Analysis . . . 663
8.6 What Do We Do if the Soil Is Layered with Varying Soil Properties? . . . 673
8.7 Checklist of Parameters to Be Looked in the Soil Report . . . 682
8.8 Epilogue . . . 684
References . . . 684

9 Earthquake Resistant Design and Analysis of Buildings 687
9.1 Introduction 687
9.2 Seismic Analysis of Building as Per Code 688
9.2.1 Pseudo-static Analysis of Buildings Constructed as a Framed Structure 690
9.3 Pseudo-static Analysis of Buildings Constructed with Shear Walls 695
9.3.1 What Is Center of Mass (CM) and Stiffness (CS)? . . . 697
9.3.2 Behavior of Buildings When CM and CS Do not Coincide 700
9.4 Analysis of Building Frames Based on the Assumed Shape Function 712
References 716

10 Earthquake Analysis of Tall Chimneys and Stack-Like Structures 717
10.1 Introduction 717
10.2 A Brief History on the Development of Dynamic Response of Chimneys 718
10.3 Analysis as Proposed in IS 1893 (2005) Part IV 724
10.3.1 Some Comments on Recommendation Made in IS 1893 (Part IV) 2005 727
10.4 Dynamic Analysis of Tall Chimneys 728
10.4.1 Transformation to IS Code Format 732
10.4.2 Calculation of Amplitude 734
10.5 Numerical Analysis of Tapered Stacks 742
10.6 Computer Analysis of Tall Chimneys 746
10.7 Should We Consider Soil–Structure Interaction for Dynamic Analysis of Chimneys? 748
10.7.1 Soil–Structure Interaction of a Chimney on Pile Foundations 753
10.7.2 Dynamic Stiffness of Partially Embedded Piles 758
References 787

11 Earthquake Analysis of Concrete Gravity Dam 789
11.1 Introduction 789
11.2 Analysis as Per IS-1893 790
11.2.1 Hydrodynamic Pressure on a Dam Face from the Reservoir (Zangar 1953) 792
11.2.2 Some Comments on the IS Code Method 793
11.3 Finite Element Analysis of a Concrete Dam 793
11.4 Chopra and Fenves's Method for Dynamic Analysis of Gravity Dam 797

11.5 Chowdhury, Singh, and Dasgupta's (CSD) Model for Dynamic Analysis of Dams 803
11.5.1 Static Analysis 805
11.5.2 Dynamic Analysis 808
References 860

12 Earthquake Response of Bridge Piers and Their Foundations . . . 863
12.1 Introduction 863
12.2 Earthquake Analysis as Recommended in Codes 863
12.2.1 Some Comments on the Procedure as Proposed in Codes 869
12.3 A Proposed Dynamic Analysis Model for Bride Piers 870
12.3.1 Shear Deformation Behavior of the Pier 872
12.4 Effect of Foundation on Response of Pier and Superstructure 882
12.4.1 Piers Resting on a Shallow Foundation 882
12.4.2 Piers Resting on Pile Foundations 892
12.4.3 Bridge Pier Resting on Caisson or Well Foundation 892
12.4.4 Generalized Model for Pier and Well Foundation 902
12.4.5 Kinematical Interaction of Well and Pier 911
12.5 A Case Study 912
References 917

13 Dynamic Response of Bunkers and Circular Silos Under Earthquake Force 919
13.1 Introduction 919
13.2 A Proposed Model to Estimate Seismic Effect on Bunker Walls (Chowdhury 2009) 920
13.2.1 Calculation of Mass Coefficient 923
13.2.2 Derivation of Stiffness Coefficient 925
13.2.3 Derivation of Fundamental Time Period 926
13.2.4 Calculation of Dynamic Amplitude of the Bunker Wall 928
13.2.5 Calculation of Dynamic Moments and Shear 929
13.2.6 Design Coefficients for Dynamic Moments and Shear 933
13.2.7 Dynamic Response of Bunkers with Walls that Is Deep 941
13.2.8 Some Comments on the Above Method 945
13.3 Analysis of Bunker with Supporting Frame Interaction and Adapting Rankine's Earth Pressure (Chowdhury and Singh 2012) 948
13.3.1 Analysis of Bunker Wall 952

13.4 Dynamic Pressure on Circular Silos Under Seismic Force (Chowdhury and Tilak 2010) . . . 962
13.4.1 A Case Study . . . 966
13.4.2 Results and Discussion . . . 970
References . . . 972

14 Analysis of Water Tanks Under Earthquake Force . . . 973
14.1 Introduction . . . 973
14.2 Rigid Tanks Resting on Ground . . . 975
14.2.1 Hydrodynamic Pressure on Tank Wall and Base . . . 977
14.3 Flexible Circular Tanks Resting on Ground . . . 985
14.3.1 Computation of Base Shear . . . 989
14.3.2 Computation of Moment at the Base of Wall . . . 989
14.4 Flexible Rectangular Tanks Resting on Ground . . . 990
14.5 Some Comments on the Method as Proposed for Flexible Tanks . . . 991
14.6 Elevated Water Tanks . . . 997
14.6.1 Simplified Method as Proposed in the Code . . . 997
14.7 Effect of Foundation Stiffness on Tanks Resting on Ground . . . 1011
14.7.1 Practices as in Trend . . . 1012
14.7.2 Dynamic Analysis of a Water Tank Constructed on Ground, Considering DSSI . . . 1014
14.7.3 Dynamic Analysis of Flexible Tank Behaving as a One-Way Slab . . . 1020
14.7.4 Dynamic Analysis of Flexible Tank Behaving as a Two-Way Slab . . . 1024
14.7.5 Estimation of Response Due to Sloshing Force and Vertical Amplitude . . . 1030
14.7.6 Tanks with Other Boundary Conditions . . . 1032
References . . . 1043

Index . . . 1045

Chapter 1
Global Seismicity and History of Earthquake Occurrence in India

1.1 Introduction

From the early dawn of civilization, human beings have been battling against nature to establish their supremacy. In this process, they have been fighting against a number of natural forces like hurricanes, tornados, flood, volcanic eruptions, and finally *the earthquake*.

A study carried out by UNESCO in 1999 shows that among all the natural phenomena, it is earthquake which has been the biggest killer of human lives and has incurred destruction of properties amounting to billions of dollars.[1]

Why do earthquakes occur in nature?

The topic has certainly fascinated human beings from Neanderthal age to modern times when geophysicists armed with state-of-the-art signal processing instruments have spent hours at observatories, trying to unravel its characteristics.

It is no wonder thus that the origin of earthquakes has treaded an interesting path in its evolution from being perceived an act of God having mythological origin to tectonic theory as adapted at present.

Tectonic processes are continuous. The most glaring examples of modern tectonics are seismic events or earthquakes. These occur at an annual rate of up to 10 thousands. Most earth tremors are very hard to detect and can only be recorded by sensitive instruments called *seismographs*.

Yet, as many as hundred earthquakes per annum are destructive and one catastrophic. This suggests that destructive earthquakes are violent movements of earth crust after a period of accumulation of stress.

It may be assumed that earthquakes are caused by major and sudden discontinuities of the crust, ruptures, and faults as well as displacements of the crust. They are associated with the physicochemical processes that are at work in earth's bowels and with changes in thermodynamic conditions in inner reaches of the earth.

[1]Refer Chap. 9 for further detail.

I. Chowdhury and S. P. Dasgupta, *Earthquake Analysis and Design of Industrial Structures and Infra-structures*, GeoPlanet: Earth and Planetary Sciences,
https://doi.org/10.1007/978-3-319-90832-8_1

The seismic impulse and vibration movements caused by it often last only a few seconds. However, during a major or strong motion event even this short-lived shock generally brings about calamitous consequences. Normal earthquakes usually last for 10–15 s or more.

An examination of accelerograms (recorded by instruments used to measure the intensity of acceleration induced in the soil due to propagating waves) shows that seismic vibrations attain greatest amplitude only after a weaker vibration has occurred. Thus, all earthquakes have an initial stage. Waves of this stage are called *precursor waves*. Seismic waves of an earthquake originate at a place in earth crust some distance from the surface, called the focus or hypocenter. The foci of earthquakes have generally been found at depths not exceeding 20–50 km. Thus, the tectonic event of earthquake takes place in deep inner reaches of the earth. A typical acceleration output for a strong motion earthquake is shown in Fig. 1.1.

Destructive earthquakes are often triggered by a single shock. Sometimes, seismic impulses may come as successive shocks occurring during several months (the Crimean earthquake, 1927). A typical example can be found in 1966 Tashkent earthquake that lasted during a year almost uninterrupted. In such a case, we can refer to a seismic period of a particular region. The down-warping of land may attain sometimes as much as several hundred meters. Sinking of the seafloor has repeatedly ruptured submarine telegraph cables. Down-warping of ocean floor caused by earthquake has sometimes caused entire islands to be engulfed. The vertical displacement of seafloor during the 1923 earthquake in Japan was a few hundred meters. Lateral displacements of individual ground surface areas are common in earthquakes. This phenomenon was caused by the major Californian earthquake of 1900 where the fault and shear zone broke for 500 km. The ledges that formed were not more than 6 m high.

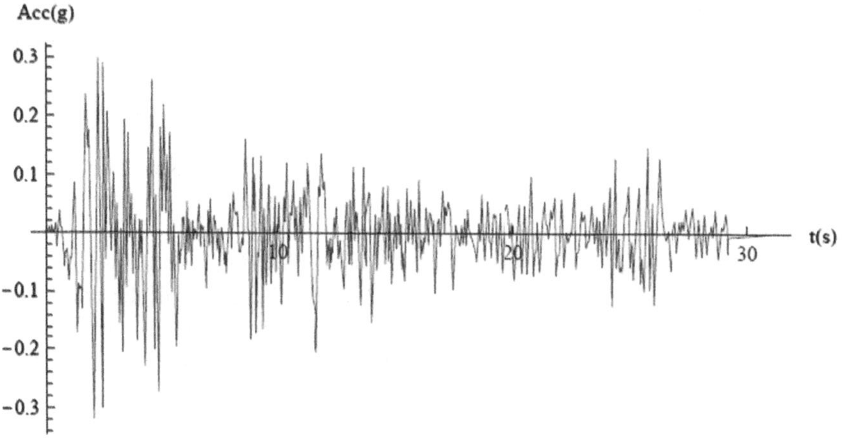

Fig. 1.1 Typical acceleration output of a strong motion earthquake

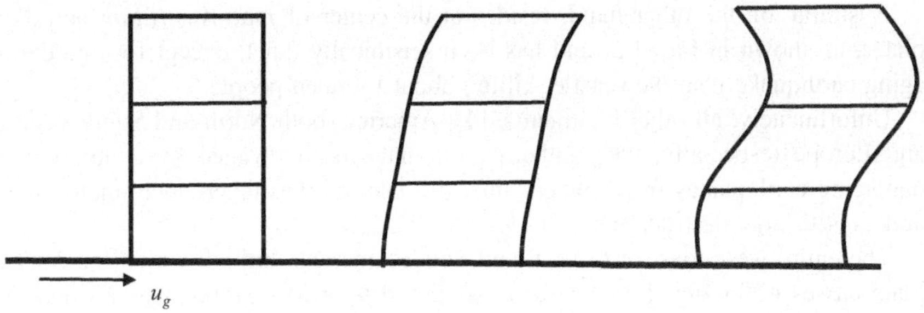

Fig. 1.2 Response of a structure subjected to ground acceleration during an earthquake

Figure 1.2 shows how earthquake affects typical structures built on ground and how they deform under ground acceleration. The job of a civil engineer is to identify the force induced in the body under various deformed shapes/modes (as shown in Fig. 1.2) and design them accordingly to ensure that under extreme case, *they might undergo reparable damages but shall not collapse*.

In this process, an engineer has to consider all the static and dynamic loads that the structure may experience during it operating life and take the worst combination to ensure its safety.

Similarly, a geotechnical engineer who is responsible for design of the sub-structure is concerned with two types of loading for his analysis and design of foundations namely for

- foundations subjected to static loads (dead load, live load, equipment load, etc);
- foundations subjected to dynamic loads (Wind load, Earthquake load, Blast load, Load due to vibration of equipment, sea waves, etc.).

Static loads are constant in magnitude and direction, e.g., self-weight of the structure. Live loads, such as weight of train on a bridge and assembly of people in a building, are also classified as static loads. The main characteristic of a dynamic load is that the loading is time dependent and gives raise an appreciable magnitude of inertial force that a designer has to take care off in his process of engineering the system.

1.1.1 Global Occurrence of Seismicity

All the five continents around the world have, to a greater or smaller extent, experienced earthquakes of strong and moderate intensities at some point of time. Occurrence of earthquakes in Africa and Australia has been minimal, with one earthquake of magnitude 5.6 that occurred in 1981 in Aswan area of Egypt. Cause of this earthquake was, however, attributed to pounding of water from Lake Nasser behind the Aswan Dam.

Australia, on the other hand, resides at the center of *Indo-Australian* and *Fiji* plates, as shown in Fig. 1.3, and has been seismically quiet, except for one damaging earthquake near Newcastle, killing about fourteen people.

Unfortunately, all other continents, like America (both North and South), Asia, and Europe (especially, the southern part), have been ravaged by a number of damaging earthquakes from ancient time till date that has taken away many lives and caused large destructions.

The entire west coast of Central and South America, through which the *Nazca* plate passes (refer Fig. 1.3), has been subjected to strong earthquakes off and on (Chile, Peru, and Mexico being the worst sufferers), causing great economic damage.

On the contrary, east coast of South America has hardly seen any earthquake and can be attributed as seismically nascent zone.

Similarly, along west coast of USA, through which passes the notorious *San Andreas* Fault, has seen a number of very strong motion earthquakes. The city of San Francisco has been shaken by a number of damaging earthquakes time and again in the past.

Other than this, Tehachapi (1952), Alaska (1964), Loma Prieta (1989), North Ridge (1994) earthquakes have incurred significant damage to property and habitats in USA.

In Europe, countries lying to its south like Portugal, Spain, Italy, Bosnia and Herzegovina (Sarajevo), Greece, Turkey, where the *Eurasian* and *Arabian* Plates clash, have seen a number of strong earthquakes in the past.

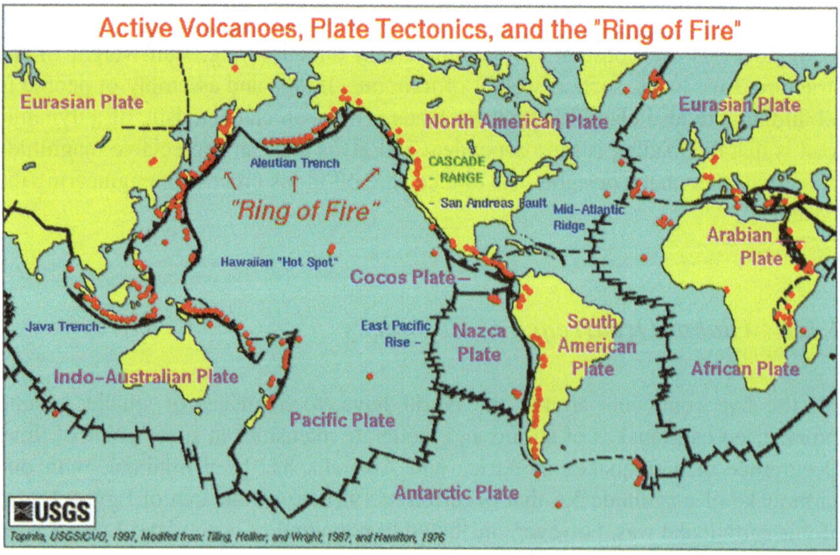

Fig. 1.3 Global mapping of tectonic plates

In 1755, a tsunami due to earthquake in seabed resulted in about 60,000 deaths in Lisbon, Portugal. The shock of the shaking was felt as far as Holland and Germany.

In 1829, an earthquake occurred in Alicante, Spain, killing about 850 people and injuring many. It also caused complete and partial destruction of many houses.

In 1908, the devastating Messina earthquake in Italy killed 120,000 people and caused widespread damage to properties and infrastructure.

The Anatolian fault that passes through Turkey and Iran has resulted in a number of earthquakes in these countries. The infamous Izmit Earthquake (1999) killed as many as 160,000 people due to an earthquake of intensity M 7.4.

Northern Europe, on the contrary, is considered as seismically more stable (countries like Germany, France, Romania). However, damaging earthquakes of moderate intensities have also occurred there in the past.

In Asia, the clash between Eurasian plate and Arabian plate has resulted in many strong motion ground shaking in the Indian subcontinent. Similarly, the clash between the Pacific and the Philippine plate has resulted in many strong earthquakes in Japan, Indonesia, etc.

Some major earthquakes around the world that have caused large fatalities in the last 5 years are listed in Table 1.1.

Table 1.1 Some major earthquakes around the world (2012–2016)

Date	Country	Fatalities	Magnitude
February 6, 2012	Philippine	113	6.7
June 11, 2012	Afghanistan	75	5.7
August 11, 2012	Azerbaijan	306	6.4
September 7, 2012	China	80	5.6
April 20, 2013	China	217	6.6
September 24, 2013	Pakistan	825	7.7
October 15, 2013	Philippine	230	7.1
August 3, 2014	China	729	6.2
April 25, 2015	Nepal	9018	7.8
May 12, 2015	Nepal	218	7.3
October 26, 2015	Afghanistan	398	7.5
February 5, 2016	Taiwan	117	6.4
April 16, 2016	Ecuador	682	7.8
August 24,2016	Italy	299	6.2
December 6, 2016	Sumatra	104	6.5

1.2 Seismic Activity in the Indian Subcontinent[2]

The Indian subcontinent has experienced a number of major ground shaking in the past. This is mainly attributed to the movement of the Indian Plate toward the Eurasian Plate which has generated seismic incidence as high as 8.0.

The entire Himalayan belt is thus highly susceptible to earthquakes, likewise other areas, including sites around Koyna Dam and some areas of Gujarat and Himachal Pradesh.

While many countries in Asia like China and Japan have maintained earthquake records of several centuries, in India it started pretty late and it was only during the British rule and creation of the Geological Survey of India that such data were generated and maintained (Oldham 1899).

1.2.1 The Kutch Earthquake of 1819

This remains one of the earliest recorded earthquakes of major magnitude. Considering that the area is remote and not very densely populated, the fatality was relatively low (about 1500–1600); however, the shock was sufficiently strong to be felt across the country as far as Kolkata. The shock created a fault scrap which was about 16 miles long and 10 feet in height. Due to its bund-like feature, this fault is popularly called *Allah Bund* (the bund created by Allah) in local lingo.

1.2.2 The Assam Earthquake 1897

This remains one of the strongest earthquakes to have devastated the subcontinent, having magnitude of the order 8.7, and is also subjected to extensive study in those early years (Oldham 1899), followed by Richter (1958) who rated in his monograph that it remained a valuable source of study in seismology.

The earthquake caused extensive damage to the area while generating two fault scraps. The earthquake was so strong that it brought significant topographical change in the landscape, including modifications in height of hills (making helio-graphic signal transmittal much easier) post-earthquake.

Observing movement of large boulders within the soil strata, Oldham came to the conclusion that peak ground acceleration (PGA) was greater than 1.0*g*.

The earthquake also caused extensive liquefaction, especially in and around banks of river Brahmaputra which consisted of clean river sand located below groundwater level.

---

[2]After Jain and Murty (1995).

Shillong area consisted of three types of buildings. One made of stone, others made of wooden frames covered with grass and plaster, while the third consisted of building with wooden frames covered with planks and resting on ground unattached.

All stone buildings collapsed under the earthquake shock, and wooden buildings that had stone chimneys mounted on top underwent severe damage; however, light wooden building resting on ground performed well under earthquake shock.

1.2.3 Kangra Earthquake 1905

The earthquake was measured at 7.8 and occurred in Kangra Valley of now Himachal Pradesh in India. This was a classic case where the Eurasian fault collided with the Indian plate (near Pamir) to generate the ground shock.

The earthquake resulted in massive destruction when about 100,000 buildings where razed to ground, killing 20,000 people and 53,000 domesticated animals.

1.2.4 Bihar Nepal Earthquake 1934

The earthquake was estimated to be of magnitude 7.8 resulting in death toll of 7253 in India and 3400 in Nepal. However, scientists estimated that the fatality could have been greater, but luckily, as the earthquake took place in the afternoon, many people were out of their home, thus being in the open space and got saved. Moreover, the strongest shaking was about 5 min with peak response occurring after about two and a half minutes, thus giving people sufficient time to run outdoor.

A large area underwent extensive liquefaction, causing heavy subsidence of roads and railway embankments. Embankments as high as 6 ft were completely flattened to ground while lakes, ponds borrow areas were further depressed. Huge fissures occurred in ground, as much as 900 ft long, 15 ft deep and 30 ft wide.

In Bihar, the town Munger was most severely damaged. Munger is a quartzite ridge emerging through alluvial soil; the deep-focused earthquake, whose focus was 100 miles from Munger, traveled through the rock and emerged at the site with negligible energy dissipation, causing extensive damage to the town.

1.2.5 The Quetta Earthquake 1935

The earthquake was of the order 7.6 and occurred in Quetta, the capital of Baluchistan (presently under Pakistan), in 1935; the quake resulted in death toll of 30,000 lives. A number of buildings made out of stone rubbles completely collapsed. Landslides from mountainsides buried and killed many people.

1.2.6 Assam–Tibet Earthquake 1950

The earthquake was estimated to be of intensity 8.7 and did more damage to property and life than the previous earthquake that took place in 1897. The earthquake caused significant changes in the topographic layout in the area, cutting off many roads and bridges. The major shock was followed by a number of aftershocks that had magnitude more than 6.0.

1.2.7 The Koyna Earthquake 1967

This earthquake has been discussed in more detail in Chap. 11, while discussing the seismic response of concrete gravity dam. It was a classic case of reservoir-induced earthquake where the water stored behind the dam-induced harmonic force to the dam causing the ground to shake killing about 200 people and injuring 1500 (Berg et al. 1969). The peak ground acceleration measured was about 0.36*g*. The dam which was designed for a nominal force of 0.05*g* performed reasonably well with only nominal damage.

1.2.8 The Bihar Nepal Earthquake 1988

The earthquake was estimated to be of magnitude 6.6. Despite its moderate intensity, it caused extensive damage, causing death to 282 people in India and 722 in Nepal and injuring more than 16,000 people. The fatality could have been much greater if it was not summer time when most people would sleep outdoor. The pattern of destruction was a replicate of the earthquake that took place in 1934, with extensive liquefaction and formation of fissures in ground, especially in Darbhanga and Madhubani district in Bihar. There was a significant destruction to embankments bridges and buildings in Bihar, Darjeeling district (West Bengal) and Sikkim.

1.2.9 Uttarkashi Earthquake 1991

The earthquake of magnitude 6.6 shook the district of Uttarkashi, Tehri, and Chamoli in the state of Uttar Pradesh. The shock killed 768 people and injuring 6000. However, the area being most thinly populated in the whole state the fatality was considered relatively low. The maximum peak ground acceleration recorded was 0.31*g* in Uttarkashi.

1.2.10 Killari (Latur) Earthquake 1993

The earthquake of magnitude 6.4 occurred in the early morning at Latur district of Maharashtra, having fatality rate of about 10,000 people. Most of the damages were concentrated in area of about 20 km × 20 km indicating that the shock was of very shallow focus. The area had no town and mostly villages that were built traditionally with stone–mud walls with wooden plank roofs topped with clay layers of depth 30–60 cm as insulation. All these buildings, being of poor earthquake resistance, were collapsed, thus killing a number of people sleeping inside. Previous to this earthquake, this area was perceived as non-seismic zone and design factors were considered for the lowest zone (Zone1 as per IS 1893–1984).

1.2.11 Bhuj Earthquake 2001

The earthquake was measured at about 7.4 in intensity and has been the most devastating earthquake in recent times creating extensive damage.

Major cities like Ahmadabad, Jamnagar, and the town of Bhuj which was only 20 km away from the epicenter experienced extensive damages. As per information collated from local newspaper, the statistics of damage is as listed in Table 1.2.

1.2.12 Indo-Nepal Earthquake 2015

The earthquake was measured at 7.3 and has made a serious dent to Nepal's fragile economy. It killed 8000 people in Kathmandu valley, injured 22,000 people, and destroyed many buildings, including few that were considered as heritage buildings of Nepal.

In India, the part adjoining Nepal saw death of 78 people and injury to 560 persons.

Table 1.2 Statistics of damage Bhuj earthquake 2001[a]

Type of damage	Extent
Number of deaths	20,000+
Number of injuries	170,000
Estimated cost of property destruction	US$ 5 billion
Buildings destroyed	~300,000
Buildings damaged	~700,000
Earth dams damaged	14
Area affected by landslide	9000 km^2
Area liquefied	~10,000 km^2

[a]Based on local newspaper reports

One of the major reasons for such large-scale destruction was that it triggered avalanches from the mountain that flattened a number of villages situated at the foothills.

A large area underwent liquefaction causing damages to roads and bridges and buildings.

1.3 Development of Earthquake-Resistant Technology Around the World and India

From days in early 1930s when engineers applied the technique of *Shindo*[3] today, 80 years down the road, technology of earthquake-resistant design has certainly taken giant strides. However, despite this laudable improvement much more needs to be done yet, as structures fail with monotonic regularity with every major occurrence of earthquake having magnitude greater than 6.5, killing people and destroying assets.

The method of SCM adapted by the structural engineers in those early days, though became a joke, where engineers talking about earthquakes often being butt of ridicule being nicknamed 10% *engineers*,[4] yet it was observed that the buildings designed by this method performed much better under earthquake than buildings designed by the traditional methods as was prevalent then (no earthquake force considered).

By the 1940s, it was well realized that force induced to the structures due to earthquake was time dependent in nature. However, no technology then existed within the domain of civil engineering based on which it was possible to quantify this force that developed due to a strong ground motion.

Thus, SCM, despite being somewhat a crude quantification of the phenomenon, enjoyed great popularity among practicing engineers as possibly the only defense line against this fearful fury of Mother Nature.

Unfortunately, unlike other engineering disciplines like mechanical, aeronautical, and shipbuilding, study of a body in motion and stress induced due to this was not a part of civil engineering curriculum (except possibly Russia) at that time. Thus, in English speaking world, civil engineers were quite in dark for a significant period of time as to how to tackle such time-dependent random forces acting on a physical system.

It was possibly in the middle 1940s when the *Massachusetts Institute of Technology* developed a multi-discipline engineering curriculum where theory of mechanical vibration was one of the courses that was being offered as an interdisciplinary optional topic by stalwarts like Professor Ray Bishplinghoff (from

[3]It is the Japanese term for seismic coefficient method (SCM).

[4]For details, refer Chap. 2.

Aerospace Department), Den Hartog (Mechanical Department), where civil engineers were allowed to take up these new fields of science (Reitherman 2012).

Students who took up this coursework certainly had a detailed exposure to study of bodies under motion as a prerequisite for study of structures subjected to harmonic and random vibration.

Among the early civil engineers who took up this course were Ray W. Clough and H. C. Martin, two civil engineering graduate students studying there. On completion of their doctoral thesis (at MIT) Martin went back to University of Washington, while Clough migrated to University of California Berkeley, and was instrumental in developing a coursework on dynamic analysis of structures, especially related to earthquake force[5], and under his able leadership, Berkeley became a leading research center for development of earthquake-resistant design, especially related to dams and buildings.

By the early 1950s, though the awareness of adapting structural dynamics as a tool to analyze earthquake force was known to many engineers, yet one of the biggest stumbling blocks to adopt this was the input acceleration data that was required for this analysis.

It was at this point that Professor George Housner from Caltech developed the famous response spectrum curve based on recorded data of several previous earthquakes and this curve remained backbone of many engineering designs (based on modal analysis) till as recent as 2002.

Advent of digital computers in the late 70s brought a great change in analytical ability of civil engineers and with some excellent computer programs developed like SAP (Bathe, Wilson, and Peterson), GTSTRUDL (Emkin), ETABS (CSI) when linear and nonlinear analysis of structures under seismic force became more or less a routine affair.

The Alaska and Niigata Earthquake (1964) in USA and Japan, respectively, brought a great change in civil engineering design. It was for the first time geotechnical engineers were brought into the fray, who till then remained just passive observers to earthquake damages of superstructures.

Huge damage to infrastructure resulting from soil liquefaction, lateral spreading, and slope failures of embankments (both manmade and natural) under seismic force opened a new vista for research in earthquake engineering. This would finally be termed as *geotechnical earthquake engineering* which is a relatively new branch in earthquake engineering that is progressively gaining importance.

Pioneering work in this area by H. B. Seed and I. M. Idriss (From Berkeley), E. Kausel, J. Roesset, R. V. Whitman (all from MIT), Professors I. Towhata, K. Ishihara et al. (from Japan) has greatly ameliorated our knowledge base on this newly evolving topic.

---

[5]The book entitled "*Dynamics of Structure*" McGraw Hill Publication, penned by Clough and Penzien, still remains one of the most read books among civil engineers studying structural dynamics and earthquake engineering.

In India, study of earthquake started as early as 1935 with Quetta earthquake when Railway Construction Board, taking lesson learned out of previous earthquakes, developed buildings specially designed to resist earthquake force.

A young engineer by the name S. L. Kumar was possibly the first engineer in the subcontinent to adopt reinforcement made out of rejected railway track sections and combine it with masonry to develop these dwellings.

It was found that the performance of these buildings to some subsequent earthquakes was very encouraging and satisfying (Kumar 1933).

In the late 1950s, especially after the Assam–Tibet earthquake, awareness dawned in India on necessity of a comprehensive earthquake engineering program both in profession and engineering education. University of Roorkee being in foothills of the Himalaya (which is highly seismic prone thus making seismic observations convenient) was selected as the center of this study.

Under the skilled guidance of Professors Jaikrishna, Anand Arya and Shamsher Prakash, it became one of the centers of excellence for study of earthquakes in India, where the Earthquake Engineering Research Center (EERC) has been instrumental in developing a number of innovative technologies (like banded buildings to resist earthquake shocks for low-cost rural housing), and providing its expertise to a number of projects of national importance.

Despite the first earthquake observatory setup in Colaba Mumbai in 1898, more observatories were certainly required for precise microzonation. EERC played a pivotal role on setting up these stations collaborating with government and other institutes and the today status of such stations is certainly impressive, as depicted in Fig. 1.4.

Other institutes, like IIT (Kanpur), IIT (Gandhinagar), IIT (Guwahati), have also contributed significantly and have developed some excellent guidelines for professionals on several topics that remain crucial in terms of earthquake engineering in practice.

Despite this laudable development, much more needs to be done in future. For instance, India is yet to develop a unified structural analysis: finite element package that can perform linear and nonlinear analysis of a system considering earthquake force catering to Indian subcontinent condition.

This particular item, with the mark "*Made in India,*" is still eluding the profession. The country is still dependent on software developed in foreign countries and paying for their upkeep in foreign currencies through the nose.[6]

In geotechnical engineering sector, the situation is still worse, as nothing exists in terms of domestically developed utility packages related to liquefaction, seismic

[6]Tragically, the country sends thousands of IIT and NIT engineers to USA to develop software on financial and hospitality sectors ruining their technological background for good. It is a shame to see brilliant structural or electrical engineers coding payroll software just for a fistful of dollars. How they contribute to "*Service of the nation*" is certainly a matter of serious sociological research as to *what ails our engineering industry that compels such bright minds to opt for other vocations* like being an IT engineer (majoring in Civil Engineering), a district magistrate or IPS officer majoring in Mechanical or Electrical Engineering.

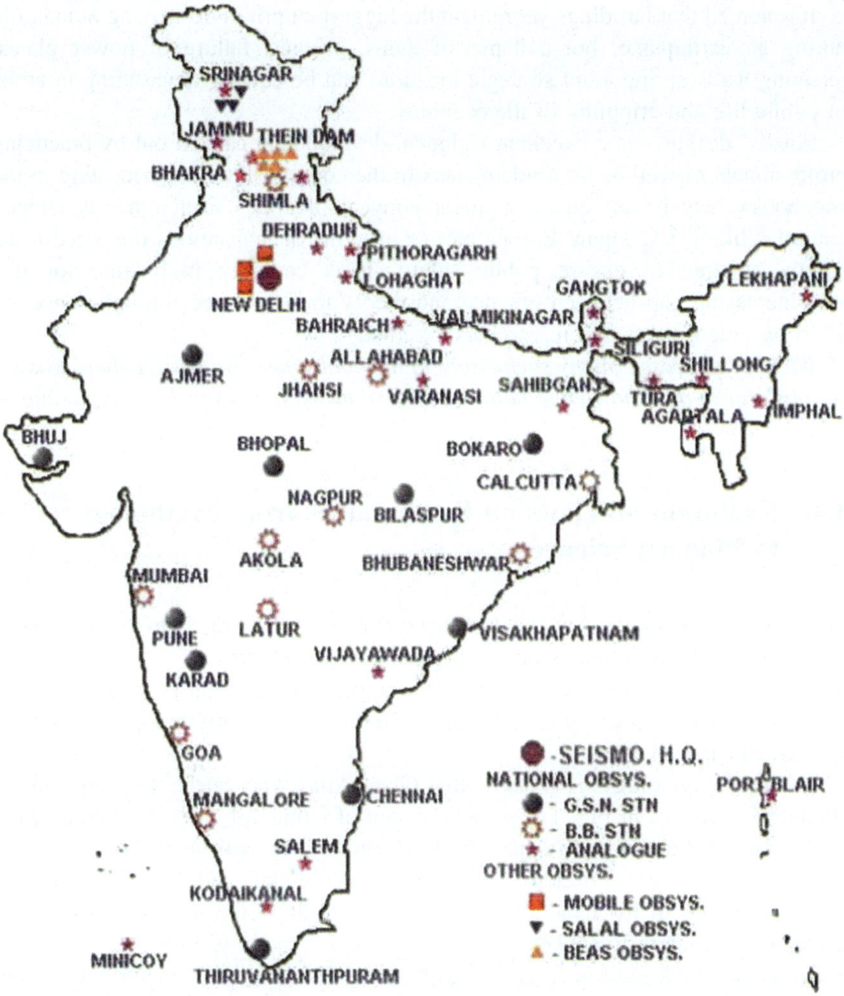

Fig. 1.4 Strong motions recording stations in India, after Madabushi and Haig (2005)

stability of embankments or coupled analysis of systems like concrete gravity dams where we are still dependent on foreign packages like ANSYS, PLAXIS or those developed by ICOLD Zurich.

Many of the technologies recommended in IS code have become obsolete (for instance, the seismic earth pressure theory of Mononobe–Okabe) but are yet to be upgraded in the code.

To the authors' perception, the focus of the earthquake fraternity in India is far too narrowed down on buildings. Other systems that can also get seriously damaged have been somewhat relegated to secondary roles.

It is agreed that buildings yet remain the biggest culprit in destroying human life during an earthquake, but collapse of dams, bridges, failure of power plants, retaining walls at important strategic locations can be equally devastating in terms of public life and crippling of the economy.

Finally, despite some excellent technical development carried out by practicing professionals as well as by academicians in the country, little is permeating in the codebooks (which yet remain a major copycat of codes from other developed countries like USA, Japan, Europe etc) or in terms of legislations that need to be strictly adhered to ensure public safety. It is certainly high time for this ever-increasing gap between practice and theory to be reduced through concerted effort by practitioners and researchers together.

We hope to present many such issues in the book where we believe there exists a serious gap in our knowledge both in terms of mechanics as well as applications.

1.4 Evolution of Views on Earthquakes from Mythology to Modern Science

Of all natural phenomena that have ravaged human civilization from its early dawn, earthquake certainly remains one of the primary perpetrators.

In ancient times when natural science was yet to take its birth, earthquake was often perceived as an act of God giving rise to a number of mythological stories as to why do earth shake.

In Greek Mythology, it is stated that Giant Atlas who carries the earth on his shoulder, as shown in Fig. 1.5, sometimes out of fatigue of carrying such a heavy load, shrugs his shoulder to adjust the load causing the earth to shake.

In another version, Poseidon, the sea God (also nicknamed *The Earth shaker*), carries a trident in his hand. When angry, he would strike the ground with the trident causing earth to shake.

In earthquake legend in Japan, it was believed that below the country lay a giant Catfish (Namazu), and when this fish moves, an earthquake takes place.

Figure 1.6 shows Japanese people who had suffered from earthquake trying to punish Namazu who is responsible for the damage.

In Indian mythology, Naga are giant serpents revered by many, but were also thought to cause earthquakes by their subterranean movements.

With the birth of natural science, a number of philosophers and scientists put forward theories pertaining to earthquakes.

According to Aristotelian philosophy, "The Earth is essentially dry but rain fills it with moisture. Then the Sun and its own fire warm the water and give rise to a quantity of wind both inside and outside it. Thus, the movement of this wind in underground caverns gives rise to earthquake."

As per Robert Hooke (famous physicist; the *Hooke's Law*), underground materials similar to gunpowder explode below the earth causing earthquakes.

Fig. 1.5 Giant Atlas carrying the world on his shoulder

Finally, with advent of *geophysics*, a special branch of science that took up seismology as a major topic of study, three theories emerged that more or less cover the major reasons why do earthquakes take place. They are: (1) molten magma theory, (2) elastic rebound theory, and (3) theory of plate tectonics.

One of the above theories or combination of them has been responsible for most of the earthquakes around the world. We explain the above theories hereafter briefly.

1.4.1 Molten Magma Theory

Figure 1.7 shows the geological compositions that constitute our Earth. The outer crust, called the lithosphere, is about 33 km thick for continents, and 10 km below oceans, as shown in Fig. 1.8.

Below the lithosphere, we have the asthenosphere, which consists of hot weak plastic material and is about 100 km thick. Below this, we have a hot but stronger layer called Mesosphere, about 350 km thick, followed by a liquid core, which is called molten magma and is under very high pressure and temperature.

Fig. 1.6 Namazu the Catfish causing earthquake in Japan

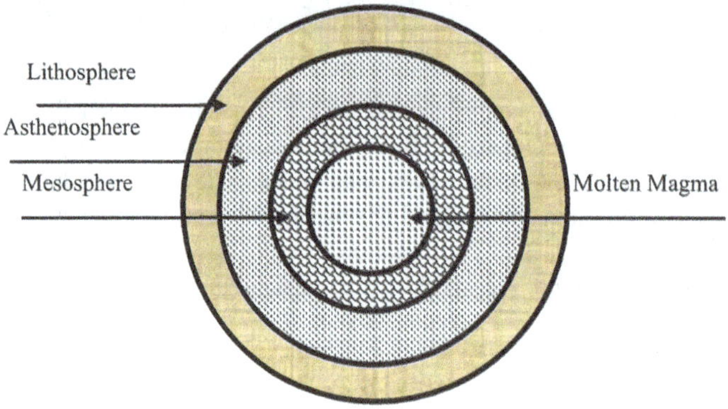

Fig. 1.7 Geological composition of the Earth

Wherever this magma finds a weak fissure or cracks, it has a tendency to push through it. When this molten magma finds its way up to the earth's surface, it spews out smoke, ashes, and lava resulting in volcanic eruptions. However, in some place, it reaches the lithosphere but cannot come out due to lack of passage, and in many cases, this huge pressure generated on the lithosphere has caused upheaval creating mountains and hills.

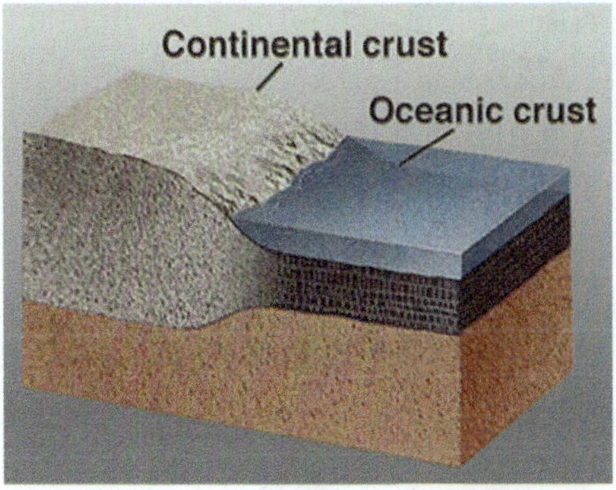

Fig. 1.8 Earth and the oceanic crust

A popular theory on birth of the Himalayan mountain range is that it has been born due to upheaval of the lithosphere due to molten magma. The huge potential energy thus created within the lithosphere is still adjusting itself that often results in earthquakes in places like South China, Nepal, and northeastern part of India.

1.4.2 Elastic Rebound Theory

Analyzing the reasons for the San Andreas earthquake (1906), based on field observation, Harry Fielding Reid, an American geologist, examined the ground displacement around the San Andreas Fault and concluded that the earthquake was a result of an elastic rebound of previously stored strain energy in the rock on either side of the fault.

In a layman's term, it can be explained by a rubber band stretched progressively when, as strain increases within the strip, so does the energy. However, beyond a limit the rubber band snaps into two pieces, releasing all energy stored in one go, creating a whiplash effect. In same way when the strain energy within rock gets released suddenly, it generates waves that dissipate all round resulting in ground to move.

Shown in Fig. 1.9 is the earth's crust having a fault where the strain generates. With time, as strain increases, the rock strata deform and finally beyond a certain stage, undergo a sudden rupture, releasing a large amount on energy that travels through surface of the earth while the rock after rupture rebounds back to its original undeformed shape.

Above concisely represents the elastic rebound theory.

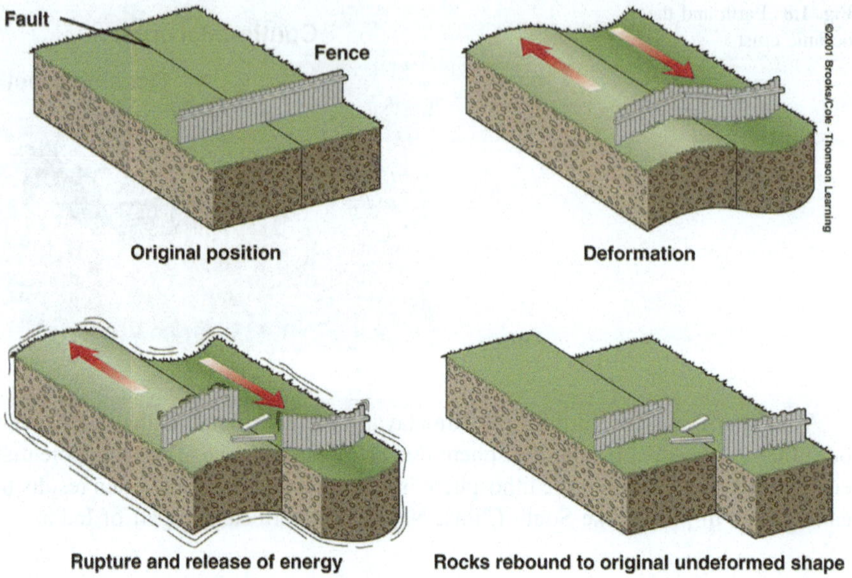

Fig. 1.9 Reid's elastic rebound theory of earthquakes

1.4.3 Theory of Plate Tectonics

To understand the basic principle of plate tectonics, let us consider Fig. 1.10 as presented hereafter.

The zigzag lines that ensemble the continents show the global lithosphere or earth surface broken up into different fragments, each one called a plate that move away or collide with each other.

A German climatologist and Arctic explorer by the name of Alfred Wegener first developed this idea.

Somewhere in the 1920s, based on evidence of samples of geological rocks and fossils collected from different continents, Wegener concluded that like the jigsaw pieces, we see the maps of our different continents today are actually near fit and once were all together as supercontinent called *Pangaea,* as shown in Fig. 1.11.

However, for some reason (that Wegener could not explain) the continents moved apart. The scientific community, though initially quite skeptical about Wegener's hypothesis, could not refute it altogether, considering the fact that his observations were based on collected samples and fossils that strongly supported it. Though no explanations at that time could be furnished in terms of mechanics as to why it should happen so, yet the hypothesis did have some very strong logic behind it.

It would be much latter (somewhere around the Second World War) that an American Naval Officer by the name of Harry Hess, while mapping the seafloor bed, came up with an evidence of *seafloor drift*. He observed that at places new

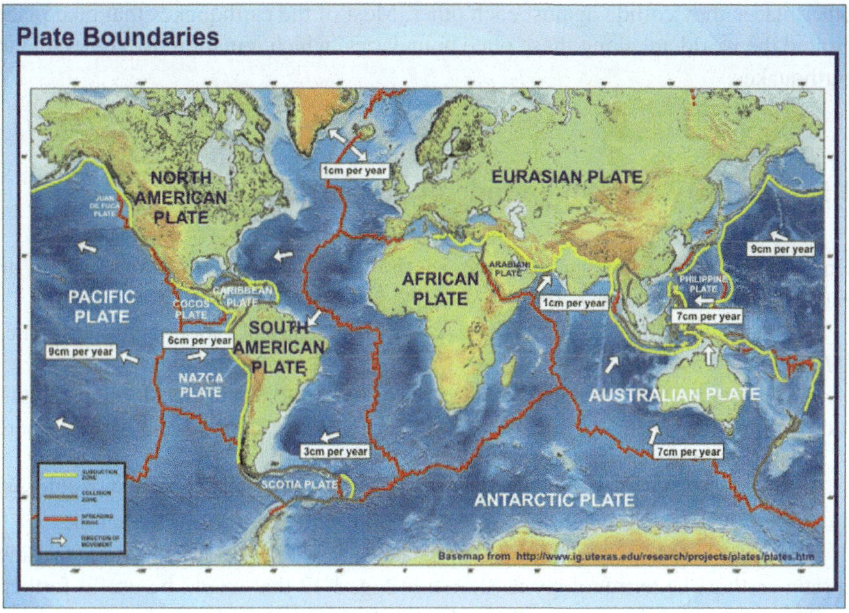

Fig. 1.10 Lithospheric plates constituting the earth surface

Fig. 1.11 Great supercontinent as per Wegener's theory

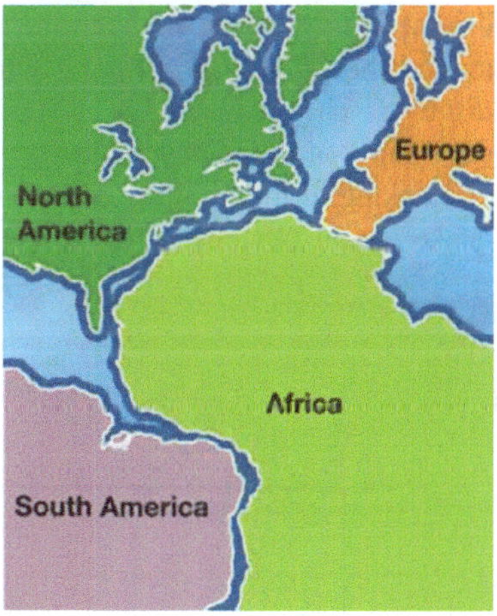

landmasses were being added to the oceanic lithosphere from below due to convecting thermodynamic cycles resulting in the seafloor to move apart. This explained the reason for the continental drift and showed that Wegener's hypothesis was indeed correct.

It is now an established fact that the seabed drifts and the lithospheric plates are under motion: At some places, they move apart (and even break apart), while at other places they collide against each other. Most of the earthquakes that take place around the world are along these plate boundaries, which remain most vulnerable to earthquakes.

1.5 Intensity and Magnitude of an Earthquake

One of the major problems seismologists initially faced was how to measure an earthquake when it happened?

The measurement of earthquake gave rise to intensity (I_0) and magnitude scales (M).

Many people (both technical and otherwise) use the term intensity and magnitude in the same sense to define the powerfulness of an earthquake. However, there is a distinct difference between the two.

The intensity scale is basically *a qualitative or perceptive scale* developed based on degree of destructiveness, or human perception of the degree of shaking.

However, before giving further explanation, it would be worth understanding certain earthquake terminology as shown in Fig. 1.12 that would be useful further.

Shown in Fig. 1.12 is a site (S) and a slip fault (point O) located at some depth and distance (usually called the source). When the fault ruptures, waves are generated. These waves travel through the soil medium and hit the site at point S as shown in Fig. 1.12. In such a case, point E, which is exactly at a point perpendicular from point O, is called the *epicenter* of the earthquake. The distance ES is termed the *epicentral distance* while the distance OS is called the *hypocentric distance*. The distance OE is termed the *focus* and can be shallow or deep, depending on the fault location.

Looking at the figure it is apparent that for a given site, the effect of an earthquake would obviously depend on a number of factors, like its location, source,

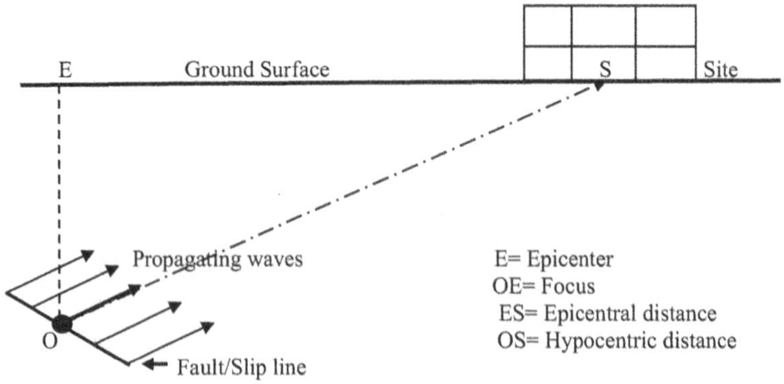

Fig. 1.12 Waves propagating from a fault to a particular site or station

focal depth, and characteristics of soil through which it propagates and on the amount of energy released due to rupture of the fault.

Some of the most common intensity scales are:

- Mercalli–Cancani–Seiberg (MCS): 12-level scale used in southern Europe
- Modified Mercalli (MM): 12-level scale proposed in 1931 by Wood and Neumann, who adapted the MCS scale to the California data set. It is used in North America and several other countries
- Medvedev–Sponheuer–Karnik (MSK): 12-level scale developed in Central and Eastern Europe and used in several other countries
- European Macro-seismic Scale (EMS): 12-level scale adopted since 1998 in Europe. It is a modification of the MM scale
- Japanese Meteorological Agency (JMA): 7-level scale used in Japan. It has been revised over the years and has recently been correlated to maximum horizontal acceleration of the ground.

These intensity scales are listed in Table 1.3a–e.

It is apparent from above that the intensity scales are qualitative in nature and are based on visual assessment of the level of destruction and human perception.

Attempts were also made by number geo-scientists to quantify the peak acceleration the ground may experience due to propagation of the waves from the above-mentioned intensity scales.

Gutenberg and Richter (1942, 1956) proposed

$$\log_{10} a_{\mathrm{h}} = \frac{I_{\mathrm{mm}}}{3} - \frac{1}{2} \tag{1.5.1}$$

Here, I_{mm} = intensity of earthquake in modified Mercalli scale, and a_{h} = ground acceleration in cm/s^2.

Kawasumi (1951) proposed

$$\log_{10} a_{\mathrm{h}} = \frac{I_{\mathrm{jma}}}{2} - 0.347 \tag{1.5.2}$$

Here, I_{jma} = intensity as per Japanese scale.

As per Medvedev and Sponheuer (1969)

$$\log_{10} a_{\max} = 0.301 I_{\mathrm{mm}} - 0.408 \tag{1.5.3}$$

Ambraseys (1974) has defined

$$\begin{aligned} \log_{10} a_{\mathrm{h}} &= 0.36 I_{\mathrm{mm}} - 0.16 \text{ and} \\ \log_{10} a_{\mathrm{v}} &= 0.38 I_{\mathrm{mm}} - 0.55 \end{aligned} \tag{1.5.4}$$

Here, a_{v} = ground acceleration in vertical direction

As per Trifunac and Brady (1975)

Table 1.3 **a** Mercalli–Cancani–Seiberg (MCS) scale. **b** Modified Mercalli (MM) scale. **c** Medvedev–Sponheuer–Karnik (MSK) scale. **d** European macro-seismic scale (EMS). **e** Japanese Meteorological Agency (JMA)

(a)

I.	Instrumental	Not felt by many people unless in favorable conditions
II.	Weak	Felt only by a few people at best, especially on the upper floors of buildings. Delicately suspended objects may swing
III.	Slight	Felt quite noticeably by people indoors, especially on the upper floors of buildings. Many do not recognize it as an earthquake. Standing motor cars may rock slightly. Vibration similar to the passing of a truck. Duration estimated
IV.	Moderate	Felt indoors by many people, outdoors by few people during the day. At night, some awakened. Dishes, windows, doors disturbed; walls make cracking sound. Sensation like heavy truck striking building. Standing motor cars rock noticeably. Dishes and windows rattle alarmingly
V.	Rather strong	Felt outside by most, may not be felt by some outside in non-favorable conditions. Dishes and windows may break and large bells will ring. Vibrations like large train passing close to house
VI.	Strong	Felt by all: many frightened and run outdoors, walk unsteadily windows, dishes, glassware broken; books fall off shelves; some heavy furniture moved or overturned, a few instances of fallen plaster. Damage slight
VII.	Very strong	Difficult to stand: furniture broken: damage negligible in building of good design and construction; slight to moderate damage in well-built ordinary structures; considerable damage in poorly built or badly designed structures; some chimneys broken. Noticed by people driving motor cars
VIII.	Destructive	Damage slight in specially designed structures; considerable in ordinary buildings with partial collapse. Damage great in poorly built structures. Fall of chimneys, factory stacks, columns, monuments, walls. Heavy furniture moved
IX.	Violent	General panic: damage considerable in specially designed structures, well-designed frame structures thrown out of plumb. Damage great in ordinary buildings, with partial collapse. Buildings shifted off foundations
X.	Intense	Some well-built wooden structures destroyed: most masonry and frame structures destroyed with foundation. Rails bent
XI.	Extreme	Few. if any masonry structures remain standing. Bridges destroyed. Rails bent greatly
XII.	Cataclysmic	Total damage—Everything is destroyed. Total destruction. Lines of sight and level distorted. Objects thrown into the air. The ground moves in waves or ripples. Large amounts of rock move position. Landscape altered, or leveled by several meters. In some cases, even the route of rivers is changed

(continued)

Table 1.3 (continued)

(b)	
I.	Not felt except by a very few under especially favorable conditions
II.	Felt only by a few persons at rest, especially on upper floors of buildings
III.	Felt quite noticeably by persons indoors, especially on upper floors of buildings. Many people do not recognize it as an earthquake. Standing motor cars may rock slightly. Vibrations similar to the passing of a truck. Duration can be estimated
V.	Felt indoors by many outdoors by few during the day At night some awakened. Dishes, windows, doors disturbed: walls make cracking sound. Sensation like heavy truck striking building. Standing motor cars rocked noticeably Felt by nearly everyone; many awakened. Some dishes, windows broken. Unstable objects overturned. Pendulum clocks may stop
VI.	Felt by all, many frightened. Some heavy furniture moved; a few instances of fallen plaster. Damage slight
VII.	Damage negligible in buildings of good design and construction; slight to moderate in well-built ordinary structures; considerable damage in poorly built or badly designed structures; some chimneys broken
VIII.	Damage slight in specially designed structures; considerable damage in ordinary buildings with partial collapse. Damage great in poorly built structures. Fall of chimneys, factory stacks, columns, monuments, walls. Heavy furniture overturned
IX.	Damage considerable in specially designed structures; well-designed frame structures thrown out of plumb. Damage great in ordinary buildings, with partial collapse. Buildings shifted off foundations.
X.	Some well-built wooden structures destroyed: most masonry and frame structures destroyed with foundations. Rails bent
XI.	Few, if any (masonry) structures remain standing. Bridges destroyed. Rails bent greatly
XII.	Damage total. Lines of sight and level are distorted. Objects thrown into the air

(c)		
I.	Not perceptible	Not felt, registered only by seismographs. No effect on objects No damage to buildings
II.	Hardly perceptible	Felt only by individuals at rest. No effect on objects. No damage to buildings
III.	Weak	Felt indoors by a few. Hanging objects swing slightly. No damage to buildings
IV.	Largely observed	Felt indoors by many and felt outdoors only by very few. A few people are awakened. Moderate vibration. Observers feel a slight trembling or swaying of the building, room, bed. chair, etc. China, glasses, windows and doors rattle. Hanging objects swing. Light furniture shakes visibly in a few cases. No damage to buildings
V.	Fairly strong	Felt indoors by most, outdoors by few. A few people are frightened and run outdoors. Many sleeping people awake Observers feel a strong shaking or rocking of the whole building, room or furniture. Hanging objects swing considerably.
VI.	Strong	Felt by most indoors and by many outdoors. A few persons lose their balance. Many people are frightened and run outdoors. Small objects may fall and furniture may be shifted. Dishes and glassware may break. Farm animals may be frightened. Visible

(continued)

Table 1.3 (continued)

		damage to masonry, structures cracks in plaster. Isolated cracks on the ground
VII.	Very strong	Most people are frightened and try to run outdoors. Furniture is shifted and may be overturned. Objects fall from shelves. Water splashes from containers. Serious damage to older buildings, masonry chimneys collapse. Small landslides
VIII.	Damaging	Many people find it difficult to stand, even outdoors. Furniture may be overturned. Waves may be seen on very soft ground. Older structures partially collapse or sustain considerable damage. Large cracks and fissures opening up, rockfalls
IX.	Destructive	General panic. People may be forcibly thrown to the ground. Waves are seen on soft ground. Substandard structures collapse. Substantial damage to well-constructed structures. Underground pipelines ruptured. Ground fracturing, widespread landslides
X.	Devastating	Masonry buildings destroyed, infrastructure crippled. Massive landslides. Water bodies may be overtopped, causing flooding of the surrounding areas and formation of new water bodies
XI.	Catastrophic	Most buildings and structures collapse. Widespread ground disturbances, tsunamis
XII.	Very catastrophic	All surface and underground structures completely destroyed. Landscape generally changed, rivers change paths, tsunamis
(d)		
I.	Not felt	Not felt, even under the most favorable circumstances
II.	Scarcely felt	Vibration is felt only by individual people at rest in houses, especially on upper floors of buildings
III.	Weak	The vibration is weak and is felt indoors by a few people. People at rest feel a swaying or light trembling
IV.	Largely observed	The earthquake is felt indoors by many people, outdoors by very few. A few people are awakened. The level of vibration is not frightening. Windows, doors and dishes rattle. Hanging objects swing
V.	Strong	The earthquake is felt indoors by most, outdoors by few. Many sleeping people awake. A few run outdoors. Buildings tremble throughout. Hanging objects swing considerably. China and glasses clatter together. The vibration is strong. Top heavy objects topple over. Doors and windows swing open or shut
VI.	Slightly damaging	Felt by most indoors and by many outdoors. Many people in buildings are frightened and run outdoors. Small objects fall. Slight damage to many ordinary buildings; for example, fine cracks in plaster and small pieces of plaster fall
VII.	Damaging	Most people are frightened and run outdoors. Furniture is shifted and objects fall from shelves in large numbers. Many ordinary buildings suffer moderate damage: small cracks in walls; partial collapse of chimneys
VIII.	Heavily damaging	Furniture may be overturned. Many ordinary buildings suffer damage: chimneys fall; large cracks appear in walls and a few buildings may partially collapse

(continued)

Table 1.3 (continued)

IX.	Destructive	Monuments and columns fall or are twisted. Many ordinary buildings partially collapse and a few collapse completely
X.	Very destructive	Many ordinary buildings collapse
XI.	Devastating	Most ordinary buildings collapse
XII.	Completely devastating	Practically all structures above and below ground are heavily damaged or destroyed

(e)	
0	Imperceptible to people. Less than 0.008 m/s^2
1	Felt by only some people in the building. 0.008–0.025 m/s^2
2	Felt by many people in the building. Some sleeping people awake. 0.025–0.08 m/s^2
3	Felt by most people in the building. Some people are frightened. 0.08–0.25 m/s^2
4	Many people are frightened. Some people try to escape from danger. Most sleeping people awake. 0.25–0.80 m/s^2
5−	Most people try to escape from danger, some finding it difficult to move. 0.80–1.40 m/s^2
5+	In many cases, unreinforced concrete block walls collapse and tombstones overturn. Many automobiles stop due to difficulty in driving. Occasionally, poorly installed vending machines fall. 1.40–2.50 m/s^2
6−	In some buildings, wall tiles and windowpanes are damaged and fall. 2.50–3.15 m/s^2
6+	In many buildings, wall tiles and windowpanes are damaged and fall. Most unreinforced concrete block walls collapse. 3.15-4.00 m/s^2
7	In most buildings, wall tiles and windowpanes are damaged and fall, In some cases, reinforced concrete block walls collapse. Peak ground acceleration greater than 4 m/s^2

$$\begin{aligned}\log_{10} a_\mathrm{h} &= 0.30 I_\mathrm{mm} + 0.014 \\ \log_{10} a_\mathrm{v} &= 0.30 I_\mathrm{mm} - 0.180\end{aligned} \tag{1.5.5}$$

One of the major problems with intensity scale is that it neither accounts for *local soil conditions* nor does it consider the effect of *epicentral distance* that can significantly amplify the response and induce damage.

Problem 1.1 A particular site near Iskenderun (Turkey) was observed to experience a peak ground acceleration of 280 cm/s^2. Compute the intensity of the earthquake as per Ambraseys and Trifunac's expression.

Solution

As per Ambraseys

$$\log_{10} a_\mathrm{h} = 0.36 I_\mathrm{mm} - 0.16, \text{here } a_\mathrm{h} = 280\ \mathrm{cm/s^2}.$$

$$\text{Or } \log_{10} 280 = 0.36 I_\mathrm{mm} - 0.16$$

$$I_{mm} = \frac{2.447 + 0.16}{0.36} = 7.242$$

Thus, the expected intensity is VII as per modified Mercalli scale.
As per Trifunac

$$\log_{10} a_h = 0.30 I_{mm} + 0.014 \text{ and } a_h = 280\ \text{cm/s}^2.$$

$$\Rightarrow I_{mm} = \frac{2.447 - 0.014}{0.30} = 8.11$$

Thus, the expected intensity is VIII as per modified Mercalli scale.

1.5.1 Magnitude of Earthquake (M_L, M_s and M_w)

Charles Richter from Caltech, based on seismographic observation of a number of earthquakes in California, first came up with a quantitative scale to measure the magnitude of an earthquake popularly known as *Richter's scale* designated by the term M_L.

Richter (1935) was first to express relationship between energy released to the magnitude of earthquake.

As per Richter

$$\log_{10} E = 11.4 + 1.5 M_L \tag{1.5.1.1}$$

Bath (1966) corrected the above to

$$\log_{10} E = 12.24 + 1.44 M_L \tag{1.5.1.2}$$

Here, E = energy in ergs and M_L = magnitude of earthquake defined by Richter as

$$M_L = \log A - 2.48 + 2.7 \log \Delta \tag{1.5.1.3}$$

Here, A is the maximum trace amplitude in microns recorded on a standard short-period seismometer, and Δ is the epicentral distance. Richter scale was the first widely used instrument-measured scale that was applied in many countries, especially USA.

The scale is based on amplitude (in millimeters) of the largest seismogram wave trace on a seismograph (having free time period of 0.8 s) which is normalized to a standard epicentral distance of 100 km.

As per Richter, M_L = 0 defines an earthquake that produces a maximum amplitude of 0.001 mm at a distance of 100 km. Each successive larger magnitude is defined as 10 times increase in amplitude beyond this base level.

Thus, an earthquake producing an amplitude of 0.01 mm at an epicentral distance of 100 km has a magnitude M_L = 1.0, and for an amplitude of 1 mm, it implies M_L = 3.0 and so on.

The Richter magnitude scale accurately reflects the amount of seismic energy released by an earthquake up to about M_L = 6.5. However, for increasingly larger earthquakes the scale progressively gets saturated and underestimates the actual energy release.

Unfortunately, all major strong ground motions that have caused destruction are invariably ≥ 6.5. Despite this limitation, it yet remains a pioneering work in the realms of seismology.

The Richter scale has also been calibrated to TNT equivalent of a bomb and some examples and equivalents are presented in Table 1.4

Richter scale has also been benchmarked qualitatively against destruction that has been observed at a particular site and is as presented in Table 1.5.

Considering that the Richter scale is benchmarked to a distance of 100 km from the source of disturbance, the effects measured are considered local.

For this reason, the Richter scale is also attributed as a measure of local magnitude, where the subscript L in the term M_L stands for the word local.

Surface wave magnitude (M_S)

Considering the fact that Richter scale saturates above 6.5, a new scale called *surface wave magnitude* (M_S) was devised to overcome this problem. The procedure of measurement is similar to Richter scale except of the fact that waves producing period 20 s or more are measured from seismographs at teleseismic distance (which means when the pickup station is located at more than 1000 km from the source). It is not necessary to measure the seismographic record within 100 km of the epicenter. Thus, teleseismic records of a number of large to moderate earthquakes globally have been assigned surface wave magnitude. Because of this,

Table 1.4 Richter scale vis-à-vis energy released

Richter scale M_L	Approximate TNT equivalent	Level and extent of damage	Example
3.0	32 metric ton	134.4 GJ	Massive ordnance air blast bomb
4.0	1000 ton	4.2 TJ	Small atomic bomb
5.0	32,000 ton	134.4 TJ	Equal to Nagasaki bomb (WW II)
6.0	10^6 ton	4.2 PJ	Killari Latur earthquake (1993)
6.5	5.6×10^6 ton	23.5 PJ	Myanmar earthquake(2015)
7.5	178×10^6 ton	748 PJ	Kashmir earthquake (2005)
8.0	1 gigaton	4.2 EJ	San Francisco (1906)
9.0	32 gigaton	134.4 EJ	Anchorage (1964)

Table 1.5 Richter scale vis-à-vis energy released

Richter scale M_L	Earthquake type	Level and extent of damage	Expectancy or frequency of occurrence
3.0–3.9	Minor	Often felt in earthquake prone area but rarely causes damage	49,000/year approximately
4.0–4.9	Light	Noticeable shaking of indoor items with rattling noises, significant damage unlikely	6200/year
5.0–5.9	Moderate	Can cause damage to poorly constructed building in small area and only slight damage to well-designed buildings	800/year
6.0–6.9	Strong	Can be destructive—up to 160-km radius from source in populated area	120/year
7–7.9	Major	Can cause serious damage over very large areas	18/year
8–8.9	Great	Can cause serious damage to several hundreds of miles around	1/year

large data set M_S is the typical magnitude used in comparisons of magnitude versus earthquake rupture length or displacement. It should, however, be noted that surface wave magnitude scale also saturates at about $M_S > 8.0$.

Moment magnitude (M_W)

The moment magnitude scale M_W is the most recently developed scale (Kanamori 1979; Hanks and Kanamori 1979) and is fundamentally different from the scales as defined above. Rather than depending on measured seismogram amplitude, the M_W scale is a function of seismic moment M_0 of an earthquake.

The seismic moment M_0 is defined as

$$M_0 = D \times A \times G \tag{1.5.1.4}$$

where D is the average displacement over the entire fault surface, A is the area of fault surface, and G is the average shear rigidity of the faulted rock. The value of D is estimated from the observed surface displacement or from displacements on the fault plane reconstructed from instrumental or geodetic modeling.

The term A is computed from the length multiplied by the estimated depth of fault plane, as revealed by surface rupture, aftershock patterns or geodetic data. The shear rigidity of a typical crustal rock is assumed to be about 3.0–3.5 × 10^{11} dyne/cm^2.

The seismic moment thus more directly represents the amount of energy released at the source rather than relying on the effects of that energy on one or more seismographs at some distance from the source.

The moment magnitude is computed from seismic moments based on the relation of Hanks and Kanamori (1979) as

$$M_W = \frac{2}{3}[\log_{10} M_0 - 16] \quad (1.5.1.5)$$

The greatest advantage of the moment magnitude is that it is capable of measuring earthquakes of magnitude $M_S > 8.0$ without any saturation, and has been effectively used to measure some very strong motion shaking around the world.

1.6 Relationship Between the Length of Fault Rupture and Earthquake Magnitude

Tocher (1958), basing on a number of earthquakes in California, proposed

$$\log_e L = 1.02M - 5.77 \quad (1.6.1)$$

where L is the length of fault rupture, measured in km, and the earthquake magnitude is represented by M. Figure 1.13 shows relation between rupture length and earthquake magnitude.

It is observed that an earthquake of magnitude of 7.5 is expected to generate a rupture length of about 5 km.

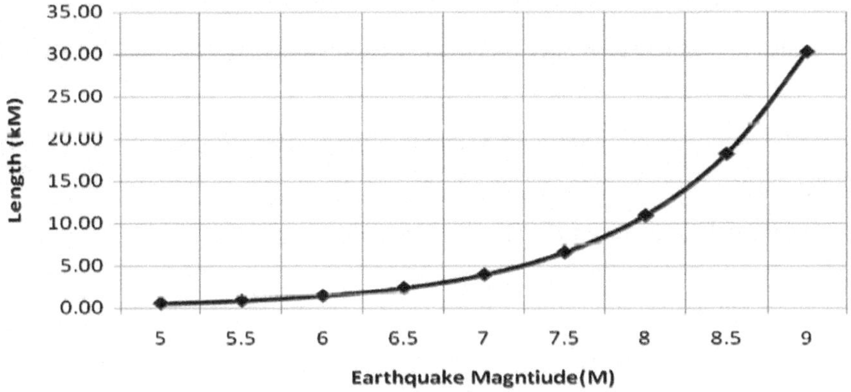

Fig. 1.13 Length of rupture for different earthquake magnitudes

1.7 Relationship Between Duration of an Earthquake and Its Magnitude

Duration of an earthquake during its happenstance is an important aspect for hazard analysis. It is evident that a very strong vibration for a few seconds and a moderate one lasting even a minute will have different consequences and the subsequent effects will be quite different. Thus, estimating duration of an earthquake that usually increases with increase in its magnitude and distance from source is quite important. Depths of bedrock and soil medium through which the waves travel also have an important bearing on its duration. The estimate is mostly probabilistic, fitted by regression analysis.

Donovan (1974) proposed a relationship based on earthquakes observed in USA and Japan as

$$D = 4 + 11(M - 5) \quad \text{for } M > 5 \tag{1.7.1}$$

where D is the duration of earthquake in seconds and M the magnitude.

Dobry et al. (1978) has proposed an expression

$$\log_{10} D = 0.432M - 1.83 \tag{1.7.2}$$

Bommer et al. (2008) based on a number of observations of earthquake in USA and Mexico have proposed for soil sites

$$\log_{10} D = 0.69M_W - 3.70 \tag{1.7.3}$$

Bullen and Bolt (1999) have proposed an expression based on observed ground acceleration of a site

$$D = 17.5 \tanh(M - 6.5) + 19 \tag{1.7.4}$$

Here, the observed acceleration was $a > 0.05g$.

Esteva and Rosenblueth (1964) gave a relation

$$D = 0.02\,\mathrm{e}^{0.74M} + 0.3\Delta \tag{1.7.5}$$

Here, Δ is the distance of the observatory from the seismic source in km.

Housner (1965) suggested the expression

$$D = 11.52M - 53 \quad \text{for } M > 5 \tag{1.7.6}$$

Variations of the duration of vibrations with some earthquake magnitudes are shown in Fig. 1.14.

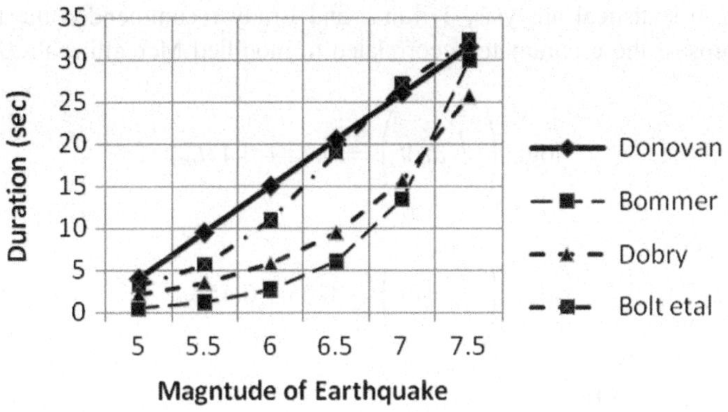

Fig. 1.14 Duration of an earthquake versus its magnitude

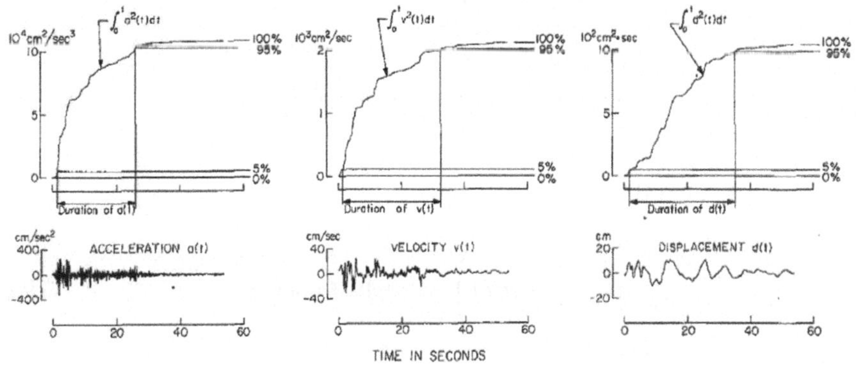

Fig. 1.15 Duration period of strong motion acceleration (*a*), velocity (*v*), and displacement (*d*), after Trifunac and Brady (1975)

Trifunac and Brady (1975) gives an interesting study on the duration of an earthquake. Using Housner's concept of mean square acceleration during rise time of strong motion, they expressed duration time as shown in Fig. 1.15.

From Fig. 1.15, it is observed that $\int_0^T a^2 \mathrm{d}t$, $\int_0^T v^2 \mathrm{d}t$ and $\int_0^T d^2 \mathrm{d}t$ increase rapidly at the start and then decays asymptotically to their final amplitudes $\int_0^\infty a^2 \mathrm{d}t$, $\int_0^\infty v^2 \mathrm{d}t$, $\int_0^\infty d^2 \mathrm{d}t$. Here, *a*, *v*, and *d* are acceleration, velocity, and displacement, respectively.

Study of a number of such integrals shows that the rapid growth corresponds to the strong motion part of the record, indicating that the first 5% of the precursor waves and last 5% have negligible response.

It is only the intermediate 90% which is considered as significant and constitute the strong motion response.

Based on statistical analysis, Trifunac and Brady recommended the following relationships of these parameters, correlated to modified Mercalli scale (I_{mm}).

$$\log_{10}\left(\int_0^T a_{\mathrm{h}}^2 \mathrm{d}t\right) = 0.72 + 0.47 I_{\mathrm{mm}} \tag{1.7.7a}$$

$$\log_{10}\left(\int_0^T a_{\mathrm{v}}^2 \mathrm{d}t\right) = 0.25 + 0.45 I_{\mathrm{mm}} \tag{1.7.7b}$$

Here, III < I_{mm} < VIII

$$\log_{10}\left(\int_0^T v_{\mathrm{h}}^2 \mathrm{d}t\right) = -0.96 + 0.47 I_{\mathrm{mm}} \tag{1.7.8a}$$

$$\log_{10}\left(\int_0^T v_{\mathrm{v}}^2 \mathrm{d}t\right) = -0.90 + 0.38 I_{\mathrm{mm}} \tag{1.7.8b}$$

where V < I_{mm} < VIII

$$\log_{10}\left(\int_0^T d_{\mathrm{h}}^2 \mathrm{d}t\right) = -1.10 + 0.45 I_{\mathrm{mm}} \tag{1.7.9a}$$

$$\log_{10}\left(\int_0^T d_{\mathrm{v}}^2 \mathrm{d}t\right) = -0.97 + 0.37 I_{\mathrm{mm}} \tag{1.7.9b}$$

And V < I_{mm} < VIII. (Refer Table 1.3b for details.)

The subscripts *h* and *v* stand for horizontal and vertical directions in all cases.

As discussed earlier, I_{mm} does not cater to epicentral distance and local soil condition, and therefore, Trifunac upgraded Eqs. (1.7.7a) through (1.7.9b) to

$$\log_{10}\left\{\int_0^T \begin{pmatrix} a^2 \\ v^2 \\ d^2 \end{pmatrix} \mathrm{d}t\right\} = as + bM + c\log_{10}\Delta + d \pm \sigma \text{ where} \tag{1.7.10}$$

s local site condition, and is 0 for soft soil, 1.0 for intermediate soil and 2 for hard soil.

M magnitude of earthquake in modified Mercalli scale.

Δ epicentral distance

Table 1.6 Values of the seismic parameters after Trifunac and Brady (1975)

Component	a	b	c	d	A	B	Σ
a_h	−0.068	1.090	−1.79	−0.66	0.681	−0.200	0.290
a_v	−0.103	1.120	−1.87	−0.14	0.613	−0.156	0.321
v_h	−0.184	0.998	−1.10	−2.66	0.657	−0.158	0.286
v_v	−0.331	1.170	−1.44	−2.60	0.556	−0.061	0.316
d_h	−0.177	0.739	−0.62	−2.05	0.663	−0.172	0.288
d_v	−0.357	0.975	−1.14	−2.16	0.511	−0.020	0.336

a, b, c, d are correlation coefficients, and σ is the standard deviation expressed as

$$\sigma = A + B + \log_{10}\Delta \pm \Sigma \tag{1.7.11}$$

The parametric values are as presented in Table 1.6.

The duration of the response D for the acceleration, velocity, and displacements has been correlated with I_{mm} as

$$\begin{aligned} D_{a_h} &= 56.3 - 4.67 I_{\mathrm{mm}} \\ D_{a_v} &= 46.5 - 3.85 I_{\mathrm{mm}} \end{aligned} \tag{1.7.10a}$$

$$\begin{aligned} D_{v_h} &= 55.5 - 2.75 I_{\mathrm{mm}} \\ D_{v_v} &= 58.4 - 3.75 I_{\mathrm{mm}} \end{aligned} \tag{1.7.10b}$$

$$\begin{aligned} D_{d_h} &= 45.2 - 0.827 I_{\mathrm{mm}} \\ D_{d_v} &= 41.8 - 0.674 I_{\mathrm{mm}} \end{aligned} \tag{1.7.10c}$$

Here, III < I_{mm} < VIII.

Considering the effect of local soil condition and epicentral distance, Trifunac et al. proposed

$$\begin{Bmatrix} D_{a_h} \\ D_{a_v} \\ D_{v_h} \\ D_{v_v} \\ D_{d_h} \\ D_{d_v} \end{Bmatrix} = as + bM + c\Delta \pm \sigma \tag{1.7.11}$$

Here,

$$\sigma = A + B \pm \Sigma \tag{1.7.12}$$

Here, the parameters are as furnished in Table 1.7.

Table 1.7 Values of the seismic parameters after Trifunac and Brady (1975)

Component	a	b	c	A	B	Σ
D_{a_h}	−6.29	2.90	0.172	4.21	0.0672	7.25
D_{a_v}	−4.88	2.33	0.149	2.92	0.0830	7.17
D_{v_h}	−6.51	4.50	0.100	8.53	0.0221	7.25
D_{v_v}	−5.60	3.55	0.141	6.94	0.0519	7.14
D_{d_h}	−5.82	5.32	0.0307	10.19	0.00397	8.78
D_{d_v}	−4.08	4.07	0.107	10.10	0.0152	8.31

Table 1.8 Values of the seismic parameters after Trifunac and Brady (1975)

Component	a	b	c	d	A	B	Σ
R_{a_h}	0.0913	1.07	−2.14	−1.36	0.873	−0.266	0.325
R_{a_v}	0.0471	1.10	−2.28	−0.661	0.866	−0.264	0.357
R_{v_h}	−0.0393	1.02	−1.30	−4.02	0.935	−0.306	0.320
R_{v_v}	−0.0197	1.21	−1.70	-3.86	0.811	−0.188	0.337
R_{d_h}	−0.0562	0.700	−0.670	-3.29	0.845	−0.283	0.287
R_{d_v}	−0.250	0.973	−1.26	−3.44	0.759	−0.144	0.351

Trifunac and Brady have also given a relationship as a rate of built-up response expressed as

$$\log_{10}\left\{\int_0^T \begin{pmatrix} a^2 \\ v^2 \\ d^2 \end{pmatrix} \mathrm{d}t \Big/ D\right\} = as + bM + c\log_{10}\Delta + d \pm \sigma = R\,(\text{say}) \qquad (1.7.13)$$

Here, D and R generically represent the duration and rate for acceleration, velocity, and displacement, respectively. The parameters are as furnished in Table 1.8.

1.8 Expected Peak Ground Acceleration of a Site

Peak ground acceleration (PGA) is another important aspect that has a significant bearing on civil engineering design for a particular site. When the earthquake waves propagate through a site, it is subjected to motion due to propagation of waves. Thus, it is evident that one would be interested to know what is the maximum acceleration it may be subjected to?

Esteva and Rosenblueth (1973) have proposed the expression

$$a = \frac{110e^{0.8M}}{R^{1.6}} \tag{1.8.1}$$

Here, M is the magnitude of an earthquake and R the *hypocentric distance* in km.

Mcguire (1974) proposed a formulation that remains quite popular in design office, expressed as

$$\log_{10} a = 2.649 + 0.278M - 1.301[\log_{10}(R + 25)] \tag{1.8.2}$$

Here, acceleration (a) is derived in cm/s^2 and R in km.

Gutenberg and Richter (1942, 1956) proposed a correlation predicting the acceleration at the epicenter as

$$\log_{10} a_0 = -2.1 + 0.81M - 0.027M^2 \tag{1.8.3}$$

Here, again the acceleration a_0 is expressed in cm/s^2.

One may note that all these expressions are valid for *acceleration in bedrock* or through soil having shear wave velocity same as that of rock, i.e., $V_s \geq 600$ m/s.

1.9 Attenuation of Earthquake with Distance

When a fault slips below ground surface, waves disperse in all directions. As waves travel through soil, the farther they are from the source, the more reduced the effect. This is known as attenuation of waves over distance. The general characteristics of the response attenuation with distance are shown in Fig. 1.16

It is apparent that as waves travel from the source, the amplitude vis-à-vis ground acceleration reduces with distance. While waves of short wavelength affect the structures near the epicenter, at a large distance it is the longer waves that render damage to the structures resting on soil.

Blume (1965) proposed an attenuation factor as

$$F_D = \frac{1}{1 + (D_e/h)^2} \tag{1.9.1}$$

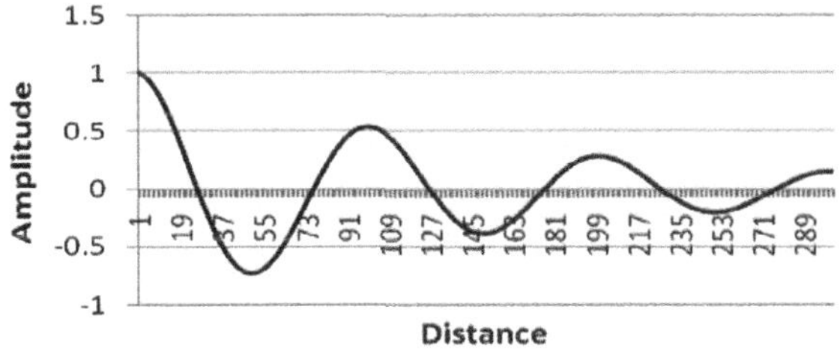

Fig. 1.16 Attenuation of waves with distance

where F_D = attenuation factor, D_e = epicentral distance, and h is the depth of focus at the epicenter.

Gutenberg and Richter (1956) have proposed that at a distance D_e from the epicenter (in miles, it shall be greater than 12 miles), the acceleration may be expressed as

$$a_D = F_D a_0 \tag{1.9.2}$$

Here, a_D = acceleration at a distance D_e, a_0 = acceleration as expressed in Eq. (1.8.3), and

$$F_D = \left[\frac{1.25}{1 + D_e/48}\right]^n \tag{1.9.3}$$

$$n = 1 + \frac{1}{2.5T_p} \tag{1.9.4}$$

Here, T_p is the predominant time period of the earthquake and is as shown in Table 1.9.

Example 1.2 A fault is located in Assam (Jorhat) India, 35 km below ground, which has a probability to generate an earthquake of magnitude 7.5. Considering Tinsukia to be 141 km from Jorhat, compute the effective peak ground acceleration expected at bedrock level.

Solution

Here, y = 35 km and D = 141 km.

Thus,

$$R = \sqrt{D^2 + y^2}$$

or

$$R = \sqrt{141^2 + 35^2} = 145\,\text{km}$$

Applying Estava and Rosenblueth's expression,

$$a = \frac{110e^{0.8M}}{R^{1.6}} = \frac{110 \times e^{7.5}}{(145)^{1.6}} = 69.24\,\text{cm/s}^2(0.07g)$$

Table 1.9 Predominant time period of earthquake versus epicentral distance

Epicentral distance in km	0–50	51–100	101–150	151–200	201–250
T_p (s)	0.25	0.3	0.4	0.5	0.6

Applying Mcguire's expression,

$$\log_{10} a = 2.649 + 0.278M - 1.301[\log_{10}(R + 25)]$$

$$\log_{10} a = 2.649 + 0.278 \times 7.5 - 1.301 \times \log_{10} 170 = 1.832 \rightarrow a = 67.92\,\text{cm/s}^2.(0.069g)$$

Applying Gutenberg and Richter's formula,

$$\log_{10} a_0 = -2.1 + 0.81M - 0.027M^2$$

or

$$\log_{10} a_0 = -2.1 + 0.81 \times 7.5 - 0.027 \times (7.5)^2 = 2.45625 \rightarrow a_0 = 285.92\,\text{cm/s}^2(0.29g)$$

This peak acceleration is expected at Jorhat, *the epicenter*.

Considering Tinsukia to be 141 km (87.6 miles) from Jorhat when predominant time period T_p = 0.4 s as per Table 1.5, we get

$$n = 1 + \frac{1}{2.5T_p} = 1 + \frac{1}{2.5 \times 0.4} = 2$$

Thus,

$$F_D = \left[\frac{1.25}{1 + D_e/48}\right]^n = \left[\frac{1.25}{1 + 87.6/48}\right]^2 = 0.1957$$

Thus, the expected peak ground acceleration at Tinsukia is

$$a_D = F_D a_0 = 0.1957 \times 285.92 = 56\,\text{cm/s}^2(0.057g)$$

Thus, taking average, the expected PGA in rock is 0.065*g*.

1.10 Summary and Conclusion

In this chapter, we have discussed in reasonable details as to why do earthquake happen in nature. We considered different major earthquakes that took place around the world, including the Indian subcontinent.

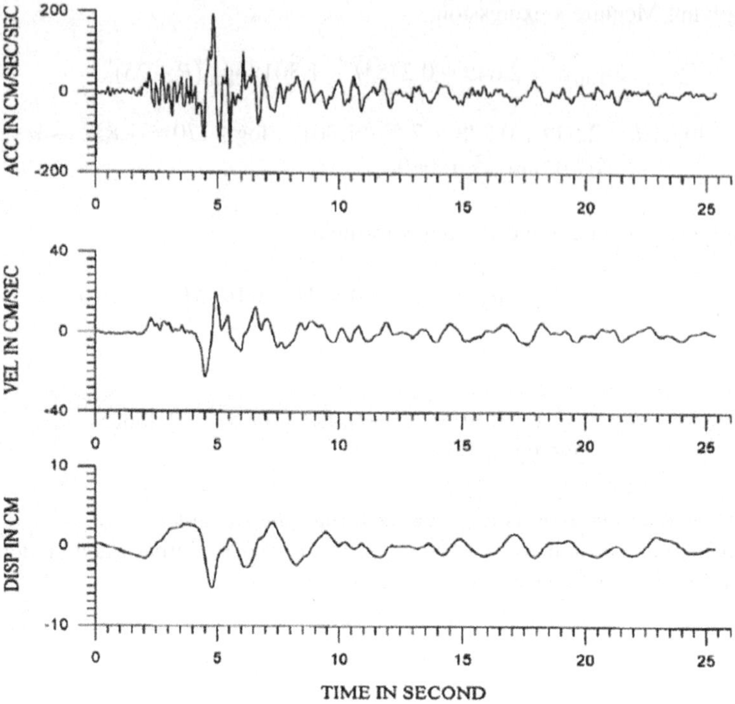

Fig. 1.17 Typical acceleration, velocity, and displacements record of Chamoli Earthquake (1991) in India

We have also discussed the evolution of an earthquake as a topic from mythology to a modern applied science where seismologists and geophysicists worked for years to quantify it both in qualitative and quantitative term.

Based on the discussion, it is apparent that slip within rock mass releases the strain energy in terms of waves and different types of waves (this will be studied further in subsequent chapters) travel through the soil inducing vibration within the soil medium. These vibrations are picked up by seismographs at different parts of the world, called observatories, and data accumulated are called strong ground motion data and can be expressed in terms of acceleration, velocity, or displacements as shown in Fig. 1.17

This particular data (ground acceleration, velocity, and displacement) are an input for design of civil engineering structures and foundations that we intend to study in some depth in subsequent chapters.

References

Ambraseys, N. N. (1974). Value of histroical records of earthquake. *Nature, 232*, 375–379.

Bath, M. (1966). Earthqauke siesmology. *Earth Science Review 1*. Elsevier Amsterdam.

Berg, G. V., Das, Y. C., Gokhale, K. V. G. K., & Setlur A. V. (1969). The Koyna India Earthquake. In *Proceedings of IVth World Conference in Earthquake Engineering* (pp. 44–57). Santiago, Chile.

Blume, J. A. (1965). Earthquake ground motion and engineering procedure for important installations near active fault. In *Proceeeding 3rd World Conference in Earthquake Engineering* (Vol. 3). New Zealand

Bommer, J. J., Ambraseys, N. N., & Straseer, F. O. (2008). Truncation of the distribution of the ground motion residuals. *Journal of Seismology, 12*, 79–105.

Bullen, K. E., & Bolt, B. (1999). *An introduction to the theory of seismology*. Cambridge, UK: Cambridge University Press.

Dobry, R., Idriss, I. M., & Ng, E. (1978). Duration characteristics of horizontal components of strong motion earthqauke records. *Bulletin of Seismological Society of America 68*(5), 1487–1520.

Donovan, N. C. (1974). Earthquake hazards for buildings. *Engineering Bulletin # 46*, 5–32 (Los Angeles: Dames and Moore).

Esteva, L., & Rosenblueth, E. (1964). Espectros de temblores a distancias moderadas y grandes. *Boletin Sociedad Mexicana de Ingenieria Sesmica, 2*(1), 1–18. Mexico.

Esteva, L., & Rosenblueth, E. (1973). Seismic risk design spectra and structural reliability. In *Proceeding Vth World Conference in Earthquake Engineering* (Vol. 2, pp. 2586–2596), Rome.

Gutenberg, B., & Richter, C. F. (1942). Earthqauke magnitude intensity, energy and acceleration. *Bulletin of Seismological Society of America*, 163–191.

Gutenberg, B., & Richter, C. F. (1956). Earthquake magnitude, intensity and acceleration. *Bulletin of Seismological Society of America, 46*, 104–155.

Hanks, T. C., & Kanamori, H. (1979). A moment magnitude scale. *Journal of Geophysics, B-84*.

Housner, G. (1965). Engineering estimate of ground shaking and maximum earthquake magnitude. *Bulletin of Sesmological Society of Amercia*, 83–102.

Jain, S. K., & Murty, C. V. R. (1995). *A short course on seismic design of reinforced concrete building*. Indian Institute of Technology. Kanpur.

Kanamori, H. (1979). The energy release in great earthquakes. *Journal of Geophysics, B-82*.

Kawasumi, H. (1951). Measures of earthquake danger and expectancy of maximum intensity throughout Japan as inferred from the seismic activity in historical times. *Bull Earthquake Research Institute Japan, 29*, 469–482.

Kumar, S. L. (1933). Theory of earthquake resistant design with a note on earthquake resistant construction in Baluchistan. Paper # 165. In *Punjab Engineering Congress* (pp. 154–189).

Madabushi, S. P. G., & Haig, S. K. (2005). *A field report on Bhuj earthquake*. London, UK: Institute of Structural Engineers.

Mcguire, R. K. (1974). Seismic structural response analysis incorporating peak response regressions response on earthquake magnitude and distance. In *Research report # R74-51*. Department of Civil Engineering. Massachusetts Institute of Technology.

Medvedev, S. V., & Sponheuer, W. (1969). Scale of seismic intensity. In *Proceedings of the 5th World Conference Earthquake Engineering* (pp. 135–138). Santiago. 13–18 January 1969.

Oldham, R. D. (1899). Report on the great earthquake 12 June 1897. *Memoir of the Geological Survey of India, 38*.

Reitherman, R. K. (2012). Five major themes in history of Earthquake Engineering. In *Proceedings of 15th World Conference in Earthquake Engineering*. Lisbon, Portugal.

Richter, C. F. (1935). A instrumental earthquake scale. *Bulletin of Seismological Society of America, 25*, 1–32.

Richter, C. F. (1958). *Elementary seismology*. San Francisco, USA: Freeman.
Tocher, D. (1958). Earthquake energy and ground breakage. *Bulletin of Seismological Society of America, 48,* 147–152.
Trifunac, M. D., & Brady, G. (1975). A study on the duration of strong earthquake ground motion. *Bulletin of Seismological Society of America, 3,* 581–626.

Chapter 2
Basic Mechanics of Earthquake-Resistant Design

2.1 Introduction

In this chapter, we will look into the ana lytical aspects of earthquake-resistant design of structures. It will show you systematically as to how the technology gradually developed from simplified equivalent static analysis that was practiced in infancy of earthquake engineering to sophisticated linear and nonlinear dynamic analysis that has become the present state of the art.

In not too distant past, engineers who would talk about earthquake in a design office were often jokingly called "*10% engineers,*" considering the fact that there usually used to be a prolonged debate among engineers as to what percentage of vertical weight was to be considered as lateral load coming on a building due to a strong motion earthquake?

In majority of the case, the consensus conclusion came to about 10% of vertical weight in places where occurrence of earthquake probability was high.

It was an era (1920s), when there were hardly any guidelines around. This percentage of vertical weight of a structure, considered as lateral force for earthquake, was dependent on the experience of the engineer concerned.

The engineer would decide on the vulnerability of a structure depending on past occurrence of earthquakes and performance of the existing structures in that particular area, taking into cognizance any damage it has caused to structures built in the vicinity.

This paucity of guideline calumniated from the major uncertainties that existed then as to how to measure an earthquake's intensity whose probability of happenstance was wrapped up in future.

However, this method of percentage of vertical weight, though appeared unsophisticated in terms of application, yet had some logic behind it.

I. Chowdhury and S. P. Dasgupta, *Earthquake Analysis and Design of Industrial Structures and Infra-structures*, GeoPlanet: Earth and Planetary Sciences,
https://doi.org/10.1007/978-3-319-90832-8_2

Beinoff (1934) elaborates …

> From past experience it has been found that buildings which are designed to withstand a constant horizontal acceleration of 0.1 times gravity are, on the whole, fairly resistant to seismic damage though we know that seismic motions do not exhibit constant accelerations; that instead they are made up of exceedingly variable oscillatory movements. A formula based on constant acceleration may thus lead to large errors especially when applied to new type of structures that has not been tested in actual earthquakes

Despite its limitation, the method started gaining acceptance among practicing engineers, for this at least gave them some tool to safeguard their structures against possible disaster against a strong motion earthquake.

Soon this came to be known as *Seismic Coefficient Method* in English-speaking countries, as *Shindo* in Japan and *Rapporto Sismico* in Italy, especially after the destructive earthquake in California in 1906, Tokyo, Japan, in 1923, and Messina Reggio, Italy, in 1908, respectively.

It was at this time, countries prone to strong motion earthquake like Japan and USA started developing seismic design codes for their countries.

The first seismic design code was introduced in Japan in 1924, while in USA, the work, though started in 1920, yet came into being only in 1927, developed as "Palo Alto Code." This was developed under the stewardship of Professor Bailey Willis and Charles D. Marx of Stanford University who recommended that buildings in California were to be designed for an equivalent horizontal force of $0.1g$, $0.15g$, and $0.2g$, when resting on hard, intermediate, and soft soil, respectively.

The first draft of *US Pacific Coast Uniform Building Code* also came out at same time (though was yet not a mandatory building law) and recommended to use 0.075, 0.075, and $0.10g$ as the equivalent horizontal force for buildings resting on hard, intermediate, and soft soils.

After Long Beach Island Earthquake in 1933, consideration of seismic force in building design became mandatory by law in the USA. Los Angeles and many other cities started to adopt 8% of g as base shear for normal building and 10% of g for schools and other important buildings that were expected to remain functional even after a strong motion ground vibration.

Thus, the *seismic coefficient* method, though simplistic in its approach, dominated the early history of earthquake-resistant design of structures for significant period of time, before the response spectrum method came into being to replace it.

2.2 Response Spectrum Method

Response spectrum method usually depicted by the term RSM is the first rational step toward scientific development of earthquake-resistant design and is a plot of absolute values of maximum response of a structure (that constitutes of displacement, velocities, accelerations, etc.) which is presented as function of natural time period of the structure/oscillator.

Originally proposed by Maurice A. Biot in 1932, it was a development that was part of Biot's Doctoral Dissertation at California Institute Technology.

In a thesis entitled "*Transient Oscillations of Elastic System,*" Biot first showed how a transient shock like that of an earthquake affects an elastic body and the response of the same is a function of its mass and stiffness. He then extrapolated it to buildings considering it as an elastic system and stated that if one was in a position to measure the ground acceleration during an earthquake, then different building will have different levels of excitation, and this will depend on individual stiffness and mass of the building.

Biot's theory on RSM can be conceptually represented as follows:

Shown in Fig. 2.1a is a cartwheel mounted by a number of single-degree oscillators that are increasingly flexible in nature (from left to right) having increasing order of time periods.

Now, if the cartwheel is subjected to a transient acceleration $\ddot{u}_g$, then it will be observed that each of the harmonic oscillators, though subjected to a common base acceleration $\ddot{u}_g$, will be subjected to different peak accelerations (S_a), though this peak for each one of them could certainly be at different time and not all of them may peak simultaneously.

Now if we can measure these peak accelerations for each of these oscillators and plot them, one would get a continuous curve, which if plotted by the chain dotted line can be attributed as the envelope of actual maximum response of the oscillators.

The concept is further elaborated through Fig. 2.1b.

Shown in Fig. 2.1b are three oscillators having different mass (m_i) and stiffness (k_i) with i = 1, 2, 3, say, such that considering $T_i = 2\pi\sqrt{m_i/k_i}$ they have natural time periods, T_1, T_2, and T_3, when $T_1 < T_2 < T_3$. One may note here that if $T_1 = 0$ (say), it means $T_1 = 2\pi\sqrt{m_1/k_1} \to 0$ as $k_1 \to \infty$; in other words, under a particular ground acceleration, the body with mass m_1 moves without any internal deformation (as $k_1 \to \infty$) as a rigid body moving in phase with the ground and experiences an acceleration that is same as the acceleration with which the ground moves. The maximum acceleration that it experienced by the rigid body m_1 is termed as the *peak ground acceleration* (*PGA*).

If all these bodies having a single degree of freedom are subjected to common ground acceleration ($\ddot{u}_g$), as shown in Fig. 1.1, then each of the bodies will have an individual total acceleration plot that is shown along axes perpendicular to their position in Fig. 2.1b.

Now, each of the *total acceleration spectrums* has one unique value that constitutes the maximum value of the acceleration depicted by Sa_{max}. Now if we plot these Sa_{max} values (for i = 1, 2, 3, …) along the axis on which the harmonic oscillators are located, we can draw a smooth curve depicted as *spectral pseudo-acceleration* curve constituting of the values Sa_{max}, Sa_2, Sa_3, etc. *This curve is not the actual response curve of the structure but an envelope constituting of maximum value of total accelerations the structures are expected* to experience during an earthquake having ground acceleration as shown in Fig. 1.1 and is actually a function of their individual time period.

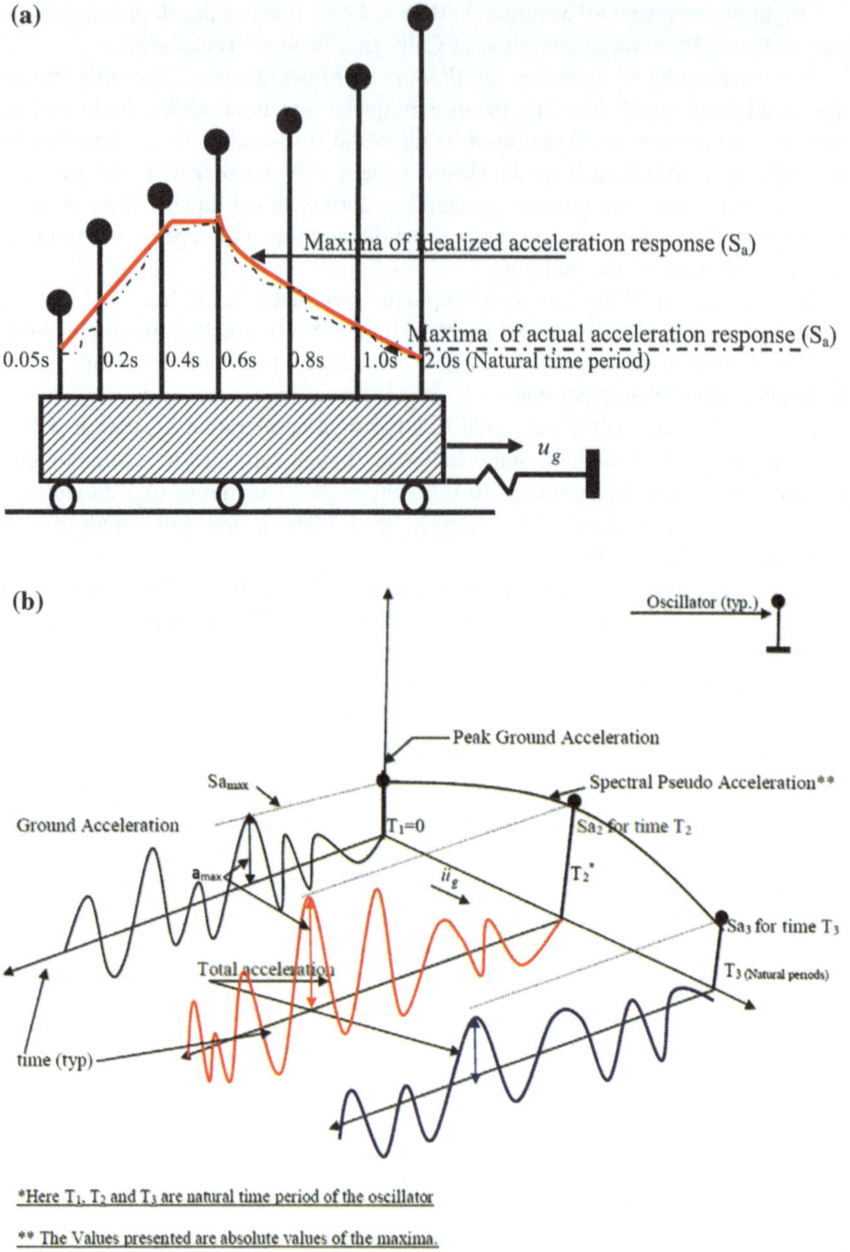

Fig. 2.1 **a** Conceptual model of response spectrum as per Biot. **b** Construction of pseudo-acceleration response spectrum

Thus, if we have a number of ground acceleration data like that in Fig. 1.1, then based on a number of such readings, one can draw an average idealized plot shown by the firm line (Fig. 2.1a, b) that represents the expected peak acceleration (or pseudo-spectral acceleration) of the harmonic oscillators when subjected to earthquake-like shocks. This averaged-out maximum envelope of the response that is drawn is called the *idealized response* and often termed as *design response spectrum* in parlance of earthquake engineering.

Generation of this design response or spectral pseudo-acceleration is an important concept in earthquake-resistant design of structures. For a structure to resist an earthquake shock, it must be built around the concept that it is capable to sustain a force equal to $m \times Sa_{max}$. Here, Sa_{max} is the maximum expected acceleration the structural body will be subjected to when experiencing a ground acceleration like one shown in Fig. 1.1 and is a function of the individual time period of the structural body.

Extending the above postulation to building systems, Biot (1932) argued that if a building system is considered as an elastic system having multiple degrees of freedom, then it can be broken down into individual components of single degree of freedom. Now provided we can measure the ground acceleration due to a transient shock like, say, an earthquake, then for each of these individual components, it is possible to measure the peak response. The overall response can then be computed by superimposition of these individual responses or modes.

Finally, based on observations of damage done to structures in previous earthquakes, Biot also predicted that though generic nature of the idealized curve would possibly remain unaltered yet, its vertical scaling can well alter, depending on a number of other engineering parameters like modal damping, strength of soil.

Biot proposed an expression that, for $T < 0.2$ s, $S_a/g = 4T + 0.2$ and for $T > 0.2$ s, $S_a/g = 0.2/T$. Here T is the time period of the single-degree oscillators and S_a/g is the ratio of spectral acceleration to acceleration due to gravity (g).

Biot's theory, despite being sound in terms of scientific basis, was difficult for immediate implementation as there was hardly any data of strong ground motion available then (1932) from which such design response spectrums could be generated.

One had to wait for another twenty years before George William Housner (1943 and 47) again from Caltech improved upon Biot's hypothesis to provide averaged-out response curve to be first used for commercial purpose.

Housner averaged out three strong motion recorded data from California, El Centro 1934 and 1940 and Tehachapi 1952, and furnished design response spectrum that remained backbone of earthquake-resistant design for many countries for a number of years to come.

Fig. 2.2 Comparison of idealized response spectra of Biot and Housner (Trifunac 2006)

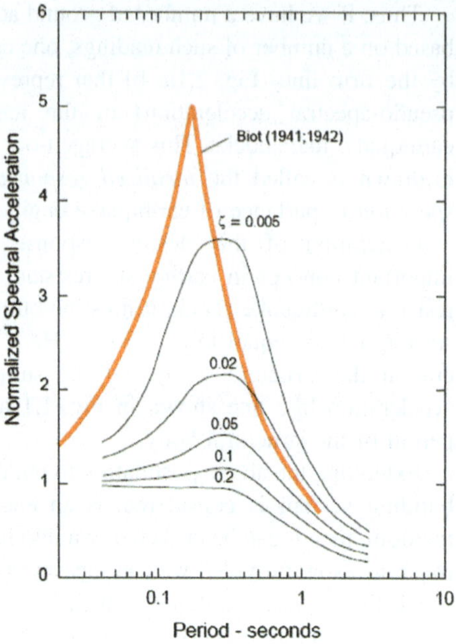

The normalized response spectra as proposed by Biot and that of Housner are as shown in Fig. 2.2.[1]

In Fig. 2.2 the firm thick line is as per Biot, while the thin lines shown are idealized spectra as per Housner for different value of damping ratio. Adjusting these curves to India's isoseismic mapping to suit the local condition, Housner's design spectra were in vogue as late as the advent of the present millennium. The response spectrum curve, adopted by IS-1893 (1984), is as shown in Fig. 2.3.

Though the theory proposed by Biot (1932) and subsequently improved upon by Housner (1947) put earthquake-resistant design on a firm footing for further progress, there still were few open issues that remained unanswered before the method of response spectrum was finally accepted as a scientific basis of earthquake-resistant design.

One of the major problems was that suppose one tries to develop a unified design spectrum curve of a country, for large countries like USA, India, not all parts of the country are equally susceptible to earthquakes. How do we cater to this, where

[1]We will park this topic for the time being here, and take it up further in detail in Chap. 3. Development of response spectrum curve is very interesting but somewhat complex technology by itself. Thus, before we have mastered a few tricks—that we are going to discuss here subsequently—we believe it would be little premature to discuss further as to how it developed, or the philosophy underlying the development of the idealized response curve.

For the time being, we will proceed with the presumption that it *exists* (like that furnish in codes as shown in Fig. 2.3) and move forward to understand how structural engineers use this spectrum curve to analyze bodies having single or multi-degrees of freedom.

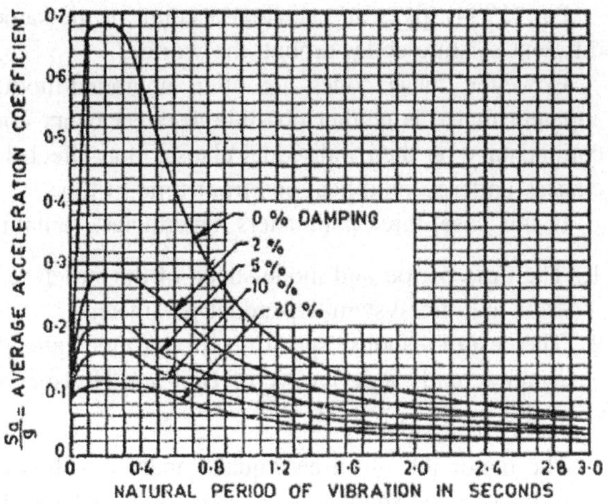

Fig. 2.3 Housner's curve as adopted by IS-1893 (1984)

structures are to be designed in areas of maximum and minimum risk and can be equally catered to?

Similarly, even in one particular area, two structures constructed can have different risk level in terms of their functional uses. How such issues can be addressed?

For example, a residential complex built in a seismic zone vis-à-vis a boiler supporting structure in a power plant built in same area will have a different level of failure risk in terms of economic and social consequences. How can one address such risks within the broad spectrum of RSM method?

We will address all these issues when we take up the explanation of individual codes.

2.2.1 The Codal Interpretation of Response Spectrum Method

In this section, a brief outline is given as to how code of practice interprets the theory of response spectrum analysis. It is apparent that the spectral acceleration curve like one shown in Fig. 2.3 would be dependent on a number of parameters that would be governed by local geology, position and depth of faults, local geological stratifications, property of soil, etc.

It will also be dependent on social issues like importance of the structure under consideration, or quality of the material used.

For this reason, most of the countries susceptible to earthquake have their own seismic codes, followed by engineers while designing structures in those countries for earthquake-resistant design.

Paz (1994) gives an excellent account of the same, giving briefly an outline of different seismic codes around the world.

However, these codes are often upgraded/modified periodically. Thus, it is apparent that it is neither possible to cover every code around the world nor give commentary on their individual clauses that affect the response, as many of these clauses undergo modifications from time to time.

In this case, three parameters nevertheless remain invariable.

1. The time period and mode shape of the structure which are independent of the code and are system-dependent parameters;
2. The design response spectrum curve (often called the backbone curve), which is again actually a derivative of Biot and Housner's curve as shown in Figs. 2.2 and 2.3.

The major job of an earthquake analyst is thus to correctly compute the time period and mode shape of the system vis-à-vis the design acceleration spectrum.

Rest of the factors, like zone factor, ductility factor, importance factor, can be considered from the respective country code to arrive at the *design base shear* (V_b).

We give a brief outline of interpretation of three codes hereafter. The main reason for choosing them, i.e., the Eurocode 8 and UBC 97, still remain most popular codes used while executing international projects (outside USA and Europe), and also the Indian code of practice, as it remains most popular in the subcontinent.

Finally, most of the real-life problems that are presented in this book have adapted one of these codes as mentioned above while computation of the response and thus are presented in some detail for better understanding of these problems by the reader.

To authors' perception, for any earthquake problem provisions of code are only about 5–10% of the issue while the rest remains basic mechanics, independent of any document or code, and this is where significant technical gaps exist and are the main focus of this book.

2.2.1.1 Response Spectrum Method as Per IS-1893 (2016)

While going through this section it is expected that *the reader has a copy of IS-1893 (2016) part 1 with him/her as a ready reference.*[2]

Since George Housner proposed the response curve in 1940 that finally came into use in the early seventies, a significant development has been made around the world on instrumentation, digitization, and real-time accumulation of strong motion data.

[2]The same is expected while we study the different clauses of Eurocode 8 and UBC 97 in subsequent sections.

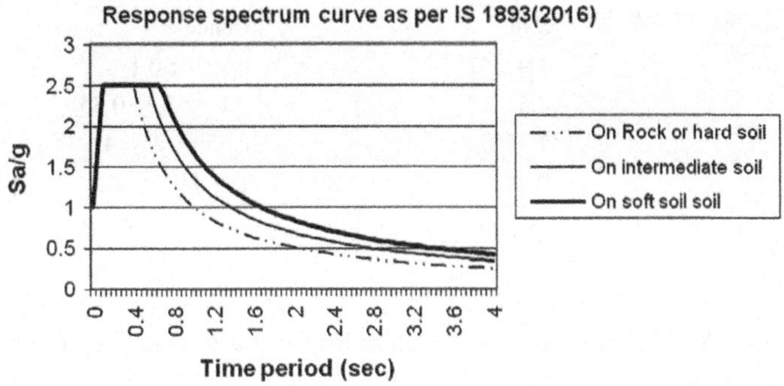

Fig. 2.4 Response spectrum curve as per IS 1893 (2016)

A number of earthquakes of significant magnitude have also occurred in India like Koyna (1967), Bihar Nepal Earthquake (1988), Uttarkashi (1991), Killari Latur (1993), Bhuj (2001) Indo-Nepal earthquake (2015) that gave a rich data bank and enabled India to modify the original response curve of Housner to suit local geological characteristics. Also, considering the fact that presently computer analysis has become the trend of the day IS code has provided the response spectrum curves in terms of expressions that can be easily digitized.

Code gives three sets of curves, which are a function of stiffness of soil supporting the structure, and is as shown in Fig. 2.4.

Observing the curves in Fig. 2.4, it is seen that curves are plotted for three types of soil: (a) rock-like or hard, (b) intermediate, and (c) soft soil. The nature of soil constituting hard, intermediate, and soft is also described in the code.

Rocky/Hard Soil

Well-graded gravel, sand gravel, and sand gravel mixtures with or without clay binder. Poorly graded clayey sand or sand clay mixtures having SPT (*N*) value greater than 30.

For this type of soil, the curve can be digitized as given in Table 2.1.

Intermediate Soil

All soils with *N* value between 10 and 30. Poorly graded sand or gravely sand, with no fines having $N > 15$. For this type of soil, the curve can be digitized as given in Table 2.2.

Table 2.1 Values of S_a/g for rocky soil

S_a/g	Remarks
$1 + 15T$	$0 < T < 0.1$
2.5	$0.1 \leq T \leq 0.4$
$1/T$	$0.4 \leq T \leq 4$
0.25	$T > 4.0$

$T =$ Modal time period of the structure

Table 2.2 Values of S_a/g for intermediate soil

S_a/g	Remarks
$1 + 15T$	$0 < T < 0.1$
2.5	$0.1 \leq T \leq 0.55$
$1.36/T$	$0.55 \leq T \leq 4$
0.34	$T > 4.0$

Soft Soil

All soils with N value less than 10. Also valid for marine clays or other sensitive clays prone to liquefy or collapse. The digitized curve can be expressed as given in Table 2.3.

It is apparent from above that once we know the time period of the structure, we can find out the value of S_a/g corresponding to that. Here, the reader is to note from Fig. 2.4 that the curve has been plotted to a scale when a rigid body placed on the ground will be subjected to an acceleration of $1g$.

This scaling has been deliberately chosen to cater to the peak ground acceleration (PGA) of different zones, having different severity of earthquake that can then be directly incorporated in lateral force estimation.

For instance, for zones subjected to *very severe shaking* (Zone V), a rigid body is expected to experience a peak ground acceleration of $0.36g$ under a maximum credible earthquake (MCE) whose probability of exceedance is only 2% in 50 years. These PGA values for different zones have been incorporated as *zone factors* (Z), as given in Table 2.4.

Figure 2.5 depicts the zone demarcation as proposed by IS 1893 (2002) based on past earthquake history.

It has been mentioned earlier that two structures built in one area may not have a same seismic risk and economic consequences. For instance, in a thermal power plant complex, we might have an administrative building, housing say 50 staffs, and also an RCC chimney 275 m tall. It is obvious that damage or collapse of these two structures will have completely different outcome on the functional operation of the plant and on the economy.

Table 2.3 Values of S_a/g for soft soil

S_a/g	Remarks
$1 + 15T$	$0 < T < 0.1$
2.5	$0.1 \leq T \leq 0.67$
$1.67/T$	$0.67 \leq T \leq 4$
0.42	$T > 4.0$

Table 2.4 Seismic zone factors (Z) as per IS code

Seismic intensity	Low	Moderate	Severe	Very severe
Seismic zone	II	III	IV	V
Z	0.1	0.16	0.24	0.36

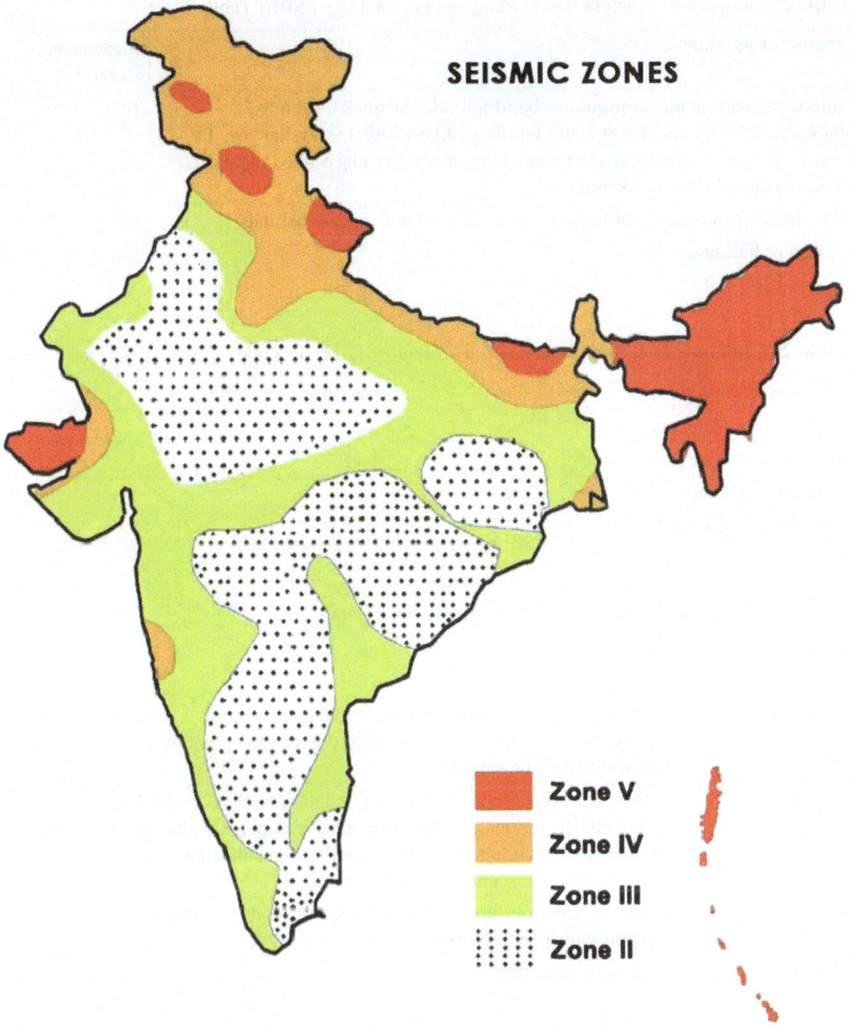

Fig. 2.5 Zone demarcations as proposed in IS 1893 (2016)

It is apparent that damage to the chimney or its collapse will have far more severe consequences than the damage to the building. Thus, it is obvious that it would be sensible to have higher margin of safety while designing the chimney against earthquake force compared to the building, despite the fact that they are constructed in the same earthquake zone. This issue has been incorporated in code by *importance factor* (*I*), which varies from structure to structure depending on its criticality.

For instance for buildings, there are two groups of classifications, as given in Table 2.5.

Table 2.5 Importance factors for buildings as per IS 1893 (2016) Table 8

Nature of building	Importance factor *I*
Important service and community buildings like hospitals, schools, monumental services. Emergency buildings like telephone exchanges, TV and radio stations, railway platforms, large assembly place like cinemas, town halls, subway, powerhouses	1.5
Residential or commercial building having greater than 200 habitats	1.2
All other buildings	1.0

Table 2.6 Importance factors for industrial structures as per IS code

Category of structures	Importance factor *I*
Structures in Category I	2.00
Structures in Category II	1.75
Structures in Category III	1.5
Structures in Category IV	1.0

Table 2.7 Definitions of structure categories as per IS code

Category of structures	Definition
Category I	Structures whose failure can cause conditions that can lead directly or indirectly to extensive loss of life/property to population at large in areas adjacent to plant complex
Category II	Structures whose failure can cause conditions that can lead directly or indirectly to serious fire hazard/extensive damage within plant complex. Structures which are required to handle immediate emergency in post-earthquake scenario are also included
Category III	Structures whose failure although expensive does not lead to serious hazard with plant complex
Category IV	All other structures

For industrial structures, code has classified it into the categories in terms of importance factors as depicted in Table 2.6.

Classifications of the categories are as defined in Table 2.7.

Code has also stated that the above is just a recommendation as a *minimum.*

Client/company or the contractor has the liberty to choose the category according to their own interpretation to ensure the safety of the structure if it is felt that it underrates its economic or social consequences.

Other than this, Table 5 in IS 1893 (2015) part IV gives extensive commentary on the categories of different types of industrial structures that are constructed in different industrial complexes. This may be referred to for further interpretation.

This brings us to the last of the parameters that affect the design earthquake force. This is called *response reduction factor* (*R*), and it is felt that it needs some explanation.

When a severe earthquake hits a structure, the design philosophy has been built around the fact that under an earthquake shock the structure is expected to undergo some damage by cracking and distortion *but shall not collapse*. It is also expected to undergo some permanent deformations.

When a structure is allowed to behave as discussed above, it reduces the response significantly.

Firstly, in reinforced concrete structures, cracking reduces the moment of inertia of a member without change in mass. This goes on to reduce the stiffness of the members and elongate the time period making it more flexible, thus attracting lesser force on it. The cracking of members also enhances damping of the members, and a significant amount of strain energy is lost through cracking by generation of kinetic energy and heat dissipation. This considerably attenuates the response. This particular behavior is embedded in design by providing adequate ductility factors in the members.

Thus, the more brittle the structure, the greater the chances that it would undergo a sudden and explosive failure; *such structures should be avoided as far as practicable.*

On the contrary, if appropriate ductility factor is incorporated in the structure, then members subjected to forces more than its capacity are expected to undergo large deformation and it would take significant time before they collapse giving adequate warning.

Response reduction factors (*R*), for some typical commonly used structural arrangements, are furnished in Table 2.8. For more detailed commentary, reader may refer to Table 9 of IS 1893 (2016) part 1.

Before we describe the evaluation of lateral force, it is reiterated that structures designed to resist an earthquake are expected to possess a minimum strength to resist a minor tremor without any damage that can occur frequently. Under moderate tremor, often called design basis earthquake (DBE) the structure is expected to

Table 2.8 Response reduction factors for various structures as per IS 1893

Structure type	Response reduction factor *R*
Ordinary reinforced concrete moment resisting frame	3.0
Special reinforced concrete moment resisting frame detailed to IS 13920	5.0
Steel frame with concentric bracings	3.0
Steel frame with eccentric bracing	5.0
Load bearing masonry wall buildings unreinforced	1.5
Reinforced with horizontal RC band	2.5
Reinforced with horizontal RC band and vertical bars at corners of room and jamb openings	3.0

Table 2.9 Scaling factors for S_a/g for different damping ratios

Damping ratio %	0	2	5	7	10	15	20	25	30
Scaling factor	3.2	1.4	1.0	0.9	0.8	0.7	0.6	0.55	0.5

Intermediate values to be linearly interpolated

withstand this without significant structural damage though some non-structural damage may occur during this and can withstand a major tremor (MCE), *without any collapse*.

The lateral force coefficient (A_h) under earthquake is expressed as

$$A_h = \left(\frac{\left(\frac{Z}{2}\right)\left(\frac{S_a}{g}\right)}{\frac{R}{I}} \right) \tag{2.2.1}$$

Provided that for any structure having $T < 0.1$ s, the value of A_h shall not be taken less than $Z/2$ whatever the value of I/R may be. The above formulation is based on DBE, which is considered as half of MCE.

In case it is decided to design the structure on MCE basis, the factor 2 in denominator of Eq. 2.2.1 is to be replaced by 1.0.

The vertical acceleration, if considered in design, shall be

- 2/3rd of peak ground acceleration (2.5) for buildings and liquid retaining structures;
- 2/3rd of design acceleration (S_a) for bridges and industrial structures, etc.

The total lateral base shear due to earthquake force may be considered as

$$V = A_h \times W \tag{2.2.2}$$

Here

V shear induced at base of the structure;
W weight of the structure;
A_h coefficient as defined in Eq. 2.2.1.

The vertical distribution of force at each floor of a framed structure is then expressed as

$$Q_i = V \frac{W_i h_i^2}{\sum_{i=1}^{n} W_i h_i^2} \tag{2.2.3}$$

Here

Q_i design lateral force at ith level;
W_i weight at ith floor;

h_i height of the ith floor from base;
n total number of floors in the frame.

It may be noted that the curve in Fig. 2.4 is plotted for 5% material damping. There could be many cases where overall damping of the system could be less or greater than 5%, for this the S_a/g value needs to be scaled up or down depending on the damping ratio considered and is as recommended by code as depicted in Table 2.9.

From the above discussion, it is apparent that key to the whole analysis is to correctly estimate the time period of the structure on which the modal spectral acceleration is dependent. Based on modal analysis, earthquake engineers spent a significant amount of time estimating this parameter. This we will certainly discuss in detail subsequently.

For simple structures, code has however given some empirical formulas for estimating time period of building-type structures in fundamental mode. These may be used to estimate at least preliminary force coming on the structure.

For a building-type framed structure, having no infill brick panel fundamental time period may be estimated as

$$\begin{aligned} T &= 0.075h^{3/4} \text{ for reinforced concrete frames} \\ &= 0.085h^{3/4} \text{ for structural steel frames} \end{aligned} \tag{2.2.4}$$

For buildings with RC structural walls

$$T = \frac{0.075h^{3/4}}{\sqrt{A_w}} \geq \frac{0.09h}{\sqrt{D}} \tag{2.2.4.1}$$

where A_w is the total effective area (m^2) of walls in the first story of the building expressed as

$$A_w = \sum_{i=1}^{N_w} \left[A_{wi} \left\{ 0.2 + \left(\frac{L_{wi}}{h} \right)^2 \right\} \right] \tag{2.2.4.2}$$

For all other types of building, including moment resistance frames with in-filled brick panels,

$$T = 0.09h/\sqrt{D} \tag{2.2.5}$$

Here

h height of building in m;
D width of frame in direction in which lateral force need to be computed;
A_{wi} effective cross-sectional area of wall i in first story of building in m^2;
L_{wi} length of structural wall in first story in the considered direction of lateral force in m;

N_w number of walls in the considered direction of earthquake shaking.

The value of L_{wi}/h to be used in Eq. (2.2.4.2) shall not exceed 0.9.

2.2.2 *Response Spectrum Method as Per Eurocode 8*

In many international projects where detailed engineering is carried out in India (low cost engineering work centers), it is observed that engineers are expected to follow codes other than IS-1893.

It is for this reason that the two most popular codes practiced internationally are briefly discussed. Of these, one is certainly Eurocode 8, that is in use in major part of Middle East and North Africa other than, of course, the whole European Union.

Unlike the IS code, that has classified the soil into three types (hard, intermediate and soft), the Eurocode has classified the soil into seven categories: A to E and S_1 and S_2.

While ground classification in IS code is somewhat left to an individual's interpretation, Eurocode is certainly more defined on this and has taken the effect of layering of soil into cognizance, which otherwise could create significant confusion.

The classification of ground type is as given in Table 2.10.

Code recommends classifying the site as per average shear wave velocity for first 30 m depth of soil, in absence of which it recommends to use the SPT (N) value.

If the soil is layered, average shear wave velocity $v_{s,30}$ is to be computed in accordance with the expression

$$v_{s,30} = \frac{30}{\sum_{i=1}^{N} h_i/v_i} \tag{2.2.6}$$

Here h_i and v_i denote thickness and shear wave velocity (at shear strain level of 10^{-5} or less) of the ith layer in a total of N layers existing in the top 30 m.

For sites with ground conditions reflecting S_1 or S_2 special studies for the definition of seismic action are required. For these types of soils and especially for soil type S_2, possibility of soil failure under seismic action is to be taken into account.

2.2.2.1 Horizontal Elastic Response Spectrum

The general characteristics curve of the spectral acceleration as per Eurocode 8 is as shown in Fig. 2.6.

As Europe is a big area and has variable susceptibility to earthquake, the response spectrum curve is controlled by peak ground acceleration (a_{gr}) for which European Union has developed extensive data from previous earthquakes. The peak ground acceleration is usually furnished for A type soil for which the scaling factor is taken as 1.0.

Table 2.10 Ground classification as Eurocode 8

Ground type	Description of profile	Parameters		
		$V_{s,30}$ (m/s)	N (blows/30 cm)	c_u (kPa)
A	Rock or other rock-like geological formation including at most 5 m of weaker material at surface	>800	–	–
B	Deposits of very dense sand, gravel or very stiff clay, at least several tens of meters in thickness, characterized by gradual increase of mechanical property with depth	360–800	>50	>250
C	Deep deposits of dense or medium-dense sand, gravel or stiff clay with thickness from several tens to many hundreds of meters	180–360	15–50	70–250
D	Deposits of loose-to-medium cohesionless soil (with or without some soft cohesive layers) or of predominantly soft to firm cohesive soil	<180	<15	<70
E	A soil profile consisting of a surface alluvium layer with v_s values of type C or D and thickness varying between about 5 and 20 m underlain by stiffer material with v_s > 800 m/s	–	–	–
S_1	Deposits consisting or containing a layer at least 10 m thick of soft clay/silts with high plasticity index (PI > 40) and high water content	<100 (indicative)	–	10–20
S_2	Deposits of liquefiable soils, of sensitive clays or any other soil profile not included in types A to E or S_1	–	–	–

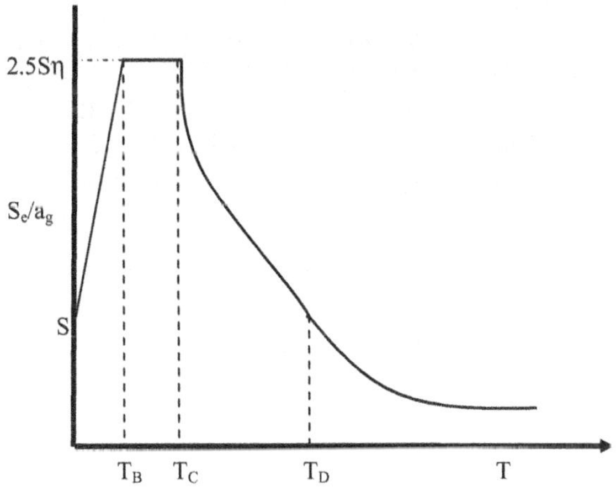

Fig. 2.6 General nature of response spectrum curve as per Eurocode 8

For other types of soil, this is multiplied by a suitable amplification factor. Depending on the structure in consideration, a_{gr} is multiplied by a reliability factor γ when design peak ground acceleration factor $a_g = \gamma \cdot a_{gr}$.

Based on above the elastic response spectrum $S_e(T)$ is expressed by the following expressions.

$$0 < T < T_B : S_e(T) = a_g S\left[1 + \frac{T}{T_B}(2.5\eta - 1)\right] \tag{2.2.7}$$

$$T_B < T < T_C : S_e(T) = a_g S(2.5\eta) \tag{2.2.8}$$

$$T_C < T < T_D : S_e(T) = a_g S(2.5\eta)\left(\frac{T_C}{T}\right) \tag{2.2.9}$$

$$T_D < T < 4.0\,\mathrm{s} : S_e(T) = a_g S(2.5\eta)\left(\frac{T_C T_D}{T^2}\right) \tag{2.2.10}$$

where

$S_e(T)$	elastic response spectrum;
T	time period of vibration of a single-degree oscillator;
a_g	design ground acceleration of type A ground ($a_g = \gamma \cdot a_{gr}$);
T_B, T_C, T_D	are explained in Fig. 2.6.
S	soil factor;
η	damping correction factor with reference value of $\eta = 1$ for 5% damping.

Further defining the seismic parameters, the Eurocode has defined two types of spectrums, Type 1 and Type 2. Type 1 is actually meant for higher seismicity area like Southern Europe, where expected intensity of earthquake is $M_s \cong 7$, while Type 2 is meant for Northern Europe, where seismic activity is not severe and M_s is not greater than 5.5.

Tables 2.11 and 2.12 furnish parameters for generating the response spectrum for both these types.

The value of damping correction factor η, if other than 5%, can be obtained from the expression

Table 2.11 Parameters defining response curve Type 1

Ground type	S	T_B (s)	T_C (s)	T_D (s)
A	1.0	0.15	0.40	2.0
B	1.2	0.15	0.50	2.0
C	1.15	0.20	0.60	2.0
D	1.35	0.20	0.80	2.0
E	1.40	0.15	0.5	2.0

Table 2.12 Parameters defining response curve Type 2

Ground type	S	T_B (s)	T_C (s)	T_D (s)
A	1.0	0.05	0.25	1.2
B	1.35	0.05	0.25	1.2
C	1.5	0.10	0.25	1.2
D	1.8	0.10	0.30	1.2
E	1.6	0.05	0.25	1.2

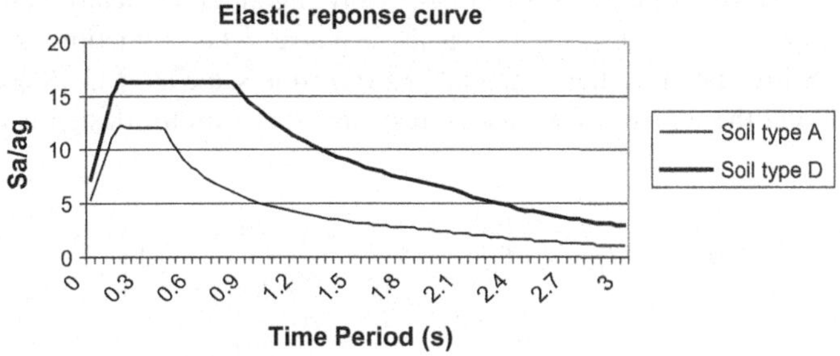

Fig. 2.7 Design response spectrum for soil type A and D

$$\eta = \sqrt{10/(5+\xi)} \leq 0.55 \tag{2.2.11}$$

Here ξ is the viscous damping ratio of the structure.

Example 2.2.1 For a particular site, the peak ground acceleration for type A soil is 3.53 m/s^2. We are building a pipe rack 12 m high, having two bays 6 m each, having importance factor of 1.5; determine the elastic response spectrum and plot the response spectrum curve for Type 1 spectrum considering the structure to have 7% damping. The soil at the site is found to be of type D.

Solution

Correction factor $\eta = \sqrt{10/(5+\xi)} = \sqrt{10/12} = 0.913$

Here $a_g = 1.5 \times 3.53 = 5.3\,\text{m/s}^2$.

Time period of the frame $T = 0.085h^{3/4} = T = 0.085 \times (12)^{3/4} = 0.548$ s. Thus it is seen that the time period lies between T_B and T_C. This gives design spectrum value as

$$S_e(T) = a_g S \cdot 2.5\eta = 5.3 \times 1.35 \times 2.5 \times 0.913 = 16.315\,\text{m/s}^2.$$

The response spectrum curves for soil type A and D and for Type 1 site are as shown in Fig. 2.7.

2.2.2.2 Design Response Spectrum

While discussing the IS code we had stated that basic philosophy of structural design during a strong motion vibration is that it is expected to undergo minor damage and some inelastic deformation.

This inelastic deformation significantly reduces the response by dissipating energy through cracking and frictional heat, damping, etc.

Thus, to avoid explicit inelastic analysis and yet allowing the structure to dissipate energy is mainly achieved through imbibing ductility in the members and/or by other mechanisms and is achieved by performing an elastic analysis based on a reduced response with respect to elastic response. This reduction is generally achieved by incorporating the behavior factor q (q is same as R in IS code).

Considering the q factor, the seismic response spectrum for design is expressed as

$$0 < T < T_B : S_e(T) = a_g S\left[\frac{2}{3} + \frac{T}{T_B}\left(\frac{2.5}{q} - \frac{2}{3}\right)\right] \quad (2.2.12)$$

$$T_B < T < T_C : S_e(T) = a_g S\left(\frac{2.5}{q}\right) \quad (2.2.13)$$

$$T_C < T < T_D : S_e(T) = a_g S\left(\frac{2.5}{q}\right)\left(\frac{T_C}{T}\right) \geq 0.2a_g \quad (2.2.14)$$

$$T_D < T < 4.0\,\mathrm{s} : S_e(T) = a_g S\left(\frac{2.5}{q}\right)\left(\frac{T_C T_D}{T^2}\right) \geq 0.2a_g \quad (2.2.15)$$

The q factor that is dependent on ductility of the frame can be broken down into three categories.

- Ductility class low (DCL), in which there is practically no ductility in the system and the resistance to earthquake loading is basically achieved through strength of the structure rather than its ductile property;
- Ductility class medium (DCM), where quite high level of plasticity is permitted and corresponding design check and detailing requirements need to be abided by;
- Ductility class high (DCH), where very large inelastic deformations are permitted, accompanied by even more onerous and complex design and detailing requirements.

It is normally the DCM and DCH categories that constitute most of the structures that resist seismic force through their ductile property and use q factor to reduce the response.

For some standard systems, recommended values of q are as given in Table 2.13.

Table 2.13 Factors q for common structural systems

Structure type	DCM	DCH
Frame system, dual system, couple wall system	3.0 (α_u/α_1)	4.5 (α_u/α_1)
Uncoupled wall system	3.0	4.0 (α_u/α_1)
Torsionally flexible system	2.0	3.0
Inverted pendulum system	1.5	2.0

The above q factors need to be reduced by 20% in case the structure is irregular in elevation.

The q factors has also been described in terms of frame configuration. Some typical frames and their q factors are described in Table 2.14.

In Table 2.14, the ratio α_u/α_1 is defined as the seismic action that causes development of full plastic mechanism (α_u) to the seismic action of first plastification in the system (α_1).

The value of α_u may be found as the ratio of the base shear on development of a full plastic mechanism according to a pushover analysis to the base shear due to the design seismic action.

As the designer may not consider it worth performing iterations of pushover analysis and design based on elastic analysis to compute the ratio α_u/α_1, code has furnished default values as given in Table 2.14.

Example 2.2.2 For the example furnished in 2.2.1, calculate the design spectral acceleration considering an appropriate q factor for DCM and DCH category.

Solution

Here $a_g = 1.5 \times 3.53 = 5.3\,\text{m/s}^2$.

Time period of the frame $T = 0.085h^{3/4} = T = 0.085 \times (12)^{3/4} = 0.548\,\text{s}$. Thus, it is seen that the time period lies between T_B and T_C. This gives the design spectrum value as

$$S_e(T) = a_g S \frac{2.5}{q}$$

Referring to Table 2.14, for ductility type DCM, $q = 4.0$ as it is a moment resistant frame having two bays

$$\therefore S_e(T) = 5.3 \times 1.35 \times \frac{2.5}{4} = 4.47\,\text{m/s}^2.$$

Thus, the structure is to be designed for $0.45g$.

For DCH category $q = 5 \times 1.3 = 6.5$ as per Table 2.14.

Table 2.14 Factors q based on structural configurations

Structure type	q	
	DCM	DCH
Moment resistant frame 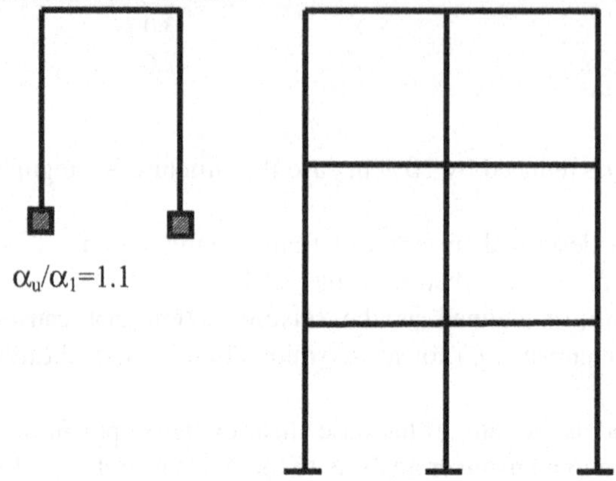	4	5 (α_u/α_1)
Braced frame 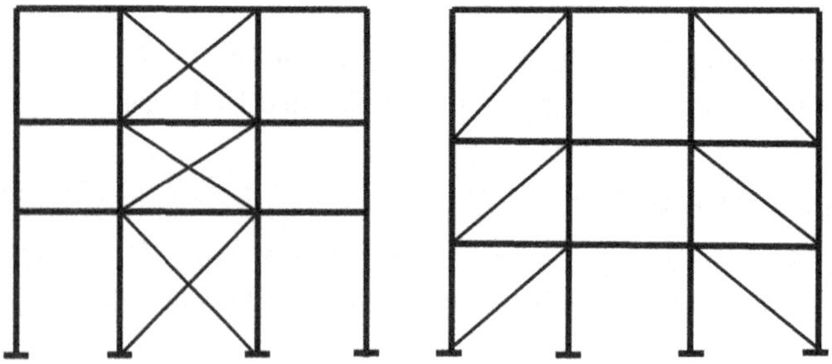	4	4
Single bay concentrically braced frame	2	2.5
Single bay eccentrically braced frame, $\alpha_u/\alpha_1 = 1.2$	4	$5(\alpha_u/\alpha_1)$

$$\therefore S_e(T) = 5.3 \times 1.35 \times \frac{2.5}{6.5} = 2.8\,\mathrm{m/s^2}.$$

Thus, the structure is to be designed for $0.28g$.

2.2.2.3 Design Response Spectrum in Vertical Direction

For design acceleration spectrum in vertical direction, the Eurocode recommends to use Eqs. 2.2.12–2.2.15 but replacing a_g by a_{vg}, where $a_{vg} = 0.9$ times a_g for Type 1 curves and 0.45 times a_g for Type 2 curves. The soil factor S is taken as 1.0 for all the cases, irrespective of soil type.

2.2.2.4 Some Recommended Fundamental Time Period Values as Per Eurocode 8

Like in IS code for some simplified structures especially building-type frames, Eurocode has given expressions for fundamental time period that may be used earthquake-resistant design.

For building-type frames up to 40 m in height

$$T = C_t H^{\frac{3}{4}} \tag{2.2.16}$$

where

- C_t 0.085 for moment resistant steel space frames and 0.075 for moment resistant space concrete frames and for eccentrically braced steel frames;
- C_t 0.050 for all other structures;
- H height of building in meter from the foundation level or top of rigid basement.

It may be noted herein that the Eurocode has not specified clearly, unlike the IS code, whether Eq. 2.2.16 is valid for frames without in-filled brick walls or not.

For shear walled structures

$$C_t = 0.075/\sqrt{A_c} \tag{2.2.17}$$

where A_c is the total effective area of shear walls in bottom story, in m^2.

2.2.2.5 Estimation of Base Shear and Its Distribution as Per Eurocode 8

For an estimated structural period T and corresponding design response spectrum S_e, the base shear is estimated as

$$V = \lambda\left(\frac{W}{g}\right)S_e \tag{2.2.18}$$

In Eq. 2.2.18 the term λ is considered as 0.85 for frames more than two stories with $T < T_c$ and is taken as 1.0 otherwise.

The nodal forces at each floor level are distributed as

$$Q_i = V\frac{W_i h_i}{\sum_{i=1}^{n} W_i h_i} \tag{2.2.19}$$

Here n is the number of floors in the frame, W_i is the weight of the ith floor, and h_i is the corresponding height from the base.

It may be noted that while IS code takes a parabolic distribution of forces along the height, Eurocode takes a linear distribution of forces along vertical direction.

2.2.3 *Response Spectrum Method as Per UBC 97*

American practice for estimation of earthquake like Eurocode 8 has well digressed beyond its geographic boundary. In many countries outside USA where important projects are executed, it is not unusual to see a client asking to carry out earthquake analysis as per American practice, especially if the client is of American origin like Chevron, Halliburton, General electric etc.

Beyond the geographic boundary of USA, it is observed that in many cases analysis based on UBC 97 is still popular, while within USA, an analysis based on ASCE-07 has been in vogue. As such, we will briefly look into UBC 97 code hereafter.

2.2.3.1 Analysis as Per UBC 97

As per Uniform Building Code 1997, like IS code, the country is broken up into different zones of seismic intensity and not dependent on the local peak ground acceleration though provision for the same has also been kept when it is felt that it is crucial to carry out the analysis as per site response spectra.

Considering that most of the countries (especially in Asia) carry out seismic analyses based on zonal demarcation and are quite comfortable with this philosophy, the design as per UBC 97 is quite popular outside USA (particularly for multinational projects), where earthquake analysis of structures as per local guidelines are in abeyance or found dubious.

As per UBC, depending on intensity of earthquake, the country is divided into five zones. These are zone 1, 2A, 2B, 3 and 4, in terms of progressive intensity. The intensity of peak ground acceleration is depicted by the zone factor (Z), as given in Table 2.15.

Based on the above, depending on the type of soil existing at site, code has demarcated the response spectrum into zones that are controlled by the ground acceleration coefficient (C_a) and ground velocity coefficient (C_v).The soil classification unlike IS code is quite specific, as given in Table 2.16.

Table 2.15 Zone demarcation as per UBC 97

Zone	1	2A	2B	3	4
Z	0.075	0.15	0.2	0.3	0.4

Table 2.16 Soil classification as per UBC 97

Soil profile type	Generic description of soil	Average shear wave velocity for upper 30 m in m/s
S_A	Hard rock	1500
S_B	Rock	760–1500
S_C	Very dense soil or soft rock	360–760
S_D	Stiff to intermediate soil	180–360
S_E	Soft soil	≤ 180
S_F	Soil requiring site-specific evaluation	

For soil type S_F, the following guidelines have been furnished by the code

- Very thick soft/medium stiff clay of thickness >120 ft;
- Clay with very high plasticity of thickness >25 ft and plasticity index (PI) >75;
- Peat and/or highly organic clay of thickness >10 ft;
- Soil vulnerable to potential failure or collapse under seismic loading, such as liquefiable soil, quick and highly sensitive clay, collapsible weakly cemented soil.

Based on the above, the acceleration-controlled part of spectrum (C_a) is shown below, in Table 2.17.

Part of the spectrum that is controlled by the ground velocity (C_v) depending on the soil type and zone is depicted in Table 2.18.

It will be observed that in Zone 4 which is the most severe, additional terms N_a and N_v come into the picture and this needs some explanation.

Zone 4 as per UBC is expected to experience earthquake of maximum severity (like in California).

Table 2.17 Seismic coefficient C_a

Soil type	Seismic zone factor				
	Z = 0.075	Z = 0.15	Z = 0.2	Z = 0.3	Z = 0.4
S_A	0.06	0.12	0.16	0.24	$0.32N_a$
S_B	0.08	0.15	0.20	0.30	$0.40N_a$
S_C	0.09	0.18	0.24	0.33	$0.40N_a$
S_D	0.12	0.22	0.28	0.36	$0.44N_a$
S_E	0.19	0.30	0.34	0.36	$0.36N_a$

Table 2.18 Seismic coefficient C_v

Soil type	Seismic zone factor				
	Z = 0.075	Z = 0.15	Z = 0.2	Z = 0.3	Z = 0.4
S_A	0.06	0.12	0.16	0.24	$0.32N_v$
S_B	0.08	0.15	0.20	0.30	$0.40N_v$
S_C	0.13	0.25	0.32	0.45	$0.56N_v$
S_D	0.18	0.32	0.40	0.54	$0.64N_v$
S_E	0.26	0.50	0.64	0.84	$0.96N_v$

In this case, the proximity of earthquake fault to the site in question, nature of fault, types of waves emanating from the source, etc., all have a bearing on the shock to be expected from such source and this is what is catered to by the factors N_a and N_v.

For this, the seismic source factor has been divided into three categories as defined in Table 2.19.

Based on the above definition of seismic source type, the acceleration controlling factor (N_a) and velocity controlling factor (N_v) are as depicted in Tables 2.20 and 2.21.

Considering the above data, the response curve is plotted according to the following guideline as shown in Fig. 2.8.

In Fig. 2.8, $T_s = C_v/2.5C_a$ and $T_0 = 0.2T_s$

The total design base shear V in a given direction is then computed from the equation

$$V = \frac{C_v I}{RT} W \tag{2.2.20}$$

Table 2.19 Definition of seismic source

Seismic source type	Description	Seismic source definition	
		Maximum moment magnitude M	Slip rate SR (mm/yr)
A	Faults that are capable of producing large magnitude events and that have a high rate of seismic activity	$M \geq 7.0$ and	SR ≥ 5
B	All faults other than type A or C	$M \geq 7.0$ and $M < 7.0$ and $M \geq 6.5$ and	SR < 5 SR > 2 SR < 2
C	Faults that are not capable of producing large magnitude earthquakes and which have a relatively low rate of seismic activity	$M < 6.5$	SR ≤ 2

Table 2.20 Seismic factor N_a

Seismic source type	Closest to known seismic source		
	≤2 km	5 km	≥10 km
A	1.5	1.2	1.0
B	1.3	1.0	1.0
C	1.0	1.0	1.0

Table 2.21 Seismic factor N_v

Seismic source type	Closest to known seismic source			
	≤2 km	5 km	≥10 km	≥15 km
A	2.0	1.6	1.2	1.0
B	1.6	1.2	1.0	1.0
C	1.0	1.0	1.0	1.0

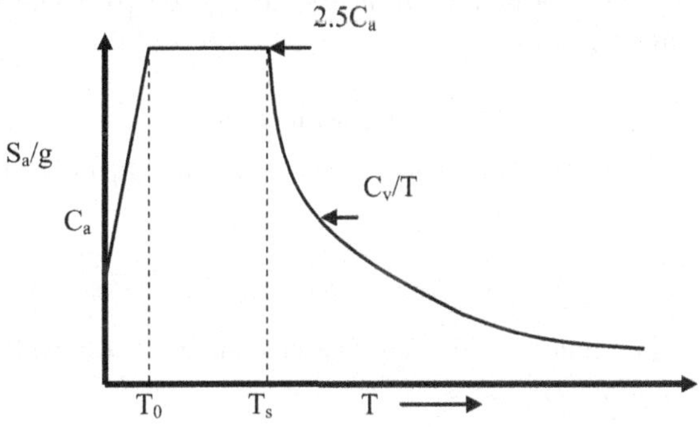

Fig. 2.8 Response spectrum curve as per UBC 97

Here

- I importance factor (having same concept as described in IS code);
- R response reduction factor (same as IS code in terms of concept);
- T fundamental time period of vibration in seconds;
- W total seismic load.

For short period, i.e., when, $T < C_v/2.5C_a$ the base shear is expressed as

$$V = \frac{2.5C_a I}{R} W \tag{2.2.21}$$

The base shear should not be less than

$$V = 0.11C_aIW \tag{2.2.22}$$

Over and above this, in zone 4 the base shear must be greater than

$$V = \frac{0.8ZN_vI}{R}W \tag{2.2.23}$$

For the value of I and R, extensive commentary is there in UBC code in Table 16K for I, and Table 16 N, O and P for response reduction factor R, and may be referred to.

2.2.3.2 Time Period of a Structure as Per UBC 97

The fundamental time period of a building-type framed structure is expressed as by the same formula as furnished in Eq. 2.2.16 where the value of C_t is as defined hereafter

$$\begin{aligned} C_t &= 0.0853 \text{ for steel resisting frame} \\ &= 0.0731 \text{ for concrete moment resisting frame or eccentrically braced frame} \\ &= 0.0488 \text{ for all other type of frames.} \end{aligned} \tag{2.2.24}$$

For frames with masonry or concrete shear walls the time period is same as Eq. (2.2.17).

In this context, UBC 97 has made an interesting restriction. If the time period of a building frame is obtained by other methods, like Rayleigh's formula or dynamic analysis, then the value of T in zone 4 cannot be over 30% greater than that determined by Eq. 2.2.16 and in zone 1, 2 and 3 this shall not be 40% more than that computed as per Eq. 2.2.16.

IS 1893 (2016) has also given recommendation in similar line in Clause 7.8.2 of the code.

It states "*Dynamic analysis may be performed by either time history or response spectrum method, however in such case the design base shear* V_{bd} *(where the subscript bd stands for base shear based on dynamic analysis shall be compared against that computed based on Eq.* 2.2.4 *which is* V_b*. If* V_b *is less than* V_{bd}*, then all responses like displacements, member forces, base reactions shall be multiplied by a factor* V_b/V_{bd}*. In other words, forces and responses are upgraded to the level of* V_b.

This particular clause has created significant confusion in many projects, especially pertaining to industrial structures.

Considering the fact that the empirical time period gives higher base shear, in many case clients engineer has insisted to take advantage of this clause and compel the contractor to design for higher base shear, even when exact dynamic response based on computer analysis has given a much lower value.

The author gives his own interpretation on this though code has yet remained taciturn on this issue.

Both UBC 97 and IS-1893 Part 1 are devised for building or building-type structures.

Majority of these structures will have masonry or hollow concrete blocks as infill that acts as internal/external walls. It has been observed that during earthquake vibration these walls, though are considered to be non-structural members, yet go on to significantly stiffen the structure when they act like strut members. However, in most of the case an analyst will ignore this strutting effect of the infill wall while designing the building frame and only considers its mass. This goes on to increase the mass however stiffness remaining unaltered makes the structure more flexible than in realty, thus giving a lower value of S_a/g than actually what would the system experience.

Thus to safeguard against such oversight or misinterpretation, code has introduced this clause. Thus, when the stiffening effect of infill panel is taken into cognizance, this clause is not valid and neither is it valid for open structures like pipe racks, boiler supporting or tech-structures.[3]

2.2.3.3 Estimation of Base Shear and Distribution of Force as Per UBC 97

The base shear is computed by Eqs. 2.2.20–2.2.23. This is distributed over the height of structure as force at each level F_i, plus and extra force F_t at top of the structure. Thus,

$$V = F_t + \sum_{i=1}^{n} F_i \qquad (2.2.25)$$

The distribution of force along the height is given by the expression.

$$Q_i = (V - F_t)\frac{W_i h_i}{\sum_{i=1}^{n} W_i h_i} \qquad (2.2.26)$$

Here

$$\begin{aligned} F_t &= 0.07TV \leq 0.25V \quad &\text{if } T > 0.7\,\text{s} \\ F_t &= 0.0 \quad &\text{if } T < 0.7\,\text{s}. \end{aligned} \qquad (2.2.27)$$

[3]Refer Chap. 9 for further detail.

Fig. 2.9 A four-storied frame with load to be analyzed for seismic load

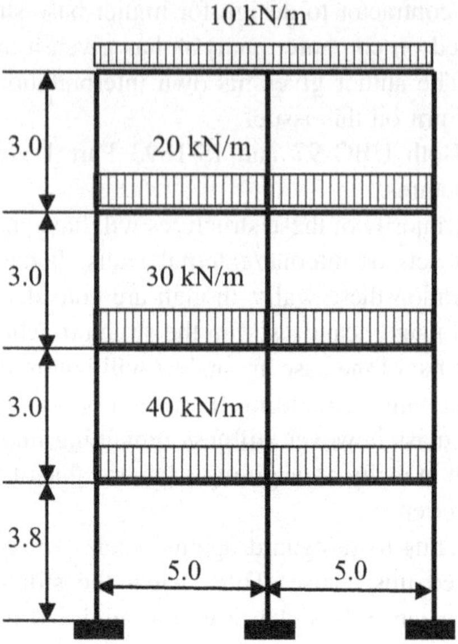

From Eq. 2.2.27 it is observed that this extra force comes into play when the structure is quite flexible. In such a case, in higher modes there is a whip lash effect that takes place in higher modes and generates an additional force at top of a structure and the force F_t caters to it. When the structure is sufficiently rigid, this is considered as zero.

It may again be observed that in UBC 97 like the Eurocode the floor force distribution is linear.

Example 2.2.3 A reinforced concrete four-storied frame having load at each floor (including structure self-weight) is as shown in Fig. 2.9. The frame is situated in zone IV as per IS code and has soil of intermediate nature having SPT value 28. Determine the base shear and floor forces as per IS 1893, Eurocode 8 and UBC 97. Consider frames to have medium ductility.

Solution
Analysis based on IS-1893 (2016) Part-1

$$T = 0.075 \times h^{3/4}, \text{here } h = 12.8\,\text{m}$$

$$\rightarrow T = 0.075 \times (12.8)^{3/4} = 0.507\,\text{s}$$

Considering intermediate soil and Zone IV $S_a/g = 2.5$, $Z = 0.24$, $I = 1.0$, $R = 3.0$ for ordinary moment resistance frame.

Thus $A_h = \left(\frac{ZI}{2R}\right)\left(\frac{S_a}{g}\right) = \left(\frac{0.24 \times 1}{2 \times 3}\right) \times 2.5 = 0.1$

Total weight of the structure = $(40 + 30 + 20 + 10) \times 10 = 1000$ kN

Thus, the base shear = $A_h \times W = 100\,\text{kN}$
Thus, the floor forces are computed as
At EL 12.8

$$F_1 = 100 \times \frac{10 \times 10 \times (12.8)^2}{100 \times (12.8)^2 + 200 \times (9.8)^2 + 300 \times (6.8)^2 + 400 \times (3.8)^2}$$
$$= 100 \times \frac{16{,}384}{55{,}240} = 29.7\,\text{kN}$$

At EL 9.8

$$F_2 = 100 \times \frac{20 \times 10 \times (9.8)^2}{100 \times (12.8)^2 + 200 \times (9.8)^2 + 300 \times (6.8)^2 + 400 \times (3.8)^2}$$
$$= 100 \times \frac{19{,}208}{55{,}240} = 34.7\,\text{kN}$$

At EL 6.8

$$F_3 = 100 \times \frac{30 \times 10 \times (6.8)^2}{100 \times (12.8)^2 + 200 \times (9.8)^2 + 300 \times (6.8)^2 + 400 \times (3.8)^2}$$
$$= 100 \times \frac{13{,}872}{55{,}240} = 25.11\,\text{kN}$$

At EL 3.8

$$F_2 = 100 \times \frac{40 \times 10 \times (3.8)^2}{100 \times (12.8)^2 + 200 \times (9.8)^2 + 300 \times (6.8)^2 + 400 \times (3.8)^2}$$
$$= 100 \times \frac{5776}{55{,}240} - 10.4\,\text{kN}$$

Analysis based on Eurocode 8

Based on Eurocode the first thing that needs to be known is what is peak ground acceleration? Keeping parity with IS code let it be $a_g = 0.24g$ (since $Z = 0.24$ in earlier case). This gives $a_g = 2.35$ m/s^2.

Considering SPT value of 28, soil classification in this case is type C. Considering Type 1 response curve $S = 1.15$, $T_B = 0.2$ s, $T_C = 0.8$ s and $T_D = 2.0$ s.

The fundamental time period of the frame remains same as IS code, i.e., $T = 0.507$ s.

Thus for this case $T_B < T < T_C$ which gives

$$S_e(T) = a_g S\left(\frac{2.5}{q}\right)$$

Referring to Table 2.14 we see that $q = 4$ for medium ductility thus

$$S_e(T) = 2.35 \times 1.15\left(\frac{2.5}{4}\right) = 1.689\,\text{m/s}^2.$$

Thus, the base shear $V = \lambda\left(\frac{W}{g}\right)S_e(T) = 0.85 \times \left(\frac{1000}{9.81}\right) \times 1.689 = 146.34\,\text{kN}$

Thus, the floor forces are computed as

At EL 12.8

$$\begin{aligned} F_1 &= 146 \times \frac{10 \times 10 \times (12.8)}{100 \times (12.8) + 200 \times (9.8) + 300 \times (6.8) + 400 \times (3.8)} \\ &= 146 \times \frac{1280}{6800} = 27.5\,\text{kN} \end{aligned}$$

At EL 9.8

$$\begin{aligned} F_2 &= 146 \times \frac{20 \times 10 \times (9.8)}{100 \times (12.8) + 200 \times (9.8) + 300 \times (6.8) + 400 \times (3.8)} \\ &= 146 \times \frac{1960}{6800} = 42.1\,\text{kN} \end{aligned}$$

At EL 6.8

$$\begin{aligned} F_3 &= 146 \times \frac{30 \times 10 \times (6.8)}{100 \times (12.8) + 200 \times (9.8) + 300 \times (6.8) + 400 \times (3.8)} \\ &= 146 \times \frac{2040}{6800} = 43.8\,\text{kN} \end{aligned}$$

At EL 3.8

$$\begin{aligned} F_2 &= 146 \times \frac{40 \times 10 \times (3.8)}{100 \times (12.8) + 200 \times (9.8) + 300 \times (6.8) + 400 \times (3.8)} \\ &= 146 \times \frac{1520}{6800} = 32.63\,\text{kN} \end{aligned}$$

Analysis based on UBC 97

Comparing the Z value we see that as per UBC the zone is somewhere between 2 and 3.

Considering that the SPT value is 28, soil classification in this case is of type S_D as per UBC.

Referring to Tables 2.17 and 2.18 we have, $C_a = 0.312$ and $C_v = 0.456$.

The fundamental time period of the frame remains same as for IS code, i.e., $T = 0.507$ s.

Thus, the base shear $V = \frac{C_v I}{RT} W$

For building concrete frame $R = 5.6$

Thus, $V = \frac{0.456 \times 1.0}{5.6 \times 0.507} \times 1000 = 161\,\text{kN}$.
For $T < C_v/2.5C_a$ the base shear is expressed as

$$V = \frac{2.5C_a I}{R} W$$

Here $\frac{C_v}{2.5C_a} = \frac{0.456}{2.5 \times 0.312} = 0.584\,\text{s}$; thus, $T = 0.507$ is less than this.
Hence

$$V = \frac{2.5 \times 0.312 \times 1.0}{5.6} \times 1000 = 139\,\text{kN}.$$

Thus, the floor forces are computed as
At EL 12.8

$$\begin{aligned} F_1 &= 139 \times \frac{10 \times 10 \times (12.8)}{100 \times (12.8) + 200 \times (9.8) + 300 \times (6.8) + 400 \times (3.8)} \\ &= 139 \times \frac{1280}{6800} = 26.2\,\text{kN} \end{aligned}$$

At EL 9.8

$$\begin{aligned} F_2 &= 139 \times \frac{20 \times 10 \times (9.8)}{100 \times (12.8) + 200 \times (9.8) + 300 \times (6.8) + 400 \times (3.8)} \\ &= 139 \times \frac{1960}{6800} = 40\,\text{kN} \end{aligned}$$

At EL 6.8

$$\begin{aligned} F_3 &= 139 \times \frac{30 \times 10 \times (6.8)}{100 \times (12.8) + 200 \times (9.8) + 300 \times (6.8) + 400 \times (3.8)} \\ &= 139 \times \frac{2040}{6800} = 41.7\,\text{kN} \end{aligned}$$

At EL 3.8

$$\begin{aligned} F_2 &= 139 \times \frac{40 \times 10 \times (3.8)}{100 \times (12.8) + 200 \times (9.8) + 300 \times (6.8) + 400 \times (3.8)} \\ &= 139 \times \frac{1520}{6800} = 31.7\,\text{kN} \end{aligned}$$

The results are tabulated below for ready reference and comparison

Elevation	IS code	Eurocode	UBC 97
12.8	30	28	26
9.8	35	42	40
6.8	25	44	42
3.8	10	32	31
Base shear	100	146	139

2.3 Dynamic Theory of Structures Related to Earthquake Engineering

Earthquake loading is time-dependent in nature. However, till now, what we have demonstrated is an analysis of structures based on pseudo-static loading, where time period of a structure was considered on an empirical basis, as suggested in various codes.

However, not all the structures in real-world fit into the category as suggested in codes, and even it is so, distribution of stiffness, mass, etc., could well be quite irregular that compels an engineer to carry out a detailed dynamic analysis.

In fact, one of the major tasks for a civil engineer carrying out an earthquake-resistant design is to correctly estimate the time period of a structure, which remains the key parameter that governs the response spectrum based modal analysis.

Readers who are conversant with basic theory of mechanical vibration/structural dynamics, can well skip this section and go to the next section. However, for those who are new to the topic (rookies), it would certainly be worth the effort to go through this section carefully to understand how a system behaves under random loads like that an earthquake motion.

For this, we start with a body having single degree of freedom (i.e., it has a single mass and stiffness).

2.3.1 Free Vibration of a Body with a Single Degree of Freedom

If a body is subjected to an external force acting for a short instant of time, and then removed, we know that the body keeps on vibrating about its mean position just like a simple pendulum. The body is then said to be undergoing a free vibration as no other external force is acting on it (Fig. 2.10).

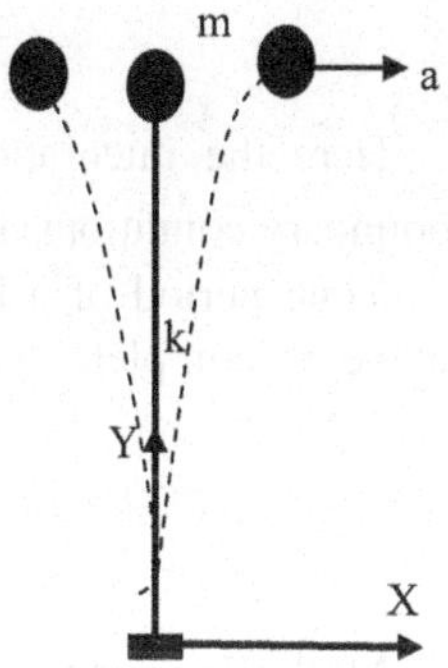

Fig. 2.10 A single-degree oscillator under free vibration

In this case, the acceleration induced to the body mass must be in equilibrium with its elastic deformation as no other external force acts on it. In other words,

$$ma + kx = 0 \tag{2.3.1}$$

Here m = mass of the body, a = acceleration induced to it, k = stiffness of the body and x = elastic deformation.

Equation 2.3.1 can be thus expressed as (Biggs 1964)

$$m\ddot{x} + kx = 0 \tag{2.3.2}$$

where

$\ddot{x}$ = acceleration of the mass, which is a second derivative of displacement with respect to time.

Considering $\omega_{\mathrm{n}}^2 = k/m$ and $D = \mathrm{d}/\mathrm{d}t$ an operator, Eq. 2.3.2 can be expressed as

$$\left(D^2 + \omega^2\right)x = 0 \tag{2.3.3}$$

$$\rightarrow D = \pm i\omega \tag{2.3.4}$$

from which it can be inferred that solution for x can be expressed as

$$x = C_1 \mathrm{e}^{i\omega t} + C_2 \mathrm{e}^{-i\omega t} \tag{2.3.5}$$

Considering $\mathrm{e}^{\pm i\theta} = \cos\theta \pm i\sin\theta$ we have

$$x = C_1(\cos\omega t + i\sin\omega t) + C_2(\cos\omega t - i\sin\omega t) \tag{2.3.6}$$

$$\rightarrow x = (C_1 + C_2)\cos\omega t + i(C_1 \quad C_2)\sin\omega t \tag{2.3.7}$$

Considering $A = C_1 + C_2$ and $B = i(C_1 - C_2)$, other sets of constants, the complete solution can be expressed as

$$x = A\cos\omega t + B\sin\omega t \tag{2.3.8}$$

Here the integration constants A and B are to determined from appropriate boundary conditions and $\omega_n = \sqrt{k/m}$ is called the *natural frequency* of the system.

Time period of a body, which is defined as the time required by the body to make on complete cycle of vibration, is thus expressed as

$$T = \frac{2\pi}{\omega_n} = 2\pi\sqrt{\frac{m}{k}} \tag{2.3.9}$$

Multiplying the numerator and the denominator within the square root by g, acceleration due to gravity, Eq. 2.3.9 can also be expressed as

$$T = 2\pi\sqrt{\frac{\delta_{st}}{g}} \tag{2.3.10}$$

where δ_{st} = static deflection of the system under the load $m \times g$.

Example 2.3.1 A simply supported beam has self-weight of 67.5 kg/m, and spans a distance of 4.0 m. A body of weight 2.56 kN drops from a height of 1.5 m by accident at its mid span. What is moment and shear induced in the body, given $I = 0.0045\ \text{m}^4$ and $E = 2.85 \times 10^7\ \text{kN/m}^2$.

Solution

Self-weight of beam = $0.675 \times 4 = 2.7\,\text{kN}$.

For the free falling body, considering, $v^2 = u^2 + 2gh$, where $u = 0$ and $h = 1.5$ m

$$v = 2\sqrt{gh} = 2\sqrt{9.81 \times 1.5} = 7.67\,\text{m/s}.$$

Now by conservation of momentum

$$\frac{W_{beam} + W_{body}}{g} V_b = \frac{W_{body}}{g} V_{body}, \text{ which gives}$$

$$V_b = \left(\frac{2.56}{5.26}\right) \times 7.67 = 3.73\,\text{m/s}.$$

Deflection due to self-weight of beam can be expressed as

$$\delta_1 = \frac{5wL^4}{384EI} = \frac{5 \times 0.675 \times (4)^4}{384 \times 2.85 \times 10^7 \times 0.0045} = 1.754 \times 10^{-5}\,\text{m}$$

Static deflection due to dropped load that moves in phase with the beam is

$$\delta_2 = \frac{PL^3}{48EI} = \frac{2.56 \times (4)^3}{48 \times 2.85 \times 10^7 \times 0.0045} = 2.66 \times 10^{-5}\,\text{m}$$

$$\text{Total static deflection } \delta_{st} = \delta_1 + \delta_2 = 4.414 \times 10^{-5}\,\text{m}$$

Considering

$$T = 2\pi\sqrt{\frac{\delta_{st}}{g}} = 2\pi\sqrt{\frac{4.414 \times 10^{-5}}{9.81}}$$
$$= 0.0133\,\text{s}$$
$$\rightarrow \omega = 2\pi/T = 471\,\text{rad/s}$$

We had seen (Eq. 2.3.8) that complete solution to deflection is expressed as

$$x - A\cos\omega t + B\sin\omega t$$

Now at $t = 0$, $x = 0 \rightarrow A = 0$ and at $t = 0$, $\dot{x} = V_b \rightarrow B = V_b/\omega$
Thus $x = \frac{V_b}{\omega}\sin\omega t = \frac{3.73}{471}\sin 471t = 7.912 \times 10^{-3}\sin 471t$
Considering maximum value of $\sin 471t = 1.0$, the maximum amplitude of dynamic amplitude is 7.912 mm.

Thus, we see that the dynamic amplitude is magnified by $\text{AF} = \frac{7.912}{4.414} \times 100 = 179$ times.

Considering $\Delta = \frac{PL^3}{48EI}$, equivalent load to generate a deflection of 7.912 mm is

$$P = \frac{48EI\Delta}{L^3} = \frac{48 \times 2.85 \times 10^7 \times 0.0045 \times 7.912 \times 10^{-3}}{(4)^3} = 760\,\text{kN}.$$

Thus, the maximum moment $M = \frac{PL}{4} = \frac{760 \times 4}{4} = 760\,\text{kN m}$
Maximum shear force = $0.5 \times 760 \cong 380\,\text{kN}$.
From the above problem we make some important conclusions

- For impulsive- or transient-type load, free vibration characteristics play an important role in estimation of stress within the body.
- Dynamic load can induce very high deformation within the body compared to static load of same magnitude.
- Though load is impulsive in nature, yet the deformation characteristics are converted into a harmonic output $x = (V_b/\omega) \times \sin\omega t$, which is quite amenable to analytical solution.

2.3.2 *Free Vibration of a Body Having Single Degree of Freedom with Damping*

Before we get into the mathematical derivation of its properties, a brief description is needed about the physical nature of damping to show that what it does is important here.

Damping can be attributed as nature's line of defense to this universe, protecting it from destruction. Our world (including other planets in Milky Way) all has certain mass and stiffness and is continuously bombarded by various forces (dynamic in nature) like wind, waves, earthquakes, gravitational pull, electromagnetic forces and sundry others. Many of them may be out of phase or even in resonance with the planets. It is the property of damping that is inherent in nature, prevents the universe from exploding and maintains the semblance of stability within the system.

Narrowing our focus from such broad philosophical level to our day-to-day world let us consider the case of a pendulum vibrating in air. At its highest position of motion (where it momentarily comes to rest), it gains a height say h from its mean position when the potential energy gained by the bob is expressed as mgh. From this position, as it descends down and the potential energy gets converted to kinetic energy, when at its mean position it gains the maximum velocity of $v = \sqrt{2gh}$.

By conservation of energy under ideal conditions, the pendulum once starts vibrating will keep on doing so till the end of time (this is called perpetual motion). However, in real world we see that it does not happen like that.

Like all good things come to an end, the vibration too progressively dies down and after certain period of time the pendulum comes to a stop.

The question is why does it happen so?

It happens so because damping prevalent in air slowly extricates energy from the body, progressively decaying its amplitude of vibration and finally brings it to rest.

In case of any dynamic load like wind or earthquake force that acts on a body, the material damping inherent within the body dissipates away the energy and attenuates the response with course of time.

There are different types of damping, like material, geometric/radiation damping, prevalent in nature. We will talk about them in more detail in various subsequent chapters. We will conclude this discussion here only by saying that this damping is usually considered as viscous in nature, mathematically represented by a dashpot in which the damping force is represented as $F_{\mathrm{d}} = c \cdot v$, where F_{d} is a damping resistive force, c is the coefficient of damping that is a function of the velocity the body gains while experiencing a dynamic load.

Referring to Fig. 2.11, presenting a body under free vibration where damping is taken into consideration, the equation of equilibrium is expressed as

$$m\ddot{x} + c\dot{x} + kx = 0 \tag{2.3.11}$$

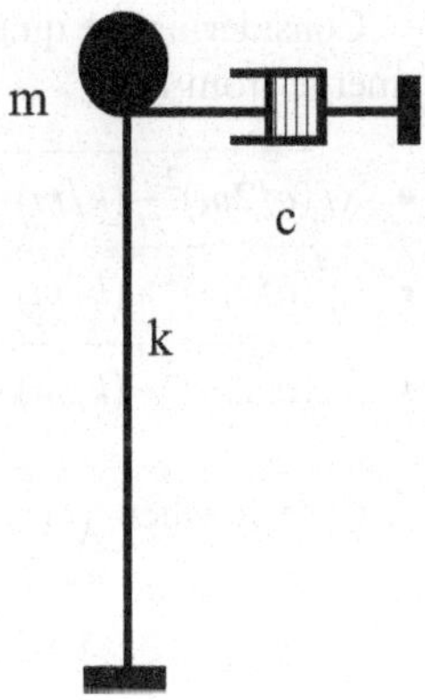

Fig. 2.11 A single-degree oscillator under-damped free vibration

Considering $x = e^{st}$, Eq. 2.3.11 can be expressed as (Mukherjee 2002)

$$ms^2 + cs + k = 0 \tag{2.3.12}$$

Solving Eq. 2.3.12 gives

$$s_{1,2} = \frac{-c \pm \sqrt{c^2 - 4km}}{2m} \tag{2.3.13}$$

Thus two roots of the solutions are

$$s_1 = \frac{-c + \sqrt{c^2 - 4km}}{2m} \tag{2.3.14}$$

and

$$s_2 = \frac{-c - \sqrt{c^2 - 4km}}{2m} \tag{2.3.15}$$

Thus, the general solution is expressed as

$$x = C_1 e^{s_1 t} + C_2 e^{s_2 t} \tag{2.3.16}$$

$$\rightarrow x = e^{-\left(\frac{c}{2m}\right)t}\left(C_1 e^{t\sqrt{\left(\frac{c}{2m}\right)^2 - \left(\frac{k}{m}\right)}} + C_2 e^{-t\sqrt{\left(\frac{c}{2m}\right)^2 - \left(\frac{k}{m}\right)}}\right) \tag{2.3.17}$$

Considering the quantities below, the square root as the discriminant, three cases emerge from it

- $\sqrt{(c/2m)^2-(k/m)} > 0$
- $\sqrt{(c/2m)^2-(k/m)} = 0$
- $\sqrt{(c/2m)^2-(k/m)} < 0$

Case-1, when $\sqrt{(c/2m)^2-(k/m)} > 0$, let this value be α. Then

$$x = e^{-\left(\frac{c}{2m}\right)t}(C_1 e^{\alpha t} + C_2 e^{-\alpha t}) \tag{2.3.18}$$

In such case, it is called an *overdamped* system and no vibration is possible.

Case-2, when $\sqrt{(c/2m)^2-(k/m)} = 0$. In such case

$$x = e^{-\left(\frac{c}{2m}\right)t}(C_1 e^{\alpha t} + C_2 t e^{-\alpha t}) \text{ and this gives}$$

$$x = x_0 e^{-\left(\frac{c}{2m}\right)t} \quad [\text{Here } x_0 = C_1 + C_2 \text{ a constant}] \tag{2.3.19}$$

Again as $\sqrt{(c/2m)^2-(k/m)} = 0$ we have

$$\frac{c^2}{4m^2} = \frac{k}{m}$$

$\rightarrow c = 2\sqrt{km}$, is defined as the critical damping c_c.

Thus, the damping ratio for a structural system is expressed as

$$\zeta = c/c_c \leq 1.0 \tag{2.3.20}$$

ζ is usually known as the damping ratio of the system, and the system is termed as *critically damped* system.

Case-3, when $\sqrt{(c/2m)^2-(k/m)} < 0$. In such case

$$x = e^{-(c/2m)t}\left[C_1 e^{i\left(\sqrt{k/m-(c/2m)^2}\right)t} + C_2 e^{-i\left(\sqrt{k/m-(c/2m)^2}\right)t}\right] \tag{2.3.21}$$

Considering $\omega_n = \sqrt{k/m}$ and $e^{\pm i\theta} = \cos\theta \pm i\sin\theta$, Eq. (2.3.21) can be finally expressed as

$$x = e^{-\zeta\omega_n t}[A \sin \omega_d t + B \cos \omega_d t] \qquad (2.3.22)$$

In Eq. (2.3.22) ω_d is called the *damped natural frequency* and is expressed as $\omega_d = \omega_n\sqrt{1 - \zeta^2}$. When the damping ratio is less than critical damping, the system is called *under-damped* system.

Example 2.3.2 Repeat the problem worked out in Example 2.3.1 considering the beam having a damping ratio of 5% of critical damping.

Solution
The general solution to the displacement is as stated in Eq. (2.3.22) and is expressed as

$$x = e^{-\zeta\omega_n t}[A \sin \omega_d t + B \cos \omega_d t].$$

Here, considering the boundary condition at t = 0, x = 0, we have B = 0. Again considering at t = 0, dx/dt = V_b we have
$A = \frac{V_b}{\omega_n\sqrt{1-\zeta^2}}$, and this gives
$x = \frac{V_b}{\omega_n\sqrt{1-\zeta^2}} e^{-\zeta\omega_n t} \sin\left(\omega_n\sqrt{1-\zeta^2}\right)t$. Substituting the appropriate values we finally have

$$x = \frac{3.73}{470.4} e^{-23.55t} \sin(470.4)t$$

The amplitude of vibration when plotted against time is as shown below (Fig. 2.12).

It is seen that due to presence of damping the amplitude decays with time.

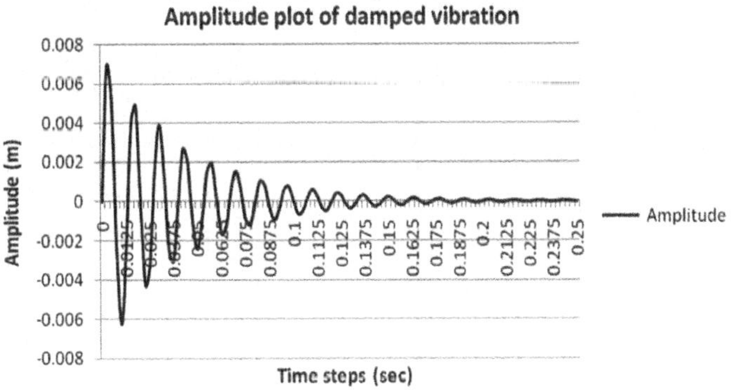

Fig. 2.12 Amplitude of vibration plotted against time

2.3.3 Characteristics and Response of a Body Under Transient or Impulse Loading

If a loading acts on a body for a very short duration of time and when the duration is significantly small compared to its natural time period, the load is called an *impulsive force or just an impulse*. The impulse $\hat{F}$ acting on a mass will result in sudden change in its velocity which is equal to $\hat{F}/m$ without any appreciable change in displacement. A mass m, which is initially at rest, will gain velocity V_0 and can be expressed in the form (Chowdhury and Dasgupta 2008)

$$\hat{F} = \int_0^{\Delta t} F(t)\mathrm{d}t = mV_0 \tag{2.3.3.1}$$

We had shown, vide Example 2.3.2, that the damped response in such case is expressed by Eq. (2.3.22). The equation implies that though the force acts for a small duration of time yet it gets converted to a sinusoidal response that decays with time.

Shown in Fig. 2.13 is a generalized force versus time plot acting on a single-degree oscillator. The force diagram can then divided into a large number of rectangular slices represented by $F(\tau)\mathrm{d}\tau$ that represents the single impulse. Summing up all such impulses (which added together gives the force time diagram), the overall response of the system can be computed.

Here τ defines the time at which a typical impulse acts and t defines the time at which the response x (this could be amplitude, velocity or acceleration) to this impulse is determined. Thus, for an un-damped system the response at an instant of time $\mathrm{d}\tau$ is expressed as

$$\mathrm{d}x = \frac{F(\tau)}{m\omega_\mathrm{n}} \sin \omega_\mathrm{n}(t - \tau)\mathrm{d}\tau \tag{2.3.3.2}$$

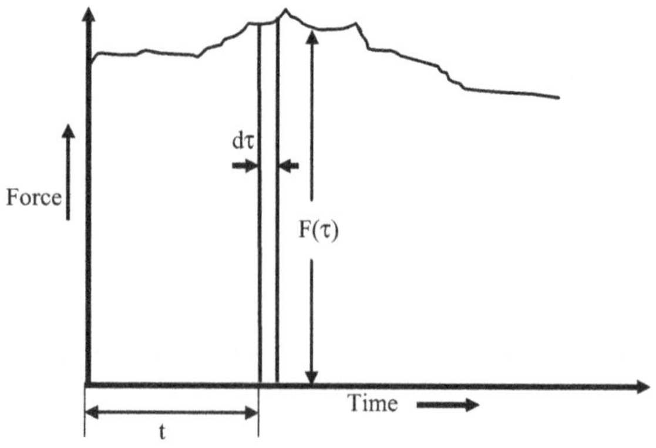

Fig. 2.13 Generalized force versus time plot

Thus, over the full time span t

$$x = \frac{1}{m\omega_n} \int_0^t F(\tau) \sin \omega_n (t - \tau) d\tau \tag{2.3.3.3}$$

The integral above will give complete solution for transient as well as forced vibration when the constants of integration are evaluated from the initial boundary conditions $x_0 = \dot{x}_0 = 0$.

If the initial boundary conditions $x_0 = \dot{x}_0 \neq 0$, the response is expressed as

$$x = x_0 \cos \omega_n t + \frac{\dot{x}_0}{\omega_n} \sin \omega_n t + \frac{1}{m\omega_n} \int_0^t F(\tau) \sin \omega_n (t - \tau) d\tau \tag{2.3.3.4}$$

If $F(\tau)$ is analytical in nature, it is possible to integrate the above equation. However if $F(\tau)$ is complicated or random in nature like that of an earthquake shock (like one shown in Fig. 1.1 in Chap. 1), one has to resort to numerical analysis to integrate the above.

For damped motion, Eq. (2.3.3.3) can be modified and expressed as

$$x = \frac{1}{m\omega_n \sqrt{1 - \zeta^2}} \int_0^t F(\tau) e^{-\zeta\omega_n (t-\tau)} \sin \omega_n \sqrt{1 - \zeta^2} (t - \tau) d\tau \tag{2.3.3.5}$$

2.3.4 Response of Single-Degree of Freedom System Under Earthquake-Type Loading

Shown in Fig. 2.14 is a single-degree frame of mass m, stiffness k, and material damping represented by a dashpot c. The frame is subjected to ground acceleration

Fig. 2.14 A single-degree frame subjected to ground acceleration ($\ddot{u}_g$)

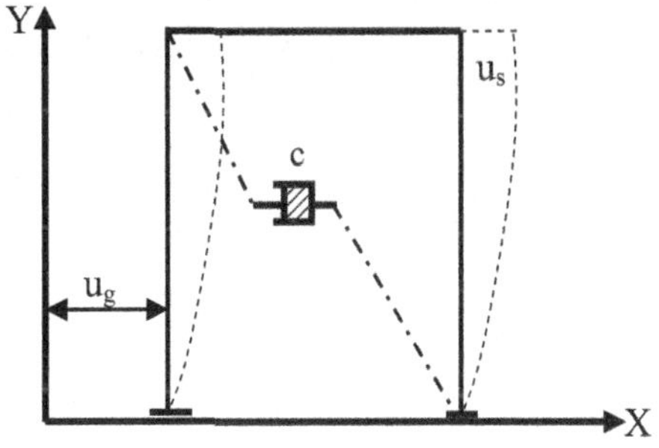

$\ddot{u}_g$ generating a ground displacement u_g. Let the acceleration, velocity and displacement induced in the frame be expressed as $\ddot{u}_s, \dot{u}_s$ and u_s, respectively.

The dynamic equation of equilibrium is thus expressed as

$$m(\ddot{u}_g + \ddot{u}_s) + c\dot{u}_s + ku_s = 0 \tag{2.3.4.1}$$

Equation (2.3.4.1) can be expressed as

$$m\ddot{u}_s + c\dot{u}_s + ku_s = F_e(t) \tag{2.3.4.2}$$

In Eq. (2.3.4.2), the term $F_e(t) = -m\ddot{u}_g$ is the earthquake force on frame. Now as this load is random in nature by analogy to the concept as explained is Sect. 2.3.3 we can express the displacement of the frame as

$$u_s = \frac{1}{m\omega_n\sqrt{1-\zeta^2}} \int_0^t F_e(t) e^{-\zeta\omega_n(t-\tau)} \sin \omega_n \sqrt{1-\zeta^2}(t-\tau) d\tau \tag{2.3.4.3}$$

$$= \frac{-1}{\omega_n\sqrt{1-\zeta^2}} \int_0^t \ddot{u}_g e^{-\zeta\omega_n(t-\tau)} \sin \omega_n \sqrt{1-\zeta^2}(t-\tau) d\tau \tag{2.3.4.4}$$

Equations (2.3.4.3) and (2.3.4.4) are usually known as the *Duhamel integral* or convolution integral and form the basis of earthquake analysis by response spectrum method.

Differentiating Eq. (2.3.4.4) we have

$$\begin{aligned}
\dot{u}_s &= \frac{-1}{w_n\sqrt{1-\zeta^2}} \int_0^t \ddot{u}_g(\tau) e^{-\zeta\omega_n(t-\tau)} \Big[-\zeta\omega_n \sin \sqrt{1-\zeta^2}\omega_n(t-\tau) d\tau \\
&\quad + \omega_n\sqrt{1-\zeta^2} \cos \sqrt{1-\zeta^2}\omega_n(t-\tau) d\tau \Big] \\
\Rightarrow \quad &= \int_0^t \ddot{u}_g(\tau) e^{-\zeta\omega_n(t-\tau)} \Big[\frac{\zeta}{\sqrt{1-\zeta^2}} \sin \sqrt{1-\zeta^2}\omega_n(t-\tau) d\tau \\
&\quad - \cos \sqrt{1-\zeta^2}\omega_n(t-\tau) d\tau \Big]
\end{aligned} \tag{2.3.4.5}$$

In Eq. (2.3.4.5) $\dot{u}_s$ is the relative velocity of the frame under earthquake force.

For a small value of ζ, the first term within the parenthesis of Eq. (2.3.4.5) can well be ignored when

$$\dot{u}_s = -\int_0^t \ddot{u}_g(\tau) e^{-\zeta\omega_n(t-\tau)} \left[\cos\sqrt{1-\zeta^2}\,\omega_n(t-\tau)\mathrm{d}\tau\right] \tag{2.3.4.6}$$

Now while discussing Biot and Housner's model at the outset of this chapter we had said that response spectrum method deals with maximum envelope of the spectrum, hence

$$|S_v|_{max} = \left| -\int_0^t \ddot{u}_g(\tau) e^{-\zeta\omega_n(t-\tau)} \left[\cos\sqrt{1-\zeta^2}\,\omega_n(t-\tau))\mathrm{d}\tau\right]\right|_{max} \tag{2.3.4.7}$$

In Eq. (2.3.4.7) the term S_v is attributed as the spectral velocity of the system, when for small damping the spectral displacement S_d is expressed as

$$S_d = \frac{S_v}{\omega_n} = \frac{S_a}{\omega_n^2} \tag{2.3.4.8}$$

In Eq. (2.3.4.8), S_a is the spectral acceleration, as depicted in Fig. 2.4.

The maximum force in the system is then expressed as

$$V = k \times S_d = k\frac{S_a}{\omega_n^2} = m \times S_a = W \times \left[\frac{S_a}{g}\right] \tag{2.3.4.9}$$

The force as expressed in Eq. (2.3.4.9) is multiplied by the factors Z (zone factor), importance factor I and response reduction factor R, etc., to finally arrive at the design seismic force.

Based on the above, it may be concluded that for any ground motion.

- The spectral acceleration is a function of natural frequency or time period of the structure.
- Intensity of spectral acceleration S_a is dependent on inherent damping property of the system.

2.3.5 *Interpretation of Duhamel Integral with Respect to Time History*

In many important structures—especially pertaining to nuclear power plants—we are often interested to know the full history of response on the structure for a specific period. This is usually carried out where site-specific spectra are used for design. In such cases we often resort to numerical analysis of the Duhamel integral as the acceleration input is random is nature. Among a number of techniques one such integral scheme that is often deployed is as elaborated hereafter

$$u_s = \frac{1}{\omega_n\sqrt{1-\zeta^2}}\int_0^t \ddot{u}_g(\tau)e^{-\zeta\omega_n(t-\tau)}\left[\sin\sqrt{1-\zeta^2}\omega_n(t-\tau)d\tau\right] \tag{2.3.5.1}$$

Considering $\omega_d = \omega_n\sqrt{1-\zeta^2}$ the damped natural frequency Eq. (2.3.5.1) can be expressed as

$$\begin{aligned} u_s &= \frac{1}{\omega_d}\int_0^t \ddot{u}_g(\tau)e^{-\zeta\omega_n t}e^{\zeta\omega_n\tau}[\sin\omega_d(t-\tau)d\tau] \\ &= \frac{e^{-\zeta\omega_n t}}{\omega_d}\int_0^t \ddot{u}_g(\tau)e^{\zeta\omega_n\tau}[\sin\omega_d t\cos\omega_d\tau - \cos\omega_d t\sin\omega_d\tau]d\tau \end{aligned} \tag{2.3.5.2}$$

Equation 2.3.5.2 can be simply expressed as

$$u_s = e^{-\zeta\omega_n t}[A\sin\omega_d t - B\cos\omega_d t] \tag{2.3.5.3}$$

In Eq. (2.3.5.3)

$$A = \frac{1}{\omega_d}\int_0^t \ddot{u}_g(\tau)e^{\zeta\omega_n\tau}\cos\omega_d\tau)\cdot d\tau \text{ and } B = \frac{1}{\omega_d}\int_0^t \ddot{u}_g(\tau)e^{\zeta\omega_n\tau}\sin\omega_d\tau)\cdot d\tau$$

For reasonably small time steps, the constants A and B can be solved by trapezoidal rule

$$\int_0^t y d\tau = \frac{\Delta t}{2}[y_0 + 2(y_1 + y_2 + y_3 + \cdots + y_{n-1}) + y_n] \tag{2.3.5.4}$$

where $\Delta t = t_{i+1} - t_i$

The acceleration and velocity can then be computed from the expression (based on finite difference technique)

$$\ddot{u}_{i+1} = \frac{1}{\Delta t^2}[u_{i+1} - 2u_i + u_{i-1}] \tag{2.3.5.5}$$

$$\dot{u}_{i+1} = \frac{1}{2\Delta t}[u_{i+1} - u_{i-1}] \tag{2.3.5.6}$$

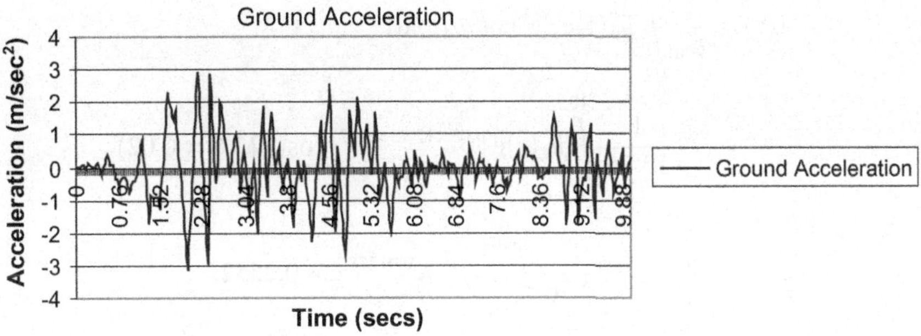

Fig. 2.15 Design site response spectra for the site

It will be observed that at $i = 0$, i.e., at start one needs to know the value of u_{-1}. This is obtained from the expression

$$u_{-1} = u_0 - \Delta t \dot{u}_0 + \frac{\Delta t^2}{2} \ddot{u}_0 \tag{2.3.5.7}$$

$u_0 = \dot{u}_0$ are usually considered zero at the start of the response analysis.

Other advanced integration schemes, like Simpson's 1/3 method or Runge–Kutta scheme, can also be deployed to derive the response.

Example 2.3.3 A water tank of weight 6000 kN is supported on a frame whose stiffness is estimated as 1×10^5 kN/m. The site-specific response spectra generated for the site is as shown below in Fig. 2.15, and whose digitized values are available. Determine the displacement, velocity and acceleration experienced by the structure. Consider 5% damping for the structure.

Solution

Weight of tank = 6000 kN

Mass of tank = 6000/9.81 = 611.6 kN s^2/m

Stiffness of frame supporting the tank = 1×10^5 kN/m

$$\omega_n = \sqrt{\frac{k}{m}} = \sqrt{\frac{1 \times 10^5}{611.6}} = 12.78\,\text{rad/s}$$

$$\omega_d = \omega_n\sqrt{1 - \zeta^2} = 12.78\sqrt{1 - (0.05)^2} = 12.76\,\text{rad/s}$$

$$A(\tau) = \frac{1}{\omega_d}\int_0^t \ddot{u}_g(\tau)e^{\zeta\omega_n\tau}\cos\omega_d\tau \cdot d\tau$$

$$\rightarrow \frac{1}{12.76}\int_0^{10} \ddot{u}_g(\tau)e^{(0.05\times 12.78\times 0.02)}\cos(12.76\times 0.02)$$

$$A(\tau) = \frac{1}{12.76}\int_0^{10} \ddot{u}_g(\tau)e^{0.01278}\cos 0.2552,$$

similarly

$$B(\tau) = \frac{1}{12.76}\int_0^{10} \ddot{u}_g(\tau)e^{0.01278}\sin 0.2552.$$

Here $\ddot{u}_g(\tau)$ is the digitized ground acceleration of figure (2.3.5.1).

It may be noted that though the values presented below are up to two seconds, the computation is actually carried out up to ten seconds for which the data was available. The numerical integration was carried out based on trapezoidal rule as elaborated above. Integrating the values form zero to ten seconds gives $A(\tau) = -8.03\times 10^{-3}$ and $B(\tau) = -2.10\times 10^{-3}$.

The values of displacement, acceleration and velocity were computed applying Eqs. (2.3.5.3), (2.3.5.5) and (2.3.5.6).

Time step (s)	Ground acceleration	$A(\tau)$	$B(\tau)$	Displacement	Acceleration	Velocity
0	0.061803	6.06E−02	1.58E−02	2.10E−03	−5.21E+00	−3.09E−04
0.02	0.035708	3.50E−02	9.14E−03	4.68E−11	1.32E−01	−5.24E−02
0.04	0.009712	9.52E−03	2.49E−03	−2.04E−03	4.57E−01	−5.11E−02
0.06	0.041987	4.11E−02	1.07E−02	−3.91E−03	7.45E−01	−4.65E−02
0.08	0.07436	7.29E−02	1.90E−02	−5.47E−03	9.77E−01	−3.91E−02
0.1	0.106635	1.05E−01	2.73E−02	−6.64E−03	1.14E+00	−2.93E−02
0.12	0.066904	6.56E−02	1.71E−02	−7.36E−03	1.23E+00	−1.79E−02
0.14	0.027174	2.66E−02	6.95E−03	−7.59E−03	1.23E+00	−5.65E−03
0.16	−0.01256	−1.23E−02	−3.21E−03	−7.32E−03	1.16E+00	6.67E−03
0.18	0.036101	3.54E−02	9.24E−03	−6.59E−03	1.01E+00	1.83E−02
0.2	0.084758	8.31E−02	2.17E−02	−5.46E−03	8.04E−01	2.84E−02
0.22	0.133416	1.31E−01	3.41E−02	−4.00E−03	5.50E−01	3.64E−02
0.24	0.071319	6.99E−02	1.83E−02	−2.32E−03	2.67E−01	4.19E−02
0.26	0.009221	9.04E−03	2.36E−03	−5.38E−04	−2.62E−02	4.46E−02
0.28	0.041202	4.04E−02	1.05E−02	1.23E−03	−3.10E−01	4.43E−02
0.3	0.02168	2.12E−02	5.55E−03	2.88E−03	−5.67E−01	4.12E−02

(continued)

(continued)

Time step (s)	Ground acceleration	$A(\tau)$	$B(\tau)$	Displacement	Acceleration	Velocity
0.32	0.00206	2.02E−03	5.27E−04	4.31E−03	−7.81E−01	3.56E−02
0.34	0.043556	4.27E−02	1.11E−02	5.42E−03	−9.39E−01	2.77E−02
0.36	0.085053	8.34E−02	2.18E−02	6.15E−03	−1.03E+00	1.83E−02
0.38	0.126549	1.24E−01	3.24E−02	6.47E−03	−1.06E+00	8.01E−03
0.4	0.168045	1.65E−01	4.30E−02	6.37E−03	−1.02E+00	−2.58E−03
0.42	−0.03365	−3.30E−02	−8.61E−03	5.86E−03	−9.08E−01	−1.27E−02
0.44	−0.23544	−2.31E−01	−6.02E−02	4.98E−03	−7.45E−01	−2.18E−02
0.46	−0.09732	−9.54E−02	−2.49E−02	3.81E−03	−5.38E−01	−2.93E−02
0.48	0.04081	4.00E−02	1.04E−02	2.43E−03	−3.02E−01	−3.47E−02
0.5	0.051797	5.08E−02	1.33E−02	9.21E−04	−5.27E−02	−3.77E−02
0.52	0.162159	1.59E−01	4.15E−02	−6.07E−04	1.94E−01	−3.82E−02
0.54	0.27262	2.67E−01	6.98E−02	−2.06E−03	4.22E−01	−3.63E−02
0.56	0.382982	3.75E−01	9.80E−02	−3.34E−03	6.17E−01	−3.20E−02
0.58	0.240247	2.35E−01	6.15E−02	−4.37E−03	7.68E−01	−2.59E−02
0.6	0.09761	9.57E−02	2.50E−02	−5.10E−03	8.65E−01	−1.82E−02
0.62	0.094274	9.24E−02	2.41E−02	−5.48E−03	9.05E−01	−9.54E−03
0.64	0.090841	8.90E−02	2.32E−02	−5.50E−03	8.85E−01	−4.99E−04
0.66	0.087505	8.58E−02	2.24E−02	−5.17E−03	8.09E−01	8.35E−03
0.68	−0.04768	−4.67E−02	−1.22E−02	−4.51E−03	6.83E−01	1.64E−02
0.7	−0.18286	−1.79E−01	−4.68E−02	−3.58E−03	5.16E−01	2.33E−02
0.72	−0.31804	−3.12E−01	−8.14E−02	−2.44E−03	3.21E−01	2.84E−02
0.74	−0.33011	−3.24E−01	−8.45E−02	−1.18E−03	1.09E−01	3.16E−02
0.76	−0.56143	−5.50E−01	−1.44E−01	1.31E−04	−1.04E−01	3.27E−02
0.78	−0.44479	−4.36E−01	−1.14E−01	1.40E−03	−3.05E−01	3.17E−02
0.8	−0.32824	−3.22E−01	−8.40E−02	2.54E−03	−4.81E−01	2.87E−02
0.82	−0.31402	−3.08E−01	−8.04E−02	3.50E−03	−6.22E−01	2.38E−02
0.84	−0.29979	−2.94E−01	−7.67E−02	4.20E−03	−7.19E−01	1.76E−02
0.86	−0.28557	−2.80E−01	−7.31E−02	4.62E−03	−7.68E−01	1.04E−02
0.88	−0.27134	−2.66E−01	−6.94E−02	4.73E−03	−7.66E−01	2.75E−03
0.9	−0.40378	−3.96E−01	−1.03E−01	4.53E−03	−7.16E−01	−4.91E−03
0.92	−0.53621	−5.25E−01	−1.37E−01	4.05E−03	−6.20E−01	−1.21E−02
0.94	−0.66865	−6.55E−01	−1.71E−01	3.32E−03	−4.87E−01	−1.83E−02
0.96	−0.80108	−7.85E−01	−2.05E−01	2.39E−03	−3.26E−01	−2.31E−02
0.98	−0.67159	−6.58E−01	−1.72E−01	1.34E−03	−1.48E−01	−2.64E−02
1	−0.5422	−5.31E−01	−1.39E−01	2.24E−04	3.45E−02	−2.79E−02
1.02	−0.4128	−4.05E−01	−1.06E−01	−8.78E−04	2.10E−01	−2.75E−02
1.04	−0.41781	−4.09E−01	−1.07E−01	−1.90E−03	3.68E−01	−2.54E−02
1.06	−0.42291	−4.14E−01	−1.08E−01	−2.77E−03	4.99E−01	−2.17E−02
1.08	−0.23819	−2.33E−01	−6.10E−02	−3.44E−03	5.94E−01	−1.68E−02
1.1	−0.05346	−5.24E−02	−1.37E−02	−3.87E−03	6.48E−01	−1.08E−02

(continued)

(continued)

Time step (s)	Ground acceleration	$A(\tau)$	$B(\tau)$	Displacement	Acceleration	Velocity
1.12	0.131258	1.29E−01	3.36E−02	−4.04E−03	6.60E−01	−4.34E−03
1.14	0.31598	3.10E−01	8.09E−02	−3.95E−03	6.29E−01	2.26E−03
1.16	0.500702	4.91E−01	1.28E−01	−3.61E−03	5.58E−01	8.55E−03
1.18	0.685425	6.72E−01	1.75E−01	−3.04E−03	4.53E−01	1.41E−02
1.2	0.870147	8.53E−01	2.23E−01	−2.30E−03	3.22E−01	1.87E−02
1.22	0.443804	4.35E−01	1.14E−01	−1.42E−03	1.73E−01	2.19E−02
1.24	0.01756	1.72E−02	4.49E−03	−4.79E−04	1.72E−02	2.36E−02
1.26	−0.40878	−4.01E−01	−1.05E−01	4.73E−04	−1.36E−01	2.38E−02
1.28	−0.83513	−8.18E−01	−2.14E−01	1.37E−03	−2.76E−01	2.24E−02
1.3	−1.26137	−1.24E+00	−3.23E−01	2.16E−03	−3.96E−01	1.97E−02
1.32	−1.68771	−1.65E+00	−4.32E−01	2.78E−03	−4.87E−01	1.57E−02
1.34	−1.26627	−1.24E+00	−3.24E−01	3.22E−03	−5.44E−01	1.08E−02
1.36	−0.84494	−8.28E−01	−2.16E−01	3.43E−03	−5.65E−01	5.39E−03
1.38	−0.87329	−8.56E−01	−2.23E−01	3.42E−03	−5.49E−01	−2.53E−04
1.4	−0.90174	−8.84E−01	−2.31E−01	3.19E−03	−4.98E−01	−5.74E−03
1.42	−0.93018	−9.12E−01	−2.38E−01	2.76E−03	−4.17E−01	−1.07E−02
1.44	−0.91468	−8.96E−01	−2.34E−01	2.17E−03	−3.11E−01	−1.49E−02
1.46	−0.89918	−8.81E−01	−2.30E−01	1.45E−03	−1.87E−01	−1.80E−02
1.48	−0.92979	−9.11E−01	−2.38E−01	6.54E−04	−5.48E−02	−1.99E−02
1.5	−0.9603	−9.41E−01	−2.46E−01	−1.63E−04	7.77E−02	−2.04E−02
1.52	−1.26569	−1.24E+00	−3.24E−01	−9.48E−04	2.02E−01	−1.96E−02
1.54	−0.75066	−7.36E−01	−1.92E−01	−1.65E−03	3.10E−01	−1.76E−02
1.56	−0.23554	−2.31E−01	−6.03E−02	−2.23E−03	3.95E−01	−1.45E−02
1.58	0.279487	2.74E−01	7.15E−02	−2.66E−03	4.53E−01	−1.06E−02
1.6	0.794512	7.79E−01	2.03E−01	−2.90E−03	4.81E−01	−6.04E−03
1.62	1.309635	1.28E + 00	3.35E−01	−2.95E−03	4.76E−01	−1.23E−03
1.64	1.82466	1.79E+00	4.67E−01	−2.81E−03	4.42E−01	3.53E−03
1.66	2.339685	2.29E+00	5.99E−01	−2.49E−03	3.79E−01	7.95E−03
1.68	2.157513	2.11E+00	5.52E−01	−2.02E−03	2.94E−01	1.17E−02
1.7	1.975244	1.94E+00	5.05E−01	−1.43E−03	1.93E−01	1.47E−02
1.72	1.792974	1.76E+00	4.59E−01	−7.67E−04	8.11E−02	1.66E−02
1.74	1.610802	1.58E+00	4.12E−01	−6.96E−05	−3.28E−02	1.74E−02
1.76	1.428532	1.40E+00	3.66E−01	6.14E−04	−1.42E−01	1.71E−02
1.78	1.583628	1.55E+00	4.05E−01	1.24E−03	−2.39E−01	1.57E−02
1.8	1.738823	1.70E+00	4.45E−01	1.77E−03	−3.18E−01	1.33E−02
1.82	1.296392	1.27E+00	3.32E−01	2.18E−03	−3.75E−01	1.01E−02
1.84	0.853961	8.37E−01	2.19E−01	2.43E−03	−4.06E−01	6.36E−03
1.86	0.411628	4.03E−01	1.05E−01	2.52E−03	−4.11E−01	2.29E−03
1.88	−0.0308	−3.02E−02	−7.88E−03	2.45E−03	−3.89E−01	−1.82E−03
1.9	−0.47323	−4.64E−01	−1.21E−01	2.22E−03	−3.42E−01	−5.70E−03

(continued)

(continued)

Time step (s)	Ground acceleration	$A(\tau)$	$B(\tau)$	Displacement	Acceleration	Velocity
1.92	−0.91567	−8.97E−01	−2.34E−01	1.86E−03	−2.75E−01	−9.13E−03
1.94	−1.358	−1.33E+00	−3.48E−01	1.38E−03	−1.92E−01	−1.19E−02
1.96	−1.80043	−1.76E+00	−4.61E−01	8.30E−04	−9.84E−02	−1.38E−02
1.98	−2.24286	−2.20E+00	−5.74E−01	2.39E−04	−1.01E−03	−1.48E−02
2	−2.68519	−2.63E+00	−6.87E−01	−3.53E−04	9.40E−02	−1.48E−02

2.3.6 Earthquake Response of Structures Having Multi-degrees of Freedom

Most of the structures related to industry or infrastructure are hardly ever modeled as a body having single degree of freedom. Leaving aside the fact that such models are construed as only approximate (if not far too crude), many of them due to there geometrical configuration demands that they be modeled as systems having multi-degrees of freedom for greater accuracy.

Thus in this section we will look into the techniques that are adapted to compute the forces due to an earthquake in system having degrees of freedom more than one. We will restrict our discussion to lumped mass discrete model herein before taking up more advanced models in subsequent sections.

2.3.6.1 Free Vibration of a System Having Two Degrees of Freedom

We start with a system having two degrees of freedom before extending it to higher degrees, as a two-degree system is quite amenable to manual or hand computation.

Figure 2.16 shows a two-storied frame having loads W_1 and W_2 on two floors, giving equivalent mass of m_1 and m_2 on each floor, respectively. The total lateral

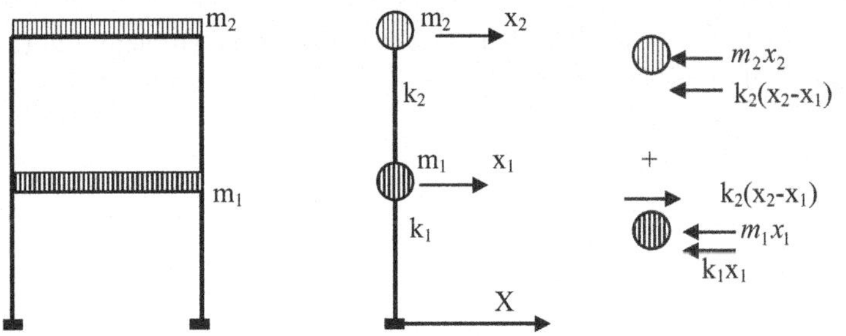

Fig. 2.16 A two-storied frame with its mathematical model and free body diagram

stiffness computed for each floors is k_1 and k_2, respectively. The free body diagram of the body vibrating freely is also shown in the figure.

The displacements at each floor level are considered as x_1 and x_2, respectively.

Assuming the body is in motion in the direction of x positive, and applying D'Alembert's equation,

we have

$$-m_1\ddot{x}_1 - k_1x_1 + k_2(x_2 - x_1) = 0 \tag{2.3.6.1}$$

$$-m_2\ddot{x}_2 - k_2(x_2 - x_1) = 0 \tag{2.3.6.2}$$

In matrix notation, Eqs. (2.3.6.1) and (2.3.6.2) can be expressed as

$$\begin{bmatrix} m_1 & 0 \\ 0 & m_2 \end{bmatrix} \begin{Bmatrix} \ddot{x}_1 \\ \ddot{x}_2 \end{Bmatrix} + \begin{bmatrix} k_1 + k_2 & -k_2 \\ -k_2 & k_2 \end{bmatrix} \begin{Bmatrix} x_1 \\ x_2 \end{Bmatrix} = 0 \tag{2.3.6.3}$$

In compact matrix notation, Eq. (2.3.6.3) can be further expressed as

$$[\mathbf{M}]\{\ddot{X}\} + [\mathbf{K}]\{X\} = 0 \tag{2.3.6.4}$$

In Eq. (2.3.6.4) matrix $[\mathbf{M}]$ and $[\mathbf{K}]$ are of the order 2×2, as depicted in Eq. (2.3.6.3).

In Sect. 2.3.1 for a body of single-degree freedom system we had shown that the amplitude of vibration is expressed as

$$x = A \sin \omega_{\rm n} t + B \cos \omega_{\rm n} t \tag{2.3.6.5}$$

Thus, for a system having multi-degrees of freedom considering $\{X\} = \{X_0\}e^{i\omega t}$, a vibration of general harmonic character, Eq. (2.3.6.4) can be expressed as

$$\left[-[\mathbf{M}]\omega^2 + [\mathbf{K}]\right]\{X_0\}e^{i\omega t} = 0 \tag{2.3.6.6}$$

Now considering $\{X_0\} \neq 0$ we have

$$[\mathbf{K}] - [\mathbf{M}]\omega^2 = 0 \tag{2.3.6.7}$$

Considering $\lambda = \omega^2$, usually called the *eigenvalue* of the problem, Eq. (2.3.6.7) can be expressed in terms of (2.3.6.3) as

$$\begin{bmatrix} k_1 + k_2 - m_1\lambda & -k_2 \\ -k_2 & k_2 - m_2\lambda \end{bmatrix} = 0 \tag{2.3.6.8}$$

Considering $k_1 + k_2 = A$ and $k_2 = B$, Eq. (2.3.6.8) can be expressed as (Chowdhury and Dasgupta 2008)

$$\begin{bmatrix} A - m_1\lambda & -B \\ -B & B - m_2\lambda \end{bmatrix} = 0 \tag{2.3.6.9}$$

Solution of Eq. (2.3.6.9) gives two roots whose values are expressed as

$$\lambda_1 = \frac{(m_1B + m_2A) - \sqrt{(m_1B - m_2A)^2 + 4m_1m_2B^2}}{2m_1m_2} \tag{2.3.6.10}$$

$$\lambda_2 = \frac{(m_1B + m_2A) + \sqrt{(m_1B - m_2A)^2 + 4m_1m_2B^2}}{2m_1m_2} \tag{2.3.6.11}$$

The natural frequencies are then computed as $\omega_1 = \sqrt{\lambda_1}$ and $\omega_2 = \sqrt{\lambda_2}$, and the corresponding time periods are obtained from the generic expression $T_i = 2\pi/\omega_i$.

Thus we see that a structure having two degrees of freedom has two eigenvalues vis-à-vis time periods and each of the degrees of freedom is attributed *as a mode of vibration* in parlance of structural dynamics. Thus, a structure having N degrees of freedom has N modes.

Considering the first mode we have

$$\begin{bmatrix} k_1 + k_2 - m_1\lambda_1 & -k_2 \\ -k_2 & k_2 - m_2\lambda_1 \end{bmatrix} \begin{Bmatrix} x_1 \\ x_2 \end{Bmatrix} = 0 \tag{2.3.6.12}$$

Considering that Eq. (2.3.6.12) is a homogenous equation, one has to find x_1 and x_2 in terms of the other; thus, considering $x_1 = 1.0$, $x_2 = (A - m_1\lambda_1)/B$.

Similarly, for the second mode we have

$$\begin{bmatrix} k_1 + k_2 - m_1\lambda_2 & -k_2 \\ -k_2 & k_2 - m_2\lambda_2 \end{bmatrix} \begin{Bmatrix} x_1 \\ x_2 \end{Bmatrix} = 0 \tag{2.3.6.13}$$

Solving the above we have $x_1 = B/(A - m_1\lambda_2)$, $x_2 = 1.0$.

The values of x_1 and x_2 computed above are known as eigenvectors of the problem and are usually depicted as

$$[\phi] = \begin{bmatrix} 1 & \frac{B}{A - m_2\lambda_2} \\ \frac{A - m_1\lambda_1}{B} & 1 \end{bmatrix} \tag{2.3.6.14}$$

The term ϕ is also sometimes termed as the *scaling vectors*. It gives us a measure of, in what ratio the displacements will take place in relation to one another under the application of an externally applied dynamic force. This is a unique property for a particular structure. Thus, for an externally applied dynamic force the amplitude of vibration is expressed as

$$\{X_i\} = \{\phi_i\}q(t) \tag{2.3.6.15}$$

where $q(t)$ is known as the displacement in generalized coordinates, where it depicts the displacement as a variable of time scaled by the factor ϕ of the ith mode.

Example 2.3.4 A two-storied frame has the following properties: m_1 = 1.0 kN s²/m, m_2 = 2.0 kN s²/m and $k_1 = k_2$ = 10 kN/m. Determine the time periods and modal vectors of the frame. Plot the deformation characteristics.

Solution

Based on Eq. (2.3.6.3), the equation of motion is expressed as

$$\begin{bmatrix} m_1 & 0 \\ 0 & m_2 \end{bmatrix} \begin{Bmatrix} \ddot{x}_1 \\ \ddot{x}_2 \end{Bmatrix} + \begin{bmatrix} k_1 + k_2 & -k_2 \\ -k_2 & k_2 \end{bmatrix} \begin{Bmatrix} x_1 \\ x_2 \end{Bmatrix} = 0$$

Substituting the numerical values we have

$$\begin{bmatrix} 1 & 0 \\ 0 & 2 \end{bmatrix} \begin{Bmatrix} \ddot{x}_1 \\ \ddot{x}_2 \end{Bmatrix} + \begin{bmatrix} 20 & -10 \\ -10 & 10 \end{bmatrix} \begin{Bmatrix} x_1 \\ x_2 \end{Bmatrix} = 0$$

Considering Eq. 2.3.6.8 we have

$$\begin{bmatrix} 20 - \lambda & -10 \\ -10 & 10 - 2\lambda \end{bmatrix} = 0$$

Expanding the above, we have a quadratic equation in λ expressed as

$$\lambda^2 - 25\lambda + 50 = 0$$

$$\rightarrow \lambda = \frac{25 \pm \sqrt{625 - 200}}{2}$$

Thus we have $\lambda_1 = 2.1922$ and $\lambda_2 = 22.807$. Considering $\lambda = \omega^2$, we have $\omega_1 = 1.48$ rad/s and $\omega_2 = 4.75$ rad/s. As $T = 2\pi/\omega$, we have $T_1 = 4.24$ s and $T_2 = 1.32$ s.

Thus for the first mode

$$\begin{bmatrix} 20 - 2.1922 & -10 \\ -10 & 10 - 2 \times 2.1922 \end{bmatrix} \begin{Bmatrix} \phi_1 \\ \phi_2 \end{Bmatrix} = 0$$

Expanding the above, we have two homogenous equations

$$17.807\phi_1 - 10\phi_2 = 0 \text{ and}$$

$$-10\phi_1 + 5.615\phi_2 = 0$$

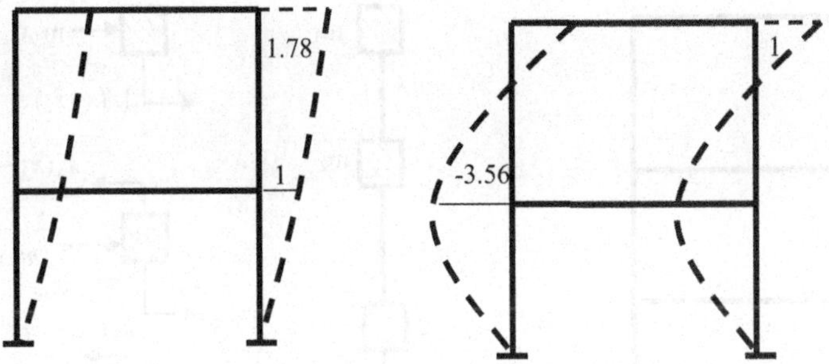

Fig. 2.17 Deformation characteristics of the two-storied frame under free vibration

From the first equation we have, considering $\phi_1 = 1.0$, $\phi_2 = 1.7807$. It may be noted that if we substitute the value of ϕ_1 and ϕ_2 in the second equation it will be automatically satisfied. Thus

$$-10 \times 1.0 + 5.615 \times 1.7807 \rightarrow 0.$$

For the second mode

$$\begin{bmatrix} 20 - 22.807 & -10 \\ -10 & 10 - 2 \times 22.807 \end{bmatrix} \begin{Bmatrix} \phi_1 \\ \phi_2 \end{Bmatrix} = 0 \text{ and this gives}$$

$$-2.807\phi_1 - 10\phi_2 = 0 \text{ and}$$

$$-10\phi_1 - 35.614\phi_2 = 0$$

Considering $\phi_2 = 1.0$, we have $\phi_1 = -3.56$

Thus, the eigenvectors for the two modes can be represented as

$$[\varphi] = \begin{bmatrix} 1.0 & -3.56 \\ 1.7807 & 1.0 \end{bmatrix}$$

The deformation characteristics of the frame are as plotted hereafter (Fig. 2.17).

2.3.6.2 Structural System Having Higher Degrees of Freedom

We extend the degree of freedom by one before extrapolating the same to a system having n degrees of freedom, where $n = 1, 2, 3, \ldots, n$.

Shown in Fig. 2.18 is a three-storied frame whose mathematical model and the free body diagram under free vibration are shown. Expressing the equations of equilibrium, we have

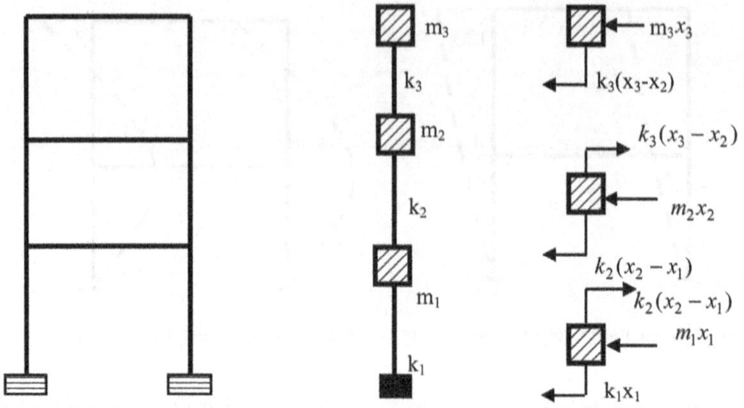

Fig. 2.18 A frame with three degrees of freedom and its free body diagram

$$-m_1\ddot{x}_1 - k_1x_1 + k_2(x_2 - x_1) = 0 \tag{2.3.6.16}$$

$$-m_2\ddot{x}_2 - k_2(x_2 - x_1) + k_3(x_3 - x_2) = 0 \tag{2.3.6.17}$$

$$-m_3\ddot{x}_3 - k_3(x_3 - x_2) = 0 \tag{2.3.6.18}$$

Expressing the above equations in matrix form, we have

$$\begin{bmatrix} m_1 & 0 & 0 \\ 0 & m_2 & 0 \\ 0 & 0 & m_3 \end{bmatrix} \begin{Bmatrix} \ddot{x}_1 \\ \ddot{x}_2 \\ \ddot{x}_3 \end{Bmatrix} + \begin{bmatrix} k_1 + k_2 & -k_2 & 0 \\ -k_2 & k_2 + k_3 & -k_3 \\ 0 & -k_3 & k_3 \end{bmatrix} \begin{Bmatrix} x_1 \\ x_2 \\ x_3 \end{Bmatrix} = 0 \tag{2.3.6.19}$$

Now considering that a frame has n degrees of freedom, Eq. (2.3.6.19) can be extrapolated to (Clough and Penzien 1982)

$$\begin{bmatrix} m_1 & 0 & 0 & \cdots & 0 & 0 \\ 0 & m_2 & 0 & \cdots & 0 & 0 \\ 0 & 0 & m_3 & \cdots & 0 & 0 \\ \cdots & \cdots & \cdots & \cdots & \cdots & \cdots \\ 0 & 0 & 0 & 0 & m_{n-1} & 0 \\ 0 & 0 & 0 & 0 & 0 & m_n \end{bmatrix} \begin{Bmatrix} \ddot{x}_1 \\ \ddot{x}_2 \\ \ddot{x}_3 \\ \cdots \\ \ddot{x}_{n-1} \\ \ddot{x}_n \end{Bmatrix} + \begin{bmatrix} k_1+k_2 & -k_2 & 0 & \cdots & 0 & 0 \\ -k_2 & k_2+k_3 & -k_3 & \cdots & 0 & 0 \\ 0 & -k_3 & k_3+k_4 & \cdots & 0 & 0 \\ \cdots & \cdots & \cdots & \cdots & \cdots & \cdots \\ 0 & 0 & 0 & \cdots & k_{n-1}+k_n & -k_n \\ 0 & 0 & 0 & \cdots & -k_n & k_n \end{bmatrix} \begin{Bmatrix} x_1 \\ x_2 \\ x_3 \\ \cdots \\ x_{n-1} \\ x_n \end{Bmatrix} = 0 \tag{2.3.6.20}$$

The eigensolution of the problem is thus expressed as

$$\begin{bmatrix} (k_1+k_2)-m_1\lambda & -k_2 & 0 & \cdots & 0 & 0 \\ -k_2 & (k_2+k_3)-m_2\lambda & -k_3 & \cdots & 0 & 0 \\ 0 & -k_3 & (k_3+k_4)-m_3\lambda & \cdots & 0 & 0 \\ \cdots & \cdots & \cdots & \cdots & \cdots & \cdots \\ 0 & 0 & 0 & \cdots & (k_{n-1}+k_n)-m_{n-1}\lambda & -k_n \\ 0 & 0 & 0 & \cdots & -k_n & k_n-m_n\lambda \end{bmatrix} = 0 \tag{2.3.6.21}$$

It will be observed that expansion of the matrix furnished by Eq. (2.3.6.21) will give a polynomial in terms of λ of power n. When $n \geq 4$ one needs special techniques to determine the eigenvalues. Some these techniques will be discussed later. We will conclude this discussion by only stating herein that the estimation of eigenvalues correctly remains the most challenging part of the computational effort in the annals of structural dynamics.

Example 2.3.5 Shown in Fig. 2.19 is a three-storied frame whose dynamic properties are furnished below. Determine the time periods and the corresponding eigenvectors for the three modes.

Solution

Based on Eq. (2.3.6.19), here m_1 = 400 kN s^2/m, m_2 = 400 kN s^2/m, and m_3 = 200 kN s^2/m and k_1 = 3000 kN, k_2 = 2000 kN/m, k_3 = 1500 kN/m.

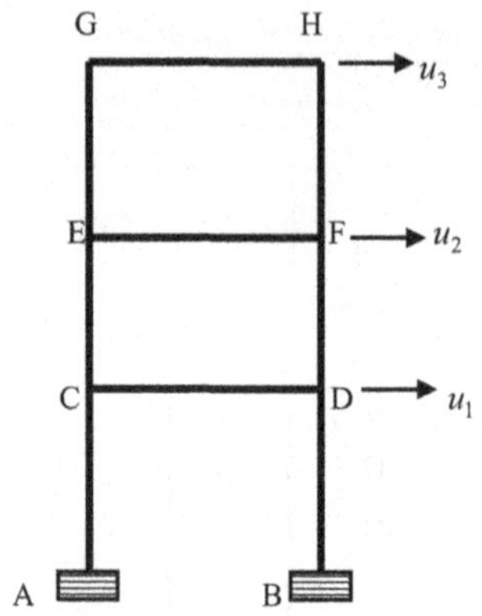

Fig. 2.19 A three-storied frame for dynamic analysis

The equilibrium equation is thus expressed as

$$\begin{bmatrix} 400 & 0 & 0 \\ 0 & 400 & 0 \\ 0 & 0 & 200 \end{bmatrix} \begin{Bmatrix} \ddot{u}_1 \\ \ddot{u}_2 \\ \ddot{u}_3 \end{Bmatrix} + \begin{bmatrix} 5000 & -2000 & 0 \\ -2000 & 3500 & -1500 \\ 0 & -1500 & 1500 \end{bmatrix} \begin{Bmatrix} u_1 \\ u_2 \\ u_3 \end{Bmatrix} = 0$$

To find the natural frequencies we have

$$\begin{bmatrix} 5000 - 400\lambda & -2000 & 0 \\ -2000 & 3500 - 400\lambda & -1500 \\ 0 & -1500 & 1500 - 200\lambda \end{bmatrix} = 0$$

On expansion we have

$$(5000 - 400\lambda) \begin{vmatrix} 3500 - 400\lambda & -1500 \\ -1500 & 1500 - 200\lambda \end{vmatrix} + 2000 \begin{vmatrix} -2000 & -1500 \\ 0 & 1500 - 200\lambda \end{vmatrix}$$
$$= 0$$

Expansion of the above gives a cubical equation

$$\lambda^3 - 28.75\lambda^2 + 215.625\lambda - 281.25 = 0$$

To determine the first root of the above cubical equation we apply the technique of Newton–Raphson method.

Let $f(\lambda) = \lambda^3 - 28.75\lambda^2 + 215.625\lambda - 281.25$ and $f'(\lambda) = 3\lambda^2 - 57.5\lambda + 215.625$

We start with the iteration

$$\lambda_{i+1} = \lambda_i - \frac{f(\lambda_i)}{f'(\lambda_i)}$$

We start with an arbitrary value of $\lambda = 50$, and this gives

$$\lambda_1 = 50 - \frac{63{,}625}{4840.6} = 35.686, \lambda_2 = 35.686 - \frac{18{,}676}{2171.5} = 28.255, \ldots,$$
$$\lambda_6 = 17.753 - \frac{83.66}{140.33} = 17.156, \lambda_7 = 17.156 - \frac{5.569}{112.14} = 17.106,$$
$$\lambda_8 = 17.106 - \frac{0.1944}{109.87} = 17.104$$

Thus, the first root of the cubical equation is 17.104.

Now as $\lambda = 17.104$ is one of the roots of the equation it must satisfy the expression

$$\lambda^3 - 28.75\lambda^2 + 215.625\lambda - 281.25 = 0$$

Thus, factorizing we have

$$(\lambda - 17.104)\lambda^2 - (\lambda - 17.104)11.646\lambda + (\lambda - 17.104)16.431816 = 0$$

$$\text{Or } (\lambda - 17.104)\left(\lambda^2 - 11.646\lambda + 16.431816\right) = 0$$

$$\rightarrow \lambda^2 - 11.646\lambda + 16.431816 = 0$$

Hence

$$\lambda = \frac{11.646 \pm \sqrt{(11.646)^2 - 4 \times 16.431816}}{2} \rightarrow 1.6426 \text{ and } 10.00$$

Thus, $\omega_1 = \sqrt{1.6426} = 1.281$ rad/s, $\omega_2 = \sqrt{10.00} = 3.162$ rad/s and $\omega_3 = \sqrt{17.104} = 4.135$ rad/s.

Thus, $T_1 = 4.9$ s, $T_2 = 1.987$ s and $T_3 = 1.52$ s.

For calculation of eigenvectors, for the first mode ($\lambda = 1.6426$)

$$\begin{bmatrix} 5000 - 400 \times 1.6426 & -2000 & 0 \\ -2000 & 3500 - 400 \times 1.6426 & -1500 \\ 0 & -1500 & 1500 - 200 \times 1.6426 \end{bmatrix} \begin{Bmatrix} \phi_1 \\ \phi_2 \\ \phi_3 \end{Bmatrix} = 0$$

$$\begin{bmatrix} 4343 & -2000 & 0 \\ -2000 & 2843 & -1500 \\ 0 & -1500 & 1171.48 \end{bmatrix} \begin{Bmatrix} \phi_1 \\ \phi_2 \\ \phi_3 \end{Bmatrix} = 0$$

The above, on simplification, gives $\phi_1 = 1.0, \phi_2 = 2.1715$ and $\phi_3 = 2.7816$
For the second mode ($\lambda = 10.0$)

$$\begin{bmatrix} 5000 - 400 \times 10.0 & -2000 & 0 \\ -2000 & 3500 - 400 \times 10.0 & -1500 \\ 0 & -1500 & 1500 - 200 \times 10.0 \end{bmatrix} \begin{Bmatrix} \phi_1 \\ \phi_2 \\ \phi_3 \end{Bmatrix} = 0$$

The above, on simplification, gives $\phi_1 = 2.0, \phi_2 = 1.0$ and $\phi_3 = -3.0$
For the third mode we have ($\lambda = 17.104$)

$$\begin{bmatrix} 5000 - 400 \times 17.104 & -2000 & 0 \\ -2000 & 3500 - 400 \times 17.104 & -1500 \\ 0 & -1500 & 1500 - 200 \times 17.104 \end{bmatrix} \begin{Bmatrix} \phi_1 \\ \phi_2 \\ \phi_3 \end{Bmatrix} = 0$$

The above, on simplification, gives $\phi_1 = 1.391, \phi_1 = -1.164$ and $\phi_3 = 1.0$.
Thus, the eigenvectors for the three modes are

$$[\boldsymbol{\varphi}] = \begin{bmatrix} 1.0 & 2.0 & 1.391 \\ 2.1715 & 1.0 & -1.164 \\ 2.7816 & -3.0 & 1.0 \end{bmatrix}$$

2.3.7 Decoupling of Matrix that Makes Life Easy

We had shown in Sect. 2.3.6 how to determine the eigenvalues of the structures having two and three degrees of freedom, and also derived the stiffness and mass matrix of order n. It is observed that computation of eigenvalues becomes progressively complex as the degree of freedom gets higher and one may well wonder as to how to solve the coupled matrix equations to finally determine the amplitude of vibration which remains the final goal of this particular exercise.

Fortunately, this becomes quite simple once the eigenvalues and eigenvectors are determined.

To this end, let us consider the equation below

$$[\mathbf{M}]\{\ddot{X}\} + [\mathbf{K}]\{X\} = 0 \tag{2.3.7.1}$$

Here [**M**] and [**K**] are mass and stiffness matrices of order $n \times n$, respectively.

Let $\omega_1, \omega_2, \omega_3, \ldots, \omega_n$, be the n number of eigenvalues, and let eigenvectors be represented by the matrix $[\boldsymbol{\varphi}]$, which is again of order n. We had stated before that the displacement amplitude is usually determined in terms eigenvectors and amplitude in generalized coordinate as $\{X_i\} = \{\phi_i\}q(t)$.Thus, Eq. (2.3.7.1) can be expressed as

$$[\mathbf{M}]\{\boldsymbol{\varphi}\}\ddot{q}_n(t) + [\mathbf{K}]\{\boldsymbol{\varphi}\}q_n(t) = 0 \tag{2.3.7.2}$$

Now multiplying Eq. (2.3.7.2) by the expression $\{\boldsymbol{\varphi}\}^{\mathrm{T}}$ we have

$$\{\boldsymbol{\varphi}\}^{\mathrm{T}}[\mathbf{M}]\{\boldsymbol{\varphi}\}\ddot{q}_n(t) + \{\boldsymbol{\varphi}\}^{\mathrm{T}}[\mathbf{K}]\{\boldsymbol{\varphi}\}q_n(t) = 0 \tag{2.3.7.3}$$

Considering $[\mathbf{K}] = [\mathbf{M}]\omega_n^2$ and substituting the same in Eq. (2.3.7.3) we have

$$\{\boldsymbol{\varphi}\}^{\mathrm{T}}[\mathbf{M}]\{\boldsymbol{\varphi}\}\ddot{q}_n(t) + \{\boldsymbol{\varphi}\}^{\mathrm{T}}[\mathbf{M}]\{\boldsymbol{\varphi}\}\omega_n^2 q_n(t) = 0 \tag{2.3.7.4}$$

$$\rightarrow \{\boldsymbol{\varphi}\}^{\mathrm{T}}[\mathbf{M}]\{\boldsymbol{\varphi}\}\left[\ddot{q}_n(t) + \omega_n^2 q_n(t)\right] = 0$$

Since $\{\boldsymbol{\varphi}\}^{\mathrm{T}}[\mathbf{M}]\{\boldsymbol{\varphi}\}$ represents the generalized mass of the system and has a nonzero value, hence

$$\ddot{q}_n(t) + \omega_n^2 q_n(t) = 0 \tag{2.3.7.5}$$

The question now is what does Eq. (2.3.7.5) signify?

It may be mentioned that when 2.2.6.7 is solved we get the eigenvalues in the form

$$[\lambda] = \begin{bmatrix} \omega_1^2 & & & \cdots & \\ & \omega_2^2 & & \cdots & \\ & & \omega_3^2 & \cdots & \\ \cdots & \cdots & \cdots & \cdots & \cdots \\ & & & \cdots & \omega_n^2 \end{bmatrix} \tag{2.3.7.6}$$

Equation (2.3.7.6) clearly shows that the eigenvalues obtained are in uncoupled form thus based on this, it may be argued that Eq. (2.3.7.5) actually represents number n of single-degree equations in uncoupled form. In other words, the system gets decoupled into n equations represented by

$$\ddot{q}_1(t) + \omega_1^2 q_1(t) = 0 \tag{2.3.7.7a}$$

$$\ddot{q}_2(t) + \omega_2^2 q_2(t) = 0 \tag{2.3.7.7b}$$

....

$$\ddot{q}_n(t) + \omega_n^2 q_n(t) = 0 \tag{2.3.7.7c}$$

Considering we know the analytical solution of single degree of freedom, it is only then a case to solve these equations in uncoupled mode and get final amplitude from the expression $\{X_i\} = \{\phi_i\}q(t)$ where i is the mode number.

To prove the above theory with numbers, we solve a problem as elaborated hereafter.

Example 2.3.6 For the problem cited in Example 2.3.4 show that the problem decouples into two uncoupled equation of single degree.

Solution

Referring to problem 2.3.4, we see that mass matrix $[M] = \begin{bmatrix} 1 & 0 \\ 0 & 2 \end{bmatrix}$ and stiffness matrix $[K] = \begin{bmatrix} 20 & -10 \\ -10 & 10 \end{bmatrix}$.

The eigenvalues computed were $\lambda_1 = 2.1922$ and $\lambda_2 = 22.807$, the corresponding eigenvectors computed were $[\phi] = \begin{bmatrix} 1 & -3.562 \\ 1.7807 & 1 \end{bmatrix}$.

$$\text{Based on the above, } [\boldsymbol{\varphi}]^{\mathrm{T}}[\mathbf{M}][\boldsymbol{\varphi}] = \begin{bmatrix} 1 & 1.7807 \\ -3.56 & 1 \end{bmatrix}\begin{bmatrix} 1 & 0 \\ 0 & 2 \end{bmatrix}\begin{bmatrix} 1 & -3.56 \\ 1.7807 & 1 \end{bmatrix} = \begin{bmatrix} 7.34 & 0 \\ 0 & 14.68 \end{bmatrix}$$

$$\text{And } [\boldsymbol{\varphi}]^{\mathrm{T}}[\mathbf{K}][\boldsymbol{\varphi}] = \begin{bmatrix} 1 & 1.7807 \\ -3.56 & 1 \end{bmatrix}\begin{bmatrix} 20 & -10 \\ -10 & 10 \end{bmatrix}\begin{bmatrix} 1 & -3.56 \\ 1.7807 & 1 \end{bmatrix} = \begin{bmatrix} 16.096 & 0 \\ 0 & 334.99 \end{bmatrix}$$

Thus, the equilibrium equation can be expressed as

$$\begin{bmatrix} 7.34 & 0 \\ 0 & 14.68 \end{bmatrix}\begin{Bmatrix} \ddot{q}_1 \\ \ddot{q}_2 \end{Bmatrix} + \begin{bmatrix} 16.096 & 0 \\ 0 & 334.99 \end{bmatrix}\begin{Bmatrix} q_1 \\ q_2 \end{Bmatrix} = 0$$

The above can thus be expressed as two equations of single degree of freedom and are expressed as $7.34\ddot{q}_1 + 16.096q_1 = 0$ and $14.68\ddot{q}_2 + 334.99q_2 = 0$, which can be simplified and expressed as $\ddot{q}_1 + 2.192q_1 = 0$ and $\ddot{q}_2 + 22.80q_2 = 0$. It may be observed that the equations are of the type as expressed by Eqs. (2.3.7.7a–2.3.7.7c) where factors of q_1 and q_2 have converged to the square of the natural frequencies.

2.3.8 Decoupling of Equations with Damping Incorporated

In this section, we try to evaluate what happens to a system having multi-degrees of freedom when damping is also considered in the analysis. In other words, how does the theory as applied in Sect. 2.3.7 work for an equation of type

$$[\mathbf{M}]\{\ddot{X}\} + [\mathbf{C}]\{\dot{X}\} + [\mathbf{K}]\{X\} = 0 \tag{2.3.8.1}$$

For a typical frame having three degrees of freedom, considering that damping also resists the motion in the same manner as stiffness, we have the equation of free vibration as

$$\begin{aligned} &\begin{bmatrix} m_1 & 0 & 0 \\ 0 & m_2 & 0 \\ 0 & 0 & m_3 \end{bmatrix} \begin{Bmatrix} \ddot{x}_1 \\ \ddot{x}_2 \\ \ddot{x}_3 \end{Bmatrix} + \begin{bmatrix} c_1 + c_2 & -c_2 & 0 \\ -c_2 & c_2 + c_3 & -c_3 \\ 0 & -c_3 & c_3 \end{bmatrix} \begin{Bmatrix} \dot{x}_1 \\ \dot{x}_2 \\ \dot{x}_3 \end{Bmatrix} \\ &\quad + \begin{bmatrix} k_1 + k_2 & -k_2 & 0 \\ -k_2 & k_2 + k_3 & -k_3 \\ 0 & -k_3 & k_3 \end{bmatrix} \begin{Bmatrix} x_1 \\ x_2 \\ x_3 \end{Bmatrix} = 0 \end{aligned} \tag{2.3.8.2}$$

Multiplying Eq. (2.3.8.1) by the expression $[\boldsymbol{\varphi}]^{\mathrm{T}}$, we have

$$[\boldsymbol{\varphi}]^{\mathrm{T}}[\mathbf{M}][\boldsymbol{\varphi}]\{\ddot{q}_n(t)\} + [\boldsymbol{\varphi}]^{\mathrm{T}}[\mathbf{C}][\boldsymbol{\varphi}]\{\dot{q}_n(t)\} + [\boldsymbol{\varphi}]^{\mathrm{T}}[\mathbf{K}][\boldsymbol{\varphi}]\{q_n(t)\} = 0 \tag{2.3.8.3}$$

Equation (2.3.8.3) can thus be expressed as

$$M_n\ddot{q}_n(t) + C_n\dot{q}_n(t) + K_nq_n(t) = 0 \tag{2.3.8.4}$$

We had seen earlier that for an undamped system the term $K_n/M_n \rightarrow \omega_{\mathrm{n}}^2$. So what happens to the term C_n/M_n here?

For a single degree of freedom having damping, the equation of equilibrium can be expressed as

$$m\ddot{x} + c\dot{x} + kx = 0 \tag{2.3.8.5}$$

This can be further expressed as

$$\ddot{x} + \frac{c}{m}\dot{x} + \omega^2 x = 0 \tag{2.3.8.6}$$

Based on the derivations carried out in Sect. 2.3.2, it can be shown that $c/m \rightarrow 2\zeta\omega$, where $\zeta =$ damping ratio of the system and $\omega =$ the natural frequency.

Thus, by same logic, Eq. (2.3.8.4) can be expressed as

$$\ddot{q}_n(t) + 2\zeta_n\omega_n\dot{q}_n(t) + \omega_n^2 q_n(t) = 0 \tag{2.3.8.7}$$

This means a system with n degrees of freedom decouples into n equations of single degree of freedom as

$$\ddot{q}_1(t) + 2\zeta_1\omega_1 + \omega_1^2 q_1(t) = 0 \tag{2.3.8.8a}$$

$$\ddot{q}_2(t) + 2\zeta_2\omega_2 + \omega_2^2 q_2(t) = 0 \tag{2.3.8.8b}$$

....

$$\ddot{q}_n(t) + 2\zeta_n\omega_n + \omega_n^2 q_n(t) = 0 \tag{2.3.8.8c}$$

Solving the above equation gives the complimentary function when the right-hand side is zero and the particular integral when the right-hand side is nonzero, which gives the complete solution of the problem.

2.3.9 Response of Multi-degree Freedom System Under Earthquake-Type Force

For a structure having n degrees of freedom, under earthquake-type force, the equation of equilibrium is expressed as (Clough and Penzien 1982)

$$[\mathbf{M}]\{\ddot{X}\} + [\mathbf{C}]\{\dot{X}\} + [\mathbf{K}]\{X\} = [\mathbf{M}]\{\mathbf{I}\}S_{\mathrm{an}} \tag{2.3.9.1}$$

In Eq. (2.3.9.1), the matrix $\{\mathbf{I}\}$ represents a column matrix having values of 1 for the degrees of freedoms that are active to acceleration S_{a}. Here, S_{a} is the spectral acceleration as furnished in code.

Considering $[X] = [\boldsymbol{\varphi}]q_n(t)$ and pre-multiplying Eq. (2.3.9.1) by $[\boldsymbol{\varphi}]^{\mathrm{T}}$, we have

$$[\boldsymbol{\varphi}]^{\mathrm{T}}[\mathbf{M}][\boldsymbol{\varphi}]\ddot{q}_n(t) + [\boldsymbol{\varphi}]^{\mathrm{T}}[\mathbf{C}][\boldsymbol{\varphi}]\dot{q}_n(t) + [\boldsymbol{\varphi}]^{\mathrm{T}}[\mathbf{K}][\boldsymbol{\varphi}]q_n(t) = [\boldsymbol{\varphi}]^{\mathrm{T}}[\mathbf{M}]\{\mathbf{I}\}S_{\mathrm{an}} \tag{2.3.9.2}$$

Equation (2.3.9.2) can thus be expressed as

$$M_n\ddot{q}_n(t) + C_n\dot{q}_n(t) + K_n q_n(t) = L_n S_{\mathrm{an}} \tag{2.3.9.3}$$

$$\ddot{q}_n(t) + 2\zeta_n\omega_n\dot{q}_n(t) + \omega_n^2 q_n(t) = \left[\frac{L_n}{M_n}\right] S_{\mathrm{an}} \tag{2.3.9.4}$$

In Eq. (2.3.9.4), $L_n = [\boldsymbol{\varphi}]^{\mathrm{T}}[\mathbf{M}]\{\mathbf{I}\}$ is called the *modal excited mass* and $M_n = [\boldsymbol{\varphi}]^{\mathrm{T}}[\mathbf{M}][\boldsymbol{\varphi}]$ is called the *generalized mass* of the system.

Thus, Eq. (2.3.9.4) gets decoupled into n number of equations of single degree of freedom as

$$\ddot{q}_1(t) + 2\zeta_1\omega_1\dot{q}_n(t) + \omega_1^2 q_n(t) = \left[\frac{L_1}{M_1}\right] S_{a1} \tag{2.3.9.5a}$$

$$\ddot{q}_2(t) + 2\zeta_2\omega_2\dot{q}_2(t) + \omega_2^2 q_2(t) = \left[\frac{L_2}{M_2}\right] S_{a2} \tag{2.3.9.5b}$$

....

$$\ddot{q}_{n-1}(t) + 2\zeta_{n-1}\omega_{\mathrm{n}}\dot{q}_{n-1}(t) + \omega_{n-1}^2 q_{n-1}(t) = \left[\frac{L_{n-1}}{M_{n-1}}\right] S_{a(n-1)} \tag{2.3.9.5c}$$

$$\ddot{q}_n(t) + 2\zeta_n\omega_{\mathrm{n}}\dot{q}_n(t) + \omega_{\mathrm{n}}^2 q_n(t) = \left[\frac{L_n}{M_n}\right] S_{\mathrm{an}} \tag{2.3.9.5d}$$

Now for modal earthquake analysis, the effect of damping is usually incorporated for the ith mode say by picking up the value of S_{ai} from the response spectrum curve and then multiplying it by the scaling factor as furnished in Table 2.9, depending on the damping ratio considered for the system. Thus, for any mode i where $1 < i < n$, the expression can be simply written as

$$\ddot{q}_i(t) + \omega_i^2 q_i(t) = \left[\frac{L_i}{M_i}\right] \overline{S}_{\mathrm{ai}} \tag{2.3.9.6}$$

Here, $L_i = [\phi_i]^{\mathrm{T}}[\mathbf{M}]\{\mathbf{I}\}$, $M_i = [\phi_i]^{\mathrm{T}}[\mathbf{M}][\phi_i]$ and $\overline{S}_{\mathrm{ai}} =$ the design response spectrum as obtained from expressions furnished in Table 2.1 through Table 2.3 scaled by the factors as furnished in Table 2.9 depending on the damping ratio for the ith mode.

For solution of Eq. (2.3.9.6), let us assume $p = [L_i/M_i]\overline{S}_{\mathrm{ai}}$.

Thus, Eq. (2.3.9.6) can be expressed as

$$\ddot{q}_i(t) + \omega_i^2 q_i(t) = p \tag{2.3.9.7}$$

It may be noted that since the damping effect was already considered while considering the spectral acceleration $\overline{S}_{\mathrm{ai}}$, in Eq. (2.3.9.7) it is not necessary to consider damping effect separately.

The complimentary function can then be expressed as

$$q_i^c(t) = A\cos\omega_i t + B\sin\omega_i t \tag{2.3.9.8}$$

In Eq. (2.3.9.8), the superscript c stands for *complementary function.*

The above thus gives $y_1(t) = \cos\omega_i t$, $y_2(t) = \sin\omega_i t$, $y_1'(t) = -\omega_i\sin\omega_i t$ and $y_2'(t) = \omega_i\cos\omega_i t$, and $g(t) = p$.

The Wronskian is thus expressed as

$$W = \begin{vmatrix} y_1(t) & y_2(t) \\ y_1'(t) & y'(t) \end{vmatrix} \rightarrow \begin{vmatrix} \cos \omega_i t & \sin \omega_i t \\ -\omega_i \sin \omega_i t & \omega_i \cos \omega_i t \end{vmatrix} = \omega_i \tag{2.3.9.9}$$

Thus,

$$A(t) = -\int \frac{g(t) y_2(t)}{W} \mathrm{d}t = -\int \frac{p \sin \omega_i t}{\omega_i} \mathrm{d}t = \frac{p \cos \omega_i t}{\omega_i^2} \tag{2.3.9.10}$$

and

$$B(t) = \int \frac{g(t) y_1(t)}{W} \mathrm{d}t = \int \frac{p \cos \omega_i t}{\omega_i} \mathrm{d}t = \frac{p \sin \omega_i t}{\omega_i^2} \tag{2.3.9.11}$$

Thus, the particular integral is expressed as

$$q_i(t) = A(t) y_1(t) + B(t) y_2(t) \rightarrow \frac{p}{\omega_i^2} \left(\cos^2 \omega_i t + \sin^2 \omega_i t \right) = \left(\frac{L_i}{M_i} \right) \left[\frac{\overline{S}_{\mathrm{ai}}}{\omega_i^2} \right] \tag{2.3.9.12}$$

Thus, in terms of provision of IS-1893, $q_i(t)$ can be expressed as

$$q_i(t) = \left(\frac{ZI}{2R} \right) \left(\frac{L_i}{M_i} \right) \left[\frac{\overline{S}_{\mathrm{ai}}}{\omega_i^2} \right] \tag{2.3.9.13}$$

The amplitude of vibration $\{X_i\}$ for the ith mode is thus expressed as

$$\{X_i\} = \left(\frac{ZI}{2R} \right) \{\phi_i\} \left(\frac{L_i}{M_i} \right) \left[\frac{\overline{S}_{\mathrm{ai}}}{\omega_i^2} \right] \tag{2.3.9.14}$$

The nodal forces $\{f_i\}$ acting at each lumped mass level are then calculated from the expression

$$\{f_i\} = [\mathbf{K}] \{\phi_i\} \left(\frac{ZI}{2R} \right) \left(\frac{L_i}{M_i} \right) \left[\frac{\overline{S}_{\mathrm{ai}}}{\omega_i^2} \right] \tag{2.3.9.15}$$

Considering $[\mathbf{K}] = [\mathbf{M}]\omega_i^2$, expression (2.3.9.15) can also be written as

$$\{f_i\} = [\mathbf{M}] \{\phi_i\} \left(\frac{ZI}{2R} \right) \left(\frac{L_i}{M_i} \right) \left[\overline{S}_{\mathrm{ai}} \right] \tag{2.3.9.16}$$

Equations (2.3.9.15) and (2.3.9.16) must give the same result, as one can be derived from the other. However, in real practice, considering that the mass matrix

[**M**] is mostly a lumped mass system, it is a diagonal matrix and thus it is usually computationally more efficient to compute the nodal force based on Eq. (2.3.9.16).

It may be noted that the term L_i/M_i is defined as the *modal participation factor* and depicted by the term $\kappa_i = L_i/M_i$ when Eq. (2.3.9.16) can be expressed as

$$\{f_i\} = \kappa_i[\mathbf{M}]\{\phi_i\}\left(\frac{ZI}{2R}\right)\left[\overline{S}_{\mathrm{ai}}\right] \tag{2.3.9.17}$$

The total base shear acting at foundation level is then expressed as

$$V_{\mathrm{b}} = \sum_{i=1}^{n}\{f_i\} \tag{2.3.9.18}$$

Equation (2.3.9.18) can also be expressed as

$$V_{\mathrm{b}} = \{\mathbf{I}\}^{\mathrm{T}}\{f_i\} \tag{2.3.9.19}$$

Here, $\{\mathbf{I}\}^{\mathrm{T}}$ is a row identity matrix of order n.

Substituting $\{f_i\}$ from Eq. (2.3.9.16) in Eq. (2.3.9.19), we have

$$V_{\mathrm{b}} = \{\mathbf{I}\}^{\mathrm{T}}[\mathbf{M}]\{\phi_n\}\left(\frac{ZI}{2R}\right)\left(\frac{L_n}{M_n}\right)\overline{S}_{\mathrm{an}} \tag{2.3.9.20}$$

It may be observed that

$$\{\mathbf{I}\}^{\mathrm{T}}[\mathbf{M}]\{\phi_n\} = [\phi_n]^{\mathrm{T}}[\mathbf{M}]\{\mathbf{I}\} \rightarrow L_n \tag{2.3.9.21}$$

Thus, Eq. (2.3.9.20) can be expressed as

$$V_{\mathrm{b}} = \left(\frac{ZI}{2R}\right)\left(\frac{L_n^2}{M_n}\right)\overline{S}_{\mathrm{an}} \tag{2.3.9.22}$$

In Eq. (2.3.9.22) the term L_n^2/M_n is called the *modal mass participation factor* and defines the part of the total mass of the structure that participates in each mode. This, when divided by the total mass (which is sum of the trace of the determinant of the mass matrix), gives the percentage of mass participating in each mode.

2.3.10 Some Other Interpretations of the Mass Participation Factor κ

We had shown in Sect. 2.3.9 that modal participation factor is expressed as

$$\kappa_n = \frac{L_n}{M_n} = \frac{\{\phi_n\}^{\mathrm{T}}[\mathbf{M}]\{\mathbf{I}\}}{\{\phi_n\}^{\mathrm{T}}[\mathbf{M}]\{\phi_n\}} \tag{2.3.10.1}$$

For a system having n degrees of freedom, the numerator of Eq. (2.3.10.1) can be expressed as

$$L_n = \{\phi_1 \quad \phi_2 \quad \dots \quad \phi_{n-1} \quad \phi_n\} \begin{bmatrix} m_1 & 0 & \dots & 0 & 0 \\ 0 & m_2 & \dots & 0 & 0 \\ \dots & \dots & \dots & \dots & \dots \\ 0 & 0 & \dots & m_{n-1} & 0 \\ 0 & 0 & \dots & 0 & m_n \end{bmatrix} \begin{Bmatrix} 1 \\ 1 \\ \dots \\ 1 \\ 1 \end{Bmatrix}$$
$$= m_1\phi_1 + m_2\phi_2 + \cdots + m_{n-1}\phi_{n-1} + m_n\phi_n \tag{2.3.10.2}$$

Similarly, the denominator can be expressed as

$$M_n = \{\phi_1 \quad \phi_2 \quad \dots \quad \phi_{n-1} \quad \phi_n\} \begin{bmatrix} m_1 & 0 & \dots & 0 & 0 \\ 0 & m_2 & \dots & 0 & 0 \\ \dots & \dots & \dots & \dots & \dots \\ 0 & 0 & \dots & m_{n-1} & 0 \\ 0 & 0 & \dots & 0 & m_n \end{bmatrix} \begin{Bmatrix} \phi_1 \\ \phi_2 \\ \dots \\ \phi_{n-1} \\ \phi_n \end{Bmatrix}$$
$$= m_1\phi_1^2 + m_2\phi_2^2 + \cdots + m_{n-1}\phi_{n-1}^2 + m_n\phi_n^2 \tag{2.3.10.3}$$

Thus, the modal participation factor can also be expressed as

$$\kappa_n = \frac{L_n}{M_n} = \frac{\sum_{i=1}^{n} m_i\phi_i}{\sum_{i=1}^{n} m_i\phi_i^2} \tag{2.3.10.4}$$

2.3.11 Modal Combination of Forces

Based on mathematical derivations as furnished in Sect. (2.3.9), it may be inferred that for system having n degrees of freedom the nodal forces are depicted as

$$f_{1n} = \left\{ f_{11} \quad f_{21} \quad \cdots \quad f_{(n-1)1} \quad f_{n1} \right\}^{\mathrm{T}} \text{ for mode 1} \tag{2.3.11.1a}$$

$$f_{2n} = \left\{ f_{12} \quad f_{22} \quad \cdots \quad f_{(n-1)2} \quad f_{n2} \right\}^{\mathrm{T}} \text{ for mode 2} \tag{2.3.11.1b}$$

....

$$f_{nn} = \left\{ f_{1n} \quad f_{2n} \quad \cdots \quad f_{(n-1)n} \quad f_{nn} \right\}^{\mathrm{T}} \text{ for mode } n \tag{2.3.11.1c}$$

The combination of all these modal forces finally gives the maximum design envelope for which the structure is finally designed.

One of the three methods as mentioned below usually carries out the modal combination.

- Absolute sum technique (ABSSUM)
- Square root of sum square (SRSS) method
- Complete quadratic combination (CQC).

2.3.11.1 Absolute Sum Technique (ABSSUM)

As the name implies, the method combines the responses of all modes by summing up all the values without considering their algebraic signs. In other words

$$\Lambda_n = \sum_{i=1}^{n} |\Lambda_i| \tag{2.3.11.2}$$

Here, $|\Lambda_i|$ represents the absolute value of the response and could be variables like amplitude, velocity, acceleration, or nodal forces. The method, though recommended by many codes (including IS-1893), is found to give far too conservative results and is thus sometimes used now, but for non-critical structures only.

2.3.11.2 Square Root of Sum Square (SRSS)

Rosenblueth (1951) originally proposed this method. In this method, summing up square of all the responses and then taking its root gives the design output. It has been observed that SRSS method gives a much more rational value than the ABBSUM technique, especially when the frequencies of the structures are widely spaced, and remains the most popular method for computation of dynamic response under earthquake force.

Thus, based on this technique

$$\Lambda_n = \sqrt{\sum_{i=1}^{n} \Lambda_i^2} \quad (2.3.11.3)$$

2.3.11.3 The Complete Quadratic Combination Method (CQC)

It was originally proposed by Der Kiureghian (1981); in this case, the response of the system is obtained from the expression

$$\Lambda_n = \sqrt{\sum_{i=1}^{n} \sum_{j=1}^{n} \Lambda_i \rho_{ij} \Lambda_j} \quad (2.3.11.4)$$

Here, n = number of modes being considered Λ_i = response quantity in mode i, Λ_j = response quantity in mode j, and ρ_{ij} = cross-modal coefficient expressed as

$$\rho_{ij} = \frac{8\sqrt{\zeta_i \zeta_j}\left(\zeta_i + \beta_{ij}\zeta_j\right)\beta_{ij}^{\frac{3}{2}}}{\left(1 - \beta_{ij}^2\right)^2 + 4\zeta_i \zeta_j \beta_{ij}\left(1 + \beta_{ij}\right)^2} \quad (2.3.11.5)$$

Here, ζ_i = modal damping ratio for mode i; ζ_j = modal damping ratio for mode j, and β_{ij} = frequency ratio $\left(\omega_i/\omega_j\right)$.

For normal dynamic analysis, the damping ratio is usually considered constant in all modes when Eq. (2.3.11.5) simplifies to

$$\rho_{ij} = \frac{8\zeta^2\left(1 + \beta_{ij}\right)\beta_{ij}^{\frac{3}{2}}}{\left(1 - \beta_{ij}^2\right)^2 + 4\zeta^2 \beta_{ij}\left(1 + \beta_{ij}\right)^2} \quad (2.3.11.6)$$

Whatever has been discussed until now will be elaborated hereafter based on two numerical examples below.

The reader is requested to go through these two examples cited thoroughly to understand the underlying concept.

Example 2.3.7 Shown below is a RCC frame where the loads at each level are depicted and this includes the self-weight of the beam. The beam size at roof level is 250 × 450 mm, and for all other floors, it is 250 × 550 mm. The column sizes are: (i) from 0 to 6.8 from ground—400 × 600 mm (ii) from 6.8 to 9.8 m—400 × 500, and (iii) from 9.8 to 12.8 m—400 × 450. The concrete grade is M30. The frame is located as per IS code in zone IV with soil of intermediate properties (average SPT value = 28). Perform a detail dynamic analysis based on modal

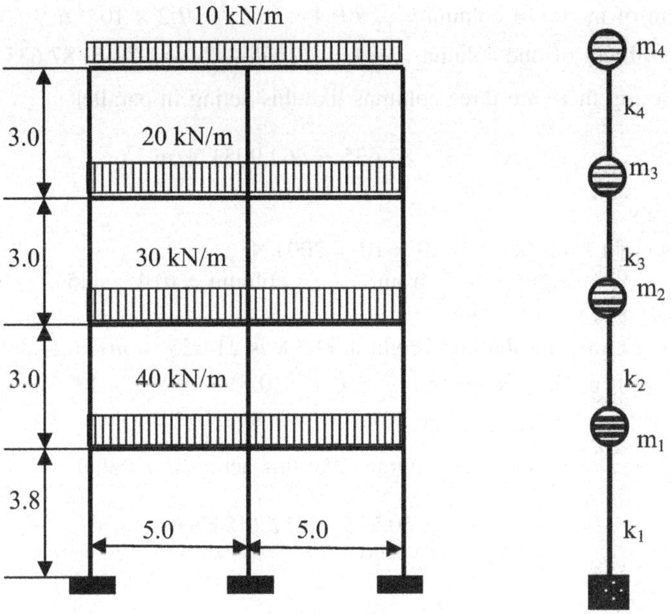

Fig. 2.20 A four-storied frame and its mathematical model

response technique to obtain the nodal forces and amplitude of displacements. Assume 5% damping for the structure (Fig. 2.20).

Solution

As a first step, we calculate the values of lumped masses and stiffness.

At EL 3.8

Total weight from beam = 40 × 10 = 400 kN

Total weight from column = 0.4 × 0.6 × 3.4 × 25 = 20.4 kN.

Thus, the total contributing weight at el 3.8 is 420.4 $\rightarrow m_1 = 42.85\,\text{kN}\,\text{s}^2/\text{m}$

Moment of inertia of column = $\frac{1}{12} \times 0.4 \times (0.6)^3 = 7.2 \times 10^{-3}\,\text{m}^4$

$$E = 5000\sqrt{f_{ck}} \rightarrow 5000 \times \sqrt{30} = 27{,}386\,\text{N/mm}^2$$

Thus, stiffness of one column = $\frac{12EI}{L^3} = \frac{12 \times 2.7386 \times 10^7 \times 7.2 \times 10^{-3}}{(3.8)^3} = 43{,}121\,\text{kN/m}$

Considering there are three columns like this acting in parallel

$$k_1 = 3 \times 43{,}121 = 129{,}364\,\text{kN/m}$$

At EL 6.8

Total weight from beam = 30 × 10 = 300 kN

Total weight from column = (0.4 × 0.6 × 1.5 + 0.4 × 0.5 × 1.5) × 25 = 16.5 kN.

Thus, the total contributing weight at el 3.8 is 316.5 $\rightarrow m_2 = 32.26\,\text{kN}\,\text{s}^2/\text{m}$.

Moment of inertia of column = $\frac{1}{12} \times 0.4 \times (0.6)^3 = 7.2 \times 10^{-3}\,\mathrm{m}^4$

Thus, stiffness of one column = $\frac{12EI}{L^3} = \frac{12\times 2.7386\times 10^7\times 7.2\times 10^{-3}}{(3.0)^3} = 87{,}635\,\mathrm{kN/m}$

Considering there are three columns like this acting in parallel

$$k_2 = 3 \times 87{,}635 = 262{,}905\,\mathrm{kN/m}$$

At EL 9.8

Total weight from beam = 20 × 10 = 200 kN

Total weight from column = (0.4 × 0.5 × 1.5 + 0.4 × 0.45 × 1.5) × 25 = 14.25 kN.

Thus, the total contributing weight at el 3.8 is 214.25 → $m_3 = 21.83\,\mathrm{kN\,s^2/m}$

Moment of inertia of column = $\frac{1}{12} \times 0.4 \times (0.5)^3 = 4.167 \times 10^{-3}\,\mathrm{m}^4$

Thus, stiffness of one column = $\frac{12EI}{L^3} = \frac{12\times 2.7386\times 10^7\times 4.167\times 10^{-3}}{(3.0)^3} = 50{,}715\,\mathrm{kN/m}$

Considering there are three columns like this acting in parallel

$$k_3 = 3 \times 50{,}715 = 152{,}145\,\mathrm{kN/m}$$

At EL 12.8

Total weight from beam = 10 × 10 = 100 kN

Total weight from column = (0.4 × 0.45 × 1.5) × 25 = 6.75 kN.

Thus, the total contributing weight at el 3.8 is 106.75 → $m_4 = 10.88\,\mathrm{kN\,s^2/m}$

Moment of inertia of column = $\frac{1}{12} \times 0.4 \times (0.45)^3 = 3.0375 \times 10^{-3}\,\mathrm{m}^4$

Thus, stiffness of one column = $\frac{12EI}{L^3} = \frac{12\times 2.7386\times 10^7\times 3.0375\times 10^{-3}}{(3.0)^3} = 36{,}971\,\mathrm{kN/m}$

Considering there are three columns like this acting in parallel

$$k_4 = 3 \times 36{,}971 = 110{,}913\,\mathrm{kN/m}$$

Having determined the stiffness and mass values, we set up the equation of motion as

$$\begin{bmatrix} m_1 & 0 & 0 & 0 \\ 0 & m_2 & 0 & 0 \\ 0 & 0 & m_3 & 0 \\ 0 & 0 & 0 & m_4 \end{bmatrix} \begin{Bmatrix} \ddot{u}_1 \\ \ddot{u}_2 \\ \ddot{u}_3 \\ \ddot{u}_4 \end{Bmatrix} + \begin{bmatrix} k_1+k_2 & -k_2 & 0 & 0 \\ -k_2 & k_2+k_3 & -k_3 & 0 \\ 0 & -k_3 & k_3+k_4 & -k_4 \\ 0 & 0 & -k_4 & k_4 \end{bmatrix} \begin{Bmatrix} u_1 \\ u_2 \\ u_3 \\ u_4 \end{Bmatrix} = 0$$

$$\begin{bmatrix} 42.85 & 0 & 0 & 0 \\ 0 & 32.26 & 0 & 0 \\ 0 & 0 & 21.83 & 0 \\ 0 & 0 & 0 & 10.88 \end{bmatrix} \begin{Bmatrix} \ddot{u}_1 \\ \ddot{u}_2 \\ \ddot{u}_3 \\ \ddot{u}_4 \end{Bmatrix} + \begin{bmatrix} 3.923\times10^5 & -2.629\times10^5 & 0 & 0 \\ -2.629\times10^5 & 4.151\times10^5 & -1.521\times10^5 & 0 \\ 0 & -1.521\times10^5 & 2.631\times10^5 & -1.109\times10^5 \\ 0 & 0 & -1.109\times10^5 & 1.109\times10^5 \end{bmatrix} \begin{Bmatrix} u_1 \\ u_2 \\ u_3 \\ u_4 \end{Bmatrix} = 0$$

The eigenproblem of the same is expressed as

$$\begin{bmatrix} 3.923\times10^5 - 42.85\lambda & -2.629\times10^5 & 0 & 0 \\ -2.629\times10^5 & 4.151\times10^5 - 32.26\lambda & -1.521\times10^5 & 0 \\ 0 & -1.521\times10^5 & 2.631\times10^5 - 21.83\lambda & -1.109\times10^5 \\ 0 & 0 & -1.109\times10^5 & 1.109\times10^5 - 10.88\lambda \end{bmatrix} = 0$$

Solving the above in computer through Mathcad, we have

$[\lambda] = \{904.4 \quad 5601 \quad 15{,}510 \quad 22{,}250\}$, considering $\omega = \sqrt{\lambda}$ and $T = 2\pi/\omega$ we have

$$[T] = \{0.209 \quad 0.084 \quad 0.05 \quad 0.042\}\ \text{s}.$$

The corresponding eigenvectors are (again computed in Mathcad)

$$[\phi] = \begin{bmatrix} -0.542 & -0.539 & -0.454 & 0.397 \\ -0.728 & -0.312 & 0.470 & -0.847 \\ -0.911 & 0.451 & 0.521 & 1 \\ -1 & 1 & -1 & -0.846 \end{bmatrix}$$

Having done the eigenvalue analysis, we proceed to determine the design forces for each mode as follows.

As per IS-1893 (2002), for zone IV $Z = 0.24$, let $I = 1.0$ and for ordinary moment resistant frame $R = 3.0$.

Thus, code factor $C_f = \frac{ZI}{2R} = \frac{0.24\times1}{2\times3} = 0.04$.

For Mode 1

$T = 0.209$ and this gives $S_a/g = 2.5$; thus, $S_a = 2.5 \times 9.81 \times 0.04 = 0.981\,\text{m/s}^2$

$$\{\phi\}_1 = \begin{Bmatrix} -0.542 \\ -0.728 \\ -0.911 \\ -1 \end{Bmatrix} \text{ and this gives}$$

$$L_1 = \{\phi\}_1^{\mathrm{T}}[\mathbf{M}]\{\mathbf{I}\} = \{-0.542 \quad -0.728 \quad -0.911 \quad -1\} \begin{bmatrix} 42.85 & 0 & 0 & 0 \\ 0 & 32.26 & 0 & 0 \\ 0 & 0 & 21.83 & 0 \\ 0 & 0 & 0 & 10.88 \end{bmatrix} \begin{Bmatrix} 1 \\ 1 \\ 1 \\ 1 \end{Bmatrix}$$

$$= -77.477$$

$$M_1 = \{\varphi\}_1^{\mathrm{T}}[\mathbf{M}]\{\varphi\}_1$$

$$= \{-0.542 \quad -0.728 \quad -0.911 \quad -1\} \begin{bmatrix} 42.85 & 0 & 0 & 0 \\ 0 & 32.26 & 0 & 0 \\ 0 & 0 & 21.83 & 0 \\ 0 & 0 & 0 & 10.88 \end{bmatrix} \begin{Bmatrix} -0.542 \\ -0.728 \\ -0.911 \\ -1 \end{Bmatrix}$$

$$= 58.682$$

The amplitude of vibration for the first mode is expressed as

$$\{u\}_1 = \{\phi\}_1 \left(\frac{L_1}{M_1}\right) \left(\frac{S_{a1}}{4\pi^2}\right) T_1^2 \quad (\text{As } T = 2\pi/\omega)$$

$$\{u\}_1 = \begin{Bmatrix} -0.542 \\ -0.728 \\ -0.911 \\ -1 \end{Bmatrix} \times \left(\frac{-77.477}{58.862}\right) \times \left(\frac{0.981}{4\pi^2}\right) \times (0.209)^2$$

$$= \begin{Bmatrix} 0.7767 \\ 1.043 \\ 1.306 \\ 1.433 \end{Bmatrix} \times 10^{-3}\ \text{m}$$

It may be noted that in the above expression the spectral acceleration S_{ai} is already multiplied by the code factor $C_f = ZI/2R$, and thus need not be taken into cognizance further.

The nodal force for the first mode is thus expressed as

$$\{f_1\} = [\mathbf{M}]\{\phi_1\}\left(\frac{L_1}{M_1}\right)[S_{a1}]$$

$$\{f_1\} = \begin{bmatrix} 42.85 & 0 & 0 & 0 \\ 0 & 32.26 & 0 & 0 \\ 0 & 0 & 21.83 & 0 \\ 0 & 0 & 0 & 10.83 \end{bmatrix} \begin{Bmatrix} -0.542 \\ -0.728 \\ -0.911 \\ -1 \end{Bmatrix} \left(\frac{-77.477}{58.862}\right)[0.981]$$

$$= \begin{Bmatrix} 30.081 \\ 30.418 \\ 25.758 \\ 14.092 \end{Bmatrix} \text{kN}$$

$$\text{Modal mass participation is } \frac{L_1^2}{M_1} = \frac{(-77.477)^2}{58.862} = 102.292\,\text{kN s}^2/\text{m}$$

$$\%\text{ of mass participated} = \frac{102.292}{42.85 + 32.26 + 21.83 + 10.88} \times 100 = 94.87\%$$

For Mode 2

$T = 0.084$ and this gives $S_a/g = 2.26$; thus, $S_a = 2.26 \times 9.81 \times 0.04 = 0.886\,\text{m/s}^2$

$$\{\phi\}_2 = \begin{Bmatrix} -0.539 \\ -0.312 \\ 0.451 \\ 1 \end{Bmatrix}, \text{ and this gives}$$

$$L_2 = \{\phi\}_2^{\mathrm{T}}[\mathbf{M}]\{\mathbf{I}\}$$

$$= \{-0.539 \quad -0.312 \quad 0.451 \quad 1\} \begin{bmatrix} 42.85 & 0 & 0 & 0 \\ 0 & 32.26 & 0 & 0 \\ 0 & 0 & 21.83 & 0 \\ 0 & 0 & 0 & 10.88 \end{bmatrix} \begin{Bmatrix} 1 \\ 1 \\ 1 \\ 1 \end{Bmatrix}$$

$$= 12.136$$

$$M_2 = \{\phi\}_2^{\mathrm{T}}[\mathbf{M}]\{\phi\}_2$$

$$= \{0.539 \quad 0.312 \quad 0.451 \quad 1\} \begin{bmatrix} 42.85 & 0 & 0 & 0 \\ 0 & 32.26 & 0 & 0 \\ 0 & 0 & 21.83 & 0 \\ 0 & 0 & 0 & 10.88 \end{bmatrix}$$

$$\begin{Bmatrix} -0.539 \\ -0.312 \\ 0.451 \\ 1 \end{Bmatrix}$$

$$= 30.909$$

The amplitude of vibration for the second mode is expressed as

$$\{u\}_2 = \{\phi\}_2\left(\frac{L_2}{M_2}\right)\left(\frac{S_{a2}}{4\pi^2}\right)T_2^2$$

$$\{u\}_2 = \begin{Bmatrix} -0.539 \\ -0.312 \\ 0.415 \\ 1 \end{Bmatrix} \times \left(\frac{-12.436}{30.909}\right) \times \left(\frac{0.886}{4\pi^2}\right) \times (0.084)^2$$

$$= \begin{Bmatrix} 3.434 \\ 1.988 \\ -2.873 \\ -6.371 \end{Bmatrix} \times 10^{-5}\ \text{m}$$

The nodal force for the second mode can thus be expressed as

$$\{f_2\} = [\mathbf{M}]\{\phi_2\}\left(\frac{L_2}{M_2}\right)[S_{a2}]$$

$$\{f_2\} = \begin{bmatrix} 42.85 & 0 & 0 & 0 \\ 0 & 32.26 & 0 & 0 \\ 0 & 0 & 21.83 & 0 \\ 0 & 0 & 0 & 10.83 \end{bmatrix} \begin{Bmatrix} -0.539 \\ -0.312 \\ 0.451 \\ 1 \end{Bmatrix} \left(\frac{-12.436}{30.909}\right)[0.886]$$

$$= \begin{Bmatrix} 8.233 \\ 3.588 \\ -3.51 \\ -3.878 \end{Bmatrix}\ \text{kN}$$

$$\text{Modal mass participation is } \frac{L_2^2}{M_2} = \frac{(-12.436)^2}{30.909} = 5.003\ \text{kN s}^2/\text{m}$$

$$\%\ \text{of mass participated} = \frac{5.003}{42.85 + 32.26 + 21.83 + 10.88} \times 100 = 4.64\%$$

For Mode 3

$T = 0.050$ and this gives $S_a/g = 1.75$; thus, $S_a = 1.75 \times 9.81 \times 0.04 = 0.686\ \text{m/s}^2$

$$\{\phi\}_3 = \begin{Bmatrix} -0.454 \\ 0.470 \\ 0.521 \\ -1 \end{Bmatrix} \text{ and this gives}$$

$$L_3 = \{\varphi\}_3^{\mathrm{T}}[\mathbf{M}]\{\mathbf{I}\}$$

$$= \{-0.454 \quad 0.470 \quad 0.521 \quad -1\}\begin{bmatrix} 42.85 & 0 & 0 & 0 \\ 0 & 32.26 & 0 & 0 \\ 0 & 0 & 21.83 & 0 \\ 0 & 0 & 0 & 10.88 \end{bmatrix}\begin{Bmatrix} 1 \\ 1 \\ 1 \\ 1 \end{Bmatrix}$$

$$= -3.798$$

$$M_3 = \{\varphi\}_3^{\mathrm{T}}[\mathbf{M}]\{\varphi\}_3$$

$$= \{-0.454 \quad 0.470 \quad 0.521 \quad -1\}\begin{bmatrix} 42.85 & 0 & 0 & 0 \\ 0 & 32.26 & 0 & 0 \\ 0 & 0 & 21.83 & 0 \\ 0 & 0 & 0 & 10.88 \end{bmatrix}$$

$$\begin{Bmatrix} -0.454 \\ 0.470 \\ 0.521 \\ -1 \end{Bmatrix}$$

$$= 32.764$$

The amplitude of vibration for the third mode is expressed as

$$\{u\}_3 = \{\phi\}_3\left(\frac{L_3}{M_3}\right)\left(\frac{S_{a3}}{4\pi^2}\right)T_3^2$$

$$\{u\}_3 = \begin{Bmatrix} -0.454 \\ 0.470 \\ 0.521 \\ -1 \end{Bmatrix} \times \left(\frac{-3.798}{32.764}\right) \times \left(\frac{0.686}{4\pi^2}\right) \times (0.050)^2 = \begin{Bmatrix} 2.286 \\ -2.367 \\ -2.624 \\ 5.036 \end{Bmatrix} \times 10^{-6}\,\mathrm{m}$$

The nodal force for the third mode can thus be expressed as

$$\{f_3\} = [\mathbf{M}]\{\phi_3\}\left(\frac{L_3}{M_3}\right)[S_{a3}]$$

$$\{f_3\} = \begin{bmatrix} 42.85 & 0 & 0 & 0 \\ 0 & 32.26 & 0 & 0 \\ 0 & 0 & 21.83 & 0 \\ 0 & 0 & 0 & 10.83 \end{bmatrix}\begin{Bmatrix} -0.454 \\ 0.470 \\ 0.521 \\ -1 \end{Bmatrix}\left(\frac{-3.798}{32.764}\right)[0.686]$$

$$= \begin{Bmatrix} 1.547 \\ -1.206 \\ -0.904 \\ 0.865 \end{Bmatrix}\mathrm{kN}$$

$$\text{Modal mass participation is } \frac{L_3^2}{M_3} = \frac{(-3.798)^2}{32.764} = 0.44\,\text{kN}\,\text{s}^2/\text{m}$$

$$\%\text{ of mass participated} = \frac{0.44}{42.85 + 32.26 + 21.83 + 10.88} \times 100 = 0.41\%$$

For Mode 4

$T = 0.042$ and this gives $S_a/g = 1.63$; thus, $S_a = 1.63 \times 9.81 \times 0.04 = 0.639\,\text{m/s}^2$

$$\{\phi\}_4 = \begin{Bmatrix} 0.397 \\ -0.847 \\ 1 \\ -0.846 \end{Bmatrix} \text{ and this gives}$$

$$L_4 = \{\phi\}_4^{\text{T}}[\mathbf{M}]\{\mathbf{I}\}$$

$$= \{0.397 \quad -0.847 \quad 1 \quad -0.846\} \begin{bmatrix} 42.85 & 0 & 0 & 0 \\ 0 & 32.26 & 0 & 0 \\ 0 & 0 & 21.83 & 0 \\ 0 & 0 & 0 & 10.88 \end{bmatrix} \begin{Bmatrix} 1 \\ 1 \\ 1 \\ 1 \end{Bmatrix}$$

$$= 2.313$$

$$M_4 = \{\varphi\}_4^{\text{T}}[\mathbf{M}]\{\varphi\}_4$$

$$= \{0.397 \quad -0.847 \quad 1 \quad -0.846\}$$

$$\begin{bmatrix} 42.85 & 0 & 0 & 0 \\ 0 & 32.26 & 0 & 0 \\ 0 & 0 & 21.83 & 0 \\ 0 & 0 & 0 & 10.88 \end{bmatrix} \begin{Bmatrix} 0.397 \\ -0.847 \\ 1 \\ -0.846 \end{Bmatrix}$$

$$= 59.514$$

The amplitude of vibration for the fourth mode is expressed as

$$\{u\}_4 = \{\phi\}_4 \left(\frac{L_4}{M_4}\right) \left(\frac{S_{a4}}{4\pi^2}\right) T_4^2$$

$$\{u\}_4 = \begin{Bmatrix} 0.397 \\ -0.847 \\ 1 \\ -0.846 \end{Bmatrix} \times \left(\frac{2.313}{59.514}\right) \times \left(\frac{0.639}{4\pi^2}\right) \times (0.042)^2$$

$$= \begin{Bmatrix} 4.405 \\ -9.398 \\ 11.1 \\ -9.387 \end{Bmatrix} \times 10^{-7}\,\text{m}$$

The nodal force for the fourth mode can thus be expressed as

$$\{f_4\} = [\mathbf{M}]\{\phi_4\}\left(\frac{L_4}{M_4}\right)[S_{a4}]$$

$$\{f_4\} = \begin{bmatrix} 42.85 & 0 & 0 & 0 \\ 0 & 32.26 & 0 & 0 \\ 0 & 0 & 21.83 & 0 \\ 0 & 0 & 0 & 10.83 \end{bmatrix} \begin{Bmatrix} 0.397 \\ -0.847 \\ 1 \\ -0.846 \end{Bmatrix} \left(\frac{2.313}{59.514}\right)[0.639]$$

$$= \begin{Bmatrix} 0.422 \\ -0.679 \\ 0.542 \\ -0.229 \end{Bmatrix} \text{kN}$$

$$\text{Modal mass participation is } \frac{L_4^2}{M_4} = \frac{(2.313)^2}{59.514} = 0.09\,\text{kN}\,\text{s}^2/\text{m}$$

$$\%\text{ of mass participated} = \frac{0.09}{42.85 + 32.26 + 21.83 + 10.88} \times 100 = 0.08\%$$

Based on above calculations, a few points need to be carefully noted

- The first or the fundamental mode has the maximum contribution to the dynamic response, while higher modes have progressively diminishing effect. It is for this, in many cases, especially for buildings and similar structures, determination of the fundamental mode, and computation of moments and shears based on the fundamental mode only suffice.
- The modal mass percentage of mass is 95% for first mode, while for higher modes it is negligible.
- Summation of percentage of mass for all four modes is 100%.

Example 2.3.8 For the problem worked out in Example 2.3.7, determine the floor and base shears and modal combination of forces by (i) ABBSUM, (ii) SRSS, and (iii) CQC technique.

Checking on the problem 2.2.7, the floor shears in kN obtained at various levels are as tabulated below

Elevation (m)	Mode 1	Mode 2	Mode 3	Mode 4
12.8	14.092	−3.878	0.865	−0.229
9.8	25.758	−3.51	−0.904	0.542
6.8	30.418	3.588	−1.206	−0.679
3.8	30.081	8.233	1.547	0.422

Floor shears are tabulated as hereafter

Elevation	Floor shear Mode 1	Floor shear Mode 2	Floor shear Mode 3	Floor shear Mode 4
12.8	14.092	−3.878	0.865	−0.229
9.8	14.092 + 25.758 = 39.85	−3.878 − 3.51 = −7.388	0.865 − 0.904 = −0.039	−0.229 + 0.542 = 0.313
6.8	39.85 + 30.418 = 70.268	−7.388 + 3.588 = −3.8	−0.039 − 1.206 = −1.245	0.313 − 0.679 = −0.366
3.8	70.268 + 30.081 = 100.349	−3.8 + 8.233 = 4.43	−1.245 + 1.547 = 0.302	−0.366 + 0.422 = 0.056
Base shear at (0.0)	100.349	4.43	0.302	0.056

Thus, based on *ABSSUM* technique, the base shear is

$$V(\text{total}) = 100.349 + 4.43 + 0.302 + 0.056 = 105.137\,\text{kN}$$

Based on *SRSS* method, the base shear is

$$V_{\text{total}} = \sqrt{(100.349)^2 + (4.43)^2 + (0.302)^2 + (0.056)^2} = 100.44\,\text{kN}.$$

Based on *CQC* method, the term β_{ij} can be expressed in matrix form as

$$\beta_{ij} = \begin{bmatrix} \omega_1/\omega_1 & \omega_1/\omega_2 & \omega_1/\omega_3 & \omega_1/\omega_4 \\ \omega_2/\omega_1 & \omega_2/\omega_2 & \omega_2/\omega_3 & \omega_2/\omega_4 \\ \omega_3/\omega_1 & \omega_3/\omega_2 & \omega_3/\omega_3 & \omega_3/\omega_4 \\ \omega_4/\omega_1 & \omega_4/\omega_2 & \omega_4/\omega_3 & \omega_4/\omega_4 \end{bmatrix}$$

Here, $T = [0.209 \quad 0.084 \quad 0.05 \quad 0.042]$, respectively. Considering $\omega = 2\pi/T$, we have

$$\omega = [30.063 \quad 74.799 \quad 125.663 \quad 149.599]\,\text{rad/s. And this gives}$$

$$\beta_{ij} = \begin{bmatrix} 1 & 22.65214 & 332.2815 & 1791.946 \\ 7.445711 & 1 & 0.105875 & 0.055046 \\ 5.644467 & 0.168559 & 1 & 1.531073 \\ 4.97619 & 2 & 1.190476 & 1 \end{bmatrix}$$

Now substituting the value of β_{ij} in Eq. (2.3.11.6) for $\zeta = 0.05$, we have the cross-modal coefficients as

$$\rho_{ij} = \begin{bmatrix} 1 & 0.07431 & 0.041717 & 0.036789 \\ 0.029863 & 1 & 0.168559 & 0.110092 \\ 0.00998 & 0.100332 & 1 & 0.618945 \\ 0.007393 & 0.055046 & 0.519914 & 1 \end{bmatrix}$$

Thus

$$V_{CQC} = \sqrt{\begin{bmatrix} 100.349 & 4.43 & 0.302 & 0.056 \end{bmatrix} \begin{bmatrix} 1 & 0.07431 & 0.041717 & 0.036789 \\ 0.029863 & 1 & 0.168559 & 0.110092 \\ 0.00998 & 0.100332 & 1 & 0.618945 \\ 0.007393 & 0.055046 & 0.519914 & 1 \end{bmatrix} \begin{bmatrix} 100.349 \\ 4.43 \\ 0.302 \\ 0.056 \end{bmatrix}}$$
$$= 100.688\,\text{kN}$$

Thus, it is observed that while ABBSUM gives the most conservative value, SRSS and CQC give comparable results, especially when the frequencies are widely spaced.

2.4 Various Computation Techniques Adapted for Eigensolutions

In the previous sections, we have shown analytical procedures followed for computation of forces in a structure based on modal response technique.

In the whole gamut of such analyses, it will be observed that the crucial issue is correct/accurate estimation of the eigenpair $(\lambda_i, \{\phi_i\})$.

Once these two parameters are obtained, rest of the computation is quite straightforward.

Presently, with a number of general-purpose utility software like Mathematica, MATLAB, Mathcad available in the market, where eigensolvers are part of built-in functions of these software, eigenvalue analysis has become almost a routine exercise.

However, it was in not too distant past, engineers, mathematicians, and analysts spent a significant amount of time and energy to devise efficient methods to compute the eigenpairs of a structural system and implement these algorithms in their computer codes. The eigenvalue analysis remains the most formidable task in the whole realm of dynamic analysis, especially when the degrees of freedom of a system are large.

In this section, we look at some of the techniques used for eigenvalue analysis for small and large structural systems.

2.4.1 Method of Visual Inspection

This can be very easily adapted for small structures where one can compute the characteristic equation, and would usually be valid for a structure having three to four degrees of freedom.

Knowing the characteristic equation, one can plot the equation graphically by any of the utility software like a spreadsheet and check the values where it gives zero values. The particular points at which the equation gives a zero value (or changes its sign) are the eigenvalues of the problem.

To elaborate further, we solve a numerical problem hereafter.

Example 2.4.1 A structure has stiffness and mass matrix as shown below; estimate the natural frequency based on visual inspection given

$$[K] = \begin{bmatrix} 5 & -2 & 0 \\ -2 & 3.5 & -1.5 \\ 0 & -1.5 & 1.5 \end{bmatrix} \times 10^6 \text{ kN/m and } [M] = \begin{bmatrix} 4 & 0 & 0 \\ 0 & 4 & 0 \\ 0 & 0 & 2 \end{bmatrix} \times 10^3 \text{ kN s}^2/\text{m}$$

Solution

Considering the expression $[\mathbf{K}] - [\mathbf{M}]\omega^2 = 0$ where $\lambda = \omega^2$, we have

$$\begin{bmatrix} 5000 - 4\lambda & -2000 & 0 \\ -2000 & 3500 - 4\lambda & -1500 \\ 0 & -1500 & 1500 - 2\lambda \end{bmatrix} = 0$$

Expanding the above, we have

$$(5000 - 4\lambda)\begin{vmatrix} 3500 - 4\lambda & -1500 \\ -1500 & 1500 - 2\lambda \end{vmatrix} + 2000\begin{vmatrix} -2000 & -1500 \\ 0 & 1500 - 2\lambda \end{vmatrix} = 0$$

Expanding and simplifying the above equation, we have $\lambda^3 - 2875\lambda^2 + 2{,}156{,}250\lambda - 281{,}250{,}000 = 0$. This is a characteristic equation of the problem.

Plotting the above equation in excel graphically, we have the plot as follows (Fig. 2.21).

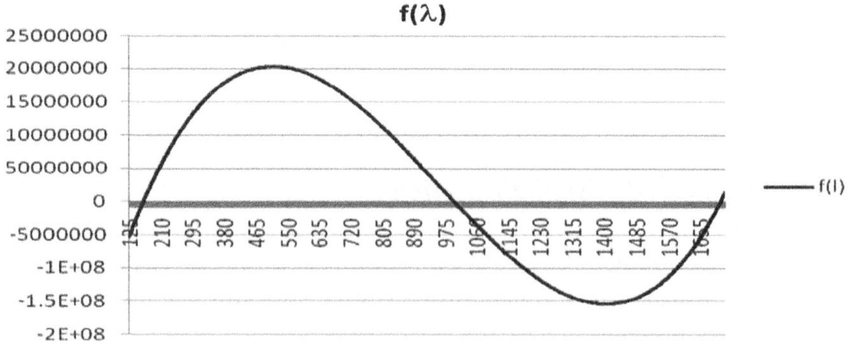

Fig. 2.21 Eigenvalue plot of the stiffness and mass matrix

The plot shows the curve cutting the x-axis at three points where it has value of zero. These three points are 164, 1000, and 1710. Thus, these points on the x-axis where the curve values tend to zero represent the eigenvalues of the problem.

2.4.2 Newton–Raphson Method

Newton–Raphson method remains possibly the most powerful numerical method to determine the roots of an algebraic equation. The underlying theory of the same is as explained hereafter.

Let there be a function

$$f(x) = Ax^3 + Bx^2 + Cx + D \tag{2.4.2.1}$$

Now let x_0 be the perceived root of the equation when the correct root is actually x_1 such that $f(x_1) = 0$.

Depending on the problem it is evident that $x_1 = x_0 \pm h$. Now, as $f(x_1) = 0 \rightarrow f(x_0 \pm h) = 0$. Expanding this by Taylor's series, we have

$$f(x_0) + hf'(x_0) + \frac{h^2}{2!}f''(x_0) + \frac{h^3}{3!}f'''(x_0) + \cdots = 0 \tag{2.4.2.2}$$

Ignoring the higher terms in Eq. (2.4.2.2), we have

$$f(x_0) + hf'(x_0) = 0 \tag{2.4.2.3}$$

$$\rightarrow h = -\frac{f(x_0)}{f'(x_0)} \tag{2.4.2.4}$$

Thus, an improved approximation is given by

$$x_1 = x_0 - \frac{f(x_0)}{f'(x_0)} \tag{2.4.2.5}$$

Successive approximation is given by $x_1, x_2, x_3, \ldots, x_n$, when

$$x_{n+1} = x_n - \frac{f(x_n)}{f'(x_n)} \text{ till the value of } x_{n+1} - x_n \rightarrow 0.$$

The value thus obtained above is one of the roots of the equation, and let it be depicted by α. Now if α is one of the roots of the equation, then $(x - \alpha) = 0$ and the right-hand side of Eq. (2.4.2.1) can be expressed as

$$(x - \alpha)\left(A'x^2 + B'x + C'\right) = 0 \tag{2.4.2.6}$$

Equation (2.4.2.6) is next considered, and we repeat the process until we find the next root β. We keep on reducing the degree of the polynomial for each successive root until all the roots are computed.

Example 2.4.2 Repeat the problem cited in Example 2.2.3 to determine eigenvalues of the problem by Newton–Raphson method.

Solution

Referring to Example 2.4.1, the characteristic third-order polynomial equation of the problem is computed as

$$\lambda^3 - 2875\lambda^2 + 2{,}156{,}250\lambda - 281{,}250{,}000 = 0$$

$$\text{Now let} f(\lambda) = \lambda^3 - 2875\lambda^2 + 2{,}156{,}250\lambda - 281{,}250{,}000$$

$$\text{Thus} f'(\lambda) = 3\lambda^2 - 5750\lambda + 2{,}156{,}250$$

We start with trial value of $\lambda_0 = 125$ (this can well be checked against the visual inspection problem. Theoretically, we should have started with 164, which would have given lower numbers of iterations)

This gives

$$\lambda_1 = 125 - \frac{f(\lambda_0)}{f'(\lambda_0)} = 125 - \left(\frac{-54{,}687{,}500}{1{,}484{,}375}\right) = 161.8421053$$

$$\lambda_2 = 161.8421053 - \frac{f(\lambda_1)}{f'(\lambda_1)} = 161.8421053 + 2.56344959 = 164.4055549$$

$$\lambda_3 = 164.4055549 - \frac{f(\lambda_2)}{f'(\lambda_2)} = 164.4176949$$

$$\lambda_4 = 164.4176949 - \frac{f(\lambda_3)}{f'(\lambda_3)} = 164.4176952$$

$$\lambda_5 = 164.4176952 - \frac{f(\lambda_4)}{f'(\lambda_4)} = 164.4176952$$

It is observed that values of λ have now become constant; thus, $\alpha = 164.4176952$ is one of the roots of the equation.

The reduction in polynomial is done as hereafter

$$(\lambda - 164.4176952)\left|\lambda^3 - 2875\lambda^2 + 2{,}156{,}250\lambda - 281{,}250{,}000\right|\lambda^2 - 2710.582305\lambda + 1{,}710{,}582.305$$

$$\frac{\lambda^3 - 164.4176952\lambda^2}{-2710.582305\lambda^2 + 2{,}156{,}250\lambda}$$

$$\frac{-2710.582305\lambda^2 + 445{,}667.6952\lambda}{1{,}710{,}582.305\lambda - 281{,}250{,}000}$$

$$\frac{1{,}710{,}582.305\lambda - 281{,}250{,}000}{\times}$$

Thus, the reduced polynomial is $\lambda^2 - 2710.582305\lambda + 1{,}710{,}582.305 = 0$.

From the above expression, roots of λ can well be computed from the expression $\lambda = \frac{-b \pm \sqrt{b^2 - 4ac}}{2a}$ where $a = 1$, $b = -2710.583205$, $c = 1{,}710{,}582.305$. However to elaborate the process, we will repeat the above for the reduced second-degree polynomial.

We start with value of $\lambda = 950$, thus

$$\lambda_1 = 950 - \frac{f(\lambda_0)}{f'(\lambda_0)} = 950 - \left(\frac{38{,}029.11525}{-810.582}\right) = 996.9158$$

$$\lambda_2 = 996.9158 - \frac{f(\lambda_1)}{f'(\lambda_1)} = 996.9158 - \left(\frac{2201.092052}{-716.751}\right) = 999.9867$$

$$\lambda_3 = 999.9867 - \frac{f(\lambda_2)}{f'(\lambda_2)} = 999.9867 - \left(\frac{9.430616866}{-710.609}\right) = 1000$$

$$\lambda_4 = 1000 - \frac{f(\lambda_3)}{f'(\lambda_3)} = 1000 - \left(\frac{0.000176124}{-710.582}\right) = 1000$$

Thus, the next root is $\lambda = 1000$; now doing a simple middle-term factorization, one gets the other value of $\lambda = 1710.6$.

2.4.3 *Matrix Deflation Method*

The method as cited above is usually adapted for small structures having three to four degrees of freedom. For structures with many degrees of freedom, the major problem is to compute the characteristic equation as it becomes an inordinately laborious exercise.

Moreover, in many cases, we are only interested to know the first few or the fundamental frequency that would principally affect the dynamic response. Thus, a more efficient method is to use matrix deflation method as elaborated hereafter.

For a structural system under free vibration, the dynamic equilibrium is expressed as

$$[\mathbf{K}]\{\phi\} = [\mathbf{M}]\omega^2\{\phi\} \tag{2.4.3.1}$$

Equation (2.4.3.1) can be expressed as (Hurty and Rubenstein 1967)

$$[\mathbf{B}]\{\phi\} = \frac{1}{\omega^2}\{\phi\} \tag{2.4.3.2}$$

In Eq. (2.4.3.2), $[\mathbf{B}] = [\mathbf{K}]^{-1}[\mathbf{M}]$, when on iteration the equation converges to the *lowest eigenvalue*.

The iteration starts with a trial value of the column vector $\{\phi\}$ which is pre-multiplied by the matrix $[\mathbf{B}]$.

The resulting column matrix is then normalized by reducing one of its element values to unity. This normalized column vector is again pre-multiplied by the matrix $[\mathbf{B}]$, and the process is repeated until the values of the normalized column vector become stationary with respect to the preceding step. The constant thus obtained gives the *lowest eigenvalue* of the system.

In case we start with an equilibrium equation

$$[\mathbf{A}]\{\phi\} = \omega^2\{\phi\} \tag{2.4.3.3}$$

where $[\mathbf{A}] = [\mathbf{K}][\mathbf{M}]^{-1}$, and following the steps as elaborated in the preceding paragraph, we converge to the *highest eigenvalue* of the system.

The method is now elaborated by a suitable numerical problem hereafter.

Example 2.4.3 A structure has stiffness and mass matrix as shown below; estimate the lowest and highest eigenvalues by matrix deflation method

$$[K] = \begin{bmatrix} 5 & -2 & 0 \\ -2 & 3.5 & -1.5 \\ 0 & -1.5 & 1.5 \end{bmatrix} \times 10^6 \text{ kN/m and } [M] = \begin{bmatrix} 4 & 0 & 0 \\ 0 & 4 & 0 \\ 0 & 0 & 2 \end{bmatrix} \times 10^3 \text{ kN s}^2/\text{m}$$

Solution

We compute the lowest eigenvalue first.

Considering $[K] = \begin{bmatrix} 5 & -2 & 0 \\ -2 & 3.5 & -1.5 \\ 0 & -1.5 & 1.5 \end{bmatrix} \times 10^6$ kN/m, we have

$$[K]^{-1} = \begin{bmatrix} 3.333 & 3.333 & 3.333 \\ 3.333 & 8.333 & 8.333 \\ 3.333 & 8.333 & 15 \end{bmatrix} \times 10^{-7}$$

Thus,

$$[B] = [K]^{-1}[M] \rightarrow \begin{bmatrix} 3.333 & 3.333 & 3.333 \\ 3.333 & 8.333 & 8.333 \\ 3.333 & 8.333 & 15 \end{bmatrix} \begin{bmatrix} 4000 & 0 & 0 \\ 0 & 4000 & 0 \\ 0 & 0 & 2000 \end{bmatrix} \times 10^{-7}$$

$$[B] = \begin{bmatrix} 1.333 & 1.333 & 0.667 \\ 1.333 & 3.333 & 1.667 \\ 1.333 & 3.333 & 3 \end{bmatrix} \times 10^{-3}$$

We start with a trial vector $\{\phi_0\} = \begin{Bmatrix} 1.0 \\ 1.0 \\ 1.0 \end{Bmatrix}$

- **Thus, for Cycle 1**

Thus,

$$[B]\{\phi_0\} = \begin{bmatrix} 1.333 & 1.333 & 0.667 \\ 1.333 & 3.333 & 1.667 \\ 1.333 & 3.333 & 3 \end{bmatrix} \begin{Bmatrix} 1 \\ 1 \\ 1 \end{Bmatrix} \times 10^{-3} = \begin{Bmatrix} 3.333 \\ 6.333 \\ 7.667 \end{Bmatrix} \times 10^{-3}$$

Taking the term 3.333×10^{-3} as common, we have the normalized eigenvector as $\{\phi_1\} = \begin{Bmatrix} 1.0 \\ 1.9 \\ 2.3 \end{Bmatrix}$.

- **Thus, for Cycle 2**

$$[B]\{\phi_1\} = \begin{bmatrix} 1.333 & 1.333 & 0.667 \\ 1.333 & 3.333 & 1.667 \\ 1.333 & 3.333 & 3 \end{bmatrix} \begin{Bmatrix} 1.0 \\ 1.9 \\ 2.3 \end{Bmatrix} \times 10^{-3} = \begin{Bmatrix} 5.4 \times 10^{-3} \\ 0.012 \\ 0.015 \end{Bmatrix}$$

Considering 5.4×10^{-3} as common, we have the normalized eigenvector as $\{\phi_2\} = \begin{Bmatrix} 1.0 \\ 2.13 \\ 2.698 \end{Bmatrix}$.

- **For Cycle 3, we have**

$$[B]\{\phi_2\} = \begin{bmatrix} 1.333 & 1.333 & 0.667 \\ 1.333 & 3.333 & 1.667 \\ 1.333 & 3.333 & 3 \end{bmatrix} \begin{Bmatrix} 1.0 \\ 2.13 \\ 2.698 \end{Bmatrix} \times 10^{-3} = \begin{Bmatrix} 5.971 \times 10^{-3} \\ 0.013 \\ 0.017 \end{Bmatrix}$$

Considering 5.971×10^{-3} as common, we have the normalized eigenvector as

$$\{\phi_3\} = \begin{Bmatrix} 1.0 \\ 2.165 \\ 2.7678 \end{Bmatrix}.$$

- **For Cycle 4, we have**

$$[B]\{\phi_3\} = \begin{bmatrix} 1.333 & 1.333 & 0.667 \\ 1.333 & 3.333 & 1.667 \\ 1.333 & 3.333 & 3 \end{bmatrix} \begin{Bmatrix} 1.0 \\ 2.165 \\ 2.7678 \end{Bmatrix} \times 10^{-3} = \begin{Bmatrix} 6.065 \times 10^{-3} \\ 0.013 \\ 0.017 \end{Bmatrix}$$

Considering 6.065×10^{-3} as common, we have the normalized eigenvector as

$$\{\phi_4\} = \begin{Bmatrix} 1.0 \\ 2.17 \\ 2.779 \end{Bmatrix}.$$

- **For Cycle 5, we have**

$$[B]\{\phi_3\} = \begin{bmatrix} 1.333 & 1.333 & 0.667 \\ 1.333 & 3.333 & 1.667 \\ 1.333 & 3.333 & 3 \end{bmatrix} \begin{Bmatrix} 1.0 \\ 2.17 \\ 2.779 \end{Bmatrix} \times 10^{-3} = \begin{Bmatrix} 6.079 \times 10^{-3} \\ 0.013 \\ 0.017 \end{Bmatrix}$$

Considering 6.079×10^{-3} as common, we have the normalized eigenvector as

$$\{\phi_4\} = \begin{Bmatrix} 1.0 \\ 2.171 \\ 2.78 \end{Bmatrix}.$$

- **For Cycle 6, we have**

$$[B]\{\phi_3\} = \begin{bmatrix} 1.333 & 1.333 & 0.667 \\ 1.333 & 3.333 & 1.667 \\ 1.333 & 3.333 & 3 \end{bmatrix} \begin{Bmatrix} 1.0 \\ 2.171 \\ 2.78 \end{Bmatrix} \times 10^{-3} = \begin{Bmatrix} 6.082 \times 10^{-3} \\ 0.013 \\ 0.017 \end{Bmatrix}$$

Considering 6.082×10^{-3} as common, we have the normalized eigenvector as

$$\{\phi_4\} = \begin{Bmatrix} 1.0 \\ 2.171 \\ 2.781 \end{Bmatrix}.$$

It is seen that the normalized eigenvector has become stationary with respect to the previous iteration. Now considering Eq. (2.4.3.2), we have

$\frac{1}{\omega^2} = \frac{1}{\lambda} = 6.082 \times 10^{-3} \rightarrow \lambda = 164.4196$. It may be noted that solving by Newton–Raphson method the same problem previously, we arrived at a value of $\lambda = 164.4177$.

The corresponding eigenvector is $\{\phi\} = \begin{Bmatrix} 1.0 \\ 2.171 \\ 2.781 \end{Bmatrix}$.

We compute the highest eigenvalue hereafter.

Referring to Eq. (2.4.3.3), we have

$$[\mathbf{A}]\{\phi\} = \omega^2\{\phi\}$$

By the problem

$$[\mathbf{A}] = [K][M]^{-1}$$

$$\text{or, } [\mathbf{A}] = \begin{bmatrix} 5 & -2 & 0 \\ -2 & 3.5 & -1.5 \\ 0 & -1.5 & 1.5 \end{bmatrix} \begin{bmatrix} 0.25 & 0 & 0 \\ 0 & 0.25 & 0 \\ 0 & 0 & 0.5 \end{bmatrix} \times 10^3$$

$$= \begin{bmatrix} 1250 & -500 & 0 \\ -500 & 875 & -750 \\ 0 & -375 & 750 \end{bmatrix}$$

We start with a trial vector $\{\phi_0\} = \begin{Bmatrix} 1.0 \\ 1.0 \\ 1.0 \end{Bmatrix}$.

- Thus, for Cycle 1, we have

$$[\mathbf{A}]\{\phi_0\} = \begin{bmatrix} 1250 & -500 & 0 \\ -500 & 875 & -750 \\ 0 & -375 & 750 \end{bmatrix} \begin{Bmatrix} 1 \\ 1 \\ 1 \end{Bmatrix} = \begin{Bmatrix} 750 \\ -375 \\ 375 \end{Bmatrix}$$

Taking 375 common from row 3, we have the normalized vector as $\{\phi_1\} = \begin{Bmatrix} 2 \\ -1 \\ 1 \end{Bmatrix}$.

- Thus, for Cycle 2, we have

$$[\mathbf{A}]\{\phi_1\} = \begin{bmatrix} 1250 & -500 & 0 \\ -500 & 875 & -750 \\ 0 & -375 & 750 \end{bmatrix} \begin{Bmatrix} 2 \\ -1 \\ 1 \end{Bmatrix} = \begin{Bmatrix} 3000 \\ -2625 \\ 1125 \end{Bmatrix}$$

Taking 1125 common from row 3, we have the normalized vector as $\{\phi_2\} = \begin{Bmatrix} 2.667 \\ -2.538 \\ 1 \end{Bmatrix}$.

- Thus, for Cycle 3, we have

$$[\mathbf{A}]\{\phi_2\} = \begin{bmatrix} 1250 & -500 & 0 \\ -500 & 875 & -750 \\ 0 & -375 & 750 \end{bmatrix} \begin{Bmatrix} 2.667 \\ -2.538 \\ 1 \end{Bmatrix} = \begin{Bmatrix} 4500 \\ -4125 \\ 1625 \end{Bmatrix}$$

Taking 1625 common from row 3, we have the normalized vector as $\{\phi_3\} = \begin{Bmatrix} 2.769 \\ -2.538 \\ 1 \end{Bmatrix}$.

- Thus, for Cycle 4, we have

$$[\mathbf{A}]\{\phi_3\} = \begin{bmatrix} 1250 & -500 & 0 \\ -500 & 875 & -750 \\ 0 & -375 & 750 \end{bmatrix} \begin{Bmatrix} 2.769 \\ -2.538 \\ 1 \end{Bmatrix} = \begin{Bmatrix} 4731 \\ -4356 \\ 1702 \end{Bmatrix}$$

Taking 1702 common from row 3, we have the normalized vector as $\{\phi_4\} = \begin{Bmatrix} 2.780 \\ -2.559 \\ 1 \end{Bmatrix}$.

- Thus, for Cycle 5, we have

$$[\mathbf{A}]\{\phi_4\} = \begin{bmatrix} 1250 & -500 & 0 \\ -500 & 875 & -750 \\ 0 & -375 & 750 \end{bmatrix} \begin{Bmatrix} 2.780 \\ -2.559 \\ 1 \end{Bmatrix} = \begin{Bmatrix} 4754 \\ -4379 \\ 1710 \end{Bmatrix}$$

Taking 1710 common from row 3, we have the normalized vector as $\{\phi_5\} = \begin{Bmatrix} 2.781 \\ -2.561 \\ 1 \end{Bmatrix}$.

- Thus, for Cycle 6, we have

$$[\mathbf{A}]\{\phi_5\} = \begin{bmatrix} 1250 & -500 & 0 \\ -500 & 875 & -750 \\ 0 & -375 & 750 \end{bmatrix} \begin{Bmatrix} 2.781 \\ -2.561 \\ 1 \end{Bmatrix} = \begin{Bmatrix} 4757 \\ -4382 \\ 1711 \end{Bmatrix}$$

Taking 1711 common from row 3, we have the normalized vector as $\{\phi_6\} = \begin{Bmatrix} 2.781 \\ -2.562 \\ 1 \end{Bmatrix}$.

- Thus, for Cycle 7, we have

$$[\mathbf{A}]\{\phi_6\} = \begin{bmatrix} 1250 & -500 & 0 \\ -500 & 875 & -750 \\ 0 & -375 & 750 \end{bmatrix} \begin{Bmatrix} 2.781 \\ -2.562 \\ 1 \end{Bmatrix} = \begin{Bmatrix} 4757 \\ -4382 \\ 1711 \end{Bmatrix}$$

Taking 1711 common from row 3, we have the normalized vector as $\{\phi_7\} = \begin{Bmatrix} 2.781 \\ -2.562 \\ 1 \end{Bmatrix}$.

The values of the normalized eigenvectors have become stationary; thus, the highest eigenvalue and its corresponding eigenvectors are $\lambda = 1711$ and $\{\phi\} = \begin{Bmatrix} 2.781 \\ -2.562 \\ 1 \end{Bmatrix}$.

It may be noted that by Newton–Raphson method we arrived at a value of $\lambda = 1710.6$.

2.4.3.1 Why Iteration Converges to the Lowest/Highest Eigenvalue?

In the previous section, we had worked out numerical problem to demonstrate that starting with a trial column vector and pre-multiplying by the dynamic matrix $[\mathbf{B}]$ or $[\mathbf{A}]$, on successive iteration, the value converges to the lowest or highest eigenvalue, depending on which matrix($[\mathbf{B}]$ or $[\mathbf{A}]$) has been used.

The question that obviously arises then is why do they converge to the lowest or highest value as demonstrated above. To this end, a mathematical proof is furnished hereafter.

For the dynamic matrix $[\mathbf{B}]$ of order $n \times n$, the number of eigenmodes shall also be of the order $n \times n$ when the trial column vector can be represented by linear superposition of the n eigenvectors of the system. This is valid because the n orthogonal vectors are linearly independent in n space.

Thus,

$$\{\phi_0\} = C_1\left\{\phi^{(1)}\right\} + C_2\left\{\phi^{(2)}\right\} + C_3\left\{\phi^{(3)}\right\} + \cdots + C_n\left\{\phi^{(n)}\right\} \tag{2.4.3.4}$$

Here

$\{\phi_0\}$	the first trial column vector
$\left\{\phi^{(1)}\right\}, \left\{\phi^{(2)}\right\}, \left\{\phi^{(3)}\right\}, \ldots, \left\{\phi^{(n)}\right\}$	the n eigenvectors
$C_1, C_2, C_3, \ldots, C_n$	constants.

As shown in the previous problem, the trial column $\{\phi_0\}$ is an arbitrary selection and is a known vector. Now according to the iteration procedure, we pre-multiply the trial vector (vide Eq. 2.4.3.4) by the dynamical matrix [**B**], say. Then

$$[B]\{\phi_0\} = C_1[B]\left\{\phi^{(1)}\right\} + C_2[B]\left\{\phi^{(2)}\right\} + C_3[B]\left\{\phi^{(3)}\right\} + \cdots + C_n[B]\left\{\phi^{(n)}\right\} \tag{2.4.3.5}$$

Now by Eq. (2.4.3.2)

$$[B]\left\{\phi^{(1)}\right\} = \frac{1}{\omega_1^2}\left\{\phi^{(1)}\right\} \tag{2.4.3.6}$$

$$[B]\left\{\phi^{(2)}\right\} = \frac{1}{\omega_2^2}\left\{\phi^{(2)}\right\} \tag{2.4.3.7}$$

....

$$[B]\left\{\phi^{(r)}\right\} = \frac{1}{\omega_r^2}\left\{\phi^{(r)}\right\} \tag{2.4.3.8}$$

....

$$[B]\left\{\phi^{(n)}\right\} = \frac{1}{\omega_n^2}\left\{\phi^{(n)}\right\} \tag{2.4.3.9}$$

Substituting the above in Eq. (2.4.3.5), we have

$$\begin{aligned}[B]\{\phi_0\} &= \{\phi_1\} \\ &= \frac{C_1}{\omega_1^2}\left\{\phi^{(1)}\right\} + \frac{C_2}{\omega_2^2}\left\{\phi^{(2)}\right\} + \frac{C_3}{\omega_3^2}[B]\left\{\phi^{(3)}\right\} + \cdots + \frac{C_n}{\omega_n^2}[B]\left\{\phi^{(n)}\right\}\end{aligned} \tag{2.4.3.10}$$

In Eq. (2.4.3.10), $\{\phi_1\}$ is the next trial vector. Now repeating the process of pre-multiplying it by the dynamical matrix [**B**], we have

$$[B]\{\phi_1\} = \{\phi_2\}$$
$$= \frac{C_1}{\omega_1^4}\left\{\phi^{(1)}\right\} + \frac{C_2}{\omega_2^4}\left\{\phi^{(2)}\right\} + \frac{C_3}{\omega_3^4}[B]\left\{\phi^{(3)}\right\} + \cdots + \frac{C_n}{\omega_n^4}[B]\left\{\phi^{(n)}\right\} \tag{2.4.3.11}$$

and so on.

Thus, if the process is repeated r times, then for the rth iteration we have

$$[B]\{\phi_r\} = \{\phi_{r+1}\}$$
$$= \frac{C_1}{\omega_1^{2r}}\left\{\phi^{(1)}\right\} + \frac{C_2}{\omega_2^{2r}}\left\{\phi^{(2)}\right\} + \frac{C_3}{\omega_3^{2r}}[B]\left\{\phi^{(3)}\right\} + \cdots + \frac{C_n}{\omega_n^{2r}}[B]\left\{\phi^{(n)}\right\} \tag{2.4.3.12}$$

Now referring to Eq. (2.4.3.12), one should recollect that modes are numbered starting with one having the lowest natural frequency such that

$$\omega_1 < \omega_2 < \omega_3 \cdots < \omega_n \tag{2.4.3.13}$$

Now if r is sufficiently large number, then it is evident that

$$\frac{1}{\omega_1^{2r}} \gg \frac{1}{\omega_2^{2r}} \gg \frac{1}{\omega_3^{2r}} \cdots \gg \frac{1}{\omega_n^{2r}} \tag{2.4.3.14}$$

Thus, the first term in Eq. (2.4.3.12) only becomes significant. The number of desired iterations depends on the accuracy that is desired, that is, the number of significant digits desired in the mode numbers and frequencies. If r is the sufficient number of iterations to achieve the necessary accuracy then

$$\{\phi_{r+1}\} = \frac{C_1}{\omega_1^{2r}}\left\{\phi^{(1)}\right\} \tag{2.4.3.15}$$

Equation (2.4.3.15) implies that r + 1th trial vector becomes identical to first natural modal vector within a multiplicative constant. An additional iteration yields

$$\{\phi_{r+2}\} = \frac{C_1}{\omega_1^{2r+2}}\left\{\phi^{(1)}\right\} \tag{2.4.3.16}$$

Now if the vectors $\{\phi_{r+1}\}$ and $\{\phi_{r+2}\}$ are normalized in the same way, the ratio of the normalizing constant for the latter column vector compared to that of the former is $1/\omega_1^2$. Thus, the iteration procedure leading to the modal column also yields the lowest natural frequency in this case.

Similarly, if one starts with the dynamical matrix $[\mathbf{A}]$, one can follow similar logic to arrive at the highest natural frequency and its corresponding eigenvectors.

2.4.4 Stodola–Vianello Method or Sweeping Matrix Technique

Stodola–Vianello method is actually a refined version of matrix iteration method where after the results converge to the lowest eigenvalue and its corresponding eigenvectors, the effect of the first eigenvectors is eliminated from the matrix based on orthogonal properties of the matrix when in effect the order of matrix is reduced by one. Thus, a matrix of order $n \times n$ gets reduced to a matrix of order $(n-1)$ $(n-1)$ when again performing second round iteration the next subsequent eigenvalues are determined. The steps of the solutions are as elaborated hereafter.

If $[\mathbf{B}]$ is the dynamical matrix as explained in Eq. (2.4.3.2), the equilibrium equation is expressed as

$$[B]\{\phi\} = \frac{1}{\omega^2}\{\phi\} \tag{2.4.4.1}$$

We had shown that starting by a trial vector $\{\phi_0\} = \{1 \quad 1 \quad 1\}^{\mathrm{T}}$ with successive iteration the value finally converges to the lowest eigenvalue ω_1 and its corresponding eigenvector $\left\{\phi^{(1)}\right\}$.

Based on orthogonal property as $\sum_{i-1,j=1}^{n} m_i\phi_i\phi_j = 0$, we can express the above as

$$m_1\left\{\phi_1^{(1)}\right\}\left\{\phi_1^{(2)}\right\} + m_2\left\{\phi_2^{(1)}\right\}\left\{\phi_2^{(2)}\right\} + \cdots + m_n\left\{\phi_n^{(1)}\right\}\left\{\phi_n^{(2)}\right\} = 0 \tag{2.4.4.2}$$

Here, the subscript defines the *number of degrees of freedom* and the superscript defines the *mode.*

Now considering that we have known m_1 and $\left\{\phi^{(1)}\right\}$ previously, Eq. (2.4.4.2) can be expressed as

$$\left\{\phi_1^{(2)}\right\} = -\frac{m_2\left\{\phi_2^{(1)}\right\}\left\{\phi_2^{(2)}\right\}}{m_1\left\{\phi_1^{(1)}\right\}} - \cdots - \frac{m_n\left\{\phi_n^{(1)}\right\}\left\{\phi_n^{(2)}\right\}}{m_1\left\{\phi_1^{(1)}\right\}} \tag{2.4.4.3}$$

Now substituting this value in Eq. (2.4.4.1), we form a new relation

$$[\mathbf{B}']\{\phi\} = \frac{1}{\omega^2}\{\phi\}$$

where $[\mathbf{B}']$ is of the order $(n-1)$ as the value of $\left\{\phi^{(1)}\right\}$ is eliminated. We proceed now with the next cycle of iteration to get the next higher frequency.

The method is now elaborated by a suitable numerical example.

Example 2.4.4 A structure has stiffness and mass matrix as shown below; estimate the eigenvalues by Stodola and Vianello method

$$[K] = \begin{bmatrix} 5 & -2 & 0 \\ -2 & 3.5 & -1.5 \\ 0 & -1.5 & 1.5 \end{bmatrix} \times 10^6 \text{ kN/m and } [M] = \begin{bmatrix} 4 & 0 & 0 \\ 0 & 4 & 0 \\ 0 & 0 & 2 \end{bmatrix} \times 10^3 \text{ kN s}^2/\text{m}$$

Solution

We had worked out in detail in Example 2.4.3 that by matrix deflation method we arrive at a value of $\frac{1}{\omega^2} = 6.082 \times 10^{-3}$ having corresponding eigenvector as $\{\phi\}^{(1)} = \begin{Bmatrix} 1.0 \\ 2.171 \\ 2.781 \end{Bmatrix}$.

We had also calculated the dynamical **[B]** and **[M]** matrices as

$$[B] = \begin{bmatrix} 1.333 & 1.333 & 0.667 \\ 1.333 & 3.333 & 1.667 \\ 1.333 & 3.333 & 3 \end{bmatrix} \times 10^{-3} \text{ and } [M] = \begin{bmatrix} 4 & 0 & 0 \\ 0 & 4 & 0 \\ 0 & 0 & 2 \end{bmatrix} \times 10^3 \text{ kN s}^2/\text{m}$$

Having calculated the first mode, by orthogonal condition

$$m_1\left\{\phi_1^{(1)}\right\}\left\{\phi_1^{(2)}\right\} + m_2\left\{\phi_2^{(1)}\right\}\left\{\phi_2^{(2)}\right\} + m_3\left\{\phi_3^{(1)}\right\}\left\{\phi_3^{(2)}\right\} = 0$$
$$\rightarrow 4000 \times 1.0 \times \left\{\phi_1^{(2)}\right\} + 4000 \times 2.171\left\{\phi_2^{(2)}\right\} + 2000 \times 2.781\left\{\phi_3^{(2)}\right\} = 0$$
$$\left\{\phi_1^{(2)}\right\} = -2.171\left\{\phi_2^{(2)}\right\} - 1.3905\left\{\phi_3^{(2)}\right\}$$

Considering $[B]\{\phi\} = \frac{1}{\omega^2}\{\phi\}$ we have

$$(1.333\phi_1 + 1.333\phi_2 + 0.667\phi_3) \times 10^{-3} = 6.082 \times 10^{-3}$$

$$(1.333\phi_1 + 3.333\phi_2 + 1.667\phi_3) \times 10^{-3} = 13.204 \times 10^{-3}$$

$$(1.333\phi_1 + 3.333\phi_2 + 3\phi_3) \times 10^{-3} = 16.914 \times 10^{-3}$$

Now substituting $\{\phi\} = -2.171\{\phi_2\} - 1.3905\{\phi_3\}$ in last two of the above equations, we eliminate $\{\phi_1\}$ to have

$$(0.43957\phi_2 - 0.1865365\phi_3) \times 10^{-3} = 13.240 \times 10^{-3} \text{ and}$$

$$(0.43957\phi_2 + 1.1464635\phi_3) \times 10^{-3} = 16.914 \times 10^{-3}$$

Thus, the modified dynamic matrix becomes $[\mathbf{B}'] = \begin{bmatrix} 0.43957 & -0.1865365 \\ 0.43957 & 1.1464635 \end{bmatrix}$ $\times 10^{-3}$, and again we start with a trial vector $\{\phi_0\} = \begin{Bmatrix} 1.0 \\ 1.0 \end{Bmatrix}$.

- Thus for first cycle, we have

$$[\mathbf{B}']\{\phi_0\} = \begin{bmatrix} 0.43957 & -0.1865365 \\ 0.43957 & 1.1464635 \end{bmatrix} \begin{Bmatrix} 1 \\ 1 \end{Bmatrix} \times 10^{-3} = 1.586 \times 10^{-3} \begin{Bmatrix} 0.16 \\ 1 \end{Bmatrix}$$
$$\rightarrow \{\phi_1\} = \begin{Bmatrix} 0.16 \\ 1 \end{Bmatrix}$$

- For the second cycle

$$[\mathbf{B}']\{\phi_1\} = \begin{bmatrix} 0.43957 & -0.1865365 \\ 0.43957 & 1.1464635 \end{bmatrix} \begin{Bmatrix} 0.16 \\ 1 \end{Bmatrix} \times 10^{-3}$$
$$= 1.217 \times 10^{-3} \begin{Bmatrix} -0.096 \\ 1 \end{Bmatrix} \rightarrow \{\phi_2\} = \begin{Bmatrix} -0.096 \\ 1 \end{Bmatrix}$$

- For the third cycle

$$[\mathbf{B}']\{\phi_2\} = \begin{bmatrix} 0.43957 & -0.1865365 \\ 0.43957 & 1.1464635 \end{bmatrix} \begin{Bmatrix} -0.096 \\ 1 \end{Bmatrix} \times 10^{-3}$$
$$= 1.104 \times 10^{-3} \begin{Bmatrix} -0.207 \\ 1 \end{Bmatrix} \rightarrow \{\phi_3\} = \begin{Bmatrix} -0.207 \\ 1 \end{Bmatrix}$$

....

- After 15th cycle

$$[\mathbf{B}']\{\phi_{14}\} = \begin{bmatrix} 0.43957 & -0.1865365 \\ 0.43957 & 1.1464635 \end{bmatrix} \begin{Bmatrix} -0.333 \\ 1 \end{Bmatrix} \times 10^{-3}$$
$$= 1.000 \times 10^{-3} \begin{Bmatrix} -0.333 \\ 1 \end{Bmatrix} \rightarrow \{\phi_{14}\} = \begin{Bmatrix} -0.333 \\ 1 \end{Bmatrix}$$

- At 16th cycle

$$[\mathbf{B}']\{\phi_{15}\} = \begin{bmatrix} 0.43957 & -0.1865365 \\ 0.43957 & 1.1464635 \end{bmatrix} \begin{Bmatrix} -0.333 \\ 1 \end{Bmatrix} \times 10^{-3}$$
$$= 1.000 \times 10^{-3} \begin{Bmatrix} -0.333 \\ 1 \end{Bmatrix} \rightarrow \{\phi_{16}\} = \begin{Bmatrix} -0.333 \\ 1 \end{Bmatrix}$$

Thus, we see that the column vectors have stabilized when $\frac{1}{\omega^2} = 1 \times 10^{-3} \rightarrow \lambda = \omega^2 = 1000$ (same value as obtained in Example 2.4.2).

The complete eigenvector is now constructed from the expression

$$\left\{\phi_1^{(2)}\right\} = -2.171\left\{\phi_2^{(2)}\right\} - 1.3905\left\{\phi_3^{(2)}\right\}$$

$$\left\{\phi_1^{(2)}\right\} = -2.171 \times (-0.333) - 1.3905 \times 1$$

$$\rightarrow \phi_1 = -0.6669$$

Thus, the complete eigenvector for the second mode is $\left\{\phi^{(2)}\right\} = \begin{Bmatrix} -0.6669 \\ -0.333 \\ 1.0 \end{Bmatrix}$

and the corresponding eigenvalue is 1000.

For the next subsequent mode, we use the following two equations.

$$m_1\left\{\phi_1^{(1)}\right\}\left\{\phi_1^{(3)}\right\} + m_2\left\{\phi_2^{(1)}\right\}\left\{\phi_2^{(3)}\right\} + m_3\left\{\phi_3^{(1)}\right\}\left\{\phi_3^{(3)}\right\} = 0$$

$$\left\{\phi_1^{(3)}\right\} = -2.171\left\{\phi_2^{(3)}\right\} - 1.3905\left\{\phi_3^{(3)}\right\} \qquad \text{(i)}$$

And $m_1\left\{\phi_1^{(2)}\right\}\left\{\phi_1^{(3)}\right\} + m_2\left\{\phi_2^{(2)}\right\}\left\{\phi_2^{(3)}\right\} + m_3\left\{\phi_3^{(2)}\right\}\left\{\phi_3^{(3)}\right\} = 0$

$$4000 \times (-0.6669)\left\{\phi_1^{(3)}\right\} + 4000 \times (-0.333)\left\{\phi_2^{(3)}\right\} + 2000 \times 1\left\{\phi_3^{(3)}\right\} = 0$$

$$\left\{\phi_1^{(3)}\right\} = -0.5\left\{\phi_2^{(3)}\right\} + 0.75\left\{\phi_3^{(3)}\right\} = 0 \qquad \text{(ii)}$$

Equating Equations (i) and (ii), we have

$$-2.171\left\{\phi_2^{(3)}\right\} - 1.3905\left\{\phi_3^{(3)}\right\} = -0.5\left\{\phi_2^{(3)}\right\} + 0.75 \times 1\left\{\phi_3^{(3)}\right\}$$

$$\rightarrow \left\{\phi_2^{(3)}\right\} = -1.280969\left\{\phi_3^{(3)}\right\}$$

Now substituting the above in the last row of matrix $[\mathbf{B}']$, we have

$$(-0.43957 \times 1.280969 + 1.464635) \times 10^{-3}\{\phi_3\} = \frac{1}{\omega_3^2}[\phi_3]$$

$$\Rightarrow \left[0.5833888396 \times 10^{-3}\right]\{\phi_3\} = \frac{1}{\omega_3^2}[\phi_3]$$

Thus, the next eigenvalue is[4] $\lambda_3 = 1/0.5833888396 \times 10^{-3} = 1714.12$ and $\left[\phi_3^{(3)}\right] = 1.0$

Thus $\left[\phi_2^{(3)}\right] = -1.280969$ and

$$\left\{\phi_1^{(3)}\right\} = -0.5 \times (-1.280969) + 0.75 \times 1 = 1.3904845$$

Thus, the required eigenvector is $\{\phi\}^{(3)} = \left\{\begin{array}{c} 1.3904845 \\ -1.280969 \\ 1.0 \end{array}\right\}$.

2.4.5 Eigenvalue Computation Based on Vector Transformation

In Sect. 2.4.4, we had shown techniques to determine the eigenvalues and vectors based on vector iteration. In this section, we show another method based on vector transformation, an algorithm that is considered one of the most dominant techniques adapted by computer software for extraction of eigenvalues for large finite element problems.

The underlying concept of the method is as elaborated hereafter.

For a system under free vibration, the equation of equilibrium is expressed as

$$[\mathbf{K}]\{\varphi\} = \omega^2[\mathbf{M}]\{\varphi\} \tag{2.4.5.1}$$

In Eq. (2.4.5.1) $[\mathbf{K}]$ = stiffness matrix of the system, $[\mathbf{M}]$ = mass matrix, $\lambda = \omega^2$ = eigenvalue of the system, and $\{\varphi\}$ = normalized eigenvector.

We had shown in Sects. 2.3.7 and 2.3.8 that if $\{\varphi\}$ is the normalized eigenvectors, then, based on operations as shown below, we have

$$\{\varphi\}^{\mathrm{T}}[\mathbf{K}]\{\varphi\} = [\lambda] \text{ and } \{\varphi\}^{\mathrm{T}}[\mathbf{M}]\{\varphi\} = [\mathbf{I}] \tag{2.4.5.2}$$

In Eq. (2.4.5.2) $[\lambda]$ = a diagonal matrix consisting of the eigenvalues of the system and $[\mathbf{I}]$ = an identity matrix.

For the eigenvector $\{\varphi\}$, as this diagonalization is unique, by transformation method we thus try to construct, based on iteration, a vector which reduces $[\mathbf{K}]$ and $[\mathbf{M}]$ into a diagonal matrix by successively pre- and post-multiplying the transformation matrices $[T_r]^{\mathrm{T}}$ and $[T_r]$, where r = number of iterations 1, 2, 3, …

[4]We had previously computed this value as 1710.6.

Thus

$$\begin{aligned}
[\mathbf{K}_2] &= [T_1]^{\mathrm{T}}[\mathbf{K}_1][T_1] \\
[\mathbf{K}_3] &= [T_2]^{\mathrm{T}}[\mathbf{K}_2][T_2] \\
&\ldots\ldots\ldots\ldots\ldots \\
[\mathbf{K}_{r+1}] &= [T_r]^{\mathrm{T}}[\mathbf{K}_r][T_{r1}]
\end{aligned} \tag{2.4.5.3}$$

And for the mass matrix

$$\begin{aligned}
[\mathbf{M}_2] &= [T_1]^{\mathrm{T}}[\mathbf{M}_1][T_1] \\
[\mathbf{M}_3] &= [T_2]^{\mathrm{T}}[\mathbf{M}_2][T_2] \\
&\ldots\ldots\ldots \\
[\mathbf{M}_{r+1}] &= [T_r]^{\mathrm{T}}[\mathbf{M}_r][T_r]
\end{aligned} \tag{2.4.5.4}$$

The transformation matrix is chosen in such a way that

$$[K_{r+1}] \to [\lambda] \text{ and } [M_{r+1}] \to [I] \text{ as } r \to \infty$$

In real-world engineering, it is not necessary for the matrix $[K_{r+1}]$ to converge to $[\lambda]$ or $[M]$ to $[I]$ as long as they converge to a diagonal form $[K_{r+1}] \to \mathrm{Diag}\left[K_{ii}^r\right]$ and $[M_{r+1}] \to \mathrm{Diag}\left[M_{ii}^r\right]$.

The eigenvalues are then computed from the expression

$$[\lambda] \to \mathrm{Diag}\left[\frac{K_{ii}^r}{M_{ii}^r}\right] \tag{2.4.5.5}$$

and

$$[\varphi] \to [T_1][T_2]\ldots[T_r]\mathrm{Diag}\left[\frac{1}{\sqrt{M_{nn}^r}}\right] \tag{2.4.5.6}$$

Here, n = order of the matrix.

In the above computation, the reader is to note that λ may not be in sequential (i.e., ascending or descending) order.

Based on the above, we now describe two techniques that have an important application in vibration problem in engineering having large degrees of freedom and finite element analysis.

2.4.5.1 Standard Jacobi Technique

Carl Gustav Jacob Jacobi (1804–1851), the German mathematician, presented this theory in 1846, and one hundred and seventy years since then it still remains one of the most powerful methods of eigensolution in the realm of applied physics and engineering. The method can be used for both standard and general eigenvalue problem. We discuss herein the standard problem to start with.

For a system under free vibration, the equation of equilibrium is expressed as

$$[\mathbf{K}]\{\varphi\} = \omega^2[\mathbf{M}]\{\varphi\} \tag{2.4.5.7}$$

Equation (2.4.5.7) can be expressed as

$$[\mathbf{A}]\{\varphi\} = \omega^2\{\varphi\} \text{ where } [A] = [\mathbf{K}][\mathbf{M}]^{-1} \tag{2.4.5.8}$$

Equation (2.4.5.8) is known as the *standard eigenvalue* problem such that on transformation

$$[T_r]^{\mathrm{T}}\ldots[T_3]^{\mathrm{T}}[T_2]^{\mathrm{T}}[T_1]^{\mathrm{T}}[\mathbf{A}][T_1][T_2][T_3]\ldots[T_r] \to \lambda \tag{2.4.5.9}$$

And

$$[T_1][T_2][T_3]\ldots[T_r] \to [I] \tag{2.4.5.10}$$

According to Jacobi, the rotation matrix $[T_r]$ is chosen in such a way that the off-diagonal element of $[\mathbf{A}_r]$ is zeroed. If element (*j*, *k*) is to be reduced to zero, the corresponding transformation matrix is chosen as

$$[T_r] = \begin{bmatrix} & & & j \text{ col} & k \text{ col} \\ 1 & 0 & 0 & 0 & 0 \\ \cdot & \cdot & \cdot & \cdot & \cdot \\ 0 & 0 & 1 & 0 & 0 \\ 0 & 0 & 0 & \cos\theta & -\sin\theta \\ & & & \sin\theta & \cos\theta \end{bmatrix} \begin{matrix} \\ \\ \\ \\ j \text{ row} \\ k \text{ row} \end{matrix} \tag{2.4.5.11}$$

Here $\tan 2\theta = \frac{2A^r_{jk}}{A^r_{jj} - A^r_{kk}}$, where $A^r_{jj} \neq A^r_{kk}$. If $A^r_{jj} = A^r_{kk}$, then $\theta = \pi/4$.

For a symmetric matrix, the transformation needs to be only applied to either lower or upper triangular matrix, including the diagonal term. For optimal solution and efficient computer implementation, a special form of the solution called *Threshold Jacobi Technique* is usually applied (Bathe 1996).

In this case, off-diagonal elements are tested column by column or row by row and transformation is only applied if the value is greater than a pre-defined threshold value for that sweep. Since the objective is to reduce the matrix into a diagonal form, the strategy adapted is to diminish the coupling effect between the *j* and *k* degrees of freedom. An estimate of this coupling effect is given by $\sqrt{A^2_{jk}/A_{jj} \cdot A_{kk}}$. This factor can be effectively used to decide if a rotation has to be applied or not. The convergence is measured based on the threshold tolerance for a finite number of iterations. Thus, for a particular finite number of iterations *q*, it is assumed to have converged to tolerance *t* if

$$\frac{\left|A_{jj}^{q+1}-A_{jj}^{q}\right|}{A_{jj}^{q+1}} \leq 10^{-t} \text{ and } \sqrt{\frac{\left[A_{jk}^{q+1}\right]^2}{A_{jj}^{q+1}A_{kk}^{q+1}}} \leq 10^{-t} \text{ for all } i<j \text{ and } i=1,2,3\ldots,n \tag{2.4.5.12}$$

Having defined the method, it would be worth examining now as to how and why the matrix gets reduced to the diagonal form.

Let us consider the matrix

$$[A]=\begin{bmatrix} A_{jj} & A_{jk} \\ A_{kj} & A_{kk} \end{bmatrix} \text{ a } 2\times 2 \text{ symmetric matrix.} \tag{2.4.5.13}$$

The transformation matrix as defined above is expressed as

$$[T]=\begin{bmatrix} \cos\theta & -\sin\theta \\ \sin\theta & \cos\theta \end{bmatrix} \tag{2.4.5.14}$$

Based on transformation operation, we have

$$[T]^{\mathrm{T}}[A][T]=\begin{bmatrix} \cos\theta & \sin\theta \\ -\sin\theta & \cos\theta \end{bmatrix}\begin{bmatrix} A_{jj} & A_{jk} \\ A_{kj} & A_{kk} \end{bmatrix}\begin{bmatrix} \cos\theta & -\sin\theta \\ \sin\theta & \cos\theta \end{bmatrix} \tag{2.4.5.15}$$

Expansion of Eq. (2.4.5.15) gives

$$[T]^{\mathrm{T}}[A][T]=\begin{bmatrix} A_{jj}\cos^2\theta+A_{jk}\sin 2\theta+A_{kk}\sin^2\theta & (A_{kk}-A_{jj})\sin\theta\cos\theta+A_{jk}\cos 2\theta \\ (A_{kk}-A_{jj})\sin\theta\cos\theta+A_{jk}\cos 2\theta & A_{jj}\sin^2\theta-A_{jk}\sin 2\theta+A_{kk}\cos^2\theta \end{bmatrix} \tag{2.4.5.16}$$

As the off-diagonal terms are to be reduced to zero, we have

$$\frac{1}{2}(A_{kk}-A_{jj})\sin 2\theta+A_{jk}\cos 2\theta=0 \tag{2.4.5.17}$$

$$\rightarrow \tan 2\theta=\frac{2A_{jk}}{A_{jj}-A_{kk}} \text{ when } A_{jj}\neq A_{kk} \tag{2.4.5.18}$$

For $A_{jj}=A_{kk}$, $\theta=\pi/4$ for $A_{jk}>0$ and $\theta=-\pi/4$ for $A_{jk}<0$.

Thus, based on the above transformation, the matrix is systematically transformed to a diagonal form. It can be proved that once the off-diagonal elements are small, the convergence is quadratic in nature and minimal cost is required to calculate the eigenpairs to high accuracy once the approximate solution is obtained (Wilkinson 1969).

Example 2.4.5 For the structure having stiffness and mass matrix as shown below, estimate the eigenpairs by standard Jacobi method

$$[K] = \begin{bmatrix} 5 & -2 & 0 \\ -2 & 3.5 & -1.5 \\ 0 & -1.5 & 1.5 \end{bmatrix} \times 10^6 \text{ kN/m and } [M] = \begin{bmatrix} 4 & 0 & 0 \\ 0 & 4 & 0 \\ 0 & 0 & 2 \end{bmatrix} \times 10^3 \text{ kN s}^2/\text{m}$$

Solution

For the given problem $[M]^{-1} = \begin{bmatrix} 0.25 & 0 & 0 \\ 0 & 0.25 & 0 \\ 0 & 0 & 0.5 \end{bmatrix} \times 10^{-3}\ \text{m/kN s}^2$

Thus,

$$A = [K][M]^{-1}$$
$$= \begin{bmatrix} 5 & -2 & 0 \\ -2 & 3.5 & -1.5 \\ 0 & -1.5 & 1.5 \end{bmatrix} \begin{bmatrix} 0.25 & 0 & 0 \\ 0 & 0.25 & 0 \\ 0 & 0 & 0.5 \end{bmatrix} \times 10^3$$
$$= \begin{bmatrix} 1250 & -500 & 0 \\ -500 & 875 & -750 \\ 0 & -375 & 750 \end{bmatrix}$$

For the first row (j = 1) and second column (k = 2)

$\theta = \frac{1}{2}\tan^{-1}\left(\frac{2A_{12}}{A_{11}-A_{22}}\right) = \frac{1}{2}\tan^{-1}\left(\frac{2\times(-500)}{1250-875}\right) = -0.606$ (here the value is presented in radians)

Thus, $\cos\theta = 0.822$ and $\sin\theta = -0.570$

Thus, the transformation matrix is expressed as

$$T_1 = \begin{bmatrix} 0.822 & 0.570 & 0 \\ -0.570 & 0.822 & 0 \\ 0 & 0 & 1 \end{bmatrix}$$

$$T_1^{\mathrm{T}} A T_1 = \begin{bmatrix} 0.822 & -0.570 & 0 \\ 0.570 & 0.822 & 0 \\ 0 & 0 & 1 \end{bmatrix} \times \begin{bmatrix} 1250 & -500 & 0 \\ -500 & 875 & -750 \\ 0 & -375 & 750 \end{bmatrix}$$
$$\times \begin{bmatrix} 0.822 & 0.570 & 0 \\ -0.570 & 0.822 & 0 \\ 0 & 0 & 1 \end{bmatrix}$$
$$= \begin{bmatrix} 1597 & 0 & 427.196 \\ 8.527\times10^{-14} & 528.5 & -616.444 \\ 213.598 & -308.222 & 750 \end{bmatrix}$$

Thus, the modified value is $[A'] = \begin{bmatrix} 1597 & 0 & 427.196 \\ 8.527 \times 10^{-14} & 528.5 & -616.444 \\ 213.598 & -308.222 & 750 \end{bmatrix}$

For $j = 1$ and $k = 3$

$$\theta = \frac{1}{2}\tan^{-1}\left(\frac{2A'_{13}}{A'_{11} - A'_{33}}\right) = \frac{1}{2}\tan^{-1}\left(\frac{2 \times (427.196)}{1597 - 750}\right) = 0.395$$

Thus, $\cos\theta = 0.923$ and $\sin\theta = 0.385$

Thus, the transformation matrix is expressed as

$$T_2 = \begin{bmatrix} 0.923 & 0 & -0.385 \\ 0 & 1 & 0 \\ 0.385 & 0 & 0.923 \end{bmatrix}$$

Thus

$$T_2^{\mathrm{T}} T_1^{\mathrm{T}} A T_1 T_2 = \begin{bmatrix} 1699 & -118.612 & 31.632 \\ -237.224 & 528.5 & -568.971 \\ -181.966 & -284.486 & 647.755 \end{bmatrix}$$

Thus $A''_2 = \begin{bmatrix} 1699 & -118.612 & 31.632 \\ -237.224 & 528.5 & -568.971 \\ -181.966 & -284.486 & 647.755 \end{bmatrix}$

For $j = 2$ and $k = 3$

$$\theta = \frac{1}{2}\tan^{-1}\left(\frac{2A''_{23}}{A''_{22} - A''_{33}}\right) = \frac{1}{2}\tan^{-1}\left(\frac{2 \times (-568.971)}{528.5 - 647.755}\right) = 0.733$$

Thus, $\cos\theta = 0.743$ and $\sin\theta = 0.669$

Thus, the transformation matrix is expressed as

$$T_3 = \begin{bmatrix} 1 & 0 & 0 \\ 0 & 0.743 & -0.669 \\ 0 & 0.669 & 0.743 \end{bmatrix} \text{hence}$$

$$T_3^{\mathrm{T}} T_2^{\mathrm{T}} T_1^{\mathrm{T}} A T_1 T_2 T_3 = \begin{bmatrix} 1699 & -66.964 & 102.844 \\ -298.047 & 157.508 & -127.417 \\ 23.551 & 157.069 & 1019 \end{bmatrix}$$

And

$$T_1 T_2 T_3 = \begin{bmatrix} 0.759 & 0.212 & -0.616 \\ -0.526 & 0.757 & -0.387 \\ 0.385 & 0.618 & 0.686 \end{bmatrix}$$

This completes the first cycle.

For the second cycle we start with

$$A = \begin{bmatrix} 1699 & -66.964 & 102.844 \\ -298.047 & 157.508 & -127.417 \\ 23.551 & 157.069 & 1019 \end{bmatrix}$$

For the first row ($j = 1$) and second column ($k = 2$)

$$\theta = \frac{1}{2}\tan^{-1}\left(\frac{2A_{12}}{A_{11} - A_{22}}\right) = \frac{1}{2}\tan^{-1}\left(\frac{2 \times (-66.964)}{1699 - 157.508}\right) = -0.043$$

Thus, $\cos\theta = 0.999$ and $\sin\theta = -0.043$

Thus, the transformation matrix is expressed as

$$T_1 = \begin{bmatrix} 0.999 & 0.043 & 0 \\ -0.043 & 0.999 & 0 \\ 0 & 0 & 1 \end{bmatrix}$$

$$T_1^{\mathrm{T}} A T_1 = \begin{bmatrix} 1712 & 0.434 & 108.308 \\ -230.649 & 144.602 & -122.84 \\ 16.724 & 157.941 & 1019 \end{bmatrix} \rightarrow [A']\text{say}$$

For $j = 1$ and $k = 3$

$$\theta = \frac{1}{2}\tan^{-1}\left(\frac{2A'_{13}}{A'_{11} - A'_{33}}\right) = \frac{1}{2}\tan^{-1}\left(\frac{2 \times (108.308)}{1712 - 1019}\right) = 0.151$$

Thus, $\cos\theta = 0.989$ and $\sin\theta = 0.151$

Thus, the transformation matrix is expressed as

$$T_2 = \begin{bmatrix} 0.989 & 0 & -0.151 \\ 0 & 1 & 0 \\ 0.151 & 0 & 0.989 \end{bmatrix}$$

Thus

$$T_2^{\mathrm{T}} T_1^{\mathrm{T}} A T_1 T_2 = \begin{bmatrix} 1715 & 24.625 & 2.086 \\ -246.546 & 144.602 & -86.624 \\ -89.498 & 156.067 & 1016 \end{bmatrix} \rightarrow [A'']\text{say}$$

For $j = 2$ and $k = 3$

$$\theta = \frac{1}{2}\tan^{-1}\left(\frac{2A''_{23}}{A''_{22} - A''_{33}}\right) = \frac{1}{2}\tan^{-1}\left(\frac{2 \times (-86.624)}{144.602 - 1016}\right) = 0.098$$

Thus, $\cos\theta = 0.995$ and $\sin\theta = 0.098$
Thus, the transformation matrix is expressed as

$$T_3 = \begin{bmatrix} 1 & 0 & 0 \\ 0 & 0.995 & -0.098 \\ 0 & 0.098 & 0.995 \end{bmatrix} \text{ hence}$$

$$\therefore T_3^T T_2^T T_1^T A T_1 T_2 T_3 = \begin{bmatrix} 1714 & 50.921 & 5.69 \\ -256.561 & 155.766 & -4.093 \\ -72.278 & 244.624 & 999.199 \end{bmatrix}$$

and

$$T_1 T_2 T_3 = \begin{bmatrix} 0.988 & 0.028 & -0.154 \\ -0.043 & 0.995 & -0.091 \\ 0.151 & 0.097 & 0.984 \end{bmatrix}$$

For the third cycle, we start with

$A = \begin{bmatrix} 1714 & 50.921 & 5.69 \\ -256.561 & 155.766 & -4.093 \\ -72.278 & 244.624 & 999.199 \end{bmatrix}$, and proceeding in identical manner, we compute

$$T_3^T T_2^T T_1^T A T_1 T_2 T_3 = \begin{bmatrix} 1705 & 2.296 & -0.117 \\ -307.337 & 164.676 & 0.062 \\ -69.058 & 248.443 & 999.223 \end{bmatrix}$$

and

$$T_1 T_2 T_3 = \begin{bmatrix} 0.999 & -0.333 & -7.785 \times 10^{-3} \\ 0.033 & 0.999 & -2.483 \times 10^{-3} \\ 7.861 \times 10^{-3} & 2.227 \times 10^{-3} & 1 \end{bmatrix} = [\phi]$$

We can stop further iteration as the eigenvector matrix $[\varphi] \to [I]$; thus, for the present problem

$$[\lambda] = \begin{Vmatrix} 164.746 \\ 999.223 \\ 1705 \end{Vmatrix} \text{ and } [\varphi] = \begin{bmatrix} 0.999 & -0.333 & -7.785 \times 10^{-3} \\ 0.033 & 0.999 & -2.483 \times 10^{-3} \\ 7.861 \times 10^{-3} & 2.227 \times 10^{-3} & 1 \end{bmatrix}$$

Comparing to previous problem, it is seen that eigenvalues are very closely matching though the off-diagonal term convergence has been poor. This is because on transformation of the matrix $[\mathbf{K}]$ and $[\mathbf{M}]$ to $[\mathbf{A}]$ through the operation $[\mathbf{K}][\mathbf{M}]^{-1}$, the matrix has lost its symmetric property. An important point to be noticed herein is that, unlike matrix deflation or Stodola's method where number of iterations increase significantly beyond the fundamental mode, in Jacobí's method the number of iteration cycles are significantly less. And this makes the technique a very attractive option for computer implementation related to dynamic analysis of systems with large degrees of freedom.

2.4.5.2 Generalized Jacobi Technique

In Sect. (2.4.5.1), we had shown solution procedure for standard eigenvalue solution of the expression

$$[\mathbf{A}][\varphi] = \lambda[\varphi] \tag{2.4.5.19}$$

where $[\mathbf{A}]$ is dynamic matrix expressed as $[\mathbf{K}][\mathbf{M}]^{-1}$.

When we consider the structural system as a discrete model consisting of lumped masses, the mass matrix $[M_{ii}]$ is diagonal, when inversion of the mass matrix is only a trivial exercise.

However, if the mass matrix is computed based on consistent mass basis or for coupled system analysis, the mass matrix would be symmetric but not diagonal and then operation $[\mathbf{A}] = [\mathbf{K}][\mathbf{M}]^{-1}$ can be a computationally expensive task for large systems. In such cases, generalized Jacobi's technique is usually adopted where the transformation is directly done on the expression

$$[\mathbf{K}][\varphi] = \lambda[\mathbf{M}][\varphi] \tag{2.4.5.20}$$

Here, we do not convert Eq. (2.4.5.20) to the standard eigenvalue form, as depicted by Eq. (2.4.5.19).

In this case, the transformation matrix is given by (for row i and column j)

$$[T_r] = \begin{bmatrix} 1 & 0 & 0 & . & . \\ 0 & 1 & 0 & a & . \\ . & . & 1 & . & . \\ . & b & . & 1 & . \\ . & . & . & . & 1 \end{bmatrix} \tag{2.4.5.21}$$

The coefficients a and b are chosen in such a way that on application it simultaneously reduces elements of row i and column j of matrix $[K_r]$ and $[M_r]$ to zero.

It is evident that, based on above statement, the coefficients a and b for the rth iteration are a function of $k_{ii}^r, k_{jj}^r, k_{ij}^r, m_{ii}^r, m_{jj}^r$ and m_{ij}^r.

Performing the matrix operation $[T_r]^T[K_r][T_r]$ and $[T_r]^T[M_r][T_r]$, and satisfying the boundary condition that k_{ij}^{r+1} and m_{ij}^{r+1} will tend to zero, will result in the following boundary conditions

$$ak_{ii}^r + (1+ab)k_{ij}^r + bk_{jj}^r = 0 \text{ and } am_{ii}^r + (1+ab)m_{ij}^r + bm_{jj}^r = 0 \qquad (2.4.5.22)$$

For solution of a and b, we express

$$\begin{aligned} k_{ii}^r &= k_{ii}^r m_{ij}^r - m_{ii}^r k_{ij}^r \\ k_{jj}^r &= k_{jj}^r m_{ij}^r - m_{jj}^r k_{ij}^r \text{ and} \\ \hat{k}^r &= k_{ii}^r m_{jj}^r - m_{ii}^r k_{jj}^r \end{aligned} \qquad (2.4.5.23)$$

where $b = -\frac{k_{ii}^r}{x}$ and $a = \frac{k_{jj}^r}{x}$, and x is defined as

$$x = \frac{\hat{k}^r}{2} + \text{sign}(\hat{k}^r)\sqrt{\left(\frac{\hat{k}^r}{2}\right) + k_{ii}^r \times k_{jj}^r} \qquad (2.4.5.24)$$

In Eq. (2.4.5.24), the function $\text{sign}(\hat{k}^r)$ means if the value of $\hat{k}^r > 0$, then we consider it as (+)1 and when $\hat{k}^r < 0$ we consider it as (−)1.

Thus, based on the above, the generalized Jacobi's technique can be structured as follows:

- Given the matrices **[K]** and **[M]**, determine for each row i and j (where $i, j = 1, 2, 3, \ldots, n$)
  - $k_{ii}^r = k_{ii}^r m_{ij}^r - m_{ii}^r k_{ij}^r$
  - $k_{ii}^r = k_{ii}^r m_{ij}^r - m_{ii}^r k_{ij}^r$
  - $\hat{k}^r = k_{jj}^r m_{ij}^r - m_{jj}^r k_{ij}^r$
- If $\hat{k}^r > 0$ then $x = \frac{\hat{k}^r}{2} + \sqrt{\left(\frac{\hat{k}^r}{2}\right) + k_{ii}^r \times k_{jj}^r}$, else
- $x = \frac{\hat{k}^r}{2} - \sqrt{\left(\frac{\hat{k}^r}{2}\right) + k_{ii}^r \times k_{jj}^r}$
- Find $a = \frac{k_{jj}^r}{x}$ and $b = -\frac{k_{ii}^r}{x}$
- Form the transformation matrix $[T_r] = \begin{bmatrix} 1 & 0 & 0 & . & . \\ 0 & 1 & 0 & a & . \\ . & . & 1 & . & . \\ . & b & . & 1 & . \\ . & . & . & . & 1 \end{bmatrix}$
- Form $[K]^{r+1} = [T_r]^T[K][T_r]$ and $[M]^{r+1} = [T_r]^T[M][T_r]$
- Repeat the cycle till the matrix is diagonalized; i.e., the off-diagonal terms get reduced below the tolerance value 10^{-t}.

- Find the eigenvalue based on the expression $[\lambda] = \text{Diag}\left[K_{ii}^r/M_{ii}^r\right]$
- Find the eigenvectors based on the expression $[\varphi] = [T_1][T_2]\ldots[T_r]\text{Diag}\left[\frac{1}{\sqrt{M_{nn}^r}}\right]$

We now elaborate the problem further by a suitable numerical problem hereafter.

Example 2.4.6 For the structure having stiffness and mass matrix as shown below, estimate the eigenpairs by generalized Jacobi method

$$[\mathbf{K}] = \begin{bmatrix} 5 & -2 & 0 \\ -2 & 3.5 & -1.5 \\ 0 & -1.5 & 1.5 \end{bmatrix} \times 10^6 \text{kN/m, and } [\mathbf{M}] = \begin{bmatrix} 4 & 0 & 0 \\ 0 & 4 & 0 \\ 0 & 0 & 2 \end{bmatrix} \times 10^3 \text{ kN s}^2/\text{m}$$

Solution

We start with the iteration with i(row) = 1 and j(col.) = 2

Thus,

$$k_{11}^1 = K_{11}^1 M_{12}^1 - M_{11}^1 K_{12}^1$$

$$\rightarrow k_{11}^1 = 5 \times 10^6 \times 0 - 4000 \times (-2 \times 10^6) = 8 \times 10^9$$

$$k_{22}^1 = K_{22}^1 M_{12}^1 - M_{22}^1 K_{12}^1$$

$$\rightarrow k_{22}^1 = 3.5 \times 10^6 \times 0 - 4000 \times (-2 \times 10^6) = 8 \times 10^9$$

$$\hat{k}_{22}^1 = k_{11}^1 m_{22}^1 - m_{11}^1 k_{22}^1$$

$$\hat{k}^1 = 5 \times 10^6 \times 4000 - 4000 \times (3.5 \times 10^6) = 6 \times 10^6 > 0$$

$$\therefore x = \frac{\hat{k}^1}{2} + \sqrt{\left(\frac{\hat{k}^1}{2}\right) + k_{11}^1 \times k_{22}^1}$$

$$x = \frac{6 \times 10^6}{2} + \sqrt{\left(\frac{6 \times 10^6}{2}\right) + (8 \times 10^9)^2} = 11{,}544{,}003{,}745$$

$$a = \frac{k_{22}^1}{x} = 0.693000468 \text{ and } b = \frac{-k_{11}^1}{x} = -0.69300468$$

$$\therefore [T_1] = \begin{bmatrix} 1 & 0.693000468 & 0 \\ -0.693000468 & 1 & 0 \\ 0 & 0 & 1 \end{bmatrix}$$

$$\therefore [T_1]^{\mathrm{T}}[\mathbf{K}][T_1] = \begin{bmatrix} 9{,}452{,}875.644 & 0 & 1{,}039{,}500.702 \\ 0 & 3{,}129{,}246.3472 & -1{,}500{,}000 \\ 1{,}039{,}500.702 & -1{,}500{,}000 & 1{,}500{,}000 \end{bmatrix} \text{ and}$$

$$[T_1]^{\mathrm{T}}[\mathbf{M}][T_1] = \begin{bmatrix} 5920.998596 & 0 & 0 \\ 0 & 5920.998596 & 0 \\ 0 & 0 & 2000 \end{bmatrix}$$

Now for $i = 1$, $i = 3$
Thus,

$$k_{11}^1 = K_{11}^1 M_{13}^1 - M_{11}^1 K_{13}^1$$

$$\rightarrow k_{11}^1 = 9{,}452{,}875.644 \times 0 - 5920.998596 \times 1{,}039{,}500.702 = -6{,}154{,}882{,}198$$

$$k_{33}^1 = K_{33}^1 M_{13}^1 - M_{33}^1 K_{13}^1$$

$$\rightarrow k_{33}^1 = 1.5 \times 10^6 \times 0 - 2000 \times 1{,}039{,}500.702 = -2{,}079{,}001{,}404$$

$$\hat{k}^1 = k_{11}^1 m_{33}^1 - m_{11}^1 k_{33}^1$$

$$\hat{k}^1 = 9{,}452{,}875.644 \times 2000 - 5920.998596 \times 1.5 \times 10^6 = 10{,}024{,}253{,}394$$

$$x = \frac{\hat{k}^1}{2} + \sqrt{\left(\frac{\hat{k}^1}{2}\right) + k_{11}^1 \times k_{33}^1}$$

$$\begin{aligned} x = {} & \frac{10{,}024{,}253{,}394}{2} \\ & + \sqrt{\left(\frac{10{,}024{,}253{,}394}{2}\right) + -(6{,}154{,}882{,}198)(-2{,}079{,}001{,}404)} \\ = {} & 11{,}169{,}839{,}159 \end{aligned}$$

$$a = \frac{k_{33}^1}{x} = -0.186126 \text{ and } b = \frac{-k_{11}^1}{x} = 0.551026932$$

$$\therefore [T_2] = \begin{bmatrix} 1 & 0 & -0.186126351 \\ 0 & 1 & 0 \\ 0.551026932 & 0 & 1 \end{bmatrix}$$

$$\therefore [T_2]^{\mathrm{T}}[T_1]^{\mathrm{T}}[\mathbf{K}][T_1][T_2] = \begin{bmatrix} 11{,}053{,}907.43 & -826{,}540.3974 & 3.49246 \times 10^{-10} \\ -826{,}540.3974 & 3{,}129{,}246.3472 & -1{,}500{,}000 \\ 0 & -1{,}500{,}000 & 1{,}440{,}519.201 \end{bmatrix} \text{ and}$$

$$[T_2]^{\mathrm{T}}[T_1]^{\mathrm{T}}[\mathbf{M}][T_1][T_2] = \begin{bmatrix} 6528.259954 & 0 & 2.27374\times10^{-13} \\ 0 & 5920.998596 & 0 \\ 2.27374\times10^{-13} & 0 & 2205.121264 \end{bmatrix}$$

Now for $i = 2, j = 3$
Thus

$$k_{22}^{1} = K_{22}^{1}M_{23}^{1} - M_{22}^{1}K_{23}^{1}$$

$$\rightarrow k_{22}^{1} = 3{,}129{,}246.3472 \times 0 - 5920.998596 \times (-1{,}500{,}000) = 8{,}881{,}497{,}893$$

$$k_{33}^{1} = K_{33}^{1}M_{23}^{1} - M_{33}^{1}K_{23}^{1}$$

$$\rightarrow k_{33}^{1} = 1{,}440{,}519.201 \times 0 - 2205.121264 \times (-1{,}500{,}000) = 3{,}307{,}681{,}896$$

$$\hat{k}^{1} = k_{22}^{1}m_{33}^{1} - m_{22}^{1}k_{33}^{1}$$

$$\begin{aligned}\hat{k}^{1} &= 3{,}129{,}246.3472 \times 2205.121264 - 5920.998596 \times 1{,}440{,}519.201 \\ &= -1{,}628{,}944{,}451 < 0\end{aligned}$$

Hence

$$x = \frac{\hat{k}^{1}}{2} - \sqrt{\left(\frac{\hat{k}^{1}}{2}\right) + k_{22}^{1} \times k_{33}^{1}}$$

$$\begin{aligned} x = &\frac{-1{,}628{,}944{,}451}{2} \\ &+ \sqrt{\left(\frac{-1{,}628{,}944{,}451}{2}\right) + (8{,}881{,}497{,}893)(3{,}307{,}681{,}896)} \\ = &-6{,}295{,}396{,}856 \end{aligned}$$

$$a = \frac{k_{33}^{1}}{x} = -0.525413 \text{ and } b = \frac{-k_{11}^{1}}{x} = 1.410792377$$

$$\therefore [T_3] = \begin{bmatrix} 1 & 0 & 0 \\ 0 & 1 & -0.525412769 \\ 0 & 1.410792377 & 1 \end{bmatrix}$$

$$\therefore [T_3]^{\mathrm{T}}[T_2]^{\mathrm{T}}[T_1]^{\mathrm{T}}[\mathbf{K}][T_1][T_2][T_3]$$
$$= \begin{bmatrix} 11{,}053{,}907.43 & -826{,}540.3974 & 434{,}274.8792 \\ -826{,}540.3974 & 1{,}763{,}985.213 & 4.65661 \times 10^{-10} \\ 34{,}274.8792 & 2.32831 \times 10^{-10} & 3{,}880{,}612.814 \end{bmatrix} \text{ and}$$

$$\therefore [T_3]^{\mathrm{T}}[T_2]^{\mathrm{T}}[T_1]^{\mathrm{T}}[\mathbf{M}][T_1][T_2][T_3]$$
$$= \begin{bmatrix} 6528.259954 & 3.20777 \times 10^{-13} & 2.27374 \times 10^{-13} \\ 3.20777 \times 10^{-13} & 10{,}309.92891 & 0 \\ 2.27374 \times 10^{-13} & 0 & 3839.663718 \end{bmatrix}$$

And

$$[T_1][T_2][T_3] = \begin{bmatrix} 1 & 0.430414831 & -0.550237646 \\ -0.693000468 & 1.18197197 & -0.396427121 \\ 0.551026932 & 1.41792377 & 1 \end{bmatrix}$$

This completes the first cycle of iteration.
For the second cycle, we start with

$$[\mathbf{K}] = \begin{bmatrix} 11{,}053{,}907.43 & -826{,}540.3974 & 434{,}274.8792 \\ -826{,}540.3974 & 1{,}763{,}985.213 & 4.65661 \times 10^{-10} \\ 34{,}274.8792 & 2.32831 \times 10^{-10} & 3{,}880{,}612.814 \end{bmatrix} \text{ and}$$

$$[\mathbf{M}] = \begin{bmatrix} 6528.259954 & 3.20777 \times 10^{-13} & 2.27374 \times 10^{-13} \\ 3.20777 \times 10^{-13} & 10309.92891 & 0 \\ 2.27374 \times 10^{-13} & 0 & 3839.663718 \end{bmatrix}$$

Following the same steps as cited for the first cycle with this updated [**K**] and [**M**], we finally have

$$\therefore [T_3]^{\mathrm{T}}[T_2]^{\mathrm{T}}[T_1]^{\mathrm{T}}[\mathbf{K}][T_1][T_2][T_3]$$
$$= \begin{bmatrix} 11{,}387{,}131 & 5811.795352 & 24.15955639 \\ 5811.795352 & 1{,}702{,}578.035 & 0 \\ 24.15955639 & -2.96495 \times 10^{-10} & 3{,}898{,}558.67 \end{bmatrix},$$

$$\therefore [T_3]^{\mathrm{T}}[T_2]^{\mathrm{T}}[T_1]^{\mathrm{T}}[\mathbf{M}][T_1][T_2][T_3]$$
$$= \begin{bmatrix} 6656.875 & 4.56535 \times 10^{-13} & 2.27374 \times 10^{-13} \\ 4.56535 \times 10^{-13} & 10355.17954 & -2.84217 \times 10^{-14} \\ 2.27374 \times 10^{-13} & -2.84217 \times 10^{-14} & 3{,}898{,}558.67 \end{bmatrix} \text{ and}$$

$$[T_1][T_2][T_3] = \begin{bmatrix} 1 & 0.083862134 & -0.094287847 \\ -0.05244 & 0.999945206 & 0.009119493 \\ 0.16159432 & -0.011041605 & 1 \end{bmatrix}$$

This completes the second cycle.
For the third cycle, the new values of [**K**] and [**M**] are

$$[\mathbf{K}] = \begin{bmatrix} 11{,}387{,}131 & 5811.795352 & 24.15955639 \\ 5811.795352 & 1{,}702{,}578.035 & 0 \\ 24.15955639 & -2.96495 \times 10^{-10} & 3{,}898{,}558.67 \end{bmatrix} \text{ and}$$

$$[\mathbf{M}] = \begin{bmatrix} 6656.875 & 4.56535 \times 10^{-13} & 2.27374 \times 10^{-13} \\ 4.56535 \times 10^{-13} & 10355.17954 & -2.84217 \times 10^{-14} \\ 2.27374 \times 10^{-13} & -2.84217 \times 10^{-14} & 3898.55867 \end{bmatrix}$$

Doing another cycle of identical iteration, one finally gets

$$\therefore [T_3]^T[T_2]^T[T_1]^T[\mathbf{K}][T_1][T_2][T_3] = \begin{bmatrix} 11{,}387{,}135 & -1.18971 \times 10^{-7} & 7.293 \times 10^{-15} \\ -1.2 \times 10^{-7} & 1{,}702{,}575.103 & -3.4694 \times 10^{-18} \\ -1.2 \times 10^{-10} & -2.96431 \times 10^{-10} & 3{,}898{,}558.67 \end{bmatrix},$$

$$\therefore [T_3]^T[T_2]^T[T_1]^T[\mathbf{M}][T_1][T_2][T_3] = \begin{bmatrix} 6656.877 & 4.56079 \times 10^{-13} & 1.5581 \times 10^{-13} \\ 4.56079 \times 10^{-13} & 10355.18167 & -2.8489 \times 10^{-14} \\ 1.16 \times 10^{-13} & -2.84893 \times 10^{-14} & 3898.55867 \end{bmatrix}$$

$$[T_1][T_2][T_3] = \begin{bmatrix} 1 & -0.000564656 & -5.1074 \times 10^{-6} \\ 0.000363 & 1 & -3.4306 \times 10^{-9} \\ 8.72 \times 10^{-6} & 4.18774 \times 10^{-9} & 1 \end{bmatrix}$$

Further iteration is not necessary as it is observed that [**K**] matrix has almost digonalized. Thus

$$\lambda_{ii} = \begin{bmatrix} \frac{\mathbf{K}_{ii}}{\mathbf{M}_{ii}} \end{bmatrix} = \begin{bmatrix} \frac{11{,}387{,}135}{6656.877} & 0 & 0 \\ 0 & \frac{1{,}702{,}575.103}{10{,}355.18167} & 0 \\ 0 & 0 & \frac{3{,}898{,}558.67}{3898.55867} \end{bmatrix}$$

$$= \begin{bmatrix} 1710.582 & 0 & 0 \\ 0 & 164.4177 & 0 \\ 0 & 0 & 1000 \end{bmatrix}$$

Based on the expression $[\varphi] = [T_1][T_2]\ldots[T_r]\text{Diag}\left[\frac{1}{\sqrt{M^r_{nn}}}\right]$, the normalized eigenvector is expressed as

$$[\varphi] = \begin{bmatrix} 0.012256451 & -5.54888 \times 10^{-6} & -8.17997 \times 10^{-8} \\ 4.44899 \times 10^{-6} & 0.009827004 & -5.49434 \times 10^{-11} \\ 1.0689 \times 10^{-7} & 4.1153 \times 10^{-11} & 0.016015775 \end{bmatrix}$$

It is again observed that the results are in excellent agreement with results of previous examples.

2.5 Modal Analysis of Structures Assuming Continuum Model

In previous sections, we had worked out in some detail modal response analysis of system under earthquake force that is discrete in nature. In other words, the system is assumed to constitute of a number of discrete elements having element mass and stiffness that are assembled to form the global stiffness and mass matrix of the overall system. And this in turn is used for free and forced vibration response under earthquake.

However, there are many cases in real-world engineering, where due to its inherent geometric or functional property it becomes much more advantageous to consider the body as a continuum, and if this concept is used judiciously, it can significantly cut down computational cost vis-à-vis a discrete element model.

Many structures in real life, like retaining walls, tall chimneys, dams, piles, shear wall, nuclear reactor buildings can be modeled by continuum technique to great computational advantage. We discuss the underlying technique of this theory pertaining to earthquake, briefly hereafter.

2.5.1 Vibration of a Flexural Beam

In this section, we examine how a flexible beam under free vibration behaves. For this, let us consider Fig. 2.22.

Shown in Fig. 2.22 is beam of length L loaded with an external load w(x,t); i.e., the load is a function of space and time.

We start with static load case such that normal stress in the section σ may be expressed as

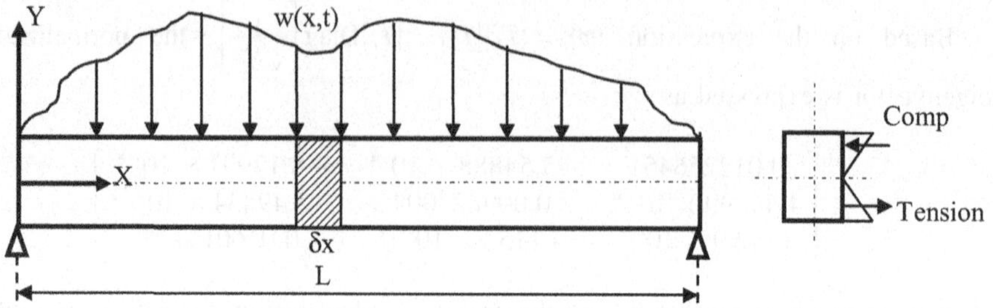

Fig. 2.22 A beam with external load $w(x, t)$

$$\sigma = Cy \tag{2.5.1}$$

Here, C = a constant and $y = 0$ is the neutral axis, where $\sigma = 0$.

Considering the sum of all internal forces equal to zero, we have

$$\int \sigma \cdot dA = 0 \tag{2.5.2}$$

Again, considering that the moment of internal force about neutral axis is equal to external bending moment M, we get

$$M = \int y \cdot \sigma \cdot dA \tag{2.5.3}$$

Substituting Eq. (2.5.1) in Eqs. (2.5.2) and (2.5.3) we have

$$C\int y \cdot dA = 0 \quad \text{and} \tag{2.5.4}$$

$$M = C\int y^2 \cdot dA \tag{2.5.5}$$

Equation (2.5.4) equals zero which implies that the neutral axis coincides with centroidal axis and

Equation (2.5.5) implies

$$C = M/\int y^2 \cdot dA \tag{2.5.6}$$

Considering $I = \int y^2 \cdot dA$, the moment of inertia of the section, Eq. (2.5.6) can be expressed as

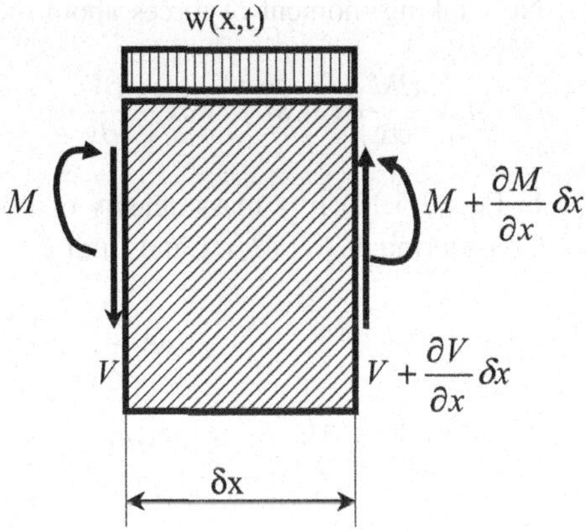

Fig. 2.23 A segment of beam δx with external load $w(x, t)$

$$C = \frac{M}{I} \tag{2.5.7}$$

Substituting this in Eq. (2.5.1), we have

$$\sigma = \frac{My}{I} \tag{2.5.8}$$

Now if ε is the strain in the outer fiber due to external load, then by Hooke's law

$$\varepsilon = \frac{\sigma}{E} = \frac{My}{EI} \tag{2.5.9}$$

Now if u is the displacement of the beam, then for small deformation

$$\frac{1}{r} = \frac{\mathrm{d}^2 u}{\mathrm{d}x^2} = \frac{-\varepsilon}{y} = \frac{-M}{EI} \tag{2.5.10}$$

In Eq. (2.5.10), r is the radius of curvature of the beam and this finally yields

$$EI\frac{\mathrm{d}^2 u}{\mathrm{d}x^2} = -M \tag{2.5.11}$$

Equation (2.5.11) is the famous Euler–Bernoulli equation for a beam under flexure.

Shown in Fig. 2.23 is the free body diagram of a segment of the beam of length δx, which is infinitesimally small subjected to external load $w(x, t)$.

Now taking moment of forces about the left face, we have

$$M + \frac{\partial M}{\partial x}\delta x - M - V\delta x - \frac{\partial V}{\partial x}\delta x^2 - w(x,t)\frac{\delta x^2}{2} + \rho A\frac{\partial^2 u}{\partial t^2}\frac{\delta x^2}{2} \tag{2.5.12}$$

In Eq. (2.5.12) ρ = mass density of the beam and A = cross-sectional area. Considering $V = \partial M/\partial x$ and limit $\delta x \to 0$, Eq. (2.5.12) can be expressed as

$$\rho A\frac{\partial^2 u}{\partial t^2} = \frac{\partial V}{\partial x} + w(x,t) \tag{2.5.13}$$

Again as $V = \partial M/\partial x$ we have

$$\rho A\frac{\partial^2 u}{\partial t^2} = \frac{\partial^2 M}{\partial x^2} + w(x,t) \tag{2.5.14}$$

Now substituting Eq. (2.5.11) and ignoring $w(x, t)$ for free vibration, we have (Rao 2007)

$$\rho A\frac{\partial^2 u}{\partial t^2} = -EI\frac{\partial^4 u}{\partial x^4} \tag{2.5.15}$$

Now considering $c^2 = \frac{EI}{\rho A}$, Eq. (2.5.15) can be expressed as

$$\frac{\partial^2 u}{\partial t^2} = -c^2\frac{\partial^4 u}{\partial x^4} \tag{2.5.16}$$

Here, the displacement $u(x, t)$ is a function of both space and time; let

$$u(x,t) = \phi(x)\cdot q(t) \tag{2.5.17}$$

Then, $\frac{\partial^2 u}{\partial t^2} = \phi(x)q''(t)$ and $\frac{\partial^4 u}{\partial x^4} = \phi''''(x)q(t)$, and substituting this in Eq. (2.5.16), we get

$$\phi(x)\cdot q''(t) = -c^2\phi''''(x)q(t) \tag{2.5.18}$$

$$\frac{q''(t)}{q(t)} = -c^2\frac{\phi''''(x)}{\phi(x)} = -\omega^2(\text{Say}) \tag{2.5.19}$$

Equation (2.5.19) can be thus expressed as two ordinary differential equations as

$$\frac{\mathrm{d}^2 q}{\mathrm{d}t^2} + \omega^2 q = 0 \text{ and} \tag{2.5.20}$$

$$\frac{\mathrm{d}^4\phi}{\mathrm{d}x^4} - \frac{\omega^2}{c^2}\phi = 0 \tag{2.5.21}$$

Solution to Eq. (2.5.20) is expressed as $q(t) = B_1 \cos \omega t + B_2 \sin \omega t$, and for Eq. (2.5.21),

considering $\mu^4 = \omega^2/c^2$, we have

$$\frac{d^4\phi}{dx^4} - \mu^4\phi = 0 \tag{2.5.22}$$

Considering $D = d/dx$, an operator, Eq. (2.5.21) can be expressed as

$$\left(D^4 - \mu^4\right)\phi = 0 \tag{2.5.23}$$

Equation (2.5.23) yields four roots, $D_1 = \mu$, $D_2 = -\mu$, $D_3 = i\mu$, $D_4 = -i\mu$, and this gives

$$\begin{aligned} &\phi(x) = A_1 e^{\mu x} + A_2 e^{-\mu x} + A_3 e^{i\mu x} + A_4 e^{-i\mu x} \\ &\rightarrow \phi(x) = C_1 \cos \mu x + C_2 \sin \mu x + C_3 \cosh \mu x + C_4 \sinh \mu x \end{aligned} \tag{2.5.24}$$

Thus, the complete solution is expressed as

$$u(x,t) = (C_1 \cos \mu x + C_2 \sin \mu x + C_3 \cosh \mu x + C_4 \sinh \mu x)(B_1 \cos \omega t + B_2 \sin \omega t) \tag{2.5.25}$$

Here, C_1, C_2, C_4, B_1, B_2, etc., are integration constants and need to be evaluated from the boundary and initial conditions of the problem.

As $\mu^4 = \omega^2/c^2$ and $c^2 = \frac{EI}{\rho A}$, we have

$$\omega^2 = \mu^4 \frac{EI}{\rho A} \tag{2.5.26}$$

$$\omega = (\mu L)^2 \sqrt{\frac{EI}{\rho A L^4}} \tag{2.5.27}$$

Here, ω is the natural frequency of the beam.

Having solved for the general case of the problem, we now take a few particular cases to see how to evaluate the natural frequencies and eigenvectors.

2.5.1.1 Beam Simply Supported at Both Ends

Shown in Fig. 2.24 is beam that is simply supported at two ends, and we need to evaluate its frequency and mode shapes under free vibration.

The four boundary conditions are

(i) At $x = 0$, $u = 0$; (ii) At $x = 0$, $EIu'' = 0$; (iii) At $x = L$, $u = 0$; and (iv) At $x = L$, $EIu'' = 0$.

Fig. 2.24 A simply supported beam of span L

Considering $\phi(x) = C_1 \cos \mu x + C_2 \sin \mu x + C_3 \cosh \mu x + C_4 \sinh \mu x$ and imposing boundary condition (i) we have $C_1 + C_3 = 0$.

Now imposing boundary condition (ii), we have $C_1 - C_3 = 0$

From these two boundary conditions, we conclude that $C_1 = C_3 = 0$ and this gives

$$\phi(x) = C_2 \sin \mu x + C_4 \sinh \mu x \tag{2.5.1.1}$$

Imposing boundary conditions (iii) and (iv), we have

$$C_2 \sin \mu L + C_4 \sinh \mu L = 0 \text{ and} \tag{2.5.1.2}$$

$$-C_2 \sin \mu L + C_4 \sinh \mu L = 0 \tag{2.5.1.3}$$

For the solution of the above two homogenous equation, we have

$$\begin{vmatrix} \sin \mu L & \sinh \mu L \\ -\sin \mu L & \sinh \mu L \end{vmatrix} = 0 \tag{2.5.1.4}$$

$$\Rightarrow \sin \mu L \sinh \mu L = 0 \tag{2.5.1.5}$$

Equation (2.5.1.5) shows that the function $\sinh \mu L$ has no admissible roots except of $\mu L = 0$ (i.e., the body is at rest) and is not an admissible function; thus, Eq. (2.5.5) is valid only when $\sin \mu L = 0$. Thus, the general solution is

$$\mu_n L = n\pi \tag{2.5.1.6}$$

Substituting Eq. (2.5.1.6) in Eq. (2.5.27), we have

$$\omega = n^2 \pi^2 \sqrt{\frac{EI}{\rho A L^4}} \tag{2.5.1.7}$$

where n = 1, 2, 3 … the mode numbers.

As we have argued above that $\sinh \mu L$ is not an admissible function, this implicitly infers that C_4 is also not valid when the displacement of the beam in different modes is expressed as

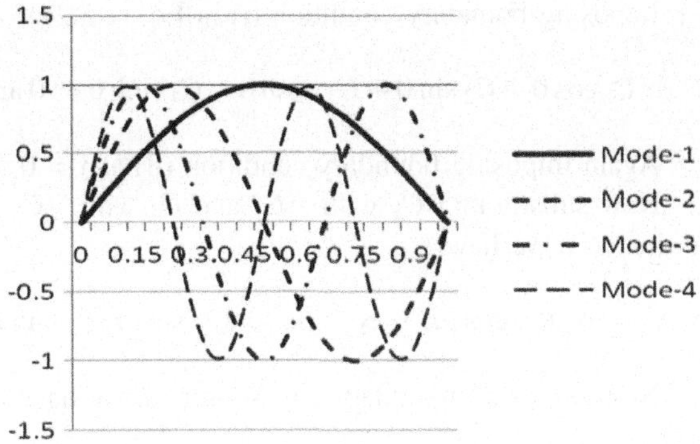

Fig. 2.25 Eigenmodes for simply supported beam for the first four modes

$$\phi_n(x) = C_n \sin\frac{n\pi x}{L} \tag{2.5.1.8}$$

Equation (2.5.1.8) can thus be expressed as

$$u_n(x) = u_0 \sin\frac{n\pi x}{L} \tag{2.5.1.9}$$

In Eq. (2.5.1.9) $\sin\frac{n\pi x}{L}$ is termed as the eigenvectors of the beam and gives the general characteristic shape as to how it is going to deform under different modes satisfying the boundary conditions, as shown in Fig. 2.25.

The complete solution is

$$u_n(x) = u_0 \sin\frac{n\pi x}{L}(B_1 \cos \omega t + B_2 \sin \omega t) \tag{2.5.1.10}$$

2.5.1.2 Beam Fixed at the Left End and Free at the Other End (Cantilever)

Shown in Fig. 2.26 is cantilever beam of span L. The boundary conditions are expressed as

(i) At $x = 0$ $u = 0$
(ii) At $x = 0$ $\mathrm{d}u/\mathrm{d}x = 0$
(iii) At $x = L$ $EIu'' = 0$
(iv) At $x = L$ $EIu''' = 0$

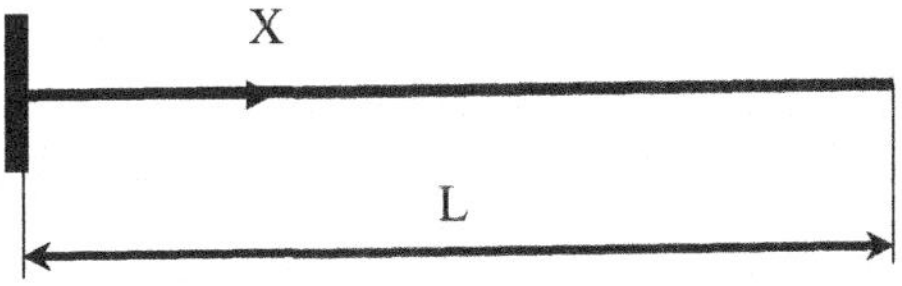

Fig. 2.26 A cantilever beam of span L

Imposing boundary condition (i) in Eq. (2.5.24), we have

$$C_1 \cos 0 + C_2 \sin 0 + C_3 \cosh 0 + C_4 \sinh 0 = 0 \text{ and this implies } C_3 = -C_1$$

Again imposing boundary condition (ii) at $x = 0$, we have $C_4 = -C_2$

Now substituting C_3 with $-C_1$ and C_4 with $-C_2$ and imposing boundary condition (iii), we have

$$C_1(\cos \mu L + \cosh \mu L) + C_2(\sin \mu L + \sinh \mu L) = 0 \tag{2.5.1.11}$$

Now imposing boundary condition (iv), we have

$$C_1(\sin \mu L - \sinh \mu L) - C_2(\cos \mu L + \cosh \mu L) = 0 \tag{2.5.1.12}$$

Solution of the two homogenous equations gives

$$\begin{vmatrix} (\cos \mu L + \cosh \mu L) & (\sin \mu L + \sinh \mu L) \\ (\sin \mu L - \sinh \mu L) & -(\cos \mu L + \cosh \mu L) \end{vmatrix} = 0 \tag{2.5.1.13}$$

$$\rightarrow (\cos \mu L + \cosh \mu L)^2 + \sin^2 \mu L - \sinh^2 \mu L = 0 \tag{2.5.1.14}$$

Expanding and simplifying Eq. (2.5.1.14) finally gives

$$\cos \mu L \cdot \cosh \mu L = -1 \tag{2.5.1.15}$$

Equation (2.5.1.15) is transcendental equation in terms of μL which gives

$$\mu_n L = 1.875,\ 4.694,\ 7.855\ldots \frac{(2n-1)\pi}{2} \quad \text{for } n \geq 4$$

Now considering $\omega = (\mu L)^2 \sqrt{\frac{EI}{\rho A L^4}}$, vide Eq. (2.5.27), we have

$\omega_1 = 3.5156\sqrt{\frac{EI}{\rho A L^4}}$, $\omega_2 = 22.034\sqrt{\frac{EI}{\rho A L^4}}$, $\omega_3 = 61.701\sqrt{\frac{EI}{\rho A L^4}}$, and so on.

For determination of displacement function, $\phi(x)$ can be expressed as

$$\begin{aligned} \phi(x) &= C_1 \cos \mu x + C_2 \sin \mu x - C_1 \cosh \mu x + - C_2 \sinh \mu x \\ &= C_2 \left[(\sin \mu x - \sinh \mu x) - \frac{C_1}{C_2}(\cos \mu x - \cosh \mu x) \right] \end{aligned} \tag{2.5.1.16}$$

Substituting the value of C_1/C_2 from Eq. (2.5.1.11), we have

$$\phi(x) = C_2\left[(\sin\mu x - \sinh\mu x) - \left(\frac{\sin\mu L + \sinh\mu L}{\cos\mu L + \cosh\mu L}\right)(\cos\mu x - \cosh\mu x)\right] \tag{2.5.1.17}$$

Thus, the general solution can be expressed as

$$u_n = C_n[(\sin\mu_n x - \sinh\mu_n x) - \alpha_n(\cos\mu_n x - \cosh\mu_n x)] \tag{2.5.1.18}$$

$$\text{where } \mu_n = \frac{1.875}{L}, \frac{4.694}{L}, \frac{7.855}{L} \ldots \frac{(2n-1)\pi}{2L} \quad \text{for } n \geq 4 \text{ and}$$

$$\alpha_n = \left(\frac{\sin\mu_n L + \sinh\mu_n L}{\cos\mu_n L + \cosh\mu_n L}\right)$$

Considering $\beta_n = \mu_n$, the eigenvectors in this case are expressed as

$$f_n(x) = \left[\left(\sin\beta_n\frac{x}{L} - \sinh\beta_n\frac{x}{L}\right) - \alpha_n\left(\cos\beta_n\frac{x}{L} - \cosh\beta_n\frac{x}{L}\right)\right] \tag{2.5.1.19}$$

The eigenfunctions for the first three modes are shown in Fig. 2.27.

Having computed the natural frequency and eigenvectors for the above two cases, we furnish hereafter the eigenshapes of beams having different boundary conditions. We will use them later for deriving stiffness and mass matrix of a system by a different technique.

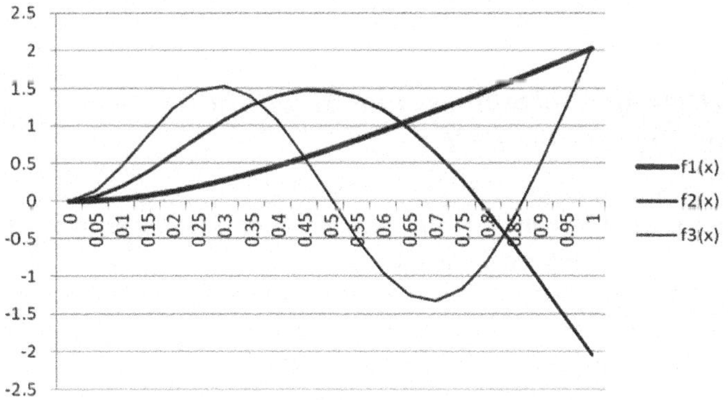

Fig. 2.27 Eigenmodes of cantilever beam for the first three modes

1. ***Left end simply supported and right end fixed.***

$$f_n(x) = \sin \beta_n \frac{x}{L} - \alpha_n \sinh \beta_n \frac{x}{L} \tag{2.5.1.20}$$

Here $\beta_n = \frac{(4n+1)\pi}{4}$, $n = 1, 2, 3, \ldots$ and $\alpha_n = \frac{\sin \beta_n}{\sinh \beta_n}$

2. ***Both ends fixed***

$$f_n(x) = \left[\left(\sin \beta_n \frac{x}{L} - \sinh \beta_n \frac{x}{L}\right) - \alpha_n \left(\cos \beta_n \frac{x}{L} - \cosh \beta_n \frac{x}{L}\right)\right] \tag{2.5.1.21}$$

Here $\beta_n = \frac{(2n+1)\pi}{2}$ $n = 1, 2, 3, \ldots.$ and $\alpha_n = \left(\frac{\sin \beta_n - \sinh \beta_n}{\cos \beta_n - \cosh \beta_n}\right)$

3. ***Both ends free***

$$f_n(x) = \left[\left(\sin \beta_n \frac{x}{L} + \sinh \beta_n \frac{x}{L}\right) - \alpha_n \left(\cos \beta_n \frac{x}{L} + \cosh \beta_n \frac{x}{L}\right)\right] \tag{2.5.1.22}$$

Here $\beta_n = \frac{(2n+1)\pi}{2}$ $n = 1, 2, 3, \ldots.$ and $\alpha_n = \left(\frac{\sin \beta_n - \sinh \beta_n}{\cos \beta_n - \cosh \beta_n}\right)$

4. ***Left end simply supported, right end free***

$$f_n(x) = \sin \beta_n \frac{x}{L} + \alpha_n \sinh \beta_n \frac{x}{L} \tag{2.5.1.23}$$

Here $\beta_n = \frac{(4n-3)\pi}{4}$ $n = 2, 3, 4 \ldots$ and $\alpha_n = \frac{\sin \beta_n}{\sinh \beta_n}$

2.5.2 Derivation of Stiffness and Mass Matrix by Eigenvector Basis

In Sect. 2.5.1, we derive the eigenfrequencies and mode shapes by solution of the fourth-order partial differential equation.

In this section, we derive the same by first deriving the stiffness and mass matrix of the system and then finding out the natural frequencies, as this is the basis which an analyst follows for doing any earthquake analysis.

For a flexural element under external load, like one cited in Fig. 2.22, the bending strain energy P_E is expressed as

$$P_E = \int_0^L \frac{M^2 \mathrm{d}x}{2EI} \tag{2.5.2.1}$$

Here, M = external moment and EI is its flexural stiffness.
Substituting Eq. (2.5.11) in the above equation, we have

$$P_E = \frac{EI}{2}\int_0^L \left[\frac{d^2u}{dx^2}\right]^2 dx \tag{2.5.2.2}$$

Now for ith mode, we had shown that

$$u_i(x,t) = \phi_i(x)q_i(t) \tag{2.5.2.3}$$

For jth mode again, we have

$$u_j(x,t) = \phi_j(x)q_j(t) \tag{2.5.2.4}$$

Substituting Eq. (2.5.2.3 and 2.5.2.4) in Eq. (2.5.2.2), we have

$$P_E = \frac{EI}{2}\int_0^L \left[\phi_i''(x)q_i(t)\phi_j''(x)q_j(t)\right] \cdot dx \tag{2.5.2.5}$$

The above can be restructured to be presented as

$$P_E = \frac{EI}{2}\int_0^L \left[\phi_i''(x)\phi_j''(x)(t)\right] \cdot dx \cdot q_i(t) \cdot q_j(t) \tag{2.5.2.6}$$

Now considering that the strain energy can also be represented by $1/2k\delta^2$, where k = stiffness and δ = displacement, we can write Eq. (2.5.2.6) as

$$\frac{1}{2}\left[k_{ij}\right]\delta_i(t)\delta_j(t) = \frac{EI}{2}\int_0^L \left[\phi_i''(x)\phi_j''(x)(t)\right] \cdot dx \cdot q_i(t) \cdot q_j(t) \tag{2.5.2.7}$$

From Eq. (2.5.2.7), we conclude that

$$\left[k_{ij}\right] = EI\int_0^L \left[\phi_i''(x)\phi_j''(x)(t)\right] \cdot dx. \tag{2.5.2.8}$$

The kinetic energy K_E of the beam is

$$K_E = \frac{1}{2}mv^2 = \frac{1}{2}\int_0^L \rho A\dot{u}(x,t)\mathrm{d}x \tag{2.5.2.9}$$

Substituting Eqs. (2.5.2.3) and (2.5.2.4), we have

$$K_E = \frac{1}{2}\left[\int_0^L \rho A\phi_i(x)\phi_j(x)\mathrm{d}x\right]\dot{q}_i(t)\dot{q}_j(t) \tag{2.5.2.10}$$

From Eq. (2.5.2.10), it can be concluded that mass matrix of the beam can be expressed as

$$\left[m_{ij}\right] = \rho A\int_0^L \phi_i(x)\phi_j(x)\mathrm{d}x \tag{2.5.2.11}$$

Thus, based on boundary condition, one can select the eigenvectors as elaborated earlier and for number n of modes can form the stiffness and mass matrix as

$$[\mathbf{K}]_{n\times n} = EI\begin{bmatrix} \int_0^L \phi_1''^2(x)\mathrm{d}x & \int_0^L \phi_1''(x)\phi_2''(x)\mathrm{d}x & \cdots & \cdots & \int_0^L \phi_1''(x)\phi_{n-1}''(x)\mathrm{d}x & \int_0^L \phi_1''(x)\phi_n''(x)\mathrm{d}x \\ \int_0^L \phi_1''(x)\phi_2''(x)\mathrm{d}x & \int_0^L \phi_2''^2(x)\mathrm{d}x & \cdots & \cdots & \int_0^L \phi_2''(x)\phi_{n-1}''(x)\mathrm{d}x & \int_0^L \phi_2''(x)\phi_n''(x)\mathrm{d}x \\ \cdots & \cdots & \cdots & \cdots & \cdots & \cdots \\ \cdots & \cdots & \cdots & \cdots & \cdots & \cdots \\ \int_0^L \phi_1''(x)\phi_{n-1}''(x)\mathrm{d}x & \int_0^L \phi_2''(x)\phi_{n-1}''(x)\mathrm{d}x & \cdots & \cdots & \int_0^L \phi_{n-1}''(x)^2\mathrm{d}x & \int_0^L \phi_{n-1}''(x)\phi_n''(x)\mathrm{d}x \\ \int_0^L \phi_1''(x)\phi_n''(x)\mathrm{d}x & \int_0^L \phi_2''(x)\phi_n''(x)\mathrm{d}x & \cdots & \cdots & \int_0^L \phi_{n-1}''(x)\phi_n''(x)\mathrm{d}x & \int_0^L \phi_n''(x)^2\mathrm{d}x \end{bmatrix} \tag{2.5.2.12}$$

$$[\mathbf{M}]_{n\times n} = \rho A\begin{bmatrix} \int_0^L \phi_1(x)^2\mathrm{d}x & 0 & \cdots & \cdots & 0 & 0 \\ 0 & \int_0^L \phi_2(x)^2\mathrm{d}x & \cdots & \cdots & 0 & 0 \\ \cdots & \cdots & \cdots & \cdots & \cdots & \cdots \\ \cdots & \cdots & \cdots & \cdots & \cdots & \cdots \\ 0 & 0 & 0 & 0 & \int_0^L \phi_{n-1}(x)^2\mathrm{d}x & 0 \\ 0 & 0 & 0 & 0 & 0 & \int_0^L \phi_n(x)^2\mathrm{d}x \end{bmatrix} \tag{2.5.2.13}$$

In Eq. (2.5.2.13), one may note that the off-diagonal terms in the mass matrix $[\mathbf{M}_{ij}]$ are zero. This is because the mass matrix is constructed from eigenvector basis, due to its orthogonal property $\phi_i(x)\phi_j(x) = 0$.

2.5.3 *Response of Flexural Beam to Earthquake-Type Load*

Based on Eqs. (2.5.2.12) and (2.5.2.13), one can derive the natural frequency of the system from the equation

$$\left[[\mathbf{K}] - [\mathbf{M}]\omega_i^2\right]\{\phi_i\} = 0 \tag{2.5.3.1}$$

Thus, the amplitude of vibration $q_i(t)$ in ith mode is expressed as

$$q_i(t) = \kappa_i\left(\frac{ZI}{2R}\right)\frac{S_{\mathrm{ai}}}{\omega^2} \tag{2.5.3.2}$$

The amplitude of displacement u_x is thus expressed

$$u_x = \kappa_i\left(\frac{ZI}{2R}\right)\frac{S_{\mathrm{ai}}}{\omega^2}\{\phi_i\}^{\mathrm{T}} f_i(x) \tag{2.5.3.3}$$

The moment and shear in the beam can be expressed as

$$M_x = -EI\kappa_i\left(\frac{ZI}{2R}\right)\frac{S_{\mathrm{ai}}}{\omega^2}\{\phi_i\}^{\mathrm{T}}\frac{\mathrm{d}^2 f_i(x)}{\mathrm{d}x^2} \quad \text{and} \tag{2.5.3.4}$$

$$V_x = -EI\kappa_i\left(\frac{ZI}{2R}\right)\frac{S_{\mathrm{ai}}}{\omega^2}\{\phi_i\}^{\mathrm{T}}\frac{\mathrm{d}^3 f_i(x)}{\mathrm{d}x^3} \tag{2.5.3.5}$$

In Eq. (2.5.3.2), κ_i = modal mass participation and, as derived in Sect. (2.3.10), can be expressed as

$$\kappa_i = \frac{\sum_{i=1}^{n} m_i\phi_i}{\sum_{i=1}^{n} m_i\phi_i^2} = \frac{\int_0^L \{\phi_i\}^{\mathrm{T}} f_i(x)\mathrm{d}x}{\int_0^L \left[\{\phi_i\}^{\mathrm{T}} f_i(x)\right]^2\mathrm{d}x} \tag{2.5.3.6}$$

2.5.4 Stiffness and Mass Matrix for Transversely Vibrating Beam Elastically Restrained at Free End with Lumped Mass

Shown in Fig. 2.28 is cantilever beam elastically restrained at end by translation and rotational springs, K_x and K_θ, respectively. It also has a mass M lumped at a distance L_m from the fixed end.

The stiffness matrix $[\mathbf{K}_{ij}]$ and mass matrix $[\mathbf{M}_{ij}]$ are expressed as

$$[\mathbf{K}_{ij}] = EI\int_0^L \phi_i''(x)\phi_j''(x)\mathrm{d}x + K_x\phi_i(L)\phi_j(L) + K_\theta\phi_i'(L)\phi_j'(L) \text{ and} \tag{2.5.4.1}$$

$$[\mathbf{M}_{ij}] = \rho A\int_0^L \phi_i(x)\phi_j(x)\mathrm{d}x + M\phi_i(L_m)\phi_j(L_m) \tag{2.5.4.2}$$

2.5.5 Vibration of a Shear Beam

In many cases, structures are short and stubby. In such cases, rather than flexure, shear deformation of the beam dominates. Examples of such shear beams are deep beams supporting walls, short and wide shear wall, an earthen dam, etc. In most of the case, for computation of seismic ground response of soil, the soil is often modeled as a one-dimensional shear beam, where shear deformation governs the response.

Shown in Fig. 2.29 is a shear beam that is subjected to free vibration. For an infinitesimally small segment δx, the equilibrium is

$$V + \frac{\partial V}{\partial x}\delta x - V - \rho A\frac{\partial^2 u}{\partial t^2}\delta x = 0 \tag{2.5.5.1}$$

Here, $V = \eta GA(\partial u/\partial x)$, G = shear modulus of the beam, A = cross-sectional area of the beam, and η = a shear correction factor.

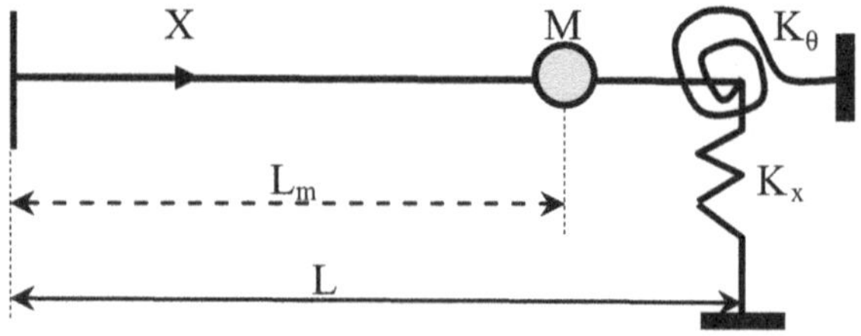

Fig. 2.28 A cantilever beam elastically restrained at edge with a lumped mass

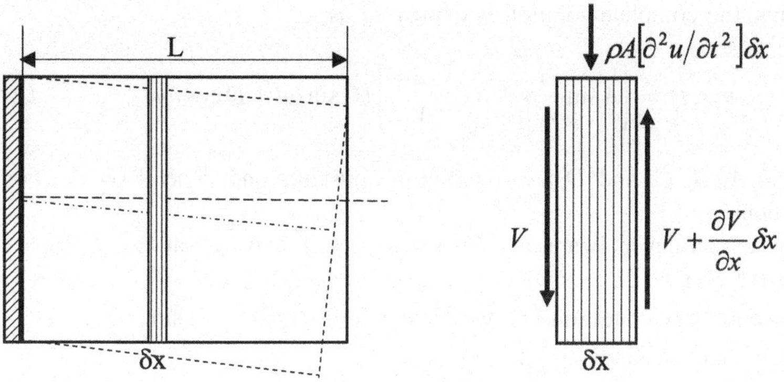

Fig. 2.29 A shear beam under free vibration with its blown-up view

Substituting the value of V in Eq. (2.5.5.1) finally yields (Rao 2007)

$$\rho\frac{\partial^2 u}{\partial t^2} = \eta G\frac{\partial^2 u}{\partial x^2} \tag{2.5.5.2}$$

Considering $V_s = \sqrt{\eta G/\rho}$, Eq. (2.5.5.2) is expressed as

$$\frac{\partial^2 u}{\partial t^2} = V_s^2\frac{\partial^2 u}{\partial x^2} \tag{2.5.5.3}$$

Now considering $u(x,t) = \phi(x)q(t)$, Eq. (2.5.5.3) can be expressed as

$$\begin{aligned} &V_s^2\phi''(x)q(t) = \phi(x)q''(t) = -\omega^2(\text{say}) \\ &\rightarrow V_s^2\frac{\phi''(x)}{\phi(x)}q(t) = \frac{q''(t)}{q(t)} = -\omega^2 \end{aligned} \tag{2.5.5.4}$$

Equation (2.5.5.4) gives two linear differential equations

$$\frac{\mathrm{d}^2 q(t)}{\mathrm{d}t^2} + \omega^2 q(t) = 0 \text{ and} \tag{2.5.5.5}$$

$$\frac{\mathrm{d}^2\phi(x)}{\mathrm{d}x^2} + \frac{\omega^2}{V_s^2}\phi(x) = 0 \tag{2.5.5.6}$$

Solution of the above two equations gives

$$q(t) = C\sin\omega t + D\cos\omega t \text{ and} \tag{2.5.5.7}$$

$$\phi(x) = A\sin\frac{\omega}{V_s}x + B\cos\frac{\omega}{V_s}x \tag{2.5.5.8}$$

Thus, the complete solution is expressed as

$$u(x,t) = \left[A \sin \frac{\omega}{V_s} x + B \cos \frac{\omega}{V_s} x\right] (C \sin \omega t + D \cos \omega t) \tag{2.5.5.9}$$

Here A, B, C, and D are integration constants and depend on the boundary conditions.

For a cantilever beam at (1) $x = 0$, $u = 0$ and (2) at $x = L$ shear stress $\tau = \eta G \frac{\partial u}{\partial x} = 0$.

For boundary condition (1), we have $B = 0 \rightarrow u(x) = A \sin \frac{\omega}{V_s} x$.

For boundary condition (2), we have

$$A \frac{\omega}{V_s} \cos \frac{\omega}{V_s} L = 0$$

$$\rightarrow \cos \frac{\omega_n}{V_s} L = 0 = \cos \frac{(2n-1)\pi}{2} \tag{2.5.5.10}$$

Thus

$$\omega_n = \frac{(2n-1)\pi V_s}{2L} = \frac{(2n-1)\pi}{2L} \sqrt{\frac{\eta G}{\rho}} \tag{2.5.5.11}$$

Thus, the time period is expressed as

$T_n = \frac{4L}{(2n-1)} \sqrt{\frac{\rho}{\eta G}}$ where n = 1, 2, 3, … the modal number

The displacement is expressed as

$$u(x) = u_0 \sin \frac{(2n-1)\pi x}{2L} \text{ where the eigenvector is}$$

$$f_n(x) = \sin \frac{(2n-1)\pi x}{2L} \tag{2.5.5.12}$$

Eigenvectors for first four modes are shown in Fig. 2.30.

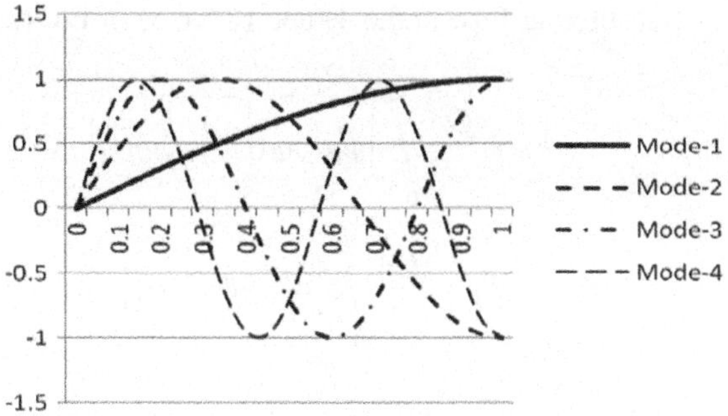

Fig. 2.30 Eigenvectors for first four modes for a cantilever shear beam

2.5.6 Derivation of Stiffness and Mass Matrix of a Shear Beam by Eigenvector Basis

For a shear beam under external load, the shear strain energy P_E is expressed as

$$P_E = \int_0^L \frac{V^2 \mathrm{d}x}{2GA} \tag{2.5.6.1}$$

Here, V = external moment, G = shear modulus and A = cross-sectional area. As

$$V = -GA\frac{\mathrm{d}u}{\mathrm{d}x} \tag{2.5.6.2}$$

substituting Eq. (2.5.6.2) in Eq. (2.5.6.1), we have

$$P_E = \frac{GA}{2}\int_0^L \left[\frac{\mathrm{d}u}{\mathrm{d}x}\right]^2 \mathrm{d}x \tag{2.5.6.3}$$

Now for ith mode, we had shown that

$$u_i(x,t) = \phi_i(x)q_i(t) \tag{2.5.6.4}$$

For jth mode again, we have

$$u_j(x,t) = \phi_j(x)q_j(t) \tag{2.5.6.5}$$

Substituting Eqs. (2.5.6.4) and (2.5.6.5) in Eq. (2.5.6.3), we have

$$P_E = \frac{GA}{2}\int_0^L \left[\phi_i'(x)q_i(t)\phi_j'(x)q_j(t)\right]\cdot \mathrm{d}x$$
$$\rightarrow P_E = \frac{GA}{2}\int_0^L \left[\phi_i'(x)\phi_j'(x)(t)\right]\cdot \mathrm{d}x\cdot q_i(t)\cdot q_j(t) \tag{2.5.6.6}$$

Considering that the strain energy can also be represented by $1/2k\delta^2$, where k = stiffness and δ = displacement, we can write Eq. (2.5.6.6) as

$$\frac{1}{2}\left[k_{ij}\right]\delta_i(t)\delta_j(t) = \frac{GA}{2}\int_0^L \left[\phi_i'(x)\phi_j'(x)(t)\right]\cdot \mathrm{d}x\cdot q_i(t)\cdot q_j(t) \tag{2.5.6.7}$$

From Eq. (2.5.6.7), we conclude that

$$\left[k_{ij}\right] = GA\int_0^L \left[\phi_i'(x)\phi_j'(x)(t)\right]\cdot \mathrm{d}x. \tag{2.5.6.8}$$

The kinetic energy of the beam is

$$K_E = \frac{1}{2}mv^2 = \frac{1}{2}\int_0^L \rho A\dot{u}(x,t)\mathrm{d}x \tag{2.5.6.9}$$

Substituting Eqs. (2.5.6.4) and (2.5.6.5), we have

$$K_E = \frac{1}{2}\left[\int_0^L \rho A\phi_i(x)\phi_j(x)\mathrm{d}x\right]\dot{q}_i(t)\dot{q}_j(t) \tag{2.5.6.10}$$

From Eq. (2.5.2.10), it is concluded that mass matrix of the shear beam can be expressed as

$$\left[m_{ij}\right] = \rho A\int_0^L \phi_i(x)\phi_j(x)\mathrm{d}x \tag{2.5.6.11}$$

2.5.7 Response of Shear Beam Under Earthquake-Type Load

Proceeding as earlier for *i*th mode, we can write the amplitude as

$$u_{ix} = \kappa_i \left(\frac{ZI}{2R}\right) \frac{S_{\mathrm{ai}}}{\omega^2} \{\phi_i\}^{\mathrm{T}} f_i(x)$$
$$\rightarrow u_{ix} = \kappa_i \left(\frac{ZI}{2R}\right) \frac{4L^2 \gamma_b}{(2i-1)^2 \pi^2 \eta G} \left(\frac{S_{\mathrm{ai}}}{g}\right) \{\phi_i\}^{\mathrm{T}} \sin \frac{(2i-1)\pi x}{2L} \tag{2.5.7.1}$$

Here, γ_b = weight density of the beam and g = acceleration due to gravity Thus, the shear force is expressed as

$$V_i = \eta G A \frac{\partial u}{\partial x} = \kappa_i \left(\frac{ZI}{2R}\right) \frac{2L\gamma_b A}{(2i-1)\pi} \left(\frac{S_{\mathrm{ai}}}{g}\right) \{\phi_i\}^{\mathrm{T}} \cos \frac{(2i-1)\pi x}{2L} \tag{2.5.7.2}$$

The modal mass participation factor κ_i is as expressed by Eq. (2.5.3.6).

Considering that this section is a preparatory section, no applications are shown herein. A number of applications of this theory with different variables, like non-homogenous elastic property, variable geometry, have been treated in different chapters of the book dealing with earthen and concrete dams, tall chimneys, retaining structures, etc.

2.6 Time History Analysis a Different Approach

Earthquake force is random in nature. We will study later in Chap. 3 in detail as to how it originates and travels through the ground to hit and excite a site. Depending on the epicentric distance, a site is subject to different intensity of excitation vis-à-vis acceleration. A typical ground acceleration spectrum of such earthquake that occurred in El Centro Earthquake 1940 is as shown in Fig. 2.31.

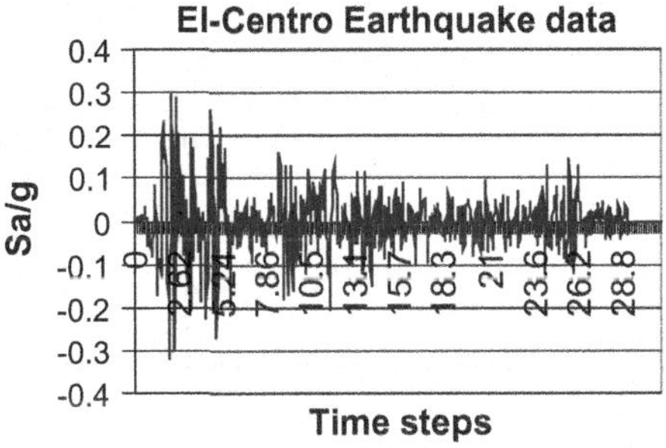

Fig. 2.31 Acceleration spectra of El Centro Earthquake 1940

Figure 2.31 shows that the acceleration is random in nature and so will be the force and amplitude induced in the structure subjected to this acceleration.

In modal response analysis, we only take the peak value of the acceleration to determine the peak modal forces and amplitude of the body. However, in many cases, especially under earthquake force, we are interested to know the complete response history of the system. Then, an engineer would adapt time history analysis to generate such results.

The method also has two important advantages. Since it is *an incremental analysis* in steps of Δt, we are in a position to determine amplitude vis-à-vis strain also at these steps.

Thus, if we are doing a *nonlinear analysis* where the stiffness varies with strain, it is very simple to incorporate this change in material property at such incremental steps and one can carry out a comprehensive nonlinear dynamic analysis under earthquake force.

There are many cases, especially for coupled analysis like soil–structure, fluid–structure interaction, in which the damping matrix becomes non-proportional. In other words, on transformation $\{\phi\}^T[\mathbf{C}]\{\phi\}$ the damping matrix does not diagonalize to $2\zeta_i\omega_i$ when off-diagonal terms of the matrix [**C**] have nonzero terms. In such cases, one cannot perform a modal analysis and the only option then left is to carry out time history analysis of the equilibrium equation directly, to arrive at a solution.

Despite such advantage, one has to concede that the analysis could be quite expensive at times, as depending on the mathematical model, volume of data generated could be significant and an engineer may find interpretation of such voluminous data an arduous task.

To understand the concept, we pose a very fundamental question: Under application of an external force, *why does a body move in space*?

Answer to the above query is that due to the application of the external force $R(t)$, the body tends to move in the same direction in which the unbalanced force acts, as cited in Fig. 2.32 and, according to Newton's first law of motion will continue to do so, as long as the force acts on the body.

Now, if we want to formulate the equation of motion, then, as per D'Alembert's principle, we apply a fictitious force in the direction opposite to the motion and state the body to be in a condition of *dynamic equilibrium* when law of statics does apply, and this gives

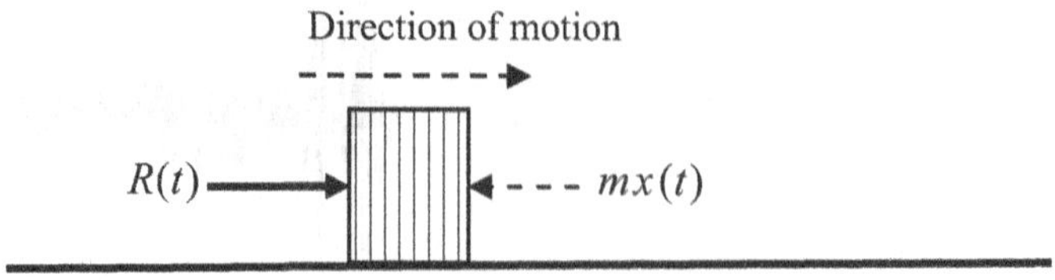

Fig. 2.32 Free body diagram of a body in motion

$$m\ddot{x}(t) = R(t) \tag{2.6.1}$$

Equation (2.6.1) is valid for a rigid body/particle under motion. When a deformable body, like say a structure, is under motion due to an external force, the equation of motion is

$$[\mathbf{M}]\{\ddot{X}\} + [\mathbf{C}]\{\dot{X}\} + [\mathbf{K}]\{X\} = \{\mathbf{R_t}\} \tag{2.6.2}$$

Thus, at an instantaneous time Δt (where Δt is a very small value) we can write Eq. (2.6.2) as

$$F_{(m)\Delta t} + F_{(c)\Delta t} + F_{(k)\Delta t} = R(t)_{\Delta t} \tag{2.6.3}$$

In Eq. (2.6.3), $F_{(m)\Delta t}$ = inertial force, $F_{(c)\Delta t}$ = damping force, and $F_{(k)\Delta t}$ = force due to stiffness of the system. Thus

$$\sum [\text{Inertial force} + \text{Damping force} + \text{Force due to stiffness}] = \text{External force at time } \Delta t$$

The above expression clearly shows that equation of motion is under static equilibrium at the time instant Δt.

Now suppose, by some mathematical manipulation we express each of the force in terms of displacement say Δx, when the equation of motion can be expressed as

$$f(m)\Delta x + f(c)\Delta x + f(k)\Delta x = R(t)\Delta t \tag{2.6.4}$$

In Eq. (2.6.4), $f(m)\Delta x$ = a function of $F_{(m)\Delta t}$ in terms of displacement Δx, $f(c)\Delta x$ = a function of $F_{(c)\Delta t}$ in terms of Δx, and $f(k)\Delta x$ = a function of $F_{(k)\Delta t}$ in terms of Δx.

From Eq. (2.6.1), we can then develop the equation

$$\Delta x = \frac{R(t)}{\Im(m, c, k)} \tag{2.6.5}$$

The value of Δx thus obtained from Eq. (2.6.5) is the displacement output at time instant Δt and becomes the initial input for finding out the displacement for the next step at an interval $2\Delta t$ and so on up to $n\Delta t$. Here, n = number of such steps, 1, 2, 3, …

The method is also known as *step-by-step integration* and is usually carried out based on finite difference equation.

We now present theoretical background of three most frequently used (and most popular) methods for time history analysis of linear and nonlinear systems under earthquake force.

2.6.1 Central Difference Technique (CDT)

As a prerequisite to this section, we expect the reader to have some background on the theory of *finite difference method.*

Central difference technique (CDT) is an *explicit integration scheme* to integrate Eq. (2.6.2) by finite difference scheme.

Equation of motion for a dynamic system having n degrees of freedom at any instant of time t is

$$[\mathbf{M}]\{\ddot{\mathbf{X}}(\mathbf{t})\} + [\mathbf{C}]\{\dot{\mathbf{X}}(t)\} + [\mathbf{K}]\{\mathbf{X}(\mathbf{t})\} = \{\mathbf{R}(t)\} \tag{2.6.1.1}$$

Applying central difference formula to $\{\ddot{X}(t)\}$ and average central difference formula to $\{\dot{X}(t)\}$ for time step increment Δt, we have

$$\{\ddot{\mathbf{X}}(\mathbf{t})\} = \frac{1}{\Delta t^2}[\mathbf{X}(\mathbf{t}+\Delta\mathbf{t}) - 2\mathbf{X}(\mathbf{t}) + \mathbf{X}(\mathbf{t}-\Delta\mathbf{t})] \text{ and}$$
$$\{\dot{\mathbf{X}}(t)\} = \frac{1}{2\Delta t}[\mathbf{X}(\mathbf{t}+\Delta\mathbf{t}) - \mathbf{X}(\mathbf{t}-\Delta\mathbf{t})]$$

For easier and more compact presentation, let us write these expressions as

$$\ddot{\mathbf{X}}_\mathbf{t} = \frac{1}{\Delta t^2}[\mathbf{X}_{\mathbf{t}+\Delta\mathbf{t}} - 2\mathbf{X}_\mathbf{t} + \mathbf{X}_{\mathbf{t}-\Delta\mathbf{t}}] \text{ and } \dot{\mathbf{X}}_\mathbf{t} = \frac{1}{2\Delta t}[\mathbf{X}_{\mathbf{t}+\Delta\mathbf{t}} - \mathbf{X}_{\mathbf{t}-\Delta\mathbf{t}}]. \tag{2.6.1.2}$$

Now substituting this in Eq. (2.6.1.1), we have

$$[\mathbf{M}]\frac{1}{\Delta t^2}[\mathbf{X}_{\mathbf{t}+\Delta\mathbf{t}} - 2\mathbf{X}_\mathbf{t} + \mathbf{X}_{\mathbf{t}-\Delta\mathbf{t}}] + [\mathbf{C}]\frac{1}{2\Delta t}[\mathbf{X}_{\mathbf{t}+\Delta\mathbf{t}} - \mathbf{X}_{\mathbf{t}-\Delta\mathbf{t}}] + [\mathbf{K}]\{\mathbf{X}(\mathbf{t})\} = \{\mathbf{R}(t)\} \tag{2.6.1.3}$$

Separating the terms of $\mathbf{X}_{\mathbf{t}+\Delta\mathbf{t}}$, $\mathbf{X}_\mathbf{t}$ and $\mathbf{X}_{\mathbf{t}-\Delta\mathbf{t}}$, we have

$$\left[\frac{\mathbf{M}}{\Delta t^2} + \frac{\mathbf{C}}{2\Delta t}\right]\mathbf{X}_{\mathbf{t}+\Delta\mathbf{t}} = \mathbf{R}_t - \left[\mathbf{K} - \frac{2\mathbf{M}}{\Delta t^2}\right]\mathbf{X}_\mathbf{t} - \left[\frac{\mathbf{M}}{\Delta t^2} - \frac{\mathbf{C}}{2\Delta t}\right]\mathbf{X}_{\mathbf{t}-\Delta\mathbf{t}} \tag{2.6.1.4}$$

Now considering $\hat{\mathbf{M}} = \frac{\mathbf{M}}{\Delta t^2} + \frac{\mathbf{C}}{2\Delta t}$, $\hat{\mathbf{K}} = \mathbf{K} - \frac{2\mathbf{M}}{\Delta t^2}$ and $\hat{\mathbf{M}}_0 = \frac{\mathbf{M}}{\Delta t^2} - \frac{\mathbf{C}}{2\Delta t}$, we can write Eq. (2.6.1.4) as

$$\hat{\mathbf{M}}\mathbf{X}_{\mathbf{t}+\Delta\mathbf{t}} = \mathbf{R}_t - \hat{\mathbf{K}}\mathbf{X}_\mathbf{t} - \mathbf{M}_0\mathbf{X}_{\mathbf{t}-\Delta\mathbf{t}} \tag{2.6.1.5}$$

It is evident that, in order to find out the value of $\mathbf{X}_{\mathbf{t}+\Delta\mathbf{t}}$, it is necessary to know, a priori, the displacements at $\mathbf{X}_\mathbf{t}$ and $\mathbf{X}_{\mathbf{t}-\Delta\mathbf{t}}$. Thus for solution at time Δt, a special condition has to be invoked. Usually, the displacement and velocity at time $t = 0$ are known. This is defined as *initial boundary condition.*

For instance, if $\mathbf{X_{t=0}} = 0$ and $\dot{\mathbf{X}}_{\mathbf{t=0}} = 0$, then we have

$$[\mathbf{M}]\{\ddot{\mathbf{X}}_{\mathbf{0}}\} + [\mathbf{C}]\{0\} + [\mathbf{K}]\{0\} = \{\mathbf{R_{t=0}}\} \rightarrow \{\ddot{\mathbf{X}}_{\mathbf{0}}\} = [\mathbf{M}]^{-1}\{\mathbf{R_{t=0}}\} \quad (2.6.1.6)$$

To find out the value of $\mathbf{X_{-\Delta t}}$, we know that by *Taylor's series*

$$f(x-h) = f(x) - h\dot{f}(x) + \frac{h^2}{2}\ddot{f}(x) \quad (2.6.1.7)$$

In same analogy to Eq. (2.6.1.7), we can write

$$\mathbf{X_{t-\Delta t}} = \mathbf{X_t} - \Delta t\dot{\mathbf{X}}_{\mathbf{t}} + \frac{\Delta t^2}{2}\ddot{\mathbf{X}}_{\mathbf{t}} \rightarrow \mathbf{X_{-\Delta t}} = \mathbf{X_0} - \Delta t\dot{\mathbf{X}}_{\mathbf{0}} + \frac{\Delta t^2}{2}\ddot{\mathbf{X}}_{\mathbf{0}} \quad (2.6.1.8)$$

Thus, having established $X_{-\Delta t}$ as in Eq. (2.6.1.8), we substitute this in Eq. (2.6.1.5) to have

$$\hat{\mathbf{M}}X_{\Delta t} = \mathbf{R}_0 - \hat{\mathbf{K}}X_0 - \mathbf{M_0}X_{-\Delta t} \quad (2.6.1.9)$$

And proceed step by step to find out the displacements at steps $\mathbf{X_{2t}}, \mathbf{X_{3t}}, \ldots \mathbf{X_{nt}}$.
The velocity and acceleration of the body at each step are computed from Eq. (2.6.1.2).

The above procedure can be structured as follows for computer implementation:

- Assemble mass matrix **M**, damping matrix **C**, and stiffness matrix **K**.
- Initialize $\mathbf{X_0}$ and $\dot{\mathbf{X}}_{\mathbf{0}}$.
- Compute $\{\ddot{\mathbf{X}}_{\mathbf{0}}\} = [\mathbf{M}]^{-1}\{\mathbf{R}_0\} - [\mathbf{C}]\{\dot{\mathbf{X}}_{\mathbf{0}}\} - [\mathbf{K}]\{\mathbf{X_0}\}$.
- Select time step Δt and calculate the integration constants
  - $\alpha_0 = 1/\Delta t^2$, $\alpha_1 = 1/(2\Delta t)$, $\alpha_2 = 2\alpha_0$ and $\alpha_3 = 1/\alpha_2$
  - Calculate $\mathbf{X_{-\Delta t}} = \mathbf{X_0} - \Delta t\dot{\mathbf{X}}_{\mathbf{0}} + \alpha_3\ddot{\mathbf{X}}_{\mathbf{0}}$
  - Form effective mass matrix as $\hat{\mathbf{M}} = \alpha_0\mathbf{M} + \alpha_1\mathbf{C}$
- For each time step compute
  - Effective load at time t $\{\hat{\mathbf{R}}_{\mathbf{t}}\} = \mathbf{R_t} - [\mathbf{K} - \alpha_2\mathbf{M}]\{X_t\} - [\alpha_0\mathbf{M} - \alpha_1\mathbf{C}]\{X_{t-\Delta t}\}$.
  - Solve for displacement at time $t+\Delta t$ as $\mathbf{X_{t+\Delta t}} = [\hat{\mathbf{M}}]^{-1}\{\hat{\mathbf{R}}_{\mathbf{t}}\}$.

Example 2.6.1 A structure has the following mass, damping, and stiffness matrix. Time history analysis is performed based on CDT for the digitized data of EL Centro 1940, for which the acceleration data up to first five steps are furnished for hand computation, while the final results have been computed for 1560 steps at an incremental steps of 0.02 s.

$$[\mathbf{K}] = \begin{bmatrix} 5 & -2 & 0 \\ -2 & 3.5 & -1.5 \\ 0 & -1.5 & 1.5 \end{bmatrix} \times 10^6\ \text{kN/m},$$

$$[\mathbf{M}] = \begin{bmatrix} 4 & 0 & 0 \\ 0 & 4 & 0 \\ 0 & 0 & 2 \end{bmatrix} \times 10^3\ \text{kN s}^2/\text{m and}$$

$$[C] = \begin{bmatrix} 15 & -10 & 0 \\ -10 & 20 & -8 \\ 0 & -8 & 55 \end{bmatrix} \times 10^3\ \text{kN m/s}$$

The digitized acceleration data for first five steps are as furnished hereafter.

Time steps	1	2	3	4	5
Time instant	0	0.02	0.04	0.06	0.08
Acceleration (m/s^2)	0.062	0.0357	9.719×10^{-3}	0.0419	0.0743

Solution
The time step $\Delta t = 0.02$ s, thus
$a_0 = 1/\Delta t^2 = 1/(0.02)^2 = 2500$, $a_1 = 1/2\Delta t = 1/(0.04) = 25$, $a_2 = 2a_0 = 5000$, $a_3 = 1/a_2 = 0.0002$

(1) **At t = 0.0**

$$R_{0.0} = [\mathbf{M}]\{\mathbf{I}\}\ddot{x}_0 = \begin{bmatrix} 4000 & 0 & 0 \\ 0 & 4000 & 0 \\ 0 & 0 & 2000 \end{bmatrix} \begin{Bmatrix} 1 \\ 1 \\ 1 \end{Bmatrix} \times 0.062 = \begin{Bmatrix} 247.212 \\ 247.212 \\ 123.606 \end{Bmatrix} \text{kN}$$

$$[\mathbf{K}] - a_2[\mathbf{M}] = \begin{bmatrix} 5 & -2 & 0 \\ -2 & 3.5 & -1.5 \\ 0 & -1.5 & 1.5 \end{bmatrix} \times 10^6 - 5000 \times \begin{bmatrix} 4000 & 0 & 0 \\ 0 & 4000 & 0 \\ 0 & 0 & 2000 \end{bmatrix}$$

$$= \begin{bmatrix} -1.5 \times 10^7 & 2 \times 10^6 & 0 \\ 2 \times 10^6 & -1.65 \times 10^7 & -1.5 \times 10^6 \\ 0 & -1.5 \times 10^6 & -8.5 \times 10^6 \end{bmatrix}$$

$$a_0[\mathbf{M}] - a_1[\mathbf{C}] = 2500 \times \begin{bmatrix} 4000 & 0 & 0 \\ 0 & 4000 & 0 \\ 0 & 0 & 2000 \end{bmatrix} - 25 \times \begin{bmatrix} 15 & -10 & 0 \\ -10 & 20 & -8 \\ 0 & -8 & 55 \end{bmatrix} \times 10^3$$

$$= \begin{bmatrix} 962{,}500 & 250{,}000 & 0 \\ 250{,}000 & 9{,}500{,}000 & 200{,}000 \\ 0 & 200{,}000 & 3{,}625{,}000 \end{bmatrix}$$

Effective mass matrix

$$[\mathbf{M_e}] = a_0[\mathbf{M}] + a_1[\mathbf{C}]$$

$$= 2500 \times \begin{bmatrix} 4000 & 0 & 0 \\ 0 & 4000 & 0 \\ 0 & 0 & 2000 \end{bmatrix}$$

$$+ 25 \times \begin{bmatrix} 15 & -10 & 0 \\ -10 & 20 & -8 \\ 0 & -8 & 55 \end{bmatrix} \times 10^3$$

$$= \begin{bmatrix} 10{,}375{,}000 & -250{,}000 & 0 \\ -250{,}000 & 1.050 \times 10^7 & -200{,}000 \\ 0 & -200{,}000 & 6{,}375{,}000 \end{bmatrix}$$

$$\{\mathbf{X}\}_{-0.02} = \{\mathbf{X}\}_0 - \Delta t\{\dot{\mathbf{X}}\}_0 + a_3\{\mathbf{X}\}_0$$

As $\{\mathbf{X}\}_0 = \{0\}$ and so is its first derivative, we have

$$\{\mathbf{X}\}_{-0.02} = \begin{Bmatrix} 0 \\ 0 \\ 0 \end{Bmatrix} - 0.02 \times \begin{Bmatrix} 0 \\ 0 \\ 0 \end{Bmatrix} + 0.0002 \times \begin{Bmatrix} 0.062 \\ 0.062 \\ 0.062 \end{Bmatrix}$$

$$= \begin{Bmatrix} 1.236 \\ 1.236 \\ 1.236 \end{Bmatrix} \times 10^{-5}\ \mathrm{m}$$

$$\{\hat{\mathbf{R}}\}_{0.0} = \{\mathbf{R}\}_{0.0} - ([\mathbf{K}] - a_2[\mathbf{M}])\{\mathbf{X}\}_{0.0} - (a_0[\mathbf{M}] - a_1[\mathbf{C}])\{X\}_{-0.02}$$

$$\rightarrow \{\hat{\mathbf{R}}\}_{0.0} = \begin{Bmatrix} 247.212 \\ 247.212 \\ 123.606 \end{Bmatrix} - \begin{bmatrix} -1.5\times10^{7} & 2\times10^{6} & 0 \\ 2\times10^{6} & -1.65\times10^{7} & -1.5\times10^{6} \\ 0 & -1.5\times10^{6} & -8.5\times10^{6} \end{bmatrix} \begin{Bmatrix} 0.0 \\ 0.0 \\ 0.0 \end{Bmatrix}$$

$$- \begin{bmatrix} 962{,}500 & 250{,}000 & 0 \\ 250{,}000 & 9{,}500{,}000 & 200{,}000 \\ 0 & 200{,}000 & 3{,}625{,}000 \end{bmatrix} \begin{Bmatrix} 1.236 \\ 1.236 \\ 1.236 \end{Bmatrix} \times 10^{-5}$$

$$\{\hat{\mathbf{R}}\}_{0.0} = \begin{Bmatrix} 125.1511 \\ 124.224 \\ 76.3267 \end{Bmatrix} \text{kN}$$

$$\text{Thus } \{\mathbf{X}\}_{0.02} = [\mathbf{M}_e]^{-1}\{\mathbf{R}\}_{0.0}$$

$$= \begin{bmatrix} 9.6\times10^{-8} & 2.3\times10^{-9} & 7.2\times10^{-11} \\ 2.3\times10^{-9} & 9.534\times10^{-8} & 2.99\times10^{-9} \\ 7.2\times10^{-11} & 2.99\times10^{-9} & 1.569\times10^{-7} \end{bmatrix} \begin{Bmatrix} 125.511 \\ 124.224 \\ 76.3267 \end{Bmatrix}$$

$$= \begin{Bmatrix} 1.2361 \\ 1.2361 \\ 1.2361 \end{Bmatrix} \times 10^{-5}\,\text{m}$$

$$\{\dot{\mathbf{X}}\}_{0.02} = a_1\left[\{\mathbf{X}\}_{0.02} - \{\mathbf{X}\}_{-0.02}\right] = \begin{Bmatrix} 0 \\ 0 \\ 0 \end{Bmatrix} \text{m/s}$$

$$\{\ddot{\mathbf{X}}\}_{0.02} = a_0\left[\{\mathbf{X}\}_{-0.02} - 2\{\mathbf{X}\}_{0.0} + \{\mathbf{X}\}_{0.02}\right]$$

$$= 2500\left[\begin{Bmatrix} 1.2361 \\ 1.2361 \\ 1.2361 \end{Bmatrix} \times 10^{-5} - 2\begin{Bmatrix} 0 \\ 0 \\ 0 \end{Bmatrix} + \begin{Bmatrix} 1.2361 \\ 1.2361 \\ 1.2361 \end{Bmatrix} \times 10^{-5}\right]$$

$$= \begin{Bmatrix} 0.0618 \\ 0.0618 \\ 0.0618 \end{Bmatrix} \text{m/s}^2$$

(2) **At t = 0.02**

$$\{\mathbf{R}\}_{0.02} = [\mathbf{M}]\{\mathbf{I}\}\ddot{x}_{0.02} = \begin{bmatrix} 4000 & 0 & 0 \\ 0 & 4000 & 0 \\ 0 & 0 & 2000 \end{bmatrix} \begin{Bmatrix} 1 \\ 1 \\ 1 \end{Bmatrix} \times 0.0357$$

$$= \begin{Bmatrix} 142.834 \\ 142.834 \\ 71.417 \end{Bmatrix} \text{kN}$$

$$\{\hat{\mathbf{R}}\}_{0.02} = \{\mathbf{R}\}_{0.02} - ([\mathbf{K}] - a_2[\mathbf{M}])\{\mathbf{X}\}_{0.02} - (a_0[\mathbf{M}] - a_1[\mathbf{C}])\{X\}_{0.0}$$

$$\{\hat{\mathbf{R}}\}_{0.02} = \begin{Bmatrix} 142.834 \\ 142.834 \\ 71.417 \end{Bmatrix} - \begin{bmatrix} -1.5 \times 10^7 & 2 \times 10^6 & 0 \\ 2 \times 10^6 & -1.65 \times 10^7 & -1.5 \times 10^6 \\ 0 & -1.5 \times 10^6 & -8.5 \times 10^6 \end{bmatrix} \begin{Bmatrix} 1.2361 \\ 1.2361 \\ 1.2361 \end{Bmatrix} \times 10^{-5} - \begin{bmatrix} 962{,}500 & 250{,}000 & 0 \\ 250{,}000 & 9{,}500{,}000 & 200{,}000 \\ 0 & 200{,}000 & 3{,}625{,}000 \end{bmatrix} \begin{Bmatrix} 0 \\ 0 \\ 0 \end{Bmatrix}$$

$$\{\hat{\mathbf{R}}\}_{0.02} = \begin{Bmatrix} 352.96 \\ 390.045 \\ 195.02 \end{Bmatrix} \text{kN; thus}$$

$$\{\mathbf{X}\}_{0.04} = [\mathbf{M}_e]^{-1}\{\mathbf{R}\}_{0.02} = \begin{bmatrix} 9.6 \times 10^{-8} & 2.3 \times 10^{-9} & 7.2 \times 10^{-11} \\ 2.3 \times 10^{-9} & 9.534 \times 10^{-8} & 2.99 \times 10^{-9} \\ 7.2 \times 10^{-11} & 2.99 \times 10^{-9} & 1.569 \times 10^{-7} \end{bmatrix} \begin{Bmatrix} 352.96 \\ 390.045 \\ 195.02 \end{Bmatrix}$$

$$= \begin{Bmatrix} 3.495 \\ 3.8585 \\ 3.1802 \end{Bmatrix} \times 10^{-5}\ \text{m}$$

$$\{\dot{\mathbf{X}}\}_{0.04} = a_1\left[\{\mathbf{X}\}_{0.04} - \{\mathbf{X}\}_{0.0}\right] = \begin{Bmatrix} 0.00087 \\ 0.0009646 \\ 0.0008 \end{Bmatrix} \text{m/s}$$

$$\{\ddot{\mathbf{X}}\}_{0.04} = a_0\left[\{\mathbf{X}\}_{0.0} - 2\{\mathbf{X}\}_{0.02} + \{\mathbf{X}\}_{0.04}\right]$$

$$= 2500\left[\begin{Bmatrix} 0 \\ 0 \\ 0 \end{Bmatrix} - 2\begin{Bmatrix} 1.2361 \\ 1.2361 \\ 1.2361 \end{Bmatrix} \times 10^{-5} + \begin{Bmatrix} 3.495 \\ 3.8585 \\ 3.1802 \end{Bmatrix} \times 10^{-5}\right]$$

$$= \begin{Bmatrix} 0.02557 \\ 0.03466 \\ 0.017702 \end{Bmatrix} \text{m/s}^2$$

(3) **At t = 0.04**

$$\{\mathbf{R}\}_{0.04} = [\mathbf{M}]\{\mathbf{I}\}\ddot{x}_{0.04} = \begin{bmatrix} 4000 & 0 & 0 \\ 0 & 4000 & 0 \\ 0 & 0 & 2000 \end{bmatrix}\begin{Bmatrix} 1 \\ 1 \\ 1 \end{Bmatrix} \times 9.719 \times 10^{-3}$$

$$= \begin{Bmatrix} 38.847 \\ 38.847 \\ 19.423 \end{Bmatrix} \text{kN}$$

$$\{\hat{\mathbf{R}}\}_{0.04} = \{\mathbf{R}\}_{0.04} - ([\mathbf{K}] - a_2[\mathbf{M}])\{\mathbf{X}\}_{0.04} - (a_0[\mathbf{M}] - a_1[\mathbf{C}])\{X\}_{0.02}$$

$$\{\hat{\mathbf{R}}\}_{0.04} = \begin{Bmatrix} 38.847 \\ 38.847 \\ 19.423 \end{Bmatrix}$$

$$- \begin{bmatrix} -1.5 \times 10^7 & 2 \times 10^6 & 0 \\ 2 \times 10^6 & -1.65 \times 10^7 & -1.5 \times 10^6 \\ 0 & -1.5 \times 10^6 & -8.5 \times 10^6 \end{bmatrix}\begin{Bmatrix} 3.495 \\ 3.8585 \\ 3.1802 \end{Bmatrix} \times 10^{-5}$$

$$- \begin{bmatrix} 962{,}500 & 250{,}000 & 0 \\ 250{,}000 & 9{,}500{,}000 & 200{,}000 \\ 0 & 200{,}000 & 3{,}625{,}000 \end{bmatrix}\begin{Bmatrix} 1.2361 \\ 1.2361 \\ 1.2361 \end{Bmatrix} \times 10^{-5}$$

$$\{\hat{\mathbf{R}}\}_{0.04} = \begin{Bmatrix} 518.212 \\ 670.118 \\ 300.342 \end{Bmatrix} \text{kN}; \text{thus}$$

$$\{\mathbf{X}\}_{0.06} = [\mathbf{M}_e]^{-1}\{\mathbf{R}\}_{0.04} = \begin{bmatrix} 9.6\times10^{-8} & 2.3\times10^{-9} & 7.2\times10^{-11} \\ 2.3\times10^{-9} & 9.534\times10^{-8} & 2.99\times10^{-9} \\ 7.2\times10^{-11} & 2.99\times10^{-9} & 1.569\times10^{-7} \end{bmatrix} \begin{Bmatrix} 518.212 \\ 670.118 \\ 300.342 \end{Bmatrix}$$

$$= \begin{Bmatrix} 5.154 \\ 6.598 \\ 4.918 \end{Bmatrix} \times 10^{-5}\ \mathrm{m}$$

$$\{\dot{\mathbf{X}}\}_{0.06} = a_1\left[\{\mathbf{X}\}_{0.06} - \{\mathbf{X}\}_{0.02}\right] = \begin{Bmatrix} 0.00097 \\ 0.00134 \\ 0.00092 \end{Bmatrix}\ \mathrm{m/s}$$

$$\{\ddot{\mathbf{X}}\}_{0.06} = a_0\left[\{\mathbf{X}\}_{0.02} - 2\{\mathbf{X}\}_{0.04} + \{\mathbf{X}\}_{0.06}\right]$$

$$= 2500\left[\begin{Bmatrix} 1.2361 \\ 1.2361 \\ 1.2361 \end{Bmatrix}\times10^{-5} - 2\begin{Bmatrix} 3.495 \\ 3.8585 \\ 3.1802 \end{Bmatrix}\times10^{-5} + \begin{Bmatrix} 5.154 \\ 6.598 \\ 4.918 \end{Bmatrix}\times10^{-5}\right]$$

$$= \begin{Bmatrix} -0.015 \\ 0.00294 \\ -0.00515 \end{Bmatrix}\ \mathrm{m/s^2}$$

and so on …

First thirty steps of the above computation are shown hereafter:

The displacement plots are shown below:

Sl no	Time steps	Input acceleration (m/s^2)	$X1$ (m)	$X2$ (m)	$X3$ (m)
1	0	0.061803	0	0	0
2	0.02	0.0357084	1.23606E−05	1.23606E−05	1.23606E−05
3	0.04	0.0097119	3.49504E−05	3.85851E−05	3.18023E−05
4	0.06	0.0419868	5.15382E−05	6.59847E−05	4.91826E−05
5	0.08	0.0743598	7.25586E−05	0.000103397	7.82245E−05
6	0.1	0.1066347	0.000107932	0.000158928	0.000126906
7	0.12	0.0669042	0.000163773	0.000239992	0.000199862
8	0.14	0.0271737	0.000212684	0.000323925	0.000276955
9	0.16	−0.0125568	0.000232017	0.000386707	0.000344972
10	0.18	0.0361008	0.000209805	0.000405423	0.000392085
11	0.2	0.0847584	0.000180418	0.000397404	0.000433675

(continued)

(continued)

Sl no	Time steps	Input acceleration (m/s^2)	$X1$ (m)	$X2$ (m)	$X3$ (m)
12	0.22	0.133416	0.00017505	0.000387037	0.000474804
13	0.24	0.0713187	0.000211891	0.000403028	0.000519572
14	0.26	0.0092214	0.000250127	0.00042796	0.000541265
15	0.28	0.041202	0.000251941	0.000437514	0.000530915
16	0.3	0.0216801	0.000232696	0.000438959	0.000516324
17	0.32	0.0020601	0.000195214	0.000418275	0.000496022
18	0.34	0.0435564	0.000146027	0.000365914	0.000464539
19	0.36	0.0850527	0.000114604	0.000304229	0.000433518
20	0.38	0.126549	0.000119143	0.00026147	0.000408864
21	0.4	0.1680453	0.000164195	0.000265456	0.00039865
22	0.42	−0.0336483	0.000244616	0.000335953	0.000416559
23	0.44	−0.23544	0.000255963	0.000383263	0.000400916
24	0.46	−0.0973152	0.000125616	0.000309848	0.000313184
25	0.48	0.0408096	−3.8983E−05	0.000161364	0.000225021
26	0.5	0.0517968	−0.00013341	4.47545E−06	0.000163134
27	0.52	0.1621593	−0.00014282	−0.00012623	9.7839E−05
28	0.54	0.2726199	−4.835E−05	−0.00015392	5.38916E−05
29	0.56	0.3829824	0.000140608	−1.9127E−05	6.88922E−05
30	0.58	0.2402469	0.000403173	0.000305097	0.000191263

Floor shear forces in frame (Mass × Output acceleration) are as shown hereafter:

Sl no	Time steps	F_1 (kN)	F_2 (m)	F_3 (m)
1	0	247.212	247.212	123.606
2	0.02	247.212	247.212	123.606
3	0.04	102.2917	138.6391	35.40563
4	0.06	−60.0196	11.75068	−10.3074
5	0.08	44.32569	100.1313	58.30868
6	0.1	143.5293	181.1739	98.19765
7	0.12	204.6779	255.3488	121.3698
8	0.14	−69.2957	28.68013	20.6878
9	0.16	−295.79	−211.518	−45.3797
10	0.18	−415.445	−440.652	−104.516
11	0.2	−71.7527	−267.346	−27.6202
12	0.22	240.1894	−23.4851	−2.3047
13	0.24	422.0937	263.575	18.19789
14	0.26	13.9529	89.41825	−115.379
15	0.28	−364.228	−153.784	−160.213

(continued)

(continued)

Sl no	Time steps	F_1 (kN)	F_2 (m)	F_3 (m)
16	0.3	−210.582	−81.082	−21.2039
17	0.32	−182.379	−221.3	−28.5565
18	0.34	−117.039	−316.764	−55.9043
19	0.36	177.6342	−93.2418	2.310041
20	0.38	359.6177	189.2595	31.83679
21	0.4	405.1304	467.4431	72.19638
22	0.42	353.6853	665.1167	140.6163
23	0.44	−690.727	−231.867	−167.759
24	0.46	−1416.95	−1207.25	−360.44
25	0.48	−342.514	−750.686	−2.16031
26	0.5	701.6751	−84.0521	131.3786
27	0.52	850.2492	261.7985	−17.038
28	0.54	1038.774	1030.189	106.7382
29	0.56	944.871	1624.866	294.7398
30	0.58	736.0665	1894.275	536.8533

The complete time history plot up to 31 s constituting 1560 steps is shown hereafter for amplitude and floor shear force (Figs. 2.33 and 2.34).

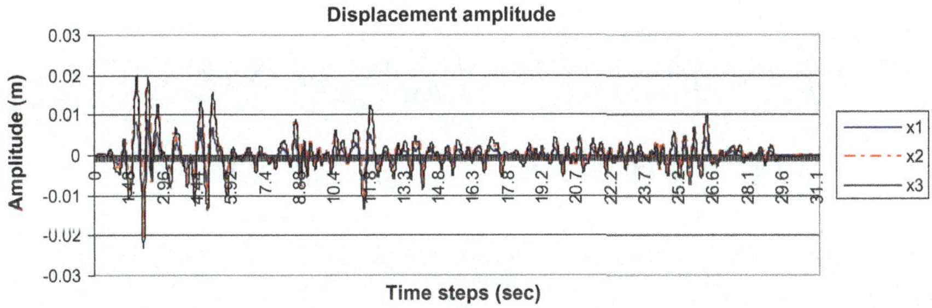

Fig. 2.33 Time history plot for amplitude in meters

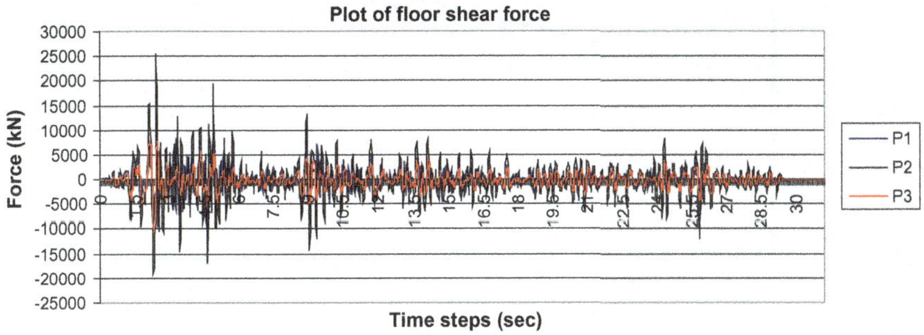

Fig. 2.34 Time history plot for floor shear force (kN)

2.6.2 Wilson-Theta Method

The method was developed by Ed Wilson (1973), at the University of California, Berkeley, and constitutes the backbone of time history analysis for the software SAP 2000.

In this technique, Wilson assumed a linear variation of acceleration with time between t and $t+\Delta t$.

Referring to Fig. 2.35 which shows linear variation of acceleration with time t to $t+\theta\Delta t$, θ is assumed to be ≥ 1.4. Let τ denote the increase in time such that $0 \leq \tau \leq \theta\Delta t$.

Then, by similar triangle

$$\frac{\ddot{\mathbf{X}}_{\mathbf{t}+\boldsymbol{\theta}\boldsymbol{\Delta}\mathbf{t}}-\ddot{\mathbf{X}}_{\mathbf{t}}}{t+\theta\Delta t-t}=\frac{\ddot{\mathbf{X}}_{\mathbf{t}+\tau}-\ddot{\mathbf{X}}_{t}}{t+\tau-t} \tag{2.6.2.1}$$

or,

$$\frac{\ddot{\mathbf{X}}_{\mathbf{t}+\tau}-\ddot{\mathbf{X}}_{t}}{\tau}=\frac{\ddot{\mathbf{X}}_{\mathbf{t}+\boldsymbol{\theta}\boldsymbol{\Delta}\mathbf{t}}-\ddot{\mathbf{X}}_{t}}{\boldsymbol{\theta}\boldsymbol{\Delta}\mathbf{t}} \rightarrow \{\ddot{\mathbf{X}}_{\mathbf{t}+\tau}\}=\{\ddot{\mathbf{X}}_{\mathbf{t}}\}+\frac{\tau}{\theta\Delta t}\{\ddot{\mathbf{X}}_{\mathbf{t}+\boldsymbol{\theta}\boldsymbol{\Delta}\mathbf{t}}-\ddot{\mathbf{X}}_{\mathbf{t}}\} \tag{2.6.2.2}$$

To find the velocity vector at time interval $t + \tau$, we integrate the L.H.S. between $t + \tau$ to t and right-hand side between τ to 0, to have

$$\begin{aligned}&\int_{t}^{t+\tau}\{\ddot{\mathbf{X}}_{\mathbf{t}+\tau}\}\mathrm{d}t=\int_{0}^{\tau}\{\ddot{\mathbf{X}}_{\mathbf{t}}\}\mathrm{d}t+\int_{0}^{\tau}\frac{\tau}{\theta\Delta t}\{\ddot{\mathbf{X}}_{\mathbf{t}+\boldsymbol{\theta}\boldsymbol{\Delta}\mathbf{t}}-\ddot{\mathbf{X}}_{\mathbf{t}}\}\mathrm{d}t\\&\rightarrow\{\dot{\mathbf{X}}_{\mathbf{t}+\tau}\}=\{\dot{\mathbf{X}}_{t}\}+\tau\{\ddot{\mathbf{X}}_{\mathbf{t}}\}+\frac{\tau^{2}}{2\theta\Delta t}\{\ddot{\mathbf{X}}_{\mathbf{t}+\boldsymbol{\theta}\boldsymbol{\Delta}\mathbf{t}}-\ddot{\mathbf{X}}_{\mathbf{t}}\}\end{aligned} \tag{2.6.2.3}$$

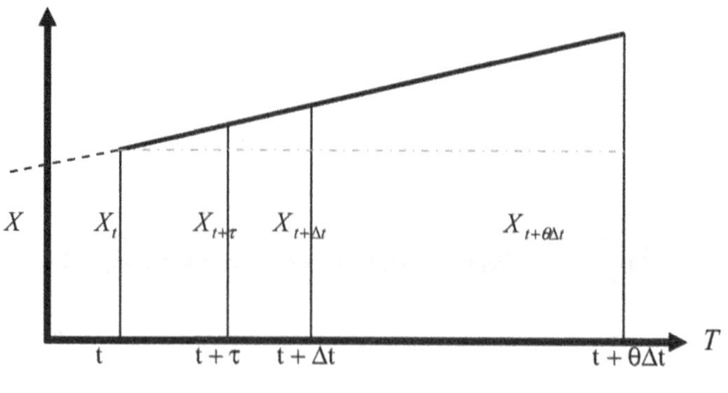

Fig. 2.35 Linear variation of time period with respect to time

To find out the displacement vector at $t+\tau$, we have on further integration of the velocity vector

$$\int_{t}^{t+\tau}\{\dot{\mathbf{X}}_{t+\tau}\}\mathrm{d}t=\int_{0}^{\tau}\{\dot{\mathbf{X}}_t\}\mathrm{d}t+\int_{0}^{\tau}\tau\{\ddot{\mathbf{X}}_t\}\mathrm{d}t+\int_{0}^{\tau}\frac{\tau^2}{2\theta\Delta t}\{\ddot{\mathbf{X}}_{t+\theta\Delta t}-\ddot{\mathbf{X}}_t\}\cdot\mathrm{d}t$$

$$\rightarrow\{\mathbf{X}_{t+\tau}\}=\{\mathbf{X}_t\}+\tau\{\dot{\mathbf{X}}_t\}+\frac{\tau^2}{2}\{\ddot{\mathbf{X}}_t\}+\frac{\tau^3}{6\theta\Delta t}\{\ddot{\mathbf{X}}_{t+\theta\Delta t}-\ddot{\mathbf{X}}_t\}\tag{2.6.2.4}$$

For $\tau=\theta\Delta t$, we have

$$\{\dot{\mathbf{X}}_{t+\theta\Delta t}\}=\{\dot{\mathbf{X}}_t\}+\theta\Delta t\{\ddot{\mathbf{X}}_t\}+\frac{\theta^2\Delta t^2}{2\theta\Delta t}\{\ddot{\mathbf{X}}_{t+\theta\Delta t}-\ddot{\mathbf{X}}_t\}$$

Thus,

$$\{\dot{\mathbf{X}}_{t+\theta\Delta t}\}=\{\dot{\mathbf{X}}_t\}+\frac{\theta\Delta t}{2}\{\ddot{\mathbf{X}}_{t+\theta\Delta t}+\ddot{\mathbf{X}}_t\}\text{and}\,\{\mathbf{X}_{t+\theta\Delta t}\}$$
$$=\{\mathbf{X}_t\}+\theta\Delta t\{\dot{\mathbf{X}}_t\}+\frac{\theta^2\Delta t^2}{6}\{\ddot{\mathbf{X}}_{t+\theta\Delta t}+2\ddot{\mathbf{X}}_t\}$$

The above two equations can be solved to obtain $\{\ddot{\mathbf{X}}_{t+\theta\Delta t}\}$ and $\{\dot{\mathbf{X}}_{t+\theta\Delta t}\}$ in terms of $\{\mathbf{X}_{t+\theta\Delta t}\}$, and this gives:

$$\begin{aligned}\{\ddot{\mathbf{X}}_{t+\theta\Delta t}\}&=\frac{6}{\theta^2\Delta t^2}\{\mathbf{X}_{t+\theta\Delta t}-\mathbf{X}_t\}-\frac{6}{\theta\Delta t}\{\dot{\mathbf{X}}_t\}-2\{\ddot{\mathbf{X}}_t\}\\\{\dot{\mathbf{X}}_{t+\theta\Delta t}\}&=\frac{3}{\theta\Delta t}\{\mathbf{X}_{t+\theta\Delta t}-\mathbf{X}_t\}-2\{\dot{\mathbf{X}}_t\}-\frac{\theta\Delta t}{2}\{\ddot{\mathbf{X}}_t\}\end{aligned}\tag{2.6.2.5}$$

To start the solution for the dynamic equation, we start with the expression

$$[\mathbf{M}]\{\ddot{\mathbf{X}}_{t+\theta\Delta t}\}+[\mathbf{C}]\{\dot{\mathbf{X}}_{t+\theta\Delta t}\}+[\mathbf{K}]\{\mathbf{X}_{t+\theta\Delta t}\}=\{\mathbf{R}_{t+\theta\Delta t}\}\tag{2.6.2.6}$$

Now, the acceleration is varying linearly; hence, the applied force within this small time step can also be assumed to vary linearly. Thus, we can draw a similar force diagram as shown in Fig. 2.36.

By a similar triangle, we have

$$\frac{\mathbf{R}_{t+\theta\Delta t}-\mathbf{R}_t}{t+\theta\Delta t-t}=\frac{\mathbf{R}_{t+\Delta t}-\mathbf{R}_t}{t+\Delta t-t}$$

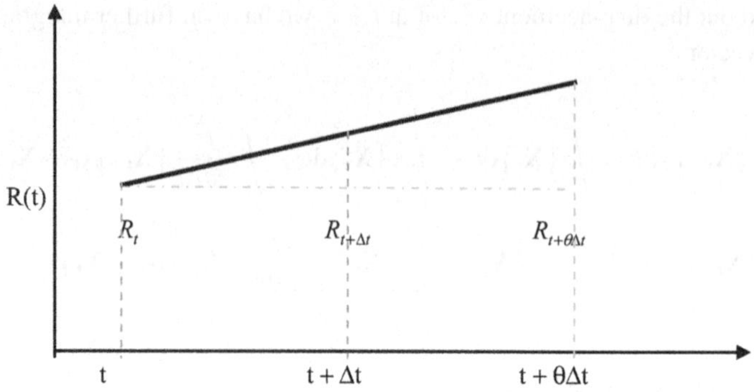

Fig. 2.36 Linear variation of *R* with time

or

$$\mathbf{R_{t+\theta\Delta t}} = \mathbf{R_t} + \theta(\mathbf{R_{t+\Delta t}} - \mathbf{R_t}) \tag{2.6.2.7}$$

Now substituting the value of $\{\ddot{\mathbf{X}}_{t+\theta\Delta t}\}$ and $\{\dot{\mathbf{X}}_{t+\theta\Delta t}\}$ in the equilibrium equation, we have

$$[\mathbf{M}]\frac{6}{\theta^2\Delta t^2}\{\mathbf{X_{t+\theta\Delta t}} - \mathbf{X_t}\} - [\mathbf{M}]\frac{6}{\theta\Delta t}\{\dot{\mathbf{X}}_t\} - 2[\mathbf{M}]\{\mathbf{X_t}\} + [\mathbf{C}]\frac{3}{\theta\Delta t}\{\mathbf{X_{t+\theta\Delta t}} - \mathbf{X_t}\} - 2[\mathbf{C}]\{\dot{\mathbf{X}}_t\}$$
$$-\frac{\theta\Delta t}{2}[\mathbf{C}]\{\ddot{\mathbf{X}}_t\} + [\mathbf{K}]\{\mathbf{X_{t+\theta\Delta t}}\} = \{\mathbf{R_t}\} + \theta\{\mathbf{R_{t+\Delta t}} - \mathbf{R_t}\}$$

The above equation can be further simplified to

$$\left[\frac{6[\mathbf{M}]}{\theta^2\Delta t^2} + \frac{3[\mathbf{C}]}{\theta\Delta t} + [\mathbf{K}]\right]\{\mathbf{X_{t+\theta\Delta t}}\} - \left[\frac{6[\mathbf{M}]}{\theta^2\Delta t^2} + \frac{3[\mathbf{C}]}{\theta\Delta t}\right]\{\mathbf{X_t}\} - \left[\frac{6[\mathbf{M}]}{\theta^2\Delta t^2} + 2[\mathbf{C}]\right]\{\dot{\mathbf{X}}_t\}$$
$$-\left[2[\mathbf{M}] - \frac{\theta\Delta t}{2}[\mathbf{C}]\right]\{\ddot{\mathbf{X}}_t\} = \{\mathbf{R_{t+\theta\Delta t}}\}$$

The above expression can now be further simplified and written in a more compact form as

$$[\hat{\mathbf{K}}]\cdot\{\mathbf{X_{t+\theta\Delta t}}\} = \{\hat{\mathbf{R}}_{t+\theta\Delta t}\} + [\hat{\mathbf{M}}_0]\{\mathbf{X_t}\} + [\hat{\mathbf{M}}_1]\{\dot{\mathbf{X}}_t\} + [\hat{\mathbf{M}}_2]\{\ddot{\mathbf{X}}_t\} \tag{2.6.2.8}$$

where

$$[\hat{\mathbf{K}}] = \left[\frac{6[\mathbf{M}]}{\theta^2\Delta t^2} + \frac{3[\mathbf{C}]}{\theta\Delta t} + [\mathbf{K}]\right];\ [\hat{\mathbf{M}}_0] = \left[\frac{6[\mathbf{M}]}{\theta^2\Delta t^2} + \frac{3[\mathbf{C}]}{\theta\Delta t}\right];$$

$$[\hat{\mathbf{M}}_1] = \left[\frac{6[\mathbf{M}]}{\theta^2\Delta t^2} + 2[\mathbf{C}]\right]\{\dot{\mathbf{X}}_t\};\ [\hat{\mathbf{M}}_2] = \left[2[\mathbf{M}] - \frac{\theta\Delta t}{2}[\mathbf{C}]\right].$$

We have

$$\{\hat{R}_{t+\theta\Delta t}\} = \{R_t\} + \theta\{R_{t+\Delta t} - R_t\} \tag{2.6.2.9}$$

from which we can find out the value of $\{X_{t+\theta\Delta t}\}$.

Once $\mathbf{X_t} + \boldsymbol{\theta\Delta}\mathbf{t}$ is obtained, we can now back substitute its values in the equations

$$\begin{aligned}\{\ddot{\mathbf{X}}_{\mathbf{t}+\theta\Delta\mathbf{t}}\} &= \frac{6}{\theta^2\Delta t^2}\{\mathbf{X}_{\mathbf{t}+\theta\Delta\mathbf{t}} - \mathbf{X_t}\} - \frac{6}{\theta\Delta t}\{\dot{\mathbf{X}}_\mathbf{t}\} - 2\{\ddot{\mathbf{X}}_\mathbf{t}\} \text{ and} \\ \{\dot{\mathbf{X}}_{\mathbf{t}+\theta\Delta\mathbf{t}}\} &= \frac{3}{\theta\Delta t}\{\mathbf{X}_{\mathbf{t}+\theta\Delta\mathbf{t}} - \mathbf{X_t}\} - 2\{\dot{\mathbf{X}}_\mathbf{t}\} - \frac{\theta\Delta t}{2}\{\ddot{\mathbf{X}}_\mathbf{t}\}\end{aligned} \tag{2.6.2.10}$$

Thus, all data at the time intervals $t + \theta\Delta t$ are evaluated.

The above method can thus be structured as follows:

- Assemble the mass matrix $[\mathbf{M}]$ the damping matrix $[\mathbf{C}]$ and stiffness matrix $[\mathbf{K}]$
- Initialize $\{\mathbf{X_0}\}$ and $\{\dot{\mathbf{X}}_\mathbf{0}\}$
- Evaluate $\{\ddot{\mathbf{X}}_\mathbf{0}\}$ (refer to central difference method to see how this is evaluated)
- Select the time step Δt and calculate the integration constant θ (this is usually taken as 1.4)
- Select the values $\alpha_0 = \frac{6}{\theta^2\Delta t^2}$, $\alpha_1 = \frac{3}{\theta\Delta t}$, $\alpha_2 = 2\alpha_1$, $\alpha_3 = \theta\Delta t/2$, $\alpha_4 = \frac{\alpha_0}{\theta}$, $\alpha_5 = \frac{-\alpha_2}{\theta}$, $\alpha_6 = 1 - \frac{3}{\theta}$, $\alpha_7 = \frac{\Delta t}{2}$, $\alpha_8 = \frac{\Delta t^2}{6}$.
- Form modified stiffness matrix $[\hat{\mathbf{K}}]$

$$[\hat{\mathbf{K}}] = [\mathbf{K}] + \alpha_0[\mathbf{M}] + \alpha_1[\mathbf{C}]$$

- Calculate the external load $\{\hat{\mathbf{R}}_{\mathbf{t}+\theta\Delta\mathbf{t}}\}$

$$\begin{aligned}\{\hat{\mathbf{R}}_{\mathbf{t}+\theta\Delta\mathbf{t}}\} = {} & \{\mathbf{R_t}\} + \theta\{\mathbf{R}_{\mathbf{t}+\Delta\mathbf{t}} - \mathbf{R_t}\} + [\mathbf{M}]\{\alpha_0\{\mathbf{X_t}\} + \alpha_2\{\dot{\mathbf{X}}_\mathbf{t}\} + 2\{\ddot{\mathbf{X}}_\mathbf{t}\}\} \\ & + [\mathbf{C}]\{\alpha_1\{\mathbf{X_t}\} + 2\{\dot{\mathbf{X}}_\mathbf{t}\} + \alpha_3\{\ddot{\mathbf{X}}_\mathbf{t}\}\}\end{aligned}$$

  - Solve for displacement at time $t + \theta\Delta t$

    - $[\hat{\mathbf{K}}]\{\mathbf{X}_{\mathbf{t}+\theta\Delta\mathbf{t}}\} = \{\hat{\mathbf{R}}_{\mathbf{t}+\theta\Delta\mathbf{t}}\}$

- Calculate the acceleration, velocity and at time displacement $t+\Delta t$
  - $\{\ddot{\mathbf{X}}_{t+\Delta t}\} = \alpha_4\{\mathbf{X}_{t+\theta\Delta t} - \mathbf{X}_t\} + \alpha_5\{\dot{\mathbf{X}}_t\} + \alpha_6\{\ddot{\mathbf{X}}_t\}$
  - $\{\dot{\mathbf{X}}_{t+\Delta t}\} = \{\dot{\mathbf{X}}_t\} + \alpha_7\{\ddot{\mathbf{X}}_{t+\Delta t} + \ddot{\mathbf{X}}_t\}$
  - $\{\mathbf{X}_{t+\Delta t}\} = \{\mathbf{X}_t\} + \{\dot{\mathbf{X}}_t\}\Delta t + \alpha_8\{\ddot{\mathbf{X}}_{t+\Delta t} + 2\ddot{\mathbf{X}}_t\}$

Example 2.6.2 Repeat the problem solved in Example 2.6.1 by Wilson-Theta method for same sets of data.

Solution

In this problem, as stated previously,

$$[\mathbf{K}] = \begin{bmatrix} 5 & -2 & 0 \\ -2 & 3.5 & -1.5 \\ 0 & -1.5 & 1.5 \end{bmatrix} \times 10^6\ \text{kN/m},$$

$$[\mathbf{M}] = \begin{bmatrix} 4 & 0 & 0 \\ 0 & 4 & 0 \\ 0 & 0 & 2 \end{bmatrix} \times 10^3\ \text{kN s}^2/\text{m and}$$

$$[C] = \begin{bmatrix} 15 & -10 & 0 \\ -10 & 20 & -8 \\ 0 & -8 & 55 \end{bmatrix} \times 10^3\ \text{kN m/s}$$

The digitized acceleration data for first five steps are as furnished hereafter.

Time steps	1	2	3	4	5
Time instant	0	0.02	0.04	0.06	0.08
Acceleration (m/s^2)	0.0618	0.0357	9.719×10^{-3}	0.0419	0.0743

The integration constants a_0 to a_8 are as computed hereafter for $\Delta t = 0.02$

α_0	α_1	α_2	α_3	α_4	α_5	α_6	α_7	α_8
7653.06	107.14	214.285	0.014	5466.47	−153.061	−1.1428	0.01	6.667×10^{-5}

$$[\hat{\mathbf{K}}] = [\mathbf{K}] + \alpha_0[\mathbf{M}] + \alpha_1[C]$$

$$[\hat{\mathbf{K}}] = \begin{bmatrix} 5 & -2 & 0 \\ -2 & 3.5 & -1.5 \\ 0 & -1.5 & 1.5 \end{bmatrix} \times 10^6 + 7653.06 \times \begin{bmatrix} 4000 & 0 & 0 \\ 0 & 4000 & 0 \\ 0 & 0 & 2000 \end{bmatrix}$$

$$+ 107.14 \times \begin{bmatrix} 15{,}000 & -10{,}000 & 0 \\ -1000 & 20{,}000 & -8000 \\ 0 & -8000 & 55{,}000 \end{bmatrix}$$

$$\rightarrow [\hat{\mathbf{K}}] = \begin{bmatrix} 3.722 \times 10^7 & -3.071 \times 10^6 & 0 \\ -3.071 \times 10^6 & 3.626 \times 10^6 & -2.357 \times 10^6 \\ 0 & -2.357 \times 10^6 & 2.27 \times 10^8 \end{bmatrix} \text{kN/m}$$

- At $t = 0.0$ s

$$\{\mathbf{X}\}_{0.0} = \begin{Bmatrix} 0 \\ 0 \\ 0 \end{Bmatrix},$$

$$\{\dot{\mathbf{X}}\}_{0.0} = \begin{Bmatrix} 0 \\ 0 \\ 0 \end{Bmatrix},$$

$$\{\ddot{\mathbf{X}}\}_{0.0} = \begin{Bmatrix} 0.0618 \\ 0.0618 \\ 0.0618 \end{Bmatrix}$$

$$\begin{aligned} \{\mathbf{R}\}_{0.0} &= [\mathbf{M}]\{\ddot{\mathbf{X}}\}_{0.0} \\ &= \begin{bmatrix} 4000 & 0 & 0 \\ 0 & 4000 & 0 \\ 0 & 0 & 2000 \end{bmatrix} \begin{Bmatrix} 0.0618 \\ 0.0618 \\ 0.0618 \end{Bmatrix} \\ &= \begin{Bmatrix} 247.212 \\ 247.212 \\ 123.606 \end{Bmatrix} \text{kN} \end{aligned}$$

- At $t = 0.02$ s

$$\{\mathbf{R}\}_{0.02} = [\mathbf{M}]\{\ddot{\mathbf{X}}\}_{0.02} = \begin{bmatrix} 4000 & 0 & 0 \\ 0 & 4000 & 0 \\ 0 & 0 & 2000 \end{bmatrix} \begin{Bmatrix} 0.0357 \\ 0.0357 \\ 0.03557 \end{Bmatrix} = \begin{Bmatrix} 142.83 \\ 142.83 \\ 71.4 \end{Bmatrix} \text{kN}$$

$$\begin{aligned}\{\hat{\mathbf{R}}_{t+\theta\Delta t}\} &= \{\mathbf{R}_t\} + \theta\{\mathbf{R}_{t+\Delta t} - \mathbf{R}_t\} + [\mathbf{M}]\{\alpha_0\{\mathbf{X}_t\} + \alpha_2\{\dot{\mathbf{X}}_t\} + 2\{\ddot{\mathbf{X}}_t\}\} \\ &\quad + [\mathbf{C}]\{\alpha_1\{\mathbf{X}_t\} + 2\{\dot{\mathbf{X}}_t\} + \alpha_3\{\ddot{\mathbf{X}}_t\}\}\end{aligned}$$

$$\{\hat{\mathbf{R}}\}_{0.02} = \{\mathbf{R}\}_{0.02} + \theta[\{\mathbf{R}\}_{0.02} - \{\mathbf{R}\}_{0.0}] + [\mathbf{M}][\alpha_0\mathbf{X}_{0.0} + \alpha_2\dot{\mathbf{X}}_{0.0} + 2\ddot{\mathbf{X}}_{0.0}]$$

$$\begin{aligned}\{\hat{\mathbf{R}}\}_{0.02} &= \begin{Bmatrix} 142.83 \\ 142.83 \\ 71.4 \end{Bmatrix} + 1.4 \times \left[\begin{Bmatrix} 142.83 \\ 142.83 \\ 71.4 \end{Bmatrix} - \begin{Bmatrix} 247.212 \\ 247.212 \\ 123.606 \end{Bmatrix}\right] \\ &\quad + \begin{bmatrix} 4000 & 0 & 0 \\ 0 & 4000 & 0 \\ 0 & 0 & 2000 \end{bmatrix} \left[7653\begin{Bmatrix} 0 \\ 0 \\ 0 \end{Bmatrix} + 214.285\begin{Bmatrix} 0 \\ 0 \\ 0 \end{Bmatrix} + 2\begin{Bmatrix} 0.0618 \\ 0.0618 \\ 0.0618 \end{Bmatrix}\right]\end{aligned}$$

$$\{\hat{\mathbf{R}}\}_{0.02} = \begin{Bmatrix} 495.421 \\ 492.826 \\ 286.176 \end{Bmatrix} \text{kN}$$

$$\begin{aligned}\{\hat{\mathbf{X}}\}_{0.02} &= [\hat{\mathbf{K}}]^{-1}\{\hat{\mathbf{R}}\}_{0.02} = \begin{bmatrix} 2.706 & 0.2308 & 0.02397 \\ 0.2308 & 2.797 & 0.2904 \\ 0.02397 & 0.2904 & 4.436 \end{bmatrix} \begin{Bmatrix} 495.421 \\ 492.826 \\ 286.176 \end{Bmatrix} \times 10^{-8} \\ &= \begin{Bmatrix} 1.461 \\ 1.576 \\ 1.424 \end{Bmatrix} \times 10^{-5}\ \text{m}\end{aligned}$$

$$\{\ddot{\mathbf{X}}\}_{0.02} = \alpha_4[\hat{\mathbf{X}}_{0.02} - \mathbf{X}_{0.0}] + \alpha_5\dot{\mathbf{X}}_{0.0} + \alpha_6\ddot{\mathbf{X}}_{0.0}$$

$$\begin{aligned}\{\ddot{\mathbf{X}}\}_{0.02} &= 5466.47\left[\begin{Bmatrix} 1.461 \\ 1.576 \\ 1.424 \end{Bmatrix} \times 10^{-5} - \begin{Bmatrix} 0 \\ 0 \\ 0 \end{Bmatrix}\right] \\ &\quad - 153.061\begin{Bmatrix} 0 \\ 0 \\ 0 \end{Bmatrix} - 1.1428\begin{Bmatrix} 0.0618 \\ 0.0618 \\ 0.0618 \end{Bmatrix} = \begin{Bmatrix} 9.243 \times 10^{-3} \\ 0.016 \\ 7.234 \times 10^{-3} \end{Bmatrix} \text{m/s}^2\end{aligned}$$

$$\{\dot{\mathbf{X}}\}_{0.02} = \dot{\mathbf{X}}_{0.0} + \alpha_7\left(\ddot{\mathbf{X}}_{0.02} + \ddot{\mathbf{X}}_{0.0}\right)$$

$$\{\dot{\mathbf{X}}\}_{0.02} = \begin{Bmatrix} 0 \\ 0 \\ 0 \end{Bmatrix} + 0.01\left[\begin{Bmatrix} -0.028 \\ -0.019 \\ -0.02 \end{Bmatrix} + \begin{Bmatrix} 0.0618 \\ 0.0618 \\ 0.0618 \end{Bmatrix}\right] = \begin{Bmatrix} 7.104 \\ 7.731 \\ 6.903 \end{Bmatrix} \times 10^{-4}\,\mathrm{m/s}$$

$$\{\mathbf{X}\}_{0.02} = \mathbf{X}_{0.0} + \Delta t\dot{\mathbf{X}}_{0.0} + \alpha_8\left(\ddot{\mathbf{X}}_{0.02} + 2\ddot{\mathbf{X}}_{0.0}\right)$$

$$\{\mathbf{X}\}_{0.02} = \begin{Bmatrix} 0 \\ 0 \\ 0 \end{Bmatrix} + 0.02\begin{Bmatrix} 0 \\ 0 \\ 0 \end{Bmatrix} + 6.6667 \times 10^{-5}$$

$$\left(\begin{Bmatrix} 7.104 \\ 7.731 \\ 6.903 \end{Bmatrix} \times 10^{-4} + 2\begin{Bmatrix} 0.0618 \\ 0.0618 \\ 0.0618 \end{Bmatrix}\right)$$

$$= \begin{Bmatrix} 8.856 \\ 9.274 \\ 8.722 \end{Bmatrix} \times 10^{-6}$$

- **At t = 0.04 s**

$$\{\mathbf{R}\}_{0.04} = [\mathbf{M}]\{\ddot{\mathbf{X}}\}_{0.04} = \begin{bmatrix} 4000 & 0 & 0 \\ 0 & 4000 & 0 \\ 0 & 0 & 2000 \end{bmatrix}\begin{Bmatrix} 0.0097119 \\ 0.0097119 \\ 0.0097119 \end{Bmatrix}$$

$$= \begin{Bmatrix} 38.847 \\ 38.847 \\ 19.4238 \end{Bmatrix}\,\mathrm{kN}$$

$$\left\{\hat{\mathbf{R}}_{\mathbf{t}+\theta\Delta\mathbf{t}}\right\} = \{\mathbf{R}_\mathbf{t}\} + \theta\{\mathbf{R}_{\mathbf{t}+\Delta\mathbf{t}} - \mathbf{R}_\mathbf{t}\} + [\mathbf{M}]\left\{\alpha_0\{\mathbf{X}_\mathbf{t}\} + \alpha_2\{\dot{\mathbf{X}}_\mathbf{t}\} + 2\{\ddot{\mathbf{X}}_\mathbf{t}\}\right\}$$
$$+ [\mathbf{C}]\left\{\alpha_1\{\mathbf{X}_\mathbf{t}\} + 2\{\dot{\mathbf{X}}_\mathbf{t}\} + \alpha_3\{\ddot{\mathbf{X}}_\mathbf{t}\}\right\}$$

$$\{\hat{\mathbf{R}}\}_{0.04} = \{\mathbf{R}\}_{0.04} + \theta\left[\{\mathbf{R}\}_{0.04} - \{\mathbf{R}\}_{0.02}\right] + [\mathbf{M}]\left[\alpha_0\mathbf{X}_{0.02} + \alpha_2\dot{\mathbf{X}}_{0.02} + 2\ddot{\mathbf{X}}_{0.02}\right]$$

$$\{\hat{\mathbf{R}}\}_{0.04} = \begin{Bmatrix} 38.847 \\ 38.847 \\ 19.4238 \end{Bmatrix} + 1.4 \times \left[\begin{Bmatrix} 38.847 \\ 38.847 \\ 19.4238 \end{Bmatrix} - \begin{Bmatrix} 142.83 \\ 142.83 \\ 71.4 \end{Bmatrix} \right]$$

$$+ \begin{bmatrix} 4000 & 0 & 0 \\ 0 & 4000 & 0 \\ 0 & 0 & 2000 \end{bmatrix} \times \left[7653 \begin{Bmatrix} 8.856 \\ 9.274 \\ 8.722 \end{Bmatrix} \times 10^{-6} \right.$$

$$\left. + 214.285 \begin{Bmatrix} 7.104 \\ 7.731 \\ 6.903 \end{Bmatrix} \times 10^{-4} + 2 \begin{Bmatrix} 9.243 \times 10^{-3} \\ 0.016 \\ 7.234 \times 10^{-3} \end{Bmatrix} \right]$$

$$\{\hat{\mathbf{R}}\}_{0.04} = \begin{Bmatrix} 857.179 \\ 974.673 \\ 515.811 \end{Bmatrix} \text{kN}$$

$$\{\hat{\mathbf{X}}\}_{0.04} = [\hat{\mathbf{K}}]^{-1} \{\hat{\mathbf{R}}\}_{0.04} = \begin{bmatrix} 2.706 & 0.2308 & 0.02397 \\ 0.2308 & 2.797 & 0.2904 \\ 0.02397 & 0.2904 & 4.436 \end{bmatrix} \begin{Bmatrix} 857.179 \\ 974.673 \\ 515.811 \end{Bmatrix} \times 10^{-8}$$

$$= \begin{Bmatrix} 2.557 \\ 3.073 \\ 2.592 \end{Bmatrix} \times 10^{-5} \text{m}$$

$$\{\ddot{\mathbf{X}}\}_{0.04} = \alpha_4 [\hat{\mathbf{X}}_{0.04} - \mathbf{X}_{0.02}] + \alpha_5 \dot{\mathbf{X}}_{0.02} + \alpha_6 \ddot{\mathbf{X}}_{0.02}$$

$$\{\ddot{\mathbf{X}}\}_{0.04} = 5466.47 \left[\begin{Bmatrix} 2.557 \\ 3.073 \\ 2.592 \end{Bmatrix} \times 10^{-5} - \begin{Bmatrix} 8.856 \\ 9.274 \\ 8.722 \end{Bmatrix} \times 10^{-6} \right]$$

$$- 153.061 \begin{Bmatrix} 7.104 \\ 7.731 \\ 6.903 \end{Bmatrix} \times 10^{-4} - 1.1428 \begin{Bmatrix} 9.243 \times 10^{-3} \\ 0.016 \\ 7.234 \times 10^{-3} \end{Bmatrix}$$

$$= \begin{Bmatrix} -0.028 \\ -0.019 \\ -0.02 \end{Bmatrix} \text{m/s}^2$$

$$\{\dot{\mathbf{X}}\}_{0.04} = \dot{\mathbf{X}}_{0.02} + \alpha_7 (\ddot{\mathbf{X}}_{0.04} + \ddot{\mathbf{X}}_{0.02})$$

$$\{\dot{\mathbf{X}}\}_{0.04} = \begin{Bmatrix} 7.104 \\ 7.731 \\ 6.903 \end{Bmatrix} \times 10^{-4} + 0.01 \left[\begin{Bmatrix} -0.028 \\ -0.019 \\ -0.02 \end{Bmatrix} + \begin{Bmatrix} 9.243 \times 10^{-3} \\ 0.016 \\ 7.234 \times 10^{-3} \end{Bmatrix} \right]$$

$$= \begin{Bmatrix} 5.233 \\ 7.408 \\ 5.632 \end{Bmatrix} \times 10^{-4}\,\mathrm{m/s}$$

$$\{\mathbf{X}\}_{0.04} = \mathbf{X}_{0.02} + \Delta t \dot{\mathbf{X}}_{0.02} + \alpha_8 \left(\ddot{\mathbf{X}}_{0.04} + 2\ddot{\mathbf{X}}_{0.02}\right)$$

$$\{\mathbf{X}\}_{0.04} = \begin{Bmatrix} 8.856 \\ 9.274 \\ 8.722 \end{Bmatrix} \times 10^{-6} + 0.02 \begin{Bmatrix} 7.104 \\ 7.731 \\ 6.903 \end{Bmatrix} \times 10^{-4}$$

$$+ 6.6667 \times 10^{-5} \left(\begin{Bmatrix} -0.028 \\ -0.019 \\ -0.02 \end{Bmatrix} + 2 \begin{Bmatrix} 9.243 \times 10^{-3} \\ 0.016 \\ 7.234 \times 10^{-3} \end{Bmatrix} \right)$$

$$= \begin{Bmatrix} 2.243 \\ 2.255 \\ 2.216 \end{Bmatrix} \times 10^{-5}\,\mathrm{m}$$

- **At t = 0.06 s**

$$\{\mathbf{R}\}_{0.06} = [\mathbf{M}]\{\ddot{\mathbf{X}}\}_{0.06} = \begin{Bmatrix} 167.947 \\ 167.947 \\ 83.973 \end{Bmatrix} \mathrm{kN}$$

$$\left\{\hat{\mathbf{R}}_{t+\theta\Delta t}\right\} = \{\mathbf{R}_t\} + \theta\{\mathbf{R}_{t+\Delta t} - \mathbf{R}_t\}$$
$$+ [\mathbf{M}]\left\{\alpha_0\{\mathbf{X}_t\} + \alpha_2\{\dot{\mathbf{X}}_t\} + 2\{\ddot{\mathbf{X}}_t\}\right\} + [\mathbf{C}]\left\{\alpha_1\{\mathbf{X}_t\} + 2\{\dot{\mathbf{X}}_t\} + \alpha_3\{\ddot{\mathbf{X}}_t\}\right\}$$

$$\{\hat{\mathbf{R}}\}_{0.06} = \{\mathbf{R}\}_{0.06} + \theta\left[\{\mathbf{R}\}_{0.06} - \{\mathbf{R}\}_{0.04}\right] + [\mathbf{M}]\left[\alpha_0 \mathbf{X}_{0.04} + \alpha_2 \dot{\mathbf{X}}_{0.04} + 2\ddot{\mathbf{X}}_{0.04}\right]$$

$$\rightarrow \{\hat{\mathbf{R}}\}_{0.06} = \begin{Bmatrix} 1267 \\ 1639 \\ 820.7311 \end{Bmatrix} \mathrm{kN}$$

$$\{\hat{\mathbf{X}}\}_{0.06} = [\hat{\mathbf{K}}]^{-1}\{\hat{\mathbf{R}}\}_{0.06} = \begin{bmatrix} 2.706 & 0.2308 & 0.02397 \\ 0.2308 & 2.797 & 0.2904 \\ 0.02397 & 0.2904 & 4.436 \end{bmatrix} \begin{Bmatrix} 1267 \\ 1639 \\ 820.7311 \end{Bmatrix} \times 10^{-8}$$

$$= \begin{Bmatrix} 3.825 \\ 5.114 \\ 4.147 \end{Bmatrix} \times 10^{-5}\,\mathrm{m}$$

$$\{\ddot{\mathbf{X}}\}_{0.06} = \alpha_4 \left[\hat{\mathbf{X}}_{0.06} - \mathbf{X}_{0.04}\right] + \alpha_5 \dot{\mathbf{X}}_{0.04} + \alpha_6 \ddot{\mathbf{X}}_{0.04}$$

$$\{\ddot{\mathbf{X}}\}_{0.04} = \begin{Bmatrix} 0.038 \\ 0.048 \\ 0.042 \end{Bmatrix} \mathrm{m/s^2}$$

$$\{\dot{\mathbf{X}}\}_{0.06} = \dot{\mathbf{X}}_{0.04} + \alpha_7 \left(\ddot{\mathbf{X}}_{0.06} + \ddot{\mathbf{X}}_{0.04}\right)$$

$$\{\dot{\mathbf{X}}\}_{0.06} = \begin{Bmatrix} 6.27 \\ 10.32 \\ 7.849 \end{Bmatrix} \times 10^{-4}\,\mathrm{m/s}$$

$$\{\mathbf{X}\}_{0.04} = \mathbf{X}_{0.02} + \Delta t \dot{\mathbf{X}}_{0.02} + \alpha_8 \left(\ddot{\mathbf{X}}_{0.04} + 2\ddot{\mathbf{X}}_{0.02}\right)$$

$$\{\mathbf{X}\}_{0.04} = \begin{Bmatrix} 3.173 \\ 4.106 \\ 3.358 \end{Bmatrix} \times 10^{-5}\mathrm{m}, \text{ and so on}\ldots$$

The time history plot of amplitude velocity and acceleration is shown for first thirty steps; here x = amplitude, V = velocity, and A = acceleration.

The time history plot of amplitude for complete 1560 steps is shown hereafter.

Time steps	X_1	V_1	A_1	X_2	V_2	A_2	X_3	V_3	A_3
0	0.000E+00	0.000E+00	6.180E−02	0.000E+00	0.000E+00	6.180E−02	0.000E+00	0.000E+00	6.18E−02
0.02	8.857E−06	7.105E−04	9.245E−03	9.274E−06	7.731E−04	1.551E−02	8.723E−06	6.905E−04	7.24E−03
0.04	2.244E−05	5.234E−04	−2.796E−02	2.556E−05	7.409E−04	−1.874E−02	2.217E−05	5.634E−04	−1.99E−02
0.06	3.173E−05	6.270E−04	3.832E−02	4.107E−05	1.032E−03	4.788E−02	3.358E−05	7.849E−04	4.21E−02
0.08	5.388E−05	1.685E−03	6.751E−02	7.356E−05	2.331E−03	8.195E−02	5.940E−05	1.882E−03	6.75E−02
0.1	1.015E−04	3.101E−03	7.410E−02	1.378E−04	4.161E−03	1.011E−01	1.114E−04	3.365E−03	8.07E−02
0.12	1.703E−04	3.377E−03	−4.652E−02	2.338E−04	5.059E−03	−1.126E−02	1.892E−04	4.123E−03	−4.96E−03
0.14	2.242E−04	1.791E−03	−1.121E−01	3.275E−04	4.057E−03	−8.901E−02	2.681E−04	3.641E−03	−4.32E−02
0.16	2.360E−04	−6.881E−04	−1.359E−01	3.869E−04	1.675E−03	−1.492E−01	3.304E−04	2.493E−03	−7.14E−02
0.18	2.049E−04	−1.942E−03	1.045E−02	3.977E−04	−2.325E−04	−4.159E−02	3.717E−04	1.920E−03	1.41E−02
0.2	1.740E−04	−8.511E−04	9.867E−02	3.899E−04	−2.955E−04	3.530E−02	4.143E−04	2.414E−03	3.53E−02
0.22	1.794E−04	1.527E−03	1.391E−01	3.956E−04	1.105E−03	1.047E−01	4.701E−04	3.194E−03	4.26E−02
0.24	2.262E−04	2.577E−03	−3.413E−02	4.302E−04	1.925E−03	−2.277E−02	5.338E−04	2.738E−03	−8.82E−02
0.26	2.640E−04	8.582E−04	−1.377E−01	4.588E−04	6.668E−04	−1.030E−01	5.682E−04	5.659E−04	−1.28E−01
0.28	2.602E−04	−9.103E−04	−3.911E−02	4.574E−04	−5.054E−04	−1.419E−02	5.613E−04	−8.733E−04	−1.49E−02
0.3	2.333E−04	−1.826E−03	−5.250E−02	4.416E−04	−1.227E−03	−5.798E−02	5.396E−04	−1.356E−03	−3.33E−02
0.32	1.870E−04	−2.762E−03	−4.110E−02	4.039E−04	−2.616E−03	−8.093E−02	5.051E−04	−2.137E−03	−4.46E−02
0.34	1.310E−04	−2.461E−03	7.125E−02	3.418E−04	−3.274E−03	1.515E−02	4.580E−04	−2.336E−03	2.47E−02
0.36	9.967E−05	−4.943E−04	1.254E−01	2.842E−04	−2.246E−03	8.768E−02	4.181E−04	−1.562E−03	5.27E−02
0.38	1.155E−04	2.107E−03	1.348E−01	2.608E−04	1.038E−04	1.473E−01	3.988E−04	−2.977E−04	7.36E−02
0.4	1.831E−04	4.579E−03	1.124E−01	2.949E−04	3.444E−03	1.867E−01	4.094E−04	1.444E−03	1.00E−01
0.42	2.696E−04	2.698E−03	−3.006E−01	3.749E−04	3.240E−03	−2.070E−01	4.391E−04	5.540E−04	−1.89E−01
0.44	2.505E−04	−5.263E−03	−4.955E−01	3.805E−04	−3.573E−03	−4.743E−01	4.038E−04	−4.501E−03	−3.15E−01

(continued)

(continued)

Time steps	X_1	V_1	A_1	X_2	V_2	A_2	X_3	V_3	A_3
0.46	7.880E−05	−1.027E−02	−5.149E−03	2.390E−04	−9.336E−03	−1.021E−01	2.746E−04	−7.220E−03	4.403E−02
0.48	−1.075E−04	−7.354E−03	2.967E−01	4.784E−05	−8.982E−03	1.374E−01	1.468E−04	−5.166E−03	1.614E−01
0.5	−1.982E−04	−1.861E−03	2.527E−01	−1.046E−04	−6.270E−03	1.338E−01	6.744E−05	−3.187E−03	3.658E−02
0.52	−1.793E−04	4.032E−03	3.365E−01	−1.892E−04	−1.498E−03	3.435E−01	1.774E−05	−1.446E−03	1.375E−01
0.54	−3.333E−05	1.046E−02	3.064E−01	−1.418E−04	6.675E−03	4.738E−01	2.237E−05	2.211E−03	2.283E−01
0.56	2.316E−04	1.575E−02	2.225E−01	8.971E−05	1.664E−02	5.225E−01	1.189E−04	7.768E−03	3.273E−01
0.58	5.590E−04	1.538E−02	−2.591E−01	4.966E−04	2.254E−02	6.738E−02	3.235E−04	1.189E−02	8.466E−02

2.6.3 *Newmark Beta Method*

Newmark (1959) developed an integration scheme based on constant average acceleration method as described below.

Considering $\ddot{\mathbf{X}}_t$ and $\ddot{\mathbf{X}}_{t+\Delta t}$ as acceleration at time t and $t + \Delta t$, he assumed that average acceleration of a body within the time interval t and $t + \Delta t$ is

$$\ddot{\mathbf{X}}_{av} = \frac{\ddot{\mathbf{X}}_{t+\Delta t} + \ddot{\mathbf{X}}_t}{2}$$

Let $\frac{dv}{dt} = \ddot{\mathbf{X}}_{av}$, we may write

$$\int_t^{t+\Delta t} dv = \int_t^{t+\Delta t} \ddot{\mathbf{X}}_{av} dt \rightarrow \dot{\mathbf{X}}_{t+\Delta t} - \dot{\mathbf{X}}_t = \ddot{\mathbf{X}}_{av}(t + \Delta t - t) \qquad (2.6.3.1)$$

i.e., $\dot{\mathbf{X}}_{t+\Delta t} = \dot{\mathbf{X}}_t + \ddot{\mathbf{X}}_{av}\Delta t$, implying $\dot{\mathbf{X}}_{t+\Delta t} = \dot{\mathbf{X}}_t + \frac{\ddot{\mathbf{X}}_{t+\Delta t} + \ddot{\mathbf{X}}_t}{2}\Delta t$

Hence, $\dot{\mathbf{X}}_{t+\Delta t} = \dot{\mathbf{X}}_t + \{(\mathbf{1} - \boldsymbol{\delta})\ddot{\mathbf{X}}_t + \boldsymbol{\delta}\ddot{\mathbf{X}}_{t+\Delta t}\}\Delta t$, where $\delta = 1/2$.

To obtain the displacement vector, we integrate the above between the limits t and $t + \Delta t$

$$\int_t^{t+\Delta t} \dot{\mathbf{X}}_{t+\Delta t} = \int_t^{t+\Delta t} \dot{\mathbf{X}}_t + \int_t^{t+\Delta t} \{(\mathbf{1} - \boldsymbol{\delta})\ddot{\mathbf{X}}_t + \delta\ddot{\mathbf{X}}_{t+\Delta t}\}\Delta t \text{ or,}$$

$$\mathbf{X}_{t+\Delta t} - \mathbf{X}_t = \dot{\mathbf{X}}_t \cdot \Delta t + \{(\mathbf{1} - \boldsymbol{\delta})\ddot{\mathbf{X}}_t + \delta\mathbf{X}_{t+\Delta t}\}\frac{\Delta t^2}{2}$$

i.e., $\mathbf{X}_{t+\Delta t} = \mathbf{X}_t + \dot{\mathbf{X}}_t \cdot \Delta t + \{(\frac{1}{2} - \boldsymbol{\beta})\ddot{\mathbf{X}}_t + \boldsymbol{\beta}\mathbf{X}_{t+\Delta t}\}\Delta t^2$, where $\beta = \frac{1}{4}$.

Here, you should note the similarity in the above procedure with Wilson-Theta method.

In Wilson-Theta method, we had expression of displacement as

$\{\mathbf{X}_{t+\Delta t}\} = \{\mathbf{X}_t\} + \{\dot{\mathbf{X}}_t\}\Delta t + \boldsymbol{\alpha}_8\{\mathbf{X}_{t+\Delta t} + 2\ddot{\mathbf{X}}_t\}$, where $\alpha_8 = \frac{\Delta t^2}{6}$ and on substitution we get

$$\{\mathbf{X}_{t+\Delta t}\} = \{\mathbf{X}_t\} + \{\dot{\mathbf{X}}_t\}\Delta t + \frac{\Delta t^2}{6}\{\mathbf{X}_{t+\Delta t} + 2\ddot{\mathbf{X}}_t\}$$

Considering $\beta = 1/6$ in Newmark's equation, we have

$$\mathbf{X}_{t+\Delta t} = \mathbf{X}_t + \dot{\mathbf{X}}_t \cdot \Delta t + \left[\left(\frac{1}{2} - \frac{1}{6}\right)\ddot{\mathbf{X}}_t + \frac{1}{6}\mathbf{X}_{t+\Delta t}\right]\Delta t^2$$

$$\rightarrow \{\mathbf{X}_{t+\Delta t}\} = \{\mathbf{X}_t\} + \{\dot{\mathbf{X}}_t\}\Delta t + \frac{\Delta t^2}{6}\{\mathbf{X}_{t+\Delta t} + 2\ddot{\mathbf{X}}_t\} \quad (2.6.3.2)$$

Thus, we see that for $\beta = 1/6$ we arrive at the same expression as in the Wilson-Theta method.

Thus, we start here with the basic equation of motion as

$$[\mathbf{M}]\{\ddot{\mathbf{X}}_{t+\Delta t}\} + [\mathbf{C}]\{\dot{\mathbf{X}}_{t+\Delta t}\} + [\mathbf{K}]\{\mathbf{X}_{t+\Delta t}\} = \{\mathbf{R}_{t+\Delta t}\} \quad (2.6.3.3)$$

and proceeding in the similar manner as shown in the Wilson-Theta method, we find out the value of $\mathbf{X}_{t+\Delta t}$ which can be systematically structured as follows

- Assemble the mass matrix $[\mathbf{M}]$ the damping matrix $[\mathbf{C}]$, and stiffness matrix $[\mathbf{K}]$
- Initialize $\{\mathbf{X}_0\}$ and $\{\dot{\mathbf{X}}_0\}$
- Evaluate $\{\ddot{\mathbf{X}}_0\}$ (*Refer to central difference method to see how this is evaluated*)
- Select time step size Δt and parameters δ and β where $\delta \geq 0.50$ and $\beta = 0.25(0.5+\delta)^2$
- Calculate integration constant …

$$\alpha_0 = \frac{1}{\beta \Delta t^2}, \alpha_1 = \frac{\delta}{\beta \Delta t}, \alpha_2 = \frac{1}{\beta \Delta t}, \alpha_3 = \frac{1}{2\beta} - 1, \alpha_4 = \frac{\delta}{\beta} - 1,$$
$$\alpha_5 = \frac{\Delta t}{2}\left(\frac{\delta}{\beta} - 2\right), \alpha_6 = \Delta t(1-\delta), \alpha_7 = \Delta t \delta$$

- Form the modified stiffness matrix as

$$[\hat{\mathbf{K}}] = [\mathbf{K}] + \alpha_0[\mathbf{M}] + \alpha_1[\mathbf{C}]$$

- Calculate modified load at time $t + \Delta t$

$$\{\hat{\mathbf{R}}_{t+\Delta t}\} = \{\mathbf{R}_{t+\Delta t}\} + [\mathbf{M}]\{\alpha_0 \mathbf{X}_t + \alpha_2 \dot{\mathbf{X}}_t + \alpha_3 \ddot{\mathbf{X}}_t\} + [\mathbf{C}]\{\alpha_1 \mathbf{X}_t + \alpha_4 \dot{\mathbf{X}}_t + \alpha_5 \ddot{\mathbf{X}}_t\}$$

- Solve for displacement vector

$$[\hat{\mathbf{K}}]\{\mathbf{X}_{t+\Delta t}\} = \{\hat{\mathbf{R}}_{t+\Delta t}\}$$

- Calculate the acceleration and velocity at time $t + \Delta t$

$$\{\ddot{\mathbf{X}}_{t+\Delta t}\} = \alpha_0\{\mathbf{X}_{t+\Delta t} - \mathbf{X}_t\} - \alpha_2\{\dot{\mathbf{X}}_t\} - \alpha_3\{\ddot{\mathbf{X}}_t\}$$
$$\{\dot{\mathbf{X}}_{t+\Delta t}\} = \{\dot{\mathbf{X}}_t\} + \alpha_6\{\ddot{\mathbf{X}}_t\} + \alpha_7\{\ddot{\mathbf{X}}_{t+\Delta t}\}$$

Example 2.6.3 Repeat the problem solved in Example 2.6.2 by Newmark beta method for same sets of data.

Solution

In this problem as stated previously

$$[\mathbf{K}] = \begin{bmatrix} 5 & -2 & 0 \\ -2 & 3.5 & -1.5 \\ 0 & -1.5 & 1.5 \end{bmatrix} \times 10^6\,\text{kN/m},$$

$$[\mathbf{M}] = \begin{bmatrix} 4 & 0 & 0 \\ 0 & 4 & 0 \\ 0 & 0 & 2 \end{bmatrix} \times 10^3\,\text{kN}\,\text{s}^2\,\text{m}, \text{ and}$$

$$[\mathbf{C}] = \begin{bmatrix} 15 & -10 & 0 \\ -10 & 20 & -8 \\ 0 & -8 & 55 \end{bmatrix} \times 10^3\,\text{kN}\,\text{m/s}.$$

The digitized acceleration data for first five steps are as furnished hereafter.

Time steps	1	2	3	4	5
Time instant	0	0.02	0.04	0.06	0.08
Acceleration (m/s^2)	0.0618	0.0357	9.719×10^{-3}	0.0419	0.0743

Here, $\delta = 0.5$, $\beta = 0.25(0.5 + \delta)^2$ and time steps $\Delta t = 0.02$ s.
The integrations constants are

α_0	α_1	α_2	α_3	α_4	α_5	α_6	α_7
10,000	100	200	1	1	0	0.01	0.01

$$[\hat{\mathbf{K}}] = [\mathbf{K}] + \alpha_0[\mathbf{M}] + \alpha_1[\mathbf{C}]$$

$$\rightarrow [\hat{\mathbf{K}}] = \begin{bmatrix} 5 & -2 & 0 \\ -2 & 3.5 & -1.5 \\ 0 & -1.5 & 1.5 \end{bmatrix} \times 10^6$$

$$+ 10000 \begin{bmatrix} 4 & 0 & 0 \\ 0 & 4 & 0 \\ 0 & 0 & 2 \end{bmatrix} \times 10^3 + 100 \begin{bmatrix} 15 & -10 & 0 \\ -10 & 20 & -8 \\ 0 & -8 & 55 \end{bmatrix} \times 10^3$$

$$= \begin{bmatrix} 4.65 \times 10^7 & -3 \times 10^6 & 0 \\ -3 \times 10^6 & 4.55 \times 10^7 & -2.3 \times 10^6 \\ 0 & -2.3 \times 10^6 & 2.7 \times 10^7 \end{bmatrix}$$

$$\text{Thus} \rightarrow [\hat{\mathbf{K}}]^{-1} = \begin{bmatrix} 2.16 \times 10^{-8} & 1.43 \times 10^{-9} & 1.218 \times 10-^{10} \\ 1.43 \times 10^{-9} & 2.217 \times 10^{-8} & 1.888 \times 10^{-9} \\ 1.218 \times 10^{-10} & 1.888 \times 10^{-9} & 3.72 \times 10^{-8} \end{bmatrix}$$

- **At t = 0.0 s**

$$\{X\}_{0.0} = \begin{Bmatrix} 0.0 \\ 0.0 \\ 0.0 \end{Bmatrix}, \{\dot{X}\}_{0.0} = \begin{Bmatrix} 0.0 \\ 0.0 \\ 0.0 \end{Bmatrix}, \{\ddot{X}\}_{0.0} = \begin{Bmatrix} 0.0618 \\ 0.0618 \\ 0.0618 \end{Bmatrix} \text{m/s}^2$$

- **At t = 0.02 s**

$$\{\mathbf{R}\}_{0.02} = \begin{Bmatrix} 142.834 \\ 142.834 \\ 71.416 \end{Bmatrix} \text{kN.}$$

Thus

$$\{\hat{\mathbf{R}}_{t+\Delta t}\} = \{\mathbf{R}_{t+\Delta t}\} + [\mathbf{M}]\{\alpha_0 \mathbf{X}_t + \alpha_2 \dot{\mathbf{X}}_t + \alpha_3 \ddot{\mathbf{X}}_t\} + [\mathbf{C}]\{\alpha_1 \mathbf{X}_t + \alpha_4 \dot{\mathbf{X}}_t + \alpha_5 \ddot{\mathbf{X}}_t\}$$

$$\{\hat{\mathbf{R}}\}_{0.02} = \begin{Bmatrix} 142.834 \\ 142.834 \\ 71.416 \end{Bmatrix} + \begin{bmatrix} 4000 & 0 & 0 \\ 0 & 4000 & 0 \\ 0 & 0 & 2000 \end{bmatrix}$$

$$\left[10{,}000\begin{Bmatrix} 0 \\ 0 \\ 0 \end{Bmatrix} + 200\begin{Bmatrix} 0 \\ 0 \\ 0 \end{Bmatrix} + 1\begin{Bmatrix} 0.0618 \\ 0.0618 \\ 0.0618 \end{Bmatrix}\right]$$

$$+ \begin{bmatrix} 15{,}000 & -10{,}000 & 0 \\ -10{,}000 & 20{,}000 & -8000 \\ 0 & -8000 & 55{,}000 \end{bmatrix}$$

$$\left[100\begin{Bmatrix} 0 \\ 0 \\ 0 \end{Bmatrix} + 1\begin{Bmatrix} 0 \\ 0 \\ 0 \end{Bmatrix} + 0\begin{Bmatrix} 0.0618 \\ 0.0618 \\ 0.0618 \end{Bmatrix}\right]$$

$$= \begin{Bmatrix} 390.012 \\ 390.012 \\ 195.006 \end{Bmatrix} \text{kN}$$

$$\{X\}_{0.02} = \left[\hat{\mathbf{K}}\right]^{-1} \times \left[\hat{\mathbf{R}}\right]_{0.02} = \begin{bmatrix} 2.16 \times 10^{-8} & 1.43 \times 10^{-9} & 1.218 \times 10-^{10} \\ 1.43 \times 10^{-9} & 2.217 \times 10^{-8} & 1.888 \times 10^{-9} \\ 1.218 \times 10^{-10} & 1.888 \times 10^{-9} & 3.72 \times 10^{-8} \end{bmatrix}$$

$$\begin{Bmatrix} 390.012 \\ 390.012 \\ 195.006 \end{Bmatrix}$$

$$= \begin{Bmatrix} 9.005 \\ 9.572 \\ 8.038 \end{Bmatrix} \times 10^{-6}\,\text{m}$$

$$\left\{\ddot{\mathbf{X}}_{\mathbf{t}+\Delta\mathbf{t}}\right\} = \alpha_0\left\{\mathbf{X}_{\mathbf{t}+\Delta\mathbf{t}} - \mathbf{X}_{\mathbf{t}}\right\} - \alpha_2\left\{\dot{\mathbf{X}}_{\mathbf{t}}\right\} - \alpha_3\left\{\ddot{\mathbf{X}}_{\mathbf{t}}\right\}$$

$$\left\{\ddot{\mathbf{X}}\right\}_{0.02} = 1000\left[\begin{Bmatrix} 9.005 \\ 9.572 \\ 8.038 \end{Bmatrix} \times 10^{-6} - \begin{Bmatrix} 0 \\ 0 \\ 0 \end{Bmatrix}\right] - 100\begin{Bmatrix} 0 \\ 0 \\ 0 \end{Bmatrix} - 1\begin{Bmatrix} 0.0618 \\ 0.0618 \\ 0.0618 \end{Bmatrix}$$

$$= \begin{Bmatrix} 0.028 \\ 0.034 \\ 0.019 \end{Bmatrix} \text{m/s}^2$$

And $\left\{\dot{\mathbf{X}}_{\mathbf{t}+\Delta\mathbf{t}}\right\} = \left\{\dot{\mathbf{X}}_{\mathbf{t}}\right\} + \alpha_6\left\{\ddot{\mathbf{X}}_{\mathbf{t}}\right\} + \alpha_7\left\{\ddot{\mathbf{X}}_{\mathbf{t}+\Delta\mathbf{t}}\right\}$

$$\{\dot{\mathbf{X}}\}_{0.02} = \begin{Bmatrix} 0 \\ 0 \\ 0 \end{Bmatrix} + 0.01 \begin{Bmatrix} 0.0618 \\ 0.0618 \\ 0.0618 \end{Bmatrix} + 0.01 \begin{Bmatrix} 0.028 \\ 0.034 \\ 0.019 \end{Bmatrix} = \begin{Bmatrix} 9.005 \\ 9.572 \\ 8.038 \end{Bmatrix} \times 10^{-4}\ \text{m/s}.$$

- **At t = 0.04 s**

$$\{\mathbf{R}\}_{0.04} = \begin{bmatrix} 4000 & 0 & 0 \\ 0 & 4000 & 0 \\ 0 & 0 & 2000 \end{bmatrix} \begin{Bmatrix} 9.712 \\ 9.712 \\ 9.712 \end{Bmatrix} \times 10^{-3} = \begin{Bmatrix} 38.848 \\ 38.848 \\ 19.4246 \end{Bmatrix} \text{kN}.$$

Thus

$$\{\hat{\mathbf{R}}_{t+\Delta t}\} = \{\mathbf{R}_{t+\Delta t}\} + [\mathbf{M}]\{\alpha_0 \mathbf{X}_t + \alpha_2 \dot{\mathbf{X}}_t + \alpha_3 \ddot{\mathbf{X}}_t\} + [\mathbf{C}]\{\alpha_1 \mathbf{X}_t + \alpha_4 \dot{\mathbf{X}}_t + \alpha_5 \ddot{\mathbf{X}}_t\}$$

$$\{\hat{\mathbf{R}}\}_{0.04} = \begin{Bmatrix} 38.848 \\ 38.848 \\ 19.424 \end{Bmatrix} + \begin{bmatrix} 4000 & 0 & 0 \\ 0 & 4000 & 0 \\ 0 & 0 & 2000 \end{bmatrix}$$

$$\left[10000 \begin{Bmatrix} 9.005 \\ 9.572 \\ 8.038 \end{Bmatrix} \times 10^{-6} + 200 \begin{Bmatrix} 9.005 \\ 9.572 \\ 8.038 \end{Bmatrix} \times 10^{-4} \right.$$

$$\left. + 1 \begin{Bmatrix} 0.028 \\ 0.034 \\ 0.019 \end{Bmatrix} \right] \begin{bmatrix} 15{,}000 & -10{,}000 & 0 \\ -10{,}000 & 20{,}000 & -8000 \\ 0 & -8000 & 55{,}000 \end{bmatrix}$$

$$\left[100 \begin{Bmatrix} 9.005 \\ 9.572 \\ 8.038 \end{Bmatrix} \times 10^{-6} + 1 \begin{Bmatrix} 9.005 \\ 9.572 \\ 8.038 \end{Bmatrix} \times 10^{-4} + 0 \begin{Bmatrix} 0.028 \\ 0.034 \\ 0.019 \end{Bmatrix} \right]$$

$$= \begin{Bmatrix} 1240 \\ 1331 \\ 611.94 \end{Bmatrix} \text{kN}$$

$$\{X\}_{0.04} = \begin{bmatrix} 2.16 \times 10^{-8} & 1.43 \times 10^{-9} & 1.218 \times 10{-}^{10} \\ 1.43 \times 10^{-9} & 2.217 \times 10^{-8} & 1.888 \times 10^{-9} \\ 1.218 \times 10^{-10} & 1.888 \times 10^{-9} & 3.72 \times 10^{-8} \end{bmatrix} \begin{Bmatrix} 1240 \\ 1331 \\ 611.94 \end{Bmatrix}$$

$$= \begin{Bmatrix} 2.876 \\ 3.242 \\ 2.543 \end{Bmatrix} \times 10^{-5}\ \text{m}$$

$$\{\ddot{\mathbf{X}}_{t+\Delta t}\} = \alpha_0\{\mathbf{X}_{t+\Delta t} - \mathbf{X}_t\} - \alpha_2\{\dot{\mathbf{X}}_t\} - \alpha_3\{\ddot{\mathbf{X}}_t\}, 10$$

$$\{\ddot{\mathbf{X}}\}_{0.04} = 1000\left[\begin{Bmatrix} 2.876 \\ 3.242 \\ 2.543 \end{Bmatrix} \times 10^{-5} - \begin{Bmatrix} 9.005 \\ 9.572 \\ 8.038 \end{Bmatrix} \times 10^{-6}\right] - 100\begin{Bmatrix} 9.005 \\ 9.572 \\ 8.038 \end{Bmatrix} \times 10^{-4}$$

$$- 1\begin{Bmatrix} 0.028 \\ 0.034 \\ 0.019 \end{Bmatrix} = \begin{Bmatrix} -0.011 \\ 3.177 \times 10^{-3} \\ -5.443 \times 10^{-3} \end{Bmatrix} \mathrm{m/s^2}$$

And $\{\dot{\mathbf{X}}_{t+\Delta t}\} = \{\dot{\mathbf{X}}_t\} + \alpha_6\{\ddot{\mathbf{X}}_t\} + \alpha_7\{\ddot{\mathbf{X}}_{t+\Delta t}\}$

$$\{\dot{\mathbf{X}}\}_{0.04} = \begin{Bmatrix} 9.005 \\ 9.572 \\ 8.038 \end{Bmatrix} \times 10^{-4} + 0.01\begin{Bmatrix} 0.028 \\ 0.034 \\ 0.019 \end{Bmatrix} + 0.01\begin{Bmatrix} -0.011 \\ 3.177 \times 10^{-3} \\ -5.443 \times 10^{-3} \end{Bmatrix}$$

$$= \begin{Bmatrix} 1.075 \\ 1.328 \\ 0.9351 \end{Bmatrix} \times 10^{-3}\,\mathrm{m/s}.$$

- **At t = 0.06 s**

$$\{\mathbf{R}\}_{0.04} = \begin{bmatrix} 4000 & 0 & 0 \\ 0 & 4000 & 0 \\ 0 & 0 & 2000 \end{bmatrix}\begin{Bmatrix} 0.0419 \\ 0.0419 \\ 0.0419 \end{Bmatrix} = \begin{Bmatrix} 167.6 \\ 167.6 \\ 83.8 \end{Bmatrix} \mathrm{kN}.$$

Thus

$$\{\hat{\mathbf{R}}_{t+\Delta t}\} = \{\mathbf{R}_{t+\Delta t}\} + [\mathbf{M}]\{\alpha_0\mathbf{X}_t + \alpha_2\dot{\mathbf{X}}_t + \alpha_3\ddot{\mathbf{X}}_t\} + [\mathbf{C}]\{\alpha_1\mathbf{X}_t + \alpha_4\dot{\mathbf{X}}_t + \alpha_5\ddot{\mathbf{X}}_t\}$$

$$\{\hat{\mathbf{R}}\}_{0.06} = \begin{Bmatrix} 167.6 \\ 167.6 \\ 83.8 \end{Bmatrix} + \begin{bmatrix} 4000 & 0 & 0 \\ 0 & 4000 & 0 \\ 0 & 0 & 2000 \end{bmatrix} \left[10{,}000 \begin{Bmatrix} 2.876 \\ 3.242 \\ 2.543 \end{Bmatrix} \times 10^{-5} \right.$$

$$\left. + 200 \begin{Bmatrix} 1.075 \\ 1.328 \\ 0.9351 \end{Bmatrix} \times 10^{-3} + 1 \begin{Bmatrix} -0.011 \\ 3.177 \times 10^{-3} \\ -5.443 \times 10^{-3} \end{Bmatrix} \right]$$

$$+ \begin{bmatrix} 15{,}000 & -10{,}000 & 0 \\ -10{,}000 & 20{,}000 & -8000 \\ 0 & -8000 & 55{,}000 \end{bmatrix} \left[100 \begin{Bmatrix} 2.876 \\ 3.242 \\ 2.543 \end{Bmatrix} \times 10^{-5} \right.$$

$$\left. + 1 \begin{Bmatrix} 1.075 \\ 1.328 \\ 0.9351 \end{Bmatrix} \times 10^{-3} + 0 \begin{Bmatrix} -0.011 \\ 3.177 \times 10^{-3} \\ -5.443 \times 10^{-3} \end{Bmatrix} \right] = \begin{Bmatrix} 2149 \\ 2564 \\ 1110 \end{Bmatrix} \text{kN}$$

$$\{X\}_{0.06} = \begin{bmatrix} 2.16 \times 10^{-8} & 1.43 \times 10^{-9} & 1.218 \times 10^{-10} \\ 1.43 \times 10^{-9} & 2.217 \times 10^{-8} & 1.888 \times 10^{-9} \\ 1.218 \times 10^{-10} & 1.888 \times 10^{-9} & 3.72 \times 10^{-8} \end{bmatrix} \begin{Bmatrix} 2149 \\ 2564 \\ 1110 \end{Bmatrix}$$

$$= \begin{Bmatrix} 5.022 \\ 6.2 \\ 4.64 \end{Bmatrix} \times 10^{-5} \text{ m}$$

$$\{\ddot{\mathbf{X}}_{t+\Delta t}\} = \alpha_0\{\mathbf{X}_{t+\Delta t} - \mathbf{X}_t\} - \alpha_2\{\dot{\mathbf{X}}_t\} - \alpha_3\{\ddot{\mathbf{X}}_t\}$$

$$\{\ddot{\mathbf{X}}\}_{0.06} = 1000 \left[\begin{Bmatrix} 5.022 \\ 6.2 \\ 4.64 \end{Bmatrix} \times 10^{-5} - \begin{Bmatrix} 2.876 \\ 3.242 \\ 2.543 \end{Bmatrix} \times 10^{-5} \right] - 100 \begin{Bmatrix} 1.075 \\ 1.328 \\ 0.9351 \end{Bmatrix} \times 10^{-3}$$

$$- 1 \begin{Bmatrix} -0.011 \\ 3.177 \times 10^{-3} \\ -5.443 \times 10^{-3} \end{Bmatrix} = \begin{Bmatrix} 0.01 \\ 0.027 \\ 0.028 \end{Bmatrix} \text{m/s}^2$$

And $\{\dot{\mathbf{X}}_{t+\Delta t}\} = \{\dot{\mathbf{X}}_t\} + \alpha_6\{\ddot{\mathbf{X}}_t\} + \alpha_7\{\ddot{\mathbf{X}}_{t+\Delta t}\}$

$$\{\dot{\mathbf{X}}\}_{0.02} = \begin{Bmatrix} 1.075 \\ 1.328 \\ 0.9351 \end{Bmatrix} \times 10^{-3} + 0.01 \begin{Bmatrix} -0.011 \\ 3.177 \times 10^{-3} \\ -5.443 \times 10^{-3} \end{Bmatrix} + 0.01 \begin{Bmatrix} 0.01 \\ 0.027 \\ 0.028 \end{Bmatrix}$$

$$= \begin{Bmatrix} 1.07 \\ 1.63 \\ 1.162 \end{Bmatrix} \times 10^{-3} \text{ m/s.}$$

And so on….

Typical time history plot for first thirty steps for amplitude is shown hereafter (Figs. 2.37 and 2.38).

Steps	X_1	X_2	X_3
1	0.000E+00	0.000E+00	0.000E+00
2	9.006E−06	9.573E−06	8.039E−06
3	2.877E−05	3.243E−05	2.543E−05
4	5.023E−05	6.202E−05	4.641E−05
5	7.578E−05	1.021E−04	7.741E−05
6	1.132E−04	1.609E−04	1.273E−04
7	1.613E−04	2.379E−04	1.956E−04
8	2.044E−04	3.171E−04	2.700E−04
9	2.227E−04	3.753E−04	3.356E−04
10	2.123E−04	4.002E−04	3.867E−04
11	1.920E−04	4.010E−04	4.301E−04
12	1.885E−04	4.005E−04	4.744E−04
13	2.103E−04	4.126E−04	5.166E−04
14	2.365E−04	4.287E−04	5.407E−04
15	2.444E−04	4.354E−04	5.399E−04
16	2.308E−04	4.298E−04	5.240E−04
17	1.989E−04	4.069E−04	4.982E−04
18	1.572E−04	3.642E−04	4.648E−04
19	1.260E−04	3.152E−04	4.324E−04
20	1.250E−04	2.826E−04	4.109E−04
21	1.637E−04	2.894E−04	4.086E−04
22	2.186E−04	3.301E−04	4.158E−04
23	2.212E−04	3.447E−04	3.912E−04
24	1.316E−04	2.801E−04	3.158E−04
25	−8.368E−07	1.554E−04	2.255E−04
26	−1.005E−04	2.056E−05	1.507E−04
27	−1.243E−04	−8.062E−05	9.512E−05
28	−4.988E−05	−9.524E−05	7.254E−05
29	1.270E−04	2.984E−05	1.137E−04
30	3.678E−04	3.019E−04	2.341E−04

Full time history plot up to 31.18 s for amplitude (Fig. 2.39) and floor shear force is shown hereafter (Fig. 2.40).

It is observed that values are closely matching with other methods both in terms of amplitude and floor shear (Fig 2.40).

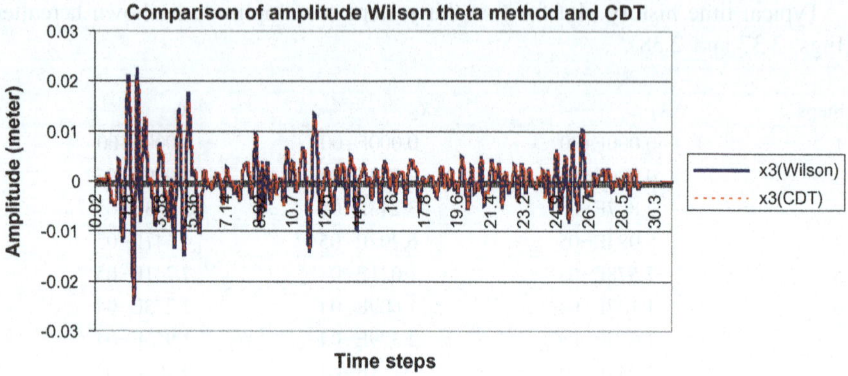

Fig. 2.37 Comparison of amplitude: Wilson-Theta versus CDT

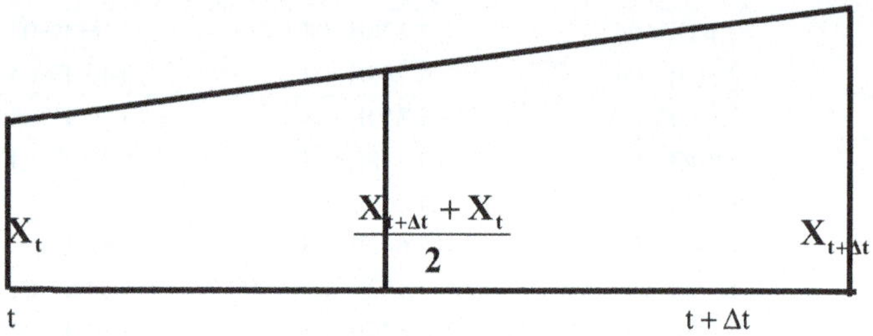

Fig. 2.38 Description of Newmark's method

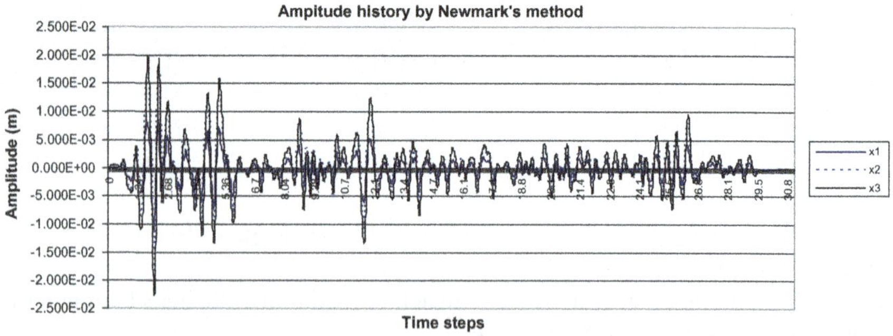

Fig. 2.39 Amplitude plot based on Newmark's method

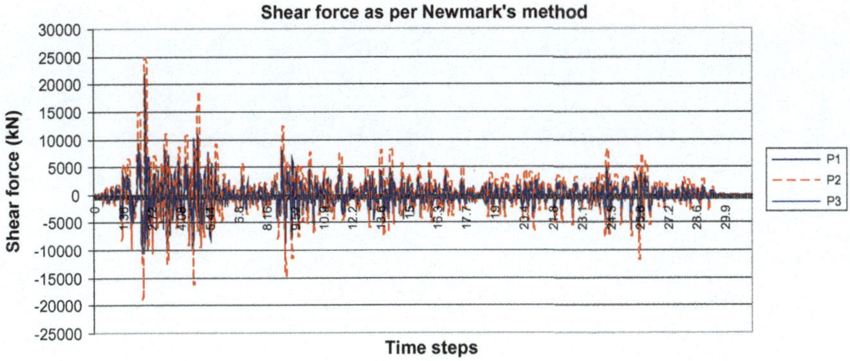

Fig. 2.40 Shear force per floor based on Newmark's method

References

Bathe, K. J. (1996). *Finite element procedures in engineering*. New Delhi, India: Prentice Hall.

Beinoff, V. H. (1934). The physical evaluation of seismic destructiveness. *Bulletin of Seismological Society of America, 24,* 398–403.

Biggs, J. M. (1964). *Introduction to structural dynamics*. NY, USA: McGraw-Hill Publications.

Biot, M. A. (1932). *Transient oscillations in elastic system* (Ph.D. thesis). Department of Aeronautics, California Institute of Technology Pasadena.

Chowdhury, I., & Dasgupta, S. P. (2008). *Dynamics of structure and foundation—A unified approach* (Vol. I & II). Leiden, Holland: Taylor and Francis.

Clough, R. W., & Penzien, J. (1982). *Dynamics of structures*. NY: McGraw-Hill Publication.

Der Kiureghian, A. (1981). A Response spectrum method for random vibration analysis for MDF systems. *Journal of Earthquake Engineering and Strucutral Dynamics, 9*, 419–435.

Housner, G. W. (1947). Characterestics of strong motion earthquake. *Bulletin of Seismological Society of America, 37*(1), 19–31.

Hurty, W., & Rubenstein, M. (1967). *Dynamics of structure*. New Delhi, India: Prentice Hall.

Mukherjee, M. (2002). *Vibration dynamics and structural systems*. New Delhi, India: Oxford IBH Publications.

Newmark, N. E. (1959). A method for computation in structural mechanics. *Journal of Engineering Mechanics ASCE, 85*, 67–94.

Paz, M., (Ed). (1994). *International handbook of earthquake engineering codes program and examples*. UK: Chapman and Hall.

Rao, S. S. (2007). *Vibration of continuous systems*. USA: Wiley.

Rosenblueth, E. (1951). *A basis for Aseismic deign*. Ph.D. thesis, University of Illinois. Urbana.

Trifunac, M. (2006). Brief history of computation of earthquake response spectra. *Journal of Soil Dynamics and Earthquake Engineering, 26,* 501–508.

Wilson, E. (1973). Non Linear dynamic analysis of complex strucutres. *Journal of Earthquake Engineering and Strucutral Dynamics, 1*, 241–252.

Chapter 3
Soil Dynamics and Earthquake Engineering

3.1 Introduction

Though soil dynamics as a formal topic of study in civil engineering was prevalent since 1936 (mostly related to machine foundations), yet its metamorphosis to a new subject termed *geotechnical earthquake engineering* that is worthy of special attention was quite late.

In fact, it was only in 1995, during the first International Conference in Geotechnical Earthquake Engineering in Tokyo, that the word itself was coined for the first time (Srbulov 2011).

Characteristics of earthquakes and the methodology to design an earthquake-resistant structure has been a topic of study by structural engineers since the 1930s.[1] Geotechnical engineers, however, came into the fray quite late and had a very limited contribution at the initial stage of growth.

One of the main reasons for this could be that, at its advent, most of the researches related to geophysical issues were carried out by geophysicists or seismologists (people from geophysics group who specialized on earthquakes).

It was only in few cases, like earth-retaining structures, earthen dams, embankments,[2] that geotechnical engineers had a limited role to play, when solutions based on simplified equivalent static method[3] were deemed sufficient.

There were, however, two earthquakes in 1964 (Niigata in Japan and Alaska) that brought geotechnical engineering into limelight when both these earthquakes resulted in spectacular geotechnical failures. These mainly consisted of

- Failure due to liquefaction of soil;
- Landslides;
- Failure of structures due to lateral movement of soil;

---

[1]Refer Chap. 2 for further details
[2]Refer Chaps. 4 and 5 for further detail.
[3]The Shindo or seismic coefficient method as discussed in Chap. 2.

I. Chowdhury and S. P. Dasgupta, *Earthquake Analysis and Design of Industrial Structures and Infra-structures*, GeoPlanet: Earth and Planetary Sciences,
https://doi.org/10.1007/978-3-319-90832-8_3

- Bearing capacity failures of foundations;
- Amplification of response due to soil overlying bedrock, that caused damage to super-structures.

These failures clearly showed that soil properties play a significant role in how waves propagate through it and affect a particular site. It also made geotechnical engineers realize that the soil characteristic strength varies significantly with earthquake-like shocks and influences its behavior (Ishihara 1993).

3.2 Soil Dynamics and Earthquake Engineering

In Chap. 1, we have discussed in detail the geophysical reasons that cause an earthquake. We have also mentioned therein how we measure the power of an earthquake and different formulae proposed by seismologists (and a few civil engineers) to measure the power, intensity, duration, etc., of an earthquake.

The slip in rock mass at the fault generates waves that travel to the surface, generating ripples on the ground causing damage to the structures located on it.

This was first pointed out by Prof. Horace Lamb in his seminal paper in 1904 in the journal of *Royal Society of London* where he showed that a transient pulse within an infinite elastic medium generates two kinds of waves namely *body waves* (P) within the soil and *shear waves* (S) at the surface. He also showed that, considering the fact that the P waves travel faster than S waves, they would appear first at a site causing a minor tremor (foreshocks), followed by S waves, which cause major tremor and disturbance at a site. Lamb's findings are shown conceptually in Fig. 3.1 that shows that a site first experiences a minor ripple due to arrival of P waves followed by major tremor due to S waves.

For an elastic half-space that is a semi-infinite medium, it was Lord Rayleigh who showed that at the ground surface, a new type of waves similar to S waves is

Fig. 3.1 Amplitude of P waves followed by S waves arriving at site during an earthquake; Lamb (1904)

generated, and the locus of soil particles vibrating under their influence follows an elliptical path. These are called *Rayleigh waves*. They are equally damaging like S waves having similar properties.

When the soil is layered, A. E. H. Love showed that earthquakes generate a special kind of waves at the soil surface that has the character of a snake wriggling across the ground. Waves of this type are called as *Love waves*.

Considering the heterogeneous character of soil, all these types of waves are generated in a soil medium creating ripples that are picked up by the seismographs in terms of amplitude, velocity, and ground acceleration. A typical ground acceleration record is as shown in Fig. 3.2.

Figure 3.2 shows the recorded ground acceleration (at an Indian observatory) for the Indo-Nepal Earthquake in 2015. Considering the earthquake waves traveling from left to right, observe the minor or precursor ripples (P waves) for the first 7 s (app.), followed by major tremors (due to S and Rayleigh waves) from 7 to 14 s. Thus, understanding the characters of these waves and how do they behave during their passage through the soil medium is of fundamental importance and would be our topic of discussion in the next section.

3.3 Characteristics of Wave Propagating Through an Elastic Medium

When waves propagate through the soil medium, considering soil as an infinite elastic medium is a convenient mathematical model that mathematicians developed to study its properties. Although, realistically speaking, soil is neither infinite nor elastic, yet, to start with, it remains a reasonably realistic model.

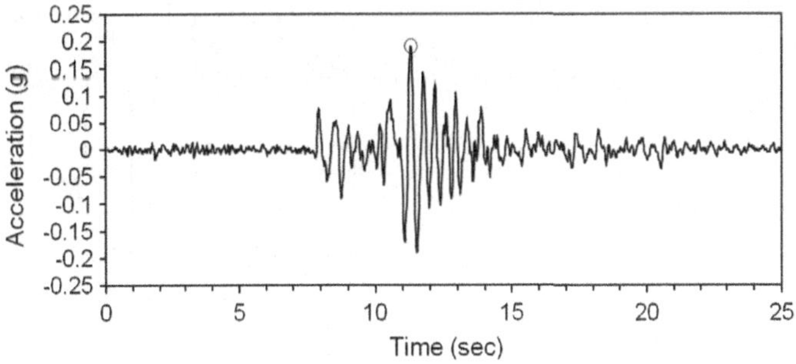

Fig. 3.2 Ground acceleration record (in an Indian observatory) of Nepal earthquake (2015)

3.3.1 Propagation of Waves in Three Dimensions in an Infinite Elastic Medium

Shown in Fig. 3.3 is an elemental soil body of magnitude Δ_x, Δ_y, and Δ_z; the stress matrix for the body at a point can be expressed as

$$[\sigma] = \begin{bmatrix} \sigma_{xx} & \tau_{xy} & \tau_{xz} \\ \tau_{yx} & \sigma_{yy} & \tau_{yz} \\ \tau_{zx} & \tau_{zy} & \sigma_{zz} \end{bmatrix} \tag{3.3.1.1}$$

If u, v, and w are displacements along X, Y, and Z coordinate, respectively, then the strain relationship is expressed as

$$\begin{aligned} &\varepsilon_{xx} = \frac{du}{dx}, \varepsilon_{yy} = \frac{dv}{dy}, \varepsilon_{zz} = \frac{dw}{dz}, \gamma_{xy} = \frac{dv}{dx} + \frac{du}{dy}, \\ &\gamma_{yz} = \frac{dw}{dy} + \frac{dv}{dz}, \gamma_{zx} = \frac{du}{dz} + \frac{dw}{dx} \end{aligned} \tag{3.3.1.2}$$

The rigid body rotation is expressed as

$$\Omega_x = \frac{1}{2}\left(\frac{dw}{dy} - \frac{dv}{dz}\right), \Omega_y = \frac{1}{2}\left(\frac{du}{dz} - \frac{dw}{dx}\right), \text{and } \Omega_z = \frac{1}{2}\left(\frac{dv}{dx} - \frac{du}{dy}\right) \tag{3.3.1.3}$$

The stress strain relationship is now expressed by the relation

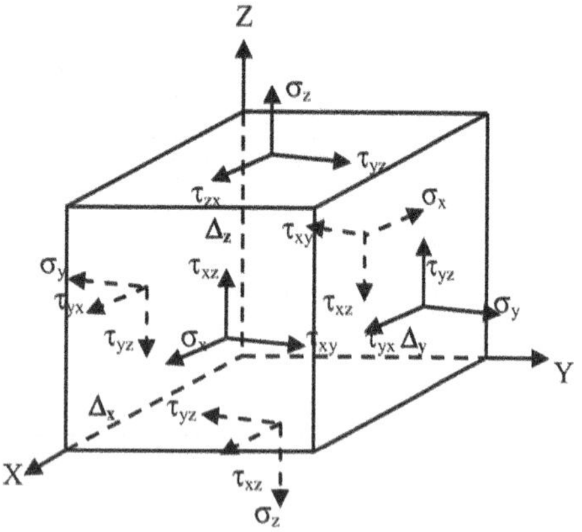

Fig. 3.3 Stresses in an elemental body in Cartesian coordinates

$$
\begin{Bmatrix} \sigma_{xx} \\ \sigma_{yy} \\ \sigma_{zz} \\ \tau_{xy} \\ \tau_{yz} \\ \tau_{zx} \end{Bmatrix} = \begin{bmatrix} \lambda+2G & \lambda & \lambda & 0 & 0 & 0 \\ \lambda & \lambda+2G & \lambda & 0 & 0 & 0 \\ \lambda & \lambda & \lambda+2G & 0 & 0 & 0 \\ 0 & 0 & 0 & G & 0 & 0 \\ 0 & 0 & 0 & 0 & G & 0 \\ 0 & 0 & 0 & 0 & 0 & G \end{bmatrix} \begin{Bmatrix} \varepsilon_{xx} \\ \varepsilon_{yy} \\ \varepsilon_{zz} \\ \gamma_{xy} \\ \gamma_{yz} \\ \gamma_{zx} \end{Bmatrix} \tag{3.3.1.4}
$$

where λ and G are Lame's constants expressed as

$$
\lambda = \frac{\nu E}{(1+\nu)(1-2\nu)} \quad \text{and} \quad G = \frac{E}{2(1+\nu)} \tag{3.3.1.5}
$$

Based on above, one can easily derive that

$$
\sigma_{xx} = \lambda e_{\text{v}} + 2G\varepsilon_{xx}, \sigma_{yy} = \lambda e_{\text{v}} + 2G\varepsilon_{yy}, \sigma_{zz} = \lambda e_{\text{v}} + 2G\varepsilon_{zz} \tag{3.3.1.6}
$$

$$
\tau_{xy} = G\gamma_{xy}, \quad \tau_{yz} = G\gamma_{yz} \quad \text{and} \quad \tau_{xz} = G\gamma_{xz} \tag{3.3.1.7}
$$

where $e_{\text{v}} = \varepsilon_{xx} + \varepsilon_{yy} + \varepsilon_{zz}$ is defined as the *volumetric strain* within the body.

The equation of motion in the X-direction can now be expressed as

$$
\begin{aligned}
\rho\Delta_x \cdot \Delta_y \cdot \Delta_z \frac{\partial^2 u}{\partial t^2} &= \left(\sigma_{xx} + \frac{\partial \sigma_{xx}}{\partial x}\Delta_x\right)\Delta_y\Delta_z - \sigma_{xx}\Delta_y\Delta_z + \left(\tau_{xy} + \frac{\partial \tau_{xy}}{\partial y}\Delta_y\right)\Delta_x\Delta_z \\
&\quad - \tau_{xy}\Delta_x\Delta_z + \left(\tau_{xz} + \frac{\partial \tau_{xz}}{\partial z}\Delta_z\right)\Delta_x\Delta_y - \tau_{xz}\Delta_x\Delta_y
\end{aligned} \tag{3.3.1.8}
$$

The above, on simplification, gives

$$
\rho\frac{\partial^2 u}{\partial t^2} - \frac{\partial \sigma_{xx}}{\partial x} + \frac{\partial \tau_{xy}}{\partial y} + \frac{\partial \tau_{zx}}{\partial z} \tag{3.3.1.9}
$$

Proceeding in identical fashion in the Y- and Z-directions, we have

$$
\rho\frac{\partial^2 v}{\partial t^2} = \frac{\partial \tau_{yx}}{\partial x} + \frac{\partial \sigma_{yy}}{\partial y} + \frac{\partial \tau_{yz}}{\partial z} \quad \text{and} \tag{3.3.1.10}
$$

$$
\rho\frac{\partial^2 w}{\partial t^2} = \frac{\partial \tau_{zx}}{\partial x} + \frac{\partial \tau_{zy}}{\partial y} + \frac{\partial \sigma_{zz}}{\partial z} \tag{3.3.1.11}
$$

Equations (3.3.1.9) thru (3.3.1.11) represent the equation of motion of an elastic body in three dimensions.

Subsequent analysis of the above equations will reveal that the waves propagating actually break up into two types of waves, namely compressive (P) waves and shear (S) waves, which we intend to discuss hereafter.

3.3.2 Compressive or P Waves Propagating Through Soil Medium

Substituting Eqs. (3.3.1.6) in (3.3.1.9), we have

$$\rho\frac{\partial^2 u}{\partial t^2}=\frac{\partial}{\partial x}(\lambda e_{\mathrm{v}}+2G\varepsilon_{xx})+\frac{\partial}{\partial y}G\gamma_{xy}+\frac{\partial}{\partial z}G\gamma_{xz} \tag{3.3.2.1}$$

Again substituting the strain displacement equations as furnished in Eq. (3.3.1.2), we can express Eq. (3.3.2.1) as

$$\rho\frac{\partial^2 u}{\partial t^2}=\frac{\partial e_{\mathrm{v}}}{\partial x}(\lambda+G)+G\nabla^2 u \tag{3.3.2.2}$$

where the Laplacian operation ∇^2 is expressed as $\nabla^2=\frac{\partial^2}{\partial x^2}+\frac{\partial^2}{\partial y^2}+\frac{\partial^2}{\partial z^2}$.

Similarly, equation of motion along the Y and Z coordinates is expressed as

$$\rho\frac{\partial^2 v}{\partial t^2}=\frac{\partial e_{\mathrm{v}}}{\partial y}(\lambda+G)+G\nabla^2 v \tag{3.3.2.3}$$

$$\rho\frac{\partial^2 w}{\partial t^2}=\frac{\partial e_{\mathrm{v}}}{\partial z}(\lambda+G)+G\nabla^2 w \tag{3.3.2.4}$$

Differentiating the equation of motion in the X-direction with respect to x, we have

$$\begin{aligned}&\rho\frac{\partial^2}{\partial t^2}\left(\frac{\partial u}{\partial x}\right)=\frac{\partial^2 e_{\mathrm{v}}}{\partial x^2}(\lambda+G)+G\nabla^2\left(\frac{\partial u}{\partial x}\right)\\ &\rightarrow\rho\frac{\partial^2\varepsilon_{xx}}{\partial t^2}=\frac{\partial^2 e_{\mathrm{v}}}{\partial x^2}(\lambda+G)+G\nabla^2\varepsilon_{xx}\end{aligned} \tag{3.3.2.5}$$

Similarly, in the Y- and Z-directions we have

$$\rho\frac{\partial^2\varepsilon_{yy}}{\partial t^2}=\frac{\partial^2 e_{\mathrm{v}}}{\partial x^2}(\lambda+G)+G\nabla^2\varepsilon_{yy} \tag{3.3.2.6}$$

$$\rho\frac{\partial^2\varepsilon_{zz}}{\partial t^2}=\frac{\partial^2 e_{\mathrm{v}}}{\partial x^2}(\lambda+G)+G\nabla^2\varepsilon_{zz} \tag{3.3.2.7}$$

Now adding Eqs. (3.3.2.5) thru (3.3.2.7), we have

$$\rho\left(\frac{\partial^2\varepsilon_{xx}}{\partial t^2}+\frac{\partial^2\varepsilon_{yy}}{\partial t^2}+\frac{\partial^2\varepsilon_{zz}}{\partial t^2}\right)=(\lambda+G)\nabla^2 e_v+G\nabla^2(\varepsilon_{xx}+\varepsilon_{yy}+\varepsilon_{zz}) \qquad (3.3.2.8)$$

Considering that $e_v = \varepsilon_{xx}+\varepsilon_{yy}+\varepsilon_{zz}$, Eq. (3.3.2.8) can be expressed as

$$\begin{aligned}\rho\left(\frac{\partial^2 e_v}{\partial t^2}\right)&=(\lambda+2G)\nabla^2 e_v\\ \left(\frac{\partial^2 e_v}{\partial t^2}\right)&=\frac{(\lambda+2G)}{\rho}\nabla^2 e_v\end{aligned} \qquad (3.3.2.9)$$

$$v_p=\sqrt{\frac{\lambda+2G}{\rho}} \qquad (3.3.2.10)$$

which is the expression of compression or primary (P) waves traveling through the medium.

Equation (3.3.2.10) can be expressed in terms of shear modulus G and the Poisson's ratio ν as

$$v_p=\sqrt{\frac{2G(1-\nu)}{\rho(1-2\nu)}} \qquad (3.3.2.11)$$

Figure 3.4 shows the nature of propagating P waves through the soil.

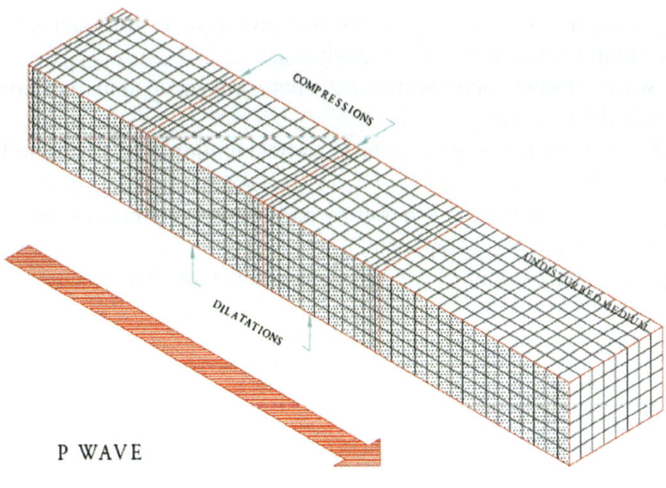

Fig. 3.4 P waves propagating through an elastic/soil medium

It is observed that as the waves travel through the medium the soil undergoes compression and dilation without any upheaval at the surface. Thus, during propagation of P waves during an earthquake the damage induced is minimal.

3.3.3 Shear or S Waves Propagating Through Soil Medium

Another kind of wave that generates in a soil medium due to propagation of waves is shear or S wave. This may be obtained by differentiating equation of motion in the *Y*-direction with respect to *z* and the equation of motion in the *Z*-direction with respect to *y* and subtracting them from each other.

Thus

$$\rho\frac{\partial^2}{\partial t^2}\left(\frac{\partial w}{\partial y}-\frac{\partial v}{\partial z}\right)=G\nabla^2\left(\frac{\partial w}{\partial y}-\frac{\partial v}{\partial z}\right) \tag{3.3.3.1}$$

Recalling the previously defined expression of rotation [vide Eq. (3.3.1.3)], Eq. (3.3.3.1) can be simplified to

$$\rho\frac{\partial^2 \Omega_x}{\partial t^2}=G\nabla^2\Omega_x \tag{3.3.3.2}$$

Equation (3.3.3.2) defines a distortional wave traveling along the *X*-axes,[4] when the waves are observed to travel with a velocity.

$$v_s=\sqrt{G/\rho} \tag{3.3.3.3}$$

This wave is commonly known as S or shear wave, which travels with velocity v_s, knows as shear wave velocity of soil and due to its distortional nature causes maximum ground damage during an earthquake.

The S waves during their motion are often broken up into two components, namely SH and SV waves.

The SH waves are those waves whose particle motion is restricted to horizontal plane only and the SV waves are those whose particle motion is restricted to vertical plane only. For a given S wave of arbitrary particle motion, this can be expressed as vector sum of SH and SV component.

From Eqs. (3.3.2.11) and (3.3.3.3), one can conclude that

$$\frac{v_p}{v_s}=\sqrt{\frac{2(1-v)}{1-2v}} \tag{3.3.3.4}$$

[4]Similar expressions can be derived in *Y*- and *Z*-direction.

For a typical value of $\nu = 0.25$, the ratio of $v_p/v_s = \sqrt{3}$, which shows that in an infinite elastic medium primary waves travel at a much faster rate than the shear waves and are first to arrive at a site during an earthquake (minor tremor) followed by a major tremor which develops due to shear waves coming at a slower speed.

Figure 3.5 shows propagation of shear waves through the soil medium during an earthquake.

It is observed in Fig. 3.5 that the earth surface is subjected to heavy upheaval and is one of the major reasons for damage during an earthquake.

3.3.4 Propagation of Rayleigh Waves Through Soil Medium

In Sects. (3.3.2) and (3.3.3), we had derived the wave equations assuming the elastic medium as an infinite domain. In real-world situation, an infinite domain is actually an idealization and the earth we live in, though big, is still a finite sphere where ignoring its spherical curvature can be considered as a semi-infinite elastic half-space.

Wave propagation through such elastic half-space develops a special kind of waves near the surface, called Rayleigh waves, that significantly affect a structure resting on earth during an earthquake.

Shown in Fig. 3.6 are waves propagating through an elastic half-space expressed by the *X-Y* plane. Let *u* and *w* be the displacements in the direction of *X* and *Z*-axes, respectively, independent of *Y*.

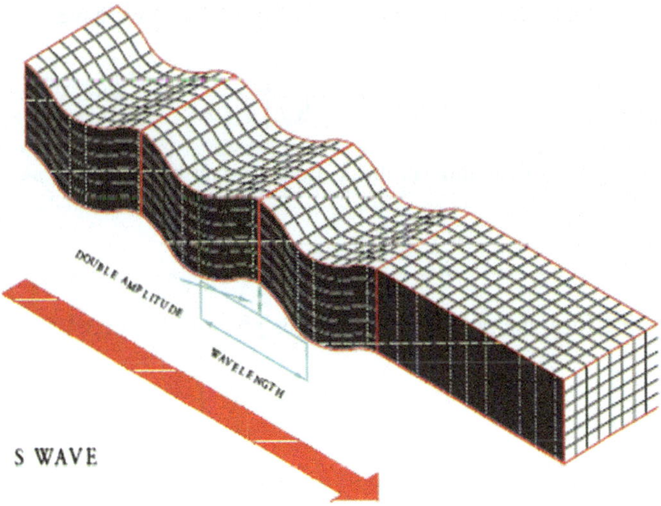

Fig. 3.5 S waves propagating through an elastic/soil medium

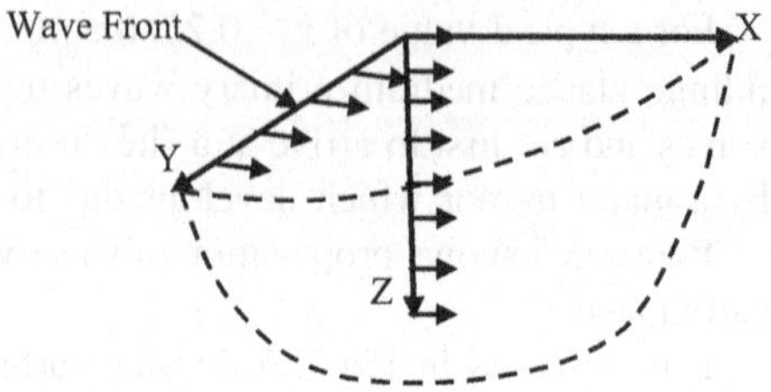

Fig. 3.6 Propagating waves in an elastic half-space

Let u and w be a function of stream potential ϕ and ψ such that

$$u = \frac{\partial \phi}{\partial x} + \frac{\partial \psi}{\partial z} \quad \text{and} \quad w = \frac{\partial \phi}{\partial z} - \frac{\partial \psi}{\partial x} \tag{3.3.4.1}$$

Considering e_v = volumetric strain where $e_v = \varepsilon_{xx} + \varepsilon_{yy} + \varepsilon_{zz}$ we have

$$e_v = \frac{\partial u}{\partial x} + \frac{\partial v}{\partial y} + \frac{\partial w}{\partial z} \tag{3.3.4.2}$$

Substituting Eq. (3.3.4.1), we have

$$e_v = \frac{\partial}{\partial x}\left(\frac{\partial \phi}{\partial x} + \frac{\partial \psi}{\partial z}\right) + \frac{\partial}{\partial y}0 + \frac{\partial}{\partial z}\left(\frac{\partial \phi}{\partial z} - \frac{\partial \psi}{\partial x}\right)$$

$$\rightarrow e_v = \frac{\partial^2 \phi}{\partial x^2} + \frac{\partial^2 \phi}{\partial z^2} = \nabla^2 \phi \tag{3.3.4.3}$$

Again considering the expression of Ω_y, as furnished in Eq. (3.3.1.3) and substituting the potential functions expressed in Eq. (3.3.4.1) we have

$$\Omega_y = \frac{1}{2}\left[\frac{\partial}{\partial z}\left(\frac{\partial \phi}{\partial x} + \frac{\partial \psi}{\partial z}\right) - \frac{\partial}{\partial x}\left(\frac{\partial \phi}{\partial z} - \frac{\partial \psi}{\partial x}\right)\right] = \frac{\partial^2 \psi}{\partial x^2} + \frac{\partial^2 \psi}{\partial z^2} = \nabla^2 \psi \tag{3.3.4.4}$$

Now the equation of motion in the X-direction can be expressed as

$$\rho \frac{\partial^2 u}{\partial t^2} = \frac{\partial e_v}{\partial x}(\lambda + G) + G\nabla^2 u \tag{3.3.4.5}$$

Substituting for u the potential functions as given in Eq. (3.3.4.1), we have

$$\rho \frac{\partial^2}{\partial t^2}\left(\frac{\partial \phi}{\partial x} + \frac{\partial \psi}{\partial z}\right) = (\lambda + G)\frac{\partial}{\partial x}\nabla^2 \phi + G\nabla^2\left(\frac{\partial \phi}{\partial x} + \frac{\partial \psi}{\partial z}\right) \tag{3.3.4.6}$$

A careful observation of Eq. (3.3.4.6) reveals that it is actually a combination of two wave equations

$$\frac{\partial^2 \phi}{\partial t^2} = \left[\frac{\lambda + G}{\rho}\right] \nabla^2 \phi \rightarrow \frac{\partial^2 \phi}{\partial t^2} = v_p^2 \nabla^2 \phi \tag{3.3.4.7}$$

and

$$\frac{\partial^2 \psi}{\partial t^2} = \left[\frac{G}{\rho}\right] \nabla^2 \psi \rightarrow \frac{\partial^2 \psi}{\partial t^2} = v_s^2 \nabla^2 \psi \tag{3.3.4.8}$$

In a similar manner for rotational case, we have

$$\rho \frac{\partial}{\partial z}\left(\frac{\partial^2 \phi}{\partial t^2}\right) - \rho \frac{\partial}{\partial x}\left(\frac{\partial^2 \psi}{\partial t^2}\right) = (\lambda + 2G)\frac{\partial}{\partial z}\nabla^2 \phi - G\frac{\partial}{\partial x}\nabla^2 \psi \tag{3.3.4.9}$$

Equation (3.3.4.9) on scrutiny again shows that it is a combination of two types of wave propagation, as expressed by Eqs. (3.3.4.7) and (3.3.4.8).

For a sinusoidal wave traveling in the X-direction, let the potential function be expressed as

$$\phi = F(z)\mathrm{e}^{[i\omega t - nx]} \quad \text{and} \quad \psi = G(z)\mathrm{e}^{[i\omega t - nz]} \tag{3.3.4.10}$$

where $F(z)$ and $G(z)$ are functions of depth and $n = 2\pi/L_w$, where L_w = wavelength.

Substituting Eq. (3.3.4.10) in Eqs. (3.3.4.7) and (3.3.4.8), we have

$$\frac{\partial}{\partial t^2} F(Z)\mathrm{e}^{[i\omega t - nx]} = v_p^2 \nabla^2 F(Z)\mathrm{e}^{[i\omega t - nx]} \tag{3.3.4.11}$$

$$\rightarrow -\omega^2 F(z) = v_p^2 \left[F''(Z) - n^2 F(Z)\right] \tag{3.3.4.12}$$

Similarly

$$\frac{\partial}{\partial t^2} G(Z)\mathrm{e}^{[i\omega t - nz]} = v_s^2 \nabla^2 G(Z)\mathrm{e}^{[i\omega t - nz]} \tag{3.3.4.13}$$

$$\rightarrow -\omega^2 G(z) = v_s^2 \left[G''(Z) - n^2 G(Z)\right] \tag{3.3.4.14}$$

Equations (3.3.4.12) and (3.3.4.13) give

$$\frac{\mathrm{d}F(Z)}{\mathrm{d}z^2} - q^2 F(Z) = 0 \tag{3.3.4.15}$$

$$\frac{\mathrm{d}G(Z)}{\mathrm{d}z^2} - s^2 G(Z) = 0 \tag{3.3.4.16}$$

where

$$q^2 = n^2 - \omega^2/v_p^2 \quad \text{and} \quad s^2 = n^2 - \omega^2/v_s^2 \tag{3.3.4.16a}$$

Solution to Eqs. (3.3.4.15) and (3.3.4.16) can be expressed as

$$F(Z) = A_1 e^{-qz} + A_2 e^{qz} \text{ and } G(Z) = B_1 e^{-sz} + B_2 e^{sz},$$

where A_1, A_2, B_1, B_2 are integration constants and are a function of the boundary condition of the problem.

Perusing the solutions, one observes that as $Z \to \infty$ the functions $F(Z)$ and $G(Z)$ also tend to infinity and are inadmissible. Thus, a realistic solution is only possible when A_2 and B_2 are zero. This gives the solution as

$$F(Z) = A_1 e^{-qz} \quad \text{and} \quad G(Z) = B_1 e^{-sz} \tag{3.3.4.17}$$

The potential function can now be expressed as

$$\phi = A_1 e^{-qz} e^{[i\omega t - nx]} \quad \text{and} \quad \psi = B_1 e^{-sz} e^{[i\omega t - nz]} \tag{3.3.4.18}$$

As the soil surface (i.e., $Z = 0$) is a free surface, hence $\sigma_{zz} = \tau_{zx} = \tau_{zy} = 0$. Thus,

$$\sigma_{zz}(Z=0) = \lambda e_v + 2G\varepsilon_{zz} = \lambda e_v + 2G\left(\frac{\partial w}{\partial z}\right) = 0 \tag{3.3.4.19}$$

for

$$\tau_{xz}(Z=0) = G\gamma_{zx} = G\left(\frac{\partial w}{\partial x} + \frac{\partial u}{\partial z}\right) = 0 \tag{3.3.4.20}$$

Using the definition of u and w and solution of ϕ and ψ as given in Eq. (3.3.4.18), Eqs. (3.3.4.19) and (3.3.4.20) can be written as

$$A_1\left[(\lambda + 2G)q^2 - \lambda n^2\right] - 2iGnsA_2 = 0$$

and

$$2iA_1 nq + A_2\left(s^2 + n^2\right) = 0 \tag{3.3.4.21}$$

Rearranging we have

$$\frac{A_1}{A_2}\left[\frac{(\lambda + 2G)q^2 - \lambda n^2}{2insG}\right] - 1 = 0 \quad \text{and} \quad \frac{A_1}{A_2}\left[\frac{2qin}{s^2 + n^2}\right] + 1 = 0 \tag{3.3.4.22}$$

Eliminating A_1 and A_2 from Eq. (3.3.4.22), we have

$$\frac{(\lambda+2G)q^2-\lambda n^2}{2insG}=-\frac{2qin}{s^2+n^2} \tag{3.3.4.23}$$

The above can be further simplified to

$$4qsn^2G=\left(s^2-n^2\right)\left[(\lambda+2G)q^2-\lambda n^2\right] \tag{3.3.4.24}$$

Squaring Eq. (3.3.4.24) and substituting the value of s and n as defined earlier one has

$$16G^2n^4\left(n^2-\frac{\omega^2}{v_p^2}\right)\left(n^2-\frac{\omega^2}{v_s^2}\right)=\left[(\lambda+2G)\left(n^2-\frac{\omega^2}{v_p^2}\right)-\lambda n^2\right]\left[n^2+\left(n^2-\frac{\omega^2}{v_s^2}\right)\right] \tag{3.3.4.25}$$

Equation (3.3.4.25) can be further reduced to

$$16\left(1-\frac{\omega^2}{n^2v_p^2}\right)\left(1-\frac{\omega^2}{n^2v_s^2}\right)=\left[2-\left(\frac{\lambda+2G}{G}\right)\left(\frac{\omega^2}{n^2v_p^2}\right)-\lambda n^2\right]^2\left[2-\frac{\omega^2}{n^2v_s^2}\right]^2 \tag{3.3.4.26}$$

Let L_R = wavelength of Rayleigh wave be expressed as $L_R=2\pi/n$ and let V_R be the Rayleigh wave velocity of the propagating waves; $n=\omega/V_R$, $R=V_R/v_s$, $\alpha R=V_R/v_p$, $V_R^2/v_p^2=\alpha^2=(1-2v)/[2(1-v)]$ and $v=\lambda/[2(\lambda+G)]$.

Substituting these values in Eq. (3.3.4.26), we have

$$16\left(1-\alpha^2R^2\right)\left(1-R^2\right)=\left(2-R^2\right)^4 \tag{3.3.4.27}$$

Expanding the above gives

$$R^6-8R^4+\left(24-16\alpha^2\right)R^2+16\left(\alpha^2-1\right)=0 \tag{3.3.4.28}$$

Equation (3.3.4.28) is actually a cubic equation in terms of R^2, from which, when solved for various values of v, the Poisson's ratio, we get the values of Rayleigh wave velocity V_R in terms of compression and shear wave velocity, v_p and v_s, respectively.

For v = 0.5, i.e., when the body is incompressible, Eq. (3.3.4.28) reduces to

$$R^6-8R^4+24R^2-16=0 \tag{3.3.4.29}$$

The real root of the above equation (other two roots are imaginary) is given by

$$V_R^2 = 0.91275v_s^2 \rightarrow V_R = 0.95538v_s \tag{3.3.4.30}$$

Similarly, for $v = 0.25$ it can be shown that three roots for Eq. (3.3.4.28) are $R^2 = 4$, $R^2 = 2 + 2\sqrt{3}$ and $R^2 = 2 - 2\sqrt{3}$. Of these, only the last root satisfies the condition of surface wave being real.

Thus for $R^2 = 2 - 2\sqrt{3}$ we have $V_R = 0.9194v_s$.

It has been shown by Richart et al. (1970) that for practical engineering problems the Poisson's ratio of soil usually varies from 0.3 to 0.4, when V_R and v_s can be considered same for all practical purpose.

The nature of propagation of Rayleigh waves is as shown in Fig. 3.7.

It is observed that Rayleigh waves and shear waves produce similar kind of distortion on the soil surface.

Example 3.3 A number of soil samples collected from a site gives an average dynamic modulus $(E) = 2.866 \times 10^5$ kN/m², and Poisson's ratio $v = 0.3$. Determine the values of v_p, v_s, and V_R. Weight density of soil = 19.0 kN/m³.

Solution

Considering $G = E/2(1 + v)$, the shear modulus of soil is estimated as

$$G = \frac{2.866 \times 10^5}{2(1 + 0.3)} = 110230.7\ \text{kN/m}^2.$$

$$\text{Therefore, } v_s = \sqrt{\frac{G}{\rho}} = \sqrt{\frac{11230.7 \times 9.81}{19.0}} = 238.56\ \text{m/sec}.$$

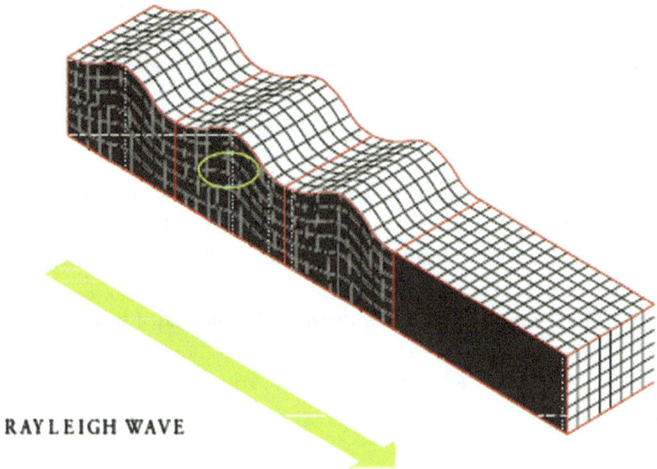

Fig. 3.7 Rayleigh waves propagating through an elastic/soil medium

$$v_p = v_s\sqrt{\frac{2(1-v)}{1-2v}} = 238.56 \times \sqrt{\frac{2 \times 0.7}{0.4}} = 446.3\ \text{m/sec.}$$

Again, $V_R = 0.95538 v_s = 0.95538 \times 238.56 = 228\ \text{m/sec.}$

3.3.4.1 Displacement Due to Rayleigh Waves

Vide Eq. (3.3.4.18), we had shown that the potential functions are

$$\phi = A_1 e^{-qz} e^{[i\omega t - nx]} \quad \text{and} \quad \psi = B_1 e^{-sz} e^{[i\omega t - nz]}$$

Substituting for amplitude u, we have

$$u = -iA_1 n e^{[-qz + i(\omega t - nx)]} + A_2 s e^{[-sz + i(\omega t - nx)]} \tag{3.3.4.1.1}$$

Considering the value of $A_2 = -\frac{2iqnA_1}{s^2 + n^2}$ as from Eq. (3.3.4.21), we have

$$\begin{aligned} u &= -A_1\left[-ine^{-qz} + \left(\frac{2iqsn}{s^2+n^2}\right)e^{-sz}\right]e^{i(\omega t - nx)} \\ \rightarrow u &= -A_1 in\left[-e^{-qz} + \left(\frac{2qs/n^2}{1+s^2/n^2}\right)e^{-sz}\right]e^{i(\omega t - nx)} \end{aligned} \tag{3.3.4.1.2}$$

Proceeding identically one can derive

$$w = A_1 n\left[\frac{2q/n}{(s^2/n^2)+1}e^{-sz} - \frac{q}{n}e^{-qz}\right]e^{i(\omega t - nx)} \tag{3.3.4.1.3}$$

Thus, variation of the amplitude with respect to depth can be expressed as

$$u(z) = -e^{-qz} + \frac{2qs/n^2}{1+(s^2/n^2)}e^{-sz} \tag{3.3.4.1.4}$$

and

$$w(z) = \frac{(2q/n)}{(s^2/n^2)+1}e^{-sz} - \frac{q}{n}e^{-qz} \tag{3.3.4.1.5}$$

By the definition as furnished in Eq. (3.3.4.16a), one can rewrite q and s as

$$\frac{q^2}{n^2} = 1 - \frac{\omega^2}{n^2 v_p^2} = 1 - \alpha^2 R^2 \quad \text{and} \quad \frac{s^2}{n^2} = 1 - \frac{\omega^2}{n^2 v_s^2} = 1 - R^2 \tag{3.3.4.1.6}$$

Thus, $u(z)$ and $w(z)$ can now be expressed by the Poisson's ratio and any wave number n.

For example, for $v = 0.25$, $u(z)$ and $w(z)$ can be represented by

$$u(z) = -e^{-0.8475zn} + 0.5773e^{-0.3993zn} \tag{3.3.4.1.7}$$

$$w(z) = 0.8475e^{-0.8475zn} - 1.469e^{-0.3993zn} \tag{3.3.4.1.8}$$

It can be shown from theoretical consideration that a wave perpendicular to the plane of motion is not possible in a homogenous half-space. However, such waves, popularly known as SH waves, are observed prominently on the earth surface with other surface waves during an earthquake.

Love (1944) developed a theory sufficient to include the SH waves by having two homogeneous layers resting over each other, which will be our topic of discussion in the next section.

3.3.5 *Propagation of Love Waves Through Soil Medium*

Shown in Fig. 3.8 is an elastic half-space overlain by another layer of elastic medium having height *H*. Love waves or SH waves propagating in positive *x*-direction would also generate a ground motion in *Y*-direction (i.e., plane perpendicular to the plane of the paper) that can be expressed as

$$v(x, z, t) = V(z)e^{i(n_L x - \omega t)} \tag{3.3.5.1}$$

where v is the displacement of ground in the y-direction, $V(z)$ is the variation of the displacement along depth Z, and n_L is the Love wave number.

The propagation of waves must satisfy the two-dimensional wave equation for both the overlain soil and the elastic half-space when

$$\frac{\partial^2 v}{\partial t^2} = \frac{G_1}{\rho_1}\left(\frac{\partial^2 v}{\partial x^2} + \frac{\partial^2 v}{\partial z^2}\right) \quad \text{valid for } 0 < z < H, \text{ and}$$
$$\frac{\partial^2 v}{\partial t^2} = \frac{G_2}{\rho_2}\left(\frac{\partial^2 v}{\partial x^2} + \frac{\partial^2 v}{\partial z^2}\right) \quad \text{valid for } z \geq H \tag{3.3.5.2}$$

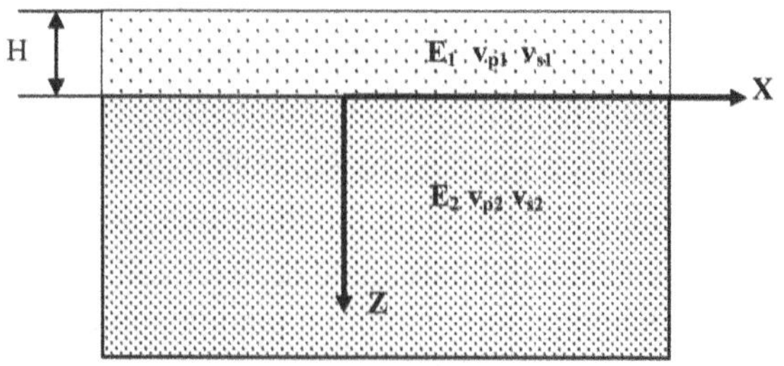

Fig. 3.8 An elastic half-space overlain by a soil of finite depth

The amplitude of vibration will vary with depth according to the expression (Aki and Richards 1980)

$$v(z) = S_1 e^{-V_1 z} + T_1 e^{V_1 z} \quad textfor\ 0 < z < H$$

and

$$v(z) = S_2 e^{-V_2 z} + T_2 e^{V_2 z} \quad \text{for } z \geq H \tag{3.3.5.3}$$

where T and S are amplitude of waves propagating up and down, respectively.

Here

$$V_1 = \sqrt{\frac{n_L^2 - \omega^2}{G_1/\rho_1}} \quad \text{and} \quad V_2 = \sqrt{\frac{n_L^2 - \omega^2}{G_2/\rho_2}} \tag{3.3.5.4}$$

As the layered half-space extends to infinity along the Z-direction no waves can reflect back due to any rebound, which implies $T_2 = 0$, and again, considering that all stresses and strains vanish at the free surface, the condition is satisfied when $\partial v/\partial z = 0$, which gives

$$(S_1 - T_1)V_1\left(e^{-V_1 z} + e^{V_1 z}\right) = 0 \tag{3.3.5.5}$$

As $V_1 \neq 0$, hence, $S_1 = T_1$, the amplitude function can now be expressed as

$$v(z) = S_1\left(e^{-V_1 z} + e^{V_1 z}\right) \quad \text{for } 0 < z < H \text{ and } v(z) = S_2 e^{-V_2 z} \quad \text{for } Z \geq H \tag{3.3.5.6}$$

At $z = H$, considering stress compatibility we get

$$2iG_1 V_1 S_1 \sin(iV_1 H) = G_2 V_2 S_2 e^{-V_2 H} \tag{3.3.5.7}$$

Similarly, the strain compatibility condition yields

$$2S_1 \cos(iV_1 H) = S_2 e^{-V_2 H} \tag{3.3.5.8}$$

Combining the above two equations, we obtain

$$v(x,z,t) = 2S_1 \cos\left[\omega\sqrt{\left(\frac{1}{V_1^2} - \frac{1}{V_L^2}\right)} \times z\right] e^{i(n_L x - \omega t)} \quad \text{for } 0 < z < H \tag{3.3.5.9}$$

and

$$v(x,z,t) = 2S_1 \cos\left[\omega\sqrt{\left(\frac{1}{V_1^2}-\frac{1}{V_L^2}\right)} \times H\right] \times \mathrm{e}^{\left[-\omega\sqrt{\left(\frac{1}{V_L^2}-\frac{1}{V_2^2}\right)}\times(z-H)\right]} \mathrm{e}^{i(n_L x-\omega t)} \quad \text{for } z \geq H. \tag{3.3.5.10}$$

Here, V_1 and V_2 are shear wave velocities of layers 1 and 2, while V_L is the Love wave velocity. Equations (3.3.5.9) and (3.3.5.10) show that the amplitude of Love wave velocity varies as a harmonic function for the top layer of depth H, while it decays as an exponential function for the lower strata of soil. That is why the Love waves remain trapped in the surface layer of the soil.

The Love wave velocity can be computed from the solution of

$$\tan \omega H\sqrt{\left(\frac{1}{V_1^2}-\frac{1}{V_L^2}\right)} = \left(\frac{G_2}{G_1}\right)\frac{\sqrt{1/V_L^2 - 1/V_2^2}}{\sqrt{1/V_1^2 - 1/V_L^2}} \tag{3.3.5.11}$$

As shown in Fig. 3.9, it is observed that Love waves are dispersive in nature.

Example 3.4 Soil strata of 22 m thickness having an average shear wave velocity of 267 m/sec are overlying the bedrock; the bedrock has a shear wave velocity of 750 m/sec. Density of soil = 18.5 kN/m^3, while that of the rock is 23 kN/m^3.

Determine the free field natural frequency of the soil and hence compute the Love wave velocity.

Solution

For this particular case, for the overlying soil v_s = 267 m/sec and H = 22 m.

This fundamental time period is expressed as $T = \sqrt{\frac{4H}{v_s}} = \sqrt{\frac{4\times22}{267}} = 0.574$ sec.

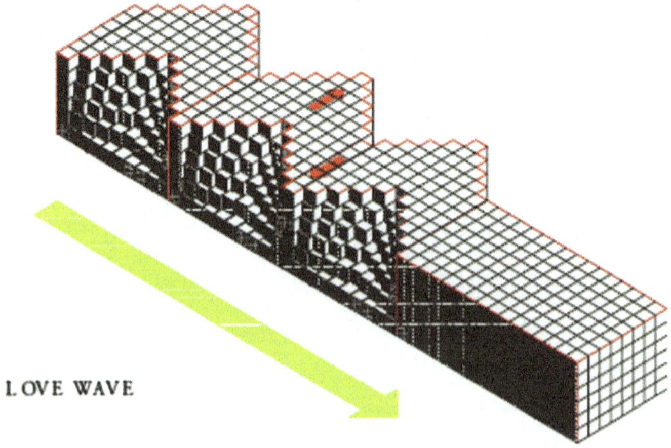

Fig. 3.9 Propagation of Love waves through elastic or soil medium

Considering

$$\omega = \frac{2\pi}{T} = \frac{2 \times \pi}{0.574} = 10.94 \text{ rad/sec.}$$
$$\rightarrow \omega H = 10.94 \times 22 = 240.7$$

Converting ωH to degrees, we have $\omega H = \frac{180}{\pi} \times 240.7 = 13791°$. This gives $\tan \omega H = -2.592$ as $\tan \omega H \sqrt{\left(\frac{1}{V_1^2} - \frac{1}{V_L^2}\right)} = \left(\frac{G_2}{G_1}\right) \frac{\sqrt{1/V_L^2 - 1/V_2^2}}{\sqrt{1/V_1^2 - 1/V_L^2}}$.

On simplification, we have

$$V_L^4 + \left(\alpha^2 V_1^4 - 2V_2^2\right) V_L^2 + \left[\left(1 - \alpha^2\right) V_1^4 V_2^2\right] = 0$$

Here, $\alpha^2 = \frac{(G_2/G_1)^2}{\tan^2 \omega H}$, where $G_2 = \left(\frac{23}{9.81}\right) \times (750)^2 = 1318807 \text{ kN/m}^2$. $G_1 = \left(\frac{18.5}{9.81}\right) \times (267)^2 = 134439 \text{ kN/m}^2$.

$$\therefore \alpha^2 = \frac{\left(\frac{1318807}{134439}\right)^2}{(-2.59)^2} = 14.34 \text{ and } V_1 = 267 \text{ m/sec and } V_2 = 750 \text{ m/sec.}$$

or,

$$V_L^4 + \left(14.34 \times (267)^4 - 2 \times (750)^2\right) V_L^2 + \left[(1 - 14.34)(267)^4 \times (750)^2\right] = 0$$
$$\Rightarrow V_L^4 + 7.2876 \times 10^{10} V_L^2 - 3.813 \times 10^{16} = 0$$
$$\Rightarrow V_L^2 = \frac{-7.2876 \times 10^{10} \pm \sqrt{(7.2876 \times 10^{10})^2 + 4 \times 3.813 \times 10^{16}}}{2} = 523214$$

Thus $\Rightarrow V_L = 723$ m/sec.

3.4 Propagation of Waves in Polar Coordinates

When wave propagation is studied in three dimensions, it is sometimes convenient to analyze the system in polar coordinates when, as shown in Fig. 3.10, we consider

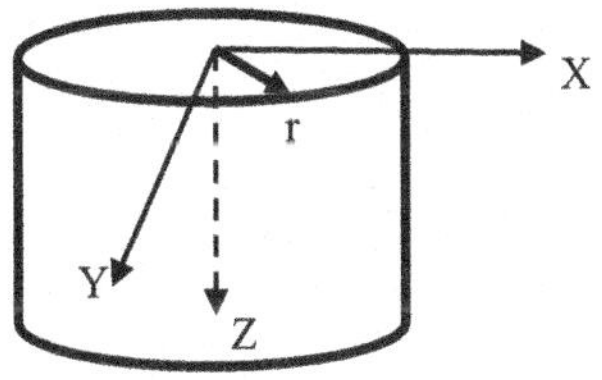

Fig. 3.10 Propagation of elastic waves in cylindrical coordinates

$r = \sqrt{x^2 + y^2}$ and the Z-axes which is vertical. The displacement function in the cylindrical coordinates (r, Z, t) can be expressed as (Das 1982)

$$u(r, z, t) = \phi(r)\psi(z)\xi(t) \tag{3.4.1}$$

The equation of wave propagation can be expressed as

$$\frac{\partial^2 u}{\partial t^2} = v_s^2\left[\frac{\partial^2 u}{\partial r^2} + \frac{1}{r}\frac{\partial u}{\partial r} + \frac{\partial^2 u}{\partial z^2}\right] \tag{3.4.2}$$

Here, the angular function θ is ignored, for the case is axisymmetric. Now, substituting Eq. (3.4.1) in Eq. (3.4.2), we have

$$\phi(r)\psi(z)\ddot{\xi}(t) = v_s^2\left[\ddot{\phi}(r)\psi(z)\xi(t) + \frac{1}{r}\dot{\phi}(r)\psi(z)\xi(t) + \phi(r)\ddot{\psi}(z)\xi(t)\right] \tag{3.4.3}$$

Dividing each of the above terms by $\phi(r)\psi(z)$ we get

$$\begin{aligned} \ddot{\xi}(t) &= v_s^2\left[\frac{\ddot{\phi}(r)}{\phi(r)}\xi(t) + \frac{1}{r}\frac{\dot{\phi}(r)}{\phi(r)}\xi(t) + \frac{\ddot{\psi}(z)}{\psi(z)}\xi(t)\right] \\ \rightarrow \frac{\ddot{\xi}(t)}{v_s^2\xi(t)} &= \left[\frac{\ddot{\phi}(r)}{\phi(r)} + \frac{1}{r}\frac{\dot{\phi}(r)}{\phi(r)} + \frac{\ddot{\psi}(z)}{\psi(z)}\right] = -k^2 \text{ (say)} \end{aligned} \tag{3.4.4}$$

Equation (3.4.4) can be separated into

$$\ddot{\xi}(t) + k^2 v_s^2 \xi(t) = 0 \rightarrow \ddot{\xi}(t) + \lambda^2\xi(t) = 0 \quad \text{where } \lambda^2 = k^2 v_s^2 \tag{3.4.5}$$

Again, let us consider

$$\frac{\ddot{\psi}(z)}{\psi(z)} = m^2 \rightarrow \ddot{\psi}(z) - m^2\psi(z) = 0 \tag{3.4.6}$$

Finally, considering $\left[\frac{\ddot{\phi}(r)}{\phi(r)} + \frac{1}{r}\frac{\dot{\phi}(r)}{\phi(r)} + m^2\right] = -k^2$ we have

$\left[\frac{\ddot{\phi}(r)}{\phi(r)} + \frac{1}{r}\frac{\dot{\phi}(r)}{\phi(r)}\right] = -k^2 - m^2$. Now considering $h^2 = k^2 + m^2$, we finally have

$$\ddot{\phi}(r) + \frac{1}{r}\dot{\phi}(r) + h^2\phi(r) = 0 \tag{3.4.7}$$

Thus, the partial differential equation [Eq. (3.4.2)] can now be separated into three linear differential equations

$$\frac{d^2\phi}{ds^2} + \frac{1}{s}\frac{d\phi}{ds} + \phi = 0 \quad \text{where } s = hr. \tag{3.4.8}$$

$$\frac{d^2\psi}{dz^2} - m^2\psi = 0 \quad \text{and} \quad \frac{d^2\xi}{dt^2} + \lambda^2\psi = 0. \tag{3.4.9}$$

The solutions are, respectively,

$$\phi(r) = C_1 J_0(hr) + C_2 K_0(r); \psi(z) = C_3 e^{mz} + C_4 e^{-mz}; \xi(t) = C_5 \cos \lambda t + C_6 \sin \lambda t \tag{3.4.10}$$

where $J_0(hr), K_0(hr)$ = Bessel's functions of first and second kind of order zero.

Thus, the complete solution is

$$u(r, z, t) = [C_1 J_0(h) + C_2 K_0(hr)][C_3 e^{mz} + C_4 e^{-mz}][C_5 \cos \lambda t + C_6 \sin \lambda t] \tag{3.4.11}$$

Equation (3.4.11) is the general solution to the partial differential Eq. (3.4.2) where C_1 to C_6 are integration constants that are boundary specific to the problem.

For instance, when $r \to \infty$, $K_0(hr) = 0$, for $z \to \infty$ as waves cannot reflect back implies $e^{mz} = 0$. For harmonic motion at $t = 0$, $u = 0$ gives $C_5 = 0$ which would finally give

$$u(r, z, t) = \hat{C} J_0(hr) e^{-mz} \sin \lambda t \tag{3.4.12}$$

3.5 Mathematical Background of Actual Ground Response, and Design Response Spectra

When waves propagate through the soil medium during an earthquake, the acceleration response picked up by the accelerometer is similar to what is depicted in Fig. 3.2. However, while designing a structure one usually follows a normalized uniform curve like one shown in Fig. 3.11.

So, what is the sanctity of it?

How does a random and irregular curve, like the one shown in Fig. 3.2, can be justified by a regularized curve as furnished in Fig. 3.11?

We discuss this issue in detail herein.

Shown in Fig. 3.12 is a single bay portal frame subjected to a ground acceleration $\ddot{u}_g(t)$,which is random in nature, like the one shown in Fig. 3.2, and acts over a duration of time t, say.

As described in Chap. 2, the equilibrium equation of the body can then be expressed as

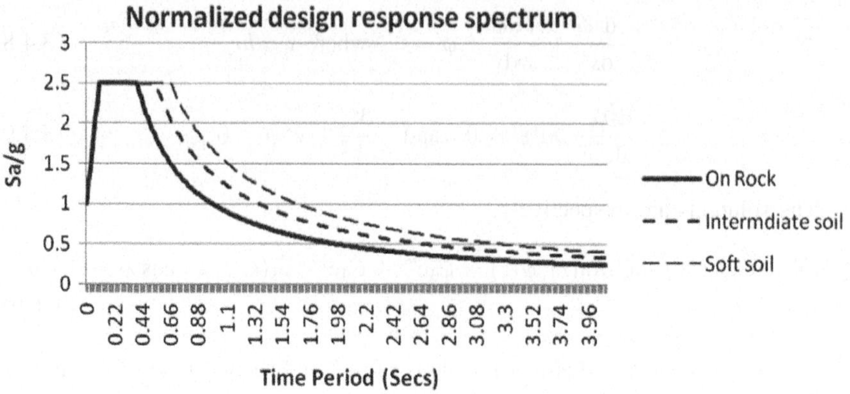

Fig. 3.11 Normalized response curve as furnished in IS-1893(2016) Part-1

Fig. 3.12 A single degree oscillator subjected ground acceleration $\ddot{u}_g$

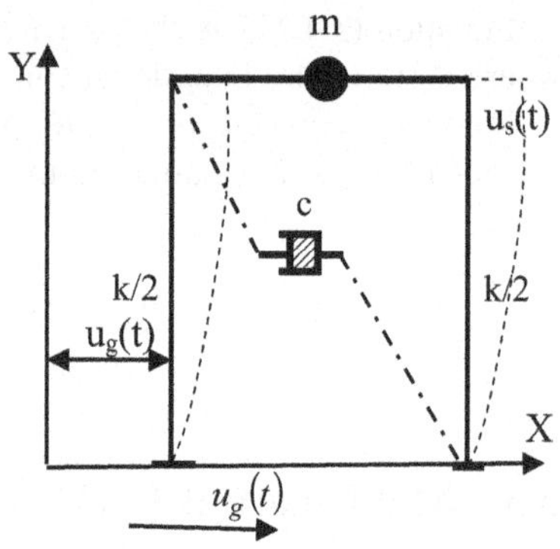

$$m\ddot{u}_s(t) + c\dot{u}_s(t) + ku_s(t) = -m\ddot{u}_g(t) \tag{3.5.1}$$

Dividing each term by the mass *m*, we have

$$\begin{aligned} &\ddot{u}_s(t) + \left(\frac{c}{m}\right)\dot{u}_s(t) + \left(\frac{k}{m}\right)u_s(t) = -\ddot{u}_g(t) \\ &\to \ddot{u}_s(t) + 2\zeta\omega_n\dot{u}_s(t) + \omega_n^2 u_s(t) = -\ddot{u}_g(t) \end{aligned} \tag{3.5.2}$$

Here, $\omega_n^2 = k/m$ is the natural frequency, $\zeta = c/c_c$ the damping ratio, $c_c = 2\sqrt{km}$ the critical damping and $u_s(t)$ is the amplitude of vibration over the time span t.

It was shown in Chap. 2 Sect. 2.3.4 that the solution to Eq. (3.5.2) can be expressed as

$$u_s(t) = \frac{1}{\omega_d}\int_0^t \ddot{u}_g(\tau)e^{-\zeta\omega_n(t-\tau)}\sin\omega_d(t-\tau)d\tau \tag{3.5.3}$$

where $\omega_d = \omega_n\sqrt{1-\zeta^2}$ is the damped natural frequency of the body, and $\ddot{u}_g(\tau)$ is the value of ground acceleration at any instant of time τ.

Differentiating Eq. (3.5.3), we have

$$\dot{u}_s(t) = \frac{1}{\omega_d}\int_0^t \ddot{u}_g(\tau)\left[-\zeta\omega_n e^{-\zeta\omega_n(t-\tau)}\sin\omega_d(t-\tau) + \omega_d e^{-\zeta\omega_n(t-\tau)}\cos\omega_d(t-\tau)\right]d\tau$$

$$= \int_0^t \ddot{u}_g(\tau)e^{-\zeta\omega_n(t-\tau)}\left[\cos\omega_d(t-\tau) - \frac{\zeta}{\sqrt{1-\zeta^2}}\sin\omega_d(t-\tau)\right]d\tau$$

For $0 \le \zeta \le 0.2$, $\sqrt{1-\zeta^2} \to 1.0$, which gives

$$\dot{u}_s(t) = \int_0^t \ddot{u}_g(\tau)e^{-\zeta\omega_n(t-\tau)}[\cos\omega_n(t-\tau) - \zeta\sin\omega_n(t-\tau)]d\tau \tag{3.5.4}$$

Again, from Eq. (3.5.2) we have

$$\rightarrow \ddot{u}_s(t) + \ddot{u}_g(t) = -2\zeta\omega_n\dot{u}_s(t) - \omega_n^2 u_s(t) \tag{3.5.5}$$

Now substituting the value of $u_s(t)$ and $\dot{u}_s(t)$ from Eqs. (3.5.3) and (3.5.4) in Eq. (3.5.5), we have

$$\begin{aligned}\ddot{u}_s(t) + \ddot{u}_g(t) &= \omega_n(2\zeta^2 - 1)\int_0^t \ddot{u}_g(\tau)e^{-\zeta\omega_n(t-\tau)}\sin\omega_n(t-\tau)d\tau \\ &\quad - 2\zeta\omega_n\int_0^t \ddot{u}_g(\tau)e^{-\zeta\omega_n(t-\tau)}\cos\omega_n(t-\tau)d\tau \\ \Rightarrow \ddot{u}_s^t(t) &= \omega_n(2\zeta^2 - 1)\int_0^t \ddot{u}_g(\tau)e^{-\zeta\omega_n(t-\tau)}\sin\omega_n(t-\tau)d\tau \\ &\quad 2\zeta\omega_n\int_0^t \ddot{u}_g(\tau)e^{-\zeta\omega_n(t-\tau)}\cos\omega_n(t-\tau)d\tau\end{aligned} \tag{3.5.6}$$

Here, $\ddot{u}_s^t(t)$ is the *total absolute acceleration* $[\ddot{u}_s(t) + \ddot{u}_g(t)]$ of the oscillator.

The absolute maximum values of the response as per Eqs. (3.5.3), (3.5.4), and (3.5.6) are termed as

- Spectral relative displacement $S_{d;}$
- Spectral relative velocity $S_{v;}$
- Spectral absolute acceleration $S_{a.}$

Considering that all these terms are a function of damping ratio ζ and natural frequency ω_n, each of the above terms is expressed as $S_d(\zeta, \omega_n)$, $S_v(\zeta, \omega_n)$ and $S_a(\zeta, \omega_n)$.

Thus,

$$S_d(\zeta, \omega_n) = \frac{1}{\omega_d}\left|\int_0^t \ddot{u}_g(\tau)e^{-\zeta\omega_n(t-\tau)} \sin \omega_d(t-\tau)d\tau\right|_{max} \tag{3.5.7}$$

$$S_v(\zeta, \omega_n) = \left|\int_0^t \ddot{u}_g(\tau)e^{-\zeta\omega_n(t-\tau)}[\cos \omega_d(t-\tau) - \zeta \sin \omega_d(t-\tau)]d\tau\right|_{max} \tag{3.5.8}$$

and

$$\begin{aligned} S_a(\zeta, \omega_n) &= \omega_n(2\zeta^2 - 1)\left|\int_0^t \ddot{u}_g(\tau)e^{-\zeta\omega_n(t-\tau)} \sin \omega_d(t-\tau)d\tau\right|_{max} \\ &\quad - 2\zeta\omega_n\left|\int_0^t \ddot{u}_g(\tau)e^{-\zeta\omega_n(t-\tau)} \cos \omega_d(t-\tau)d\tau\right|_{max} \end{aligned} \tag{3.5.9}$$

For a low value of damping ($\zeta \leq 0.20$),[5] from Eq. (3.5.7) we have

$$S_d(\zeta, \omega_n) \times \omega_n = \left|\int_0^t \ddot{u}_g(\tau)e^{-\zeta\omega_n(t-\tau)} \sin \omega_n(t-\tau)d\tau\right|_{max} \tag{3.5.10}$$

$$\rightarrow S_{pv}(\zeta, \omega_n) = \left|\int_0^t \ddot{u}_g(\tau)e^{-\zeta\omega_n(t-\tau)} \sin \omega_n(t-\tau)d\tau\right|_{max} \tag{3.5.11}$$

Here, S_{pv} is called as *pseudo-spectral velocity*.

It should be clear at this juncture that $S_{pv}(\zeta, \omega_n)$ *is not equal to* $\dot{u}_s(t)$, but has the unit of velocity, as such is demarcated by the word *pseudo* to differentiate with the

---

[5]As $\sqrt{1-\zeta^2} \rightarrow 1.0$.

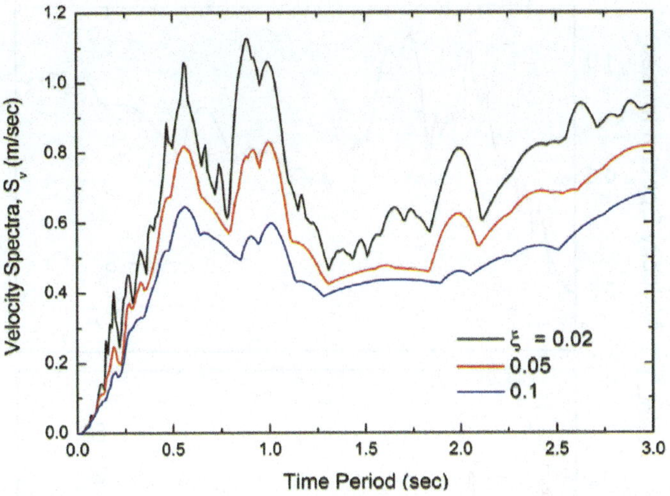

Fig. 3.13 Typical velocity spectra plot for different values of damping and time periods

actual velocity $\dot{u}_s(t)$. It is a function of the damping ratio and natural frequency/time period of the oscillator and is actually an envelope of maximum output that varies with different values of damping (ζ) and time period (T_n). Figure 3.13 shows plots of velocity spectra for different values of ζ and T_n.

Equations (3.5.10) and (3.5.11) give

$$S_d(\zeta, \omega_\mathrm{n}) = \frac{S_{pv}}{\omega_\mathrm{n}} \tag{3.5.12}$$

From Eq. (3.5.8), we have

$$S_v(0, \omega_\mathrm{n}) = \left| \int_0^t \ddot{u}_g(\tau)[\cos \omega_\mathrm{n}(t-\tau) -]\mathrm{d}\tau \right|_{\max} \tag{3.5.13}$$

while from Eq. (3.5.11), we get

$$S_{pv}(0, \omega_\mathrm{n}) = \left| \int_0^t \ddot{u}_g(\tau) \sin \omega_\mathrm{n}(t-\tau)\mathrm{d}\tau \right|_{\max} \tag{3.5.14}$$

It has been shown by Hudson (1956, 1962) that Eqs. (3.5.13) and (3.5.14) give very closely matched results except when the time period is too long. This is shown in Fig. 3.14.

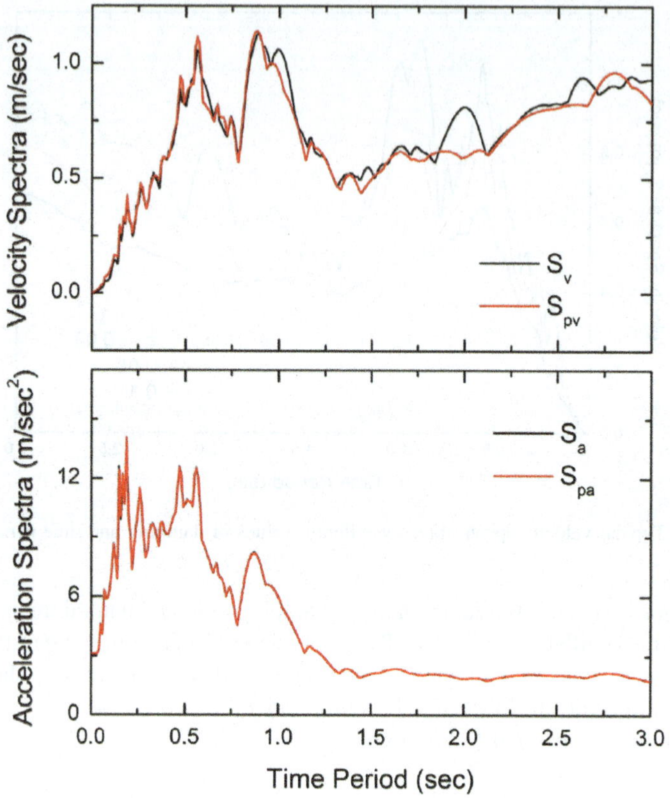

Fig. 3.14 Comparison between actual and pseudo-velocity and acceleration spectra

Putting $\zeta = 0$ in Eq. (3.5.9), we have ignored the negative sign.[6]

$$S_a(0, \omega_\mathrm{n}) = \omega_\mathrm{n} \left| \int_0^t \ddot{u}_g(\tau) \sin \omega_\mathrm{n}(t - \tau) \mathrm{d}\tau \right|_{\max} \tag{3.5.15}$$

Comparing the value within parenthesis with Eq. (3.5.14), one can write

$$S_a(0, \omega_\mathrm{n}) = \omega_\mathrm{n} S_{pv}(0, \omega_\mathrm{n}) \tag{3.5.16}$$

Numerical analysis has shown that Eq. (3.5.16) is almost satisfied for damping values within the range $0 \le \zeta \le 0.2$. Thus, for normal structures that usually have low damping values one may write

[6]For earthquake +/(−) sign does not count.

$$S_a(\zeta, \omega_n) = \omega_n S_{pv}(\zeta, \omega_n) \tag{3.5.17}$$

Considering the right-hand side of Eq. (3.5.17) as $S_{pa}(\zeta, \omega_n)$, the *pseudo-spectral acceleration,* we find that the value has one to one equivalence with total spectral acceleration [S_a (ζ, ω_n)]. The concept is figuratively presented in Fig. 3.14.

Computation of this maximum value of response (displacement, velocity, and acceleration) has an important bearing on structural analysis and design, simply because, while carrying out an earthquake-resistant design, a design engineer is interested to know the maximum absolute value of these parameters the structure will be able to withstand rather than the full response history.

For instance, as per Eq. (3.5.7), $S_d(\zeta, \omega_n)$ gives the maximum value of relative amplitude of the structure under a particular ground acceleration $\ddot{u}_g(t)$.

Thus, the maximum strain energy of the of the body can be expressed as

$$E = \frac{1}{2} k S_d^2 \tag{3.5.18}$$

From Eq. (3.5.12), we can express this as

$$E = \frac{1}{2} k \left(\frac{S_{pv}}{\omega_n^2} \right)^2 \tag{3.5.19}$$

Considering $k = m\omega_n^2$, one can express Eq. (3.5.19) as $E = \frac{1}{2} k S_d^2 = \frac{1}{2} m S_{pv}^2 =$ kinetic energy. The energy of the system is thus conserved.

Thus, maximum force the body will experience can well be represented by

$$V = \frac{S_a(\zeta, \omega_{n)})}{g} W = \frac{S_{pa}(\zeta, \omega_n)}{g} W \tag{3.5.20}$$

Here, W = weight of the structure and g = acceleration due to gravity.

Thus, if we have sufficient number of earthquake data, then one can take an average envelope of the maximum response and for different time periods and damping can plot them and finally form a smoothened design spectra curve, as shown in Fig. (3.15).

As observed in Fig. 3.15, the magnitude of peak acceleration (i.e., the flat portion) varies with damping ratio. An increase in damping ratio reduces the magnitude of peak acceleration, as shown in Fig. 3.15.

The normalized curve, as furnished in Fig. 3.11, is usually drawn for standard 5% damping. For other values of damping ratio, the normalized curve can be adjusted by a scaling factor [Eurocode 8]

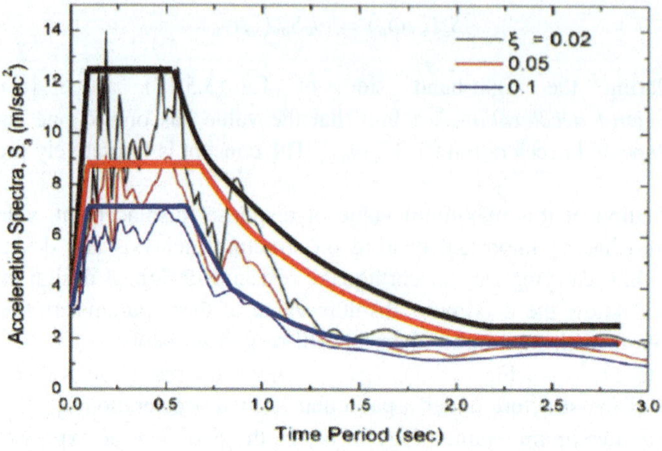

Fig. 3.15 Smoothened design spectra (acceleration) for structural design

$$SF = \sqrt{\frac{10}{5+\zeta}} \tag{3.5.21}$$

Here, SF = scaling factor and ζ = damping ratio of the system.

The response curve, especially the peak ground acceleration (PGA), i.e., a rigid body (having $T_n = 0$) moving with same acceleration as that of the ground is dependent on a number of issues like location of fault from a particular site, geological and soil condition, magnitude of earthquake.[7]

Thus, for big countries like USA, India, European countries, Iran, it is convenient to generate a normalized unified curve having $\mathrm{Lt}_{T_n \to 0} Sa_0 = 1g$. Here, Sa_0 is the peak spectral ground acceleration in the normalized curve. The local effects are then catered through the zone factor Z (refer Table 2.4) to arrive at the elastic response a particular structure having a given time period T_n at this site.

Thus, the S_a/g value is obtained from normalized curve when multiplied by zone factor Z, and the importance factor I (as explained in Chap. 2) is known as the *design spectral acceleration* of structure at a particular location.

Thus, net lateral response under *maximum considered earthquake* (MCE)[8] is expressed as

$$A_h = \frac{ZI}{R}\left(\frac{S_a}{g}\right) \tag{3.5.22}$$

[7]Refer Chap. 1 for details.

[8]For definition of MCE refer Chap. 2 Sect. 2.2.1.

Thus, for a moderate zone where $Z = 0.16$ considering $I/R = 1.0$, the acceleration to be expected by a rigid body resting on ground is $0.16g$. This value often amplifies significantly due to the presence of soft soil overlying the rock. This is known as local ground amplification and remains an important study in the domain of geotechnical earthquake engineering.

3.5.1 Plotting the Responses in Log Scale

The three response spectra, S_d, $S_{pa,}$ and S_{pv}, essentially give the same information. The three spectra are just different ways of presenting the structural response of a body and knowing one of them, the other spectra can easily be derived from the expression

$$S_d = \frac{S_{pv}}{\omega_\mathrm{n}} = \frac{S_{pa}}{\omega_\mathrm{n}^2} \tag{3.5.1.1}$$

In Eq. (3.5.1.1), each spectrum directly provides a physically meaningful quantity.

The deformation response spectrum (S_d) provides the peak deformation or displacement of a system.

The pseudo-velocity response spectrum (S_{pv}) is directly related to peak strain energy stored in the system during an earthquake, while the pseudo-acceleration response spectrum (S_{pa}) is directly related to the peak value of the equivalent lateral force and base shear-induced in the structure.

The three responses can be drawn together in a single plot (called tripartite plot) and remain a compact way of representing the spectra over a wide range of natural frequency and damping values, so that it provides peak response of all possible structures.

From Eq. (3.5.1.1), we have

$$S_{pv} = \frac{S_{pa}}{\omega_\mathrm{n}} \tag{3.5.1.2}$$

Considering $\omega_\mathrm{n} = 2\pi/T_n$, where T_n is time period of the system, we have

$$\frac{T_n}{2\pi} S_{pa} = S_{pv} \tag{3.5.1.3}$$

Taking log on both sides, we have

$$\log T_n + \log S_{pa} - \log 2\pi = \log S_{pv} \tag{3.5.1.4}$$

Considering $\log T_n = x$ and $\log S_{pv} = y$, we can express Eq. (3.5.1.4) as $y = x + \log S_{pa} - \log 2\pi \rightarrow y = x + \text{constant}$, which gives

$$\frac{\mathrm{d}y}{\mathrm{d}x} = +1 \tag{3.5.1.5}$$

This is the equation of a straight line having gradient +1 (or +45°). The intercept on Y-axis depends on S_{pa}. The S_{pa} line is at an inclination of +45° with the X-axis.

Again, from Eq. (3.5.1.1) we can write

$$S_{pv} = \frac{2\pi}{T_n} S_d \tag{3.5.1.6}$$

$$\log S_{pv} = \log 2\pi + \log S_d - \log T_n \tag{3.5.1.7}$$

Considering $\log T_n = x$, $\log S_{pv} = y$ and $\log S_d = \text{constant}$, we have

$$y = \log 2\pi + \log S_d - x$$

$y = -x + \text{constant}$, and this gives

$$\frac{\mathrm{d}y}{\mathrm{d}x} = -1 \tag{3.5.1.8}$$

This represents a straight line with slope -1, which means that the constant S_d line makes an angle of $-45°$ with the X-axis.

Figure 3.16 shows a typical tripartite plot of the spectrums for different damping values normalized to $1g$.

3.5.2 Computation of the Integral Having Random Acceleration Response $\ddot{u}_g(\tau)$

For an oscillator subjected to ground acceleration $\ddot{u}_g(\tau)$, like the one shown in Fig. 3.2, the equation of equilibrium is given by Eq. (3.5.2), whose solution is expressed as

$$u_s(t) = \frac{1}{\omega_\mathrm{d}} \int_0^t \ddot{u}_g(\tau) e^{-\zeta\omega_\mathrm{n}(t-\tau)} \sin \omega_\mathrm{d}(t-\tau) \mathrm{d}\tau \tag{3.5.2.1}$$

Considering that the ground acceleration $\ddot{u}_g(\tau)$ is random in nature, it is evident that the integration cannot be carried out analytically, and one has to resort to numerical analysis to find the values.

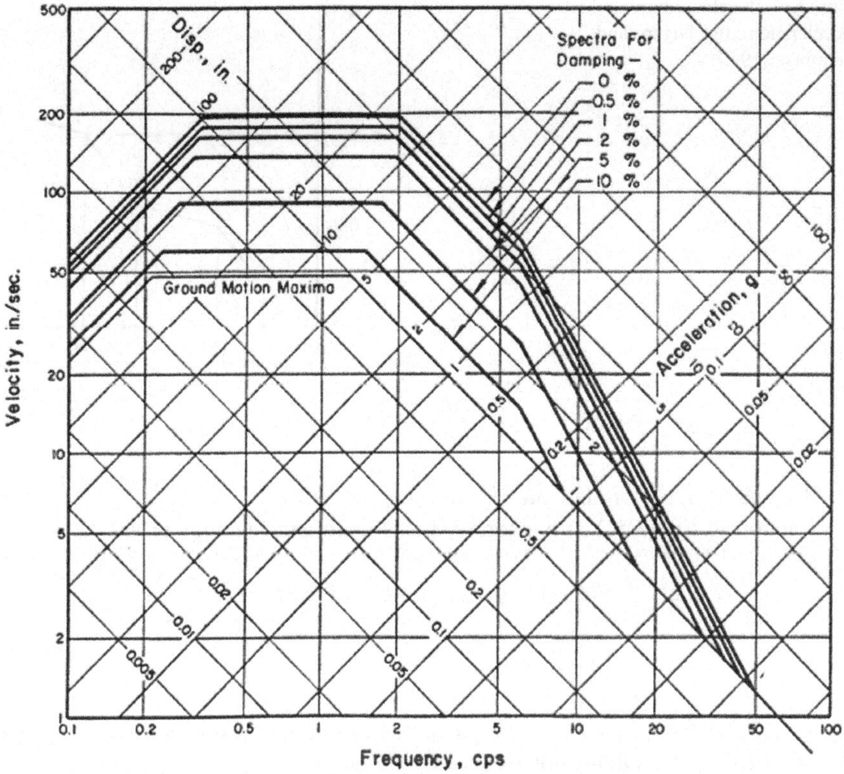

Fig. 3.16 Typical tripartite plot of the response spectrums after Newmark and Hall (1973)

This can be carried out either by any of the time history methods, as presented in Sect. 2.6 of Chap. 2, readjusted for a body having single degree of freedom, or the algorithm as proposed by Nigam and Jennings (1969) as elaborated below may be adopted.

For a body of single degree freedom subjected to acceleration $a(t)$, the equation of equilibrium can be expressed as

$$\ddot{x} + 2\zeta\omega_n\dot{x} + \omega_n^2 x = -a(t) \tag{3.5.2.2}$$

Here, $\zeta =$ damping ratio and ω_n = natural frequency of the body and $\omega_d = \omega_n\sqrt{1-\zeta^2}$ is the damped natural frequency of the body.

Assuming that $a(t)$ can be approximated by a segmental linear function, as shown in Fig. (3.17), Eq. (3.5.1.1) may be expressed as

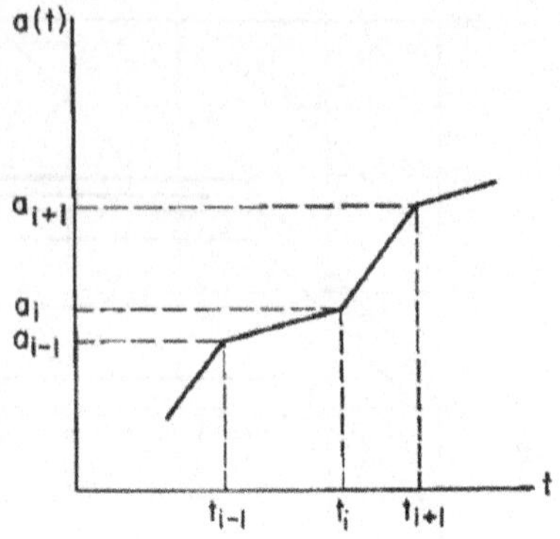

Fig. 3.17 Idealized base acceleration after Nigam and Jennings (1969)

$$\ddot{x}+2\zeta\omega_n\dot{x}+\omega_n^2x=-a_i+\frac{\Delta a_i}{\Delta t_i}(t-t_i)\quad \text{for } t\leq_i t\leq t_{i+1} \tag{3.5.2.3}$$

Here, $\Delta t_i = t_{i+1} - t_i$ and $\Delta a_i = a_{i+1} - a_i$.
Solution to Eq. (3.5.2.3) is expressed as Nigam and Jennings (1969)

$$\begin{aligned} x &= e^{\zeta\omega_n(t-t_i)}[C_1\sin\omega_d(t-t_i)+C_2\cos\omega_d(t-t_i)] \\ &\quad -\frac{a_i}{\omega_n^2}+\frac{2\zeta}{\omega_n^3}\frac{\Delta a_i}{\Delta t_i}-\frac{1}{\omega_n^2}\frac{\Delta a_i}{\Delta t_i}(t-t_i) \end{aligned} \tag{3.5.2.4}$$

In Eq. (3.5.2.4), C_1 and C_2 are integration constants. Considering $x = x_i$ and $\dot{x} = \dot{x}_i$ at $t = t_i$ and solving for C_1 and C_2 we get

$$C_1=\frac{1}{\omega_n\sqrt{1-\zeta^2}}\left(\zeta\omega_n x_i+\dot{x}_i-\frac{2\zeta^2-1}{\omega_n^2}\frac{\Delta a_i}{\Delta t_i}+\frac{\zeta}{\omega_n}a_i\right) \tag{3.5.2.5}$$

$$C_2=x_i-\frac{2\zeta}{\omega_n^3}\frac{\Delta a_i}{\Delta t_i}+\frac{a_i}{\omega_n^2} \tag{3.5.2.6}$$

Substituting the values of C_1 and C_2 in Eq. (4.5.2.4), it can be demonstrated that x and $\dot{x}$ at $t = t_{i+1}$ is given by

$$\hat{x}_{i+1}=A(\zeta,\omega_n,\Delta t_i)\hat{x}_i+B(\zeta,\omega_n,\Delta t_i)\hat{a}_i \tag{3.5.2.7}$$

where

$\hat{x}_i = \begin{Bmatrix} x_i \\ \dot{x}_i \end{Bmatrix}$ and $\hat{a}_i = \begin{Bmatrix} a_i \\ a_{i+1} \end{Bmatrix}$; the matrices A and B are expressed as

$$A=\begin{bmatrix} a_{11} & a_{12} \\ a_{21} & a_{22} \end{bmatrix} \quad \text{and} \quad B=\begin{bmatrix} b_{11} & b_{12} \\ b_{21} & b_{22} \end{bmatrix}$$

The elements of matrices A and B are given by

$$
\begin{aligned}
a_{11} &= \mathrm{e}^{-\zeta\omega_{\mathrm{n}}\Delta t_i}\left(\frac{\zeta}{\sqrt{1-\zeta^2}}\sin\omega_{\mathrm{d}}\Delta t_i + \cos\omega_{\mathrm{d}}\Delta t_i\right)\\
a_{12} &= \frac{\mathrm{e}^{-\zeta\omega_{\mathrm{n}}\Delta t_i}}{\omega_{\mathrm{n}}\sqrt{1-\zeta^2}}\sin\omega_{\mathrm{d}}\Delta t_i\\
a_{21} &= -\left(\frac{\omega_{\mathrm{n}}\mathrm{e}^{-\zeta\omega_{\mathrm{n}}\Delta t_i}}{\sqrt{1-\zeta^2}}\right)\sin\omega_{\mathrm{d}}\Delta t_i\\
a_{22} &= \mathrm{e}^{-\zeta\omega_{\mathrm{n}}\Delta t_i}\left(\frac{-\zeta}{\sqrt{1-\zeta^2}}\sin\omega_{\mathrm{d}}\Delta t_i + \cos\omega_{\mathrm{d}}\Delta t_i\right)
\end{aligned}
\tag{3.5.2.8}
$$

$$
\begin{aligned}
b_{11} &= \mathrm{e}^{-\zeta\omega_{\mathrm{n}}\Delta t_i}\left[\left(\frac{2\zeta^2-1}{\omega_{\mathrm{n}}^2\Delta t_i}+\frac{\zeta}{\omega_{\mathrm{n}}}\right)\frac{\sin\omega_{\mathrm{d}}\Delta t_i}{\omega_{\mathrm{n}}\sqrt{1-\zeta^2}}+\left(\frac{2\zeta}{\omega_{\mathrm{n}}^3\Delta t_i}+\frac{1}{\omega_{\mathrm{n}}^2}\right)\cos\omega_{\mathrm{d}}\Delta t_i\right]-\frac{2\zeta}{\omega_{\mathrm{n}}^3\Delta t_i}\\
b_{12} &= -\mathrm{e}^{\ \zeta\omega_{\mathrm{n}}\Delta t_i}\left[\left(\frac{2\zeta^2-1}{\omega_{\mathrm{n}}^2\Delta t_i}\right)\frac{\sin\omega_{\mathrm{d}}\Delta t_i}{\omega_{\mathrm{n}}\sqrt{1-\zeta^2}}+\frac{2\zeta}{\omega_{\mathrm{n}}^3\Delta t_i}\cos\omega_{\mathrm{d}}\Delta t_i\right]-\frac{1}{\omega_{\mathrm{n}}^2}+\frac{2\zeta}{\omega_{\mathrm{n}}^3\Delta t_i}\\
b_{21} &= \mathrm{e}^{-\zeta\omega_{\mathrm{n}}\Delta t_i}\left[\begin{array}{l}\left(\frac{2\zeta^2-1}{\omega_{\mathrm{n}}^2\Delta t_i}+\frac{\zeta}{\omega_{\mathrm{n}}}\right)\left(\cos\omega_{\mathrm{d}}\Delta t_i-\frac{\zeta}{\sqrt{1-\zeta^2}}\sin\omega_{\mathrm{d}}\Delta t_i\right)\\ -\left(\frac{2\zeta}{\omega_{\mathrm{n}}^3\Delta t_i}+\frac{1}{\omega_{\mathrm{n}}^2}\right)\left(\omega_{\mathrm{n}}\sqrt{1-\zeta^2}\sin\omega_{\mathrm{d}}\Delta t_i+\zeta\omega_{\mathrm{n}}\cos\omega_{\mathrm{d}}\Delta t_i\right)\end{array}\right]+\frac{1}{\omega_{\mathrm{n}}^2\Delta t_i}\\
b_{22} &= \mathrm{e}^{-\zeta\omega_{\mathrm{n}}\Delta t_i}\left[\begin{array}{l}\frac{2\zeta^2-1}{\omega_{\mathrm{n}}^2\Delta t_i}\left(\cos\omega_{\mathrm{d}}\Delta t_i-\frac{\zeta}{\sqrt{1-\zeta^2}}\sin\omega_{\mathrm{d}}\Delta t_i\right)\\ -\left(\frac{2\zeta}{\omega_{\mathrm{n}}^3\Delta t_i}\right)\left(\omega_{\mathrm{n}}\sqrt{1-\zeta^2}\sin\omega_{\mathrm{d}}\Delta t_i+\zeta\omega_{\mathrm{n}}\cos\omega_{\mathrm{d}}\Delta t_i\right)\end{array}\right]-\frac{1}{\omega_{\mathrm{n}}^2\Delta t_i}
\end{aligned}
\tag{3.5.2.9}
$$

From Eq. (3.5.2.3), one can see that the absolute acceleration

$$
\ddot{x}_i^{\mathrm{abs}} = \ddot{x}_i + a_i = -\left(2\zeta\omega_{\mathrm{n}}\dot{x}_i + \omega_{\mathrm{n}}^2 x_i\right) \tag{3.5.2.10}
$$

Thus, if the velocity and displacement of the body is known at some time t_0, then the state of the body at all subsequent time moments t_i can be computed exactly by following the mathematical steps as elaborated above.

The main computational advantage with this method is that the values of A and B are functions of ω_{n}, ζ and Δt_i. Considering that ω_{n} and ζ are constant during calculation of each spectrum value and if Δt_i is also a constant (that is, the accelerometer reads the input at equal spacing) then x_i, $\dot{x}_i$ and $\ddot{x}_i^{\mathrm{abs}}$ can be computed most efficiently.

Thus, the response spectra are furnished by

$$S_d(\zeta, \omega_n) = \text{Max}_{i=1,N}[x_i(\zeta, \omega_n)] \tag{3.5.2.11}$$

$$S_v(\zeta, \omega_n) = \text{Max}_{i=1,N}[\dot{x}_i(\zeta, \omega_n)] \tag{3.5.2.12}$$

$$S_a(\zeta, \omega_n) = \text{Max}_{i=1,N}[\ddot{x}_i^{\text{abs}}(\zeta, \omega_n)] \tag{3.5.2.13}$$

Here, S_d, S_v, S_a are the spectral values of displacement velocity and acceleration, respectively, for selected values of damping and natural frequency and N is the total number of discrete points at which the response is obtained. Nigam recommends a value of $\Delta t_i \leq T_n/20$ (T_n = natural time period) for discretization error to be less than 1.2%.

3.6 Amplification of Response Due to Soil Overlying Bedrock

For important structures, like nuclear power plants, bridges of national importance, studies are also carried out to determine site-specific spectra rather than following the generalized recommendations of code as furnished in Fig. 3.11.

Unless a particular site has an unusual configuration, like being a valley or confined by hills/ridges, one-dimensional analysis of wave propagation through elastic media gives an excellent estimate of ground motion at a particular site.

We present hereafter a few models that are popular around the world.

3.6.1 The SHAKE Method

Originally developed by Schnabel (1972), the technique is implemented in software called SHAKE, developed at University of California Berkeley, Department of Earthquake Engineering, and remains till date a popular technique adopted for linear analysis of ground response.

We elaborate the same hereafter for both a simple case and subsequently for a more complex one.

Shown in Fig. 3.18 is a soil layer extending to infinity horizontally and having depth H, resting on bedrock. The soil layer is considered to be un-damped. The differential equation of motion in this case is thus

$$\rho \frac{\partial^2 u}{\partial t^2} = G \frac{\partial^2 u}{\partial z^2} \tag{3.6.1.1}$$

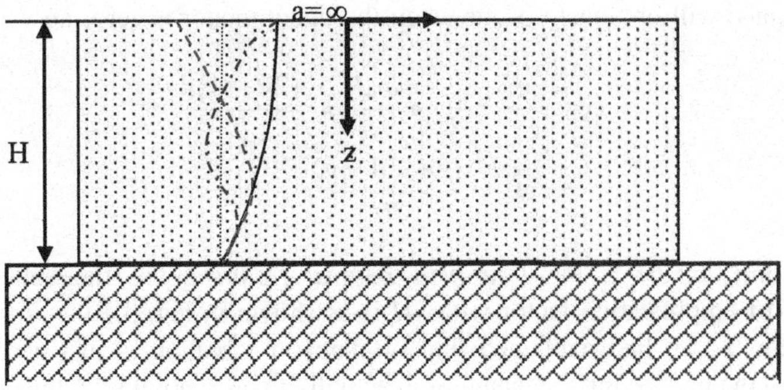

Fig. 3.18 Horizontal soil layer of depth H resting on a rigid rock

The solution to the above in a complex form is expressed as

$$u(z,t) = X_1 e^{i(\omega t + kz)} + X_2 e^{-i(\omega t - kz)} \tag{3.6.1.2}$$

where ω = circular frequency of ground motion, k = wave number ω/v_s, and v_s = shear wave velocity of the soil.

At the free surface, as the shear stress is equal to zero, we have

$$\tau(0,t) = G\frac{\partial u}{\partial z} = 0 \rightarrow Gik\left(X_1 e^{ik0} - X_2 e^{ik0}\right) = 0$$

It is evident that this can only be zero when $X_1 = X_2$; thus

$$u(z,t) = 2X_1\left(\frac{e^{ikz} + e^{-ikz}}{2}\right)e^{i\omega t} \tag{3.6.1.3}$$

Considering $X_1 = X$, Eq. (3.6.1.3) can be written as

$$u(z,t) = 2X_1 \cos kz e^{i\omega t} \tag{3.6.1.4}$$

Equation (3.6.1.4) can be used to define a transfer function that describes the ratio of displacement amplitude at any two points in the soil layer. Choosing these two points at top and bottom of the soil layer, we have the transfer function as

$$H(\omega) = \frac{u(0,t)}{u(H,t)} = \frac{1}{\cos kH} \tag{3.6.1.5}$$

Since the denominator in Eq. (3.6.1.5) cannot be more than 1 it shows that the surface motion at worst can be equal to the amplitude of the bedrock and for other values more than 1.

Resonance will obviously occur when the denominator is zero, i.e.,

$$\cos\frac{\omega_n H}{v_s} = 0 \rightarrow \frac{\omega_n H}{v_s} = \frac{(2n-1)\pi}{2}$$
$$\Rightarrow \omega_n = \frac{(2n-1)\pi v_s}{2H} \tag{3.6.1.6}$$

Equation (3.6.1.6) is the *free field natural frequency* of the soil medium. Variation of amplification factor with kH is as shown in Fig. 3.19.

In Chap. 2, we had explained that un-damped vibration is just a concept to understand the underlying mechanics and actually there is nothing in this world that is un-damped. In this case, also the soil medium does have its own damping that affects the response considerably.

The damped equation of motion is expressed as

$$\rho\frac{\partial^2 u}{\partial t^2} = G\frac{\partial^2 u}{\partial z^2} + \eta\frac{\partial^3 u}{\partial z^2 \partial t} \tag{3.6.1.7}$$

Considering $u = \phi e^{i\omega t}$, substitution in Eq. (3.6.1.7) gives

$$-\rho\omega^2\phi = (G + i\omega\eta)\frac{d^2\phi}{dt^2} \rightarrow -\rho\omega^2\phi = G^*\frac{d^2\phi}{dt^2}$$

Here, G^* is known as *complex shear modulus of soil*. Solution to the differential equation can thus be expressed as

$$u(z,t) = X_1 e^{i(\omega t + k*z)} + X_2 e^{-i(\omega t - k*z)} \tag{3.6.1.8}$$

Here, k^* is the complex wave number expressed as $k^* = \omega/v_s^*$, where $v_s^* = \sqrt{G^*/\rho}$ is the complex shear wave velocity of the soil.

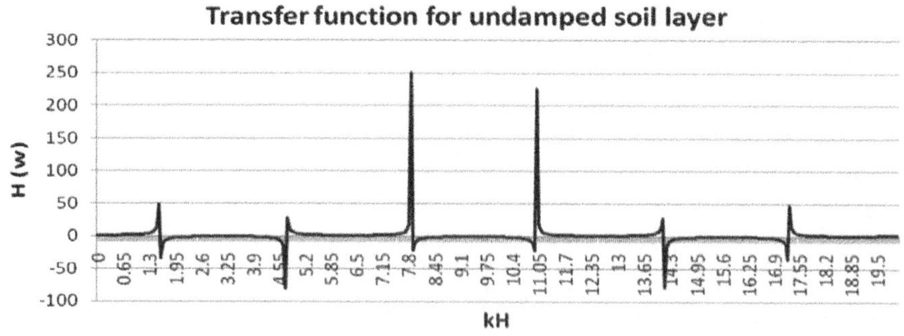

Fig. 3.19 Amplification of un-damped soil medium overlying the rock

It can be shown that considering the soil constitutive model as Kelvin and Voigt type, the viscosity factor $\eta = 2GD/\omega$, where D is the material damping ratio of the soil. Substituting this value of D we have $G^* = G(2 + iD)$. Now considering the expression $v_s^* = \sqrt{G^*/\rho}$, this can be approximated to

$$v_s^* = v_s(1 + iD) \tag{3.6.1.9}$$

At the free surface, as the shear stress must be zero, we have

$$\tau(0,t) = G\frac{\partial u}{\partial z} = 0 \rightarrow G^* ik^* \left(X_1 e^{ik0} - X_2 e^{ik0}\right) = 0$$

This results in $X_1 - X_2 = 0$ or $X_1 = X_2 = X$, which gives

$$u(z,t) = 2X \ \cos k^* z e^{i\omega t} \tag{3.6.1.10}$$

The transfer function is thus expressed as

$$H(\omega) = \frac{u(0,t)}{u(H,t)} = \frac{1}{\cos k^* H} \tag{3.6.1.11}$$

where

$$k^* = \frac{\omega}{v_s^*} = \frac{\omega}{v_s(1 + iD)} \tag{3.6.1.12}$$

$$\rightarrow k^* = \frac{\omega}{v_s}(1 + iD)^{-1} \approx \frac{\omega}{v_s}(1 - iD) = k(1 - iD) \tag{3.6.1.13}$$

The transfer function can now be expressed as

$$H(\omega) == \frac{1}{\cos k(1 - iD)H} = \frac{1}{\cos(kH - iDkH)} \tag{3.6.1.14}$$

Considering $a = kH$ and $b = -DkH$, we have $\cos(a + ib) = \cos^2 a + \sinh^2 b$ which gives

$$H(\omega) = \frac{1}{\sqrt{\cos^2 kH + \sinh^2 DkH}} \tag{3.6.1.15}$$

The variation of amplification factor for different damping ratios and kH is shown in Fig. 3.20.

From Fig. 3.20, it will be observed that when the bedrock is overlain by soil, there can be a significant amplification of acceleration at the ground level and,

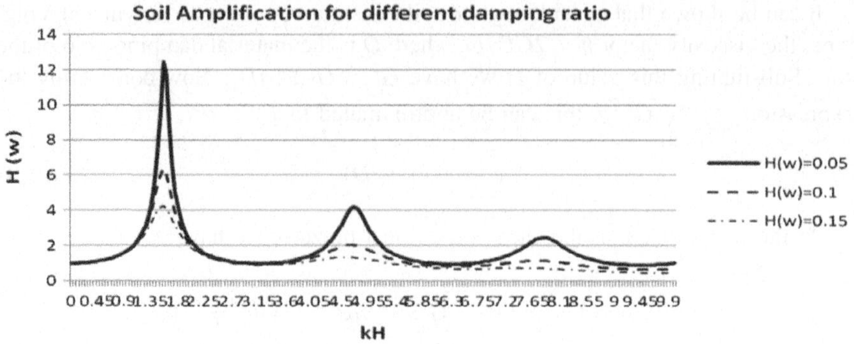

Fig. 3.20 Damped amplification of soil resting on a bedrock

therefore, even for a moderate earthquake at a site, structures located on hillocks or at a height are susceptible to more damage.

At resonance, due to inherent damping of soil, although there is magnification of response, yet the value remains finite.

The theory can well be extended to layered soil and the reader may refer to Kramer (1996) or SHAKE manual for further details.

Considering the analysis as presented above is in frequency domain, it is restricted to linear response only. As soil behavior is notoriously nonlinear, especially under large strain, some engineers have remained quite skeptical about its output, despite its popularity in industry. Many researchers perceive that the above technique just remains a qualitative analysis of the soil behavior as it does take into cognizance the stiffness degradation and damping augmentation of soil that is usually observed when soil undergoes large strains under strong ground motion.

Example 3.5 A soil stratum of 25 m thickness that lies over bedrock has an average shear wave velocity of 245 m/s. Compute the amplification at the ground level for maximum considered earthquake for the first three modes. Consider material damping of soil as 15%.

Solution

The general equation for time period is expressed as $T = \frac{4H}{(2n-1)v_s}$, $\omega = 2\pi/T$ and $k = \omega/v_s$, which gives for the first three modes the values listed in the table below.

Mode (n)	1	2	3
T (s)	0.408	0.136	0.082
ω (rad/s)	15.394	46.181	76.969
k	0.063	0.188	0.314
kH	1.571	4.712	7.854

Substituting the values in Eq. (3.6.1.15), considering $D = 0.15$, we have $H(15.4) = 4.32$, $H(46.2) = 1.54$ and $H(76.9) = 1.082$.

3.6.2 *Idriss and Seed's (1968) Model*

This method is based on time domain analysis, when the equation of equilibrium is expressed as

$$\rho(z)\frac{\partial^2 u}{\partial t^2} + c(z)\frac{\partial u}{\partial t} - G(z)\frac{\partial^2 u}{\partial z^2} = -\rho(z)\frac{\partial^2 u_g}{\partial t^2} \quad (3.6.2.1)$$

where $\rho(z)$ = mass density of soil in vertical z-direction, $c(z)$ = damping of soil in z-direction, $G(z)$ = dynamic shear modulus of soil varying with depth as Gz^{α}, u = relative displacement of ground with respect to bedrock, u_g = displacement at bedrock level.

Considering $u(z,t)\sum_{n=1}^{\infty}\phi_n(z)\psi_n(t)$ and applying the technique of separation of variables, the above second-order hyperbolic partial differential equation can be broken up into two linear differential equations whose solutions are given by

$$\phi_n(z) = \left(\frac{\beta_n}{2}\right)^b \Gamma(1-b)\left(\frac{z}{H}\right)^{b/\theta} J_{-b}\left[\beta_n\left(\frac{z}{H}\right)^{1/\theta}\right] \quad \text{and} \quad (3.6.2.2)$$

$$\ddot{\psi}_n(t) + 2D\omega_n\dot{\psi}_n(t) + \omega_n^2\psi_n(t) = -P_n\ddot{u}_g \quad (3.6.2.3)$$

Here, J_{-b} is Bessel's function of first kind of order $-b$, β_n represents the roots of the equation and $\beta_n = 1.7510, 4.8785, 8.0166$ and 11.1570 for $n = 1, 2, 3$ and 4, respectively. Γ represents the Gamma function.

Here, the circular natural frequency ω_n is given by

$$\omega_n = \frac{\beta_n\sqrt{G/\rho}}{(\theta H)^{1/\theta}} \quad (3.6.2.4)$$

$$D = \frac{c}{2\rho\omega_n} \quad \text{and} \quad P_n = \left[\left(\frac{\beta_n}{2}\right)^{1+b}\Gamma(1-b)J_{1-b}(\beta_n)\right]^{-1} \quad (3.6.2.5)$$

Here, $\theta = 2/(2-\alpha)$ and $b = (\alpha-1)/(\alpha-2)$.

For calculating the displacement, the steps to be followed can be summarized as follows:

- Find out α, G, H, D and c.
- Determine ϕ_n for various modes—usually the fundamental mode suffices.
- Perform time history response analysis for a given bedrock earthquake data to find out $\psi_n(t)$. This may be carried out by adopting any of the methods as presented in Sect. 2.6 of Chap. 2.
- The time history analysis will give $u(z,t), \dot{u}(z,t)$ and $\ddot{u}(z,t)$
- The absolute total displacement, velocity, and acceleration are obtained as $u_{tot}(z,t) = u(z,t) + u_g(z,t)$, $\dot{u}_{tot}(z,t) = \dot{u}(z,t) + \dot{u}_g(z,t)$ and $\ddot{u}_{tot}(z,t) = \ddot{u}(z,t) + \ddot{u}_g(z,t)$

For cohesionless soil, considering $\alpha = 1/3$, Idriss and Seed have given the following expression for sandy soil.

$$\phi_n(z) = \left(\frac{\beta_n}{2}\right)^{0.4} \Gamma(0.6) \left(\frac{z}{H}\right)^{1/3} J_{-0.4}\left[\beta_n \left(\frac{z}{H}\right)^{5/6}\right] \tag{3.6.2.6}$$

$$\ddot{\psi}_n(t) + 2D\omega_n\dot{\psi}_n(t) + \omega_n^2\psi_n(t) = -\ddot{u}_g\left[\left(\frac{\beta_n}{2}\right)^{1.4} \Gamma(0.6) J_{0.6}(\beta_n)\right]^{-1} \tag{3.6.2.7}$$

where

$$\omega_n = \frac{\beta_n v_s}{1.2H^{5/6}} \tag{3.6.2.8}$$

For a pure cohesive soil, when $\alpha = 0$, the above gets modified to

$$\phi_n(z) = \cos\left[\frac{(2n-1)\pi z}{2H}\right] \tag{3.6.2.9}$$

when

$$\ddot{\psi}_n(t) + 2D\omega_n\dot{\psi}_n(t) + \omega_n^2\psi_n(t) = (-1)^n \ddot{u}_g\left[\frac{4}{(2n-1)\pi}\right] \tag{3.6.2.10}$$

where $\omega_n = \frac{(2n-1)\pi v_s}{2H}$.

The advantage with time domain analysis is that it is possible to carry out both linear and nonlinear analyses within the same framework.

3.6.3 Chowdhury and Dasgupta's (2007, 2013) Model

Though developed for a different application (Refer Chaps. 4, 6 and 8), a mathematical model developed by Chowdhury and Dasgupta (2007, 2013) can also be used to compute the ground response and local soil amplification.

When vertically propagating S_v waves travel through the soil from bedrock level, it is apparent that the soil medium over it will get excited and starts vibrating. The general character of such mode shapes of vibrations is as shown in Fig. 3.18 by the curves ϕ_1, ϕ_2, ϕ_3, etc.

Referring to Fig. 3.18 in two dimensions, the propagation of S_v waves through the medium can be expressed as

$$\frac{\partial^2 u(x,z,t)}{\partial x^2} + \frac{\partial^2 u(x,z,t)}{\partial z^2} = \frac{1}{Vs^2}\frac{\partial^2 u(x,z,t)}{\partial t^2} \tag{3.6.3.1}$$

where V_s = shear wave velocity of the soil medium; $u(x, z, t)$ = the displacement function and can be considered as $u = H(x) \cdot Q(z) \cdot P(t)$ [H, P, Q are the three independent functions of x, z, and t, respectively].

It can be shown that Eq. (3.6.3.1) can be broken up into three ordinary differential equations of second order, given by

$$\frac{d^2 P}{dt^2} + \lambda^2 P = 0 \tag{3.6.3.2}$$

where $\lambda = iVs$, with i being a constant

$$\frac{d^2 H(x)}{dx^2} + k^2 H(x) = 0, \tag{3.6.3.3}$$

where k is another constant

$$\frac{d^2 Q(z)}{dz^2} + p^2 Q(z) = 0 \tag{3.6.3.4}$$

where p, i, and k are related through $p^2 = i^2 - k^2$.

The solution of Eqs. (3.6.3.3) and (3.6.3.4) is given by

$$H(x) = A\cos kx + B\sin kx \tag{3.6.3.5}$$

$$Q(z) = C\cos pz + D\sin pz \tag{3.6.3.6}$$

Imposing the boundary conditions

At $x = 0$, $u = 0 \rightarrow H(x) = 0$, which implies $A = 0$.

At $x = a$ (where a may be very large), $u = 0 \rightarrow H(a) = 0$ which implies $H(a) = B\sin ka = 0$

$$\rightarrow k = \frac{m\pi}{a} \tag{3.6.3.7}$$

and hence

$$H_m(x) = \sin\frac{m\pi x}{a}. \tag{3.6.3.8}$$

At the free surface, i.e., where $z = 0$ (Fig. 3.18), the boundary conditions are
At $z = 0$, the shear strain $\frac{\partial u}{\partial z} = 0$ or $\frac{\mathrm{d}Q(z)}{\mathrm{d}z} = 0$, which implies $D = 0$.
At $z = H$, displacement $u = 0$, i.e. $Q(H) = 0$.
It implies that

$$p = \frac{(2n-1)\pi}{2H} \tag{3.6.3.9}$$

and hence

$$Q(z) = \cos\frac{(2n-1)\pi z}{2H} \tag{3.6.3.10}$$

The eigenvectors of the problem can thus be established as

$$\phi(x,z) = H(x)Q(z) = \sin\frac{m\pi x}{a}\cos\frac{(2n-1)\pi z}{2H} \tag{3.6.3.11}$$

where m, n = 1, 2, 3,…
Again, from the description of Eqs. (3.6.3.2) and (3.6.3.3)

$$\lambda = V_s\sqrt{p^2 + k^2}$$

Substituting the value of p and k from Eq. (3.6.3.9) and (3.6.3.7), one can have $\lambda = V_s\pi\sqrt{\frac{m^2}{a^2} + \frac{(2n-1)^2}{4H^2}}$.
For the fundamental mode, considering that $m, n = 1$ and lim $a \rightarrow \infty$, the value of λ reduces to

$$\lambda = \omega = \pi V_s\sqrt{0 + \frac{1}{4H^2}} \quad \text{and}$$
$$\omega = \frac{\pi V_s}{2H} \tag{3.6.3.12}$$

The period T can be derived from Eq. (3.6.3.12) as $T = 4H/V_s$ which is basically the free field time period in one dimension for the site.
For lim $a \rightarrow \infty$, the first term of eigenfunction (in the x-direction) can be dropped in Eq. (3.6.3.11), which gives the eigenfunction in one dimension as

$$\phi(z) = \cos\frac{(2n-1)\pi z}{2H} \tag{3.6.3.13}$$

In this case, the eigenvalues vis-à-vis free field time period and eigenvectors are derived by direct solution of the differential equation of motion of the wave propagation in two dimensions, for G constant with depth.

However, the same can also be derived from energy equation as furnished hereafter.

The strain energy equation of a soil body, in general, is given by Timoshenko and Goodier (1983)

$$U - \frac{Ae^2}{2} + G\left(\varepsilon_x^2 + \varepsilon_y^2 + \varepsilon_z^2\right) + \frac{G}{2}\left(\gamma_{xy}^2 + \gamma_{yz}^2 + \gamma_{xz}^2\right) \tag{3.6.3.14}$$

where U = strain energy density of the soil body; $\Lambda = 2Gv/(1-2v)$; G = dynamic shear modulus of the soil medium and v its Poisson's ratio; $e = \varepsilon_x + \varepsilon_y + \varepsilon_z$; $\varepsilon_{x,y,z}$ = strain in the x-, y- and z-direction, and $\gamma_{xy,yz,zx}$ = shear strains in the xy, yz, and zx planes, respectively.

With reference to Fig. 3.18 and assuming the condition of plane strain, Eq. (3.6.3.14) can be rewritten as

$$U = \frac{Gv}{1-2v}(\varepsilon_x + \varepsilon_z)^2 + G\left(\varepsilon_x^2 + \varepsilon_z^2\right) + \frac{G}{2}\left(\gamma_{xz}^2\right) \tag{3.6.3.15}$$

For impulsive seismic response, $\varepsilon_z = 0$ which reduces Eq. (3.6.3.15) further to

$$U = \frac{G(1-v)}{1-2v}\varepsilon_x^2 + \frac{G}{2}\gamma_{xy}^2 \tag{3.6.3.16}$$

Considering $u(x, z) = \phi(x, z), q(t)$ and $\varepsilon_r = \partial u/\partial x$ and $\gamma_{xy} = \partial u/\partial z$ one can have

$$\frac{\partial U}{\partial q_r} = \frac{2G(1-v)}{1-2v}\frac{\partial u}{\partial x}\frac{\partial}{\partial q_r}\left(\frac{\partial u}{\partial x}\right) + G\frac{\partial u}{\partial z}\frac{\partial}{\partial q_r}\left(\frac{\partial u}{\partial z}\right)$$

That is

$$\frac{\partial U}{\partial q_r} = \frac{2G(1-v)}{1-2v}\frac{\partial \phi_i}{\partial x}\frac{\partial \phi_r}{\partial x}q_i q_r + G\frac{\partial \phi_i}{\partial z}\frac{\partial \phi_r}{\partial z}q_i q_r \tag{3.6.3.17}$$

Here

$\phi(x, z)$ generalized shape function with respect to x and z coordinate;
$q(t)$ displacement function with respect to time in generalized coordinate.

From which it can be proved that stiffness and mass matrices can be written as

$$K_{ir} = \int_0^H \int_0^a \left[\frac{2G(1-2v)}{1-2v} \frac{\partial \phi_i}{\partial x} \frac{\partial \phi_r}{\partial x} + G \frac{\partial \phi_i}{\partial z} \frac{\partial \phi_r}{\partial z} \right] \mathrm{dx} \cdot \mathrm{dz} \tag{3.6.3.18}$$

and

$$M_{ir} = \frac{\gamma_s}{g} \int_0^H \int_0^a \phi_i \phi_r \mathrm{dx} \cdot \mathrm{dz} \tag{3.6.3.19}$$

where K = stiffness matrix of the soil medium; M = mass matrix of the soil medium; i and r are different modes 1,2,3…

K and M for the fundamental mode are then given by

$$K_{11} = \int_0^H \int_0^a \left[\frac{2G(1-v)}{1-2v} \left(\frac{\partial \phi}{\partial x} \right)^2 + G \left(\frac{\partial \phi}{\partial z} \right)^2 \right] \mathrm{dx} \cdot \mathrm{dz} \tag{3.6.3.20}$$

$$M_{11} = \frac{\gamma_s}{g} \int_0^H \int_0^a (\phi)^2 \mathrm{dx} \cdot \mathrm{dz} \tag{3.6.3.21}$$

It was shown earlier that when lim $a \to \infty$, the first term can be dropped and Eqs. (3.6.3.20) and (3.6.3.21) reduced to

$$K_{11} = \int_0^H \left[G \left(\frac{\partial \phi}{\partial z} \right)^2 \right] \cdot \mathrm{dz} \tag{3.6.3.22}$$

$$M_{11} = \frac{\gamma_s}{g} \int_0^H (\phi)^2 dz \tag{3.6.3.23}$$

Considering the shape function as given in Eq. (3.6.3.13) as $\phi(z) = \cos \frac{\pi z}{2H}$ (i.e. for n = 1) and substituting it in Eqs. (3.6.3.22) and (3.6.3.23) for a constant G value and by integrating we have

$$K_{11} = \frac{\pi^2 G}{8H} \tag{3.6.3.24}$$

and

$$M_{11} = \frac{\gamma_s H}{2g} \tag{3.6.3.25}$$

Considering $T = 2\pi\sqrt{M/K}$ substituting Eqs. (3.6.3.24) and (3.6.3.25), one can arrive at the same expression as $T = 4H/V_s$ derived earlier. This shows the correctness of the energy principle as adopted herein.

Thus, for multi-degrees of freedom, the stiffness and mass matrix can be expressed as

$$[K] = \int_0^H \left[G\left(\frac{\partial\phi_i}{\partial z}\right)\left(\frac{\partial\phi_r}{\partial z}\right)\right] \cdot \mathrm{d}z \tag{3.6.3.26}$$

$$[M] = \frac{\gamma_s}{g}\int_0^H (\phi_i)(\phi_r)\mathrm{d}z \tag{3.6.3.27}$$

Thus, for the first four modes, expanding Eqs. (3.6.3.26) and (3.6.3.27) gives

$$[K] = \frac{G}{H}\begin{bmatrix} \frac{\pi^2}{8} & 0 & 0 & 0 \\ 0 & \frac{9\pi^2}{8} & 0 & 0 \\ 0 & 0 & \frac{25\pi^2}{8} & 0 \\ 0 & 0 & 0 & \frac{49\pi^2}{8} \end{bmatrix} \tag{3.6.3.28}$$

$$[M] = \frac{\gamma_s H}{g}\begin{bmatrix} \frac{1}{2} & 0 & 0 & 0 \\ 0 & \frac{1}{2} & 0 & 0 \\ 0 & 0 & \frac{1}{2} & 0 \\ 0 & 0 & 0 & \frac{1}{2} \end{bmatrix} \tag{3.6.3.29}$$

Solving for the eigenvalue λ from $[K] - \lambda[M] = 0$ and knowing $\lambda = \omega^2$, where $T = 2\pi/\omega$, we finally have

$$\{T\} = \frac{H}{V_s}\{4 \quad 1.333 \quad 0.8 \quad 0.571\}^{\mathrm{T}} \tag{3.6.3.30}$$

The corresponding eigenvectors $[\phi]$ are obtained as

$$[\phi] = \begin{bmatrix} 1 & 0 & 0 & 0 \\ 0 & 1 & 0 & 0 \\ 0 & 0 & 1 & 0 \\ 0 & 0 & 0 & 1 \end{bmatrix} \tag{3.6.3.31}$$

Considering modal analysis, the amplitude of displacement can be expressed as

$$S_d = \frac{S_a}{\omega^2} \tag{3.6.3.32}$$

Based on codal definition, the amplitude of vibration may be expressed as

$$u_n = \kappa_n \beta \frac{S_{an}}{\omega^2} \cos(2n - 1) \frac{\pi z}{2H} \tag{3.6.3.33}$$

Thus

$$u_n \times \omega^2 = \ddot{u}_n = \kappa_n \beta S_{an} \tag{3.6.3.34}$$

Here $\ddot{u}_n =$ acceleration at ground surface ($z = 0$) and S_{an} = acceleration at bedrock level ($z = H$)

Here, κ_n is the modal mass participation factor and is expressed as $\kappa_n = \sum_{i=1}^{n} m_i\phi_i / \sum_{i=1}^{n} m_i\phi_i^2$.

For the first four modes, this is expressed as shown in Table 3.1.

Here, β is a code factor expressed as $\beta = ZI/2R$, where Z = zone factor, I = importance factor, and R = response reduction factor.

Thus, the ground amplification (Ω_n) can be expressed as

$$\Omega_n = \frac{\ddot{u}_n}{S_{an}} = \kappa_n \tag{3.6.3.35}$$

Properties of Gibson-Type Soil

As soil is notoriously heterogeneous in nature, it is only in case of normally consolidated clay that G remains constant with depth. For cohesionless soil, G is often found to vary with depth as $G = G_0(z/H)$, where G varies from 0 at $z = 0$ to G_0 at depth H. This type of soil is often called as Gibson-type soil.

Table 3.1 Modal mass participation first four modes

Mode	1	2	3	4
κ	$8/(\pi + 2)$	$-8(3\pi - 2)$	$8/(5\pi + 2)$	$-8/(7\pi - 2)$

Adopting this type of soil in differential equation, as furnished in Eq. (4.7.1), changes the properties of the partial differential equation giving rise to Bessel's function of first type of order 0, where $J_0(\beta_n z)$ are the eigenvectors of the problem. Though this is solvable quite easily, there are other types of models, like $G = G_0(z/H)^2$, $G = G_0(1 + z/H)$, $G = G_0(1 + z/H)^2$, $G = G_0(z/H)^{0.5}$, where it becomes extremely difficult to solve the equations analytically as they become inordinately complex or even not possible to handle (as no analytical solution exists for some of them).

Thus, to circumvent this problem and yet arrive at a realistic result, we approach the problem as elaborated hereafter.

Considering Eq. (3.6.3.13) as the assumed shape function will generically satisfy the static equation of a cantilever shear beam expressed as

$$G_0\left(\frac{z}{H}\right)A\left(\frac{\mathrm{d}^2u}{\mathrm{d}z^2}\right) = q \tag{3.6.3.36}$$

$$\Rightarrow GA\left(\frac{\mathrm{d}^2u}{\mathrm{d}z^2}\right) = q \tag{3.6.3.37}$$

where $G = G_0(z/H)$ and A = area of cross-section of the shear beam and q = the externally applied load.

If ideally by Eq. (3.6.3.13) $u = u_0 \cdot \cos(2n - 1)\pi z/2H$ is the exact solution of Eq. (3.6.3.37), it will yield

$$GA\left(\frac{\mathrm{d}^2u}{\mathrm{d}z^2}\right) - q = 0 \tag{3.6.3.38}$$

However, since it is not, it will have a residual error Re expressed as

$$\mathrm{Re} = GA\left(\frac{\mathrm{d}^2u}{\mathrm{d}z^2}\right) - q \tag{3.6.3.39}$$

Now as per Galerkinian basis of weighted residual method, we minimize the error over the domain that will give

$$\int_0^H GA\left(\frac{\mathrm{d}^2u}{\mathrm{d}z^2}\right)\phi_j(z)\mathrm{d}z - \int_0^H q\phi_j(z)\mathrm{d}z = 0 \tag{3.6.3.40}$$

where $u = u_0 \cdot \phi_i(z)$ and $\phi_i(z)$ is as defined in Eq. (3.6.3.13).

Integrating Eq. (3.6.3.40) by parts, we have

$$\int_0^H GA\frac{\mathrm{d}}{\mathrm{d}z}\left(\frac{\mathrm{d}u}{\mathrm{d}z}\right)\phi_j(z)\mathrm{d}z - \int_0^H q\phi_j(z)\mathrm{d}z = 0$$
$$\rightarrow \int_0^H GA\frac{\mathrm{d}u}{\mathrm{d}z}\phi_j(z) - \int_0^H \frac{\mathrm{d}}{\mathrm{d}z}\left(GA\frac{\mathrm{d}u}{\mathrm{d}z}\right)\phi_j(z)\mathrm{d}z = \int_0^H q\phi_j(z)\mathrm{d}z \tag{3.6.3.41}$$

The first term in Eq. (3.6.3.41) depicts the shear V_b in beam, which gives

$$-\int_0^H \frac{\mathrm{d}}{\mathrm{d}z}\left(GA\frac{\mathrm{d}u}{\mathrm{d}z}\right)\phi_j(z)\mathrm{d}z = \int_0^H q\phi_j(z)\mathrm{d}z - V_b \tag{3.6.3.42}$$

Now considering $u = u_0 \cdot \phi_i(z)$, we have

$$-\int_0^H \frac{\mathrm{d}}{\mathrm{d}z}\left(GA\frac{\mathrm{d}\phi_i(z)}{\mathrm{d}z}\right)\phi_j(z)\mathrm{d}z \cdot u_0 = \int_0^H q\phi_j(z)\mathrm{d}z - V_b \tag{3.6.3.43}$$

From the above, it is apparent that based on Galerkinian basis of weighted residual method, the stiffness of the shear beam can be expressed as

$$[K] = -\int_0^H \frac{\mathrm{d}}{\mathrm{d}z}\left(GA\frac{\mathrm{d}\phi_i(z)}{\mathrm{d}z}\right)\phi_j(z)\mathrm{d}z \tag{3.6.3.44}$$

$$\rightarrow [K] = -\int_0^H [GA\phi_i'(z)]'\phi_j(z)\mathrm{d}z \tag{3.6.3.45}$$

It may be observed that the stiffness matrix derived herein is different from Eq. (3.6.3.26) which is based on the Rayleigh–Ritz basis.

For mass matrix based on Galerkinian basis, the matrix remains same as expressed by Eq. (3.6.3.27).

For soil type $G = G_0(z/H)$, considering $\phi = \cos(2n-1)\frac{\pi z}{2H}$ as the assumed shape function, for $n = 1, 2, 3\ldots$, also satisfies the boundary condition of the equation (it is to be remembered that the actual solution is $\phi = J_0(\beta_n z/H)$, as discussed above).

$$G_0\left(\frac{z}{H}\right)\left(\frac{\mathrm{d}^2u}{\mathrm{d}z^2}\right) = \rho\frac{\partial^2 u}{\partial t^2} \tag{3.6.3.46}$$

Now applying Galerkin's basis to derive the stiffness matrix, vide Eq. (3.6.3.45), we have

$$[K] = \frac{G}{H}\begin{bmatrix} \frac{\pi^2}{16}+\frac{1}{4} & \frac{-3}{4} & \frac{5}{36} & \frac{-7}{36} \\ \frac{-3}{4} & \frac{9\pi^2}{16}+\frac{1}{4} & \frac{-15}{4} & \frac{21}{100} \\ \frac{5}{36} & \frac{-15}{4} & \frac{25\pi^2}{16}+\frac{1}{4} & \frac{-35}{4} \\ \frac{-7}{36} & \frac{21}{100} & \frac{-35}{4} & \frac{49\pi^2}{16}+\frac{1}{4} \end{bmatrix} \quad (3.6.3.47)$$

The mass matrix remains same as Eq. (3.6.3.29).

Solving for the eigenvalue λ from $[K]-\lambda[M]=0$ and knowing $\lambda = \omega^2$, where $T = 2\pi/\omega$, we finally have

$$\{T\} = \frac{H}{V_s}\{5.19 \quad 2.133 \quad 1.228 \quad 0.755\}^{\mathrm{T}} \quad (3.6.3.48)$$

The scaling factors for the eigenvectors are expressed as

$$\{\phi\} = \begin{bmatrix} -0.971 & -0.534 & 0.078 & 0.021 \\ -0.217 & 0.723 & -0.701 & -0.125 \\ -0.086 & 0.39 & 0.562 & 0.668 \\ -0.046 & 0.199 & 0.432 & -0.733 \end{bmatrix} \quad (3.6.3.49)$$

Considering that Eq. (4.7.59) was derived based on assumed shape function $\phi = \cos(2n-1)\frac{\pi z}{2H}$ in lieu of $\phi = J_0(\beta_n z/H)$ (which is exact), it would be enlightening to compare the values of time period coefficient $T = C_T(H/V_s)$ by both the methods, as shown hereafter (Table 3.2).

The results are found to be in excellent agreement, having error less than 0.5% for first three modes. Considering fundamental mode to be most critical, we can surely argue that this can well be acceptable for practical design work. Advantage with approaching this problem based on Galerkin's weighted residual technique is that for other types of soil ($G = G_0(1 + z/H)$, $G = G_0(z/H)^{0.5}$, $G = G_0(z/H)^2$ etc.), for which no exact solution exists, we can come to a realistic solution.

As elaborated in Eq. (3.6.3.33), the amplitude of displacement can be expressed in this case as

Table 3.2 Error in value C_T exact versus Galerkin's

Mode	1	2	3	4
$\phi = J_0(\beta_n z/H)$	5.19	2.14	1.232	0.769
$\phi = \cos(2n-1)\pi z/2H$	5.19	2.133	1.228	0.755
Error (%)	0	0.32%	0.32%	1.8%

$$u_n = \kappa_n \beta \frac{S_{an} T_n^2}{4\pi^2} [\phi_n]^{\mathrm{T}} \cos(2n-1) \frac{\pi z}{2H} \tag{3.6.3.50}$$

Here, κ_n does not change and remains the same as the values derived earlier for G constant with depth; $[\phi_n]$ is the scale factor of the eigenvectors, as derived in Eq. (3.6.3.49).

Equation (3.6.3.50) can be further expressed for the fundamental mode as

$$u_1 = \frac{8}{\pi + 2} \gamma_s \beta \left(\frac{S_{a1}}{g} \right) \frac{(5.19)^2 H^2}{4\pi^2 G} \begin{bmatrix} -0.971 \cdot \cos \dfrac{\pi z}{2H} \\ -0.217 \cdot \cos \dfrac{3\pi z}{2H} \\ -0.086 \cdot \cos \dfrac{5\pi z}{2H} \end{bmatrix} \tag{3.6.3.51}$$

Proceeding in identical fashion as stated in Eq. (3.6.3.35) one has $\Omega_n = 1.274\left(\frac{8}{2+\pi}\right)$ for the fundamental mode.

Proceeding in identical fashion, considering $T = C_T(H/Vs)$, values of C_T for different soil types are presented in Table 3.3.

Computation of Damping Matrix

For dynamic equilibrium, we have shown in Chap. 2 that

$$\begin{aligned} &[\mathbf{K}] - [\mathbf{M}]\omega_i^2 = 0 \\ &\rightarrow [\mathbf{K}] = [\mathbf{M}]\omega_i^2 \end{aligned} \tag{3.6.3.52}$$

Now considering $c = Dc_c$, where c = damping of soil and c_c is critical damping = $2\sqrt{km}$, we have

$$[\mathbf{C}] = 2D[\mathbf{M}]\omega_i \tag{3.6.3.53}$$

Now, substituting the value of $[\mathbf{M}]$ from Eq. (3.6.3.29) we have

$$[\mathbf{C}] = 2D \frac{\gamma_s H}{2g} [\mathbf{I}] \omega_i \tag{3.6.3.54}$$

where $[\mathbf{I}]$ = identity matrix having diagonal elements as 1 and all off-diagonal terms as zero.

Example 3.6 A sandy soil stratum of 32 m thickness has an average shear wave velocity of 450 m/sec, which is overlying the bedrock. The shear modulus of soil is found to vary as $G(z) = G_0(z/H)$. The recorded acceleration at bedrock level (whose digitized data is available) is as shown in the figure below. Determine the response

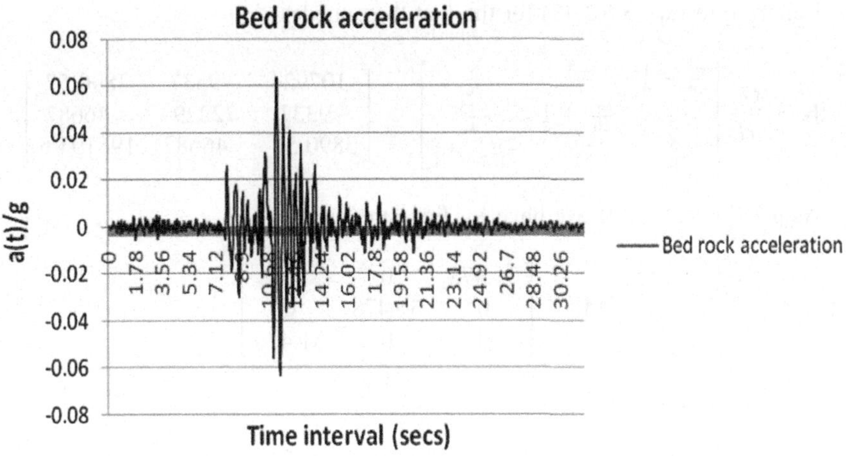

Fig. 3.21 Recorded acceleration at bedrock level

Table 3.3 Values of C_T for different soils for the first four modes

Mode				
	1	2	3	4
$G = G_0(z/H)$	5.19	2.133	1.228	0.755
$G = G_0(z/H)^{0.5}$	4.486	1.668	1.005	0.679
$G = G_0(z/H)^2$	7.826	3.301	1.66	0.867
$G = G_0(1 + z/H)$	3.094	1.095	0.66	0.462
$G = G_0(1 + z/H)^2$	2.421	0.906	0.545	0.361

at the ground level considering density of soil as 19.3 kN/m^3 and soil damping ratio as 15% (Fig. 3.21).

Solution

Considering that G varies linearly with depth and referring to Table 3.3 we have C_T = 5.19, 2.133, 1.228 for the first three modes.

Thus, $G_0 = \left(\frac{19.3}{9.81}\right) \times 450^2 = 398395\ \text{kN/m}^2$.

Thus, time period ($C_T H/V_s$) and natural frequency ($2\pi/T$) for the first three modes are as expressed below

Mode	1	2	3
Time period (T)	0.369	0.152	0.087
Natural frequency (ω)	17.02	41.42	71.95

Referring to Eq. (3.6.3.47) for the first three modes

$$[\mathbf{K}] = \frac{G_0}{H}\begin{bmatrix} \frac{\pi^2}{16}+\frac{1}{4} & \frac{-3}{4} & \frac{5}{36} \\ \frac{-3}{4} & \frac{9\pi^2}{16}+\frac{1}{4} & \frac{-15}{4} \\ \frac{5}{36} & \frac{-15}{4} & \frac{25\pi^2}{16}+\frac{1}{4} \end{bmatrix} = \begin{bmatrix} 10792 & -9337 & 3890.57 \\ -9337 & 72229 & -46687 \\ 3890.57 & -46687 & 195104.5 \end{bmatrix}$$

As per Eq. (3.6.3.29), we have for first three modes

$$[\mathbf{M}] = \begin{bmatrix} 31.478 & 0 & 0 \\ 0 & 31.478 & 0 \\ 0 & 0 & 31.478 \end{bmatrix}$$

Referring to Eq. (3.6.3.53)

$$[\mathbf{C}] = \begin{bmatrix} 160.7 & 0 & 0 \\ 0 & 391 & 0 \\ 0 & 0 & 679 \end{bmatrix}$$

Now applying the Newmark-β method as elaborated in Chap. 2, the acceleration plot at top of ground is plotted hereafter (Fig. 3.22).

It is observed that in this case the peak acceleration of 0.062g at the bedrock level amplifies to 1.76g at the ground level.

3.6.4 *Equivalent Linear and Nonlinear Constitutive Models*

Analysis shown in the previous section is a linear model where the shear modulus G and damping ratio D remain invariant with strain. Actually, soil is essentially a

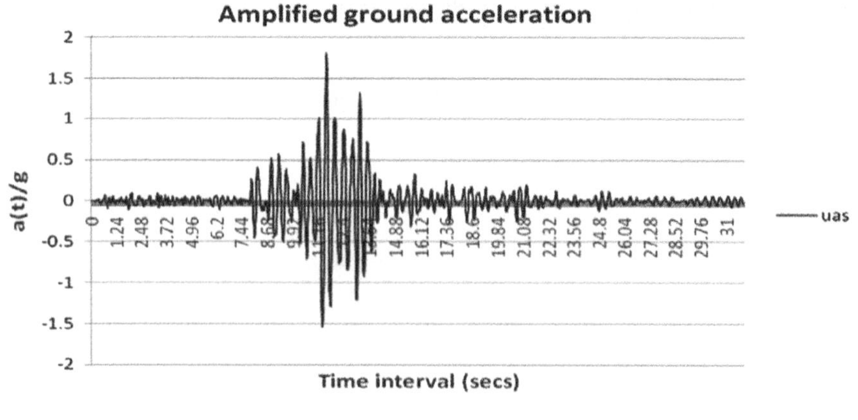

Fig. 3.22 Amplified ground acceleration at top of ground

heterogeneous material having profound nonlinear properties that show variation in stiffness and damping as the strain increases. It is only at a very low strain (like that in case of a machine foundation) that the shear modulus G and damping ratio D remain constant with strain. For most of the earthquakes ($M > 6.0$) that are considered significant, strain induced in the soil body is large, for which both G and D values vary with the strain level (Chowdhury et al. 2015).

Thus, while performing the time history analysis, at every step the amplitude vis-à-vis strain is computed, that gives the change in G and D at every step from the constitutive model. The stiffness and damping matrices are thus upgraded for the next step and the analysis continues.

We have discussed this phenomenon in detail in Chap. 8 and would like to only mention here that a number of models, like hyperbolic model, Ramberg, and Osgood's model (1943), Hardin and Drnevich (1972), Cundall and Pyke's (1979) model (HDCP), are available for nonlinear dynamic analysis of soil.

3.7 Liquefaction of Soil Under Earthquake

Liquefaction, unlike other materials' properties, is unique to soil, when it can fail under a transient shock, like that of an earthquake force.

Unless taken care off properly, it can result in spectacular failure of structure, when the rehabilitation cost could be astronomical.

Figure 3.23 shows such a failure of buildings in Niigata Earthquake in Japan 1964 due to soil liquefaction.

Therefore, before we get into the analytical aspect of liquefaction, it would perhaps be worth the effort to understand the physics of the problem. For this, let us look at Fig. 3.24.

Shown in Fig. 3.24 is a cohesionless soil sample which is saturated and enclosed in a rubber sheath, so that no drainage from the soil is possible. Let it be in equilibrium under an overburden pressure σ_v. If we now look at the sample through an electron microscope, the magnified vision of the sample will look something like Fig. 3.24, where the voids existing within the soil sample will be full of water and the overburden pressure σ_v will be taken up both by the soil particles and fluid existing in the void. Let us now subject this sample to a cyclic load at a rapid pace. As an impervious rubber sheath encloses the sample, the additional stress induced due to the cyclic load will be transferred to the soil as well as pore water that cannot dissipate out. It is obvious then that, with repeating of each cycle of load, the pore water pressure will go on increasing when at any stage the shear stress induced in soil may be expressed as

$$\tau = (\sigma_v - u) \tan \phi \tag{3.7.1}$$

where τ = shear stress in soil. σ_v = overburden stress, u = induced pore pressure and ϕ = angle of internal friction of soil.

Fig. 3.23 Toppling of a building due to soil liquefaction; Niigata Earthquake (1964), Japan

Fig. 3.24 A cohesionless saturated soil sample in magnified form

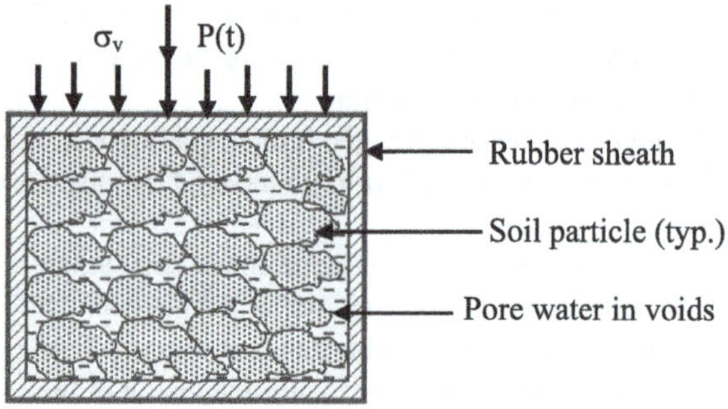

When the pore pressure u progressively increases with each cycle, there will come a stage when $\sigma_v = u$ and the *soil loses its shear strength and starts flowing like a liquid.*

This phenomenon, when, under repetitive cyclic loading, the soil loses its strength due to sudden increase in its pore pressure, is called *liquefaction of soil* and is a very common occurrence during an earthquake around the world.

Once liquefied under seismic load, soil starts flowing like a liquid and can create devastating damage to structures resting on it. Sand boils are usually observed on ground, which starts flowing freely (especially if the ground is sloped) resulting in cracks on the ground or crests of embankments.

Fig. 3.25 Typical sand boil, 6–8 m in diameter and 20 cm in height, due to liquefaction in Bhuj Earthquake, India

Figure 3.25 shows typical sand boils observed in ground due to liquefaction during Bhuj Earthquake in India, while Fig. 3.26 shows cracks at the crest of earth dam due to lateral spreading of ground as a secondary effect due to liquefaction.

It is thus apparent that it is necessary to carefully evaluate the liquefaction potential of a soil, especially when the soil is cohesionless and below groundwater level in a seismic prone area. Even a general c-ϕ soil, if susceptible to liquefaction, it will lose a significant load carrying capacity and could undergo unacceptable settlement, resulting in secondary damage to structure.

3.7.1 Liquefaction Analysis of Cohesionless Sandy Soil

Based on explanations furnished in the previous section, it is apparent that saturated cohesionless soil existing below groundwater table is most susceptible to liquefaction. Considering that, for such sandy soil, the Standard Penetration Test (SPT) values have been extensively used to evaluate its engineering properties, it was but natural that researchers have tried to correlate the liquefaction property of soil to the number of blows (N), the index of measurement of SPT test.

Fig. 3.26 Cracks at crest of embankment due to lateral spreading of ground; Bhuj Earthquake in India

Pioneering work on this was carried out by Seed et al. (1984) and has been an accepted design methodology that is practiced globally. The procedure is presented hereafter.

When earthquake waves propagate through soil, it is subjected to cyclic shear stress. Seed et al. defined the term *cyclic stress ratio* (CSR) expressed as

$$\mathrm{CSR} = \frac{\tau_{av}}{\sigma'_v} = 0.65\left(\frac{a_{\max}}{g}\right)\left(\frac{\sigma_v}{\sigma'_v}\right) r_\mathrm{d} \qquad (3.7.1.1)$$

Here, $a_{\max}$ = maximum acceleration at ground level, g = acceleration due to gravity, σ_v = total overburden pressure at the depth where liquefaction analysis is carried out, σ'_v = effective overburden pressure at the depth where liquefaction analysis is carried out, and r_d = reduction factor that varies with depth and is given by

$$\begin{aligned} r_\mathrm{d} &= 1.0 - 0.000765z \quad \text{for } z \leq 9.15 \text{ m} \\ r_\mathrm{d} &= 1.174 - 0.0267z \quad \text{for } 9.15 \leq z \leq 23 \text{ m} \end{aligned} \qquad (3.7.1.2)$$

$$r_\mathrm{d} = 0.744 - 0.008z \quad \text{for } 23 \leq z \leq 30\ textm$$

$$r_\mathrm{d} = 0.5 \quad \text{for } z \geq 30 \text{ m}$$

For ease of electronic computation, r_d can also be expressed (Blake et al. 2002)

$$r_d = \frac{1.000 - 0.4113z^{0.5} + 0.04052z + 0.001753z^{1.5}}{1.000 - 0.4177z^{0.5} + 0.05729z - 0.006205z^{1.5} + 0.00121z^2} \quad (3.7.1.3)$$

Figure 3.27 gives comparative plot of r_d as recommended by Seed et al. (1984) and Blake et al. (2002).

It is observed that the values are very closely matching and can well be used if one wishes to develop a computer program on this procedure.

The next important parameter that controls the value of CSR is a_{max} and its choice, while performing a liquefaction analysis, has remained somewhat ambiguous.

The authors had seen a number of cases where the designer used the expressions as proposed in Eqs. (3.3.1.1) and (3.3.1.11). This, to our perception is not realistic, as these values attribute to ground acceleration at the bedrock level and not at the ground surface.

Swaran (1999) has proposed an expression

$$a_{max} = 0.184 \times 10^{0.320M} D^{-0.8} g \quad (3.7.1.4)$$

Here, a_{max} = maximum ground acceleration, M = expected moment magnitude of earthquake and D = maximum distance of the site from the epicenter in km.

The background of this formula has, however, not been furnished by Swaran (1999), though the expression is often used in many design offices in India.

One disadvantage with the formula is that one needs to know priori the fault location D, which is always not available for a given project. Secondly, one can certainly assume a moment magnitude for computing the acceleration; however, that could overestimate or underrate the response, as it remains a guesswork at best.

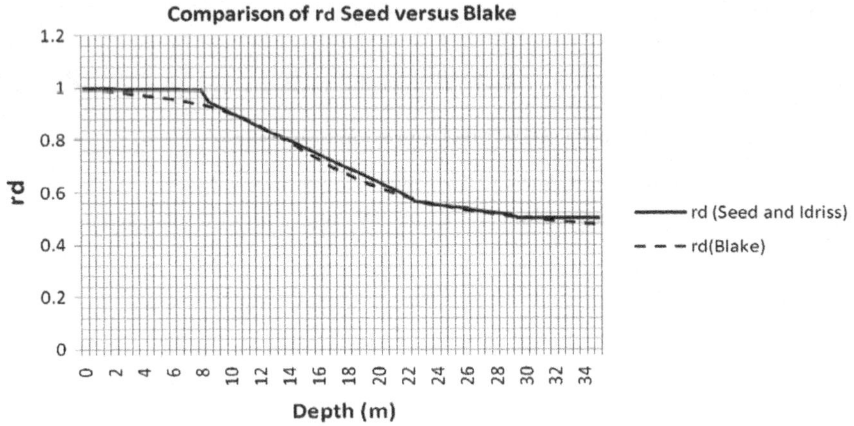

Fig. 3.27 Comparative plot of r_d according to Seed et al. (1984) versus Blake et al. (2002)

An alternative method one adopts is to follow the code considering zero period acceleration. Referring to Eq. (3.5.1) and explanations as furnished in Sect. 2.2.1 of Chap. 2 for $T = 0$ the effective acceleration becomes

$$a_{max} = Z\left(\frac{I}{R}\right)g \quad (3.7.1.5)$$

Thus for 5% damping, as furnished in the code for zone V, considering $I/R = 1.0$ in this case and referring to Table 2.4 of Chap. 2, we get $a_{max} = 0.36g$.

Considering that the damping for soil is usually greater than 5% (usually about 10–15%), for 10% damping peak ground acceleration, vide Table 2.9, it is computed as $a_{max} = 0.36 \times 0.8 \times g = 0.29g$.

The above technique is often adapted by designers to estimate the peak acceleration at ground surface during an earthquake as per IS-1893(2016). However, the recommendation of code being generic in nature, the results can significantly vary in reality because of the local geological condition of the site, which can significantly affect the result.

A more realistic approach would be to assume the soil strata as a shear beam resting on a bedrock and then compute the time period from the expressions as furnished in Table 3.3. Once the time period is computed assuming the beam is resting on rock, the maximum spectral acceleration can be computed.

The amplitude of vibration can then be computed as

$$u_z = \frac{8}{2+\pi}\left(\frac{ZI}{R}\right)\frac{S_a}{\omega^2}\cos\frac{(2n-1)\pi z}{2H} \quad (3.7.1.6)$$

At the ground surface consider $z = 0$, and for sandy cohesionless soil, as shown by Chowdhury and Dasgupta (2007), take $R = 2$ and $R = 3.0$ for soil with fine content greater than 10%.

Once the amplitude is computed for each mode, the ground acceleration can be computed as

$$S_a(z=0) = u_0\omega^2 \quad (3.7.1.7)$$

The procedure is now explained through a suitable example.

Example 3.7 For the problem 3.5 furnished earlier determine the peak ground acceleration given the site is Zone IV as per IS code. All other data remain same.

Solution

As computed in Problem 3.5, the time period vis-à-vis spectral acceleration (S_a/g) is shown hereafter in tabular form.

Mode	1	2	3
Time period (*T*)	0.369	0.152	0.087
Sa/*g*	2.5	2.5	2.31
Natural frequency (ω)	17.02	41.42	71.95

Considering $Z = 0.24$ for Zone 4 and $R = 2$
$u_z = \frac{8}{2+\pi}\left(\frac{ZI}{R}\right)\frac{S_a}{\omega^2}\cos\frac{(2n-1)\pi z}{2H}$ we have

$$u_1 = \frac{8}{2+\pi}\left(\frac{0.24}{2}\right)\frac{2.5\times 9.81}{17.02^2}\cos\frac{\pi\times 0}{2\times 32} = 0.01575 \text{ m for mode 1}$$

Thus, the acceleration at ground level = $0.01575 \times 17.02^2 = 4.562\ \text{m/sec}^2 \cdot$ $(0.465g)$ for 5% damping. Considering 15% damping.

$$a_{\max}(\text{design}) = \sqrt{\frac{10}{5+D}} \times a_{\max}(5\%) = \sqrt{\frac{10}{20}} \times 0.465g = 0.328g$$

Observe here that if we would have ignored the local site effect and carried out further computation based on $0.24g$, we would have underrated the response by almost 37%.

From the above clarifications, it is apparent that CSR is an index of force induced in the body due to action of propagating waves under earthquake. This is resisted by another parameter, the *cyclic resistance ratio* (CRR), which is a function of *N*, the number of blow count for an SPT test.

As a first step, to compute CRR the field observed SPT value N_0 is subjected to certain corrections as mentioned below.

$$(N_1)_{60} = N_0 C_N C_E C_B C_R C_S \tag{3.7.1.8}$$

Here, N_0 = observed *N* value in field. C_N = correction factor for overburden pressure; C_E = correction factor for hammer energy ratio, C_B = correction factor related to borehole diameter, C_R = correction factor for rod length C_S = correction factor for sample with or without liner, $(N_1)_{60}$ = corrected SPT value with 60% hammer efficiency.

The correction factors as mentioned above have been further elaborated in Table 3.4.

In Table 3.4, P_0 = standard atmospheric pressure or 100 kPa and σ'_v = effective overburden pressure at the depth of the standard penetration sample.

Having established the design SPT value $(N_1)_{60}$, the CRR value is expressed for clean sand (i.e. having fine content (FC < 5%) as Blake (1997),

$$\text{CRR} = \frac{a + by + cy^2 + dy^3}{1 + ey + fy^2 + gy^3 + hy^4} \tag{3.7.1.9}$$

Table 3.4 Correction factors to the observed SPT values

Factor	Equipment parameter	Term	Correction factor
Overburden pressure	Independent of equipment	C_N	$\sqrt{P_0/\sigma'_v}$
Energy ratio	Safety hammer	C_E	0.6–1.17
	Donut hammer		0.45–1.0
Rod length	3–4 m	C_R	0.75
	4–6 m		0.85
	6–10 m		0.95
	10–30 m		1.0
	>30 m		>1.0
Borehole diameter	65–115 mm	C_B	1.0
	150 mm		1.05
	200 mm		1.15
Sampling method	Standard sampler	C_S	1.0
	Sampler without liners		1.2

where $a = 0.048$, $b = -0.004721$, $c = 0.0006136$, $d = -1.673 \times 10^{-5}$, $e = -0.1248$, $f = 0.009578$, $g = -0.0003285$, $h = 3.714 \times 10^{-6}$ and $y = (N_1)_{60}$.

Equation (3.7.1.9) is valid when $(N_1)_{60}$ *is less than 30.*

For $(N_1)_{60} > 30$, the soil is deemed too dense and is usually considered as *non-liquefiable*.

An alternate formula proposed by Rausch (1998) is also popular in design office practice and is expressed as

$$\mathrm{CRR}_{7.5} = \frac{1}{34 - (N_1)_{60}} + \frac{(N_1)_{60}}{135} + \frac{50}{\left[10(N_1)_{60} + 45\right]^2} - \frac{1}{200} \tag{3.7.1.10}$$

In Eq. (3.7.1.10), $\mathrm{CRR}_{7.5}$ represents the cyclic resistance ratio at an earthquake of magnitude 7.5.

Comparative values of CRR, based on the above two expressions, are as shown in Fig. 3.28.

From Fig. 3.28, it is observed that both the expressions are in excellent agreement and either of them can be used to compute CRR for liquefaction analysis.

3.7.1.1 Effect of Fine Content (FC) on CRR

The above development by Seed et al. is based on clean sand under saturated conditions. However, in reality it will be rare when at a site one would get clean sand only. In most cases, some amount of silt and clay always remains mixed with it. It has been observed by Seed et al. (1984) that with presence of fine content, there

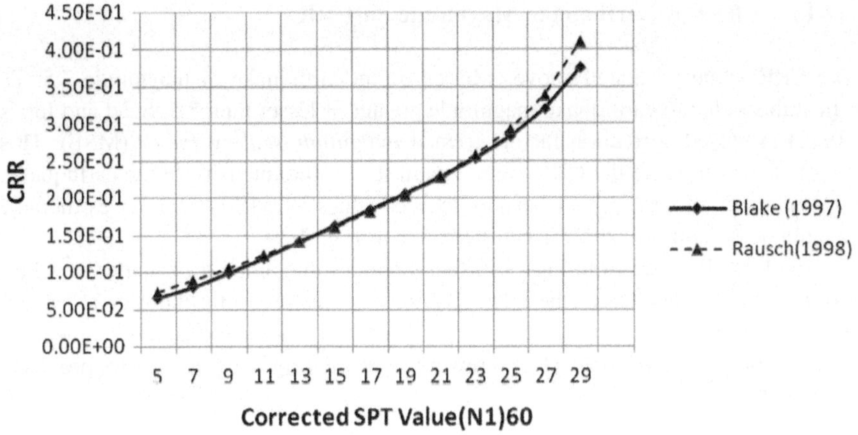

Fig. 3.28 CRR values as per Blake (1997) and Rausch (1998)

is an apparent increase in value of CRR. Whether this increase is due to increase in resistance or decrease in penetration is not clear.

To cater to this observation, corrections to SPT value have been recommended. It may be noted that other grain characteristics, like plasticity index (PI), may also affect the liquefaction resistance as well but extent to which they affect the soil behavior is not yet well defined. As such, corrections based on fine content only are used and should be mellowed *with caution and judgment.*

Seed and Idriss (1982) have proposed correction to $(N_1)_{60}$—to an equivalent clean sand value $(N_1)_{60CS}$ given by

$$(N_1)_{60CS} = \alpha + \beta \times (N_1)_{60} \tag{3.7.1.11}$$

Here, α and β are constants which are functions of the % of fine content FC present in the soil sample and are as furnished in Table 3.5.

Table 3.5 Values of α and β with FC

Values of α and β	Fine content %
$\alpha = 0$	For FC $\leq$ 5%
$\alpha = e^{\left[1.76-\left(190/FC^2\right)\right]}$	$5\% \leq FC \leq 35\%$
$\alpha = 5.0$	$FC \geq 35\%$
$\beta = 1.0$	For FC $\leq$ 5%
$\beta = \left[0.99 + \frac{(FC)^{1.5}}{1000}\right]$	$5\% \leq FC \leq 35\%$
$\beta = 1.2$	$FC \geq 35\%$

3.7.1.2 Effect of Earthquake Magnitude on CRR

The CRR value evaluated above is based on an earthquake of magnitude 7.5. To adjust this value to earthquake magnitude greater or lesser than 7.5, Seed and Idriss (1982) proposed correction factors termed *magnitude-scaling factor* (MSF). This factor is used to scale the CRR value to adjust it to the intensity of the earthquake.

A number of researchers have proposed values of MSF vis-à-vis earthquake magnitude M and are as furnished hereafter in Fig. 3.29.

It will be observed from Fig. 3.29 that there is some amount of scatter in values at the two extreme ends of the interval. Moreover, as presently industry is leaning more and more to electronic computation rather than graphical data (more amenable to hand computation), Idriss (1999) reworked on the original data and has proposed an expression

$$\mathrm{MSF} = \frac{10^{2.24}}{M_w^{2.56}} \tag{3.7.1.12}$$

Equation (3.7.1.12) has more-or-less global acceptance in computing the magnitude-scaling factor.

The factor of safety (FOS) against liquefaction is finally computed from the expression

$$\mathrm{FOS} = \left[\frac{\mathrm{CRR}_{7.5}}{\mathrm{CSR}}\right] \times \mathrm{MSF} \geq 1.0 \tag{3.7.1.13}$$

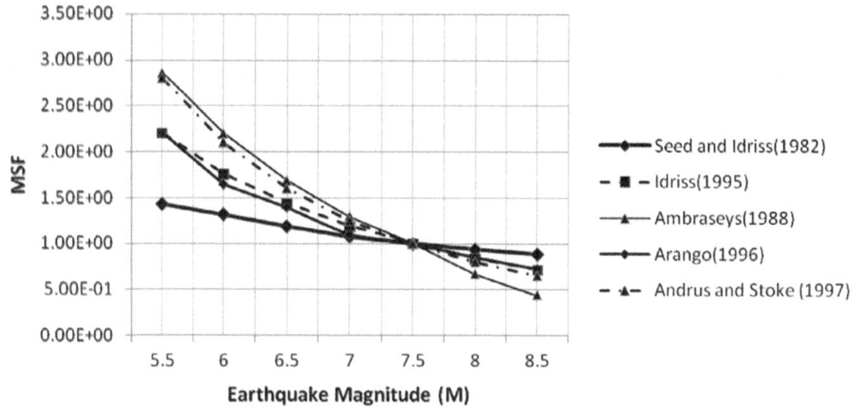

Fig. 3.29 MSF values compared to different earthquake magnitudes

If the factor of safety (FOS) *is greater than 1.0, it is perceived that the soil is not susceptible to liquefaction.* On the contrary, when FOS is less than 1.0, liquefaction failure is expected.

Despite the fact that the use of the above technology is very popular in practice, uncertainties prevailing are quite significant; as such, a FOS of 1.2 or more is usually generally targeted as a minimum.

Example 3.8 A water treatment plant site is in South East Asia; its plot plan showing level contours and borehole locations are shown in Fig. 3.30. The boreholes show consistent soil strata in the site and one of the detailed bore logs with N value is as shown hereafter. The site shows medium stiff-to-stiff brown silty clay for the first 6.0 m and then has a deep stratum of medium-dense to very-dense brown to greyish-brown silty sand with about 20% fine content. The average standing water level is 8.9 m below the existing ground level. The SPT values obtained are presented in tabular form below. Compute the liquefaction potential of the soil (Fig. 3.31).

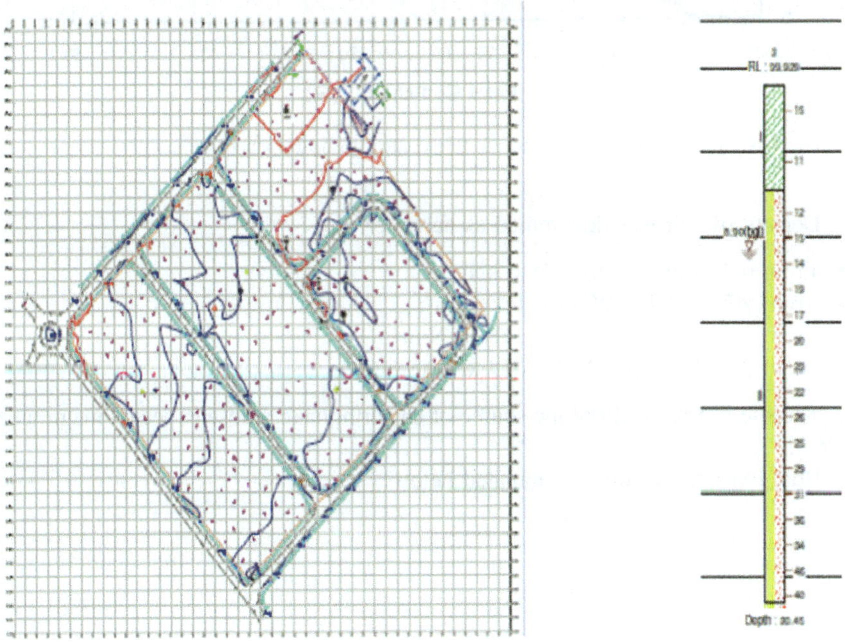

Fig. 3.30 Plot plan and representative bore log for the given site

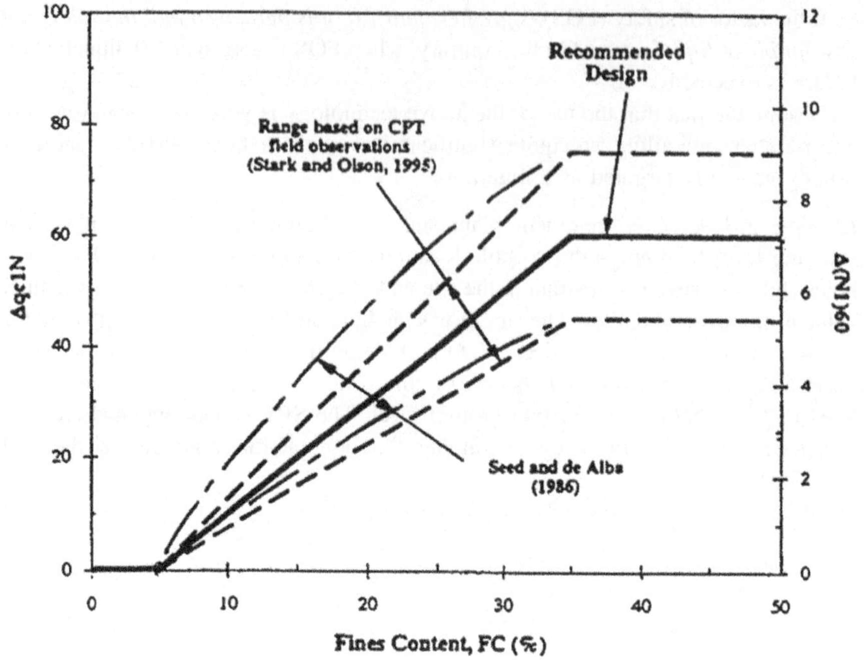

Fig. 3.31 Graph to compute correction to fine content to tip resistance

Density of soil was determined as hereafter:

- From 0.25 to 6.225 m = 18.6 kN/m^3.
- From 7.725 to 16.725 m = 18 kN/m^3.
- From 18.225 to 21.225 m = 18.5 kN/m^3.
- From 22.725 to 30.225 m = 19 kN/m^3.

The site is 150 km from the source of an earthquake whose expected magnitude is 6.8.

Bore log data of soil are tabulated below.

Depth	Avg depth	*N* values BH-1	*N* values BH-2	*N* values BH-3	*N* values BH-4	*N* values BH-5	*N* values BH-6	*N* values BH-7	Average *N* values
0.0–0.5	0.25								15
1.0–1.45	1.225	17			13			15	15
1.5–1.95	1.725		13	15					14
2.0–2.45	2.225						15		15
2.5–3.0	2.75								16
3.0–3.45	3.225	14			17	18		17	17
4–4.95	4.475		6	11			7		8
5.0–6.0	5.5								7
6.0–6.45	6.225	5			5	5	8	5	6
7.5–7.95	7.725	14	13	12	13	14	15	13	13
9–9.45	9.225	13	11	15	15	15	17	12	14
10.5–10.95	10.725	12	14	14	16	13	17	19	15
12–12.45	12.225	14	10	19	14	15	15	15	15
13.5–13.95	13.725	16	13	17	16	15	19	18	16
15–15.45	15.225	18	17	20	18	19	17	20	18
16.5–16.95	16.725	19	19	23	21	18	26	25	22
18–18.45	18.225	20	25	22	25	21	24	24	23
19.5–19.95	19.725	23	25	26	28	25	34	28	27
21–21.45	21.225	30	23	25	30	26	31	29	28
22.5–22.95	22.725	36	32	29	32	31	36	31	32
24–24.45	24.225	38	32	31	37	25	32	35	33
25.5–25.95	25.725	35	30	36	32	32	36	41	35
27–27.45	27.225	39	42	34	31	37	35	38	37
28.5–28.95	28.725	42	38	46	40	46	43	40	42

Average SPT values and other relevant soil properties are tabulated hereafter.

Depth (m)	N_0 (avg)	Wt density	ϕ	FC (%)
0.25	15	18.6		
1.225	15	18.6		
1.725	14	18.6		
2.225	15	18.6		
2.75	16	18.6		
3.225	17	18.6		
4.475	8	18.6		
5.5	7	18.6		
6.225	6	18.6		
7.725	13	18	31	20
9.225	14	18	31	20
10.725	15	18	31	20
12.225	15	18	34	20
13.725	16	18	34	20
15.225	18	18	34	20
16.725	22	18	34	20
18.225	23	18.5	34	20
19.725	27	18.5	34	20
21.225	28	18.5	34	20
22.725	32	19	34	20
24.225	33	19	34	20
25.725	35	19	34	20
27.225	37	19	34	20
28.725	42	19	34	20

As a first step, we correct the observed average N_0 values to $(N_1)_{60}$, as shown hereafter. While doing this, considering that the water table is at 8.9 m, we only take into account the correction to SPT values below the water table, because it is only the saturated sand that is subjected to liquefaction.

Depth (m)	N (avg)	Wt density	FC	σ_v	u	σ'_v	C_N	C_E	C_R	C_B	C_S	$(N_1)_{60}$
0.25	15	18.6		4.65								
1.225	15	18.6		22.785								
1.725	14	18.6		32.085								
2.225	15	18.6		41.385								
2.75	16	18.6		51.15								
3.225	17	18.6		59.985								
4.475	8	18.6		83.235								
5.5	7	18.6		102.3								
6.225	6	18.6		115.785								
7.725	13	18	20	139.05								
9.225	**14**	**18**	**20**	**166.05**	**3.25**	**162.80**	**0.784**	**0.885**	**0.95**	**1**	**1**	**9**
10.725	15	18	20	193.05	18.25	174.80	0.756	0.885	1	1	1	10
12.225	15	18	20	220.05	33.25	186.80	0.732	0.885	1	1	1	9
13.725	16	18	20	247.05	48.25	198.80	0.709	0.885	1	1	1	10
15.225	18	18	20	274.05	63.25	210.80	0.689	0.885	1	1	1	11
16.725	22	18	20	301.05	78.25	222.80	0.670	0.885	1	1	1	13
18.225	23	18.5	20	337.1625	93.25	243.91	0.640	0.885	1	1	1	13
19.725	27	18.5	20	364.9125	108.25	256.66	0.624	0.885	1	1	1	15
21.225	28	18.5	20	392.6625	123.25	269.41	0.609	0.885	1	1	1	15
22.725	32	19	20	431.775	138.25	293.53	0.584	0.885	1	1	1	17
24.225	33	19	20	460.275	153.25	307.03	0.571	0.885	1	1	1	17
25.725	35	19	20	488.775	168.25	320.53	0.559	0.885	1	1	1	17
27.225	37	19	20	517.275	183.25	334.03	0.547	0.885	1	1	1	18
28.725	42	19	20	545.775	198.25	347.53	0.536	0.885	1	1	1	20
30.225	47	19	20	574.275	213.25	361.03	0.526	0.885	1	1	1	22

For clarity of calculation, we show in detail the computation at $z = 9.225$ m, i.e., the layer just below the groundwater level (marked bold in above table). For rest of the layers below it, the values are autogenerated through a computer program and presented in tabular form.

Considering the epicentric distance is 150 km from the site, according to Eq. (3.7.1.4) we get

$$a_{\max} = 0.184 \times 10^{0.320M} D^{-0.8} g = 0.184 \times 10^{0.320 \times 6.8} \times 150^{-0.8} g = 0.501g$$

Thus, $CSR = \frac{\tau_{av}}{\sigma'_v} = 0.65\left(\frac{a_{\max}}{g}\right)\left(\frac{\sigma_v}{\sigma'_v}\right) r_d$ here $r_d = 1.174 - 0.0267z$ (Refer Eq. 3.7.1.2)

$$\Rightarrow \text{CSR} = 0.65\left(\frac{0.501g}{g}\right)\left(\frac{166}{163}\right) \times (1.174 - 0.0267 \times 9.225) = 0.307$$

Considering FC = 20%

$$\alpha = e^{\left[1.76 - \left(190/FC^2\right)\right]} = e^{\left[1.76 - \frac{190}{400}\right]} = 3.614 \quad \text{and}$$

$$\beta = \left[0.99 + \frac{(FC)^{1.5}}{1000}\right] = \left[0.99 + \frac{(20)^{1.5}}{1000}\right] = 1.08$$

Considering $(N_1)_{60CS} = \alpha + \beta \times (N_1)_{60}$ as the correction to SPT due to fine content, here $(N_1)_{60} = 9$, we have

$$(N_1)_{60CS} = 3.614 + 1.08 \times 9 \cong 13$$

Thus, applying Rausch's (1998) expression, i.e., Eq. (3.7.1.10)

$$\begin{aligned} \text{CRR}_{7.5} &= \frac{1}{34 - (N_1)_{60}} + \frac{(N_1)_{60}}{135} + \frac{50}{\left[10(N_1)_{60} + 45\right]^2} - \frac{1}{200}\text{CRR}_{7.5} \\ &= \frac{1}{34 - 13} + \frac{13}{135} + \frac{50}{[10 \times 13 + 45]^2} - \frac{1}{200} \\ &= 0.0476 + 0.096 + 1.632 \times 10^{-3} - 5 \times 10^{-3} = 0.146 \end{aligned}$$

Considering that the design earthquake magnitude is $M = 6.8$, we get

$$\text{MSF} = \frac{10^{2.24}}{M_w^{2.56}} = \frac{10^{2.24}}{6.8^{2.56}} = 1.284$$

$$\text{FOS} = \left[\frac{CRR_{7.5}}{CSR}\right] \times \text{MSF} = \frac{0.146 \times 1.284}{0.307} = 0.608 < 1.0$$

Thus, the layer can liquefy under an earthquake of magnitude 6.8.

Computation of cyclic stress ratio (CSR) for all layers.

Z (m)	γ_{soil} (kN/m^3)	FC (%)	σ_v (kN/m^2)	u (kN/m^2)	σ'_v (kN/m^2)	r_d	CSR
9.225	18	20	166.05	3.25	162.80	0.927693	0.308134
10.725	18	20	193.05	18.25	174.80	0.887643	0.31924
12.225	18	20	220.05	33.25	186.80	0.847593	0.325149
13.725	18	20	247.05	48.25	198.80	0.807543	0.326802
15.225	18	20	274.05	63.25	210.80	0.767493	0.324926
16.725	18	20	301.05	78.25	222.80	0.727443	0.320091
18.225	18.5	20	337.16	93.25	243.91	0.687393	0.309429
19.725	18.5	20	364.91	108.25	256.66	0.647343	0.299717
21.225	18.5	20	392.66	123.25	269.41	0.607293	0.288238
22.725	19	20	431.775	138.25	293.53	0.567243	0.271727
24.225	19	20	460.275	153.25	307.03	0.5502	0.268606
25.725	19	20	488.775	168.25	320.53	0.5382	0.267265
27.225	19	20	517.275	183.25	334.03	0.5262	0.265365
28.725	19	20	545.775	198.25	347.53	0.5142	0.262973
30.225	19	20	574.275	213.25	361.03	0.5022	0.260142

Computation of cyclic resistance ratio (CRR) for all layers.

Z (m)	$(N_1)_{60}$	α	β	$(N_1)_{60CS}$	$CRR_{7.5}$
9.225	9	3.614668	1.079443	14	0.146022
10.725	10	3.614668	1.079443	14	0.15461
12.225	9	3.614668	1.079443	14	0.148216
13.725	10	3.614668	1.079443	15	0.15655
15.225	11	3.614668	1.079443	16	0.167579
16.725	13	3.614668	1.079443	17	0.185397
18.225	13	3.614668	1.079443	18	0.188291
19.725	15	3.614668	1.079443	20	0.211889
21.225	15	3.614668	1.079443	20	0.212258
22.725	17	3.614668	1.079443	22	0.237724
24.225	17	3.614668	1.079443	22	0.23539
25.725	17	3.614668	1.079443	22	0.242894
27.225	18	3.614668	1.079443	23	0.252785
28.725	20	3.614668	1.079443	25	0.296087
30.225	22	3.614668	1.079443	27	0.146022

Computation of factor of safety (FOS) at $M = 6.8$ and inference.

Z (m)	$CRR_{7.5}$	MSF	CSR	$FOS_{6.8}$	Conclusion
9.225	0.146022	1.284	0.308134	**0.608**	Liquefied
10.725	0.15461	1.284	0.31924	**0.622**	Liquefied
12.225	0.148216	1.284	0.325149	**0.585**	Liquefied
13.725	0.15655	1.284	0.326802	**0.615**	Liquefied
15.225	0.167579	1.284	0.324926	**0.662**	Liquefied
16.725	0.185397	1.284	0.320091	**0.744**	Liquefied
18.225	0.188291	1.284	0.309429	**0.781**	Liquefied
19.725	0.211889	1.284	0.299717	**0.908**	Liquefied
21.225	0.212258	1.284	0.288238	**0.946**	Liquefied
22.725	0.237724	1.284	0.271727	1.123	Not Liquefied
24.225	0.23539	1.284	0.268606	1.125	Not Liquefied
25.725	0.242894	1.284	0.267265	1.167	Not Liquefied
27.225	0.252785	1.284	0.265365	1.223	Not Liquefied
28.725	0.296087	1.284	0.262973	1.446	Not Liquefied
30.225	0.146022	1.284	0.260142	1.694	Not Liquefied

Thus, we see that from top 9.225 to 21 m the soil is susceptible to liquefaction. Liquefaction in this case is deep-seated in nature.

3.7.2 *Liquefaction Based on CPT Value*

In many projects around the world, the Cone penetration test (CPT) is presently being given precedence over SPT, especially if the soil is silty or semi-cohesionless in nature.

Primary advantage of CPT is that a nearly continuous profile of penetration resistance is obtained for stratigraphic interpretation.

CPT results are usually more continuous than SPT data and allow for a more explicit definition of the soil layers. However, unlike SPT, visual verification of soil samples is not possible which is somewhat disadvantageous in terms of liquefaction analysis. It is for this, Idriss (1999) advocates that for liquefaction analysis, CPT tests results shall be supplemented by some well-chosen bore logs carrying SPT test to backup the results obtained from CPT values.

Three methods are usually popular around the world for estimation of liquefaction potential of soil based on CPT values; these are

- Seed et al. (1986) method;
- Suzuki et al.'s (1997) method;
- Robertson and Wride's (1997) method.

3.7.2.1 Method Proposed by Seed et al. (1986)

In this case, as a first step, the measured tip resistance q_c for the cone has to be corrected for overburden pressure expressed as

$$q_{cl} = C_q \cdot q_c \tag{3.7.2.1.1}$$

Here, C_q is given by

$$C_q = \frac{1.8}{0.8 + \frac{\sigma'_v}{P_a}} \tag{3.7.2.1.2}$$

were σ'_v = vertical effective overburden pressure, P_a = atmospheric pressure in the same units as σ'_v (in SI units it is considered as 100 kPa).

The value of q_{cl} now needs to be converted to ton per square feet (tsf) to match Seed et al.' curve and unit.

Next, correction due to fine content is to be carried out, being usually read out from Fig. 3.31.

The corrected value of tip resistance due to fine content is given by

$$q_{clf} = q_{cl} + \Delta q_{cl} \tag{3.7.2.1.3}$$

Idriss and Boulanger (2008) have proposed an expression for Δq_{cl} as

$$\Delta q_{cl} = \left(11.9 + \frac{q_{cl}}{14.6}\right) e^{\left(1.63 - \left(\frac{9.7}{FC+2}\right) - \left(\frac{15.7}{FC+2}\right)^2\right)} \tag{3.7.2.1.4}$$

Now $CRR_{7.5}$ is read off from the graph as presented in Fig. 3.32.

Alternatively, Idriss and Boulangar (2008) have proposed an expression

$$CRR_{7.5} = e^{\left[\frac{q_{clf}}{113} + \left(\frac{q_{clf}}{1000}\right)^2 - \left(\frac{q_{clf}}{140}\right)^3 + \left(\frac{q_{clf}}{137}\right)^4 - 2.80\right]} \tag{3.7.2.1.5}$$

that may be used for direct computation of the value of $CRR_{7.5}$.

Computation of CSR and factor of safety remains same as explained in Sect. 3.7.1.1.

3.7.2.2 Suzuki et al.'s (1997) Method

In this procedure, the measured tip resistance is corrected for the overburden pressure as

$$q_{cl} = \frac{q_c}{P_a}\sqrt{\frac{P_a}{\sigma'_v}} \tag{3.7.2.2.1}$$

Fig. 3.32 Chart to determine $CRR_{7.5}$ from corrected q_{clf} value

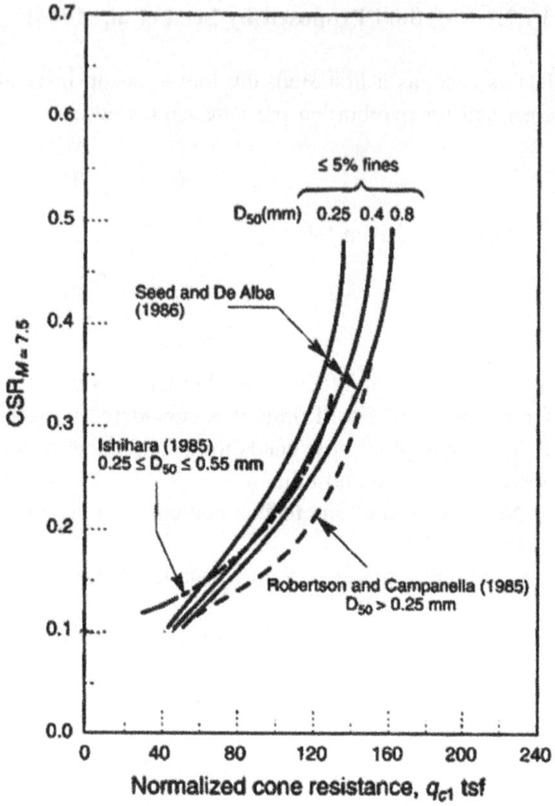

Here, q_c = observed tip resistance in field, q_{cl} = corrected tip resistance due to overburden, $\sigma'_v =$ effective overburden pressure, P_a = atmospheric pressure (100 kPa).

Equation (3.7.2.2.1) is corrected for fine content as presented in Eqs. (3.7.2.1.3) and (3.7.2.1.4) to compute the value of q_{clf}.

Next, one needs to estimate the soil behavior index I_c expressed as

$$I_c = \sqrt{(3.47 - \log_{10} Q)^2 + (\log_{10} R_f + 1.22)^2} \quad (3.7.2.2.2)$$

where

$$Q = \frac{q_{clf} - \sigma_v}{\sigma'_v} \quad (3.7.2.2.3)$$

$$R_f = \left[\frac{f_s}{q_{clf} - \sigma_v}\right] \times 100 \quad (3.7.2.2.4)$$

Here, q_{clf} = corrected tip resistance in tsf, σ'_v = effective overburden pressure σ_v = gross overburden pressure, f_s = measure sleeve friction, Q = normalized tip resistance, R_f = sleeve friction ratio.

The corrected tip resistance q_{clf} is further adjusted to q_{ca} based on the index property as

$$q_{ca} = q_{clf} \times f(I_c) \tag{3.7.2.2.5}$$

where $f(I_c)$ is as furnished hereafter in Table 3.6.

The CRR 7.5 is finally obtained from the chart as furnished in Fig. 3.33.

3.7.2.3 Robertson and Wride's (1997) Method

Like Suzuki's method explained in previous section Robertson and Wride's method is also soil index (I_c) driven. However, the evaluation of I_c is iterative that makes the procedure a little tedious in terms of hand computation; however, it can very well be fitted into a computer program which makes it a very attractive choice in today's age of cyber computation.

Table 3.6 Variation of $f(I_c)$ with I_c

I_c	1.6	1.8	1.9	2.0	2.1	2.2	2.3	≥2.4
$f(I_c)$	1.0	1.2	1.3	1.5	1.7	2.1	2.6	3.5

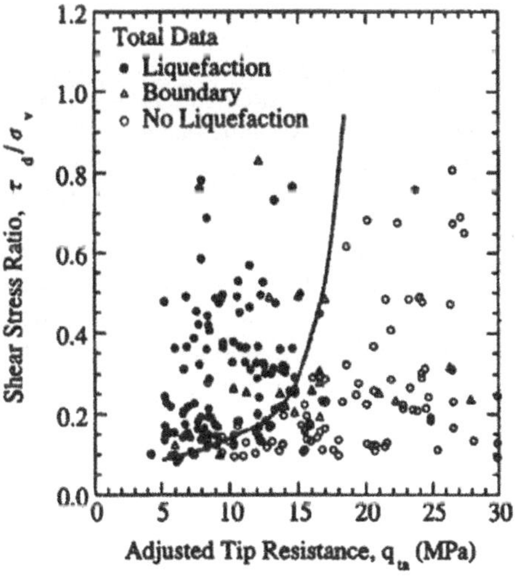

Fig. 3.33 $CRR_{7.5}$ versus q_{ca} as per Suzuki et al. (1997)

It starts with computation of I_c as furnished in Eq. (3.7.2.2.2) where

$$Q = \frac{q_c - \sigma_v}{P_a}\left(\frac{P_a}{\sigma'_v}\right)^n \quad (3.7.2.3.1)$$

and

$$R_f = \left[\frac{f_s}{q_c - \sigma_v}\right] \times 100 \quad (3.7.2.3.2)$$

In Eq. (3.7.2.3.1), the nomenclatures are same as furnished in Suzuki's method. We start the calculation with n = 1.0. If the computed value of $I_c > 2.6$, then one concludes that the soil is clayey in nature and non-liquefiable, and *the calculations stop there.*

In case $I_c < 2.6$, it means that assuming n = 1 was wrong in the first place, and it is necessary to re-calculate I_c assuming n = 0.5 when

$$Q = \frac{q_c}{P_a}\left(\frac{P_a}{\sigma'_v}\right)^n \text{ (i.e. considering } n = 0.5) \quad (3.7.2.3.3)$$

R_f remains same as shown in Eq. (3.7.2.3.2). If I_c is now <2.6, it means that the assumption n = 0.5 is correct and the soil is probably non plastic and granular, and we go to the next step of normalizing the tip resistance q_c.

In case I_c is yet >2.6, it shows that again the assumption of n = 0.5 was wrong and the soil is probably silty in nature; then we assume n = 0.7 where Q is same as given by Eq. (3.7.2.3.3.), but with n = 0.7, and subsequently I_c is computed.

As a next step we normalize the tip resistance q_c to q_{cN} which is given by

$$q_{cN} = \frac{q_c}{P_a} \times C_Q \quad (3.7.2.3.4)$$

Here, q_{cN} is a dimensionless number and

$$C_Q = \left(\frac{P_a}{\sigma'_v}\right)^n \quad (3.7.2.3.5)$$

Here, the value of n is the same as the one adopted to compute I_c (i.e., either 0.5 or 0.7), and other abbreviations are as defined previously.

As a next step, we need to do fine corrections to the normalized tip resistance q_{cN}.

Namely

$$(q_{cN})_f = K_c q_{cN} \quad (3.7.2.3.6)$$

If $I_c < 1.64$ the $K_c = 1.0$; else:

$$K_c = -0.403I_c^4 + 5.581I_c^3 - 21.63I_c^2 + 33.75I_c - 17.88 \qquad (3.7.2.3.7)$$

In a recent research, Robertson has given an updated version of K_c that may also be used; it is given by

$$(q_{cN})_f = q_{cN} + \Delta q_{cN} \qquad (3.7.2.3.8)$$

where

$$\Delta q_{cN} = \left[\frac{K_c}{1 - K_c}\right] q_{cN} \qquad (3.7.2.3.9)$$

Here, K_c is a function of fine content FC (%).

When $FC \leq 5\%$, $K_c = 0.0$; for $5 < FC < 35$, $K_c = 0.0267 \times (FC - 5)$; and for $FC > 35$, $K_c = 0.8$.

The fine content FC is related to I_c by the expression

$$FC = 1.75I_c^{3.25} - 3.7 \qquad (3.7.2.3.10)$$

The CRR value at magnitude 7.5 is finally calculated from the expressions:

If $(q_{cN})_f < 50$ then

$$CRR_{7.5} = 0.833\left[\frac{(q_{cN})_f}{1000}\right] + 0.05 \qquad (3.7.2.3.11)$$

If $50 < (q_{cN})_f < 160$ then

$$CRR_{7.5} = 93\left[\frac{(q_{cN})_f}{1000}\right]^3 + 0.08 \qquad (3.7.2.3.12)$$

Robertson has further recommended to correct $CRR_{7.5}$ due to vertical overburden pressure when

$$CRR_V = K_\sigma CRR_{7.5} \qquad (3.7.2.3.13)$$

Here, K_σ is obtained from Fig. 3.34 as shown.

In the chart, the confining pressure σ_{cp} is in tsf and can be expressed by

$$\sigma_{cp} = \left[\frac{1 + 2K_0}{3}\right]\sigma_v' \approx 0.65\sigma_v' \qquad (3.7.2.3.14)$$

Here, K_0 is the lateral earth pressure and σ_v' is the effective vertical overburden pressure.

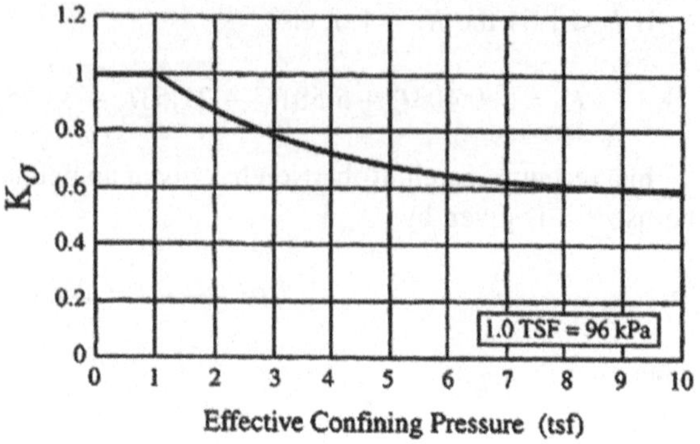

Fig. 3.34 Graph representing the correction facto K_σ

The factor of safety FOS is finally computed as

$$\text{FOS}_M = \left(\frac{\text{CRR}_V}{\text{CSR}}\right) \times \text{MSF}$$

where MSF is as defined by Eq. (3.7.1.12).

Example 3.9 A petrochemical complex in a CIS country having typical CPT boring data is as shown in Fig. 3.35. Compute the liquefaction potential of the site, given $a_{\max} = 0.40g$. Given $\gamma_{\text{soil}}(\text{sat}) = 18.2\ \text{kN/m}^3$ and dry density of soil $\gamma_{\text{d}} = 17.4\ \text{kN/m}^3$. The groundwater table is 1.0 m below the existing ground level. Average FC = 10% for the site. Expected magnitude of earthquake $M = 7.0$, with target FOS = 1.2.

Solution

In this case, to elaborate the problem we work out in detail the problem at 6.0 m below the existing ground level where CPT value is observed minimum. For the rest of the soil profile, we present the results in a graphical form.

The average tip coefficient $q_{c(\text{avg})}$ and friction resistance f_s as computed from two boreholes are shown hereafter in tabular form.

Z (m)	1	2	3	4	5	6	7	8	9	10
$q_{c(\text{avg})}$ (kPa)	12000	10000	10000	12000	10000	3000	500	10000	8000	9500
f_s (kPa)	100	190	250	175	40	15	10	35	20	35

(i) **Seed et al.'s** (1986) **method:**

At a depth of 6.0 ms below GL, $q_c = 3000\ \text{kN/m}^2$ and $fs = 15\ \text{kN/m}^2$.
Thus, the gross overburden pressure $\sigma_v = 17.4 \times 1 + 18.2 \times 5 = 108.4\ \text{kN/m}^2$.
The pore pressure considering water 1.0 m below GL = $10 \times (6 - 1) = 50\ \text{kN/m}^2$.
The net effective pressure $\sigma'_v = 108.4 - 50 = 58.4\ \text{kN/m}^2$.

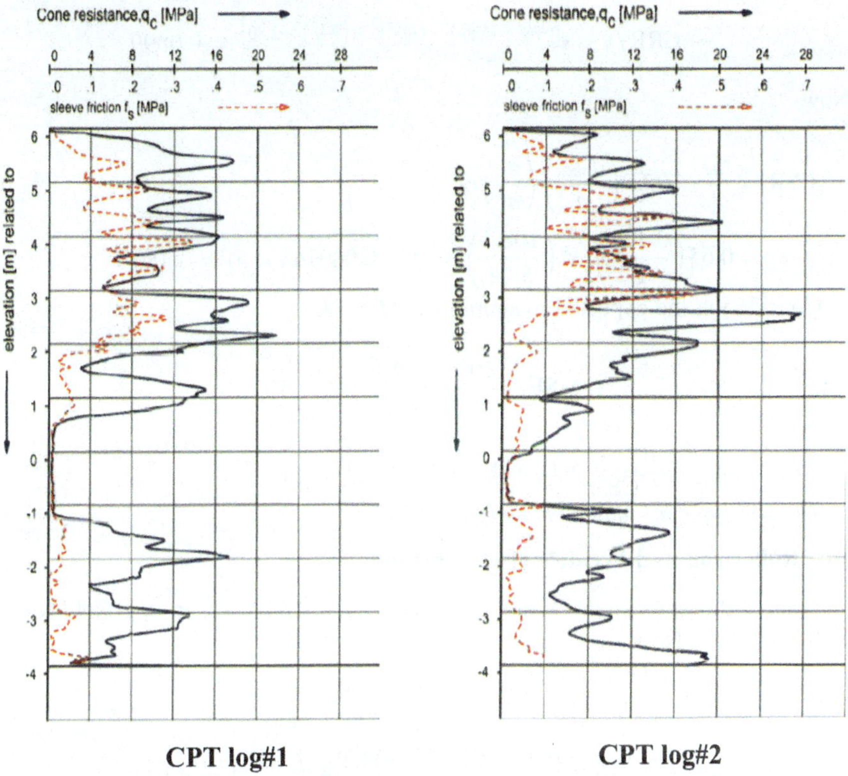

Fig. 3.35 CPT bore log of the given site

Now $C_q = \dfrac{1.8}{0.8 + \frac{\sigma'_v}{P_a}} = \dfrac{1.8}{0.8 + \frac{58.4}{100}} = 1.30$

Thus, the corrected tip pressure due to the overburden $q_{cl} = C_q \cdot q_c = 1.30 \times 3000 = 3900$ kPa $\rightarrow 40.64$ tsf.

The average fineness content FC = 10% when the correction due to fineness is

$$\Delta q_{cl} = \left(11.9 + \frac{q_{cl}}{14.6}\right) e^{\left(1.63 - \left(\frac{9.7}{FC+2}\right) - \left(\frac{15.7}{FC+2}\right)^2\right)}$$

$$= \left[11.9 + \frac{40.64}{14.6}\right] e^{\left(1.63 - \frac{9.7}{12} - \left(\frac{15.7}{12}\right)^2\right)} = 6.03\text{tsf}$$

$\therefore q_{clf} = q_{cl} + \Delta q_{cl} = 40.64 + 6.03 = 46.67$ tsf.

Now $\text{CRR}_{7.5} = e^{\left[\frac{q_{clf}}{113} + \left(\frac{q_{clf}}{1000}\right)^2 - \left(\frac{q_{clf}}{140}\right)^3 + \left(\frac{q_{clf}}{137}\right)^4 - 2.80\right]}$

$$\rightarrow \text{CRR}_{7.5} = \text{e}^{\left[\frac{46.67}{113} + \left(\frac{46.67}{1000}\right)^2 - \left(\frac{46.67}{140}\right)^3 + \left(\frac{46.67}{137}\right)^4 - 2.80\right]} = 0.0899$$

Now

$$\text{CSR} = \frac{\tau_{av}}{\sigma'_v} = 0.65\left(\frac{a_{\max}}{g}\right)\left(\frac{\sigma_v}{\sigma'_v}\right)r_\text{d}$$
$$= 0.65\left(\frac{0.40g}{g}\right) \times \left(\frac{108.4}{58.4}\right) \times (1 - 0.000765 \times 6) = 0.480$$

Considering the expected magnitude as $M = 7.0$

$$\text{MSF} = \frac{10^{2.24}}{M_w^{2.56}} = \frac{10^{2.24}}{7^{2.56}} = 1.19$$

Thus, $\text{FOS} = \frac{\text{CRR} \times \text{MSF}}{\text{CST}} = \frac{0.0899 \times 1.19}{0.480} = 0.223 < 1.0$

The soil liquefies under the shock.

(ii) **Robertson and Wride's** (1997) **method:**

Here, as shown previously, $q_c = 3000$ kN/m^2, $f_s = 15$ kN/m^2, $\sigma_v = 108.4$ kN/m^2, and $\sigma'_v = 58.4$ kN/m^2.

Here, $Q = \frac{q_c - \sigma_v}{P_\text{a}}\left(\frac{P_\text{a}}{\sigma'_v}\right)^n$ where $n = 1.0$ as a start.

$$Q = \frac{q_c - \sigma_v}{P_\text{a}}\left(\frac{P_\text{a}}{\sigma'_v}\right)^n = \frac{3000 - 108.4}{100} \times \left(\frac{100}{58.4}\right)^1 = 49.51$$
$$R_\text{f} = \left[\frac{f_s}{q_c - \sigma_v}\right] \times 100 = \left[\frac{15}{3000 - 108.4}\right] \times 100 = 0.518$$
$$I_c = \sqrt{(3.47 - \log_{10} Q)^2 + (\log_{10} R_\text{f} + 1.22)^2}$$

$I_c = \sqrt{(3.47 - \log_{10} 49.51)^2 + (\log_{10} 0.518 + 1.22)^2} = 1.79 < 2.6$, thus the assumption $n = 1$ is *not correct* and the soil could be liquefiable. So assume $n = 0.5$ when

$Q = \frac{q_c}{P_\text{a}}\left(\frac{P_\text{a}}{\sigma'_v}\right)^n = \frac{3000}{100} \times \sqrt{\frac{100}{58.4}} = 39.256$ and $R_\text{f} = 0.518$ that remains unchanged.

Thus, $I_c = \sqrt{(3.47 - \log_{10} 39.26)^2 + (\log_{10} 0.518 + 1.22)^2} = 1.89 < 2.6$, thus the assumption $n = 0.5$ is *correct,* hence

$$C_Q = \left(\frac{P_\text{a}}{\sigma'_v}\right)^n = \sqrt{\frac{100}{58.4}} = 1.308$$

Thus, the normalized tip resistance is given by

$$q_{cN} = \frac{q_c}{P_a} \times C_Q = \frac{3000}{100} \times 1.308 = 39.256$$

As $I_c > 1.64$

$$K_c = -0.403 I_c^4 + 5.581 I_c^3 - 21.63 I_c^2 + 33.75 I_c - 17.88$$
$$K_c = -0.403 \times 1.79^4 + 5.581 \times 1.79^3 - 21.63 \times 1.79^2 + 33.75 \times 1.79 - 17.88 = 1.18$$

Thus, the value of normalized tip resistance corrected for fineness is

$$(q_{cN})_f = K_c q_{cN} = 1.18 \times 39.256 = 46.356$$

As $(q_{cN})_f < 50$, hence

$$CRR_{7.5} = 0.833 \left[\frac{(q_{cN})_f}{1000}\right] + 0.05 = 0.833 \left[\frac{46.356}{1000}\right] + 0.05 = 0.089$$

The confining pressure σ_{cp} is expressed as

$\sigma_{\text{cp}} = \left[\frac{1+2K_0}{3}\right]\sigma'_v \approx 0.65\sigma'_v = \frac{0.65 \times 58.4}{96} = 0.395$ tsf; thus, from Fig. 3.37 we see that $K_\sigma = 1.0$.

Thus, $\text{CRR}_V = K_\sigma \text{CRR}_{7.5} = 1 \times 0.089 = 0.089$

$$\text{FOS} = \frac{\text{CRR} \times \text{MSF}}{\text{CST}} = \frac{0.089 \times 1.19}{0.480} = 0.220 < 1.0$$

Thus, the soil liquefies in this layer.

It is observed that FOS computed by Seed's method and that by Robertson's method match quite well. The rest of the FOS data are presented graphically hereafter in Fig. 3.36.

It is observed that beyond 5.0 m below GL the soil is highly susceptible to liquefaction and falls below the target FOS of 1.20.

3.7.3 Settlement of Soil Due to Liquefaction

It is apparent that if a soil stratum undergoes liquefaction, it would also undergo large settlement. How much would this settlement be has been a topic of significant research by earthquake engineers.

A number of researchers have sought solution to this problem, of which Tokimatsu and Seed (1987), Robertson (1997), Ishihara and Yoshimine (1990) has been most significant.

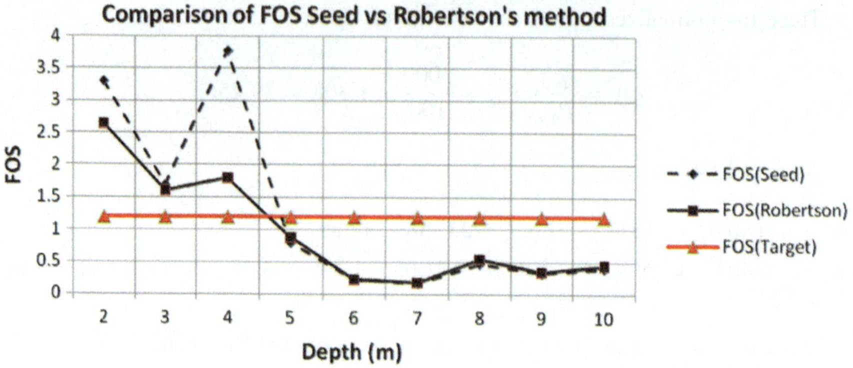

Fig. 3.36 Comparison of FOS Seed et al.'s (1986) versus Robertson and Wride's (1997) method

Of all the above methods, Ishihara's method, being most popular in practice and easy to apply, is presented below.

3.7.3.1 Ishihara and Yoshimine's (1990) Method of Settlement Calculation

For estimation of settlement in soil due to liquefaction, Ishihara and Yoshimine (1990) developed a chart as shown Fig. 3.37.

The chart can be used to compute settlement of soil based on both CPT and SPT tests, when $(N_1)_{60}$ value is correlated to q_{cl} and D_r, as shown in Table 3.7.

Thus, knowing the FOS for a particular layer and the corresponding value of $(N_1)_{60}$ (For CPT test, q_{cl} is the value of tip resistance to be converted to an equivalent $(N_1)_{60}$ value), from Fig. 3.63 one can estimate the post-liquefaction volumetric strain (ε_v%). Once the strain is known, multiplying it by the thickness of the layer one can compute the estimated settlement. Like in the problem as cited in Examples 3.8 and 3.9, if the FOS is estimated for individual layers, then for each layer the settlements are computed separately and summation of all these settlements would give the net effective settlement for the full layer of soil that has liquefied.

Example 3.10 For the problem as cited in Example 3.8, compute the effective settlement for the liquefied strata.

Table 3.7 Correlation between $(N_1)_{60}$ and relative density D_r and CPT tip resistance q_{cl}

$(N_1)_{60}$	3	6	10	14	20	25	30
D_r (%)	30	40	50	60	70	80	90
q_{cl} (N/mm^2)	3.3	4.5	6	8	11	14.7	20

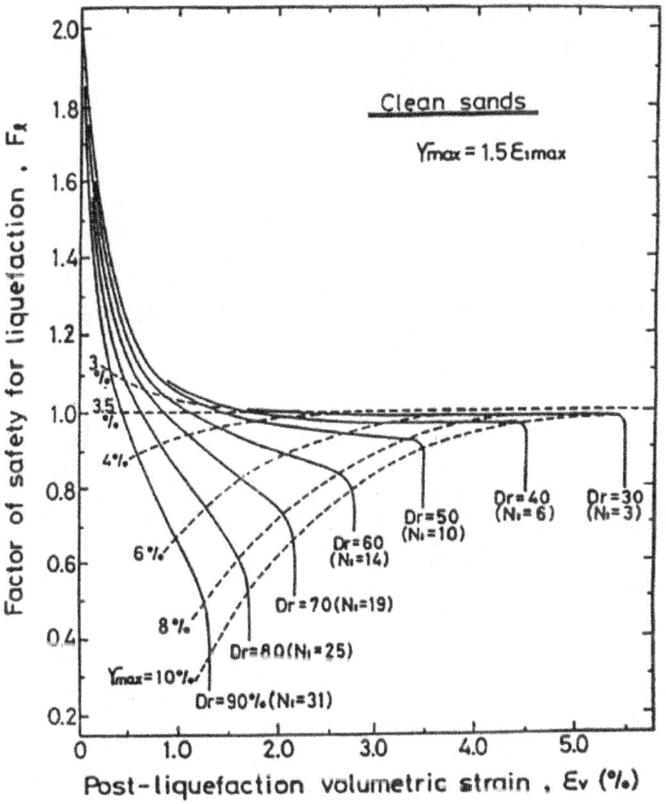

Fig. 3.37 Chart developed by Ishihara and Yoshimine (1990) for settlement estimation

Solution

From the example cited in Example 3.9, we have the following data

Z (m)	$(N_1)_{60}$	$FOS_{6.8}$	ε_v (%) [Read off from Fig. 3.37]
7.725	8	>1.0	1.5
9.225	9	0.608	3.3
10.725	10	0.622	3.2

(continued)

(continued)

Z (m)	$(N_1)_{60}$	$FOS_{6.8}$	ε_v (%) [Read off from Fig. 3.37]
12.225	9	0.585	3.7
13.725	10	0.615	3.5
15.225	11	0.662	3.2
16.725	13	0.744	3.0
18.225	13	0.781	2.8
19.725	15	0.908	1.2
21.225	15	0.946	1.0
22.725	17	1.123	0.9
24.225	17	1.125	0.8
25.725	17	1.167	0.78
27.225	18	1.223	0.75
28.725	20	1.446	0.4
30.225	22	1.694	0.3

Thus, the thickness of each layer and average volumetric strain and settlement for each soil layer are as shown below.

Z (m)	Volumetric strain ε_v (%)	Thickness of each layer (mm)	Average strain ε_{avg} (%)	Settlement (mm) Col 3 × 4
7.725	1.5	1500	2.4	36
9.225	3.3			
10.725	3.2	1500	3.45	51.75
12.225	3.7			
13.725	3.5	1500	3.35	50.25
15.225	3.2			
16.725	3	1500	2.9	43.5
18.225	2.8			
19.725	1.2	1500	1.1	16.5
21.225	1			
22.725	0.9	1500	0.85	12.75
24.225	0.8			
25.725	0.78	1500	0.765	11.475
27.225	0.75			
28.725	0.4	1500	0.35	5.25
30.225	0.3			

Thus, taking summation of all the individual layers, the total estimated settlement = 227.475 mm.

Considering the uncertainty prevailing in the problem, like the error in estimation from the graph and soil data, a (±) 20% uncertainty in the above-estimated value of settlement could well be expected. Thus, the predicted settlement can well be anywhere between 182 and 273 mm.

3.8 Do Clays Liquefy Under Earthquake?

Normally a soil sample, cohesive in nature, is deemed as a non-liquefiable. For c-ϕ soil, the c component is usually not affected by fluctuating pore pressure.

However, there are exceptions.

For instance, in China it has been observed that there are certain types of clay that do flow like a liquid under transient shock.

Based on the above, the Chinese researchers have developed certain criteria under which clay is susceptible to liquefaction, and these are

- Weight of fine particles (FC) with diameter less than 0.005 mm is less than 15% of the dry weight of the soil sample.
- The Liquid limit (LL) of the soil is less than 35%.
- The moisture content of the soil is less than 0.9 times the liquid limit of the soil.

Clay soils not meeting all the above criteria are usually considered as non-liquefiable.

3.9 Effect of Liquefaction on Bearing Capacity of Soil

Normally, for foundations resting on soil which is non-liquefiable or has a FOS against liquefaction greater than 1.0, the seismic effect on foundation is usually ignored.

In fact, in practice, many codes even permit augmentation of the allowable bearing capacity to the extent of 25% in its design capacity.

Recent research has shown (refer Chap. 6 for details) that the hypothesis of increasing this bearing capacity by 25% may not be a prudent idea. Even when the soil is sufficiently strong to withstand the seismic shock without liquefaction, depending on foundation geometry and its location with respect to the bedrock level, there can be a significant reduction in the available FOS and can result in large deformation of the footing and even trigger a failure.

When the soil liquefies, it is thus evident that a foundation can undergo significant deformation and make the superstructure unsafe.

In case liquefaction is near the surface, there can be many remedial measures, like ground improvement locally, supporting the whole structure on common raft foundation, etc.

For example, in Fig. 3.38, we show a structure resting on a soil which can liquefy followed by another layer that is non-liquefiable. It is apparent that if layer 1

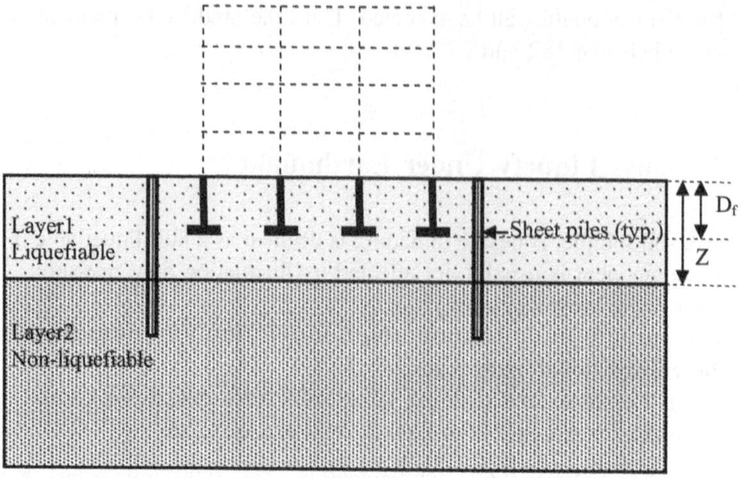

Fig. 3.38 Foundation soil confined by sheet pile all round and compacted

liquefies the foundations can undergo heavy random settlement, damaging the structure.

One solution to overcome this risk would be to surround the structure with sheet piles thus confining the potentially liquefiable soil (thus preventing its lateral flow) and locally improving the confined soil's property by vibro-compaction or preloading of soil, etc.

One can also use piles as a foundation and consider the portion within the liquefiable layer as freestanding while carrying out the design. Cost, convenience of construction, local practice, and importance of the structure built, many issues like this will influence the decision of managing this risk.

On the contrary, if the liquefaction failure is deep-seated, the problem becomes far more complex and with little reflection, one can realize that all the solutions proposed above would fail miserably.

One of the worst things that can happen to a shallow foundation is failure of the same by local punching when, due to heavy load from the column, it punches through the supporting layer and fails as soil layer below it has undergone liquefaction.

3.9.1 *Local Punching Failure of Foundation Due to Soil Liquefaction*

Shown in Fig. 3.39 are foundations supported on a soil layer which is non-liquefiable (Layer 1) which is followed by another soil layer (Layer 2) which is liquefiable. Now, during a seismic shock, if the bottom layer liquefies, it is apparent

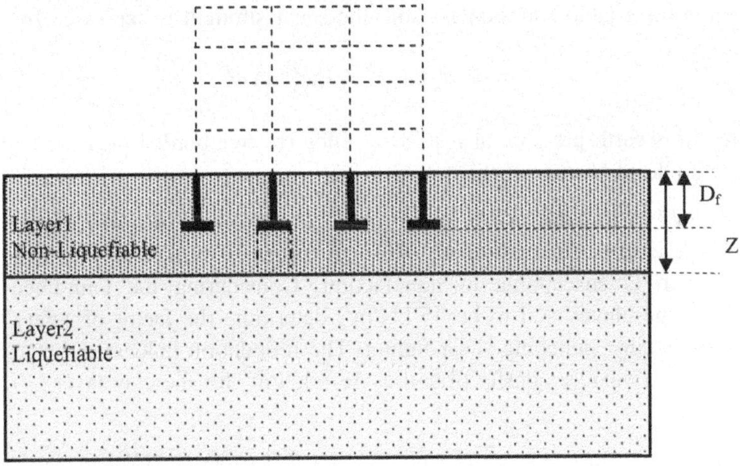

Fig. 3.39 Local punching failure of foundation due to liquefaction of the bottom layer

that the footing would have a tendency to punch through the upper, stiffer layer and fail (this is shown by chain-dotted line in Fig. 3.39.). This is similar to a column that tries to punch through footing under heavy load for which we check its structural integrity.

To prevent this happen, we check the factor of safety to footing.

For isolated footing

$$FS = \frac{2(L+B) \times (Z - D_f)\tau_f}{P} \tag{3.9.1.1}$$

For strip footing

$$FS = \frac{2(Z - D_f)\tau_f}{P} \tag{3.9.1.2}$$

Here, L = length of footing. B = width of footing, Z, D_f are dimensions as shown in Fig. 3.39, P = load from superstructure.

For layer 1, to be cohesive in nature, $\tau_f = S_u$, where S_u is the undrained shear strength of soil.

For the soil layer to be c-ϕ soil, the shear strength is defined as

$$\tau_f = c + \sigma_h \tan \phi \tag{3.9.1.3}$$

Here, $\sigma_h =$ horizontal stress induced in soil and for cohesive soil it is usually considered $0.5\sigma_v$, where $\sigma_v =$ overburden stress.

For non-liquefiable cohesionless soil, the shear strength is expressed by

$$\tau_f = \sigma'_h \tan\phi' = k_0 \sigma'_v \tan\phi' \quad (3.9.1.4)$$

Here, k_0 is earth pressure at rest, σ'_v = effective overburden pressure and ϕ' = effective angle of friction of the cohesive soil.

Example 3.11 A foundation of size 3.5 m × 3.0 m carries a load of 750 kN. It rests on a cohesive soil of overall depth of 3.0 m followed by a sandy layer of 1.5 m depth, which is susceptible to liquefaction. Considering the undrained shear strength of the clayey soil to be 45 kN/m^2, determine the factor of safety of the foundation against punching shear failure. The foundation is located 1.6 m below ground level. Consider density of soil as 18.8 kN/m^3 for the clay layer.

Solution
Here, the overburden pressure of soil = 18.8 × 1.6 = 30.08 kN/m^2.

$$\therefore \tau_f = c + 0.5 \times \sigma_v = 45 + 0.5 \times 30.08 = 60.04 \text{ kN/m}^2.$$

Considering $FS = \frac{2(L+B) \times (Z - D_f)\tau_f}{P} = \frac{2 \times (3.5+3) \times (3.0 - 1.6) \times 60.08}{750} = 1.45$

Taking into account the uncertainty in soil, the FS value should be at least 2.0.

3.9.2 General Shear Failure of Foundation Due to Soil Liquefaction

Collapse or failure of structures due to failure of shallow foundations under earthquake has been a very common observation. A typical failure of buildings due to such foundation failure is as shown in Fig. 3.40.

These types of failure are most common when foundations are resting on clayey stratum followed by a layer of loosely saturated cohesionless soil below the groundwater level.

As per Terzaghi (1943), the ultimate bearing capacity of soil under vertical load is expressed as

$$q_{\text{ult}} = cN_c + qN_q + \frac{1}{2}\gamma B N_\gamma \quad (3.8.2.1)$$

For rectangular foundations of size $L \times B$, Eq. (3.8.2.1) can be expressed as

$$q_{\text{ult}} = cN_c\left[1 + 0.3\frac{B}{L}\right] + qN_q + \frac{1}{2}\gamma B N_\gamma \quad (3.9.2.2)$$

In Eq. (3.9.2.2), c = cohesive property of soil q = overburden pressure on footing $\gamma \times D_f$, and N_c, N_q, N_γ = bearing capacity factors. γ = weight density of soil. D_f = depth of foundation below ground.

Fig. 3.40 Failure of shallow foundation under earthquake in Turkey, after Prakash and Puri (2013)

For clayey soil $\phi = 0 \rightarrow N_\gamma = 0$ and $N_q = 1$, which gives

$$q_{\text{ult}} = cN_c\left[1 + 0.3\frac{B}{L}\right] + \gamma D_{\text{f}} \qquad (3.9.2.3)$$

The above can be further expressed by

$$q_{\text{ult}} = S_u N_c\left[1 + 0.3\frac{B}{L}\right] + \gamma D_{\text{f}} \qquad (3.9.2.4)$$

where S_u = undrained shear strength of soil.

For foundations resting at shallow depth, the second term in Eq. (3.9.2.4) has a marginal effect and for all practical purposes it can be represented as

$$q_{\text{ult}} = S_u N_c\left[1 + 0.3\frac{B}{L}\right] \qquad (3.9.2.5)$$

For a bottom layer of liquefiable soil, there is a reduction in the value of N_c, as shown in Table 3.8.

Example 3.12 For the Example 3.11, as cited previously, determine the reduced bearing capacity of soil, given $S_u = 45$ kN/m^2.

Table 3.8 Reduced values of N_c

$(Z - D_f)/B^*$	0	0.25	0.5	1.0	1.5	∞
N_c	0	0.7	1.3	2.5	3.8	5.5

*Refer Fig. 3.39 for definition of Z and D_f

Solution

Under un-liquefied condition we have

$$q_{\text{ult}} = S_u N_c \left[1 + 0.3\frac{B}{L}\right] + \gamma D_f$$

Here, $B = 3.0$ m and $L = 3.5$ m, $D_f = 1.6$ m, $N_c = 5.5$ and $\gamma = 18.8$ kN/m^3.
Thus, $q_{\text{ult}} = 45 \times 5.5 \times \left[1 + 0.3\frac{3}{3.5}\right] + 18.8 \times 1.6 = 341.2$ kN/m^2.
Thus, the ultimate load carrying capacity is expressed as

$$P_{\text{ult}} = q_{\text{ult}} \times L \times B = 341.2 \times 3.5 \times 3 = 3582.6 \text{ kN}.$$

Therefore, factor of safety before any earthquake = $\frac{3582.6}{750} = 4.77$.

During liquefaction, $\frac{Z - D_f}{B} = \frac{3 - 1.6}{3} = 0.466 \approx 0.5$(say); this gives $N_c = 1.3$ (Refer Table 3.8).

Thus $q_{\text{ult}} = S_u N_c \left[1 + 0.3\frac{B}{L}\right] = 45 \times 1.3 \times \left[1 + 0.3 \times \frac{3}{3.5}\right] = 73.54$ kN/m^2.
Hence, the ultimate load carrying capacity during liquefaction is expressed as

$$P_{\text{ult}} = q_{\text{ult}} \times L \times B = 73.54 \times 3.5 \times 3 = 772.2 \text{ kN}.$$

Thus, the reduced factor of safety = $\frac{772}{750} = 1.03$.

3.10 Ground Subsidence Due to an Earthquake

During liquefaction, if the ground is sloping, or terminates in a canal, due to softening of soil it starts flowing in the lateral direction, creating ground subsidence.

A number of such failures have occurred to ground in different earthquakes around the world, like Loma Prieta (USA), Kobe (Japan).

Figure 3.41 shows failures of highway due to ground subsidence in the Loma Prieta earthquake.

A number of researchers have tried to model this phenomenon; however, parameters involved are so numerous that a theoretical mathematical model to predict correctly this phenomenon is still eluding the profession.

Nonetheless, Bartlett and Youd (1992) have developed an empirical mathematical model based on observations of historical data collected from six earthquakes in USA and two in Japan, which are often used for practical engineering work to estimate the lateral flow hazard. However, authors of the present book

Fig. 3.41 Ground subsidence due to liquefaction, Loma Prieta Earthquake

would recommend them to be used with caution, as local geological condition may strongly affect the actual behavior.

According to Bartlett and Youd (1992), when the land terminates to a canal or a gorge having free end face

$$\log_{10} D_H = -16.3658 + 1.1782M - 0.9275 \log_{10} R - 0.0133R + 0.6572W + 0.3483 \log_{10} T_{15} + 4.5270 \log_{10}(100 - F_{15}) - 0.9224D50_{15} \tag{3.10.1}$$

When the ground is generally sloping

$$\log_{10} D_H = -15.787 + 1.1782M - 0.9275 \log_{10} R - 0.0133R + 0.4293S + 0.3483 \log_{10} T_{15} + 4.527 \log_{10}(100 - F_{15}) - 0.9224D50_{15} \tag{3.10.2}$$

where D_H = estimated average ground displacement in meters; $D50_{15}$ = average mean grain size of the liquefiable layers included in T_{15} in mm; M = moment magnitude of earthquake; R = epicenter distance in km; F_{15} = average fine content (passing ASTM 200 sieve) for liquefiable layer in % included in T_{15}, T_{15} = cumulative thickness in meters of the saturated granular layer having SPT blow count <15. Here, S = ground slope in percent, W = ratio of height (H) of the free face to the distance L from the base of the free face to the point in question.

3.11 Concluding Remarks

It is apparent from the above illustration that soil dynamics as applied to earthquake engineering is a vast subject, having widely diversified nature constituting of geophysics, seismology, geotechnical engineering, and even digital signal processing.

Compared to other topics in civil and earthquake engineering, soil dynamics as a topic is relatively new and there are still a number of issues on which we are in the dark and certainly envisage seeing further development.

In the present chapter, some basic applications to geotechnical earthquake engineering are presented. Further applications of the subject on different topics, like retaining structures, earthen dams, pile foundations, are dealt with subsequent chapters.

Despite significant development in different mathematical models to analyze different aspect of geotechnical phenomena one should remember that in this case we are trying to model nature and, as such, it can never be 100% accurate as many idealizations are made to fit the problem into our known domain of analysis. For this, the design related to geotechnical earthquake engineering should always be mellowed with judgment, conservativeness, and advice from experienced engineers should always be sought if we want to arrive at a realistic result.

References

Aki, P., & Richards, P. (1980). *Quantitative seismology*. USA: University Science Books.

Bartlett, S. F., & Youd, T. L. (1992). *Empirical analysis of horizontal ground displacement generated by liquefaction induced lateral spread*. Technical Report #NCEER-92-0021. National Center for Earthquake Engineering Research.

Blake, T. F. (1997). Summary Report of the procceding of the NCEER workshop on evaluation of liquefaction potential of soil. *Technical report # 0022*. USA: NCEER.

Blake, T. F., Hollingworth, R. A., & Stewart, J. P. (2002). *Recommended procedure for implementation of DMG*. Special Publication #117. Ventura California: Fugro West Inc.

Chowdhury, I., & Dasgupta, S. P. (2007). Dynamic earth pressure on rigid unyielding walls under earthquake forces. *Indian Geotechnical Journal*, *37*(2), 81–93.

Chowdhury, I., & Dasgupta, S. P. (2013). Dynamic response of substructures under earthquake force. *Electronic Journal of Geotechnical Engineering*, *17L*. Oklahoma, USA.

Chowdhury, I., Ghosh, A., & Dasgupta, S. P. (2015). Stiffness degradation and damping augmentation of soil under earthquake loading. *Electronic Journal of Geotechnical Engineering*, *20*(9). Oklahoma, USA.

Cundall, S., & Pyke, R. (1979). Nonlinear soil models for irregular cyclic loading. *Journal of the Geotechnical Engineering Division ASCE*, *105*(6), 715–725.

Das, B. M. (1982). *Fundamentals of soil dynamics*. USA: Elsevier Publication.

Hardin, B., & Drnevich, V. P. (1972). Shear modulus and damping in soil, design equations and curves. *Journal of Soil Mechanics and Foundation Engineering ASCE*, *98*(7), 667–692.

Hudson, D. E. (1956). Response spectrum techniques in engineering seismology. In *Proceedings World Conference on Earthquake Engineering*. California: Berkley.

Hudson, D. E. (1962). Some problems in application of spectrum techniques to strong motion earthquake analysis. *Bulletin of the Seismological Society of America, 52*(2), 417–430.

Idriss, I. (1999). An update to the Seed-Idriss simplified procedure for evaluation of liquefaction potential. In *Proceeding TRB Workshop on New Approaches to Liquefaction*. Publication #FHWA-RD-99-165 Federal Highway Adminstration.

Idriss, I. M., & Boulanger, R. W. (2008). *Soil liquefaction during earthquakes* (261 pp). Monograph MNO-12. Oakland, CA: Earthquake Engineering Research Institute.

Idriss, I. M., & Seed, H. B. (1968). Analysis of liquefaction to Nigata Earthquake. *ASCE 93(SM3)*.

Ishihara, K. (1993). *Soil behavior in earthquake geotechniques*. Oxford, United Kingdom: Clarendon Press.

Ishihara, K., & Yoshimine, Y. (1990). Settlement of buildings on saturated sand during earthquake. *Journal of Soils and Foundations, 17*(1), 23–38.

Kramer, S. L. (1996). *Geotechnical earthquake engineering*. Upper Saddle River, New Jersey, USA: Pearson Education Inc.

Lamb, H. (1904). On propagation of tremors over the surface of an elastic solid. *Philosophical Transactions*. Royal Society of London.

Love, A. E. H. (1944). *A Treatise on Mathmatical Theory in Elasticity*. New York: Dover Publications.

Newmark, N. M., & Hall, W. J. (1973). Procedures and criteria for earthquake resistant design. Building practices for disaster mitigation. *Building Research Series, 46,* 209–236. National Bureau of Standards, US department of commerce.

Nigam, N. C., & Jennings, P. C. (1969). Calculation of response spectra from strong motion earthquake records. *Bulletin of the Seismological Society of America, 59*(2), 909–922.

Prakash, S., & Puri, V. (2013). Shallow foundations for seismic loads. In *7th International Conference in Case Histories in Geotechnical Engineering*. USA: Missouri Institute of Science and Technology.

Ramberg, W., & Osgood, W. (1943). *Description of stress strain curves by three parameters*. Technical note #902. Washington, DC: National Adivisory Committee of Aeronautics.

Rausch, A. F. (1998). An expression for cyclic resistance ratio. Report # 98/3; Department of Civil Engineering. Austin: University of Texas.

Richart, Jr., F. E., Hall, J. R., & Woods, R. D. (1970). *Vibrations of soils and foundations*. N.J.: Prentice Hall Englewood Cliff.

Robertson, P. K., & Wride's, C. E. (1997). Cyclic liquefaction and its evaluation based on SPT and CPT. In *Proceedings NCEER Workshop on Liquefaction Resistance of Soil*.

Schnabel, P. B., Lysmer, J., & Seed, H. B. (1972). *SHAKE a computer program for earthquake response analysis of horizontally layered site*. Research Report #72 12. Earthquake Engineering Research Center, University of California Berkeley.

Seed, H. B., & Idriss, I. M. (1982). *Ground motion and soil liquefaction during earthquake*. Earthquake Engineering Research Institute, University of California Berkeley.

Seed, H. B., Tokimatsu, K., Harder, Jr., L. F., Chung, R. (1984). *Influence of SPT procedures in soil liquefaction resistance evaluations*. Research Report # 84–15. Earthquake Engineering Research Center, University of California Berkeley.

Seed, H. B., Idriss, I. M., & De Abla's, P. (1986). Use of SPT and CPT test for evaluating the liquefaction resistance of soil. *Journal of Soil Mechanics and Foundation Engineering ASCE*.

Srbulov, M. (2011). Practical soil dynamics. In *Case studies in earthquake and geotechnical engineering*. Springer Publications.

Swaran, S. (1999). *Soil dynamics and machine foundation*. New Delhi, India: Galgotia Publication.

Suzuki, Y., Koyamada, K., & Tokimatsu, K. (1997). Prediction of liquefaction resistance based on CPT tip resistance and sleeve friction. In *14th International Conference in Soil Mechanics* (vol. 1 603–606). Hamburg Germany.

Terzaghi, K. (1943). *Theoretical soil mechanics*. USA: John Wiley Publication.

Timoshenko, S., & Goodier, J. N. (1983). *Theory of elasticity*. USA: McgrawHill Publication.

Tokimatsu, K., & Seed, H. B. (1987). Evaluation of settlement in sand due to earthquake shaking. *Journal of Geotechnical Engineering ASCE, 13*(8).

Chapter 4
Earthquake Analysis of Earth Retaining Structures

4.1 Introduction

Retaining walls supporting earth make an important component in infrastructure works like highways, roads, ports, bridge abutment.

It has not been uncommon that during post-earthquake relief operations (especially in mountainous region), relief could not be sent in time, for damage to retaining wall itself has cut off the link road or tilting of the same has resulted in secondary collapse of bridge girders, making some part of relief area inaccessible, thus compounding the problems of the relief workers who have anyway a tough job to execute.

Thus, understanding the behavior of such structures under earthquake load is of paramount importance to civil engineers undertaking such tasks. It is important at times that such structures remain functional even after a severe earthquake maintaining the vital link among two places.

One of the earliest methods for earthquake analysis of such retaining walls has been proposed by Mononobe (1926) and Okabe (1929) from Japan, which still remains the backbone of analysis in almost all design offices and codes of practice around the world.

4.2 Mononobe and Okabe's (M–O) Method (1926) for Analysis of a Gravity-Type Retaining Wall

Mononobe developed a pseudo-static analysis for estimation of earth pressure on a gravity-type retaining wall during an earthquake.

He basically extended the original Coulomb's theory to develop the soil pressure behind a retaining wall catering to the horizontal and vertical component of the ground acceleration.

I. Chowdhury and S. P. Dasgupta, *Earthquake Analysis and Design of Industrial Structures and Infra-structures*, GeoPlanet: Earth and Planetary Sciences,
https://doi.org/10.1007/978-3-319-90832-8_4

Coulomb (1776) derived the equation of active and passive earth pressure as

$$P_a = \frac{1}{2} K_A \gamma H^2 \quad \text{and} \quad P_p = \frac{1}{2} K_P \gamma H^2$$

where

$$K_A = \frac{\cos^2(\phi - \beta)}{\cos^2\beta \cos(\delta + \beta)\left[1 + \left\{\frac{\sin(\delta+\phi)\sin(\phi - i)}{\cos(\delta+\beta)\cos(\beta - i)}\right\}^{\frac{1}{2}}\right]^2} \tag{4.2.1}$$

$= \text{coefficient of active earth pressure, and}$

$$K_P = \frac{\cos^2(\phi + \beta)}{\cos^2\beta \cos(\delta - \beta)\left[1 - \left\{\frac{\sin(\delta+\phi)\sin(\phi + i)}{\cos(\delta-\beta)\cos(i - \beta)}\right\}^{\frac{1}{2}}\right]^2} \tag{4.2.2}$$

$= \text{coefficient of passive earth pressure,}$

γ weight density of soil;
H height of the retaining wall;
ϕ friction angle of the soil;
δ angle of friction between wall and soil;
β angle subtended with the back of the wall with vertical;
i slope of soil with respect to the horizontal.

Mononobe and Matsuo (1926) and Okabe (1929) modified the above coefficient of earth pressure considering the horizontal and vertical seismic coefficient of earthquake force α_h and α_v to

$$K_{AE} = \frac{\cos^2(\phi - \beta - \theta)}{\cos\theta\cos^2\beta \cos(\delta + \beta + \theta)\left[1 + \left\{\frac{\sin(\delta+\phi)\sin(\phi - \theta - i)}{\cos(\delta+\beta+\theta)\cos(\beta - i)}\right\}^{\frac{1}{2}}\right]^2} \tag{4.2.3}$$

where $\theta = \tan^{-1}\left[\frac{\alpha_h}{1-\alpha_v}\right]$; similarly, the coefficient of passive earth pressure under earthquake is given by

$$K_{PE} = \frac{\cos^2(\phi + \beta - \theta)}{\cos\theta\cos^2\beta \cos(\delta - \beta + \theta)\left[1 - \left\{\frac{\sin(\delta+\phi)\sin(\phi - \theta + i)}{\cos(\delta-\beta+\theta)\cos(i - \beta)}\right\}^{\frac{1}{2}}\right]^2} \tag{4.2.4}$$

Mononobe's solution was based on the following assumptions:
Referring to Fig. 4.1

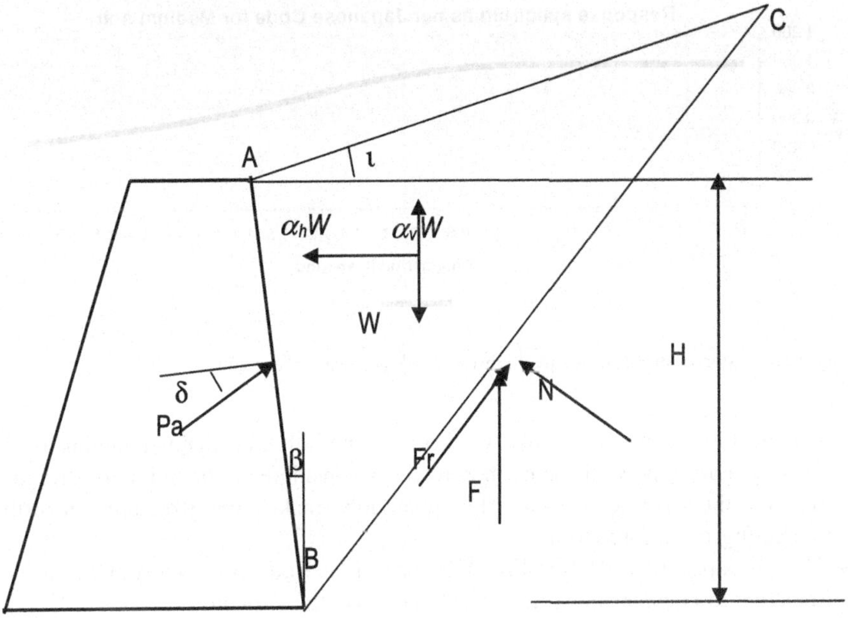

Fig. 4.1 Mononobe's (1926) force diagram for gravity-type retaining wall under earthquake

- The failure in soil takes place along the plane BC.
- The movement of the wall is sufficient to produce an active plane.
- At failure full shear strength along the failure plane BC is mobilized.
- The soil mass behind the wall behaves as a rigid body.
- α_h and α_v are maximum peak acceleration coefficient (*Sa*/*g*) in horizontal and vertical directions, respectively.

It will be observed that in the above calculation no reference is made to time period, which shows the analysis to be pseudo-static in nature.

The value of peak ground acceleration a_0 may be calculated empirically from Guttenberg and Richter's (1956) expressions as

$$\log_{10} a_0 = -2.1 + 0.81M - 0.027M^2 \tag{4.2.5}$$

Here a_0 is in cm/sec^2 and $\alpha_h = a_0/g$.

As per IS-1893(2007), this can be considered as

$$\alpha_h = (ZI/2R)\left(\frac{S_{a\max}}{g}\right) \tag{4.2.6}$$

and, $\alpha_v = 0.5\alpha_h$ half of the lateral response.

The α_h and α_v values are considered as peak acceleration, which is quite justified in this case because of the following two reasons

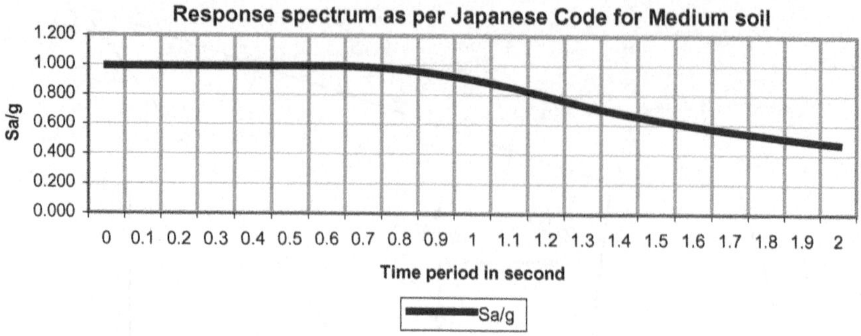

Fig. 4.2 Response spectrum for medium-type soil as per Japanese code

- At the time when Mononobe worked out the solution, major retaining walls were gravity-type walls being massive, was considered to be infinitely stiff, i.e., having time period $T \rightarrow 0$. This invariably makes the structure very stiff attracting more force to it.
- As per Japanese code (see Fig. 4.2) for time period up to nearly 0.8 s, it is a practice to consider maximum acceleration on the structure.

Mononobe (1926) considered the dynamic pressure distribution as parabolic in nature, and for active case, it was assumed that the force is acting at a height $H/2$ from the bottom of wall. While for passive case, again considering a parabolic distribution, the force was assumed to be acting at a height of $2/3H$ from the base of wall.

The above procedure has almost universally dominated the design office practice in almost all countries (including India), though characteristics of retaining wall has undergone significant change with advent of reinforced concrete structures.

Nowadays, the common practice is to go for a cantilever retaining wall for a height up to 6.0 m, beyond which engineers normally provide counterfort retaining wall which would provide overall economy in design. Typical sketch of such walls is as shown in Fig. 4.3.

It is obvious from Fig. 4.3 that the present-day retaining walls have become much more flexible as such considering maximum acceleration (i.e., for $T = 0$ as shown in Japanese code in Fig. 4.2) inducing design pressure may not be true in all cases. Considering the wall is flexible, the time period of the walls will surely play a dominant role in dynamic response of the system where a pseudo-static analysis only could well become insufficient.

Example 4.2.1 A 6-m-high-gravity retaining wall has the following properties:

Density of soil is 20 kN/m^3. The soil is inclined at $\iota = 10°$ with horizontal. The internal angle of friction is $\phi = 32°$. Friction between the wall and soil is $\delta = 18°$. The earthquake zone is zone 4 as per IS-1893, when $Z = 0.24$, $I = 1.2$, and $R = 1.5$. Determine the active and passive earth pressure on the wall and find out the base moment based on M–O method. Assume the wall soil interface as perfectly vertical, i.e., $\beta = 0°$.

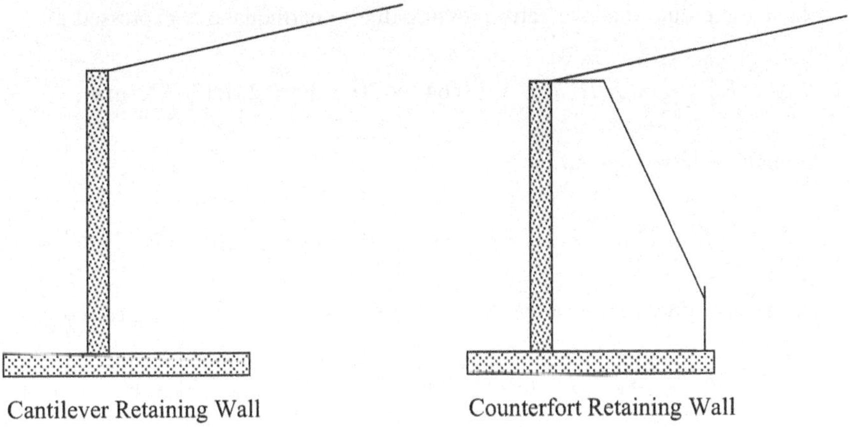

Fig. 4.3 Present-day cantilever and counterfort retaining walls

Peak ground acceleration $Sa/g = 2.5$ as per IS-1893. Vertical acceleration is ½ the horizontal acceleration = 1.25.

$$\alpha_h = (ZI/2R) \times (Sa/g) = \tfrac{0.24\times 1.2}{2\times 1.5} \times 2.5 = 0.24\text{, similarly } \alpha_v = 0.12.$$

$$\theta = \tan -1\left(\frac{\alpha_h}{1-\alpha_v}\right) = \tan -1\left(\frac{0.24}{1-0.12}\right) = 15.24^\circ.$$

$$K_{AE} = \frac{\cos^2(\phi-\beta-\theta)}{\cos\theta\cos^2\beta\cos(\delta+\beta+\theta)\left[1+\left\{\frac{\sin(\delta+\phi)\sin(\phi-\theta-i)}{\cos(\delta+\beta+\theta)\cos(\beta-i)}\right\}^{\frac{1}{2}}\right]^2}$$

$$\Rightarrow K_{AE} = \frac{\cos^2(32-0-15.24)}{\cos 15.24\cos^2 0\cos(18+0+15.24)\left[1+\left\{\frac{\sin(18+32)\sin(32-15.24-10)}{\cos(18+0+15.24)\cos(0-10)}\right\}^{\frac{1}{2}}\right]^2} = 0.642$$

Similarly

$$K_{PE} = \frac{\cos^2(\phi+\beta-\theta)}{\cos\theta\cos^2\beta\cos(\delta-\beta+\theta)\left[1-\left\{\frac{\sin(\delta+\phi)\sin(\phi-\theta+i)}{\cos(\delta-\beta+\theta)\cos(i-\beta)}\right\}^{\frac{1}{2}}\right]^2}$$

$$\Rightarrow K_{PE} = \frac{\cos^2(32+0-15.24)}{\cos 15.24\cos^2 0\cos(18-0+15.24)\left[1-\left\{\frac{\sin(18+32)\sin(32-15.24+10)}{\cos(18-0+15.24)\cos(10-0)}\right\}^{\frac{1}{2}}\right]^2} = 2.465$$

Shear force due to active earth pressure due to earthquake is expressed as

$$P_{ae} = \frac{1}{2} K_{ae} \gamma_s H^2 = 0.5 \times 0.642 \times 20 \times 36 = 231.136\ \text{kN/m}$$

Moment at base is given by

$$M_{ae} = \frac{1}{2} K_{ae} \gamma_s H^2 \times \frac{H}{2} = 0.5 \times 0.642 \times 20 \times 36 \times \frac{6}{2} = 643.409\ \text{kN m/m}$$

Shear force due to passive earth pressure due to earthquake is expressed as

$$P_{pe} = \frac{1}{2} K_{pe} \gamma_s H^2 = 0.5 \times 2.465 \times 20 \times 36 = 887.4\ \text{kN/m}$$

Moment at base is given by

$$M_{pe} = \frac{1}{2} K_{pe} \gamma_s H^2 \times \frac{2H}{3} = 887.4 \times \frac{2 \times 6}{3} = 3549.6\ \text{kN m/m}$$

Observe here that time period of the wall does not come into play. The formula is valid only for cohesionless soil with no overburden.

4.2.1 Some Comments on M–O Method and Its Validity

Despite its popularity in design office and research, it is now an established truth that M–O method underrates the dynamic response significantly.

Major lacunae that exist in M–O method can be summarized as hereafter

- M–O method assumes the wall to be rigid and only considers the soil subjected to peak ground acceleration. This is possibly not realistic, as because the wall may have high stiffness, it is surely not infinite and can affect the overall seismic response. Structural influence of the wall itself on the overall dynamic response is completely ignored in the M–O method.
- It fails to predict the point of action of the lateral thrust, which at best remains an approximation.
- M–O method was actually derived for dry cohesionless backfill and really does not cater to general c-ϕ soil or soil with additional surcharge over the backfill, though have been rampantly used or abused in many such cases.
- It does not address the dynamic soil–structure interaction (DSSI) effect with the foundation soil that can well amplify the response.

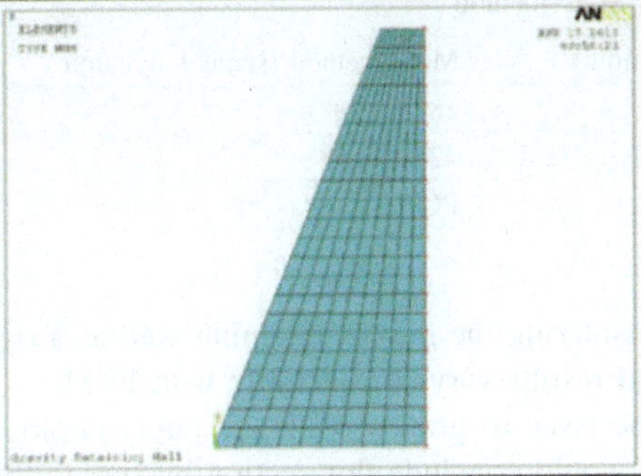

Details of mesh

Type of plane stress element: PLANE183

Type of mass element: MASS21

No. of plane stress elements: 220

No. of mass elements: 23

Boundary condition: bottom nodes fixed

FE mesh of fixed base gravity retaining wall

Fig. 4.4 Finite element model of gravity retaining wall after Chowdhury and Singh (2015)

An investigation on this account was carried out by Chowdhury and Singh (2015)[1] where a detailed finite element analysis and a method proposed by them were compared with conventional M–O method.

The data used were as follows

(1) Properties of gravity retaining wall

Height (H) of 10 m, base width (B_b) of 5 m, top width (B_t) of 0.5 m, modulus of elasticity of retaining wall (E) 1.2×10^7 kN/m^2, Poisson's ratio (v) 0.15, and unit weight (γ) of 24 kN/m^3,

(2) Properties of backfilled soil

Angle of internal friction (ϕ) of 31.19°, unit weight of soil (γ_s) 17 kN/m^3, cohesion (c) 1 kN/m^2, surcharge (q) 10 kN/m^2, depth of water table from top 1 m, angle of internal friction below water table (ϕ) 25°, submerged unit weight of soil below water table (γ_s) 7 kN/m^3, angle of inclination ($\imath$) of 10°.

(1) Seismic code factors

Peak ground acceleration was considered as 0.36g, having importance factor = 3.0 and response reduction factor $R = 1.5$. Vertical acceleration (a_v) considered was half of the horizontal (a_h).

The analysis comprehensively showed that M–O method consistently underrates the response.

The typical finite element model of the gravity wall, as analyzed by ANSYS, is as shown in Fig. 4.4 under seismic loading and compared with conventional M–O method. The comparative results are as shown in Table 4.1.

---

[1]Chowdhury and Singh (2015); Behavior of gravity-type retaining wall under earthquake load under generalized earth fill—*Journal of Earthquake Engineering* December Taylor and Francis USA.

Table 4.1 Comparative values FEM versus M–O method

Forces	ANSYS (static + dynamic)	M–O method (static + dynamic)
Moment	2821	1825
Shear	559	424
Acceleration	Sa(T)	PGA

It is evident that the concept of considering the gravity retaining wall as a rigid body (M–O method) is not correct and results vary considerably with FEM.

We will not further elaborate on the issue as gravity walls are rarely deployed nowadays as retaining structures. But would conclude this section by stating that, despite the above limitations, Mononobe's pseudo-static method is in extensive use, irrespective of whether the wall is a cantilever or a counterfort one or if it is thin or thick in dimension.

Before we try to tackle and circumvent the above limitations as posed in M–O method, we describe herein other methods which are also in practice for evaluation of pressure due to earthquake force in different countries.

4.3 Seed and Whitman's Method

Seed and Whitman (1970) developed an expression which may also be used to determine the horizontal pseudo-static force acting on a retaining wall.

As per their calculations, the active earth pressure may be expressed as

$$P_{AE} = \frac{3}{8}\frac{a_{\max}}{g}\gamma_s H^2 \tag{4.3.1}$$

According to this work, the location pseudo-static force is assumed to act at height of $0.6H$ above the base of the structure.

Example 4.2.2 A 6-m-high-gravity retaining wall has the following properties:

Density of soil of 20 kN/m^3. The soil is inclined at $\imath = 10°$ with horizontal. The internal angle of friction is $\phi = 32°$. Friction between wall and soil is $\delta = 18°$. The earthquake zone is zone 4 as per IS-1893, when $Z = 0.24$, $I = 1.2$, and $R = 1.5$. Determine the active and passive earth pressure on the wall and find out the base moment based on Seed and Whitman's method. Assume the wall soil interface as perfectly vertical, i.e., $\beta = 0°$.

Solution

Peak ground acceleration $Sa/g = 2.5$ as per IS-1893. Vertical acceleration is ½ the horizontal acceleration = 1.25.

$\alpha_h = (ZI/2R) \times (Sa/g) = \frac{0.24\times1.2}{2\times1.5} \times 2.5 = 0.24$, similarly $\alpha_v = 0.12$.

Dynamic shear on wall, vide Eq. 4.3.1, is $(3/8) \times 0.24 \times 20 \times 36 = 64.8$ kN

As per Coulomb's equation

$$K_{AE} = \frac{\cos^2(\phi - \beta)}{\cos^2\beta\,\cos(\delta+\beta)\left[1+\left\{\frac{\sin(\delta+\phi)\sin(\phi-i)}{\cos(\delta+\beta)\cos(\beta-i)}\right\}^{\frac{1}{2}}\right]^2}$$

Considering ϕ = 32, ι = 10, and δ = 18, we have K_{AE} = 0.313.
Thus, static shear $P_{AE} = (1/2)K_{AE}\gamma_s H^2$

$$P_{AE} = 0.5 \times 0.313 \times 24 \times 36 = 135.216 \text{ kN}.$$

Thus, the total static plus dynamic shear = 135.216 + 64.8 = 200 kN.

As per Seed and Whitman, the dynamic force acts at a height of 0.6H above base.

Thus, maximum moment at base $= 135.216 \times 6/3 + 64.8 \times 0.6 \times 6 =$ 503.71 kN m.

4.4 Steedman and Zeng's (1990) Method

For the fixed base cantilever, as shown above, if subjected to ground acceleration a_h, the acceleration at depth z is given by (Fig. 4.5)

$$a(z,t) = a_h \sin\left[\omega\left(t - \frac{H-z}{v_s}\right)\right] \tag{4.4.1}$$

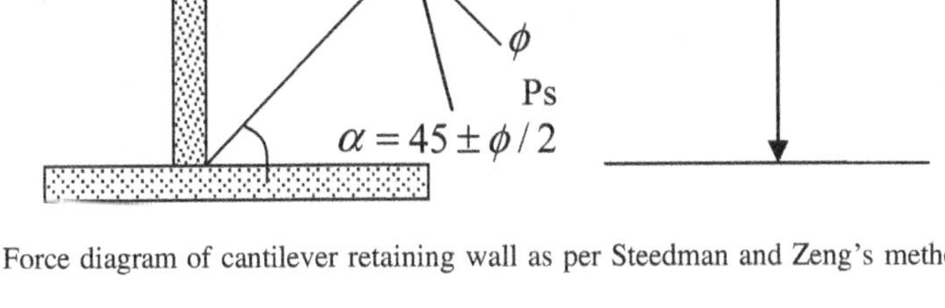

Fig. 4.5 Force diagram of cantilever retaining wall as per Steedman and Zeng's method

Considering that the pressure on the wall is resulting from the triangular wedge only, which is at a state of incipient failure, mass of a thin strip of depth dz within the soil wedge is given by $m(z,t) = \frac{\gamma H - z}{g \tan\alpha} dz$ where γ is weight density of the soil.

The total inertial force P_h acting on the wall can thus be expressed as

$$P_h(t) = \int_0^H m(z)a(z,t) = \frac{\lambda\gamma a_h}{4\pi^2 g \tan\alpha}[2\pi H \cos\omega\zeta + \lambda(\sin\omega\zeta - \sin\omega t)] \quad (4.4.2)$$

Here $\lambda = \frac{2\pi}{v_s\omega}$ is the vertically propagating shear wave length and $\zeta = t - \frac{H}{v_s}$.

For a special case when the failed wedge acts as a rigid block (i.e., $v_s \to \infty$), we have

$$(P_h(\max))_{\lim v_s \to \infty} = \frac{\gamma H^2 a_h}{2g \tan\alpha} = \frac{a_h}{g} W = \alpha_y W \quad (4.4.3)$$

The above can be stated to be equivalent force as assumed in the Mononobe's method.

The total static plus dynamic thrust can then be obtained by resolving the force in the wedge and is given by

$$P_{AE} = \frac{P_h(t)\cos(\alpha - \phi) + W \sin(\alpha - \phi)}{\cos(\delta + \phi - \alpha)} \quad (4.4.4)$$

The total earth pressure is obtained by differentiating the above expression with respect to z; thus,

$$p_{AE} = \frac{\partial P_{AE}}{\partial z} = \frac{\gamma z}{\tan\alpha} \frac{\sin(\alpha - \phi)}{\cos(\delta + \phi - \alpha)} + \frac{\alpha_h \gamma z}{\tan\alpha} \frac{\cos(\alpha - \phi)}{\cos(\delta + \phi - \alpha)} \sin\left[\omega\left(t - \frac{z}{v_s}\right)\right] \quad (4.4.5)$$

The first term in the above equation represents the static pressure acting at a height of $H/3$ from the base while the second term represents the dynamic pressure where the thrust point is found to be varying with time and is given by

$$h_d = H - \frac{2\pi^2 H^2 \cos\omega\zeta + 2\pi\lambda H \sin\omega\zeta - \lambda^2(\cos\omega\zeta - \cos\omega t)}{2\pi H \cos\omega\zeta + \pi\lambda(\sin\omega\zeta - \sin\omega t)} \quad (4.4.6)$$

For very low-frequency motion, the dynamic thrust is found to be at $H/3$ (this is when H/λ is small and the backfill moves in the same phase). For higher frequency motions, the point of application is found to move higher up on the wall.

4.5 Dynamic Analysis of Cantilever Retaining Wall

So far we have explained the various pseudo-static and pseudo-dynamic methods available for analysis of retaining walls under earthquake force. We have also explained briefly some of the limitations of the pseudo-static method which is perhaps more appropriate for gravity type of wall. We now present herein a procedure for dynamic response of cantilever and counterfort type of retaining wall.

While studying the dynamic response of structure previously we had seen that the acceleration it would be subjected to, depends on the time period of the structure. Then considering a suitable damping ratio, we can find out the earthquake response based on code prescribed response spectrum.

Based on the above philosophy, we proceed to explain the investigation which has been carried out by Chowdhury and Dasgupta (2011).

To start with, we take the simplest case as shown below.

Shown in Fig. 4.6 is a cantilever retaining wall with dry sandy backfill and the ground with no inclination. It is to be noted that the response can also be derived from a more generalized soil condition but has been considered first for the sake of brevity and also to use it as a benchmark for more generalized cases that will be taken up subsequently.

While performing the analysis, it is assumed here that

- The soil profile under active case is at incipient failure when the failure line makes angle $\alpha = \tan(45 + \phi/2)$, as shown in the below figure.
- Since soil profile is already under failed condition under static load, it will not induce any stiffness to the overall dynamic response but will only contribute to the inertial effect.

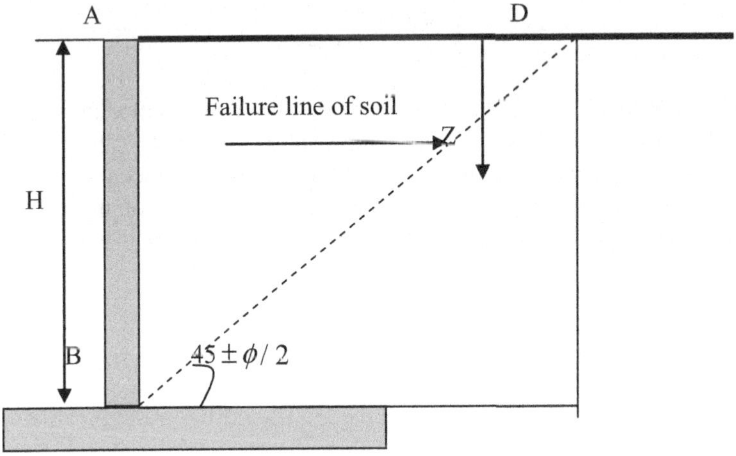

Fig. 4.6 A cantilever retaining wall with dry cohesionless backfill ($c = 0$)

- Since the cantilever wall is relatively thin, mass contribution of the wall itself may be ignored, compared to that of the soil. The wall thus contributes only to stiffness of the overall soil–structure system.
- The retaining wall is fixed at base, and foundation compliance has been ignored for the present analysis.

It will be observed that assumptions made above are identical to what Mononobe (1926) or Steedman and Zeng (1990) have assumed in their analysis. Based on above assumptions, the analysis is carried out as elaborated hereafter.

4.5.1 *Dynamic Response of Dry Cohesionless Backfill* $(C = 0)$

Considering ϕ as the internal angle of friction, the active and passive coefficients of pressure K_A and K_P may be expressed as

$$K_A = \frac{1 - \sin\phi}{1 + \sin\phi}; \text{ and } K_P = \frac{1 + \sin\phi}{1 - \sin\phi} \tag{4.5.1}$$

The static pressures acting on the wall can be expressed as

$$p_{az} = K_A \cdot \gamma_s \cdot z \tag{4.5.2}$$

$$p_{pz} = K_P \cdot \gamma_s \cdot z \tag{4.5.3}$$

For the wall considered as a cantilever beam fixed at the base slab, the differential equation of static equilibrium under active soil condition can be expressed as

$$EI\frac{d^4u}{dz^4} = K_A \cdot \gamma_s \cdot z \tag{4.5.4}$$

Here u is displacement of the retaining wall, E = Young's modulus of the RCC wall and I is moment of inertia of the wall considered, $I = 1/12 \cdot (B \times t^3)$; here t is the thickness of the wall (can be considered as an average thickness for variation between top and bottom thickness of the wall), B = width of the wall usually considered as 1.0 m as the analysis is usually carried out per meter width of the wall.

On successive integration of Eq. (4.5.4), we have

$$EI\frac{d^3u}{dz^3} = \frac{K_A \cdot \gamma_s z^2}{2} + C_1 \tag{4.5.5}$$

$$EI\frac{d^2u}{dz^2} = \frac{K_A \cdot \gamma_s z^3}{6} + C_1 z + C_2 \tag{4.5.6}$$

$$EI\frac{du}{dz} = \frac{K_A \cdot \gamma_s z^4}{24} + C_1\frac{z^2}{2} + C_2 z + C_3 \tag{4.5.7}$$

$$EIu = \frac{K_A \cdot \gamma_s z^5}{120} + C_1\frac{z^3}{6} + C_2\frac{z^2}{2} + C_3 z + C_4 \tag{4.5.8}$$

For the given wall (Fig. 4.6), we consider the boundary conditions

(1) $$At\ z = 0\ \frac{d^3u}{dz^3} = 0 \Rightarrow C_1 = 0 \tag{4.5.8a}$$

(2) $$\text{At } z = 0\ \frac{d^2u}{dz^2} = 0 \Rightarrow C_2 = 0 \tag{4.5.8b}$$

(3) $$\text{At } z = H\ \frac{du}{dz} = 0 \Rightarrow C_3 = -\frac{K_A \cdot \gamma_s \cdot H^4}{24} \tag{4.5.8c}$$

(4) $$\text{At } z = H\ u = 0 \Rightarrow C_4 = \frac{K_A \cdot \gamma_s \cdot H^5}{30} \tag{4.5.8d}$$

Thus, Eq. (4.5.8) can be expressed as

$$EIu = \frac{K_A \cdot \gamma_s z^5}{120} - \frac{K_A \cdot \gamma_s H^4 z}{24} + \frac{K_A \gamma_s H^5}{30} \tag{4.5.9}$$

Equation (4.5.9) can be further expressed in generalized form as

$$u = \frac{K_A \cdot \gamma_s \cdot H^5}{30EI}\left[\frac{\xi^5}{4} - \frac{5\xi}{4} + 1\right] \tag{4.5.10}$$

where $\xi = z/H$ is a non-dimensional term that varies between 0 and 1.

Static deflection of wall under the given condition can thus be expressed as

$$u_{\text{static}} = \frac{K_A \cdot \gamma_s \cdot H^5}{30EI} \text{ at } \xi = 0 \tag{4.5.11}$$

Time period of the wall can then be expressed as

$$T = 2\pi\sqrt{\frac{u_{\text{static}}}{g}} \tag{4.5.12}$$

Substituting Eqs. (4.5.11) in (4.5.12) and considering $I = t^3/12$, we have

$$T_A = 3.97\sqrt{\frac{K_A \gamma_s H^5}{Et^3 g}} \tag{4.5.13}$$

For modal analysis, maximum amplitude (S_d) is expressed as

$$S_d = \frac{S_a}{\omega^2} \tag{4.5.14}$$

Here S_a = spectral acceleration and $\omega = 2\pi/T$ is the natural frequency of the wall.

In terms of code, Eq. (4.5.14) can be expressed as

$$S_d = \kappa\beta \frac{S_a}{\omega^2} \tag{4.5.15}$$

Here $\kappa =$ modal mass participation factor and is expressed as $\sum_{i=1}^{n} m_i \varphi_i / \sum_{i=1}^{n} m_i \varphi_i^2$. β = code factor expressed as $ZI/2R$, where Z = zone factor, I = Importance factor, and R = Response reduction factor.

Thus, based on Eq. (4.5.15) the dynamic amplitude of the wall can be expressed as

$$u = \kappa\beta \frac{S_a}{4\pi^2} T^2 f(\xi) \tag{4.5.16}$$

where $f(\xi) = \frac{\xi^5}{4} - \frac{5\xi}{4} + 1$.

Equation (4.5.16) can be finally expressed as

$$u = \kappa\beta \frac{K_A \gamma_s H^5}{30EI} \left(\frac{S_a}{g}\right) \left[\frac{\xi^5}{4} - \frac{5\xi}{4} + 1\right] \tag{4.5.17}$$

Considering $M = EI\ \mathrm{d}^2u/\mathrm{d}z^2$ and $V = EI\ \mathrm{d}^3u/\mathrm{d}z^3$, we have

$$M_\xi = \kappa\beta \frac{K_A \gamma_s H^3}{6} \left(\frac{S_a}{g}\right) [\xi^3] \tag{4.5.18}$$

$$M_z = \kappa\beta \frac{K_A \gamma_s z^3}{6} \left(\frac{S_a}{g}\right) \tag{4.5.19}$$

Similarly,

$$V_z = \kappa\beta \frac{K_A \gamma_s z^2}{2} \left(\frac{S_a}{g}\right) \tag{4.5.20}$$

Equations (4.5.17), (4.5.19), and (4.5.20) are exact and give the dynamic displacement, moment, and shear for a cantilever retaining wall under earthquake force in fundamental mode for cohesionless dry backfill.

The modal participation factor κ can be expressed in this case as

$$\kappa = \frac{\sum_{i=1}^{n} m_i \varphi_i}{\sum_{i=1}^{n} m_i \varphi_i^2} = \frac{\int_0^1 \gamma_s H^2 \xi f(\xi) \mathrm{d}\xi}{\int_0^1 \gamma_s H^2 \xi f(\xi)^2 \mathrm{d}\xi} \tag{4.5.21}$$

$$\kappa = \frac{\int_0^1 \xi \left(\frac{\xi^5}{4} - \frac{5\xi}{4} + 1\right) \mathrm{d}\xi}{\int_0^1 \xi \left(\frac{\xi^5}{4} - \frac{5\xi}{4} + 1\right)^2 \mathrm{d}\xi} \tag{4.5.22}$$

Equation (4.5.22) on integration gives $\kappa = 2.133$.

4.5.2 Effect of Vertical Acceleration S_v

From Eq. (4.5.4), we have seen that

$$p = EI \frac{\mathrm{d}^4 u}{\mathrm{d}z^4} = K_A \cdot \gamma_s \cdot z \tag{4.5.23}$$

Thus, for the present case, dynamic pressure on wall is expressed as

$$p_{\mathrm{dyn}} = EI \frac{\mathrm{d}^4 u}{\mathrm{d}z^4} = \frac{EI}{H^4} \frac{\mathrm{d}}{\mathrm{d}\xi^4} \kappa\beta \frac{K_A \gamma_s H^5}{30EI} \left(\frac{S_a}{g}\right) \left[\frac{\xi^5}{4} - \frac{5\xi}{4} + 1\right] \tag{4.5.24}$$

$$\Rightarrow p_{\mathrm{dyn}} = \kappa\beta K_A \gamma_s \left(\frac{S_a}{g}\right) z \tag{4.5.25}$$

Now if S_v is the vertical acceleration corresponding to time period T_A, then the dynamic pressure in vertical direction can be expressed as

$$p_{V\,\mathrm{dyn}} = \pm \kappa\beta \gamma_s \left(\frac{S_V}{g}\right) z \tag{4.5.26}$$

In horizontal direction, the effect of this pressure can be expressed as

$$p_{H\,\mathrm{dyn}} = \pm \kappa\beta K_A \gamma_s \left(\frac{S_V}{g}\right) z \tag{4.5.27}$$

Thus, the total dynamic pressure considering the vertical component of acceleration can now be expressed as

$$p_{\mathrm{dyn}} = \kappa\beta K_A\gamma_s\left(\frac{S_a}{g}\right)z \pm \kappa\beta K_A\gamma_s\left(\frac{S_v}{g}\right)z \tag{4.5.28}$$

For maximum pressure, we must take the positive sign that gives

$$p_{\mathrm{dyn}} = \kappa\beta K_A\gamma_s\left(\frac{S_a}{g}\right)z + \kappa\beta K_A\gamma_s\left(\frac{S_v}{g}\right)z \tag{4.5.29}$$

As per IS-1893(2016) considering $S_v = 2S_a/3$ Eq. (4.5.29) can thus be expressed as

$$p_{\mathrm{dyn}} = \kappa\beta K_A\gamma_s\left(\frac{5S_a}{3g}\right)z \tag{4.5.30}$$

Thus, considering the effect of vertical acceleration the dynamic displacement, moment, and shear can be expressed as

$$u = \kappa\beta\frac{K_A\gamma_s H^5}{18EI}\left(\frac{S_a}{g}\right)\left[\frac{\xi^5}{4} - \frac{5\xi}{4} + 1\right] \tag{4.5.31}$$

$$M_z = \kappa\beta\frac{5K_A\gamma_s z^3}{18}\left(\frac{S_a}{g}\right) \tag{4.5.32}$$

$$V_z = \kappa\beta\frac{5K_A\gamma_s z^2}{6}\left(\frac{S_a}{g}\right) \tag{4.5.33}$$

Equations (4.5.31) to (4.5.33) show that the displacement, moments, and shears get amplified by 67% when effect of vertical acceleration is considered *and should not be ignored.*

4.5.3 Soil Inclined at an Angle i with the Vertical

In this case, expressions presented in Eqs. (4.5.31), (4.5.32), and (4.5.33) remain valid except that in this case the active and passive earth pressures are expressed as (Fig. 4.7)

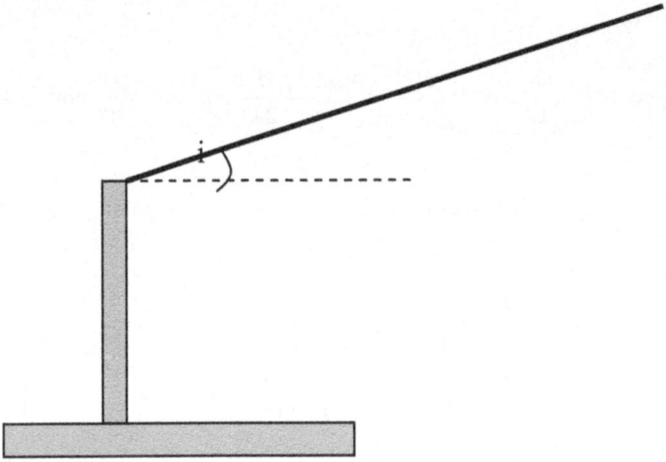

Fig. 4.7 Retaining wall with inclined backfill at an angle i

$$K_A = \cos i \times \frac{\cos i - \sqrt{\cos^2 i - \cos^2 \phi}}{\cos i + \sqrt{\cos^2 i - \cos^2 \phi}} \tag{4.5.34}$$

$$K_P = \cos i \times \frac{\cos i + \sqrt{\cos^2 i - \cos^2 \phi}}{\cos i - \sqrt{\cos^2 i - \cos^2 \phi}} \tag{4.5.35}$$

4.5.4 Dynamic Response of Wall with c-ϕ Backfill

For a general c-ϕ soil, the active earth pressure is expressed as

$p_a = \gamma_s z \cot^2 \alpha - 2c \cot \alpha$, where $\alpha = 45 + \phi/2$
Substituting this in Eq. (4.5.4), we have

$$EI \frac{d^4 u}{dz^4} = \gamma_s \cdot z \cdot \cot^2 \alpha - 2c \cot \alpha \tag{4.5.36}$$

Proceeding in identical fashion as explained earlier and imposing the boundary conditions as stated in Eqs. 4.5.8a–4.5.8d, we have $C_1 = C_2 = 0$ and

$$C_3 = \frac{cH^3 \cot \alpha}{3} - \frac{\gamma_s H^4}{24} \cot^2 \alpha \tag{4.5.37}$$

$$C_4 = \frac{\gamma_s H^5}{30} \cot^2 \alpha - \frac{cH^4 \cot \alpha}{4} \tag{4.5.38}$$

This gives

$$EIu = \frac{\gamma_s z^5}{120}\cot^2\alpha - \frac{cz^4}{12}\cot\alpha + \frac{cH^3 z\cot\alpha}{3} - \frac{\gamma_s H^4 z}{24}\cot^2\alpha + \frac{\gamma_s H^5}{30}\cot^2\alpha - \frac{cH^4}{4}\cot\alpha \tag{4.5.39}$$

Equation (4.5.39) after some simple algebraic manipulation can be finally expressed as

$$u = \frac{\gamma_s H^5\cot^2\alpha}{30EI}\left[\frac{\xi^5}{4} - \frac{5\xi}{4} + 1 - \psi\frac{\xi^4}{3} + \psi\frac{4\xi}{3} - \psi\right] \tag{4.5.40}$$

Here $\psi = \frac{15H_c}{4H}\tan\alpha$, a dimensionless parameter, and $H_c = \frac{2c}{\gamma_s}$ the freestanding height of soil.

Thus, u_{static} at $\xi = 0$ is expressed as

$$u_{\text{static}} = \frac{\gamma_s H^5\cot^2\alpha}{30EI}[1-\psi] \tag{4.5.41}$$

Substituting the above in Eq. (4.5.12), we have

$$T_A = 3.97\cot\alpha\sqrt{\frac{\gamma_s H^5}{Et^3 g}(1-\psi)} \tag{4.5.42}$$

Thus, for modal analysis the dynamic amplitude is expressed as

$$u = \kappa\cdot\beta\cdot\frac{\gamma_s H^5\cot^2\alpha}{30EI}\left(\frac{S_a}{g}\right)(1-\psi)\left[\frac{\xi^5}{4} - \frac{5\xi}{4} + 1 - \psi\frac{\xi^4}{3} + \psi\frac{4\xi}{3} - \psi\right] \tag{4.5.43}$$

The dynamic moment and shear can be expressed as

$$M_z = \kappa\beta\gamma_s\cot^2\alpha\left(\frac{S_a}{g}\right)(1-\psi)\left[\frac{z^3}{6} - \frac{H_c z^2\tan\alpha}{2}\right] \tag{4.5.44}$$

$$V_z = \kappa\beta\gamma_s\cot^2\alpha\left(\frac{S_a}{g}\right)(1-\psi)\left[\frac{z^2}{2} - H_c z\tan\alpha\right] \tag{4.5.45}$$

Considering the effect of vertical acceleration $S_v = 2S_a/3$, we have

$$u = \kappa\cdot\beta\cdot\frac{\gamma_s H^5\cot^2\alpha}{18EI}\left(\frac{S_a}{g}\right)(1-\psi)\left[\frac{\xi^5}{4} - \frac{5\xi}{4} + 1 - \psi\frac{\xi^4}{3} + \psi\frac{4\xi}{3} - \psi\right] \tag{4.5.46}$$

$$M_z = \kappa\beta\gamma_s \cot^2\alpha\left(\frac{S_a}{g}\right)(1-\psi)\left[\frac{5z^3}{18} - \frac{2H_c z^2 \tan\alpha}{9}\right] \tag{4.5.47}$$

and

$$V_z = \kappa\beta\gamma_s \cot^2\alpha\left(\frac{S_a}{g}\right)(1-\psi)\left[\frac{5z^2}{6} - \left(\frac{4}{9}\right)H_c z \tan\alpha\right] \tag{4.5.48}$$

The modal mass participation may be expressed as

$$\kappa = \frac{\int_0^1 \xi\left(\frac{\xi^5}{4} - \frac{5\xi}{4} + 1 - \psi\frac{\xi^4}{3} + \psi\frac{4\xi}{3} - \psi\right)\mathrm{d}\xi}{\int_0^1 \xi\left(\frac{\xi^5}{4} - \frac{5\xi}{4} + 1 - \psi\frac{\xi^4}{3} + \psi\frac{4\xi}{3} - \psi\right)^2\mathrm{d}\xi}$$
$$\rightarrow \kappa = \frac{5/42 - \psi/9}{25/448 + 16\psi^2/315 - 295\psi/2772} \tag{4.5.49}$$

The above derivation is for a general soil that has finite value of c and ϕ.

When the soil is purely cohesionless, i.e., $c = 0$, $\psi \rightarrow 0$, Eqs. (4.5.46) to (4.5.48) degenerate to Eqs. (4.5.31) to (4.5.33) and Eq. (4.5.22). This shows the correctness of the derivation of the above expressions.

In Eq. (4.5.42), it will be observed that for limiting value of $\psi \rightarrow 1$, the time period tends to zero and for $\psi > 1$ the solution collapses. The physical significance of this is as explained hereafter.

The above solution is valid when the value of c is low so that the soil is adhering to the wall. For high values of c ($\psi > 1$), the negative pressure will be sufficiently high to develop tension cracks and loose contact over the wall for a height $(2c/\gamma_s)\tan\alpha$.

In such a case, for evaluation of static pressure it is usual practice to neglect the cracked portion and consider the wall to be partially loaded by the positive pressure to a height ($H - 2c/\gamma_s\tan\alpha$) from the base of the wall.

This is a special case and requires separate treatment. This has been dealt later in Sect. 4.5.6.

For passive case, we have

$$EI\frac{\mathrm{d}^4u}{\mathrm{d}z^4} = \gamma_s \cdot z.\tan^2\alpha + 2c\tan\alpha \tag{4.5.50}$$

As before, after successive integration and imposing the boundary conditions as cited in Eqs. (4.5.8a)–(4.5.8d), we have $C_1 = 0$, $C_2 = 0$.

$$C_3 = -\frac{\gamma_s H^4}{24}\tan^2\alpha - \frac{cH^3}{3}\tan\alpha \tag{4.5.51}$$

$$C_4 = \frac{\gamma_s H^5}{30} \tan^2 \alpha + \frac{cH^4}{4} \tan \alpha \tag{4.5.52}$$

Imposing the above integration constants, we have

$$EIu = \frac{\gamma_s z^5}{120} \tan^2 \alpha + \frac{cz^4}{12} \tan \alpha - \frac{cH^3 z \tan \alpha}{3} - \frac{\gamma_s H^4 z}{24} \tan^2 \alpha + \frac{\gamma_s H^5}{30} \tan^2 \alpha + \frac{cH^4}{4} \tan \alpha \tag{4.5.53}$$

Equation (4.5.53) on simplification gives

$$u = \frac{\gamma_s H^5 \tan^2 \alpha}{30EI} \left[\frac{\xi^5}{4} - \frac{5\xi}{4} + 1 + \psi_p \frac{\xi^4}{3} - \psi_p \frac{4\xi}{3} + \psi_p \right] \tag{4.5.54}$$

Here $\psi_p = \frac{15H_c}{4H} \cot \alpha$, a dimensionless a parameter, and H_c is as defined earlier = $2c/\gamma_s$.

Based on the above

$$u_{\text{static}} = \frac{\gamma_s H^5 \tan^2 \alpha}{30EI} \left[1 + \psi_p \right] \tag{4.5.55}$$

Substituting this in Eq. (4.5.12), we finally have the time period for passive case as

$$T_P = 3.97 \tan \alpha \sqrt{\frac{\gamma_s H^5}{Et^3 g} \left(1 + \psi_p \right)} \tag{4.5.56}$$

Thus, based on modal analysis, dynamic amplitude, moments, and shears are expressed as

$$u_p = \kappa \cdot \beta \cdot \frac{\gamma_s H^5 \tan^2 \alpha}{30EI} \left(\frac{S_a}{g} \right) \left(1 + \psi_p \right) \left[\frac{\xi^5}{4} - \frac{5\xi}{4} + 1 + \psi_p \frac{\xi^4}{3} - \psi_p \frac{4\xi}{3} + \psi_p \right] \tag{4.5.57}$$

$$M_{pz} = \kappa \beta \gamma_s \tan^2 \alpha \left(\frac{S_a}{g} \right) \left(1 + \psi_p \right) \left[\frac{z^3}{6} + \frac{H_c z^2 \cot \alpha}{2} \right] \tag{4.5.58}$$

$$V_{pz} = \kappa \beta \gamma_s \tan^2 \alpha \left(\frac{S_a}{g} \right) \left(1 + \psi_p \right) \left[\frac{z^2}{2} + H_c z \cot \alpha \right] \tag{4.5.59}$$

Considering the effect of vertical acceleration, we have

$$u_p = \kappa \cdot \beta \cdot \frac{\gamma_s H^5 \tan^2 \alpha}{18EI} \left(\frac{S_a}{g}\right) \left(1 + \psi_p\right) \left[\frac{\xi^5}{4} - \frac{5\xi}{4} + 1 + \psi_p \frac{\xi^4}{3} - \psi_p \frac{4\xi}{3} + \psi_p\right] \tag{4.5.60}$$

$$M_{pz} = \kappa\beta\gamma_s \tan^2 \alpha \left(\frac{S_a}{g}\right) \left(1 + \psi_p\right) \left[\frac{5z^3}{18} + \frac{2H_c z^2 \cot \alpha}{9}\right] \tag{4.5.61}$$

$$V_{pz} = \kappa\beta\gamma_s \tan^2 \alpha \left(\frac{S_a}{g}\right) \left(1 + \psi_p\right) \left[\frac{5z^2}{6} + \frac{4}{9} H_c z \cot \alpha\right] \tag{4.5.62}$$

The modal mass participation factor is expressed as

$$\kappa = \frac{\int_0^1 \xi \left[\frac{\xi^5}{4} - \frac{5\xi}{4} + 1 + \psi_p \frac{\xi^4}{3} - \psi_p \frac{4\xi}{3} + \psi_p\right] d\xi}{\int_0^1 \xi \left[\frac{\xi^5}{4} - \frac{5\xi}{4} + 1 + \psi_p \frac{\xi^4}{3} - \psi_p \frac{4\xi}{3} + \psi_p\right]^2 d\xi} \tag{4.5.63}$$

$$\rightarrow \kappa = \frac{5/42 - \psi_p/9}{25/448 + 16\psi_p^2/315 - 295\psi_p/2772}$$

4.5.5 *Dynamic Response of Wall with Pure Intact Clay ($\phi = 0$) as Backfill*

This case can be easily derived from the previous general case considering cotα & tanα = 1 which gives

$$T_A = 3.97 \sqrt{\frac{\gamma_s H^5}{Et^3 g}(1 - \psi_c)} \tag{4.5.64}$$

$$u = \kappa \cdot \beta \cdot \frac{\gamma_s H^5}{20EI} \left(\frac{S_a}{g}\right) (1 - \psi_c) \left[\frac{\xi^5}{4} - \frac{5\xi}{4} + 1 - \psi_c \frac{\xi^4}{3} + \psi_c \frac{4\xi}{3} - \psi_c\right] \tag{4.5.65}$$

$$M_z = \kappa\beta\gamma_s \left(\frac{S_a}{g}\right) (1 - \psi_c) \left[\frac{z^3}{4} - \frac{3H_c z^2}{4}\right] \tag{4.5.66}$$

$$V_z = \kappa\beta\gamma_s \left(\frac{S_a}{g}\right) (1 - \psi_c) \left[\frac{3z^2}{4} - \left(\frac{3}{2}\right) H_c z\right] \tag{4.5.67}$$

here $\psi_c = \frac{15H_c}{4H}$

Similarly, for passive case, considering the vertical acceleration effect can be expressed as

$$T_P = 3.97\sqrt{\frac{\gamma_s H^5}{Et^3 g}\left(1+\psi_{pc}\right)} \text{ where, } \psi_{pc} = \psi_c \tag{4.5.68}$$

$$u_p = \kappa \cdot \beta \cdot \frac{\gamma_s H^5}{18EI}\left(\frac{S_a}{g}\right)\left(1+\psi_{pc}\right)\left[\frac{\xi^5}{4} - \frac{5\xi}{4} + 1 + \psi_{pc}\frac{\xi^4}{3} - \psi_{pc}\frac{4\xi}{3} + \psi_{pc}\right] \tag{4.5.69}$$

$$M_{pz} = \kappa\beta\gamma_s\left(\frac{S_a}{g}\right)\left(1+\psi_{pc}\right)\left[\frac{5z^3}{18} + \frac{2H_c z^2}{9}\right] \tag{4.5.70}$$

$$V_{pz} = \kappa\beta\gamma_s\left(\frac{S_a}{g}\right)\left(1+\psi_{pc}\right)\left[\frac{5z^2}{6} + \frac{4}{9}H_c z\right] \tag{4.5.71}$$

Here κ is same as Eq. 4.5.63 where ψ_p is replaced by either ψ_c or ψ_{pc} as the case will be.

4.5.6 *Dynamic Response of Wall with* c-ϕ *Backfill and Overburden Surcharge* q

This type of problem is often faced by engineers in industrial plants and construction in dense urban region and remains a serious problem under earthquake. No solution exists till date for this problem under earthquake force, and engineers have to often resort to finite element analysis (FEM) to arrive at a workable result.

Shown in Fig. 4.8 is a retaining wall with generic c-ϕ soil as the backfill and also having a surcharge q at top. While it is possible to derive, the combined pressure due to this, including the overburden and then performing successive integration as described earlier, makes the analysis tedious. Considering that we are performing modal analysis, we can argue that the problem is linear, and, as such, a superimposition of displacement is permissible.

Thus, in this case, we derive the static displacement of the wall for the surcharge load q only and finally add it to Eq. (4.5.40) to arrive at the final static displacement.

Hence, we start with the expression

$$EI\frac{d^4u}{dz^4} = q \cdot \cot^2\alpha \tag{4.5.72}$$

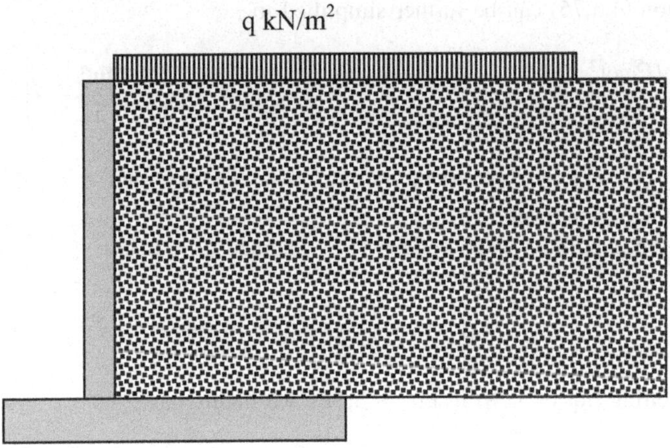

Fig. 4.8 A cantilever retaining wall with c-ϕ backfill and surcharge load q

On successive integration as explained earlier and imposing the boundary conditions as cited in Eqs. (4.5.8a)–(4.5.8d), we have $C_1 = 0$, $C_2 = 0$

$$C_3 = -\frac{q \cot^2 \alpha H^3}{6} \tag{4.5.73}$$

$$C_4 = -\frac{q \cot^2 \alpha H^4}{8}$$

Substituting these values, we finally have

$$EIu = \frac{qz^4 \cot^2 \alpha}{24} - \frac{q \cot^2 \alpha H^3 z}{6} + \frac{q \cot^2 \alpha H^4}{8} \tag{4.5.74}$$

Equation (4.5.74) can be finally written as

$$u = \frac{q \cot^2 \alpha H^4}{8EI}\left[\frac{\xi^4}{3} - \frac{4\xi}{3} + 1\right] \tag{4.5.75}$$

Equation (4.5.75) will be now added to Eq. (4.5.40) to arrive at the total displacement of the system.

Thus, the total static displacement may now be expressed as

$$u = \frac{\gamma_s H^5 \cot^2 \alpha}{30EI}\left[\frac{\xi^5}{4} - \frac{5\xi}{4} + 1 - \psi\frac{\xi^4}{3} + \psi\frac{4\xi}{3} - \psi\right] + \frac{q \cot^2 \alpha H^4}{8EI}\left[\frac{\xi^4}{3} - \frac{4\xi}{3} + 1\right] \tag{4.5.76}$$

Equation (4.5.76) can be further simplified to

$$u = \frac{\gamma_s H^5 \cot^2 \alpha}{30EI}\left[\frac{\xi^5}{4} - \frac{5\xi}{4} + 1 - \psi\frac{\xi^4}{3} + \psi\frac{4\xi}{3} - \psi + \frac{\eta\xi^4}{3} - \frac{4\eta\xi}{3} + \eta\right] \tag{4.5.77}$$

In Eq. (4.5.77), as mentioned before, $\psi = \frac{15H_c}{4H}\tan\alpha$ and $\eta = \frac{15q}{4\gamma_s H}$ are both dimensionless parameters.

Thus, the maximum static deflection is obtained ($\xi = 0$) as

$$u_{\text{static}} = \frac{\gamma_s H^5 \cot^2 \alpha}{30EI}[1 - \psi + \eta] \tag{4.5.78}$$

Substituting Eq. (4.5.78) in Eq. (4.5.12), we finally have

$$T_A = 3.97 \cot\alpha \sqrt{\frac{\gamma_s H^5}{Et^3 g}(1 - \psi + \eta)} \tag{4.5.79}$$

Thus, for modal analysis the dynamic amplitude, moments, and shears can be expressed as

$$u = \frac{\kappa\beta\gamma_s H^5 \cot^2 \alpha}{30EI}\left(\frac{S_a}{g}\right)(1 - \psi + \eta)\left[\frac{\xi^5}{4} - \frac{5\xi}{4} + 1 - \psi\frac{\xi^4}{3} + \psi\frac{4\xi}{3} - \psi + \frac{\eta\xi^4}{3} - \frac{4\eta\xi}{3} + \eta\right] \tag{4.5.80}$$

$$M_z = \kappa\beta\gamma_s \cot^2\alpha\left(\frac{S_a}{g}\right)(1 - \psi + \eta)\left[\frac{z^3}{6} - \frac{z^2}{2}\left(H_c \tan\alpha - \frac{q}{\gamma_s}\right)\right] \tag{4.5.81}$$

$$V_z = \kappa\beta\gamma_s \cot^2\alpha\left(\frac{S_a}{g}\right)(1 - \psi + \eta)\left[\frac{z^2}{2} - z\left(H_c \tan\alpha - \frac{q}{\gamma_s}\right)\right] \tag{4.5.82}$$

Equations (4.5.80)–(4.5.82) give the displacement, moment, and shear under seismic loading for the most general condition of soil.

When there is no overburden, i.e., $\eta \to 0$, the formulas converge to Eqs. (4.5.46)–(4.5.48) and represents the case of c-ϕ soil only. When again, $\psi \to 0$, the equations converge to the case of pure cohesionless soil (c = 0).

Finally, an interesting case is observed when $\psi \to 0$, $\eta \to 0$ and even $\phi = 0$ (i.e., $\cot^2\alpha \to 1$). The values obtained in Eqs. (4.5.80)–(4.5.82) are not zero but finite, i.e., it converges to a hydrostatic pressure. Thus, if we take density of soil in this particular case ($\gamma_{\text{sat}} - \gamma_w$), it reflects a case when the soil is in a liquefied state.

The modal mass participation factor in this case is expressed as

$$\kappa = \frac{\int_0^1 \xi\left[\frac{\xi^5}{4} - \frac{5\xi}{4} + 1 - \psi\frac{\xi^4}{3} + \psi\frac{4\xi}{3} - \psi + \frac{\eta\xi^4}{3} - \frac{4\eta\xi}{3} + \eta\right]\mathrm{d}\xi}{\int_0^1 \xi\left[\frac{\xi^5}{4} - \frac{5\xi}{4} + 1 - \psi\frac{\xi^4}{3} + \psi\frac{4\xi}{3} - \psi + \frac{\eta\xi^4}{3} - \frac{4\eta\xi}{3} + \eta\right]^2\mathrm{d}\xi} \tag{4.5.83}$$

$$\rightarrow \kappa = \frac{5/42 + (\eta - \psi)/9}{25/448 + 16(\psi^2 + \eta^2)/315 + 295(\eta - \psi)/2772 - 32\eta\psi/315}$$

Considering the effect of vertical acceleration, we have

$$u = \frac{\kappa\beta\gamma_s H^5 \cot^2\alpha}{18EI}\left(\frac{S_a}{g}\right)(1 - \psi + \eta)\left[\frac{\xi^5}{4} - \frac{5\xi}{4} + 1 - \psi\frac{\xi^4}{3} + \psi\frac{4\xi}{3} - \psi + \frac{\eta\xi^4}{3} - \frac{4\eta\xi}{3} + \eta\right] \tag{4.5.84}$$

$$M_z = \kappa\beta\gamma_s \cot^2\alpha\left(\frac{S_a}{g}\right)(1 - \psi + \eta)\left[\frac{5z^3}{18} - \frac{2z^2}{9}\left(H_c \tan\alpha - \frac{q}{\gamma_s}\right)\right] \tag{4.5.85}$$

$$V_z = \kappa\beta\gamma_s \cot^2\alpha\left(\frac{S_a}{g}\right)(1 - \psi + \eta)\left[\frac{5z^2}{6} - \frac{4}{9}z\left(H_c \tan\alpha - \frac{q}{\gamma_s}\right)\right] \tag{4.5.86}$$

4.5.7 *Dynamic Response of Wall with c-ϕ Backfill Partially Submerged Below Water*

Shown in Fig. 4.9 is a retaining wall having partially submerged c-ϕ soil as backfill. The soil is partially submerged for a height H_2. In this case, considering the complex nature of the pressure diagram, it would become difficult to perform the integration with appropriate boundaries. As such, we approach the problem differently for this particular case. Considering that the first step in the process is to determine the static deflection and this is obtained based on the loading on the retaining wall, the net horizontal loading on the wall can be expressed as

$$P_A = \frac{1}{2}\cot^2\alpha\left[\gamma_s H_1^2 + (\gamma_{sat} - \gamma_w)H_2^2\right] - 2c\cot\alpha H + \gamma_s H_1 \cdot H_2 \cot^2\alpha + \frac{2c^2}{\gamma_s} \tag{4.5.87}$$

Here

γ_s dry density of soil of height H_1;
γ_{sat} saturated density of soil of height H_2;
γ_w density of water.

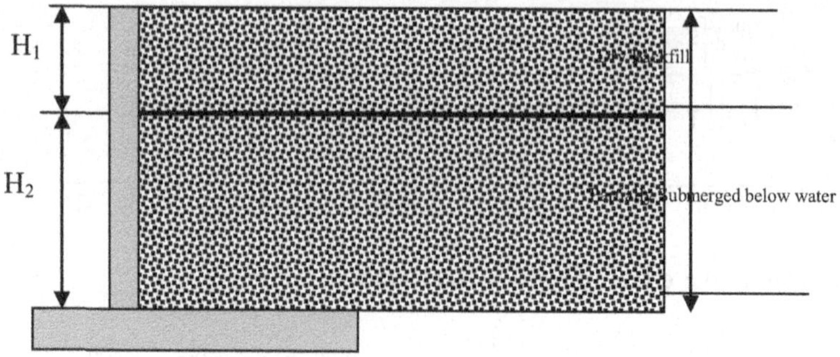

Fig. 4.9 A cantilever retaining wall with *c*-ϕ backfill partially submerged

Now if we consider an equivalent dry backfill of density γ_s which imposes the same load on the wall, the displacement of the wall will be same as that which would be induced by load as expressed in Eq. (4.5.87).

Thus, considering

$$\frac{1}{2}K_{AE}\gamma_s H^2 = \frac{1}{2}\cot^2\alpha\left[\gamma_s H_1^2 + (\gamma_{sat} - \gamma_w)H_2^2\right] - 2c\,\cot\alpha H + \gamma_s H_1 \cdot H_2 \cot^2\alpha + \frac{2c^2}{\gamma_s}$$

we have

$$K_{AE} = \cot^2\alpha \frac{\left[\gamma_s H_1^2 + (\gamma_{sat} - \gamma_w)H_2^2\right]}{\gamma_s (H_1 + H_2)^2} - \frac{4c\,\cot\alpha}{\gamma_s \cdot (H_1 + H_2)} + \frac{2\gamma_s H_1 \cdot H_2}{(H_1 + H_2)^2} + \frac{4c^2}{\gamma_s^2 (H_1 + H_2)^2} \tag{4.5.88}$$

Here K_{AE} = an equivalent coefficient of active earth pressure when considering a pressure diagram of $p_{az} = K_{AE} \cdot \gamma_s \cdot z$ over the height H will give same deflection as that produced by P_A in Eq. (4.5.87).

Thus, for the present case the problem now gets simplified considerably when we have

$$T_A = 3.97\sqrt{\frac{K_{AE}\gamma_s H^5}{Et^3 g}} \tag{4.5.89}$$

$$u = \kappa\beta \frac{K_{AE}\gamma_s H^5}{30EI}\left(\frac{S_a}{g}\right)\left[\frac{\xi^5}{4} - \frac{5\xi}{4} + 1\right] \tag{4.5.90}$$

$$M_z = \kappa\beta \frac{K_{AE}\gamma_s z^3}{6}\left(\frac{S_a}{g}\right) \tag{4.5.91}$$

$$V_z = \kappa\beta \frac{K_{AE}\gamma_s z^2}{2}\left(\frac{S_a}{g}\right) \tag{4.5.92}$$

Considering vertical acceleration, we have

$$u = \kappa\beta \frac{K_{AE}\gamma_s H^5}{18EI}\left(\frac{S_a}{g}\right)\left[\frac{\xi^5}{4} - \frac{5\xi}{4} + 1\right] \tag{4.5.93}$$

$$M_z = \kappa\beta \frac{5K_{AE}\gamma_s z^3}{18}\left(\frac{S_a}{g}\right) \tag{4.5.94}$$

$$V_z = \kappa\beta \frac{5K_{AE}\gamma_s z^2}{6}\left(\frac{S_a}{g}\right) \tag{4.5.95}$$

Here κ remains same as Eq. 4.5.22

4.5.8 Special Case of c-ϕ Soil When It Loses Contact for Some Portion at Top

For this case, the maximum load on the wall may be expressed as

$$P_a = \frac{\gamma_s H^2}{2}\cot^2\alpha - 2cH\cot\alpha + \frac{2c^2}{\gamma_s} \tag{4.5.96}$$

Based on the above, for a partially loaded beam, static deflection can be expressed based on the fundamentals of beam theory as

$$u_{\text{static}} = \frac{P_a\left(H - \frac{2c}{\gamma_s}\tan\alpha\right)^3}{15EI}\left[1 + \frac{5}{4}\frac{2c\tan\alpha}{\gamma_s\left[H - \frac{2c}{\gamma_s}\tan\alpha\right]}\right] \tag{4.5.97}$$

Here P_a is as expressed in Eq. (4.5.96).
Thus, the time period, based on Eq. (4.5.12), is expressed as

$$T_A = 2\pi\sqrt{\frac{P_A\left(H - \frac{2c}{\gamma_s}\tan\alpha\right)^3}{15EIg}\left(1 + \frac{5}{4}\frac{2c\tan\alpha}{\gamma_s\left(H - \frac{2c}{\gamma_s}\tan\alpha\right)}\right)} \tag{4.5.98}$$

Maximum dynamic amplitude, moments, and shears along the depth of the wall are expressed as

$$u = \kappa\beta\left(\frac{S_a}{g}\right)\frac{P_a\left(H - \frac{2c}{\gamma_s}\tan\alpha\right)^3}{15EI}\left[1 + \frac{5}{4}\frac{2c\,\tan\alpha}{\gamma s\left[H - \frac{2c}{\gamma_s}\tan\alpha\right]}\right] \quad (4.5.99)$$

$$M_z = \kappa\beta\left(\frac{S_a}{g}\right)P_a\frac{z^3}{3\left(H - \frac{2c}{\gamma_s}\tan\alpha\right)^2} \quad (4.5.100)$$

$$V_z = \kappa\beta\left(\frac{S_a}{g}\right)P_a\frac{z^2}{\left(H - \frac{2c}{\gamma_z}\tan\alpha\right)^2} \quad (4.5.101)$$

It is to be noted that in this case $z = 0$ where pressure is zero, that is, at a depth $(H - 2c/\gamma_s \tan\alpha)$ from the top of wall. Here κ remains same as in Eq. 4.5.22.

4.5.9 Damping of Soil and that of the Wall

For the wall considering it as RCC, we can consider 5% material damping as the usual practice.

During earthquake, the failed soil wedge ABD (vide Fig. 4.6) will move to and fro during vibration. When the soil body moves toward the wall, it will generate an active earth pressure and a passive pressure when moving in the opposite direction. It is evident that the soil body during its motion will generate a friction force along slope BD with the soil below the failed surface, and this friction force will oppose the motion and generate a damping that will attenuate the excitation.

The damping force can be approximately estimated as follows.

Considering the free body diagram of the failed wedge ABD, the friction acting along the slope BD may be expressed as

$$F_R = W(\sin\alpha - \mu\cos\alpha) \quad (4.5.102)$$

Here

F_R friction force along the surface BD that resists the motion;
W weight of the soil body ABD ($W = m \cdot g$);
α $45 + \phi/2$ for active case;
μ internal angle of friction of soil @tan ϕ.

Considering the resistive friction force F_R as the damping force, Eq. (4.5.102) can be expressed as

$$C \cdot v = W(\sin\alpha - \mu\cos\alpha) \quad (4.5.103)$$

Here C = damping of the system, velocity $v = S_a/\omega$ where S_a is the seismic acceleration, and ω the natural frequency @ $\sqrt{k/m}$.

Equation (4.5.103) on simple algebraic manipulation can be expressed as

$$\frac{C}{C_c} = \frac{\sin\alpha - \mu\cos\alpha}{2\left(\frac{S_a}{g}\right)} \tag{4.5.104}$$

Here C_c = critical damping $= 2\sqrt{k/m}$
Considering

$\zeta = \frac{C}{C_c}$ the damping ratio, we have

$$\zeta_A = \frac{\sin\alpha - \mu\cos\alpha}{2\left(\frac{S_a}{g}\right)} \tag{4.5.105}$$

The above gives the damping ratio of the soil in active case. For passive case

$$\zeta_p = \frac{\sin\alpha_p + \mu\cos\alpha_p}{2\left(\frac{S_a}{g}\right)}, \text{ where } \alpha_p = 45 - \phi/2 \tag{4.5.106}$$

For a conservative estimate, we should consider ζ to be minimum, but having a finite rational value. This is valid either when the numerator in Eqs. (4.5.104) and (4.5.105) is minimum or the denominator is maximum.

For numerator to be minimum, it must be zero which gives $\phi = 90$, which is impossible to achieve. Thus, the condition is denominator is to be maximum. In other words, Sa/g is to be maximum.

For instance, maximum value of Sa/g as per IS 1893(2016) is 2.5. Applying this value, we have

$$\zeta_A = \frac{\sin\alpha - \mu\cos\alpha}{5} \tag{4.5.107}$$

For sandy soil, the ϕ value usually varies from 15° for very loose sand to 40° for very dense sand. Considering the above values, variations in damping ratio, for active and passive cases, are as shown in Fig. 4.10.

It will be observed that variation of damping ratio with respect to friction angle ϕ is not widely varying; thus, an estimated value of 15% in active case and 20% for passive case would cater for soil with almost all levels of in situ compaction. The damping ratio values also look quite reasonable and match the data that are usually

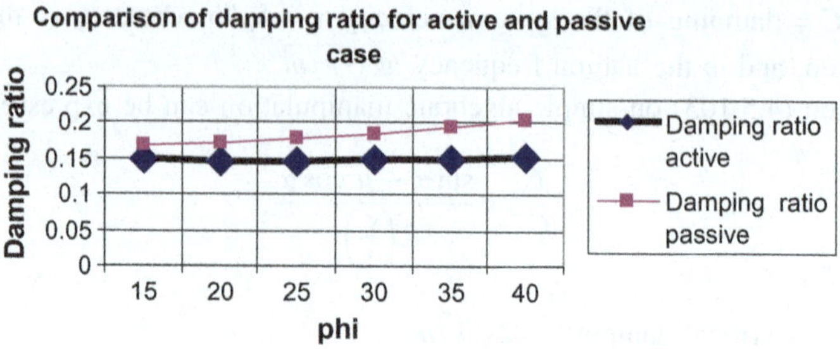

Fig. 4.10 Variation of damping ratio of soil with friction angle of soil

considered from experience for practical seismic design of such walls, based on FEM.

For c-ϕ soil the damping ratios can be expressed as

$$\zeta_A = \frac{\sin\alpha - \mu\cos\alpha - \frac{cH\,\mathrm{cosec}\alpha}{W}}{5} \text{ and} \tag{4.5.108}$$

$$\zeta_p = \frac{\sin\alpha_p + \mu\cos\alpha_p + \frac{cH\,\mathrm{cosec}\alpha_p}{W}}{5} \tag{4.5.109}$$

for general c-ϕ soil (Fig. 4.6) $W = 0.5\gamma_s H^2\cot\alpha$ and for c-ϕ soil with overburden $W = 0.5\gamma_s H^2\cot\alpha + q\cot\alpha$. Replace α by α_p, vide Eq. (4.5.106) for passive case.

The 5% material damping of wall may be added to the above, to arrive at the design damping ratio.

Example 4.5.1 A 6.0-m-high retaining wall with the following soil properties is analyzed under earthquake force.

Height of wall = 6.0 m, top thickness = 0.25 m, bottom thickness = 0.4 m grade of concrete = M30, foundation width = 4.0 m, foundation thickness = 0.6 m density of backfill = 22 kN/m^3, angle of internal friction = 28°, cohesive strength (c) = 5 kN/m^2, overburden (q) = 50 kN/m^2, earthquake zone = Zone IV as per IS-1893. Damping of soil considered (average) = 15%. Determine the dynamic moment and shear under active pressure case.

Considering that this is a generic c-ϕ soil with overburden, Mononobe's method cannot be applied,

$$H_c = \frac{2c}{\gamma s} = \frac{2\times 5}{22} = 0.454m$$

$\alpha = 45 + \phi/2 = 45 + 14 = 59°$. Here H = 6.0 m

$$\psi_c = \frac{15H_c}{4H}\tan\alpha = \frac{15\times 0.454}{24}\tan 59 = 0.473$$
$$\eta = \frac{15q}{4\gamma_s H} = \frac{15\times 50}{4\times 22\times 6} = 1.42$$
$$E_{\text{conc}} = 5000\sqrt{f_{ck}} = 5000\sqrt{30}\times 1000 = 2.74\times 10^7\ \text{kN/m}^2.$$

Average thickness = 0.5(250 + 400) = 325 mm = 0.325 m
Moment of inertia $I = \frac{1}{12}\times 1\times(0.325)^3 = 0.00286\ \text{m}^4$
As per Eq. 4.5.79

$$T_A = 3.97\cot\alpha\sqrt{\frac{\gamma_s H^5}{Et^3 g}(1-\psi+\eta)}$$

$$\rightarrow T_A = 3.97\cot 59\sqrt{\frac{22\times 6^5}{2.74\times 10^7\times 0.325^3\times 9.81}(1-0.473+1.42)} = 0.4532s$$

For medium soil site, Sa/g = 2.5. Considering 15% damping ratio scaling factor is 0.7 as per IS-1893.
Thus, design Sa/g = 0.7 × 2.5 = 1.75.
For zone IV, Z = 0.24, I = 1.2 and R = 3.0; thus, $\beta = \frac{ZI}{2R} = \frac{0.24\times 1.2}{6} = 0.05$
Modal mass participation factor is expressed as

$$\kappa = \frac{\int_0^1 \xi\left[\frac{\xi^5}{4} - \frac{5\xi}{4} + 1 - \psi\frac{\xi^4}{3} + \psi\frac{4\xi}{3} - \psi + \frac{\eta\xi^4}{3} - \frac{4\eta\xi}{3} + \eta\right]d\xi}{\int_0^1 \xi\left[\frac{\xi^5}{4} - \frac{5\xi}{4} + 1 - \psi\frac{\xi^4}{3} + \psi\frac{4\xi}{3} - \psi + \frac{\eta\xi^4}{3} - \frac{4\eta\xi}{3} + \eta\right]^2 d\xi}$$

As per Eq. 4.5.83, κ = 1.111.
The displacement amplitude is expressed as

$$u = \frac{\kappa\beta\gamma_s H^5\cot^2\alpha}{18EI}\left(\frac{S_a}{g}\right)(1-\psi+\eta)\left[\frac{\xi^5}{4} - \frac{5\xi}{4} + 1 - \psi\frac{\xi^4}{3} + \psi\frac{4\xi}{3} - \psi + \frac{\eta\xi^4}{3} - \frac{4\eta\xi}{3} + \eta\right]$$
$$u = \frac{1.111\times 0.05\times 20\times 6^5\cot^2 31}{18\times 2.74\times 10^7\times 0.00286}(1.75)(1-0.473+1.42)$$
$$\times\left[\frac{\xi^5}{4} - \frac{5\xi}{4} + 1 - \psi\frac{\xi^4}{3} + \psi\frac{4\xi}{3} - \psi + \frac{\eta\xi^4}{3} - \frac{4\eta\xi}{3} + \eta\right]$$

Similarly for moment and shear

$$M_z = \kappa\beta\gamma_s \cot^2\alpha\left(\frac{S_a}{g}\right)(1-\psi+\eta)\left[\frac{5z^3}{18} - \frac{2z^2}{9}\left(H_c\tan\alpha - \frac{q}{\gamma_s}\right)\right]$$

$$V_z = \kappa\beta\gamma_s \cot^2\alpha\left(\frac{S_a}{g}\right)(1-\psi+\eta)\left[\frac{5z^2}{6} - \frac{4}{9}z\left(H_c\tan\alpha - \frac{q}{\gamma_s}\right)\right]$$

Substituting the appropriate values, we get

z	Displacement (mm)	Moment (kN m/m)	Shear (kN/m)
0	1.897124	0	0
0.6	1.652324	0.31417	1.2205232
1.2	1.408029	1.672574	3.4807779
1.8	1.16571	4.69905	6.7807642
2.4	0.928164	10.01744	11.120482
3	0.699801	18.25157	16.499932
3.6	0.48694	30.0253	22.919113
4.2	0.298099	45.96246	30.378025
4.8	0.144286	66.68688	38.876669
5.4	0.039297	92.82241	48.415045
6	0	124.9929	58.993153

For catering to effect of vertical acceleration, increase the above values by 67%.

4.6 Dynamic Analysis of Counterfort Retaining Walls

Walls used to retain soil of height of more than 6.0 ms are often provided with counterforts to economize their design. Behavior of such stiffened walls under earthquake force is still uncertain and practically no solution exists till date in terms of their dynamic behavior, except of the dynamic pressure concept as proposed by Mononobe and Matsuo (1926) and Okabe's (1929) (M–O method) based on pseudo-static analysis. The seismic coefficient method that forms the backbone of M–O method, though obsolete for other types of structures (where more rational techniques like modal and time history analysis have been in vogue) has, however, continued to be in practice for this type of walls for the last 86 years, though found to give a considerably lower bound solution to the problem (Ostadan 2004). The present method is thus an attempt to revisit this long-standing practical problem and propose a more realistic analysis based on modal response technique considering a generalized backfilled soil having both c and ϕ values with overburden—an oft faced condition in practice for which again M–O method is deficient (Chowdhury et al. 2012).

To start with, we take the simplest the case as shown below.

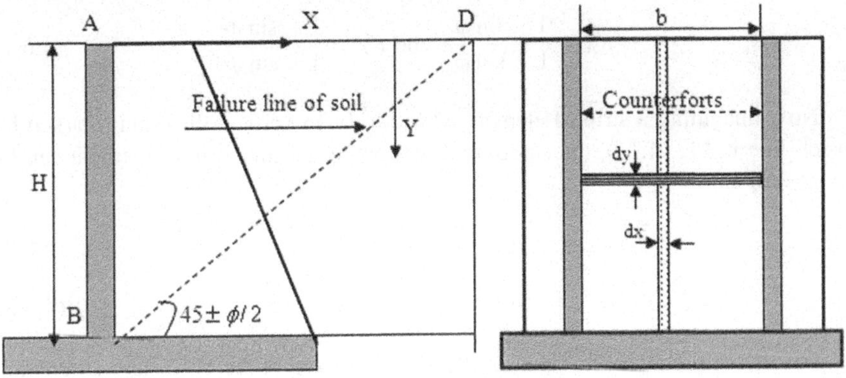

Fig. 4.11 A counterfort retaining wall with dry cohesionless backfill ($c = 0$)

Shown in Fig. 4.11 is a counterfort retaining wall with dry sandy backfill and the ground with no inclination. Like thc cantilever retaining wall, we start with this simple benchmark problem before digressing to a more generalized case.

While performing the analysis, it is assumed here that

- Soil profile under active case is at incipient failure when the failure line makes angle $\alpha = \tan\,(45 + \phi/2)$, as shown in Fig. 4.11.
- Since soil profile is already under failed condition under static load, it will not induce any stiffness to the overall dynamic response but will only contribute to inertial effect.
- Since the counterfort wall is relatively thin, mass contribution of the wall itself may be ignored compared to that of failed soil. The wall thus contributes only to stiffness of overall soil–structure system.
- The retaining wall is fixed at base and at the edge of counterforts.
- Foundation compliance has been ignored for the present analysis.
- The wall essentially behaves as a plate having three sides fixed and one side free as its boundary condition.

4.6.1 *Dynamic Response of Wall Having Dry Cohesionless Backfill* ($C = 0$)

Consider the wall behaving as a two-way slab of aspect ratio $r = H/b$, which essentially behaves as a plate element. Considering ϕ as internal angle of friction, active and passive coefficients of pressure, K_A and K_P, may be expressed as

$$K_A = \frac{1-\sin\phi}{1+\sin\phi}; \text{ and } K_P = \frac{1+\sin\phi}{1-\sin\phi} \tag{4.6.1}$$

Now imagining a strip of element of width dx, in vertical direction (marked by hatch line in Fig. 4.11), the static pressures acting on this strip of element can be expressed as

$$p_{ay} = K_A \cdot \gamma_s \cdot y \cdot \mathrm{d}x \tag{4.6.2}$$

$$p_{py} = K_P \cdot \gamma_s \cdot y \cdot \mathrm{d}x \tag{4.6.3}$$

For the strip considered as a cantilever beam fixed at the base slab, differential equation of static equilibrium under active soil condition can be expressed as

$$EI\frac{\mathrm{d}^4u}{\mathrm{d}y^4} = K_A \cdot \gamma_s \cdot y \cdot \mathrm{d}x \tag{4.6.4}$$

Here u is displacement of the strip element, E = Young's modulus of the RCC wall, and I is moment of inertia of the strip considered $I = (1/12) \cdot (\mathrm{d}x \times t^3)$, where t is thickness of the wall (can be considered as average thickness for variation between top and bottom thickness of the wall).

On successive integration of Eq. (4.6.4), as shown earlier for the cantilever wall, we have

$$EIu = \frac{K_A \cdot \gamma_s y^5}{120}\mathrm{d}x + C_1\frac{y^3}{6} + C_2\frac{y^2}{2} + C_3 y + C_4 \tag{4.6.5}$$

For the given strip (Fig. 4.11), we consider the boundary conditions

(5) $$\text{At } y = 0 \quad \frac{\mathrm{d}^3u}{\mathrm{d}y^3} = 0 \Rightarrow C_1 = 0 \tag{4.6.6}$$

(6) $$\text{At } y = 0 \quad \frac{\mathrm{d}^2u}{\mathrm{d}y^2} = 0 \Rightarrow C_2 = 0 \tag{4.6.7}$$

(7) $$\text{At } y = H \quad \frac{\mathrm{d}u}{\mathrm{d}y} = 0 \Rightarrow C_3 = -\frac{K_A \cdot \gamma_s \cdot H^4}{24}\mathrm{d}x \tag{4.6.8}$$

(8) $$\text{At } y = H \; u = 0 \Rightarrow C_4 = \frac{K_A \cdot \gamma_s \cdot H^5}{30}\mathrm{d}x \tag{4.6.9}$$

Thus, Eq. (4.6.5) can be expressed as

$$EIu = \frac{K_A \cdot \gamma_s y^5}{120} \mathrm{d}x - \frac{K_A \cdot \gamma_s H^4 y}{24} \mathrm{d}x + \frac{K_A \gamma_s H^5}{30} \mathrm{d}x \tag{4.6.10}$$

Equation (4.6.10) can be further expressed in generalized form as

$$u = \frac{2K_A \cdot \gamma_s \cdot H^5}{5Et^3} \left[\frac{\eta^5}{4} - \frac{5\eta}{4} + 1 \right] \tag{4.6.11}$$

Here $\eta = y/H$ a non-dimensional term that varies between 0 and 1.

From Eq. (4.6.11), we can conclude that for the strip element having base fixed and the other end free, the generic shape function of displacement can be expressed as

$$f(\eta) = \frac{\eta^5}{4} - \frac{5\eta}{4} + 1 \tag{4.6.12}$$

Proceeding in identical fashion, it can be shown that in horizontal direction, considering a strip element of width dy having both ends fixed and subjected to a uniformly distributed load w, the generic shape function for displacement in natural coordinates can be expressed as

$$f(\xi) = 16\xi^4 - 32\xi^3 + 16\xi^2 \tag{4.6.13}$$

Here $\xi = x/b$ a non-dimensional term that varies between 0 and 1.

The shape functions derived in Eqs. (4.6.12) and (4.6.13) will generically satisfy the plate equilibrium equation having boundary conditions of three sides fixed and one side free but may still have residual error (R_e) as they are not derived from exact analysis of the fourth order partial differential of an isotropic plate expressed as

$$\frac{\partial^4 u}{\partial x^4} + 2\frac{\partial^4 u}{\partial x^2 \partial y^2} + \frac{\partial^4 u}{\partial y^4} = \frac{q}{D} \tag{4.6.14}$$

Here u is the displacement of the plate under a pressure load q, and D is the flexural stiffness of plate expressed as $D = Et^3/12(1 - \nu^2)$.

In natural coordinates, as expressed in Eqs. (4.6.12) and (4.6.13), Eq. (4.6.14) can be expressed as

$$\frac{1}{b^4}\frac{\partial^4 u}{\partial \xi^4} + \frac{2}{H^2 b^2}\frac{\partial^4 u}{\partial \xi^2 \partial \eta^2} + \frac{1}{H^4}\frac{\partial^4 u}{\partial \eta^4} = \frac{q}{D} \tag{4.6.15}$$

where displacement u is expressed as

$$u = \Delta \cdot f(\xi) \cdot f(\eta) \tag{4.6.16}$$

Substituting Eq. (4.6.16) in (4.6.15), we have

$$\frac{\Delta}{b^4}\frac{\partial^4}{\partial \xi^4}f(\xi)\cdot f(\eta) + \frac{2\Delta}{H^2b^2}\frac{\partial^4}{\partial \xi^2 \partial \eta^2}f(\xi)\cdot f(\eta) + \frac{\Delta}{H^4}\frac{\partial^4}{\partial \eta^4}f(\xi)\cdot f(\eta) - \frac{q}{D} = 0 \tag{4.6.17}$$

The shape functions $f(\xi)$ and $f(\eta)$ will generically satisfy Eq. (4.6.17) as they conform to the boundary condition of the given plate; however, they will have residual error R_e which may be expressed as

$$R_e = \frac{\Delta}{b^4}\frac{\partial^4}{\partial \xi^4}f(\xi)\cdot f(\eta) + \frac{2\Delta}{H^2b^2}\frac{\partial^4}{\partial \xi^2 \partial \eta^2}f(\xi)\cdot f(\eta) + \frac{\Delta}{H^4}\frac{\partial^4}{\partial \eta^4}f(\xi)\cdot f(\eta) - \frac{q}{D} \tag{4.6.18}$$

Residual error R_e is now minimized by Galerkin's weighted residual method based on which

$$\int_0^1\int_0^1 \begin{bmatrix} \frac{\Delta}{b^4}\frac{\partial^4}{\partial \xi^4}f(\xi)\cdot f(\eta) + \frac{2\Delta}{H^2b^2}\frac{\partial^4}{\partial \xi^2 \partial \eta^2}f(\xi)\cdot f(\eta) \\ + \frac{\Delta}{H^4}\frac{\partial^4}{\partial \eta^4}f(\xi)\cdot f(\eta) - \frac{q}{D} \end{bmatrix} f(\xi)f(\eta)\mathrm{d}\xi\mathrm{d}\eta = 0 \tag{4.6.19}$$

Equation (4.6.19) can thus be expressed as

$$\begin{aligned} &\frac{\Delta}{b^4}\int_0^1\int_0^1 f''''(\xi)f(\xi)f(\eta)^2\mathrm{d}\xi\cdot\mathrm{d}\eta + \frac{2\Delta}{H^2b^2}\int_0^1\int_0^1 f''(\xi)f''(\eta)f(\xi)f(\eta)\,\mathrm{d}\xi\cdot\mathrm{d}\eta \\ &+ \frac{\Delta}{H^4}\int_0^1\int_0^1 f(\xi)^2 f(\eta)f''''(\eta)\mathrm{d}\xi\mathrm{d}\eta - \frac{1}{D}\int_0^1\int_0^1 qf(\xi)f(\eta)\mathrm{d}\xi\cdot\mathrm{d}\eta = 0 \end{aligned} \tag{4.6.20}$$

Considering $q = K_A \cdot \gamma_s \cdot y$, Eq. (4.6.20) can be expressed in a simplified form as

$$\Delta\left[\frac{H^4}{b^4}X_1 + \frac{2H^2}{b^2}X_2 + X_3\right] = \frac{K_A\cdot\gamma_s\cdot H^5}{D}X_4 \tag{4.6.21}$$

Considering $r = H/b$, the aspect ratio of the plate, Eq. (4.6.21) can be expressed as

Table 4.2 Values of integral functions X_1 to X_5 for cohesionless soil ($c = 0$)

Integral functions	Values
$X_1 = \int_0^1 \int_0^1 \left[\frac{\partial^4}{\partial \xi^4} f(\xi)\right] f(\xi) f(\eta)^2 \, d\xi d\eta$	55.4113
$X_2 = \int_0^1 \int_0^1 \left[\frac{\partial^2}{\partial \xi^2} f(\xi)\right] \left[\frac{\partial^2}{\partial \eta^2} f(\eta)\right] f(\xi) f(\eta) \, d\xi d\eta$	−0.6772
$X_3 = \int_0^1 \int_0^1 \left[\frac{\partial^4}{\partial \eta^4} f(\eta)\right] f(\eta) f(\xi)^2 \, d\xi d\eta$	1.4512
$X_4 = \int_0^1 \int_0^1 \eta f(\eta) f(\xi) \, d\xi d\eta$	0.0635

$$\Delta = \frac{\frac{K_A \cdot \gamma_s \cdot H^5}{D} X_4}{r^4 X_1 + 2r^2 X_2 + X_3} \tag{4.6.22}$$

Here X_1, X_2, X_3, and X_4 are integral functions that can be solved numerically or explicitly and are as expressed in Table 4.2.

The displacement u can thus be expressed as

$$u = \frac{K_A \gamma_s \cdot H^5}{D} \left[\frac{X_4}{r^4 X_1 + 2r^2 X_2 + X_3}\right] f(\xi) \cdot f(\eta) \tag{4.6.23}$$

Here $f(\xi)$ and $f(\eta)$ are as expressed in Eqs. (4.6.13) and (4.6.12), respectively.
The maximum static deflection will occur at $f(\xi) = 0.5$ and $f(\eta) = 0$, which gives

$$u_{\text{static}} = \frac{K_A \gamma_s \cdot H^5}{D} \left[\frac{X_4}{r^4 X_1 + 2r^2 X_2 + X_3}\right] \tag{4.6.24}$$

The time period of the wall can then be expressed as

$$T = 2\pi \sqrt{\frac{u_{\text{static}}}{g}} \tag{4.6.25}$$

Substituting Eqs. (4.6.24) in (4.6.25) under active failure, we have

$$T_A = 2\pi \sqrt{\frac{K_A \gamma_s H^5}{Dg} \left[\frac{X_4}{r^4 X_1 + 2r^2 X_2 + X_3}\right]} \tag{4.6.26}$$

For modal analysis, maximum amplitude (S_d) is expressed as

$$S_d = \frac{S_a}{\omega^2} \tag{4.6.27}$$

Here S_a = spectral acceleration, and $\omega = 2\pi/T$, natural frequency of the wall.

In terms of code, Eq. (4.6.27) can be expressed

$$S_d = \kappa\beta\frac{S_a}{\omega^2} \tag{4.6.28}$$

Here $\kappa =$ modal mass participation factor and is expressed as $\sum_{i=1}^{n} m_i\varphi_i / \sum_{i=1}^{n} m_i\varphi_i^2$, $\beta =$ a code factor expressed as $ZI/2R$ where $Z =$ zone factor, $I =$ importance factor, and $R =$ response reduction factor.

Thus, based on Eq. (4.6.28) the dynamic amplitude of the wall can be expressed as

$$u_d = \kappa\beta\frac{S_a}{4\pi^2}T_A^2 f(\xi)\cdot f(\eta) \tag{4.6.29}$$

Equation (4.6.32) can be finally expressed as

$$u_d = \kappa\beta\frac{K_A\gamma_s H^5}{D}\left(\frac{S_a}{g}\right)\left[\frac{X_4}{r^4X_1 + 2r^2X_2 + X_3}\right]\left[16\xi^4 - 32\xi^3 + 16\xi^2\right]\left[\frac{\eta^5}{4} - \frac{5\eta}{4} + 1\right] \tag{4.6.30}$$

Equation (4.6.30) gives the complete dynamic displacement profile of the wall between limits 0–1 in both *X*- and *Y*- direction.

4.6.2 Dynamic Bending Moment and Shear Force

Bending moment and shear force induced in the wall can be expressed as (Timoshenko Woinowsky-Kreiger 1987)

$$M_x = -D\left[\frac{\partial^2 u}{\partial x^2} + v\frac{\partial^2 u}{\partial y^2}\right] \tag{4.6.31}$$

$$M_y = -D\left[v\frac{\partial^2 u}{\partial x^2} + \frac{\partial^2 u}{\partial y^2}\right] \tag{4.6.32}$$

$$Q_x = -D\left[\frac{\partial^3 u}{\partial x^3} + \frac{\partial^3 u}{\partial x.\partial y^2}\right] \tag{4.6.33}$$

$$Q_y = -D\left[\frac{\partial^3 u}{\partial x^2\partial y} + \frac{\partial^3 u}{\partial y^3}\right] \tag{4.6.34}$$

Transferring the above equations in natural coordinates and substituting in Eq. (4.6.30), dynamic moments and shears can be calculated using the following equations. To find out moments and shear, η is replaced with $(1-\eta)$ to keep the reference coordinate identical to that of the displacement, which gives

$$M_x = \kappa \cdot \beta \cdot K_A \cdot \gamma_s \cdot H^3 \left[\frac{S_a}{g}\right] \left[\frac{X_4}{r^4 X_1 + 2r^2 X_2 + X_3}\right] \times \left[r^2 f''(\xi) f(1-\eta) + v f(\xi) f''(1-\eta)\right] \tag{4.6.35}$$

$$M_y = \kappa \cdot \beta \cdot K_A \cdot \gamma_s \cdot H^3 \left[\frac{S_a}{g}\right] \left[\frac{X_4}{r^4 X_1 + 2r^2 X_2 + X_3}\right] \times \left[f(\xi) f''(1-\eta) + v r^2 f''(\xi) f(1-\eta)\right] \tag{4.6.36}$$

$$Q_x = \kappa \cdot \beta \cdot K_A \cdot \gamma_s \cdot H^2 \left[\frac{S_a}{g}\right] \left[\frac{X_4}{r^4 X_1 + 2r^2 X_2 + X_3}\right] \times \left[r^3 f'''(\xi) f(1-\eta) + r f'(\xi) f''(1-\eta)\right] \tag{4.6.37}$$

$$Q_y = \kappa \cdot \beta \cdot K_A \cdot \gamma_s \cdot H^2 \left[\frac{S_a}{g}\right] \left[\frac{X_4}{r^4 X_1 + 2r^2 X_2 + X_3}\right] \times \left[r^2 f''(\xi) f'(1-\eta) + f(\xi) f'''(1-\eta)\right] \tag{4.6.38}$$

The modal participation factor κ can be expressed in this case as

$$\kappa = \frac{\sum_{i=1}^{n} m_i \varphi_i}{\sum_{i=1}^{n} m_i \varphi_i^2} = \frac{\int_0^1 K_A \gamma_s H^2 \eta f(\xi) f(\eta)\, d\xi d\eta}{\int_0^1 K_A \gamma_s H^2 \eta f(\xi)^2 f(\eta)^2 d\xi\, d\eta} \tag{4.6.39}$$

$$\kappa = \frac{\int_0^1 \int_0^1 \eta \left(16\xi^4 - 32\xi^2 + 16\xi^2\right)\left(\frac{\eta^5}{4} - \frac{5\eta}{4} + 1\right) d\xi \cdot d\eta}{\int_0^1 \int_0^1 \eta \cdot \left(16\xi^4 - 32\xi^2 + 16\xi^2\right)^2 \left(\frac{\eta^5}{4} - \frac{5\eta}{4} + 1\right)^2 d\xi \cdot d\eta} \tag{4.6.40}$$

Equation (4.6.40) on computation gives a value of $\kappa = 2.8$.

For passive pressure case, replace K_A by K_P in Eqs. (4.6.35) through (4.6.38) as expressed in Eq. (4.6.1).

Dynamic amplitude, bending moment, and shear can also be expressed in a simplified form for quick computation as

$$u_d = \text{Coeff}_d \times \kappa \cdot \beta \frac{K_A \gamma_s H^5}{D} \left[\frac{S_a}{g}\right] \tag{4.6.41a}$$

$$M_x = \text{Coeff}_{mx} \times \kappa \cdot \beta \cdot K_A \cdot \gamma_s \cdot H^3 \left[\frac{S_a}{g}\right] \tag{4.6.41b}$$

$$M_y = \text{Coeff}_{my} \times \kappa \cdot \beta \cdot K_A \cdot \gamma_s \cdot H^3 \left[\frac{S_a}{g}\right] \tag{4.6.41c}$$

$$Q_x = \text{Coeff}_{qx} \times \kappa \cdot \beta \cdot K_A \cdot \gamma_s \cdot H^2 \left[\frac{S_a}{g}\right] \tag{4.6.41d}$$

$$Q_y = \text{Coeff}_{qy} \times \kappa \cdot \beta \cdot K_A \cdot \gamma_s \cdot H^2 \left[\frac{S_a}{g}\right] \tag{4.6.41e}$$

Typical values of such design coefficients for aspect ratio $H/b = 1.5$ and Poisson's ratio $\nu = 0.25$ are furnished in Tables 4.3, 4.4, 4.5, 4.6, and 4.7.

Similar design charts can be developed for various values of aspect ratio (H/b) for quick computation of design moments and shears in a design office environment (Fig. 4.12).

Table 4.3 Coeff_d for estimation of amplitude for various values of η(vert.) and ξ(horz.)

η, ξ	0.0	0.2	0.4	0.5	0.6	0.8	1
0.0	0.00000	0.00009	0.00021	0.00023	0.00021	0.00009	0.00000
0.2	0.00000	0.00007	0.00016	0.00017	0.00016	0.00007	0.00000
0.4	0.00000	0.00005	0.00011	0.00011	0.00011	0.00005	0.00000
0.5	0.00000	0.00004	0.00008	0.00009	0.00008	0.00004	0.00000
0.6	0.00000	0.00003	0.00006	0.00006	0.00006	0.00003	0.00000
0.8	0.00000	0.00001	0.00002	0.00002	0.00002	0.00001	0.00000
1.0	0.00000	0.00000	0.00000	0.00000	0.00000	0.00000	0.00000

Table 4.4 Coeff_{mx} for estimation of Mx for various values of η and ξ

η, ξ	0	0.2	0.4	0.5	0.6	0.8	1
0.0	0.00000	0.00012	0.00026	0.00028	0.00026	0.00012	0.00000
0.2	0.00134	0.00011	−0.00046	−0.00053	−0.00046	0.00011	0.00134
0.4	0.00442	0.00020	−0.00189	−0.00215	−0.00189	0.00020	0.00442
0.5	0.00000	0.00012	0.00026	0.00028	0.00026	0.00012	0.00000
0.6	0.00824	0.00034	−0.00361	−0.00410	−0.00361	0.00034	0.00824
0.8	0.01230	0.00049	−0.00541	−0.00615	−0.00541	0.00049	0.01230
1.0	0.01639	0.00066	−0.00721	−0.00820	−0.00721	0.00066	0.01639

Table 4.5 Coeff$_{my}$ for estimation of *My* for various values of η and ξ

η, ξ	0	0.2	0.4	0.5	0.6	0.8	1
0.0	0.00000	0.00047	0.00105	0.00114	0.00105	0.00047	0.00000
0.2	0.00034	0.00025	0.00039	0.00041	0.00039	0.00025	0.00034
0.4	0.00110	0.00014	-0.00026	-0.00031	-0.00026	0.00014	0.00110
0.5	0.00000	0.00047	0.00105	0.00114	0.00105	0.00047	0.00000
0.6	0.00206	0.00011	-0.00084	-0.00096	-0.00084	0.00011	0.00206
0.8	0.00307	0.00013	-0.00134	-0.00153	-0.00134	0.00013	0.00307
1.0	0.00410	0.00016	-0.00180	-0.00205	-0.00180	0.00016	0.00410

Table 4.6 Coeff$_{qx}$ for estimation of *Qx* for various values of η and ξ

η, ξ	0	0.2	0.4	0.5	0.6	0.8	1
0.0	0.00000	0.00525	0.00262	0.00000	−0.00262	−0.00525	0.00000
0.2	−0.01209	−0.01423	−0.02041	−0.02417	−0.02793	−0.03411	−0.03626
0.4	−0.03975	−0.05452	−0.07098	−0.07950	−0.08801	−0.10448	−0.11925
0.5	0.00000	0.00525	0.00262	0.00000	−0.00262	−0.00525	0.00000
0.6	−0.07414	−0.10346	−0.13328	−0.14828	−0.16328	−0.19310	−0.22242
0.8	−0.11066	−0.15488	−0.19916	−0.22131	−0.24346	−0.28775	−0.33197
1.0	−0.14752	−0.20653	−0.26554	−0.29505	−0.32455	−0.38356	−0.44257

Table 4.7 Coeff$_{qy}$ for estimation of *Qy* for various values of η and ξ

η, ξ	0.0	0.2	0.4	0.5	0.6	0.8	1
0.0	0.00000	0.00140	0.00315	0.00341	0.00315	0.00140	0.00000
0.2	−0.01210	0.00041	0.00734	0.00823	0.00734	0.00041	−0.01210
0.4	−0.01783	−0.00021	0.00898	0.01015	0.00898	−0.00021	−0.01783
0.5	0.00000	0.00140	0.00315	0.00341	0.00315	0.00140	0.00000
0.6	−0.01996	−0.00057	0.00929	0.01053	0.00929	−0.00057	−0.01996
0.8	−0.02046	−0.00076	0.00913	0.01036	0.00913	−0.00076	−0.02046
1.0	−0.02049	−0.00082	0.00902	0.01024	0.00902	−0.00082	−0.02049

4.6.3 Soil Inclined at an Angle *i* with the Horizontal

In this case, as shown in Fig. 4.13, expressions presented vide Eqs. (4.5.35)–(4.5.40) remain valid except that in this case the active and passive earth pressure coefficients are expressed as

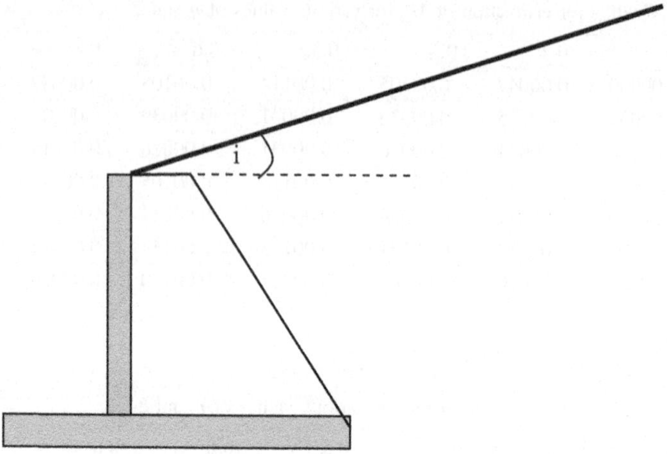

Fig. 4.12 Retaining wall with inclined backfill at an angle i

Fig. 4.13 Retaining wall with general c-ϕ soil as backfill

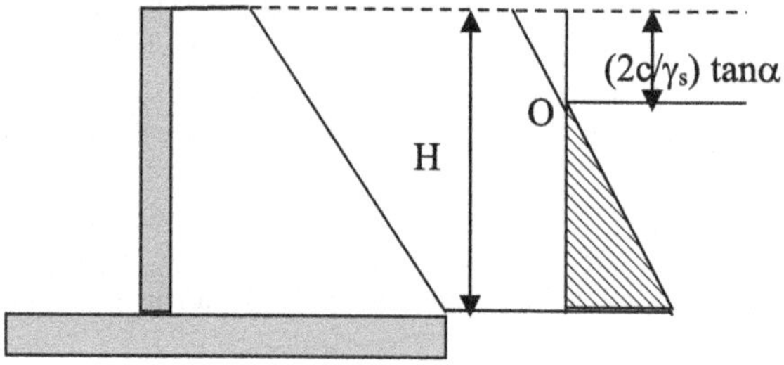

$$K_A = \cos i \times \frac{\cos i - \sqrt{\cos^2 i - \cos^2 \phi}}{\cos i + \sqrt{\cos^2 i - \cos^2 \phi}} \tag{4.6.42}$$

$$K_P = \cos i \times \frac{\cos i + \sqrt{\cos^2 i - \cos^2 \phi}}{\cos i - \sqrt{\cos^2 i - \cos^2 \phi}} \tag{4.6.43}$$

4.6.4 *Dynamic Response of Wall with c-ϕ Backfill*

For a general c-ϕ soil, the active earth pressure is expressed as

$$p_a = \gamma_s z \cot^2 \alpha - 2c \cot \alpha \text{ where } \alpha = 45 + \phi/2. \tag{4.6.44}$$

Shown in Fig. 4.13 is a retaining wall with pressure induced on it as per Eq. (4.6.44) for generalized c-ϕ soil.

It is apparent from the figure that due to cohesion a portion of soil of height $(2c/\gamma_s)\tan\alpha$ cracks and separates away from the wall when, as per practice, the static pressure due to this portion is ignored and the wall pressure is derived based on the net area shown by the hatched line in Fig. 4.13.

For the above case, displacement at point O in Fig. 4.13, where the pressure is zero, is expressed as

$$u_o = \frac{K_A\gamma_s \cdot \left(H - \frac{2c}{\gamma_s}\tan\alpha\right)^5}{D}\left[\frac{X_4}{r^4X_1 + 2r^2X_2 + X_3}\right] \tag{4.6.45}$$

It is to be noted that in this case, $f(\xi)$ and $f(\eta)$ are as expressed in Eqs. (4.6.12) and (4.6.13).

The integral functions X_1, X_2, X_3, and X_4 are as furnished in Table 4.2.

Considering a dimensionless term

$$\mu = \frac{2c\,\tan\alpha}{\gamma_s H} \tag{4.6.46}$$

the maximum deflection at the tip of the wall is expressed as

$$u_{\max} = \frac{K_A\gamma_s \cdot H^5(1-\mu)^5}{D}\left[\frac{X_4}{r^4X_1 + 2r^2X_2 + X_3}\right]\left[1 + \frac{5}{4}\left\{\frac{\mu}{(1-\mu)}\right\}\right] \tag{4.6.47}$$

Time period of the wall can now be expressed as

$$T_A = 2\pi\sqrt{\frac{K_A\gamma_s H^5 \cdot (1-\mu)^5}{Dg}\left[\frac{X_4}{r^4X_1 + 2r^2X_2 + X_3}\right]\left[1 + \frac{5}{4}\left\{\frac{\mu}{(1-\mu)}\right\}\right]} \tag{4.6.48}$$

Considering Eq. (4.6.29), the dynamic amplitude can now be expressed as

$$u_d = \frac{\kappa\beta K_A\gamma_s H^5(1-\mu)^5}{D}\left[\frac{S_a}{g}\right]\left[\frac{X_4}{r^4X_1 + 2r^2X_2 + X_3}\right]\left[1 + \frac{5}{4}\left\{\frac{\mu}{(1-\mu)}\right\}\right]f(\xi)f(\eta) \tag{4.6.49}$$

Dynamic bending moment and shear force are expressed as

$$M_x = \kappa\beta K_A\gamma_s H^3(1-\mu)^3\left[\frac{S_a}{g}\right]\left[\frac{X_4}{r^4X_1 + 2r^2X_2 + X_3}\right]\left[1 + \frac{5}{4}\left\{\frac{\mu}{(1-\mu)}\right\}\right] \times \left[r^2f''(\xi)f(1-\eta) + vf(\xi)f''(1-\eta)\right] \tag{4.6.50}$$

$$M_y = \kappa\beta K_A\gamma_s H^3(1-\mu)^3\left[\frac{S_a}{g}\right]\left[\frac{X_4}{r^4X_1+2r^2X_2+X_3}\right]\left[1+\frac{5}{4}\left\{\frac{\mu}{(1-\mu)}\right\}\right]$$
$$[f(\xi)f''(1-\eta)+vr^2f''(\xi)f(1-\eta)] \tag{4.6.51}$$

$$Q_x = \kappa\beta K_A\gamma_s H^2(1-\mu)^2\left[\frac{S_a}{g}\right]\left[\frac{X_4}{r^4X_1+2r^2X_2+X_3}\right]\left[1+\frac{5}{4}\left\{\frac{\mu}{(1-\mu)}\right\}\right]$$
$$[r^3f'''(\xi)f(1-\eta)+rf'(\xi)f''(1-\eta)] \tag{4.6.52}$$

$$Q_y = \kappa\beta K_A\gamma_s H^2(1-\mu)^2\left[\frac{S_a}{g}\right]\left[\frac{X_4}{r^4X_1+2r^2X_2+X_3}\right]\left[1+\frac{5}{4}\left\{\frac{\mu}{(1-\mu)}\right\}\right]$$
$$[r^2f''(\xi)f'(1-\eta)+f(\xi)f'''(1-\eta)] \tag{4.6.53}$$

In this case, design coefficients furnished in Table 4.3 also could be used, provided $r = 1.5$ and $v = 0.25$.

The modal mass participation may be expressed as

$$\kappa = \frac{\int_\mu^1\int_\mu^1 \eta\left(16\xi^4-32\xi^2+16\xi^2\right)\left(\frac{\eta^5}{4}-\frac{5\eta}{4}+1\right)\mathrm{d}\xi\cdot\mathrm{d}\eta}{\int_\mu^1\int_\mu^1 \eta\cdot\left(16\xi^4-32\xi^2+16\xi^2\right)^2\left(\frac{\eta^5}{4}-\frac{5\eta}{4}+1\right)^2\mathrm{d}\xi\cdot\mathrm{d}\eta} \tag{4.6.54}$$

Values of modal participation factor κ for various values of $\mu = \frac{2c\tan\alpha}{\gamma_s H}$ varying from 0 to 0.6 are as furnished in Table 4.8. Intermediate values can be linearly interpolated.

The above derivation is for a general soil that has finite values of c and ϕ. When the soil is purely cohesionless, i.e., $c = 0$, $\mu \to 0$, Eqs. (4.6.49)–(4.6.54) degenerate to Eqs. (4.6.35)–(4.6.40). This shows the correctness of the derivation of the above expressions.

For a passive case, the earth pressure coefficient K_A is to be replaced by K_P. For the case of pure intact clay ($\phi = 0$), the value $\mu = 2c/\gamma_s H$ when $\alpha = 45°$, and all other formulations above remain the same.

Table 4.8 Values of modal participation Factor

Sl. no	μ	κ
1	0	2.8
2	0.1	2.89
3	0.2	3.086
4	0.3	3.45
5	0.4	4.205
6	0.5	5.841
7	0.6	9.785

4.6.5 Dynamic Response of Wall with c-ϕ Backfill and Surcharge q

This type of problem is often faced by engineers in brownfield industrial plants and construction in dense urban region and remains a serious problem under earthquake. No solution exists till date for this problem, and engineers have to often resort to finite element analysis (FEM) to arrive at a workable result.

Shown in Fig. 4.14 is a retaining wall with generic c-ϕ soil as the backfill, and also having a surcharge q on the top.

In horizontal direction, the shape function $f(\xi)$ remains as $f(\xi) = 16\xi^4 - 32\xi^3 + 16\xi^2$, while in vertical direction it has been shown, vide Eq. (4.5.77), that the shape function is expressed as

$$f(\eta) = \left[\frac{\eta^5}{4} - \frac{5\eta}{4} + 1 - \psi\frac{\eta^4}{3} + \psi\frac{4\xi}{3} - \psi + \frac{\zeta\eta^4}{3} - \frac{4\zeta\eta}{3} + \zeta\right] \tag{4.6.55}$$

In Eq. (4.6.54), $\psi = \frac{15H_c}{4H}\tan\alpha$ and $\zeta = \frac{15q}{4\gamma_s H}$ are both dimensionless parameters and $H_c = 2c/\gamma_s$.

For analysis of this case, it is assumed that the overburden pressure is sufficiently high so that the top portion of soil does not separate away from the wall and remains in contact with the wall, in other words $\zeta > \psi$.

Based on above, the modified Galerkin's equation can be expressed for this case as

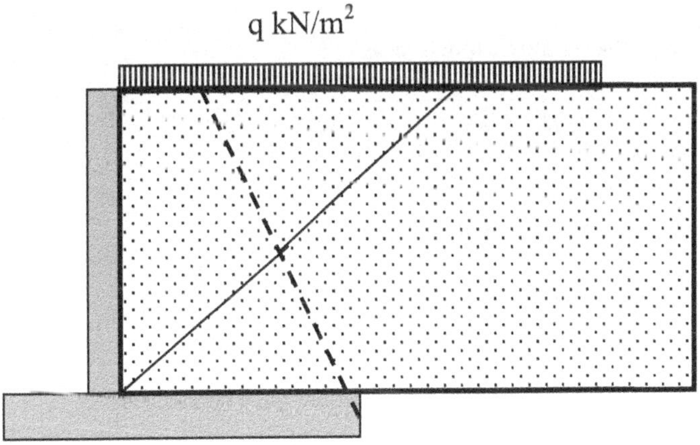

Fig. 4.14 A counterfort retaining wall with c-ϕ backfill and surcharge load q

$$\frac{\Delta}{b^4}\int_0^1\int_0^1 f''''(\xi)f(\xi)f(\eta)^2\mathrm{d}\xi\cdot\mathrm{d}\eta+\frac{2\Delta}{H^2b^2}\int_0^1\int_0^1 f''(\xi)f''(\eta)f(\xi)f(\eta)\mathrm{d}\xi\cdot\mathrm{d}\eta$$

$$+\frac{\Delta}{H^4}\int_0^1\int_0^1 f(\xi)^2f(\eta)f''''(\eta)\mathrm{d}\xi\mathrm{d}\eta-\frac{K_A\gamma_sH^5}{D}\int_0^1\int_0^1 \eta f(\xi)f(\eta)\mathrm{d}\xi\cdot\mathrm{d}\eta$$

$$-\frac{K_AqH^4}{D}\int_0^1\int_0^1 f(\xi)f(\eta)\mathrm{d}\xi\mathrm{d}\eta=0 \tag{4.6.56}$$

$$\Rightarrow\Delta\left[\frac{H^4}{b^4}X_1+\frac{2H^2}{b^2}X_2+X_3\right]=\frac{K_A\cdot\gamma_s\cdot H^5}{D}\left[X_4+\frac{q}{\gamma_sH}X_5\right] \tag{4.6.57}$$

Considering $u=\Delta f(\xi)\cdot f(\eta)$, the static deflection is thus expressed as

$$u=\frac{K_A\gamma_s\cdot H^5}{D}\left[\frac{X_4+\lambda X_5}{r^4X_1+2r^2X_2+X_3}\right]f(\xi)\cdot f(\eta) \tag{4.6.58}$$

where $\lambda=\frac{q}{\gamma_sH}$ and $r=H/b$ the aspect ratio.

In this case, the integral functions X_1, X_2, X_3, X_4, and X_5 are not unique, but are function of both ψ and ζ. The explicit values of these integrals are furnished in Table 4.9.

The maximum static deflection is expressed at $f(\xi)=0.5$ and $f(\eta)=0$ as

$$u_{\max}=\frac{K_A\gamma_s\cdot H^5}{D}\left[\frac{X_4+\lambda X_5}{r^4X_1+2r^2X_2+X_3}\right](1-\psi+\zeta) \tag{4.6.59}$$

The time period is thus expressed as

$$T_A=2\pi\sqrt{\frac{K_A\gamma_sH^5}{Dg}\left[\frac{X_4+\lambda X_5}{r_c^4X_1+2r_c^2X_2+X_3}\right][1-\psi+\zeta]} \tag{4.6.60}$$

The dynamic amplitude is then expressed as

$$u_{\text{dyn}}=\frac{\kappa\beta K_A\gamma_s\cdot H^5}{D}\left[\frac{S_a}{g}\right]\left[\frac{X_4+\lambda X_5}{r^4X_1+2r^2X_2+X_3}\right](1-\psi+\zeta)f(\xi)f(\eta) \tag{4.6.61}$$

Dynamic bending moment and shear force are expressed as

$$M_x=\kappa\beta K_A\gamma_sH^3(1-\psi+\zeta)\left[\frac{S_a}{g}\right]\left[\frac{X_4+\lambda X_5}{r^4X_1+2r^2X_2+X_3}\right]\left[r^2f''(\xi)f(1-\eta)+vf(\xi)f''(1-\eta)\right] \tag{4.6.62}$$

Table 4.9 Explicit values of integrals X_1 through X_5 for general c-ϕ soil with overburden

Integral functions	Values
$X_1 = \int_0^1 \int_0^1 \left[\frac{\partial^4}{\partial \xi^4} f(\xi)\right] f(\xi) f(\eta)^2 \mathrm{d}\xi \mathrm{d}\eta$	$52.59(\psi^2 + \zeta^2) + 107.93(\zeta - \psi) - 105.18\psi \cdot \zeta + 55.41$
$X_2 = \int_0^1 \int_0^1 \left[\frac{\partial^2}{\partial \xi^2} f(\xi)\right] \left[\frac{\partial^2}{\partial \eta^2} f(\eta)\right] f(\xi) f(\eta)\, \mathrm{d}\xi \mathrm{d}\eta$	$-0.677 - 0.93(\psi^2 + \zeta^2) + 1.858\psi \cdot \zeta + 1.625(\psi - \zeta)$
$X_3 = \int_0^1 \int_0^1 \left[\frac{\partial^4}{\partial \eta^4} f(\eta)\right] f(\eta) f(\xi)^2 \mathrm{d}\xi \mathrm{d}\eta$	$1.451 - 2.6\psi \cdot \zeta + 2.709(\zeta - \psi) + 1.3(\psi^2 + \zeta^2)$
$X_4 = \int_0^1 \int_0^1 \eta f(\eta) f(\xi)\, \mathrm{d}\xi \mathrm{d}\eta$	$0.0635 + 0.06(\zeta - \psi)$
$X_5 = \int_0^1 \int_0^1 f(\eta) f(\xi)\, \mathrm{d}\xi \mathrm{d}\eta$	$0.222 + 0.213(\zeta - \psi)$

$$M_y = \kappa\beta K_A\gamma_s H^3(1-\psi+\zeta)\left[\frac{S_a}{g}\right]\left[\frac{X_4+\lambda X_5}{r^4X_1+2r^2X_2+X_3}\right] \times \left[f(\xi)f''(1-\eta)+vr^2f''(\xi)f(1-\eta)\right] \tag{4.6.63}$$

$$Q_x = \kappa\beta K_A\gamma_s H^2(1-\psi+\zeta)\left[\frac{S_a}{g}\right]\left[\frac{X_4+\lambda X_5}{r^4X_1+2r^2X_2+X_3}\right]\left[r^3f'''(\xi)f(1-\eta)+rf'(\xi)f''(1-\eta)\right] \tag{4.6.64}$$

$$Q_y = \kappa\beta K_A\gamma_s H^2(1-\psi+\zeta)\left[\frac{S_a}{g}\right]\left[\frac{X_4+\lambda X_5}{r^4X_1+2r^2X_2+X_3}\right] \left[r^2f''(\xi)f'(1-\eta)+f(\xi)f'''(1-\eta)\right] \tag{4.6.65}$$

Equations (4.6.61)–(4.6.65) give the displacement, moment, and shear under seismic loading for the most general condition of soil. When there is no overburden, i.e., $\eta \to 0$, the formulas converge to the case of c-ϕ soil only. When again, $\psi \to 0$, the equations converge to the case of pure cohesionless soil ($c = 0$).

Finally, an interesting case is observed: when $\psi \to 0$, $\eta \to 0$ and even $\phi = 0$ (i.e., $K_A \to 1$), the values obtained in Eqs. (4.6.61)–(4.6.65) are not zero and finite and converge to a hydrostatic case. Thus, if we take density of soil in this particular case ($\gamma_{sat} - \gamma_w$), it reflects a case when the soil has undergone liquefaction.

The modal mass participation factor in this case is expressed as

$$\kappa = \frac{\int_0^1\int_0^1 \eta f(\xi)f(\eta)\,d\xi d\eta + \int_0^1\int_0^1 \lambda f(\xi)f(\eta)d\xi d\eta}{\int_0^1\int_0^1 \eta f(\xi)^2f(\eta)^2 + \int_0^1\int_0^1 \lambda f(\xi)^2f(\eta)^2 d\xi d\eta} \tag{4.6.66}$$

Explicit value of the above integral in terms of ψ, ζ, and λ is expressed as

$$\kappa = \frac{0.0635+0.0593(\zeta-\psi)+0.222\lambda+0.213\lambda(\zeta-\psi)}{0.0226+0.0432(\zeta-\psi)+0.0206(\psi^2-\zeta^2)-0.0412\psi\cdot\zeta+0.1099\lambda-0.2086\lambda\psi\zeta+0.1043\lambda(\psi^2+\zeta^2)+0.214\lambda(\zeta-\psi)} \tag{4.6.67}$$

4.6.6 Dynamic Response of Wall with c-ϕ Backfill Partially Submerged Below Water

Shown in Fig. 4.15 is a retaining wall having partially submerged c-ϕ soil as backfill. The soil is partially submerged for a height H_2. In this case, considering the complex nature of the pressure diagram it would become difficult to perform the integration with appropriate boundaries. As such, we approach the problem differently for this particular case. Considering that the first step in the process is to

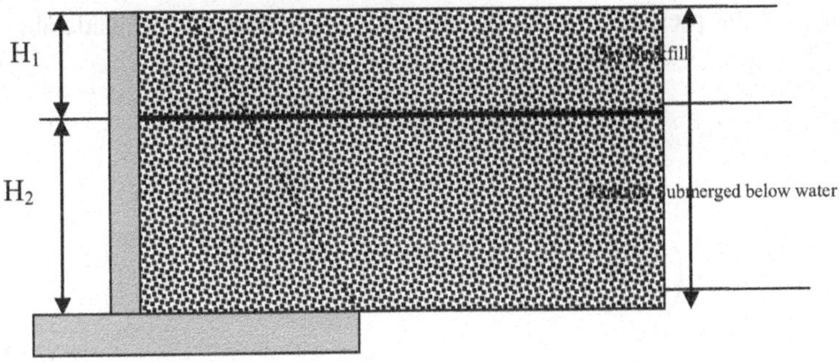

Fig. 4.15 A counterfort retaining wall with c-ϕ backfill partially submerged

determine the static deflection, and this is obtained based on the loading on the retaining wall, the net horizontal loading on the wall can be expressed as

$$P_A = \frac{1}{2}\cot^2\alpha\left[\gamma_s H_1^2 + (\gamma_{sat} - \gamma_w)H_2^2\right] - 2c\cot\alpha H + \gamma_s H_1 \cdot H_2 \cot^2\alpha + \frac{2c^2}{\gamma_s} \tag{4.6.68}$$

Here

γ_s dry density of soil of height H_1;
γ_{sat} saturated density of soil of height H_2;
γ_w density of water.

Now if we consider an equivalent dry backfill of density γ_s, which imposes the same load on the wall, the displacement of the wall will be same as that which would be induced by load, as expressed in Eq. (4.6.68).

Thus considering

$$\frac{1}{2}K_{AE}\gamma_s H^2 = \frac{1}{2}\cot^2\alpha\left[\gamma_s H_1^2 + (\gamma_{sat} - \gamma_w)H_2^2\right] - 2c\cot\alpha H + \gamma_s H_1 \cdot H_2 \cot^2\alpha + \frac{2c^2}{\gamma_s} \tag{4.6.69}$$

we have

$$K_{AE} = \cot^2\alpha \frac{\left[\gamma_s H_1^2 + (\gamma_{sat} - \gamma_w)H_2^2\right]}{\gamma_s (H_1 + H_2)^2} - \frac{4c\cot\alpha}{\gamma_s \cdot (H_1 + H_2)} + \frac{2\gamma_s H_1 \cdot H_2 \cot^2\alpha}{(H_1 + H_2)^2} + \frac{4c^2}{\gamma_s^2 (H_1 + H_2)^2} \tag{4.6.70}$$

Here K_{AE}= an equivalent coefficient of active earth pressure when considering a pressure diagram of $p_{ay} = K_{AE} \cdot \gamma_s \cdot y$ over the height H will give same deflection as that produced by P_A in Eq. (4.6.68).

Thus, for the present case the problem now gets simplified considerably when we have

$$T_A = 2\pi\sqrt{\frac{K_{AE}\gamma_s H^5}{Dg}\left[\frac{X_4}{r^4X_1 + 2r^2X_2 + X_3}\right]} \tag{4.6.71}$$

$$u_d = \kappa\beta\frac{K_{AE}\gamma_s H^5}{D}\left(\frac{S_a}{g}\right)\left[\frac{X_4}{r^4X_1 + 2r^2X_2 + X_3}\right]\left[16\xi^4 - 32\xi^3 + 16\xi^2\right]\left[\frac{\eta^5}{4} - \frac{5\eta}{4} + 1\right] \tag{4.6.72}$$

$$M_x = \kappa\cdot\beta\cdot K_{AE}\cdot\gamma_s\cdot H^3\left[\frac{S_a}{g}\right]\left[\frac{X_4}{r^4X_1 + 2r^2X_2 + X_3}\right]$$
$$\left[r^2 f''(\xi)f(1-\eta) + \nu f(\xi)f''(1-\eta)\right] \tag{4.6.73}$$

$$M_y = \kappa\beta K_{AE}\gamma_s H^3\left[\frac{S_a}{g}\right]\left[\frac{X_4}{r^4X_1 + 2r^2X_2 + X_3}\right]$$
$$\left[f(\xi)f''(1-\eta) + \nu r^2 f''(\xi)f(1-\eta)\right] \tag{4.6.74}$$

$$Q_x = \kappa\cdot\beta\cdot K_{AE}\cdot\gamma_s\cdot H^2\left[\frac{S_a}{g}\right]\left[\frac{X_4}{r^4X_1 + 2r^2X_2 + X_3}\right]$$
$$\left[r^3 f'''(\xi)f(1-\eta) + r f'(\xi)f''(1-\eta)\right] \tag{4.6.75}$$

$$Q_y = \kappa\cdot\beta\cdot K_{AE}\cdot\gamma_s\cdot H^2\left[\frac{S_a}{g}\right]\left[\frac{X_4}{r^4X_1 + 2r^2X_2 + X_3}\right]$$
$$\left[r^2 f''(\xi)f'(1-\eta) + f(\xi)f'''(1-\eta)\right] \tag{4.6.76}$$

The modal participation factor κ will remain same as expressed by Eq. (4.6.40) cited earlier.

4.6.7 Validation of the Plate Theory

Using a polynomial shape function that conforms to the boundary condition of a differential equation (linear or partial) and minimizing the residual error for solution of this equation based on various weighted residual techniques (like sub-domain, Galerkin, least square) has been an acceptable and established technique in realms of numerical methods in engineering, especially by finite element method.

For rectangular plate elements, assuming polynomial functions based on Pascal triangle (Szilard 2004) or an approximate assumed boundary satisfying shape function and applying weighted residual technique has been often used to derive approximate solution of a plate element (Chandrashekara 2001).

Tocher (1962) first developed an approximate solution based on non-conforming polynomial shape function, the stiffness of a rectangular plate element. This was followed by Adini and Clough (1964) who derived an explicit expression of stiffness matrix for a general plate element for finite element analysis.

The authors are, however, yet to come across any work where shape functions have been derived based on explicit boundary conditions of plates considering a strip of beam element having similar boundary conditions and then subsequently using it in a plate equation to arrive at a solution based on Galerkin's technique of minimizing the error as presented herein.

So naturally the question arises as to how accurate the technique is?

To this end, a simply supported plate subjected to a load w is solved by the present technique and the displacement coefficient thus obtained is compared to the analytical solution, as cited in Timoshenko and Kreiger (1987).

For a simply supported plate considering the shape functions as (shape functions derived from solution of differential equation of a simply supported beam by identical procedure as cited at the outset)

$$f(\xi) = \frac{16}{5}\xi^4 - \frac{32}{5}\xi^3 + \frac{16}{5}\xi \tag{4.6.77}$$

$$f(\eta) = \frac{16}{5}\eta^4 - \frac{32}{5}\eta^3 + \frac{16}{5}\eta \tag{4.6.78}$$

The static deflection is expressed as

$$u = \frac{w \cdot H^4}{D}\left[\frac{X_4}{r^4 X_1 + 2r^2 X_2 + X_3}\right] f(\xi) \cdot f(\eta) \tag{4.6.79}$$

The above can be expressed as

$$u = \text{Coeff} \times \frac{w \cdot H^4}{D} \tag{4.6.80}$$

where

$$\text{Coeff} = \left[\frac{X_4}{r^4 X_1 + 2r^2 X_2 + X_3}\right] f(\xi) \cdot f(\eta) \tag{4.6.81}$$

The coefficient thus derived above is compared to Timoshenko's data at $f(\xi) = 0.5$ and $f(\eta) = 0.5$ and are shown hereafter for different aspect ratios of the plate.

From Fig. 4.16, it is observed that the values are in excellent agreement. Maximum variation is about 5% at $r = 1.9$ and minimum is 1.8% at $r = 1.0$. Thus,

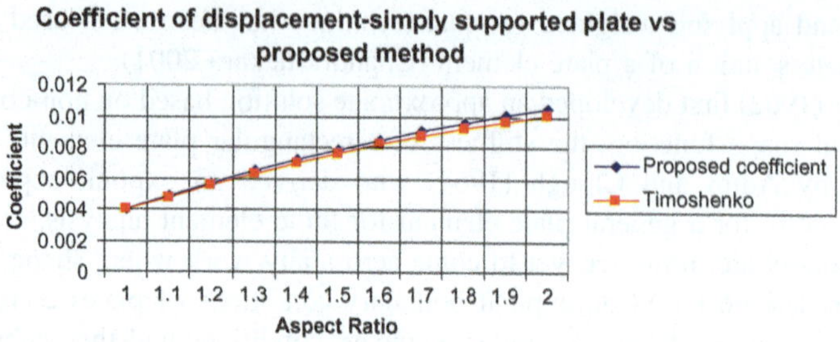

Fig. 4.16 Comparison of coefficient of displacement proposed versus analytical solution by Timoshenko and Kreiger (1987)

it can be concluded that considering its accuracy for this benchmark problem the proposed technique should also be valid for other boundary conditions like free, fixed simply supported and suffice for practical application, as presented in this case.

Efficiency of the method lies in using such higher-order polynomial is the minimum computational effort required to arrive at an acceptable solution of sufficient accuracy without having to resort to finite element analysis or tedious computation with trigonometric functions that are encountered in classical analysis. The solution can well be carried out in a spreadsheet or generic utility software like Mathcad which are common computational tools available in any design office environment.

Example 4.6.1 A 8.0-m-high counterfort retaining wall with the following structural and soil properties is analyzed under earthquake force.

Height of wall = 8.0 m, thickness of wall = 0.25 m, counterfort spacing = 6.0 m thickness of counterfort = 0.4 m, grade of concrete = M39, foundation width = 4.0 m, foundation thickness = 0.6 m, density of backfill = 22 kN/m^3. Angle of internal friction = 10°, cohesive strength (c) = 5kN/m^2, overburden (q) = 50 kN/m^2. Earthquake zone = Zone V as per IS-1893. Damping of soil considered (average) = 15%. Determine the displacement amplitude, moments, and shear on the wall due to seismic force.

The soil being a generic c-ϕ soil with overburden, as such Mononobe's formula cannot be used.

Here as the earthquake zone is V $Z = 0.36$, $I = 1.0$, $R = 3.0$

$$\beta = ZI/2R = \frac{0.36 \times 1}{6} = 0.06$$

$$\alpha = 45 + \phi/2 = 45 + 10/2 = 50$$

$$H_c = 2c/\gamma_s = \frac{2 \times 5}{22} = 0.4545 \text{ m}$$

$$\psi = \frac{15H_c}{4H} \tan\alpha = \frac{15 \times 0.454}{4 \times 8} \tan 50 = 0.25239$$

$$\zeta = \frac{15q}{4\gamma_s H} = \frac{15 \times 50}{4 \times 22 \times 8} = 1.0653$$

Here $E_{\text{conc}} = 5000\sqrt{f_{ck}} = 3.122 \times 10^7\ \text{kN/m}^2$, here $f_{ck} = 39\ \text{N/mm}^2$.

$$D = \frac{E_{\text{conc}} t^3}{12(1-\nu^2)} = \frac{3.122 \times 10^7 \times (0.25)^3}{12(1-0.25^2)} = 4.3361 \times 10^4\ \text{kN/m}$$

$$r_c = H/b = 8/6 = 1.333$$

$$K_A = \frac{1-\sin\phi}{1+\sin\phi} = \frac{1-\sin 10}{1+\sin 10} = 0.704$$

$$\lambda = \frac{q}{\gamma_s H} = \frac{50}{22 \times 8} = 0.284$$

Referring to Table 4.9

$$X_1 = 52.59(\psi^2+\zeta^2) + 107.93(\zeta-\psi) - 105.18\psi\cdot\zeta + 55.41 = 177.6$$
$$X_2 = -0.677 - 0.93(\psi^2+\zeta^2) + 1.858\psi\cdot\zeta + 1.625(\psi-\zeta) = -2.6076$$
$$X_3 = 1.451 - 2.6\psi\cdot\zeta + 2.709(\zeta-\psi) + 1.3(\psi^2+\zeta^2) = 4.5055$$
$$X_4 = 0.0635 + 0.06(\zeta-\psi) = 0.1116$$
$$X_5 = 0.222 + 0.213(\zeta-\psi) = 0.3953$$

$$T_A = 2\pi\sqrt{\frac{K_A\gamma_s H^5}{Dg}\left[\frac{X_4+\lambda X_5}{r_c^4X_1+2r_c^2X_2+X_3}\right][1-\psi+\zeta]}$$

$$T_A = 2\pi\sqrt{\frac{0.704\times 22\times 8^5}{4.3361\times 10^4\times 9.81}\left[\frac{0.1116+0.284\times 0.3953}{1.333^4\times 177.6+2\times 1.333^2\times(-2.607)+4.505}\right][1-0.252+1.065]}$$

$T_A = 0.185$ sec.

As per IS-1893, the corresponding $Sa/g = 2.5$. Now considering 15% average damping scale down factor is 0.7.

This design $Sa/g = 1.75$.

$$\kappa = \frac{0.0635+0.0593(\zeta-\psi)+0.222\lambda+0.213\lambda(\zeta-\psi)}{0.0226+0.0432(\zeta-\psi)+0.0206(\psi^2-\zeta^2)-0.0412\psi\cdot\zeta+0.1099\lambda\quad 0.2086\lambda\psi\zeta+0.1043\lambda(\psi^2+\zeta^2)+0.214\lambda(\zeta-\psi)}$$
$$= 1.305$$

The amplitude of vibration is given by

$$u_{\text{dyn}} = \frac{\kappa\beta K_A\gamma_s\cdot H^5}{D}\left[\frac{S_a}{g}\right]\left[\frac{X_4+\lambda X_5}{r^4X_1+2r^2X_2+X_3}\right](1-\psi+\zeta)f(\xi)f(\eta)$$

here $f(\eta) = \left[\frac{\eta^5}{4} - \frac{5\eta}{4} + 1 - \psi\frac{\eta^4}{3} + \psi\frac{4\xi}{3} - \psi + \frac{\zeta\eta^4}{3} - \frac{4\zeta\eta}{3} + \zeta\right]$ & $f(\xi) = 16\xi^4 - 32\xi^3 + 16\xi^2$, varying from 0 to 1, substituting the values we have

$$u(0.5,0) = 2.1182 \times 10^{-3}\text{m}$$

The distribution of the amplitude is shown below, as plotted in Mathcad.

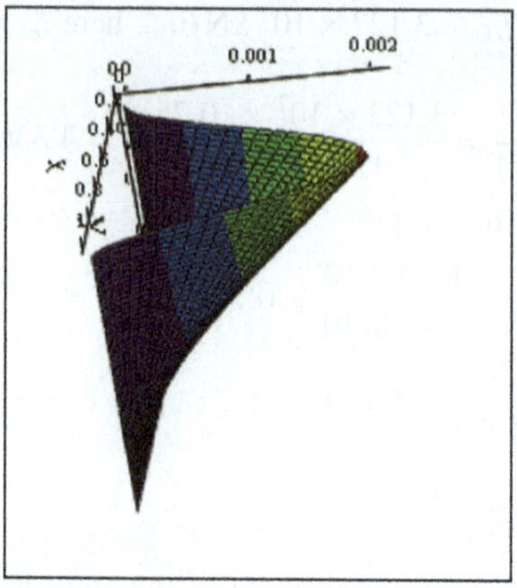

Plotu

Displacement amplitude of the wall under shear force

Dynamic bending moment and shear force are expressed as

$$M_x = \kappa\beta K_A\gamma_s H^3(1-\psi+\zeta)\left[\frac{S_a}{g}\right]\left[\frac{X_4+\lambda X_5}{r^4X_1+2r^2X_2+X_3}\right]\left[r^2f''(\xi)f(1-\eta)+vf(\xi)f''(1-\eta)\right]$$

$$M_y = \kappa\beta K_A\gamma_s H^3(1-\psi+\zeta)\left[\frac{S_a}{g}\right]\left[\frac{X_4+\lambda X_5}{r^4X_1+2r^2X_2+X_3}\right]\left[f(\xi)f''(1-\eta)+vr^2f''(\xi)f(1-\eta)\right]$$

$$Q_x = \kappa\beta K_A\gamma_s H^2(1-\psi+\zeta)\left[\frac{S_a}{g}\right]\left[\frac{X_4+\lambda X_5}{r^4X_1+2r^2X_2+X_3}\right]\left[r^3f'''(\xi)f(1-\eta)+rf'(\xi)f''(1-\eta)\right]$$

$$Q_y = \kappa\beta K_A\gamma_s H^2(1-\psi+\zeta)\left[\frac{S_a}{g}\right]\left[\frac{X_4+\lambda X_5}{r^4X_1+2r^2X_2+X_3}\right]\left[r^2f''(\xi)f'(1-\eta)+f(\xi)f'''(1-\eta)\right]$$

Substituting the appropriate values, the moment, and shear diagrams are shown hereafter as obtained by Mathcad.

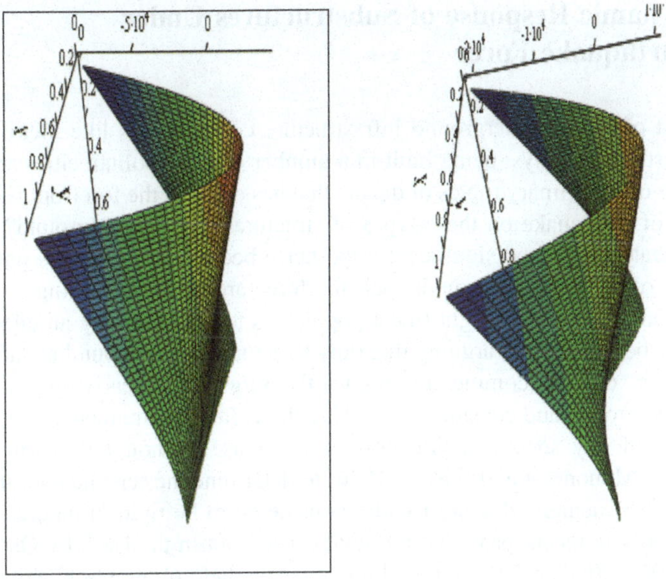

Plotmx

Moment *Mx* and *My* of the wall due to seismic force

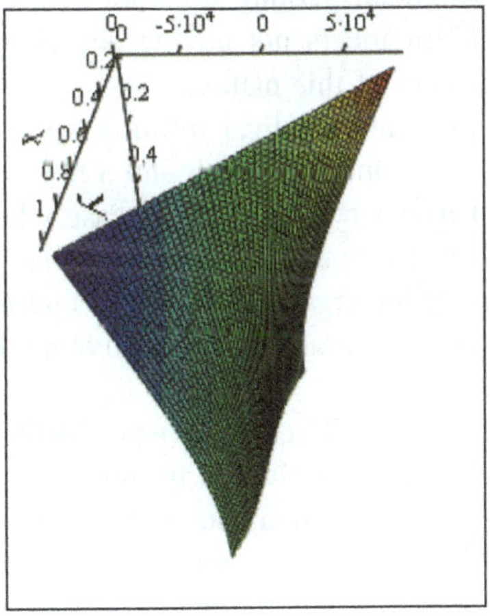

Plotqx

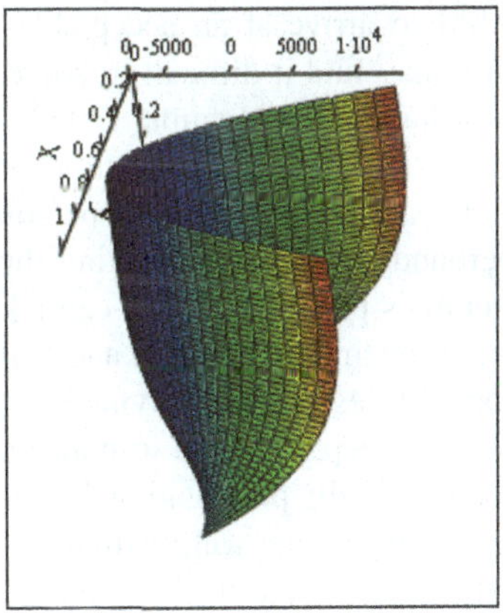

Plotqy

Shear V_x and V_y of the wall due to seismic force

4.7 Dynamic Response of Substructures Under Earthquake Force

With spurt of recent underground infrastructure constructions like metro rail tunnels, pedestrian subways, being built in a number of metropolitan cities around the world, one of the primary topics of debate that has come to the forefront is—what is the effect of earthquake on these types of structures built below ground?

The debate has gained significant importance because philosophies put forward for design of such structures under seismic force are highly diverging.

While one school of thought (most popular) is to ignore totally an effect of any earthquake below ground, arguing that only structures above ground are affected by seismic force, others recommend to reduce the Sa/g value linearly up to a depth of 30 m below ground and consider $0.5 \times Sa/g$ thereafter. Again, there are others who in absence of any specific guidelines prefer to use Mononobe's earth pressure coefficients (Mononobe and Matsuo 1926) to determine the seismic pressure on the wall of such structures, that are usually considered to be rigid in nature.

It was only in recent past, that research carried out in the USA by Ostadan and White (1997), Ostadan (2004), based on extensive field observation and then back checking the same by finite element method (FEM) through software like SHAKE and SASSI, came up with a pressure equation showing a best fit over a number of soil data. This formula is now adapted by NEHRP and also by US Nuclear Regulatory Authority for all their substructures for nuclear power plants.

However, one of the major limitations of this method is that the expression is valid only for G constant with depth and would still require software like SHAKE or SASSI to arrive at an acceptable result. Engineers not having access to such software may find it difficult to solve a problem of this nature.

Chowdhury and Dasgupta (2007) presented an analytical solution for determination of such dynamic pressure, having G constant with depth and a few selected cases like sandy- or gravelly-type soil. But it was restricted to the case where the underground structure was resting directly on the bedrock itself. A similar to this philosophy is presented which can take care of underground structures that may be floating or resting on bedrock and soil having a generalized constitutive model can be adapted to estimate the dynamic pressure.

As a first step, we need to understand the free field characteristics of the site.

When vertically propagating S_v waves travel through the soil from bedrock level, it is apparent that the soil medium over it will get excited and start vibrating. The general character of such mode shapes of vibrations is as shown in Fig. 4.17 by the curves ϕ_1, ϕ_2, ϕ_3, etc.

Referring to Fig. 4.17 in two dimensions, the propagation of S_v waves through the medium can be expressed as

$$\frac{\partial^2 u(x,z,t)}{\partial x^2} + \frac{\partial^2 u(x,z,t)}{\partial z^2} = \frac{1}{Vs^2}\frac{\partial^2 u(x,z,t)}{\partial t^2} \tag{4.7.1}$$

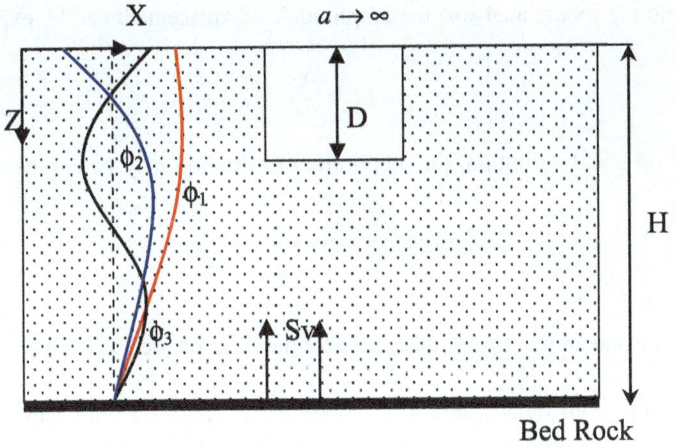

Fig. 4.17 An underground structure built below ground

Equation (4.7.1) has already been solved in detail in Chap. 3, Sect. 3.5.3, wherein we have shown that the stiffness and mass matrix for the soil can be expressed for the first four modes for G constant with depth as

$$[K] = \frac{G}{H}\begin{bmatrix} \frac{\pi^2}{8} & 0 & 0 & 0 \\ 0 & \frac{9\pi^2}{8} & 0 & 0 \\ 0 & 0 & \frac{25\pi^2}{8} & 0 \\ 0 & 0 & 0 & \frac{49\pi^2}{8} \end{bmatrix} \tag{4.7.2}$$

$$[M] = \frac{\gamma_s H}{g}\begin{bmatrix} \frac{1}{2} & 0 & 0 & 0 \\ 0 & \frac{1}{2} & 0 & 0 \\ 0 & 0 & \frac{1}{2} & 0 \\ 0 & 0 & 0 & \frac{1}{2} \end{bmatrix} \tag{4.7.3}$$

Solving for eigenvalue λ from $[\mathbf{K}] - \lambda[\mathbf{M}] = 0$ and knowing $\lambda = \omega^2$, where $T = 2\pi/\omega$, we finally have

$$\{T\} = \frac{H}{V_s}\{4 \quad 1.333 \quad 0.8 \quad 0.571\}^T \tag{4.7.4}$$

The corresponding eigenvectors $[\phi]$ are obtained as

$$[\phi] = \begin{bmatrix} 1 & 0 & 0 & 0 \\ 0 & 1 & 0 & 0 \\ 0 & 0 & 1 & 0 \\ 0 & 0 & 0 & 1 \end{bmatrix} \tag{4.7.5}$$

Considering modal analysis, the amplitude of displacement can be expressed as

$$S_d = \frac{S_a}{\omega^2} \tag{4.7.6}$$

Based on codal definition, the amplitude of vibration may be expressed as

$$u_n = \kappa_n \beta \frac{S_{an}}{\omega^2} \cos(2n-1)\frac{\pi z}{2H} \tag{4.7.7}$$

Here κ_n is modal mass participation factor and is expressed as $\kappa_n = \sum_{i=1}^{n} m_i \phi_i / \sum_{i=1}^{n} m_i \phi_i^2$

For the first four modes, this is as expressed in Table 3.13 in Chap. 3.

Here β is a code factor expressed as $\beta = ZI/2R$ where Z = zone factor I = importance factor R = response reduction factor.

The strain within the soil body is expressed as

$\varepsilon_{xx} = 0$ and $\varepsilon_{zz} = \frac{\partial u}{\partial z}$ which gives

$$\varepsilon_{zz\,n} = -\kappa_n \frac{(2n-1)\pi}{2H} \beta \frac{S_{an} T_n^2}{4\pi^2} \sin(2n-1)\frac{\pi z}{2H} \tag{4.7.8}$$

Equation (4.7.8) on simplification for first four modes finally gives

$$\varepsilon_{zz\,1} = -\frac{16}{\pi(\pi+2)} \beta \gamma_s \left(\frac{S_{a1}}{g}\right) \frac{H}{G} \sin\frac{\pi z}{2H} \tag{4.7.9}$$

$$\varepsilon_{zz\,2} = \frac{16}{3\pi(3\pi-2)} \beta \gamma_s \left(\frac{S_{a2}}{g}\right) \frac{H}{G} \sin\frac{3\pi z}{2H} \tag{4.7.10}$$

$$\varepsilon_{zz\,3} = -\frac{16}{5\pi(5\pi+2)} \beta \gamma_s \left(\frac{S_{a2}}{g}\right) \frac{H}{G} \sin\frac{5\pi z}{2H} \tag{4.7.11}$$

$$\varepsilon_{zz\,4} = \frac{16}{7\pi(7\pi-2)} \beta \gamma_s \left(\frac{S_{a4}}{g}\right) \frac{H}{G} \sin\frac{7\pi z}{2H} \tag{4.7.12}$$

For plain strain condition

$$\begin{Bmatrix} \sigma_{xx} \\ \sigma_{zz} \\ \tau_{xz} \end{Bmatrix} = \frac{2G}{(1-2v)} \begin{bmatrix} 1-v & v & 0 \\ v & 1-v & 0 \\ 0 & 0 & \frac{1-2v}{2} \end{bmatrix} \begin{Bmatrix} \varepsilon_{xx} \\ \varepsilon_{zz} \\ \gamma_{xz} \end{Bmatrix} \tag{4.7.13}$$

This gives

$$\sigma_{xx} = \frac{2G(1-v)}{1-2v}\varepsilon_{xx} + \frac{2Gv}{1-2v}\varepsilon_{zz} \tag{4.7.14}$$

Now considering $\varepsilon_{xx} = 0$, we have

$$\sigma_{xx} = \frac{2Gv}{1-2v}\varepsilon_{zz} \tag{4.7.15}$$

Now considering that σ_{xx} is the dynamic pressure on the wall and we have for the first four modes

$$p_1 = \frac{-32}{\pi(\pi+2)}\left(\frac{v}{1-2v}\right)\beta\left(\frac{S_{a1}}{g}\right)\gamma_s H \sin\frac{\pi z}{2H} \tag{4.7.16}$$

$$p_2 = \frac{32}{3\pi(3\pi-2)}\left(\frac{v}{1-2v}\right)\beta\left(\frac{S_{a2}}{g}\right)\gamma_s H \sin\frac{3\pi z}{2H} \tag{4.7.17}$$

$$p_3 = \frac{-32}{5\pi(5\pi+2)}\left(\frac{v}{1-2v}\right)\beta\left(\frac{S_{a3}}{g}\right)\gamma_s H \sin\frac{5\pi z}{2H} \tag{4.7.18}$$

$$p_4 = \frac{32}{7\pi(7\pi-2)}\left(\frac{v}{1-2v}\right)\beta\left(\frac{S_{a4}}{g}\right)\gamma_s H \sin\frac{7\pi z}{2H} \tag{4.7.19}$$

Equations (4.6.42)–(4.6.45) can also be generically expressed as

$$p = \text{coeff} \times \beta \times \psi \times \left(\frac{S_a}{g}\right) \times \gamma_s \times H \tag{4.7.20}$$

Here $\psi = \left(\frac{v}{1-2v}\right)$

In this formula, coeff is a coefficient and is a SRSS value of the above pressures for different modes, as shown hereafter in Fig. 4.18.

Based on the above, the following are to be noted

- For G remaining constant with depth, $\cos(2n-1)\pi z/2H$ is an exact solution to the differential equation of motion and gives identical solution to the problem based on Rayleigh–Ritz basis(energy principle) for obtaining the stiffness and mass matrix of the system.
- The response for fundamental mode is most critical. The higher-order modes have significantly reduced response (Fig. 4.18).
- The SRSS values of pressure for first three modes vis-à-vis fundamental mode—there is a negligible difference (Fig. 4.18).
- Thus, if one works with fundamental mode only, for practical engineering problem, it is more than adequate.

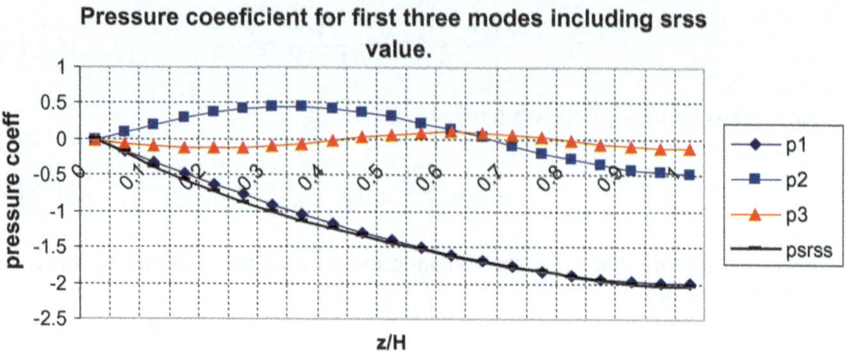

Fig. 4.18 Dynamic pressure coefficient on wall for G constant with depth

- **Properties of Gibson-Type Soil**

Again, as shown in Chap. 3 vide Eq. (3.6.3.47), the stiffness matrix is expressed as

$$[K] = \frac{G}{H}\begin{bmatrix} \frac{\pi^2}{16}+\frac{1}{4} & \frac{-3}{4} & \frac{5}{36} & \frac{-7}{36} \\ \frac{-3}{4} & \frac{9\pi^2}{16}+\frac{1}{4} & \frac{-15}{4} & \frac{21}{100} \\ \frac{5}{36} & \frac{-15}{4} & \frac{25\pi^2}{16}+\frac{1}{4} & \frac{-35}{4} \\ \frac{-7}{36} & \frac{21}{100} & \frac{-35}{4} & \frac{49\pi^2}{16}+\frac{1}{4} \end{bmatrix} \tag{4.7.21}$$

The mass matrix remains same as Eq. (4.7.3).

Solving for the eigenvalue λ from $[\mathbf{K}] - \lambda[\mathbf{M}] = 0$ and knowing $\lambda = \omega^2$ where $T = 2\pi/\omega$ we finally have

$$\{T\} = \frac{H}{V_s}\{\,5.19 \quad 2.133 \quad 1.228 \quad 0.755\,\}^T \tag{4.7.22}$$

The scaling factors for the eigenvectors are expressed as

$$\{\phi\} = \begin{bmatrix} -0.971 & -0.534 & 0.078 & 0.021 \\ -0.217 & 0.723 & -0.701 & -0.125 \\ -0.086 & 0.39 & 0.562 & 0.668 \\ -0.046 & 0.199 & 0.432 & -0.733 \end{bmatrix} \tag{4.7.23}$$

The amplitude of displacement can be expressed in this case [refer Eq. (3.6.3.50)] as

$$u_n = \kappa_n \beta \frac{S_{an} T_n^2}{4\pi^2}[\phi_n]^T \cos(2n-1)\frac{\pi z}{2H} \tag{4.7.24}$$

Here κ_n does not change and remains same as values derived earlier for G constant with depth, and $[\phi_n]$ is the scale factor of the eigenvectors as derived in Eq. (4.7.23).

Equation (4.7.24) can be further expressed for fundamental mode as

$$u_1 = \frac{8}{\pi+2}\gamma_s\beta\left(\frac{S_{a1}}{g}\right)\frac{(5.19)^2H^2}{4\pi^2G}\begin{bmatrix}-0.971\cdot\cos\frac{\pi z}{2H}\\-0.217\cdot\cos\frac{3\pi z}{2H}\\-0.086\cdot\cos\frac{5\pi z}{2H}\end{bmatrix} \tag{4.7.25}$$

$$\rightarrow u_1 = 1.0616\gamma_s\beta\left(\frac{S_{a1}}{g}\right)\frac{H^2}{G}\begin{bmatrix}-0.971\cdot\cos\frac{\pi z}{2H}\\-0.217\cdot\cos\frac{3\pi z}{2H}\\-0.086\cdot\cos\frac{5\pi z}{2H}\end{bmatrix} \tag{4.7.26}$$

Considering $\varepsilon_{xx} = 0$ and $\varepsilon_{zz} = \frac{\partial u}{\partial z}$, we have

$$\varepsilon_{zz1} = 1.0616\beta\left(\frac{S_{a1}}{g}\right)\frac{H}{G}\begin{bmatrix}1.525\sin\frac{\pi z}{2H}+\\1.022\sin\frac{3\pi z}{2H}+\\0.675\sin\frac{5\pi z}{2H}\end{bmatrix} \tag{4.7.27}$$

Considering $\sigma_{xx} = p = \frac{2Gv}{1-2v}\varepsilon_{zz}$

$$p_1 = 2.123\beta\gamma_s H\left(\frac{S_{a1}}{g}\right)\left(\frac{v}{1-2v}\right)\begin{bmatrix}1.525\sin\frac{\pi z}{2H}+\\1.022\sin\frac{3\pi z}{2H}+\\0.675\sin\frac{5\pi z}{2H}\end{bmatrix} \tag{4.7.28}$$

Higher modes can well be ignored as it has been shown earlier (Fig. 4.18) that their effect is insignificant.

Proceeding in identical fashion, considering $T = C_T(H/Vs)$, values of C_T for different soil type are presented in Table 3.3.

Now following the same steps as shown from Eqs. (4.7.25)–(4.7.28), the dynamic pressure for soils with different characteristics is as shown hereafter:

(i) For soil of type $G = G_0(z/H)^{0.5}$, the dynamic pressure is expressed as

$$p = 1.586\beta\gamma_s H\left(\frac{S_a}{g}\right)\left(\frac{v}{1-2v}\right)(1.566A_z + 0.372B_z + 0.104C_z) \tag{4.7.29}$$

(ii) For soil of type $G = G_0(z/H)^2$

$$p = 4.826\beta\gamma_s H\left(\frac{S_a}{g}\right)\left(\frac{v}{1-2v}\right)(1.424A_z + 1.822B_z + 1.26C_z) \tag{4.7.30}$$

(iii) For soil of type $G = G_0(1 + z/H)$

$$p = 0.754\beta\gamma_s H\left(\frac{S_a}{g}\right)\left(\frac{v}{1-2v}\right)(1.569A_z + 0.240B_z + 0.013C_z) \quad (4.7.31)$$

(iv) For soil of type $G = G_0(1 + z/H)^2$

$$p = 0.462\beta\gamma_s H\left(\frac{S_a}{g}\right)\left(\frac{v}{1-2v}\right)(1.561A_z + 0.51B_z + 0.075C_z) \quad (4.7.32)$$

Here $A_z = \sin(\pi z/2H)$, $B_z = \sin(3\pi z/2H)$ and $C_z = \sin(5\pi z/2H)$.

Equations (4.6.65)–(4.6.68) can be generically expressed as

$$p = \text{coeff} \times \beta \times \psi \times \left(\frac{S_a}{g}\right) \times \gamma_s \times H \quad (4.7.33)$$

The variation of pressure profile for different types of soil over the depth of the soil to bedrock is as shown (Figs. 4.19 and 4.20).

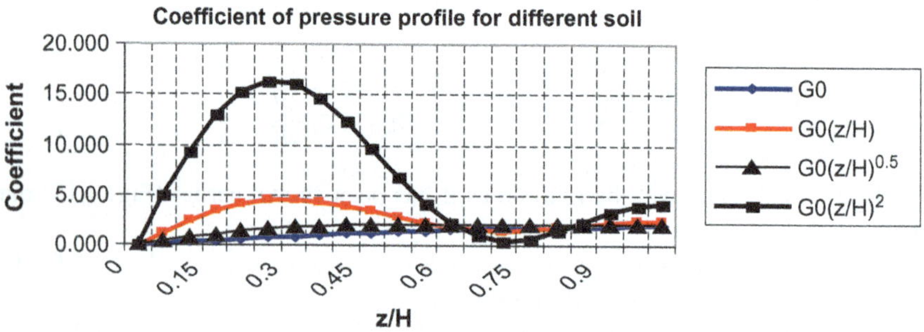

Fig. 4.19 Variation of dynamic pressure over the soil depth

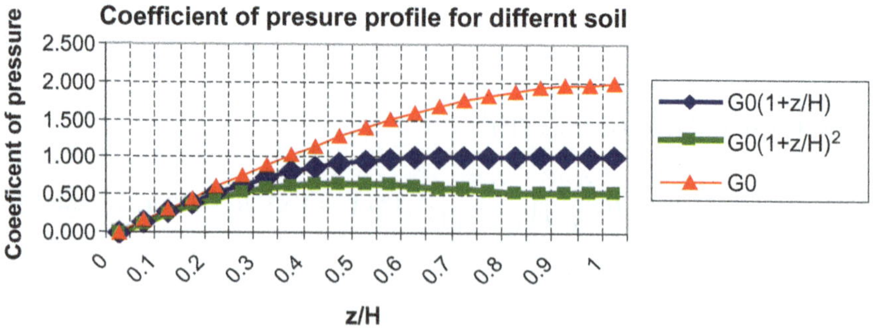

Fig. 4.20 Variation of dynamic pressure over the soil depth

For quick computation, the pressure coefficients for different types of soil are as shown in Table 4.10.

Based on the above table, the steps for estimation of dynamic pressure on the wall of the substructure can be summarized as follows.

1. Determine the depth of bedrock from where waves can emanate (in the absence of bedrock this can be taken as depth where SPT value $N \geq 50$).
2. Select the variation profile of dynamic shear modulus of soil body to the depth of bedrock which closely resembles expressions like $G_0(z/H)$, $G_0(1 + z/H)$.
3. Estimate Poisson's ratio v vis-à-vis the value ψ.
4. Compute free field time period of the site from expression $T = C_T(H/V_s)$, where the coefficient C_T is as given in Table 3.3 Estimate the response reduction factor of soil, consider $R = 3.0$ for soft soil and $R = 2.0$ for stiff soil.
5. Estimate the code factor $\beta = ZI/2R$.
6. From the value of T as computed in step 4, determine the spectral acceleration Sa/g from code.
7. To be on conservative side, consider 15% damping for soft soil and 10% damping for stiff soil while estimating the value Sa/g.
8. Now referring to Table 4.10 extract the pressure profile from Eq. (4.7.20) and (4.7.33).

Table 4.10 Pressure coefficient for different soil

$\xi = z/H$	Coeff G_0	Coeff $G_0(z/H)$	Coeff $G_0(z/H)^{0.5}$	Coeff $G_0(z/H)^2$	Coeff $G_0(1 + z/H)$	Coeff $G_0(1 + z/H)^2$
0	0.000	0.00	0.000	0.000	0.000	0.000
0.05	0.155	1.30	0.396	4.919	0.139	0.125
0.1	0.310	2.50	0.773	9.367	0.274	0.244
0.15	0.462	3.48	1.115	12.933	0.403	0.353
0.2	0.612	4.18	1.410	15.318	0.522	0.448
0.25	0.758	4.56	1.648	16.371	0.629	0.526
0.3	0.899	4.62	1.827	16.104	0.723	0.585
0.35	1.035	4.40	1.949	14.684	0.802	0.625
0.4	1.164	3.96	2.021	12.402	0.867	0.648
0.45	1.287	3.40	2.053	9.633	0.919	0.656
0.5	1.401	2.81	2.057	6.777	0.958	0.652
0.55	1.506	2.27	2.044	4.202	0.985	0.639
0.6	1.603	1.85	2.027	2.196	1.003	0.622
0.65	1.689	1.60	2.012	0.932	1.014	0.601
0.7	1.765	1.53	2.004	0.448	1.019	0.581
0.75	1.830	1.61	2.006	0.657	1.020	0.563
0.8	1.884	1.80	2.015	1.367	1.019	0.547
0.85	1.926	2.04	2.030	2.323	1.016	0.535
0.9	1.957	2.27	2.044	3.253	1.014	0.527
0.95	1.975	2.44	2.055	3.919	1.012	0.522
1	1.981	2.50	2.059	4.160	1.012	0.520

It is to be noted that the dynamic pressure profile will vary with location of the substructure below ground, that is, whether it is near the surface or deeply embedded inside ground—for this refer to the worked out example cited below.

To check validity of the proposed method, the formulation has been compared with Ostadan (2004) having G constant with depth and a recommended value of $v = 1/3$ and expressed as

$$p(z) = -0.0015 + 5.05z - 15.84z^2 + 28.25z^3 - 24.59z^4 + 8.14z^5$$

It is observed in Fig. 4.21 that the values are well matching within the acceptable limit of civil engineering design.

While the method proposed in Ostadan (2004) and NEHRP requires estimation of natural frequency based on SHAKE/SASSI and then computation of the pressure, the method as proposed herein does not require any software to be used either to estimate the free field time period or the pressure for any type of soil that fits in the profiles as mentioned above.

Example 4.7.1 A metro tunnel, 35 m wide, 20 m deep, is located at two sites as shown below (i) at the surface, and (ii) 16 m below ground. We need to determine the pressure on the wall when $V_s = 200$ m/s (varying linearly with depth z/H) having $v = 0.3$ and weight density of 20 kN/m^3. The site is seismic Zone III as per IS code. The bedrock level is at 40 m below ground surface (Fig. 4.22).

As V_s varies as z/H, the time period of site is expressed as $T = 5.19\frac{H}{V_s} = \frac{5.19 \times 40}{200} = 1.038$ s

Considering the soil as soft, the damping ratio chosen is 15%.

For $T = 1.038$ s, $Sa/g = 1.67/T = 1.608$. For damping ratio of 15%, scaling factor = 0.7.

Thus, design $Sa/g = 0.7 \times 1.608 = 1.126$.

For Zone III $Z = 0.16$, $I = 1.5$, and $R = 3 \rightarrow \beta = \frac{ZI}{2R} = \frac{0.16 \times 1.5}{6} = 0.04$.

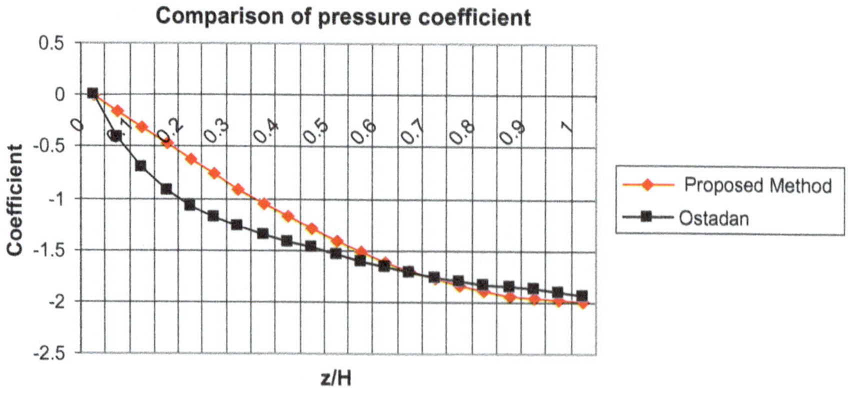

Fig. 4.21 Comparison of pressure coefficients for $v = 1/3$

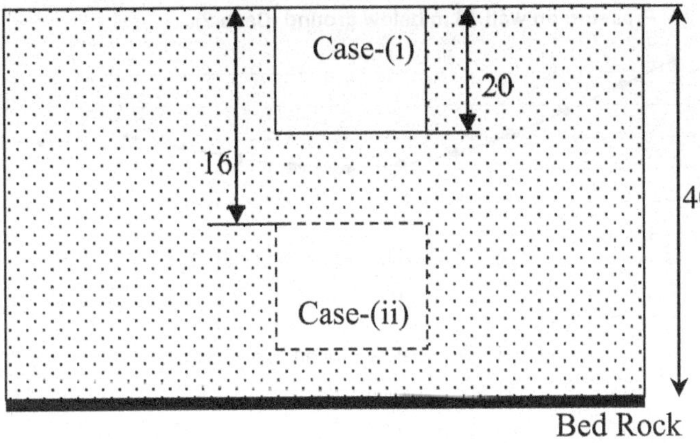

Fig. 4.22 A 35 × 20-m metro railway tunnel below ground

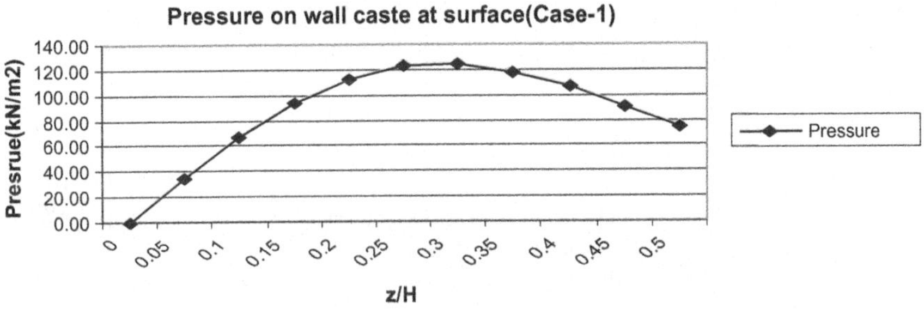

Fig. 4.23 Pressure on tunnel wall at ground surface (Case-1)

$$\psi = \frac{v}{1 - 2v} = \frac{0.3}{1 - 0.6} = 0.75$$

For Case (i) $\xi_1 = 0.0$ and $\xi_2 = \frac{20}{40} = 0.5$; thus, the pressure distribution is as shown in Fig. 4.23.

For Case (ii) $\xi_1 = \frac{16}{40} = 0.4$ and $\xi_2 = \frac{36}{40} = 0.9$ the pressure distribution is as shown in Fig. 4.24.

The pressures are computed from the expression $p = \text{coeff} \times \beta \times \psi \times (S_a/g) \times \gamma_s \times H$ where the coefficients are extracted for the corresponding ξ values from Table 4.10 (Table 4.11).

Figures 4.23 and 4.24 clearly reflect that ignoring seismic effect altogether below ground is not a prudent decision. Structures built below ground are significantly affected by seismic force. This is logical, as because the waves propagating through the soil, interact with the rigid substructure and the frequency content of the wave excites the structure generating pressure on the wall. The depth of bedrock

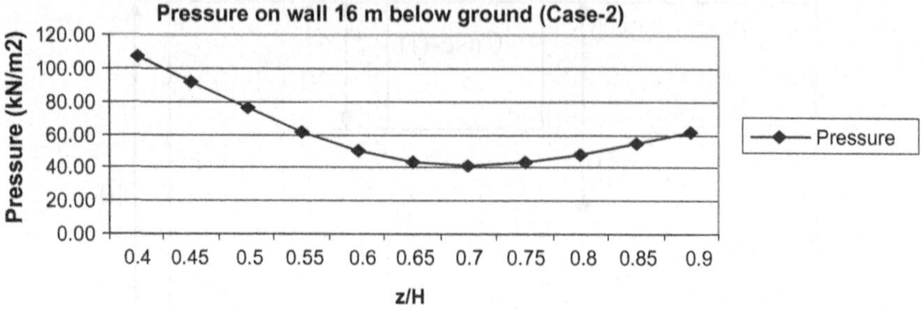

Fig. 4.24 Pressure on tunnel wall 16 m below ground surface (Case-2)

Table 4.11 Dynamic pressure on tunnel wall

$\xi = z/H$	Pressure coefficient as per Table 4.10	Dynamic pressure Case-1 (kN/m^2)	$\xi = z/H$	Pressure coefficient as per Table 4.10	Dynamic pressure Case-2 (kN/m^2)
0	0	0.00	0.4	3.967	107.20
0.05	1.309	35.37	0.45	3.404	91.99
0.1	2.505	67.70	0.5	2.81	75.94
0.15	3.489	94.29	0.55	2.272	61.40
0.2	4.189	113.20	0.6	1.857	50.18
0.25	4.567	123.42	0.65	1.607	43.43
0.3	4.626	125.01	0.7	1.532	41.40
0.35	4.403	118.99	0.75	1.612	43.56
0.4	3.967	107.20	0.8	1.804	48.75
0.45	3.404	91.99	0.85	2.047	55.32
0.5	2.81	75.94	0.9	2.278	61.56

level with respect to the position of the substructure has a significant effect on the magnitude of dynamic pressure. The shallower the bedrock level, the more intense the pressure on the wall.

4.8 Ostadan's Method

Ostadan (1997, 2004) conducted extensive study on the subject based on analysis carried out in SASSI 2000 with various soil properties and Poisson's ratio value and came with a normalized dynamic pressure coefficient curve given by the expression

$$p(z) = -0.0015 + 5.05z - 15.84z^2 + 28.25z^3 - 24.59z^4 + 8.14z^5. \quad (4.8.1)$$

The above-normalized pressure coefficient when compared with the analytical solution proposed herein gives quite closely matched value, as shown in Fig. 4.21.

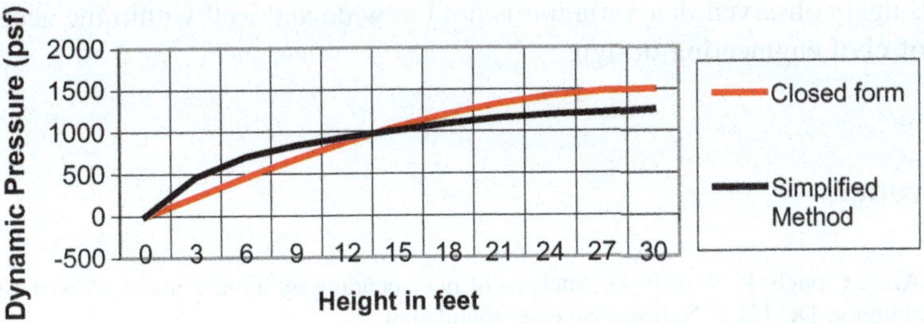

Fig. 4.25 Comparison of dynamic pressure for the 30 ft basement wall with Vs = 1000 ft/sec

Based on the above equation, Ostadan proposed a simplified method for determination of dynamic pressure on such unyielding walls whose steps are as mentioned hereunder:

- Perform free field soil column analysis and obtain the ground response at the depth corresponding to the base of the wall in the free field. The response motion in terms of acceleration response spectrum at 30% damping should be obtained. The free field soil column analysis may be performed using a computer program like SHAKE with input motion specified either at the ground surface or at the depth of foundation base-mat. The choice for location of control motion is an important decision that needs to be made consistent with development of the design motion. The location of input motion may significantly affect the dynamic response of the building and seismic soil pressure amplitude.
- Obtain total mass of the soil body as $m = 0.5 \times \rho \times H^2 \times \psi_v$, for the present case $\psi_v = 2/[(1 - v)(2 - v)]^{0.5}$, where ρ is mass density of the soil and v is Poisson's ratio.
- Obtain the lateral seismic force from the product of the total mass obtained above and the acceleration value of the free field response at the soil column obtained at the depth of the bottom of the wall.
- Obtain the maximum seismic soil pressure at the ground surface level by dividing the lateral force obtained above by a factor $0.744H$ (which is actually the area under curve for the equation furnished by him as mentioned above).
- Obtain the pressure profile by multiplying the peak pressure from the above step by the pressure distribution relationship as furnished earlier.

Ostadan's method[2] has been compared with the analytical method proposed herein for a basement wall which is 30-feet deep having shear wave velocity of soil as 1000 ft/sec. The density of soil is considered as 125 lbs/ft^3.

The wall is considered to be in a zone subjected to severe earthquake where $Z = 0.24$, $I = 1.2$, and $R = 2.0$. The outcome of the results is shown in Fig. 4.25.

[2]Personal communications with Dr. Ignacio Arango and Dr. Farhang Ostadan April 2005.

It is again observed that variation is not too wide and well within the acceptable limit of civil engineering design.

References

Adini, A., & Clough, R. W. (1964). Analysis of plate bending by finite element. *Report # G7337*. Washington DC USA: National Science foundation.

Chandrashekhara, K. (2001). *Theory of plates*. New Delhi, India: University Press of India.

Chowdhury, I., & Dasgupta, S. P. (2007). Dynamic pressure on rigid unyielding wall under earthquake pressure. *Indian Geotechnical Journal, 37*(2).

Chowdhury, I., & Dasgupta, S. P. (2011). An analytical solution to dynamic response of cantilever retaining wall with generalized backfill. *Electronic Journal of Geotechnical Engineering, 16C*. USA.

Chowdhury, I., & Singh, J. P. (2015). Behavior of gravity retaining wall under earthquake force. *Journal of Earthquake Engineering, 19*(1), 563–591. Illinois, USA.

Chowdhury, I., & Singh, J. P. (2014). Behaviour of gravity type retaining wall under earthquake load with generalized back fill. *Journal of Earthquake Engineering (T&F) USA, 19*, 563–591.

Chowdhury, I., Singh, J. P., & Dasgupta, S. P. (2012). Seismic response of counterfort retaining wall with generalized backfilled soil. *Electronic Journal of Geotechnical Engineering, 16R*, 1329–1357. USA.

Gutenberg, B., & Richter, C. F. (1956). Earthquake magnitude intensity, energy and acceleration. *Bulletin of Seismological Society of America*, 104–155.

Mononobe, N., & Matsuo, H. (1926). On determination of earth pressures during earthquakes. *Proceedings on World Engineering Congress*.

Okabe, S. (1929). General theory of earth pressure. *Journal of Japanese society of civil engineering, 12*(1).

Ostadan, F. (2004). Seismic soil pressure on building walls an updated approach. In: *11th international conference in soil dynamics and earthquake engineering*. University of California Berkeley.

Ostadan, F., & White, W. H. (1997). Lateral seismic soil pressure an updated approach. *Bechtel Technical Grant Report*. LA, USA.

Seed, H. B., & Whitman, R. V. (1970). Design of earth retaining structures under dynamic loads. In *ASCE speciality conference on lateral stress on ground and earth retaining structures* (pp. 103–147).

Steedman, R. S., & Zeng, X. (1990). The seismic response of waterfront retaining walls. In *Proceedings on specialty conference on design and performance of earth retaining structures*. Ithaca New York: Cornell University.

Szilard, R. (2004). *Theory and application of plate analysis*. USA: John Wiley Publication.

Timoshenko, S., & Woinowsky-Kreiger, S. (1987). *Theory of plates and shells*. NewYork: McgrawHill Publication.

Tocher, J. L. (1962). *Analysis of plate bending using triangular element*. Ph.D. thesis. Berkeley: University of California.

Chapter 5
Dynamic Analysis of Earth Dams Under Earthquakes

5.1 Introduction

Earthen dams and embankments play a very important role in water management and infrastructure development where embankments built out of soil make the bases of highways and rail tracks.

For earthen dams used especially as water reservoirs for agricultural purpose, the system usually consists of carefully chosen natural soil with a clay core (to control the seepage) and has been successfully utilized for effective utilization of water stored thereby.

We deal with the earthquake analysis of such dams herein.

Unlike concrete dams, which are far more flexible mostly due to their geometry (bending and shear deformation both control their behavior), earthen dams because of their large base width compared to their height are usually assumed to behave as pure shear beams where shear deformation controls its dynamic characteristics.

5.2 Mononobe's (1936) Method for Analysis of Earth Dam

One of the earliest analyses of such earth dam was proposed by Mononobe in his classic paper in 1936, wherein the dam is considered as an isosceles triangle having uniform mass density.

As width of the base of the dam is considered as far greater then the height of the dam, it is assumed that the shear deformation is predominant and bending deformation which is secondary in nature may be ignored.

With respect to Fig. 5.1 for a strip dz, the horizontal shear force and the inertial shear is given by

I. Chowdhury and S. P. Dasgupta, *Earthquake Analysis and Design of Industrial Structures and Infra-structures*, GeoPlanet: Earth and Planetary Sciences,
https://doi.org/10.1007/978-3-319-90832-8_5

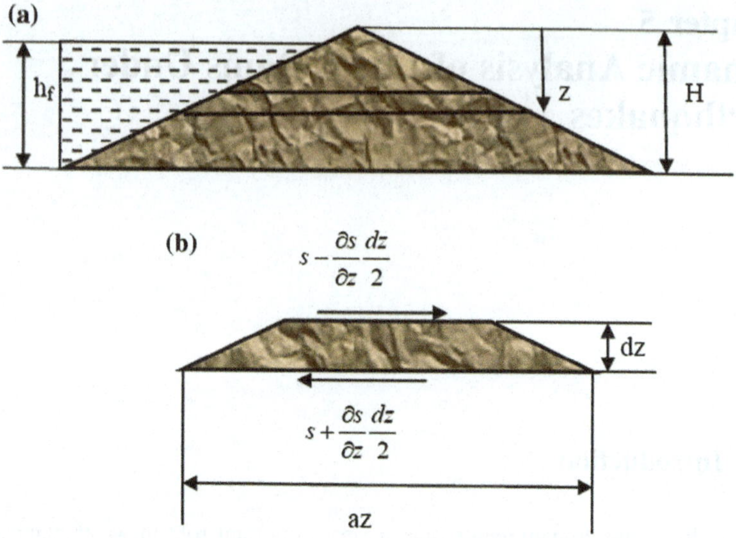

Fig. 5.1 Shear force in a triangular wedge-shaped earth dam

$$\rho(az)\mathrm{d}z\left(\frac{\partial^2 X}{\partial t^2}\right)=\frac{\partial s}{\partial z}\mathrm{d}z \tag{5.2.1}$$

where X is the horizontal amplitude of displacement, and s is the shear force or

$$\rho az\left(\frac{\partial^2 X}{\partial t^2}\right)=\frac{\partial s}{\partial z} \tag{5.2.2}$$

The shear modulus[1] of the soil medium is given by

$$G=\frac{\text{Shear Stress}}{\text{Shear Strain}}=\frac{s/az}{\mathrm{d}X/\mathrm{d}z}$$

which gives

$$s=Gaz\frac{\mathrm{d}X}{\mathrm{d}z} \tag{5.2.3}$$

Differentiating the above with respect to Z, we have

$$\frac{\partial s}{\partial z}=Ga\frac{\partial X}{\partial z}+Gaz\frac{\partial^2 X}{\partial z^2}$$

[1]This is considered constant over the depth in this case.

from which we have

$$\rho z \frac{\partial^2 X}{\partial t^2} = Ga \frac{\partial X}{\partial z} + Gaz \frac{\partial^2 X}{\partial z^2} \tag{5.2.4}$$

which is the basic equation of shear vibration of the triangular wedge.

For solution of the above equation, considering $z = z'\mathrm{H}$ where H is the height of the dam, and considering

$$X = x(z') \sin \omega t$$

we have

$$\frac{\partial^2 X}{\partial t^2} = -x\omega^2 \sin \omega t$$

and

$$\frac{\partial X}{\partial z} = \frac{\partial x}{\partial z} \sin \omega t \tag{5.2.5}$$

or

$$\frac{\partial X}{\partial z} = \frac{1}{H} \frac{\partial x}{\partial z'} \sin \omega t$$

and

$$\frac{\partial^2 X}{\partial z^2} = \frac{1}{H^2} \frac{\partial^2 x}{\partial z'^2} \sin \omega t$$

Substituting the above in the basic equation of shear vibration, we have

$$G\left(\frac{\partial x}{\partial z'} + z' \frac{\partial^2 x}{\partial z'^2}\right) + \rho H^2 z' x\omega^2 = 0 \tag{5.2.6}$$

The above is the Bessel's equation, whose solution is given by the expression

$$x = AJ_0\left[\left(H\omega\sqrt{\frac{\rho}{G}}\right) z'\right] \tag{5.2.7}$$

where A is a constant, and J_0 is the Bessel function of order zero.

Implementing the boundary condition at the base of the dam ($z' = 1$) $x = 0$, we have

$$J_0\left[\left(H\omega\sqrt{\frac{\rho}{G}}\right)\right] = 0 \tag{5.2.8}$$

The above can be expanded as

$$\left[\left(H\omega\sqrt{\frac{\rho}{G}}\right)\right]_n \frac{1}{\pi} = n - \frac{1}{4} + \frac{0.0507}{4n-1} + \cdots \tag{5.2.9}$$

where $n = 1, 2, 3, \ldots$.

or,

$$\omega_n = \frac{v_s}{H}\pi\left(n - \frac{1}{4} + \frac{0.0507}{4n-1}\right) \tag{5.2.10}$$

Considering $T = \frac{2\pi}{\omega_n}$, we have

$$T = \frac{2H}{v_s\left(n - \frac{1}{4} + \frac{0.0507}{4n-1}\right)} \tag{5.2.11}$$

Thus, for the first three modes we have the time periods as shown in Table 5.1 IS-1893 only furnishes the first fundamental mode as mentioned above for calculation of the time period.

We started the derivation of the time period equation based on the equilibrium of the elemental strip given by Eq. (5.2.1) and $\frac{\partial^2 X}{\partial t^2} = -x\omega^2 \sin \omega t$ which gives maximum value of $\frac{\partial^2 X}{\partial t^2}$ as

$$\frac{\partial^2 X}{\partial t^2} = -x\omega^2 \tag{5.2.12}$$

Again considering $\omega_n = \frac{v_s}{H}\pi\left(n - \frac{1}{4} + \frac{0.0507}{4n-1}\right)$ for $n = 1$ $\omega_n = 0.767\frac{v_s}{H}\pi$ which gives

Table 5.1 Time period of earthen dams for the first three modes

Mode	1	2	3
Time period	$2.607\frac{H}{v_s}$	$1.138\frac{H}{v_s}$	$0.726\frac{H}{v_s}$

$$\left(\frac{\partial^2 X}{\partial t^2}\right)_{\max} = -x(0.767)^2 \pi^2 \frac{v_s^2}{H^2} \tag{5.2.13}$$

Substituting the value of x and ω_n derived previously, we have

$$\left(\frac{\partial^2 X}{\partial t^2}\right)_{\max} = -AJ_0\left\{H\sqrt{\frac{\rho}{G}} \cdot z' \cdot 0.767\frac{\pi}{H}\sqrt{\frac{G}{\rho}}\right\}(0.767)^2\pi^2\frac{v_s^2}{H^2} \tag{5.2.14}$$

or proportional to $\frac{v_s^2}{H^2} J_0(0.767\pi z')$, the numerical constants being considered within the constant proportionality.

For a particular dam, as the $\frac{v_s^2}{H^2}$ is a constant quantity, hence the maximum acceleration $\left(\frac{\partial^2 X}{\partial t^2}\right)_{\max}$ is proportional to $J_0(2.41z')$.

The values of $J_0(2.41z')$ for various values of z' are as shown hereafter in Table 5.2.

The above values may be used to determine the inertial force at different heights of the dam.

When resonance occurs, the deformations would tend to infinite value without any internal friction; however, due to internal friction the maximum deflection is restricted to a finite value.

Mononobe calculated the ratio of top and bottom deflection for G and ρ and also when the parameters vary linearly with depth. Considering linear variation with depth, the acceleration is expressed as

$$\alpha = \alpha_0\left(1 + \frac{d_0 - d}{d}\frac{H - z}{H}\right) \tag{5.2.15}$$

where α is the acceleration at any depth z below top, α_0 is the ground level acceleration, and d_0 is the maximum displacement at the top. The ratio of top and bottom displacement came to 2.5 and 3.5, respectively, from which Mononobe concluded that the maximum acceleration at the top may be 2.5–3.5 times the acceleration at the base.

Table 5.2 Variation of acceleration along the height of the dam as per Mononobe (1936)

$z' = z/H$	0.0	0.1	0.3	0.5	0.7	0.9	1.0
Maximum acceleration $\alpha, J_0(2.41z')$	1.00	0.986	0.874	0.668	0.406	0.127	0.00

5.3 Gazetas's (1982) Method for Analysis of Earth Dam

Gazetas (1982) developed solutions to the shear beam wave equation for the case where the shear modulus value varies with the depth of the dam, given by the expression

$$G(z) = G_{\mathrm{av}}\left(\frac{z}{H}\right)^{\alpha} \tag{5.3.1}$$

where G_{av} is the average shear modulus of material constituting the dam.

The nth natural frequency of the dam considering the dam to be of triangular shape (i.e., $h/H = 1$) is given by

$$\omega_n = \frac{v_s}{H}\frac{\beta_n}{8}(4+\alpha)(2-\alpha) \tag{5.3.2}$$

where v_s = average shear wave velocity of the soil in the dam, and β_n is a function given in Table 5.3 for the first three modes.

Here, h is the height of the dam to its crest, and H is full height of the triangle.

The mode shape for the nth natural frequency is given by

$$\phi_n(z) = \left(\frac{z}{H}\right)^{-\frac{\alpha}{2}} J_q\left[\beta_n\left(\frac{z}{H}\right)^{1-\frac{\alpha}{2}}\right] \tag{5.3.3}$$

where J_q = Bessel's function of first kind of order $q = \alpha/(2-\alpha)$ and can be evaluated from the expression

$$J_q(x) = \sum_{k=0}^{\infty}\frac{(-1)^k}{k!\Gamma(q+k+1)}\left(\frac{x}{2}\right)^{q+2k} \tag{5.3.4}$$

where $\Gamma(x)$ is the gamma function and is given by the expression

$$\Gamma(x) = \int_0^{\infty} \mathrm{e}^{-x}x^{n-1}\mathrm{d}x \tag{5.3.5}$$

Table 5.3 Frequency coefficient for earth dam for first three modes as per Gazetas (1982)

α	β_1	β_2	β_3
0	2.404	5.520	8.654
1/2	2.903	6.033	9.171
4/7	2.999	6.133	9.273
2/3	3.142	6.283	9.525
1	3.382	7.106	10.174

5.4 Makdisi and Seed's (1977) Method for Analysis of an Earth Dam

A simplified procedure proposed by Makdisi and Seed (1977), taking into consideration the strain-dependent degradation of shear modulus and damping of soil, possibly remains the most popular method for dam analysis in design offices other than finite element analysis and was developed in similar line as Mononobe's method, as explained earlier.

Based on the shear beam equation as derived earlier, the authors found that the acceleration at any level z as a function of time is given by

$$\ddot{u}(z,t) = \sum_{n=1}^{\infty} \frac{2J_0(\beta_n z/H)}{\beta_n J_1 \beta_n} \omega_n V_n(t) \tag{5.4.1}$$

where J_0, J_1 = Bessel's functions of first kind of orders zero and one, β_n = the zero value of frequency equation $J_0(\omega H\sqrt{\rho/G}) = 0$, and ω_n = natural frequency of the nth mode = $\beta_n V_s/H$.

$$V_n(t) = \int_0^t \ddot{u}_g e^{-\lambda_n \omega_n (t-\tau)} \sin[\omega_{dn}(t-\tau)] d\tau \tag{5.4.2}$$

where $\omega_{dn} = \omega_n\sqrt{1-\lambda^2}$ the damped natural frequency and λ_n = critical damping ratio.

Thus,

$$\ddot{u}(z,t) - \sum_{n=1}^{\infty} \phi_n(z)\omega_n V_n(t) \tag{5.4.3}$$

where modal participation factor is

$$\phi_n(z) = \sum_{n=1}^{\infty} \frac{2J_0(\beta_n z/H)}{\beta_n J_1 \beta_n} \tag{5.4.4}$$

Considering the first three modes of vibration, the corresponding values of β_n are always $\beta_1 = 2.4$, $\beta_2 = 5.52$, and $\beta_3 = 8.65$, which gives the first three natural frequencies as in Table 5.4.

At the crest of the dam ($z = 0$), the corresponding values of mode participation factors $\phi_n(0)$ for the first three modes are listed in Table 5.5.

Thus, the crest acceleration at each mode is given by

Table 5.4 Natural frequencies for first three modes for earth dams as per Makdisi and Seed (1977)

Mode	1	2	3
Natural frequency	$\omega_1 = 2.4\frac{v_s}{H}$	$\omega_2 = 5.52\frac{v_s}{H}$	$\omega_3 = 8.65\frac{v_s}{H}$

Table 5.5 Modal participation factor for first three modes for earth dams as per Makdisi and Seed (1977)

Mode	1	2	3
Modal participation factor	$\phi_1 = 1.6$	$\phi_2 = 1.06$	$\phi_3 = 0.86$

$$\ddot{u}(0,t) = \sum_{n=1}^{\infty} \phi_n(0)\omega_n V_n(t); \tag{5.4.5}$$

when based on response spectrum analysis, the maximum expected acceleration is given by

$$\ddot{u}(0,t) = \phi_n(0)S_{an} \tag{5.4.6}$$

Thus, for the first three modes the maximum acceleration is given by

$$\ddot{u}_1\max = 1.6S_{a1}, \quad \ddot{u}_2\max = 1.06S_{a2}, \quad \ddot{u}_3\max = 0.86S_{a3}$$

The maximum SRSS values of the acceleration are thus given by the expression

$$\ddot{u}_n\max = \sum_{n=1}^{3} \sqrt{(\ddot{u}_n\max)^2} \tag{5.4.7}$$

To estimate the strain compatible material properties, an expression for the average shear strain over the entire section should be determined. From the shear slice theory, the expression for shear strain at any level in the dam is given by

$$\gamma(z,t) = \sum_{n=1}^{\infty} \frac{2J_1(\beta_n z/H)}{H\omega_n J_1(\beta_n)} V_n(t) \tag{5.4.8}$$

$$\begin{aligned} &= \frac{H}{v_s^2} \sum_{n=1}^{\infty} \frac{2J_1(\beta_n z/H)}{\beta_n^2 J_1(\beta_n)} \omega_n V_n(t) \\ &= \frac{H}{v_s^2} \sum_{n=1}^{\infty} \bar{\phi}_n(z)\omega_n V_n(t) \end{aligned} \tag{5.4.9}$$

where $\bar{\phi}_n(z) = \frac{2J_1\left(\beta_n \frac{z}{H}\right)}{\beta_n^2 J_1(\beta_n)}$ = shear strain mode participation factor.

It is recommended that the contribution of higher modes being small, it is sufficient to consider the contribution of the first mode only over the entire depth of the dam for calculation of the average shear strain.

Thus, the maximum average shear strain is obtained as

$$\gamma_{\max}(z) = \frac{H}{v_s^2}\bar{\phi}_1 S_{a1} \tag{5.4.10}$$

It has been shown that the average value of the first factor is given by $\bar{\phi}_{1_{\text{avg}}} = 0.3$.

Assuming an equivalent cyclic shear strain as approximately 65% of the average shear strain,

$$\gamma_{\text{avg}}(\max) = 0.65 \times 0.3 \times \frac{H}{v_s^2} S_{a1} \tag{5.4.11}$$

Having obtained a new value for the average shear strain, a new set of moduli and damping values can be obtained from the expressions[2]

$$G = \frac{G_{\max}}{\left(1 + \frac{\psi}{\psi_r}\right)} \quad \text{and} \quad \frac{D_c}{D_m} = \frac{\psi/\psi_r}{1 + \psi/\psi_r} \tag{5.4.12}$$

The iterations are carried out till the values become constant with respect to the earlier cycle. It has been observed that the system generally converges by three cycles.

5.5 Chowdhury and Dasgputa's (2010) Model for Dynamic Analysis of a Homogeneous and Clay-Cored Dam

In the previous section, we have explained the major analytical techniques that are popular for earthquake analysis of earthen dams, as developed by Mononobe (1936), Makdisi and Seed (1977), and Gazetas (1982). Numerical analysis, based on finite element method as proposed by Clough and Chopra (1966) is also in practice but requires extensive computer analysis backed by up by detailed and exhaustive in situ geotechnical data.

Of all these techniques, the first three methods are analytical in nature, where the dam is considered as a triangular homogenous shear beam and a solution is sought,

[2]For further explanation of how to modify the damping and shear modulus with respect to shear strain, refer Chap. 8 titled Geotechnical Consideration for Dynamic Soil–Structure Interaction.

based on its equation of motion in one dimension under the seismic force. Gazetas' formulation based on the above theory is usually considered to be most rigorous (Kramer 1984), while Seed's method can cater to the nonlinear deformation of the dam based on hyperbolic stiffness degradation of the soil which is a function of induced strain level due to vibration (Seed and Idriss 1971).

Finite element method obviously gives the most exhaustive solution but is not without its limitation for other than exhaustive data preparation and input requirement—that makes the analysis expensive; if proper constitutive model cannot be fitted into the mathematical model, it can often yield unrealistic result. In many cases, it is difficult to fit an earth dam problem into general-purpose finite element software, thus necessitating the development of special purpose software which is not readily available.

It is for this, the analytical methods as mentioned above have remained more popular around the world because of their simplicity in application, though they suffer from deficiencies in many cases. For instance, all the methods cited above assume the dam to be a perfect triangle (i.e., crest area is zero) while in reality it is not. As a matter fact, the United States Bureau of Reclamation of small dams suggests an empirical formula for the crest width (B_t) and height H (feet in FPS units) as $B_t = H/5 + 10$. A dam always has some finite dimension at the crest for maintenance and inspection. How much these data affect its dynamic response is not very apparent.

Both Mononobe (1936) and Makdisi and Seed (1977) gave solution assuming the dynamic shear modulus (G) constant with depth, while this would hardly be the case in real field condition. It has been found that G varies with height (H) of the dam. Gazetas (1982) proposed a formulation where he considered G to vary with depth (z) as $G(z/H)^{\alpha}$, where $\alpha = 0$, 1, 2, etc., depict variation of G as constant, linear, and parabolic with depth, respectively. However, many designers feel this could be unrealistic, especially for linear and parabolic variation, because it depicts $G = 0$ at the crest of the dam, which is unrealistic.

Finally, none of the analytical methods takes care of the practical situation of impermeable clay core which is often deployed in an earth dam of large magnitude to control the seepage through it.

In the present section, we present a solution where a number of such deficiencies as cited above are overcome to arrive at a more realistic solution.

Shown in Fig. 5.2 is a typical triangular shear beam usually considered for a conventional linear analysis of an earth dam.

The equation of motion of such dams is usually expressed as

$$\frac{\partial^2 u}{\partial t^2} = \frac{G}{\rho}\left[\frac{\partial^2 u}{\partial z^2} + \frac{1}{z}\frac{\partial u}{\partial z}\right] \tag{5.6.1}$$

in which soil properties are: G = dynamic shear modulus; ρ = mass density (γ/g); γ = unit weight; g = acceleration due to gravity; and u = displacement amplitude of the soil in x-direction.

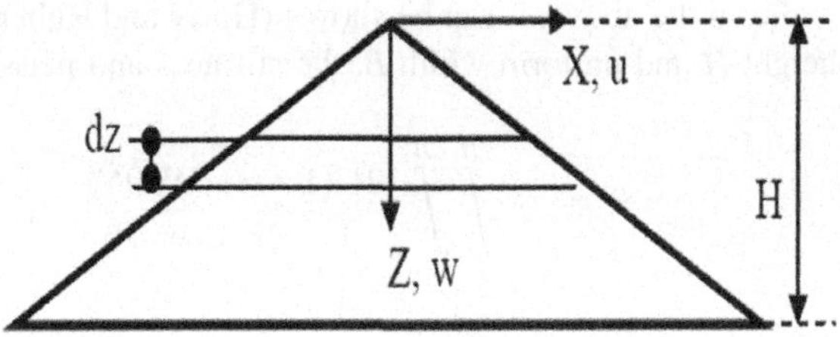

Fig. 5.2 Typical triangular shear beam considered as earth dam

For the above type of earth dams, the boundary conditions considered are

(a) At $z = 0$, $\partial u/\partial z = 0$ for all values of t;
(b) At $z = H$, $u = 0$ for all values of t.

Imposing the above boundary conditions, Eq. (5.6.1) may be written as

$$u(z,t) = \sum_{n=1}^{\infty} [A_n \sin \omega_n t + B_n \cos \omega_n t] J_0(\beta_n z/H) \tag{5.6.2}$$

in which $A_n = B_n$ = constants that are functions of the initial conditions; J_0 = Bessel's function of first kind of order 0; β_n = a constant having values $\beta_1 = 2.404$, $\beta_2 = 5.52$, $\beta_3 = 8.65$, etc. (Jeffrey 2005).

Based on the above, it may be stated that the generalized shape function of a triangular shear beam vibrating under its own inertia can be expressed as

$$N_j = J_0(\beta_j z/H) \tag{5.6.3}$$

where j = 1, 2, 3, 4, … are the number of modes considered for analysis.

For a two-dimensional plane strain body under deformation, the shear strain energy (Timoshenko and Young 1982) may be expressed as

$$V = \frac{vG}{1-2v}(\varepsilon_x + \varepsilon_z)^2 + G(\varepsilon_x^2 + \varepsilon_z^2) + \frac{G}{2}(\gamma_{xz}^2) \tag{5.6.4}$$

in which V = strain energy density; v = Poisson's ratio; ε_x, ε_z and γ_{xz} = strain in the body.

Considering $u(x, z, t) = N_r(x, z) \cdot q(t)$, we have

$$\frac{\partial V}{\partial q_r} = \frac{2G(1-v)}{1-2v} \frac{\partial N_i}{\partial x} \frac{\partial N_r}{\partial x} q_i q_r + G \frac{\partial N_i}{\partial z} \frac{\partial N_r}{\partial z} q_i q_r \tag{5.6.5}$$

where $q(t)$ is exclusively a time function.

From the above, it can be shown (Hurty and Rubenstein 1967) that for a body of height H and uniform width B, the stiffness and mass matrices can be expressed as

$$K_{ir} = \int_0^H \int_0^B \left[\frac{2G(1-v)}{1-2v} \frac{\partial N_i}{\partial x} \frac{\partial N_r}{\partial x} + G \frac{\partial N_i}{\partial z} \frac{\partial N_r}{\partial z} \right] \mathrm{d}x\mathrm{d}z \qquad (5.6.6)$$

and

$$M_{ir} = \frac{\gamma}{g} \int_0^H \int_0^B N_i N_r \, \mathrm{d}x\mathrm{d}z \qquad (5.6.7)$$

For shear waves propagating in the vertical direction, it has been shown (refer Chap. 3) that in one dimension, a soil body of depth H and extending to infinity in horizontal direction, Eqs. (5.6.6) and (5.6.7) can be expressed as

$$K_{ir} = \int_0^H \left[G \frac{\partial N_i}{\partial z} \frac{\partial N_r}{\partial z} \right] \mathrm{d}z \qquad (5.6.8)$$

$$M_{ir} = \frac{\gamma}{g} \int_0^H N_i N_r \mathrm{d}z \qquad (5.6.9)$$

For a typical trapezoidal section of an earth dam shown in Fig. 5.3, Eqs. (5.6.8) and (5.6.9) can be modified and expressed as

$$K_{ir} = GA_t \int_0^H \left[\left(1 + \psi \frac{z}{H}\right) \frac{\partial N_i}{\partial z} \frac{\partial N_r}{\partial z} \right] \mathrm{d}z \qquad (5.6.10)$$

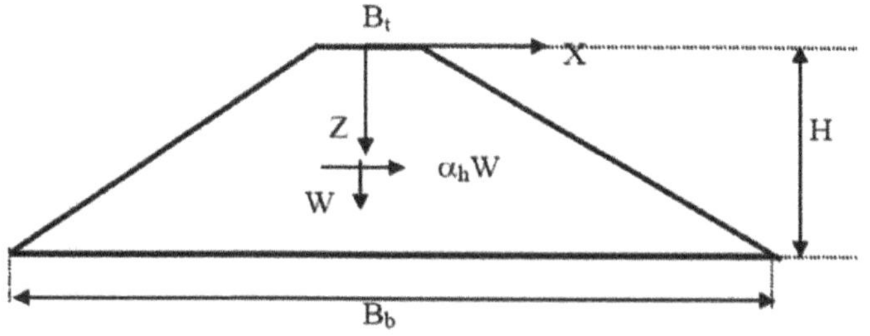

Fig. 5.3 A typical earth dam with trapezoidal section

and

$$M_{ir} = \frac{\gamma A_t}{g} \int_0^H \left(1 + \psi \frac{z}{H}\right) N_i N_r \, dz \tag{5.6.11}$$

where A_t = cross-sectional area of the dam at $z = 0$ and for 1.0 m depth (perpendicular to the plane of the paper) is equal to B_t; ψ is dimensionless constant expressed as $\psi = (B_b - B_t)/B_t$; N_i and N_r are expressed in Eq. (5.6.3).

Considering $\xi = z/H$ when $z \to 0$ $\xi \to 0$ and when $z \to H$, $\xi \to 1$, Eqs. (5.6.10) and (5.6.11) can be expanded and expressed for the first three modes

$$[K] = \frac{GA_t}{H} \begin{bmatrix} \beta_1^2 \int_0^1 (1+\psi\xi) J_1^2(\beta_1\xi) d\xi & & \text{Symmetrical} \\ \beta_1\beta_2 \int_0^1 (1+\psi\xi) J_1(\beta_1\xi) J_1(\beta_2\xi) d\xi & \beta_2^2 \int_0^1 (1+\psi\xi) J_1^2(\beta_2\xi) d\xi & \\ \beta_1\beta_3 \int_0^1 (1+\psi\xi) J_1(\beta_1\xi) J_1(\beta_3\xi) d\xi & \beta_2\beta_3 \int_0^1 (1+\psi\xi) J_1(\beta_2\xi) J_1(\beta_3\xi) d\xi & \beta_3^2 \int_0^1 (1+\psi\xi) J_1^2(\beta_3\xi) d\xi \end{bmatrix} \tag{5.6.12}$$

$$[M] = \frac{\gamma A_t H}{g} \begin{bmatrix} \int_0^1 (1+\psi\xi) J_0^2(\beta_1\xi) d\xi & & \text{Symmetrical} \\ \int_0^1 (1+\psi\xi) J_0(\beta_1\xi) J_0(\beta_2\xi) d\xi & \int_0^1 (1+\psi\xi) J_0^2(\beta_2\xi) d\xi & \\ \int_0^1 (1+\psi\xi) J_0(\beta_1\xi) J_0(\beta_3\xi) d\xi & \int_0^1 (1+\psi\xi) J_0(\beta_2\xi) J_0(\beta_3\xi) d\xi & \int_0^1 (1+\psi\xi) J_0^2(\beta_3\xi) d\xi \end{bmatrix} \tag{5.6.13}$$

Equations (5.6.12) and (5.6.13) are generalized expression for stiffness and mass matrices for a trapezoidal dam shown in Fig. 5.3, having G invariant with depth. The individual elements of **[K]** and **[M]** matrices are obtained by numerical integration of the respective terms between limits 1 and 0.

For the particular case of trapezoidal section is a triangle, i.e., limit $A_t \to 0$ and $\psi \to \infty$, Eqs. (5.6.12) and (5.6.13) can be expressed as

$$[K] = \frac{GA_t}{H} \begin{bmatrix} 3.894 \times 10^6 & -736.854 & -6.425 \times 10^3 \\ -736.854 & 8.82 \times 10^6 & 1.582 \times 10^4 \\ -6.425 \times 10^3 & 1.582 \times 10^4 & 1.378 \times 10^7 \end{bmatrix} \tag{5.6.14}$$

and

$$[M] = \frac{\gamma A_t H}{g} \begin{bmatrix} 6.742 \times 10^5 & 156.128 & -18.743 \\ -156.128 & 2.895 \times 10^5 & 207.419 \\ -18.473 & 207.419 & 1.844 \times 10^5 \end{bmatrix} \tag{5.6.15}$$

Consider the eigenvalue problem of the above (Bathe 1996) as

$$[\mathbf{K}][\phi] = \lambda[\mathbf{M}][\phi] \tag{5.6.16}$$

where λ and ϕ are the eigenvalues and eigenvectors of matrix [K] and [M], respectively.

We have then

$$[\lambda] = \begin{bmatrix} 5.775 & 0 & 0 \\ 0 & 30.469 & 0 \\ 0 & 0 & 74.758 \end{bmatrix} \frac{V_s^2}{H^2} \tag{5.6.17}$$

where $V_s = \sqrt{Gg/\gamma}$, the shear wave velocity of the soil.

Since $\omega^2 = \lambda$, one can write

$$[\omega] = \begin{bmatrix} 2.403 & 0 & 0 \\ 0 & 5.52 & 0 \\ 0 & 0 & 8.646 \end{bmatrix} \frac{V_s}{H} \tag{5.6.18}$$

$$\Rightarrow [C_{\omega i}] \frac{V_s}{H}$$

The values of $C_{\omega i}$ obtained based on above method are compared with other established methods as stated earlier in Table 5.6.

It is observed that $C_{\omega t}$ based on the proposed method is in excellent agreement with established methods for limit $A_t \to 0$.

Having established the correctness of the basic condition of derivation for the particular case of limit $A_t \to 0$, we present the variance of time period of the dam for various values of ψ considering $T = 2/\omega$. This is given in Table 5.7. The time period is expressed as $T_i = C_{Ti}(H/V_s)$.

Table 5.6 Comparison of $C_{\omega t}$ by various methods

Mode	Mononobe	Makdisi and Seed	Gazetas	Proposed
1	2.409	2.4	2.404	2.403
2	5.521	5.52	5.52	5.52
3	8.654	8.65	8.654	8.646

Table 5.7 Variation C_T—first three modes

ψ	C_{T1}	C_{T2}	C_{T3}
∞	2.615	1.138	0.727
99	2.64	1.149	0.733
49	2.666	1.159	0.739
24	2.715	1.178	0.749
15	2.763	1.193	0.756
12	2.809	1.207	1.762
9	2.854	1.219	0.766

It is observed that the time periods get elongated with increase in crest width though the variation for fundamental time period (which is most critical) is not large and is of the order of 9% only (from $\psi \to \infty$ to 9). As such, the proposed method, though more realistic, does not give any significant advantage over the established methods using triangular homogeneous sections. However, the distinct advantage of the proposed method will emerge subsequently when we take up the case of dam with impermeable cores which is more a reality in the field than a homogenous section assumed for dynamic analysis of such earth dams.

The first three mode shapes/eigenvectors for the dam based on Eq. (5.6.3) are shown in Fig. 5.4.

Effect of variation of *G* with depth

Gazetas (1982) derived solution of the dynamic response of an earth dam considering $G(z) = G(z/H)$. It is apparent from the above that while it gives a G value constant with depth ($\alpha = 0$), for linear and parabolic variation it gives $G = 0$ at the crest of the dam. Since this will invariably have a finite value, we propose to consider $G(z)$ as

$$G(z) = G(1 + z/H)^{\alpha} \tag{5.6.19}$$

For solution to this for which there are no established answers, Eq. (5.6.12) gets modified to

$$[K] = \frac{GA_t}{H}\begin{bmatrix} \beta_1^2\int_0^1 (1+\xi)^{\alpha}(1+\psi\xi)J_1^2(\beta_1\xi)\mathrm{d}\xi & & \text{Symmetrical} \\ \beta_1\beta_2\int_0^1 (1+\xi)^{\alpha}(1+\psi\xi)J_1(\beta_1\xi)J_1(\beta_2\xi)\mathrm{d}\xi & \beta_2^2\int_0^1 (1+\xi)^{\alpha}(1+\psi\xi)J_1^2(\beta_2\xi)\mathrm{d}\xi & \\ \beta_1\beta_3\int_0^1 (1+\xi)^{\alpha}(1+\psi\xi)J_1(\beta_1\xi)J_1(\beta_3\xi)\mathrm{d}\xi & \beta_2\beta_3\int_0^1 (1+\xi)^{\alpha}(1+\psi\xi)J_1(\beta_2\xi)J_1(\beta_3\xi)\mathrm{d}\xi & \beta_3^2\int_0^1 (1+\xi)^{\alpha}(1+\psi\xi)J_1^2(\beta_3\xi)\mathrm{d}\xi \end{bmatrix}. \tag{5.6.20}$$

Here, $\alpha = 1$ and 2, respectively, for linear and parabolic variation.

Based on the above, variations of C_T for linear and parabolic profile for different values of ψ are as presented in Table 5.8.

The intermediate values can well be linearly interpolated considering that the variation of C_T with respect to ψ is almost linear.

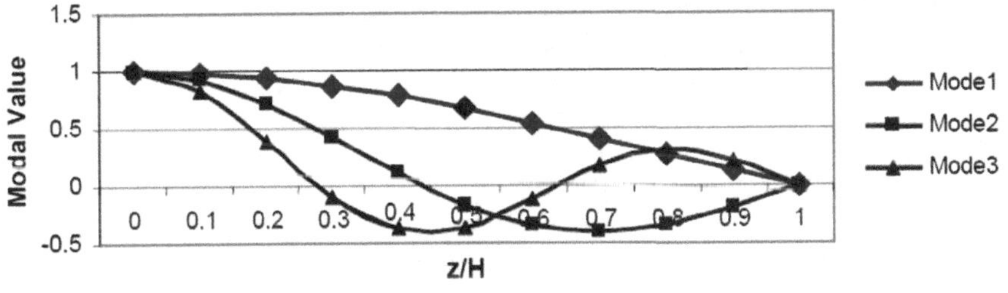

Fig. 5.4 Eigenvectors of the dam in first three mode

Table 5.8 C_T value for shear modulus having linearly and parabolic variation with depth

ψ	Linear variation			Parabolic variation		
	C_{T1}	C_{T2}	C_{T3}	C_{T1}	C_{T2}	C_{T3}
∞	2.005	0.924	0.585	1.554	0.754	0.46
99	2.026	0.934	0.59	1.573	0.763	0.463
49	2.048	0.943	0.594	1.591	0.776	0.466
24	2.089	0.959	0.601	1.626	0.785	0.47
15	2.134	0.975	0.607	1.664	0.798	0.475
12	2.163	0.984	0.61	1.688	0.806	0.477
9	2.203	0.996	0.614	1.722	0.816	0.48

It is again observed here that while the variation in time period is about 25% for linear variation of G with respect to G constant with depth, the variation due to tapering of the section is of the order of 10% only.

5.5.1 *Effect of Inner Core on Dynamic Response of the Dam*

No analytical methods exist for this case in terms of dynamic response. It is in such a situation that one resorts to finite element method for a comprehensive analysis. However, considering the dynamic response remaining linear, the problem can be approached as hereunder.

Shown in Fig. 5.5 is a typical earth dam with an internal clay core.

Let the width of the core at crest be B_{tc}, and B_{bc} be the width at the bottom. Here, the suffix c denotes the clay core.

Maintaining mathematical similarity, let us define

$$\psi_c = (B_{bc} - B_{tc})/B_{tc} \tag{5.6.1.1}$$

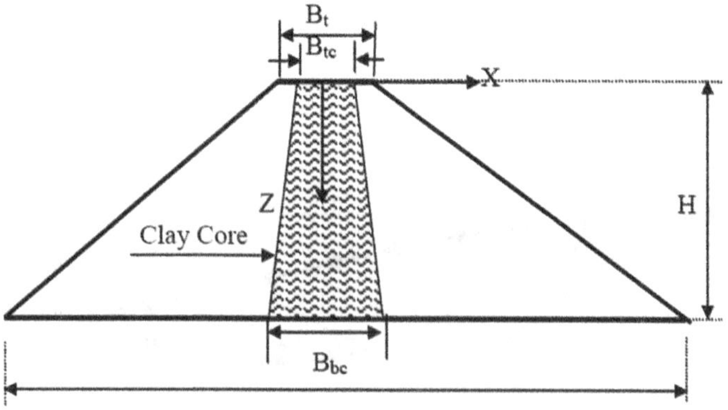

Fig. 5.5 An earth dam with central clay core to arrest seepage

in which G_c = dynamic shear modulus of the clay core and let $m = G_c/G$ the modular ratio of the clay core and the outer material.

The stiffness matrix for the dam section can be expressed as

$$[K] = GA_t \int_0^H \left[\left(1+\psi\frac{z}{H}\right)\frac{\partial N_i}{\partial z}\frac{\partial N_r}{\partial z}\right]\mathrm{d}z - GA_{tc}\int_0^H \left[\left(1+\psi_c\frac{z}{H}\right)\frac{\partial N_i}{\partial z}\frac{\partial N_r}{\partial z}\right]\mathrm{d}z + G_cA_{tc}\int_0^H \left[\left(1+\psi_c\frac{z}{H}\right)\frac{\partial N_i}{\partial z}\frac{\partial N_r}{\partial z}\right]\mathrm{d}z \tag{5.6.1.2}$$

$$\rightarrow \quad [K] = GA_t \int_0^H \left[\left(1+\psi\frac{z}{H}\right)\frac{\partial N_i}{\partial z}\frac{\partial N_r}{\partial z} + (m-1)\left(\frac{A_{tc}}{A_t}\right)\int_0^H \left(1+\psi_c\frac{z}{H}\right)\frac{\partial N_i}{\partial z}\frac{\partial N_r}{\partial z}\right]\mathrm{d}z \tag{5.6.1.3}$$

in which A_{tc} is the area of inner core at the crest of the dam = $B_{tc} \times 1$.

Substituting Eq. (5.6.3), the above can be expanded for the first three modes as

$$[K] = \begin{bmatrix} k_{11} & \text{Symm.} & \\ k_{12} & k_{22} & \\ k_{13} & k_{32} & k_{33} \end{bmatrix} \tag{5.6.1.4}$$

where

$$k_{11} = \frac{GA_t}{H}\left[\int_0^1 (1+\psi\xi)\beta_1^2 J_1^2(\beta_1\xi) + (m-1)\left(\frac{A_{tc}}{A_t}\right)\int_0^1 (1+\psi_c\xi)\beta_1^2 J_1^2(\beta_1\xi)\right]\mathrm{d}\xi \tag{5.6.1.5}$$

$$k_{12} = \frac{GA_t}{H}\left[\int_0^1 (1+\psi\xi)\beta_1\beta_2 J_1(\beta_1\xi)J_1(\beta_2\xi) + (m-1)\left(\frac{A_{tc}}{A_t}\right)\int_0^1 (1+\psi_c\xi)\beta_1\beta_2 J_1(\beta_1\xi)J_1(\beta_2\xi)\right]\mathrm{d}\xi \tag{5.6.1.6}$$

$$k_{13} = \frac{GA_t}{H}\left[\int_0^1 (1+\psi\xi)\beta_1\beta_3 J_1(\beta_1\xi)J_1(\beta_3\xi) + (m-1)\left(\frac{A_{tc}}{A_t}\right)\int_0^1 (1+\psi_c\xi)\beta_1\beta_3 J_1(\beta_1\xi)J_1(\beta_3\xi)\right]\mathrm{d}\xi \tag{5.6.1.7}$$

$$k_{22} = \frac{GA_t}{H}\left[\int_0^1 (1+\psi\xi)\beta_2^2 J_1^2(\beta_2\xi) + (m-1)\left(\frac{A_{tc}}{A_t}\right)\int_0^1 (1+\psi_c\xi)\beta_2^2 J_1^2(\beta_2\xi)\right]\mathrm{d}\xi \tag{5.6.1.8}$$

$$k_{23} = \frac{GA_t}{H}\left[\int_0^1 (1+\psi\xi)\beta_3\beta_2 J_1(\beta_3\xi)J_1(\beta_2\xi) + (m-1)\left(\frac{A_{tc}}{A_t}\right)\int_0^1 (1+\psi_c\xi)\beta_3\beta_2 J_1(\beta_3\xi)J_1(\beta_2\xi)\right]\mathrm{d}\xi \tag{5.6.1.9}$$

$$k_{33} = \frac{GA_t}{H}\left[\int_0^1 (1+\psi\xi)\beta_3^2 J_1^2(\beta_3\xi) + (m-1)\left(\frac{A_{tc}}{A_t}\right)\int_0^1 (1+\psi_c\xi)\beta_3^2 J_1^2(\beta_3\xi)\right]\mathrm{d}\xi \tag{5.6.1.10}$$

Similarly, the mass matrix is given by the expression

$$[M] = \begin{bmatrix} m_{11} & \text{Symmetrical} & \\ m_{12} & m_{22} & \\ m_{13} & m_{23} & m_{33} \end{bmatrix} \tag{5.6.1.11}$$

Considering $p = \gamma_c/\gamma$, the unit weight ratio of core soil and the outer soil, the mass matrix coefficients can be expressed as

$$m_{11} = \frac{\gamma A_t H}{g}\int_0^1 \left[(1+\psi\xi)J_0^2(\beta_1\xi) + (p-1)\left(\frac{A_{tc}}{A_t}\right)(1+\psi_c\xi)J_0^2(\beta_1\xi)\right]\mathrm{d}\xi \tag{5.6.1.12}$$

$$m_{12} = \frac{\gamma A_t H}{g} \int_0^1 \left[(1 + \psi\xi) J_0(\beta_1 \xi) J_0(\beta_2 \xi) + (p-1)\left(\frac{A_{tc}}{A_t}\right)(1 + \psi_c \xi) J_0(\beta_1 \xi) J_0(\beta_2 \xi) \right] \mathrm{d}\xi \tag{5.6.1.13}$$

$$m_{13} = \frac{\gamma A_t H}{g} \int_0^1 \left[(1 + \psi\xi) J_0(\beta_1 \xi) J_0(\beta_3 \xi) + (p-1)\left(\frac{A_{tc}}{A_t}\right)(1 + \psi_c \xi) J_0(\beta_1 \xi) J_0(\beta_3 \xi) \right] \mathrm{d}\xi \tag{5.6.1.14}$$

$$m_{22} = \frac{\gamma A_t H}{g} \int_0^1 \left[(1 + \psi\xi) J_0^2(\beta_2 \xi) + (p-1)\left(\frac{A_{tc}}{A_t}\right)(1 + \psi_c \xi) J_0^2(\beta_2 \xi) \right] \mathrm{d}\xi \tag{5.6.1.15}$$

$$m_{23} = \frac{\gamma A_t H}{g} \int_0^1 \left[(1 + \psi\xi) J_0(\beta_2 \xi) J_0(\beta_3 \xi) + (p-1)\left(\frac{A_{tc}}{A_t}\right)(1 + \psi_c \xi) J_0(\beta_2 \xi) J_0(\beta_3 \xi) \right] \mathrm{d}\xi \tag{5.6.1.16}$$

$$m_{33} = \frac{\gamma A_t H}{g} \int_0^1 \left[(1 + \psi\xi) J_0^2(\beta_3 \xi) + (p-1)\left(\frac{A_{tc}}{A_t}\right)(1 + \psi_c \xi) J_0^2(\beta_3 \xi) \right] \mathrm{d}\xi \tag{5.6.1.17}$$

Once the stiffness and mass matrices are obtained based on above formulation, one can follow Eq. (5.6.16) for standard eigenvalue solution of the problem to derive the time periods and vibration modes.

Stiffness coefficients derived through Eqs. (5.6.1.5)–(5.6.1.10) are based on soil having G constant with depth. However, considering the generality of the solution it is not difficult to incorporate the variation of G with depth. For instance, if we have a case where the outer shell has a soil that varies linearly with depth while the inner core has stiff clay whose dynamic shear modulus is constant with depth, the stiffness matrix gets modified as follows

$$k_{11} = \frac{GA_t}{H}\left[\int_0^1 (1+\psi\xi)(1+\xi)\beta_1^2 J_1^2(\beta_1\xi) + (m-1)\left(\frac{A_{tc}}{A_t}\right)\int_0^1 (1+\psi_c\xi)\beta_1^2 J_1^2(\beta_1\xi)\right]\mathrm{d}\xi \tag{5.6.1.18}$$

and so on.

It is apparent from the above that where established methods do not provide an answer to this problem in terms of *propagation of waves through a homogeneous shear beam of varying cross section,* the proposed method comes up with a rational solution of this practical problem.

5.5.2 Estimation of Dynamic Amplitude and Stress

For a body subjected to an earthquake force, the maximum amplitude is expressed as (Clough and Penzien 1982)

$$S_d = S_a/\omega^2 \tag{5.6.2.1}$$

In terms of provision as furnished in code, Eq. (5.6.2.1) can be finally expressed as

$$S_{di} = \kappa_i\left(\frac{ZI}{2R}\right)\frac{S_{ai}}{\omega^2} \tag{5.6.2.2}$$

where Z = zone factor; I = importance factor; R = ductility factor, and κ_i = modal mass participation factor usually expressed as $\sum_{i=1}^n m_i\phi_i/\sum_{i=1}^n m_i\phi_i^2$.

It is to be noted that IS-1893 (2016) [draft code for earthen dam] does not have any provision for R, the ductility factor in terms of earthen dam, though it implicitly assumes a value of $R = 1.5$ in a semi-empirical approach proposed therein. However, it has been shown by Chowdhury and Dasgupta (2007) that for R = 2–3, it will mostly give a reasonable result. Surely, more research is needed to define this value with more clarity for earth dams.

The modal mass participation factor based on the above formulation can be expressed in this case as

$$\kappa_i = \frac{\int_0^1 (1+\psi\xi)J_0(\beta_i\xi)}{\int_0^1 (1+\psi\xi)J_0(\beta_i\xi)^2} \tag{5.6.2.3}$$

It is again observed here instead of a unique value as expressed by Makdisi and Seed (1977), the modal participation factor is a function of the top and bottom width in terms of . Different values of κ_i with respect to ψ are furnished hereafter in Table 5.9.

It is observed from Table 5.9 that modal participation factor varies significantly, especially for the second mode with respect to the top and bottom width of the dam. The value of κ_i as obtained for $\psi \to \infty$ matches exactly the value as proposed by Makdisi and Seed (1977).

The displacement of the dam over the height H can be thus expressed as

$$u_i(z,t) = \kappa_i \left(\frac{ZI}{2R}\right) \frac{S_{ai}}{\omega_i^2} J_0\left(\beta_i \frac{z}{H}\right) \tag{5.6.2.4}$$

Substituting $T = 2\pi/\omega$, Eq. (5.6.2.4), after some simplification, can be expressed as

$$u_i(z,t) = \kappa_i \left(\frac{ZI}{2R}\right) \frac{C_{Ti}^2 \gamma H^2}{4\pi^2 G} \left(\frac{S_{ai}}{g}\right) J_0\left(\beta_i \frac{z}{H}\right) \tag{5.6.2.5}$$

The shear strain is expressed as

$$\gamma_{xzi} = \frac{\partial u}{\partial z} = \kappa_i \left(\frac{ZI}{2R}\right) \frac{C_{Ti}^2 \beta_i \gamma H}{4\pi^2 G} \left(\frac{S_{ai}}{g}\right) J_1\left(\beta_i \frac{z}{H}\right) \tag{5.6.2.6}$$

The shear stress that varies along the depth is finally expressed as

$$\tau_{xzi} = \kappa_i \left(\frac{ZI}{2R}\right) \frac{C_{Ti}^2 \beta_i \gamma H}{4\pi^2} \left(\frac{S_{ai}}{g}\right) J_1\left(\beta_i \frac{z}{H}\right) \tag{5.6.2.7}$$

Table 5.9 Variation of modal mass participation factor (κ_i) with ψ

ψ	κ_1	κ_2	κ_3
∞	1.602	−1.065	0.851
99	1.591	−1.00	0.85
49	1.581	−0.94	0.849
24	1.562	−0.827	0.847
15	1.543	−0.718	0.845
12	1.532	−0.655	0.844
9	1.515	−0.561	0.843

5.5.3 Earthquake Stability and Estimation of Lateral Seismic Coefficient (α_H)

Estimation of lateral stability of earthen dams under earthquake is still marred by some uncertainties, and a number of theories have been proposed, e.g.,

- Slip circle analysis based on pseudo-static method;
- Sliding block method;
- Dynamic finite element analysis.

Of all the above methods, the slip circle method based on pseudo-static analysis is the most popular, though it has been shown that the method can at times give unsatisfactory results. Terzaghi (1950) who first proposed it for stability of embankments under earthquake conceded the method to be crude, while Seed (1979) showed a number of dams in USA that had undergone failure in earthquake had a factor of safety (FOS) greater than 1.0 based on pseudo-static analysis.

Newmark (1965) proposed a sliding block technique which, though found to be theoretically sound, essentially requires time history response of a site for which site response data are always not available. Moreover, the movement of the failed mass as a rigid block may not be always realistic when the soil is not reasonably stiff.

Dynamic finite element analysis (Chopra and Clough 1966; Duncan 1992) though gives the most comprehensive result, requires significant engineering effort and very careful evaluation of the in situ soil parameters of the dam which is surely an expensive exercise and can possibly be justified only for very large dam of utmost importance.

Thus, in spite of its limitation, the pseudo-static analysis based on slip circle method, due to its simplicity in application, continues to be the most popular method in practice where the earthquake force is simulated by an equivalent lateral seismic coefficient α_h (Fig. 5.3).

One of the major difficulties and limitation that arises in estimation of α_h for dams is that, unlike a normal structure, where the lateral seismic coefficient can be considered as constant over the height of a structure, an earth dam is a far more flexible system and as such α_h varies with height. Hence, in many cases what would be the effective lateral seismic coefficient becomes a matter of judgment.

As per IS-1893 (2016) [draft code for earthen dam], for instance, α_h is expressed by an empirical formula

$$\alpha_h = \frac{1}{3} \times Z \times I \times S \tag{5.6.3.1}$$

where Z and I are as explained earlier and S is soil factor that varies from max 2.0 for soft soil to 1.0 minimum.

The peak horizontal ground acceleration (PHGA) is considered as $a_{\max} = Z \times I \times S$, while the net lateral force due to earthquake is given by

$$\mathrm{F} = \frac{1}{3} \times Z \times I \times S \times W \tag{5.6.3.2}$$

It is obvious that, irrespective of the time period of the dam, this value remains constant for a particular zone (II, III, IV, etc.) though it is well known that how much acceleration a body is subjected to is dependent on the time periods of the body in different modes.

Since the value of α_h proposed is empirical in nature, there could be cases when the response could be overestimated and vice versa and matching with the above value can just be a coincidence.

Based on the proposed dynamic theory, the lateral coefficient α_h can be estimated as mentioned hereafter. Typical slope failure profile of an earth dam, based on wedge theory, is shown in Fig. 5.6.

Following Makdisi and Seed (1977), the total horizontal force acting on this wedge is expressed as

$$F(z,t) = \text{Mass of triangular wedge} \times \ddot{u}_{\text{av}} \tag{5.6.3.3}$$

where $\ddot{u}_{\text{av}}$ = average acceleration over the dam height.

Based on Fig. 5.6, Eq. (5.6.3.3) can be expressed as

$$\ddot{u}_{\text{av}} = \frac{2F(z,t)g}{B_z \times z \times \gamma} = \frac{2\tau_{xz}g}{z\gamma} \tag{5.6.3.4}$$

or,

$$\alpha_{hi} = \frac{\ddot{u}}{g} = \frac{2\tau_{xzi}}{z\gamma} = 2\kappa_i \left(\frac{ZI}{2R}\right) \frac{C_{Ti}^2 \beta_i H}{4\pi^2 z} \left(\frac{S_{ai}}{g}\right) J_1\left(\beta_i \frac{z}{H}\right) \tag{5.6.3.5}$$

where i = 1, 2, 3, … is the number of modes.

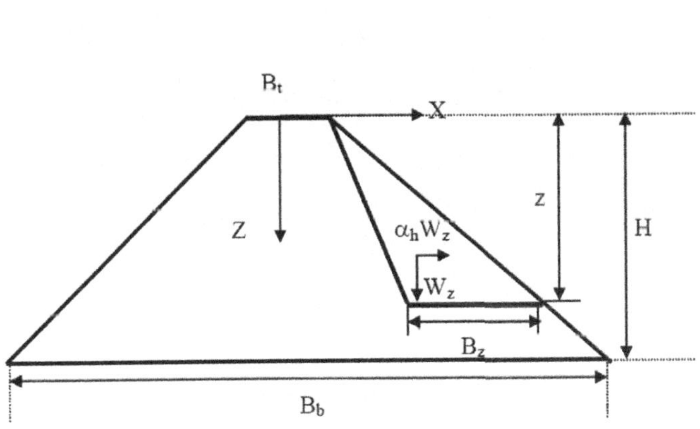

Fig. 5.6 Assumed failure profile of an earth dam

It is evident from Eq. (5.6.3.5) that α_h is a function of the depth z and shape function, as described in Eq. (5.6.3).

Now, let as assume that this lateral seismic coefficient is effective over the whole cross section of the dam.

Thus, over full height H of the dam, combining Eqs. (5.6.3.5) and (5.6.11), we have

$$F_{ei} = \alpha_{hi} W = 2\kappa_i \left(\frac{ZI}{2R}\right) \frac{C_{Ti}^2 \beta_i H \gamma A_t}{4\pi^2} \left(\frac{S_{ai}}{g}\right) \int_0^H \frac{1}{z} \left(1 + \psi \frac{z}{H}\right) J_1\left(\beta_i \frac{z}{H}\right) \mathrm{d}z \quad (5.6.3.6)$$

where

$$W = \gamma A_t \int_0^1 \left(1 + \psi \frac{z}{H}\right) \mathrm{d}z \quad (5.6.3.7)$$

Thus, the average lateral seismic coefficient over the full height of the dam can be expressed as

$$\alpha_{hi} = \frac{F_{ei}}{W} = \frac{2\kappa_i \left(\frac{ZI}{2R}\right) \frac{C_{Ti}^2 \beta_i}{4\pi^2} \left(\frac{S_{ai}}{g}\right) \int_0^1 \frac{1}{\xi} (1 + \psi\xi) J_1(\beta_i \xi) \mathrm{d}\xi}{\int_0^1 (1 + \psi\xi) \mathrm{d}\xi} \quad (5.6.3.8)$$

It may be observed that α_{hi} obtained using Eq. (5.6.3.8), as expected, is a dimensionless quantity.

Eqn (5.6.3.8) can be further simplified to

$$\alpha_{hi} = \mathrm{Coeff}_i \times \left(\frac{ZI}{2R}\right) \left(\frac{S_{ai}}{g}\right) \quad (5.6.3.9)$$

where

$$\mathrm{Coeff}_i = \frac{\left[2\kappa_i \frac{C_{Ti}^2 \beta_i}{4\pi^2} \int_0^1 \frac{1}{\xi} (1 + \psi\xi) J_1(\beta_i \xi) \mathrm{d}\xi\right]}{\int_0^1 (1 + \psi\xi)\, \mathrm{d}\xi} \quad (5.6.3.10)$$

Values of the coefficients in Eq. (5.6.3.10) for various values of ψ are furnished in Table 5.10.

The design lateral coefficient based on Eq. (5.6.3.9) can finally be considered for first three modes as

$$\alpha_{hr} = \sqrt{\alpha_{h1}^2 + \alpha_{h2}^2 + \alpha_{h3}^2} \quad (5.6.3.11)$$

Table 5.10 Values of seismic coefficient for different values of ψ for first three modes

ψ	*G* constant with depth Modes			*G* linear with depth Modes			*G* parabolic with depth Modes		
	First	Second	Third	First	Second	Third	First	Second	Third
99	1.11	0.14	0.05	0.65	0.09	0.03	0.39	0.06	0.02
49	1.13	0.14	0.05	0.66	0.09	0.03	0.40	0.06	0.02
24	1.14	0.14	0.05	0.68	0.09	0.03	0.41	0.06	0.02
15	1.18	0.14	0.06	0.70	0.09	0.04	0.42	0.06	0.02
12	1.21	0.14	0.07	0.72	0.09	0.04	0.44	0.06	0.03
9	1.26	0.14	0.08	0.74	0.09	0.05	0.45	0.06	0.03

5.6 Stability Analysis of a Dam

As mentioned earlier, the most popular method for stability analysis of dams under static load is based on slip circle methods, like Swedish slip circle method, Bishops simplified analysis. That soil fails along an arc was observed in a number of landslides in the Alps, from where Fellenius (1936) originally developed the method. In spite of its sound scientific background, it has been observed to over-estimate FOS under an actual earthquake (Seed 1979). One of the possible reasons for this could be that for a homogenous isotropic medium under transient dynamic load, the failure surface could be different from a circular arc as assumed under static load.

Without going into much of details, it can be argued that under a transient shock, if a body fails, the failure profile would follow the path of least resistance. Since the resisting force within the soil body is developed by the friction and cohesion in a particular plane, it is apparent that for least resistance, the surface should be a minimum. In other words, it should be a straight line rather than a circular arc. Failure of dams along a straight line path is not uncommon, and Whitman and Lambe (1979) point out that a number of dams resting on firm ground have been observed to undergo failure along a straight line even under static loads. Since most of the earthen dams are built on relatively firm ground, it would thus not be illogical to assume a straight line failure profile of a dam surface under an earthquake. Finally, under earthquake a number of frames which are essentially treated as shear frames the infill panels have been observed to undergo cracks that follow a straight line rather than a circular arc (Agarwal and Shrikhande 2006). Similarly, deep beams and squat shear walls (where shear strain energy dominates) failing under earthquake have been observed to follow more or less a straight line (Park and Pauley 1975).

Based on the above logic, we propose to evaluate the dynamic stability of the dam as mentioned hereunder.

Shown in Fig. 5.7 is an earth dam subjected to lateral earthquake force.

It is evident that we can draw infinite number of planes having an angle θ that can vary from 0 to angle α (the slope of the dam) through which the dam can

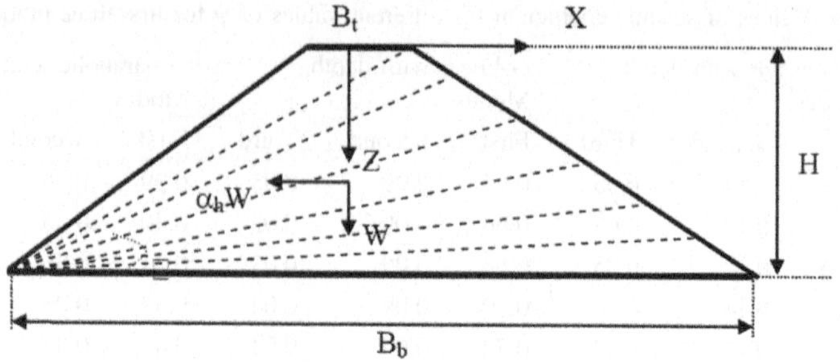

Fig. 5.7 An earth dam with various possible failure planes

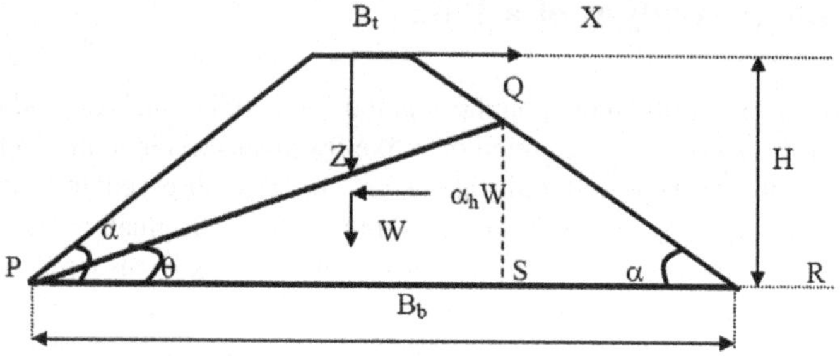

Fig. 5.8 A typical trapezoidal section of an earth dam

potentially slide. Among all these planes, there exists one unique plane at an angle θ which is the weakest and has the maximum probability of failure along that plane. In other words, FOS against failure on this plane is a minimum.

Shown in Fig. 5.8 is an earth dam section where the critical failure profile plane is at an angle θ with the horizontal.

Based on geometry, it can then be shown that area of triangle PQR (Δ) is given by

$$\Delta = \frac{1}{2} B_b^2 \frac{\sin\alpha \sin\theta}{\sin(\alpha + \theta)} \tag{5.6.4.1}$$

Thus, weight of the dam above failure plane PQ can be expressed as

$$W = \frac{\gamma}{2}\left[(B_b + B_t)H - B_b^2 \frac{\sin\alpha \sin\theta}{\sin(\alpha + \theta)}\right] \tag{5.6.4.2}$$

The weight W is then expected to slide down the plane PQ which is the critical plane of failure.

As shown in Fig. 5.9, the weight W slides down a critical plane at an angle θ with horizontal.

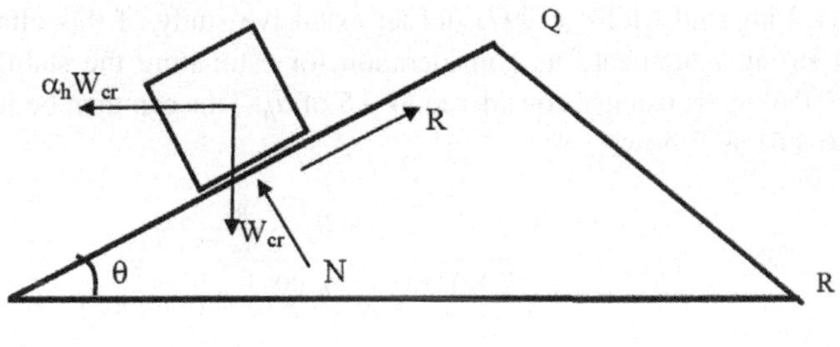

Fig. 5.9 Sliding of weight W_{cr} along the sloped surface

The resistive force R may be expressed as

$$R = \mu N + c \times \text{PQ} \tag{5.6.4.3}$$

where μ = angle of internal friction of soil usually expressed as tan ϕ; c = cohesion of soil, and PQ = length of the failure plane = $B_b[\sin\alpha/\sin(\alpha+\theta)]$.

From free body diagram, one can write

$$N = W(\cos\theta - \alpha_h \sin\theta) \tag{5.6.4.4}$$

The force which drives the body down is given by $F_D = W(\sin\theta + \alpha_h \cos\theta)$. The resisting force is thus given by

$$F_R = \mu W(\cos\theta - \alpha_h \sin\theta) + c \times B_b \frac{\sin\alpha}{\sin(\alpha+\theta)} \tag{5.6.4.5}$$

The factor of safety (FOS) is then expressed as

$$\text{FOS} = \frac{F_R}{F_D} = \frac{\mu(\cos\theta - \alpha_h \sin\theta) + \frac{cB_b}{W}\frac{\sin\alpha}{\sin(\alpha+\theta)}}{\sin\theta + \alpha_h \cos\theta} \tag{5.6.4.6}$$

where W is the weight expressed by Eq. (5.6.4.2).

To determine the minimum factor of safety, one can now vary θ from 0 to α and evaluate the plane which is critical and FOS is the minimum.

5.6.1 *Effect of Vertical Seismic Coefficient α_V on FOS*

The normal perception is that vertical component of an earthquake has no effect on the overall stability of the dam, though Mononobe (1924) showed that a combination of horizontal and vertical acceleration led to severe damage in earth retaining structures in Kanto earthquake in Japan as early as 1920. Chopra (1966) through finite element analysis showed that vertical acceleration affects seismic response of earth dams. Ling and Mohri (1997) did an extensive study of this effect on earth dams and strongly advocate its consideration for estimating the stability. As per IS-1893 (2016), α_v is usually considered as 0.5 of α_h. This can then be incorporated in Eq. (5.6.4.6) as follows

$$\mathrm{FOS} == \frac{\mu((1 \pm \alpha_v)\cos\theta - \alpha_h \sin\theta) + \frac{cB_b}{W_{cr}}\frac{\sin\alpha}{\sin(\alpha+\theta)}}{(1 \pm \alpha_v)\sin\theta + \alpha_h \cos\theta} \tag{5.6.5.1}$$

5.6.2 *Stability of a Dam with Impermeable Core*

Stability of dam including the impermeable core induces significant computational difficulties. Firstly, due to the heterogeneity of the soil medium, shear stress changes between the outer and inner shell material. Moreover, as there is a significant variation in stiffness of the outer and inner core material, distribution of forces between the two parts also varies. The usual practice is to use FEM for estimation of the stability, where uncertainties prevail for if the meshing of the soil material is not sufficiently refined, one can well arrive at an upper bound results.[3]

We propose to approach the problem as given hereafter.

Normally, the cores constitute of stiff compacted silty clay/clayey silt (little friction is preferable for it significantly enhances the shear strength with depth along with cohesion c) having minimum void ratio to reduce seepage. From strength point of view, also they are usually stiffer than the outer shell material in the order of 3–4 times. Thus, if one looks at the overall response of the heterogeneous earth dam under earthquake, it can be concluded that the system acts analogous to a shear wall-framed structure where the inner core acts as a stiff shear wall and outer shell acts like the connected frame.

In other words, distribution of the lateral force among the outer and inner shell material is dependent on the relative stiffness of these two materials.

For a homogeneous isotropic body of area A and height H, the static stiffness of the soil is simply expressed as

$$K_{st} = GA/H \tag{5.6.6.1}$$

For the earth dam with variable cross section having clay core, Eq. (5.6.6.1) can be expressed as

[3]Means result slightly in the higher side with respect to exact solution.

$$K_{st} = \frac{GA_t}{H}\int_0^1 (1+\psi\xi)\mathrm{d}\xi - \frac{GA_{tc}}{H}\int_0^1 (1+\psi_c\xi)\mathrm{d}\xi + \frac{G_cA_{tc}}{H}\int_0^1 (1+\psi_c\xi)\mathrm{d}\xi \tag{5.6.6.2}$$

On simplification, Eq. (5.6.6.2) can be expressed as

$$K_{st} = \frac{GA_t}{H}\left[\left(1+\frac{\psi}{2}\right) + (m-1)r_c\left(1+\frac{\psi_c}{2}\right)\right] \tag{5.6.6.3}$$

where $m = G_c/G$ and $r_c = A_{tc}/A_t$.

The stiffness of the inner core may be expressed as

$$K_c = \frac{G_cA_{tc}}{H}\left(1+\frac{\psi_c}{2}\right) \tag{5.6.6.4}$$

Thus, if $\alpha_h W$ is the total force on the dam the force resisted by the inner core is given by

$$F_{hc} = \frac{K_c}{K_{st}}(\alpha_h W) \quad \Rightarrow F_{hc} = \left[\frac{mr_c\left(1+\frac{\psi_c}{2}\right)}{1+\frac{\psi}{2}+(m-1)r_c\left(1+\frac{\psi_c}{2}\right)}\right](\alpha_h W) \tag{5.6.6.5}$$

$$\Rightarrow F_{hc} = (\alpha_{hc} W) \quad \text{where } \alpha_{hc} = \alpha_h\left[\frac{mr_c\left(1+\frac{\psi_c}{2}\right)}{1+\frac{\psi}{2}+(m-1)r_c\left(1+\frac{\psi_c}{2}\right)}\right] \tag{5.6.6.6}$$

Proceeding in identical manner, it can be shown that the force induced in outer shell can be expressed as

$$F_{hS} = \left(\frac{\alpha_h W}{2}\right)\left[\frac{1+\frac{\psi}{2}-r_c\left(1+\frac{\psi_c}{2}\right)}{1+\frac{\psi}{2}+(m-1)r_c\left(1+\frac{\psi_c}{2}\right)}\right] \tag{5.6.6.7}$$

Equation (5.6.6.7) can be rewritten as

$$F_{hS} = \alpha'_{hs} W$$

where

$$\alpha'_{hs} = \frac{\alpha_h}{2}\left[\frac{1 + \frac{\psi}{2} - r_c\left(1 + \frac{\psi_c}{2}\right)}{1 + \frac{\psi}{2} + (m-1)r_c\left(1 + \frac{\psi_c}{2}\right)}\right] \tag{5.6.6.8}$$

Here, the term ½ comes because of the fact that the outer shell constitutes of two parts, and as such, the total force on the outer shell is thus divided equally into the two sections (provided, of course, the outer shell is symmetric to the core).

Based on the time period obtained earlier, the seismic lateral force coefficient α_h is obtained using Eq. (5.6.3.11).

For a composite dam, the total weight W is expressed as

$$W = \gamma A_t H\left[1 + \frac{\psi}{2} + r_c(p-1)\left(1 + \frac{\psi_c}{2}\right)\right] \tag{5.6.6.9}$$

For calculation of FOS, let us consider Fig. 5.10.

It is evident that there can be two failures within the dam section:

(1) Slope failure of the outer shell at a critical angle θ (the PQ plane);
(2) Shearing off of the inner core due to the force F_{hc}; vide Eq. (5.6.6.5).

Proceeding in the same line as explained earlier if θ be the critical plane at which the FOS is considered minimum. Then, based on Fig. 5.10, the weight of soil, W_b, below the critical PQ plane is expressed as

$$W_b = \frac{\gamma(B_b - B_{bc})^2}{8}\frac{\sin\theta\sin\delta}{\sin(\delta-\theta)} \tag{5.6.6.10}$$

The gross weight of one outer shell is given by

$$W_s = \frac{\gamma H}{2}[B_{bs} + B_{ts}] \tag{5.6.6.11}$$

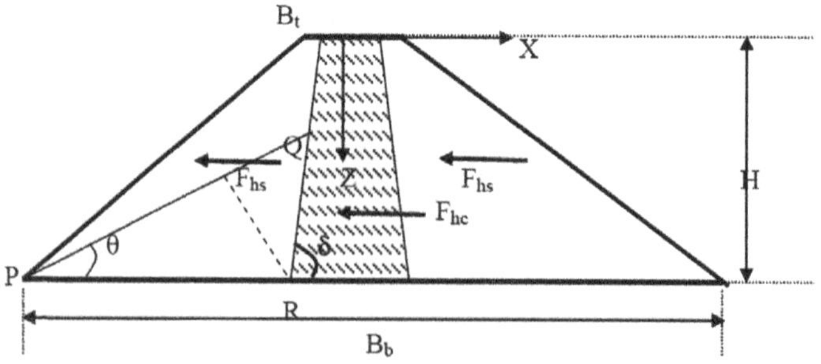

Fig. 5.10 Lateral seismic force on clay-cored earth dam

where $B_{bs} = 0.5(B_b - B_{bc})$ and $B_{ts} = 0.5(B_t - B_{tc})$.

Subtracting Eq. (5.6.6.13) from Eq. (5.6.6.11), we get the critical weight W_{cr} that slides along the plane PQ.

The factor of safety for the outer shell is now expressed as

$$\text{FOS} = \frac{\mu\left(\left(1 \pm \alpha_v'\right)\cos\theta - \alpha_h'\sin\theta\right) + \dfrac{c(B_b - B_{bc})}{2W_{cr}}\dfrac{\sin\delta}{\sin(\delta-\theta)}}{\left(1 \pm \alpha_v'\right)\sin\theta + \alpha_h'\cos\theta} \quad (5.6.6.12)$$

where α_{hs}' is as expressed by Eq. (5.6.6.8) and α_v' is 0.5 times α_{hs}'.

To arrive at the minimum FOS, we now vary θ from 0 to α at some predefined increment of say 1° to obtain the minimum value.

For the inner core, the force F_{hc} as per Eq. (5.6.6.5) basically acts as the base shear like in a continuous shear wall. Then, this can be distributed over the height of the dam as

$$Q_i = F_{hc}\frac{W_i \times h_i}{\sum W_i \times h_i^2} \quad (5.6.6.13)$$

where W_i = weight of core at a particular depth h_i given by $\gamma_c A_t\left(1 + \psi_c \frac{z_i}{H}\right)(z_i - z_{i-1})$; Q_i = nodal force at depth h_i.

Summing up of all these nodal forces Q_i up to a particular depth h_i gives the shear force acting at that particular level.

Thus,

$$V_{i(z=hi)} = \sum_0^{h_i} Q_i \quad (5.6.6.14)$$

The shear stress developed at any section is given by

$$\tau_{7i} = \frac{V_{zi}}{B_{tc}\left(1 + \psi_c \frac{z}{H}\right) \times 1} \le c + \gamma_c \times z \tan\varphi_c \quad (5.6.6.15)$$

where $c + \gamma_c \times z \tan\varphi_c$ is the allowable shear stress at a particular depth z from crest, c is cohesion of soil in core, and φ_c is its internal angle of friction.

5.6.3 *Effect of Hydrodynamic Pressure on Stability*

Hydrodynamic pressure on the dam will be most critical when it creates a suction pressure on the sloped surface.

It has been shown by Chowdhury and Dasgupta (2008) that the hydrodynamic pressure due to the earthquake on a vertical surface for a fluid medium extending to infinity can be expressed as

$$p = \frac{12}{\pi(\pi+2)} \frac{S_a \gamma_w H_w}{g} \sin \frac{\pi z}{2H_w} \tag{5.6.7.1}$$

in which H_w = vertical height of water in dam; S_a = acceleration corresponding to free field time period of the water expressed as $4H_w/v_w$; v_w = velocity of sound in water = 1149 m/s, and γ_w = unit weight of water.

Thus, for a surface at an inclination α the pressure normal (p_n) to the surface can be expressed as

$$p_n = \frac{12}{\pi(\pi+2)} \frac{S_a \gamma_w H_w}{g} \sin \frac{\pi z}{2H_w} \sin \alpha \tag{5.6.7.2}$$

Thus, the suction force V which reduces the normal force N is given by

$$V = \int_0^{H_w} p_n dz = 0.473 \frac{\gamma_w S_a H_w^2}{g} \sin \alpha \tag{5.6.7.3.}$$

The factor of safety against stability considering the suction effect of hydrodynamic pressure can then be expressed as

$$\mathrm{FOS} = \frac{\mu\left((1 \pm \alpha_v)\cos\theta - \alpha_h \sin\theta - \frac{V\cos(\alpha-\theta)}{W_{cr}}\right) + \frac{cB_b}{W_{cr}} \frac{\sin\alpha}{\sin(\alpha+\theta)}}{(1 \pm \alpha_v)\sin\theta + \alpha_h \cos\theta + \frac{V\sin(\alpha-\theta)}{W_{cr}}} \tag{5.6.7.4}$$

For clay-cored dam, replace α_v and α_h by α'_{v_s} and α'_{h_s}, respectively, and shall affect the outer shell material only.

Example 5.1 A typical data of a dam as shown in Fig. 5.11 are considered. The dam is analyzed for two cases:

(1) Homogeneous section having v_s = 125 m/s, ϕ = 32°, c = 50 kN/m^2, unit weight γ_c = 20 kN/m^3, height of water = 40 m.

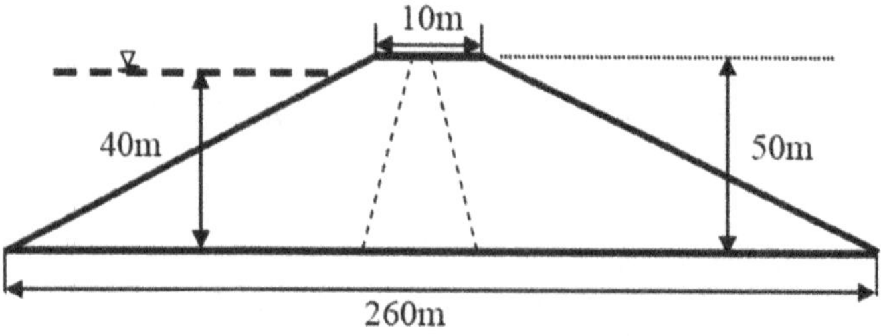

Fig. 5.11 A typical cross-section of an earth dam

(2) The same dam with a silty clay core (shown by dotted line) the core having width at crest of 4.0 m and width at base of 65 m having unit weight of $\gamma_c = 21\ \text{kN/m}^3$ and shear wave velocity of 220 m/s, $c = 85\ \text{kN/m}^2$ and $\phi_c = 12°$.

Maximum height of water in the reservoir = 40 m.

The dam is in Zone IV as per IS-1893 (2016) having importance factor I as 2.0 and R is taken as 2.0 for this case.

Solution Comparison of time periods (in seconds) is shown in Table 5.11.

It is observed that the time periods obtained by the proposed method are in excellent agreement with other established methods for $B_t = 0$. However, the top width having finite dimension ($B_t = 10.0$ m) elongates the time period by about 4% for the fundamental mode, which would be most critical. The clay core being much stiffer than the outer shell has an overall stiffening effect on the dam response when the time period gets reduced by 25% than the established expression—this is logical.

Accelerations induced in different modes in dam in terms of g are given in Table 5.12.

The acceleration induced for first three modes by proposed method is as shown above; it is evident that depending on the time period value, the acceleration varies, unlike recommendation of IS code that suggests a constant value. Here, it is again observed that the clay core stiffening the overall response attracts more acceleration than the homogeneous dam.

Modal displacement of dam crest is given in Table 5.13.

The modal response is found to be quite high, though within acceptable limit. Here, it is observed that stiffening effect of impermeable core induces more acceleration and also induces a higher displacement in fundamental mode though the SRSS values are almost the same.

Table 5.11 Comparison of time periods

Mode	IS-1893 (1984)	Mononobe (1936)	Makdisi and Seed (1977)	Gazetas (1982)	Proposed $B_t = 0.0$	Proposed $B_t = 10.0$	With clay core
1	1.044	1.044	1.046	1.046	1.046	1.085	0.783
2	–	0.455	0.455	0.455	0.455	0.471	0.338
3	–	0.291	0.3	0.29	0.30	0.299	0.214

Table 5.12 Induced acceleration of the earth dam

Mode	S_a (homogeneous dam) proposed (g)	S_a with clay core (proposed) (g)	S_a as per IS-1893 (2016) (g)
1	0.15	0.21	0.19
2	0.3	0.3	NA
3	0.3	0.3	NA

Table 5.13 Maximum crest amplitude of the earth dam

Mode	u (mm): homogeneous dam	u (mm): with clay core
1	675	683
2	−135	−98
3	55	40
SRSS value	690	692

Table 5.14 Lateral seismic coefficient of the earth dam

Dam type	α_h
Homogeneous section	0.183
With clay core	0.25
As per IS code	0.19

The lateral seismic coefficient for the dam is given in Table 5.14.

Following the same trend as expressed earlier, the lateral seismic coefficient for the heterogeneous dam is higher—this reflects the importance of considering the heterogeneous effect rather than ignoring it, as is the practice of now. It may however be noted that the values were obtained based on damping value $D = 5\%$ for all cases. For clay core with high plasticity, damping is usually high; thus, considering 7% damping for the homogeneous dam and 10% for the dam with clay core value of $\alpha_{h,}$ we will come to 0.16 and 0.20, respectively.

Force transmitted to outer and inner shell based on stiffness distribution is outlined in Table 5.15.

The minimum factor of safety of the dam under various load combinations is given in Table 5.16.

Variation of shear stress in the internal core is given in Fig. 5.12.

Table 5.15 Lateral seismic coefficient of the clay-cored dam and lateral force

Dam section	Lateral seismic coefficient (α_h)	Force (kN)
Outer core	0.053	7268
Inner core	0.118	16,228

Table 5.16 Factor of safety of the homogenous and clay-cored earth dam

Load case	FOS (reservoir full)	FOS (reservoir empty)
Under static load	–	3.47*
Under earthquake—homogenous dam	1.60	2.01
Under earthquake—with clay core	1.71	1.97

*Carried out by Swedish slip circle method

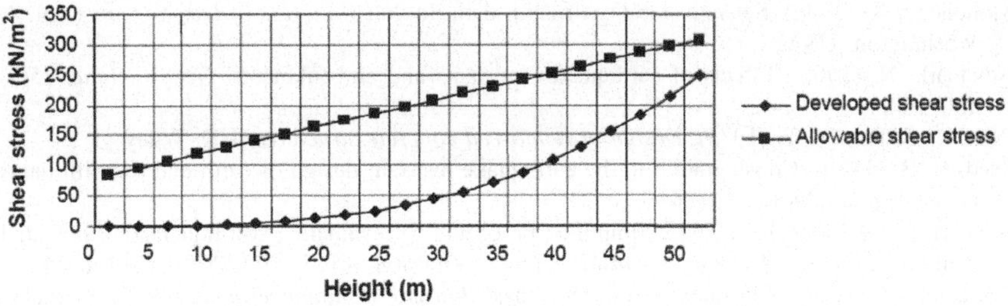

Fig. 5.12 Variation of shear stress in inner core of dam having c = 85 kN/m^2 and ϕ = 12°

References

Agarwal, P., & Shrikhande, M. (2006). *Earthquake resistant design of structures*. New Delhi, India: Prentice Hall.

Bathe, K. J. (1996). *Finite element procedures in engineering*. New Delhi, India: Prentice Hall.

Chopra, A. K. (1966). The importance of vertical component of earthquake motion. *Bulletin of Seismological Society of America, 56*, 1163–1175.

Chopra, A. K., & Clough, R. W. (1966). Earthquake stress analysis in earth dams. *Journal of the Engineering Mechanics, ASCE, 92*(EM2), 197–212.

Chowdhury, I., & Dasgupta, S. P. (2007). Dynamic earth pressures on rigid unyielding wall under earthquake force. *Indian Geotechnical Journal, 37*(2), 81–93.

Chowdhury, I., & Dasgupta, S. P. (2008). *Dynamics of structure and foundation- A unified approach* (Vol. II). London, UK: Taylor and Francis.

Chowdhury, I., & Dasgupta, S. P. (2010). Dynamic analysis of earth dams and embankments under earthquake force. *Indian Geotechnical Journal, 40*(3), 143–158.

Clough, R. W., & Penzien, J. (1982). *Dynamics of structures*. New York: McGraw-Hill Publication.

Duncan, J. M. (1992). State of the art: Static stability and deformation analysis. In *Proceedings of the Specialty Conference on Stability and Performance of slopes and Embankments, ASCE* (Vol. 1, pp. 222–266), New York.

Fellenius, W. (1936). Calculation of the stability of earth dams. In *2nd Congress of Large Dams* (Vol. 4, p. 445), Washington, USA.

Gazetas, G. (1982). Shear vibration of vertically inhomogeneous earth dams. *International Journal for Numerical and Analytical Methods in Geomechanics, 6*(1), 219–241.

Hurty, W., & Rubenstein, M. (1967). *Dynamics of structure*. New Delhi, India: Prentice Hall.

IS-1893. (1984). *Code of earthquake resistant design*. New Delhi, India: Bureau of Indian Standard.

IS-1893. (2016). *Code for earthquake resistant design of structures*. New Delhi, India: Bureau of Indian Standard Institution.

Jeffrey, A. (2005). *Handbook of mathematical formula and integrals*. New York: Elsevier Publication.

Kramer, S. (1984). *Geotechnical earthquake engineering*. India: Pearson Education.

Ling, H., Leschinsky, D., & Mohri, Y. (1997). Soil slopes under combined horizontal and vertical seismic accelerations. *Journal of Earthquake Engineering and Structural Dynamics, 26*, 1231–141.

Makdisi, F. I., & Seed, H. B. (1977). A simplified procedure for estimating earthquake induced deformation in dams and embankments. *Report # UCB/EERC-77/19*, University of California, Berkeley, USA.

Mononobe, N. (1924). Effects of vertical acceleration and theory of vibration. *Proceedings of the Japanese Society of Civil Engineering, 10*(5), 1063–1094.

Mononobe, N. (1936). Seismic stability of earth dam. In *2nd Congress of Large Dams* (Vol. 4), Washington, USA.
Newmark, N. (1965). Effects of earthquakes on dams and embankments. *Geotechnique, 15*(2), 130–160.
Park, R., & Pauley, T. (1975). *Design of reinforced concrete structures*. UK: Wiley.
Seed, H. B. (1979). Consideration in the earthquake resistant design of earth and rock fill dams. *Geotechnique, 29*(3), 215–263.
Seed, H. B., & Idriss, M. (1971). Simplified procedure for evaluating soil liquefaction potential. *Journal of the Soil Mechanics and Foundations Division, ASCE, 107*(SM9), 1249–1274.
Terzaghi, K. (1950). Mechanism of Landslides. *Journal of Engineering Geology*. Berkeley: Geological Society of America.
Timoshenko, S. P., & Young, D. H. (1982). *Theory of elasticity*. New York: McGraw Hill publication.
Whitman, R. V., & Lambe, T. W. (1979). *Soil mechanics*. New York: Wiley.

Chapter 6
Seismic Design of Shallow and Deep Foundations

6.1 Introduction

This chapter deals with the design of shallow and deep foundation under seismic load. However, those who are conversant with design of structures under seismic load might wonder as to what is so special about it? Most of the codes around the world recommend *an increase in allowable bearing capacity of soil* by about 25% under seismic load combination for design of shallow foundations. Thus, the question naturally arises as to what makes it be given special attention?

6.2 Basis of Increment in Bearing Capacity Under Seismic Force

To start with, we look at the basis on which codes recommend an increase in allowable bearing capacity.

Under static loading, the general shear failure expression for a strip footing in c–ϕ soil, as given by Terzaghi (1943), is expressed as

$$q_{\mathrm{ult}} = cN_{\mathrm{c}} + \gamma D_{\mathrm{f}} N_{\mathrm{q}} + \frac{1}{2}\gamma B N_{\gamma} \tag{6.2.1}$$

In Eq. (6.2.1)

q_{ult}	ultimate bearing capacity of soil, under general shear failure condition;
c	cohesive strength of soil;
γ	weight density of soil;
D_{f}	depth of foundation below ground level;
B	width of footing; and
N_{c}, N_{q}, N_{γ}	bearing capacity factors.

I. Chowdhury and S. P. Dasgupta, *Earthquake Analysis and Design of Industrial Structures and Infra-structures*, GeoPlanet: Earth and Planetary Sciences,
https://doi.org/10.1007/978-3-319-90832-8_6

Equation (6.2.1) has been further generalized by Brinch Hansen (1970), as furnished in IS-6403 (1997) as

$$q_{\mathrm{ult}} = cN_c s_c d_c i_c + \gamma D_f N_q s_q d_q i_q + \frac{1}{2}\gamma B N_\gamma s_\gamma d_\gamma i_\gamma W' \quad (6.2.2)$$

In Eq. (6.2.2)

s_c, s_q, s_γ = shape factors; d_c, d_q, d_γ = depth factors; i_c, i_q, i_γ = inclination factors; and W' = water table correction factor.

The shape factors s_c, s_q, and s_γ are defined as hereafter.

For continuous footing

$$s_c = s_q = s_\gamma = 1.0 \quad (6.2.3)$$

For rectangular footing

$$s_c = s_q = 1 + 0.2B/L \quad \text{and} \quad s_\gamma = 1.0 - 0.4B/L \quad (6.2.4)$$

For square footing

$$s_c = 1.3, s_q = 1.2, s_\gamma = 0.8 \quad (6.2.5)$$

For circular footing

$$s_c = 1.3, s_q = 1.2, s_\gamma = 0.6 \quad (6.2.6)$$

The depth factors, d_c, d_q, and d_γ, are defined as

$$d_c = 1 + 0.2(D_f/B)\sqrt{N_\phi} \quad (6.2.7)$$

Here, D_f = depth of foundation below ground level and B = width of foundation and $N_\phi = \tan^2(45 + \phi/2)$.

$$d_q = d_\gamma = 1.0 \quad \text{for } \phi < 10^\circ$$
$$d_q = d_\gamma = 1 + 0.1(D_f/B)\sqrt{N_\phi} \quad \text{for } \phi > 10^\circ. \quad (6.2.8)$$

Inclination factors are expressed as

$$i_c = i_q = (1 - \alpha/90)^2 \quad (6.2.9)$$

and

$$i_\gamma = (1 - \alpha/\phi)^2 \tag{6.2.10}$$

Here, α = angle subtended with the horizontal.
The effect of water table is considered as hereafter:

- If depth of water table is likely to permanently remain at or below a depth of $(D_f + B)$ beneath the ground level surrounding the footing, then $W' = 1.0$.
- If the water table is located at a depth D_f or likely to rise to the base of footing or above, then the value of W' shall be taken as 0.5.
- If the water table is likely to be permanently located at depth $D_f \leq D_w \leq (D_f + B)$, then the value of W' is to be obtained by linear interpolation.

Equation (6.2.2), when divided by a suitable factor of safety (FOS) [usually considered between 2.5 to 3], gives the allowable bearing capacity (q_{all}) as

$$q_{all} = \frac{q_{ult}}{\text{FOS}} \tag{6.2.11}$$

In most of the codes around the world, the return period of an earthquake occurrence to the expected intensity is once in fifty years. For this reason, for an average structure having a life span of say 25–30 years, *it is considered once in a lifetime event.*

Thus, in order to economize on the foundation, one may increase the allowable bearing capacity by 25% when, without endangering the safety, one can cut down on the cost. The fact is that during its occurrence, the factor of safety gets reduced to 2.0–2.4, but still has a sufficient margin as to not to trigger a failure of the foundation.

The logic as proposed above looks quite rational, but yet is inconsistent!

To understand the lacunae, we would briefly look into the basis of Eq. (6.2.1) hereafter.

6.3 Terzaghi's (1943) Basis of Ultimate Bearing Capacity of a Shallow Foundation Under Static Loading

Shown in Fig. 6.1 is the soil failure profile as perceived by Terzaghi (1943) for a footing of width B.

The soil under influence of vertical load from the foundation will generate a failure profile of edcfg under general shear failure, as shown in the figure where zone abc is a triangular elastic zone located immediately below the foundation with the inclinations of the side ac and bc of the wedge with base of foundation being ϕ, the angle of friction of the soil.

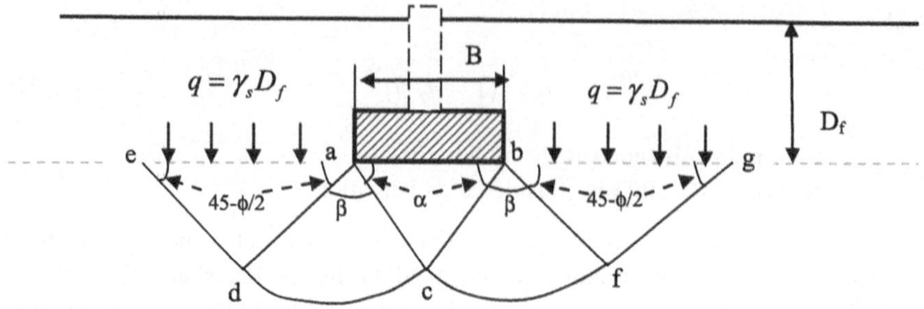

Fig. 6.1 Terzaghi's basis for determination of ultimate bearing capacity of a shallow foundation

Zones acd and bcf are known as *Prandtl's radial shear zones.*

Zones ade and bfg are *Rankine's passive pressure zones.* The slip lines in these zones intersect the ground surface at an angle $(45 - \phi/2)°$.

Lines cd and cf are arcs of a log spiral defined by the equation $r = r_0 e^{\theta \tan\phi}$.

The lines ad, de, bf, and fg are straight lines. Lines de and fg actually extend up to the ground surface. Terzaghi assumed that the soil located above foundation can be replaced by a surcharge loading $q = \gamma_s D_f$.

Ultimate bearing capacity q_u of the foundation can now be determined if we consider the faces ac and bc of the triangular wedge abc and determine the passive force on each face required to cause failure.

The total passive pressure will be the *superimposition* of three cases, i.e.,

1. Due to surcharge $q = \gamma_s D_f$
2. Due to cohesion c
3. Due to unit weight of soil γ_s and friction angle ϕ.

Terzaghi assumed that under vertical load the foundation forms the failure wedge abc making an angle $\alpha = \phi$ with the horizontal. The Rankine's passive pressure zone makes an angle $(45 - \phi/2)°$ with the horizontal, as shown in Fig. 6.1. Thus, angle cbf = $[(3/4)\pi - \phi/2]° = \beta$ (say).

6.3.1 Determination of Passive Pressure Due to Surcharge Load Q

Let us consider the free body diagram footing (symmetrical about the centerline of footing) under surcharge load. The different forces and their locations are as shown in Fig. 6.2.

Based on the figure and as defined above, $\alpha = \phi, \delta = 45 - \phi/2 \rightarrow \beta = (3\pi/4 - \varphi/2)$.

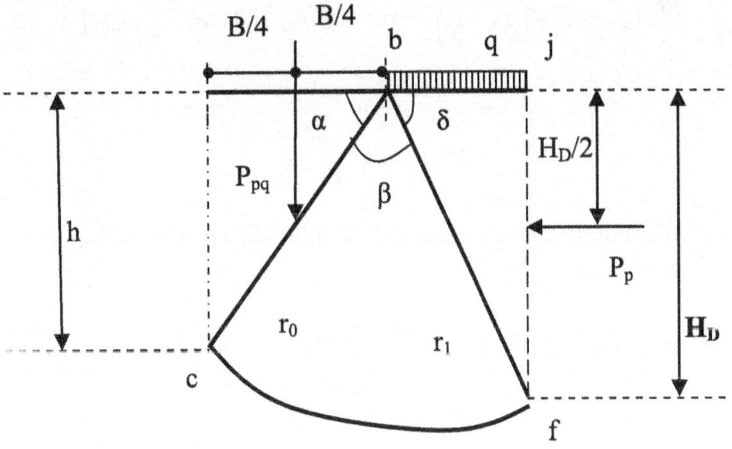

Fig. 6.2 Free body diagram of the soil under surcharge load including seismic force

From geometry of figure

$$r_1 = r_0 e^{\beta \tan \phi} \tag{6.3.1.1}$$

$$r_1 \cos(45 - \varphi/2) = r_0 e^{\beta \tan \phi} \cos(45 - \varphi/2) \tag{6.3.1.2}$$

$$H_D = r_0 e^{\beta \tan \phi} \sin(45 - \phi/2) \tag{6.3.1.3}$$

$$h = \frac{B}{2} \tan \phi, \quad r_0 = \frac{B}{2} \sec \phi \tag{6.3.1.4}$$

and

$$W = \frac{1}{2} \gamma_s Bh \tag{6.3.1.5}$$

Now taking moment of all forces about point b in Fig. 6.2, one gets

$$P_{pq} \times \frac{B}{4} - q\frac{(bj)^2}{2} - P_p \times \frac{H_D}{2} = 0 \tag{6.3.1.6}$$

In Eq. (6.3.1.6)

$$P_p = qK_p H_D \rightarrow qH_D \tan^2(45 + \varphi/2) \tag{6.3.1.7}$$

Now substituting Eqs. (6.3.1.2) to (6.3.1.5) and (6.3.1.7) in Eq. (6.3.1.6), one may get

$$P_{pq} \times \frac{B}{4} = \frac{qB^2}{8}\left[\sec^2\phi\cos^2\left(45 - \frac{\phi}{2}\right)e^{2\beta\tan\phi} + \sec^2\phi\,\sin^2\left(45 - \frac{\phi}{2}\right)\tan^2\left(45 + \frac{\phi}{2}\right)e^{2\beta}\right] \tag{6.3.1.8}$$

$$\rightarrow 2P_{pq} = qB\left[\frac{e^{2\beta\tan\phi}}{1 - \sin\phi}\right] \tag{6.3.1.9}$$

As shown in Fig. 6.3, equilibrium of the wedge abc under surcharge load can be expressed as

$$q_q B = 2P_{pq} \tag{6.3.1.10}$$

$$q_q B = qB\left[\frac{e^{2\beta\tan\phi}}{1 - \sin\phi}\right] \tag{6.3.1.11}$$

Considering $(1 - \sin\phi) = 2\cos^2(45 + \varphi/2)$, Eq. (6.3.1.11) can be finally expressed as

$$q_q = q\left[\frac{e^{2\left(\frac{3\pi}{4} - \frac{\phi}{2}\right)\tan\phi}}{2\cos^2(45 + \varphi/2)}\right] \tag{6.3.1.12}$$

$$\rightarrow q_q = qN_q \tag{6.3.1.13}$$

As per Terzaghi and Peck (1982), the term in bracket in Eq. (6.3.1.12) is the static bearing capacity factor N_q.

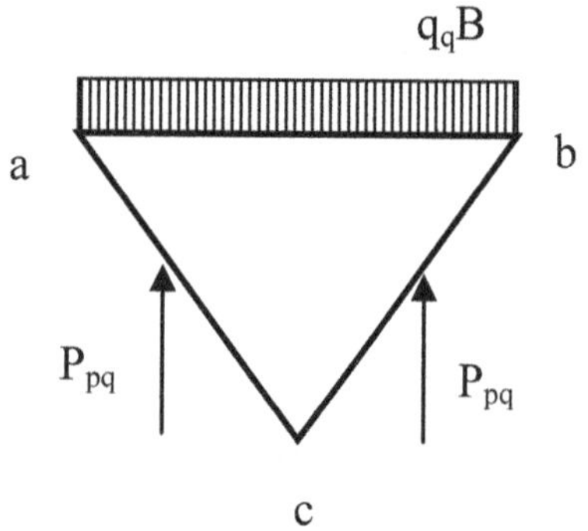

Fig. 6.3 Equilibrium of wedge under surcharge load

6.3.2 Determination of Passive Force P_{pc} for Cohesion c

Shown in Fig. 6.4 is the free body diagram of forces acting on the soil body under cohesive stress *c*.

Thus, taking moment of all forces about point b, we have

$$P_{pc} \times \frac{B}{4} = P_p \times \frac{H_D}{2} + M_c \tag{6.3.2.1}$$

In Eq. (6.3.2.1)

$$P_p = 2cH_D \tan(45 + \varphi/2) \tag{6.3.2.2}$$

and

$$M_c = \frac{c}{2\tan\phi\left(r_1^2 - r_0^2\right)} \tag{6.3.2.3}$$

Substituting Eqs. (6.3.2.2) and (6.3.2.3) in Eq. (6.3.2.1), we have

$$P_{pc} \times \frac{B}{4} = \frac{1}{2} c r_0^2 \cos\phi \times e^{2\beta\tan\phi} + \frac{c r_0^2}{2\tan\phi}\left(e^{2\beta\tan\phi} - 1\right) \tag{6.3.2.4}$$

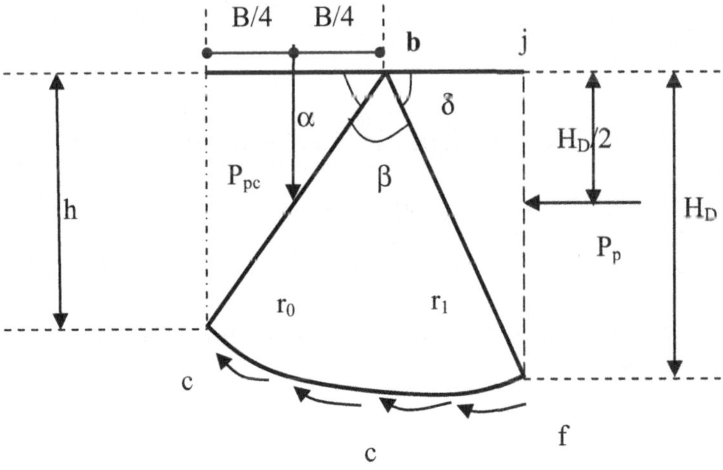

Fig. 6.4 Free body diagram of the soil under cohesive force including seismic force

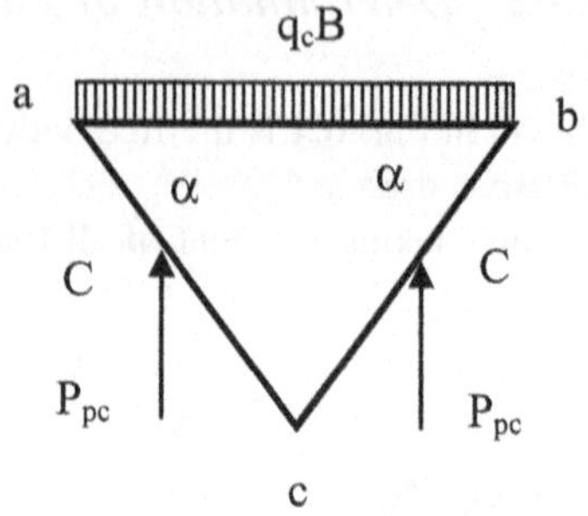

Fig. 6.5 Equilibrium of wedge under cohesive force

Substituting Eq. (6.3.1.4) above, we have

$$\rightarrow 2P_{\mathrm{pc}} = cB \sec\phi \times e^{2\beta \tan\phi} + \frac{cB}{2\tan\phi}\left(e^{2\beta\tan\phi} - 1\right)\sec^2\phi \qquad (6.3.2.5)$$

$$2P_{\mathrm{pc}} = cB\left[\sec\phi \times e^{2\beta\tan\phi} + \frac{\sec^2\phi}{2\tan\phi}\left(e^{2\beta\tan\phi} - 1\right)\right] \qquad (6.3.2.6)$$

Now considering equilibrium of the wedge as shown in Fig. 6.5, we have

$$q_{\mathrm{c}}B = 2C\sin\phi + 2P_{\mathrm{pc}} \qquad (6.3.2.6)$$

$$\rightarrow q_{\mathrm{c}}B = cB\sin\phi\sec\phi + cB\left[e^{2\beta\tan\phi}\sec\phi + \frac{\sec^2\phi}{\tan\phi}\left(e^{2\beta\tan\phi} - 1\right)\right] \qquad (6.3.2.7)$$

This finally gives

$$\begin{aligned} q_{\mathrm{c}} &= c\cot\phi\left[\frac{e^{2\left(\frac{3\pi}{4}-\frac{\phi}{2}\right)\tan\phi}}{2\cos^2(45+\varphi/2)} - 1\right] \\ &= c\cot\phi\left[N_q - 1\right] = cN_{\mathrm{c}} \end{aligned} \qquad (6.3.2.8)$$

6.3.3 *Determination of Passive Pressure Due to Weight of Soil* $P_{p\gamma}$

Figure 6.6 shows the free body diagram of the soil wedge bcfj. In this case, the center of log spiral (of which cf is an arc) lies at point O that lies somewhere along the line bf. For this, the minimum value of $P_{\mathrm{p}\gamma}$ has to be determined by several trials. The forces that need to considered here are

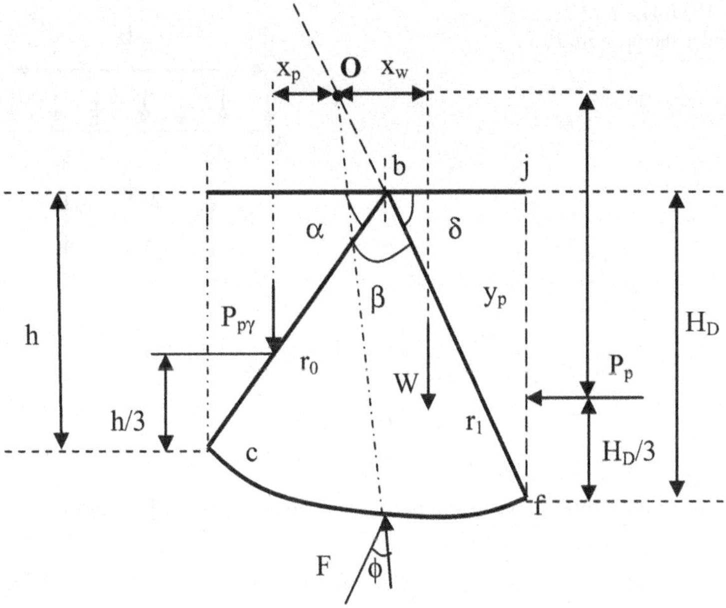

Fig. 6.6 Free body diagram of soil body under its own weight

- weight W of wedge bcfj;
- passive force $P_{p\gamma}$;
- Rankine's passive force P_p;
- friction force acting along arc cf and passing through O.

Taking moment of all the forces about O, we have

$$P_{p\gamma} \times x_p = W \times x_w + P_p \times y_p \tag{6.3.3.1}$$

$$\rightarrow P_{p\gamma} = \frac{1}{x_p}\left[W \times x_w + P_p \times y_p\right] \tag{6.3.3.2}$$

In Eq. (6.3.3.2)

$$P_p = \frac{1}{2}\gamma_s H_D^2 \tan^2\left(45 + \frac{\phi}{2}\right) \tag{6.3.3.3}$$

Considering the equilibrium of wedge as shown in Fig. 6.7, we have

$$q_\gamma B = 2P_{p\gamma} + W \tag{6.3.3.4}$$

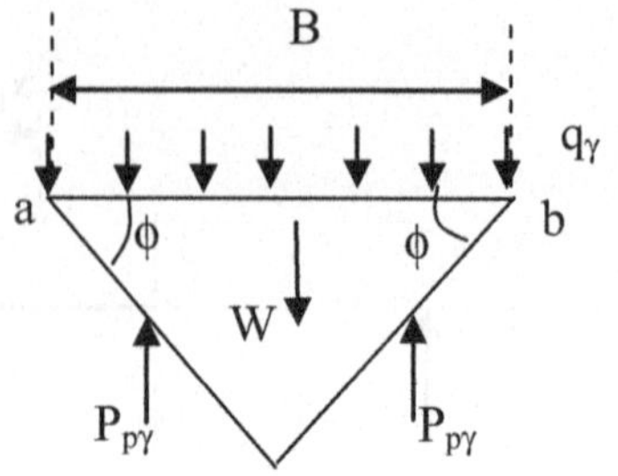

Fig. 6.7 Equilibrium of wedge under the pressure $P_{p\gamma}$

In Eq. (6.3.3.4),

$$W = \frac{B^2}{4}\gamma_s \tan\phi \tag{6.3.3.5}$$

The minimum value of $P_{p\gamma}$ obtained based on a number of trials is

$$P_{p\gamma} = \frac{1}{8}\gamma_s B^2 K_{p\gamma} \tan^2\phi \tag{6.3.3.6}$$

In Eq. (6.3.3.5) $K_{p\gamma}$ = coefficient of passive earth pressure.
Substituting Eqs. (6.3.3.5) and (6.3.3.6) in Eq. (6.3.3.4), we finally have

$$q_\gamma = \frac{1}{2}\gamma_s B\left[\frac{\tan\phi}{2}\left(\frac{K_{p\gamma}}{\cos^2\phi}\right) - 1\right] \to \frac{1}{2}\gamma_s B N_\gamma \tag{6.3.3.7}$$

Now based on superimposition which was assumed to be valid for the above

$$q_u = q_q + q_c + q_\gamma = qN_q + cN_c + \frac{1}{2}\gamma_s B N_\gamma \tag{6.3.3.8}$$

Equation (6.3.3.8) is Terzaghi's expression of ultimate bearing capacity for a strip footing subjected to a vertical load.

Terzaghi modified the above for *square footing* of size $B \times B$ as

$$q_u = qN_q + 1.3cN_c + 0.4\gamma_s B N_\gamma \tag{6.3.3.9}$$

For a *circular footing* of diameter B, the ultimate stress is expressed as

$$q_u = qN_q + 1.3cN_c + 0.3\gamma_s B N_\gamma \tag{6.3.3.10}$$

From the above derivation, it is observed that despite the problem being non-linear in nature, Terzaghi computed the ultimate bearing capacity by superimposing the values q_q, q_c, and q_g, as shown in Eq. (6.3.3.8).

Table 6.1 Comparative value of bearing capacity (q_u) by superimposition and coupled analysis in kN/m^2 (after Manoharan and Dasgupta 1995)

Friction angle	Superposition method		Coupled FEM (Manoharan and Dasgupta)	
	Terzaghi	Chen	Associated	Non-associated
0	139	127.8	139.5	–
10	271.5	251.8	281.3	277.7
20	623.5	574.2	636.6	609.5
25	1001.0	929.1	1020.0	939.0
30	1715.0	1597.3	1713.3	1501.6

The validity of the assumption of superimposition has been investigated by a number of researchers, like Griffith (1982), Chen and Davidson (1973), Manoharan and Dasgupta (1995). While Griffith used nonlinear finite element (FEM) to arrive at the ultimate bearing capacity of the soil, Chen and Davidson used limit equilibrium theory, and Manoharan and Dasgupta used FEM adapting viscoelastic theory with associated and non-associated flow rule.

All these researches show that Terzaghi's assumption of superimposition is slightly conservative and the end result of ultimate bearing capacity (q_u) varies by 3–12%. Considering that we divide the value of q_u by an adequate factor of safety, the net allowable bearing capacity has negligible error and can well be adapted for practical engineering problem while keeping the mathematical derivation simplified.

Manoharan and Dasgupta (1995) compared the q_u for different values of ϕ (angle of friction of soil), and the same are presented hereafter for ready reference in Table 6.1.

The values clearly show that the variation is not large and for practical engineering problems can be well adopted without any significant errors.

Choudhury and Rao (2005) have also done a comparison on the bearing capacity computed by different researchers, and the conclusions they have drawn are identical in nature.

6.4 The Mechanics of Equilibrium of a Surface Footing Under Earthquake Force

When an earthquake hits a site, as shown in Fig. 6.8, the soil starts vibrating along with the foundation in different modes ϕ_1, ϕ_2, ϕ_3, etc. This generates an inertial force in soil expressed as $\alpha_h W$ in horizontal direction and $\alpha_v W$ in vertical direction within the soil that modifies the equilibrium condition of the wedge abc below the foundation.

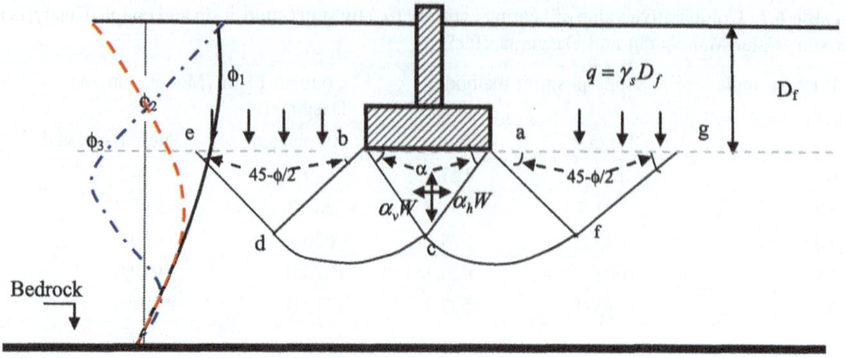

Fig. 6.8 Soil behaviour under an earthquake force

It is apparent that under this force, seeking the equilibrium condition of the wedge abc will modify the values of N_c, N_q, and N_γ compared to their values as derived for the static case.

Based on the above idea, a number of researchers have tried to develop the *dynamic bearing capacity factors* (N_{cd}, N_{qd}, $N_{\gamma d}$) under seismic loading, endeavoring to establish a realistic mathematical model for estimation of ultimate bearing capacity of a foundation under seismic loading.

Meyerhoff (1953, 1963) was one of the earliest, who attempted to develop the bearing capacity factors considering lateral force generated due to earthquake load. However, in his work, effect of seismic acceleration on inertia of the supporting soil was ignored.

This was followed by a number of researchers like Sarma and Iossifelis (1990), Richard et al. (1993), Budhu and Al Karni (1993), who all addressed the problem by considering the seismic force on superstructure and supporting soil mass.

Others, namely Dormieux and Pecker (1995), Soubra (1997, 1999), Kumar and Rao (2002), Choudhury and Rao (2005), have all used different techniques like upper bound limit equilibrium theory, stress characteristics to compute the dynamic bearing capacity factors for the soil.

All these researchers point out the fact that, under seismic force, the bearing capacity factors are something different than what is computed under static force—despite suffering from three significant lacunae like

- The seismic force is considered as pseudo-static in nature.
- The acceleration remains invariant with depth.
- The specific value of seismic coefficients α_h and α_v needs to be considered for the foundation design—which has remained somewhat undefined, when many use the same values as that of the superstructure.

Pecker (1996) points out that considering same seismic coefficient for the structure and foundation is not rational and could provide a misleading picture.

This has also been pointed out by Chowdhury and Tilak (2012) where, considering the free field time period of the site and variation in acceleration along the depth to the bedrock, it was shown that the depth of bedrock level strongly affects the dynamic bearing capacity of the soil.

One common expression deployed by geotechnical engineers to determine the peak ground acceleration is after Richter and Gutenberg (1956) and is expressed as

$$\log_{10} a_0 = -2.1 + 0.81M - 0.027M^2 \tag{6.3.1}$$

In Eq. (6.3.1), a_0 = peak ground acceleration in cm/s^2 and M is the moment magnitude of earthquake in Richter scale.

The seismic coefficient $\alpha_h = a_0/g$ and α_v (coefficient of vertical acceleration) are usually considered as (1/2 to 2/3)α_h. It is apparent in this case that the acceleration is *pseudo-static* in nature.

Of all the research available, we present hereafter two mathematical models that are easy to apply and amenable to design office application.

6.4.1 Model for Dynamic Bearing Capacity Factors Under Seismic Load

Richard et al. (1993) developed a bearing capacity theory for a strip footing resting on cohesionless soil under seismic load based on Coulomb's active and passive earth pressure theory.

The theory, unlike Terzaghi or Meyerhof, assumes a simplified failure profile of the soil, as shown in Fig. 6.9.

As per Richard et al. (1993), the angle α_A made by wedge I (vide Fig. 6.9a) with horizontal is expressed as

$$\alpha_A = \phi + \tan^{-1}\left\{\frac{\tan\phi(\tan\phi + \cot\phi)\sqrt{(1 + \tan\delta\cot\phi)} - \tan\phi}{1 + \tan\delta(\tan\phi + \cot\phi)}\right\} \tag{6.4.1.1}$$

Similarly for wedge II

$$\alpha_P = -\phi + \tan^{-1}\left\{\frac{\tan\phi(\tan\phi + \cot\phi)\sqrt{(1 + \tan\delta\cot\phi)} - \tan\phi}{1 + \tan\delta(\tan\phi + \cot\phi)}\right\} \tag{6.4.1.2}$$

In Eqs. (6.4.1.1) and (6.4.1.2) ϕ = soil friction angle and δ = friction along the plane BC.

The equilibrium of the soil wedge is as shown in Fig. 6.9b where

P_A Coulomb's active pressure;
P_P Coulomb's passive pressure;

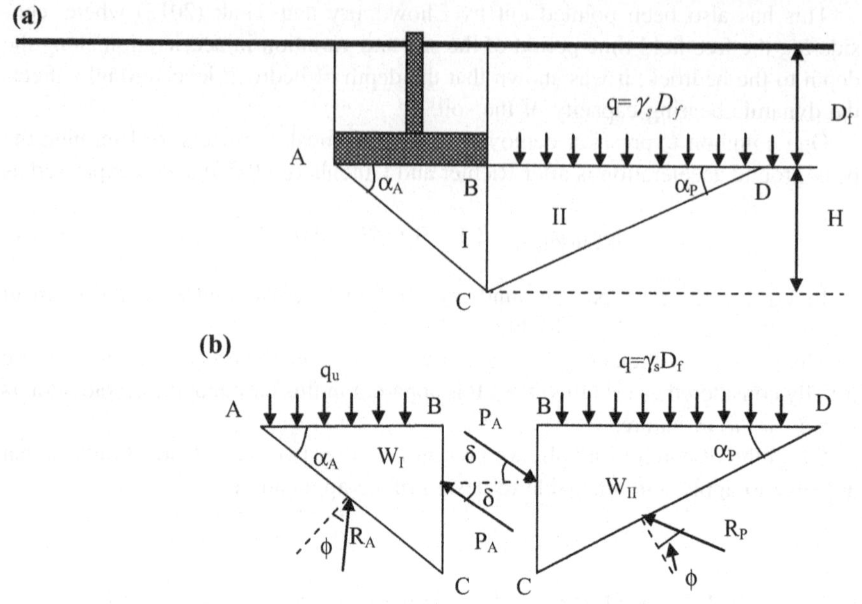

Fig. 6.9 Richards et al. (1993) model for general shear failure of soil

R_A resultant force (of shear and normal force) along plane AC;
R_P resultant force along CD;
W_I weight of soil wedge I; and
W_{II} weight of soil wedge II.

For $\phi \neq 0, q \neq 0$ and $\gamma_s = 0$ we have

$$q_u = q'_u \tag{6.4.1.3}$$

and

$$P_A \cos \delta = P_P \cos \delta \tag{6.4.1.4}$$

but

$$P_A \cos \delta = q'_u K_A H \tag{6.4.1.5}$$

Here, K_{A} = horizontal component of Coulomb's active earth pressure where

$$K_{\mathrm{A}} = \frac{\cos^2\varphi}{\cos\delta\left[1+\sqrt{\frac{\sin(\phi+\delta)\sin\phi}{\cos\delta}}\right]^2} \tag{6.4.1.6}$$

Similarly $P_{\mathrm{P}}\cos\delta = q'_{\mathrm{u}}K_{\mathrm{P}}H$ where

K_{P} horizontal component of Coulomb's passive earth pressure where

$$K_{\mathrm{P}} = \frac{\cos^2\varphi}{\cos\delta\left[1-\sqrt{\frac{\sin(\phi-\delta)\sin\phi}{\cos\delta}}\right]^2}. \tag{6.4.1.7}$$

Combing Equations from (6.4.1.3) to (6.4.1.7), we finally have

$$q_{\mathrm{u}} = q\left(\frac{K_{\mathrm{P}}}{K_{\mathrm{A}}}\right) \rightarrow qN_{\mathrm{q}}. \tag{6.4.1.8}$$

In Eq. (6.4.1.8), N_{q} is the bearing capacity factor.
Now for $\phi \neq 0, q = 0$ and $\gamma_{\mathrm{s}} \neq 0$, we have $q_{\mathrm{u}} = q''_{\mathrm{u}}$ and

$$P_{\mathrm{A}}\cos\delta = q''_{\mathrm{u}}K_{\mathrm{A}}H + \frac{1}{2}K_{\mathrm{A}}\gamma_{\mathrm{s}}H^2 \tag{6.4.1.9}$$

Also

$$P_P\cos\delta = \frac{1}{2}K_P\gamma_{\mathrm{s}}H^2 \tag{6.4.1.10}$$

For equilibrium

$$\begin{aligned} &q''_{\mathrm{u}}K_{\mathrm{A}}H + \frac{1}{2}K_{\mathrm{A}}\gamma_{\mathrm{s}}H^2 - \frac{1}{2}K_{\mathrm{P}}\gamma_{\mathrm{s}}H^2 \\ &\rightarrow q''_{\mathrm{u}} = \frac{\gamma_{\mathrm{s}}H}{2}\left(\frac{K_{\mathrm{P}}-K_{\mathrm{A}}}{K_{\mathrm{A}}}\right) = \frac{\gamma_{\mathrm{s}}H}{2}\left(\frac{K_{\mathrm{P}}}{K_{\mathrm{A}}}-1\right) \end{aligned} \tag{6.4.1.11}$$

Now, as $H = B\tan\alpha_{\mathrm{A}}$, we have

$$\rightarrow q''_{\mathrm{u}} == \frac{\gamma_{\mathrm{s}}B\tan\alpha_A}{2}\left(\frac{K_{\mathrm{P}}}{K_{\mathrm{A}}}-1\right) \rightarrow \frac{1}{2}\gamma_{\mathrm{s}}BN_{\gamma}$$

where,

$$N_\gamma = \tan\alpha_A\left(\frac{K_P}{K_A} - 1\right) \tag{6.4.1.12}$$

Now superimposing the above two values for granular soil, we have

$$q_u = qN_q + \frac{1}{2}\gamma_s BN_\gamma \tag{6.4.1.13}$$

where N_q and N_γ are depicted by Eqs. (6.4.1.8) and (6.4.1.13).

For computation of the bearing capacity factors, Richard recommends to consider $\delta = \phi/2$.

For pure clayey soil when $\phi = 0$, it can be shown that bearing capacity factor converges to the value of $N_c = 6.0$. For this, following other researchers (for, e.g., Meyerhoff, Brinch Hansen), Richard recommends

$$N_c = (N_q - 1)\cot\phi \tag{6.4.1.14}$$

The values computed are as shown in Table 6.2 vis-à-vis values computed by other researchers.

Thus, the ultimate bearing capacity can be finally expressed as

$$q_u = cN_c + qN_q + \frac{1}{2}\gamma_s BN_\gamma \tag{6.4.1.15}$$

For earthquake case, the free body diagram is as shown in Fig. 6.9b and gets modified as shown hereafter.

In Fig. 6.10, the angles α_{AE} and α_{PE} are angles made by the soil wedge with horizontal under Coulomb's active and passive earth pressure condition. Here

$$\alpha_{AE} = -\alpha + \tan^{-1}\left[\frac{\sqrt{(1+\tan^2\alpha)\{1+\tan(\delta+\theta)\cot\alpha\}} - \tan\alpha}{1+\tan(\delta+\theta)(\tan\alpha + \cot\alpha)}\right] \tag{6.4.1.16}$$

Table 6.2 Bearing capacity factors as derived by Richard et al. vis-à-vis other researchers

ϕ	Richard et al. (1993)			Terzaghi (1943)			Meyerhof (1974)		
	N_c	N_q	N_γ	N_c	N_q	N_γ	N_c	N_q	N_γ
0	6	1	0	5.7	1	0	5.14	1	0
10	7.77	2.37	1.38	9.6	2.7	1.2	8.35	2.47	0.4
20	13.46	5.9	6.06	17.7	7.4	4.0	14.8	6.4	3.95
30	26.86	16.51	23.76	37.2	22.5	19.7	30.1	18.4	15.7
40	58.43	59.04	111.9	95.7	81.4	100.4	75.3	64.2	93.7

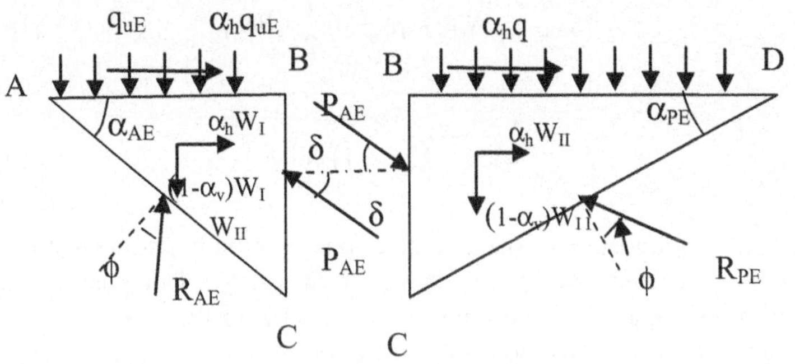

Fig. 6.10 Equilibrium condition of soil under seismic force

In Eq. (6.4.1.16)

$$\alpha = \phi - \theta,\ \theta = \tan^{-1}\frac{\alpha_h}{1-\alpha_v},$$

α_h = coefficient of horizontal seismic acceleration, and α_v = coefficient of vertical seismic acceleration.

Proceeding in the same way as that for the static case, it can be shown that

$$N_{qE} = \frac{K_{PE}}{K_{AE}} \tag{6.4.1.17}$$

and

$$N_{\gamma E} = \tan\alpha_{AE}\left(\frac{K_{PE}}{K_{AE}} - 1\right) \tag{6.4.1.18}$$

where

$$K_{AE} = \frac{\cos^2(\phi-\theta)}{\cos\theta\cos(\delta+\theta)\left[1+\sqrt{\frac{\sin(\phi+\delta)\sin(\phi-\theta)}{\cos(\delta+\theta)}}\right]^2}; \tag{6.4.1.19}$$

$$K_{PE} = \frac{\cos^2[\phi-\theta]}{\cos\theta\cos(\delta+\theta)\left[1-\sqrt{\frac{\sin(\phi+\delta)\sin(\phi-\theta)}{\cos(\delta+\theta)}}\right]^2}. \tag{6.4.1.20}$$

As cited earlier, considering that $\delta = \phi/2$, the values of K_{AE} and K_{PE} can be computed to subsequently derive the values of N_{qE} and $N_{\gamma E}$, when the ultimate bearing capacity of a strip footing under earthquake load can finally be computed from the expression:

$$q_{\mathrm{uE}} = cN_{\mathrm{cE}} + qN_{\mathrm{qE}} + \frac{1}{2}\gamma_{\mathrm{s}}BN_{\gamma\mathrm{E}}, \tag{6.4.1.21}$$

when

$$N_{\mathrm{cE}} = \left(N_{\mathrm{qE}} - 1\right)\cot\phi. \tag{6.4.1.22}$$

Example 6.1 A foundation of size 2 m × 2 m is to be built in a site that is expected to experience an earthquake of $M = 6.8$ in Richter scale. Compute the static and seismic bearing capacity of the foundation considering the soil properties $c = 30$ kN/m^2, $\phi = 10°$. The foundation is expected to be located at 1.5 m below ground level. Density of soil = 18 kN/m^3.

Solution

We compute the static bearing capacity based on Terzaghi's as well as Richard's expression keeping the shaper factors as cited in Eq. (6.3.3.9).

Computation of ultimate bearing capacity based on Terzaghi's solution:
For $\phi = 10°$, $N_{\mathrm{c}} = 9.6$, $N_{\mathrm{q}} = 2.7$, $N_{\gamma} = 1.3$
Now considering $q_{\mathrm{u}} = qN_{\mathrm{q}} + 1.3cN_{\mathrm{c}} + 0.4\gamma_{\mathrm{s}}BN_{\gamma}$

$$q_{\mathrm{u}} = 18 \times 1.5 \times 2.7 + 1.3 \times 30 \times 9.7 + 0.4 \times 18 \times 2 \times 1.3 = 469.92$$
$$\cong 470\,\mathrm{kN/m^2}, \text{say}.$$

Computation based on Richard's equation
For $\phi = 10°$, $N_{\mathrm{c}} = 7.77$, $N_q = 2.37$, $N_{\gamma} = 1.38$
Now considering $q_{\mathrm{u}} = qN_{\mathrm{q}} + 1.3cN_{\mathrm{c}} + 0.4\gamma_{\mathrm{s}}BN_{\gamma}$

$$q_{\mathrm{u}} = 18 \times 1.5 \times 2.37 + 1.3 \times 30 \times 7.77 + 0.4 \times 18 \times 2 \times 1.38 = 386.89$$
$$\cong 387\,\mathrm{kN/m^2}, \text{ say}.$$

Seismic ultimate bearing capacity of foundation
Considering $\log_{10} a_0 = -2.1 + 0.81M - 0.027M^2$ where $M = 6.8$ gives $a_0 = 144.38$ cm/s^2. Thus $\alpha_{\mathrm{h}} = \frac{144.38}{981} = 0.15$ and $\alpha_{\mathrm{v}} = \frac{2}{3}\alpha_{\mathrm{v}} = 0.1$

$$\theta = \tan^{-1}\frac{\alpha_h}{1 - \alpha_{\mathrm{v}}} = \tan^{-1}\frac{0.15}{1 - 0.1} = 9.46°$$

$$\alpha = \phi - \theta = 0.538°.$$

Now considering $\delta = 5°$, i.e., $\phi/2$

$$\alpha_{AE} = -\alpha + \tan^{-1}\left[\frac{\sqrt{(1+\tan^2\alpha)\{1+\tan(\delta+\theta)\cot\alpha\}} - \tan\alpha}{1+\tan(\delta+\theta)(\tan\alpha+\cot\alpha)}\right]$$

$$\alpha_{AE} = -0.538 + \tan^{-1}\left[\frac{\sqrt{(1+\tan^2 0.538)\{1+\tan(14.46)\cot 0.538\}} - \tan 0.538}{1+\tan(14.46)(\tan 0.538+\cot 0.538)}\right] = 10.05°$$

Now

$$K_{AE} = \frac{\cos^2(\phi-\theta)}{\cos\theta\,\cos(\delta+\theta)\left[1+\sqrt{\frac{\sin(\phi+\delta)\sin(\phi-\theta)}{\cos(\delta+\theta)}}\right]^2}$$

$$K_{AE} = \frac{\cos^2(0.538)}{\cos 9.46\,\cos 14.46\left[1+\sqrt{\frac{\sin(15)\sin(0.538)}{\cos(14.46)}}\right]^2} = 0.95$$

and

$$K_{PE} = \frac{\cos^2[\phi-\theta]}{\cos\theta\,\cos(\delta+\theta)\left[1-\sqrt{\frac{\sin(\phi+\delta)\sin(\phi-\theta)}{\cos(\delta+\theta)}}\right]^2}$$

$$\rightarrow K_{PE} = \frac{\cos^2[0.538]}{\cos 9.46\cos 14.46\left[1-\sqrt{\frac{\sin(15)\sin(0.538)}{\cos(14.46)}}\right]^2} = 1.160$$

Considering $N_{qE} = \frac{K_{PE}}{K_{AE}} = \frac{1.160}{0.95} = 1.22$ and

$$N_{\gamma E} == \tan\alpha_{AE}\left(\frac{K_{PE}}{K_{AE}} - 1\right) = \tan 10.05 \times \left(\frac{1.16}{0.95} - 1\right) = 0.04$$

$$N_{cE} = \left(N_{qE} - 1\right)\cot\phi = (1.22 - 1)\cot 10 = 1.247$$

Now considering

$$q_{uE} = qN_{qE} + 1.3cN_{cE} + 0.4\gamma_s BN_{\gamma E} \text{ we have}$$
$$q_{uE} = 18 \times 1.5 \times 1.22 + 1.3 \times 30 \times 1.247 + 0.4 \times 18 \times 2 \times 0.04 = 82.149 \cong 82\,\text{kN/m}^2.$$

Thus, it is observed that there is a drastic reduction in bearing capacity when under service load, i.e., considering a FOS = 3.0 and designing the foundation for the same, the foundation will yet undergo failure under seismic force.

6.4.2 Some Comments on the Above Method

Based on the above analysis, it is apparent that the bearing capacity factors do not remain same as that under static load, and, under worst condition, can significantly reduce the load carrying capacity of the soil.

Thus, increasing the allowable bearing capacity by 25% or more, as suggested by a number of codes around the world, may not be an apt decision and can well aggravate the failure.

However, the method as proposed suffers from three serious deficiencies that can be summarized as hereafter.

1. The computation of α_h and α_v considered is pseudo-static in nature.
2. The linear failure profile is assumed under ultimate load which is certainly an oversimplification of the problem.
3. Deriving the ultimate bearing capacity based on reduced bearing capacity factors N_{cd}, N_{qd}, and $N_{\gamma d}$.

Considering uniform peak ground acceleration as depicted above is not realistic. Namely, based on the theory of soil dynamics, it can be shown (e.g., refer Chap. 3, Sect. 3.4.3) that acceleration induced in the soil is dependent on a number of factors like mode of vibration, material damping of soil, bedrock level, sundry, and other factors and is a variable parameter and needs careful evaluation.

Considering a linear failure profile is an oversimplification, for the experimental evidence shows that the failure profile is actually a log spiral curve.

For this, the bearing capacity factors under static load match reasonably well with Terzaghi's and Meyerhof's up to $\phi = 30°$; however, beyond this, the variation is quite significant, as seen from Table 6.2.

Apparently, reducing the static bearing capacity due to seismic force looks attractive but is possibly not logical. The reason for this has been explained hereafter.

The ultimate bearing capacity is a state when the soil undergoes failure. Thus, it is the ultimate state in soil, which is never actually mobilized (unless the design engineer heavily fouls up somewhere).

During service condition of the foundation, this ultimate bearing capacity at failure is divided by an adequate factor of safety, when the foundation's stress is restricted to this serviceability limit.

Thus, when we compute the dynamic bearing capacity factors N_{cd}, N_{qd}, and $N_{\gamma d}$, we will compute the bearing capacity of soil based on Eq. (6.3.1.2.1) as stated hereafter as

$$q_{aE} = \frac{q_{uE}}{FOS} \tag{6.4.2.1}$$

Here, q_{aE} = allowable bearing capacity of the foundation under seismic load.

In this process, if one considers Richard's technique, the effect of earthquake force is also divided by the factor of safety (FOS) which is not conceptually correct *as the earthquake force is induced under the service load condition of the foundation and not under failure.*

Equation (6.4.2.1) can thus seriously underrate the earthquake effect, if we pursue this line of *reduction of dynamic bearing capacity factor* (followed by Richard and a number of other researchers).

Nonetheless, as a pioneering effort on this particular topic, Richard's method remains an excellent research work in the realms of geotechnical earthquake engineering.

6.5 A Proposed Model to Compute Bearing Capacity of Foundation Under Seismic Force

In this section, we present a model (Chowdhury and Dasgupta 2016) that overcomes a number of lacunae, which, as cited in the earlier section, have been evaded.

To start with, the original Terzaghi's derivation is extended with additional seismic load for a strip foundation supported on soil which is considered homogenous over a large depth.

Figure 6.11 is a foundation located at a depth D_f below ground level. The depth of bedrock (i.e., the depth where shear wave velocity $V_s \geq 600$ m/s or SPT value $N \geq 50$) is at a depth H from top of soil.

Thus, under the effect of seismic force, the soil will undergo free field vibration when free field time period can be expressed as (Kramer 1996),

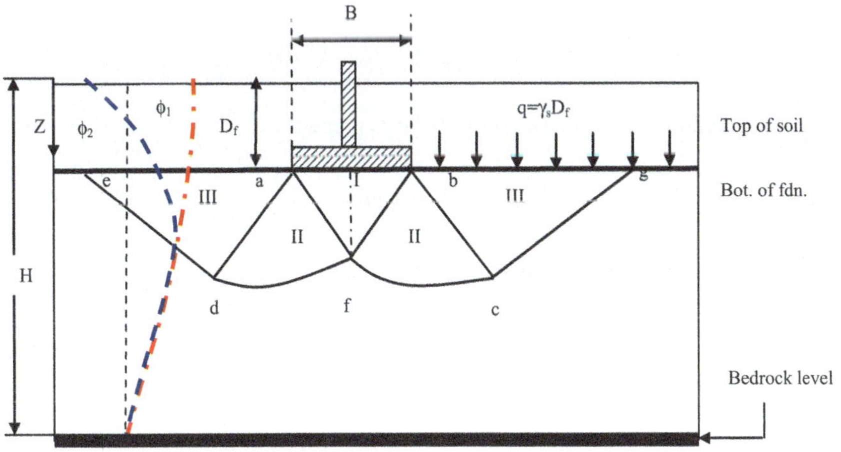

Fig. 6.11 Strip foundation resting on soil subjected to earthquake force

$$T_n = \frac{4H}{(2n-1)V_s} \tag{6.5.1}$$

Then, the site will start vibrating, having deformation modes $\phi_1, \phi_2, \ldots, \phi_n$, etc., where eigenvector shapes can be expressed as $\phi_n = \cos(2n-1)\pi z/2H$. Here, n = mode number → 1, 2, 3, …, ∞.

Thus, as per modal analysis, the amplitude of vibration for the nth mode can be expressed as

$$S_{dn} = \kappa_n C_f \frac{S_{an}}{\omega_n^2} \tag{6.5.2}$$

Here, S_{dn} = peak spectral amplitude. S_{an} is the spectral acceleration obtained from Fig. 2.5 of Chap. 2 which is related to the time period, as expressed by Eq. (6.5.1).

C_f = code factor, expressed as $C_f = ZI/2R$, here Z = zone factor as cited in Table 2.4 of Chap. 2.

I = importance factor, R = response reduction factor and can be considered as R = 2.0 for stiff clayey soil having high plasticity index and R = 1–1.5 for loose-to-medium cohesionless soil.

The modal participation factor κ_n can be expressed as shown in Table 6.3.

The amplitude of vibration that varies with depth z can be expressed as

$$u_{zn} = \kappa_n C_f \frac{S_{an}}{\omega_n^2} \cos(2n-1)\frac{\pi z}{2H} \tag{6.5.3}$$

Thus, at foundation depth $z = D_f$, the amplitude is expressed as

$$u_{zn} = \kappa_n C_f \frac{S_{an}}{\omega_n^2} \cos(2n-1)\frac{\pi D_f}{2H} \tag{6.5.4}$$

The acceleration at depth D_f can thus be expressed as

$$\ddot{u}_{zn} = \kappa_n C_f S_{an} \cos(2n-1)\frac{\pi D_f}{2H} \tag{6.5.5}$$

And coefficient of lateral seismic force α_{hn} can be expressed as

$$\alpha_{hn} = \kappa_n C_f \cos\left[(2n-1)\frac{\pi D_f}{2H}\right]\left(\frac{S_{an}}{g}\right) \tag{6.5.6}$$

Table 6.3 Modal participation factor of soil vibration (Chowdhury and Dasgupta 2015)

Modal participation factor	Mode 1	Mode 2	Mode 2
κ_n	$8/(\pi+2)$	$8/(3\pi-2)$	$8/(5\pi+2)$

As suggested in IS-1893 (2016), the vertical component of acceleration coefficient is now expressed as

$$\alpha_{vn} = \frac{2}{3}\alpha_{hn} \tag{6.5.7}$$

6.5.1 Determination of Passive Force P_{pq} Under Surcharge Load

The free body diagram of the soil below the foundation under the effect of surcharge load is as shown in Fig. 6.12.

As elaborated in Fig. 6.12, $\alpha = \phi$, $\delta = 45 - \phi/2 \rightarrow \beta = \frac{3\pi}{4} - \frac{\phi}{2}$.

From geometry of Fig. 6.12

$$r_1 = r_0 e^{\beta \tan\phi} \tag{6.5.1.1}$$

$$\mathrm{bj} = r_1 \cos\left(45 - \frac{\phi}{2}\right) = r_0 e^{\beta \tan\phi} \cos\left(45 - \frac{\phi}{2}\right) \tag{6.5.1.2}$$

$$H_D = r_0 e^{\beta \tan\phi} \sin\left(45 - \frac{\phi}{2}\right) \tag{6.5.1.3}$$

$$h = \frac{B}{2}\tan\phi, \quad r_0 = \frac{B}{2}\sec\phi \tag{6.5.1.4}$$

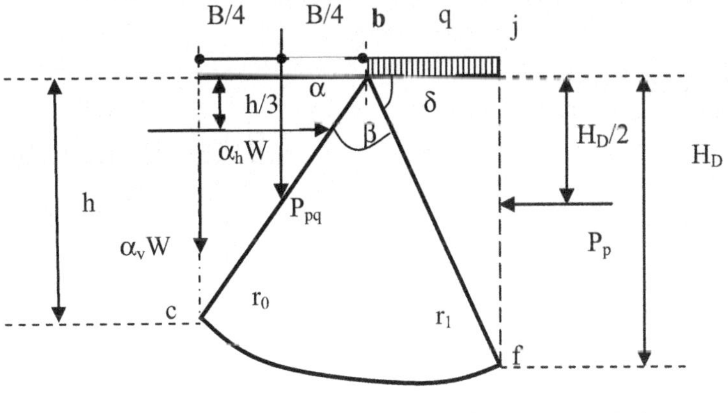

Fig. 6.12 Free body diagram of soil under surcharge load including seismic force

and

$$W = \frac{1}{2}\gamma_s B h \tag{6.5.1.5}$$

Now, taking the moment of all forces about point b in Fig. 6.12, one gets:

$$P_{pq} \times \frac{B}{4} \pm \alpha_h W \times \frac{h}{3} \pm \alpha_v W \times \frac{B}{2} - q\frac{(bj)^2}{2} - P_p \times \frac{H_D}{2} = 0 \tag{6.5.1.6}$$

In Eq. (6.5.1.6)

$$P_p = qK_p H_D \rightarrow qH_D \tan^2\left(45 + \frac{\phi}{2}\right) \tag{6.5.1.7}$$

Now substituting Eqs. (6.5.1.2) to (6.5.1.5) and (6.5.1.7) in Eq. (6.5.1.6), one gets

$$P_{pq} \times \frac{B}{4} = \frac{qB^2}{8}\left[\begin{array}{l}\sec^2\phi\cos^2\left(45 - \frac{\phi}{2}\right)e^{2\beta\tan\phi} + \sec^2\phi\sin^2\left(45 - \frac{\phi}{2}\right)\tan^2\left(45 + \frac{\phi}{2}\right)e^{2\beta\tan\phi} \\ \pm \frac{\alpha_h \gamma_s B}{3q}\tan^2\phi \pm \frac{\alpha_v \gamma_s B}{q}\tan\phi\end{array}\right] \tag{6.5.1.8}$$

$$\rightarrow 2P_{pq} = qB\left[\frac{e^{2\beta\tan\phi}}{1 - \sin\phi} - \frac{\alpha_h}{3}\left(\frac{\gamma_s B}{q}\right)\tan^2\phi \pm \frac{2\alpha_h}{3}\left(\frac{\gamma_s B}{q}\right)\tan\phi\right] \tag{6.5.1.9}$$

As shown in Fig. 6.13, the equilibrium of the wedge abc under surcharge load and vertical seismic force can be expressed as

$$q_q B \pm \alpha_v W = 2P_{pq} \tag{6.5.1.10}$$

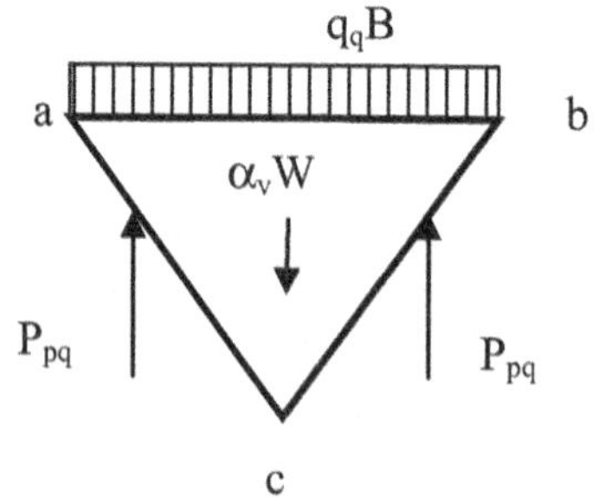

Fig. 6.13 Equilibrium of wedge under a surcharge load

$$q_q B = qB\left[\frac{e^{2\beta \tan\phi}}{1-\sin\phi} \pm \frac{\alpha_{hn}}{3}\left(\frac{\gamma_s B}{q}\right)\tan^2\phi \pm \frac{2\alpha_{hn}}{3}\left(\frac{\gamma_s B}{q}\right)\tan\phi\right] \pm \frac{\alpha_{hn}}{6}\gamma_s B^2 \tan\phi \tag{6.5.1.11}$$

Considering $1-\sin\phi = 2\cos^2\left(45+\frac{\phi}{2}\right)$, Eq. (6.5.1.11) can be finally expressed as

$$q_q = q\left[\frac{e^{2\left(\frac{3\pi}{4}-\frac{\phi}{2}\right)\tan\phi}}{2\cos^2\left(45+\frac{\phi}{2}\right)} \pm \alpha_{hn}\left(\frac{\gamma_s B}{q}\right)\left(\frac{1}{3}\tan^2\phi + \frac{5}{6}\tan\phi\right)\right] \tag{6.5.1.12}$$

$$\rightarrow q_q = qN_{qd} \tag{6.5.1.13}$$

Thus

$$N_{qd} = \left[\frac{e^{2\left(\frac{3\pi}{4}-\frac{\phi}{2}\right)\tan\phi}}{2\cos^2\left(45+\frac{\phi}{2}\right)} \pm \alpha_{hn}\left(\frac{\gamma_s B}{q}\right)\left(\frac{1}{3}\tan^2\phi + \frac{5}{6}\tan\phi\right)\right] \tag{6.5.1.14}$$

As per Terzaghi and Peck (1982), the first term in Eq. (6.5.1.14) is the static bearing capacity factor N_q. Hence,

$$N_{qd} = \left[N_q \pm \alpha_{hn}\left(\frac{\gamma_s B}{q}\right)\left(\frac{1}{3}\tan^2\phi + \frac{5}{6}\tan\phi\right)\right] \tag{6.5.1.15}$$

Thus, from Eq. (6.5.1.15) we see that the static bearing capacity factor under worst combination of load gets alternatively enhanced or reduced by a factor (depending on the vector direction) represented by the second expression in Eq. (6.5.1.15) and will vary with modal deformation of soil, where n is the mode number.

Equation (6.5.1.15) can be finally expressed as

$$N_{qd} = \left[N_q \pm \alpha_{hn}\lambda_q\left(\frac{1}{3}\tan^2\phi + \frac{5}{6}\tan\phi\right)\right] \tag{6.5.1.16}$$

where $\lambda_q = B/D_f$ a dimensionless number.

6.5.2 Determination of Passive Force P_{pc} for Cohesion C

Shown in Fig. 6.14 is the free body diagram of forces acting on the soil body under cohesive stress c.

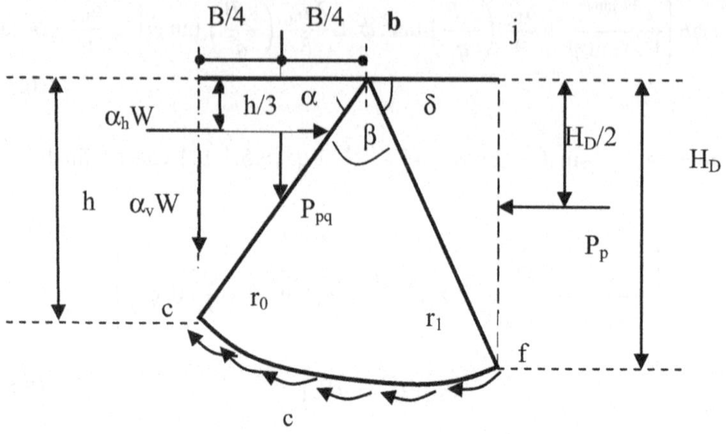

Fig. 6.14 Free body diagram of the soil under cohesive force including seismic force

Thus, taking moment of all forces about point b, we have

$$P_{pc} \times \frac{B}{4} + \alpha_v W \times \frac{B}{2} \pm \alpha_h W \times \frac{h}{3} = P_p \times \frac{H_D}{2} + M_c \tag{6.5.2.1}$$

In Eq. (6.5.2.1)

$$P_p = 2cH_D \tan(45 + \phi/2) \tag{6.5.2.2}$$

And

$$M_c = \frac{c}{2\tan\phi\left(r_1^2 - r_0^2\right)} \tag{6.5.2.3}$$

Substituting Eq. (6.5.2.2) and (6.5.2.3) in Eq. (6.5.2.1), we have

$$\begin{aligned} P_{pc} \times \frac{B}{4} = \frac{1}{2} c r_0^2 \cos\phi \times e^{2\beta\tan\phi} + \frac{c r_0^2}{2\tan\phi}\left(e^{2\beta\tan\phi} - 1\right) \\ \pm \frac{\alpha_v}{4}\gamma_s B^2 r_0 \sin\phi \pm \frac{\alpha_h}{6}\gamma_s B r_0^2 \sin^2\phi \end{aligned} \tag{6.5.2.4}$$

Now substituting Eq. (6.5.1.5) in Eq. (6.5.2.4), one finally has

$$2P_{\text{pc}} = cB \sec\phi \times e^{2\beta \tan\phi} + \frac{cB}{2\tan\phi}\left(e^{2\beta \tan\phi} - 1\right)\sec^2\phi \pm \alpha_{\text{v}}\gamma_{\text{s}}B^2 \sin\phi$$
$$\pm \frac{\alpha_{\text{h}}}{3}\gamma_s B \tan^2\phi \tag{6.5.2.5}$$

$$2P_{\text{pc}} = cB\left[\sec\phi \times e^{2\beta \tan\phi} + \frac{\sec^2\phi}{2\tan\phi}\left(e^{2\beta \tan\phi} - 1\right)\right] \pm \alpha_{\text{v}}\gamma_{\text{s}}B^2 \sin\phi$$
$$\pm \frac{\alpha_{\text{h}}}{3}\gamma_{\text{s}} B \tan^2\phi. \tag{6.5.2.6}$$

Now considering equilibrium of the wedge as shown in Fig. 6.15, we have

$$q_{\text{c}}B = 2C \sin\phi + 2P_{\text{pc}} - \alpha_{\text{v}}W \tag{6.5.2.7}$$

$$q_{\text{c}}B = cB \sin\phi \sec\phi + cB\left[e^{2\beta \tan\phi}\sec\phi + \frac{\sec^2\phi}{\tan\phi}\left(e^{2\beta \tan\phi} - 1\right)\right]$$
$$\pm \alpha_{\text{v}}\gamma_s B^2 \tan\phi \pm \frac{\alpha_{\text{h}}}{3}\gamma_s B^2 \tan^2\phi \tag{6.5.2.8}$$

Substituting Eq. (6.5.7) in (6.5.2.8), we finally have

$$q_{\text{c}} = c\cot\phi\left[\frac{e^{2\left(\frac{3\pi}{4}-\frac{\phi}{2}\right)\tan\phi}}{2\cos^2\left(45+\frac{\phi}{2}\right)} - 1\right] \pm \frac{5}{6}\alpha_{\text{hn}}\gamma_s B \tan\phi \pm \frac{\alpha_{\text{hn}}}{3}\gamma_{\text{s}} B \tan^2\phi$$
$$= c\cot\phi\left[N_{\text{q}} - 1\right] \pm \alpha_{\text{hn}}\gamma_s B\left[\frac{5}{6}\tan\phi + \frac{1}{3}\tan^2\phi\right]$$
$$= c\left[\cot\phi\left[N_{\text{q}} - 1\right] \pm \alpha_{\text{hn}}\left(\frac{\gamma_s B}{c}\right)\left\{\frac{5}{6}\tan\phi + \frac{1}{3}\tan^2\phi\right\}\right]$$
$$= c\left[\cot\phi\left(N_{\text{q}} - 1\right) \pm \alpha_{\text{hn}}\lambda_c\left\{\frac{5}{6}\tan\phi + \frac{1}{3}\tan^2\phi\right\}\right] \tag{6.5.2.9}$$

In Eq. (6.5.2.9), $\lambda_{\text{c}} = \gamma_{\text{s}}B/c$ is a dimensionless number.

As per Terzaghi and Peck (1982), the first term in Eq. (6.5.2.9) is the static bearing capacity factor N_{c}. Hence,

$$q_{\text{c}} = c\left[N_{\text{c}} \pm \alpha_{\text{hn}}\lambda_c\left\{\frac{5}{6}\tan\phi + \frac{1}{3}\tan^2\phi\right\}\right] \rightarrow cN_{\text{cd}} \tag{6.5.2.10}$$

Thus, we see that the static bearing capacity factor again gets modified due to seismic force by the second term in Eq. (6.5.2.10) and it is mode-dependent.

6.5.3 Determination of Passive Force $P_{p\gamma}$ for Weight Density γ_S

In this case, one can proceed in identical steps as elaborated in previous two cases to finally arrive at

$$q_\gamma = \frac{1}{2}\gamma_s B\left[\frac{\tan\phi}{2}\left(\frac{K_{p\gamma}}{\cos^2\phi}\right) - 1\right] \pm \frac{5}{6}\alpha_h\gamma_s B\tan\phi \pm \frac{\alpha_h}{3}\gamma_s B\tan^2\phi$$

$$\rightarrow q_\gamma = \frac{1}{2}\gamma_s B\left[\frac{\tan\phi}{2}\left(\frac{K_{p\gamma}}{\cos^2\phi} - 1\right) \pm 2\alpha_h\left(\frac{5}{6}\tan\phi + \frac{1}{3}\tan^2\phi\right)\right] \quad (6.5.3.1)$$

Equation (6.5.3.2) can be finally expressed as

$$q_\gamma = \frac{1}{2}\gamma_s B\left[N_\gamma \pm 2\alpha_{hn}\left(\frac{5}{6}\tan\phi + \frac{1}{3}\tan^2\phi\right)\right] \rightarrow \frac{1}{2}\gamma_s B N_{\gamma d} \quad (6.5.3.2)$$

In Eq. (6.5.3.2), N_γ is the static bearing capacity factor which gets altered by the second term due to seismic combination.

6.5.4 Dynamic Ultimate Bearing Capacity of Soil

Thus, following Terzaghi's equation, the ultimate bearing capacity of the soil under seismic loading can be expressed as

$$q^n_{ult} = cN^n_{cd} + qN^n_{qd} + \frac{1}{2}\gamma_s B N^n_{\gamma d} \quad (6.5.4.1)$$

In Eq. (6.5.4.1), the superscript n stands for the modal number and reflects that bearing capacity is not unique but will vary with each mode.

Though the above is derived based on equilibrium condition as per laws of mechanics, it suffers from a fundamental fallacy. Referring to Eq. (6.5.2.10), it is observed that for $\phi = 0$, i.e., pure cohesive soils, under seismic loading the bearing capacity factor $N_{cd} \cong N_c$; i.e., the factor remains unchanged. However, as cited in the previous method of Richard et al., real field observations show that N_c value does undergo reduction during earthquake.

This apparent anomaly can be explained considering the fact that in the mathematical model as perceived by Terzaghi, angle $\alpha = \phi$ (as per Fig. 6.15). But experimental evidence shows that for c–ϕ soil the angle α is actually $45° + \phi/2$ (Das 1987).

This prompted Meyerhof and others to rework on the problem to arrive at a more rational bearing capacity factors expressed as

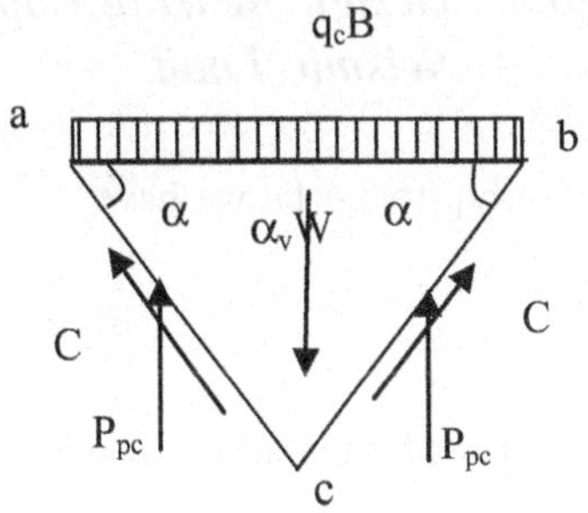

Fig. 6.15 Equilibrium of wedge under cohesive force

$$N_{qM} = \tan^2\left(45^\circ + \frac{\phi}{2}\right)e^{\pi \tan \phi} \quad (6.5.4.2)$$

$$N_{cM} = \cot \phi\left(N_{qM} - 1\right) \text{ and} \quad (6.5.4.3)$$

$$N_{\gamma M} = \left(N_{qM} - 1\right)\tan(1.4\phi) \quad (6.5.4.4)$$

In Eqs. (6.5.4.2) to (6.5.4.4), the subscript M stands for *Meyerhof's expression.*

6.5.5 Dynamic Bearing Capacity Factors as Per Meyerhof's Expression

Proceeding in identical fashion as cited above for Terzaghi's case considering $\alpha = 45^\circ + \phi/2$, it is easy to arrive at the expressions based on principle of superimposition that

$$N_{qd} = N_{qM} \pm \alpha_{hn}\left(\frac{B}{D_f}\right)\left[\frac{5}{6}\tan\alpha + \frac{1}{3}\tan^2\alpha\right] \quad (6.5.5.1)$$

$$N_{cd} = N_{cM} \pm \alpha_{hn}\lambda_c\left[\frac{5}{6}\tan\alpha + \frac{1}{3}\tan^2\alpha\right] \quad (6.5.5.2)$$

$$N_{\gamma d} = N_{\gamma M} \pm 2\alpha_{hn}\left[\frac{5}{6}\tan\alpha + \frac{1}{3}\tan^2\alpha\right] \quad (6.5.5.3)$$

6.5.6 Design Bearing Capacity of Foundation Under Seismic Load

From Eq. (6.5.4.1), we have

$$q^n_{\text{ult}} = cN^n_{\text{cd}} + qN^n_{\text{qd}} + \frac{1}{2}\gamma_s BN^n_{\gamma\text{d}} \tag{6.5.6.1}$$

Expanding the above, one gets, ignoring the mode number n,

$$q_{\text{ult}} = c\left[N_c \pm \alpha_{\text{hn}}\left(\frac{\gamma_s B}{c}\right)f(\alpha)\right] + q\left[N_q \pm \alpha_{\text{hn}}\left(\frac{\gamma_s B}{q}\right)f(\alpha)\right] + \frac{1}{2}\gamma_s B\left[N_\gamma \pm \alpha_{\text{hn}}\left(\frac{\gamma_s B}{\frac{1}{2}\gamma_s B}\right)f(\alpha)\right] \tag{6.5.6.2}$$

where

$$f(\alpha) = \frac{5}{6}\tan\alpha + \frac{1}{3}\tan^2\alpha \tag{6.5.6.3}$$

Equation (6.4.6.4) can be expressed as

$$q_{\text{ult}} = cN_c + qN_q + \frac{1}{2}\gamma_s BN_\gamma \pm 3\alpha_{\text{hn}}\gamma_s Bf(\alpha) \tag{6.5.6.4}$$

In Eq. (6.5.6.4) the first three terms express the static ultimate bearing capacity under general shear failure and are actually never simulated, as these terms are divided by an adequate factor of safety to arrive at the serviceable loading condition. The earthquake part is a force that would thus alternatively enhance or reduce this serviceable load condition considering it is a force that is expected under service condition of the foundation in its operative state.

Hence, the allowable design bearing capacity under seismic force can be expressed as

$$q_{\text{ad}} = \frac{\left[cN_c + qN_q + \frac{1}{2}\gamma_s BN_\gamma\right]}{FOS} \pm 3\alpha_{\text{hn}}\gamma_s Bf(\alpha) \tag{6.5.6.5}$$

$$q_{\text{ad}} = q_a(\text{static}) \pm 3\kappa_n\left[\frac{ZI}{2R}\right]\cos\left[(2n-1)\frac{\pi D_f}{2H}\right]\left(\frac{S_{\text{an}}}{g}\right)\gamma_s B\left(\frac{5}{6}\tan\left(45+\frac{\phi}{2}\right) + \frac{1}{3}\tan^2\left(45+\frac{\phi}{2}\right)\right) \tag{6.5.6.6}$$

$$\rightarrow q_{ad} = \left[\frac{P}{A} \pm \frac{M}{Z_f}\right] \pm 3\kappa_n \left[\frac{ZI}{2R}\right] \cos\left[(2n-1)\frac{\pi D_f}{2H}\right]\left(\frac{S_{an}}{g}\right)\gamma_s B\left(\frac{5}{6}\tan\left(45 + \frac{\phi}{2}\right)\right.$$
$$\left.+\frac{1}{3}\tan^2\left(45 + \frac{\phi}{2}\right)\right) \quad (6.5.6.7)$$

Here, P = load from super structure, M = moment on foundation from super structure A = area of foundation. Z_f = section modulus of foundation.

In Eq. (6.5.6.7), q_{ad} = allowable bearing capacity of foundation under seismic/dynamic load and q_a = allowable bearing capacity under static load.

Looking at Eq. (6.5.6.7), it is observed that q_a can be either derived based on general shear or local shear failure or even from settlement criteria which gets augmented/reduced by the second term in Eq. (6.5.6.7), i.e., the seismic force due to soil inertia. Thus, in one case it enhances the stress induced in foundation which in turn reduces the factor of safety to the ultimate stress, and in other case, it just acts in reverse sense.

The factor of safety under worst seismic load is thus considered as

$$FOS = \frac{q_{ult}}{\left[\frac{P}{A} + \frac{M}{Z}\right] + 3\kappa_n \left[\frac{ZI}{2R}\right] \cos\left[(2n-1)\frac{\pi D_f}{2H}\right]\left(\frac{S_{an}}{g}\right)\gamma_s B\left(\frac{5}{6}\tan\left(45 + \frac{\phi}{2}\right) + \frac{1}{3}\tan^2\left(45 + \frac{\phi}{2}\right)\right)} \quad (6.5.6.8)$$

6.5.7 *Computation of Free Field Time Period of Soil Having G Varying with Depth*

The derivation above was done based on G constant with depth. But in majority of cases G is found to vary with depth. This variation would primarily affect the time period and this may be computed from the expression as cited below in Table 6.3, when the eigenvectors and the modal mass participation factor remain same as cited earlier.

For variable shear modulus with depth, the time period can be expressed as

$$T_n = C_T\left(\frac{H}{V_s}\right) \quad (6.5.7.1)$$

The value of C_T for soil having G varying with depth is as furnished in Table 6.4

Table 6.4 Values of C_T for Gibson-type soil

Mode	1	2	3
$G_z = G_0(z/H)$	5.2	2.13	1.23
$G_z = G_0\sqrt{z/H}$	4.5	1.66	1.00
$G_z = G_0(z/H)^2$	7.8	3.30	1.66

6.5.8 Selection of Appropriate Value of Damping Ratio

For this, guidelines may be sought from Sect. 8.4.1.3 of Chap. 8 where detailed discussion is carried out on this topic.

Example 6.2 A foundation of size 2 m × 2 m is to be built in a site that is considered to be in Zone IV as per IS-1893 (2002). The SPT value N was observed to be greater than 50 at a depth of 28 m from ground surface.

Soil being predominantly clayey in nature, G is assumed constant with depth.

Compute the static and seismic bearing capacity of the foundation considering the soil properties as average $c = 30\ \text{kN/m}^2$ and $\phi = 10°$. The foundation is expected to be located at 1.5 m below ground level. Density of soil = 20 kN/m^3. Shear wave velocity of soil at foundation level is 220 m/sec.

Solution

For the computation of static bearing capacity, we use Meyerhof's expression.

Here, $\phi = 10°$. Referring to Table 6.2, $N_q = 2.47$, $N_c = 8.35$, $N_\gamma = 0.4$.

$$q_{\text{ult}} = cN_c s_c d_c i_c + \gamma D_f N_q s_q d_q i_q + \frac{1}{2}\gamma B N_\gamma s_\gamma d_\gamma i_\gamma$$

$d_q = d_\gamma = 1 + 0.1\sqrt{N_\phi}\left(\frac{D_f}{B}\right)$, where $N_\phi = \tan^2\left(45 + \frac{\phi}{2}\right) \rightarrow 1.42$

$D_f = 1.5$ m and $B = 2.0$ give $d_q = d_\gamma = 1.089$

$$s_q = s_\gamma = 1 + 0.1N_\phi\left(\frac{B}{L}\right) \rightarrow 1 + 0.1 \times 1.42 \times 1 = 1.142$$

$$d_c = 1 + 0.2\sqrt{N\phi}\left(\frac{D_f}{B}\right) \rightarrow 1 + 0.2 \times \sqrt{1.42} \times \left(\frac{1.5}{2.0}\right) = 1.178$$

$$s_c = 1 + 0.2N_\phi\left(\frac{B}{L}\right) \rightarrow 1 + 0.2 \times 1.42 \times 1 = 1.284$$

$$i_q = i_c = i_\gamma = 1$$

Thus

$$q_{\text{ult}} = 30 \times 8.35 \times 1.284 \times 1.178 \times 1 + 20 \times 1.5 \times 2.47 \times 1.142 \times 1.089 \times 1 + \frac{1}{2}20 \times 2 \times 0.4 \times 1.142 \times 1.089 \times 1$$

$\rightarrow q_{\text{ult}} = 378.89 + 80.69 + 9.9 = 469.5\,\text{kN/m}^2.$

Taking factor of safety of 3.0, $q_{\text{all}} = 469.5/3 = 156$ kN/m^2.

Depth to bedrock = 28 m, shear wave velocity = 220 m/s.

For Mode 1

$$T_1 = 4H/V_s \rightarrow 4 \times 28/220 = 0.509\,\mathrm{s}$$

For $T_1 = 0.51$ s, $S_a/g = 2.5$. For zone IV, $Z = 0.24$, take $I = 1.5$ $R = 2$.
Referring to Table 6.3 $\kappa_1 = 8/(2+\pi)$

$$V_{EQ} = 3\kappa_n \left[\frac{ZI}{2R}\right] \cos\left[\frac{\pi D_f}{2H}\right]\left(\frac{S_{an}}{g}\right)\gamma_s B\left(\frac{5}{6}\tan\left(45+\frac{\phi}{2}\right)+\frac{1}{3}\tan^2\left(45+\frac{\phi}{2}\right)\right)$$

$$V_{EQ} = 3\left(\frac{8}{2+\pi}\right)\left[\frac{0.24\times 1.5}{4}\right]\cos\left[\frac{\pi\times 1.5}{2\times 28}\right](2.5)\times 20$$
$$\times 2\left(\frac{5}{6}\tan\left(45+\frac{10}{2}\right)+\frac{1}{3}\tan^2\left(45+\frac{10}{2}\right)\right) = 61.36\,\mathrm{kN/m^2}$$

Here, V_{EQ} is the additional stress induced in the foundation due to earthquake.
Thus, the total stress induced in the foundation = 156 + 61 = 217 kN/m^2.
Thus, the factor of safety of the foundation under earthquake = $\frac{470}{217} = 2.16 < 3.0$.

For Mode 2

$$T_1 = 4H/3V_s \rightarrow 4 \times 28/3 \times 220 = 0.17\,\mathrm{s}$$

For $T_1 = 0.17$ s, $S_a/g = 2.5$. For zone IV, $Z = 0.24$, take $I = 1.5$ $R = 2$.
Referring to Table 6.3 $\kappa_1 = 8/(3\pi - 2)$

$$V_{EQ} = 3\kappa_n \left[\frac{ZI}{2R}\right] \cos\left[\frac{3\pi D_f}{2H}\right]\left(\frac{S_{an}}{g}\right)\gamma_s B\left(\frac{5}{6}\tan\left(45+\frac{\phi}{2}\right)+\frac{1}{3}\tan^2\left(45+\frac{\phi}{2}\right)\right)$$

$$V_{EQ} = 3\left(\frac{8}{3\pi-2}\right)\left[\frac{0.24\times 1.5}{4}\right]\cos\left[\frac{3\pi\times 1.5}{2\times 28}\right](2.5)\times 20$$
$$\times 2\left(\frac{5}{6}\tan\left(45+\frac{10}{2}\right)+\frac{1}{3}\tan^2\left(45+\frac{10}{2}\right)\right) = 41.3\,\mathrm{kN/m^2}$$

Thus, the total stress induced in the foundation = 156 + 41 = 197 kN/m^2.
Thus, the factor of safety of the foundation under earthquake = $\frac{470}{197} = 2.38 \leq 3.0$.

For Mode 3

$$T_1 = 4H/5V_s \rightarrow 4 \times 28/5 \times 220 = 0.101\,\mathrm{s}$$

For $T_1 = 0.10$ s, $S_a/g = 2.5$. For zone IV, $Z = 0.24$, take $I = 1.5$ $R = 2$.
Referring to Table 6.3 $\kappa_1 = 8/(5\pi + 2)$

$$V_{EQ} = 3\kappa_n \left[\frac{ZI}{2R}\right] \cos\left[\frac{5\pi D_f}{2H}\right]\left(\frac{S_{an}}{g}\right)\gamma_s B\left(\frac{5}{6}\tan\left(45+\frac{\phi}{2}\right)+\frac{1}{3}\tan^2\left(45+\frac{\phi}{2}\right)\right)$$

$$V_{EQ} = 3\left(\frac{8}{5\pi+2}\right)\left[\frac{0.24\times 1.5}{4}\right]\cos\left[\frac{5\pi\times 1.5}{2\times 28}\right](2.5)\times 20$$
$$\times 2\left(\frac{5}{6}\tan\left(45+\frac{10}{2}\right)+\frac{1}{3}\tan^2\left(45+\frac{10}{2}\right)\right) = 17\,\text{kN/m}^2$$

Thus, the total stress induced in the foundation = 156 + 17 = 173 kN/m^2.
Thus, the factor of safety of the foundation under earthquake = $\frac{470}{173} = 2.72$.

6.6 What Do We Do When Structures Are Supported on Pile Foundations?

We actually do nothing, except increasing the static capacity of pile by 25%, like we do for a shallow foundation!

The conventional concept is that we estimate the static capacity of pile under vertical and lateral load under serviceability condition based on geotechnical data. For DL + LL + EL load combination[1], we increase the pile load carrying capacity by 25% (which is P_p say). Based on the axial load, shear, and moment coming on the foundation from the superstructure, we estimate the number of piles, as shown in Fig. 6.16 and Eq. 6.6.1.

Shown in Fig. 6.16 is a group of piles below a pile cap subjected to axial force (*P*), moment (*M*), and shear (*V*) due to dead load, live load, and seismic load combination.

The load per pile is then computed from the expression

$$P_p = \frac{P}{n} \pm \frac{(M + V \times D_p)\times r}{\sum_i^n r^2} \tag{6.6.1}$$

In Eq. (6.6.1)

- P_p load per pile that should be $\leq$ static design capacity augmented by 25% under seismic load combination;
- n number of pile in the pile group;
- M, V moments and shear at top of pile cap as shown in Fig. 6.16a;
- D_p depth of pile cap;
- r distance of pile in extreme row from center of loading.

Specific to Fig. 6.16b, Eq. (6.6.1) can be expressed as

[1]Here, DL = dead load, LL = live load, EL = earthquake load.

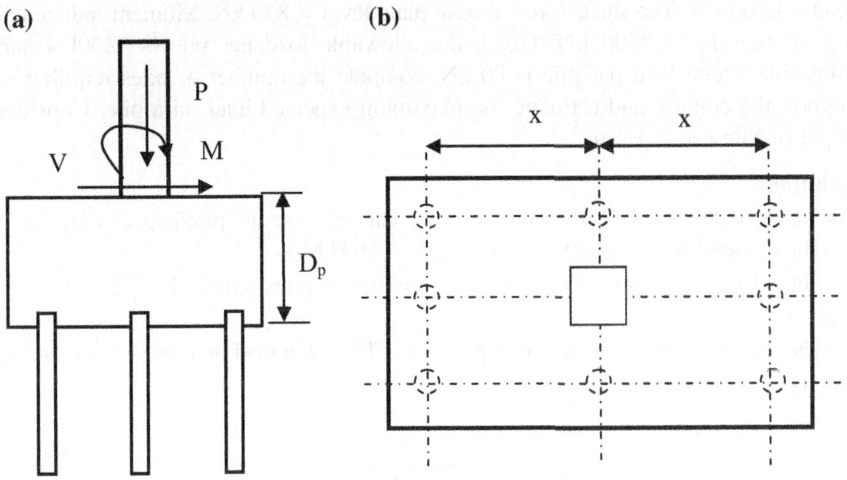

Fig. 6.16 A pile cap with piles subjected to seismic load combination

Fig. 6.17 Layout of powerhouse pile cap and pile

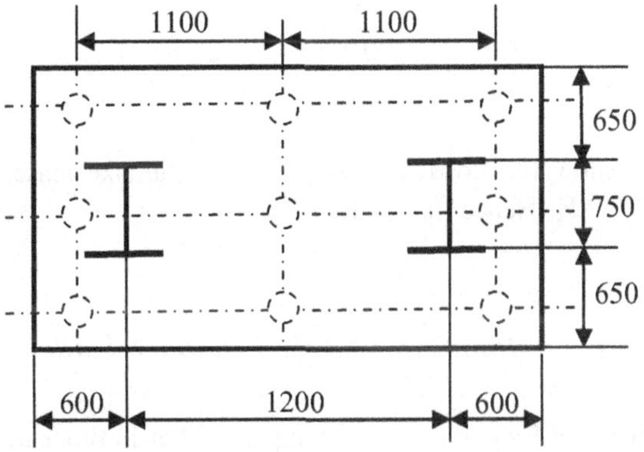

$$P_{\mathrm{p}} = \frac{P}{8} \pm \frac{\left(M + V \times D_{\mathrm{p}}\right) \times x}{2 \times 3 \times x^2} \tag{6.6.2}$$

The above should be less than the static capacity of pile multiplied by the factor 1.25 in vertical direction.

The base shear V acting on top of pile head, when divided by number of piles in pile group, gives the individual shear each pile is subjected to, and this should be less than design static capacity again enhanced by 25%.

The above remains the standard procedure for estimating the pile loads in a pile cap supporting loads from superstructure.

Example 6.5 A typical powerhouse column is as shown in Fig. 6.17, supported on piles. The load carried by the column under dead Load + crane load + seismic load = 6500 kN. The shear force at base plate level = 850 kN. Moment induced at top of pile cap is 2500 kN. Given the allowable load per pile is 1200kN and allowable lateral load per pile is 80 kN, compute the number of piles required to support the column and estimate the maximum expected load on a pile. Consider depth of pile cap = 1.2 m.

Solution

As we are considering earthquake case, we can increase the pile capacity by 25%.

Thus, lateral load capacity = 80 × 1.25 = 100 kN.

Thus, based on lateral load capacity, number of piles required = $\frac{850}{100} = 8.5 \cong 9$ (say).

The layout of pile is as shown in Fig. 6.17. Thus, the load on pile on the extreme row is

$$P_p = \frac{P}{n} \pm \frac{(M + V \times D_p) \times r}{\sum_i^n r^2}$$

Self-weight of pile cap = 2.4 × 2.05 × 1.2 × 25 = 147.6 kN
Total weight = 6500 + 147.6 = 6647.6

$$P_p = \frac{6647.6}{9} \pm \frac{(2500 + 850 \times 1.2) \times 1.1}{2 \times 3 \times (1.1)^2} = 739 \pm 533 = 1272\,\text{kN or } 206\,\text{kN}$$

Allowable load per pile for earthquake case = 1200 × 1.25 = 1500 kN > 1272 kN. Hence, it is safe.

6.6.1 *Estimation of Pile Capacity Under Vertical Load*

Based on Example 6.5, it is apparent that in order to compute the number of piles below column, we need to know *priori* the vertical and lateral load carrying capacity of a pile under a particular set of geotechnical data.

Estimation of these capacities strictly does not fall under purview of earthquake engineering, yet, for completeness of the same, we elaborate the same hereafter briefly, and this we believe would give a better appreciation of subsequent development that has taken place in the realms of dynamic response of piles under seismic loading.

A pile under vertical load transfers the load to the surrounding soil

- By friction only usually called *friction piles*.
- By bearing only termed as *bearing piles*.
- By both bearing and friction (the most common case)

Shown in Fig. 6.18 is a single pile embedded in ground and the mechanism of load transfer to the surrounding soil.

As depicted in Fig. 6.18 for a pile subjected to vertical load, a part of the load is transferred through friction (Q_f) along the shaft with surrounding soil, while the rest is transferred through bearing (Q_b) at base of the pile.

When both Q_f and Q_b reach the ultimate state, the pile undergoes failure. Thus, for the pile to carry a vertical load effectively, one takes

$$Q_a = \frac{Q_b + Q_f}{\text{FOS}} \tag{6.6.1.1}$$

In Eq. (6.6.1.1)

Q_a allowable pile capacity;
FOS factor of safety, usually considered as 2.5;
Q_b ultimate bearing capacity at pile base; and
Q_f ultimate frictional load transferred through pile shaft.

An alternate expression which is also considered is

$$Q_a = \frac{Q_b}{3.0} + \frac{Q_f}{1.5} \tag{6.6.1.2}$$

Lower of the above two values is usually considered as design capacity of the pile.

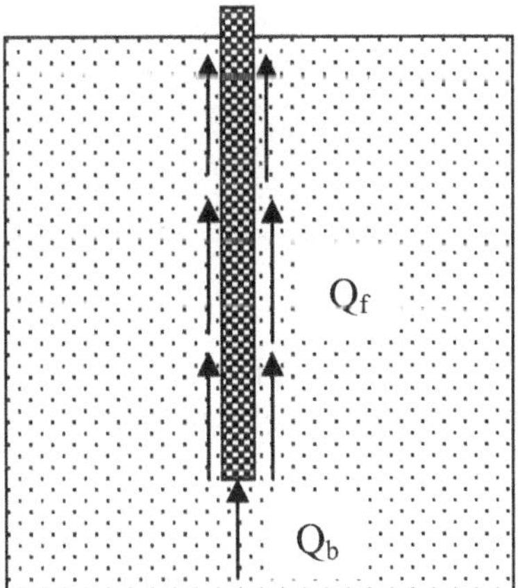

Fig. 6.18 Load transfer mechanism of a pile under vertical load

6.6.1.1 Estimation of Pile Load Capacity for c–ϕ Soil Which Is Predominantly Granular

For general c–ϕ soil, ultimate baring capacity at the pile base is considered as

$$Q_b = \left(cN_c + qN_q + \frac{1}{2}\gamma_s DN_\gamma\right)\left(\frac{\pi D^2}{4}\right) \quad (6.6.1.3)$$

In Eq. (6.6.1.3), the last term within first parenthesis has negligible effect, as N_γ and D, the diameter of pile, are of small magnitude and are usually neglected, which gives

$$Q_b = \left(cN_c + qN_q\right)\left(\frac{\pi D^2}{4}\right) \quad (6.6.1.4)$$

In Eq. (6.6.1.4), c is the cohesion of soil and is usually considered as the undrained shear strength at pile tip.

N_c value as per Meyerhof is usually considered as 9.0, and this gives

$$Q_b = \left(9c + \gamma_{av} LN_q\right)\left(\frac{\pi D^2}{4}\right) \quad (6.6.1.5)$$

In Eq. (6.6.1.5), L is the length of the pile shaft embedded in ground and

$$\gamma_{av} = \frac{\gamma_1 L_1 + \gamma_2 L_2 + \cdots + \gamma_n L_n}{\sum_{i=1}^{n} L_i} \quad (6.6.1.6)$$

The N_q value as per Meyerhof is as expressed in Table 6.5.

The friction force along the shaft is expressed as

$$Q_f = \pi D K_s \tan\delta \sum_{L=0}^{L} \sigma'_v \Delta L \quad (6.6.1.7)$$

In Eq. (6.6.1.7)

D diameter of pile;
K_s coefficient of earth pressure;

Table 6.5 Design N_q values after Meyerhof (1959)

ϕ	20	25	28	30	32	34	36	38	40	42	45	Remarks
N_q	8	12	20	25	35	45	60	80	120	160	230	Driven piles
N_q	4	5	8	12	17	22	30	40	60	80	115	Bored piles

δ friction angle between pile and contact soil;
σ'_v effective overburden pressure ($\gamma_{sub} \times z$), and
ΔL pile shaft length for the given layer.

For layered soil with variable ϕ, Q_f is expressed as

$$Q_f = \pi D K_s \sum_{i=0}^{n} \tan \delta_i \sigma'_v \Delta Li \quad (6.6.1.8)$$

Based on tests carried out on full-scale pile, it has been observed that the effective overburden pressure does not increase linearly with depth, as depicted in Eqs. (6.6.1.7) and (6.6.1.8), and becomes constant after a certain depth.

For practical design purpose, it is usually assumed that the effective overburden increases linearly to a depth of 10–20 times the diameter of pile, beyond which it becomes constant. This approach is found to be adequate for ordinary foundations where the penetration depths of piles are restricted to 10–20 D when refusals are usually encountered for closed-end piles driven in cohesionless soil.

Tomlinson (2012), however, reports that when open-ended tubular piles are driven in soil that transfers the load mainly through friction, like that in transmission tower lines or offshore platforms, the above gives an extremely conservative value and does not look realistic. Other researchers have also seriously contested the above restriction, especially for long slender piles, raising doubts on its sanctity.

Various researchers have proposed different values of δ based on field observed data as well as material of construction, when δ is usually considered as $\delta = (2/3)\phi$.

If the pile is a concrete pile, some researchers recommend $\delta = (3/4)\phi$. For steel piles, $\delta = 20°$. IS 2911 (Part 1 to 4), Indian Code, for design of pile foundation has recommended a value of $\delta = \phi$.

K_s, the coefficient of earth pressure, has also been recommended by a number of researchers, which is a function of the method of installation and has also been corelated to the pile material, as furnished hereafter in Table 6.6.

Other researchers have recommended a value of K_s = 0.5–1.0 for steel piles and K_s = 1.0–2.0 for concrete piles.

Estimation of Friction Angle ϕ of Soil for Piling Design

Friction property of the soil, ϕ, is determined prior to installation of pile either based on field test like SPT or carrying out laboratory test on the undisturbed sample. Unfortunately, during piling installation this value undergoes significant change due to installation disturbance.

Table 6.6 Design values of K_s based on method of installation

Piling installation type	Bored pile	Driven H pile	Driven displacement type
K_s	0.5	0.5–1.0	1.0–2.0

While piles are driven, the soil gets compacted, enhancing the value. For bored piles, it is just the opposite when the value usually gets reduced.

No theoretical model exists by which this can be derived. However, based on field observation and field load test data of piles, the following values are usually adopted to estimate the design ϕ value.

For driven piles, Kishida (1967) recommends

$$\phi_{\mathrm{D}} = 15 + 20\sqrt{N} \tag{6.6.1.9}$$

Here, ϕ_{D} = design friction angle of soil for pile design and N = design SPT value (number of blows).

For bored piles, Tomlinson recommends

$$\phi_{\mathrm{D}} = \phi_i - 3 \tag{6.6.1.10}$$

Here, ϕ_i = friction angle of soil prior to pile installation.

As per Meyerhoff

$\phi_{\mathrm{D}} = 0.75\phi_i + 10°$ for driven piles and $\phi_{\mathrm{D}} = \phi_i - 3$ for bored piles.

Estimation of Pile Capacity Directly from SPT Value

For a non-cohesive soil, in most cases the SPT test is carried out to determine its geotechnical properties. Thus, the pile capacity has also been directly corelated to N values, as expressed hereafter.

For driven piles, where soil gets displaced and compacted:

$$Q_{\mathrm{u}} = 400N_{\mathrm{b}} \times \left(\frac{\pi D^2}{4}\right) + 2\bar{N} \times \pi DL \tag{6.6.1.11}$$

For H piles:

$$Q_{\mathrm{u}} = 400N_{\mathrm{b}} \times (\text{Area at base}) + \bar{N} \times \text{Surface area} \times \mathrm{L} \tag{6.6.1.12}$$

For bored piles:

$$Q_{\mathrm{u}} = 133N_b \times \left(\frac{\pi D^2}{4}\right) + 0.6\bar{N} \times \pi DL \tag{6.6.1.13}$$

In Eqs. (6.6.1.11) to (6.6.1.13)

Q_{u} ultimate load carrying capacity of the pile;

N_{b} SPT value at base of pile (i.e., pile tip);

$\bar{N}$ average SPT value along pile shaft;
D diameter of pile; and
L length of pile.

The factors of safety adapted to compute the allowable load are FOS = 4.0 for driven piles and FOS = 2.5 for bored piles.

6.6.1.2 Estimation of Pile Load Capacity for Clayey Soil

For cohesive-type soil, the ultimate load capacity of pile, Q_u, is computed from the expression

$$Q_u = c_b N_c \times \left(\frac{\pi D^2}{4}\right) + \alpha \times \bar{c} \times \pi DL \tag{6.6.1.14}$$

In Eq. (6.6.1.14)

c_b average cohesive value at pile tip;
N_c bearing capacity factor considered as 9.0;
$\bar{c}$ average cohesive factor along pile shaft of length L; and
α reduction factor which is dependent on consistency of the soil and is usually considered as shown in Table 6.7.

Example 6.6 Shown in Fig. 6.19 is the geotechnical profile of a site where a pile of 300 mm diameter and 20 m long is to be installed. The pile is a precast driven pile. The net overburden pressure as computed is also shown in the right-hand side of the figure. Considering that the groundwater table exists at 2.0 m below ground level, compute the vertical capacity of the pile. Estimate also the pile capacity based on direct SPT value.

Solution
The pile here sustains load through both bearing and friction.
The ultimate bearing load is expressed as

$$Q_b = (cN_c + qN_q)\left(\frac{\pi D^2}{4}\right)$$

Table 6.7 Reduction factor *a* for pile design

Consistency	SPT value (N)	Reduction factor (*a*)
Soft to very soft	≤ 4	0.7
Medium	4–8	0.5
Stiff	8–15	0.4
Stiff to hard	>15	0.3

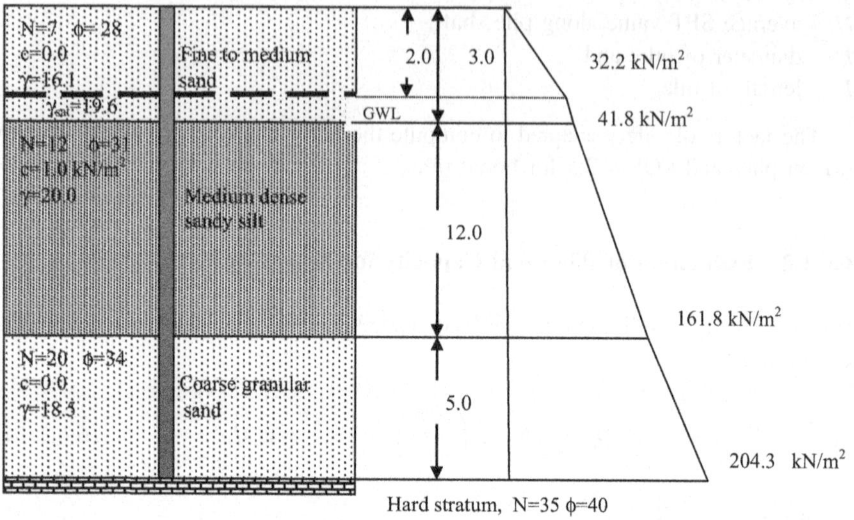

Fig. 6.19 Geotechnical profile of soil and pile driven in it

Here, c = 0, q = 204.3 (as per pressure diagram drawn) and N_q = 120 (after Meyerhoff's data) for driven piles.

Thus

$$Q_b = (0 \times 9 + 204.3 \times 120)\left(\frac{\pi \times 0.3^2}{4}\right) = 1733\,\text{kN}.$$

Computation of shaft resistance

$$Q_f = \pi D K_s \sum_{i=0}^{n} \tan \delta_i \sigma'_v \Delta Li$$

<u>**From 0 to 2 m**</u>

$\phi = 28°$, K_s = 1.5 for driven piles. $\phi_D = 0.75 \times 28° + 10 = 31$, $\delta = \frac{2}{3} \times 31° = 21°$

Thus, $Q_1 = \pi \times 0.3 \times 1.5 \times \tan 21 \times \frac{32.2}{2} \times 2 = 17.47\,\text{kN}$

<u>**From 2 to 3 m**</u>

$\phi = 28°$, K_s = 1.5 for driven piles. $\phi_D = 31$, $\delta = 21°$

Thus, $Q_2 = \pi \times 0.3 \times 1.5 \times \tan 21 \times \frac{32.2 + 41.8}{2} \times 1 = 20.0\,\text{kN}$

From 3 to 15 m

$\phi = 31°$, $K_s = 1.5$ for driven piles. $\phi_D = 0.75 \times 31° + 10 = 33.2$, $\delta = \frac{2}{3} \times 33.2° = 22°$

Thus, $Q_3 = \pi \times 0.3 \times 1.5 \times \tan 22 \times \frac{(41.8+161.8)}{2} \times 12 = 697.7$ kN

From 15 to 20 m

$\phi = 34°$, $K_s = 1.5$ for driven piles. $\phi_D = 0.75 \times 34° + 10 = 36$, $\delta = \frac{2}{3} \times 36° = 24°$

Thus, $Q_4 = \pi \times 0.3 \times 1.5 \times \tan 24 \times \frac{(161.8+204.3)}{2} \times 5 = 576$ kN

Thus, the total shaft resistance $Q_f = Q_1 + Q_2 + Q_3 + Q_4 = 17.47 + 20 + 698 + 576 = 1311.47$ kN

Thus, the allowable load on pile

$$Q_a = \frac{Q_b + Q_f}{FOS} = \frac{1733 + 1311}{3} = 1014.66\,\text{kN, or}$$

$$Q_a = \frac{Q_b}{3.0} + \frac{Q_f}{1.5} = \frac{1733}{3} + \frac{1311}{1.5} = 1451.6\,\text{kN}$$

Considering the lower value, estimated allowable capacity of the pile = 1015 kN.

Computation based on SPT value

$$Q_u = 400N_b \times \left(\frac{\pi D^2}{4}\right) + 2\bar{N} \times \pi DL$$

Here, $N_b = 35$ and

$$\bar{N} = \frac{7 \times 3 + 12 \times 12 + 20 \times 5}{20} = 13.25$$

Thus,

$Q_u = 400 \times 35 \times \left(\frac{\pi \times 0.3^2}{4}\right) + 2 \times 13 \times \pi \times 0.3 \times 20 = 989.6 + 490 = 1479.6$ kN

Thus $Q_a = \frac{Q_u}{4} = 370$ kN.

The value is very conservative.

Considering a FOS of 2.5 as recommended in IS-2911 (part 1)

$$Q_a = \frac{Q_u}{2.5} = \frac{1479.6}{2.5} = 592\,\text{kN}$$

Computation restricting overburden pressure till 20 times diameter of pile

The critical depth for pile here is considered as $20 \times 300 = 6000$ mm (Fig. 6.20).

Effective overburden at this level is = $41.8 + (20 - 10) \times 3 = 71.8$ kN/m^2.

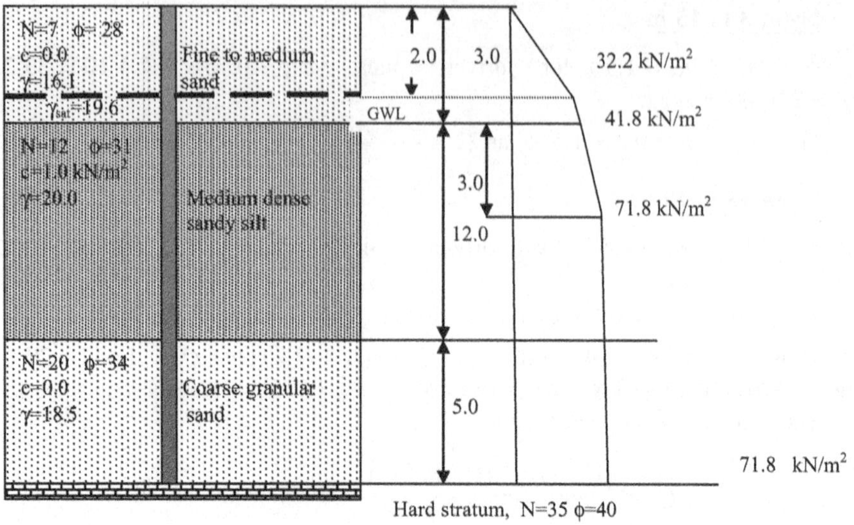

Fig. 6.20 Geotechnical profile of soil and pile driven in it

The ultimate bearing load is expressed as

$$Q_b = (cN_c + qN_q)\left(\frac{\pi D^2}{4}\right)$$

Here, $c = 0$, $q = 71.8$ (as per pressure diagram drawn) and $N_q = 120$ (after Meyerhoff's data) for driven piles.

Thus, $Q_b = (0 \times 9 + 71.8 \times 120)\left(\frac{\pi \times 0.3^2}{4}\right) = 609$ kN.

Computation of shaft resistance

$$Q_f = \pi D K_s \sum_{i=0}^{n} \tan \delta_i \sigma'_v \Delta Li$$

From 0 to 2 m

$\phi = 28°$, $K_s = 1.5$ for driven piles. $\phi_D = 0.75 \times 28° + 10 = 31$, $\delta = \frac{2}{3} \times 31° = 21°$

Thus, $Q_1 = \pi \times 0.3 \times 1.5 \times \tan 21 \times \frac{32.2}{2} \times 2 = 17.47$ kN

From 2 to 3 m

$\phi = 28°$, $K_s = 1.5$ for driven piles. $\phi_D = 31$, $\delta = 21°$

Thus, $Q_2 = \pi \times 0.3 \times 1.5 \times \tan 21 \times \frac{32.2 + 41.8}{2} \times 1 = 20.0$ kN

From 3 to 6 m

$\phi = 31°$, $K_s = 1.5$ for driven piles. $\phi_D = 0.75 \times 31° + 10 = 33.2$, $\delta = \frac{2}{3} \times 33.2° = 22°$

Thus, $Q_3 = \pi \times 0.3 \times 1.5 \times \tan 22 \times \frac{(41.8 + 71.8)}{2} \times 3 = 97.33$ kN

From 6 to15 m

$\phi = 31°$, $K_s = 1.5$ for driven piles. $\phi_D = 0.75 \times 31° + 10 = 33.2$, $\delta = \frac{2}{3} \times 33.2° = 22°$

Thus, $Q_4 = \pi \times 0.3 \times 1.5 \times \tan 22 \times 71.8 \times 9 = 369$ kN

From 15 to 20 ms

$\phi = 34°$, $K_s = 1.5$ for driven piles. $\phi_D = 0.75 \times 34° + 10 = 36$, $\delta = \frac{2}{3} \times 36° = 24°$

Thus, $Q_5 = \pi \times 0.3 \times 1.5 \times \tan 24 \times 71.8 \times 5 = 225.96$ kN

Thus, the total shaft resistance $Q_f = 17.47 + 20 + 97.33 + 369 + 225.96 = 729.76$ kN

Thus, the allowable load on pile

$$Q_a = \frac{Q_b + Q_f}{FOS} = \frac{609 + 730}{3} = 444\,\text{kN, or}$$

$$Q_a = \frac{Q_b}{3.0} + \frac{Q_f}{1.5} = \frac{609}{3} + \frac{730}{1.5} = 690\,\text{kN}$$

Considering the lower value, the estimated allowable capacity of the pile = 444 kN.

6.6.2 Conventional Methods for Estimation of Lateral Load Capacity of Piles

Under earthquake force (as elucidated in Chap. 2), the major force that a structure experiences is in lateral direction. Thus, it is apparent that if a structure is supported on pile foundation, the piles will also be subjected to lateral force that it should effectively sustain for the structure to remain functional.

At the advent of piles as a foundation component (in the early 1940–50), it was considered as a member that was capable of taking vertical load only. When the pile was subjected to significant lateral force, like that in a jetty or transmission line towers batter piles were adopted to cater for the lateral force as shown in Fig. 6.21, which shows typical batter piles carrying compressive load $R = \sqrt{P^2 + V^2}$ which is purely compressive in nature.

However, in many practical engineering situations, such piles are not feasible as neither boring nor driving is easy keeping the alignment intact and the execution

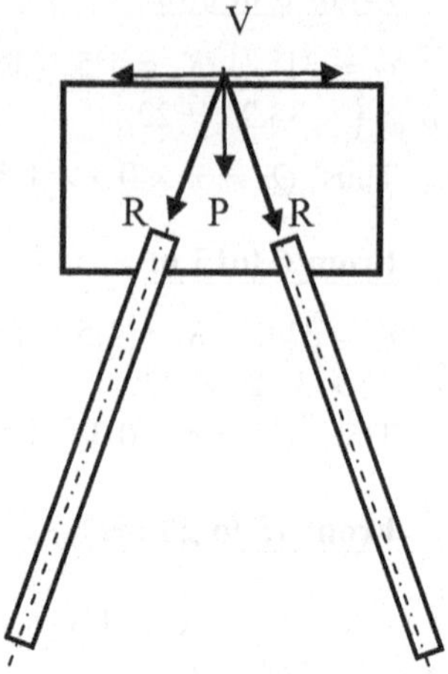

Fig. 6.21 Batter piles carrying compressive load under bi-directional force

cost becomes expensive. Thus, research was focused on estimation of lateral load capacity of piles that were driven vertically.

Brinch Hansen (1961) and Broms (1964) were the first to develop the lateral load capacity of vertical piles subjected to horizontal force based on ultimate load analysis.

The underlying theory is explained hereafter.

6.6.2.1 Brinch Hansen's (1961) Method for Estimation of Lateral Load of Piles

Based on earth pressure theory, Brinch Hansen(1961) developed a procedure for computation of lateral load capacity of a pile under ultimate load that is applicable for c–ϕ soil and also can be applied for layered soil strata.

The technique is as explained below.

Shown in Fig. 6.22 is a pile embedded in ground having length L and having a freestanding height of e subjected to a force Q_u that induces a moment $M_u = Q_u \cdot e$ at the pile head, as shown in Fig. 6.22b. Due to application of this load, the pile rotates about a point C when the pressure diagram is as shown in Fig. 6.22b.

Considering that the sum of all forces in horizontal direction is zero, we have

$$Q_u - \int_0^{x_c} p_{xu} D dx + \int_{x_c}^{L} p_{xu} D dx = 0 \qquad (6.6.2.1)$$

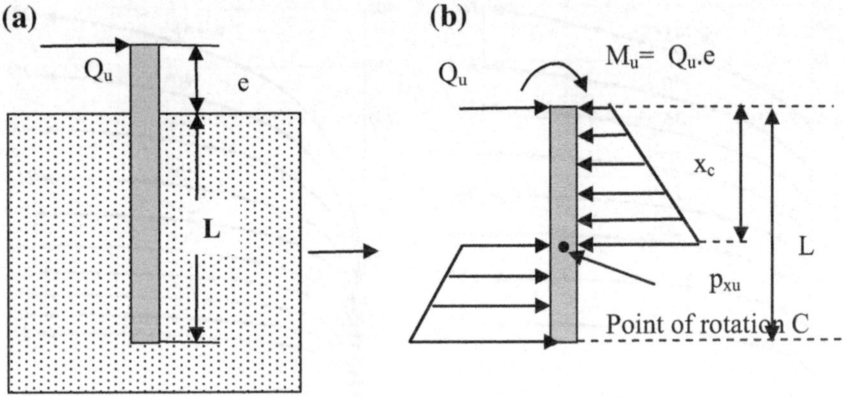

Fig. 6.22 Pressure distribution in a free headed laterally loaded piles after Brinch Hansen

Again taking summation of moments = 0, we have

$$Q_u e + \int_0^{x_c} p_{xu} x D dx - \int_{x_c}^{L} p_{xu} D x dx = 0 \tag{6.6.2.2}$$

In Eqs. (6.6.2.1) and (6.6.2.2), D is the diameter of the pile and x_c is the point of rotation.

For a particular pile, if one can find out p_{xu} and x_c, then it is possible to compute Q_u from Eqs. (6.6.2.1) and (6.6.2.2).

The ultimate pressure p_{xu} is expressed as

$$p_{xu} = \bar{\sigma}_v K_q + c K_c \tag{6.6.2.3}$$

Here

$\bar{\sigma}_v$	effective overburden pressure;
c	cohesion of soil;
K_q and K_c	factors that are a function of x/D and friction angle ϕ, as shown in Fig. 6.23.

It may be noted that Brinch Hansen considered width of pile as B, which as per notation as derived herein is D. The term x_c is actually obtained by trial and error, and the steps to be adapted to compute the ultimate load capacity Q_u are summarized hereafter.

- Discretize the soil profile into number of layers.
- Compute $\bar{\sigma}_v$, K_p, and K_c for each thickness h of the soil layer.
- Calculate the pressure p_{xu} for each of the soil layers, and plot it with depth.

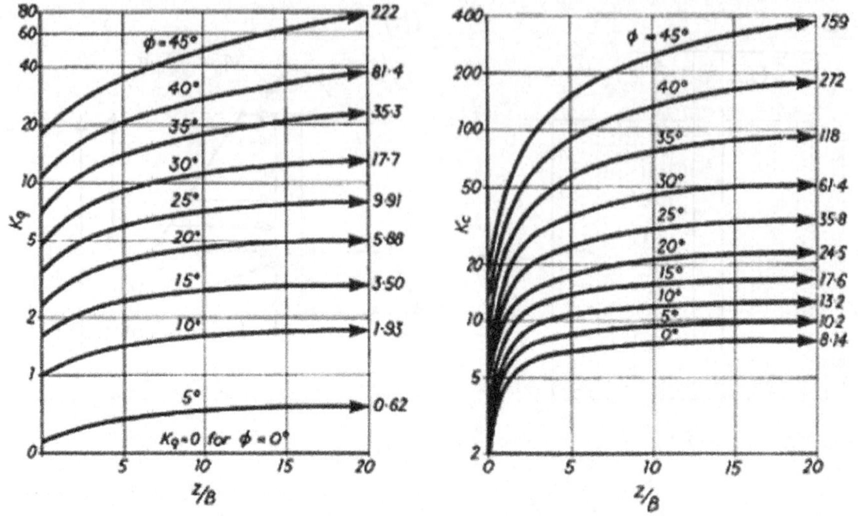

Fig. 6.23 Charts of K_p and K_c as per Brinch Hansen (1961)

- Assume depth of rotation at a depth x_c below soil line, and take moment about the application point Q_u (refer to Fig. 6.22).
- If this moment is of small magnitude or zero, the assumed depth x_c is correct; else modify x_c to achieve a lower value till it converges.
- Once x_c is computed, take moment about x_c as depicted in Eq. (6.6.2.2) to compute the value of Q_u.

The method is now elaborated by a suitable numerical problem.

Example 6.7 Shown in Fig. 6.24 is the geotechnical profile of a site where a pile of 1200 mm diameter and 16.2 m long is to be installed. The pile is a precast driven pile. The net overburden pressure as computed is also shown in the right hand-side of the figure. The pile is freestanding above ground for a height of 1.2 m. Considering that the groundwater table exists at 2.0 m below ground level, compute the lateral load capacity of the pile by Brinch Hansen's method. Considering a factor of safety FOS = 2.5, compute the allowable lateral capacity of pile.

Solution

X(m)	x/D	ϕ_D[a]	c	K_p	K_c	$\bar{\sigma}_v$	p_{xu}
0	0	31	0	5.5	5	0	0
2	1.64	31	0	7	13	32.2	225.4
3	2.5	33	0	7.5	30	41.8	313.5
15	12.5	33	1	12	49	161.8	1990.6

[a]For computation of ϕ_D, refer to Example 6.6.

The average pressure on pile shaft is as shown in Fig. 6.25.

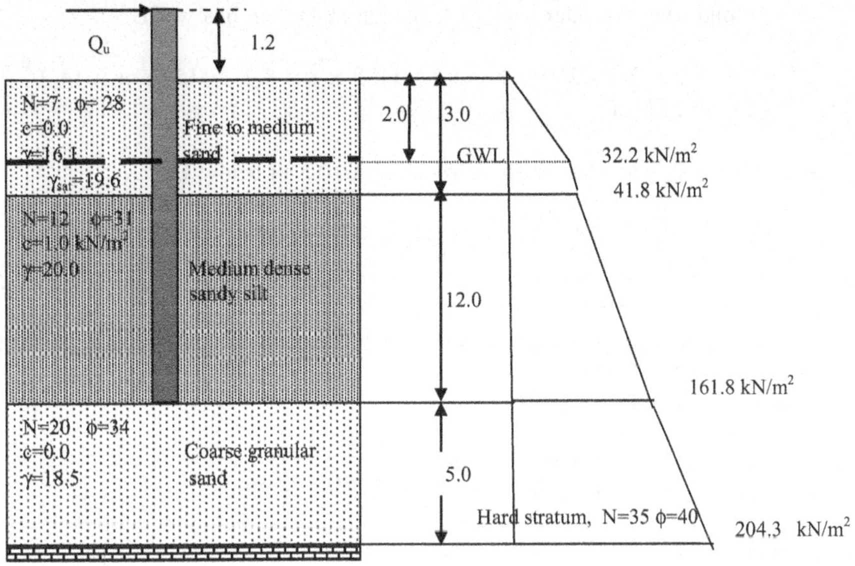

Fig. 6.24 Geotechnical profile of soil and pile driven in it

Fig. 6.25 Average pressures on pile shaft under lateral load

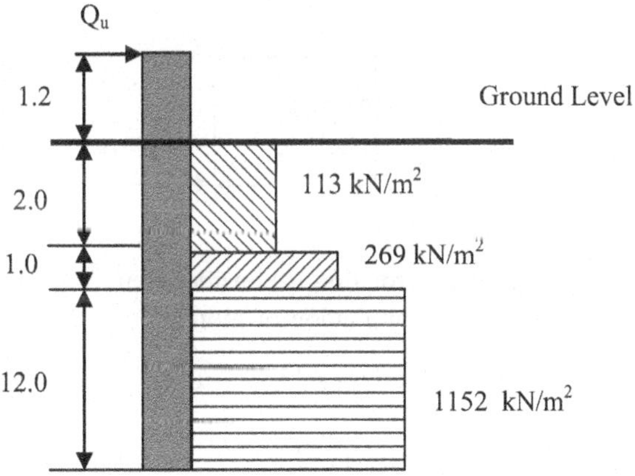

As a first trial, let us assume the center of rotation at 10.0 m below ground level. Thus taking moment about the free head, we have

$$\begin{aligned} M &= 113 \times 2 \times 2.2 + 269 \times 1 \times 3.7 + 1152 \times 7 \times 7.7 - 1152 \times 5 \times 13.7 \\ &= -15326.7\,\text{kN m} \end{aligned}$$

As a second trial, consider center of rotation at 11.0 m below GL

$$M = 113 \times 2 \times 2.2 + 269 \times 1 \times 3.7 + 1152 \times 8 \times 8.2 - 1152 \times 4 \times 14.2 = 11630.1\,\mathrm{kN\,m}$$

As a third trial, consider center of rotation at 10.5 m below GL

$$M = 113 \times 2 \times 2.2 + 269 \times 1 \times 3.7 + 1152 \times 7.5 \times 7.95 - 1152 \times 4.5 \times 13.95 = -2136.3\,\mathrm{kN\,m}$$

Considering that the moments change sign between 11 and 10.5 m, we can plot this as shown below

From the figure, we get

$$\frac{11630}{0.5 - x} = \frac{2136.3}{x} \rightarrow x = 0.0776\,\mathrm{m}$$

where moment is zero.

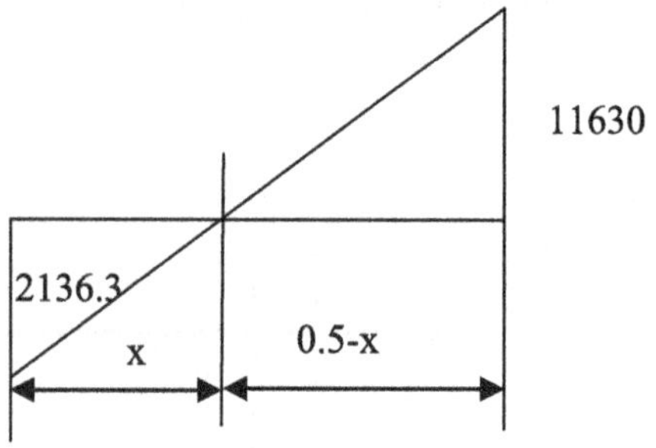

Thus the center of rotation is considered at x_c = 10.5 +0.0776 = 10.5776 m below GL. Considering summation of moment about the point x_c = 0

$$Q_u(10.5776 + 1.2) - 113 \times 2 \times 9.5776 \times 1.2 - 269 \times 1 \times 8.0776 \times 1.2 - 1152 \times \frac{7.5776^2}{2} \times 1.2 + 1152 \times \frac{4.4424^2}{2} \times 1.2 = 0$$

In the above equation, diameter of pile is considered as 1200 mm.

Solving the above, we have $Q_u = \frac{(2164.53 + 2172.87 + 33073.93 - 11367.31) \times 1.2}{11.7776} = 2654\,\mathrm{kN}$

Thus, $Q_{all} = \frac{2654}{2.5} = 1061$ kN.

6.6.2.2 Broms' (1964) Method for Estimation of Lateral Load of Piles

Broms (1964) developed a method analogous in approach to that of Brinch Hansen, however with some simplifying assumptions. While Brinch Hansen developed the procedure for generalized c–ϕ soil, Broms developed separate analysis for cohesionless soil ($c = 0$) and pure cohesive soil ($\phi = 0$).

However, one advancement Broms achieved was: Brinch Hansen's method is essentially valid for short piles, while Broms proposed the formulation for both long and short piles.

As per Broms, a pile is considered to be short and rigid when $L/T \leq 2$ or $L/R \leq 2$.

Here

$$T = [EI/\hat{n}_h]^{\frac{1}{5}} \text{ and } R = [EI/n_h]^{\frac{1}{4}} \tag{6.6.2.4}$$

E Young's modulus of pile material;
I moment of inertia of pile material;
$\hat{n}_{\mathrm{h}}$ linearly increasing soil modulus with depth z; and
n_{h} constant of modulus of sub-grade reaction.

Broms' Analysis of Short Piles in Cohesive Soil

Figure 6.26 shows the soil reaction and bending moment for a pile of length L and width B when it is free headed and restrained.

The lateral load (H_{u}) induced on the pile is computed from the graph as shown in Fig. 6.27.

Once the lateral load is determined, the point of zero shear or maximum moment is computed from the expression

$$f = \frac{H_{\mathrm{u}}}{9c_{\mathrm{u}}B} \tag{6.6.2.5}$$

$$M_{\max} = H_{\mathrm{u}}(e + 1.5B + 0.5f) = 2.25c_{\mathrm{u}}Bg^2 \tag{6.6.2.6}$$

For short, rigid, *fixed-headed* pile (Fig. 6.28b), as per Broms, the maximum moment is expressed as

$$M_{\max} = \frac{1}{2}\left[9c_{\mathrm{u}}B\left(L^2 - 2.25B^2\right)\right] \tag{6.6.2.7}$$

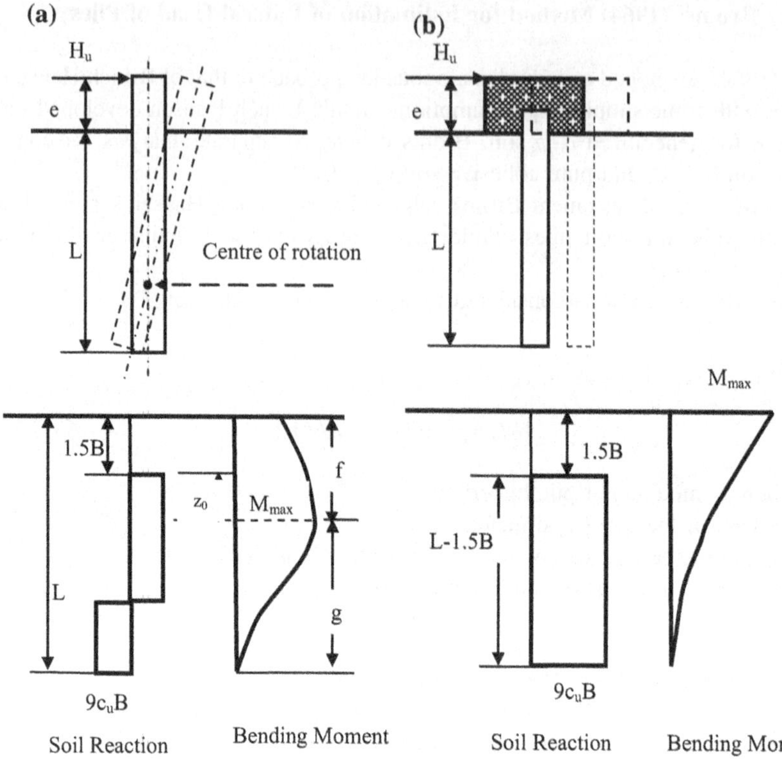

Fig. 6.26 Rotational (**a**) and translational (**b**) movement under ultimate load for short pile in cohesive soil

Analysis of Short Piles in Cohesionless Soil ($c = 0$)

For cohesionless soil for free-headed and fixed-headed short piles, the soil reactions are as shown in Fig. 6.28a, b. At any depth z, the soil reaction is expressed as

$$p_z = 3B\sigma'_v K_p \tag{6.6.2.8}$$

Here

- B width of pile;
- σ'_v effective overburden pressure; and
- K_p passive earth pressure when $K_p = 1 + \sin\phi/1 - \sin\phi$. For uniform cohesionless soil, Broms represented the ultimate lateral force graphically, as shown in Fig. 6.29.

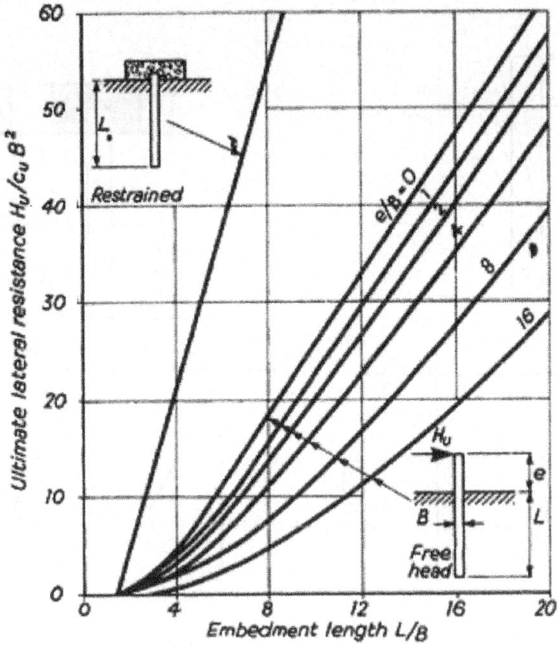

Fig. 6.27 Ultimate lateral load resistances of short piles in cohesive soil (ϕ = 0) after Broms (1964)

The ultimate lateral resistance can also be expressed as

$$H_u = \frac{\gamma K_p B L^3}{2(e + L)} \tag{6.6.2.9}$$

For short rigid fixed-headed piles where failure takes place due to translational movement is shown in Fig. 6.28b

$$H_u = \frac{3\gamma K_p B L^2}{2} \tag{6.6.2.10}$$

The maximum moment at pile head (due to soil failure) is expressed as

$$M_{max} = \gamma K_p B L^3 \tag{6.6.2.11}$$

Equation (6.6.2.11) is only valid when the value is less than structural moment capacity of the pile.

Example 6.8 For Example 6.7, as furnished earlier, ignore the small value of c for middle layer and determine the ultimate lateral resistance and moment for the pile

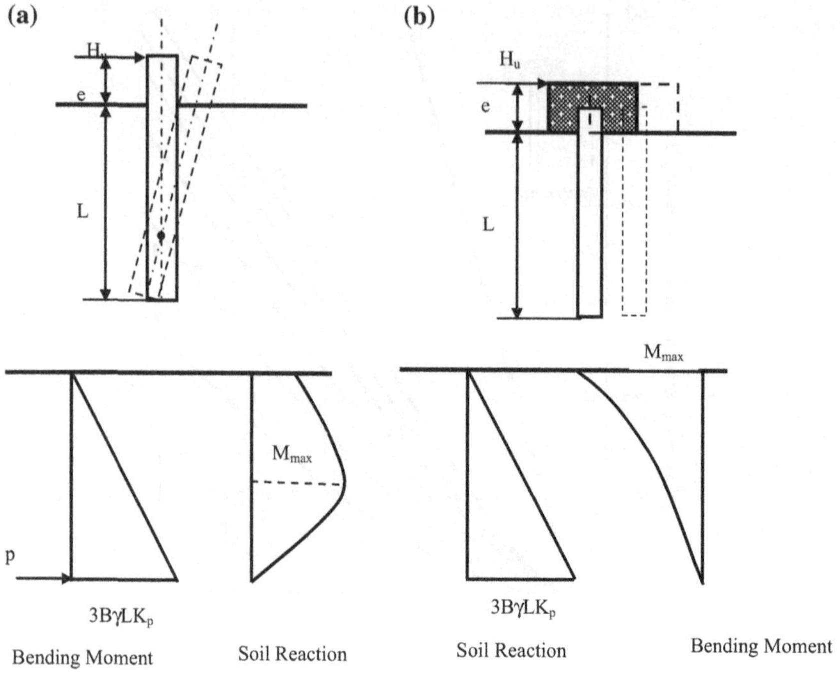

Fig. 6.28 Rotational (**a**) and translational (**b**) movement under ultimate load for short pile in cohesionless soil

Fig. 6.29 Ultimate lateral load resistances of short piles in cohesionless soil ($c = 0$) after Broms (1964)

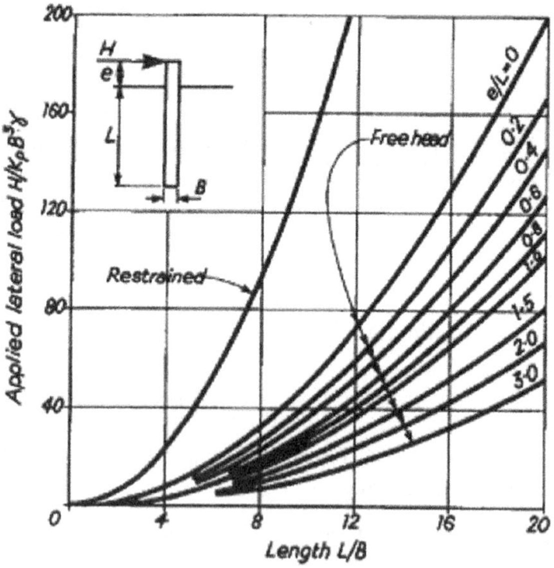

as per Broms' method. Consider free head of 1.2 m, and also compute the lateral shear for restrained head pile.

Solution

As per Example 6.6 design, ϕ is for 0–2 m 31° and for 2–15 m it is 33°. Thus, the weighted average value of ϕ is

$$\phi_w = \frac{31 \times 2 + 33 \times 13}{15} = 31.73$$

Thus, $K_p = \frac{1 + \sin\phi_w}{1 - \sin\phi_w} = \frac{1 + \sin 32.73}{1 - \sin 32.73} = 3.35.$

The weighted average value of soil density is expressed as

$$\gamma_{av} = \frac{16.1 \times 2 + 9.6 \times 1 + 10 \times 12}{15} = 10.78\,\text{kN/m}^3$$

Consider $H_u = \frac{\gamma K_p B L^3}{2(e+L)} = \frac{10.78 \times 3.35 \times 1.2 \times (15)^3}{2(1.2+15)} = 4514$ kN for the pile being free-headed.

For restrained head pile $H_u = \frac{3\gamma K_p B L^2}{2} = \frac{3 \times 10.78 \times 3.35 \times 1.2 \times (15)^2}{2} =$ 14625 kN.

$$M_{max} = \gamma K_p B L^3 = 10.78 \times 3.35 \times 1.2 \times (15)^3 = 146258\,\text{kN m}$$

Analysis of Long Piles in Cohesive Soil (ϕ = 0)

For long piles, the behavior of free-headed piles restrained at head is as shown in Fig. 6.30. The maximum moment for the free-headed pile is expressed as

$$M_{max} = H_u(e + 1.5B + 0.5f) \qquad (6.6.2.12)$$

where f is as expressed in Eq. (6.6.2.5).

If M_u is the ultimate moment capacity of pile (this is the structural moment of the pile shaft), then H_u can be derived from the graph as proposed by Broms, as shown in Fig. 6.31.

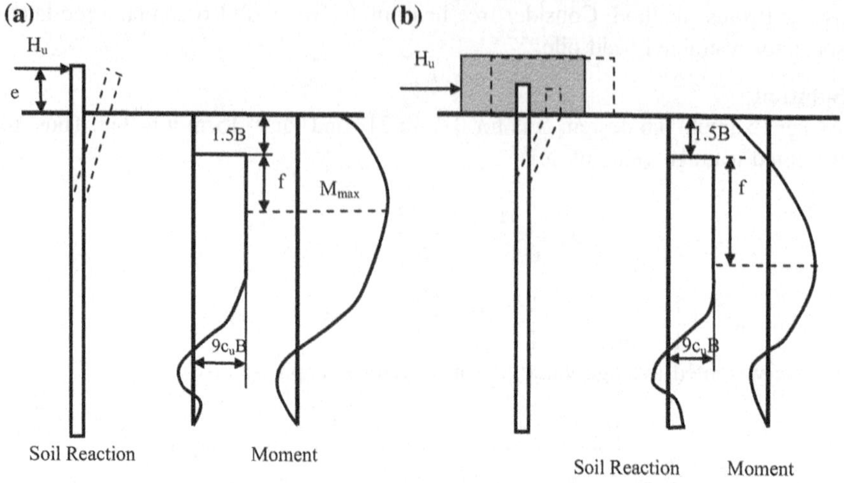

Fig. 6.30 Rotational and translational movement under ultimate load for long pile in cohesive soil

Fig. 6.31 Ultimate lateral resistance of long piles in cohesive piles after Broms (1964)

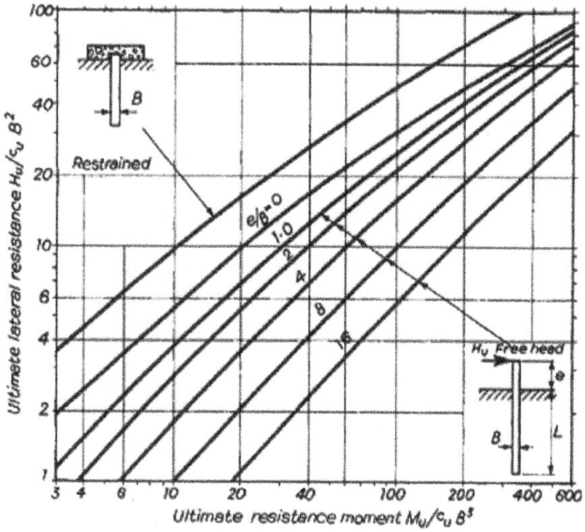

Analysis of Long Piles in Cohesionless Soil (c = 0)

The behavior of long piles in cohesionless soil is as shown in Fig. 6.32. In this case

$$f = 0.82\sqrt{\frac{H}{\gamma B K_p}} \tag{6.6.2.13}$$

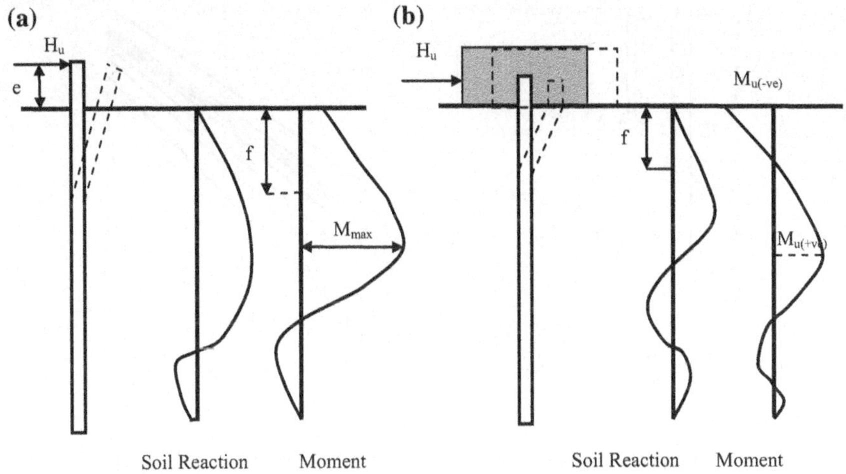

Fig. 6.32 Rotational (**a**) and translational (**b**) movement under ultimate load for long pile in cohesionless soil

The maximum positive bending moment for free-headed pile is expressed as

$$M_{\max} = H\left(e + \frac{2}{3}f\right) \tag{6.6.2.14}$$

For zero bending moment at pile head, the ultimate lateral load is expressed as

$$H_u = \frac{M_u}{e + 0.54\sqrt{\frac{H_u}{\gamma B K_p}}} \tag{6.6.2.15}$$

Broms has also established design charts from which the ultimate lateral resistance can also be established, as shown in Fig. 6.33.

In Fig. 6.33, the moment M_u is the ultimate structural moment capacity of the pile shaft.

Some Comments on the Methods as Cited Above

As observed, Broms' method is the quickest and easy for computation, while Brinch Hansen's method, though more general, is more tedious as it requires trial and error for estimation of the center of rotation.

One of the major disadvantages with the ultimate load approach is that one cannot estimate the pile head deflection that becomes critical for earthquake analysis, especially when piles support tall structures or buildings.

Unacceptable deflection at pile head can further aggravate the deflection at top of the structure and can also induce secondary damage to its members.

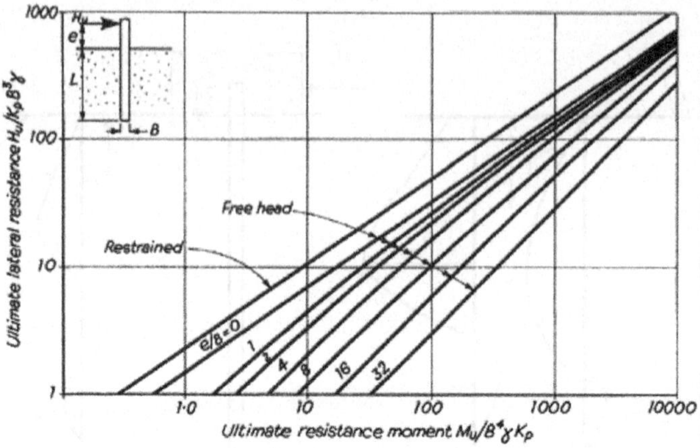

Fig. 6.33 Ultimate lateral resistances of long piles in cohesionless piles after Broms (1964)

Subsequent research and comparison with finite element method (FEM) has also shown (Ruigrok 2010) that the conventional methods are not very accurate and in many cases the outcome widely varies with field observation and FEM analysis.

Thus, most of the codes of practice that adopts such type of analysis for estimation of lateral load capacity of a pile usually advocate load testing to finally vindicate the theoretical calculations.

Length of Fixity Approach as Adopted by IS-2911, Indian Code of Practice IS-2911 code of practice for design and installation of pile foundation by Bureau of Indian Standards uses the depth of fixity concept for estimation of lateral capacity of piles.

The depth of fixity can be conceptually explained as hereafter.

Shown in Fig. 6.34 is a long slender free-headed pile embedded in ground and subjected to lateral load. As the load gets transferred to the soil, there is a particular depth beyond which the load quickly dies down and beyond this point the pile shaft remains more or less undeformed. This particular depth beyond which the pile remains undeformed is known as the depth of fixity of pile (L_f) where the pile is assumed to be idealized as fixed and acts as a cantilever beam, as shown in Fig. 6.34.

As per IS-2911, the depth of fixity for a long and slender pile fixed or free headed is determined from the diagram as presented in Fig. 6.35 hereafter.

In Fig. 6.35, $T = [EI/K_1]^{\frac{1}{5}}$ and $R = [EI/K_2]^{\frac{1}{4}}$. Here, E and I are the Young's modulus of pile and moment of inertia of pile cross section, respectively. The values of K_1 and K_2 are furnished in Tables 6.8 and 6.9, respectively.

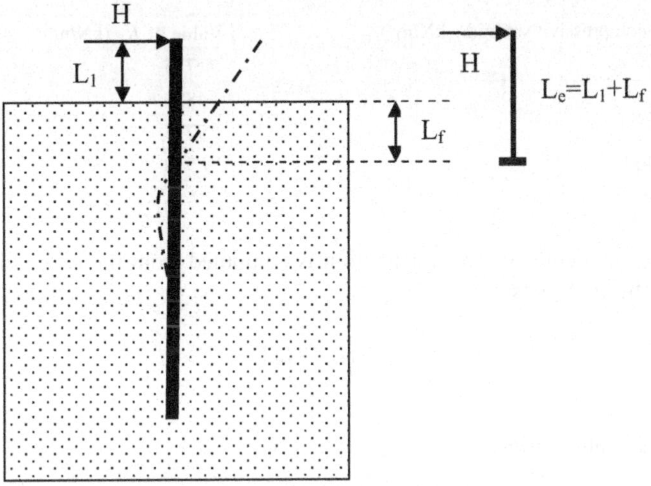

Fig. 6.34 Deformation of long slender pile embedded in ground

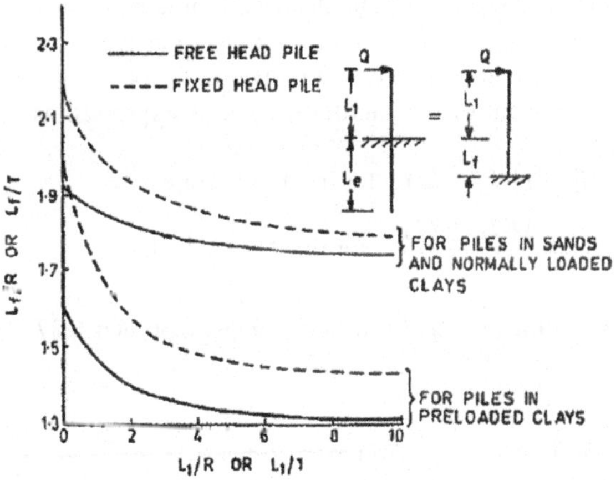

Fig. 6.35 Determination of depth of fixity as per IS 2911

Table 6.8 Values of K_1 as recommended in IS-2911

Soil type	Value of K_1 (kN/m^3)	
	Dry	Submerged
Loose sand	2600	1460
Medium sand	7750	5250
Dense sand	20,750	12,450
Very loose sand under repeated loading or normally loaded clays	–	400

Table 6.9 Values of K_2 as recommended in IS-2911

Unconfined compressive strength (kN/m^2)	Value of K_2 (kN/m^2)
20–40	775
100–200	4880
200–400	9775
More than 400	19,550

Knowing the depth of fixity, deflection is computed from the expressions.
For free-headed piles

$$y = \frac{Q(L_1 + L_f)^3}{3EI} \tag{6.6.2.16}$$

For fixed-headed piles

$$y = \frac{Q(L_1 + L_f)^3}{12EI} \tag{6.6.2.17}$$

The moment at the fixed end of the idealized cantilever beam (M_f) is higher than the actual moment induced in the pile (M_p); thus, to arrive at the value of M_p, the fixed-end moment M_f is multiplied by a reduction factor m, when $M_p = m \cdot M_f$.

The fixed-end moment of the cantilever beam is expressed as

$$\begin{aligned} M_f &= Q(L_1 + L_f) \quad \text{for free headed piles} \\ &= \frac{Q(L_1 + L_f)}{2} \quad \text{for fixed headed piles} \end{aligned} \tag{6.6.2.18}$$

The reduction factor m is as furnished in Figs. 6.36 and 6.37.

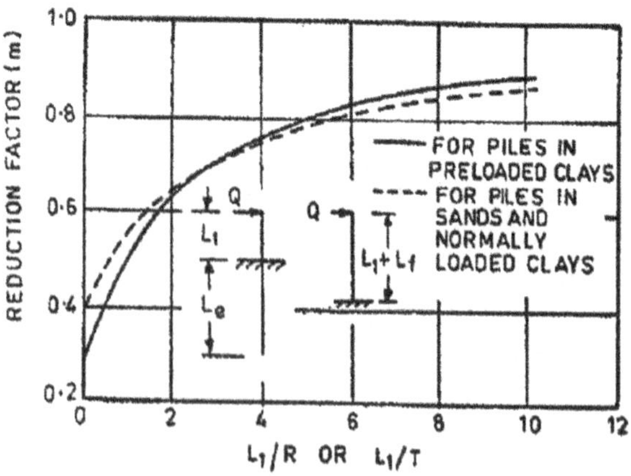

Fig. 6.36 Reduction factor m for a free-headed pile as per IS 2911

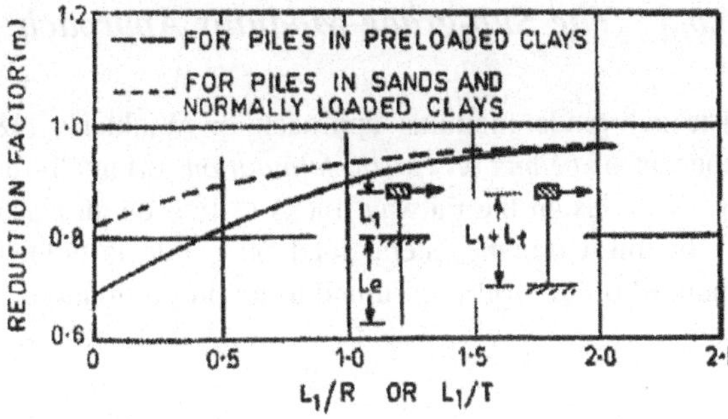

Fig. 6.37 Reduction factor m for fixed-headed pile as per IS 2911

As per the author's experience, the IS codes procedure is highly simplified and varies widely with field observations and other sophisticated analyses like finite difference and FEM analysis and *should be used with caution*. The results should preferably be corroborated against field load tests.

Example 6.9 For Example 6.7 as cited earlier, determine the allowable lateral load of the pile when the lateral displacement is restricted to 20 mm. What moment is induced in the pile?

Solution

As cited in Example 6.8, the average design friction angle $\phi = 32.73°$ that shows the soil to be medium dense. Thus, referring to Table 6.8, assume $K_1 = 6000\ \text{kN/m}^3$.

$E = 2.5 \times 10^7\ \text{kN/m}^2$, and $I = \frac{\pi D^4}{64} = \frac{\pi \times (1.2)^4}{64} = 0.10178\ \text{m}^4 \rightarrow EI = 2544690$ kN m^2.

$$T = \left[\frac{EI}{K_1}\right]^{\frac{1}{5}} = \left[\frac{2544690}{6000}\right]^{0.2} = 3.35$$

Thus $L_1/T = 1.2/3.35 = 0.357$. From Fig. 6.35 $\rightarrow L_f/T = 1.8$ which gives $L_f = 6.03$ m.

For free headed piles

$$y = \frac{Q(L_1 + L_f)^3}{3EI}$$

or

$$Q = \frac{3EIy}{(L_1 + L_f)^3} = \frac{3 \times 2544690 \times 20}{(1.2 + 6.03)^3 \times 1000} = 404\ \text{kN}$$

$$M_f = Q(L_1 + L_f) = 404 \times 7.23 = 2921\ \text{kN m}$$

Referring to Fig. 6.6, for $L_1/T = 0.357$, reduction factor $m = 0.48$; thus, the expected moment in pile is $M_p = 0.48 \times 2921 = 1402$ kN m.

6.6.3 The Sub-grade Modulus Approach

The sub-grade modulus approach or Winkler's (1867) model is based on the concept of *beams on elastic foundation*, originally used by Russian railway engineers to design the railway tracks resting on an elastic base.

In this case, the lateral load on a pile is embedded in ground; the pressure induced on the soil is assumed to be proportional to deflection, which is expressed as

$$p = k_s B u \qquad (6.6.3.1)$$

Here

p pressure induced in soil per unit width;
k_s sub-grade modulus of soil having unit kN/m^3;
B width of pile; and
u deflection.

For a beam element, the Euler–Bernoulli's equation of equilibrium is expressed as

$$EI\frac{d^2u}{dz^2} = -M \qquad (6.6.3.2)$$

$$\rightarrow EI\frac{d^3u}{dz^3} = -V \qquad (6.6.3.3)$$

and

$$EI\frac{d^4u}{dz^4} = -p = -k_s B u \qquad (6.6.3.4)$$

Equation (6.6.3.4) can be expressed as

$$EI\frac{d^4u}{dz^4} + k_s B u = 0 \qquad (6.6.3.5)$$

$$\frac{d^4u}{dz^4} + \frac{k_s B}{EI}u = 0 \qquad (6.6.3.6)$$

Let $\lambda^4 = \frac{k_s B}{EI}$, and considering $D = d/dx$, an operator, we have

$$(D^4 + \lambda^4)u = 0 \qquad (6.6.3.7)$$

As

$$u \neq 0 \rightarrow (D^4 + \lambda^4) = 0 \tag{6.6.3.8}$$

This gives

$$D^4 + \lambda^4 + 2D^2\lambda^2 - 2D^2\lambda^2 = 0 \tag{6.6.3.9}$$

$$\rightarrow (D^2 + \lambda^2)^2 - 2D^2\lambda^2 = 0 \tag{6.6.3.10}$$

or

$$\left(D^2 + D\lambda\sqrt{2} + \lambda^2\right)\left(D^2 - D\lambda\sqrt{2} + \lambda^2\right) = 0 \tag{6.6.3.11}$$

Equation (6.6.3.11) gives

$$\left(D^2 + D\lambda\sqrt{2} + \lambda^2\right) = 0 \text{ or } \left(D^2 - D\lambda\sqrt{2} + \lambda^2\right) = 0 \tag{6.6.3.12}$$

For $\left(D^2 + D\lambda\sqrt{2} + \lambda^2\right) = 0$ we have

$$D = \frac{-\lambda\sqrt{2} \pm i\lambda\sqrt{2}}{2} = -\frac{\lambda}{\sqrt{2}} \pm i\frac{\lambda}{\sqrt{2}} \tag{6.6.3.13}$$

For $\left(D^2 - D\lambda\sqrt{2} + \lambda^2\right) = 0$ we have

$$D = \frac{\lambda\sqrt{2} \pm i\lambda\sqrt{2}}{2} = \frac{\lambda}{\sqrt{2}} \pm i\frac{\lambda}{\sqrt{2}} \tag{6.6.3.14}$$

Thus, Eq. (6.6.3.8) gives four roots

$$(1)\ -\tfrac{\lambda}{\sqrt{2}} + i\tfrac{\lambda}{\sqrt{2}} \quad (2)\ -\tfrac{\lambda}{\sqrt{2}} - i\tfrac{\lambda}{\sqrt{2}} \quad (3)\ \tfrac{\lambda}{\sqrt{2}} + i\tfrac{\lambda}{\sqrt{2}} \quad (4)\ \tfrac{\lambda}{\sqrt{2}} - i\tfrac{\lambda}{\sqrt{2}}$$

Let $p = \frac{\lambda}{\sqrt{2}} = \sqrt{4\frac{k_s B}{4EI}}$ which gives the solution as

$$u = Ae^{(-p+ip)z} + Be^{(-p-ip)z} + Ce^{(p+ip)z} + De^{(p-ip)z} \tag{6.6.3.15}$$

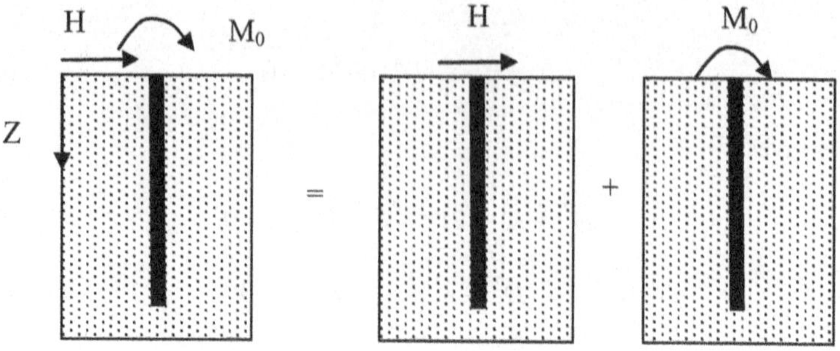

Fig. 6.38 A pile subjected to lateral load (H) and moment (M_0) embedded in ground

Considering $e^{\pm ipz} = \cos pz \pm i \sin pz$, Eq. (6.6.3.15) can be finally expressed as

$$u = e^{-pz}[A_0 \cos pz + B_0 \sin pz] + e^{pz}[C_0 \cos pz + D_0 \sin pz] \tag{6.6.3.16}$$

Here, A_0, B_0, C_0, and D_0 are integration constants and need to be evaluated from the boundary condition of the problem in hand.

While describing previously the length of fixity approach as adopted by IS-2911, Indian code of practice, it is observed that the response dies down quickly beyond a certain distance. Thus, u (displacement) cannot increase as z increases, and this shows that the second portion of Eq. (6.6.3.15) is not valid, which finally gives the solution as

$$u = e^{-pz}[A_0 \cos pz + B_0 \sin pz] \tag{6.6.3.17}$$

Considering the analysis is in elastic domain, a pile embedded in ground subjected to load and moment is as shown in Fig. 6.38.

Solution for developed bending moment and shear force in the pile is as presented below (Den Hartog 1952).

For lateral load H

$$u = \frac{2pH}{k_s B} e^{-pz} \cos pz \tag{6.6.3.18a}$$

$$\theta = -\frac{-2p^2 H}{k_s B} e^{-pz} (\cos pz + \sin pz) \tag{6.6.3.18b}$$

$$M_z = \frac{H}{p} e^{-pz} \sin pz \tag{6.6.3.18c}$$

$$V_z = He^{-pz} (\cos pz - \sin pz) \tag{6.6.3.18d}$$

For applied moment M_0

$$u = \frac{2p^2 M_0}{k_s B} e^{-pz} (\cos pz - \sin pz) \quad (6.6.3.19a)$$

$$\theta = -\frac{-4p^3 M_0}{k_s B} e^{-pz} \cos pz \quad (6.6.3.19b)$$

$$M_z = M_0 e^{-pz} (\cos pz + \sin pz) \text{ and } V_z = -2pM_0 e^{-pz} \sin pz \quad (6.6.3.19c)$$

Considering the analysis is elastic in nature, the values (for lateral load H and applied moment M_0) can now be superimposed to determine the final response.

Example 6.10 A pile of 300 mm diameter and 16.2 m long is embedded in soil whose average sub-grade modulus is 16,000 kN/m^3. Determine the allowable lateral load of the pile when the lateral displacement is restricted to 20 mm.

The pile has freestanding length of 1.2 m above ground and the remaining 15 m embedded in ground. What is the moment induced in the pile? Consider grade of concrete used for pile as M25.

Solution

Diameter of pile = 300 mm.

Thus

$$I = \frac{\pi D^4}{64} = \frac{\pi \times (0.3)^4}{64} = 0.000398 \, \text{m}^4.$$

$$E_{\text{conc}} = 2.5 \times 10^7 \, \text{kN/m}^2 \rightarrow EI = 9.94 \times 10^3 \, \text{kN/m}^2.$$

$$p = \sqrt[4]{\frac{k_s B}{4EI}} = \sqrt[4]{\frac{16000 \times 0.3}{4 \times 9.94 \times 10^3}} = 0.589449$$

The formulation as cited in Sect. 6.6.3 was developed in a computer program, and the outputs are as shown hereafter.

Z (m)	y(H) m	y(M) m	y(T) m	$V_z(H)$ kN	$V_z(M)$ kN	$V_z(T)$ kN**	$M_z(H)$ kN m	M_z (M) kN m	M_z (T) kN m[a]
0	1.18E−02	8.34E−03	2.01E−02	48	0	48	0	57.6	57.6
1	5.44E−03	1.27E−03	6.71E−03	7.33	−20.94	−13.61	25.11	44.32	69.42
2	1.39E−03	−1.39E−03	−5.83E−06	−8.01	−19.30	−27.31	23.15	23.14	46.29
3	−3.95E−04	−1.67E−03	−2.07E−03	−9.64	−11.36	−21.00	13.62	7.71	21.33
4	−7.90E−04	−1.12E−03	−1.91E−03	−6.42	−4.54	−10.96	5.44	−0.01	5.43
5	−6.07E−04	−5.14E−04	−1.12E−03	−2.96	−0.69	−3.65	0.83	−2.38	−1.56
6	−3.17E−04	−1.31E−04	−4.47E−04	−0.75	0.76	0.01	−0.91	−2.19	−3.11
7	−1.05E−04	3.77E−05	−6.76E−05	0.22	0.91	1.13	−1.10	−1.29	−2.38
8	3.39E−07	7.49E−05	7.52E−05	0.43	0.61	1.04	−0.73	−0.51	−1.24
9	3.27E−05	5.75E−05	9.02E−05	0.33	0.28	0.61	−0.34	−0.08	−0.41
10	3.01E−05	3.00E−05	6.00E−05	0.17	0.07	0.24	−0.09	0.09	0.00
11	1.76E−05	9.94E−06	2.76E−05	0.06	−0.02	0.04	0.02	0.10	0.13
12	7.03E−06	−4.81E−08	6.98E−06	0.00	−0.04	−0.04	0.05	0.07	0.12
13	1.05E−06	−3.10E−06	−2.05E−06	−0.02	−0.03	−0.05	0.04	0.03	0.07
14	−1.19E−06	−2.85E−06	−4.04E−06	−0.02	−0.02	−0.03	0.02	0.01	0.03
15	−1.42E−06	−1.67E−06	−3.09E−06	−0.01	−0.01	−0.02	0.01	0.00	0.00

[a]Here, V_z (T) and M_z (T) represent the sum of shears and moments due to horizontal load and applied moment

The graphical plots of displacement, shear, and moment are as shown hereafter (Figs. 6.39, 6.40, and 6.41).

Values in the table show that for displacement to be restricted to 20 mm, the maximum shear force is 48 kN and the bending moment is 69.4 kN m.

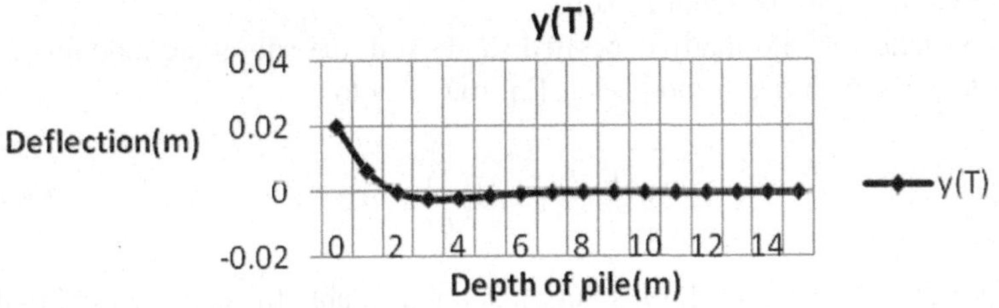

Fig. 6.39 Displacement plot of a pile

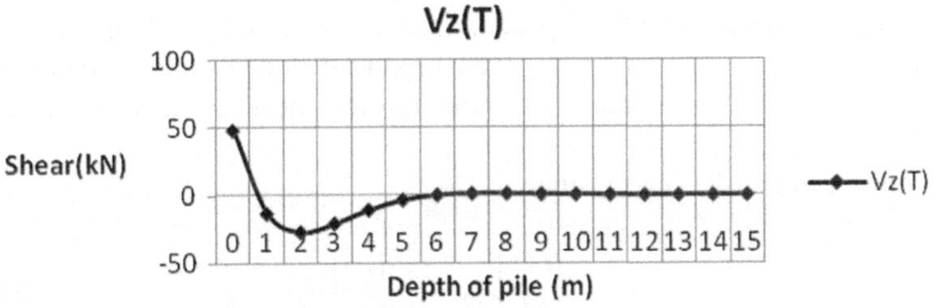

Fig. 6.40 Shear plot of a pile

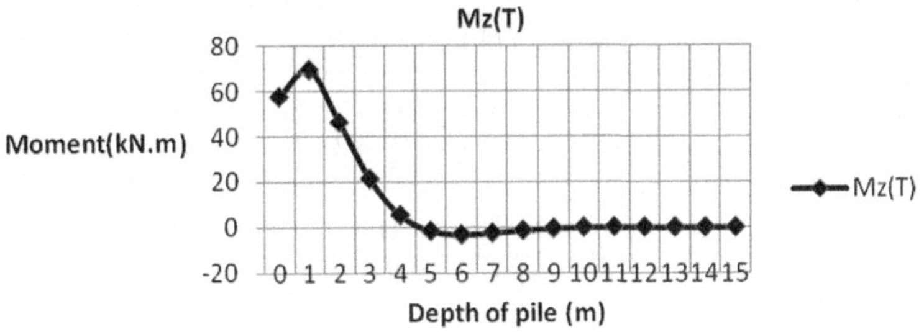

Fig. 6.41 Bending moment diagram of the pile

6.6.4 Matlock and Reese' (1960) Method for Lateral Load Analysis of Long Slender Piles

The method proposed earlier is mathematically rigorous but suffers from one significant lacunae; that is, the sub-grade modulus of the soil is assumed invariable with depth. Except for normally/over consolidated stiff clay, sub-grade modulus usually varies with the depth of soil.

For instance, for normally deposited sandy soil, the sub-grade modulus usually varies linearly with depth modifying Eq. (6.6.3.5) to

$$EI\frac{d^4u}{dz^4} + k_s B\left(\frac{z}{L}\right)u = 0 \tag{6.6.4.1}$$

Unfortunately, a closed-form solution is not available for such a case. Finally, in many cases, soil is layered, where each layer will have different sub-grade moduli. It is apparent that only possible way to arrive at solutions to these cases is based on numerical analysis.

Matlock and Reese (1960), based on finite difference method, arrived at a solution for free-headed and fixed-headed pile where the sub-grade modulus varies with depth. Based on this analysis, they produced dimensionless coefficients to determine deflection (y), rotation (S), moment (M), shear (V), and pressure (p) as follows.

For pile head that is free to rotate

$$y = \left[\frac{HT^3}{EI}\right]A_y + \left[\frac{MT^2}{EI}\right]B_y \tag{6.6.4.2}$$

$$S = \left[\frac{HT^2}{EI}\right]A_s + \left[\frac{MT}{EI}\right]B_s \tag{6.6.4.3}$$

$$M = [HT]A_m + [M_0]B_m \tag{6.6.4.4}$$

$$V = [H]A_v + \left[\frac{M_0}{T}\right]B_v \tag{6.6.4.5}$$

$$p = \left[\frac{H}{T}\right]A_p + \left[\frac{M_0}{T^2}\right]B_p \tag{6.6.4.6}$$

In Eqs. (6.6.4.2) to (6.6.4.6), T is the relative stiffness factor expressed as

$$T = \left[\frac{EI}{n_h}\right]^{\frac{1}{5}} \tag{6.6.4.7}$$

Here, n_h = coefficient of sub-grade modulus that varies with depth.

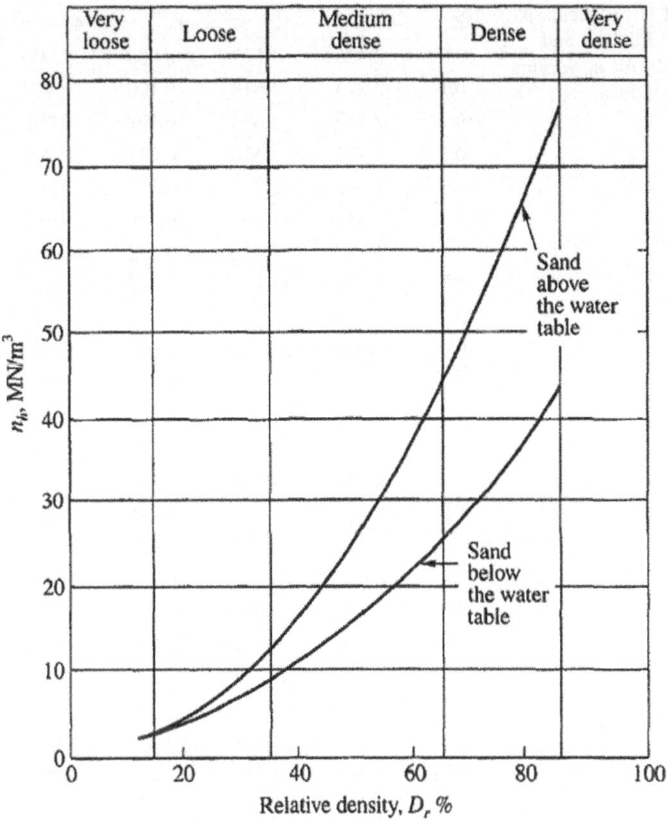

Fig. 6.42 Values of n_h as recommended by Matlock and Reese (1960)

Values of n_h, as proposed by Matlock and Reese (1960) correlating it to relative density of sand D_r, are furnished in Fig. 6.42.

Typical values of n_h for cohesive soil are furnished in Table 6.10 (after Poulos and Davis 1980) in SI units.

The values of A_y, A_s, A_M, A_v, A_p and B_y, B_s, B_M, B_v, B_p are furnished in Table 6.11.

Table 6.10 Typical values of n_h for cohesive oil

Soil type	n_h (kN/m³)	After
Soft normally consolidated clay	166–3515	Reese and Matlock (1956)
	276.8–553.6	Davisson and Prakash (1963)
Normally consolidated organic clay	111–276.8	Peck and Davisson (1962)
	111–830.4	Davisson (1970)
Loess	8027–11,072	Bowles (1968)
Peat	55.4	Davisson (1970)
	27.7–110.7	Wilson and Hills (1967)

Table 6.11 Coefficients A and B after Matlock and Reese (1960) for n_h varying linearly with depth

Z	A_y	A_s	A_m	A_v	A_p
0.0	2.435	−1.623	0.000	1.000	0.000
0.1	2.273	−1.618	0.100	0.989	−0.227
0.2	2.112	−1.603	0.198	0.966	−0.422
0.3	1.952	−1.578	0.291	0.906	−0.586
0.4	1.796	−1.545	0.379	0.840	−0.718
0.5	1.644	−1.503	0.459	0.764	−0.822
0.6	1.496	−1.454	0.532	0.677	−0.897
0.7	1.353	−1.397	0.595	0.585	−0.947
0.8	1.216	−1.335	0.649	0.489	−0.973
0.9	1.086	−1.268	0.693	0.392	−0.977
1.0	0.962	−1.197	0.727	0.295	−0.962
1.2	0.738	−1.047	0.767	0.109	−0.885
1.4	0.544	−0.893	0.772	−0.056	−0.761
1.6	0.381	−0.741	0.746	−0.193	−0.609
1.8	0.247	0.596	0.696	−0.298	−0.445
2.0	0.142	−0.464	0.628	−0.371	−0.283
3.0	−0.075	0.040	0.225	0.349	0.226
4.0	−0.050	0.052	0.000	−0.016	0.201
5.0	0.009	0.025	−0.033	0.013	0.046
Z	B_y	B_s	B_m	B_v	B_p
0.0	1.623	−1.750	1.000	0.000	0.000
0.1	1.453	−1.650	1.000	−0.007	−0.145
0.2	1.293	−1.550	0.999	−0.028	−0.259
0.3	1.143	−1.450	0.994	−0.058	−0.343
0.4	1.003	−1.351	0.987	−0.095	−0.401
0.5	0.873	−1.253	0.976	−0.137	−0.436
0.6	0.752	−1.156	0.960	−0.181	−0.451
0.7	0.642	−1.061	0.939	−0.226	−0.449
0.8	0.540	−0.968	0.914	−0.270	−0.432
0.9	0.448	−0.878	0.885	−0.312	−0.403
1.0	0.364	−0.792	0.852	−0.350	−0.364
1.2	0.223	−0.629	0.775	−0.414	−0.268
1.4	0.112	−0.482	0.668	−0.456	−0.157
1.6	0.029	−0.354	0.594	−0.477	0.047
1.8	−0.030	−0.245	0.498	−0.476	0.054
2.0	−0.070	−0.155	0.404	−0.456	0.140
3.0	0.089	0.057	0.059	−0.0213	0.268
4.0	−0.028	0.049	0.042	0.017	0.112
5.0	0.000	0.011	0.026	0.029	−0.002

In Table 6.11, the term $Z = z/T$, where z is any depth of pile below ground level.

When the piles support a heavy structure or embedded in a heavy pile cap, the pile head becomes fixed against rotation.

The deflection in such case can be computed from the previous case, considering $S = 0$, which finally gives

$$y = \left[A_y - \left(\frac{A_s}{B_s}\right)B_y\right]\frac{HT^3}{EI} \qquad (6.6.4.8)$$

The deflection at soil line ($z = 0$) is thus $y = 0.93\frac{HT^3}{EI}$

Identically, the moment equation can be expressed as

$$M = \left[A_m - \left(\frac{A_s}{B_s}\right)B_m\right]HT \to 0.93HT \text{ at } z = 0. \qquad (6.6.4.9)$$

Example 6.11 A pile of 450 mm diameter and 20 m long is embedded in a sandy soil having $D_r = 60\%$. The groundwater level is 600 mm below soil line. Compute the deflection, moment, and shear at the pile head if the applied load at top of pile cap is 50 kN. The pile cap has a depth of 1.5 m. Grade of concrete = M30.

Solution

For M30 grade concrete $E = 2.74 \times 10^7$ kN/m^2.

Moment of inertia of the pile $= \frac{\pi D^4}{64} = \frac{\pi \times (0.45)^4}{64} = 2.013 \times 10^{-3}$ m^4.

For soil submerged below water with $D_r = 60\%$, $n_h = 25{,}000$ kN/m^3 (refer Fig. 6.42).

Now $T = \left[\frac{EI}{n_h}\right]^{\frac{1}{5}} = \left[\frac{2.74 \times 10^7 \times 2.013 \times 10^{-3}}{25000}\right]^{\frac{1}{5}} = 1.17.$

Here $H = 50$ kN and $M - 50 \times 1.5 - 75$ kN m.

At $z = 0$ from Table 6.10: $A_y = 2.435$ $A_s = -1.623$, $A_m = 0.0$, $A_v = 1.0$ and $B_y = 1.623$, $B_s = -1.750$, $B_m = 1.00$, $B_v = 0.0$.

Thus, the deflection

$$y = \left[\frac{HT^3}{EI}\right]A_y + \left[\frac{MT^2}{EI}\right]B_y = \left[\frac{50 \times 1.17^3}{2.74 \times 10^7 \times 2.013 \times 10^{-3}}\right] \times 2.435 + \left[\frac{75 \times 1.17^2}{2.74 \times 10^7 \times 2.013 \times 10^3}\right] \times 1.623$$

$$\to y = 3.535 \times 10^{-3} + 3.02 \times 10^{-3} = 6.556 \times 10^{-3}\text{m}$$

Moment

$$M = [HT]A_m + [M_0]B_m = [50 \times 1.17] \times 0.0 + [75] \times 1 = 75\,\text{kN m}$$

Shear

$$V = [H]A_v + \left[\frac{M_0}{T}\right]B_v = [50] \times 1.0 + \left[\frac{75}{1.15}\right] \times 0 = 50\,\text{kN}.$$

In case the pile head is assumed fixed

$$y = 0.93\frac{HT^3}{EI} = \frac{0.93 \times 50 \times 1.17^3}{2.74 \times 10^7 \times 2.013 \times 10^{-3}} = 1.35 \times 10^{-3}\,\text{m}$$
$$M = 0.93HT = 54.4\,\text{kN. m}$$

6.6.5 *Numerical Methods for Lateral Load Analysis of Piles*

In most of the cases, a pile embedded in soil has heterogeneous soil profile, as shown in Fig. 6.43. For a pile embedded in soil heterogeneous in nature, it is apparent that magnitude of sub-grade modulus varies with each of the layers and an analytical solution of the same based on Eq. (6.6.3.5) is not possible.

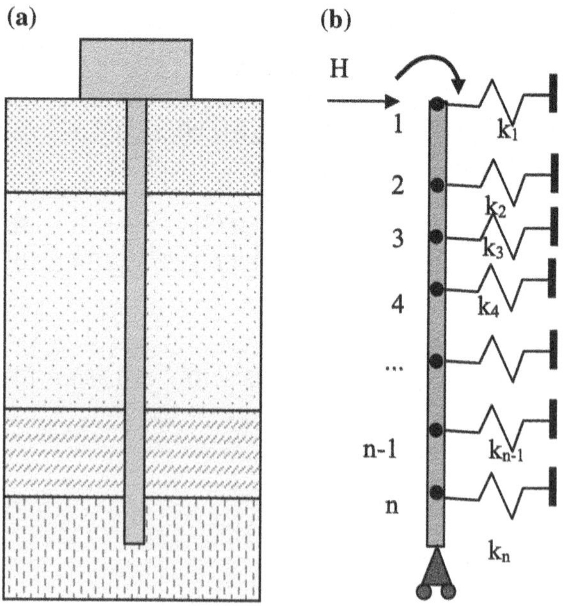

Fig. 6.43 A pile embedded in heterogeneous soil layer and its equivalent mathematical model

Matlock and Reese's method will also be not valid in this case, if the sub-grade modulus of layer 2 say is weaker than layer 1. Moreover, under earthquake in many cases it is not a rarity that one of the layers, especially the one near surface or the one below that, undergoes liquefaction, when the stress distribution is completely different.

It is in these cases that a numerical method based on finite difference (FDM) or finite element method (FEM) can be conveniently deployed to arrive at a solution.

Before getting into the mathematical aspects of the method, the model conceived is described hereafter.

In this case, the pile is discretized into n convenient nodes and the soil pressure is assumed to be modeled as linear spring having magnitude

$$k_i = k_s \mathrm{D} h_i \tag{6.6.5.1}$$

Here, D = diameter of pile and h_i = distance between i + 1th and ith node.

Once the spring values are computed, based on equilibrium equations, by FDM or FEM, one may arrive at the desired solution.

The procedures are as explained hereafter.

6.6.5.1 Finite Difference Analysis of Pile Under Lateral Load

Shown in Fig. 6.44 is a pile which is broken into 11 discrete points called nodes, marked 1–11 when we have 10 beam elements as a whole constituting the pile. Points zero and twelve are nodes outside the pile called *phantom nodes*. The soil is represented by springs of magnitude k_sDh, where k_s = sub-grade modulus of the soil segment, D the diameter of pile, and h the length of each pile element segment as shown in the figure.

With reference to Eq. (6.6.3.5), the equilibrium equation at any node i is expressed as

$$EI\frac{\mathrm{d}^4 y_i}{\mathrm{d}z^4} + k_i y_i = 0 \tag{6.6.5.2}$$

Here, k_i = k_{si}Dh. The finite difference equation can thus be expressed as[2]

$$y_{i+2} - 4y_{i+1} + \left[6 + \frac{k_i h^4}{EI}\right] y_i - 4y_{i-1} + y_{i-2} = 0 \tag{6.6.5.3}$$

Equation (6.6.5.3) is applied successively from node 2 to 11 to yield nine equations.

[2]Here, we have presumed that the reader has some background on finite difference method.

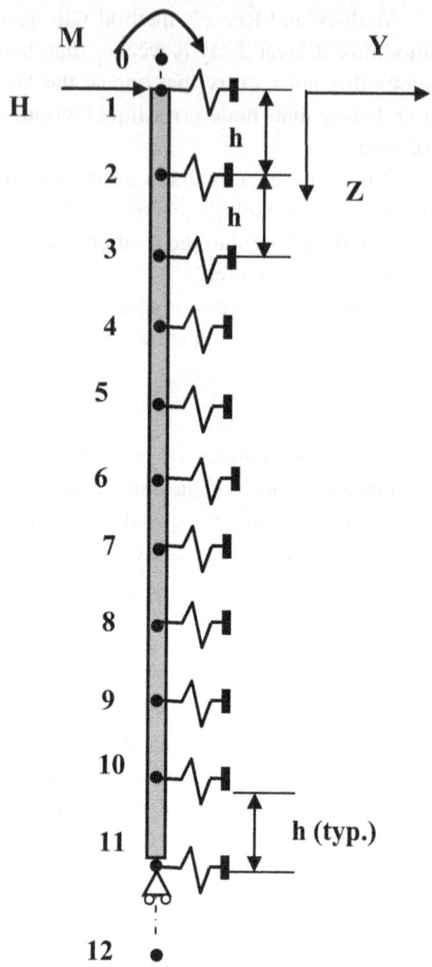

Fig. 6.44 Mathematical model of pile with soil springs with lateral load and moment

These nine equations will constitute of displacement from node 1 to 11 as well as the two phantom nodes, 0 and 12.

At node 1 considering the equation $EI\frac{d^2y}{dx^2} = -M_x = \frac{y_0 - 2y_1 + y_2}{h^2} = -\frac{M}{EI} \rightarrow y_0 = 2y_1 - y_2 - \frac{Mh^2}{EI}$

Similarly, applying the moment equation at node 11, we have $\frac{y_{10} - 2y_{11} + y_{12}}{h^2} = 0 \rightarrow y_{12} = 2y_{11} - y_{10}$.

We back substitute these values in the nine equations as derived above to eliminate the phantom nodes y_0 and y_{12}. However, now we have nine equations and eleven unknowns, y_1 to y_{11}. Thus, to make the problem determinate, we get additional two equations from the conditions $\sum F_y = 0$ and $\sum M = 0$, which gives

$$k_1y_1 + k_2y_2 + k_3y_3 + \cdots + k_{11}y_{11} = H \tag{6.6.5.4}$$

$$10k_1y_1h + 9k_2y_2h + 8k_3y_3h + \cdots + k_{10}y_{10}h = 10Hh + M \tag{6.6.5.5}$$

Thus, expressing these eleven equations in matrix notation, we have

$$[\mathbf{A}]\{\mathbf{Y}\} = \{\mathbf{F}\} \tag{6.6.5.6}$$

$$\begin{bmatrix} -2 & 5+\frac{k_2h^4}{EI} & -4 & 1 & 0 & 0 & 0 & 0 & 0 & 0 & 0 \\ 1 & -4 & 6+\frac{k_3h^4}{EI} & -4 & 1 & 0 & 0 & 0 & 0 & 0 & 0 \\ 0 & 1 & -4 & 6+\frac{k_4h^4}{EI} & -4 & 1 & 0 & 0 & 0 & 0 & 0 \\ 0 & 0 & 1 & -4 & 6+\frac{k_5h^4}{EI} & -4 & 1 & 0 & 0 & 0 & 0 \\ 0 & 0 & 0 & 1 & -4 & 6+\frac{k_6h^4}{EI} & -4 & 1 & 0 & 0 & 0 \\ 0 & 0 & 0 & 0 & 1 & -4 & 6+\frac{k_7h^4}{EI} & -4 & 1 & 0 & 0 \\ 0 & 0 & 0 & 0 & 0 & 1 & -4 & 6+\frac{k_8h^4}{EI} & -4 & 1 & 0 \\ 0 & 0 & 0 & 0 & 0 & 0 & 1 & -4 & 6+\frac{k_9h^4}{EI} & -4 & 1 \\ 0 & 0 & 0 & 0 & 0 & 0 & 0 & 1 & -4 & 5+\frac{k_{10}h^4}{EI} & -2 \\ k_1 & k_2 & k_3 & k_4 & k_5 & k_6 & k_7 & k_8 & k_9 & k_{10} & k_{11} \\ 10k_1 & 9k_2 & 8k_3 & 7k_4 & 6k_5 & 5k_6 & 4k_7 & 3k_8 & 2k_9 & k_{10} & 0 \end{bmatrix}$$

$$\begin{Bmatrix} y_1 \\ y_2 \\ y_3 \\ y_4 \\ y_5 \\ y_6 \\ y_7 \\ y_8 \\ y_9 \\ y_{10} \\ y_{11} \end{Bmatrix} = \begin{bmatrix} \frac{\frac{Mh^2}{EI}}{0} \\ 0 \\ 0 \\ 0 \\ 0 \\ 0 \\ 0 \\ 0 \\ 0 \\ H \\ 10H + \frac{M}{h} \end{bmatrix} \tag{6.6.5.7}$$

Having set up the equilibrium equation, the displacements of the node can be computed from the expression

$$\{\mathbf{Y}\} = [\mathbf{A}]^{-1}[\mathbf{F}] \tag{6.6.5.8}$$

Bending moment at any node i can be determined from the expression

$$M_i = EI\frac{y_{i-1} - 2y_i + y_{i+1}}{h^2} \tag{6.6.5.9}$$

The soil reaction is obtained from the expression

$$R_i = k_i \times y_i \tag{6.6.5.10}$$

Shear force is obtained from the algebraic summation of the lateral shear and soil reaction.

Equation (6.6.5.7) shows that the matrix is of modest size and can very easily be solved in a spreadsheet or general-purpose utility engineering software like Mathcad, MATLAB.

Example 6.12 A concrete pile of 500 mm diameter and 17.6 m long is embedded in a layered soil having sub-grade modulus as shown in Fig. 6.45. The pile is subjected to a horizontal shear of 50 kN at top pile cap along with a moment of 350 kN m due to earthquake. Compute the displacement moment and shear in pile under these loads. What will be the shear and moment in pile if the first layer of 4.0 m gets liquefied due to an earthquake shock? Consider the pile cap depth as 1.5 m and grade of concrete for pile = M30.

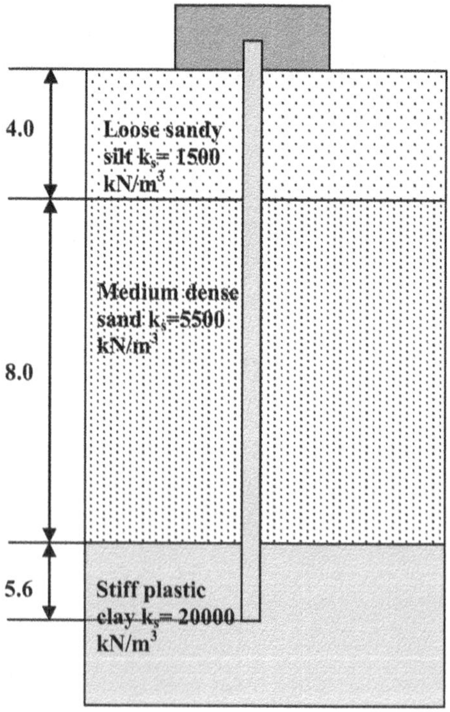

Fig. 6.45 Pile embedded in layered soil

Solution
Shown in the table below are the input and the results in tabular form

Node	i-coordinate (m)	K_s (kN/m^3)	Soil spring (k) kN/m	Displacement (m)	Shear (kN)	Moment (kN m)
1	0.00	1500	283	1.02717E−01	20.95	425.00
2	0.84	1500	566	7.27039E−02	−20.18	436.44
3	1.68	1500	566	4.82517E−02	−47.48	418.98
4	2.51	1500	566	2.91380E−02	−63.96	382.35
5	3.35	1500	566	1.48964E−02	−72.39	334.15
6	4.19	1500	2074	4.91244E−03	−82.58	280.02
7	5.03	5500	2074	−1.50345E−03	−79.46	218.74
8	5.87	5500	2074	−5.13213E−03	−68.82	159.65
9	6.70	5500	2074	−6.72655E−03	−54.86	108.04
10	7.54	5500	2074	−6.94437E−03	−40.46	66.22
11	8.38	5500	2074	−6.31838E−03	−27.35	34.53
12	9.22	5500	2074	−5.25244E−03	−16.46	12.04
13	10.06	5500	2074	−4.03312E−03	−8.09	−2.80
14	10.90	5500	2074	−2.84947E−03	−2.18	−11.76
15	11.73	5500	2074	−1.81569E−03	1.58	−16.57
16	12.57	20,000	7543	−9.93045E−04	9.07	−18.73
17	13.41	20,000	7543	−4.09113E−04	12.16	−15.64
18	14.25	20,000	7543	−2.44275E−05	12.34	−10.37
19	15.09	20,000	7543	2.28095E−04	10.62	−4.98
20	15.92	20,000	7543	4.17187E−04	7.48	−0.79
21	16.76	20,000	7543	5.96184E−04	2.98	1.18
22	17.60	20,000	3771	7.90258E−04	0.00	425.00

Solving the expression (6.6.5.8), we compute the displacement and finally compute the bending moment and shear force. The respective diagrams are plotted in Figs. 6.46, 6.47, and 6.48.

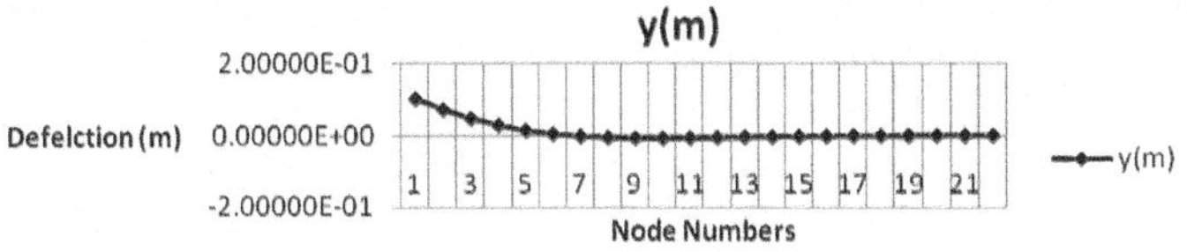

Fig. 6.46 Displacement plot of pile under given load

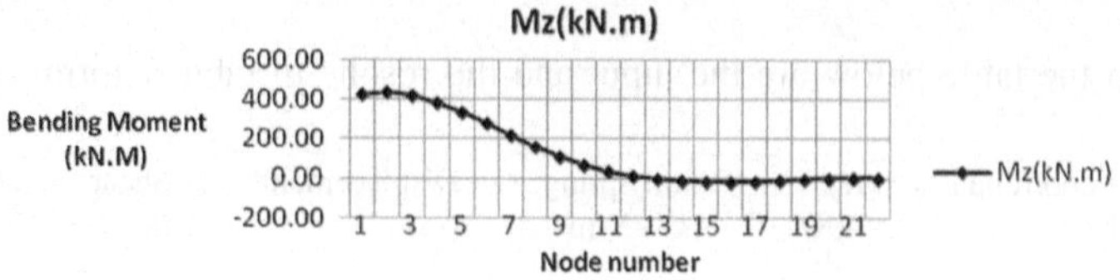

Fig. 6.47 Bending moment plot of pile under given load

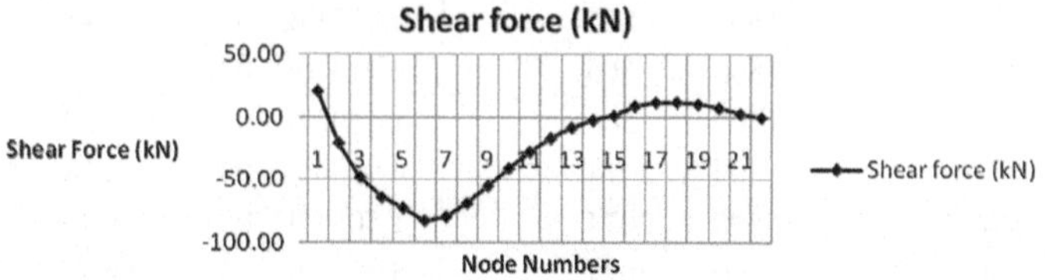

Fig. 6.48 Shear force diagram of under given load

In case of liquefaction of soil, data get modified as hereafter

Node	Z-coordinate (m)	K_s (kN/m^3)	Soil spring (k) kN/m	Displacement	Shear (kN)	Moment (kN m)
1	0.00	0	0	2.72782E−01	50.00	425.00
2	0.84	0	0	2.15898E−01	50.00	456.84
3	1.68	0	0	1.64835E−01	50.00	488.69
4	2.51	0	0	1.19999E−01	50.00	520.53
5	3.35	0	0	8.17953E−02	50.00	552.37
6	4.19	0	0	5.06300E−02	−55.02	584.22
7	5.03	5500	2074	2.69089E−02	−110.84	542.29
8	5.87	5500	2074	1.00976E−02	−131.78	461.16
9	6.70	5500	2074	−8.37528E−04	−130.05	365.32
10	7.54	5500	2074	−7.11773E−03	−115.28	270.70
11	8.38	5500	2074	−9.94867E−03	−94.65	186.45
12	9.22	5500	2074	−1.04039E−02	−73.06	116.69
13	10.06	5500	2074	−9.37215E−03	−53.62	62.10
14	10.90	5500	2074	−7.54918E−03	−37.97	21.15
15	11.73	5500	2074	−5.45667E−03	−26.65	−8.79
16	12.57	20,000	7543	−3.47614E−03	−0.43	−30.78
17	13.41	20,000	7543	−1.88784E−03	13.81	−34.36
18	14.25	20,000	7543	−7.37319E−04	19.37	−27.93
19	15.09	20,000	7543	5.72893E−05	18.94	−17.60
20	15.92	20,000	7543	6.27642E−04	14.21	−7.57
21	16.76	20,000	7543	1.10153E−03	5.90	−0.87
22	17.60	20,000	3771	1.56436E−03	0.00	0.00

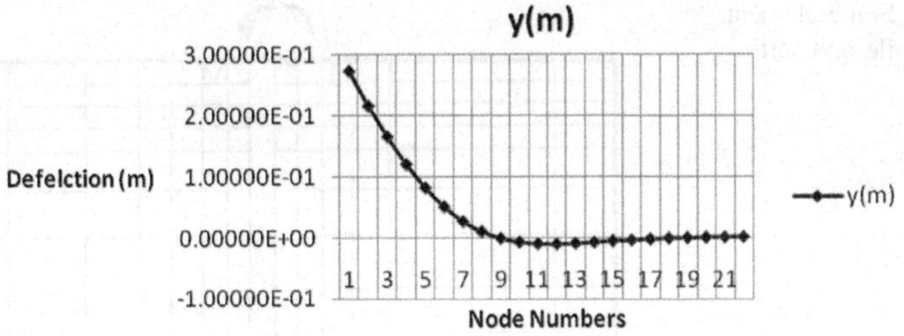

Fig. 6.49 Displacement diagram of pile under given load when the first layer liquefies

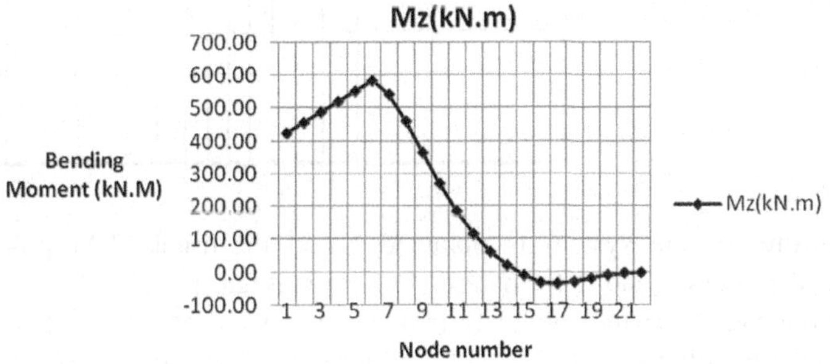

Fig. 6.50 Moment diagram of pile under given load when the first layer liquefies

Fig. 6.51 Shear force diagram of pile under given load when the first layer liquefies

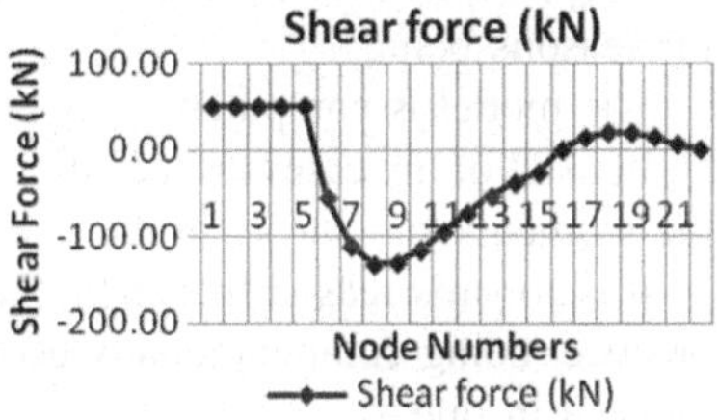

Displacement, moment, and shear force diagram are as shown in Figs. 6.49, 6.50, and 6.51.

6.6.5.2 Finite Element Analysis of Pile Embedded in Ground

Two most popular models used in FEM are the one like shown in Fig. 6.44, and the other as shown in Fig. 6.52.

As shown in Fig. 6.52, the soil is modeled as plain strain element and the pile could be modeled either as beam elements or a plain stress element which is subjected to lateral load H and moment *M*.

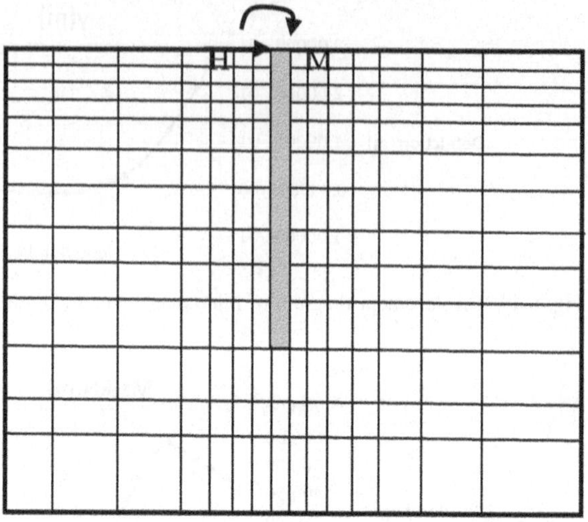

Fig. 6.52 Finite element model of pile and soil

Equilibrium of the system is obtained based on usual FEM procedure to determine the stresses and deformations in both soil and pile.

The advantage with the model is that it can take cognizance of soil heterogeneity and can solve both linear and nonlinear problems, provided the soil constitutive model is available. The model gives an accurate output of stress induced in soil and pile.

Despite the above advantages, the model is seldom used in professional practice, the reasons being:

The model is computationally expensive and sometimes difficult to interpret.

In most of the cases, especially if it is a long pile, one is more interested to know the moments and shears in the pile rather than the stress induced in soil.

It is for this reason that model as shown in Fig. 6.44 remains more popular in practice, being computationally economic and equally proficient in terms of stress induced in pile.

We describe below the FEM technique adopted for Winkler-type model.

Considering that this book is not exclusively devoted to finite element method, we only give a brief description of the mathematics underlying it. For detailed treatment, reader may refer to a number of standard references on FEM, like that by Bathe (1996), Zienkiewicz (1970).

As shown in Fig. 6.44, let us consider the free body diagram of the element between node 2 and 3, as shown in Fig. 6.53. Shown is a beam element having two degrees of freedom per node (one translation and one rotation). The degrees of freedom are marked $1(\delta_1)$, $2(\theta_1)$, $3(\delta_2)$, and $4(\theta_2)$.

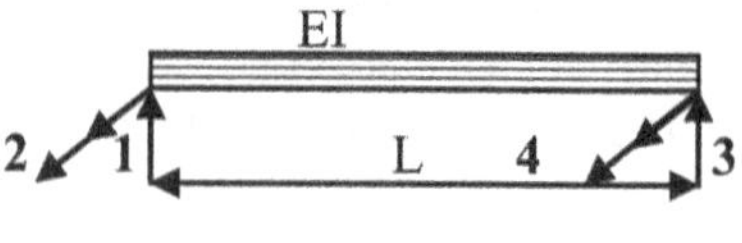

Fig. 6.53 Finite element model of the beam element

Since the total number of degrees of freedom is four, we choose the displacement function as

$$f = \alpha_1 + \alpha_2 x + \alpha_3 x^2 + \alpha_4 x^3 \tag{6.6.6.2.1}$$

Now as $\{f\} = [\mathrm{M}]\,\{\alpha\}$ we have

$$\{f\} = \langle 1 \quad x \quad x^2 \quad x^3 \rangle \langle \alpha_1 \; \alpha_2 \quad \alpha_3 \quad \alpha_4 \rangle^T \tag{6.6.6.2.2.}$$

Now, at $x = 0, \{f\} = \delta_1 = \alpha_1$ and again at $x = L, \{f\} = \delta_2 = \alpha_1 + \alpha_2 L + \alpha_3 L^2 + \alpha_4 L^3$.

At $x = 0$, $\frac{\mathrm{d}f}{\mathrm{d}x} = \theta_1 = \alpha_2$, and at $x = L, \frac{\mathrm{d}f}{\mathrm{d}x} = \theta_2 = \alpha_2 + 2\alpha_3 L + 3\,\alpha_4 L^2$.

The above can be expressed in matrix form as

$$\{f\} = \begin{bmatrix} 1 & 0 & 0 & 0 \\ 0 & 1 & 0 & 0 \\ 1 & L & L^2 & L^3 \\ 0 & 1 & 2L & 3L^2 \end{bmatrix} \begin{Bmatrix} \alpha_1 \\ \alpha_2 \\ \alpha_3 \\ \alpha_4 \end{Bmatrix} = \begin{Bmatrix} \delta_1 \\ \theta_1 \\ \delta_2 \\ \theta_2 \end{Bmatrix} \rightarrow \begin{Bmatrix} \alpha_1 \\ \alpha_2 \\ \alpha_3 \\ \alpha_4 \end{Bmatrix}$$

$$= \begin{bmatrix} 1 & 0 & 0 & 0 \\ 0 & 1 & 0 & 0 \\ \frac{-3}{L^2} & \frac{-2}{L} & \frac{3}{L^2} & \frac{-1}{L} \\ \frac{2}{L^3} & \frac{1}{L^2} & \frac{-2}{L^2} & \frac{1}{L^2} \end{bmatrix} \begin{Bmatrix} \delta_1 \\ \theta_1 \\ \delta_2 \\ \theta_2 \end{Bmatrix} \tag{6.6.6.2.3}$$

i.e. $\{\alpha\} = [\mathbf{C}]^{-1}\{\delta\}$

Considering $\{f\} = [\mathrm{M}]\,[\mathrm{C}]^{-1}\{\delta\}$, we have

$$\{f\} = \langle 1 \quad x \quad x^2 \quad x^3 \rangle \begin{bmatrix} 1 & 0 & 0 & 0 \\ 0 & 1 & 0 & 0 \\ \frac{-3}{L^2} & \frac{-2}{L} & \frac{3}{L^2} & \frac{-1}{L} \\ \frac{2}{L^3} & \frac{1}{L^2} & \frac{-2}{L^2} & \frac{1}{L^2} \end{bmatrix} \begin{Bmatrix} \delta_1 \\ \theta_1 \\ \delta_2 \\ \theta_2 \end{Bmatrix} \tag{6.6.6.2.4}$$

$$\rightarrow \{f\} = [\mathbf{N}][\boldsymbol{\delta}]$$

where

$$[\mathbf{N}] = \langle 1 \quad x \quad x^2 \quad x^3 \rangle \begin{bmatrix} 1 & 0 & 0 & 0 \\ 0 & 1 & 0 & 0 \\ \frac{-3}{L^2} & \frac{-2}{L} & \frac{3}{L^2} & \frac{-1}{L} \\ \frac{2}{L^3} & \frac{1}{L^2} & \frac{-2}{L^2} & \frac{1}{L^2} \end{bmatrix}$$

$$\rightarrow [N] = \left\langle \left(1 - \frac{3x^2}{L^2} + \frac{2x^3}{L^3}\right) \quad \left(x - \frac{2x^2}{L} + \frac{x^3}{L^2}\right) \quad \left(\frac{3x^2}{L^2} - \frac{2x^3}{L^3}\right) \quad \left(-\frac{x^2}{L^2} + \frac{x^3}{L^3}\right) \right\rangle$$

Thus

$$\{f\} = \left\langle \left(1 - \frac{3x^2}{L^2} + \frac{2x^3}{L^3}\right) \quad \left(x - \frac{2x^2}{L} + \frac{x^3}{L^2}\right) \quad \left(\frac{3x^2}{L^2} - \frac{2x^3}{L^3}\right) \quad \left(\frac{-x^2}{L^2} + \frac{x^3}{L^3}\right) \right\rangle \begin{Bmatrix} \delta_1 \\ \theta_1 \\ \delta_2 \\ \theta_2 \end{Bmatrix} \tag{6.6.6.2.5}$$

For a beam element, we know that $EI\frac{d^2f}{dx^2} = -M_x$ which is same as $\{\sigma\} = [D]\{\varepsilon\}$, where $\{\sigma\} = M_x$. $[\mathbf{D}] = EI$ and $\{\varepsilon\} = \frac{d^2f}{dx^2} = [\mathbf{B}]\,\{\delta\}$.

Thus, $[\sigma] = EI\left\langle \left(\frac{6}{L^2} - \frac{12x}{L^3}\right) \quad \left(\frac{4}{L} - \frac{6x}{L^2}\right) \quad \left(\frac{-6}{L^2} + \frac{12x}{L^3}\right) \quad \left(\frac{2}{L} - \frac{6x}{L^2}\right) \right\rangle < \delta_1\ \theta_1\ \delta_2\ \theta_2 >^T$ which gives

$$[B] = \left\langle \left(\frac{6}{L^2} - \frac{12x}{L^3}\right) \quad \left(\frac{4}{L} - \frac{6x}{L^2}\right) \quad \left(\frac{-6}{L^2} + \frac{12x}{L^3}\right) \quad \left(\frac{2}{L} - \frac{6x}{L^2}\right) \right\rangle \tag{6.6.6.2.6}$$

as $[\mathbf{K}] = \int [\mathbf{B}]^T[\mathbf{D}][\mathbf{B}] \cdot dv$ for beam element we have $[\mathbf{K}] = EI\int_0^L [\mathbf{B}]^{\mathrm{T}}[\mathbf{B}]\,dx$

$$[\mathbf{K}] = EI\begin{bmatrix} \frac{12}{L^3} & \frac{6}{L^2} & -\frac{12}{L^3} & \frac{6}{L^2} \\ \frac{6}{L^2} & \frac{4}{L} & -\frac{6}{L^2} & \frac{2}{L} \\ -\frac{12}{L^3} & -\frac{6}{L^2} & \frac{12}{L^3} & -\frac{6}{L^2} \\ \frac{6}{L^2} & \frac{2}{L} & -\frac{6}{L^2} & \frac{4}{L} \end{bmatrix} \tag{6.6.6.2.7}$$

This is the element stiffness matrix for the beam element.

It will be observed that in this case the FEM approximation converges to the exact value as obtained usually by classical slope deflection method (Weaver and Gere 1980).

Assembly of global matrix

Shown in Fig. 6.54 is a beam broken up into two beam elements having two degrees of freedom per node.

For beam elements 1 and 2, the element stiffness matrix can be expressed as

$$[\mathbf{K}_{11}] = EI\begin{bmatrix} \frac{12}{L^3} & \frac{6}{L^2} & -\frac{12}{L^3} & \frac{6}{L^2} \\ \frac{6}{L^2} & \frac{4}{L} & -\frac{6}{L^2} & \frac{2}{L} \\ -\frac{12}{L^3} & -\frac{6}{L^2} & \frac{12}{L^3} & -\frac{6}{L^2} \\ \frac{6}{L^2} & \frac{2}{L} & -\frac{6}{L^2} & \frac{4}{L} \end{bmatrix}^{\;1\;2\;3\;4} \quad \text{and} \quad [\mathbf{K}_{22}] = EI\begin{bmatrix} \frac{12}{L^3} & \frac{6}{L^2} & -\frac{12}{L^3} & \frac{6}{L^2} \\ \frac{6}{L^2} & \frac{4}{L} & -\frac{6}{L^2} & \frac{2}{L} \\ -\frac{12}{L^3} & -\frac{6}{L^2} & \frac{12}{L^3} & -\frac{6}{L^2} \\ \frac{6}{L^2} & \frac{2}{L} & -\frac{6}{L^2} & \frac{4}{L} \end{bmatrix}^{\;3\;4\;5\;6} \tag{6.6.6.2.8}$$

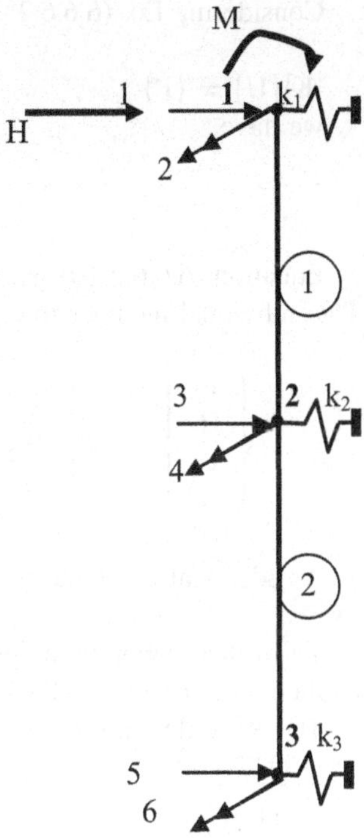

Fig. 6.54 FEM of beams connected to springs

The global stiffness matrix including the soil spring is thus expressed as

$$[\mathbf{K_G}] = \begin{array}{c} \begin{matrix} 1 & 2 & 3 & 4 & 5 & 6 \end{matrix} \\ \begin{bmatrix} \frac{12EI}{L^3}+\mathbf{k_1} & \frac{6EI}{L^2} & -\frac{12EI}{L^3} & \frac{6EI}{L^2} & 0 & 0 \\ \frac{6EI}{L^2} & \frac{4EI}{L} & -\frac{6EI}{L^2} & \frac{2EI}{L} & 0 & 0 \\ -\frac{12EI}{L^3} & -\frac{6EI}{L^2} & \frac{24EI}{L^3}+\mathbf{k_2} & 0 & -\frac{12EI}{L^3} & \frac{6EI}{L^2} \\ \frac{6EI}{L^2} & \frac{2EI}{L} & 0 & \frac{8EI}{L} & \frac{-6EI}{L^2} & \frac{2EI}{L} \\ 0 & 0 & -\frac{12EI}{L^3} & \frac{-6EI}{L^2} & \frac{12EI}{L^3}+\mathbf{k_3} & -\frac{6EI}{L^2} \\ 0 & 0 & \frac{6EI}{L^2} & \frac{2EI}{L} & -\frac{6EI}{L^2} & \frac{4EI}{L} \end{bmatrix} \end{array} \tag{6.6.6.2.8}$$

The equilibrium equation is thus expressed as

$$\begin{bmatrix} \frac{12EI}{L^3}+\mathbf{k_1} & \frac{6EI}{L^2} & -\frac{12EI}{L^3} & \frac{6EI}{L^2} & 0 & 0 \\ \frac{6EI}{L^2} & \frac{4EI}{L} & -\frac{6EI}{L^2} & \frac{2EI}{L} & 0 & 0 \\ -\frac{12EI}{L^3} & -\frac{6EI}{L^2} & \frac{24EI}{L^3}+\mathbf{k_2} & 0 & -\frac{12EI}{L^3} & \frac{6EI}{L^2} \\ \frac{6EI}{L^2} & \frac{2EI}{L} & 0 & \frac{8EI}{L} & \frac{-6EI}{L^2} & \frac{2EI}{L} \\ 0 & 0 & -\frac{12EI}{L^3} & \frac{-6EI}{L^2} & \frac{12EI}{L^3}+\mathbf{k_3} & -\frac{6EI}{L^2} \\ 0 & 0 & \frac{6EI}{L^2} & \frac{2EI}{L} & -\frac{6EI}{L^2} & \frac{4EI}{L} \end{bmatrix} \begin{Bmatrix} \delta 1 \\ \theta 1 \\ \delta 2 \\ \theta 2 \\ \delta 3 \\ \theta_3 \end{Bmatrix} = \begin{Bmatrix} H \\ M \\ 0 \\ 0 \\ 0 \\ 0 \end{Bmatrix} \tag{6.6.6.2.9}$$

Considering Eq. (6.6.6.2.9) as

$[\mathbf{K}]\{\mathbf{U}\} = \{\mathbf{P}\}$
we have

$$\{\mathbf{U}\} = [\mathbf{K}]^{-1}\{\mathbf{P}\} \quad (6.6.6.2.10)$$

Equation (6.6.6.2.10) gives the global displacement of the structure at nodes. The individual member forces are then expressed as

$$\begin{Bmatrix} V_1 \\ M_1 \\ V_2 \\ M_2 \end{Bmatrix} = EI \begin{bmatrix} \frac{12}{L^3} & \frac{6}{L^2} & -\frac{12}{L^3} & \frac{6}{L^2} \\ \frac{6}{L^2} & \frac{4}{L} & -\frac{6}{L^2} & \frac{2}{L} \\ -\frac{12}{L^3} & -\frac{6}{L^2} & \frac{12}{L^3} & -\frac{6}{L^2} \\ \frac{6}{L^2} & \frac{2}{L} & -\frac{6}{L^2} & \frac{4}{L} \end{bmatrix} \begin{Bmatrix} \delta_1 \\ \theta_1 \\ \delta_2 \\ \theta_2 \end{Bmatrix} \quad \text{for member 1.}$$

An identical equation can be written for member 2 to determine the member forces.

To further elaborate the problem, we repeat the problem 6.12 by finite element method carried out in STAAD. The FEM is shown hereafter (Fig. 6.55).

The STAAD input model of the problem is presented hereafter.

INPUT:-

```
STAAD PLANE
START JOB INFORMATION
ENGINEER DATE 26-May-16
END JOB INFORMATION
INPUT WIDTH 79
UNIT METER KN
JOINT COORDINATES
1 0 0 0;
2 0-0.5 0;
3 0-1 0;
4 0-1.5 0;
5 0-2 0;
6 0-2.5 0;
7 0-3 0;
8 0-3.5 0;
9 0-4 0;
10 0-5 0;
11 0-6 0;
12 0-7 0;
13 0-8 0;
14 0-9 0;
15 0-10 0;
16 0-11 0;
```

```
17 0-12 0;
18 0-13.4 0;
19 0-14.8 0;
20 0-16.2 0;
21 0-17.6 0;
MEMBER INCIDENCES
1 1 2; 2 2 3; 3 3 4; 4 4 5; 5 5 6; 6 6 7; 7 7 8; 8 8 9; 9 9 10; 10 10 11;
11 11 12; 12 12 13; 13 13 14; 14 14 15; 15 15 16; 16 16 17; 17 17 18; 18 18 19;
19 19 20; 20 20 21;
DEFINE MATERIAL START
ISOTROPIC CONCRETE
E 2.74e + 007
POISSON 0.17
DENSITY 25
ALPHA 1e-005
DAMP 0.05
END DEFINE MATERIAL
MEMBER PROPERTY INDIAN
1 TO 20 PRIS YD 0.5
CONSTANTS
MATERIAL CONCRETE ALL
SUPPORTS
1 FIXED BUT FY FZ MX MY MZ KFX 187.5
2 FIXED BUT FY FZ MX MY MZ KFX 375
3 FIXED BUT FY FZ MX MY MZ KFX 375
4 FIXED BUT FY FZ MX MY MZ KFX 375
5 FIXED BUT FY FZ MX MY MZ KFX 375
6 FIXED BUT FY FZ MX MY MZ KFX 375
7 FIXED BUT FY FZ MX MY MZ KFX 375
8 FIXED BUT FY FZ MX MY MZ KFX 375
9 FIXED BUT FY FZ MX MY MZ KFX 1563
10 FIXED BUT FY FZ MX MY MZ KFX 2750
11 FIXED BUT FY FZ MX MY MZ KFX 2750
12 FIXED BUT FY FZ MX MY MZ KFX 2750
13 FIXED BUT FY FZ MX MY MZ KFX 2750
14 FIXED BUT FY FZ MX MY MZ KFX 2750
15 FIXED BUT FY FZ MX MY MZ KFX 2750
16 FIXED BUT FY FZ MX MY MZ KFX 2750
17 FIXED BUT FY FZ MX MY MZ KFX 8375
18 FIXED BUT FY FZ MX MY MZ KFX 14000
19 FIXED BUT FY FZ MX MY MZ KFX 14000
20 FIXED BUT FY FZ MX MY MZ KFX 14000
21 FIXED BUT FZ MX MY MZ KFX 7000
*21 PINNED
LOAD 1 LOADTYPE None TITLE LOAD CASE 1
```

```
JOINT LOAD
1 FX 50 MZ -425
PERFORM ANALYSIS
PRINT SUPPORT REACTION ALL
PRINT JOINT DISPLACEMENTS ALL
FINISH
```

The displacement plot of the problem by FEM is as shown in Fig. 6.56.

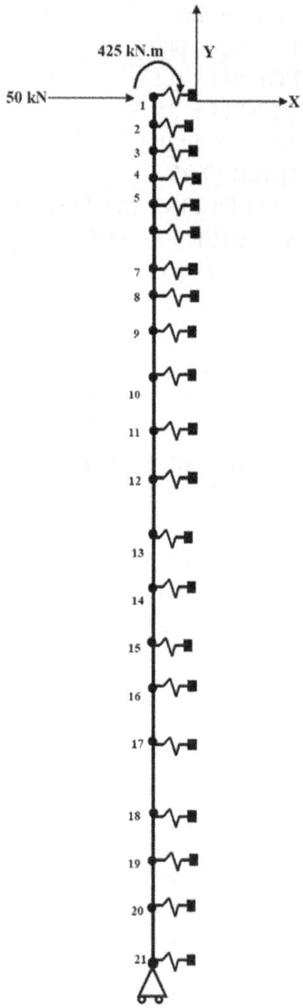

Fig. 6.55 Finite element model of the pile

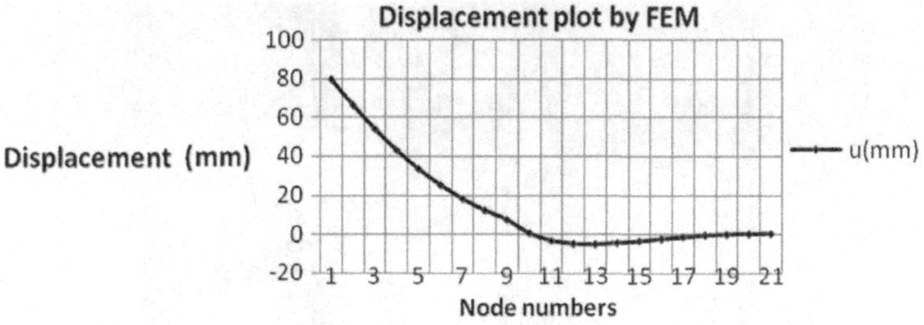

Fig. 6.56 Displacement plot of pile by finite element method

The maximum values of displacement moment and shear are compared with finite difference method below.

Method	Displacement (mm)	Moment (kN m)	Shear (kN)
Finite difference method	102	436	82.6
Finite element method	80	456	78.3

The techniques as mentioned above remain standard techniques adapted for computation of load in a pile due to seismic force.

Yet, piles have failed with regular monotony under earthquake loads. And some classic cases are presented pictorially in Figs. 6.57, 6.58, and 6.59.

Study on history of pile failures has shown that in most of the cases the structural failures have occurred in piles due to liquefaction of soil. However, for many other structures (resting on pile foundations), which have survived an earthquake, one can never say for sure if the piles have incurred any damage or not, for, unlike the

Fig. 6.57 Pile failures below NHK building; Niigata Earthquake (1964)

Fig. 6.58 Failure of structure supported on piles during Bhuj Earthquake (2001)

Fig. 6.59 Failure of a pile supported building during earthquake in Mexico (1985)

building shown in Fig. 6.57, one never really carries out such exhaustive investigation to see what is actually the health of pile after an earthquake.

It could so happen that some of the structures could well have survived an earthquake and are yet being supported on damaged piles. Thus, that the present design practice is not a full proof is now amply evident to the earthquake researchers.

The argument is further substantiated by Figs. 6.58 and 6.59 where superstructures supported on pile foundations were heavily damaged.

The question that then naturally arises is *where is the* snag *that makes the conventional design vulnerable*?

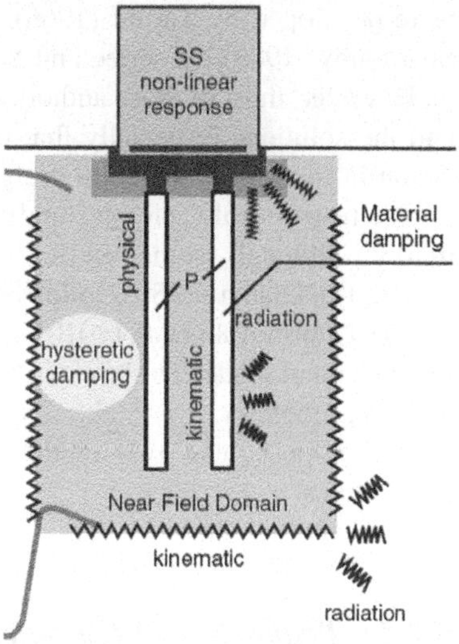

Fig. 6.60 Behavior of piles supporting the structure under an earthquake force

Perusal of previous sections shows that even if a detailed dynamic analysis is carried out for the superstructure, it is usually considered fixed at base and whatever the force is induced in the structure is transferred to the pile which is assumed to have taken no part in the vibration.

This simplifying assumption could have some semblance of reality for a frame supporting rotating machine but would certainly not be true in case of an earthquake.

Shown in Fig. 6.60 is a pile embedded in soil supporting a structure.

During an earthquake, the waves traveling through soil will excite the pile first and this will go on to transfer the load to the superstructure. How much load is transferred to the superstructure or what would be the excitation of the pile will depend on a number of factors, like relative stiffness of pile, superstructure and surrounding soil, material damping of pile, and also the hysteretic and radiation damping of soil.

Thus, it is seen that the actual behavior of pile supporting a structure under an earthquake is actually a complex soil structure interaction problem and the behavior is quite different than what happens under static load.

6.7 Dynamic Response of Pile Foundations Under Earthquake Force

Based on the argument cited above, a number of researchers and engineers had studied the behavior of piles under dynamic load. The development actually started in the early 1970s in the realm of machine foundations when a number of solutions were developed by Tajimi (1966), Penzien (1970), Novak (1974), Novak and El Sharnouby (1983), Banerjee and Sen (1987), Dobry and Gazetas (1988), etc.

However, most of those authors assumed the forcing function as a harmonic load and the solutions (especially that of Novak and Gazetas) have had a wide application in machine foundation design.

The response of such piles under earthquake-type force has remained somewhat murky, although some earlier attempts were carried out by Prakash and Chandrashekharan (1980), followed by a number of researchers like Bhattacharya (2003), Cubrinovski et al. (2010), and many others. Finally, the dynamic analysis of finite element model, as shown in Fig. 6.55, has also been popular as a professional analytical tool.

We propose here two mathematical models that are amenable to practical applications.

6.7.1 Prakash and Chandrashekharan's (1980) Model for Dynamic Analysis of Piles

This remains to be one of the earliest models for dynamic response of piles under earthquake force. Though typically chart oriented,[3] it is a useful tool to estimate at least preliminarily the effect of earthquake on a pile foundation and is quite acquiescent to hand calculation.

The steps involved are summarized hereafter.

1. Evaluate the value of k_s or n_h from the soil report. These values are to be correlated to shear wave velocity of the soil and not the static value.
2. Based on data obtained in step 1, compute T or R as furnished in Eq. (6.6.2.4). Calculate the maximum depth factor $Z_{max} = L/T$ or L/R. Here, L is length of pile and Z_{max} shall be greater than 5.0.
3. For the computed value of Z_{max}, determine frequency factor from Fig. 6.61a–c.
4. Estimate the allowable vertical load the pile can sustain (W) which gives the lumped mass at pile head.
5. Estimate the natural frequency from the equations

$$\omega_{n1} = F_{CL1} / \sqrt{\frac{W}{g k_s R}} \quad \text{for soil modulus constant with depth} \qquad (6.7.1.1)$$

[3]A typical geotechnical engineer's approach.

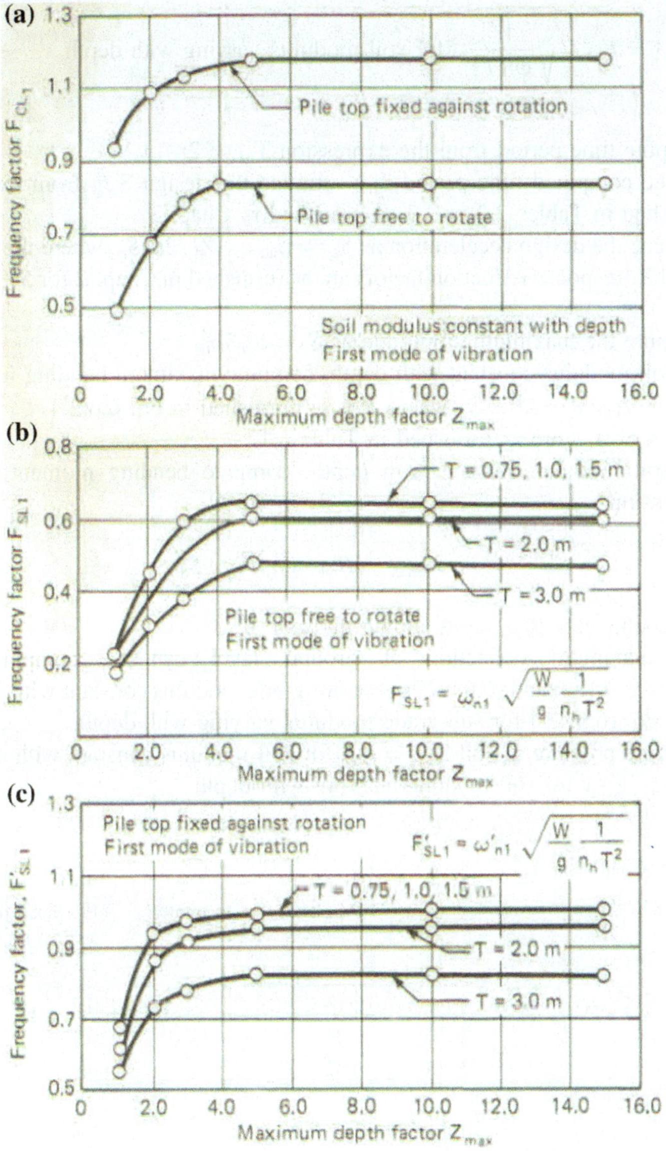

Fig. 6.61 **a** Frequency factor or soil having modulus constant with depth after Prakash and Chandrashekaran (1980). **b** Frequency factor or soil modulus varying linearly with depth; pile head free. **c** Frequency factor or soil modulus varying linearly with depth; pile head fixed

$$\omega_{n1} = F_{SL1} \Big/ \sqrt{\frac{W}{g n_h T^2}} \quad \text{for soil modulus varying with depth} \tag{6.7.1.2}$$

6. Compute time period from the expression $T_{n1} = 2\pi/\omega_{n1}$.
7. For the computed time period T_{n1}, estimate the design S_a/g from expressions furnished in Tables 2.2 or 2.3 as furnished in Chap. 2.
8. Estimate the design acceleration as $S_{ad} = S_{ad} = (ZI/2R)S_a$, where the values Z, I, and R (response reduction factor) are as furnished in Chap. 2 for 5% damping ratio.
9. Compute the maximum amplitude as $S_d = S_{ad}/\omega_{n1}^2$.
10. For soil modulus constant with depth, compute maximum bending moment as $M_{max} = A_{me} \times n_h R^2 \times S_d$ where R is as furnished in Eq. 6.6.2.4. Values of A_{me} are as furnished in Table 6.12.
11. For soil modulus varying with depth, compute bending moment from the expression

$$M_{max} = B_{me} \times n_h T^3 \times S_d \tag{6.7.1.3}$$

The coefficients B_{me} are as shown in Table 6.13.

12. The maximum deflection at ground level can be computed from Eqs. (6.6.3.18a) and (6.6.3.19a) for sub-grade modulus constant with depth and from Eq. (6.6.4.2) for sub-grade modulus varying with depth.
13. Compute pressure in soil as $p = n_h y$ for soil modulus constant with depth and $p = n_h \cdot z \cdot y$ for soil modulus varying with depth.

Table 6.12 Coefficient A_{me} for different Z_{max}

Z_{max}	A_{me} Pile head free to rotate	A_{me} Pile head fixed against rotation	
		(−ve)	(+ve)
2	0.13	0.9	0.00
3	0.24	0.9	0.04
5–15	0.32	0.9	0.18

Table 6.13 Coefficient B_{me} for different values of Z_{max}

Z_{max}	B_{me} Pile head free to rotate	B_{me} Pile head fixed against rotation	
		(−ve)	(+ve)
2	0.100	0.93	0.00
3	0.255	0.93	0.10
5–15	0.315	0.90	0.28

14. The pressure in soil is to be checked against allowable pressure

$$p_{\text{all}} = \left[\frac{1+\sin\phi}{1-\sin\phi}\right]\gamma_s Dz \tag{6.7.4}$$

Example 6.13 A reinforced concrete pile of 20 m length and 450 mm diameter is embedded in a cohesive soil having sub-grade modulus = 1560 kN/m^3. The allowable vertical capacity of the pile is 1205 kN. Compute the maximum amplitude, moment, and shear the pile is subjected to if it is situated in Zone IV as per IS1893 (2002). Consider the importance factor $I = 1.5$. Consider grade of concrete as M35.

Solution

Grade of concrete = M35.

Thus, as per IS-456, $E = 5000\sqrt{f_{\text{ck}}}$ N/mm^2. $\rightarrow \frac{5000\times\sqrt{35}}{1000} \times 10^6 = 2.96 \times 10^7$ kN/m^2.

$$\text{I} = \frac{\pi \times (0.45)^4}{64} = 2.013 \times 10^{-3}\,\text{m}^4.$$

$$R = \sqrt[4]{\frac{EI}{n_h}} = \sqrt[4]{\frac{2.96\times10^7\times2.013\times10^{-3}}{1560}} = 1.4\,\text{m}$$

$Z_{\max} = L/R = 20/1.4 = 14 > 5.0.$

For $Z_{\max} = 14.0$ $F_{\text{CL1}} = 0.85$ from Fig. 6.61a,

Thus $\omega_{n1} = F_{\text{CL1}}/\sqrt{\frac{W}{gk_sR}}$

$\omega_{n1} = \frac{0.85}{\sqrt{\frac{1205}{9.81\times1560\times1.4}}} = 3.6$ rad/s. $\rightarrow T = 2\pi/\omega_{n1} = 1.75$ s

As per IS-1893 $S_{\text{ad}} = \left[\frac{ZI}{2R}\right] \times \left(\frac{1.67}{T}\right) \times 9.81 = \left[\frac{0.24\times1.5}{6}\right] \times \left(\frac{1.67}{1.75}\right) \times 9.81 = 0.56\,\text{m/s}^2$

$$\text{Thus, the maximum amplitude} = \frac{S_{\text{ad}}}{\omega_{n1}^2} = \frac{0.56}{(3.6)^2} \times 1000 = 43\,\text{mm}$$

Maximum moment

$$M_{\max} = A_{\text{me}} \times n_h R^2 \times S_d = 0.32 \times 1560 \times (1.4)^2 \times 43 \times 10^{-3} = 42.07\,\text{kN. m}$$

Considering the pile is free headed, the deflection is expressed by

$$u = \frac{2pH}{k_s B} e^{-pz} \cos pz \text{ where } p = \sqrt[4]{\frac{k_s B}{4EI}}$$

Here, $p = \sqrt[4]{\frac{1560 \times 0.45}{4 \times 5958}} = 0.586$

Now at $z = 0$ $u = \frac{2pH}{k_s B} \rightarrow H = \frac{k_s B u}{2p} = \frac{1560 \times 0.45 \times 43 \times 10^{-3}}{2 \times 0.586} = 25.75$ kN.

6.7.1.1 Some Comments on Prakash and Chandrashekharan's Method

The greatest advantage with the above method is that with least calculation it gives a reasonable estimate of the pile response under seismic load.

In the early and mid-seventies when computer was not so common feature in a design office, this was what one could do at best to estimate the response.

However, with advent of finite element analysis it is possible to use a much more sophisticated model for dynamic response through computer but of course at an additional cost.

Notwithstanding its utility Prakash's model suffers from certain lacunae as mentioned hereafter.

1. It ignores the mass of pile that can significantly affect the final response.
2. The model considers overall damping as 5% which may or not be true in all cases.
3. The analysis is silent on the phenomenon that has contributed to failure of a number of piles around the world, which is *liquefaction* and effect of lateral spreading.

We show in Fig. 6.62 a classic case of structural collapse due to pile failure under liquefaction.

We will certainly discuss more on this issue in the subsequent model presented hereafter.

6.7.2 *Chowdhury and Dasgupta's (2012) Model for Seismic Response of Pile*

As a first step, we need to understand the free field characteristics of soil in which the pile is embedded. Referring to Fig. 6.63 in two dimensions, the propagation of S_V waves through the medium can be expressed as

$$\frac{\partial^2 u(x,z,t)}{\partial x^2} + \frac{\partial^2 u(x,z,t)}{\partial z^2} = \frac{1}{V_s^2} \frac{\partial^2 u(x,z,t)}{\partial t^2} \tag{6.7.2.1}$$

For this, it has already been established, vide Sect. 3.4.3 of Chap. 3, that stiffness and mass matrix for the soil can be expressed as

Fig. 6.62 Collapse of Showa Bridge due to piling failure, Niigata Earthquake (1964)

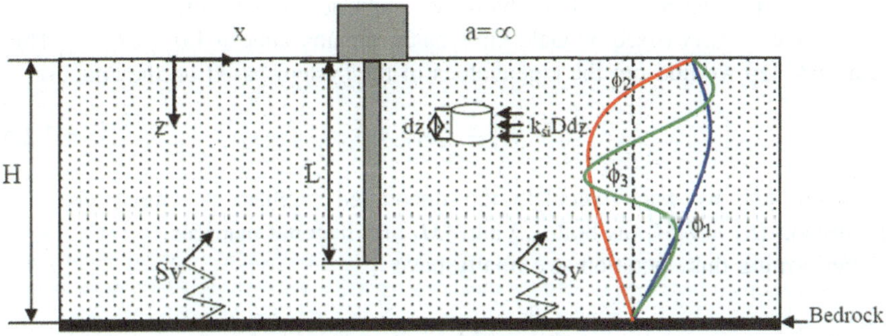

Fig. 6.63 Free field motion of site with piles embedded in it

$$[K] = \frac{G}{H}\begin{bmatrix} \frac{\pi^2}{8} & 0 & 0 \\ 0 & \frac{9\pi^2}{8} & 0 \\ 0 & 0 & \frac{25\pi^2}{8} \end{bmatrix} \quad (6.7.2.2)$$

$$[M] = \frac{\gamma_s H}{g}\begin{bmatrix} \frac{1}{2} & 0 & 0 \\ 0 & \frac{1}{2} & 0 \\ 0 & 0 & \frac{1}{2} \end{bmatrix} \quad (6.7.2.3)$$

Equation (6.7.2.2) that depicts the stiffness of soil medium has an interesting characteristic. On close examination, we see that it has got a unit of kN/m^3.

This means that this stiffness of soil actually works as a dynamic pressure and can be attributed as a dynamic sub-grade modulus of the soil. The dynamic sub-grade modulus is found to be varying with each mode of the stratum's vibration. This we depict as k_{si} where i represents number of mode like $i = 1, 2, 3, \ldots, \infty$.

We also notice that for G constant with depth, fortunately the dynamic sub-grade modulus remains uncoupled with respect to their modes.

6.7.2.1 Properties of Gibson-Type Soil

Again, as depicted in Sect. 3.4.3 Chap. 3, we have shown that the stiffness matrix of the soil can be expressed as

$$[K] = \frac{G}{H}\begin{bmatrix} \frac{\pi^2}{16} + \frac{1}{4} & -\frac{3}{4} & \frac{5}{36} \\ -\frac{3}{4} & \frac{9\pi^2}{16} + \frac{1}{4} & -\frac{15}{4} \\ \frac{5}{36} & -\frac{15}{4} & \frac{25\pi^2}{16} + \frac{1}{4} \end{bmatrix} \quad (6.7.2.4)$$

The problem here complicates as the stiffness matrix becomes coupled due to the factor z/H. To circumvent this problem, we approach it as follows.

The mass matrix based on Galerkin's basis remains same as Eq. (6.7.2.3). Thus, as a first step we find out the eigenvalues of the system based on the equation

$$[K] - \lambda[M] = 0 \quad (6.7.2.5)$$

Knowing $[\lambda]$, we back calculate the uncoupled stiffness per mode from the expression $[K] = \lambda[M]$, as both $[\lambda]$ and $[M]$ are diagonal matrices.

Performing this operation, we finally have

$$[K] = \frac{G}{H}\begin{bmatrix} 0.733 & 0 & 0 \\ 0 & 4.337 & 0 \\ 0 & 0 & 13.099 \end{bmatrix} \quad (6.7.2.6)$$

Table 6.14 gives the dynamic sub-grade modulus of the soil for first three modes for different types of soils based on the technique as proposed above.

From Table 6.14, it may be noted that depth of bedrock H has an important bearing on the stiffness and significantly affects its properties under seismic loading.

Dynamic modulus as derived above will be used subsequently to compute the dynamic stiffness of the pile that is embedded in a half-space.

Table 6.14 Dynamic sub-grade modulus for different soils for the first three modes

G (varying type)	K_{S1}	K_{S2}	K_{S3}
$G = G_0$	1.234 G_0/H	11.103 G_0/H	30.843 G_0/H
$G = G_0(z/H)$	0.733 G_0/H	4.337 G_0/H	13.099 G_0/H
$G = G_0(z/H)^2$	0.322 G_0/H	1.812 G_0/H	7.162 G_0/H
$G = G_0(1 + z/H)$	2.062 G_0/H	16.464 G_0/H	45.32 G_0/H
$G = G_0(1 + z/H)^2$	3.366 G_0/H	24.024 G_0/H	66.367 G_0/H
$G = G_0\sqrt{z/H}$	0.981 G_0/H	7.093 G_0/H	19.544 G_0/H

6.7.2.2 Dynamic Stiffness of Pile

For this, as a first step, we start with static equilibrium equation of pile of length L and diameter D embedded in a soil of sub-grade modulus k_s, as shown in Fig. 6.63.

This is expressed as

$$EI\frac{d^4u}{dz^4} = -k_s \cdot D \cdot u \qquad (6.7.2.7)$$

As explained in Sect. 6.5.3, the solution to Eq. (6.7.2.7) can be expressed as

$$u = e^{-pz}(C_0 \cdot \cos pz + C_1 \cdot \sin pz) + e^{pz}(C_2 \cos pz + C_3 \sin pz) \qquad (6.7.2.8)$$

where $p = \sqrt{4}k_s.D/4EI$, E = Young's modulus of pile material, I = moment of inertia of the pile cross-sectional area.

For a long pile subjected to load or moment at its head, it is reasonable to assume that at a significant distance from the pile head (where the load is applied) the curvature along the pile axis vanishes. This condition can only be satisfied when C_2 and C_3 in Eq. (6.7.2.43) are considered insignificant and the deflection equation becomes

$$u = e^{-pz}(C_0 \cdot \cos pz + C_1 \cdot \sin pz) \qquad (6.7.2.9)$$

Now considering $u = u_0$ at $z = 0$, we have $C_0 = u_0$, which gives

$$u = e^{-pz}(u_0 \cdot \cos pz + C_1 \cdot \sin pz) \qquad (6.7.2.10)$$

Again, considering $du/dz = \theta_0$ at $z = 0$ we have

$$C_1 = u_0 + \frac{\theta_0}{p} \qquad (6.7.2.11)$$

Thus,

$$u = e^{-pz}\left[u_0 \cdot \cos pz + \left(u_0 + \frac{\theta_0}{p}\right) \cdot \sin pz\right] \tag{6.7.2.12}$$

For the magnitude of rotation being small, $\theta_0 \to u_0/L$ gives

$$u = u_0 e^{-pz}\left[\cos pz + \left(1 + \frac{1}{pL}\right) \cdot \sin pz\right] \tag{6.7.2.13}$$

Or,

$$\frac{u}{u_0} = e^{-pz}\left[\cos pz + \left(1 + \frac{1}{pL}\right). \sin pz\right] \tag{6.7.2.14}$$

Now considering $\beta = pL$, a dimensionless term, the generic shape function of displacement for a long pile can be expressed as

$$\phi(z) = e^{-\beta\frac{z}{L}}\left[\cos\beta\frac{z}{L} + \eta \cdot \sin\beta\frac{z}{L}\right] \tag{6.7.2.15}$$

Here $\beta = \sqrt{4}k_s \cdot DL^4/4EI$ and $\eta = 1 + 1/\beta$.

For dynamic condition, as the sub-grade modulus k_s varies from mode to mode (vide Table 6.14), the shape function $\phi(z)$ also varies and is expressed as

$$\phi_i(z) = e^{-\beta_i\frac{z}{L}}\left[\cos\beta_i\frac{z}{L} + \eta_i \cdot \sin\beta_i\frac{z}{L}\right] \tag{6.7.2.16}$$

Here, as defined earlier, i = modal number 1, 2, 3, …, ∞.

Equation (6.7.2.16) will now be used to derive the dynamic stiffness of the pile.

Referring to Fig. 6.62, if we now take a cylindrical body of the pile element of depth dz, where dz is infinitesimally small, the strain energy of this body along with the reactive soil pressure can be expressed as (Shames and Dym 1995)

$$\mathrm{d}\Pi_i = \frac{EI}{2}\left[\frac{\mathrm{d}^2u}{\mathrm{d}z^2}\right]^2 + \frac{1}{2}k_{si}D \cdot \mathrm{d}z \cdot u^2 \tag{6.7.2.17}$$

Thus, over the whole length L of the pile

$$\Pi_i = \frac{EI}{2}\int_0^L\left[\frac{\mathrm{d}^2u}{\mathrm{d}z^2}\right]^2\mathrm{d}z + \frac{1}{2}k_{si}D \cdot \int_0^L u^2\mathrm{d}z \tag{6.7.2.18}$$

Now considering $u = \phi(z) \cdot q(t)$, it can be shown by Hurty and Rubenstein (1967) that

$$[K]_{ij}= EI \int_0^L \frac{d^2\phi_i(z)}{dz^2}\frac{d^2\phi_j(z)}{dz^2} dz + k_{si} \cdot D \int_0^L \phi_i(z) \cdot \phi_j(z) dz \tag{6.7.2.19}$$

Considering $\phi_i(z)$ as expressed in Eq. (6.7.2.16) and also

$$\frac{d^2\phi_i(z)}{dz^2} = \frac{2\beta_i^2}{L^2}\left[e^{-\beta_i \frac{z}{L}}\left(\sin\frac{\beta_i z}{L} - \eta_i \cos\frac{\beta_i z}{L}\right)\right],$$

where

$$f_i(z) = \left[e^{-\beta_i \frac{z}{L}}\left(\sin\frac{\beta_i z}{L} - \eta_i \cos\frac{\beta_i z}{L}\right)\right] \tag{6.7.2.20}$$

Equation (6.7.2.19) can be expanded and expressed for first three modes as

$$[K]_{3\times 3} = \frac{4EI}{L^3}\begin{bmatrix} k_{11} & k_{12} & k_{13} \\ k_{21} & k_{22} & k_{23} \\ k_{31} & k_{32} & k_{33} \end{bmatrix} \tag{6.7.2.21}$$

Considering $\xi = z/L$, the stiffness coefficients k_{ij} can be expressed as

$$k_{11} = \int_0^1 \beta_1^4 f_1(\xi)^2 d\xi + \int_0^1 \beta_1^4 \phi_1(\xi)^2 d\xi \tag{6.7.2.22}$$

$$k_{12} = \int_0^1 \beta_1^2\beta_2^2 f_1(\xi) f_2(\xi) d\xi + \int_0^1 \beta_1^2\beta_2^2 \phi_1(\xi)\phi_2(\xi) d\xi \tag{6.7.2.23}$$

$$k_{13} = \int_0^1 \beta_1^2\beta_3^2 f_1(\xi) f_3(\xi) d\xi + \int_0^1 \beta_1^2\beta_3^2 \phi_1(\xi)\phi_3(\xi) d\xi \tag{6.7.2.24}$$

$$k_{21} = k_{12} \tag{6.7.2.25}$$

$$k_{22} = \int_0^1 \beta_2^4 f_2(\xi)^2 d\xi + \int_0^1 \beta_2^4 \phi_2(\xi)^2 d\xi \tag{6.7.2.26}$$

$$k_{23} = \int_0^1 \beta_2^2\beta_3^2 f_2(\xi)f_3(\xi)\mathrm{d}\xi + \int_0^1 \beta_2^2\beta_3^2\phi_2(\xi)\phi_3(\xi)\mathrm{d}\xi \tag{6.7.2.27}$$

$$k_{31} = k_{13} \tag{6.7.2.28}$$

$$k_{32} = k_{23} \tag{6.7.2.29}$$

$$k_{33} = \int_0^1 \beta_3^4 f_3(\xi)^2\mathrm{d}\xi + \int_0^1 \beta_3^4\phi_3(\xi)^2\mathrm{d}\xi \tag{6.7.2.30}$$

The integrals in Eqs. (6.7.2.22)–(6.7.2.30) can very easily be solved by numerical integration like Simpson's or trapezoidal rule. In this case, the steps between 0 and 1 are taken ≤ 0.02, even a simple summation produces values of sufficient accuracy for design office applications.

6.7.2.3 Mass Matrix of Pile

In this context, leaving aside the self-weight of pile, the major question arises as to what is the total lumped mass at the pile head?

If we ponder over the issue, it will be realized that under no case can this load be more than static design capacity as recommended in the soil report. This is the load to which a designer would always restrict his load to.

Considering P_d as the allowable vertical load on pile, mass matrix can now be expressed as (Meirovitch 1967)

$$[M]_{ij} = \frac{\gamma_c A}{g}\int_0^L \phi_i(z)\phi_j(z)dz + \frac{P_\mathrm{d}}{g}\phi_i(0)\phi_j(0) \tag{6.7.2.31}$$

Equation (6.7.2.31) can be expanded and expressed for first three modes as

$$[M]_{3\times3} = \frac{P_\mathrm{d}}{g}\begin{bmatrix} 1 + r\int_0^1 \phi_1(\xi)^2\mathrm{d}\xi & 1 + r\int_0^1 \phi_1(\xi)\phi_2(\xi)\mathrm{d}\xi & 1 + r\int_0^1 \phi_1(\xi)\phi_3(\xi)\mathrm{d}\xi \\ 1 + r\int_0^1 \phi_2(\xi)\phi_1(\xi)\mathrm{d}\xi & 1 + r\int_0^1 \phi_2(\xi)^2\mathrm{d}\xi & 1 + r\int_0^1 \phi_2(\xi)\phi_3(\xi)\mathrm{d}\xi \\ 1 + r\int_0^1 \phi_3(\xi)\phi_1(\xi)\mathrm{d}\xi & 1 + r\int_0^1 \phi_3(\xi)\phi_2(\xi)\mathrm{d}\xi & 1 + r\int_0^1 \phi_3(\xi)^2\mathrm{d}\xi \end{bmatrix} \tag{6.7.2.32}$$

In Eq. (6.7.2.32), r is a weight ratio and is expressed $r = \gamma_c \cdot \mathrm{A} \cdot \mathrm{L/P_d}$. Here, γ_c = weight density of the pile material and A = area of cross section of the pile.

The integrals can again be obtained numerically or by simple summation, taking intermittent steps ≤ 0.02 between boundaries 0 to 1.

6.7.2.4 Modal Damping Ratio of Pile

The damping of pile under seismic load consists of two parts: (i) material damping and (ii) radiation damping.

While the material damping ratio is usually taken as 5% for RCC piles and 2–3% for steel piles, for pile foundations it is the radiation damping which invariably governs the response.

It has been shown by Chowdhury and Dasgupta (2008, 2010) that for fundamental mode, the radiation pile damping ratio can be expressed as

$$\zeta = \frac{0.43L}{V_s}\left(\int_0^1 \phi(z)^2 \mathrm{d}z\right)\cdot\omega \tag{6.7.2.33}$$

Here, V_s = shear wave velocity of the soil and ω is natural frequency of pile in fundamental mode.

Equation (6.7.2.33) can be extended to estimate the modal damping ratio for a system with multi-degree of freedom as

$$[\zeta]_{3\times3} = \frac{0.43L}{V_s}\begin{bmatrix} \int_0^1 \phi_1(z)^2 \mathrm{d}z & 0 & 0 \\ 0 & \int_0^1 \phi_2(z)^2 \mathrm{d}z & 0 \\ 0 & 0 & \int_0^1 \phi_3(z)^2 \mathrm{d}z \end{bmatrix}\begin{bmatrix}\omega_1\\ \omega_2\\ \omega_3\end{bmatrix} \tag{6.7.2.34}$$

It will be observed in many cases that in second and third mode the damping value would be high. Considering we are analyzing a single pile, while in reality there will be multiple piles under a pile cap, the present analysis does not take into cognizance the reflection of waves from other piles that goes on to attenuate the damping in higher modes. As such, damping ratios for higher modes should always be reduced. ACI-351.3R-04 recommends a reduction of 50% in damping ratio to the values obtained analytically.

It is suggested, that to be on conservative side, under no case shall pile damping ratio be considered to be more than 20% in any mode—this includes summation of both material and radiation damping.

6.7.2.5 Dynamic Modal Analysis

Having established the stiffness, mass, and damping matrix, we proceed to perform the dynamic modal analysis as mentioned hereafter.

As a first step, we obtain the eigenvalues from the expression

$$[K] - \lambda[M] = 0 \tag{6.7.2.35}$$

Considering that the matrix is of order 3 × 3, the solution is simple and can be easily carried out by standard methods like polynomial solution, Stodola–Vinello method.

Having obtained the eigenvalues λ_i and knowing $\lambda = \omega^2$ and $T_i = 2\pi/\omega_i$, we finally find out the time period of the pile–soil system for the first three modes.

Based on modal analysis (Manos and Clough 1982), maximum amplitude (S_{di}) can be expressed as

$$S_{\text{di}} = \frac{S_{\text{ai}}}{\omega_i^2} \tag{6.7.2.36}$$

Expressed in terms of code, Eq. (6.7.2.36) can be expressed as

$$S_{\text{di}} = \kappa_i C_{\text{F}} \frac{S_{\text{ai}}}{\omega_i^2} \tag{6.7.2.37}$$

Here, κ_i = modal mass participation factor and can be expressed as

$$\kappa_i = \frac{\sum_{i=1}^{3} m_i \phi_i}{\sum_{i=1}^{3} m_i \phi_i^2} \tag{6.7.2.38}$$

Equation (6.7.2.38) in terms of the present problem can be expressed as

$$\kappa_i = \frac{\left(\frac{\gamma_c AL}{g}\right) \int_0^1 \phi_i(z)\mathrm{d}z + \frac{P_{\text{d}}}{g} \phi_i(0)}{\left(\frac{\gamma_c AL}{g}\right) \int_0^1 \phi_i(z)^2\mathrm{d}z + \frac{P_{\text{d}}}{g} \phi_i(0)^2} \tag{6.7.2.39}$$

As P_{d} will usually be a large value compared to the self-weight of the pile, in most of the case $\kappa_i \to 1.0$ and can be considered the same without any significant error.

C_{F} is a code factor expressed as $C_{\text{F}} = \frac{ZI}{2R}$, where Z is the zone factor, I the importance factor, and R the response reduction factor. S_{ai}/g is obtained from code corresponding to the time period T_i.

Thus, Eq. (6.7.2.37) can now be expressed as

$$S_{\mathrm{di}} = C_{\mathrm{F}} \frac{S_{\mathrm{ai}} T_i^2}{4\pi^2} \tag{6.7.2.40}$$

The dynamic amplitude of the pile for first three modes are then expressed as

$$u_i = C_F \frac{S_{\mathrm{ai}} T_i^2}{4\pi^2} e^{-\beta_i \frac{z}{L}} \left[\cos \beta_i \frac{z}{L} + \eta_i \sin \beta_i \frac{z}{L}\right] \tag{6.7.2.41}$$

Considering $EI \frac{\mathrm{d}^2 u}{\mathrm{d}z^2} = -M_z$, the dynamic moment in pile can be expressed as

$$M_{\mathrm{zi}} = -EIC_{\mathrm{F}} \beta_i^2 \frac{S_{\mathrm{ai}} T_i^2}{2\pi^2 L^2} \left(\sin \beta_i \frac{z}{L} - \eta_i \cos \beta_i \frac{z}{L}\right) \tag{6.7.2.42}$$

Dynamic shear force ($V = \mathrm{d}M_z/\mathrm{d}z$) can be expressed as

$$V_{\mathrm{zi}} = -EIC_{\mathrm{F}} \beta_i^3 \frac{S_{\mathrm{ai}} T_i^2}{2\pi^2 L^3} \left((1 + \eta_i) \cos \beta_i \frac{z}{L} + (1 - \eta_i) \sin \beta_i \frac{z}{L}\right) \tag{6.7.2.43}$$

6.7.2.6 Kinematical Interaction with Ground

What we have discussed till now is the inertial interaction of the pile with superstructure load incorporated in the analysis as P_{d}.

Other than this, the free field motion of the ground will also pull the pile along with it. Especially, a long slender pile will move along with ground motion following the same deformation geometry of the ground.

The free field time period of the ground, referring to Fig. 6.62, can be expressed as

$$T_n = \frac{4H}{(2n - 1)V_{\mathrm{s}}} \tag{6.7.2.44}$$

Here, H = depth of soil to the bedrock, V_{s} = shear wave velocity of the soil, n = the mode number → n = 1, 2, 3 ….

Let S_{an}/g be spectral acceleration corresponding to time period T_n.

The ground displacement u_{gn} can be expressed as

$$u_{\mathrm{gn}} = \kappa_n C_{\mathrm{F}} \frac{S_{\mathrm{an}} T_n^2}{4\pi^2} \cos(2n - 1) \frac{\pi z}{2H} \tag{6.7.2.45}$$

Equation (6.7.2.45) can be simplified and expressed as

$$u_{\mathrm{gn}} = \kappa_n C_F \left(\frac{S_{\mathrm{an}}}{g}\right)\left(\frac{4\gamma_s H^2}{\pi^2 (2n-1)^2 G}\right)\cos(2n-1)\frac{\pi z}{2H} \tag{6.7.2.46}$$

Considering

$$EI\frac{\mathrm{d}^2 u_{\mathrm{gn}}}{\mathrm{d}z^2} = -M_z \tag{6.7.2.47}$$

$$\Rightarrow M_z = \kappa_n C_{\mathrm{F}}\gamma_{\mathrm{s}}\left(\frac{S_{\mathrm{an}}}{g}\right)\left(\frac{EI}{G}\right)\cos(2n-1)\frac{\pi z}{2H} \tag{6.7.2.48}$$

and

$$V_z = -\kappa_n C_{\mathrm{F}}\gamma_{\mathrm{s}}\left(\frac{S_{\mathrm{an}}}{g}\right)\left(\frac{EI}{G}\right)\left(\frac{(2n-1)\pi}{2H}\right)\sin(2n-1)\frac{\pi z}{2H} \tag{6.7.2.49}$$

Here, κ_n is modal participation factor for first three modes, as expressed in Table 6.3.

Moments and shears obtained, vide Eqs. (6.7.2.48) and (6.7.2.49), now need to be added to Eqs. (6.7.2.42) and (6.7.2.43) to get the complete response of the pile under seismic load.

For a long slender pile, considering that the ratio (*EI*/*G*) is not large, it will usually yield a negligible value and may be ignored from the analysis. However, for short rigid piles which can resist the free field ground motion, this can contribute significantly to the overall response.

6.7.2.7 Effect of Liquefaction of Soil Below Pile Cap

As explained in Chap. 3, liquefaction is a phenomenon that is peculiar to earthquake and has been responsible for damage to a number of pile foundations around the world. In cohesionless soil, rapid buildup of pore pressure due to transient shock often makes the soil lose its shear strength, making a portion of pile embedded in soil freestanding. This sudden change of stiffness in pile can make it unstable and induce a structural failure of the pile, either by formation of plastic hinges or even by buckling of the member.

When the liquefiable layer is overlain by non-liquefying soil (like clayey soil), as the bottom layer liquefies, the overlying layer often spreads out and induces additional drag to the pile and pile cap, inducing significant damage to them.

Thus, liquefaction of soil compounded by lateral spreading remains a major challenge to an earthquake engineer undertaking design of such pile foundations.

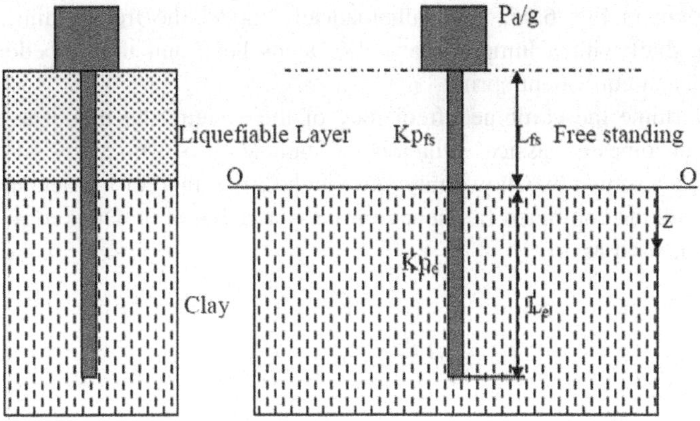

Fig. 6.64 Pile embedded in partially liquefiable soil layer

Fig. 6.65 Mathematical model of the embedded pile in liquefied soil

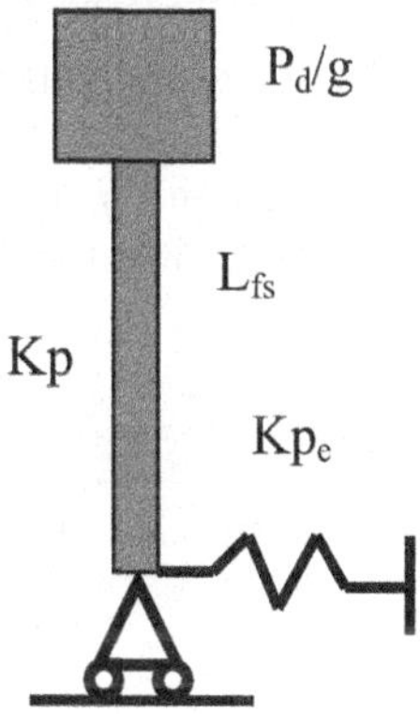

In Fig. 6.64, we show a pile driven in a soil whose top portion is susceptible to liquefaction. We try to seek a solution to this problem where a part of soil just below the pile cap has been assumed to have lost its shear strength due to liquefaction.

It is obvious that the simplest yet most rational model that can be perceived is that the pile becomes freestanding for a part of its length L_{fs}. The subscript fs means *freestanding,* and it is same as the depth of soil that is liquefiable. The depth of pile that still remains embedded in soil is denoted by L_e. Here, the subscript e stands for *embedded* and is used for mathematical notations where appropriate subsequently.

We try to seek solution to this problem for the fundamental mode that will be most critical in terms of its stability/failure.

If K_{pfs} and K_{pe} are stiffness of pile for freestanding and the embedded portion, respectively, then the mathematical model can perceived to be the same, as shown in Fig. 6.65.

As shown in Fig. 6.65, we mathematically model the freestanding part as a cantilever shaft with a lumped mass P_d/g at its head and the embedded part is modeled as an equivalent spring.

To determine the combined frequency of the system, the model as shown in Fig. 6.65 becomes a classical situation for application of Dunkerley's Equation.

It has been shown by Chowdhury and Singh (2010) that for mathematical model of this nature the combined frequency of the system based on Dunkerley's formula can be expressed as

$$\frac{1}{\omega_c^2} = \frac{1}{\omega_{fs}^2} + \frac{1}{\omega_e^2} \tag{6.7.2.50}$$

Here

ω_c combined natural frequency of the system;
ω_{fs} natural frequency of freestanding part considering its base as fixed;
ω_e natural frequency of the portion of pile that is embedded in soil.

Thus, to study the dynamic characteristics, as a first step we determine K_{pfs} and K_{pe}.

Shown in Fig. 6.66 is a beam under unit deflection at node 1. It can be shown (Paz-1987) that the geometric curvature of the beam can be expressed as

$$\psi(z) = 1 - 3\left(\frac{z}{L}\right)^2 + 2\left(\frac{z}{L}\right)^3 \tag{6.7.2.51}$$

The stiffness of the beam is then expressed as

$$K = \int_0^L EI\psi''(z)^2 dz \tag{6.7.2.52}$$

Considering $L = L_{fs}$, we have

$$K_{pfs} = \int_0^{L_{fs}} EI\left[-\left(\frac{6}{L_{fs}}\right) + 12\left(\frac{z}{L_{fs}^3}\right)\right]^2 dz \rightarrow \frac{12EI}{L_{fs}^3} \tag{6.7.2.53}$$

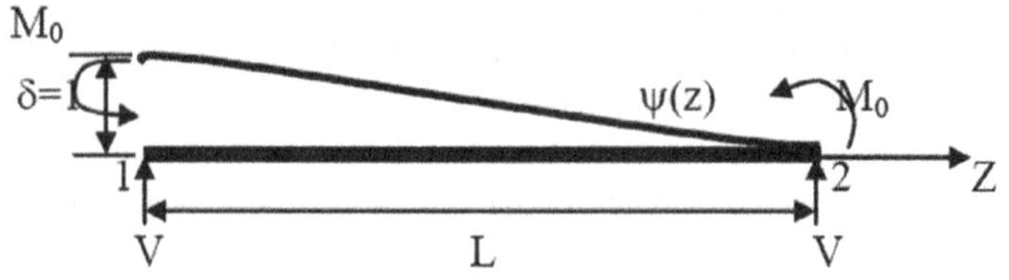

Fig. 6.66 Static deflection of beam subjected to unit deflection at node 1

The distributed mass matrix for the self-weight of pile shaft is expressed as

$$m_{fs} = \frac{\gamma_c A}{g} \int_0^{L_{fs}} \psi(z)^2 dz \rightarrow \frac{13}{35} \frac{\gamma_c A L_{fs}}{g} \tag{6.7.2.54}$$

Thus

$$\omega_{fs} = 3.464 \sqrt{\frac{EI}{L_{fs}^3 \left(\frac{P_d}{g} + \frac{13}{35} \frac{\gamma_c A L_{fs}}{g} \right)}} \tag{6.7.2.55}$$

For the embedded portion, the stiffness for the fundamental mode is expressed as

$$K_{pe} = \frac{4EI}{L_e^3} \left[\int_0^1 \beta_e^4 f_e(\xi)^2 d\xi + \int_0^1 \beta_e^4 \phi_e(\xi)^2 d\xi \right] \tag{6.7.2.56}$$

Here $\beta_e = \sqrt[4]{k_{se}.DL_e^4/4EI}$, $\eta_e = 1 + 1/\beta_e$

$$f_e(\xi) = \left[e^{-\beta_e \xi} (\sin \beta_e \xi - \eta_e \cos \beta_e \xi) \right] \tag{6.7.2.57}$$

and

$$\phi_e(\xi) = e^{-\beta_e \xi} [\cos \beta_e \xi + \eta_e \cdot \sin \beta_e \xi] \tag{6.7.2.58}$$

It is to be noted that $z = 0$ will be at the top of non-liquefiable soil layer, as shown in Figure 6, and $\xi = z/L_e$.

The mass matrix of the embedded part can be expressed as

$$M_e = \frac{\gamma_c A L_e}{g} \int_0^1 \phi_e(\xi)^2 d\xi + \frac{P_d}{g} \tag{6.7.2.59}$$

The natural frequency of the embedded part, ω_e, can be w obtained from the expression

$$\omega_e = \sqrt{\frac{\frac{4EI}{L_e^3} \left[\int_0^1 \beta_e^4 f_e(\xi)^2 d\xi + \int_0^1 \beta_e^4 \phi_e(\xi)^2 d\xi \right]}{\frac{\gamma_c A L_e}{g} \int_0^1 \phi_e(\xi)^2 d\xi + \frac{P_d}{g}}} \tag{6.7.2.60}$$

Having obtained ω_e and ω_{fs}, one can now substitute the same in Eq. (6.7.2.50) to obtain the combined natural frequency ω_c of the system.

Thus, as per modal analysis, deflection at pile head due to structural deformation of the freestanding part can be established as

$$\Delta_{fs} = C_F \frac{S_{ac} T_c^2}{4\pi^2} \tag{6.7.2.61}$$

Here, $T_c = 2\pi/\omega_c$ and S_{ac} = acceleration corresponding to code for the combined time period T_c.

The moment and shear at pile head can be expressed as

$$M_0 = \frac{6EI\Delta_{fs}}{L_{fs}^2}, \; V_0 = \frac{12EI\Delta_{fs}}{L_{fs}^3} \tag{6.7.2.62}$$

while at the soil–air interface (section O–O, vide Fig. 6.64)

$$M_0 = -\frac{6EI\Delta_{fs}}{L_{fs}^2}, \text{ and } V_0 = \frac{12EI\Delta_{fs}}{L_{fs}^3} \tag{6.7.2.63}$$

Now to determine the deflection and distribution of moments and shear at the soil line O–O, we consider Fig. 6.67 as shown hereafter.

To determine the deflection at soil line O–O, we start with Eq. (6.7.2.15) and rewrite it as

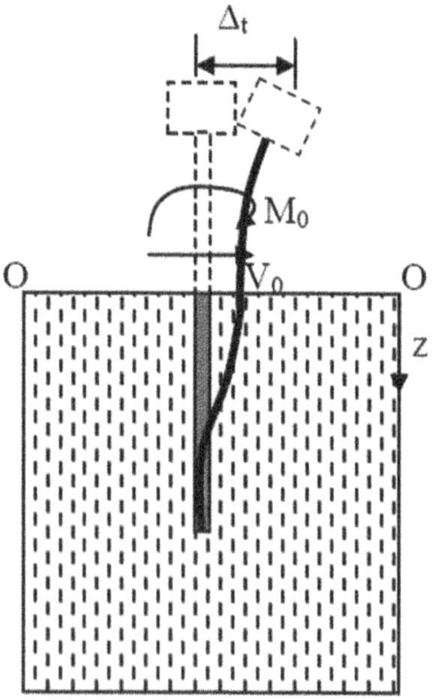

Fig. 6.67 Boundary condition for embedded portion of pile with probable displacement profile

$$\Delta_e = e^{-\beta_e \frac{z}{L_e}} \left[C_0 \cos\left(\frac{\beta_e z}{L_e}\right) + C_1 \sin\left(\frac{\beta_{ez}}{L_e}\right) \right] \tag{6.7.2.64}$$

Now imposing the boundary condition at $z = 0$, $EI\frac{d^2\Delta_e}{dz^2} = -M_0$, and at $z = 0$ $EI\frac{d^3\Delta_e}{dz^3} = -V_0$, we finally have

$$\Delta_e = \frac{L_e^3\left(e^{-\beta_e \frac{z}{L_e}}\right)}{2\beta_e^3 EI} \left[\left(V_0 - \frac{M_0\beta_e}{L_e}\right) \cos\left(\frac{\beta_e z}{L_e}\right) + \frac{\beta_e}{L_e} M_0 \sin\left(\frac{\beta_e z}{L_e}\right) \right] \tag{6.7.2.65}$$

The distribution of moment in pile below soil line O-O is expressed as

$$\begin{aligned} M_z &= -EI\frac{d^2\Delta_e}{dz^2} \\ \rightarrow M_z &= \frac{-L_e\left(e^{-\beta_e \frac{z}{L_e}}\right)}{\beta_e} \left[\left(V_0 - \frac{M_0\beta_e}{L_e}\right) \sin\beta_e \frac{z}{L_e} - \frac{\beta_e M_0}{L_e} \cos\beta_e \frac{z}{L_e} \right] \end{aligned} \tag{6.7.2.66}$$

The shear force distribution V_z (dM_z/dz) can be expressed as

$$V_z = e^{-\beta_e \frac{z}{L_e}} \left[V_0 \cos\frac{\beta_e z}{L_e} + \left(\frac{2\beta_e M_0}{L_e} - V_0\right) \sin\beta_e \frac{z}{L_e} \right] \tag{6.7.2.67}$$

The total deflection at top of the pile cap is thus given by

$$\Delta_t = \Delta_{fs} + \Delta_e \pm L_{fs} \cdot \theta_e \tag{6.7.2.68}$$

When $V_0 \geq \frac{2\beta_e M_0}{L_e}$, take (−) $L_{fs}\theta_e$ and vice versa for maximum amplitude. Here, θ_e is the rotation of pile at soil line O–O

$$\rightarrow \Delta_t = C_F \frac{S_{ac} T_c^2}{4\pi^2} + \frac{L_e^3}{2\beta_e^3 EI}\left(V_0 - \frac{M_0\beta_e}{L_e}\right) \pm L_{fs} \cdot \left(\frac{L_e^2}{2\beta_e^2 EI}\left(\frac{2\beta_e M_0}{L_e} - V_0\right)\right) \tag{6.7.2.69}$$

A fundamental point to be noted here is that while the pile becomes partially embedded, the portion below soil (of length L_e) should be sufficiently capable to carry the design load P_d without the freestanding part. Thus, while designing the pile for vertical capacity, this has to be clearly ensured.

6.7.2.8 Effect of Deep-Seated Liquefaction with Non-liquefiable Crust at Top

Shown in Fig. 6.68 is a pile embedded in soil, where the top crust does not liquefy (depth L_c) while the layer below it (depth L_l) is susceptible to liquefaction.

Dynamic behavior of piles under such situation has been a topic of significant controversy and recent research. Based on a number of failure case studies in different parts of the world it is now an established fact that this type of geological stratification is the most potential source of pile damage during ground shaking and in its aftermath during an earthquake (Hamada 2000; Tokimatsu et al. 1999, 2009; Ishihara et al. 2009).

A number of researchers have tried to explain the phenomenon adapting advanced numerical modeling and sophisticated laboratory test based on centrifuge and shake table testing (Cubrinovski et al. 2006, 2010; Finn and Thavaraj 2001; Dobry and Abdoun 2001). However, the parametric variations are so wide and diversified that it is extremely difficult if not impossible to identify all the parameters that affect the phenomenon and predict with confidence the actual behavior. And yet the topic continues to be a serious hindrance and zone of uncertainty to practicing engineers dealing with piles in earthquake zone.

Except for Japanese Code (JRA) and recent Euro Code (Euro Code 8) & NEHRP, no codes of any other country (including IS-2911/1893) address this issue.

Bhattacharya (2003) points out that even the recommendations as furnished in JRA (1996), Euro code, & NEHRP are at best a simplistic idealization and do not necessarily depict the actual picture.

In post failure analysis of Showa bridge foundation, Bhattacharya reports that as per JRA's design criteria for piles, Showa Bridge pile foundations on analysis were found to be sufficiently safe, and yet they failed miserably under the seismic force!

Thus, as a first step it would be worthwhile to see what does JRA recommends.

As shown in Fig. 6.69, the JRA recommends the portion of structure within the top crust to be designed for passive earth pressure (q_{NL}) and to design the pile for

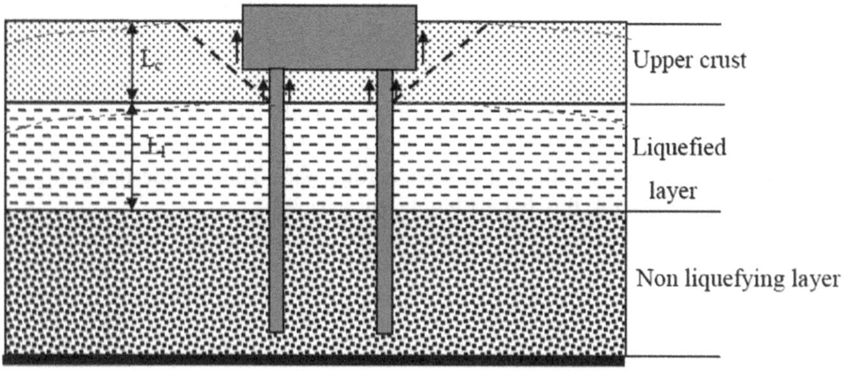

Fig. 6.68 Pile and pile cap embedded in soil susceptible to deep-seated failure

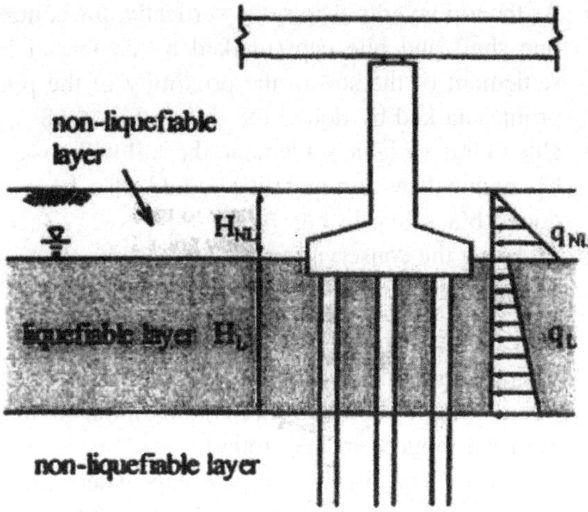

Fig. 6.69 Recommendation of JRA (1996) for piles under deep-seated liquefaction

30% of overburden pressure (q_L) within the liquefied zone. The background of this pressure diagram is based on instrumented test data of some bridges supported on pile foundations in Hanshin expressway that performed well under earthquake (Yokoyama et al. 1997).

However, JRA recommends that this force be considered in isolation and need not be considered acting simultaneously with inertial force.

This recommendation has been subject to criticism (Haigh and Madabhushi 2002), who pointed out based on centrifuge test that JRA recommendation under-predicts the force on pile by almost 300%, especially during the transient phase.

Berril et al. (2001) based on their research concluded that main threat to the pile emanates from force induced on it from non-liquefied crust rather than the force from the liquefied soil layer.

Thus, in spite of all these researches, a rational and practical mathematical model amenable to design office application is still a serious quest in the industry.

We intend to seek solution to the problem as hereafter.

Before we delve into the mathematics of the problem, let us first try to physically conceptualize the problem as to what happens to the system when an earthquake hits the foundation?

- The pile and pile cap along with the soil start vibrating.
- The vibration induces two conditions in the interaction within the pile and the soil.
  - The pile pushes the soil, generating a passive earth pressure case.
  - The soil pushes on the pile, generating an active earth pressure case.
- As the soil vibrates, there is a rapid increase in pore pressure in the second layer that induces degradation in shear strength and it liquefies.
- As the second layer loses its shear strength and flows, the top layer will try to settle.

- As the top layer tries to settle vertically, the contact friction between the soil and pile shaft and pile cap (marked by arrows in Fig. 6.68 will try to resist the settlement of the soil in the proximity of the pile cap, inducing a deformation profile marked by dotted red line in Fig. 6.68.
- Due to this variable settlement, the soil will possibly rupture along a line marked by an idealized straight line (could also be a logarithmic spiral) marked by dotted black line in Fig. 6.68.
- To be on the conservative side we assume (like in JRA) that the failure is along the passive pressure profile, i.e., makes an angle $45 - \phi/2$ with the horizontal.

It is apparent that this mass then clings to pile and pile cap and induces an additional lateral force. Moreover, as the top crust moves and rides along with the liquefied bottom soil, the acceleration induced in it will be same as free field acceleration induced in the ground.

Shown in Fig. 6.70 is a pile cap where the piles are centered at 2.5D. Considering the passive pressure failure, the width of the clinging soil can be estimated as $L = B = 3.5D + 2L_c \tan(45 + \phi/2)$ at ground level and at the interface of the first and second layer this is estimated as 3.5D.

Then, considering surface area at ground level,

$$A_1 = [3.5D + 2L_c \tan(45 + \phi/2)]^2$$

and at top liquefied layer

$$A_2 = [3.5D]^2 \tag{6.7.2.70}$$

The volume of soil clinging to the pile cap may be estimated as

$$V_{\text{soil}} = \left[\frac{A_1 + A_2}{2}\right] \times L_c \tag{6.7.2.71}$$

Thus, the suspended weight from the pile at the bottom of the top crust level can be expressed as

$$\begin{aligned} W_d = {} & (V_{\text{soil}} - \text{Vol. of pile cap} + \text{pile up to depth } L_c) \\ & \times \gamma_s + \text{Vol. of pile cap} + \text{pile up to depth } L_c \times \gamma_c \end{aligned}$$

The dynamic drag force on the pile V_{dr} can then be expressed as

$$V_{dr} = \left(\frac{8}{2+\pi}\right)\left(\frac{ZI}{2R}\right)\left(\frac{S_{ag}}{g}\right) W_d \tag{6.7.2.72}$$

Here S_{ag} is the acceleration due to free field time period of the site in fundamental mode calculated as

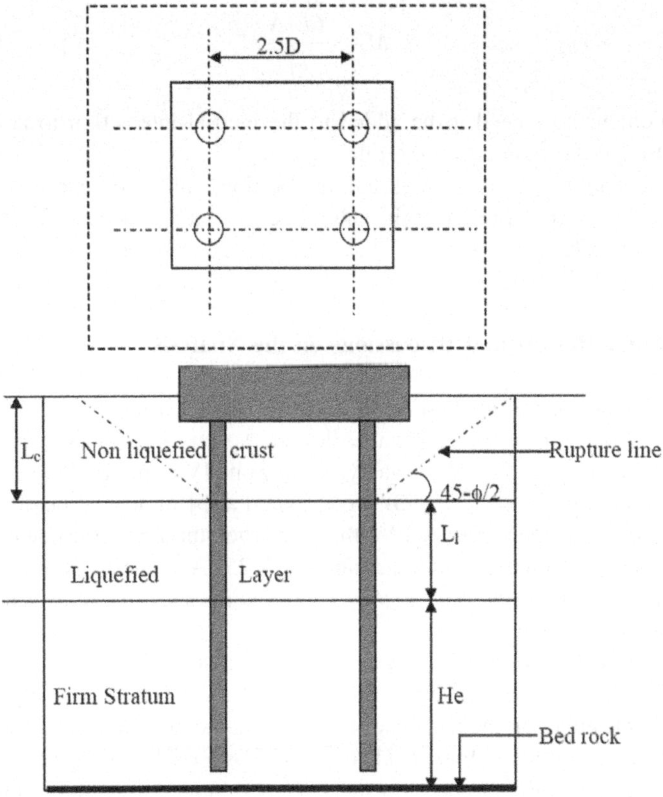

Fig. 6.70 Pile cap with soil clinging to it

$$T_{ag} = \frac{4H_e}{V_s} \tag{6.7.2.73}$$

Dividing Eq. (6.7.2.72) by the number of piles N (as per Fig. 6.70, $N = 4$ in this case), we estimate the additional force on each pile due to spreading of the soil.

Thus, the mathematical model becomes same as Fig. 6.64 except that here $L_{fs} = L_l$, the depth of soil layer that liquefies.

Thus, the displacement at top of liquefied layer can be estimated as

$$\Delta = \frac{V_{dr} L_l^3}{N \times 12EI} \tag{6.7.2.74}$$

Moment at top of the third layer (firm bottom layer) is expressed as

$$M_z = \frac{6EI\Delta}{L_1^2} \tag{6.7.2.75}$$

This moment now need to be added to the inertial interaction to arrive at the maximum expected moment in the pile.

For calculation of the freestanding portion, all the expressions from Eqs. (6.7.2.53) to (6.7.2.69) remain valid, except that L_{fs} is considered as L_1, as shown in Fig. 6.70.

6.7.2.9 Some Results and Discussions on the Analysis

To further elaborate the method an RCC bored pile of a length of 24 ms of and diameter of 900 mm, having allowable design capacity of 2500 kN, and average shear wave velocity of 186 m/s, is analyzed for Zone IV as per IS-1893 (2016). The bedrock (where SPT value $N \geq 50$) is considered at 24 m, where the pile is rested.

The same pile is then analyzed for the case when the first 6.0 m below the pile cap get liquefied. Concrete grade assumed is M25. Average density of soil = 17 kN/m^3. Density of concrete = 25 kN/m^3. The dynamic shear modulus is assumed to vary as $G = G_0 \cdot (z/H)$.

The salient design properties are tabulated in Table 6.15.

It is observed that the fundamental mode is the most critical, and higher modes have negligible effect on the response. The bending moment and shear force on pile are as presented hereafter (Figs. 6.71, 6.72, 6.73 and 6.74).

We also compare the moment for the same soil having different variation of G, namely

(i) $G = G_0\ (z/H)$—sandy soil, (ii) G constant—clayey soil, and (iii) $G = G_0\ (1 + z/H)$ - c–ϕ soil.

Next we analyze the same pile for 6.0-m depth of soil below the pile cap that has undergone liquefaction. In this context, a very peculiar property has been observed:

Table 6.15 Design parameters for fully embedded pile under earthquake force

Mode	1	2	3
G (kN/m^2)	59,950	59,950	59,950
Dynamic sub-grade modulus (kN/m^3)	1831	10,830	32,720
Time period (s)	1.303	0.099	0.024
Damping ratio (%)	10.3	57.3[a]	108[a]
S_a (m/s^2) (g)	0.103	0.134	0.073
Modal participation factor (κ_i)	1.012	1.007	1.005
Displacement at pile head (mm)	4.5	0.033	0.0011
Moment at pile head (kN m)	175	3.172	0.176
Shear at pile head (kN)	58	1.6	0.11

[a]Restricted to 20% for second and third modes

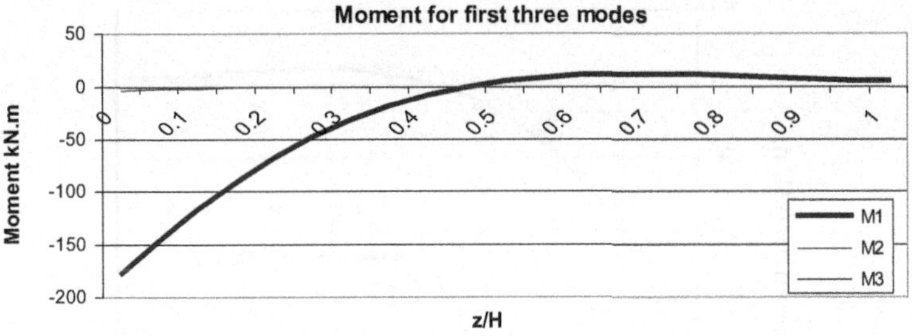

Fig. 6.71 Bending moment diagram for the pile under earthquake

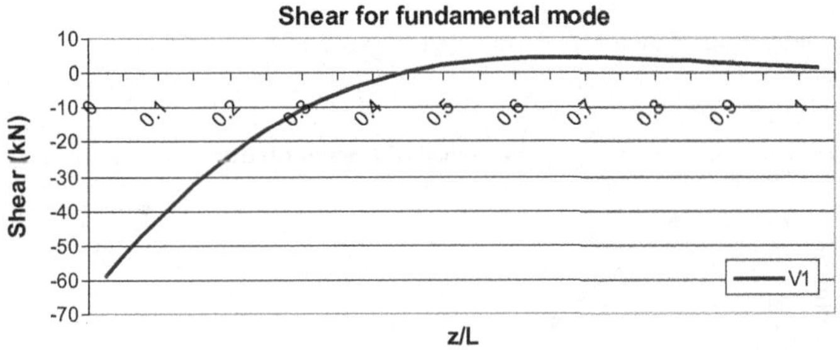

Fig. 6.72 Shear force diagram for the pile under earthquake

Fig. 6.73 Displacement diagram for the pile under earthquake

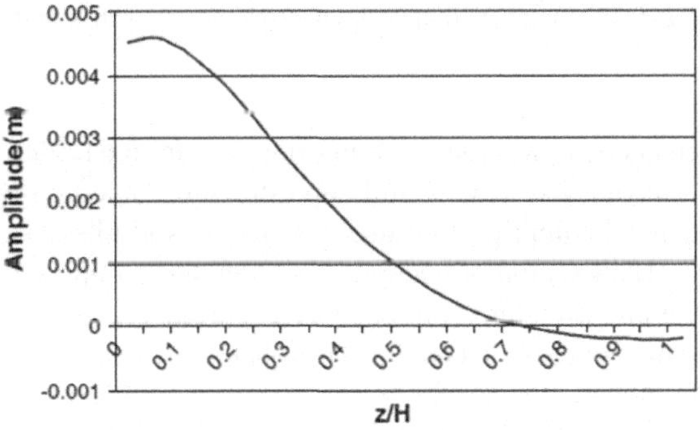

The shallower the liquefaction depth, the greater the displacement. Up to a certain depth, the displacement of the pile at top is very high and beyond this the displacement reduces significantly and then changes sign.

The salient design parameters are as tabulated below for progressive freestanding depths.

The apparent displacement anomaly can be explained by the fact that when the freestanding part is very short, it acts like a very stiff stub, compared to the portion

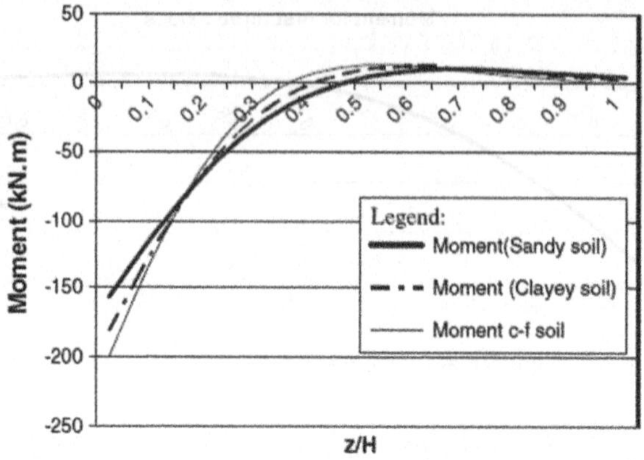

Fig. 6.74 Comparison of moments for different types of soil

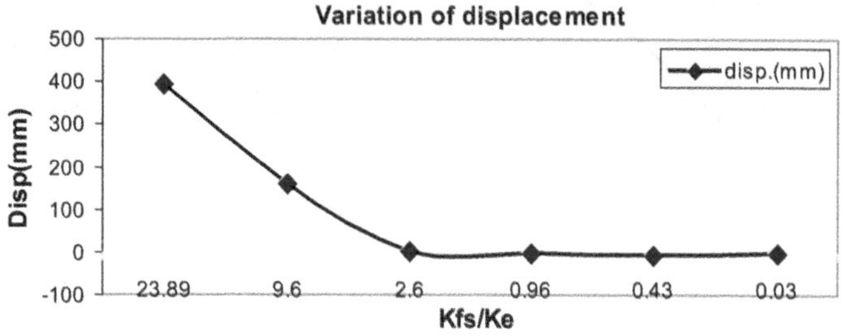

Fig. 6.75 Variation of pile head displacement with relative freestanding and embedded pile stiffness ratio

embedded in ground. Referring to Fig. 6.64, the freestanding part then starts behaving like a very stiff superstructure underlain by soft foundation resulting in dynamic amplification due to dynamic soil pile structure interaction.

The location of bedrock level here has an important bearing. For in this case, as the pile is resting on it at 24 m, for 6-m free-stand the displacement is 26 mm. The same bedrock, if taken to 60 m below the pile head (i.e. presuming the pile to be frictional), the displacement amplifies to 124 mm (5 times) while the moments jump to 4314 kN m (1.53 times) and shear becomes 1438 KN (1.53 times).Thus, the softer the pile embedment zone, the higher the soil structure amplification under earthquake can be expected.

Figure 6.75 clearly explains this phenomenon. As the ratio K_{pfs} and K_{pe} reduces progressively, there is marked reduction to pile head displacement. The change in sign to the displacement can be attributed to the depth when flexural behavior predominates over shear, giving an opposite slope to the pile at the soil line, making the pile bend in a direction opposite to the direction in which the shear force acts.

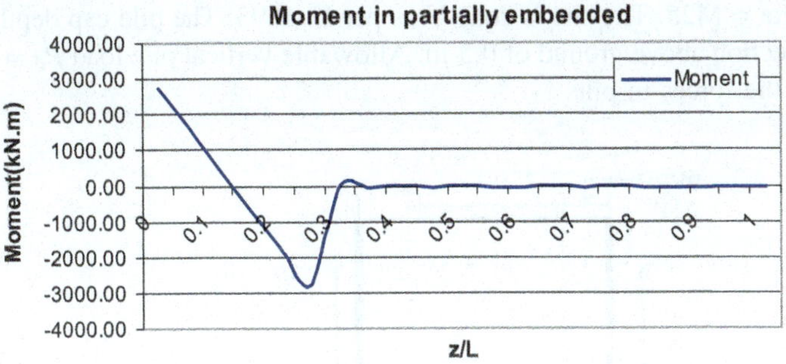

Fig. 6.76 Moment in pile with 6.0 m of top portion liquefied

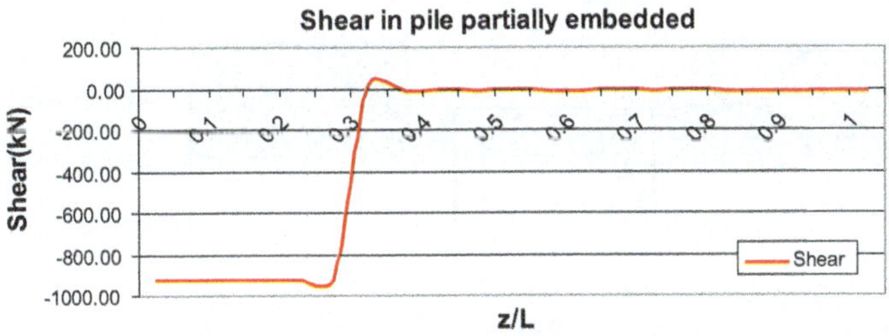

Fig. 6.77 Shear in pile with 6.0 m of top portion liquefied

It is apparent that for this case, if the soil liquefies to a depth up to 4 m, a huge stress would be induced in the top portion of the pile (e/D ratio = 6.23 and 3.33 for 3 and 4 m free stand, respectively; here $e = M_0/P_d$). Thus, an engineer designing such a pile has to give careful consideration as to how to sustain it. Either the ground itself needs to be locally tamped (dynamic compaction) to increase the SPT value so that the soil does not liquefy, or the pile shaft may have to be locally strengthened by providing annular steel jackets when the combined RCC and the annular steel jacket should be strong enough to sustain the load.

A shallow liquefaction below and around the pile cap can be far more dangerous than a deeper liquefaction.

The bending moment and shear force diagram of the pile after soil liquefaction (up to 6 m depth) are shown hereafter (Figs. 6.76 and 6.77).

The lateral soil spreading issue obviously requires more study; however, comprehensive design steps to estimate the moment and shear in a pile are furnished in the example hereafter.

Example 6.14 Shown is a 24-m RCC pile of 900 mm diameter. The top 4 m are of non-liquefiable clay ($\phi = 0$), with next 6 m of liquefiable sand. The pile penetrates 14 m inside non-liquefiable dense sand and finally rests on the bedrock. Average density of soil is 17 kN/m^3. Shear wave velocity 186 m/s is varying linearly. Grade of concrete = M25. The site is zone V as per IS-1893. The pile cap depth is 1.5 m with projection above ground of 0.5 m. Allowable vertical pile load $P_d = 2500$ kN. Estimate the forces in pile.

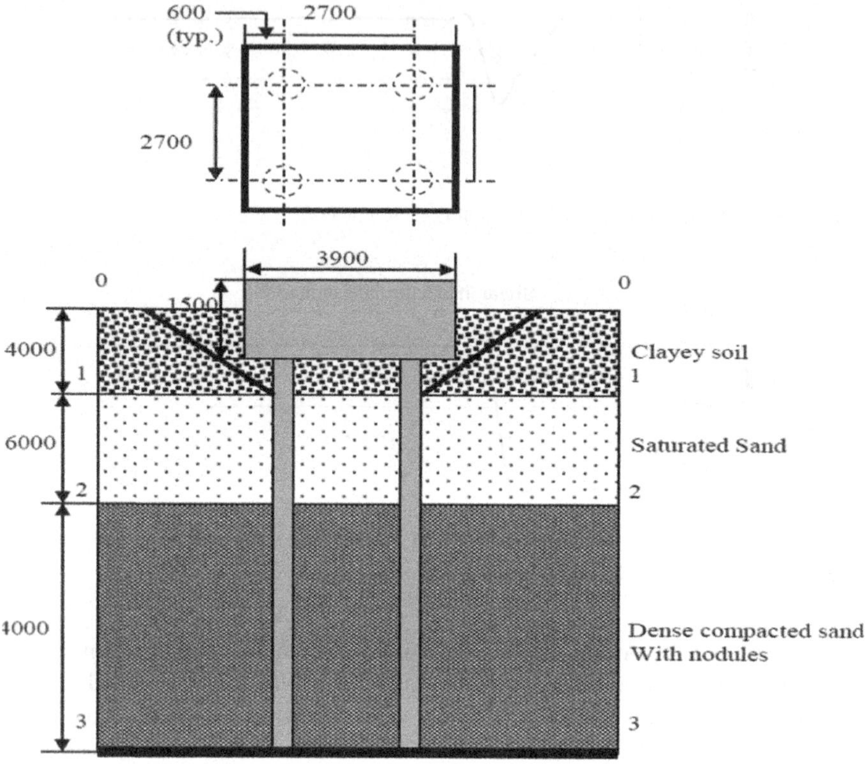

Solution

Young's modulus of pile = $5700\sqrt{f_{ck}}$ N/mm^2 $\rightarrow 2.85 \times 10^7$ kN/m^2

Diameter of pile = 900 mm

$I = \frac{\pi D^4}{64} = 0.0322$ m^4. $A = \frac{\pi D^2}{4} = 0.636$ m^2.

Dynamic shear modulus = $\rho V_s^2 = \frac{17}{9.81} \times (186)^2 = 59950$ kN/m^2.

Calculation of drag force on pile from non-liquefiable layer

Influence area at soil surface (0-0) = $(3.5 + 2 \times 4\tan 45°)^2 = 132.25$ m^2.

Area at interface 1–1 = $(3.5)^2 = 12.25$ m^2.

Gross volume of prismoid = $\left(\frac{132.25 + 12.25}{2}\right) \times 4 = 289$ m^3.

Vol. of pile cap below soil = $3.9 \times 3.9 \times 1 = 15.21$ m^3.

Vol. of 4 piles below soil = $\frac{\pi D^2}{4} \times 3 \times 4 = 7.63$ m^3.

∴ Net volume of soil = 289 − 15.21 − 7.63 = 266.15 m^3.

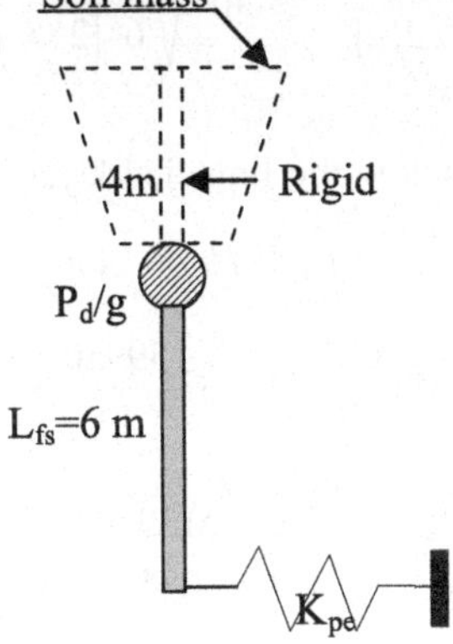

Thus, the total influence load = 266.15 × 17 + (15.21 + 7.63) × 25 = 5096 kN.
Free field time period = $\frac{4H_e}{V_s} = \frac{4 \times 14}{186} = 0.301$ s.
⇒ $\frac{S_a}{g} = 2.5$ as per IS-1893. For zone V, zone factor $Z = 0.36$, Let $I = 1.5$, $R = 3$

$$\therefore C_F = \frac{ZI}{2R} = \frac{0.36 \times 1.5}{2 \times 3} = 0.09$$

Thus, the drag force $V_{dr} = 0.09 \times 5096 \times 2.5 = 1147$ kN.
Hence, the drag force per pile = $\frac{1147}{4} = 287$ kN.

$$K_{fs} = \frac{12EI}{L_{fs}^3} = \frac{12 \times 2.85 \times 10^7 \times 0.0322}{(6)^3} = 50984\,\text{kN/m}.$$

Deflection $\Delta = \frac{287}{50984} = 5.629 \times 10^{-3}\,\text{m} \cong 5.63\,\text{mm}$.

$$M_{fs} = \frac{6EI\Delta}{L_{fs}^2} = \frac{6 \times 2.85 \times 10^7 \times 0.0322 \times 5.629 \times 10^{-3}}{(6)^2} = 861\,\text{kN m}$$

It is to be noted that the pile length embedded in the top 4 m of clayey soil will act rigidly and compared to the 6.0 m shaft within liquefiable soil will not deform (see the figure above). The plastic hinge may thus either develop at 4.0 m or 10.0 m below the pile head

Dynamic loading due to vertical load P_d = 2500 kN.

Lumped mass M = $\frac{P_d}{g} = \frac{2500}{9.81} = 255$ kN s^2/m

$$\omega_{\text{fs}} = 3.464\sqrt{\frac{EI}{L_{\text{fs}}^3\left[\frac{P_{\text{d}}}{g} + \frac{13}{35}\frac{\gamma_c A L_{\text{fs}}}{g}\right]}} = 3.464\sqrt{\frac{2.85\times10^7\times0.0322}{6^3\left[255+\frac{13}{35}\frac{25\times0.636\times6}{9.81}\right]}} = 14.04\,\text{rad/s}$$

For embedded portion below the liquefied layer

$$L_{\text{e}} = H_{\text{e}} = 14\,\text{m}$$

$$k_{\text{se}} = 0.733\frac{G_0}{H_{\text{e}}} = 0.733\frac{59950}{14} = 3139\,\text{kN/m}$$

$$\beta_{\text{e}} = 4\sqrt{\frac{k_{\text{se}}DL_{\text{e}}^4}{4EI}} = 4\sqrt{\frac{3139\times0.9\times14^4}{4\times2.85\times10^7\times0.0322}} = 2.33,$$

$$\eta_{\text{e}} = 1 + 1/\beta_{\text{e}} = 1.43.$$

$$K_{\text{pe}} = \frac{4EI}{L^3}\left[\int_0^1 \beta_{\text{e}}^4 e^{-2\beta_{\text{e}}\xi}(\sin\beta_{\text{e}}\xi - \eta_{\text{e}}\cos\beta_{\text{e}}\xi)^2 d\xi + \int_0^1 \beta_{\text{e}}^4 e^{-2\beta_{\text{e}}\xi}[\cos\beta_{\text{e}}\xi + \eta_{\text{e}}\sin\beta_{\text{e}}\xi]^2 d\xi\right]$$

$$\rightarrow K_{\text{pe}} = \frac{4EI}{L^3}[I_1 + I_2]$$

$$M = \frac{\gamma_c A L_{\text{e}}}{g}\int_0^1 e^{-2\beta_{\text{e}}\xi}[\cos\beta_{\text{e}}\xi + \eta_{\text{e}}\sin\beta_{\text{e}}\xi]^2 d\xi \rightarrow M_{\text{e}} = \frac{\gamma_c A L_{\text{e}}}{g} I_3$$

The integrals I_1, I_2 and I_3 are obtained as shown below.

$\xi = z/L_{\text{e}}$	I_1	I_2	I_3
0	60.20053	29.4730	1.0000
0.05	39.65476	31.3804	1.0647
0.1	24.87194	31.3988	1.0653
0.15	14.66085	29.9142	1.0150
0.2	7.946469	27.3638	0.9284
0.25	3.805498	24.1657	0.8199
0.3	1.478432	20.6790	0.7016
0.35	0.365311	17.1868	0.5831
0.4	0.011111	13.8929	0.4714
0.45	0.085514	10.9286	0.3708
0.5	0.360609	8.3642	0.2838

(continued)

(continued)

$\xi = z/L_e$	I_1	I_2	I_3
0.55	0.688977	6.2224	0.2111
0.6	0.983713	4.4912	0.1524
0.65	1.201245	3.1357	0.1064
0.7	1.327262	2.1083	0.0715
0.75	1.365719	1.3558	0.0460
0.8	1.330639	0.8252	0.0280
0.85	1.240322	0.4674	0.0159
0.9	1.113522	0.2393	0.0081
0.95	0.967136	0.1045	0.0035
1	0.815031	0.0339	0.0011
Sum→	6.69834	12.448887	0.422383

In the above table, summation is carried out by trapezoidal rule

$$I = \frac{h}{2}[y_0 + 2(y_1 + y_2 + y_3 + \ldots y_{n-1}) + y_n]; \quad \text{here } h = 0.05$$

$$K_{pe} = \frac{4 \times 2.85 \times 10^7 \times 0.0322}{14^3}[6.698 + 12.448] = 25666\,\text{kN/m}$$

$$M_e = \frac{25 \times 0.636 \times 14}{9.81} \times 0.422 + 255 = 265\,\text{kN s}^2/\text{m}$$

$$\omega_e = \sqrt{\frac{K_{pe}}{M_e}} = \sqrt{\frac{25666}{265}} = 9.85\,\text{rad/s}.$$

The combined frequency is computed as

$$\frac{1}{\omega_c^2} = \frac{1}{\omega_{fs}^2} + \frac{1}{\omega_e^2}$$

Substituting values of ω_{fs} and ω_e we have $\omega_c^2 = 65 \rightarrow \omega_c = 8.063$ rad/s $\rightarrow$ $T_c = 0.779\,\text{s}$

For $T_c = 0.779$ s, as per IS 1893, $S_a/g = 1.67/0.779 = 2.143$.

$$\therefore \Delta = C_F \frac{S_a}{\omega_c^2} = 0.09 \frac{2.143 \times 9.81}{65} = 0.029\,\text{m} \cong 29\,\text{mm}$$

Maximum dynamic shear force

$$V = \frac{12EI\Delta}{L_{fs}^3} = \frac{12 \times 2.85 \times 10^7 \times 0.0322 \times 0.029}{(6)^3} = 1479\,\text{kN}$$

Table 6.16 Design parameters for partially embedded pile, with progressive liquefaction at top

G (kNm^2)	59,950	59,950	59,950	59,950	59,950	59,950
Freestanding length L_{fs} (m)	3	4	6	8	10	18
Embedded length L_e (m)	21	20	18	16	14	6
Dynamic k_s (kN/m^3)	1831	1831	1831	1831	1831	1831
Depth of bedrock (m)	24	24	24	24	24	24
Time period (s)	0.798	0.802	0.852	0.972	1.159	2.392
Damping ratio (%)	11	12	12	13	13	19
S_a (m/s^2) (g)	0.151	0.143	0.141	0.121	0.101	0.044
K_{pfs}/K_{pe}	23.89	9.6	2.6	0.96	0.43	0.03
Max. moment (kN m)	14,020	7488	2822	1207	565	31
Max. shear (kN)	9350	3744	941	302	113	3
Total displacement at pile head (mm)	395	160	26	−0.57	−4.5	−0.33

Maximum dynamic moment

$$M_{fs} = \frac{6EI\Delta}{L_{fs}^2} = \frac{6 \times 2.85 \times 10^7 \times 0.0322 \times 0.029}{(6)^2} = 4435\,\text{kN m}$$

Thus, the total maximum moment on pile = 4435 + 861 = 5296 kN m.

The total shear force = 1479 + 287 = 1766 kN.

Comparing to 10.0-m freestanding pile as well as 6.0-m one (vide Table 6.16), it will be observed that forces are far higher than in both cases. This clearly reflects the potential risk for this case.

References

ACI-351.3R-04. (2011). Foundation for dynamic equipments. *American Concrete Institute Committee # 351.*

Banerjee, P. K., & Sen, R. (1987). Dynamic behavior of axially and laterally loaded piles and pile groups, Chapter 3. In P. K. Banerjee & R. Butterfield (Eds.), *Developments in soil mechanics and foundation engineering* (3rd ed.). London: Elsevier, Applied Science.

Bathe, K. J. (1996). *Finite element method in engineering procedure*. New Delhi: Prentice Hall Pvt. Limited.

Berrill, J. B., Christensen, S. A., Kennan, R. P., Okada, W., & Pettinga, J. R. (2001). Case study of lateral spreading forces on a piled foundation. *Geotechnique, 51*(6), 501–517.

Bhattacharya, S. (2003). *Pile Instability during earthquake liquefaction*. Ph.D. thesis of Subhamoy Bhattacharya, Cambridge University.

Bowles, J. E. (1968). *Foundation analysis and design*. NY: McGraw-Hills.

Brinch Hansen, J. (1961). A general formula for bearing capacity. A revised and extended formula for bearing capacity. The *Danish Geotechnical Institute*. Bulletin No. 11. Copenhagen.

Brinch Hansen, J. (1970). A revised and extended formula for bearing capacity *Danish Geotechnical Institute Bulletin # 98*, pp. 5–11.

Broms, B. B. (1964). Lateral resistance of piles in cohesive soils. *Journal of the Soil Mechanics and Foundations Division ASCE, 90*(2), 27–63.

Budhu, M., & Al Karni, A. (1993). Seismic bearing capacity of soil. *Geotechnique, 43*(1), 181–187.

Chandrashekharan, V. (1974). *Analysis of Pile foundation under earthquake and dynamic loads.* Ph.D. Thesis, University of Roorkee, Roorkee, UP.

Chen, W. F., & Davidson, H. L. (1973). Bearing capacity determination by limit analysis. *Journal of the Soil Mechanics and Foundations Division ASCE, 99,* 443–449.

Chowdhury, I., & Dasgupta, S. P. (2008). *Dynamics of structures and foundation a unified approach* (Vol. II). Leiden, Holland: Taylor and Francis.

Chowdhury, I., & Dasgupta, S. P. (2010). Estimation of lateral load capacity of short piles under earthquake force (May 24). In *International Conferences on Recent Advances in Geotechnical Earthquake Engineering and Soil Dynamics.* Paper 20. http://scholarsmine.mst.edu/icrageesd/05icrageesd/session05/20.

Chowdhury, I., & Dasgupta, S. P. (2012). Seismic response of piles under earthquake force. *Indian Geotechnical Journal, 42*(2).

Chowdhury, I., & Dasgupta, S. P. (2015). Dynamic bearing capacity of shallow foundation under earthquake force. *Indian Geotechnical Journal* (May).

Chowdhury, I., & Dasgupta, S. P. (2016). Dynamic bearing capacity of shallow foundation under earthquake force. *Indian Goetechnical Journal, 47*(1), 36–45.

Choudhury, D., & Subba Rao, K. S. (2005). Seismic bearing capacity of shallow strip footings. *Journal of Geotechnical and Geological Engineering, 23,* 403–418.

Chowdhury, I., & Tilak, R. (2012). Effect of bedrock level on Dynamic bearing capacity of Shallow Foundation under earthquake load. In *International Symposium on Engineering Uncertainty Safety Assessment and Management Bengal Engineering College*, Jan 4–6, 2012.

Chowdhury, I., & Singh, J. P. (2010). Do DSSI attenuate dynamic response of building? In *Proceedings 14th symposium on Earthquake Engineering IIT Roorkee* (Vol. 1, pp. 644–654).

Chowdhury, I., Ghosh, A., & Dasgupta, S. P. (2015). Stiffness degradation and damping augmentation of soil under earthquake loading. *Electronic Journal of Geotechnical Engineering* (Vol. 20), Bundle 9, 2015 Oklahama, USA.

Christchurch City Council. (2010). *Construction standard specification*, Part 4: Water Supply (p. 59) (Cubrinovski M, Hughes M, Bradley B, McCahon I, McDonald Y, Simpson H, Cameron R, Christison).

Cubrinovski, M., Haskell, J. J. M., & Bradley, B. (2010). Soil pile interaction in liquefying soils—modeling issues. In: R. P. Orense, N. Chuow, & M. J. Pender (Eds.), *Soil foundation structure interaction.* London: Taylor and Francis.

Cubrinovski, M., Kokusho, T., & Ishihara, K. (2006). Interpretation from large-scale shake table tests on piles undergoing lateral spreading in liquefied soils. *Soil Dynamics and Earthquake Engineering, 26,* 275–286.

Das, B. M. (1987). *Theoretical foundation engineering developments in geotechnical engineering* (Vol. 47). Amsterdam: Elsevier Publications.

Davisson, M. T. & Prakash, S. (1963). A review of soil pole behavior. *Highway Research Record No. 39* (pp. 25–48).

Davisson, M. T. (1970). *Lateral load capacity of piles.* Washington, DC: Highway Research Record, Transportation Research Board.

Den Hartog, J. P. (1952). *Advanced strength of materials.* New York: Dover Publication.

Dobry, R., & Gazetas, G. (1988). Simple method for dynamic stiffness and damping of floating pile groups. *Geotechnique, 38*(4), 557–574.

Dobry, R., & Abdoun, T. (2001). Recent studies on seismic centrifuge modeling of liquefaction and its effect on deep foundations. State-of-the-Art Report (SOAP3). In *Proceedings of the 4th International Conference on Recent Advances in Geotechnical Earthquake Engineering and Soil Dynamics* (Vol. 2). San Diego, CA, March 26–31.

Dormieux, L., & Pecker, A. (1995). Seismic bearing capacity of cohesion less soil. *Journal of Geotechnical Engineering ASCE, 121*(3), 300–303.

Dunkerley, S. (1894). On the whirling and vibration of shafts. *Philosophical Transactions of the Royal Society, 185*, 279–360 (London, UK).
Finn, W. D. L., & Thavaraj, T. (2001). Deep foundations in liquefiable soil, SOAP 1. In *Proceedings of the 4th International Conference on Recent Advances in Geotechnical Earthquake Engineering and Soil Dynamics*, San Diego.
Griffith, D. V. (1982). Computation of bearing capacity factors using finite elements. *Geotechnique, 32*, 195–202.
Gutenberg, B., & Richter, C. F. (1956). Magnitude and energy of earthquakes. *Annali di Geofisica, 9*, 1–15.
Haigh, S. K., & Madabhushi, S. P. G. (2002). Centrifuge modelling of lateral spreading past pile foundations. In *International Conference on Physical Modelling in Geotechnics*, St John's, Newfoundland, Canada, July 2002.
Hamada, M. (2000). Performance of foundations against liquefaction-induced permanent ground displacement. In *Proceedings of the 12th World Conference on Earthquake Engineering, Auckland, New Zealand*, Paper No. 1754.
Hurty, W. C., & Rubinstein, M. F. (1967). *Dynamics of structures*. New Delhi: Prentice-Hall of India.
Ishihara, T., Midorikawa, M., & Azuhata, T. (2009). Modal properties and free vibration of uplifting behavior of multistory buildings modeled as uniform shear-beam. *Journal of Structural and Construction Engineering, AIJ, 74*(640), 1055–1061. (in Japanese).
IS-2911/1893. *Bureau of Indian Standards*. New Delhi, India.
IS 1893 Part1. (2002). Criteria for earthquake resistant design of structures. *Bureau of Indian Standards*. New Delhi, India.
IS 2911(Part 1 to 4). Design and construction of pile foundations—code of practice. *Bureau of Indian Standards*. New Delhi, India.
IS-6403. (1997). Code of practice for determination of bearing capacity of shallow foundations. *Bureau of Indian Standards*. New Delhi, India.
Karnovsky, I., & Olga, L. (2001). *Formulas for structural dynamics* (p. 535). New York, N.Y.: McGraw-Hill, Inc.
Kishida, H. (1967). Ultimate bearing capacity of piles driven into loose sand. *Soils and Foundations, 7*(3), 20–29.
Kramer, S. (1996). *Geotechnical earthquake engineering*. New Delhi, India: Pearson Education.
Kreyszig. (1969). *Advanced engineering mathematics* (2nd ed.). Wiley Eastern: New Delhi.
Kumar, J., & Rao, V. B. K. M. (2002). Seismic bearing capacity of spread foundations. *Geotechnique, 52*(2), 79–88.
Manoharan, N., & Dasgupta, S. P. (1995). Bearing capacity of surface footings by finite elements. *Journal of Computers and Strucutres, 4*, 563–586.
Manos, G. C., & Clough, R. W. (1982). Further study of the earthquake response of a broad cylindrical tank Model. *EERC Report No 82/07*. UBC, California.
Meyerhof, G. G. (1951). The ultimate bearing capacity of foundations. *Geotechnique*, 301–332.
Meyerhof, G. G. (1953). The bearing capacity of foundation under eccentric and inclined load. In *Proceedings 3rd International Conference in Soil Mechanics and Foundation Engineering* (Vol. 1, pp 440–445) Zurich.
Meyerhof, G. G. (1959). Compaction of sands and bearing capacity of piles. *ASCE Journal of Soil Mechanics and Foundation Division* (pp. 1–29). SM6.
Meyerhof, G. G. (1963). Some recent research on bearing capacity of foundations. *Canadian Geotechnical Journal, 1*(1), 16–26.
Meyerhof, G. G. (1974). Ultimate bearing capacity of footings on sand layer overlying clay. *Candian Geotechnical Journal, 11*(2), 224–229.
Matlock, H., & Reese, L. C. (1960). Generalized solutions for laterally loaded piles. *Journal of the Soil Mechanics and Foundations Division, 86*(5), 91–97.
Meirovitch, L. (1967). *Analytical methods in vibration*. London: Macmillan Publication.
Novak, M. (1974). Dynamic stiffness and damping of piles. *Canadian Geotechnical Journal, 11*, 574–598.

Novak, M., & El Sharnouby, B. (1983). Stiffness and damping constants for single piles. *Journal of Geotechnical Engineering Division, ASCE, 109*, 961–974.

Paz, M. (1987). *Structural dynamics*. New Delhi: CBS Publishers Ltd.

Pecker, A. (1996). Seismic bearing capacity of shallow foundations. In *Proceedings XIth World Conference in Earthquake Engineering Mexico*, Paper # 2076.

Poulos, H. G. & Davis, E. H. (1980). *Pile foundation analysis and design*. Hoboken: Wiley.

Peck, R. B., & Davisson, M. T. (1962). *Discussion of "Design and Stability of Pile Groups Subjected to Lateral Loads"*. Ph.D. thesis, University of Illinois.

Penzien, J. (1970). *Soil pile foundation interaction in earthquake engineering*, R. L. Wiegel (Ed.), Englewood Cliff, New Jersey: Prentice Hall.

Prakash, S., & Chandrashekharan, V. (1980). Analysis of piles in clay against earthquakes. *Spring Convention ASCE* (Portland, Oregon, April).

Reese, L. C., & Matlock, H. (1956). *Non-dimensional solutions for laterally loaded piles with soil modulus*. Publication # 29. Bureau of Engineering Research, University of Texas Austin.

Richards, R., Elms, D. G., & Budhu, M. (1993). Seismic bearing capacity and settlement of foundations. *Journal of Geotechnical Engineering Division ASCE, 119*(4), 662–674.

Ruigork, J. A. T. (2010). *Laterally loaded piles models and measurements*. Ph.D. thesis. Holland: Technical University of Delft.

Sarma, S. K., & Iossefelis, I. S. (1990). Seismic bearing capacity factors of shallow strip footings. *Geotechniqe, 40*(2), 265–273.

Shames, I., & Dym, C. L. (1995). *Energy and finite element method in structural mechanics*. New Delhi, India: New Age Publisher.

Soubra, A. (1997). Seismic bearing capacity of shallow strip footings. *Proceedings Institution of Civil Engineers Geotechnical Engineering, 125*(4), 230–241.

Soubra, A. H. (1999). Upper bound solutions for bearing capacity of foundations. *Journal of Geotechnical Division ASCE, 125*(1), 59–69.

Tajimi, H. (1966). Earthquake response of Foundation Structures (in Japanese). *Report, Faculty of Science and Engineering* (pp. 1.1–3.5). Nihon University Tokyo.

Terzaghi, K. (1943). *Theoretical soil mechanics*. USA: John Wiley Publication.

Terzaghi, K., & Peck, R. B. (1982). *Soil mechanics and foundation engineering*. USA: John Wiley Publication.

Timoshenko and Goodier. (1983). *Theory of elasticity*. NY: McGrawHill Book Company.

Tokimatsu, K., & Aska, Y. (1998). Effects of liquefaction-induced ground displacements on pile performance in the 1995 Hyogoken-Nambu earthquake. In *Soils and Foundations, Special Issue on Geotechnical Aspects of the January 17, 1995 Hyogoken-Nambu Earthquake* (No. 2, pp. 163–178) (September).

Tokimatsu, K., Ibaraki, Y., & Arai, H. (2009). Dynamic soil properties back-calculated from strong motions recorded at two downhole arrays during the 2007 Niigata-ken Chuetsu-oki earthquakes. In *Proceedings of the 6th International Conference on Urban Earthquake Engineering, Tokyo, Japan* (pp. 485–489), March 3–4, 2009.

Tomlinson, M. (2012). *Pile design and construction practice*. UK: Taylor and Francis Publication.

Verruijt, A. (2010). *An introduction to soil dynamics*. Berlin: Springer.

Wilson, S. D., & Hills, D. E. (1967). How to determine lateral load capacity of piles. In *Pile-Foundation Know How* (41–45). Washington: Amer. Wood Press. Inst.

Winkler, E. (1867). *Die Lehre von der Elastizitat und Festigate*. Prague, Czechaslovakia.

Weaver, W., & Gere, J. M. (1980). *Matrix analysis of framed structures* (2nd ed.), New York: Van Nostrand Reinhold Company Inc.

Yokoyama, K., Tamura, K., & Matsuo, O. (1997). Design methods of bridge foundations against soil liquefaction and liquefaction-induced ground flow. In *Proceedings of the Second Italy–Japan Workshop on Seismic Design and Retrofit of Bridges* (pp. 297–319).

Zienkiewicz, O. C. (1970). *Finite element method in engineering science*. London: McGraw-Hill Inc.

Chapter 7
Dynamic Soil–Structure Interaction in Earthquake Analysis

7.1 Introduction

In the last thirty years or so, the dynamic soil–structure interaction (DSSI) has undertaken giant leaps in terms of application. From being perceived initially as a utopian exercise in the realms of theoretical physics (understood by few), in the early 1960s it has stormed into the hard nosed arena of professional design and has possibly cut itself a permanent niche, rather than being perceived as just a passing fad.

However, despite its graduation from being considered a theoretical conjecture in hallowed domain of academics to an applied science and that which should justifiably fit into the daily chore of design office, the topic is yet to gain universal acceptance.

The reason is that it unfortunately suffers from a number of misplaced perceptions and myths that have created a lot of misinterpretations, especially among many professionals, in terms of its behavior, vis-à-vis its application.

In India, research on DSSI started with great impetus and fanfare in the late 1970s and some significant contribution was made both by academicians (Anandakrishnan and Krishnaswamy 1973; Dasgupta 1976; Dasgupta and Kameswara Rao 1976, 1978; Rao 1977) and industry (Ghosh et al. 1984; Ghosh and Batavayal 1985; Chowdhury and Som 1993; Chowdhury 1984), but unfortunately in between for some reasons that are circumstantial and otherwise, the enthusiasm to undertake research on this topic somehow got waned.

From 1990 to 2000, not much development took place in India till the infamous Bhuj (2001) and Indo-Nepal (2015) earthquake made a severe dent in the country's socioeconomic fabric, when interest and compulsion to work on this topic rekindled again.

On retrospection, reason for decline in its popularity as a research topic and application can perhaps be attributed to the following reasons:

I. Chowdhury and S. P. Dasgupta, *Earthquake Analysis and Design of Industrial Structures and Infra-structures*, GeoPlanet: Earth and Planetary Sciences,
https://doi.org/10.1007/978-3-319-90832-8_7

- India stopped building major nuclear power plants (where, as per IAEA safety norms, considering DSSI is mandatory) after the late 1980s (KAIGA and RAPP).
- India never suffered from a major earthquake creating havoc since 1970–2000, possibly bringing in some sense of complacency in the profession (IS 1893, code for "Earthquake Resistant Design of Structures" never got revised or updated since 1984 till 2002, i.e., only after the Bhuj disaster).
- Some shortsighted policy in our education system and misplaced perceptions in our professional framework, *deplorably existing till date.*

While the first two reasons cited above are circumstantial, for which one can do very little, the third point is a major issue, which needs some careful evaluation.

Any development in DSSI needs an equally competent background in soil and structural mechanics. Unfortunately, this combination is difficult to achieve (if not impossible) under the present educational framework we have in many countries, including India.

It is unfortunate that structural engineering and geotechnical engineering are mostly treated as two diverging sciences in civil engineering, where the two disciplines hardly interact or function in tandem.

This has also brought about some misconception among structural engineers that soil mechanics par se is an empirical science, which consists of limited mathematics, more laboratory and field observations that are finally fitted into charts and graphs for design. Rather than blaming only our structural colleagues for this misplaced perception, perhaps geotechnical engineers are to be equally blamed for aggravating this notion. For, not too long back, they did have this inclination to resort to charts and graphs at every drop of hat, rather than providing a distinct mathematical formula that can be fitted into a computer model.

Finally, trying to find solace behind the famous quotation of Dr. Karl Terzaghi and oft chanting the mantra—"*Soil Mechanics is more an art than a science*" has possibly not helped much to gain the subject a reputation that it is indeed a science (if not something more) than being an art only.

In profession, while structural engineers would hardly bother beyond knowing the bearing capacity of soil—to size the foundation or the pile capacity to decide on the number of piles to be provided below a column—geotechnical engineers, on the other hand, would never raise their nose above the ground level or venture beyond their *Hansenian world* of N_c, N_q, N_γ, etc., to understand what is happening above the ground level. This attitude of "*This is structural mechanics and, as such, none of our business*" and vice versa is surely not a very healthy atmosphere where research and development in DSSI can thrive.

Perhaps the educational course content at postgraduate level prevalent in the majority of engineering colleges in many countries is one of the main culprits for development of this uncalled for syndrome.

Looking at the curriculum, especially in geotechnical engineering at postgraduate level in most of the premier institutes, it appears that the curriculum is designed with an implicit assumption that people graduating would work only for a company whose main line of business is purely soil investigation, or activities related to soil only. A scenario that clearly reflects a myopic vision, in imparting engineering education, is surely not very conducive for growth and development of DSSI as a technology.

The author can recall his own experience in the early 1980s when *Finite Element Method* (FEM) was making in roads in the majority of engineering institutes in India and engulfing all branches of applied science.

It was amazing to see the resistance it received from traditional geotechnical academicians, who perceived it as almost an impolite intrusion of structural engineering into their time-honored domain of experimental and field observation, and was something definitely not to their liking.

However, like among any controversy, the truth finally prevails.

Happily, FEM did make at last a grudging entry into the domain of geotechnical engineering and has established itself since then as a powerful tool for analysis.

Today the situation is distinctly better (though much more need to be done yet), and people started realizing that this distinct rightist and leftist attitude (where each has a demarcation line not to be trespassed under any circumstance) will not work. One does have to form a united front, where one science would overlap the other for that is the only way forward to be in the front line of technology—a realization it is hoped would bring about changes for the better in the days to come.

7.2 A Short History on Development of Dynamic Soil–Structure Interaction (DSSI)

Though DSSI as an applied science has been in vogue for last thirty years or so, the roots of its development are more than a century old. It was applied by mathematicians—who first saw the possibility when Lamb (1904) first published the pioneering paper in journal of the *Philosophical Transactions of* the *Royal Society of London*. However, it was hardly of any immediate influence, for civil engineers were not conscious about its significance then and Lamb, being a hydrodynamist and not an engineer, never explained the possibility of how the theory could be applied.

Moreover, the solution—being analytically far too convoluted—eluded the mathematical perception of most of the civil engineers of his time. It was left to Pekeris (1955), Pekeris and Lifson (1957) to further ameliorate on Lamb's theory, but the solution still remained complex enough to be far away from the slide rule dominated domain of civil engineers.

Seismologists and geophysicists, however, took notice of this development and realized the fact as to why during an earthquake a major tremor (propagation of S and Rayleigh waves) is always preceded by minor tremors (P waves).

The first structured and scientific development of DSSI actually started in realms of machine foundations, when DEGEBO (The German Society of Soil mechanics), in 1935, commissioned Reissner (1936) to find solution to a footing vibrating under harmonic load.[1]

Erich Reissner was certainly a maverick among civil engineers. Born 1913, in Aachen, Germany, he received his doctorate in civil engineering from Technical University of Berlin in 1935 (he received a second doctorate in *mathematics* from the Massachusetts Institute of Technology in 1938). The topic mentioned above was his subject for doctoral dissertation.

Reissner, however, was not alone in his quest for an answer to this problem; Karl Margurre, between 1931 and 1933, presented some very interesting papers dealing with harmonically loaded soil and plate, including soil having layered media. He was, however, possibly overwhelmed by complexities of wave propagation and his developments petered out after some time. Thus, his work now remains only as a historical anecdote and not of much practical significance.

Reissner used Lamb's analysis to develop a solution for the problem. However, Reissner's solutions matched poorly with the field observation. The reason was that Reissner made some simplifying assumptions, like the plate to have frictionless contact with soil, the displacement at center of the load to be equal to displacement of plate, and the contact stress between soil and plate to be uniform. There was also a sign error in his solution that was later corrected and improved upon by Y. O. Shekhtar (Barkan 1962).

Irrespective of the simplifying assumptions he made in terms of distribution of contact stress, etc., his paper of 1936 and a subsequent paper (Reissner 1937) gave excellent insight into the problem sans a computer regarding radiation damping and to the equivalent mass–spring–damper analog system that can be adapted for this difficult mixed boundary-value problem.

In this context, Reissner can truly be treated as the forefather of modern DSSI.

Further research on this topic was, however, abandoned then for the time being, more possibly due to the reason that Germany was preparing for imminent war and the focus of scientific development got digressed somewhere else.

In between, in Germany, an engineer by the name of G. Ehlers published a paper (Ehlers 1942) which was possibly the first distinctive step toward development of DSSI, but got very little attention then, as the time and place for this publication was wrong. The paper was first of all far ahead of its time, and Germany, at that point of time, was completely immersed in running over Europe (Second World War), that resulted in the paper not getting due attention it really deserved.

It was not until the 1950s when Sung (1953) and Quinlan (1953) independently submitted solutions to Reissner's problem, when interest on the topic got reinstated again.

This was followed by a number of significant papers, like the seminal paper of Bycroft (1956), Warburton (1957), Hsieh (1962), Thomson and Kobori (1963), Awojobi and Grootenhius (1965), Gladwell (1968) that significantly enhanced our

[1]An excellent exposition of the historical development of soil–structure interaction right from its humble beginning has been put forward by Kausel (2010).

knowledge on dynamic behavior of rigid circular, rectangular, and strip footing under dynamic loads.

However, all these solutions were still perceived as far too complicated to be applied in daily and routine design office activity and remained only known to a few academicians and researchers.

After the Second World War, USA was going through a great industrial metamorphosis when demand for power was growing leaps and bounds. To meet this spiraling demand, power companies like General Electric, Westinghouse were churning out turbo machineries for power generation of progressively higher capacities. But many of them during operation misbehaved. The turbines tripped due to excessive vibration or developed cracks in the foundations as there was no scientific technology available then to study their behavior properly.

It was at this time a Russian engineer Gregory P. Teschbatarioff, who had migrated to USA (and was in the faculty of Princeton University then), translated, at his own initiative, the works of one of his countrymen by the name of Barkan (1962).

The work demonstrated in a systematic way how vibration analysis of such frames and foundations can be considered for rotating equipments, within the framework of analytical tools then available in a design office—the western world sat up and started to take notice and the *modern soil dynamics possibly saw its first light on birth.*

Barkan's seminal work (though presently outdated) remains one of the most influencing memoirs that has ever been written within the realms of civil engineering and had a far-reaching influence in a number of countries (including India) that were going through rapid industrialization.

After this, a number of contributions to this end had been made by researchers like Richart (1966), Richart et al. (1970), Whitman (1970, 1972), Lysmer (1965, 1966), Novak (1974a, b), Novak and Beredugo (1972a, b, c), Novak and Hifnawy (1983), Novak and Sharnouby (1983), Gazetas (1983), Gazzetas and Tassoulas (1987), Kausel (1974). But most of these solutions were restricted to loads that were harmonic in nature, where foundations were considered to be rigid having infinite stiffness.

In structural engineering side, the 1952 Tehachapi Earthquake in the USA clearly brought the structural engineering community out of their slumber. Devastation it caused clearly made the structural engineers realize that traditional tools (pseudo-static analysis) they used for designing buildings and other structures were distinctly inadequate and the development started in frantic pace to consider the effect of time-dependent load and its behavior on the structure, giving birth to structural dynamics pioneered by engineers like Clough and Penzien (1975), Hurty and Rubenstein (1967), Biggs (1964), Newmark (1959), Newmark and Rosenblueth (1971).

By 1970, the technology of dynamic analysis considering the structure as fixed at ground level has developed significantly and the development of computers by IBM in the USA brought about great enhancement in computational ability of an average engineer and catapulted civil engineering into a complete different era of cyber technology (especially in the USA) that saw a rapid development in matrix analysis and numerical methods in civil engineering—an era when finite difference method (FDM) and FEM started making rapid progress.

Soil dynamics and structural dynamics were still two different branches of technology that were yet to integrate.

The integration came into being in the late 1970s when USA started the development of nuclear power plant commercially.

However, to be historically correct, it was Martel (1940) who possibly offered the earliest commentaries on plausible interaction between structure and soil. Based on his observations on a number of structures' behavior under earthquake in the USA and Japan, he concluded that damage to buildings resting on soft soils, deep alluvia, or high elevations can be expected to be more intense than buildings resting on a level ground. Also, citing relatively lower seismic excitations observed in tunnels, he speculated on possible reduction in motion intensity with depth.

In a nuclear power plant, the reactor building is especially massive and sufficiently rigid. It did not take long for engineers to figure out that the conventional theory of structural dynamics was inadequate, at least in this case. For, normal structures are usually more flexible compared to soil, when, considering soil–structure boundary as fixed is valid- especially for earthquake analysis.

But the reactor building is sufficiently rigid, with stiffness comparable to soil (if not greater), and thus the concept of perceiving soil boundary as fixed and ignoring its deformation would possibly give a wrong picture.

It was at this juncture that Anestis (Andy) Veletsos and J. W. Meek (1974) wrote that legendry paper on dynamic behavior of building-foundation system that first gave insight to how superstructures behave under earthquake load when the foundation and soil get coupled, and even after 40 years since its publication, this paper along with another published by Jennings and Bielak (1973) still remains the backbone of significant research to this day.

That structures built over soft soil modify the response significantly became an established fact, and this was further confirmed by research carried out by Luco and Hadijian (1974), Hall and Kissenpfennig (1976), Wolf (1985, 1988) and was the advent of DSSI of the modern era and which is where it is presently.

The technology has obviously cascaded to all types of structures, including building engineering design (especially for high rises with basements), bridges, underground facilities, but in many countries, including India, it is yet to have a universal acceptance in the industry as a practice, unfortunately.

7.3 The Difference in Approach Between Soil and Structural Dynamics

The major stumbling block to application of DSSI has been perhaps the difference in approach in analysis of structures and soil under earthquake load. The structural dynamics, which involves analysis of second-order linear differential equations, whose solutions are relatively simple and most of the interpretations are more physical than mathematical, has been a very favorite analytical tool especially to professionals in industry.

On the contrary, the soil dynamics theories are based on propagation of waves through an elastic media and generate fear-evoking second- and fourth-order partial differential equations, whose solutions involve Bessel's functions of first and second kind, special functions like Hankel's function, Bateman–Pekeris integrals (just to name some notorious few), with which the majority of professional engineers are not comfortable at all. And, no wonder, people shun the topic and resort to easy way-out of considering the structural boundary as fixed.

Initially, one of the major challenges in DSSI was how to bridge the discrete (structural dynamics) and the continuum (soil dynamics) theory. It was only after Lysmer (1966) made the landmark contribution of developing equivalent springs based on elastic half-space theory that people started considering DSSI as a serious tool for application. Of course, development of FEM has also contributed significantly to bring this tool into the grip of an average engineer in the industry.

7.4 Why Are We Reluctant to Apply DSSI as an Analytical Tool?

Despite DSSI being an established technology now, people are still reluctant to apply it in their day-to-day design practices. Even those who are conversant with the theory are acutely aware of the complexity it could bring in the analysis and modeling the soil and structure based on FEM, and the fact that the analysis certainly becomes expensive and laborious.

It develops a lot of conflicting queries in one's mind, like

1. The structure I am designing—how much will DSSI affect its response?
2. Is it for the good or worse? Meaning: Will the structural response attenuate or amplify?
3. What risk I take if I ignore DSSI?
4. Considering the soil effect (especially FEM analysis) makes the analysis more laborious and time consuming, thus more costly, *is it worth the effort*?
5. My boss is a traditionalist and is under the time pressure; can I convince him it is worth the effort and can I really do some valuable addition by considering DSSI?

Very relevant questions, but not always very easy to answer, as will be examined hereafter.

7.4.1 How Do We Know Whether DSSI Effect Is Significant or not?

Veletsos and Meek (1974) suggested that when the wave parameter,

$$V_s/fh \leq 20 \tag{7.4.1}$$

the DSSI effect could be significant.

Here, V_s is the shear wave velocity of the soil, f is the fundamental frequency of the fixed-base structure, and h is height of the structure. The wave parameter may be looked upon as a measure of the relative stiffness of the foundation and the structure.

Thus for a building, say, $f = 1/T$ and $T = 0.1n$ (the rule of thumb we often apply to estimate fundamental time period of a building) where n is the number of stories, one can modify Eq. (7.4.1) to $V_s n/h \leq 200$.

Now considering $h/n = 3.3$ m for any average building, we arrive at an expression $V_s \leq 660\,\text{m/s}$. Thus, one may conclude that for $V_s \leq 660\,\text{m/s}$, the DSSI effect could be significant. Considering the fact that $V_s = 660$ m/s is the shear wave velocity in a rock, we can surely say that any building in India which is being built on the Gangetic Plane, or anywhere else, except the Himalayan belt, is susceptible to the DSSI effect.

Based on the explanation furnished above, Fig. 7.1 shows limiting stories versus shear wave velocity, above which the DSSI effect is significant.

One major lacuna with DSSI is that the analysis and its outcome are available only at a stage when the project is in quite an advanced stage of detail engineering. The architectural planning, including functional space arrangement, height of building, etc., all get fixed by the time detailed soil–structure analysis may be carried out.

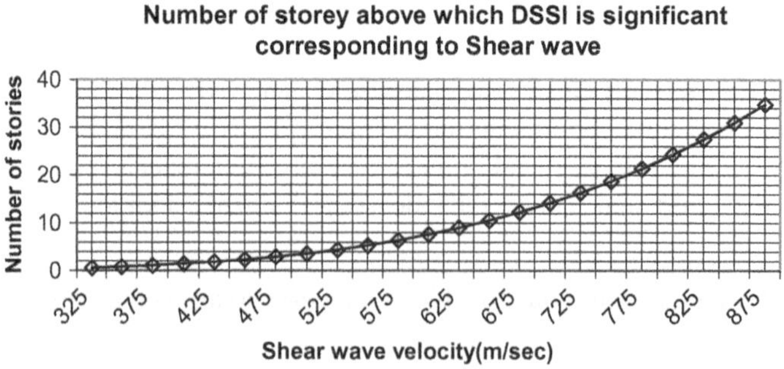

Fig. 7.1 Limiting story height versus shear wave velocity (V_s) below which the DSSI effect takes place

Now, if the analyst comes up with a suggestion at such an advanced stage that the planning needs complete reorientation, for the DSSI effect could have an adverse influence on the structure—the analyst is surely not to be considered very smart.

But, on the contrary, how can one say from the outset what could be the limiting boundaries that should not be trespassed so that amplification effect of DSSI, if any, is nullified at planning stage, i.e., before any detailed analysis has been carried out?

Difficult question, yet not impossible to answer, provided one does some intelligent assessment.

To answer this riddle, one has to understand how the soil overlying bedrock affects the earthquake response.

For those who are conversant with the theory of soil dynamics, would be aware that free field time period of a site increases with increasing depth of overlying soil over bedrock. This helps to determine the frequency of the waves amplified and filtered out of the soil and is also related to the amount of soil–structure interaction that will take place during an earthquake.

Dowrick (2003) reports that in Mexico earthquake in 1957 an extensive damage occurred to buildings that were tall and were found to be resting on alluvium soil of depth >1000 m.

In 1967, the Caracas earthquake showed identical results, where tall structures that were resting on deep alluvium soil overlying bedrock underwent extensive damage.

In 1970 earthquake at Gediz, in Turkey, a part of a factory was demolished in a town about 140 km from the epicenter, while no other buildings in the town underwent any damage!

Subsequent investigation revealed that fundamental period of the building matched the free field time period of the site.

The Caracas earthquake, as cited earlier, also showed a distinctive pattern where medium-rise buildings (5–9 stories) underwent extensive damage, where depth to bedrock was less than 100 m, while buildings over 14 stories were damaged where the depth to bedrock was greater than 150 m.

The bottom line of the above phenomenon is that *if the fundamental time period of structure matches the free field time period of the site—it is the surest way to courting a disaster.*

Two-dimensional equation of wave propagation through soil is given by the equation

$$\frac{\partial^2 u}{\partial x^2} + \frac{\partial^2 u}{\partial z^2} = \frac{1}{V_s^2} \frac{\partial^2 u}{\partial t^2} \tag{7.4.2}$$

Solution of the above partial differential equation gives free field time period of a site extending to infinity in horizontal direction as

$$T_n = \frac{4H}{(2n-1)V_s} \tag{7.4.3}$$

Here, H = depth of soil over bedrock, V_s = shear wave velocity of the soil, and n = modal number 1, 2, 3, ….

The fundamental time period of site (for n = 1) is thus given by $T = 4H/V_s$. From the preceding paragraph, it is apparent that if this matches the fixed-base time period of the structure, it is expected to undergo damage and thus should be avoided at the planning stage.

For instance, for moment resisting reinforced concrete frames, the fundamental time period of the building as per IS-1893 (2016) is given by

$$T = 0.075h^{3/4} \tag{7.4.4}$$

Here, h = height of the building above its foundation.

Equating Eq. (7.4.4) with the fundamental time period of the site we have, considering each floor height to be 3.3 m, approximately,

$$n = 61(H/V_s)^{4/3} \tag{7.4.5}$$

Here, n = number of stories of the building for which damage can occur due to resonance for a particular H/V_s ratio.

Figure 7.2 reflects the limiting stories that should be avoided for such soil–structure resonance.

Based on the above figure, for instance, for a site having shear wave velocity of 200 m/s, where depth of soil is 100 m, the overlying bedrock a 24-storied building is likely to undergo damage due to resonance. So either we should construct a building of height 20 stories or beyond 29 stories in such a site to avoid resonance.

If the depth of overlying soil is, say, 45 m, the limiting story that should be avoided is about 8 stories. The concept is a simple but very effective tool to avoid adverse effect of DSSI at planning stage.

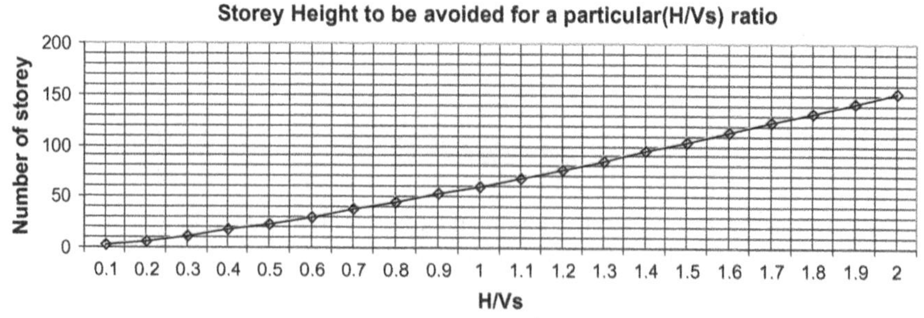

Fig. 7.2 Resonance story height to be avoided at planning stage

Thus, based on the above concept even during geotechnical investigation stage, it is possible to predict the story/height danger to be avoided and it is possibly the right time, for the design is still in drafting and planning stage when changes are quite possible.

The concept as above, though presented in terms of buildings, can well be adopted for any structure whose fixed-base fundamental period can be identified.

The authors are, however, yet to see a geotechnical report in India and abroad where geotechnical engineers have made such recommendations to an architect.

7.5 Coupling of Soil and Structure

Having established some rational basis for avoiding resonance at the planning stage, as cited above, one would still like to know how the foundation compliance affects the overall response of the structure. This is usually termed as *inertial interaction*.

For this, the most convenient method for modeling the soil and connecting it to the building superstructure would be either by providing boundary springs and dashpots or modeling the soil itself as finite elements. However, in most of the cases the use of equivalent soil springs is the technology in vogue for buildings modeled as three-dimensional (3D) space structures.

Except for nuclear power plants, and steam turbine foundation for power plants having capacity of 500 MW or more, resting on complicated soil, solutions available are scanty, where both the superstructure and the soil are modeled by FEM in three dimensions, for the cost of computation becomes far too expensive and laborious and is prohibitive for any normal commercial/industrial structure.

The use of equivalent soil spring and dampers to model the foundation compliance effect for DSSI analysis is increasingly becoming popular for analysis of structures in 3D, because the modeling is simple, convenient, and can be very easily fitted into a commercially available structural analysis packages that are available in the market.

The coupling of soil and structure is very conveniently done by adding the spring stiffness to diagonal element (k_{ii}) of structural element based on displacement compatibility principle.

The system as shown in Fig. 7.3 has 6 degrees of freedom. The global stiffness matrix of the system is expressed as

$$[\mathbf{K}] = \begin{bmatrix} k_{11} & k_{12} & k_{13} & k_{14} & 0 & 0 \\ k_{21} & k_{22} & k_{23} & k_{24} & 0 & 0 \\ k_{31} & k_{32} & k_{33}+k_{11} & k_{34}+k_{12} & k_{13} & k_{14} \\ k_{41} & k_{42} & k_{43}+k_{21} & k_{44}+k_{22} & k_{23} & k_{24} \\ 0 & 0 & k_{31} & k_{32} & k_{33}+\mathbf{K_x} & k_{34} \\ 0 & 0 & k_{41} & k_{42} & k_{43} & k_{44}+\mathbf{K_\theta} \end{bmatrix} \tag{7.5.1}$$

In Eq. (7.5.1), k_{11}, k_{22}, etc., are the element stiffness of the individual members in translational and rotational modes assembled to form the global stiffness matrix of the system. This is a standard equation and has been elaborated in a number of textbooks of FEM and matrix analysis of structures. It will be observed that boundary conditions at degree 5 and 6 are enforced by adding the foundation/soil springs K_x and K_θ directly to the diagonal element of the 5th and 6th degrees of freedom. For a case when the foundation base is considered fixed, either the 5th and 6th rows and columns are eliminated in Eq. (7.5.1), or K_x and K_θ are replaced by a very high value (say of the order of 10^{20}) to enforce the fixed-base boundary condition. Performing the eigensolution based on the expression $[\mathbf{K}] - [\mathbf{M}]\omega^2 = 0$ will show an elongation in time period as the springs go softer. There is no ambiguity as such. Here, $[\mathbf{K}]$ = global stiffness matrix of the system, $[\mathbf{M}]$ = mass matrix of the system, and ω = natural frequency of the system.

Structural engineers have been using this technique for the last thirty years or more to interpret the behavior of structure under DSSI and also standing by the argument that fixed-base analysis is conservative as the time period elongates.

Based on the explanation cited above, Eq. (7.5.1) certainly looks mathematically sound and irrefutable. However, putting on the hat of a theoretical physicist and probing the equation a bit more deeply, one comes up with some very fascinating outcome.

For Eq. (7.5.1) to be true and sacrosanct, it should be valid for all cases, irrespective of the material of choice, time, and place.

Let us assume that 100 years down the road scientists have developed a material whose Young's modulus value is $E = 2 \times 10^{30}\ \mathrm{kN/m^2}$ and build the superstructure with this material. The structure is put on a foundation having same/earlier values of K_x and K_θ, as shown in Fig. 7.2. In other words, the structural stiffness is now significantly high ($K_{\mathrm{struct}} \to \infty$) compared to the foundation stiffness. Now one can well pose the queries as to how does the structure behave under a base acceleration $\ddot{u}_g$? What is the characteristic of its vibration?

Without performing any mathematics, by mere inspection of the system one could safely conclude that the body will certainly vibrate; however, considering the superstructure to be very stiff, in this case it will undergo a rigid body mode, as there will be no/negligible deformation within the body itself and the first two fundamental modes will be governed by the foundation springs ($\sqrt{K_x/M}$ and $\sqrt{K_\theta/J}$) that will govern the response. The response is as shown in the right part of Fig. 7.3.

Now the question is, does Eq. (7.5.1), which is most often adopted for such analyses, furnish us with what has been explained above? *Surprisingly not!*

For considering now the values of element stiffness, k_{11}, k_{22}, etc., being of very high order of 1×10^{20} or 1×10^{25}, will simply gobble up K_x and K_θ whose order will be of mere 10^6 or 10^7 only. Thus, on applying the eigensolution one would arrive at a time period $T \to 0$. Thus, it is seen that formulating the DSSI phenomenon based on the principle as adopted in Eq. (7.5.1) could be mathematically convenient, but certainly not realistic, as it fails to invoke the fundamental soil

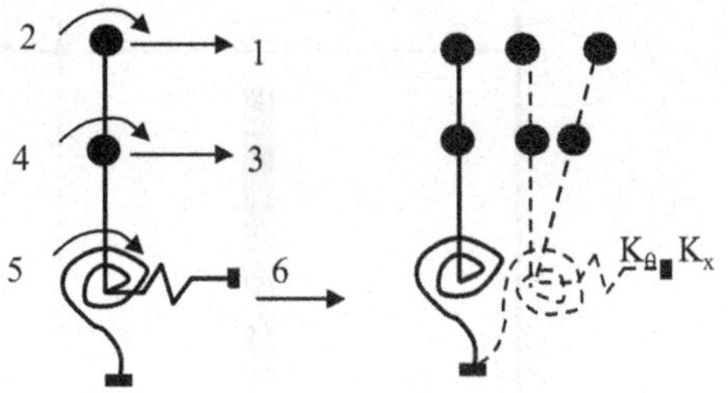

Fig. 7.3 A structure connected to foundation springs and deformation as a rigid body

modes in case the structure is far too stiff. Thus, practicing engineers who believe that by merely adding soil springs at the base of a structure's foundation and getting some time period output of an eigenfrequency run in a computer, is what it all takes to do a DSSI analysis based on a commercially available software, they can well retrospect the outcome of their results they have obtained till date.

The major mathematical flaw that exists in Eq. (7.5.1) is adding soil springs directly to the diagonal element of the matrix, which implicitly makes it a parallel spring connection, meaning thereby that for externally applied load, the nodal compatibility in terms of displacement is satisfied. This is certainly true for a structure or its internal nodes, but not valid for a soil–structure boundary where each would deform depending on its own inherent stiffness.

Like patch test in FEM is a test to check the sanctity of a stiffness formulation, the logic proposed above may be called an RB test (short of Rigid Body), to check the soil–structure coupling logic, and we see that formulation shown in Eq. (7.5.1) fails this test with parallel spring connection, especially when the structure has got significant stiffness compared to soil.

Thus, sitting before a computer if one blindly starts plugging soil spring below a foundation raft supporting a structure in a commercially available software and starts believing that it is DSSI, that he/she is carrying out—*think twice as to what you are doing?*

7.5.1 The Correct Modeling of Foundation and Structure

So what is the correct way of modeling the foundation coupled to a super structure?

To evaluate this, let us consider Fig. 7.4 as shown hereafter.

Figure 7.4 shows a body having single degree of freedom having mass m and stiffness $K_{\text{fixed}} = K_s$; the fixed-base stiffness of the system is connected to a foundation having translational and rotational stiffness expressed as $\bar{k}_x$ and $\bar{k}_\theta$, respectively. Let the body be given acceleration $= \ddot{u}_g$ at ground level.

Here, subscript g stands for ground.

It is apparent from Fig. 7.4 that under the ground acceleration $\ddot{u}_g$, the system moves to the right and undergoes deformation, when total deformation u_e of the system can be expressed as

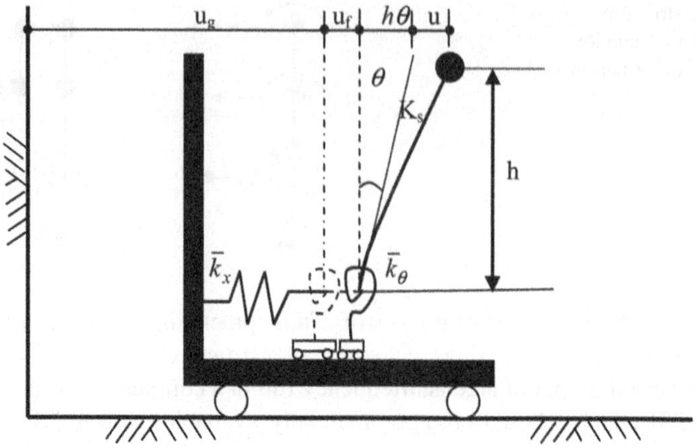

Fig. 7.4 A single-degree oscillator supported on foundation springs under base acceleration

$$u_e = u + u_f + h\theta \tag{7.5.2}$$

Here, u = deformation of the structure;
u_f = deformation of foundation in translation;
θ = rotation of foundation giving a displacement $h\theta$ at top;
u_e = effective displacement of the system.
Now, considering $P = m \times \ddot{u}_g = Ku$, Eq. (7.5.2) can be expressed as

$$P/K_e = P/K_s + P/\bar{k}_x + Ph^2/\bar{k}_\theta \tag{7.5.3}$$

Now taking out the common term $\ddot{u}_g$ in Eq. (7.5.3), we have

$$m/K_e = m/K_s + m/\bar{k}_x + mh^2/\bar{k}_\theta \tag{7.5.4}$$

Taking the first term common in the right-hand side of Eq. (7.5.4), we have

$$m/K_e = m/K_s\left[1 + K_s/\bar{k}_x + K_s h^2/\bar{k}_\theta\right] \tag{7.5.5}$$

Taking square root on both sides and multiplying by the term 2π, we have

$$2\pi\sqrt{m/K_e} = \left[2\pi\sqrt{m/K_s}\right] \times \sqrt{\left[1 + K_s/\bar{k}_x + K_s h^2/\bar{k}_\theta\right]} \tag{7.5.6}$$

Equation (7.5.6) can be conveniently expressed as

$$T_e = T\sqrt{1 + \frac{K_s}{\bar{k}_x}\left(1 + \frac{\bar{k}_x h^2}{\bar{k}_\theta}\right)} \tag{7.5.7}$$

Here, T_e = modified time period of the structure due to soil stiffness, T = time period of fixed-base structure, K_s = stiffness of fixed-base structure, $\bar{k}_x, \bar{k}_\theta$ = horizontal and rotational spring constant of the foundation-soil system, h = effective height or inertial centroid of the system.

Equation (7.5.7) is actually the celebrated expression of Veletsos and Meek's (1974) and is extensively used and referred in almost all references pertaining to DSSI earthquake analyses and also recommended by Federal Emergency Management Agency (FEMA) of USA and ASCE 7-02.

Studying Eq. (7.5.7), we conclude two things:

1. The fixed-base time period of a building gets elongated by foundation compliance.
2. The effect is dependent on relative stiffness of the foundation and superstructure and its height.

We also conclude from the above that if we take out the common mass term m from Eq. (7.5.4), we have

$$\frac{1}{K_e} = \frac{1}{K_s} + \frac{1}{\bar{k}_x} + \frac{h^2}{\bar{k}_\theta} \tag{7.5.8}$$

where K_e = equivalent stiffness of the soil–structure system having single degree of freedom.

Equation (7.5.8) clearly shows that the stiffness combination is *in series and not in parallel*, as is usually perceived. Thus plugging in soil springs in the way as shown in Eq. (7.5.1) in any commercially available structural software could be convenient, but is certainly not the correct way to model the DSSI phenomenon.

Based on the above, we study the effect of DSSI has on the frame based on a suitable numerical problem as worked out below.

Example 7.1 A concrete frame having elevation dimensions is as shown above. The structure is being constructed at a place considered zone III as per IS code. Columns are of dimension 300 × 450 and beams of size 250 × 400. Foundation F1 has size of 2 × 2 m and F2 has size of 2.75 × 2.75 m. Concrete grade is M25. The bearing capacity of soil is 150 kN/m^2, shear wave velocity is 250 m/s, Poisson's ratio of soil = 0.32, and density of soil = 19 kN/m^3. Weight due to DL and live load contribution at each floor are as shown in the figure. Determine the base shear and story force, considering 5% damping of the structure. Compute for

- the frame as fixed-base;
- considering the foundation compliance effect (Fig. 7.5).

(1) **Fixed-base analysis**

We do a simple response spectrum analysis based on IS-1893 for fundamental period only.

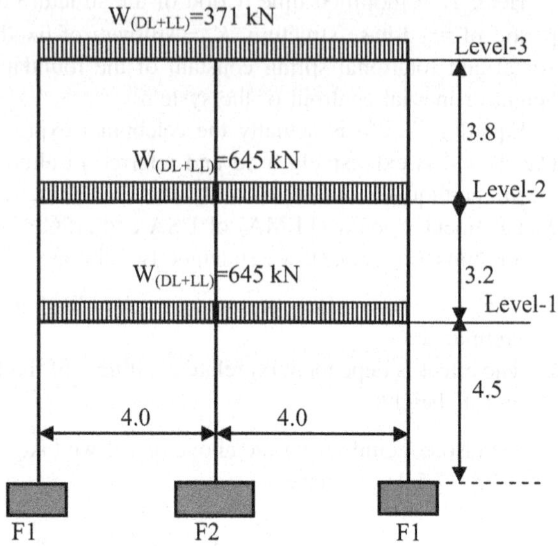

Fig. 7.5 A three-storied frame supported on foundations assuming a fixed-base

As per IS code, fundamental time period of a concrete building is expressed as $T = 0.075h^{0.75}$, where h = height of the building and is 11.5 m in this case.

Thus, $T = 0.468$ s as the fundamental time period, which gives $S_a/g = 2.5$ from the response spectrum curve.

For zone 3, as per IS code, $Z = 0.16$, $I = 1.0$, $R = 3.0$.

This gives $A_h = (ZI/2R) \times (S_a/g) = 0.066$.

Thus, the base shear is given by $V_{textb} = 0.066 \times (371 + 2 \times 645) = 109.63$ kN.

Thus, the floor lateral force at Level 1 is expressed as

$$P_1 = V_{textb} \times \left[\frac{W_1 h_1^2}{\sum_1^3 W_i h_i^2}\right] = 109.63 \times \left[\frac{645 \times (4.5)^2}{371 \times (11.5)^2 + 645 \times (7.7)^2 \times 645 \times (4.5)^2}\right]$$

$$= 109.63 \times \left(\frac{13{,}061}{100{,}368}\right) = 14.26 \text{ kN}$$

At Level 2

$$P_2 = V_{textb} \times \left[\frac{W_2 h_2^2}{\sum_1^3 W_i h_i^2}\right] = 109.63 \times \left[\frac{645 \times (7.7)^2}{371 \times (11.5)^2 + 645 \times (7.7)^2 \times 645 \times (4.5)^2}\right]$$

$$= 109.63 \times \left(\frac{38{,}242}{100{,}368}\right) = 41.77 \text{ kN}$$

At Level 3

$$P_3 = V_b \times \left[\frac{W_3 h_3^2}{\sum_1^3 W_i h_i^2}\right] = 109.63 \times \left[\frac{371 \times (11.5)^2}{371 \times (11.5)^2 + 645 \times (7.7)^2 \times 645 \times (4.5)^2}\right]$$

$$= 109.63 \times \left(\frac{49{,}065}{100{,}368}\right) = 53.6 \text{ kN}$$

(2) **Analysis considering soil–structure interaction**

In this case, we first identify the foundation spring stiffness.

Dynamic shear modulus of soil is expressed as $G = (\gamma_s/g) \times V_s^2$.

Here, γ_s = 19 kN/m³, V_s = 250 m/s and g = 9.81 m/s². This gives G = 121,050 kN/m².

For foundation $F1$

Area = 2.0 × 2.0 = 4.0 m² and $I = (1/12) \times 2.0 \times (2.0)^3 = 1.33$ m⁴.

This gives $r_x = \sqrt{\frac{4}{\pi}} = 1.27$ m and $r_\theta = \sqrt{4}\frac{4\times 1.33}{\pi} = 1.14$ m.

Thus $k_x = \frac{8Gr_x}{2-v} = \frac{8\times 121{,}050\times 1.27}{1.68} = 732{,}064$ kN/m and

$$k_\theta = \frac{8Gr_\theta^3}{3(1-v)} = \frac{8 \times 121{,}050 \times (1.14)^3}{2.04} = 703{,}298 \text{ kN/m}$$

For foundation F2

Area = 2.75 × 2.75 = 7.56 m² and $I = (1/12) \times 2.75 \times (2.75)^3 = 4.76$ m⁴.

This gives $r_x = \sqrt{\frac{7.56}{\pi}} = 1.55$ m and $r_\theta = \sqrt{4}\frac{4\times 4.76}{\pi} = 1.185$ m.

Thus $k_x = \frac{8Gr_x}{2-v} = \frac{8\times 121{,}050\times 1.55}{1.68} = 893{,}464$ kN/m and

$$k_\theta = \frac{8Gr_\theta^3}{3(1-v)} = \frac{8 \times 121{,}050 \times (1.185)^3}{2.04} = 789{,}914 \text{ kN/m}$$

Thus, the total stiffness of the foundation system is $2 \times F_1 + F_2$, and this gives

$$\bar{k}_x = 2 \times 732{,}064 + 893{,}464 = 2{,}357{,}592 \text{ kN/m}$$
$$\bar{k}_\theta = 2 \times 703{,}298 + 789{,}914 = 2{,}196{,}510 \text{ kN/m}$$

The total structural stiffness of the frame, as per Veletsos and Meek (1974), can be expressed

$$K_s = \frac{4\pi^2 W}{gT^2}$$

Here, W = total weight on frame = $2 \times 645 + 371 = 1661$ kN, T = fundamental time period.

Thus, $K_s = \frac{4\pi^2 \times 1661}{9.81 \times 0.468^2} = 30,519$ kN/m.

Here, $\bar{h} = \frac{\sum_{i=1}^{3} W_i h_i}{\sum_{i=1}^{3} W_i} = \frac{371 \times 11.5 + 645 \times 7.7 + 645 \times 4.5}{1661} = 7.306$ m from base.

Now considering

$$T_e = T\sqrt{1 + \frac{K_s}{k_x}\left(1 + \frac{\bar{k}_x h^2}{k_\theta}\right)},$$ we have

$$T_e = 0.468\sqrt{1 + \frac{30,519}{2,357,592}\left(1 + \frac{2,357,592 \times 7.3^2}{2,196,510}\right)} = 0.62\ \text{s},$$ which gives $\rightarrow S_a/g = 2.2$.

Thus, the base shear can now be calculated by proportion as $V_{textb} = (109.63/2.5) \times 2.2 = 97.47$ kN.

Similarly, the floor lateral force can be estimated by proportion as

$$P_1 = 12.6\ \text{kN}, \quad P_2 = 36.97\ \text{kN} \quad \text{and} \quad P_3 = 47.44\ \text{kN}$$

We tabulate the floor shear values below for ready comparison

Level	1	2	3
Floor shear (fixed-base)	109.63	95.37	53.6
Floor shear (DSSI)	97.6	84.7	47.6

The calculation clearly shows that there is attenuation in response. Soil–structure interaction has reduced the response by about 12% compared to the fixed-base.

7.6 Unraveling the Popular Myth

Based on Example 7.1, we see that due to addition of foundation springs, there is an elongation in the time period the response got reduced. This has given rise to the popular belief that the soil–structure interaction is beneficial and if we ignore the soil in the analysis and consider the frame as fixed at its base, the analysis is on *conservative* side.

This logic is certainly convenient, and many traditional analysts would ignore DSSI believing that what is computed above is true.

As per the authors, *this is the biggest misinterpretation engineers suffer from regarding DSSI* and brings us to our first model of interpreting Veletsos's equation in a different perspective (Chowdhury and Singh 2010).

If we look at Eq. (7.5.4), we will see that it can also be conveniently expressed as

$$\frac{1}{\omega_e^2} = \frac{1}{\omega_s^2} + \frac{1}{\omega_x^2} + \frac{1}{\omega_\theta^2} \tag{7.6.1}$$

Equation (7.6.1) has also been derived by Kramer (1998) and Wolf and Zhang (1999), wherein both confirm that coupling of soil and structure takes place like in Eq. (7.6.1) during an earthquake.

In fact, careful examination of Eq. (7.6.1) reveals that it is nothing else but the classical Dunkerley's equation derived by Stanley Dunkerley in 1894.

In a seminal paper published in the *Philosophical Transactions of the Royal Society of London*, Stanley Dunkerley derived the expression for determining the natural frequency of a vibrating shaft mounted with a number of disks. Dunkerley showed by considering one disk at a time and estimating there individual frequencies as $\omega_1, \omega_2, \ldots, \omega_n$, the combined frequency of the system can be expressed as

$$\frac{1}{\omega_e^2} = \frac{1}{\omega_1^2} + \frac{1}{\omega_2^2} + \frac{1}{\omega_3^2} + \cdots + \frac{1}{\omega_n^2} \tag{7.6.2}$$

Here, n = number of disks mounted on the shaft. It may also be mentioned that the experiment carried out by him and others showed only a variation of 2–5% from the expression he derived as per Eq. (7.6.2), and these variations could mostly be attributed to small experimental errors that creep in while carrying such elaborate experiments.

In the same spirit as Dunkerley, let us consider Fig. 7.6.

It is apparent from Fig. 7.6 that the single-degree oscillator can be broken down into three systems and, on finding out individual frequencies of each of the system, the coupled natural frequency may be determined as per Eq. (7.6.1) by adding the individual frequencies.

Multiplying Eq. (7.5.4) by the factor g, from which we derived Eq. (7.6.1), we have

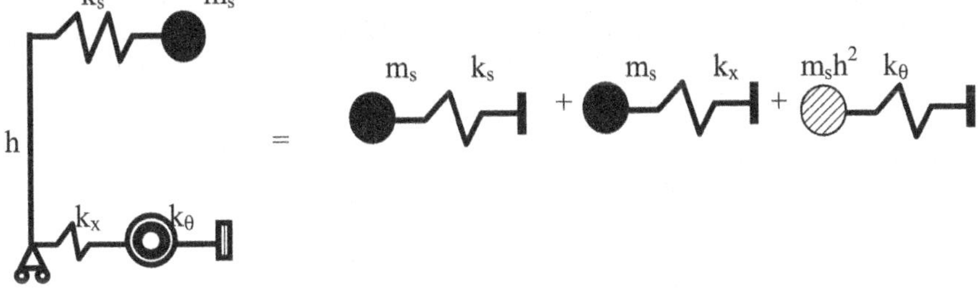

Fig. 7.6 Single-degree of freedom body broken up into three equivalent systems

$$\frac{m_s g}{k_e} = \frac{m_s g}{k_s} + \frac{m_s g}{k_x} + \frac{m_s h^2 g}{k_\theta} \tag{7.6.3}$$

from which we can go back to our fundamental Eq. (7.5.2) expressed as

$$\delta_e = \delta_s + \delta_x + \delta_\theta \tag{7.6.4}$$

Equation (7.6.4) shows that displacement of these three separate systems can be superimposed to arrive at the total displacement of the system, provided of course the system is linear.

Veletsos considered massless foundation, the inertial effect of foundation can be now considered based on Dunkerley's formulation as elaborated in Fig. 7.7.

Following Dunkerley's principle and Eq. (7.6.3), we have

$$\frac{1}{\omega_e^2} = \frac{m_s}{k_s} + \frac{m_0 + m_s}{k_x} + \frac{J_0 + m_s h^2}{k_\theta} \tag{7.6.5}$$

$$\Rightarrow \frac{m_s + m_0}{k_e} = \frac{m_s}{k_s} + \frac{m_0 + m_s}{k_s} + \frac{J_0 + m_s h^2}{k_\theta} \tag{7.6.6}$$

$$\Rightarrow \frac{1}{k_e} = \frac{\mu_x}{k_s} + \frac{1}{k_x} + \frac{\mu_\theta h^2}{k_\theta} \tag{7.6.7}$$

Here, $\mu_x = \frac{m_s}{m_0 + m_s}$, and $\mu_\theta = \frac{\left(\frac{J_0}{h^2} + m_s\right)}{m_0 + m_s}$ are dimensionless parameters.

It is thus observed in Eq. (7.6.7) that the flexibility of the structure and rotational flexibility of foundation are affected by the terms μ_x and μ_θ considering inertial effect of the foundation.

The above concept can be extended to modal analysis for earthquakes, as given hereafter.

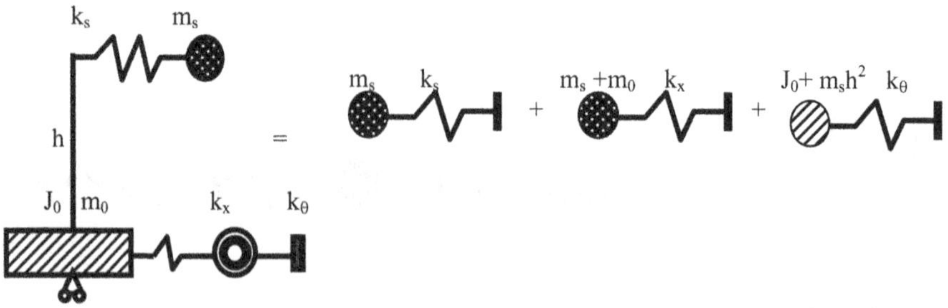

Fig. 7.7 A structure with single degree of freedom with a foundation mass

Let us start with Eq. (7.6.4). In terms of modal analysis, this can be expressed as

$$S_{de} = \frac{S_{as}}{\omega_s^2} + \frac{S_{ax}}{\omega_x^2} + \frac{S_{a\theta}}{\omega_\theta^2} \tag{7.6.8}$$

Thus, in terms of code, the above can be expressed as

$$u_e = \kappa_i \left(\frac{ZI}{2R}\right)\left(\frac{S_{as}}{\omega_s^2} + \frac{S_{ax}}{\omega_x^2} + \frac{S_{a\theta}}{\omega_\theta^2}\right) \tag{7.6.9}$$

$$\Rightarrow u_e = \kappa_i \left(\frac{ZI}{2R}\right)\left(\frac{m_s S_{as}}{k_s} + \frac{(m_0 + m_s)S_{ax}}{k_x} + \frac{\left(\frac{J_0}{h^2} + m_s\right)h^2 S_{a\theta}}{k_\theta}\right) \tag{7.6.10}$$

Equation (7.6.10) can be further expressed as

$$\begin{aligned} \Rightarrow u_e = \kappa_i \left(\frac{ZI}{2R}\right)\frac{m_s S_{as}}{k_s}\Bigg(1 + \left(\frac{m_0 + m_s}{m_s}\right)\left(\frac{k_s}{k_x}\right)\left(\frac{S_{ax}}{S_{as}}\right) \\ + \left(1 + \frac{J_0}{m_s h^2}\right)\left(\frac{S_{a\theta}}{S_{as}}\right)\left(\frac{k_s}{k_\theta}\right)h^2\Bigg) \end{aligned} \tag{7.6.11}$$

In Eq. (7.6.11), it will be observed that expression within the parenthesis is a dimensionless term, and when k_x and $k_\theta \to \infty$, the amplitude of displacement converge to the fixed response of the structure in fundamental mode.

Thus, from Eq. (7.6.11) we can clearly say that due to soil–foundation effect the fixed-base spectral acceleration (S_{as}) gets modified by the factor within the parenthesis which is nothing but the amplification factor (AF) due to DSSI, or

$$\Rightarrow \text{A.F.} = \left(1 + \left(\frac{m_0 + m_s}{m_s}\right)\left(\frac{k_s}{k_x}\right)\left(\frac{S_{ax}}{S_{as}}\right) + \left(1 + \frac{J_0}{m_s h^2}\right)\left(\frac{S_{a\theta}}{S_{as}}\right)\left(\frac{k_s}{k_\theta}\right)h^2\right) \tag{7.6.12}$$

In case the inertial effect of foundation is ignored, the amplification factor can be simply expressed as

$$\text{A.F.} = \left(1 + \left(\frac{k_s}{k_x}\right)\left(\frac{S_{ax}}{S_{as}}\right) + \left(\frac{S_{a\theta}}{S_{as}}\right)\left(\frac{k_s}{k_\theta}\right)h^2\right) \tag{7.6.13}$$

Equations (7.6.12) and (7.6.13) clearly show that fixed-base spectral acceleration gets amplified when DSSI is considered unless the structure is resting on rock (i.e., when k_x and $k_\theta \to \infty$). Thus, softer the soil and stiffer the structure, effect of foundation compliance on this soil amplification would be more profound.

In the above expressions, the acronyms have the following definitions:

h height of lumped mass m_s from c.g. of the foundation;
I importance factor of the structure;
J_0 mass moment of inertia of foundation;
k_s structural stiffness of the structure;
k_x translational stiffness of foundation;
k_θ rocking stiffness of foundation;
m_0 mass of foundation;
m_s mass of structure;
R response reduction factor;
S_{as} acceleration of the fixed-base structure corresponding to material damping of ζ_s;
S_{ax} acceleration of foundation in translation mode, corresponding to radiation damping ζ_x;
$S_{a\theta}$ acceleration of foundation in rocking mode, corresponding to radiation damping ζ_θ;
Z zone factor as per IS code.

It is, thus, observed that all plausible parameters that affect the DSSI phenomenon are incorporated in the amplification factor. Thus, the effective time period and effective damping do not really have any meaning and possibly give a murky picture.

There is one idealization in the above model though, namely, all responses (for both structure and foundation) peak simultaneously. It is apparent that when waves travel through the structure and foundation, based on their inherent properties, there will be some time lag and phase difference in the peak response for each of the mode; however, as modal analysis usually ignores this part and considers the maximum envelope to arrive at a conservative result, Eq. (7.6.12) may be considered as a conservative estimate of DSSI amplification factor for which the structure needs to be designed.

So, we have possibly broken the myth to perhaps great consternation of traditionalists in structural engineering and it would be worth evaluating how much it affects the problem we have solved earlier in Sect. 7.5.

Example 7.2 Determine the base and floor shear for the frame as solved in Example 7.1. Consider 10% damping of soil in translation and 15% in rotation. Consider footing thickness for F_1 and F_2 as 450 mm. Determine the response considering

- Ignoring inertial effect of foundation;
- Considering inertial effect of foundation.

Ignoring the effect of foundation inertia

The total weight from superstructure $= 2 \times 645 + 371 = 1661$ kN.

Thus, dividing the weight by $g = 9.81$ m/s^2, mass of superstructure = 169.3 kN s^2/m.

The total stiffness of foundation calculated previously is

$$\bar{k}_x = 2,357,592\,\text{kN/m}$$
$$\bar{k}_\theta = 2,196,510\,\text{kN/m}$$

The structural stiffness calculated previously K_s = 30,519 kN/m.

Height $\bar{h} = 7.306$ m.

Thus, $T_x = 2\pi\sqrt{\frac{m}{k_x}} = 2 \times \pi \times \sqrt{\frac{169.3}{2,357,592}} = 0.053$ s, which gives $S_a/g = (1 + 15 \times 0.053) \times 0.8 = 1.436$.

Multiplying by code factors, we have $A_h = (ZI/2R) \times (S_a/g) = \frac{0.16 \times 1}{2 \times 3} \times 1.436 = 0.038$. It may be noted that S_a/g value is multiplied by a factor of 0.8 to cater to 10% damping in translation as per code. Thus, S_{ax} = 0.038g.

$$T_\theta = 2\pi\sqrt{\frac{mh^2}{k_\theta}} = 2 \times \pi \times \sqrt{\frac{169.3 \times (7.3)^2}{2,196,510}} = 0.3953\text{ s} \rightarrow S_a/g = 2.5 \times 0.7$$
$$= 1.75$$

which gives A_h = 0.047.

And this shows that $S_{a\theta}$ = 0.047g. From the previous calculation S_{as} = 0.066, g = 0.65 m/s^2.

Considering A.F. $= \left(1 + \left(\frac{k_s}{k_x}\right)\left(\frac{S_{ax}}{S_{as}}\right) + \left(\frac{S_{a\theta}}{S_{as}}\right)\left(\frac{k_s}{k_\theta}\right)h^2\right)$.

$$\text{AF} = \left(1 + \left(\frac{30,519}{2,357,592}\right)\left(\frac{0.038}{0.066}\right) + \left(\frac{0.047}{0.066}\right)\left(\frac{30,519}{2,196,510}\right) \times (7.3)^2\right)$$
$$= 1 + 7.453 \times 10^{-3} + 0.53 = 1.534$$

Thus, the base shear value gets amplified by the above number and V_b(DSSI) = 109.63 × 1.534 = 168.25 kN.

The lateral force at each floor and shear per floor are tabulated below and compared with the fixed-base case

Level	1	2	3
Floor force (fixed-base)	14.26	41.77	53.60
Floor force (DSSI)	21.87	64.07	82.22
Shear/floor (fixed-base)	109.63	95.37	53.60
Shear/floor (DSSI)	168.16	146.29	82.22

It is apparent that the forces have amplified quite significantly. This is however certainly a conservative estimate as it will be shown later that, although the amplification is high for higher modes, the modal mass participation significantly reduces higher mode amplification and the increase in force will be not so profound finally. Nonetheless, as a conceptual model, it clearly depicts the mechanics uderlying the phenomenon.

Considering the effect of foundation inertia

Weight of footing $F1 = 2.0 \times 2.0 \times 0.45 \times 25 = 45$ kN.
Weight of footing $F2 = 2.75 \times 2.75 \times 0.45 \times 25 = 85$ kN.
Thus, the total weight o foundation = $2 \times 45 + 85 = 175$ kN.
Thus, the mass of foundation = 17.84 kN s^2/m.

Mass moment of inertia of footing $F1 = \frac{4.587}{12}\left[(2)^2 + (0.45)^2\right] = 1.606 \text{ kN m}^2/\text{s}^2$.

Mass moment of inertia of footing $F2 = \frac{8.66}{12}\left[(2.75)^2 + (0.45)^2\right] = 5.603 \text{ kN m}^2/\text{s}^2$.

Thus, $J_\theta = 2 \times 1.606 + 5.603 = 8.815 \text{ kN m}^2/\text{s}^2$.

Thus, $T_x = 2\pi\sqrt{\frac{m + m_f}{k_x}} = 2 \times \pi \times \sqrt{\frac{169.3 + 17.84}{2{,}357{,}592}} = 0.056\text{s} \rightarrow S_a/g = (1 + 15 \times 0.056) \times 0.8 = 1.47$.

This gives $S_{ax} = 0.038g$.

$$T_\theta = 2\pi\sqrt{\frac{mh^2 + J_\theta}{k_\theta}} = 2 \times \pi \times \sqrt{\frac{169.3 \times (7.3)^2 + 8.815}{2,196,510}}$$
$$= 0.4028\text{ s} \rightarrow S_a/g = 2.5 \times 0.7 = 1.75$$

This gives $S_{a\theta} = 0.374$ m/s^2. From the previous calculation $S_{as} = 0.066g$. Now considering

$$\text{A.F.} = \left(1 + \left(\frac{m_0 + m_s}{m_s}\right)\left(\frac{k_s}{k_x}\right)\left(\frac{S_{ax}}{S_{as}}\right) + \left(1 + \frac{J_0}{m_s h^2}\right)\left(\frac{S_{a\theta}}{S_{as}}\right)\left(\frac{k_s}{k_\theta}\right)h^2\right)$$

$$\Rightarrow \text{A.F.} = \left(1 + \left(\frac{17.84 + 169.3}{169.3}\right)\left(\frac{30,519}{2,357,592}\right)\left(\frac{0.039}{0.066}\right) + \left(1 + \frac{8.815}{169.3 \times (7.3)^2}\right)\left(\frac{0.047}{0.066}\right)\left(\frac{30519}{2196510}\right) \times (7.3)^2\right)$$

$$\Rightarrow \text{A.F.} = \left(1 + 8.45 \times 10^{-3} + 0.527\right) = 1.535$$

Thus, in this case the inertial effect of foundation is negligible.

Going through the above problem, we can draw some very important conclusions:

- The DSSI always amplifies the response and will only approach fixed-base case when foundation stiffness is significantly high, like when the footings are resting on rock. Thus, the concept of considering fixed-base analysis as a conservative case is *certainly a myth* and not a reality.
- In case of framed structure on isolated footings, though the inertial effect of foundation on the overall response is negligible, in other cases, like frames on raft or pile foundations, it can significantly affect the response.

- The foundation inertial effect is possibly more profound for inverted pendulum-type structures, like elevated water tanks resting on a raft foundation or tall chimneys supported on circular or octagonal pads.

The above phenomenon can also be explained physically, which many of us have possibly felt, but have never realized. If one runs at certain speed on a hard road, it would be much easier to do so than running at same speed, over soft soil, like say a sandy beach.

While running on a soft soil, one would see that body muscles would start aching in a very short time, compared to that on the road. Aching in the muscles clearly shows that they are far more stressed, compared to the situation when somebody runs on a hard road for same distance. Stress within the body is nothing but a manifestation of dynamic bending moment and shear force the body is experiencing when one is on the move, and we can surely say that for a man running on beach, the dynamic bending moment and shears experienced by his muscles are more than what he will experience while running on the road.

By same analogy, one may conclude that a body under motion on soft soil will experience more stress/forces than when resting over hard soil. That is, the dynamic forces will amplify compared to a fixed-base case and do not attenuate as many believe.

7.7 Extension of Veletsos's Equation to Multi-degree of Freedom

Veletsos's equation can be extended further (Chowdhury and Dasgupta 2002) to systems having multi-degree of freedom of order $N \times N$. This can be expressed as

$$\frac{[\mathrm{I}]}{[\mathrm{K_e}]} = \frac{[\mathrm{I}]}{[\mathrm{K}]} + \frac{[\mathrm{I}]}{K_x} + \frac{[\mathrm{h}^2]}{K_\theta} \tag{7.7.1}$$

Here, [I] = identity matrix of order N having its diagonal element as 1.

Equation (7.7.1) can be further expressed as

$$[\mathbf{I}][\mathbf{Ke}]^{-1} = [\mathbf{I}][\mathbf{K}]^{-1} + [\mathbf{I}/K_x] + [\mathbf{h}^2/K_\theta] \tag{7.7.2}$$

$$\rightarrow \quad [\mathbf{Fe}] = [\mathbf{F}] + [\mathbf{F_x}] + [\mathbf{F_\theta}] \tag{7.7.3}$$

Inversion of the combined flexibility matrix as furnished in Eq. (7.7.3) gives combined stiffness matrix $[\mathbf{K_e}]$ of the soil and structure. Here, h would be the position vector of each of the lumped masses to the center of the springs.

Now, if we put Eq. (7.7.2) to RB test, we find that it passes the test with flying colors for, as limit of $K \rightarrow \infty$, the first term in the right-hand side of Eq. (7.7.2) approaches zero and we are left with the soil springs values only, based on which the body vibrates and satisfies RB test conditions posed earlier.

In Eqs. (7.7.2) and (7.7.3), it is clearly seen that the soil flexibility gets directly added to the diagonal and then, on inversion, affects all the terms of the $[K_e]$ and gives the true interaction, unlike a parallel spring which affects only locally the interaction effects and does not possibly give a true picture when the stiffness of the superstructure becomes quite high compared to that of the soil.

For estimation of damping, the effective damping ratio of soil–structure system can be expressed as (Chowdhury and Dasgupta 2002)

$$[\zeta_e] = [\mathbf{K_e}]\{[\zeta][\mathbf{F}] + [\zeta_x][\mathbf{F}_x] + [\zeta_\theta][\mathbf{F_\theta}]\} \tag{7.7.4}$$

In Eq. (7.7.4):

$[\zeta_e]$	equivalent damping matrix of the soil–structure system;
$[\mathbf{K_e}]$	combined stiffness matrix of the frame and soil;
$[\zeta]$	modal damping ratio of the structure;
$[\zeta_x]$	damping ratio of foundation in translational mode;
$[\zeta_\theta]$	damping ratio of foundation in rocking mode;
$[\mathbf{F}]$, $[\mathbf{F}_x]$, $[\mathbf{F}_\theta]$	flexibility matrices of the structure and foundation in translation and rocking mode.

It may be noted that the matrix $[\zeta_e]$ here is non-diagonal, considering that the soil damping is of radiation type, while the structural damping is material-dependent; however, computation will show that the non-diagonal terms are of small magnitude and the diagonal elements dominate. Thus, considering the modal damping ratios based on the diagonal element only would give a conservative estimation of modal damping ratios and may well be used for practical analyses.

7.7.1 Why Does Some Commercially Available Software Show Attenuation of Response Considering DSSI?

1. If what has been stated above is true, then why does commercial software show lower values of base shear for they also show an elongation in time period?
2. An increase in time period generally shows a reduced spectral acceleration; why does spectral acceleration amplify as shown above?

Answers to these paradoxes are elucidated hereafter.

For a fixed-base structure, after the eigensolution is carried out, the nodal forces are usually computed from the expression

$$\{P_i\} = [\mathbf{M}]\{\phi_i\}\left(\frac{L_i}{M_i}\right)\left(\frac{ZI}{2R}\right)S_{ai} \tag{7.7.5}$$

Here, $[\mathbf{M}]$ = mass matrix of the structure and is a diagonal matrix considering masses are lumped at nodes. $\{\phi_i\}$ = scaling or eigenvectors for ith mode. $ZI/2R$ = code factor as per IS-1893 (Part-1), S_{ai} = spectral acceleration as obtained from response spectrum curve for fixed-base time period T_i, $L_i = \{\phi_i\}^T[\mathbf{M}]\{I\}$ and $M_i = \{\phi_i\}^T[\mathbf{M}]\{\phi_i\}$, $\{I\}$ = identity matrix.

In Eq. (7.7.5), the external nodal forces $\{P_i\}$ have been derived by computing them from inertial side. Same can also be computed from the stiffness side as follows:

$$\{u_i\} = \{\phi_i\}\left(\frac{L_i}{M_i}\right)\left(\frac{ZI}{2R}\right)\left(\frac{S_{ai}}{\omega_i^2}\right) \tag{7.7.6}$$

Here, u_i = modal displacement of structure in ith mode and ω_i is natural frequency in ith mode for the fixed-base structure.

Equation (7.7.6) can thus be expressed as

$$\{u_i\} = \{\phi_i\}\left(\frac{L_i}{M_i}\right)\left(\frac{ZI}{2R}\right)\left(\frac{S_{ai}T_i^2}{4\pi^2}\right) \tag{7.7.7}$$

Multiplying Eq. (7.7.7) by $[\mathbf{K}]$, global stiffness matrix of the fixed-base structure, one gets

$$\{P_i\} = [\mathbf{K}]\{\phi_i\}\left(\frac{L_i}{M_i}\right)\left(\frac{ZI}{2R}\right)\left(\frac{S_{ai}T_i^2}{4\pi^2}\right) \tag{7.7.8}$$

Now considering $[\mathbf{K}] = [\mathbf{M}]\omega_i^2 \to [\mathbf{M}]\left(4\pi^2/T_i^2\right)$ and substituting the above in Eq. (7.7.8), it will be observed that one can arrive at Eq. (7.7.5).

Thus, it is seen that irrespective of whether one approaches the nodal force computation from inertial side or stiffness side, the outcome is same. However, in most of the case as the mass matrix is lumped and diagonal, it is computationally more efficient to compute the nodal forces from inertial side and the same is adopted by most of the commercially used software in their coding.

Let us now see what happens when DSSI effect is considered.

The nodal load is expressed as

$$\{P_i\} = [\mathbf{M}]\{\phi_i\}\left(\frac{L_i}{M_i}\right)\left(\frac{ZI}{2R}\right)\widehat{S}_{ai} \tag{7.7.9}$$

In Eq. (7.7.9), the term $\widehat{S}_{ai}$ is the spectral acceleration from the response curve corresponding to time period as expressed by Eq. (7.5.7) and will certainly show a reduced response due to DSSI. However, if this is true, like derived earlier, it should match the nodal force if one comes from the stiffness side. Deformation amplitude of the fixed-base structure due to the effect of DSSI is now expressed as

$$\{u_i\} = \{\phi_i\}\left(\frac{L_i}{M_i}\right)\left(\frac{ZI}{2R}\right)\left(\frac{\widehat{S}_{ai}\widehat{T}_i^2}{4\pi^2}\right) \quad (7.7.10)$$

Multiplying Eq. (7.7.10) by [**K**], the fixed-base stiffness of the structure (as the major interest is to know the forces induced in the structure), structural nodal forces can be computed as

$$\{P_i\} = [\mathbf{K}]\{\phi_i\}\left(\frac{L_i}{M_i}\right)\left(\frac{ZI}{2R}\right)\left(\frac{\widehat{S}_{ai}\widehat{T}_i^2}{4\pi^2}\right) \quad (7.7.11)$$

Again considering the expression $[\mathbf{K}] = \rightarrow [\mathbf{M}]\left(4\pi^2/T_i^2\right)$ and substituting it in Eq. (7.7.11), one has

$$\{P_i\} = [\mathbf{M}]\{\phi_i\}\left(\frac{L_i}{M_i}\right)\left(\frac{ZI}{2R}\right)\widehat{S}_{ai}\left(\frac{\widehat{T}_i}{T_i}\right)^2 \quad (7.7.12)$$

Comparing Eq. (7.7.12) with that furnished in Eq. (7.7.9), it is seen that for the nodal force from inertial side to be compatible with one computed from stiffness side it has to be multiplied by a factor $\left(\widehat{T}_i/T_i\right)^2$ and one can safely predict *that this has not been looked into and missing in most of the software coding and this is what produces a reduced response and generates the popular myth "DSSI on soft soil is beneficial to the structure."*

With a little reflection, one would realize that the term $\left(\widehat{T}_i/T_i\right)^2$, which is again a dimensionless number, is nothing but is the dynamic amplification factor AF, same as derived earlier by which the fixed-base response is amplified. Thus, despite $\widehat{S}_{ai}$ being less than S_{ai}, because it gets multiplied by the term $\left(\widehat{T}_i/T_i\right)^2$, the final nodal force would show an increase in magnitude compared to fixed-base case.

Example 7.3 For the frame cited in Fig. 7.8, carry out a detailed modal analysis to determine the base shear and floor shears considering

- fixed-base;
- Considering soil–structure interaction as per Sect. 7.7.

All dimensions and data remain same as mentioned before.

Solution

The frame is redrawn for convenience again.

Column size in this case is 300 × 450 and this gives $I_{col} = (1/12) \times 0.3 \times (0.45)^3 = 2.28 \times 10^{-3}\ m^4$

Thus, at each floor effective $I_{eff} = 3 \times 2.28 \times 10^{-3} = 6.834 \times 10^{-3}\ m^4$.

Young's modulus of concrete = $5000\sqrt{f_{ck}}\ N/mm^2 \rightarrow 2.5 \times 10^7\ kN/m^2$.

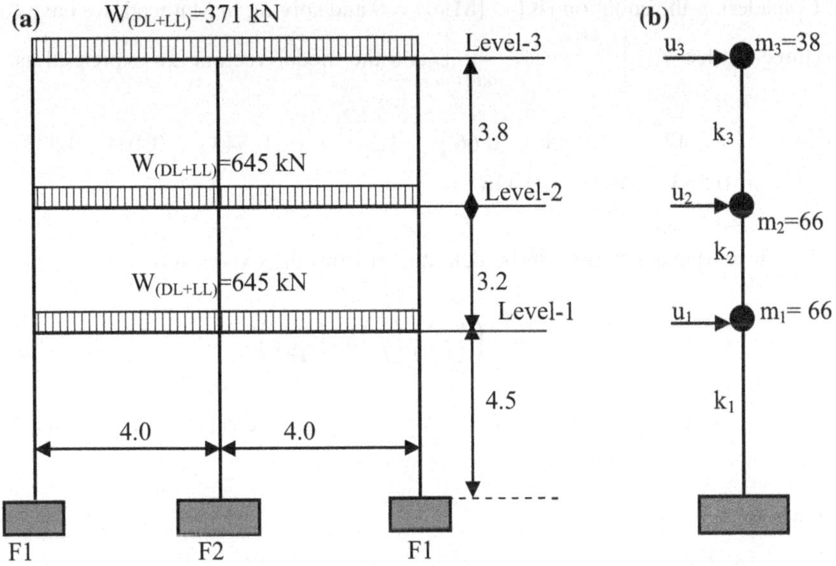

Fig. 7.8 Three-storied fixed-base frame and its mathematical model for modal analysis

Thus, as per the problem,

$$k_1 = \frac{12EI_{\text{eff}}}{(L_1)^3} = \frac{12 \times 2.5 \times 10^7 \times 6.834 \times 10^{-3}}{(4.5)^3} = 22,499\ \text{kN/m}$$

$$k_2 = \frac{12EI_{\text{eff}}}{(L_2)^3} = \frac{12 \times 2.5 \times 10^7 \times 6.834 \times 10^{-3}}{(3.2)^3} = 62,567\ \text{kN/m}$$

and

$$k_3 = \frac{12EI_{\text{eff}}}{(L_3)^3} = \frac{12 \times 2.5 \times 10^7 \times 6.834 \times 10^{-3}}{(3.8)^3} = 37,363\ \text{kN/m}$$

Based on the theory as explained in Chap. 2, the stiffness matrix considering the structure is fixed at the foundation top and the mass matrix is expressed as

$$[K] = \begin{bmatrix} k_1 + k_2 & -k_2 & 0 \\ -k_2 & k_2 + k_3 & -k_3 \\ 0 & -k_3 & k_3 \end{bmatrix}, \quad [M] = \begin{bmatrix} m_1 & 0 & 0 \\ 0 & m_2 & 0 \\ 0 & 0 & m_3 \end{bmatrix}$$

This gives $[K] = \begin{bmatrix} 85066 & -62567 & 0 \\ -62567 & 99939 & -37363 \\ 0 & -37363 & 37363 \end{bmatrix}$ and $[M] = \begin{bmatrix} 66 & 0 & 0 \\ 0 & 66 & 0 \\ 0 & 0 & 38 \end{bmatrix}$.

Considering the equation $[\mathbf{K}] - [\mathbf{M}]\omega^2 = 0$ and solving in Mathcad, we have for the three modes $[T] = \begin{bmatrix} 0.594 & & \\ & 0.189 & \\ & & 0.124 \end{bmatrix}$, and the modal vectors are expressed as

$\{\phi_1\}^{\mathrm{T}} = \{-0.471 \quad -0.585 \quad -0.66\}$, $\{\phi_2\}^{\mathrm{T}} = \{-0.543 \quad -0.104 \quad 0.833\}$,
$\{\phi_3\}^{\mathrm{T}} = \{0.533 \quad -0.713 \quad 0.446\}$

The floor displacements can be determined from the expression

$$u_i = \{\phi_i\}\left(\frac{L_i}{M_i}\right)\left(\frac{ZI}{2R}\right)S_{\mathrm{a}i}\left(\frac{T_i^2}{4\pi^2}\right)$$

Here, $L_i = \{\phi_i\}^{\mathrm{T}}[\mathbf{M}]\{I\}$ and $M_i = \{\phi_i\}^{\mathrm{T}}[\mathbf{M}]\{\phi_i\}$, where $\{\mathrm{I}\}$ = identity matrix.

For Mode 1

$Z = 0.16$, $I = 1.0$ and $R = 3.0$ this gives $ZI/2R = 0.027$.

$$\{\phi_1\}^{\mathrm{T}} = \{-0.471 \quad -0.585 \quad -0.66\}$$

Substituting, we get

$$L_i = \{-0.471 \quad -0.585 \quad -0.66\}\begin{bmatrix} 66 & 0 & 0 \\ 0 & 66 & 0 \\ 0 & 0 & 38 \end{bmatrix}\begin{Bmatrix} 1 \\ 1 \\ 1 \end{Bmatrix} = -94.776$$

$$M_i = \{-0.471 \quad -0.585 \quad -0.66\}\begin{bmatrix} 66 & 0 & 0 \\ 0 & 66 & 0 \\ 0 & 0 & 38 \end{bmatrix}\begin{Bmatrix} -0.471 \\ -0.585 \\ -0.66 \end{Bmatrix} = 53.781$$

For the first mode $T_1 = 0.594$ s, which gives $S_a = 24.525\ \mathrm{m/s^2}$. (Here $S_a/g = 2.5$)

Thus, $u_1 = \{\phi_1\}\left(\frac{L_1}{M_1}\right)\left(\frac{ZI}{2R}\right)S_{\mathrm{a}1}\left(\frac{T_1^2}{4\pi^2}\right)$

$$u_1 = \begin{Bmatrix} -0.471 \\ -0.585 \\ -0.660 \end{Bmatrix} \times \left(\frac{-94.776}{53.781}\right) \times (0.027) \times 24.525\left(\frac{(0.594)^2}{4\pi^2}\right)$$
$$= \begin{Bmatrix} 4.852 \\ 6.026 \\ 6.798 \end{Bmatrix} \times 10^{-3}\ \mathrm{m}$$

Floor lateral forces can now be computed from the expression

$$\{P_1\} = [K] \times \{u_1\} = \begin{bmatrix} 85066 & -62567 & 0 \\ -62567 & 99939 & -37363 \\ 0 & -37363 & 37363 \end{bmatrix} \times \begin{Bmatrix} 4.852 \\ 6.026 \\ 6.798 \end{Bmatrix} \times 10^{-3}$$

$$= \begin{Bmatrix} 35.685 \\ 44.660 \\ 28.865 \end{Bmatrix} \text{kN}$$

Floor lateral forces can also be calculated from the inertial expression

$$\{P_1\} = [\mathbf{M}] \times \{\phi_1\} \times \left(\frac{L_1}{M_1}\right) \times \left(\frac{ZI}{2R}\right) \times S_{a1} = \begin{bmatrix} 66 & 0 & 0 \\ 0 & 66 & 0 \\ 0 & 0 & 38 \end{bmatrix}$$

$$\times \begin{Bmatrix} -0.471 \\ -0.585 \\ -0.660 \end{Bmatrix} \times \left(\frac{-94.776}{53.781}\right) \times 0.027 \times 24.525 = \begin{Bmatrix} 35.827 \\ 44.499 \\ 28.905 \end{Bmatrix}$$

The reader is to carefully note that irrespective of whether the computation is done based on stiffness basis [**K**] or inertial basis [**M**], the force must match at each floor and also the base shear force. Because the inertial basis is an external force and the stiffness basis is an internal force, *this must match at all nodes at any time to maintain equilibrium*.

However, most of the commercially available programs would calculate the floor forces based on inertial basis because the lumped mass matrix is diagonal, and it is computationally more efficient to derive from inertial basis, as deriving the force from stiffness basis requires computationally more involved process in terms of matrix operations. It may be observed that in the above calculations the P_1 values match well from both stiffness and inertial basis. The miniscule difference it shows is due to numerical error arising out of matrix operation and is very small: actually of the order of 0.1%.

The floor shears in first mode are expressed by

$$\{V_1\} = \begin{bmatrix} 28.865 + 44.66 + 35.685 \\ 28.865 + 44.66 \\ 28.865 \end{bmatrix} = \begin{Bmatrix} 109.39 \\ 73.525 \\ 28.865 \end{Bmatrix} \text{kN}$$

For Mode 2

$ZI/2R = 0.027$

$$\{\phi_1\}^{\mathrm{T}} = \{-0.543 \quad -0.104 \quad 0.833\}$$

Substituting, we get

$$L_i = \{-0.543 \quad -0.104 \quad 0.833\}\begin{bmatrix} 66 & 0 & 0 \\ 0 & 66 & 0 \\ 0 & 0 & 38 \end{bmatrix}\begin{Bmatrix} 1 \\ 1 \\ 1 \end{Bmatrix} = -11.048$$

$$M_i = \{-0.543 \quad -0.104 \quad 0.833\}\begin{bmatrix} 66 & 0 & 0 \\ 0 & 66 & 0 \\ 0 & 0 & 38 \end{bmatrix}\begin{Bmatrix} -0.543 \\ -0.104 \\ -0.833 \end{Bmatrix} = 46.542$$

Here, $T_2 = 0.189$ s, which gives $S_a = 24.525$ m/s^2. (Here $Sa/g = 2.5$).
Thus, $u_2 = \{\phi_2\}\left(\frac{L_2}{M_2}\right)\left(\frac{ZI}{2R}\right)S_{a2}\left(\frac{T_2^2}{4\pi^2}\right)$.
The above expression finally gives

$$u_2 = \begin{Bmatrix} 7.682 \\ 1.461 \\ -11.7 \end{Bmatrix} \times 10^{-5}\,\mathrm{m}$$

The floor lateral forces can now be computed from the expression

$$\{P_2\} = [K] \times \{u_2\} = \begin{bmatrix} 85066 & -62567 & 0 \\ -62567 & 99939 & -37363 \\ 0 & -37363 & 37363 \end{bmatrix} \times \begin{Bmatrix} 7.682 \\ 1.461 \\ -11.7 \end{Bmatrix} \times 10^{-5}$$

$$= \begin{Bmatrix} 5.574 \\ 1.06 \\ -4.918 \end{Bmatrix}\,\mathrm{kN}$$

From the inertial expression, we have

$$\{P_2\} = [\mathbf{M}] \times \{\phi_2\} \times \left(\frac{L_2}{M_2}\right) \times \left(\frac{ZI}{2R}\right) \times S_{a2} = \begin{Bmatrix} 5.564 \\ 1.066 \\ -4.914 \end{Bmatrix}\,\mathrm{kN}$$

The floor shear for Mode 2 are expressed as

$$\{V_2\} = \begin{bmatrix} -4.918 + 1.06 + 5.574 \\ -4.918 + 1.06 \\ -4.918 \end{bmatrix} = \begin{Bmatrix} 1.716 \\ -3.858 \\ -4.918 \end{Bmatrix}\,\mathrm{kN}$$

For Mode 3

Proceeding in identical fashion as for modes 1 and 2, we have

$\{P_3\} = \begin{Bmatrix} 1.94 \\ -2.615 \\ 0.934 \end{Bmatrix}$ kN, and the amplitudes are

$\{u_3\} = \begin{Bmatrix} 1.149 \\ -1.538 \\ 0.962 \end{Bmatrix} \times 10^{-5}$ m.

The floor shear for Mode 3 is expressed as

$$\{V_3\} = \begin{bmatrix} 0.934 - 2.615 + 1.94 \\ 0.934 - 2.615 \\ 0.934 \end{bmatrix} = \begin{Bmatrix} 0.259 \\ -1.681 \\ 0.934 \end{Bmatrix} \text{kN}$$

The SRSS values of floor forces can now be calculated as
At Level 3

$$V_3 = \sqrt{(28.865)^2 + (-4.914)^2 + (0.934)^2} = 29.3\,\text{kN}$$

At Level 2

$$V_2 = \sqrt{(73.525)^2 + (-3.858)^2 + (-1.681)^2} = 73.74\,\text{kN}$$

At Level 1

$$V_3 = \sqrt{(109.39)^2 + (1.716)^2 + (0.259)^2} = 109.4\,\text{kN}$$

The reader is to note that based on IS code provisions we got a base shear $V_b = 109.63$ kN; thus, we see that both values commendably agree, though in terms of distribution of lateral force per floor, the IS code gives conservative values.

Consideration of soil–structure interaction

*Before we solve for this portion we would request the reader to carefully go through the comments after certain steps marked by asterisk**. These are very important commentaries to help the reader get insight into the problem solved herein.*

Based on previous calculations, vide Example 7.2,

$$[K_x] = \begin{bmatrix} 2357592 & 0 & 0 \\ 0 & 2357592 & 0 \\ 0 & 0 & 2357592 \end{bmatrix} \quad \text{and} \quad [K_\theta] = \begin{bmatrix} 2196510 & 0 & 0 \\ 0 & 2196510 & 0 \\ 0 & 0 & 2196510 \end{bmatrix}$$

$$[K] = \begin{bmatrix} 85066 & -62567 & 0 \\ -62567 & 99939 & -37363 \\ 0 & -37363 & 37363 \end{bmatrix} \quad \text{and} \quad [h] = \begin{bmatrix} 4.5 & 0 & 0 \\ 0 & 7.7 & 0 \\ 0 & 0 & 11.5 \end{bmatrix}$$

Thus, $[F_e] = [K]^{-1} + [K_x]^{-1} + [h^2] \times [K_\theta]$

$$\rightarrow [F_e] = \begin{bmatrix} 5.407 & 4.442 & 4.442 \\ 4.442 & 8.781 & 6.04 \\ 4.442 & 6.04 & 14.78 \end{bmatrix} \times 10^{-5} \rightarrow [K_e] = [F_e]^{-1}$$

$$= \begin{bmatrix} 33570 & -13970 & -4382 \\ -13970 & 21650 & -4650 \\ -4382 & -4650 & 9983 \end{bmatrix}$$

**Please observe the [K_e] matrix; it is completely different from the fixed-base stiffness matrix and also different from the case when the springs would have been locally added to the foundation node. One may also note that the matrix is symmetric but all degrees of freedom are affected, showing a *true soil–structure interaction*.

The [M] matrix remains the same as previously computed; thus, solving the expression $[\mathbf{K_e}] - [\mathbf{M}]\omega^2 = 0$ in Mathcad, we have

$$[T_e] = \begin{bmatrix} 0.648 & 0 & 0 \\ 0 & 0.33 & 0 \\ 0 & 0 & 0.246 \end{bmatrix},$$ which shows elongation in time period in all modes.

Considering 5% damping, the accelerations for the three modes are $S_{ae1} = 20.58\ \text{m/s}^2$, $S_{ae2} = 24.525\ \text{m/s}^2$ and $S_{ae3} = 24.525\ \text{m/s}^2$ as per IS-1893 (2002).

Thus, structural deformation based on the above time periods can be determined as

$$u_1 = \{\phi_1\} \left(\frac{L_1}{M_1}\right) \left(\frac{ZI}{2R}\right) S_{ae1} \left(\frac{T_{e1}^2}{4\pi^2}\right)$$

**The reader is to note that since we are interested in finding out the deformation of structure, the scaling vectors and modal mass participation factors to be considered are for fixed-base case. The only terms that change in this case are *acceleration* and time periods due to foundation compliance. Thus,

$$u_1 = \begin{Bmatrix} -0.471 \\ -0.585 \\ -0.660 \end{Bmatrix} \times \left(\frac{-94.776}{53.781}\right) \times (0.027) \times 20.58\left(\frac{(0.648)^2}{4\pi^2}\right)$$
$$= \begin{Bmatrix} 4.847 \\ 6.02 \\ 6.792 \end{Bmatrix} \times 10^{-3}\,\mathrm{m}$$

Forces at each floor in the first mode can be expressed as

$$\{P_1\} = [K] \times \{u_1\} = \begin{bmatrix} 85066 & -62567 & 0 \\ -62567 & 99939 & -37363 \\ 0 & -37363 & 37363 \end{bmatrix} \times \begin{Bmatrix} 4.847 \\ 6.02 \\ 6.792 \end{Bmatrix} \times 10^{-3}$$
$$= \begin{Bmatrix} 35.652 \\ 44.619 \\ 28.838 \end{Bmatrix} \mathrm{kN}$$

**The reader is to note here that since we are trying to find out the structural deformation, we need to multiply u_1 by [K], i.e., the fixed-base stiffness, and not by [K_e].

Now we check the lateral forces from the inertial expression to see if this matches the above value

$$\{P_1\} = [\mathbf{M}] \times \{\phi_1\} \times \left(\frac{L_1}{M_1}\right) \times \left(\frac{ZI}{2R}\right) \times S_{ae1} = \begin{bmatrix} 66 & 0 & 0 \\ 0 & 66 & 0 \\ 0 & 0 & 38 \end{bmatrix}$$
$$\times \begin{Bmatrix} -0.471 \\ -0.585 \\ -0.660 \end{Bmatrix} \times \left(\frac{-94.776}{53.781}\right) \times 0.027 \times 24.525 = \begin{Bmatrix} 30.07 \\ 37.36 \\ 24.26 \end{Bmatrix} \mathrm{kN}$$

**On comparison we see that the forces mismatch and coming from inertial side we see a reduced response!! We have however said at the outset that elastic force and inertial force must match—*so what went wrong*?

To this end, let us consider

$$\{P_i\} = [K] \times \{u_i\} = [M]\omega_i^2 \times \{\phi_1\}\left(\frac{L_i}{M_i}\right)\left(\frac{ZI}{2R}\right)S_{ae1}\left(\frac{T_{ei}^2}{4\pi_2}\right)$$

In the above expression [K] = $M\omega_i^2$, where ω_i is the natural fixed-base frequency of the system. Now, considering $T_i = 2\pi/\omega_i$, we have

$$\{P_i\} = [K] \times \{u_i\} = [M] \times \{\phi_1\}\left(\frac{L_i}{M_i}\right)\left(\frac{ZI}{2R}\right)S_{ae1}\left(\frac{T_{ei}^2}{T_i^2}\right)$$

Thus, from the above we see that if we come from inertial side, the fixed-base response must be multiplied by a factor $(T_{ei}/T_i)^2$ to arrive at the correct value of DSSI. In case one runs this problem in software where the forces are calculated from inertial side and does not have this amplification factor incorporated in the source, code will show a reduced response and create the myth that DSSI reduces the response.

Thus, multiplying the lateral force obtained from inertial mass by the factor $(T_{ei}/T_i)^2$, we now have

$$\{P_1\} = [\mathbf{M}] \times \{\phi_1\} \times \left(\frac{L_1}{M_1}\right) \times \left(\frac{ZI}{2R}\right) \times S_{ae1} \times \left(\frac{T_{ei}}{T_i}\right)^2 = \begin{bmatrix} 66 & 0 & 0 \\ 0 & 66 & 0 \\ 0 & 0 & 38 \end{bmatrix}$$

$$\times \begin{Bmatrix} -0.471 \\ -0.585 \\ -0.660 \end{Bmatrix} \times \left(\frac{-94.776}{53.781}\right) \times 0.027 \times 24.525 \times \left(\frac{0.648}{0.594}\right)^2 = \begin{Bmatrix} 35.79 \\ 44.45 \\ 28.88 \end{Bmatrix} \text{kN}$$

The above is in close agreement with the force obtained from stiffness side and should be OK.

The reader is also to note that in this case that the amplification due to DSSI is actually $\text{AF} = (T_{ei}/T_i)^2$, and this shows that due to DSSI the fixed-base time period gets elongated by a factor $\sqrt{\text{AF}}$.

The floor shears are expressed as

$$\{V_1\} = \begin{bmatrix} 28.85 + 44.45 + 35.79 \\ 28.88 + 44.45 \\ 28.88 \end{bmatrix} = \begin{Bmatrix} 109.12 \\ 73.33 \\ 28.88 \end{Bmatrix} \text{kN}$$

Similarly, for the second mode

$$u_2 = \{\phi_2\}\left(\frac{L_2}{M_2}\right)\left(\frac{ZI}{2R}\right)S_{ae2}\left(\frac{T_{e2}^2}{4\pi^2}\right)$$

The above expression finally gives

$$u_2 = \begin{Bmatrix} 23.68 \\ 4.535 \\ -36.32 \end{Bmatrix} \times 10^{-5}\ \text{m when}$$

$$\{P_2\} = [K] \times \{u_2\} = \begin{bmatrix} 85066 & -62567 & 0 \\ -62567 & 99939 & -37363 \\ 0 & -37363 & 37363 \end{bmatrix} \times \begin{Bmatrix} 23.68 \\ 4.535 \\ -36.32 \end{Bmatrix} \times 10^{-5}$$

$$= \begin{Bmatrix} 17.305 \\ 3.289 \\ -15.266 \end{Bmatrix} \text{kN}$$

This gives $\{V_2\} = \begin{Bmatrix} 5.33 \\ -11.97 \\ -15.26 \end{Bmatrix}$ kN

Similarly, for the third mode

$$u_3 = \begin{Bmatrix} 4.524 \\ -6.052 \\ -3.785 \end{Bmatrix} \times 10^{-5}\,\text{m} \quad \text{and} \quad \{P_3\} = \begin{Bmatrix} 7.635 \\ -10.293 \\ -32.836 \end{Bmatrix} \text{kN} \quad \text{and}$$

$$\{V_3\} = \begin{Bmatrix} -35.49 \\ -43.12 \\ -32.836 \end{Bmatrix} \text{kN}$$

The SRSS values at Levels 1, 2, and 3 are expressed as $\{V\} = \begin{Bmatrix} 115 \\ 86 \\ 46 \end{Bmatrix}$ kN.

We compare the values arrived at due to DSSI and fixed-base below

Level	1	2	3
Shear per floor (fixed-base)	109	73	29
Shear per floor due to DSSI	115	86	46
% increase	5.5	18	58

**One should observe here that amplification of force is modest in this case. The reason being $V_s/(\text{f} \times \bar{h})$ in this case is 20.3 ≅ 20; thus, as suggested by Veletsos and Meek (1974), the DSSI effect is not so profound.

In case the soil is soft, and time period gets further prolonged (T_e), the DSSI effect would surely be stronger. For stiff structure and soft soil, the system will have significant amplification due to DSSI. This remains one of the critical reasons why DSSI is important for nuclear power plants. However, there is one redeeming aspect in this issue—the radiation damping of soil, which we are going to discuss a little bit latter.

7.7.2 Extending the Proposed Model for a Space Frame

Extending the model as proposed in Sect. 7.7 for a three-dimensional space frame is not difficult.

As shown in Fig. 7.9, let the fixed-base stiffness of the superstructure be $[K]_{N\times N}$. Here, N is the number of active degrees of freedom of the frame. Let $[K_x]_{N\times N}$ and $[K_\theta]_{N\times N}$ be the representative stiffness of the foundation written as an $N \times N$ diagonal matrix. Then, extending Eq. 7.7.3, we can write

$$[\mathbf{F_e}] = [\mathbf{F}] + [\mathbf{F_x}] + [\Re][F_\theta] \quad (7.7.13)$$

Here

[F]	flexibility matrix of the fixed-base structure of order $N \times N$;
$[F_x]$	flexibility of foundation matrix of order $N \times N$ in translational mode;
$[F_\theta]$	flexibility of foundation matrix of order $N \times N$ in rocking mode;
$[\Re]$	square of radius vector of each of the lumped masses expressed as $(x_i - x_f)^2 + (y_i - y_f)^2 + (z_i - z_f)^2$;
(x_i, y_i, z_i)	coordinates of the lumped mass at ith node;
(x_f, y_f, z_f)	coordinates of the center of foundation where the representative foundation spring is connected.

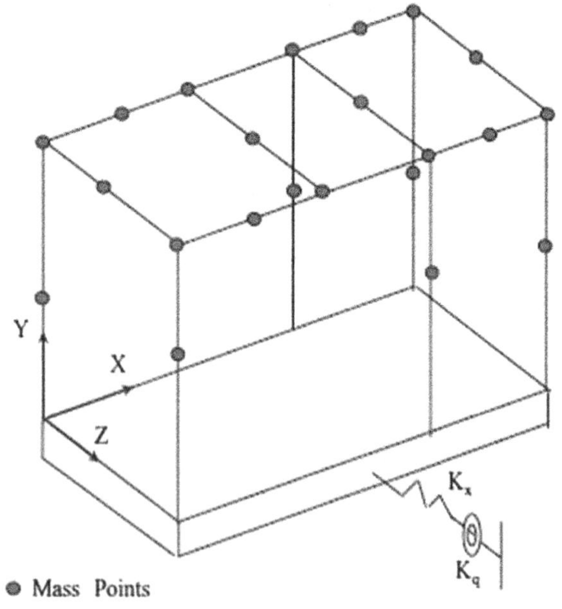

Fig. 7.9 A 3D frame with multi-degrees of freedom and representative foundation springs

7.8 Alternate Model for DSSI Based on Lagrange's Equation

The model was developed way back in the mid-1980s (Chowdhury 1984), originally to determine the DSSI effect on steam turbine framed foundations but can also be effectively used for studying response of frames under earthquake.

Consider the frame as shown in Fig. 7.10, where the frame is coupled to a foundation having translation spring K_x and rotational spring K_θ. The masses at each floor are m_1, m_2, m_3, etc., and the foundation mass and mass moment of inertia are m_f and J_f, respectively.

We choose translation and rotation of foundation as u_f and θ. Displacement at each floor is expressed as u_i, where i is the number of floors or levels in the frame.

The kinetic energy of the system is expressed as

$$T = \frac{1}{2} m_f \dot{u}_f^2 + \frac{1}{2} J_f \dot{\theta}^2 + \frac{1}{2} \sum_{i=1}^{3} m_i \left(u_f + h_i \theta + u_i\right)^2 \tag{7.8.1}$$

The potential energy of the system is expressed as

$$U = \frac{1}{2} K_x u_f^2 + \frac{1}{2} K_\theta \theta^2 + \sum_{i=0}^{3} \frac{1}{2} k_{i+1} (u_{i+1} - u_i)^2 \tag{7.8.2}$$

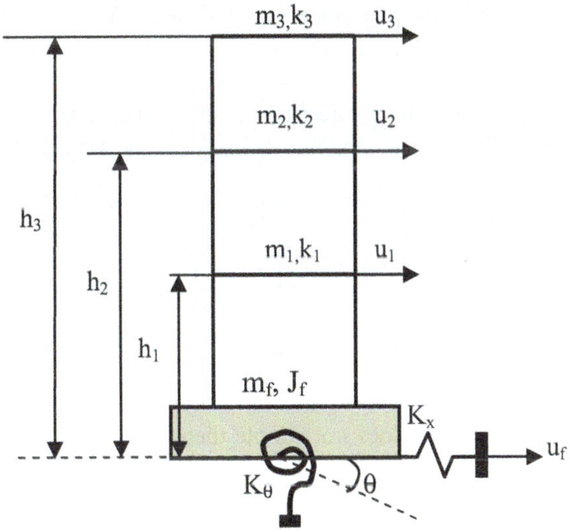

Fig. 7.10 Frame resting on representative foundation springs in translation and rotation

Now, considering Lagrange's expression,

$$d(T+U)=\sum_{i=1}^{N}\left[\frac{\mathrm{d}}{\mathrm{d}t}\left(\frac{\partial T}{\partial \dot{q}_i}\right)-\frac{\partial T}{\partial q_i}+\frac{\partial U}{\partial q_i}\right]=0 \tag{7.8.3}$$

Equations (7.8.1) and (7.8.2) finally give

$$\begin{bmatrix} m_f+\sum\limits_{i}^{3} m_i & \sum\limits_{i=1}^{3} m_ih_i & m_1 & m_2 & m_3 \\ \sum\limits_{1}^{3} m_ih_i & J_f+\sum\limits_{i=1}^{3} m_ih_i^2 & m_1h_1 & m_2h_2 & m_3h_3 \\ m_1 & m_1h_1 & m_1 & 0 & 0 \\ m_2 & m_2h_2 & 0 & m_2 & 0 \\ m_3 & m_3h_3 & 0 & 0 & m_3 \end{bmatrix} \begin{Bmatrix} \ddot{u}_f \\ \ddot{\theta} \\ \ddot{u}_1 \\ \ddot{u}_2 \\ \ddot{u}_3 \end{Bmatrix} + \begin{bmatrix} K_x & 0 & 0 & 0 & 0 \\ 0 & K_\theta & 0 & 0 & 0 \\ 0 & 0 & k_1+k_2 & -k_2 & 0 \\ 0 & 0 & -k_2 & k_2+k_3 & -k_3 \\ 0 & 0 & 0 & -k_3 & -k_3 \end{bmatrix} \begin{Bmatrix} u_f \\ \theta \\ u_1 \\ u_2 \\ u_3 \end{Bmatrix} = 0 \tag{7.8.4}$$

Equation (7.8.4) gives the free vibration equation of the frame including foundation compliance for the frame shown in Fig. 7.10. In this case, compared to the earlier model, matrix order for structure having N degrees of freedom gets increased to $N + 2$, as additional translational (u_f) and rotation (θ) displacements of foundation come into play. One may also note that in this case the bottom corner matrix gives the fixed-base stiffness and mass of the structure.

Coupling of this nature, when the mass matrix is full and symmetric, is called dynamic coupling and the greatest advantage in this case is the damping matrix [C] remains uncoupled, considering that its basis is same as the stiffness matrix, and is expressed as

$$[C]=\begin{bmatrix} C_x & 0 & 0 & 0 & 0 \\ 0 & C_\theta & 0 & 0 & 0 \\ 0 & 0 & c_1+c_2 & -c_2 & 0 \\ 0 & 0 & -c_2 & c_2+c_3 & -c_3 \\ 0 & 0 & 0 & -c_3 & c_3 \end{bmatrix} \tag{7.8.5}$$

Considering the foundation damping is diagonal, in this case the orthogonal transformation does not couple them with other modes when we have

$$\{\phi\}^{\mathrm{T}}[C]\{\phi\}=\begin{bmatrix} 2\zeta_x\omega_x & 0 & 0 & 0 & 0 \\ 0 & 2\zeta_\theta\omega_\theta & 0 & 0 & 0 \\ 0 & 0 & 2\zeta_1\omega_1 & 0 & 0 \\ 0 & 0 & 0 & 2\zeta_2\omega_2 & 0 \\ 0 & 0 & 0 & 0 & 2\zeta_3\omega_3 \end{bmatrix} \tag{7.8.6}$$

Thus, considering $\{U\} = [\phi]\{\xi\}$ where ξ is the generalized coordinate, the uncoupled equation of motion can be expressed as

$$\ddot{\xi}_x + 2\zeta_x\omega_x\dot{\xi}_x + \omega_x^2\xi_x = -S_{ax}(\omega_x) \tag{7.8.7a}$$

$$\ddot{\xi}_\theta + 2\zeta_\theta\omega_\theta\dot{\xi}_\theta + \omega_\theta^2\xi_\theta = -S_{a\theta}(\omega_\theta) \tag{7.8.7b}$$

$$\ddot{\xi}_1 + 2\zeta_1\omega_1\dot{\xi}_1 + \omega_1^2\xi_1 = -S_{a1}(\omega_1) \tag{7.8.7c}$$

$$\ddot{\xi}_2 + 2\zeta_2\omega_2\dot{\xi}_2 + \omega_2^2\xi_2 = -S_{a2}(\omega_2) \tag{7.8.7d}$$

$$\ddot{\xi}_3 + 2\zeta_3\omega_3\dot{\xi}_3 + \omega_3^2\xi_3 = -S_{a3}(\omega_3) \tag{7.8.7e}$$

In Eqs. (7.8.7a)–(7.8.7e), S_a is the spectral acceleration corresponding to the system frequency ω with respect to the damping ratio ζ. Either modal or a time history response can be carried out. Once the value of ξ is known, [U] is determined from the expression $\{U\} = [\phi]\{\xi\}$. Now, knowing u_1, u_2, and u_3, one can substitute them in fixed-base stiffness of the frame depicted by the 3×3 bottom corner of the matrix in Eq. (7.8.4) as P = [K]{U} to determine the forces in each of the structure nodes due to DSSI.

Example 7.4 A tech structure steel frame system is as shown in Fig. 7.11. It is used for dual purpose of carrying pipes at top having load of 3 kN/m^2. Lower story carries a heat exchanger of 750 kN weight. The live load at heat exchanger level is 2 kN/m^2. The vessel is supported on secondary beams. The structure has bolted connection. The frame is supported on a reinforced concrete raft of size 16.5 m × 8.2 m and 750 mm thick. Base raft carries various equipments and other sundry loads which are estimated as 10 kN/m^2. Shear wave velocity of soil is 180 m/s and Poisson's ratio ν = 0.4 having weight density of 18 kN/m^3. The structure is to be built in a location identified as Zone IV as per IS-1893. Carry out the earthquake modal analysis.

- Considering the frame as fixed at top of raft;
- Considering dynamic soil–structure interaction using model as per Sect. 7.8.

Section properties of the members used are as tabulated hereafter

Members[a]	Member designation	I_{xx} (cm^4)	I_{yy} (cm^4)	A (cm^2)	Unit weight (kg/m)	Remarks
Columns	UC 203 × 203 × 60	6125	2065	76.4	60	EL 14.2 to 0.0
Beam on row A	UB 457 × 191 × 67	–	–	85	67	EL +14.2
Beam on row B	Do	–	–	85	67	EL +14.2
Beam on row 1	UB 533 × 210 × 82	–	–	105	82.2	Do
Beam on row 2	Do	–	–	105	82.2	Do
Beam on row 3	Do	–	–	105	82.2	Do
Beam on row 1	UB 610 × 229 × 101	–	–	129	101	El +6.2

(continued)

(continued)

Members[a]	Member designation	I_{xx} (cm^4)	I_{yy} (cm^4)	A (cm^2)	Unit weight (kg/m)	Remarks
Beam on row 2	UB 610 × 229 × 125	–	–	159	125	Do
Beam on row 3	UB 610 × 229 × 101	–	–	129	101	Do
Beam on row A	Do	–	–	129	101	Do
Beam on row B	Do	–	–	129	101	Do
Secondary beams	UB 406 × 178 × 54	–	–	69	54	EL +14.2
Secondary beams	UB 533 × 210 × 101	–	–	129	101	El +6.2
Floor bracings	Ch 102 × 51	–	–	13.3	14.2	El +14.2
Floor bracings	Ch 127 × 64	–	–	19	14.9	El +6.2
Longitudinal bracings	Ch 203 × 76	–	–	30.3	23.82	All levels

[a]Members used are British section

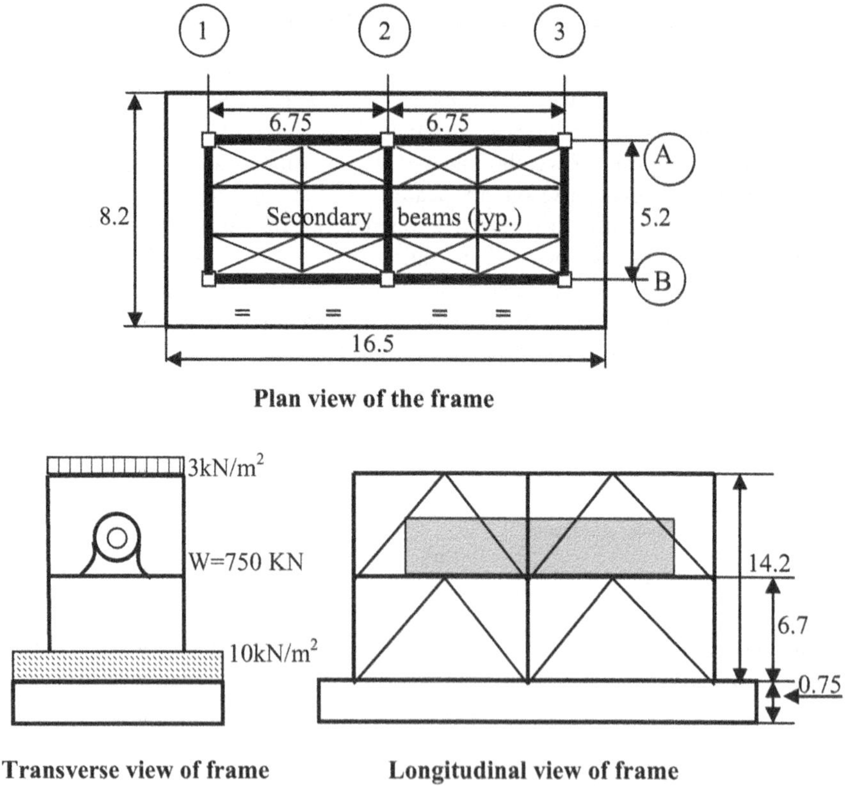

Fig. 7.11 A tech structure frame with plan view and orthogonal sections

Solution

We calculate the load on each level based on data furnished in the table above

- **At elevation 14.2**

Weight of beams on row $A = 67 \times 6.75 \times 2 = 904\,\text{kg}$.
Weight of beam on row B = 904 kg.
Weight of transverse beam on rows 1, 2, and 3 = $82 \times 5.2 \times 3 = 1279\,\text{kg}$.
Weight of secondary beams = 1458 kg.
Weight from plan bracings = 3101 kg
Load from piping = $300 \times 6.75 \times 2 \times 5.2 = 21,060\,\text{kg}$.
Contributing weight from column = 2066 kg.
Contributing weight of longitudinal bracings = 789 kg.
Total weight = 31,561 kg ≅ 316 kN.
Thus, mass m_2 = 32 kN s^2/m.

At elevation 6.7

Weight of beams on row $A = 101 \times 6.75 \times 2 = 1364\,\text{kg}$.
Weight of beam on row B = 1364 kg.
Weight of transverse beam on rows 1 and 3 = 1050 kg.
Weight of transverse beam on row 2 = 650 kg.
Weight of secondary beams = 2727 kg.
Weight from floor bracings = 3891 kg.
Live load = 14,040 kg. (50% of this considered for seismic load)
Contributing weight from column = 6665 kg.
Weight of exchanger = 75,000 kg.
Contributing weight of longitudinal bracings = 1509 kg.
Total weight = 101,230 kg ≅ 1012 kN.
Mass m_1 = 103 kN s^2/m.

At elevation 0.0

Wt of raft = $16.5 \times 8.2 \times 0.75 \times 25 = 2537\,\text{kN}$.
Equipment load on raft = 1353 kN.
Load from bracing (50% contribution) and sundry other loads = 720.
Total load at foundation level = 4610 kN.
Mass at foundation = 470 kN m^2/s.
Mass moment of inertia in transverse direction $= \frac{m}{12}(B^2 + d^2) = \frac{470}{12}\left[(8.2)^2 + (0.75)^2\right] = 2656$ kN m/s^2.

Calculation of stiffness in transverse direction

At EL 14.2
I_{xx} of columns = 6125 cm^2.

$$k_2 = \frac{12EI_{xx}}{L^3} \times N = \frac{12 \times 2 \times 10^8 \times 6.125 \times 10^{-5}}{(7.5)^3} \times 6 = 2091\,\text{kN/m}$$

At El 6.7

$$k_1 = \frac{12EI_{xx}}{L^3} \times N = \frac{12 \times 2 \times 10^8 \times 6.125 \times 10^{-5}}{(6.7)^3} \times 6 = 2933\,\text{kN/m}$$

Considering the critical damping as $c_c = 2\sqrt{km}$ and ζ the damping ratio of 5% for each mode, we have $c_1 = 55\,\text{kN m/s}$ and $c_2 = 26\,\text{kN m/s}$ which gives

Thus, the fixed-base problem can be represented as shown in Fig. 7.12.

The free vibration equation can be expressed as

$$\begin{bmatrix} m_1 & 0 \\ 0 & m_2 \end{bmatrix} \begin{Bmatrix} \ddot{u}_1 \\ \ddot{u}_2 \end{Bmatrix} + \begin{bmatrix} c_1 + c_2 & -c_2 \\ -c_2 & -c_2 \end{bmatrix} \begin{Bmatrix} \dot{u}_1 \\ \dot{u}_2 \end{Bmatrix} + \begin{bmatrix} k_1 + k_2 & -k_2 \\ -k_2 & k_2 \end{bmatrix} \begin{Bmatrix} u_1 \\ u_2 \end{Bmatrix} = 0$$

$$\begin{bmatrix} 103 & 0 \\ 0 & 32 \end{bmatrix} \begin{Bmatrix} \ddot{u}_1 \\ \ddot{u}_2 \end{Bmatrix} + \begin{bmatrix} 81 & -26 \\ -26 & 26 \end{bmatrix} \begin{Bmatrix} \dot{u}_1 \\ \dot{u}_2 \end{Bmatrix} + \begin{bmatrix} 5024 & -2091 \\ -2091 & 2091 \end{bmatrix} \begin{Bmatrix} u_1 \\ u_2 \end{Bmatrix} = 0$$

For $[K] - [M]\omega^2 = 0$ and considering $T = 2\pi/\omega$, we finally have $T_1 = 1.415$ s and $T_2 = 0.647$ s, the corresponding eigenvectors are

$$[\phi] = \begin{bmatrix} 0.698 & -0.445 \\ 1 & 1 \end{bmatrix}$$

For the site to be zone IV, $Z = 0.24$, $I = 1.5$, and for steel frame $R = 4.0$

$$[\beta] = ZI/2R = 0.045$$

Fig. 7.12 Fixed mathematical model of the structure

For the first mode

$$\{\phi_1\} = \begin{Bmatrix} 0.698 \\ 1 \end{Bmatrix}, \quad T_1 = 1.415 \text{ s}, \quad \beta = 0.045$$

$$L_1 = \{\phi_1\}^{\mathrm{T}}[M]\{I\} = \{0.698 \quad 1\}\begin{bmatrix} 103 & 0 \\ 0 & 32 \end{bmatrix}\begin{Bmatrix} 1 \\ 1 \end{Bmatrix} = 103.894$$

$$M_1 = \{\phi_1\}^{\mathrm{T}}[M]\{\phi_1\} = \{0.698 \quad 1\}\begin{bmatrix} 103 & 0 \\ 0 & 32 \end{bmatrix}\begin{Bmatrix} 0.698 \\ 1 \end{Bmatrix} = 82.182$$

$$\zeta_1 = \left[\frac{1}{M_1}\right]\frac{\{\phi_1\}[C]\{\phi_1\}}{2\omega_1} \rightarrow 0.039.$$

Thus, for 4% damping $S_{a1} = (1.67/T_1) \times 1.1 \times 9.81 = 12.736\,\text{m/s}^2$.

$$u_1 = \{\phi_1\} \times \left[\frac{L_1}{M_1}\right] \times \beta \times S_{a1} \times \left(\frac{T_1^2}{4\pi^2}\right) = \begin{Bmatrix} 0.026 \\ 0.037 \end{Bmatrix} \text{m}$$

Thus, $\{P_1\} = [K]\{u_1\} = \begin{bmatrix} 5024 & -2091 \\ -2091 & 2091 \end{bmatrix}\begin{Bmatrix} 0.026 \\ 0.037 \end{Bmatrix} = \begin{Bmatrix} 52.02 \\ 23.20 \end{Bmatrix} \text{kN}.$

For the second mode

$$\{\phi_2\} = \begin{Bmatrix} -0.445 \\ 1 \end{Bmatrix}, \quad T_2 = 0.647 \text{ s}, \quad \beta = 0.045$$

$$L_2 = \{\phi_2\}^{\mathrm{T}}[M]\{I\} = \{-0.445 \quad 1\}\begin{bmatrix} 103 & 0 \\ 0 & 32 \end{bmatrix}\begin{Bmatrix} 1 \\ 1 \end{Bmatrix} = -13.835$$

$$M_2 = \{\phi_2\}^{\mathrm{T}}[M]\{\phi_2\} = \{-0.445 \quad 1\}\begin{bmatrix} 103 & 0 \\ 0 & 32 \end{bmatrix}\begin{Bmatrix} -0.445 \\ 1 \end{Bmatrix} = 52.397$$

$$\zeta_2 = \left[\frac{1}{M_2}\right]\frac{\{\phi_2\}[C]\{\phi_2\}}{2\omega_2} \rightarrow 0.055$$

Thus, for 5% damping, $S_{a2} = 2.5 \times 9.81 = 24.525\,\text{m/s}^2$.

$$u_2 = \{\phi_2\} \times \left[\frac{L_2}{M_2}\right] \times \beta \times S_{a2} \times \left(\frac{T_2^2}{4\pi^2}\right) = \begin{Bmatrix} 1.375 \times 10^{-3} \\ -3.09 \times 10^{-3} \end{Bmatrix} \text{m}$$

Thus,

$$\{P_2\} = [K]\{u_2\} = \begin{bmatrix} 5024 & -2091 \\ -2091 & 2091 \end{bmatrix}\begin{Bmatrix} 1.375 \times 10^{-3} \\ -3.09 \times 10^{-3} \end{Bmatrix} = \begin{Bmatrix} 13.37 \\ -9.336 \end{Bmatrix} \text{kN}.$$

Considering soil–structure interaction in transverse direction

The equivalent radius of the raft in translation = 6.56 m.
The equivalent radius of the raft in rocking = 5.57 m.
Dynamic shear modulus of soil $G = (18/9.81) \times (180)^2 = 59,450\,\text{kN/m}^2$.
Poisson's ratio of soil = 0.4.

$$k_x = \frac{8Gr_x}{2-\nu} = \frac{8 \times 59,450 \times 6.56}{2-0.4} = 1,949,960\,\text{kN/m}$$

$$k_\theta = \frac{8Gr_\theta^3}{3(1-\nu)} = \frac{8 \times 59,450 \times (5.57)^3}{1.8} = 4.566 \times 10^7\,\text{kN/rad.}$$

$$c_x = 3.054\sqrt{\frac{G\rho r_x^4}{1-\nu}} = 5.064 \times 10^4\,\text{kN s/m}$$

$$c_\theta = 1.33\sqrt{\frac{G\rho r_\theta^8}{2-\nu}} = 3.343 \times 10^5\,\text{kN s/rad.}$$

Vide Eq. (7.8.4), and the equation of equilibrium can be expressed as

$$\begin{bmatrix} m_f + \sum_{i=1}^{2} m_i & \sum_{i=1}^{2} m_i h_i & m_1 & m_2 \\ \sum_{i=1}^{2} m_i h_i & J_\theta + \sum_{i=1}^{2} m_i h_i^2 & m_1 h_1 & m_2 h_2 \\ m_1 & m_1 h_1 & m_1 & 0 \\ m_2 & m_2 h_2 & 0 & m_2 \end{bmatrix} \begin{Bmatrix} \ddot{u}_f \\ \ddot{\theta} \\ \ddot{u}_1 \\ \ddot{u}_2 \end{Bmatrix} + \begin{bmatrix} c_x & 0 & 0 & 0 \\ 0 & c_\theta & 0 & 0 \\ 0 & 0 & c_1 + c_2 & -c_2 \\ 0 & 0 & -c_2 & c_2 \end{bmatrix} \begin{Bmatrix} \dot{u}_f \\ \dot{\theta} \\ \dot{u}_1 \\ \dot{u}_2 \end{Bmatrix}$$
$$+ \begin{bmatrix} k_x & 0 & 0 & 0 \\ 0 & k_\theta & 0 & 0 \\ 0 & 0 & k_1 + k_2 & -k_2 \\ 0 & 0 & -k_2 & k_2 \end{bmatrix} \begin{Bmatrix} u_f \\ \theta \\ u_1 \\ u_2 \end{Bmatrix} = 0$$

This gives

$$\begin{bmatrix} 605 & 1246 & 103 & 32 \\ 1246 & 15520 & 767.35 & 478.4 \\ 103 & 767.35 & 103 & 0 \\ 32 & 478.4 & 0 & 32 \end{bmatrix} \begin{Bmatrix} \ddot{u}_f \\ \ddot{\theta} \\ \ddot{u}_1 \\ \ddot{u}_2 \end{Bmatrix} + \begin{bmatrix} 56040 & 0 & 0 & 0 \\ 0 & 334300 & 0 & 0 \\ 0 & 0 & 81 & -26 \\ 0 & 0 & -26 & 26 \end{bmatrix} \begin{Bmatrix} \dot{u}_f \\ \dot{\theta} \\ \dot{u}_1 \\ \dot{u}_2 \end{Bmatrix}$$
$$+ \begin{bmatrix} 1.95 \times 10^6 & 0 & 0 & 0 \\ 0 & 4.566 \times 10^7 & 0 & 0 \\ 0 & 0 & 5024 & -2091 \\ 0 & 0 & -2091 & 2091 \end{bmatrix} \begin{Bmatrix} u_f \\ \theta \\ u_1 \\ u_2 \end{Bmatrix} = 0$$

For $[K] - [M]\omega^2 = 0$ and considering $T = 2\pi/\omega$, we finally have

$T = \{1.42, 0.647, 0.097, 0.048\}$ s and the corresponding eigenvectors are

$$[\varphi] = \begin{bmatrix} -1.055 \times 10^{-3} & -6.905 \times 10^{-4} & -0.998 & 2.362 \times 10^{-4} \\ -4.383 \times 10^{-3} & 2.844 \times 10^{-4} & 6.257 \times 10^{-4} & 0.067 \\ -0.698 & -0.449 & 1 & -0.498 \\ -1 & 1 & 0.988 & -1 \end{bmatrix}$$

For the first mode

$$T_1 = 1.42\,\text{s, and } \beta = 0.045$$
$$\{\phi_1\}^{\mathrm{T}} = \{-1.055 \times 10^{-3} \quad -4.383 \times 10^{-3} \quad -0.698 \quad -1\}$$

$$L_1 = \{\phi_1\}^{\mathrm{T}}[M]\{I\} = \{-1.055 \times 10^{-3} \quad -4.383 \times 10^{-3} \quad -0.698 \quad -1\}$$
$$\times \begin{bmatrix} 605 & 1246 & 103 & 32 \\ 1246 & 15520 & 767.35 & 478.4 \\ 103 & 767.35 & 103 & 0 \\ 32 & 478.4 & 0 & 32 \end{bmatrix} \begin{Bmatrix} 1 \\ 1 \\ 1 \\ 1 \end{Bmatrix} = -1.232 \times 10^3$$
$$M_1 = \{\phi_1\}^{\mathrm{T}}[M]\{\phi_1\} = 83.295$$

$D_1 = \frac{\{\phi_1\}^{\mathrm{T}}[C]\{\phi_1\}}{2 \times M_1 \times \omega_1} = 0.039$ here $\omega_1 = 2\pi/T_1$.

Thus, $S_{a1} = (1.67/T_1) \times 9.81 \times 1.1 = 12.7\,\text{m/s}^2$.

$$\{U_1\} = \{\phi_1\} \times \left[\frac{L_1}{M_1}\right] \times \beta \times S_{a1} \times \left(\frac{T_1^2}{4\pi^2}\right) = \begin{Bmatrix} 4.551 \times 10^{\quad 1} \\ 1.891 \times 10^{-4} \\ 0.301 \\ 0.431 \end{Bmatrix}$$

Thus, the structural deformation in the frame is $u_1 = \begin{Bmatrix} 0.301 \\ 0.431 \end{Bmatrix}$ m.

Thus, the nodal force in frame is

$$P_1 = \begin{bmatrix} 5024 & -2091 \\ -2091 & 2091 \end{bmatrix} \begin{Bmatrix} 0.301 \\ 0.431 \end{Bmatrix} = \begin{Bmatrix} 611 \\ 272 \end{Bmatrix} \text{kN.}$$

If we go from the inertial side, as stated in problem 7.3,

$$P_1 = [M]\{\phi_1\}\left(\frac{L_1}{M_1}\right)\beta S_{a1}\left(\frac{T_{\text{DSSI}}}{T_{\text{fixed}}}\right)^2$$

Substituting the appropriate values, we have $P_1 = \begin{Bmatrix} 893.711 \\ 8693 \\ 615 \\ 274 \end{Bmatrix}$.

Three issues need to be observed here.

1. The last two rows give the nodal force in frames which is very closely matching the stiffness basis force, and this should again match to maintain equilibrium.
2. The first row gives sum of nodal forces, i.e., the base shear getting transferred to the foundation.
3. The second row gives moment transferred to foundation as $\sum_1^2 P_i h_i$.

So extracting the response based on stiffness equation does not alter the problem basis and is more efficient computationally.

For the second mode

$T_2 = 0.67$ s, and $\beta = 0.045$

$$\{\phi_2\}^{\mathrm{T}} = \{-6.905 \times 10^{-4} \quad -2.844 \times 10^{-4} \quad -0.449 \quad 1\}$$

$$L_2 = \{\phi_2\}^{\mathrm{T}}[M]\{I\} = 109.119$$
$$M_2 = \{\phi_2\}^{\mathrm{T}}[M]\{\phi_2\} = 52.862$$
$$D_2 = \frac{\{\phi_2\}^{\mathrm{T}}[C]\{\phi_2\}}{2 \times M_2 \times \omega_2} = 0.055$$

Thus, $S_{a2} = 2.5 \times 9.81 = 24.525\ \mathrm{m/s^2}$.

$$\{U_2\} = \{\phi_2\} \times \left[\frac{L_2}{M_2}\right] \times \beta \times S_{a2} \times \left(\frac{T_2^2}{4\pi^2}\right) = \begin{Bmatrix} -1.668 \times 10^{-5} \\ 6.87 \times 10^{-6} \\ -0.011 \\ 0.024 \end{Bmatrix}$$

Thus, the structural deformation in the frame is $u_2 = \begin{Bmatrix} -0.011 \\ 0.024 \end{Bmatrix}$ m.

Thus, the nodal force in the frame is

$$P_2 = \begin{bmatrix} 5024 & -2091 \\ -2091 & 2091 \end{bmatrix} \begin{Bmatrix} -0.011 \\ 0.024 \end{Bmatrix} = \begin{Bmatrix} -105.45 \\ 73.185 \end{Bmatrix} \mathrm{kN}.$$

For the third mode

$T_3 = 0.097$ s, and $\beta = 0.045$

$$\{\phi_3\}^{\mathrm{T}} = \{-0.998 \quad 6.257 \times 10^{-4} \quad 1 \quad 0.988\}$$

$$L_3 = \{\phi_3\}^{\mathrm{T}}[M]\{I\} = -461.264$$
$$M_3 = \{\phi_3\}^{\mathrm{T}}[M]\{\phi_3\} = 468.13$$
$$D_3 = \frac{\{\phi_3\}^{\mathrm{T}}[C]\{\phi_3\}}{2 \times M_3 \times \omega_3} = 0.926$$

Thus, $S_{a3} = 2.4555 \times 9.81 \times 0.5 = 12.042\,\mathrm{m/s^2}$.

$$\{U_3\} = \{\phi_3\} \times \left[\frac{L_3}{M_3}\right] \times \beta \times S_{a3} \times \left(\frac{T_3^2}{4\pi^2}\right) = \begin{Bmatrix} 2.032 \times 10^{-4} \\ -1.274 \times 10^{-7} \\ -2.036 \times 10^{-4} \\ -2.012 \times 10^{-4} \end{Bmatrix}$$

Thus, the structural deformation in the frame is $u_3 = \begin{Bmatrix} -2.036 \times 10^{-4} \\ -2.012 \times 10^{-4} \end{Bmatrix}$ m.

Thus, nodal force in the frame is

$$P_3 = \begin{bmatrix} 5024 & -2091 \\ -2091 & 2091 \end{bmatrix} \begin{Bmatrix} -2.036 \times 10^{-4} \\ -2.012 \times 10^{-4} \end{Bmatrix} = \begin{Bmatrix} -0.602 \\ 5.018 \times 10^{-3} \end{Bmatrix} \mathrm{kN}$$

For the fourth mode

Proceeding in identical manner as shown above for first three modes, we have

$$\{U_4\} = \begin{Bmatrix} 7.92 \times 10^{-8} \\ 2.247 \times 10^{-5} \\ -1.67 \times 10^{-4} \\ -3.353 \times 10^{-4} \end{Bmatrix}, \text{ and this gives } P_4 = \begin{Bmatrix} 0.617 \\ -0.666 \end{Bmatrix} \mathrm{kN.}$$

Frame in long direction, fixed-base analysis

Stiffness of each bracing $= \frac{AE}{L}\cos^2\theta = \frac{3.03 \times 10^{-3} \times 2 \times 10^8}{8.2} = 12,447\,\mathrm{kN/m}$ at EL 14.2.

I_{yy} for main beam = 2065 cm^2 at EL 14.2.
Thus, the stiffness for the top panel is

$$\begin{aligned} k_2 &= \frac{12EI_{yy}}{L^3} \times 6 + \frac{AE}{L}\cos^2\theta \times 8 \\ &= \frac{12 \times 2 \times 10^8 \times 2.065 \times 10^{-5}}{(7.5)^3} + 8 \times 12,447 \\ &= 100,281\,\mathrm{kN/m} \end{aligned}$$

Similarly, for the bottom panel

$$
\begin{aligned}
k_1 &= \frac{12EI_{yy}}{L^3} \times 6 + \frac{AE}{L}\cos^2\theta \times 8 \\
&= \frac{12 \times 2 \times 10^8 \times 2.065 \times 10^{-5}}{(6.7)^3} + 8 \times 16,358 \\
&= 131,853\,\text{kN/m}
\end{aligned}
$$

Considering 5% damping for structure

$$
\begin{aligned}
c_1 &= 2\sqrt{k_1 m_1} \times 0.05 = 369\,\text{kN s/m} \quad \text{and} \\
c_2 &= 2\sqrt{k_2 m_2} \times 0.05 = 179\,\text{kN s/m}
\end{aligned}
$$

Thus, the free vibration equation is expressed as

$$
\begin{bmatrix} 103 & 0 \\ 0 & 32 \end{bmatrix} \begin{Bmatrix} \ddot{v}_1 \\ \ddot{v}_2 \end{Bmatrix} + \begin{bmatrix} 548 & -179 \\ -179 & 179 \end{bmatrix} \begin{Bmatrix} \dot{v}_1 \\ \dot{v}_2 \end{Bmatrix} + \begin{bmatrix} 232100 & -100281 \\ -100281 & 100281 \end{bmatrix} \begin{Bmatrix} v_1 \\ v_2 \end{Bmatrix} = 0
$$

For $[K] - [M]\omega^2 = 0$ and considering $T = 2\pi/\omega$, we finally have $T_1 = 0.21$ s and $T_2 = 0.094$ s, and the corresponding eigenvectors are

$$
[\phi] = \begin{bmatrix} -0.582 & 0.398 \\ -0.813 & -0.917 \end{bmatrix}
$$

For the site to be zone IV, $Z = 0.24$, $I = 1.5$ and for steel frame $R = 4.0$

$$
[\beta] = ZI/2R = 0.045
$$

For the first mode

$$
\begin{aligned}
\{\phi_1\} &= \begin{Bmatrix} 0.582 \\ 0.813 \end{Bmatrix}, \quad T_1 = 0.21\,\text{s}, \quad \beta = 0.045 \\
L_1 &= \{\phi_1\}^{\mathrm{T}}[M]\{I\} = -85.962 \\
M_1 &= \{\phi_1\}^{\mathrm{T}}[M]\{\phi_1\} = 56.04 \\
\zeta_1 &= \left[\frac{1}{M_1}\right] \frac{\{\phi_1\}[C]\{\phi_1\}}{2\omega_1} \to 0.04
\end{aligned}
$$

Thus, for $S_{a1} = 2.5 \times 9.81 = 24.525\,\text{m/s}^2$.

$$u_1 = \{\phi_1\} \times \left[\frac{L_1}{M_1}\right] \times \beta \times S_{a1} \times \left(\frac{T_1^2}{4\pi^2}\right) = \begin{Bmatrix} 1.101 \times 10^{-3} \\ 1.537 \times 10^{-3} \end{Bmatrix} \text{m}$$

Thus,

$$\{P_1\} = [K]\{u_1\} = \begin{bmatrix} 232100 & -100281 \\ -100281 & 100281 \end{bmatrix} \begin{Bmatrix} 1.101 \times 10^{-3} \\ 1.537 \times 10^{-3} \end{Bmatrix} = \begin{Bmatrix} 101.3 \\ 43.807 \end{Bmatrix} \text{kN.}$$

For the second mode

$$\{\phi_2\} = \begin{Bmatrix} 0.398 \\ -0.917 \end{Bmatrix}, \quad T_1 = 0.094\text{s}, \quad \beta = 0.045$$

$$L_2 = \{\phi_2\}^{\mathrm{T}}[M]\{I\} = 11.65$$

$$M_2 = \{\phi_2\}^{\mathrm{T}}[M]\{\phi_{12}\} = 43.224$$

$$\zeta_2 = \left[\frac{1}{M_2}\right] \frac{\{\phi_2\}[C]\{\phi_2\}}{2\omega_2} \rightarrow 0.064$$

Thus, for 5% damping $S_{a2} = 2.41 \times 9.81 = 23.64\,\text{m/s}^2$.

$$u_2 = \{\phi_2\} \times \left[\frac{L_2}{M_2}\right] \times \beta \times S_{a2} \times \left(\frac{T_2^2}{4\pi^2}\right) = \begin{Bmatrix} 2.554 \times 10^{-5} \\ -5.885 \times 10^{-5} \end{Bmatrix} \text{m}$$

Thus, $\{P_1\} = [K]\{u_1\} = \begin{bmatrix} 232100 & -100281 \\ -100281 & 100281 \end{bmatrix} \begin{Bmatrix} 2.554 \times 10^{-5} \\ -5.885 \times 10^{-5} \end{Bmatrix} =$

$$\begin{Bmatrix} 11.831 \\ -8.463 \end{Bmatrix} \text{kN}$$

Considering soil–structure interaction in longitudinal direction

Equivalent radius of the raft in translation = 6.56 m.
Equivalent radius of the raft in rocking = 7.907 m.
The dynamic shear modulus of soil $G = 59,450\,\text{kN/m}^2$.
Poisson's ratio of soil = 0.4.

$$k_x = \frac{8Gr_x}{2 - v} = \frac{8 \times 59,450 \times 6.56}{2 - 0.4} = 1,949,960\,\text{kN/m}$$

$$k_\theta = \frac{8Gr_\theta^3}{3(1 - v)} = \frac{8 \times 59,450 \times (7.907)^3}{1.8} = 2.612 \times 10^8\,\text{kN/rad.}$$

$$c_x = 3.054\sqrt{\frac{G\rho r_x^4}{1-\nu}} = 5.064 \times 10^4 \text{ kN s/m}$$

$$c_\theta = 1.33\sqrt{\frac{G\rho r_\theta^8}{2-\nu}} = 1.357 \times 10^6 \text{ kN s/rad.}$$

As earlier, the equation of equilibrium can be expressed as

$$\begin{bmatrix} 605 & 1246 & 103 & 32 \\ 1246 & 23550 & 767.35 & 478.4 \\ 103 & 767.35 & 103 & 0 \\ 32 & 478.4 & 0 & 32 \end{bmatrix} \begin{Bmatrix} \ddot{u}_f \\ \ddot{\theta} \\ \ddot{u}_1 \\ \ddot{u}_2 \end{Bmatrix} + \begin{bmatrix} 50640 & 0 & 0 & 0 \\ 0 & 1.357 \times 10^6 & 0 & 0 \\ 0 & 0 & 548 & -179 \\ 0 & 0 & -179 & 179 \end{bmatrix} \begin{Bmatrix} \dot{u}_f \\ \dot{\theta} \\ \dot{u}_1 \\ \dot{u}_2 \end{Bmatrix}$$
$$+ \begin{bmatrix} 1.95 \times 10^6 & 0 & 0 & 0 \\ 0 & 2.612 \times 10^8 & 0 & 0 \\ 0 & 0 & 232100 & -100300 \\ 0 & 0 & -100300 & 100300 \end{bmatrix} \begin{Bmatrix} u_f \\ \theta \\ u_1 \\ u_2 \end{Bmatrix} = 0$$

For $[K] - [M]\omega^2 = 0$ and considering $T = 2\pi/\omega$, we finally have

$T = \{0.222,\ \ 0.098,\ \ 0.09,\ \ 0.039\}$ s and the corresponding eigenvectors are

$$[\varphi] = \begin{bmatrix} 0.049 & -0.231 & -0.316 & -5.547 \times 10^{-3} \\ 2.947 \times 10^{-3} & 3.297 \times 10^{-3} & -9.816 \times 10^{-4} & -0.056 \\ 0.582 & -0.057 & 0.796 & 0.427 \\ 0.812 & 0.971 & -0.517 & 0.903 \end{bmatrix}$$

For the first mode

$$T_1 = 0.222\,\text{s}, \quad \text{and} \quad \beta = 0.045$$
$$\{\phi_1\}^{\mathrm{T}} = \{0.049 \quad -2.947 \times 10^{-3} \quad 0.582 \quad 0.812\}$$
$$L_1 = \{\phi_1\}^{\mathrm{T}}[M]\{I\} = 1.181 \times 10^3 M_1 = \{\phi_1\}^T[M]\{\phi_1\} = 71.35$$
$$D_1 = \frac{\{\phi_1\}^{\mathrm{T}}[C]\{\phi_1\}}{2 \times M_1 \times \omega_1} = 0.07$$

Thus, $S_{a1} = 2.5 \times 9.81 \times 0.9 = 22.073\,\text{m/s}^2$.

$$\{U_1\} = \{\phi_1\} \times \left[\frac{L_1}{M_1}\right] \times \beta \times S_{a1} \times \left(\frac{T_1^2}{4\pi^2}\right) = \begin{Bmatrix} 1.006 \times 10^{-3} \\ 6.049 \times 10^{-5} \\ 0.012 \\ 0.017 \end{Bmatrix}$$

Thus, the structural deformation in the frame is $\{u_1\} = \begin{Bmatrix} 0.012 \\ 0.017 \end{Bmatrix}$ m.

Thus, the nodal force in frame is

$$P_1 = \begin{bmatrix} 232100 & -100300 \\ -100300 & 100300 \end{bmatrix} \begin{Bmatrix} 0.012 \\ 0.017 \end{Bmatrix} = \begin{Bmatrix} 1081 \\ 501.4 \end{Bmatrix} \text{kN}$$

.

For the second mode

$$\begin{aligned} T_2 &= 0.098\,\text{s}, \quad \text{and} \quad \beta = 0.045 \\ \{\phi_2\}^{\mathrm{T}} &= \{-0.231 \quad -3.927 \times 10^{-3} \quad -0.057 \quad 0.971\} \\ L_2 &= \{\phi_2\}^{\mathrm{T}}[M]\{I\} = 114.761 \\ M_2 &= \{\phi_2\}^{\mathrm{T}}[M]\{\phi_2\} = 52.554 \\ D_2 &= \frac{\{\phi_2\}^{\mathrm{T}}[C]\{\phi_2\}}{2 \times M_2 \times \omega_2} = 0.474 \end{aligned}$$

Thus, $S_{a2} = 2.47 \times 9.81 \times 0.5 = 11.53\,\text{m/s}^2$.

$$\{U_2\} = \{\phi_2\} \times \left[\frac{L_2}{M_2}\right] \times \beta \times S_{a2} \times \left(\frac{T_2^2}{4\pi^2}\right) = \begin{Bmatrix} -6.69 \times 10^{-5} \\ 1.137 \times 10^{-6} \\ -1.651 \times 10^{-5} \\ 2.812 \times 10^{-4} \end{Bmatrix}$$

Thus, the structural deformation in the frame is $\{u_2\} = \begin{Bmatrix} -1.651 \times 10^{-5} \\ 2.812 \times 10^{-4} \end{Bmatrix}$ m.

Thus, the nodal force in the frame is

$$P_2 = \begin{bmatrix} 232100 & -100300 \\ -100300 & 100300 \end{bmatrix} \begin{Bmatrix} -1.651 \times 10^{-5} \\ 2.812 \times 10^{-4} \end{Bmatrix} = \begin{Bmatrix} -32.03 \\ 29.855 \end{Bmatrix} \text{kN}$$

.

For the third mode

$$T_3 = 0.09\text{s}, \quad \text{and} \quad \beta = 0.045$$
$$\{\phi_3\}^{\mathrm{T}} = \{-0.316 \quad -9.816 \times 10^{-4} \quad 0.796 \quad 0.517\}$$
$$L_3 = \{\phi_3\}^{\mathrm{T}}[M]\{I\} = -158.697$$
$$M_3 = \{\phi_3\}^{\mathrm{T}}[M]\{\phi_3\} = 92.95$$
$$D_3 = \frac{\{\phi_3\}^{\mathrm{T}}[C]\{\phi_3\}}{2 \times M_3 \times \omega_3} = 0.474$$

Thus, $S_{a3} = 2.35 \times 9.81 \times 0.5 = 11.53\,\text{m/s}^2$.

$$\{U_3\} = \{\phi_3\} \times \left[\frac{L_3}{M_3}\right] \times \beta \times S_{a3} \times \left(\frac{T_3^2}{4\pi^2}\right) = \begin{Bmatrix} 4.197 \times 10^{-5} \\ -7.136 \times 10^{-7} \\ -1.036 \times 10^{-5} \\ -1.764 \times 10^{-4} \end{Bmatrix}$$

Thus, the structural deformation in the frame is $\{u_3\} = \begin{Bmatrix} 1.036 \times 10^{-5} \\ -1.764 \times 10^{-4} \end{Bmatrix}$ m.

Thus, the nodal force in the frame is
$$P_2 = \begin{bmatrix} 232100 & -100300 \\ -100300 & 100300 \end{bmatrix} \begin{Bmatrix} 1.036 \times 10^{-5} \\ -1.764 \times 10^{-4} \end{Bmatrix} = \begin{Bmatrix} 20.094 \\ -18.728 \end{Bmatrix} \text{kN}.$$

For the fourth mode

Proceeding in identical manner as shown, we have

$$\{U_4\} = \begin{Bmatrix} 1.254 \times 10^{-6} \\ 1.266 \times 10^{-5} \\ -9.652 \times 10^{-5} \\ -2.041 \times 10^{-4} \end{Bmatrix}, \text{ and this gives } P_4 = \begin{Bmatrix} -1.938 \\ -10.788 \end{Bmatrix} \text{kN}.$$

We tabulate hereafter nodal force in frames with fixed-base case and soil–structure interaction

Frame along	Row 1 and 3		Row 2		Row A and B		
	Fixed-base	DSSI	Fixed-base	DSSI	Fixed-base	DSSI	Mode
El 14.2	5.8	69	11.6	137	22	251	1
EL 6.7	13.0	154	26.01	307	51	540	1
El 14.2	−2.33	18.3	−4.66	37	−4.2	15	2
El 6.7	3.34	−26.5	6.68	−53	5.9	−16	2
El 14.2		0.001		0.002		−9.35	3
El 6.7		−0.150		−0.301		10	3
El 14.2		0.165		0.333		−5.39	4
El 6.7		0.15		−0.301		−1	4

7.9 DSSI Model for Foundations that Are Massive and Embedded in Ground

Till now, we were discussing foundations that are mostly isolated or of strip footing type or rafts of shallow depth. For heavy structures, in many cases we have massive foundations that are embedded within the soil, when dynamic soil–structure interaction gets significantly complicated. Some the examples are as shown in Fig. 7.13.

Shown in Fig. 7.13 are examples of foundations that are massive and embedded in ground:

- A high-rise building, with basement extending well below ground;
- A 275-m-tall chimney, supported on a raft, whose thickness would be of about 3–4 m;
- A reactor building of a nuclear power plant, resting on a raft, 2.5–3 m in depth, embedded in ground. There are also cases when a part of the external wall, including the foundation, is embedded within the ground usually up to the height of breeder reactor located on the building floor.

In such cases, the geometry of the foundation cannot be ignored as it plays a significant role in the interaction and usually generates coupled mode in translation and rocking. Modeling complete soil, foundation, and superstructure by FEM makes the analysis prohibitively expensive. For leaving aside modeling the foundation in 3D or 2D, the soil has to be taken to a substantial distance away from the foundation to ensure waves are not reflected back resulting in spurious modes. It is not at all surprising to arrive at such DSSI finite element models, having half a

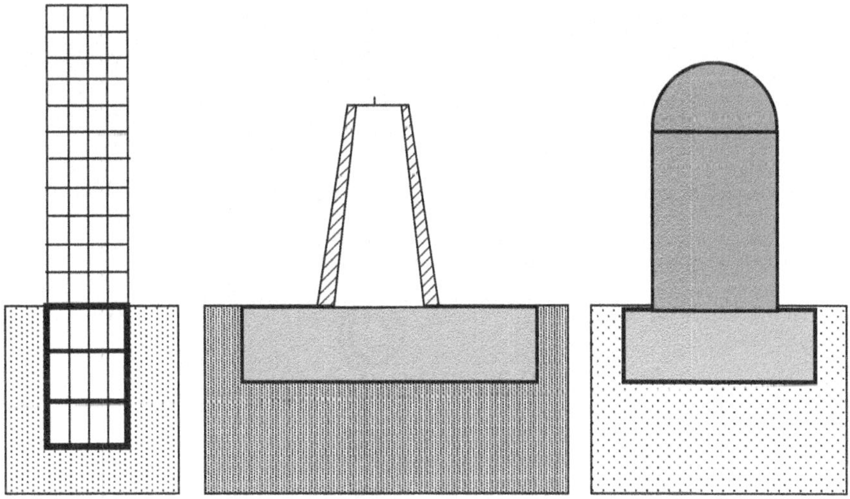

Fig. 7.13 Examples of massive foundations embedded in ground

million degrees of freedom. This makes the analysis certainly expensive and laborious, and an engineer under project time pressure would definitely be reluctant to carry out such analysis. Easy way-out, of course, would be to ignore soil and carry out a fixed-base analysis.

Finally, research has shown that when a foundation is embedded in soil, both stiffness and damping become highly frequency-dependent. Now, what is this frequency guiding the dimensionless number $a_0 = (\omega \times r)/V_s$, there is a significant confusion among many engineers while solving an earthquake problem.

To circumvent all these difficulties and also to clear up many uncertainties, we propose hereafter a model (Chowdhury et al. 2015a) which is basically based on substructure technique.

7.9.1 Model 1—Shallow Foundation

Shown in Fig. 7.14 is raft foundation resting on ground whose stiffness and damping in translational and rotational mode are expressed as K_x, C_x, K_θ, C_θ, respectively. Let the mass and mass moment of inertia of the foundation be expressed by m_f and J_θ. The foundation is subjected to a shear V_i and moment M_i, due to earthquake and vertical load P form superstructure where i is the mode number of vibrations, i = 1, 2, 3, …, etc.

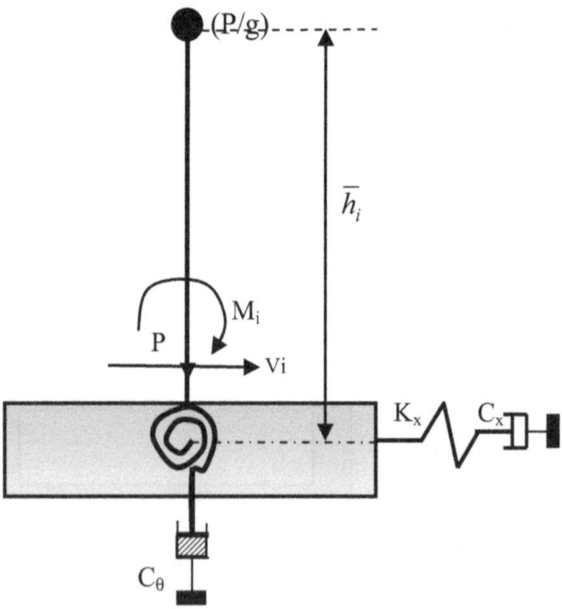

Fig. 7.14 A shallow foundation subjected to earthquake load

We consider the foundation as large, but its depth is shallow, say, of about 1–1.2 m, resting on ground. We ignore the coupling effect foundation in translation and rocking mode, assuming the depth is shallow compared to its length and width.

The equilibrium equations in such a case can obviously be expressed as

$$\left(m_f + P/g\right)\ddot{x} + C_x\dot{x} + K_x x = V_i \tag{7.9.1}$$

and

$$\left[J_\theta + (P/g)\bar{h}^2\right]\ddot{\theta} + C_\theta\dot{\theta} + K_\theta x = M_i \tag{7.9.2}$$

Solution to Eq. (7.9.1) consists of a complimentary function and a particular integral, and this is expressed as

$$x_i = \mathrm{e}^{-\zeta_x\omega_x t}\left(A\ \cos\omega_x\sqrt{1-\zeta_x^2} + B\ \sin\omega_x\sqrt{1-\zeta_{x2}t}\right) + V_i/K_x \tag{7.9.3}$$

Now considering the boundary conditions (i) at $t = 0$, $x = 0$ and (ii) at $t = 0$ $\mathrm{d}x/\mathrm{d}t = S_d\omega_x$, it can be shown that

$$x_i = \frac{V_i}{K_x}\left(1 + \mathrm{e}^{-\zeta_x\omega_x t}\left\{\sqrt{\frac{1-\zeta_x}{1+\zeta_x}}\sin\omega_x\sqrt{1-\zeta_x^2 t} - \cos\omega_x\sqrt{1-\zeta_x^2 t}\right\}\right) \tag{7.9.4}$$

Now, one should remember here that V_i *is actually a frequency-dependent term and is a function of modal frequency of the superstructure.* Based on this, Eq. (7.9.4) can be expressed as

$$x_i = \frac{V_i}{K_x\sqrt{\left(1-r_x^2\right)^2 + \left(2\zeta_x r_x\right)^2}}\left(1 + \mathrm{e}^{-\zeta_x\omega_x t}\left\{\sqrt{\frac{1-\zeta_x}{1+\zeta_x}}\sin\omega_x\sqrt{1-\zeta_x^2 t} - \cos\omega_x\sqrt{1-\zeta_x^2 t}\right\}\right) \tag{7.9.5}$$

Here, $r_x = T_x/T_i$, where T_x = time period of foundation in translation $T_x = 2\pi\sqrt{\left(m_f + P_i/g\right)/K_x}$ and T_i = fixed-base time period of the structure in *i*th mode; $\zeta_x = C_x/2\sqrt{K_x\left(m_f + P_i/g\right)}$.

In earthquake engineering, considering that we compute everything in time domain, Eq. (7.9.5) can be expressed as

$$x_i = \frac{V_i}{K_x\sqrt{\left(1-r_x^2\right)^2 + \left(2\zeta_x r_x\right)^2}}\left(1 + \mathrm{e}^{-\zeta_x\frac{2\pi}{T_x}t}\left\{\sqrt{\frac{1-\zeta_x}{1+\zeta_x}}\sin 2\pi\sqrt{1-\zeta_x^2}\frac{t}{T_x} - \cos 2\pi\sqrt{1-\zeta_x^2}\frac{t}{T_x}\right\}\right) \tag{7.9.6}$$

Proceeding in identical fashion, it can be shown that

$$\theta_i = \frac{M_i}{K_\theta\sqrt{(1-r_\theta^2)^2+(2\zeta_x r_\theta)^2}}\left(1+e^{-\zeta_\theta\frac{2\pi}{T_\theta}t}\left\{\sqrt{\frac{1-\zeta_\theta}{1+\zeta_\theta}}\sin 2\pi\sqrt{1-\zeta_\theta^2}\frac{t}{T_\theta}-\cos 2\pi\sqrt{1-\zeta_\theta^2}\frac{t}{T_\theta}\right\}\right) \tag{7.9.7}$$

Here, $r_\theta = T_\theta/T_i$ and $T_\theta = 2\pi\sqrt{[J_\theta+(P_i/g)\bar{h}^2]/K_\theta}$ and $\zeta_\theta = C_\theta/2\sqrt{K_\theta(J_\theta+(P_i/g)\bar{h}^2)}$.

Now substituting the above in Eq. (7.5.2), we have

$$\begin{aligned} u_{Ti} = u_{si}\Bigg[1 &+ \frac{V_i}{\widehat{K}_x u_{si}}\left(1+e^{-\zeta_x\frac{2\pi}{T_x}t}\left\{\sqrt{\frac{1-\zeta_x}{1+\zeta_x}}\sin 2\pi\sqrt{1-\zeta_x^2}\frac{t}{T_x}-\cos 2\pi\sqrt{1-\zeta_x^2}\frac{t}{T_x}\right\}\right) \\ &+ \frac{M_i\bar{h}_i}{\widehat{K}_\theta u_{si}}\left(1+e^{-\zeta_\theta\frac{2\pi}{T_\theta}t}\left\{\sqrt{\frac{1-\zeta_\theta}{1+\zeta_\theta}}\sin 2\pi\sqrt{1-\zeta_\theta^2}\frac{t}{T_\theta}-\cos 2\pi\sqrt{1-\zeta_\theta^2}\frac{t}{T_\theta}\right\}\right)\Bigg] \end{aligned} \tag{7.9.8}$$

Here, $\widehat{K}_x = K_x\sqrt{(1-r_x^2)^2+(2\zeta_x r_x)^2}$ and $\widehat{K}_\theta = K_\theta\sqrt{(1-r_\theta^2)^2+(2\zeta_\theta r_\theta)^2}$, and u_{si} = amplitude of the fixed-base structure at height $\bar{h}_i$, Thus, the dynamic amplification AF due to DSSI is expressed as

$$\begin{aligned} \mathrm{AF}_i = \Bigg[1 &+ \frac{V_i}{\widehat{K}_x u_{si}}\left(1+e^{-\zeta_x\frac{2\pi}{T_x}t}\left\{\sqrt{\frac{1-\zeta_x}{1+\zeta_x}}\sin 2\pi\sqrt{1-\zeta_x^2}\frac{t}{T_x}-\cos 2\pi\sqrt{1-\zeta_x^2}\frac{t}{T_x}\right\}\right)+ \\ &+ \frac{M_i\bar{h}_i}{\widehat{K}_\theta u_{si}}\left(1+e^{-\zeta_\theta\frac{2\pi}{T_\theta}t}\left\{\sqrt{\frac{1-\zeta_\theta}{1+\zeta_\theta}}\sin 2\pi\sqrt{1-\zeta_\theta^2}\frac{t}{T_\theta}-\cos 2\pi\sqrt{1-\zeta_\theta^2}\frac{t}{T_\theta}\right\}\right)\Bigg] \end{aligned} \tag{7.9.9}$$

The modal c.g. of the structure can be expressed as

$$\bar{h}_i = \frac{\sum_{i=1}^{N} m_i h_i}{\sum_{i=1}^{N} m_i} \tag{7.9.10}$$

Here, $\{m_i\} = \{P_i/(\beta\times S_{ai})\}$ without sign, as mass is basically a scalar.

Example 7.5 A 400 m^3 capacity reinforced concrete bunker is supported on a three-stage framing, as shown in Fig. 7.15. The weight of empty bunker is 225 kN. The bunker is supported on four 750 mm × 750 mm columns, while the staging beams of 750 × 1200 mm are supporting the bunker and 500 × 750 mm at lower levels. Concrete grade used for bunker is M30. The bunker is resting on a

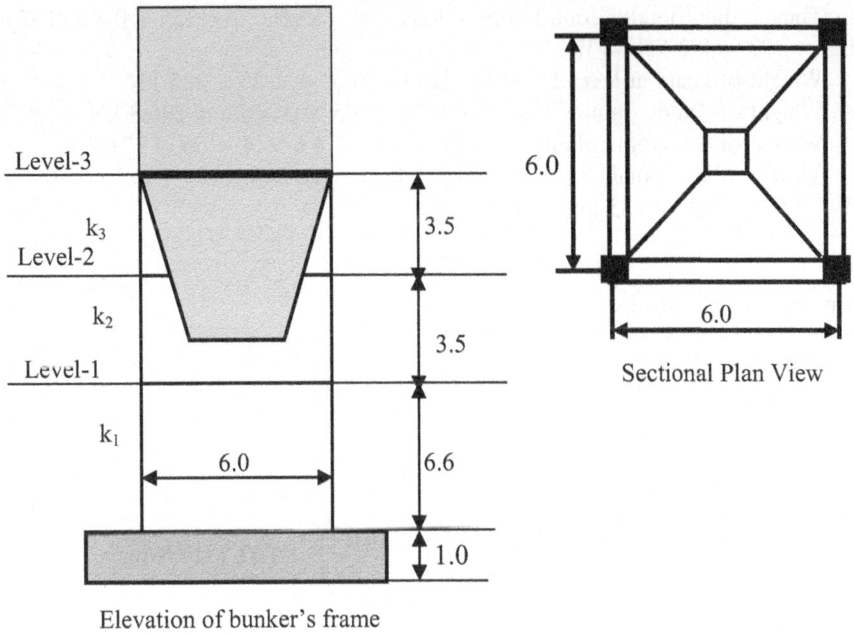

Fig. 7.15 Schematic details of the RCC bunker for analysis

reinforced concrete raft of size 7.5 m × 7.5 m having 1.0 m depth. The shear wave velocity of soil is 155 m/s. Poisson's ratio $\nu = 0.4$. The weight density of soil is 20 kN/m^3. The weight density of material to be stored in bunker is 19.5 kN/m^3. The bunker is constructed at a location which is zone V as per IS code. Carry out modal response analysis, considering:

- Frame as a fixed-base;
- Considering DSSI based on model as proposed in Sect. 7.9.1.

Solution

Calculation of weight

Volume of bunker = 400 m^3, and density of material = 19.5 kN/m^3.

Total storage weight = 400 × 19.5 = 7800 kN.

Weight of 4 supporting beams = 0.75 × 1.2 × 6 × 4 × 25 = 540 kN.

Weight of 4 supporting columns = 0.75 × 0.75 × 3.5 × 4 × 25 = 196.8.

Thus, the total contributing load at level 3 = 7800 + 225 (Self-weight of bunker) + 540 + 0.3 × 197 = 8624 kN. $\approx W_3$

Weight of beam at level 2 = 0.5 × 0.75 × 6 × 4 × 25 = 225 kN.

Weight of 4 supporting columns = 0.75 × 0.75 × 3.5 × 4 × 25 = 196.8 kN.

Thus, the total contributing load at level 2 = 225 + 0.7 × 197 + 0.3 × 197 = 422 kN. $\approx W_2$

Weight of beam at level 1 = 0.5 × 0.75 × 6 × 4 × 25 = 225 kN.

Weight of 4 top columns = 0.75 × 0.75 × 3.5 × 4 × 25 = 196.8 kN.

Weight of 4 bottom columns = 0.75 × 0.75 × 6.6 × 4 × 25 = 371 kN.

Thus, the total contributing load at level 1 = 225 + 0.7 × 197 + 0.3 × 371 = 533 kN. $\approx W_1$

Calculation of stiffness

Moment of inertia of column = $(0.75)^4/12 = 0.0264\,\mathrm{m}^4$.

This $k_3 = 4 \times \left(\frac{12\times 2.74\times 10^7\times 0.0264}{(3.5)^3}\right) = 809,826\,\mathrm{kN/m}$.

$$k_2 = k_3$$

$$k_1 = 4 \times \left(\frac{12 \times 2.74 \times 10^7 \times 0.0264}{(6.6)^3}\right) = 120,771\,\mathrm{kN/m}$$

Formation of stiffness and mass matrix

Thus, the stiffness matrix $[K] = \begin{bmatrix} k_1 + k_2 & -k_2 & 0 \\ -k_2 & k_2 + k_3 & -k_3 \\ 0 & -k_3 & k_3 \end{bmatrix}$ and mass matrix $[M] = \begin{bmatrix} W_1/g & 0 & 0 \\ 0 & W_2/g & 0 \\ 0 & 0 & W_3/g \end{bmatrix}$.

$$\rightarrow [K] = \begin{bmatrix} 930597 & -809826 & 0 \\ -809826 & 1619652 & -809826 \\ 0 & -809826 & 809826 \end{bmatrix} \text{ and } [M] = \begin{bmatrix} 54 & 0 & 0 \\ 0 & 43 & 0 \\ 0 & 0 & 879 \end{bmatrix}.$$

Performing the operation $[K] - [M]\omega^2 = 0$ in Mathcad, we have

$\{T\} = \{0.633 \quad 0.068 \quad 0.029\}$; the corresponding eigenvectors are expressed as

$\{\phi\}_1^{\mathrm{T}} = \{-0.782 \quad -0.893 \quad -1\}$, $\{\phi\}_2^{\mathrm{T}} = \{-1 \quad -0.597 \quad 0.073\}$, $\{\phi\}_3^{\mathrm{T}} = \{0.508 \quad -1 \quad 0.02\}$

Analysis for Mode 1

$$L_1 = \{\phi\}_1^{\mathrm{T}}[M]\{I\} = \{-0.782 \quad -0.893 \quad -1\}\begin{bmatrix} 54 & & \\ & 43 & \\ & & 879 \end{bmatrix}\begin{Bmatrix} 1 \\ 1 \\ 1 \end{Bmatrix} = -958.845$$

$$M_1 = \{\phi\}_1^{\mathrm{T}}[M]\{\phi\}_1 = \{-0.782 \quad -0.893 \quad -1\}\begin{bmatrix} 54 & & \\ & 43 & \\ & & 879 \end{bmatrix}\begin{Bmatrix} -0.782 \\ -0.893 \\ -1 \end{Bmatrix} = 945.701$$

For $T = 0.633$, $S_{a1} = 24.525\ \mathrm{m/s^2}$, $Z = 0.36$, $I = 1.5$, $R = 3.0$.

$$\beta = ZI/2R = 0.09$$

Thus, $u_1 = \{\phi\}_1 \times \left(\frac{L_1}{M_1}\right) \times \beta \times S_{a1} \times \left[\frac{T_1^2}{4\pi^2}\right] = \begin{Bmatrix} 0.018 \\ 0.02 \\ 0.023 \end{Bmatrix}$ m.

Nodal force $P_1 = [K] \times u_1 = \begin{bmatrix} 930597 & -809826 & 0 \\ -809826 & 1619652 & -809826 \\ 0 & -809826 & 809826 \end{bmatrix}$

$$\times \begin{Bmatrix} 0.018 \\ 0.02 \\ 0.023 \end{Bmatrix} = \begin{Bmatrix} 103.4 \\ 73.6 \\ 1968 \end{Bmatrix}.$$

Thus, the floor shear $V_1 = \begin{Bmatrix} 1968 + 73.6 + 103.4 \\ 1968 + 73.6 \\ 1968 \end{Bmatrix} = \begin{Bmatrix} 2145 \\ 2042 \\ 1968 \end{Bmatrix}$ kN.

Moment at base = $1968 \times 13.6 + 73.6 \times 10.1 + 103.4 \times 6.6 = 28{,}191$ kN m.

Analysis for Mode 2

For second mode, considering that $\{\phi\}_2^{\mathrm{T}} = \{-1 \quad -0.597 \quad 0.073\}$, $L_2 = \{\phi\}_2^{\mathrm{T}}[M]\{I\} = -14.404$, $M_2 = \{\phi\}_2^{\mathrm{T}}[M]\{\phi\}_1 = 73.01$.

For $T = 0.068$, $S_{a2} = 19.82\ \mathrm{m/s^2}$, and $\beta = 0.09$.

$$u_2 = \{\phi\}_2 \times \left(\frac{L_2}{M_2}\right) \times \beta \times S_{a2} \times \left[\frac{T_2^2}{4\pi^2}\right] = \begin{Bmatrix} 4.151 \\ 2.478 \\ -0.303 \end{Bmatrix} \times 10^{-5}\ \mathrm{m}$$

This gives $P_2 = [K] \times u_2 = \begin{Bmatrix} 18.86 \\ 8.98 \\ -22.52 \end{Bmatrix}$ kN and $V_2 = \begin{Bmatrix} 5.32 \\ -13.54 \\ -22.52 \end{Bmatrix}$ kN.

Moment at base = $-22.52 \times 13.6 + 8.98 \times 10.1 + 18.86 \times 6.6 = -91.1$ kN m.

Analysis for Mode 3

Proceeding in identical manner, we get

$$u_3 = \begin{Bmatrix} 3.616 \\ -7.118 \\ 0.142 \end{Bmatrix} \times 10^{-7}\,\text{m},\quad P_3 = \begin{Bmatrix} 0.913 \\ -1.457 \\ 0.588 \end{Bmatrix}\,\text{kN} \quad \text{and} \quad V_3 = \begin{Bmatrix} 0.044 \\ -0.869 \\ 0.588 \end{Bmatrix}$$

Moment at base $0.588 \times 13.6 - 1.457 \times 10.1 + 0.913 \times 6.6 = -0.69\,\text{kN m}$.

Consideration of soil–structure interaction

Dynamic shear modulus of soil $G = (\gamma_s/g) \times V_s^2 = (20/9.81) \times (155)^2 = 48,980\,\text{kN/m}^2$.

The area of raft = 7.5 × 7.5 = 56.25 m^2 and $I_{xx} = (7.5)^4/12 = 263.7\ \text{m}^4$.

Thus, $r_x = \sqrt{\frac{56.25}{\pi}} = 4.23$ m and $r_\theta = \sqrt{4}{\frac{4 \times 56.25}{\pi}} = 2.909$ m.

$$K_x = \frac{8Gr_x}{2-v} = \frac{8 \times 48,980 \times 4.23}{2 - 0.4} = 1,035,927\,\text{kN/m}$$

$$K_\theta = \frac{8Gr_\theta^3}{3(1-v)} = \frac{8 \times 48,980 \times (2.91)^3}{3 \times (1 - 0.4)} = 1,788,109\,\text{kN/rad}$$

$$C_x = 3.054\sqrt{\frac{G\rho r_x^4}{2-v}} = 3.054\sqrt{\frac{48,980 \times 2.04 \times (4.23)^4}{2 - 0.4}} = 13,656\,\text{kN s/m}$$

$$C_\theta = 1.33\sqrt{\frac{G\rho r_\theta^8}{2-v}} = 1.33\sqrt{\frac{48,980 \times 2.04 \times (2.909)^8}{2 - 0.4}} = 23,800\,\text{kN s/rad}$$

The mass of raft = 148 kN s^2/m.

The mass from bunker = 54 + 43 + 879 = 976 kN s^2/m (Add diagonal terms of the mass matrix).

The total mass = 148 + 976 = 1124 kN s^2/m.

Thus, $T_x = 2\pi\sqrt{M/K_x} = 2\pi\sqrt{1124/1,035,927} = 0.206$ s.

Damping ratio $\zeta_x = C_x/2\sqrt{K_x \times M} = 13,656/\left(2\sqrt{1,035,927 \times 1124}\right) = 0.20$.

Mass moment of inertia of foundation = $J_\theta = (m/12) \times (L^2 + t^2) = (148/12) \times \left[(7.5)^2 + (1)^2\right] = 706\,\text{kN s}^2/\text{m}$.

Contribution from the bunker = 879 × (13.6)^2 + 43 × (10.1)^2 + 54 × (6.6)^2 = 169,318 kN s^2/m.

Thus, the total mass moment on inertia = 706 + 169,318 = 170,025 kN s^2/m.

Thus, $T_\theta = 2\pi\sqrt{(J_\theta + \sum mh^2)/K_\theta} = 2\pi\sqrt{170,025/1,788,109} = 1.937$ s.

Damping ratio $\zeta_\theta = C_\theta/2\sqrt{K_\theta \times (J_\theta + \sum mh^2)} = 23,800/(2\sqrt{1,788,109 \times 170,025}) = 0.0215.$

For the first mode

$$r_x = T_x/T_i = 0.206/0.633 = 0.325$$

$$\bar{K}_x = K_x\sqrt{\left(1 - r_x^2\right)^2 + (2\zeta r_x)^2} = 1,035,927 \times \sqrt{\left(1 - (0.325)^2\right)^2 + (2 \times 0.2 \times 0.325)^2} = 936,243\ \text{kN/m}$$

$$r_\theta = T_\theta/T_i = 1.937/0.633 = 3.06$$

$$\bar{K}_\theta = K_\theta\sqrt{\left(1 - r_\theta^2\right)^2 + (2\zeta r_\theta)^2} = 1,788,109 \times \sqrt{\left(1 - (3.06)^2\right)^2 + (2 \times 0.0215 \times 3.06)^2} = 14,956,879\ \text{kN/rad.}$$

$$\text{AF}_i = \left[1 + \frac{V_i}{\widehat{K}_x u_{si}}\left(1 + \mathrm{e}^{-\zeta_x \frac{2\pi t}{T_x}}\left\{\sqrt{\frac{1-\zeta_x}{1+\zeta_x}}\sin 2\pi\sqrt{1-\zeta_x^2}\frac{t}{T_x} - \cos 2\pi\sqrt{1-\zeta_x^2}\frac{t}{T_x}\right\}\right) + \frac{M_i\bar{h}_i}{\widehat{K}_\theta u_{si}}\left(1 + \mathrm{e}^{-\zeta_\theta \frac{2\pi t}{T_\theta}}\left\{\sqrt{\frac{1-\zeta_\theta}{1+\zeta_\theta}}\sin 2\pi\sqrt{1-\zeta_\theta^2}\frac{t}{T_\theta} - \cos 2\pi\sqrt{1-\zeta_\theta^2}\frac{t}{T_\theta}\right\}\right)\right]$$

$$V_1 = 2145\ \text{kN} \quad M_1 = 28,191\ \text{kN} \quad u_{s1} = \begin{Bmatrix} 0.018 \\ 0.02 \\ 0.023 \end{Bmatrix}, \quad \{m_1\} = \left|\frac{P_1}{\beta \times S_{a1}}\right| = \begin{Bmatrix} 47 \\ 33 \\ 889 \end{Bmatrix}\ \text{kN s}^2/\text{m}$$

$$h_1 = \frac{889 \times 13.63 + 33 \times 10.1 + 47 \times 6.6}{889 + 33 + 47} = 13.16\ \text{m}$$

$$u_1 = 0.02 + \frac{(0.023 - 0.02)}{3.5} \times (13.16 - 10.1) = 0.0226\ \text{m} \quad \text{at } 13.16\ \text{m}$$

$$\text{AF}_1 = \left[1 + \frac{2145}{936,243 \times 0.0226}\left(1 + \mathrm{e}^{-6.1t}\left\{\sqrt{\frac{1-0.2}{1+0.2}}\sin 29.88t - \cos 29.88t\right\}\right) + \frac{28191 \times 13.16}{14,956,879 \times 0.0226}\left(1 + \mathrm{e}^{-0.07t}\left\{\sqrt{\frac{1-0.0215}{1+0.0215}}\sin 3.24t - \cos 3.24t\right\}\right)\right]$$

$$\text{AF}_1 = \left[1 + 0.1014\left(1 + \mathrm{e}^{-6.1t}\{0.816\sin 29.88t - \cos 29.88t\}\right) + 1.096\left(1 + \mathrm{e}^{-0.07t}\{0.978\sin 3.24t - \cos 3.24t\}\right)\right]$$

We show the amplification plot in Fig. 7.16.

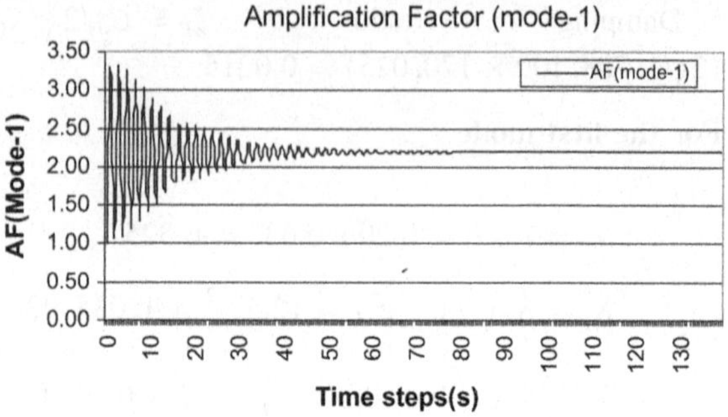

Fig. 7.16 Amplification plot for Mode 1

For the second mode

$$r_x = T_x/T_i = 0.206/0.068 = 3.03$$

$$\bar{K}_x = K_x\sqrt{\left(1 - r_x^2\right)^2 + (2\zeta r_x)^2} = 1,035,927$$

$$\times\sqrt{\left(1 - (3.03)^2\right)^2 + (2 \times 0.2 \times 3.03)^2} = 8,567,315\,\text{kN/m}$$

$$r_\theta = T_\theta/T_i = 1.937/0.068 = 28.5$$

$$\bar{K}_\theta = K_\theta\sqrt{\left(1 - r_\theta^2\right)^2 + (2\zeta r_\theta)^2}$$

$$= 1,788,109 \times \sqrt{\left(1 - (28.5)^2\right)^2 + (2 \times 0.0215 \times 28.5)^2}$$

$$= 1,450,605,081\,\text{kN/rad}$$

$$V_2 = 5.32\text{kN} \quad M_2 = -91.1\text{kN} \quad u_{s2} = \begin{Bmatrix} 4.151 \\ 2.478 \\ -0.303 \end{Bmatrix} \times 10^{-5}\text{m},$$

$$\{m_2\} = \left|\frac{P_2}{\beta \times S_{a2}}\right| = \begin{Bmatrix} 10.6 \\ 5.04 \\ 12.65 \end{Bmatrix} \text{kN s}^2/\text{m}$$

$$h_2 = \frac{12.65 \times 13.6 + 5.04 \times 10.1 + 10.6 \times 6.6}{12.65 + 5.04 + 10.6} = 10.353\text{m}$$

$$u_2 = 4.151 \times 10^{-5} - \frac{(1.673 \times 10^{-5})}{3.5} \times (13.6 - 10.353) = 2.6 \times 10^{-5} \quad \text{at } 10.353\text{m}$$

$$\text{AF}_2 = \left[1 + \frac{5.32}{8567315 \times 2.6 \times 10^{-5}}\left(1 + e^{-6.1t}\left\{\sqrt{\frac{1-0.2}{1+0.2}}\sin 29.88t - \cos 29.88t\right\}\right) - \right.$$

$$\left. - \frac{91.1 \times 10.353}{1,450,605,081 \times 2.6 \times 10^{-5}}\left(1 + e^{-0.07t}\left\{\sqrt{\frac{1-0.0215}{1+0.0215}}\sin 3.24t - \cos 3.24t\right\}\right)\right]$$

$$\mathrm{AF}_2 = \left[1 + 0.024\left(1 + \mathrm{e}^{-6.1t}\{0.816 \sin 29.88t - \cos 29.88t\}\right) \right.$$
$$\left. -0.025\left(1 + \mathrm{e}^{-0.07t}\{0.978 \sin 3.24t - \cos 3.24t\}\right)\right]$$

Plot of amplification factor is shown in Fig. 7.17.

For the third mode

$$r_x = T_x/T_i = 0.206/0.029 = 7.1$$

$$\bar{K}_x = K_x\sqrt{\left(1 - r_x^2\right)^2 + (2\zeta r_x)^2} = 1{,}035{,}927 \times \sqrt{\left(1 - (7.1)^2\right)^2 + (2 \times 0.2 \times 7.1)^2}$$
$$= 51{,}269{,}635\ \text{kN/m}$$

$$r_\theta = T_\theta/T_i = 1.937/0.029 = 67$$

$$\bar{K}_\theta = K_\theta\sqrt{\left(1 - r_\theta^2\right)^2 + (2\zeta r_\theta)^2} = 1{,}788{,}109$$
$$\times \sqrt{\left(1 - (67)^2\right)^2 + (2 \times 0.0215 \times 67)^2} = 8{,}025{,}034{,}845\ \text{kN/rad}$$

$$V_3 = 0.044\ \text{kN}, \quad M_3 = -0.69\ \text{kN}, \quad u_{s3} = \begin{Bmatrix} 3.616 \\ -7.118 \\ 0.142 \end{Bmatrix} \times 10^{-7}\text{m},$$

$$\{m_3\} = \left|\frac{P_3}{\beta \times S_{a3}}\right| = \begin{Bmatrix} 0.72 \\ 1.14 \\ 0.463 \end{Bmatrix} \text{kN s}^2/\text{m},$$

$$h_3 = \frac{0.463 \times 13.6 + 1.14 \times 10.1 + 0.72 \times 6.6}{0.463 + 1.14 + 0.72} = 9.713\ \text{m}.$$

Interpolation gives $u_3 = -5.93 \times 10^{-7}$m at 9.713 m.

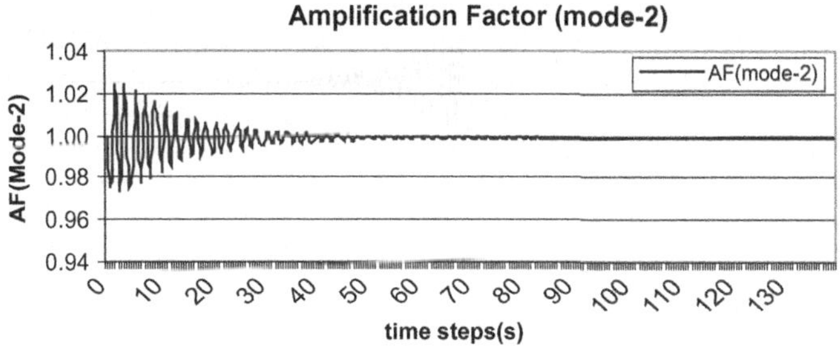

Fig. 7.17 Amplification plot for Mode 2

$$AF_3 = \left[1 - \frac{0.044}{51,269,635 \times 5.93 \times 10^{-7}}\left(1 + e^{-6.1t}\left\{\sqrt{\frac{1-0.2}{1+0.2}}\sin 29.88t - \cos 29.88t\right\}\right)\right.$$
$$\left. + \frac{0.69 \times 9.71}{8,025,034,845 \times 5.93 \times 10^{-7}}\left(1 + e^{-0.07t}\left\{\sqrt{\frac{1-0.0215}{1+0.0215}}\sin 3.24t - \cos 3.24t\right\}\right)\right]$$

$$AF_3 = \left[1 - 0.0014\left(1 + e^{-6.1t}\{0.816 \sin 29.88t - \cos 29.88t\}\right)\right.$$
$$\left. + 0.0014\left(1 + e^{-0.07t}\{0.978 \sin 3.24t - \cos 3.24t\}\right)\right]$$

We plot the AF_3 in Fig. 7.18.

The amplification in the third mode is negligible.

Data of amplification for the first 5 s, which show the rate of changes, are presented in the table below:

Time steps	AF (Mode 1)	AF (Mode 2)	AF (Mode 3)
0	1.00	1.00	1.000
0.5	3.29	0.98	1.001
1	3.12	0.98	1.001
1.5	1.10	1.02	0.999
2	1.45	1.02	0.999
2.5	3.29	0.97	1.001
3	2.79	0.99	1.001
3.5	1.12	1.02	0.999
4	1.74	1.01	0.999
4.5	3.25	0.98	1.001
5	2.52	0.99	1.000

Observing the above table, one sees that the maximum amplification, amounting to 3.29, is in Mode 1 at 0.5 and 2.5 s, when there is attenuation in the second mode and negligible amplification in the third mode.

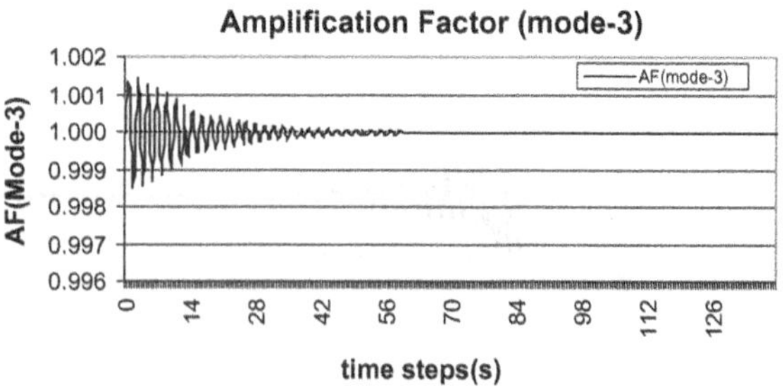

Fig. 7.18 Amplification plot for Mode 3

Thus, based on amplification due to DSSI

$V_1 = 3.29 \times \begin{Bmatrix} 2145 \\ 2042 \\ 1968 \end{Bmatrix} = \begin{Bmatrix} 7057 \\ 6718 \\ 6475 \end{Bmatrix}$ kN, $\quad V_2 = 0.98 \times \begin{Bmatrix} 5.32 \\ -13.54 \\ -22.52 \end{Bmatrix} =$ $\begin{Bmatrix} 5.21 \\ -13.27 \\ -22 \end{Bmatrix}$ kN and $V_3 = \begin{Bmatrix} 0.044 \\ -0.869 \\ 0.588 \end{Bmatrix}$ kN, having no amplification.

Thus, the SRSS values are

$$V_{(\text{Level3})} = \sqrt{(6475)^2 + (22)^2 + (0.588)^2} = 6475\,\text{kN}$$

$$V_{(\text{Level2})} = \sqrt{(6718)^2 + (13.2)^2 + (0.869)^2} = 6718\,\text{kN}$$

$$V_{(\text{Level1})} = \sqrt{(7057)^2 + (5.32)^2 + (0.044)^2} = 7057\,\text{kN}.$$

For Mode 1, the moment at base = 28,191 × 3.29 = 92,748.
For Mode 2, the moment at base = (−91.1) × 0.98 = −89.27.
For Mode 3, the moment at base = −0.69 × = −0.69.
Thus, the SRSS value of moment at base $= M_{\text{SRSS}} = \sqrt{(92,748)^2 + (89.27)^2 + (0.69)^2} = 92,748\,\text{kN m}$.

Thus, based on the above problem, we can arrive at the following conclusions:

- The first mode response is most important and gets greatest amplification.
- In this case, the soil was relatively soft, so the amplification was significant, 3.3 times.
- In higher modes, the dynamic stiffness of soil gets significantly enhanced due to frequency separation and this reduces amplification considerably, as it starts to behave almost like a fixed-base structure.
- The height of structure has significant influence on rocking mode and considerably reduces the damping ratio of foundation.
- For a normal structure, considering a fixed-base SRSS moment and shear and then multiplying the same by amplification factor obtained by considering static spring of soil would be sufficient for estimation of DSSI in the majority of structures.

7.9.2 Model 2—Foundation with Large Depth Supporting Heavy/High Structure

Based on Lagrange's equation as shown in Eq. (7.8.3), it has been shown by Chowdhury et al. (2015b) that for a block foundation, the coupled equation of equilibrium can be expressed as

$$\begin{bmatrix} m_f & m_f Z_c \\ m_f Z_c & J_\theta + m_f Z_c^2 \end{bmatrix} \begin{Bmatrix} \ddot{x} \\ \ddot{\theta} \end{Bmatrix} + \begin{bmatrix} C_x & 0 \\ 0 & C_\theta \end{bmatrix} \begin{Bmatrix} \dot{x} \\ \dot{\theta} \end{Bmatrix} + \begin{bmatrix} K_x & 0 \\ 0 & K_\theta \end{bmatrix} \begin{Bmatrix} x \\ \theta \end{Bmatrix} = \begin{Bmatrix} V_i \\ M_i \end{Bmatrix} \tag{7.9.11}$$

Extending the above for the system shown in Fig. 7.19, we have

$$\begin{aligned} &\begin{bmatrix} m_f + P/g & \left[m_f Z_c + \left(\frac{P}{g}\right)(h_i + D_f)\right] \\ \left[m_f Z_c + \left(\frac{P}{g}\right)(h_i + D_f)\right] & J_\theta + m_f Z_c^2 + (P/g)(\bar{h}_i + D_f)^2 \end{bmatrix} \begin{Bmatrix} \ddot{x} \\ \ddot{\theta} \end{Bmatrix} \\ &+ \begin{bmatrix} C_x & 0 \\ 0 & C_\theta \end{bmatrix} \begin{Bmatrix} \dot{x} \\ \dot{\theta} \end{Bmatrix} + \begin{bmatrix} K_x & 0 \\ 0 & K_\theta \end{bmatrix} \begin{Bmatrix} x \\ \theta \end{Bmatrix} = \begin{Bmatrix} V_i \\ M_i \end{Bmatrix} \end{aligned} \tag{7.9.12}$$

The eigenvalue expression can thus be depicted as

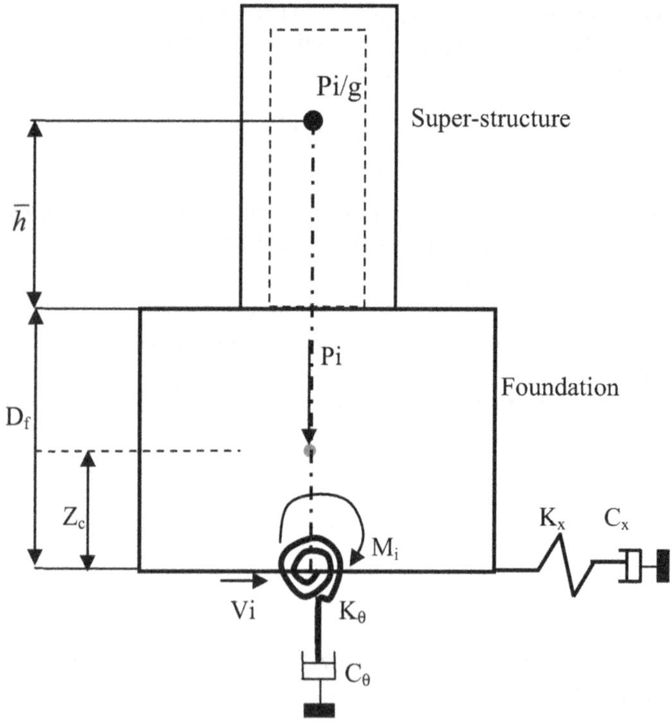

Fig. 7.19 Foundation having large depth subjected to earthquake load

$$\begin{bmatrix} K_x - (m_f + P/g)\lambda & -\left[m_f Z_c + \left(\frac{P}{g}\right)(h_i + D_f)\right] \\ -\left[m_f Z_c + \left(\frac{P}{g}\right)(h_i + D_f)\right] & K_\theta - \left[J_\theta + m_f Z_c^2 + (P/g)(\bar{h}_i + D_f)^2\right]\lambda \end{bmatrix} = 0 \tag{7.9.13}$$

Based on Eq. (7.9.13), let ω_x and ω_θ be the natural frequencies of the system, and the normalized eigenvectors are

$$[\varphi] = \begin{bmatrix} \phi_{xx} & \phi_{\theta x} \\ \phi_{x\theta} & \phi_{\theta\theta} \end{bmatrix} \tag{7.9.14}$$

For orthogonal transformation, considering the operation

$$[\varphi]^{\mathrm{T}}[\mathbf{M}][\varphi]\left\{\ddot{\xi}\right\} + [\varphi]^{\mathrm{T}}[\mathbf{C}][\varphi]\{\xi\} + [\varphi]^{\mathrm{T}}[\mathbf{K}][\varphi]\{\xi\} = [\varphi]^{\mathrm{T}}\{\mathbf{F}\} \tag{7.9.15}$$

Here

$$\{\mathbf{X}\} = \{u, \theta\} = [\phi]\{\xi\} \tag{7.9.16}$$

This gives two uncoupled equations of a single degree of freedom

$$\ddot{\xi}_x + 2\zeta_x\omega_x + \omega_x^2 = f_x \tag{7.9.17}$$

and

$$\ddot{\xi}_\theta + 2\zeta_\theta\omega_\theta + \omega_\theta^2 = m_x \tag{7.9.18}$$

$$\zeta_x = \frac{f_x/\omega_x^2}{\sqrt{(1 - s_x^2) + (2\zeta_x s_x)^2}} \quad \text{and} \quad \xi_\theta = \frac{m_x/\omega_\theta^2}{\sqrt{(1 - s_\theta^2) + (2\zeta_\theta s_\theta)^2}} \tag{7.9.19}$$

Here, $s_x = \omega_i/\omega_x$ and $s_\theta = \omega_i/\omega_\theta$, where ω_i = natural frequency of the fixed-base structure in ith mode.

Having obtained the uncoupled displacements in generalized coordinates, the global displacement is obtained from Eq. (7.9.16).

This gives u_x and θ at the base of footing, the point where spring is connected. The translational amplitude at the top of foundation is expressed as

$$u_x^t = u_x + D_f\theta \tag{7.9.20}$$

The amplification factor AF can now be obtained as

$$\mathrm{AF}_{ix} = \left[1 + \left(\frac{u_x^t}{u_{si}} + \frac{\theta \times h_i}{u_{si}}\right)\left(1 + e^{-\zeta_x \frac{2\pi}{T_x} t}\left\{\sqrt{\frac{1-\zeta_x}{1+\zeta_x}} \sin 2\pi\sqrt{1-\zeta_x^2}\frac{t}{T_x} - \cos 2\pi\sqrt{1-\zeta_x^2}\frac{t}{T_x}\right\}\right)\right] \tag{7.9.21}$$

One can also carry out a time history response analysis of Eq. (7.9.12) to determine the value of u_x and θ

$$\mathrm{AF}_{i\theta} = \left[1 + \left(\frac{u_x^t}{u_{si}} + \frac{\theta \times h_i}{u_{si}}\right)\left(1 + e^{-\zeta_\theta \frac{2\pi}{T_\theta} t}\left\{\sqrt{\frac{1-\zeta_\theta}{1+\zeta_\theta}} \sin 2\pi\sqrt{1-\zeta_\theta^2}\frac{t}{T_\theta} - \cos 2\pi\sqrt{1-\zeta_\theta^2}\frac{t}{T_\theta}\right\}\right)\right]. \tag{7.9.22}$$

Example 7.6 A nuclear reactor building has the following geometric and engineering data:

The total height of building is 63.09 m, the height of the cylindrical part is 43.83 m, internal radius of building = 18.89 m, thickness of shell for cylindrical part = 1.067 m, thickness of dome part = 0.762 m, radius of circular raft = 22.89 m, thickness of raft 2.0 m, Young's modulus of concrete = 3.304 × 10^7 kN/m^2, weight density of concrete = 24 kN/m^3, Poisson's ratio of concrete = 0.278, shear wave velocity of soil = 567.4 m/s, Poisson's ratio of soil = 0.33, weight density of soil 21 kN/m^3. The building is assumed to be situated in Zone IV as per IS code.

The building, based on fixed-base analysis, has been observed to have the following design data

Mode	1	2	3
Moment at base (kN m)	7,487,000	3,038,000	841,600
Base shear (kN)	230,100	235,200	99,900
Amplitude (mm) at centroid of building	16	0.32	0.065
Time period (s)	0.189	0.052	0.03

Determine the amplification factor of the building in coupled mode of foundation due to soil–structure interaction.

Solution

For the cylindrical part:
Internal radius = 18.89 m.
External radius = 19.957 m.
Height of cylindrical part = 43.83 m.
Volume of cylinder = $\pi(19.957^2 - 18.89^2) \times 43.83 = 5707.46\,\mathrm{m}^3$.
For the hemispherical part:
Internal radius = 18.89 m.
External radius 19.652 m.

Volume of hemispherical shell = $(2\pi/3) \times (19.652^3 - 18.89^3) = 1778\text{m}^3$.
Weight of building = (5707 + 1778) × 24 = 179,640 kN.
Mass of structure = 18,312 kN s^2/m.
Weight of foundation = $\pi \times 22.89^2 \times 2 \times 24 = 79{,}010\,\text{kN}$.
Mass of foundation = 8054 kN s^2/m.
Mass moment of inertia = $J_\theta = (1/4)mr^2 = (1/4) \times 8054 \times 22.89^2 = 1{,}054{,}978\,\text{kN s}^2/\text{m}$.

Dynamic shear modulus of soil $G = (\gamma_s/g)V_s^2 = (21/9.81) \times (567.4)^2 = 689{,}174\,\text{kN/m}^2$.

$$K_x = \frac{8Gr_0}{2-\nu} = \frac{8 \times 689{,}174 \times 22.89}{2-0.33} = 75{,}569{,}798\,\text{kN/m}$$

$$K_\theta = \frac{8Gr_\theta^3}{3(1-\nu)} = \frac{8 \times 689{,}174 \times 22.89^3}{3 \times (1-0.33)} = 3.289 \times 10^{10}\,\text{kN/rad}$$

$$C_x = 3.054\sqrt{\frac{G\rho r_x^4}{2-\nu}} = 3.054 \times \sqrt{\frac{689{,}174 \times 21 \times 22.89^4}{9.81 \times (2-0.33)}} = 1{,}503{,}980\,\text{kN s/m}$$

$$C_\theta = 1.33\sqrt{\frac{G\rho r_\theta^8}{2-\nu}} = 1.33 \times \sqrt{\frac{689{,}174 \times 21 \times 22.89^8}{9.81 \times (2-0.33)}} = 3.432 \times 10^8\,\text{kN s/rad}$$

The embedment factor for stiffness of soil is expressed as

$$\eta_x = 1 + 0.55(2-\upsilon)(h/r_x) = 1 + 0.55 \times (2-0.33) \times \left(\frac{2}{22.89}\right) = 1.08 \quad \text{(and)}$$

$$\begin{aligned}\eta_\theta &= 1 + 1.2(1-\upsilon)(h/r_\theta) + 0.2(2-\upsilon)(h/r_\theta)^3 \\ &= 1 + 1.2 \times (1-0.33) \times (2/22.89) + 0.2 \times (2-0.33) \times (2/22.89)^3 = 1.07\end{aligned}$$

$$\therefore K_x^e = K_x \times \eta_x = 81{,}615{,}382\,\text{kN/m}$$
$$\therefore K_\theta^e = K_\theta \times \eta_e = 3.52 \times 10^{10}\,\text{kN/rad}$$

The embedment factor for damping of soil is expressed as

$$\alpha_x = [1 + 1.9(1-\nu)(h/r_x)]/\sqrt{\eta_x} \rightarrow 1.07 \quad \text{(and)}$$

$$\alpha_\theta = \left[1 + 0.7(1-\nu)(h/r_\theta) + 0.62(2-\nu)(h/r_\theta)^3\right]/\sqrt{\eta_\theta} \rightarrow 1.007$$

$$\therefore C_x^e = C_x \times \alpha_x = 1{,}609{,}258\,\text{kN s/m}$$
$$C_\theta^e = C_\theta \times \alpha_\theta = 3.432 \times 10^8\,\text{kN s/rad}$$

For the first three modes, the height of centroid of the reactor building is

$$h_i = M_i/V_i \rightarrow h_1 = 32.53\,\mathrm{m}, \quad h_2 = 12.916\,\mathrm{m} \quad \text{and} \quad h_3 = 8.42\,\mathrm{m}$$

Analysis for the first mode

As per Eq. (7.9.12), equation of motion is expressed as

$$\begin{bmatrix} m_f + P/g & \left[m_f Z_c + \left(\frac{P}{g}\right)(h_i + D_f)\right] \\ \left[m_f Z_c + \left(\frac{P}{g}\right)(h_i + D_f)\right] & J_\theta + m_f Z_c^2 + (P/g)(\bar{h}_i + D_f)^2 \end{bmatrix} \begin{Bmatrix} \ddot{x} \\ \ddot{\theta} \end{Bmatrix} + \begin{bmatrix} C_x & 0 \\ 0 & C_\theta \end{bmatrix} \begin{Bmatrix} \dot{x} \\ \dot{\theta} \end{Bmatrix} + \begin{bmatrix} K_x & 0 \\ 0 & K_\theta \end{bmatrix} \begin{Bmatrix} x \\ \theta \end{Bmatrix} = \begin{Bmatrix} V_i \\ M_i \end{Bmatrix}$$

$$\begin{bmatrix} 26366 & 640367 \\ 640367 & 22896812 \end{bmatrix} \begin{Bmatrix} \ddot{x} \\ \ddot{\theta} \end{Bmatrix} + \begin{bmatrix} 1609258 & 0 \\ 0 & 3.432 \times 10^8 \end{bmatrix} \begin{Bmatrix} \dot{x} \\ \dot{\theta} \end{Bmatrix} + \begin{bmatrix} 81615382 & 0 \\ 0 & 3.52 \times 10^{10} \end{bmatrix} \begin{Bmatrix} x \\ \theta \end{Bmatrix} = \begin{Bmatrix} 230100 \\ 7487000 \end{Bmatrix}$$

Free vibration analysis gives

$$\begin{bmatrix} 81615382 - 26366\lambda & -640367 \\ -640367 & 3.52 \times 10^{10} - 22896812\lambda \end{bmatrix} = 0$$

Solution of the above gives $T_x = 1.572\,\mathrm{s}$ and $T_\theta = 0.064\,\mathrm{s}$.
The corresponding eigenvectors are expressed as

$$[\phi] = \begin{bmatrix} -0.126 & '-1 \\ '-1 & 0.028 \end{bmatrix}$$

For translational mode

$$M_1 = \sqrt{\phi_1^T[M]\phi_1} = \sqrt{\{-0.126 \quad -1\} \begin{bmatrix} 26366 & 640367 \\ 640367 & 22896812 \end{bmatrix} \begin{Bmatrix} -0.126 \\ -1 \end{Bmatrix}}$$
$$= 2.306 \times 10^7$$

For translational mode, load can now be expressed as

$$\{p\} = \frac{[\phi]_n^{\mathrm{T}}[P]}{M_1} = \frac{1}{2.306\times 10^7}\times\begin{bmatrix}-0.126\\-1\end{bmatrix}^{\mathrm{T}}\begin{Bmatrix}230100\\7487000\end{Bmatrix} = -0.326$$

$$\zeta_x = \frac{\{\phi_1\}^{\mathrm{T}}[C]\phi_1}{2M_1\omega_x} = 1.862$$

The above shows a very high damping ratio; thus, as recommended in ACI-351-3R-04, we restrict the damping ratio to $\zeta_x = 0.25$ in translational mode.

$$\omega_x = 2\pi/1.572 = 3.99\,\text{rad/s}, \quad \text{and} \quad \omega_1 = 2\pi/0.188 = 33.746\,\text{rad/s} \quad \text{and}$$
$$s_x = \omega_1/\omega_x = 8.317$$

$$\xi_1 = \frac{\frac{p_0}{\omega_x^2}}{\sqrt{\left(1-s_x^2\right)^2+(2\zeta_x s_x)^2}} = \frac{-0.326/(3.99)^2}{\sqrt{(1-8.317^2)^2+(2\times 0.25\times 8.317)^2}}$$
$$= -2.988\times 10^{-4}\,\text{m}$$

Thus, amplitudes in global coordinates are expressed as

$$[U] = \begin{Bmatrix}u_x\\\theta\end{Bmatrix} = [\phi]_1\{\xi\} = \begin{bmatrix}-0.126\\-1\end{bmatrix}\times -2.988\times 10^{-4} = \begin{Bmatrix}3.764\times 10^{-5}\\2.988\times 10^{-4}\end{Bmatrix}$$

The maximum fixed-base amplitude at centroid of building is 16 mm, and in the first mode the center of gravity of the building is at h = 32.53 m.

Hence,

$$\mathrm{AF}_x = \left[1+\left(\frac{u_x^t}{u_{si}}+\frac{\theta\times h_i}{u_{si}}\right)\left(1+\mathrm{e}^{-\zeta_x\frac{2\pi t}{T_x}}\left\{\sqrt{\frac{1-\zeta_x}{1+\zeta_x}}\sin 2\pi\sqrt{1-\zeta_x^2}\frac{t}{T_x}-\cos 2\pi\sqrt{1-\zeta_x^2}\frac{t}{T_x}\right\}\right)\right]$$

Substituting the appropriate values, we finally have the amplification factor varying with time, as shown hereafter for the first mode.

Plot in Fig. 7.20 shows that the maximum amplification factor for translational mode is @ 2.066.

Analysis for the rocking mode

For the rocking mode, $T_\theta = 0.064$ s and $Ts = 0.188$ s, the fixed-base time period of the structure $\rightarrow \omega_\theta = 98.17$ rad/s and $\omega_s = 33.42$ rad/s, and the corresponding eigenvectors as obtained above are $\phi_2 = [-1 \quad 0.028]^{\mathrm{T}}$.

$$\therefore M_2 = \sqrt{\phi_2^{\mathrm{T}}[M]\phi_2} = \sqrt{\{-1 \quad 0.028\}\begin{bmatrix}26366 & 640367\\640367 & 22896812\end{bmatrix}\begin{Bmatrix}-1\\0.028\end{Bmatrix}} = 8457$$

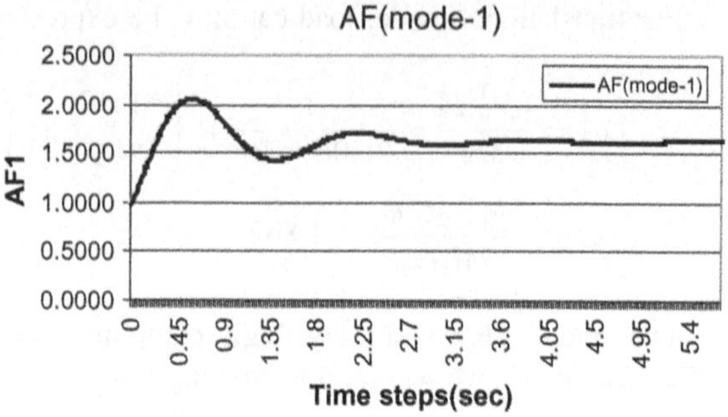

Fig. 7.20 Amplification plot for translational mode

$$\zeta_\theta = \frac{\{\phi_2\}^{\mathrm{T}}[C]\phi_2}{2M_2\omega_\theta} = 1.129$$

The damping value is high, considering 50% of damping as per ACI 351-3R-04. The modal load for the rocking mode can now be expressed as

$$p_2 = \frac{[\phi]_2^{\mathrm{T}}[P]}{M_2} = \left(\frac{1}{8457}\right) \times \begin{bmatrix} -1 \\ 0.028 \end{bmatrix}^{\mathrm{T}} \begin{Bmatrix} 230100 \\ 7487000 \end{Bmatrix} = -2.408, \quad \text{and}$$

$$s_\theta = \frac{\omega_1}{\omega_\theta} = 0.338$$

$$\xi_1 = \frac{\frac{p_2}{\omega_\theta^2}}{\sqrt{\left(1 - s_\theta^2\right)^2 + \left(2\zeta_\theta s_\theta\right)^2}} = \frac{-2.408/(98.17)^2}{\sqrt{(1 - 0.338^2)^2 + (2 \times 0.50 \times 0.338)^2}} = -2.756 \times 10^{-4}\,\mathrm{m}$$

Thus, the amplitudes in global coordinates are expressed as

$$[U] = \begin{Bmatrix} u_x \\ \theta \end{Bmatrix} = [\phi]_2\{\xi\} = \begin{bmatrix} -1 \\ 0.028 \end{bmatrix} \times -2.756 \times 10^{-4} = \begin{Bmatrix} 2.756 \times 10^{-4} \\ -7.722 \times 10^{-6} \end{Bmatrix}$$

Considering

$$\mathrm{AF}_\theta = \left[\left(1 + \frac{u_x^t}{u_{si}} + \frac{\theta \times h_i}{u_{si}}\right)\left(1 + \mathrm{e}^{-\zeta_\theta \frac{2\pi}{T_\theta} t}\left\{\sqrt{\frac{1-\zeta_\theta}{1+\zeta_\theta}} \sin 2\pi\sqrt{1-\zeta_\theta^2}\frac{t}{T_\theta} - \cos 2\pi\sqrt{1-\zeta_\theta^2}\frac{t}{T_\theta}\right\}\right)\right]$$

we finally have the amplification factor with respect to time as $\mathrm{AF}_\theta = 1.0$. That is, there is no amplification in the rocking mode.

Thus, for the fundamental time period of the building, we have the values as follows:

	Fixed-base case	Amplification translational mode	Rocking mode	SRSS value
Moment	7,487,000	15,722,700	7,487,000	15,740,516
Shear	230,100	483,210	230,100	535,199

Thus, the SRSS amplification is 2.32 times the fixed-base case.

The steps shown above now can be repeated for other fixed-base modes of the building to determine the amplification factor for the corresponding modes.

7.10 Kinematical Interaction with Foundation and Structure

What we have discussed till now is normally termed as inertial interaction of soil and structure. However, for earthquakes there is an additional interaction between soil and foundation that influences the response of the overall system and is dependent on propagation of waves through soil medium, and this is known as *kinematical interaction.*

The total DSSI analysis for any system usually consists of combined effect of kinematical and inertial interaction.

So what is the kinematical interaction?

To understand this, let us have a look at Fig. 7.21.

As shown in Fig. 7.21, let us consider a site where seismic shear waves are propagating in vertical direction. It is apparent that for a small, shallow foundation it will hardly affect wave patterns (as the foundation stiffness and dimensions are far too small compared to wavelengths of the waves), and it will follow the same deformation pattern of the soil and will be subjected to the same ground acceleration as the free field.

The foundation in this case can be visualized as a small boat floating in a sea, and, when the wave travels, it also moves with it in the same phase.

However, when the foundation is large and embedded to a substantial depth within ground, it is apparent, due to its significant geometrical presence as well as its large stiffness, that there will be an interaction between the waves and foundation and peak ground response will certainly be something different than free field response, i.e., when the foundation was not there.

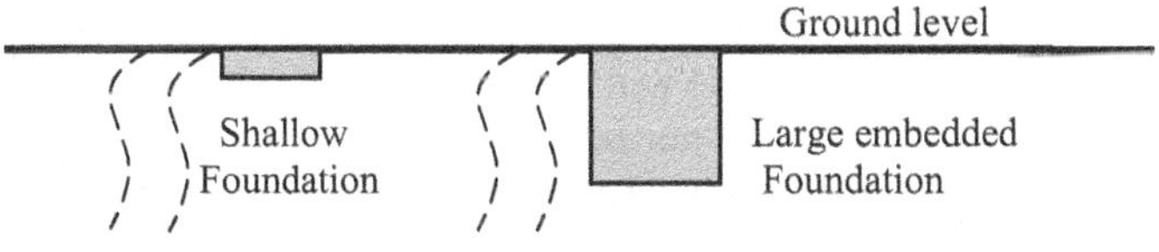

Fig. 7.21 Vertically propagating shear wave in a soil medium

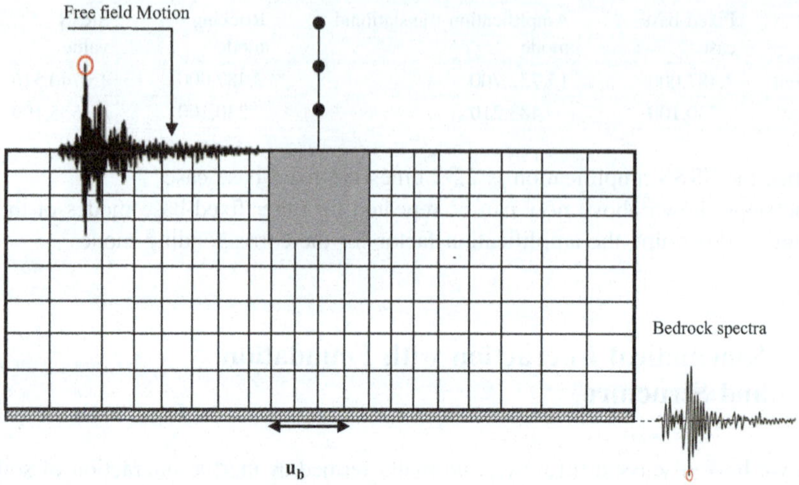

Fig. 7.22 FEM model for superstructure and foundation along with surrounding soil

This modification of ground response of embedded foundation vis-à-vis interaction with superstructure is called kinematical interaction under DSSI.

While performing a FEM analysis (Fig. 7.22), this is automatically taken care of when the input at the base level is provided to the model.

Shown in Fig. 7.22 is 2D FEM model for DSSI analysis. In this case, provided the input at bedrock level is known and is furnished as an input, the analysis becomes quite straightforward. It is apparent that waves travel through the soil and interact with foundation (also modeled as FEM) and give the total response that includes both kinematical and inertial interaction.

But unless and until the structure is sufficiently important, like a reactor building or an intake pump house with building above, housing important machineries at top, such expensive analysis will rarely be done, not to overlook the fact that an analyst may find serious difficulty in modeling this problem satisfying the appropriate boundary conditions.

Some of the practical difficulties often encountered while modeling such problems are as summarized hereafter.

- It is obvious that in this case the bedrock motion data are required. However, in most of the case, where site-specific response spectra are used, they are all furnished at the ground level and not at the base rock level. So one has to transfer the same at the base rock level. If the soil is far too layered, the results could well be dubious, as the material damping and stiffness of soil input can be tricky.
- At some sites, the bedrock level will be far below, 500–1000 m, when soil data to such a depth will definitely not be available.
- Some sites may not have specifically any bedrock at all, and then, where should be the boundary of the FEM model that would realistically reflect the free field

behavior is completely a matter of opinion and may vary from person to person (one common practice is to cut off the boundary where SPT value encountered, $N > 50$).

- Cutting off the boundary at near field will result in reflection of waves generating a spurious mode. On the other hand, taking it sufficiently far makes the analysis expensive and, most importantly, geotechnical data may not be available to such a great depth.
- Modeling the soil to a depth to which geotechnical data is known, and to ensure that waves are not reflected back, one may use infinite finite element or paraxial boundaries; however, if the software in hand does not have such options, the analysis would certainly be difficult to execute.
- Finally, analyzing a structure as per a particular code like, say, IS 1893 or UBC, etc., and also within the framework of modal analysis (which is the most common case) incorporating kinematical interaction is difficult.

For this reason, it is not unusual to see that while engineers are somewhat comfortable with inertial interaction analysis, they often ignore this phenomenon, despite the foundation having significant dimension.

So the question that naturally arises: is it possible to devise a rational procedure where kinematical interaction can be considered within modal analysis framework often deployed for industrial structures or bridges or high-rise buildings where sophistication as warranted in nuclear power industry based on FEM is not essential and yet the analysis comes up with a reasonable result?

For this, let us consider Fig. 7.23.

As shown in Fig. 7.23, let there be a site which is extending to infinity in the horizontal direction and has a soil layer of depth H over bedrock or has encountered $N > 50$ at depth H. The free field time period of the site can then be expressed as

$$T_n = 4H/[(2n - 1)V_0] \tag{7.10.1}$$

Here, n = mode number 1, 2, 3, …, [this has been derived in detail in Sect. 4.7 of Chap. 4].

It has also been shown therein, vide Eq. 4.7.33, that

$$u_n = \kappa_n \beta \frac{S_{an}}{\omega^2} \cos(2n - 1)\frac{\pi z}{2H} \tag{7.10.2}$$

From Eq. (7.10.2) as per definition of modal analysis

$$\ddot{u}_n = \kappa_n \beta S_{an} \cos(2n - 1)\frac{\pi z}{2H} \tag{7.10.3}$$

Thus, at $z = 0$

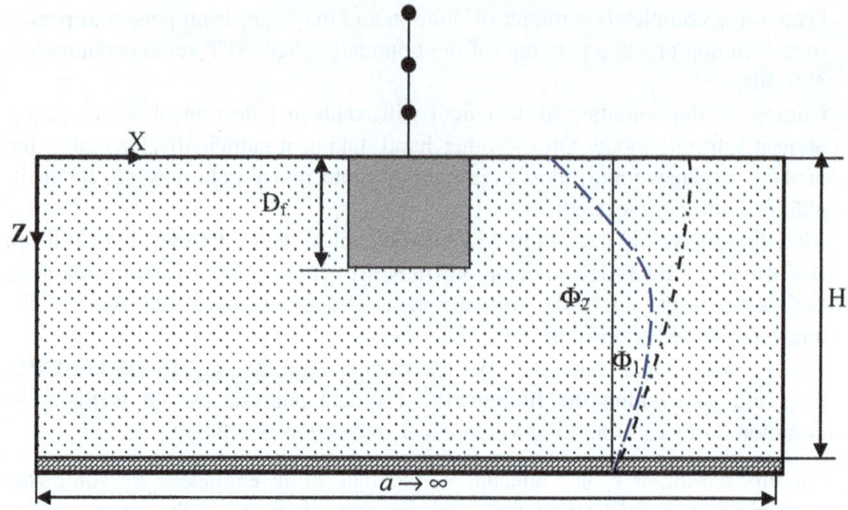

Fig. 7.23 A two-dimensional site extending to infinity with an embedded foundation

$$\ddot{u}_n = \kappa_n \beta S_{an} \tag{7.10.4}$$

At $z = D_f$,

$$\ddot{u}_n = \kappa_n \beta S_{an} \cos(2n-1)\frac{\pi D_f}{2H} \tag{7.10.5}$$

Thus, the average acceleration experienced by the basement for height from 0 to D_f is expressed as

$$\ddot{u}_n(\text{avg}) = \frac{1}{2}\kappa_n \beta S_{an}\left(1 + \cos(2n-1)\frac{\pi D_f}{2H}\right) \tag{7.10.6}$$

Thus, if we now isolate the foundation and draw the free body diagram of the system, we can develop a diagram as shown in Fig. 7.24.

It is thus observed that this average acceleration induces a lateral thrust $V_f = (M + m_s) \times \ddot{u}_n(\text{avg})$ and $M_f = [M \times Z_c + m_s(h + D_f)]\ddot{u}_{n(\text{avg})}$ at the base of the foundation at point S, which is the center of stiffness.

Here, Z_c is the centroidal height of c.g. of the basement and its load from the center of stiffness S.

The equilibrium equation of the system can now be expressed about its center of stiffness S as

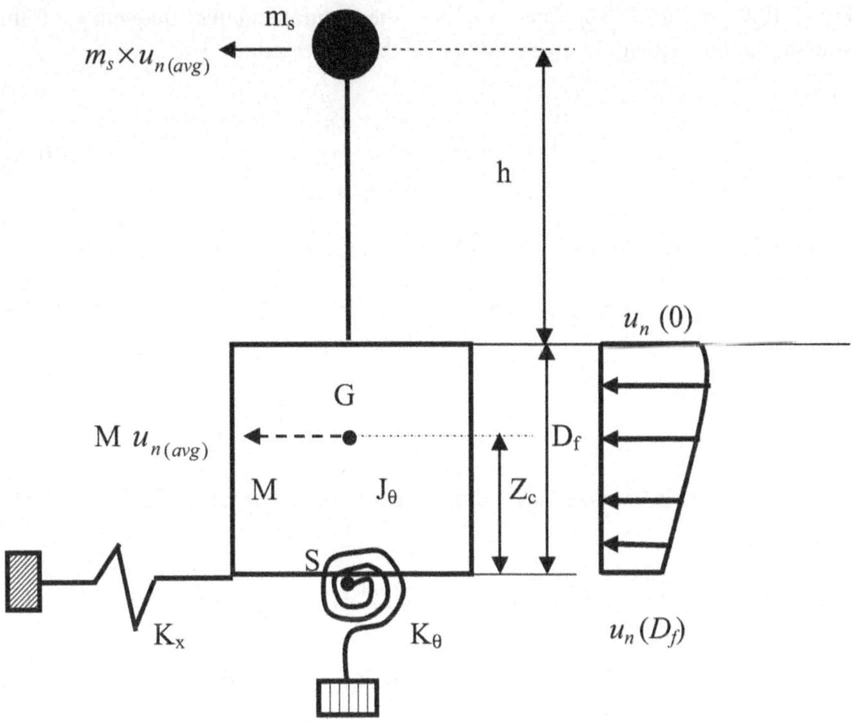

Fig. 7.24 Free body diagram of the embedded foundation under seismic force

$$\begin{bmatrix} M+m_s & MZ_c+m_s(h+D_f) & m_s \\ MZ_c+m_s(h+D_f) & J_\theta+MZ_c^2+m_s(h+D_f)^2 & m_s(h+D_f) \\ m_s & m_s(h+D_f) & m_s \end{bmatrix} \begin{Bmatrix} \ddot{u}_f \\ \ddot{\theta} \\ \ddot{u}_s \end{Bmatrix} + \begin{bmatrix} C_f & 0 & 0 \\ 0 & C_\theta & 0 \\ 0 & 0 & C_s \end{bmatrix} \begin{Bmatrix} \dot{u}_f \\ \dot{\theta} \\ \dot{u}_s \end{Bmatrix} + \begin{bmatrix} K_f & 0 & 0 \\ 0 & K_\theta & 0 \\ 0 & 0 & K_s \end{bmatrix} \begin{Bmatrix} u_f \\ \theta \\ u_s \end{Bmatrix} = \begin{Bmatrix} V_f \\ M_f \\ 0 \end{Bmatrix} \quad (7.10.8)$$

In Eq. (7.10.8), the stiffness and damping values are estimated considering the embedment effect of the foundation, as cited in Sect. 7.11.

K_s, the stiffness of the fixed-base frame, can be obtained from the expression

$$K_s = 4\pi^2 W / g\widehat{T}^2 \quad (7.10.9)$$

Here, W = total weight of the frame including appropriate dead and live load at each floor, $\widehat{T}$ = fixed-base time period of the frame @ $0.09H/\sqrt{D}$ or $C_T H^{3/4}$, etc., as furnished in IS-1893 (2002) Part 1. Now for un-damped vibration of

Eq. (7.10.8), if $[\boldsymbol{\omega}] = [\omega_f \quad \omega_\theta \quad \omega_s]^{\mathrm{T}}$ is the natural circular frequency of the system, the corresponding eigenvectors can be expressed as

$$[\varphi] = \begin{bmatrix} \phi_{ff} & \phi_{f\theta} & \phi_{fs} \\ \phi_{\theta f} & \phi_{\theta\theta} & \phi_{\theta s} \\ \phi_{sf} & \phi_{s\theta} & \phi_{ss} \end{bmatrix} \tag{7.10.10}$$

Now if we express Eq. (7.10.8) in matrix notation, we get

$$[\mathbf{M}]\{\ddot{U}\} + [\mathbf{C}]\{\dot{U}\} + [\mathbf{K}]\{U\} = \{\mathbf{P}_t\} \tag{7.10.11}$$

Considering

$$\{U_i\} = \{\phi_i\}\{\xi_i\} \tag{7.10.12}$$

where $\{\xi\}$ is the generalized co ordinate, and multiplying Eq. (7.10.11) by $\{\varphi\}^{\mathrm{T}}$ we have

$$\{\varphi\}^{\mathrm{T}}[\mathbf{M}]\{\varphi\}\{\ddot{\xi}_i\} + \{\varphi\}^{\mathrm{T}}[\mathbf{C}]\{\varphi\}\{\dot{\xi}_i\} + \{\varphi\}^{\mathrm{T}}[\mathbf{K}]\{\varphi\}\{\xi\} = \{\varphi\}^{\mathrm{T}}\{P\} \tag{7.10.13}$$

$$\rightarrow M_i\{\ddot{\xi}_i\} + C_i\{\dot{\xi}_i\} + K_i\{\xi\} = \{\varphi\}^{\mathrm{T}}\{P\} \tag{7.10.14}$$

Dividing each term of Eq. (7.10.14) by M_i, we have

$$\{\ddot{\xi}_i\} + 2\zeta_i\omega_i\{\dot{\xi}_i\} + \omega_i^2\{\xi\} = \frac{\{\varphi\}^{\mathrm{T}}\{P\}}{M_i} = q_i \quad \text{(say)} \tag{7.10.15}$$

Equation (7.10.15) thus gives the three uncoupled equations

$$\ddot{\xi}_f + 2\zeta_f\omega_f\dot{\xi}_f + \omega_f^2\xi_f = q_f \tag{7.10.16}$$

$$\ddot{\xi}_\theta + 2\zeta_\theta\omega_\theta\dot{\xi}_\theta + \omega_\theta^2\xi_\theta = q_\theta \tag{7.10.17}$$

$$\ddot{\xi}_s + 2\zeta_s\omega_s\dot{\xi}_s + \omega_s^2\xi_s = q_s \tag{7.10.18}$$

The general solution to Eqs. (7.10.16)–(7.10.18) can be expressed as

$$\xi_i = \frac{q_i}{\omega_i^2}\left[1 + \mathrm{e}^{-\zeta_i\omega_i t}\left\{\sqrt{\frac{1-\zeta_i}{1+\zeta_i}}\sin\left(\omega_i\sqrt{1-\zeta_i^2 t}\right) - \cos\left(\omega_i\sqrt{1-\zeta_i^2 t}\right)\right\}\right] \tag{7.10.19}$$

where the boundary conditions are at (1) $t = 0$, $u_i = 0$; and (2) $t = 0$ $\dot{u}_i = S_{vi} = S_{ai}/\omega_i$.

In Eq. (7.10.19), the subscript i stands for $i = f$, θ and s, respectively, as shown in Eq. (7.10.8).

In Eqs. (7.10.16)–(7.10.18), q_i is actually a function of natural frequency/time period of the site as expressed by Eq. (7.10.1).

Thus considering this effect, Eq. (7.10.19) can be expressed as

$$\xi_i = \frac{q_i}{\omega_i^2\sqrt{(1 - r_i^2)^2 + (2\zeta_i r_i)^2}}\left[1 + e^{-\zeta_i\omega_i t}\left\{\sqrt{\frac{1-\zeta_i}{1+\zeta_i}}\sin\left(\omega_i\sqrt{1-\zeta_i^2 t}\right) - \cos\left(\omega_i\sqrt{1-\zeta_i^2 t}\right)\right\}\right] \tag{7.10.20}$$

Here, $r_i = \omega_1^{si}/\omega_i$, where 1 stands for the fundamental natural mode of site. It is obvious in this case that, if the natural frequency of the site is in proximity to the natural frequency of the system when $r_i \rightarrow 1.0$, there will be a significant excitation to the system.

On deriving Eq. (7.10.20) in uncoupled modes, they are finally converted to the structural coordinate by the relation as expressed by Eq. (7.10.12).

The structural deformation of the building, u_s, for all modes can finally be expressed as

$$u_s^f = \sqrt{\sum_{i=1}^{3}\left(u_s^1\right)^2} \tag{7.10.21}$$

The base shear due to kinematical interaction can be obtained from the expression

$$V_{textb}^{KI} = K_s \times u_s^f \tag{7.10.22}$$

Though the above has been conceptually elaborated with the fundamental mode of the structure, the same can also be represented by the modal stiffness considering multi-degrees of freedom for the structure.

The problem is now analyzed with a suitable numerical example to further elaborate the issue.

Example 7.7 Shown in Fig. 7.25 is a three-storied commercial complex with basements. The basement consists of two floors below ground that is used as car parking.

The dimensions of the building are as shown in Fig. 7.25. Dead and live loads are considered for the building is as per IS-875. Other than this, the roof carries equipment loads, load like cooling plants and sundry other functional assets of load = 500 kN.

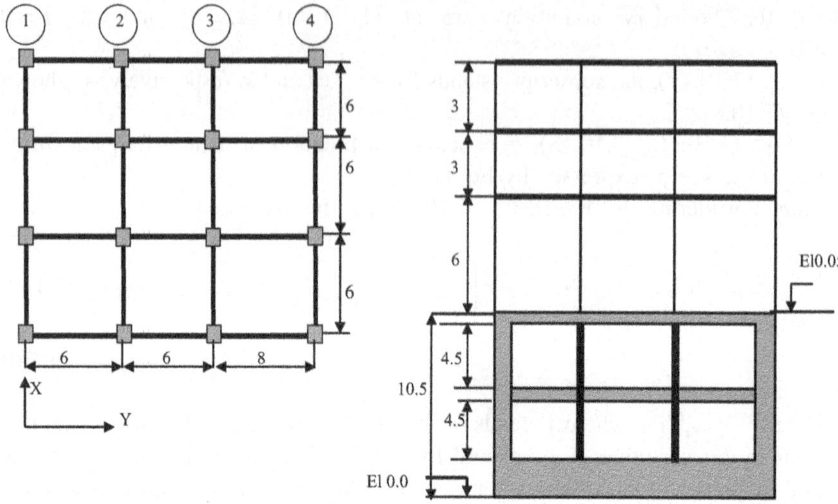

Fig. 7.25 Plan and elevation of the commercial complex analyzed for soil–structure interaction

Roof and ground floor slabs are 200 mm thick, while for other floors they are 125 mm thick. External walls are 250-mm-thick masonry walls, and internal partition walls are 100 mm thick. The basement slab of parking is 250 mm thick. Walls of basement are 450 mm thick. Base slab thickness is 600 mm. All beams along the Y-direction = 300 × 700 mm. All beams in the X-direction = 300 × 600 mm. All columns are 400 × 600 mm up to ground floor. In the basement, the columns are circular having diameter of 750 mm. Grade of concrete is M25.

Shear wave velocity of soil = 310 m/s. Poisson's ratio of soil = 0.4, density of soil is 19 kN/m^3.

The bedrock level is at −35.0 m below ground level (EL 0.0). Earthquake zone is Zone V as per IS code. Determine the

- Fixed-base response as per IS 1893 2016;
- Fixed-base response as per modal analysis;
- Kinematical response;
- Kinematical response if free field time period matches the system response;
- Inertial interaction response.

The earthquake is assumed to be acting in the X-direction; vide Fig. 7.25.

Solution

The loads calculated at each floor are as furnished in Table 7.1.

The soil stiffness and damping values thus computed are shown in Table 7.2.

The centroid of building and that of the basement were computed as EL +9.0 and EL −2.95 m, respectively.

Table 7.1 Computed weight of each floor of the building

Elevation	12.0	9.0	6.0	0.0	−4.50	−9.0
W(DL + LL) kN	5150	4997	5141	5594	12,893	18,215

Table 7.2 Stiffness and damping values considered for the soil and structure

K_x	K_q	C_x	C_θ
1.82×10^7	1.906×10^9	2.546×10^5	1.26×10^7

Fixed-base response as per IS-1893 (2016)

As per IS-1893 (2016) time period of the building is expressed as

$T = 0.09H/\sqrt{D}$, here $H = 12.0$ m height of building above basement. $D = 18.0$ m.

Thus, $T = 0.09 \times 12/\sqrt{18} = 0.254$ s.

Now, considering the medium soil site, $S_a/g = 2.5$ as per IS-1893.

For zone IV, $Z = 0.24$, $I = 1.0$ and $R = 3.0$ for ordinary moment resistant frame.

Thus $A_h = (ZI/2R) \times (S_a/g) = \frac{0.24}{6} \times 2.5 = 0.1$.

The total weight of building as obtained from Table 7.1 is

$$W_b = 5150 + 4997 + 5141 = 15,228\,\text{kN}$$

Thus, the base shear $V_b = 0.1 \times 15{,}228 = 1523$ kN.

Fixed-base response as per modal analysis

All columns till the ground floor are 400 × 600.

$$I_{col} = (1/12) \times 0.4 \times (0.6)^3 = 7.2 \times 10^{-3}\,\text{m}^4.$$

Thus, the stiffness at ground floor $k_1 = 16 \times \frac{12EI}{L^3} = 16 \times \frac{12 \times 2.5 \times 10^7 \times 7.2 \times 10^{-3}}{(6)^3} = 160,000\,\text{kN/m}$.

Stiffness at other floors $k_2 = k_3 = 16 \times \frac{12EI}{L^3} = 16 \times \frac{12 \times 2.5 \times 10^7 \times 7.2 \times 10^{-3}}{(3)^3} = 1,280,000\,\text{kN/m}$.

The stick model assumed can be expressed as shown in Fig. 7.26.

Thus, the stiffness matrix is expressed as

$$[K] = \begin{bmatrix} k_1 + k_2 & -k_2 & 0 \\ -k_2 & k_2 + k_3 & -k_3 \\ 0 & -k_3 & k_3 \end{bmatrix} = \begin{bmatrix} 1440000 & -1280000 & 0 \\ -1280000 & 2560000 & -128000 \\ 0 & -1280000 & 1280000 \end{bmatrix} \text{kN/m}$$

The mass matrix is computed as

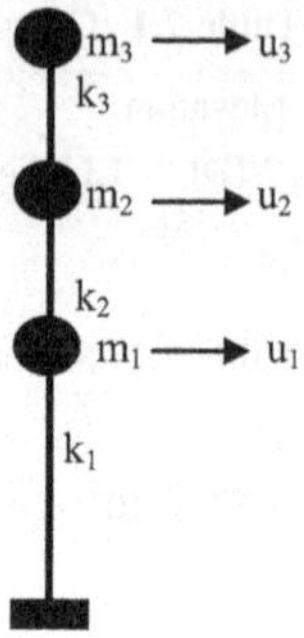

Fig. 7.26 Fixed-base model for the building

$$[M] = \begin{bmatrix} m_1 & 0 & 0 \\ 0 & m_2 & 0 \\ 0 & 0 & m_3 \end{bmatrix} = \begin{bmatrix} 524 & 0 & 0 \\ 0 & 509 & 0 \\ 0 & 0 & 525 \end{bmatrix} \text{kN s}^2/\text{m}$$

Eigensolution of the above gives

$$\begin{bmatrix} 1440000 - 524\lambda & -1280000 & 0 \\ -1280000 & 2560000 - 509\lambda & -128000 \\ 0 & -1280000 & 1280000 - 525\lambda \end{bmatrix} = 0$$

The above finally gives

$T = \langle 0.642 \quad 0.123 \quad 0.072 \rangle$ s and the corresponding eigenvectors as

$$[\phi] = \begin{bmatrix} -0.885 & -1 & 0.512 \\ -0.961 & -0.062 & -1 \\ -1 & 0.94 & 0.48 \end{bmatrix}$$

Analysis for the first mode

For the first mode, $T_1 = 0.642$ s and $\phi_1 = \{-0.885 \quad -0.961 \quad -1\}^{\mathrm{T}}$. This gives

$$L_1 = \{\phi_1\}^{\mathrm{T}}[M]\{I\} = \{-0.885 \quad -0.961 \quad -1\}\begin{bmatrix} 524 & 0 & 0 \\ 0 & 509 & 0 \\ 0 & 0 & 0 \end{bmatrix}\begin{Bmatrix} 1 \\ 1 \\ 1 \end{Bmatrix} = -1478$$

$$M_1 = \{\phi_1\}^{\mathrm{T}}[M]\{\phi_1\} = \{-0.885 \quad -0.961 \quad -1\}\begin{bmatrix} 524 & 0 & 0 \\ 0 & 509 & 0 \\ 0 & 0 & 0 \end{bmatrix}\begin{Bmatrix} -0.885 \\ -0.961 \\ -1 \end{Bmatrix}$$

$$= 1405$$

For $T = 0.642$ s, $S_{a1} = \left(\frac{1.36}{0.642}\right) \times 9.81 = 20.784\,\text{m/s}^2$ and $\left(\frac{ZI}{2R}\right) = 0.04$.

$$\text{Thus } u_1 = \{\phi_1\}\left(\frac{L_1}{M_1}\right)S_{a1}\left(\frac{ZI}{2R}\right)\left(\frac{T_1^2}{4\pi^2}\right) = \begin{Bmatrix} 8.075 \times 10^{-3} \\ 8.769 \times 10^{-3} \\ 9.124 \times 10^{-3} \end{Bmatrix} \text{m.}$$

Thus, the floor forces are

$$P_1 = \begin{bmatrix} 1440000 & -1280000 & 0 \\ -1280000 & 2560000 & -128000 \\ 0 & -1280000 & 1280000 \end{bmatrix} \begin{Bmatrix} 8.075 \times 10^{-3} \\ 8.769 \times 10^{-3} \\ 9.124 \times 10^{-3} \end{Bmatrix}$$
$$= \begin{Bmatrix} 405.396 \\ 427.608 \\ 458.949 \end{Bmatrix} \text{kN}$$

Thus, the shear per floor is

$$V_1 = \begin{Bmatrix} 405.396 + 427.608 + 458.949 \\ 427.608 + 458.949 \\ 458.949 \end{Bmatrix} = \begin{Bmatrix} 1292 \\ 887 \\ 459 \end{Bmatrix} \text{kN.}$$

Analysis for the second mode

For the second mode, $T_2 = 0.123$ s and $\phi_2 = \{-1 \quad -0.062 \quad 0.94\}^{\mathrm{T}}$. This gives

$$L_2 = \{\phi_2\}^{\mathrm{T}}[M]\{I\} = \{-1 \quad -0.062 \quad 0.94\} \begin{bmatrix} 524 & 0 & 0 \\ 0 & 509 & 0 \\ 0 & 0 & 0 \end{bmatrix} \begin{Bmatrix} 1 \\ 1 \\ 1 \end{Bmatrix} = -62.06$$

$$M_2 = \{\phi_2\}^{\mathrm{T}}[M]\{\phi_2\} = \{-1 \quad -0.62 \quad 0.94\} \begin{bmatrix} 524 & 0 & 0 \\ 0 & 509 & 0 \\ 0 & 0 & 0 \end{bmatrix} \begin{Bmatrix} -1 \\ -0.62 \\ 0.94 \end{Bmatrix}$$
$$= 989.847$$

For $T_2 = 0.123$ s, $S_{a2} = (2.5) \times 9.81 = 24.525\,\text{m/s}^2$ and $\left(\frac{ZI}{2R}\right) = 0.04$.

$$\text{Thus, } u_2 = \{\phi_2\}\left(\frac{L_2}{M_2}\right)S_{a2}\left(\frac{ZI}{2R}\right)\left(\frac{T_2^2}{4\pi^2}\right) = \begin{Bmatrix} 2.368 \times 10^{-5} \\ 1.468 \times 10^{-6} \\ -2.226 \times 10^{-5} \end{Bmatrix} \text{m.}$$

Thus, the floor forces are

$$P_2 = \begin{bmatrix} 1440000 & -1280000 & 0 \\ -1280000 & 2560000 & -128000 \\ 0 & -1280000 & 1280000 \end{bmatrix} \begin{Bmatrix} 2.368 \times 10^{-5} \\ 1.468 \times 10^{-6} \\ -2.226 \times 10^{-5} \end{Bmatrix}$$
$$= \begin{Bmatrix} 32.213 \\ 1.94 \\ -30.366 \end{Bmatrix} \text{kN}$$

Thus, the shear per floor is

$$V_2 = \begin{Bmatrix} 32.213 + 1.94 - 30.366 \\ 1.94 - 30.366 \\ -30.366 \end{Bmatrix} = \begin{Bmatrix} 3.8 \\ -28.4 \\ -30.366 \end{Bmatrix} \text{kN.}$$

Analysis for the third mode

For third mode, $T_3 = 0.072$ s and $\phi_1 = \{0.512 \quad -1 \quad 0.48\}^{\mathrm{T}}$.
This gives
For $T_3 = 0.072$ s $S_{a1} = (1 + 15T_3) \times 9.81 = 20.47\,\mathrm{m/s^2}$ and $\left(\frac{ZI}{2R}\right) = 0.04$, $L_3 = 11.288$ and $M_3 = 767.323$.

$$\rightarrow u_3 = \begin{Bmatrix} 8.199 \times 10^{-7} \\ -1.601 \times 10^{-6} \\ 7.685 \times 10^{-7} \end{Bmatrix} \text{m}$$

Thus, the floor forces are

$$P_2 = \begin{bmatrix} 1440000 & -1280000 & 0 \\ -1280000 & 2560000 & -128000 \\ 0 & -1280000 & 1280000 \end{bmatrix} \begin{Bmatrix} 8.199 \times 10^{-7} \\ -1.601 \times 10^{-6} \\ 7.685 \times 10^{-7} \end{Bmatrix}$$
$$= \begin{Bmatrix} 3.23 \\ -6.133 \\ 3.034 \end{Bmatrix} \text{kN}$$

Thus, the shear per floor

$$V_3 = \begin{Bmatrix} 3.23 - 6.133 + 3.034 \\ 3.034 - 6.133 \\ 3.034 \end{Bmatrix} = \begin{Bmatrix} 0.131 \\ -3.1 \\ -3.034 \end{Bmatrix} \text{kN.}$$

Thus, the SRSS base shear = $\sqrt{(1292)^2 + (3.8)^2 + (0.131)^2} \cong 1292\,\text{kN}$.

Based on the above, it is observed that the fundamental modal analysis is adequate for this case.

Analysis for kinematical interaction

Depth of bedrock (*H*) = 35 m. Shear wave velocity = 310 m/s.

Thus, the fundamental free field time period of the site is $T_s = 4H/V_s = 4 \times 35/310 = 0.452\,\text{s}$.

For 10% material damping of soil, this gives $S_{a1}/g = 2.0$ code factor $\beta = 0.06$.

The depth of foundation D_f = 10.05 m.

Thus, the average acceleration the basement is subjected to is

$$\ddot{u}_{n(\text{avg})} = \frac{1}{2}\kappa_n \beta S_{an}\left[1 + \cos(2n-1)\frac{\pi D_f}{2H}\right]$$

$$\ddot{u}_{n(\text{avg})} = \frac{1}{2} \times \left(\frac{8}{2+\pi}\right) \times 0.06 \times 2.0 \times \left(1 + \cos\left(\frac{\pi \times 10.5}{2 \times 35}\right)\right) \times 9.81$$
$$= 1.739\,\text{m/s}^2$$

The total weight of basement = 36,702 kN.

The total weight of superstructure = $5150 + 4997 + 5141 = 15,228\,\text{kN}$.

Mass of basement = 3741.3 kN s^2/m.

Mass of superstructure = 1558 kN s^2/m.

$$V_b = (3741.3 + 1558) \times 1.739 = 9215\,\text{kN}$$

Here, $Z_c = \frac{5594 \times 10.05 + 12893 \times 5.35 \times + 18215 \times 0.6}{5594 + 12893 + 18215} = 3.709\,\text{m}$ above from center of stiffness *S*.

$$M_b = 3741.3 \times 3.709 \times 1.739 + \left[\frac{5150}{9.81} \times 22.05 + \frac{4997}{9.81} \times 19.05 + \frac{5141}{9.81} \times 16.05\right] \times 1.739$$
$$= 24,131 + (11,576 + 9704 + 8411) \times 1.739\, M_b$$
$$= 75,764\,\text{kN m}$$

Thus, the equation of equilibrium can be expressed as

$$\begin{bmatrix} M_f+\sum_{i=1}^{3} m_{si} & M_fZ_c+\sum_{i=1}^{3} m_{si}(h_i+D_f) & m_1 & m_2 & m_3 \\ M_fZ_c+\sum_{i=1}^{3} m_{si}(h_i+D_f) & J_\theta+M_fZ_c^2+\sum_{i=1}^{3} m_i(h_i+D_f)^2 & m_1(h_1+D_f) & m_2(h_2+D_f) & m_3(h_3+D_f) \\ m_1 & m_1(h_1+D_f) & m_1 & 0 & 0 \\ m_2 & m_2(h_2+D_f) & 0 & m_2 & 0 \\ m_3 & m_3(h_3+D_f) & 0 & 0 & m_3 \end{bmatrix} \begin{Bmatrix} \ddot{u}_x \\ \ddot{\theta} \\ \ddot{u}_1 \\ \ddot{u}_2 \\ \ddot{u}_3 \end{Bmatrix}$$
$$+\begin{bmatrix} C_x & 0 & 0 & 0 & 0 \\ 0 & C_\theta & 0 & 0 & 0 \\ 0 & 0 & c_1+c_2 & -c_2 & 0 \\ 0 & 0 & -c_2 & c_2+c_3 & -c_3 \\ 0 & 0 & 0 & -c_3 & c_3 \end{bmatrix} \begin{Bmatrix} \dot{u}_x \\ \dot{\theta} \\ \dot{u}_1 \\ \dot{u}_2 \\ \dot{u}_3 \end{Bmatrix}$$
$$+\begin{bmatrix} K_x & 0 & 0 & 0 & 0 \\ 0 & K_\theta & 0 & 0 & 0 \\ 0 & 0 & k_1+k_2 & -k_2 & 0 \\ 0 & 0 & -k_2 & k_2+k_3 & -k_3 \\ 0 & 0 & 0 & -k_3 & k_3 \end{bmatrix} \begin{Bmatrix} u_x \\ \theta \\ u_1 \\ u_2 \\ u_3 \end{Bmatrix} = \begin{Bmatrix} V_{textb} \\ M_b \\ 0 \\ 0 \\ 0 \end{Bmatrix}$$

Mass of foundation $M_f = \frac{5594+12{,}893+18{,}215}{9.81} = 3741\,\text{kN s}^2/\text{m}$.
Mass moment of inertia about the X-direction $J_\theta = \frac{3741}{12}(18^2+10.05^2) = 132{,}495\,\text{kN s}^2/\text{m}$.
Considering 5% damping of structure, $c = \zeta \times 2\sqrt{k.m}$.
Thus,

$$c_1 = 0.05 \times 2\sqrt{k_1 \times m_1} = 0.05 \times 2\sqrt{160{,}000 \times 524} = 916$$
$$c_2 = 0.05 \times 2\sqrt{k_2 \times m_2} = 0.05 \times 2\sqrt{1{,}280{,}000 \times 509} = 2552$$
$$c_3 = 0.05 \times 2\sqrt{k_3 \times m_3} = 0.05 \times 2\sqrt{1{,}280{,}000 \times 525} = 2592$$

Here, D_f = 10.05 m, h_1 = 6.0 m, h_2 = 9.0 m, h_3 = 12.0 m and Z_c = 3.709 m. Substituting the above values and those furnished in Table 7.2, we have

$$
\begin{bmatrix} 5299 & 43560 & 524 & 509 & 525 \\ 43560 & 758900 & 8410 & 9696 & 11580 \\ 524 & 8410 & 524 & 0 & 0 \\ 509 & 9696 & 0 & 509 & 0 \\ 525 & 11580 & 0 & 0 & 525 \end{bmatrix} \begin{Bmatrix} \ddot{u}_x \\ \ddot{\theta} \\ \ddot{u}_1 \\ \ddot{u}_2 \\ \ddot{u}_3 \end{Bmatrix}
+ \begin{bmatrix} 2.546\times 10^5 & 0 & 0 & 0 & 0 \\ 0 & 1.26\times 10^7 & 0 & 0 & 0 \\ 0 & 0 & 3468 & -2552 & 0 \\ 0 & 0 & -2552 & 5144 & -2592 \\ 0 & 0 & 0 & -2592 & 2592 \end{bmatrix} \begin{Bmatrix} \dot{u}_x \\ \dot{\theta} \\ \dot{u}_1 \\ \dot{u}_2 \\ \dot{u}_3 \end{Bmatrix}
+ \begin{bmatrix} 1.82\times 10^7 & 0 & 0 & 0 & 0 \\ 0 & 1.906\times 10^9 & 0 & 0 & 0 \\ 0 & 0 & 1440000 & -1280000 & 0 \\ 0 & 0 & -1280000 & 2560000 & -1280000 \\ 0 & 0 & 0 & -1280000 & 1280000 \end{bmatrix} \begin{Bmatrix} u_x \\ \theta \\ u_1 \\ u_2 \\ u_3 \end{Bmatrix}
= \begin{Bmatrix} 9215 \\ 75764 \\ 0 \\ 0 \\ 0 \end{Bmatrix}
$$

The eigenvalue analysis of the above equation gives

$$
\begin{aligned} \omega &= \langle 9.609 \quad 50.806 \quad 64.89 \quad 86.735 \quad 131.71 \rangle \rightarrow T \\ &= \langle 0.654 \quad 0.124 \quad 0.097 \quad 0.072 \quad 0.048 \rangle \end{aligned}
$$

The corresponding eigenvectors are

$$
[\varphi] = \begin{bmatrix} 8.031\times 10^{-3} & 6.22\times 10^{-3} & -0.479 & 0.011 & 0.294 \\ 1.443\times 10^{-3} & -2.983\times 10^{-3} & -0.022 & -2.504\times 10^{-4} & -0.057 \\ 0.885 & 1 & 0.824 & -0.523 & 0.612 \\ 0.961 & 0.112 & 0.95 & 1 & 0.797 \\ 1 & -0.839 & 1 & -0.488 & 1 \end{bmatrix}
$$

Analysis for the first mode

Time period $T_1 = 0.653$ s $\rightarrow \omega_1 = 9.609\,\text{rad/s}$.
The corresponding scaling vectors are

$$\{\phi_1\} = \left\{ 8.031 \times 10^{-3} \quad 1.443 \times 10^{-3} \quad 0.885 \quad 0.961 \quad 1 \right\}^{\mathrm{T}}$$

$$M_1 = \{\varphi_1\}^{\mathrm{T}}[M]\{\varphi_1\} = \left\{ 8.031 \times 10^{-3} \quad 1.443 \times 10^{-3} \quad 0.885 \quad 0.961 \quad 1 \right\}$$
$$\times \begin{bmatrix} 5299 & 43560 & 524 & 509 & 525 \\ 43560 & 758900 & 8410 & 9696 & 11580 \\ 524 & 8410 & 524 & 0 & 0 \\ 509 & 9696 & 0 & 509 & 0 \\ 525 & 11580 & 0 & 0 & 525 \end{bmatrix} \begin{Bmatrix} 8.031 \times 10^{-3} \\ 1.443 \times 10^{-3} \\ 0.885 \\ 0.961 \\ 1 \end{Bmatrix}$$

$$\rightarrow M_1 = 1513 \zeta_1 = \frac{\{\phi_1\}^{\mathrm{T}}[C]\{\phi_1\}}{2\omega_1 M_1} \rightarrow 0.027$$

$$p_1 = \frac{\{\phi_1\}^{\mathrm{T}}[P]}{M_1} \rightarrow \left\{ 8.031 \times 10^{-3} \quad 1.443 \times 10^{-3} \quad 0.885 \quad 0.961 \quad 1 \right\}$$
$$\times \begin{Bmatrix} 4164 \\ 15444 \\ 0 \\ 0 \\ 0 \end{Bmatrix} \left(\frac{1}{1513} \right) = 0.121$$

The free field time period of site $T_s = 0.452\,\text{s} \rightarrow \omega_s = 13.9\,\text{rad/s}$.

$$r_1 = 13.9/9.609 = 1.449$$

Thus,

$$\xi_i = \frac{q_i}{\omega_i^2 \sqrt{(1 - r_i^2)^2 + (2\zeta_i r_i)^2}} \left[1 + \mathrm{e}^{-\zeta_i \omega_i t} \left\{ \sqrt{\frac{1 - \zeta_i}{1 + \zeta_i}} \sin\left(\omega_i \sqrt{1 - \zeta_i^2 t} \right) - \cos\left(\omega_i \sqrt{1 - \zeta_i^2 t} \right) \right\} \right]$$

$$\rightarrow \xi_i = \frac{0.051}{(9.609)^2 \sqrt{(1 - 1.449^2)^2 + (2 \times 0.027 \times 1.444)^2}} \times \left[1 + \mathrm{e}^{-0.259t} \{ 0.973 \sin 9.605t - \cos 9.605t \} \right]$$

Plot of the above is as shown in Fig. 7.27.

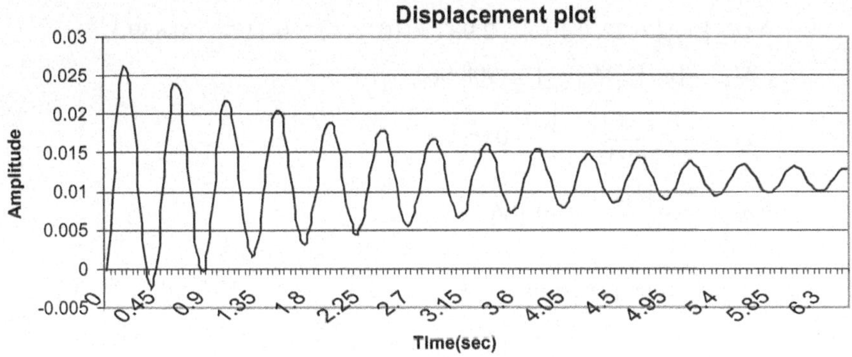

Fig. 7.27 Amplitude plot for the first mode

This shows the maximum amplitude in uncoupled mode as 0.00274 m.
Thus, the amplitude of vibration

$$\{U_1\} = \{\phi_1\}\xi_1 = \begin{Bmatrix} 0.015 \\ 9.093 \times 10^{-4} \\ 0.885 \\ 0.961 \\ 1 \end{Bmatrix} \times 0.00274 = \begin{Bmatrix} 2.197 \times 10^{-5} \\ 3.945 \times 10^{-6} \\ 2.42 \times 10^{-3} \\ 2.627 \times 10^{-3} \\ 2.735 \times 10^{-3} \end{Bmatrix} \text{m}$$

Now knowing the displacement in frame, to cut down computation we can determine the force in frame by multiplying the nodal displacement by structural part of the stiffness matrix only.

$$P_1 = \begin{bmatrix} 1440000 & -1280000 & 0 \\ -1280000 & 2560000 & -128000 \\ 0 & -1280000 & 1280000 \end{bmatrix} \begin{Bmatrix} 2.42 \times 10^{-3} \\ 2.627 \times 10^{-3} \\ 2.735 \times 10^{-3} \end{Bmatrix} = \begin{Bmatrix} 121.2 \\ 128 \\ 137.8 \end{Bmatrix} \text{kN}$$

Thus, the shear per floor

$$V_1 = 121 + 128 + 137 = 386\,\text{kN}$$
$$V_2 = 128 + 138 = 266\,\text{kN}$$
$$V_3 = 138\,\text{kN}$$

Thus, the base shear in the frame at top of basement is 386 kN.

Analysis for the second mode

Time period $T_2 = 0.124\,\text{s} \rightarrow \omega_2 = 50.806\,\text{rad/s}$.
The corresponding scaling vectors are

$$\{\phi_2\} = \{6.22 \times 10^{-3} \quad 2.983 \times 10^{-3} \quad 1 \quad 0.112 \quad -0.839\}^{\mathrm{T}}$$
$$M_2 = \{\phi_2\}^{\mathrm{T}}[M]\{\phi_2\} = 908.063$$
$$\zeta_2 = \frac{\{\phi_2\}^{\mathrm{T}}[C]\{\phi_2\}}{2\omega_2 M_2} \rightarrow 0.058$$
$$p_2 = \frac{\{\phi_2\}^{\mathrm{T}}[P]}{M_2} = -0.186$$

The free field time period of site $T_s = 0.452\,\mathrm{s} \rightarrow \omega_s = 13.9\,\mathrm{rad/s}$.

$$r_2 = 13.9/50.806 = 0.274$$

Thus

$$\xi_i = \frac{q_i}{\omega_i^2\sqrt{(1-r_i^2)^2 + (2\zeta_i r_i)^2}}\left[1 + \mathrm{e}^{-\zeta_i\omega_i t}\left\{\sqrt{\frac{1-\zeta_i}{1+\zeta_i}}\sin\left(\omega_i\sqrt{1-\zeta_i^2}t\right) - \cos\left(\omega_i\sqrt{1-\zeta_i^2}t\right)\right\}\right]$$

For $i = 2$

$$\rightarrow \xi_i = \frac{-0.186}{(50.806)^2\sqrt{(1-0.274^2)^2 + (2\times 0.058\times 0.274)^2}} \times \left[1 + \mathrm{e}^{-2.9467t}\{0.9435\sin 50.18t - \cos 50.18t\}\right]$$

Plot of the above is shown hereafter in Fig. 7.28.
This shows maximum amplitude as 0.00017 m.
Thus, the amplitude of vibration is

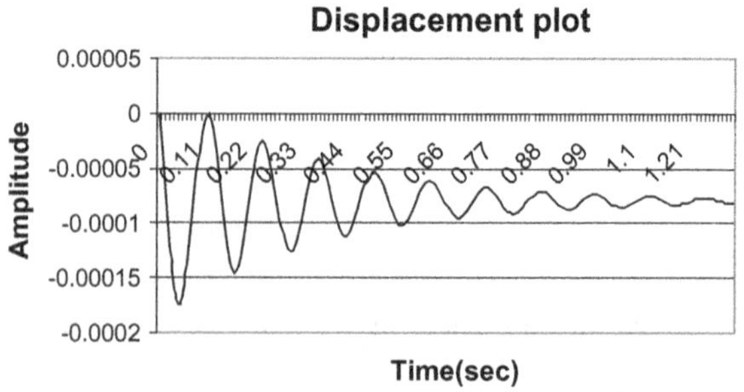

Fig. 7.28 Amplitude plot for the second mode

$$\{U_2\} = \{\phi_2\}\xi_2 = \begin{Bmatrix} -1.057 \times 10^{-6} \\ 5.072 \times 10^{-7} \\ 1.70 \times 10^{-4} \\ -1.909 \times 10^{-5} \\ -1.426 \times 10^{-4} \end{Bmatrix} \text{m}$$

Thus,

$$P_2 = \begin{bmatrix} 1440000 & -1280000 & 0 \\ -1280000 & 2560000 & -128000 \\ 0 & -1280000 & 1280000 \end{bmatrix} \begin{Bmatrix} 1.70 \\ -0.191 \\ -1.426 \end{Bmatrix} \times 10^{-4}$$
$$= \begin{Bmatrix} -220.36 \\ -13.782 \\ 206.94 \end{Bmatrix} \text{kN}$$

The shear per floor

$$V1 = -220 - 14 + 207 = -27$$
$$V_2 = 207 - 14 = -193\,\text{kN}$$
$$V_3 = -220\text{kN}$$

The base shear in the frame at the top of basement is 27kN in the second mode.

Analysis for the third mode

Time period $T_2 = 0.0.097\,\text{s} \rightarrow \omega_2 = 64.89\,\text{rad/s}$
The corresponding scaling vectors are

$$\{\phi_1\} = \{-0.479 \quad -0.022 \quad 0.824 \quad 0.950 \quad 1\}^{\text{T}}$$

$$M_3 = \{\phi_3\}^{\text{T}}[M]\{\phi_3\} = 1250$$
$$\zeta_3 = \frac{\{\phi_3\}^{\text{T}}[C]\{\phi_3\}}{2\omega_3 M_3} \rightarrow 0.404$$
$$p_3 = \frac{\{\phi_3\}^{\text{T}}[P]}{M_3} = -4.887$$

The free field time period of the site $T_s = 0.452\,\text{s} \rightarrow \omega_s = 13.9\,\text{rad/s}$.

$$r_2 = 13.9/64.89 = 0.214$$

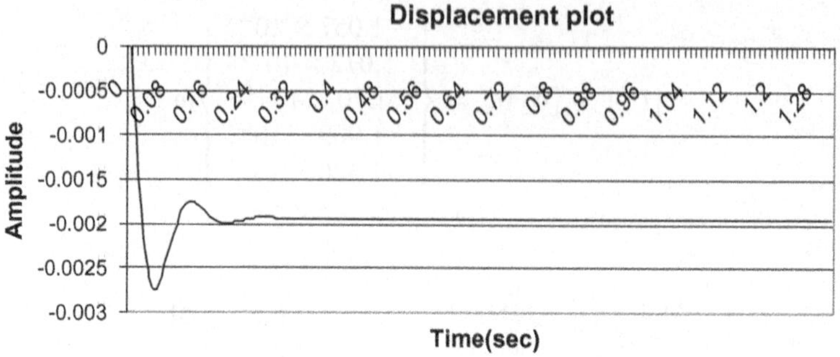

Fig. 7.29 Amplitude plot for the third mode

Thus,

$$\xi_i = \frac{q_i}{\omega_i^2\sqrt{(1-r_i^2)^2+(2\zeta_i r_i)^2}}\left[1+e^{-\zeta_i\omega_i t}\left\{\sqrt{\frac{1-\zeta_i}{1+\zeta_i}}\sin\left(\omega_i\sqrt{1-\zeta_i^2 t}\right)-\cos\left(\omega_i\sqrt{1-\zeta_i^2 t}\right)\right\}\right]$$

$$\to \xi_i = \frac{-1.494}{(64.89)^2\sqrt{(1-0.214^2)^2+(2\times 0.404\times 0.214)^2}} \times\left[1+e^{-26.215t}\{0.651\sin 46.72t-\cos 46.72t\}\right]$$

For $i = 3$

A plot of the above is shown in Fig. 7.29.

This shows that the maximum amplitude is −0.00273 m.

Thus, the amplitude of vibration

$$\{U_3\} = \{\phi_3\}\xi_3 == \begin{Bmatrix} 1.309\times 10^{-3} \\ 6.089\times 10^{-5} \\ -2.249\times 10^{-3} \\ -2.595\times 10^{-3} \\ -2.73\times 10^{-3} \end{Bmatrix} \text{m}$$

Hence,

$$P_3 = \begin{bmatrix} 1440000 & -1280000 & 0 \\ -1280000 & 2560000 & -128000 \\ 0 & -1280000 & 1280000 \end{bmatrix}\begin{Bmatrix} -2.249 \\ -2.595 \\ -2.73 \end{Bmatrix}\times 10^{-3}$$

$$= \begin{Bmatrix} 82.894 \\ -269.317 \\ -173.37 \end{Bmatrix}\text{kN}$$

The shear per floor

$$V_1 = -173 - 269 + 83 = -13\,\text{kN}$$
$$V_2 = -173 - 269 = -96\,\text{kN}$$
$$V_3 = 83\,\text{kN}$$

Thus, the base shear in the frame at the top of basement is −13 kN in the third mode.

We ignore the higher modes as the base shear progressively reduces and the first mode basically dominates.

The SRSS values of the first three modes are $V_{\text{SRSS}} = \sqrt{(386)^2 + (27)^2 + (13)^2} = 387\,\text{kN}$.

The moment and shear in the basement due to kinematical interaction can be obtained from the expression

$$\begin{Bmatrix} V_{bi} \\ M_{bi} \end{Bmatrix} = \begin{bmatrix} K_x & 0 \\ 0 & K_\theta \end{bmatrix} \begin{Bmatrix} u_x \\ \theta \end{Bmatrix}$$

These values are tabulated as hereafter

Mode	1	2	3
Shear in base of basement (kN)	400	19	23,820
Moment at base of basement (kN m)	7520	968	1.161×10^5

The SRSS values of shear and moment experienced by the base for the first three modes are V_b = 23,823 kN, M_b = 116,347 kN m. It may be observed that in the third mode the mass participation of the basement is maximum and gets excited significantly.

Analysis in case of kinematical resonance

This is a particular case to make the reader aware how dangerous can be the situation when the free field time period matches the fundamental frequency of the system.

Let us assume that the depth and soil properties are such that free field time period is within 10% of the first fundamental frequency of the system, i.e., r = 1.0.

Applying the equation for first mode

$$\xi_i = \frac{q_i}{\omega_i^2\sqrt{(1 - r_i^2)^2 + (2\zeta_i r_i)^2}}\left[1 + e^{-\zeta_i\omega_i t}\left\{\sqrt{\frac{1-\zeta_i}{1+\zeta_i}}\sin\left(\omega_i\sqrt{1-\zeta_i^2 t}\right) - \cos\left(\omega_i\sqrt{1-\zeta_i^2 t}\right)\right\}\right]$$

We finally have the maximum amplitude = 0.0262 m.
This gives structural deformation as

$$\{U_s\} = \begin{Bmatrix} 0.023 \\ 0.025 \\ 0.026 \end{Bmatrix} m$$

Thus, the nodal forces are

$$P_R = \begin{bmatrix} 1440000 & -1280000 & 0 \\ -1280000 & 2560000 & -128000 \\ 0 & -1280000 & 1280000 \end{bmatrix} \begin{Bmatrix} 0.023 \\ 0.025 \\ 0.026 \end{Bmatrix} = \begin{Bmatrix} 1161 \\ 1227 \\ 1321 \end{Bmatrix} \text{kN}$$

Thus,

$$V_1 = 1161 + 1227 + 1321 = 3709\,\text{kN}$$
$$V_2 = 1321 + 1227 = 2548\,\text{kN}$$
$$V_3 = 1321\,\text{kN}$$

The base shear for the fundamental mode in this case is 3709 kN, almost 9.6 times that of the present case. Thus, for framing with basements, it is very important to check that free field time period of the site does not match fundamental time period of the system, as the kinematical interaction gets amplified significantly.

Analysis for inertial interaction

In this case, as we have seen earlier that basically the fundamental mode is dominating, we do the analysis for the fundamental mode only.

Fixed-base time period = 0.642 s.

Total weight of building = 15,228 kN.

Total base shear = 1292 kN.

Thus equivalent structural stiffness $K_s = 4\pi^2 W/g\widehat{T}^2 = \frac{4\times\pi^2\times 15{,}228}{9.81\times(0.642)^2} = 148,683\,\text{kN/m}$.

Thus, the amplitude of structure $u_{st} = 1292/148,683 = 8.68 \times 10^{-3}$ m.

The centroid of building is 9.0 m above the top of basement.

Thus, the shear at the base of basement = 1292 kN.

Moment at the base of basement = 1292(9 + 10.05) = 24,613 kN m.

Thus, the equilibrium equation is

$$\begin{bmatrix} M_f + ms & M_f Z_c + m_s(h + D_f) \\ M_f Z_c + m_s(h + D_f) & J_\theta + M_f Z_c^2 + m_s(h + D_f)^2 \end{bmatrix} \begin{Bmatrix} \ddot{u}_f \\ \ddot{\theta} \end{Bmatrix} + \begin{bmatrix} C_f & 0 \\ 0 & C_\theta \end{bmatrix} \begin{Bmatrix} \dot{u}_f \\ \dot{u}_\theta \end{Bmatrix}$$
$$+ \begin{bmatrix} K_f & 0 \\ 0 & K_\theta \end{bmatrix} \begin{Bmatrix} u_x \\ \theta \end{Bmatrix} = \begin{Bmatrix} V_s \\ M_s \end{Bmatrix}$$

$$\begin{bmatrix} 5299 & 43560 \\ 43560 & 749400 \end{bmatrix} \begin{Bmatrix} \ddot{u}_f \\ \ddot{\theta} \end{Bmatrix} + \begin{bmatrix} 254600 & 0 \\ 0 & 1.26 \times 10^7 \end{bmatrix} \begin{Bmatrix} \dot{u}_f \\ \dot{u}_\theta \end{Bmatrix}$$
$$+ \begin{bmatrix} 1.82 \times 10^7 & 0 \\ 0 & 1.906 \times 10^9 \end{bmatrix} \begin{Bmatrix} u_x \\ \theta \end{Bmatrix} = \begin{Bmatrix} 1292 \\ 24613 \end{Bmatrix}$$

Eigenvalue analysis of the above equation gives

$$T = \langle 0.152 \quad 0.064 \rangle s \rightarrow \omega = \langle 41.469 \quad 98.626 \rangle \text{rad/s}$$

The corresponding eigenvectors are

$$\varphi = \begin{bmatrix} -1 & -1 \\ -0.121 & 0.079 \end{bmatrix}$$

The first mode analysis

$$T_1 = 0.152\,\text{s}$$

$$\{\phi_1\} = \{-1 \quad -0.121\}$$

$$M_1 = \{\phi_1\}^{\text{T}}[M]\{\phi_1\} = \{-1 \quad -0.121\} \begin{bmatrix} 5299 & 43560 \\ 43560 & 749400 \end{bmatrix} \begin{Bmatrix} -1 \\ -0.121 \end{Bmatrix}$$
$$= 26{,}900$$

$$\text{Load vector } Q = \begin{Bmatrix} 1292 \\ 24613 \end{Bmatrix}$$

$$q_1 = \frac{\{\phi_1\}^{\text{T}}\{Q\}}{M_1} = -0.159$$

$$\zeta_1 = \frac{\{\phi_1\}^{\text{T}}[C]\{\phi_1\}}{2\omega_1 M_1} \rightarrow 0.197$$

The fundamental frequency of the fixed-base structure is $\omega_f = 9.78\,\text{rad/s}$. Substituting the above values in the equation

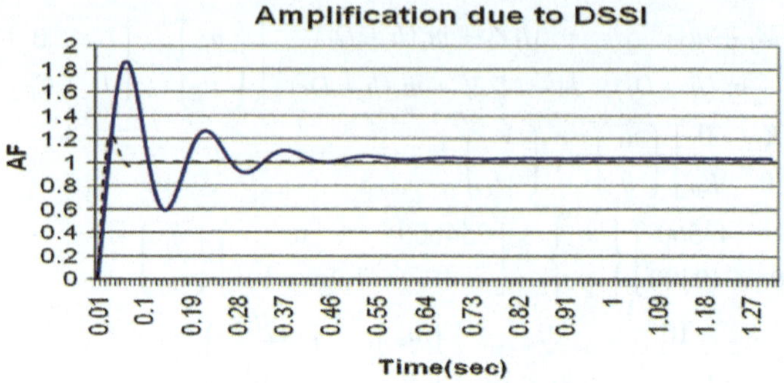

Fig. 7.30 Amplification factor due to soil–structure interaction

$$\mathrm{AF}_i = 1 + \left[\frac{u_{fi}}{u_s} + \frac{(h + D_\mathrm{f})\theta_i}{u_s}\right]\left[1 + \mathrm{e}^{-\zeta_i\omega_i t}\left\{\sqrt{\frac{1-\zeta_i}{1+\zeta_i}}\sin\left(\omega_i\sqrt{1-\zeta_i^2 t}\right) - \cos\left(\omega_i\sqrt{1-\zeta_i^2 t}\right)\right\}\right]$$

we get the computer plot as shown in Fig. 7.30.

Thus, it is observed in the first mode that the amplification due to DSSI is 1.856, while for the second mode, marked by the dotted line (whose calculations follow same steps), the amplification is AF = 1.22.

Thus, the maximum amplified base shear due to DSSI = 1292 × 1.856 = 2398 kN.

Thus, based on different methods, the values obtained for base shear are as tabulated hereafter

Base shear (kN)	IS-1893 (2016)	Modal analysis	Kinematical interaction (KI)	Inertial interaction (II)	Total shear (KI + II)
	1523	1292	387	2398	2795

Thus, we see that with all the conservativeness built in the code, the base shear still exceeds the design value by roughly 83%. Hence, while planning such structures, an engineer should give careful attention to both these interactions and consider the same in the analysis to design and construct a safe structure against seismic force.

7.11 Different Mechanical Analogs of Foundation and Soil

We furnish hereafter some standard spring and damping values of soil usually adopted for modeling the soil.

Stiffness values in Table 7.3, when multiplied by factors furnished in Table 7.4, give the stiffness of foundation embedded to a depth h below ground.

Damping values in Table 7.3, when multiplied by factors furnished in Table 7.5, give the effective damping of foundation embedded to a depth h below ground.

Here, $C = (r/V_s)k\gamma_0$ and $m = (r/V_s)^2 k\mu_0$, where r = equivalent radius and shall be r_0, r_θ, r_ψ as the case may be; G = dynamic shear modulus of soil; ρ = mass density of soil; V_s = shear wave velocity of soil; m = mass of soil participating in vibration with foundation; and C = damping of soil.

In Table 7.9, $\alpha = V_p/V_s$, V_p = velocity of compression(longitudinal) waves in soil, V_s = shear wave velocity in soil, $\aleph$ = embedment factor as furnished Table 7.8, and $a_0 = \omega r/V_s$ a dimensionless number, $b_\theta = 2/[1 + (e/r_\theta)]$, $b_\psi = 1/\left[0.37 + 0.87(e/r_\theta)^{\frac{2}{3}}\right]$, K_z^s, K_x^s, etc., represent static springs in vertical, horizontal, rocking, etc., as expressed within the third parenthesis in the first column of Table 7.9. Here superscript s stands for the word static.

For above $C = (b/V_s)k\gamma_0$ and $m = (b/V_s)^2 k\mu_0$.

In Table 7.11, λ = aspect ratio of the foundation ($2l/2b$), and e is the embedment depth. Here $C = (b/V_s)k\gamma_0$.

It is observed from the above that provided we are in a position to correctly model the soil as springs and dashpots, we can realistically predict the amplification due to dynamic soil–structure interaction for a superstructure under earthquake loading.

However, to the authors' experience, there exists a serious knowledge gap among many users as to what should be the correct values of spring to be used, what are the limitations and which spring to use where?

Like the finite element analysis, there are a number of formulations available for use, and without understanding their background, if one blindly follows them, the result could vary significantly from the reality where one may overestimate the response (which is still better though) but can also seriously underestimate the same and can well encroach in the danger zone of damage.

This area still remains a twilight zone to many, because the knowledge pertaining to behavior of soil under high strain seismic load remains deficient, and not surprisingly the hesitation is there to take up a detailed soil–structure interaction analysis.

The spring and dashpot values are actually a function of dynamic shear modulus G on Poisson's ratio ν.

The mechanical spring and dashpot analogs were mostly developed in the realms of machine foundation where the strain induced in soil is low and usually varies between $10^{-4\%}$ to a maximum of $10^{-3\%}$. For an earthquake, even a moderately strong one will induce a much higher strain in soil, varying from $10^{-2\%}$ to even $10^{-1\%}$ for strong motion records.

It has been observed that dynamic shear modulus G and damping of soil are strongly dependent on the strain range. While G gets reduced significantly with

Table 7.3 Values of soil springs for rigid circular footing resting on an elastic half-space

Sl No	Direction	Spring values	Damping values	Added mass	Equivalent radius for a rectangular footing
1	Vertical	$K_z = 4Gr_z/(1-v)$	$3.58\sqrt{G\rho r_z^4/(1-v)}$	$1.5\rho r_z^3$	$r_z = \sqrt{LB/\pi}$
2	Horizontal	$K_x = 8Gr_x/(2-v)$	$3.054\sqrt{G\rho r_x^4/(2-v)}$	$0.28\rho r_x^3$	$r_x = \sqrt{LB/\pi}$
3	Rocking	$K_{\varphi x} = 8Gr_{\varphi x}^3 \big/ 3(1-v)$	$1.33\sqrt{G\rho r_{\phi x}^8/(2-v)}$	$0.49\rho r_x^5$	$r_{\varphi x} = \sqrt{4}LB^3/3\pi$
3.1	Rocking	$K_{\varphi y} = 8Gr_{\varphi y}^3/3(1-v)$	$1.33\sqrt{G\rho r_{\phi y}^8/(2-v)}$	$0.49\rho r_y^5$	$r_{\varphi y} = \sqrt{4}L^3B/3\pi$
4	Twisting	$K_\psi = 16Gr_\psi^3/3$	$2.56\sqrt{G\rho r_{\phi y}^8}$	$0.70\rho r_\psi^5$	$r_\psi = \sqrt[4]{L^3B+BL^3/3\pi}$

Table 7.4 Embedment coefficients for spring constants (Whitman 1972)

Sl No	Direction	Coefficient
1	Vertical	$\eta_z = 1 + 0.6(1-\upsilon)(h/r_z)$
2	Horizontal	$\eta_x = 1 + 0.55(2-\upsilon)(h/r_x)$
3	Rocking	$\eta_{\varphi x} = 1 + 1.2(1-\upsilon)\left(h/r_{\varphi x}\right) + 0.2(2-\upsilon)\left(h/r_{\varphi x}\right)^3$
3.1	Rocking	$\eta_{\varphi y} = 1 + 1.2(1-\upsilon)\left(h/r_{\varphi y}\right) + 0.2(2-\upsilon)\left(h/r_{\varphi y}\right)^3$
4	Twisting	None available

Table 7.5 Embedment coefficients for soil damping ratio (Whitman 1972)

Sl No.	Direction	Coefficient
1	Vertical	$\alpha_z = [1 + 1.9(1-v)(h/r_z)]/\sqrt{\eta_z}$
2	Horizontal	$\alpha_x = [1 + 1.9(1-v)(h/r_x)]/\sqrt{\eta_x}$
3	Rocking	$\alpha_{\phi r} = \left[1 + 0.7(1-v)\left(h/r_{\phi x}\right) + 0.62(2-v)\left(h/r_{\phi x}\right)^3\right]/\sqrt{\eta_{\phi x}}$
3.1	Rocking	$\alpha_{\phi y} = \left[1 + 0.7(1-v)\left(h/r_{\phi y}\right) + 0.62(2-v)\left(h/r_{\phi y}\right)^3\right]/\sqrt{\eta_{\phi y}}$
4	Twisting	None available

Table 7.6 Soil spring constants (Wolf 1988)

Mode	Spring Stiffness (k)	γ_0 (Damping coeff.)	μ_0 (Mass coeff.)
Vertical	$\frac{4Gr_0}{1-\upsilon}$	0.85	0.27
Horizontal	$\frac{8Gr_0}{2-\upsilon}$	0.78–0.4v	0.095
Rocking	$\frac{8Gr_\theta^3}{3(1-\upsilon)}$	$\frac{0.3}{1+\frac{3(1-\upsilon)m}{8r_\theta^5\rho}}$	0.24
Torsion	$\frac{16Gr_\psi^3}{3}$	$\frac{0.433}{1+\frac{2m}{r_\psi^5\rho}}\left(\frac{m}{r_\psi^5\rho}\right)$	0.045

higher strain level (called cyclic stiffness degradation), radiation damping also reduces, while the material damping is enhanced (Seed and Idriss 1970a, b).

Thus, the first stumbling block is what G value should we take to design for a future earthquake whose intensity vis-à-vis strain range induced on soil is unknown to us?

The majority of the spring expressions were developed where the soil has been assumed to be an isotropic homogenous medium, but when one executes a project, it is rarely one would encounter soil that is homogenous and would be layered, so what G value is to be considered in the first place?

What happens if there is a soft soil of very shallow depth that is underlain by bedrock when waves get reflected from the same, and can significantly amplify the response?

Table 7.7 Values of soil springs for flexible circular footing resting on elastic half-space

Sl No.	Direction	Spring values	Damping values[a]	Added mass	Equivalent radius for a rectangular footing
1	Vertical	$K_z = \pi G r_z/(1-v)$	$3.4\sqrt{G\rho r_z^4/(2-v)}$	$1.5\rho r_z^3$	$r_z = \sqrt{LB/\pi}$
2	Horizontal	$K_x = 2\pi G r_x/(2-v)$	$3.054\sqrt{G\rho r_x^4/(2-v)}$	$0.28\rho r_x^3$	$r_x = \sqrt{LB/\pi}$
3	Rocking	$K_{\varphi x} = \pi G r_{\varphi x}^3/[2(1-v)]$	$1.33\sqrt{G\rho r_{\phi x}^8/(2-v)}$	$0.49\rho r_x^5$	$r_{\varphi x} = \sqrt{4}LB^3/3\pi$
3.1	Rocking	$K_{\varphi y} = \pi G r_{\varphi y}^3/[2(1-v)]$	$1.33\sqrt{G\rho r_{\phi y}^8/(2-v)}$	$0.49\rho r_y^5$	$r_{\varphi y} = \sqrt{4}L^3B/3\pi$
4	Twisting	$K_\psi = \pi G r_\psi^3/[2(1-v)]$	$2.56\sqrt{G\rho r_{\phi y}^8}$	$0.70\rho r_\psi^5$	$r_\psi = \sqrt{4}L^3B + BL^3/3\pi$

[a]No data exists; rigid foundation values are usually used

Table 7.8 Values of dynamic soil springs for rigid cylindrical footing embedded in an elastic half-space (Kausel and Pais 1998)

Mode	Spring stiffness (k) without embedment	Embedment factor ($\aleph$)	Remarks
Vertical	$K_z = 4Gr_z/(1-v)$	$1 + 0.54(e/r_z)$	e = Embedment depth
Horizontal	$K_x = 8Gr_x/(2-v)$	$1 + e/r_x$	
Rocking	$K_{\varphi x} = 8Gr_\theta^3/3(1-v)$	$1+(2/3)(e/r_\theta)+0.58(e/r_\theta)^3$	
Torsion	$K_\psi = 16Gr_\psi^3/3$	$1+2.67(e/r_\psi)$	

Table 7.9 Values of dynamic soil springs for rigid cylindrical footing resting on elastic half-space after Kausel and Pais (1998)

Mode	Dynamic spring stiffness	k	c
Vertical	$K_z = [4Gr_z\aleph/(1-\nu)](k + ia_0c)$	1.0	$\pi Gr_z[\alpha + 2(e/r_z)]/K_z^s$
Horizontal	$K_x = [8Gr_x\aleph/(2-\nu)](k + ia_0c)$	1.0	$\pi Gr_x[1 + (1+\alpha)(e/r_x)]/K_x^s$
Rocking	$K_\theta = \left[8Gr_\theta^3\aleph/3(1-\nu)\right](k + ia_0c)$	$1 - 0.35a_0^2/(1 + a_0^2)$	$\pi Gr_\theta^3\left[(\alpha/4) + \frac{(1+\alpha)}{3}(e/r_x)^3\right]\left(\frac{a_0^2}{b_\theta + a_0^2}\right) + 0.84(1+\alpha)(e/r_\theta)^3\left(\frac{b_\theta}{b_\theta + a_0^2}\right) \Big/ K_\theta^s$
Torsion	$K_\psi = 16Gr_\psi^3/3$	Do	$\frac{\pi}{2}Gr_\psi^3\left[1 + 4(e/r_\psi)\right]\left(\frac{a_0^2}{b_\psi + a_0^2}\right)\Big/K_\psi^s$

Table 7.10 Spring stiffness for strip foundation of width $2b$ after Wolf (1994)

Mode	Static spring stiffness (k) without embedment	γ_0 (Damping coeff.)	μ_0 (Mass coeff.)
Vertical	$K_z = G(1+4v^2)$	$3.5–2v$	–
Horizontal	$K_x = G(1+5v^2)$	$2–2.2v$	0 for $v < 0.33$,$4.5(v - 1/3)$ for $v > 0.33$
Rocking	$K_\theta = Gb^2(1.8+5.2v^2)$	$0.14–0.24v^2$	0 for $v < 0.33$, $0.25(v - 1/3)$ for $v > 0.33$

Table 7.11 Spring stiffness for rectangular foundation of length $2l$ and width $2b$, after Wolf (1988)

Mode	Static spring stiffness (k) without embedment	γ_0 (Damping coeff.)
Vertical	$K_z = \frac{Gb}{1-v}\left[3\lambda^{0.75} + 1.6\right]\left[1 + (0.25 + 0.25/\lambda)(e/b)^{0.8}\right]$	$0.9 + 0.4(\lambda - 1)^{2/3}$
Horz. translation in x-direction	$K_x = \frac{Gb}{2-v}\left[0.68(\lambda)^{0.65} + 0.8\lambda + 1.6\right]\left[1 + \left(0.33 + 1.34b/(b+1)(e/b)^{0.8}\right)\right]$	$0.75 + 0.2(\lambda - 1)^{2/3}$
Horz. translation in y-direction	$K_y = \frac{Gb}{2-v}\left[0.68(\lambda)^{0.65} + 2.4\right]\left[1 + \left(0.33 + 1.34b/(b+1)(e/b)^{0.8}\right)\right]$	$0.75 + 0.2(\lambda - 1)^{2/3}$
Rocking about x-axis	$K_{\theta x} = \frac{Gb^3}{1-v}(3.2\lambda + 0.8)\left[1 + (e/b) + \{1.6/(0.35 + \lambda)\}(e/b)^2\right]$	–
Rocking about y-axis	$K_{\theta y} = \frac{Gb^3}{1-v}(3.73\lambda + 0.27)\left[1 + (e/b) + \{1.6/(0.35 + \lambda)\}(e/b)^2\right]$	–
Torsion	$K_\psi = Gb^3\left(4.25(\lambda)^{2.45} + 4.06\right)\left[1 + (1.3 + 1.32/\lambda)(e/b)^{0.9}\right]$	–

Finally, most of the spring values developed for machine foundation consider the foundation to be rigid (majority of these equipments are mounted on block-type foundation), while for foundation for any ordinary industrial or building structure the footing could be flexible, or semi rigid. Are the springs recommended still valid for these cases?

There are a number of such issues, a designer must have absolute clarity on, before he should proceed with a DSSI analysis.

Some of these very important, yet oft neglected issues will be examined in the next chapter.

References

Anandakrishnan, M., & Krishnaswamy, N. R. (1973). Response of embedded footings to vertical vibrations. *Journal of the Soil Mechanics and Foundations Engineering, ASCE, 99*(10), 863–883.

Barkan, D. D. (1962). *Dynamics of bases and foundations*. New York: McGraw-Hill Book Co.

Biggs, J. M. (1964). *Introduction to structural dynamics*. New York: McGraw Hill Publication.

Chowdhury, I. (1984). *Dynamic soil structure interaction of turbo-genrator foundation*. M.Tech thesis. Kharagpur: Indian Institute of Technlogy.

Chowdhury, I., & Dasgupta, S. P. (2002). Earthquake response of soil-structure system. *Indian Geotechnical Journal, 32*(2), 309–328.

Chowdhury, I., & Dasgupta, S. P. (2007). Dynamic earth pressure on rigid unyielding walls under earthquake forces. *Indian Geotechnical Journal, 37*(2), 81–93.

Chowdhury, I., & Singh, J. P. (2010). Do DSSI attenuate dynamic response of Buildings? In *Proceedings 14th International Symposium on Earthquake Engineering*. Roorkee, India.

Chowdhury, I., & Som, P. K. (1993). Dynamic pile structure interaction of boiler feed pump frame foundation. *Indian Geotechnical Conference, 1*, 411–414.

Chowdhury, I., Ghosh, A., & Dasgupta, S. P. (2015a, May). Dynamic soil structure interaction of structures under earthquake motion. *Journal of Indian Geotechnical Society*.

Chowdhury, I., Tarafdar, R., & Ghosh, A. (2015b). An analytical solution to kinematic andinertial interaction of building with deep basements. In *Proceedings, 23rd International Conference in Structural Mechanics in Reactor Technology*. Manchester, U.K.

Clough, R. W., & Penzien, J. (1975). *Dynamics of structure*. New York: McGraw-Hill Kogakusha Ltd.

Dasgupta, S. P. (1976). Finite element solutions to some problems in the dynamics of footings. Ph. D Thesis, Indian Institute of Technology, Kanpur, India, August.

Dasgupta, S. P., & Kameswara Rao, N. S. V. (1976). Some finite element solutions in the dynamics of circular footings. In *Proceedings of the 2nd International Conference on Numerical Methods in Geo-mechanics*, Blacksburg, USA.

Dasgupta, S. P., & Kameswara Rao, N. S. V. (1978). Dynamics of rectangular footings by finite elements. *Journal of GT Division, ASCE, 104*(5).

Dowrick, D. J. (2003). *Earthquake risk reduction*. UK: Willey.

Ehlers, G. (1942). The effect of soil flexibility on vibrating systems. *Beton und Eisen, 41*(21/22), 197–203.

Ghosh, D., Chakroborty, D., Batavyal, H. (1984). Dynamic response and static analysis of RCC frames supporting high speed centrifugal machines with Soil Structure Interaction. In *International Conference on Case Histories in Geo-technical Engineering St. Louis University of Missouri Rolla*, USA.

Ghosh, D. K., & Batavayal, H. N. (1985). Analysis of structural response to earthquake for 150M high RCC Chimney with soil-structure interaction. In *Proceedings of National Seminar on Tall Chimneys Vigyan Bhavan*, New Delhi.

Gazetas, G. (1983). Analysis of machine foundation a state of the art. *Journal of Soil Dynamics and Earthquake Engineering, 2*(1), 2–42.

Gazetas, G., & Tassoulas, A. L. (1987). Horizontal stiffness of arbitrarily shaped embedded foundation. *Journal of GT Division ASCE, 113*(5).

Hall, J. R., & Kissenpfennig, J. F. (1976). Special topics on soil-structure interaction. *Nuclear Engineering Design, 38*, 273–287.

Hurty, W. C., & Rubenstein, M. F. (1967). *Dynamics of structure*. New Delhi: Prentice-Hall of India Pvt. Ltd.

Jennings, P. C., & Bielak, J. (1973). Dynamics of building-soil interaction. *Bulletin of Seismological Society of America, 63*(1), 9–48.

Kameswara Rao, N. S. V. (1977). Dynamic soil structure system—a brief review. *Journal of Structural Engineering India, 4*.

Kausel, E. (1974). Forced vibrations of circular foundations on layered media. *Research Report*. USA: MIT.

Kausel, E. (2010). Early history of soil-structure interaction. *Soil Dynamics and Earthquake Engineering, 30*, 822–832.

Kausel, E., & Pais, A. (1998). Formulas for dynamics stiffness of rigid foundation. *Journal of Soil Dynamics andEarthquake Engineering, 7*.

Kramer, S. (1998). *Geotechnical Earthquake Engineering*. New Delhi, India: Pearson Education.

Lamb, H. (1904). On propagation of tremors over the surface of an elastic solid. *Philosophical Transactions of the Royal Society of London, 203,* 1–42.

Luco, J. E., & Hadijian, A. H. (1974). Two dimensional approximation to three dimensional soil-structure interaction problem. *Nuclear Engineering Design, 31*, 195–23.

Lysmer, J. (1965). Vertical motion of rigid footings, Contract Report No. 3-115, U.S. Army Engineer, WES, Vicksburg, Miss., S. 1–130.

Lysmer, J., & Richart, F. E. Jr. (1966, January). Dynamic response of footings to vertical loading. *Journal of Soil Mechanics and Foundations Division, ASCE, 92*(SM1), 65–91.

Marguerre, K. (1931). Druckverteilung durch eine elastische Schicht auf starrer rauer Unterlage. *Ingenieur Archiv* Band II, S108–117.

Margurre, K. (1933). Spannungsvertilung und Wellenausbreitung in der kontinuierlich gestutzten Platte. *Ingenieur Archiv* Band IV, S332–353.

Martel, R. R. (1940). Effect of foundation on earthquake motion. *Civil Engineering ASCE, 10*(1), 7–10.

Newmark, N. M. (1959). A method of computation for structural dynamics. *Journal of the engineering mechanics division, ASCE, 85*, 67–94.

Newmark, N. M., & Rosenblueth, E. (1971). *Fundamentals of earthquake engineering*. Englewood Cliffs, NJ: Prentice-Hall.

Novak, M. (1974, November). Dynamic stiffness and damping of piles. *Canadian Geotechnical Journal, 11*(4), 574–598.

Novak, M., & Beredugo, Y. O. (1972a). Coupled horizontal and rocking vibration of embedded footings. *Canadian Geotechnical Journal, 9*, 477–497.

Novak, M., & Beredugo, Y. O. (1972b). Vertical vibration of embedded footings. *Journal of Soil Mechanics and Foundation Divisons, ASCE, 12,* 1291–1310.

Novak, M., & El Sharnouby, B. (1983). Stiffness and damping constants for single piles. *Journal of Geotechnical Engineering Division, ASCE, 109*, 961–974.

Novak, M., & Hifnawy, L. E. (1983). Vibration of hammer foundation. *Journal of Soil Dynamics and Earthquake Engineering, 2*(1), 43–53.

Pekeris, C. L. (1955). The seismic surface pulse. *Proceedings of the National Academy of Science USA, 41,* 469–480.

Pekeris, C. L., & Lifson, H. (1957). Motion of the surface of a uniform elastic half space produced by a Buried Pulse. *Journal of Acoustical Society of America, 29*(11), 1233–1238.

Quinlan, P. M. (1953). Elastic theory of soil dynamics. In *ASTM Special Technical Publication No 156, Symposium on Soil Dynamics*, pp. 3–34.
Reissner, E. (1936, December). Staionare axilasymmetriche durch eine schuterende Masse erregte Schwingungen eines homogenen elastischen Halbraumes. *Ingenieur Archiv, 22*, 381–396.
Reissner, E. (1937). Frie und erzwungene Torsionschwingungen des elastischen Halbraume. *Ingenieur Archiv* VIII Band 4 Heft S229–245.
Richart, F. E., Jr. (1966). Foundation vibration. *Transactions ASCE, 127*, Part 1, 863–898.
Richart, F. E. Jr., Hall, Jr. J. R., & Woods, R. D. (1970). *Vibrations of soils and foundations.* Englewood Cliffs, NJ: Prentice-Hall.
Seed, H. B., & Idriss, I. M. (1970a). *Soil moduli and damping factors for dynamic response analysis*. Report No. 70-1, EERC Berkley, California.
Seed, H. B., & Idriss, I. M. (1970b). *Soil moduli and damping factors for dynamic response analysis*. University of California Berkley; Earthquake Engineering Research Center Report No EERC 70-10.
Sung, T. Y. (1953, July). Vibrations in semi infinite solids due to periodic surface loading. ASTM Special Technical Publication No.156, *Symposium on soil dynamics*, pp. 35–64.
Veletsos, A. S., & Meek, J. W. (1974). Dynamic behavior of building-foundation systems. *Earthquake Engineering and Structural Dynamics, 3*(2), 121–138.
Whitman, R. V. (1972). *Analysis of soil structure interaction-a state of the art review*. Soil Publication No 300 M.I.T.
Wolf, J. P. (1985). *Dynamic soil-structure interaction*. Englewood-Cliffs: Prentice-Hall.
Wolf, J. P. (1988). *Dynamic soil structure interaction in time domain*. New Jersey, USA: Prentice Hall Publication.
Wolf, J. P. (1994). *Foundation vibration analysis, using simple physical model*. Englewood-Cliffs: Prentice-Hall.
Wolf, J., & Zhang, C. (Ed.). (1999). *Dynmaic soil structure interaction*. USA: Elsevier Publication.
Whitman, R. V. (1970). *Soil structure interaction, seismic design for nuclear power plants*. Cambridge: The MIT Press.

Chapter 8
Geotechnical Considerations for Seismic Analysis

8.1 Introduction

In this chapter, we will discuss the geotechnical considerations that are relevant to earthquake-resistant design of a structure and also the pertinent parameters that are part of a successful dynamic soil–structure interaction (DSSI) analysis.

At the very outset, we beseech readers having a strong structural bias, not to skip over this section as our experience shows that misunderstanding and wrong interpretation of dynamic soil parameters have led to significant errors and subsequent ill behavior of many structures, resulting in expensive remedial measures and delay in projects.

For industrial plants, geotechnical investigation of dynamic properties of soil is usually carried out by default, as these plants usually constitute of a number of rotating or reciprocating machines. However, for building industry or infrastructure development like bridges and highways, ports and harbors, people rarely appreciate the significance and importance of these properties and it has not been uncommon in the past that, in many cases, engineers have paid dearly for this expensive oversight.

As such, before launching into a linear or nonlinear finite element analysis of soil–structure system, the conceptual aspect of the influencing soil parameters, its limitations, and its effects should be clearly understood.

8.2 Geotechnical Parameters that Influence Earthquake-Resistant Design Including DSSI

The readers having some background on this issue may find the heading intriguing. The answer to this would be *dynamic shear modulus* (G) and *Poisson's ratio* (ν) of the soil.

I. Chowdhury and S. P. Dasgupta, *Earthquake Analysis and Design of Industrial Structures and Infra-structures*, GeoPlanet: Earth and Planetary Sciences,
https://doi.org/10.1007/978-3-319-90832-8_8

The obvious query that subsequently comes to mind then is, does it require a full chapter to address this issue?

The answer would surely be an emphatic *yes*, for in our opinion the values adopted from a geotechnical report are often misunderstood/abused in many a case and often make the analysis questionable or unrealistic.

Moreover, under seismic loading under large strain, soil behaves quite differently than how it would behave while one would design a machine foundation (low strain problem) and the qualitative and quantitative difference should be clearly understood by the designer undertaking the task.

The reasons why this topic needs special emphasis may be summarized as follows:

- Geotechnical tests (laboratory or field) based on which data are evaluated are not understood properly in many cases. As the limitation of such data is not understood clearly, it often results in incorrect interpretation.
- Data considered from soil report often needs readjustment in terms of real situation in the field, especially for a layered soil.
- Insufficient data and or lack of knowledge on the strain level to which the foundation–structure system will be subjected to–*especially during a strong motion earthquake*.
- Lack of dynamic test data and improperly corelated value from static soil parameters which could be widely varying with actual reality.
- Finally, often forgetting the bottom line that, unlike man-made material like concrete and steel, soil is far more heterogeneous and unpredictable; thus, for a real soil–structure interaction analysis it is unfair to have an analysis on an absolute scale. It should preferably be done for a particular range of values and the best estimate is to be made out of it—and this is where the engineering judgment would count to a great extent.

Having made the above statements, let us evaluate various aspects of dynamic property of soil, which are important for an integrated soil–structure interaction analysis.

Before even looking at the soil report, the engineer should be clear with himself on

- The type of structure he is dealing with.
- The type of foundation that is anticipated, like shallow foundation (could be isolated or combined footing), raft or piles.
- What analysis is he looking for? Like, is it an analysis for machine-induced load, earthquake, blast force.

Understanding of the above criterions will not only help him in understanding the data obtained from different tests but could also possibly make him realize their interpretation in a more realistic perspective.

The engineering parameters we look for in soil report for developing the soil model either for finite element or linear/nonlinear spring–dashpot model are

1. Dynamic shear modulus (G) or shear wave velocity (V_s), corelated by the expression $G = \rho V_s^2$ where ρ = mass density of soil;
2. Poisson's ratio (v);
3. Damping value of soil (both geometric and material damping).

The values are usually obtained either from field test, laboratory test, or from theoretical corelation with other engineering soil parameters. Before we step further into the topic, it would possibly be worthwhile to understand how soil behaves under cyclic loading and what its characteristics are.

It should be remembered that even under low strain, soil behavior is essentially nonlinear though at low strain it does show some kind of linearity.

Shown in Fig. 8.1 is the shear stress–strain curve of soil under cyclic loading. It is evident from the above figure that shear strain varies with stress and goes on increasing with the number of cycles of loading.

Thus, before an analysis is being carried out, one has to have an idea about the average strain range to which the soil will be subjected to under the induced dynamic loading.

The characteristic curve which shows the variation of shear modulus and damping ratio with respect to shear strain is as shown in Fig. 8.2.

The curve shown in Fig. 8.2 is otherwise known as Seed and Idriss's (1970) curve, which shows the variation of dynamic shear modulus ratio (G/G_{max}) and damping ratio (D/D_{max}) of soil with shear strain.

Soil subjected to stress by machine foundation usually develops low strain, varying from 10^{-5} to 10^{-4}.

However, for an earthquake of even moderate magnitude this will be much higher—having strain range varying from 10^{-3} to even 10^{-1} for very severe earthquake.

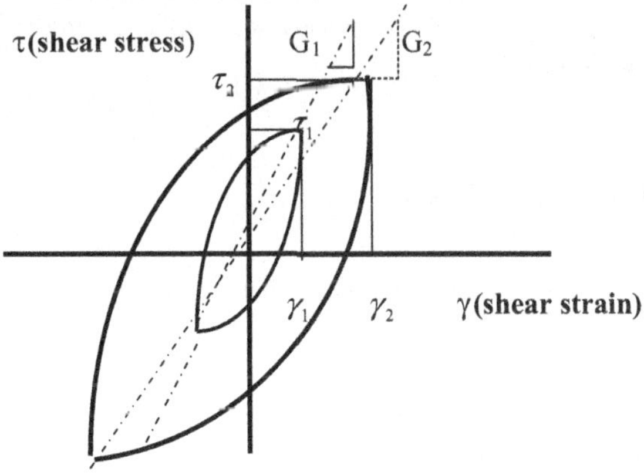

Fig. 8.1 Shear stress–strain curve of soil under cyclic loading

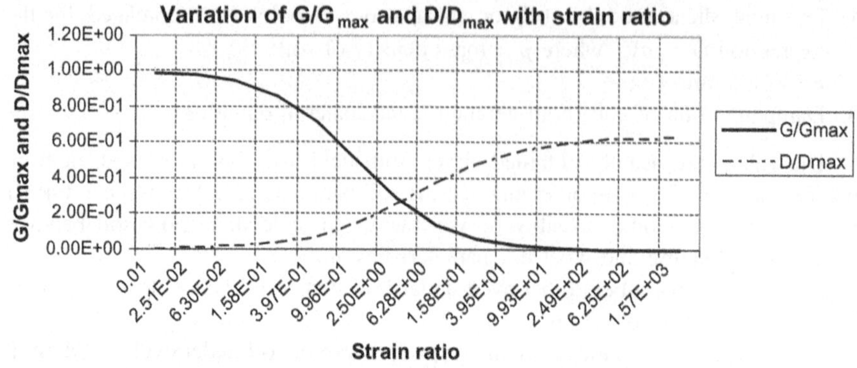

Fig. 8.2 Variation of shear modulus (G/G_{max}) and damping ratio (D/D_{max})

Since it is difficult to gauge at the outset of an analysis how much strain soil will be subjected to, the correction factor to be used to modify the data as obtained in the soil report becomes difficult to quantify.

On the contrary, rendering no correction would result in assuming a more stiff soil and the result obtained based on this could be significantly different from the reality.

Fortunately or unfortunately, most of the tests carried out in the field or in the laboratory for determination of the dynamic shear modulus are based on low strain range having values restricted to 10^{-4}.

Thus, it should be clearly understood that dynamic shear modulus data furnished in a soil report is commonly valid for low strain range and can be only used directly for analysis where the strain induced in the soil is significantly low, like in design of machine foundations only. For earthquake analysis where the site is situated in an area of moderate to severe earthquake zone, *direct use of such soil dynamic data is not valid for design of normal structures, as the strain induced in soil is usually much higher.*

8.3 Field Tests

The most common field tests that are carried out at site for evaluation of dynamic shear modulus or shear wave velocity are

1. Block vibration test;
2. Seismic cross-hole;
3. Cyclic plate load test.

8.3.1 Block Vibration Test

In block vibration test, as shown in Fig. 8.3, an oscillator is placed on a concrete block resting on ground and induces dynamic loading on the soil. Two geophones are placed at a distance to pick up the signal from the oscillator.

Once the oscillator induces dynamic force on the soil, the geophones pick up this signal and transfer it to an oscilloscope which shows an elliptical figure of Lissajous. The operating speed of the oscillator is varied till the time the natural frequency of the soil and the operating frequency of the oscillator matches (the Lissajous' figure in the oscilloscope becomes a perfect circle).

The shear wave velocity of the site is then given by

$$V_s = 4fL_x \tag{8.3.1.1}$$

where V_s = shear wave velocity of the soil; f = operating frequency of the oscillator in c.p.s; L_x = distance between the two geophones.

For arriving at meaningful results, usually high-frequency oscillators (>100 cps) are used, for which the waves generated are of the order of 0.6–1.2 m.

Thus, the results obtained from this test only influence soil of depth 0.6–1.2 m below the depth of foundation and should not be used where piles or other types of deep foundations, having influence area propagating much deeper, are used.

Trying to induce lower frequency calls for much heavier oscillators, which makes the test uneconomical compared to other types of tests.

IS-5249 (1992) and ACI 351.3R-04 (2011) Code of practice for determination of dynamic properties of soil of Bureau of Indian Standards and analysis and design of machine foundation (American Concrete Institute) recommend two methods for determination of G value based on block vibration test. The test setup is as shown in Fig. 8.4.

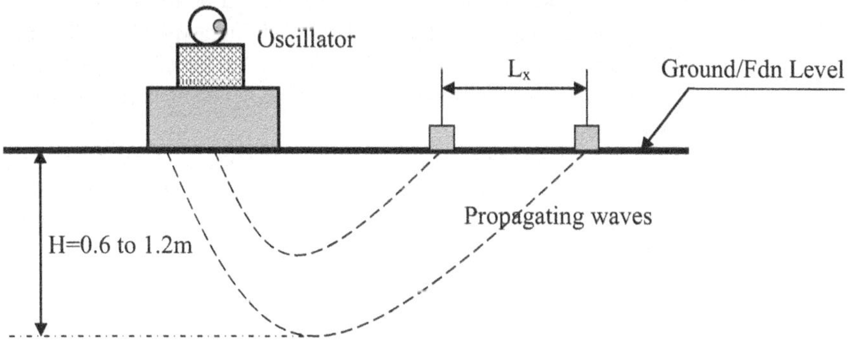

Fig. 8.3 Schematic diagram for block vibration test

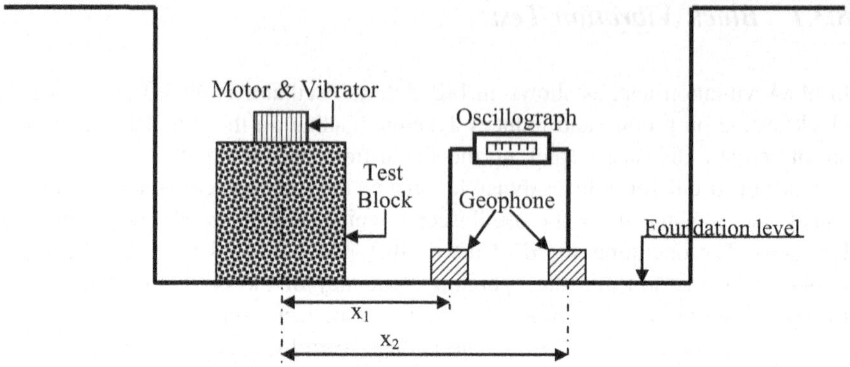

Fig. 8.4 Typical test setup for block vibration test

Forced vibration test

A plain cement concrete block of M15/20 grade concrete should be constructed in test pit, as shown in Fig. 8.4. The test pit size is usually 3 m × 6 m in plane and depth, preferably equal to the proposed depth of foundation. The test should be conducted above the groundwater table. In ordinary soils, the test block is usually of size 1 m × 1 m × 1.5 m and, in dense soils, the size adopted is 0.75 m 0.75 m × 1 m. The block size should be so adjusted that the mass ratio $B_z = \frac{(1-\nu)m_f}{\rho_s r^3}$ is always more than unity. Here m_f = mass of foundation r = equivalent radius of the footing ρ_s = mass density of soil and ν = Poisson's ratio of soil. The concrete block should be cured for at least 15 days before testing.

As shown in Fig. 8.4, the vibration exciter is mounted on the block such that it generates purely vertical sinusoidal vibrations and the line of action of vibrating force passes through the center of gravity of the block. The exciter is operated at a constant frequency. The signals are picked up by the geophones and fed into an oscillograph where the amplitudes are measured for the particular operating frequency of the exciter. The frequency of the exciter is fed into steps of small values (1–4 cycles/s) up to maximum frequency of the exciter and for each steps amplitudes are recorded. Based on the data recorded, a frequency versus amplitude curve is plotted, as shown in Fig. 8.5.

It is apparent from Fig. 8.5 that the amplitude will be *maximum* when the operating frequency is in resonance with the natural frequency of the foundation. Thus, the natural frequency of the foundation can be extricated from the recorded data.

Once the natural frequency f_n of the footing is known, C_z (coefficient of uniform compression having unit of kN/m^3) is computed from the expression Barkan (1962),

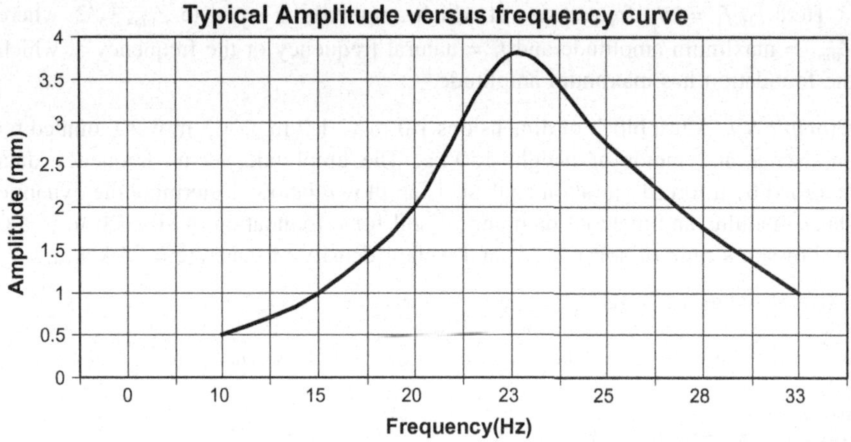

Fig. 8.5 Typical amplitude versus frequency curve for a block vibration test

$$C_z = \frac{4\pi^2 f_n^2 M}{A_f} \tag{8.3.1.2}$$

Here M = mass of foundation plus exciter and motor mounted on the block, A_f = contact area of the block with soil.

The design C_{za} value is finally computed for the actual foundation (A_1) from the expression

$$C_{za} = C_z \sqrt{\frac{A_f}{A_1}} \tag{8.3.1.3}$$

Young's modulus of soil, E_s, is finally corelated to C_{za} by the expression

$$C_{za} = \frac{1.13}{\sqrt{A_1}} \left(\frac{E_s}{1 - \nu^2} \right) \tag{8.3.1.4}$$

The dynamic shear modulus of soil, G, is finally computed from the expression

$$G = \frac{E_s}{2(1 + \nu)} \tag{8.3.1.5}$$

The material damping of soil ζ is computed from the expression

$$\zeta = \frac{f_2 - f_1}{2 f_n} \tag{8.3.1.6}$$

Here f_2, f_1 = two frequencies at which amplitude is equal to $Z_{\max}/\sqrt{2}$, where $Z_{\max}$ = maximum amplitude and f_n = natural frequency or the frequency at which the foundation has maximum amplitude.

Example 8.1 A test block of dimensions 1.0 m × 1.0 m × 1.5 m was mounted by an exciter and motor of weight 150 kg. The amplitude versus frequency data recorded by a forced vibration analysis is as shown below. Determine the dynamic shear modulus and material damping of soil for a foundation of size 3.0 m × 3.0 m, Poisson's ratio of soil = 0.35, and weight density of concrete = 24 kN/m^3.

Frequency (Hz)	50	75	100	123	125	130	175
Amplitude (m)	0.5×10^{-4}	1.0×10^{-4}	2.0×10^{-4}	3.8×10^{-4}	2.75×10^{-4}	1.0×10^{-4}	0.6×10^{-4}

Mass of test block = 1 × 1 × 1.5 × 24 = 36 kN
Weight of exciter and motor = 1.5 kN
Total weight = 37.5 kN
Mass (M) = 3.822 kN s^2/m

Based on the table furnished it is observed that resonant frequency = natural frequency = 123 Hz, as the maximum amplitude is observed at this frequency. Thus,

$$C_z = \frac{4\pi^2 f_n^2 M}{A_f} = \frac{4 \times \pi^2 \times 123 \times 123 \times 3.822}{1 \times 1} = 2,282,762\ \mathrm{kN/m^3}$$

Thus, for a foundation of size 3 m × 3 m the design C_{za} is expressed as

$$C_{za} = C_z\sqrt{\frac{A_f}{A_1}} = 2,282,762 \times \sqrt{\frac{1}{9}} = 76,0921\ \mathrm{kN/m^3}$$

Thus, Young's modulus is expressed as

$$E_s = \frac{C_{za}(1 - \nu^2)\sqrt{A_1}}{1.13} = \frac{760,921 \times \left[1 - (0.35)^2\right]\sqrt{9}}{1.13} = 1,772,677\ \mathrm{kN/m^2}.$$

Thus, $G = \frac{E_s}{2(1+\nu)} = \frac{1,772,677}{2\times 1.35} = 656,547\ \mathrm{kN/m^2}$.

The maximum computed amplitude = 3.8×10^{-4} m.

Thus, $Z_{\max}/\sqrt{2} = 2.687 \times 10^{-4}$ m.

Interpolating linearly, we have f_2 = 125.2 Hz and f_1 = 108.7 Hz, where the amplitude is 2.687×10^{-4} m.

Thus, $\zeta = \frac{f_2 - f_1}{2f_n} = \frac{125.2 - 108.7}{2 \times 123} = 0.067$.

Thus, the estimated material damping of soil is 6.7%.

Free vibration test

In this case, the test block is excited to vertical vibration by impact of a sledgehammer or any suitable device located as near to the centre of the top face of the block as possible. The vibrations shall be recorded in a pen recorder or suitable device to measure the frequency and amplitude of vibration. The value of C_z is computed from Eq. 8.3.1.2. The damping coefficient ζ is computed from the expression.

$$\zeta = \frac{1}{2\pi} \log_e \left[\frac{Z_1}{Z_2} \right] \tag{8.3.1.7}$$

Here Z_1 and Z_2 are the magnitudes of two successive amplitudes of vibration.

8.3.2 Seismic Cross-Hole Test

As shown in the schematic sketch in Fig. 8.6, a probe is placed in a borehole to the desired depth and a shear wave is generated in the soil by hitting it hard with a hammer.

The waves are picked up by a geophone entrenched firmly to the casing of another borehole located at a known distance (L_x) from the first hole.

The time taken to pick up the signal is measured by the oscilloscope.

The dynamic shear modulus G_{dyn} is then obtained from the expression

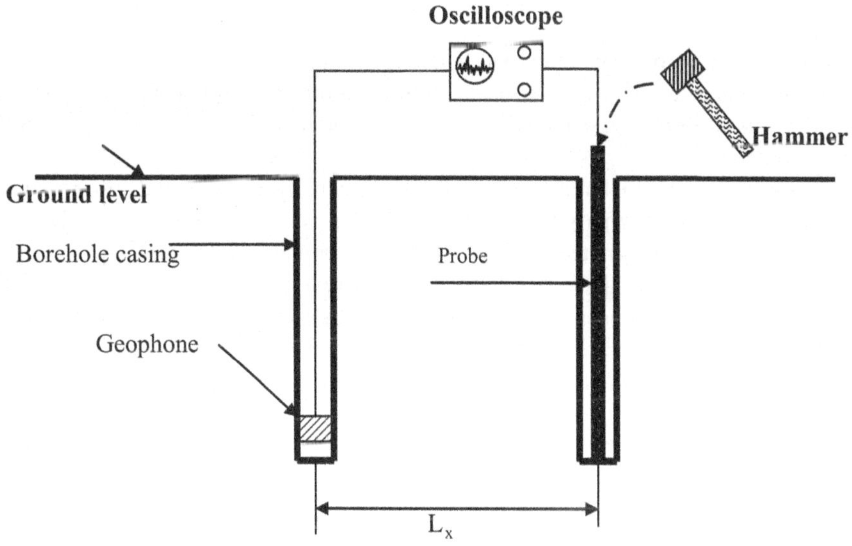

Fig. 8.6 Schematic diagram for seismic cross-borehole test

$$G_{\mathrm{dyn}} = \frac{\gamma_{\mathrm{s}}}{g}\left(\frac{L_x}{t}\right)^2 \tag{8.3.2.1}$$

where G_{dyn} = dynamic shear modulus of the soil; γ_s = weight density of soil; g = acceleration due to gravity; L_x = distance between the two boreholes; and t = elapsed time.

One of the major advantages with this test is that the dynamic shear modulus can be measured to any desired depth and can very well be an integrated part of a SPT program.

The test is very effective in case that the soil is layered in nature where visual inspection of each layer is possible based on SPT test.

However, the strain range for test is again restricted to 10^{-4} which is normally less than the strain range experienced by machine foundations and earthquake analysis and needs to be corrected to arrive at the design value of G.

8.3.3 *Cyclic Plate Load Test*

As the name suggests, this test is a modified version of the standard plate load test often used to determine allowable bearing capacity or sub-grade modulus of soil. The test is conducted using a standard plate of size 300 m × 300 mm subjected to cyclic loading that involves several loading cycles (i.e., loading, unloading and reloading as per standard practice.). This facilitates in separating the recoverable or elastic part of the deformation from the irrecoverable or the plastic part, which in turn can be related to Young's modulus of soil.

After the test setup has been arranged, the initial dial gauge readings are noted and the first increment of static load is applied to the plate. Thus, the load shall be maintained constant throughout for a period till no further settlement occurs or the rate of settlement becomes negligible. The final readings of the dial gauge shall then be recorded.

The entire load is then removed quickly but gradually, and the plate is allowed to rebound. When no further rebound occurs or it becomes negligible, the reading of the dial gauge is again noted. The load shall then be increased progressively till its magnitude acquires a value equal to the proposed next higher stage of loading which shall be maintained constant and again the final dial gauge readings are noted.

This cycle of loading and unloading is continued as shown in Fig. 8.7 till the estimated ultimate load has been reached when the final values of dial gauge are noted each time.

The coefficient of elastic compression C_z is finally computed from the expression

$$C_z = \frac{P}{\delta_e}\ \mathrm{kN/m^3} \tag{8.3.3.1}$$

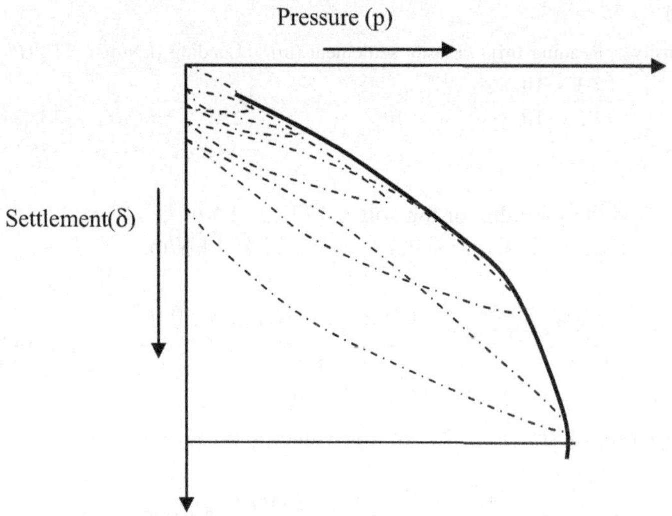

Fig. 8.7 Typical loading unloading curve under cyclic plate load test

Here

P load intensity on plate @ kN/m^2;
δ_e elastic rebound corresponding to load P in m.

Example 8.2 Following are the results of a cyclic plate load test. The size of plate used for test was 300 mm × 300 mm. Calculate the sub-grade modulus spring constant and shear modulus of soil. What would be the value of dynamic shear modulus for a foundation of size 2 m × 2 m. Assume Poisson's ratio = 0.35.

Load intensity (kN/m^2)	0.0	70	0.0	140	0.0	210	0.0	280	0.0
Settlement (mm)	0.0	2.1	1.6	3.9	2.8	5.7	4.2	7.3	5.2

Solution

Load intensity	Reading (m)	Elastic settlement (m)	Loading (kN/m^2)	C_z (P/d) kN/m^3
0.0	0.0			
70	2.1×10^{-3}			
0.0	1.6×10^{-3}	0.5×10^{-3}	70	140,000
140	3.9×10^{-3}			
0.0	2.8×10^{-3}	1.1×10^{-3}	140	127,272
210	5.7×10^{-3}			
0.0	4.2×10^{-3}	1.5×10^{-3}	210	140,000

(continued)

(continued)

Load intensity	Reading (m)	Elastic settlement (m)	Loading (kN/m^2)	C_z (P/d) kN/m^3
280	7.3×10^{-3}			
0.0	5.2×10^{-3}	2.1×10^{-3}	280	133,333
			Average ⟶	135,151

Thus sub-grade modulus of the soil = 135,151 kN/m^3.
Spring stiffness = 135,151 × 0.3 × 0.3 = 12,164 kN/m.

$$E_s = \frac{C_{za}(1 - \nu^2)\sqrt{A_f}}{1.13} = \frac{135,151 \times \left[1 - (0.35)^2\right]\sqrt{0.09}}{1.13} = 31,485\ \text{kN/m}^2.$$

Thus, $G = \frac{E_s}{2(1+\nu)} = \frac{31,485}{2 \times 1.35} = 11,661\ \text{kN/m}^2$.

For a footing of size 2 m × 2 m, the value of C_z is expressed as

$$C_{za} = C_z\sqrt{\frac{A_f}{A_1}} = 135,151 \times \sqrt{\frac{0.09}{4}} = 20,273\ \text{kN/m}^3.$$

Thus, $E_s = \frac{C_{za}(1-\nu^2)\sqrt{A_f}}{1.13} = \frac{20,273 \times \left[1-(0.35)^2\right]\sqrt{4}}{1.13} = 31,486\ \text{kN/m}^2$

Thus, G = 11,661 kN/m^2; the dynamic shear modulus is not affected by foundation size.

8.4 Techniques to Compute Dynamic Shear Modulus When No Dynamic Soil Test Has Been Carried Out

A not so uncommon phenomenon that even puts an experienced engineer under difficult situation while modeling a soil–structure interaction problem.

In many cases in a project, especially related to buildings or infrastructure development, it has been observed that no dynamic test has been carried out during the geotechnical investigation. Though not unusual, it should not happen as a rule, for this only shows the lack of foresight on the part of the engineer while submitting the technical and commercial proposal for a project.

At the advent of a project, location of a particular site is usually known, and one can easily find out from the codes how active this zone is seismically and, if felt reasonable, all he has to do is to include this additional cost of dynamic geotechnical investigation in his commercial bid.

People suffer from misnomer that dynamic tests are expensive—which is actually not true, for an average dynamic test in international market takes roughly US$ 2000–25,000 which would, however, be 0.25% of a small petrochemical refinery and possibly 0.1% of a combined cycle 350 MW power plant.

The lack of these tests can have serious consequences in an event of an earthquake and can cause serious damages to the equipments operating in such projects whose cost itself would constitute 30–40% of the whole project cost!

So one has to decide on the risk involved and come to a conclusion of its worth. Though theoretical corelation exist for evaluation of dynamic shear modulus of soil from static soil tests (which has been successfully used in project works), it is always preferable to have these dynamic tests carried out at site, for not only does it imbibe more confidence in the design process but the engineer should also be aware that "theoretically corelated values have also varied widely with respect to actual field data and should be mellowed with judgment." Considering the uncertainty prevalent in soil, it is surely not an easy task to accomplish.

8.4.1 Theoretical Corelation from Other Soil Parameters

The most outstanding work in establishing theoretical corelation for evaluating the dynamic property of soil has been done by Hardin and Richart (1963), Seed and Idriss (1970), and Ohsaki and Iwasaki (1973), to name some of the pioneering few.

Engineers, for many real projects in the past, have successfully used the expressions suggested by them. We are going to have a look at some of them hereafter and understand their limitations, if any

8.4.1.1 Corelation for Sandy and Gravelly Soil

Hardin and Richart's (1963) Formula

For rounded grained soil having void ratio less or equal to 0.8, the dynamic shear modulus is given by

$$G = \frac{6908(2.17 - e)^2}{1 + e}\sqrt{\sigma_0}\ \mathrm{kN/m^2} \tag{8.4.1.1}$$

For angular grained soil, the dynamic shear modulus is given by

$$G = \frac{3230(2.97 - e)^2}{1 + e}\sqrt{\sigma_0}\ \mathrm{kN/m^2}. \tag{8.4.1.2}$$

where G = dynamic shear modulus of the soil, e = in situ void ratio of the soil sample, σ_0 = mean effective stress = 0.333 $\sigma_v(1 + 2K_0)$, σ_v = vertical effective stress, σ_h = horizontal effective stress = $K_0\sigma_v$, K_0 = earth pressure at rest and can be considered as $1 - \sin\phi$, where ϕ is the angle of friction of the soil.

Seed and Idriss'* (1970) *Formula

The formula for dynamic modulus in this case has been related to relative density of sand which can usually be quantified from SPT test and is given by

$$G = 83.3K_2\sqrt{\sigma_0}\ \text{in psi} \tag{8.4.1.3}$$

Here K_2 is a function of the relative density of sand, which can again be estimated from the SPT value.

The relationship between SPT value and the relative density is as given in Tables 8.1 and 8.2.

For computer implementation, K_2 can also be expressed as

$$K_2 = 0.6D_r + 16 \tag{8.4.1.4}$$

It is to be noted that in this case, in order to determine the relative density, the observed SPT value has to be corrected for the overburden pressure and dilatancy to arrive at the design SPT value before it is corelated with the above table.

Corrections to SPT value

Though available in standard textbooks of soil mechanics and foundation engineering, for the sake of completeness and clarity, we present the correction expressions as mentioned hereafter.

For dilatancy correction, if the observed SPT value (N_0) is greater than 15, then the corrected SPT value N' is given by Terzaghi and Peck (1967).

$$N' = 15 + \frac{1}{2}(N_0 - 15) \tag{8.4.1.5}$$

Table 8.1 Soil properties with SPT values

SPT—value	Compactness	Relative density D_r	Angle of friction (ϕ)
0–4	Very loose	0–15	<28
4–10	Loose	15–35	28–30
10–30	Medium	35–65	30–36
30–50	Dense	65–80	36–41
>50	Very dense	>85	>41

Table 8.2 Values of K_2 vs relative density D_r at strain of 10^{-6}

Relative density D_r	90	75	60	45	40	30
K_2	70	61	52	43	40	34

The overburden correction as per Terzahi and Peck (1967) is given by

$$N'' = 0.77N' \log_{10}\frac{2000}{p'} \text{ for } p' \geq 25 \text{ kPa} \tag{8.4.1.6}$$

For $p \leq 25$ kPa, as per Murthy (1991)

$$N'' = \frac{4N'}{2 + 0.034p'} \tag{8.4.1.7}$$

in which $\mathbf{N}''$ = corrected SPT value for overburden, $\mathbf{N}'$ = corrected SPT value for dilatancy, p' = gross overburden pressure in kN/m^2.

Correlation between dynamic shear modulus and SPT data

Ohsaki and Iwasaki (1973) have given correlation for dynamic shear modulus and SPT values, expressed as

$$G = 12{,}000N^{0.8} \text{ kPa} \tag{8.4.1.8}$$

Here N = SPT value recorded at the site. As per Ishihara (2003), the N value proposed in Eq. 8.4.1.8 is as per Japanese standard and is to be multiplied by a factor of 0.833 to calibrate it to international practice.

Example 8.3 A site is having dimensions 48 m × 36 m, which would be supporting a water treatment plant.

Soil investigation was carried out at five boreholes at four corners and one at centre of the site. The soil was found to be cohesionless in nature and SPT values observed at the five boreholes are as tabulated hereafter

Depth (m)	BH1 (SPT value)	BH2 (SPT value)	BH3 (SPT value)	BH4 (SPT value)	BH-5 (SPT value
2	4	6	4	3	6
4	8	8	10	12	14
6	12	9	11	12	12
8	15	12	16	11	14
10	20	18	24	16	22
14	22	24	28	20	20

Based on laboratory and field analyses, the following parameters were further established:

Groundwater table = 1.6 m, below finished grade level

Saturated density of soil = 22 kN/m^3. Dry density = 19 kN/m^3.

Void ratio e_0 = 0.60. Plasticity index = 0.0. Poisson's ratio = 0.35.

Determine the best estimate of dynamic shear modulus (G) of soil at 5 ms below ground level presuming no dynamic soil test was done during geotechnical investigation.

Solution

Average observed SPT value at a depth of 4.0 m = $\frac{8+8+10+12+14}{5}=10.4=10(\text{say})$.

Average observed SPT value at a depth of 6 m = $\frac{12+9+11+12+12}{4}=23.5=11.2=11(\text{say})$.

At a depth of 5 m below ground level, based on linear interpolation, the average observed SPT.

$$\text{value}=\frac{11-10}{2}\times 1+10=10.5\cong 11(\text{say}).$$

The above-observed SPT value has now to be corrected for dilatancy and overburden pressure.

(1) **Correction for dilatancy**

As per Terzaghi and Peck (1967), the corrected SPT (N') value is given by $N'=15+\frac{1}{2}(N_0-15)$ for $N>15$.

As $N<15$, no dilatancy correction is required; thus, $N'=11$.

(2) **Correction for overburden pressure**

As per Terzaghi and Peck (1967).

$$N''=0.77N'\log_{10}\frac{2000}{p'}\ \text{for}\, p'\geq 25\,\text{kPa}$$

N'' = corrected SPT value for overburden; N' = corrected SPT value for dilatancy ($N'=11$); p' = gross overburden pressure in kN/m^2.

Here p' = 22 × 3.4 + 19 × 1.6 = 105.2 kN/m^2.

Substituting the above in Peck's formula, we have

$$N''=0.77N'\log_{10}\frac{2000}{105.2}=10.83$$

Thus, the corrected design SPT value = 11.

Referring to the table as mentioned earlier for $N=11$, Dr (relative density) = 37% and $\phi=32°$

Net overburden pressure at the 5 m level is expressed as

$$\sigma_v=(22-10)\times 3.4+19\times 1.6=71.2\ \text{kN/m}^2(10.326\ \text{p.s.i})$$

$$K_0=1-\sin 32=0.470$$

Considering the confining pressure

$$\sigma_o = 0.333\sigma_v(1 + 2K_0);$$
$$\text{we have } \sigma_0 = \frac{71.2}{3}(1 + 2 \times 0.47) = 46.04\ \text{kN/m}^2(6.677\ \text{p.s.i})$$

(1) As per **Hardin and Richart's formula**

$$G = \frac{6908(2.17 - e)^2}{1 + e}\sqrt{\sigma_0}$$

$$G = \frac{6908(2.17 - 0.60)^2}{1 + 0.60}\sqrt{46.04} = 72{,}210\ \text{kN/m}^2$$

(2) As per **Seed and Idriss'** formula

Referring to the chart given above for $D_r = 37\%$ and strain in the range of $10^{-4}\%$ $K_2 = 38$.

And as $G = 83.3K_2\sqrt{\sigma_0}$, we have

$$G = 83.3 \times 38 \times \sqrt{6.677} = 8179\ \text{p.s.i.}(56{,}392\ \text{kN/m}^2).$$

Thus, taking an average value of G based on **Hardin and Seed's** method.

$$\text{Average } G = \frac{72{,}210 + 56{,}392}{2} = 64{,}301\ \text{kN/m}^2$$

(3) As per **Ohsaka and Iwasaki's** formula

$$G = 12{,}000N^{0.8}\ \text{in kPa}$$

Here as recommended by Ishihara N (design) = $0.833 \times N = 0.833 \times 11 = 9.16$

Thus, $G = 12{,}000N^{0.8} = 12{,}000 \times (9)^{0.8} = 69{,}594\ \text{kN/m}^2$.

Thus, it will be observed that variation with average G obtained based on the three formulae is not significant.

What we suggest here is not to go by one formula, but check with possibly all of them and comparing them to arrive at result which would possibly be best fitting and hopefully the most realistic.

Here again it is to be noted that we had not used the angular sand formula of Hardin; if the soil description does not reflect it or the soil has both rounded and angular grains, an intermediate value has to be chosen judiciously.

8.4.1.2 Corelation for Saturated Clay

Hardin and Drnevich's Formula

Hardin and Drnevich (1972) have given the following formula applicable to clayey soil as

$$G_{max} = \frac{3230(2.97 - e)^2}{1+e}(\mathrm{OCR})^k \sqrt{\sigma_0} \text{ in kPa} \tag{8.4.1.9}$$

where e = void ratio; OCR = overconsolidation ratio; σ_0 = mean effective stress = $0.333(\sigma_v + 2\sigma_h)$; σ_v = vertical effective stress; σ_h = horizontal effective stress = $K_0\sigma_v$, K_0 = earth pressure at rest and is a function of the plasticity index and the overconsolidation ratio.

k = is a function of the plasticity index of the soil and is given as follows

Plasticity index	0	20	40	60	80	≥ 100
k	0	0.18	0.30	0.41	0.48	0.51

It is to be noted that G_{max} as obtained above corresponds to a shear strain range of 0.25×10^{-4}% and needs to be modified for the appropriate strain range as appropriate for a problem in hand based on the expression.

$$G = \frac{G_{max}}{(1 + \psi/\psi_r)} \tag{8.4.1.10}$$

Here ψ = desired strain range; ψ_r = reference strain range and is expressed as

$$\psi_r = \frac{\tau_{max}}{G_{max}} \times 100$$

and

$$\tau_{max} = \left[\left\{\left(\frac{1+K_0}{2}\right)(\sigma_v - u)\sin\phi + c\cos\phi\right\}^2 - \left\{\left(\frac{1-K_0}{2}\right)(\sigma_v - u)\right\}^2\right]^{0.5} \tag{8.4.1.11}$$

in which σ_v = total vertical stressing in soil; u = pore pressure; c = cohesion of soil; ϕ = angle of friction of soil; and K_0 = coefficient of earth pressure at rest. As per Booker and Ireland (1965) for normally consolidated clay, we have

$K_0 = 0.4 + 0.007\,PI$ for $0 < PI < 40\%$ and

$K_0 = 0.68 + 0.001(PI - 40)$ for $40 \leq PI \leq 80\%$
For overconsolidated clay

$$K_{0(\mathrm{OCR})} = K_0\sqrt{\mathrm{OCR}} \tag{8.4.1.12}$$

Example 8.5 It has been decided to place foundation of an industrial structure at 2.0 m below the existing ground level.

Based on laboratory and field tests, it has been found that the groundwater table is at a depth of 1.0 m below GL.

Unconsolidated undrained triaxial tests reveal the sample to have the following values:

- Cohesion value c = 0.21 kg/cm^2;
- Angle of resistance = 20°;
- Pore pressure = 0.1 kg/cm^2.

Consolidation tests reveal that it had a history of pre-consolidation pressure of 100 kN/m^2:

- Initial void ratio = 0.61;
- Plasticity limit PI = 35;
- Saturated unit weight of soil = 22 kN/m^3.

The site has a history of moderate to severe earthquake when from previous record it is observed to generate a strain range up to 0.1%.

Calculate the dynamic shear modulus of soil for this predicted strain range.

Solution

For foundation located at 2.0 m below the ground level, the net vertical pressure is

$$\sigma_v = 22 \times 1.0 + (22 - 10) \times 1 = 34\ \mathrm{kN/m^2}.$$

$\mathrm{OCR} = \frac{100}{34} = 2.94$, for plasticity index of 35 as per Brooker and Ireland's
$K_0 = 0.4 + 0.007 \times 35 = 0.645$

$$\therefore\quad K_{0(\mathrm{OCR})} = K_0\sqrt{\mathrm{OCR}} = 0.645 \times \sqrt{2.94} = 1.105$$

Thus, considering **that** $\sigma_o = 0.333\sigma_v\ (1 + 2K_0)$, we have

$$\sigma_o = 0.333 \times 34(1 + 2 \times 1.105) = 36.40\ \text{kN/m}^2.$$

$$G_{max} = \frac{3230(2.97 - e)^2}{1 + e}(\text{OCR})^k\sqrt{\sigma_0}$$

Here $k = 0.27$ for $PI = 35$, as per the table furnished above.
Thus, substituting the values we have

$$G_{max} = 3230 \times \left[\frac{(2.973 - 0.61)^2}{(1 + 0.61)}\right] \times (2.94)^{0.27} \times (36.4)^{0.5} = 90429\ \text{kN/m}^2.$$

Calculation for shear stress

$$\tau_{max} = \left[\left\{\frac{1 + K_0}{2}(\sigma_v - u)\sin\phi + c\cos\phi\right\}^2 - \left\{\frac{1 - K_0}{2}(\sigma_v - u)\right\}^2\right]^{0.5}$$

or

$$\tau_{max} = \left[\left\{\frac{1 + 1.105}{2}(34 - 10)\sin 20 + 21\cos 20\right\}^2 - \left\{\frac{1 - 1.105}{2}(34 - 10)\right\}^2\right]^{0.5}$$
$$= 28.34\ \text{kN/m}^2$$

ψ_r = reference strain range expressed as

$$\psi_r = \frac{\tau_{max}}{G_{max}} \times 100 = \frac{28.34}{90,429} \times 100 = 0.031\%$$

Thus for 0.1% strain

$$G = \frac{G_{max}}{(1 + \frac{\psi}{\psi_r})} \rightarrow G = \frac{90,429}{(1 + \frac{0.1}{0.031})} = 21,580\ \text{kN/m}^2$$

It is thus observed that the dynamic shear modulus is 23.8% of the theoretically calculated data.

Based on the above example, it would perhaps be not difficult to realize that how important role does the strain range play on the design value of dynamic shear modulus of soil.

8.4.1.3 Estimation of Material Damping of Soil

Damping plays a significant part in the overall response of soil–structure system. While for structural members the material damping plays a significant part (mostly considered as Rayleigh damping), for soil, the two types of damping are basically involved.

- Geometric damping;
- Material damping.

Geometric damping of a soil foundation system is a mean by which the energy is dissipated by means of radiation from the source and is a function of distance from source from where the disturbing waves generates.

Material damping of the soil foundation system is a mean by which the energy is dissipated by hysteresis and is an inherent property of the constituting material of the soil.

This can very well be found from resonant column test in the laboratory when, after the soil has been vibrated, the exciter is stopped and successive amplitudes are measured.

The total damping ratio of a soil foundation system is a sum of geometric and material damping. It is generally observed that material damping has a significant magnitude relative to geometric damping, especially in rotational modes. In such cases, the total damping rather than geometric damping should be used to obtain the response of the structural foundation system.

For translational mode, on the contrary, material damping plays an insignificant role and may be neglected in the analysis. Thus, for tall narrow structures like chimneys, boiler structures, tall buildings where the coupled horizontal and rocking mode could play a significant role, it would perhaps be realistic to consider the material damping of soil in order to have a meaningful response.

Hardin's Formula

Hardin[1] has expressed material damping of sandy soil by the expression

$$D_m = \frac{0.985\psi_r^{0.2}}{\sqrt{\sigma_0}} \tag{8.4.1.13}$$

Here notations are same as expressed earlier except of the fact that the confining pressure σ_0 is expressed in kPa. The equation is valid for a shear strain amplitude of 10^{-6} to 10^{-4} with a confining pressure of 24–144 kPa.

For a particular strain range, the value obtained above can be corrected based on the expression

$$\frac{D_c}{D_m} = \frac{\psi/\psi_r}{1+\psi/\psi_r} \tag{8.4.1.14}$$

[1]Hardin.B.O.—The nature of damping in sand JSMFD ASCE 91(SM1) 63-97 Part1.

Example 8.6 For the example as shown in Example 8.3, estimate the damping ratio of the soil as per Hardin's formula. The soil properties remain same as given in Example 8.3.

Solution

Based on the solution furnished in Example 8.3, the value of dynamic shear modulus is given by $G = 72,210$ kN/m^2; $\sigma_0 = 46.04$ kN/m^2; $K_0 = 0.47$

For design SPT value $N = 11$, $\phi = 32°$, and $\sigma_v = 71.2$ kN/m^2

Considering

$$\tau_{\max} = \left[\left\{\frac{1+K_0}{2}(\sigma_v - u)\sin\phi + c\cos\phi\right\}^2 - \left\{\frac{1-K_0}{2}(\sigma_v - u)\right\}^2\right]^{0.5},$$

we have

$$\tau_{\max} = \left[\left\{\frac{1+0.47}{2}71.2\sin 32\right\}^2 - \left\{\frac{1-0.48}{2}71.2\right\}^2\right]^{0.5} = 20.64 \text{ kN/m}^2$$

Here the pore pressure $u = 0.0$

As $\psi = \frac{\tau}{G} \times 100$, we have $\psi_r = \frac{20.64}{72{,}210} \times 100 = 0.0285\%$

Considering $D_m = \frac{0.985\psi_r^{0.2}}{\sqrt{\sigma_0}}$, we have

$$D_m = \frac{0.985(0.028)^{0.2}}{\sqrt{46.04}} = 0.071$$

Thus, the material damping ratio is estimated as 7%.

Ishibashi and Zhang's formula

Ishibashi and Zhang (1993) have proposed an expression for the damping ratio of plastic and non-plastic soil, namely

$$\zeta = 0.333\frac{1+\exp(-0.0145\text{PI}^{1.3})}{2}\left[0.586\left(\frac{G}{G_{\max}}\right)^2 - 1.547\left(\frac{G}{G_{\max}}\right) + 1\right] \tag{8.4.1.15}$$

The notations for the above expression are already explained in earlier formulas.

It will be observed (Fig. 8.2) that as $G/G_{\max}$ reduces, the damping ratio goes on increasing, meaning thereby that as strain increases, the damping ratio goes on increasing too.

Example 8.7 For the clayey soil sample as shown in Example 8.5, determine the damping ratio for the strain range level of 0.1%, based on Ishibashi and Zhang's (1993) formula. Consider all soil properties same as Example 8.5.

Solution
Based on earlier examples, we have seen that plasticity index $PI = 35$
In example 8.5, we have already calculated that for 0.1% strain $G/G_{max} = 0.238$
Substituting the above in Ishibashi and Zhang's formula, we have

$$\zeta = 0.333\frac{1 + \exp(-1.47)}{2}\left[0.586(0.238)^2 - 1.547 \times 0.238 + 1\right] = 0.1361$$

Thus, the estimated damping ratio is 13.6%.

8.5 Procedure to Estimate Large Strain in Soil Under Earthquake Force

We acknowledge at the very outset that the solution to this problem is not very easy. The uncertainties involved are so widely varying that it would be difficult to give a precise answer to this issue. To the best of our knowledge, there is no straightforward answer to this problem and the best one can achieve is a reasonable estimate or can possibly study a range of values and try to predict the overall behavior.

For earthquake, the strain would invariably be much larger than that measured during test.

Even for a moderate earthquake, as the strain range increases, degradation in soil stiffness becomes significant and has a major contribution to overall response.

It is obvious that the strain induced in soil will depend upon the strength of dynamic loading, the geological condition of the site, stress history of soil, and a number of other factors. So the point remains that, if there are no previous records of strain from previously occurring earthquake data, how does one rationalize the strain?

We discuss below some of the techniques which could be used for evaluation of the strain induced in the soil.

8.5.1 Estimation of Soil Strain for Earthquake Analysis

For earthquake analysis, things are surely more complicated for not only the forces induced in the soil are much more complex, the behavior itself is different from machine foundations.

While in machine foundation the force is induced in the soil from the structure, in case of an earthquake the force is induced within the soil where the soil first starts vibrating based on the waves propagating through it. Thus, the acceleration, it is excited to, depends on the free field vibration of the site. This acceleration induced in the soil generates shear strain on which the stiffness degradation and damping ratio would depend.

Shown in Fig. 8.8 is a schematic diagram of an industrial site with propagating earthquake waves. The depth of the site (*H*) is considered to the bedrock level from where the waves are presumed to be propagating. For a site having no bedrock, this level is usually considered at the depth where shear wave velocity of the site is greater or equal to 600 m/s. Based on SPT value, this can be considered as the depth where the design SPT value is greater than 50.

The waves propagating at bedrock level travel upward and hit the surface ($z = 0$) when the site surface undergoes a motion. However, as the surface is free, it is free to shake, and, as such, no strain energy develops at the surface.

The motion of such elastic waves propagating through an elastic medium can be defined by the partial differential equation, as cited in Chap. 3 (Eq. 3.5.3.1).

For the above problem, as explained in Sect. 3.6.3 of Chap. 3, the stiffness and mass matrices of soil can be expressed as

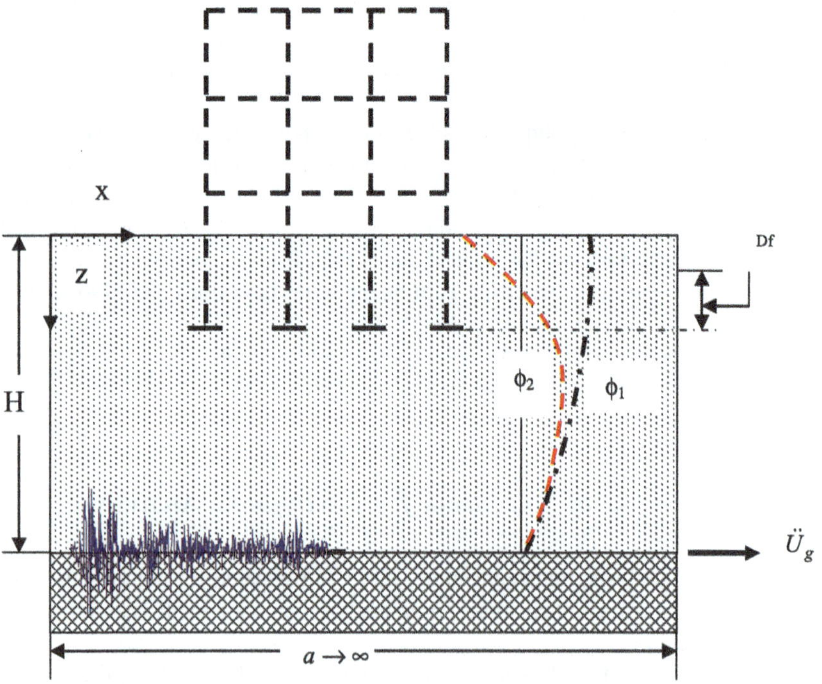

Fig. 8.8 A site with structure subjected to base motion at bed **rock level**

$$[K] = \frac{G}{H}\begin{bmatrix} \frac{\pi^2}{8} & 0 & 0 & 0 \\ 0 & \frac{9\pi^2}{8} & 0 & 0 \\ 0 & 0 & \frac{25\pi^2}{8} & 0 \\ 0 & 0 & 0 & \frac{49\pi^2}{8} \end{bmatrix} \text{ and } [M] = \frac{\gamma_s H}{g}\begin{bmatrix} \frac{1}{2} & 0 & 0 & 0 \\ 0 & \frac{1}{2} & 0 & 0 \\ 0 & 0 & \frac{1}{2} & 0 \\ 0 & 0 & 0 & \frac{1}{2} \end{bmatrix} \tag{8.5.1.1}$$

Now, performing the operation $[\mathbf{K}] - [\mathbf{M}]\omega^2 = 0$, where $T = 2\pi/\omega$, we get $\{T\} = \frac{H}{V_s}\{4 \quad 1.333 \quad 0.8 \quad 0.571\}^{\mathrm{T}}$ and the corresponding eigenvectors are

$$[\phi] = \begin{bmatrix} 1 & 0 & 0 & 0 \\ 0 & 1 & 0 & 0 \\ 0 & 0 & 1 & 0 \\ 0 & 0 & 0 & 1 \end{bmatrix} \tag{8.5.1.2}$$

Based on codal definition, the amplitude of vibration as explained in Chap. 3, Sect. 3.5.4, may be expressed as

$$u_n = \kappa_n \beta \frac{S_{an}}{\omega^2} \cos(2n-1)\frac{\pi z}{2H} \tag{8.5.1.3}$$

Here κ_n is the modal mass participation factor and is expressed as $\kappa_n = \sum_{i=1}^{n} m_i\phi_i \Big/ \sum_{i=1}^{n} m_i\phi_i^2$ and as elaborated in Table 3.1.

The strain within the soil body, considering that it acts as a one-dimensional shear beam, is expressed as

$\varepsilon_{xx} = 0$ and $\gamma_{xz} = \frac{\partial u}{\partial z}$, which gives

$$\gamma_{xz}^n = -\kappa_n \frac{(2n-1)\pi}{2H} \beta \frac{S_{an} T_n^2}{4\pi^2} \sin(2n-1)\frac{\pi z}{2H} \tag{8.5.1.4}$$

Here the superscript $n = 1, 2, 3, 4\ldots$ indicates the numbers of modes considered. Equation (8.5.1.4), on simplification for the first four modes, finally gives

$$\gamma_{xz}^1 = -\frac{16}{\pi(\pi+2)} \beta\gamma_s \left(\frac{S_{a1}}{g}\right) \frac{H}{G} \sin\frac{\pi z}{2H} \tag{8.5.1.5}$$

$$\gamma_{xz}^2 = \frac{16}{3\pi(3\pi-2)} \beta\gamma_s \left(\frac{S_{a2}}{g}\right) \frac{H}{G} \sin\frac{3\pi z}{2H} \tag{8.5.1.6}$$

$$\gamma_{xz}^{3} = -\frac{16}{5\pi(5\pi+2)}\beta\gamma_{s}\left(\frac{S_{a2}}{g}\right)\frac{H}{G}\sin\frac{5\pi z}{2H} \tag{8.5.1.7}$$

$$\gamma_{xz}^{4} = \frac{16}{7\pi(7\pi-2)}\beta\gamma_{s}\left(\frac{S_{a4}}{g}\right)\frac{H}{G}\sin\frac{7\pi z}{2H} \tag{8.5.1.8}$$

For the sake of computation, though the strain in soil for the first four modes has been shown in Eqs. (8.5.1.5)–(8.5.1.8), it has been shown by Chowdhury and Dasgupta (2012) that the effects of higher modes are negligible and it is sufficiently accurate to work with fundamental mode only for practical engineering problems.

It may be noted at this point that the strain induced in soil is dependent on the spectral acceleration factor S_a/g, which in turn is dependent on the time period, which is a function of the soil height H over bedrock and shear modulus G.

The IS 1893 gives the response spectrum curve for 5% damping; however, for damping having other values it has furnished a scaling table that is used to scale up or down the spectral acceleration factor. The table has been presented in Chap. 2 (Table 2.9).

While the initial value of G can be computed from the shear wave velocity furnished in the soil report, quantifying damping ratio is a little confusing as a number of formulas exist to identify the same. Having tinkered around with a few of them, the authors recommend the unified expression proposed by Ishibashi and Zhang Eq. (8.4.1.15) for both clay- and sandy-type soil, which is felt to be most suitable for this purpose.

To determine the design dynamic shear modulus (G) and damping ratio (ζ), we start with an iterative process where one first needs to compute the time period from the expression $4H/V_s$. Here V_s is the shear wave velocity corresponding to the value G_{max}, as obtained from soil report. Considering $G/G_{max} = 1.0$ one can compute the damping ratio of the soil from Eq. (8.4.1.15).

Now, based on the free field time period as computed above, one can find out the spectral acceleration ratio S_a/g from the curve/expression as furnished in IS-1893 (Fig. 2.5 of Chap. 2). The value then needs to be scaled up or down depending on the damping value obtained from Eq. (8.4.1.15), considering $G/G_{max} = 1.0$.

Now suppose, one is interested to compute the design G value at a depth D_f below the free surface where the foundation is to be located (refer Fig. 8.11), consider $z = D_f$ and compute the strain from Eq. (8.5.1.4). This strain in the first cycle is the reference strain γ_r (corresponding to the shear modulus having the value G_{max}). The G value for subsequent cycle is computed from the hyperbolic nonlinear relationship as proposed by Seed and Idriss (1970), Ishihara (1996), Desai and Zaman (2014).

$$\frac{G}{G_{max}} = \frac{1}{\left(1+\frac{\gamma_a}{\gamma_r}\right)} \tag{8.5.1.9}$$

In Eq. (8.5.1.9) γ_a = actual strain in soil at depth D_f, and γ_r = reference strain in soil at depth D_f. For the first cycle, $\gamma_r = \gamma_a$; in other words, G becomes 0.5 G_{max} after the first shock. With this reduced value, we go back to repeat the steps as described above for the second cycle and again put it in Eq. (8.5.1.9) to get a new value of G, keeping γ_r as the reference strain, as obtained in the first cycle. It will be observed that after two or three cycles the values of both G and D will stabilize and become constant. This constant value thus obtained is a correct value of G and D that is compatible with the strain induced in soil due to earthquake and should be used to compute the stiffness and impedance values for the foundation in a DSSI analysis. A complete example has been solved later to further elaborate the process and may be referred to.

To just cite an example here, a 30-m-soil strata having uniform G value with depth of 154897 kN/m^2 having weight density = 19 kN/m^3 was subjected to this analysis for a site at zone V as per IS code. The typical values of design, the G and D values at depth 2.5 m below ground level are as shown in Figs. 8.9 and 8.10.

It may be observed that low strain shear modulus of 154,897 kN/m^2 got degraded to 102,924 kN/m^2. The convergence is quite fast and takes in about 6th cycle. Similarly, for G/G_{max} value = 1.0 the material damping ratio is very low, <1%. This gets enhanced and finally stabilizes to a value of about 14% as soil is subjected to large deformation under seismic load. The 14% material damping ratio looks certainly a very reasonable value.

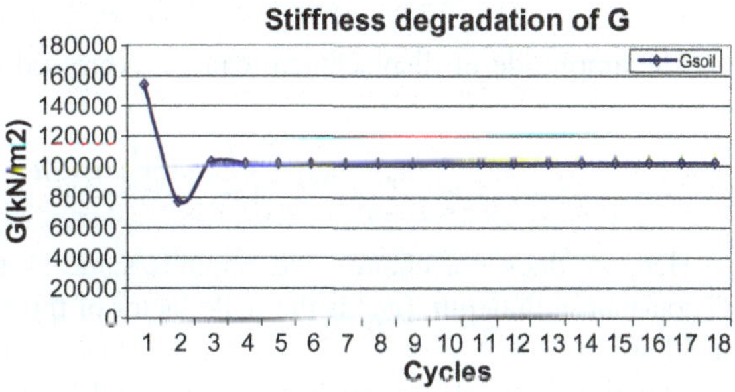

Fig. 8.9 Degradation in value of G with strain under earthquake load

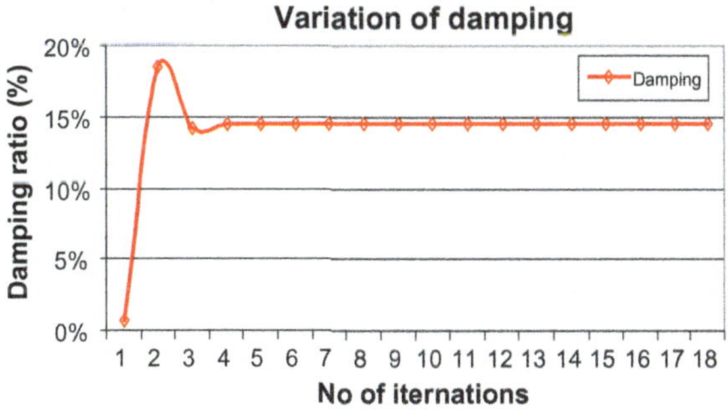

Fig. 8.10 Inrease in material damping ratio with strain under earthquake load

8.5.1.1 Soil Where Value of G Varies with Depth (Gibson-Type Soil)

For this, it has been already proved in Chap. 3, Sect. 3.6.3, that stiffness and mass matrix of the soil with varying properties are expressed as

$$[K]=\frac{G_H}{H}\begin{bmatrix}\frac{\pi^2}{16}+\frac{1}{4} & \frac{-3}{4} & \frac{5}{36} & \frac{-7}{36}\\ \frac{-3}{4} & \frac{9\pi^2}{16}+\frac{1}{4} & \frac{-15}{4} & \frac{21}{100}\\ \frac{5}{36} & \frac{-15}{4} & \frac{25\pi^2}{16}+\frac{1}{4} & \frac{-35}{4}\\ \frac{-7}{36} & \frac{21}{100} & \frac{-35}{4} & \frac{49\pi^2}{16}+\frac{1}{4}\end{bmatrix} \tag{8.5.2.1}$$

The mass matrix remains same as Eq. (8.5.1.1).

Solving for the eigenvalue λ from $[\mathbf{K}]-\lambda[\mathbf{M}]=0$ and knowing $\lambda=\omega^2$, where $T=2\pi/\omega$, we finally have

$$\{T\}=\frac{H}{V_s}\{5.19\quad 2.133\quad 1.228\quad 0.755\}^{\mathrm{T}} \tag{8.5.2.2}$$

The scaling factors for the eigenvectors are expressed as

$$\{\phi\}=\begin{bmatrix}-0.971 & -0.534 & 0.078 & 0.021\\ -0.217 & 0.723 & -0.701 & -0.125\\ -0.086 & 0.39 & 0.562 & 0.668\\ -0.046 & 0.199 & 0.432 & -0.733\end{bmatrix} \tag{8.5.2.3}$$

The amplitude of displacement can be expressed in this case as

$$u_n=\kappa_n\beta\frac{S_{an}T_n^2}{4\pi^2}[\phi_{\mathrm{n}}]^{\mathrm{T}}\cos(2n-1)\frac{\pi z}{2H} \tag{8.5.2.4}$$

Here κ_n does not change and remains same as the values derived earlier for G constant with depth; $[\phi_n]$ is the scale factor of the eigenvectors, as derived earlier in Chap. 3.

Equation (8.5.2.4) can be further expressed for fundamental mode as

$$u_1=\frac{8}{\pi+2}\gamma_s\beta\left(\frac{S_{a1}}{g}\right)\frac{(5.19)^2H^2}{4\pi^2G_H}\begin{bmatrix}-0.971\cdot\cos\frac{\pi z}{2H}\\ -0.217\cdot\cos\frac{3\pi z}{2H}\\ -0.086\cdot\cos\frac{5\pi z}{2H}\end{bmatrix} \tag{8.5.2.5}$$

$$\rightarrow u_1=1.0616\gamma_s\beta\left(\frac{S_{a1}}{g}\right)\frac{H^2}{G_H}\begin{bmatrix}-0.971\cdot\cos\frac{\pi z}{2H}\\ -0.217\cdot\cos\frac{3\pi z}{2H}\\ -0.086\cdot\cos\frac{5\pi z}{2H}\end{bmatrix}$$

Considering $\varepsilon_{xx}=0$ and $\gamma_{xz}=\frac{\partial u}{\partial z}$ we have

$$\gamma_{xz1} = 1.0616\beta\gamma_s\left(\frac{S_{a1}}{g}\right)\frac{H}{G_H}\begin{bmatrix} 1.525\sin\frac{\pi z}{2H} + \\ 1.022\sin\frac{3\pi z}{2H} + \\ 0.675\sin\frac{5\pi z}{2H} \end{bmatrix} \quad (8.5.2.6)$$

Higher modes can well be ignored, as elaborated earlier, as their effect is insignificant.

Proceeding in identical fashion, considering $T = C_T(H/V_s)$, values of C_T for different soil types are presented hereafter. The corresponding V_s values that need to be computed from the G_{Df} value are also furnished in Table 8.3.

Expressions of strain for different types of soil in fundamental mode are as furnished hereafter.

For $G = G_H\ (z/H)^{0.5}$

$$\gamma_{xz1} = 1.243\beta\gamma_s\left(\frac{S_{a1}}{g}\right)\frac{H}{G_H}[1.0A_z + 0.237B_z + 0.065C_z] \quad (8.5.2.7)$$

For $G = G_H\ (z/H)^2$

$$\gamma_{xz1} = 3.79\beta\gamma_s\left(\frac{S_{a1}}{g}\right)\frac{H}{G_H}[1.0A_z + 1.281B_z + 0.885C_z] \quad (8.5.2.8)$$

For $G = G_0\ (1 + z/H)$

$$\gamma_{xz1} = 0.592\beta\gamma_s\left(\frac{S_{a1}}{g}\right)\frac{H}{G_0}[1.0A_z + 0.153B_z + 0.00804C_z] \quad (8.5.2.9)$$

For $G = G_0\ (1 + z/H)^2$

$$\gamma_{xz1} = 0.363\beta\gamma_s\left(\frac{S_{a1}}{g}\right)\frac{H}{G_0}[1.0A_z + 0.327B_z + 0.0047C_z] \quad (8.5.2.10)$$

In Eqs. (8.5.2.7) through (8.5.2.10), the coefficients $A_z = \sin(\pi z/2H)$, $B_z = \sin(3\pi z/2H)$ and $C_z = \sin(5\pi z/2H)$.

Depending on the nature of soil, one may use the particular strain equation to compute the degraded stiffness and augmented value of damping in an earthquake.

The design steps based on the above can thus be summarized as hereafter.

- Determine the nature of soil as to whether $G = G_0$, $G = G_H\ (z/H)$ or $G = G_H\ (z/H)^{0.5}$, etc.
- Determine the initial value of shear wave velocity from the G_{max} value, as furnished in geotechnical report.
- Compute the free field time period from the expression, selecting the value of C_T from Table 8.3.
- Considering $G/G_{max} = 1.0$, determine the value of D from Eq. (8.4.1.15).

Table 8.3 Values of C_T for different soils for fundamental mode (Chowdhury et al. 2015)

Value of G	$G = G_H\ (z/H)$	$G = G_H\ (z/H)^{0.5}$	$G = G_H\ (z/H)^2$	$G = G_0\ (1 + z/H)$	$G = G_0\ (1 + z/H)^2$
C_T	5.19	4.486	7.826	3.094	2.421
V_s (Design)	$V_{sH} = \sqrt{\frac{G_{D_f}}{\rho}\left(\frac{H}{D_f}\right)}$	$V_{sH} = \sqrt{\frac{G_{D_f}}{\rho}\sqrt{\frac{H}{D_f}}}$	$V_{sH} = \sqrt{\frac{G_{D_f}}{\rho}\left(\frac{H}{D_f}\right)^2}$	$V_{s0} = \sqrt{\frac{G_{D_f}}{\rho}\left(\frac{2}{D_f+H}\right)}$	$V_{s0} = \sqrt{\frac{G_{D_f}}{\rho} \times \left(\frac{2}{D_f+H}\right)^2}$

- Choose the scaling factor (SF) for the above-computed value of D from Table 2.9 of Chap. 2.
- Select the spectral acceleration ratio (S_a/g) from code and multiply it by the scaling factor (SF) as obtained in the previous step, to compute the design spectral ratio.
- Compute the reference strain γ_r from Eq. (8.5.1.9) or Eqs. (8.5.2.7) through (8.5.2.10), as the case may be considering $z = D_f$, where D_f is the foundation where the design values are sought.
- Find the reduced value of G from Eq. (8.5.1.9). To start with $\gamma_r = \gamma_a \rightarrow G/G_{max} = 0.5$
- With this new value of G obtained at depth D_f, back-calculate Vs at depth H or 0, as the model deems fit from Table 8.3, and then the corresponding time period T and damping ratio D.
- Reselect the spectral value from code and multiply by new SF corresponding to new value of D, to obtain new design value of S_a/g.
- Recalculate the strain γ_a
- Reduce G, vide Eq. (8.5.1.9), for the next cycle.

Keep repeating the above steps till the values of G and D converge to become constant.

Example 8.8 A soil layer having depth of 30 ms has shear wave velocity of 450 m/s at 30.0 m level. Soil property varies as the expression $G_0 = G_H(z/H)$. The weight density of soil is 20 kN/m^3 and $PI = 5.0$. The site is Zone IV as per IS-1893[22]. It is required to determine the corrected G value at 3.0 m below ground level for DSSI analysis which is the found in level of a structure to be built on it.

Solution

Considering that the shear wave velocity at 30.0 m is 450 m/s,

$$G_{30} = \left(\frac{20}{9.81}\right) \times (450)^2 = 412,844 \text{ kN/m}^2.$$

Time period $T = C_T\left(\frac{H}{V_{sH}}\right) = 5.19 \times \frac{30}{450} = 0.346$ s

As a first step, considering $G/G_{max} = 1.0$,

$$D = \frac{1}{6}\left(1 + e^{-0.0145(PI)^{1.3}}\right)\left[1 + 0.586\left(\frac{G}{G_{max}}\right) - 1.547\left(\frac{G}{G_{max}}\right)^2\right]$$

$$D = \frac{1}{6}\left(1 + e^{-0.0145(5)^{1.3}}\right)[1 + 0.586 - 1.547] = 0.012$$

This gives the scaling factor = 3.2 (Refer Table 2.9 of Chap. 2), and S_a/g for $T = 0.346$ s is 2.5.

This gives the design $S_a/g = 2.5 \times 3.2 = 8.0$

For zone IV $\beta = \frac{ZI}{2R} = \frac{0.24 \times 1}{2 \times 3} = 0.04$

Thus, the strain at 3.0 m level is expressed as

$$\gamma_{xz1} = 1.0616\beta\gamma_s\left(\frac{S_{a1}}{g}\right)\frac{H}{G_H}[1.525A_z + 1.022B_z + 0.675C_z]$$

Taking $z = 3.0$ m, we get

$$\begin{aligned}\gamma_{xz1} &= 1.0616 \times 0.04 \times 20 \\ &\times 8\frac{30}{412,844}[1.525 \times 0.156 + 1.022 \times 0.454 + 0.675 \times 0.707] \\ &= 5.57 \times 10^{-4}\end{aligned}$$

Thus, the reference strain $\gamma_r = 5.57 \times 10^{-4}$

At 3.0 m, $G_{Df} = 412,844 \times \left(\frac{3}{30}\right) = 41284.4\ \text{kN/m}^2$

Thus, the reduced G in the first cycle at $z = 3.0$ is[2]
$\frac{G}{G_{max}} = \frac{1}{\left(1+\frac{\gamma_a}{\gamma_r}\right)} = \frac{41284.4}{\left(1+\frac{1}{1}\right)} = 20,642\ \text{kN/m}^2$ at 3 m depth

This gives the velocity at 30 m depth as $V_{sH} = \sqrt{\frac{G_{D_f}}{\rho}\left(\frac{H}{D_f}\right)} = \sqrt{\frac{20,642}{2.038}\left(\frac{30}{3}\right)}$
$= 318$ m/s

Thus, for the next cycle $T = C_T\left(\frac{H}{V_{sH}}\right) = 5.19 \times \frac{30}{318} = 0.489$ s

and

$$D = \frac{1}{6}\left(1 + e^{-0.0145(\mathrm{PI})^{1.3}}\right)\left[1 + 0.586\left(\frac{G}{G_{max}}\right) - 1.547\left(\frac{G}{G_{max}}\right)^2\right]$$

$$D = \frac{1}{6}\left(1 + e^{-0.0145(5)^{1.3}}\right)\left[1 + 0.586 \times (0.5) - 1.547 \times (0.5)^2\right] = 0.29$$

Thus, for $T = 0.489$ s, $S_a/g = 2.5$ and the scaling factor is 0.55 from Table 3. This gives the design $S_a/g = 1.38$.

Thus,

$$\begin{aligned}\gamma_{xz2} &= 1.0616 \times 0.04 \times 1.38\frac{30}{412,844}[1.525 \times 0.156 + 1.022 \times 0.454 + 0.675 \times 0.707] \\ &= 9.58 \times 10^{-5}\ \text{at}\ z = 3.0.\end{aligned}$$

This gives the reduce G values as $\frac{G}{G_{max}} = \frac{1}{\left(1+\frac{\gamma_a}{\gamma_r}\right)} = \frac{41284.4}{\left(1+\frac{9.58\times10^{-5}}{5.57\times10^{-4}}\right)} = 35,229\ \text{kN/m}^2$

[2]Note: For the first cycle, $\gamma_r = \gamma_a = 5.57 \times 10^{-4}$.

The steps are further shown in tabulated form for a total of 8 cycles.

Cycle	0	1	2	3	4	5	6	7	8
G_{soil}	41,284	20,642	35,229	33,028	33,874	33,028	33,874	33,028	33,874
V_s at 30 m	450	318	416	402	408	402	408	402	408
T	0.3460	0.489	0.375	0.387	0.382	0.387	0.382	0.387	0.382
Beta	0.04	0.04	0.04	0.04	0.04	0.04	0.04	0.04	0.04
Damping (%)	1	29	12	15	14	15	14	15	14
Sa/g (5%)	2.5	2.5	2.5	2.5	2.5	2.5	2.5	2.5	2.5
Sa/g (design)	8.00	1.38	2.00	1.75	2.00	1.75	2.00	1.75	2.00
Strain	5.57E −04	9.58E-05	1.39E −04	1.22E −04	1.39E −04	1.22E −04	1.39E −04	1.22E −04	1.39E −04

Thus, it is observed that reduced G value at 3.0 m below ground level is 33,450 kN/m^2 (avg.), and the material damping ratio is 14.5% (avg). The reduction factor to G_{max} is 0.80 in this case.

8.6 What Do We Do if the Soil Is Layered with Varying Soil Properties?

Till now, theories we have presented assume soil as a homogenous isotropic medium or varying with depth in a way that can be modeled analytically. But in many cases, the soil encountered at a particular site will be layered in nature.

Shown in Fig. 8.11 is a typical stratified soil profile where the shear modulus, density of soil, and Poisson's ratio vary for these stratified layers.

For most of the cases, engineers would take a weighted average where the average dynamic property may be taken as

$$G_{av} = \frac{G_1 \cdot H_1 + G_2 \cdot H_2 + G_3.H_3 + G_4 \cdot H_4}{H_1 + H_2 + H_3 + H_4} \tag{8.6.1}$$

And same principle will be applied for mass density and Poisson's ratio. Eq. (8.6.1) may be valid where the difference in magnitude in shear wave velocity for each of the layers vis-à-vis G is not large.

However, for soil having layers with widely varying properties or for soil–structure interaction of very important structures where the site is susceptible to major earthquakes, methods based on discrete element analysis may also be applied to arrive at a design dynamic modulus and damping value.

Shown in Fig. 8.12 is the mathematical model of a site having layered soil properties. We had shown earlier that for a soil site extending to infinity in

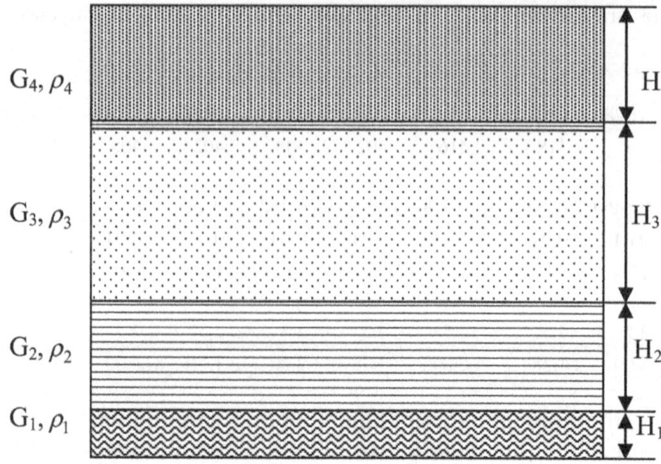

Fig. 8.11 Layered soil profile with varying soil properties

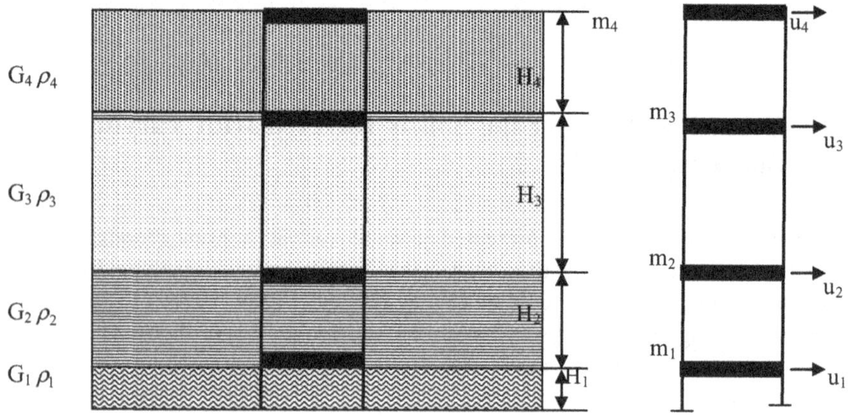

Fig. 8.12 Layered site with equivalent shear frame model

horizontal direction and having a depth H to the bedrock, it can well be modeled as a one-dimensional shear beam. In this case, as the soil is heavily layered having variable properties it can well be modeled as a shear frame, as depicted in Fig. 8.12.

In this case, the element stiffness matrix is expressed as Roesset and Whitman (1969)

$$[\mathbf{K_i}] = \frac{G_i}{H_i}\begin{bmatrix} 1 & -1 \\ -1 & 1 \end{bmatrix} \tag{8.6.2}$$

Thus, for the four-layer soil strata, as shown in Fig. 8.12, the global stiffness matrix can be expressed as

$$[\mathbf{K}] = \begin{bmatrix} \frac{G_1}{H_1} + \frac{G_2}{H_2} & -\frac{G_2}{H_2} & 0 & 0 \\ -\frac{G_2}{H_2} & \frac{G_2}{H_2} + \frac{G_3}{H_3} & -\frac{G_3}{H_3} & 0 \\ 0 & \frac{-G_3}{H_3} & \frac{G_3}{H_3} + \frac{G_4}{H_4} & -\frac{G_4}{H_4} \\ 0 & 0 & -\frac{G_4}{H_4} & \frac{G_4}{H_4} \end{bmatrix} \tag{8.6.2}$$

The lumped mass matrix [M] of the soil is expressed as

$$[\mathbf{M}] = \frac{1}{2}\begin{bmatrix} \rho_1 H_1 + \rho_2 H_2 & 0 & 0 & 0 \\ 0 & \rho_2 H_2 + \rho_3 H_3 & 0 & 0 \\ 0 & 0 & \rho_3 H_3 + \rho_4 H_4 & 0 \\ 0 & 0 & 0 & \rho_4 H_4 \end{bmatrix} \tag{8.6.3}$$

The damping matrix [C] is expressed as

$$[\mathbf{C}] = \begin{bmatrix} c_1 + c_2 & -c_2 & 0 & 0 \\ -c_2 & c_2 + c_3 & -c_3 & 0 \\ 0 & -c_3 & c_3 + c_4 & -c_4 \\ 0 & 0 & -c_4 & c_4 \end{bmatrix} \tag{8.6.4}$$

Here $c_i = \zeta_i \times 2\sqrt{k_i m_i}$. Here ζ_i can be computed from Eq. (8.4.1.15).

In Eqs. (8.7.2) and (8.7.3), G_i = shear modulus of ith layer, where i = 1, 2, 3, 4. H_i = depth of soil strata of the ith layer, and ρ_i = mass density of soil of the ith layer.

As explained in Chap. 2, considering that $[\mathbf{K}] - [\mathbf{M}]\omega^2 = 0$, we can express the amplitude of vibration for the ith mode as

$$\{U_i\} = \left(\frac{ZI}{2R}\right)\{\phi_i\}\left(\frac{L_i}{M_i}\right)\left[\frac{S_{ai}}{\omega_i^2}\right] \tag{8.6.5}$$

In Eq. (8.6.5), $\{U_i\} = \{u_1 \quad u_2 \quad u_3 \quad u_4\}^{\mathrm{T}}$. As discussed earlier, it suffices to work with the fundamental mode when

$$\{U_1\} = \left(\frac{ZI}{2R}\right)\{\phi_1\}\left(\frac{L_1}{M_1}\right)\left[\frac{S_{a1}}{\omega_1^2}\right] \tag{8.6.6}$$

Having computed the amplitude of vibration, the strain in different layers of soil can be represented by

$$\gamma_1 = \frac{u_1}{H_1}, \gamma_2 = \frac{u_2 - u_1}{H_2}, \gamma_3 = \frac{u_3 - u_2}{H_3} \text{ and } \gamma_4 = \frac{u_4 - u_3}{H_4}.$$

For the first cycle iteration, as discussed previously, $\gamma_a/\gamma_r = 1.0$, which gives G/G_{max} = 0.5. For subsequent cycles, the steps as explained in Sect. 8.6.2.1 may be followed to determine the reduced stiffness and augmented value of damping.

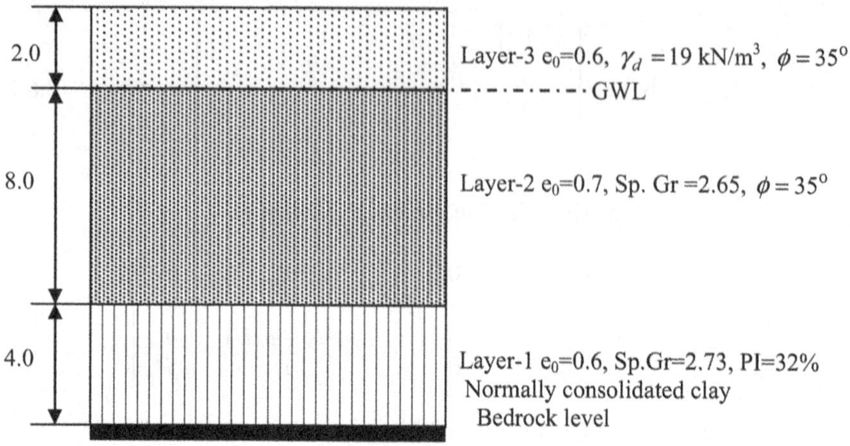

Fig. 8.13 Layered soil site with their engineering soil property

Calculation of shear modulus of soil based on free field time period is an effective tool for assessing the dynamic shear modulus of soil; however, there is a possibility that the time period obtained by this method could be higher and then the reality unless proper consideration is taken for the confining effect of the surrounding soil and proper judgment of the depth is not made.

The ATC[3] has defined H_{max} as the depth limited to 183 m, having low strain shear wave velocity of 760 m/s.

Example 8.9 Shown above in Fig. 8.13 is a layered soil site having properties as shown in the right-hand side of the figure.

The site is located in zone *V* as per IS 1893(2016). Determine the degraded stiffness and enhanced damping material of the soil.

Solution

As a first step, we determine the low strain dynamic shear modulus G_{max} and damping ratio (ζ) from the above soil property.

For layer-3

Angle of friction $\phi = 35°$.

Thus, $K_0 = 1 - \sin\phi = 0.426$

The net overburden pressure at the centre of layer 3 is

$$\sigma_v = 19 \times 1 = 19\ \text{kN/m}^2.$$

Now $\sigma_0 = \frac{\sigma_v}{3}[1 + 2K_0] = \frac{19}{3}[1 + 2 \times 0.426] = 11.73\ \text{kN/m}^2$.

[3]Applied Technology Council "Tentative provisions for the development of seismic regulations for buildings" ATC 3-06, NBS SP-510, NSF 78-8, National Bureau of Standards USA(1982).

$$G_{\max} = \frac{6908(2.17 - e_0)^2}{1 + e_0}\sqrt{\sigma_0},$$

here $e_0 = 0.6$; thus,

$$\rightarrow G_{\max} = \frac{6908(2.17 - 0.6)^2}{1 + 0.6}\sqrt{11.73} = 36,448\ \text{kN/m}^2.$$

As per Ishibashi and Zhang

$$\zeta = \frac{1}{6}\left(1 + \mathrm{e}^{-0.0145(\mathrm{PI})^{1.3}}\right)\left[1 + 0.586\left(\frac{G}{G_{\max}}\right) - 1.547\left(\frac{G}{G_{\max}}\right)^2\right]$$

For $PI = 0.0$ and $G/G_{\max} = 1$, $\zeta \rightarrow 0.013$

For layer-2

Angle of friction $\phi = 30°$.

Thus, $K_0 = 1 - \sin\phi = 0.50$,

here $e_0 = 0.7$; thus

$$\gamma_b \frac{G_s - 1}{1 + e}\gamma_w = \left(\frac{2.65 - 1}{1 + 0.7}\right) \times 10 = 9.74\ \text{kN/m}^3.$$

The net overburden pressure at the centre of layer 2 is

$$\sigma_v = 19 \times 2 + 9.74 \times 4 = 76.8\ \text{kN/m}^2.$$

Now $\sigma_0 = \frac{\sigma_v}{3}[1 + 2K_0] = \frac{76.8}{3}[1 + 2 \times 0.50] = 51.2\ \text{kN/m}^2$.

$$G_{\max} = \frac{6908(2.17 - e_0)^2}{1 + e_0}\sqrt{\sigma_0},$$

$$\rightarrow G_{\max} = \frac{6908(2.17 - 0.7)^2}{1 + 0.7}\sqrt{51.2} = 62,830\ \text{kN/m}^2.$$

As per Ishibashi and Zhang (1993)

$$\zeta = \frac{1}{6}\left(1 + \mathrm{e}^{-0.0145(PI)^{1.3}}\right)\left[1 + 0.586\left(\frac{G}{G_{\max}}\right) - 1.547\left(\frac{G}{G_{\max}}\right)^2\right]$$

For $PI = 0.0$ and $G/G_{\max} = 1$, $\zeta \rightarrow 0.013$

For layer-1

Angle of friction $\phi = 0°$.

$PI = 32\%$

Thus, $K_0 = 0.4 + 0.007PI = 0.624$,

Here $e_0 = 0.6$; thus

$$\gamma_b \frac{G_s - 1}{1+e} \gamma_w = \left(\frac{2.73 - 1}{1+0.6}\right) \times 10 = 10.8\ \text{kN/m}^3.$$

The net overburden pressure at the centre of layer 2 is

$$\sigma_v = 19 \times 2 + 9.74 \times 8 + 10.8 \times 2 = 137.52\ \text{kN/m}^2.$$

Now $\sigma_0 = \frac{\sigma_v}{3}[1+2K_0] = \frac{137.52}{3}[1+2\times 0.624] = 103\ \text{kN/m}^2$.

$G_{max} = \frac{3230(2.97-e_0)^2}{1+e_0}(\text{OCR})^k\sqrt{\sigma_0}$, Here $k = 0.252$ and OCR = 1 as the clay is normally consolidated

$$\rightarrow G_{max} = \frac{3230(2.97-0.6)^2}{1+0.6}\sqrt{103} = 115{,}079\ \text{kN/m}^2.$$

As per Ishibashi and Zhang

$$\zeta = \frac{1}{6}\left(1+e^{-0.0145(PI)^{1.3}}\right)\left[1+0.586\left(\frac{G}{G_{max}}\right) - 1.547\left(\frac{G}{G_{max}}\right)^2\right]$$

For PI = 32% and G/G_{max} = 1, $\zeta \rightarrow 0.0082$
Thus, $k_1 = G_1/H_1 = 28{,}770\ \text{kN/m}^2$, $k_2 = G_2/H_2 = 7854\ \text{kN/m}^2$, $k_3 = G_3/H_3 = 18{,}224\ \text{kN/m}^2$
Thus, the stiffness matrix is expressed as

$$[K] = \begin{bmatrix} k_1+k_2 & -k_2 & 0 \\ -k_2 & k_2+k_3 & -k_3 \\ 0 & -k_3 & k_3 \end{bmatrix} = \begin{bmatrix} 36620 & -7854 & 0 \\ -7854 & 26080 & -18224 \\ 0 & -18224 & 18224 \end{bmatrix}$$

The mass matrix is expressed as

$$[M] = \frac{1}{2}\begin{bmatrix} \rho_1H_1+\rho_2H_2 & 0 & 0 \\ 0 & \rho_2H_2+\rho_3H_3 & 0 \\ 0 & 0 & \rho_3H_3 \end{bmatrix} = \begin{bmatrix} 13.09 & 0 & 0 \\ 0 & 10.63 & 0 \\ 0 & 0 & 2.07 \end{bmatrix}$$

Considering $c_i = \zeta_i \times 2\sqrt{k_i m_i}$ and substituting the appropriate values, we have

$c_1 = 22.565$ $c_2 = 10.624$ and $c_3 = 4.505$. Thus,

$$[C] = \begin{bmatrix} c_1+c_2 & -c_2 & 0 \\ -c_2 & c_2+c_3 & -c_3 \\ 0 & -c_3 & c_3 \end{bmatrix} = \begin{bmatrix} 33.189 & -10.624 & 0 \\ -10.624 & 15.129 & -4.505 \\ 0 & -4.505 & 4.505 \end{bmatrix}$$

Considering $[K] - [M]\omega^2 = 0$, we have

$$\begin{bmatrix} 36620 - 13.09\lambda & -7854 & 0 \\ -7854 & 26080 - 10.63\lambda & -18224 \\ 0 & -18224 & 18224 - 2.07\lambda \end{bmatrix} = 0$$

Considering $\omega^2 = \lambda$ and $T = 2\pi/\omega$, we have

$T = \{0.294 \quad 0.116 \quad 0.061\}^{\mathrm{T}}$ and the corresponding eigenvectors are

$$[\varphi] = \begin{bmatrix} 0.243 & 1 & 0.016 \\ 0.948 & -0.241 & -0.021 \\ 1 & -0.362 & 1 \end{bmatrix}$$

We consider the effect of first mode only.
Thus, $T = 0.294$ s and $\{\phi_1\} = \{0.243 \quad 0.948 \quad 1\}^{\mathrm{T}}$

$$2\zeta_1\omega_1 = \{\phi_1\}^{\mathrm{T}}[C]\{\phi_1\} \rightarrow 6.625$$

$$\zeta_1 = \frac{6.625}{2 \times 21.352} = 0.155$$

Considering the site resting on a hard rock, $S_a = 2.5 \times 9.81 = 24.525$ m/s^2.
Considering $\zeta_1 = 0.155$ the design $S_{ad} = 0.7 \times 24.525 = 17.16$ m/s^2.

$$L_1 = \{\phi_1\}T[M]\{I\} = 15.328$$

$$M_1 = \{\psi_1\}T[M]\{\psi_1\} = 12.396$$

For zone V, as per IS-1893, $Z = 0.36$, $I = 1.5$, and $R = 2.0$

$$\{U\} = \left(\frac{ZI}{2R}\right)\{\phi_1\}\left(\frac{L_1}{M_1}\right)\left(\frac{S_{ad}}{\omega_1^2}\right) = \begin{Bmatrix} 1.524 \\ 5.946 \\ 6.272 \end{Bmatrix} \times 10^{-3}\ \mathrm{m}$$

Thus,

$$\gamma_1 = \frac{u_1}{H_1} = \frac{1.524 \times 10^{-3}}{4} = 3.81 \times 10^{-4}$$

$$\gamma_2 = \frac{u_2 - u_1}{H_2} = \frac{5.946 - 1.524}{8} \times 10^{-3} = 5.527 \times 10^{-4}$$

and

$$\gamma_3 = \frac{u_3 - u_2}{H_3} = \frac{6.272 - 5.946}{2} \times 10^{-3} = 1.631 \times 10^{-4}$$

Thus, for the first cycle, actual and reference strains are same; thus, the reference strain and actual strain are as shown hereafter

Reference strain	3.81×10^{-4}	5.527×10^{-4}	1.631×10^{-4}
Actual strain	3.81×10^{-4}	5.527×10^{-4}	5.527×10^{-4}

Thus, for all three cases, $G/G_{max} = 0.5$
This gives $G_1 = 57,540$ kN/m^2, $G_2 = 31,415$ kN/m^2 $G_3 = 18,224$ kN/m^2.
This gives the reduced stiffness matrix as follows

$$[K] = \begin{bmatrix} k_1 + k_2 & -k_2 & 0 \\ -k_2 & k_2 + k_3 & -k_3 \\ 0 & -k_3 & k_3 \end{bmatrix} = \begin{bmatrix} 18310 & -3927 & 0 \\ -3927 & 13040 & -9112 \\ 0 & -9112 & 9112 \end{bmatrix}$$

For layers 3 and 2

$$\zeta = \frac{1}{6}\left(1 + e^{-0.0145(PI)^{1.3}}\right)\left[1 + 0.586\left(\frac{G}{G_{max}}\right) - 1.547\left(\frac{G}{G_{max}}\right)^2\right]$$

For $G/G_{max} = 0.5$, $\zeta = 0.151$
For layer 1, for $G/G_{max} = 0.5$ and $PI = 32\%$, $\zeta = 0.383$

$$\text{Thus, } [C] = \begin{bmatrix} 557.845 & -87.283 & 0 \\ -87.283 & 145.956 & -58.672 \\ 0 & -58.672 & 58.672 \end{bmatrix}$$

The mass matrix remains same as before, and considering $\omega^2 = \lambda$ and $T = 2\pi/\omega$, we have

$T = \{0.416 \quad 0.164 \quad 0.086\}^{T}$ and the corresponding eigenvectors are

$$[\varphi] = \begin{bmatrix} 0.243 & 1 & 0.016 \\ 0.948 & -0.241 & -0.021 \\ 1 & -0.362 & 1 \end{bmatrix}$$

$$2\zeta_1\omega_1 = \{\phi_1\}^{T}[C]\{\phi_1\} \rightarrow 71.286$$

$$\zeta_1 = \frac{6.625}{2 \times 15.098} = 2.361$$

Considering $\zeta_1 = 2.361$, the design $S_{ad} = 0.55 \times 24.525 = 13.48$ m/s^2.

$$\{U\} = \left(\frac{ZI}{2R}\right)\{\phi_1\}\left(\frac{L_1}{M_1}\right)\left(\frac{S_{ad}}{\omega_1^2}\right) = \begin{Bmatrix} 2.399 \\ 9.358 \\ 9.871 \end{Bmatrix} \times 10^{-3}\ \text{m}$$

Thus,

$$\gamma_1 = \frac{u_1}{H_1} = \frac{2.399 \times 10^{-3}}{4} = 5.977 \times 10^{-4}$$

$$\gamma_2 = \frac{u_2 - u_1}{H_2} = \frac{9.358 - 2.399}{8} \times 10^{-3} = 8.699 \times 10^{-4}$$

and

$$\gamma_3 = \frac{u_3 - u_2}{H_3} = \frac{9.871 - 9.358}{2} \times 10^{-3} = 2.657 \times 10^{-4}$$

Soil layer	3	2	1
Reference strain	3.81×10^{-4}	5.527×10^{-4}	1.631×10^{-4}
Actual strain	5.977×10^{-4}	8.699×10^{-4}	2.657×10^{-4}
γ_a/γ_r	1.56	1.573	1.63
G/G_{max}	0.390	0.388	0.380
$G=$	44,881	24,378	13,850
ζ	0.166	0.167	0.42

With the above values, we reconstitute the stiffness and damping matrix, which gives revised time period as follows

$T = \{0.63 \quad 0.249 \quad 0.055\}^T$, and the corresponding eigenvectors are

$$[\varphi] = \begin{bmatrix} 0.58 & 1 & 3.709 \times 10^{-3} \\ 0.991 & -0.596 & -0.199 \\ 1 & -0.663 & 1 \end{bmatrix}$$

$$2\zeta_1\omega_1 = \{\phi_1\}^T[C]\{\phi_1\} \to 71.286$$

$$\zeta_1 = \frac{99.28}{2 \times 9.971} = 4.979$$

Considering $\zeta_1 = 4.979$, the design $S_{ad} = 0.55 \times \left(\frac{1}{0.63}\right) \times 9.81 = 8.562\ \text{m/s}^2$.

$$L_1 = \{\phi_1\}T[M]\{I\} = 20.191$$

$$M_1 = \{\phi_1\}T[M]\{\phi_1\} = 16.905$$

$$\{U\} = \left(\frac{ZI}{2R}\right)\{\phi_1\}\left(\frac{L_1}{M_1}\right)\left(\frac{S_{ad}}{\omega_1^2}\right) = \begin{Bmatrix} 8.05 \times 10^{-3} \\ 0.014 \\ 0.014 \end{Bmatrix} \times 10^{-3}\ \text{m}$$

Thus,

$$\gamma_1 = \frac{u_1}{H_1} = 2.013 \times 10^{-3}$$

$$\gamma_2 = \frac{u_2 - u_1}{H_2} = 7.137 \times 10^{-4}$$

and

$$\gamma_3 = \frac{u_3 - u_2}{H_3} = 6.367 \times 10^{-5}$$

Soil layer	3	2	1
Reference strain	3.81×10^{-4}	5.527×10^{-4}	1.631×10^{-4}
Actual strain	2.013×10^{-3}	7.137×10^{-4}	6.367×10^{-5}
γ_a/γ_r	5.28	1.29	0.39
G/G_{max}	0.16	0.43	0.72
G=	5831	27,017	82,857
ζ	0.176	0.161	0.262

And so on.

We continue the iteration for two more cycles, and it will be observed that the values have stabilized.

8.7 Checklist of Parameters to Be Looked in the Soil Report

Based on our discussion above, the parameters which require particular attention in a soil report from the engineer are summarized as follows

Field test

- **Has SPT test been carried out?**

The obvious intention is find out *N* on which *G* value depends. This can also be utilized to cross-check the field observed dynamic data.

- **If SPT values are furnished, are they observed data or corrections need to be done?**

A point to be checked for field observed data, as shown earlier, needs to be corrected.

While the soil consultant will do this correction during his own calculation of bearing capacity of soil for foundation recommendations, usually furnishes observed field data while furnishing the bore log detail in the report. So for your calculation this data needs to be corrected. If you are not too sure, you can back-calculate it from the recommended ϕ value.

- **Has the groundwater table been established during boring?**

Usually provided in a soil report, but better to check for this has significant effect on the net vertical stress.

- Has any dynamic field test been carried out?
  - Block vibration test;
  - Seismic cross-hole test.

One of the tests should be a part of the soil report. But do not take the values furnished sacrosanct. Back-check with theoretical corelation to establish if the order is close; if not—you do have the right to ask your soil consultant why there is this discrepancy. There could be a special geological condition which could result in such discrepancy and you should be clear about it.

- **If the above tests are carried out, what is the strain range induced in the soil during the test?**

This is something usually not supplied by the soil consultant who usually would recommend a unique G value. This should not be acceptable to you.

You should clear it at the very outset when providing him the specification for geotechnical investigation that this is an input you are looking for, and it should be a part of his report. It is more realistic to start with this value rather than guessing a theoretical value of $10^{-3}/10^{-4}$.

Laboratory tests

- **Check Atterberg's limit—gives values of liquid limit, plastic limit, plasticity index, etc.**

Generally, the structural engineers/analysts ignore this topic. Our suggestion would be do not disregard this for this is the basic data which gives you the first insight into the fact as to how the soil behaves. Moreover, the plasticity index, being an important property, it is all the more one should pay attention to.

- **Triaxial test gives values of c, ϕ and pre-consolidation history.**

Again, given a backhanded treatment by the analysts who, without going thru the test data, would prefer to pick up the numbers which concern him (c_a and ϕ), we suggest to go thru the test and develop enough skill to interpret the pre-consolidation stress. Make sure to ask the consultant during enquiry to supply this data. *The overconsolidation ratio plays a very important role in arriving at the correct value of* G_{dyn}.

During interpretation, if need be, seek help of a geotechnical specialist to make sure that what has been understood is correct—this will save a lot of headache in the long run.

- Weight density and void ratio of soil;
- Grain distribution to check if the soil is gap graded, uniformly graded, or poorly graded. Relative density of cohesionless soil is highly dependent on this.

8.8 Epilogue

The technology described in this chapter is still an evolving technology and hopefully, with time and research that is being carried out, all around the world we shall be in a better position in future to predict more realistically the dynamic properties of soil which affect the response of a structure.

Topics herein have been selectively chosen, which we believe are simple to apply, provide reasonably realistic results and are practical for day to day design office practice.

Most of the techniques developed herein are based on research papers (and typical practices followed in some design offices).

The idea to present this chapter is to make reader aware of the limitations prevalent with soil and also to caution him on the fact that, without these values, a realistic estimation of the whole seismic analysis could become a questionable exercise.

So one should be cautious and use his best judgment to furnish a meaningful design.

References

ACI-351.3R-04. (2011). *Foundation for dynamic equipments*. American Concrete Institute committee, p. 351.

Chowdhury, I., Ghosh, A. & Dasgupta, S. P. (2015). Stiffness degradation and damping augmentation of soil under earthquake loading. *Electronic Journal of Geotechnical Engineering, 20*, 9 (Oklahoma USA).

Desai, C. S., & Zaman, M. (2014). *Advanced geotechnical engineering-soil structure interaction using computer and material model*. London U.K: CRC press.

Hardin, B. O., & Drnevich, V. P. (1972, July). Shear modulus and damping in soils: Design equations and curves. *JSMFD. ASCE, 98* (SM7).

Hardin, B. O., & Richart, F. E., Jr. (1963). Elastic wave velocities in granular media. *JSMFD ASCE, 89* (SM1).
IS-5249. (1992). *Measurement of dynamic properties of soil.* New Delhi India: Bureau of Indian Standard.
Ishibashi, I., & Zhang, K. (1993). Unified dynamic shear modulii and damping ratios of sand and clay. *Journal of Soils and Foundations, 33*, 1.
Murthy, V. N. S. (1991). *Soil mechanics and foundation engineering*. Bangalore India: Sai Kripa Publications.
Ohsaki, Y., & Iwasaki, R. (1973). On dynamic shear modulus and Poisson's ratio of soil deposites. *Journal of Soils and Foundations, 13,* 4.
Roesset, J. M., & Whitman, R. V. (1969). *Theoretical background for amplification studies* (Research Report # 69–16. Department of Civil Engineering MIT). USA: School of Engineering, Massachusetts Institute of Technology.
Seed, H. B., & Idriss, I. M. (1970). *Soil modulii and damping factors for dynamic response analysis* (Report # 70–1 EERC). Berkeley California: The College of Engineering, University of Californaia.
Terzaghi, K., & Peck, R. B. (1967) *Soil mechanics in engineering practice*. New York USA: Wiley.

Chapter 9
Earthquake Resistant Design and Analysis of Buildings

9.1 Introduction

In this chapter, we take up perhaps the most researched topic in the realm of earthquake engineering—*the buildings*.

Despite all the researches, if we look at the history of earthquake it is observed that buildings, especially of residential type, undergo maximum damage during an earthquake, thus remaining responsible for killing maximum human lives and causing significant economic damage.

Federation of Emergency Management Agency (FEMA), USA, carried out a study dating back from 1950 to 1999, trying to assess the fatality rate and loss of property in billions of dollars across the world due to natural catastrophes. This included hurricanes, floods, earthquakes, and others. The study is pictorially presented in Fig. 9.1.

It is apparent from Fig. 9.1 that of all the natural disasters, it is the earthquake that has created maximum havoc and has been the major perpetrator.

In most of the countries around the world (including India), it is seen that ill-planned residential buildings are producing the biggest casualties compared to structurally well-planned buildings and have been the biggest killers of human life.

One should always remember the priceless saying that "*Earthquakes do not kill, but ill-engineered buildings do.*"

Figure 9.2 shows a typical residential building that underwent severe damage, while an office complex in the same city (Ahmedabad) survived the earthquake with only some minor cracks, testifying the same.

I. Chowdhury and S. P. Dasgupta, *Earthquake Analysis and Design of Industrial Structures and Infra-structures*, GeoPlanet: Earth and Planetary Sciences,
https://doi.org/10.1007/978-3-319-90832-8_9

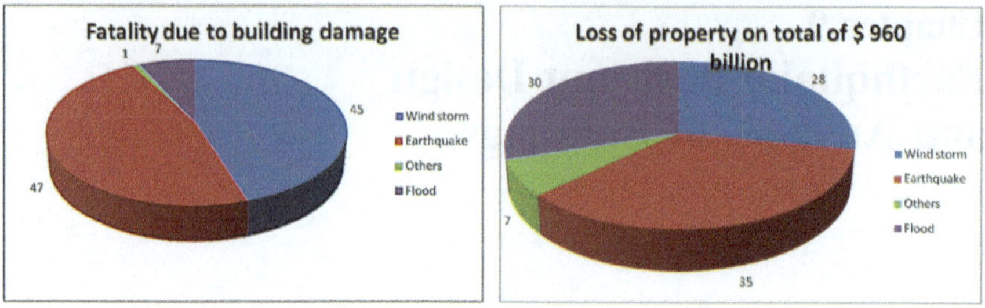

Fig. 9.1 Loss of life and property due to natural calamities

Fig. 9.2 Damage to residential and office complex; Bhuj earthquake

9.2 Seismic Analysis of Building as Per Code

Buildings come in different configurations and shapes. However, while planning buildings in areas prone to earthquake, it is recommended that the building has a shape that is regular and robust, having a clear load path to foundation. Stiffness of the members should be distributed regularly over the structure and should not have any abrupt change.

Some standard structural configurations common for buildings are as shown hereafter.

Figure 9.3a shows a building that consists of standard reinforced concrete beam column in structural arrangement commonly deployed and are either ordinary moment resistant frames (OMRF) or may constitute Special Moment Resistant Frames (SMRF), depending on what the structural engineer chooses to design.

Figure 9.3b shows flat slabs supported by shear walls usually deployed in high-rise multi-story buildings. The lift well (marked S_3) located in the center of building is usually made of reinforced concrete and acts as a shear wall too.

The shear walls have large stiffness in lateral direction and are designed to resist the seismic force induced in the building. Taking into account the fact that the floor

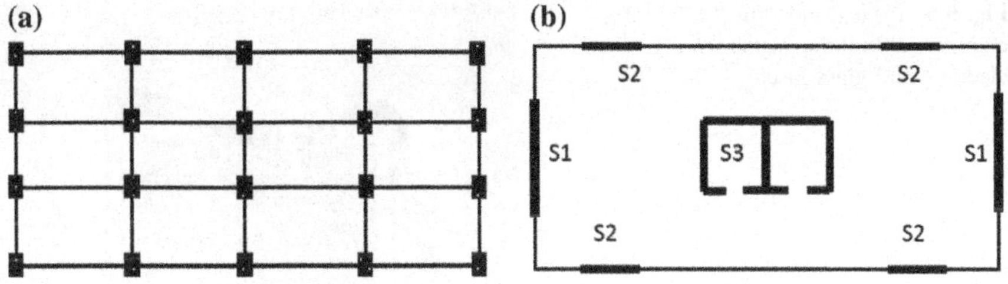

Fig. 9.3 Buildings supported on beam columns system and shear wall

slabs have infinite stiffness in their own plane, and they are usually considered as rigid diaphragms that transfer the seismic force to the shear walls without any or with negligible deformation.

When the floor area is large, like that of an office building, sometimes, to bring economy in design, the buildings are supported on shear walls and columns, as shown in Fig. 9.4a. In this case, the lateral loads are all transferred to the shear walls while the columns are designed to sustain the gravity load only. Connected by span beams, it significantly reduces the depth of floor slabs and cuts down on its overall thickness.

In many countries (especially Middle East and USA), building high rises with structural steel members are quite popular. With colored sun-reflecting glass fascia, they look quite spectacular and aesthetically pleasing. Fig. 9.5 shows such a typical building with glass fascia.

Figure 9.4b shows typical floor plan of such buildings that are often constructed nowadays. In such buildings, the transverse (shorter) direction usually consists of frames with moment connection, while the longitudinal or long direction usually consists of braced frames.

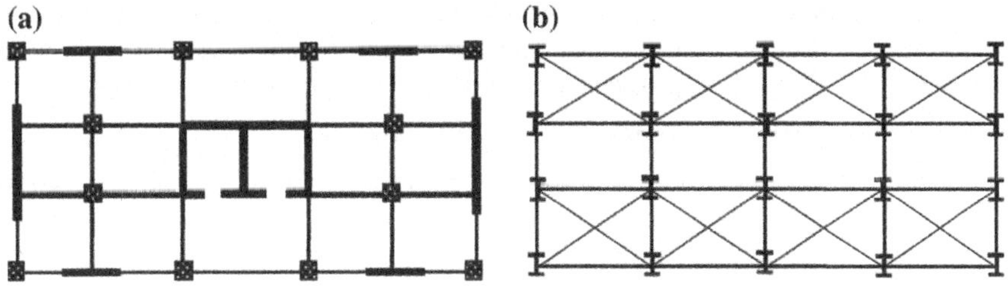

Fig. 9.4 Buildings supported by combination of shear wall, column and steel column; beams with bracing system

Fig. 9.5 Typical modern commercial building with cladding and glass fascia

9.2.1 Pseudo-static Analysis of Buildings Constructed as a Framed Structure

Design of buildings encompassing all national or international codes of practice follows a unique trait. It follows a simplest design based on equivalent lateral force extracted from an empirical time period expression.

Based on a study conducted by Dayaratnam and Jain (1991) and Jain et al. (1995), it has been observed that filled brick panels contribute significantly to stiffness of the overall structure and thus equation as proposed in codes gives time period closer to field observed figures (refer Eqs. 2.1.4 and 2.1.5 in Chap. 2).

These equations were developed based on real-time vibrations of buildings recorded during the San Fernando Earthquake in USA.

Once the time period is derived, based on building geometry, the design spectral acceleration is derived from the code and a pseudo-static analysis is carried out to compute the base shear and moments in the building frame.

For dynamic analysis, the Indian code recommends

- A shear frame-type *stick model analysis* for buildings greater than 40 m (roughly 12 story)—in Zone IV (severe), and Zone V (very severe), and for building height greater than 90 m (30 storied) in Zone II (low) or zone III (moderate), dynamic analysis (stick model) is obligatory—*provided the building is of regular shape or having nominal irregularity.*
- It is only in case of buildings with high geometric irregularities in vertical or horizontal plane (plan), the code recommends a dynamic two- or three-dimensional frame analysis.

While carrying out dynamic analysis, the code has imposed a particular restriction to base shear as per Clause 7.7.3 of IS-1893 (2016). In this particular clause, it has been stated that *if the base shear force computed based on dynamic analysis* $V_{dyn} \leq V_B$, *where* V_B *is as depicted by the code based on empirical expression, then one has to follow the empirical equation as the design basis.*

Thus, it implicitly suggests that time period computed based on empirical basis is sufficiently conservative to ensure a safe structure and shall *not be violated.*

The design procedure for such moment resistant frame based on response spectrum method adopting pseudo-static analysis has already been explained in Sect. 2.1.1 of Chap. 2. The same is further elucidated hereafter posing a suitable problem.

Example 9.1 A four-storied building is as shown in Fig. 9.3, which is a residential complex. The depth of slab at each floor is 125 mm. Beams in long direction have size 300 mm × 600 mm and that in short direction are 300 mm × 400 mm. The column size is considered to be 300 mm × 300 mm at all floors. Consider live load in roof to be 1.5 kN/m^2 and that at other floors to be 2.0 kN/m^2. The building consists of 250 mm thick peripheral walls and has 125 mm thick internal partition walls, considered at the rate of 1.0 kN/m^2. The building is located in zone IV as per IS code. Compute the nodal force and analyze the frame. The foundation is located 1.0 m below the existing ground level (Fig. 9.6).

Solution
Calculation of roof loads:

Weight of slab = 24 × 11 × 0.125 × 25 = 825 kN.
Weight of beam in long direction = 0.3 ×(0.6 − 0.125) × 24 × 4 × 25 = 342 kN.
Weight of beam in short direction = 0.3 ×(0.4 − 0.125) × 11 × 5 × 25 = 113 kN.
Weight of roof finish (avg 100 mm thick) = 24 × 11 × 0.1 × 22.5 = 594 kN.
Weight of ceiling plaster (25 mm thick) = 24 × 11 × 0.025 × 20.4 = 135 kN.

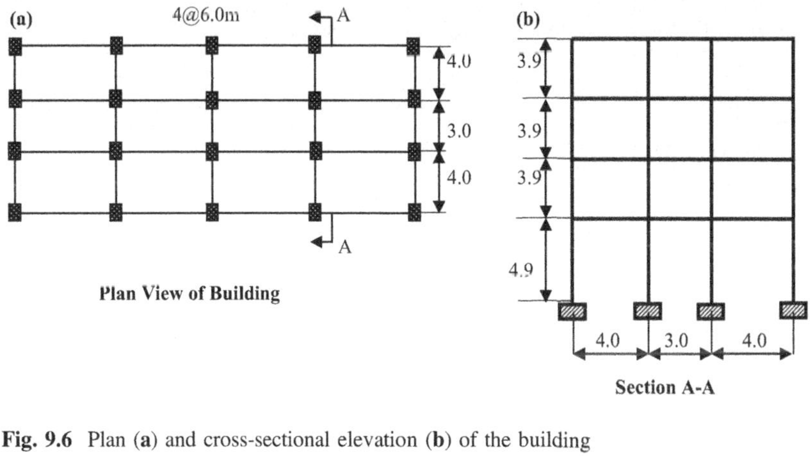

Fig. 9.6 Plan (**a**) and cross-sectional elevation (**b**) of the building

Weight of peripheral parapet wall 250 mm thick = 2 × (24 + 11) × 1.5 × 0.25 ×18 = 473 kN.
Live load = 24 × 11 × 1.5 = 396 kN.
Weight of column = 0.3 × 0.3 × (3.9/2) × 20 × 25 = 88 kN.
Total dead load = 825 + 342 + 113 + 594 + 135 + 473 + 88 = 2570 kN.

Calculation of floor loads:

Weight of slab = 825 kN.
Weight of beam in long direction = 342 kN.
Weight of beam in short direction = 113 kN.
Weight of floor tiles = 24 × 11 × 1.35 = 356 kN.
Weight of ceiling plaster = 135 kN.
Weight of external wall assuming 25% opening for windows = 0.75 × 2 ×(24 + 11) × 3.9 × 0.25 × 18 = 922 kN.
Weight of internal partition wall = 24 × 11 × 1.0 = 264 kN.
Live load = 24 × 11 × 2.0 = 528 kN.
Weight of column = 176 kN.
Total dead load = 825 + 342 + 113 + 356 + 135 + 922 + 264 + 176 = 3133 kN.

Thus, the load to be considered for each floor is as given in tabular form below

Sl No	Floor	Dead load (kN)	Live load** (kN)	Remarks
1	1st	3133	396	**Reduced by 25%
2	2nd	3133	396	
3	3rd	3133	396	
4	Roof	2570	0*	Live load on roof is ignored
	Total	11,969	1188	

Earthquake force in transverse direction:
The time period of building is expressed as $T = 0.09\frac{H}{\sqrt{D}}$
Here H = 16.6 m and D = 11 m
Thus, $T = 0.09\frac{16.6}{\sqrt{11}} = 0.45$ s, for medium soil site $S_a/g = 2.5$.

$\therefore A_h = \left(\frac{ZI}{2R}\right)\left(\frac{S_a}{g}\right)$, here $Z = 0.24$ for zone IV, $I = 1.0$ and $R = 3.0$ for ordinary moment resistant frame.

$$\rightarrow A_h = \left(\frac{0.24 \times 1.0}{2 \times 3}\right) \times 2.5 = 0.1$$

Thus, the lateral force induced on the building is $V_b = 0.1 \times (11{,}969 + 1188) = 1315.7$ kN.

Nodal force per floor:
This is given by the expression

$$Q_i = V_b \frac{W_i h_1^2}{\sum_{i=1}^{4} W_i h_i^2}$$

At the roof level,

$$Q_i = 1316 \frac{2570 \times (16.6)^2}{2570 \times (16.6)^2 + 3529 \times (12.7)^2 + 3529 \times (8.8)^2 + 3529 \times (4.9)^2}$$
$$= 1316 \times \left(\frac{708{,}189.2}{1{,}635{,}398.66}\right) = 569.87\ \text{kN}.$$

At the third floor,

$$Q_i = 1316 \frac{3529 \times (12.7)^2}{2570 \times (16.6)^2 + 3529 \times (12.7)^2 + 3529 \times (8.8)^2 + 3529 \times (4.9)^2}$$
$$= 1316 \times \left(\frac{569{,}192.41}{1{,}635{,}398.66}\right) = 458\ \text{kN}.$$

At the second floor,

$$Q_i = 1316 \frac{3529 \times (8.8)^2}{2570 \times (16.6)^2 + 3529 \times (12.7)^2 + 3529 \times (8.8)^2 + 3529 \times (4.9)^2}$$
$$= 1316 \times \left(\frac{273{,}286}{1{,}635{,}398.66}\right) = 220\ \text{kN}.$$

At the first floor:

$$Q_i = 1316 \frac{3529 \times (4.9)^2}{2570 \times (16.6)^2 + 3529 \times (12.7)^2 + 3529 \times (8.8)^2 + 3529 \times (4.9)^2}$$
$$= 1316 \times \left(\frac{84{,}731.29}{1{,}635{,}398.66}\right) = 68.2\ \text{kN}$$

Computation of nodal load per frame:
Two types of approaches are used in this case

- The influence area approach.
- Displacement compatible approach.

In case of first method, the total lateral load at roof level due to earthquake is 570 kN.

Thus, the whole frame is subjected to a uniformly distributed load of 23.75 kN/m, as shown in Fig. 9.7.

Fig. 9.7 Load distributed on each floor in plan view

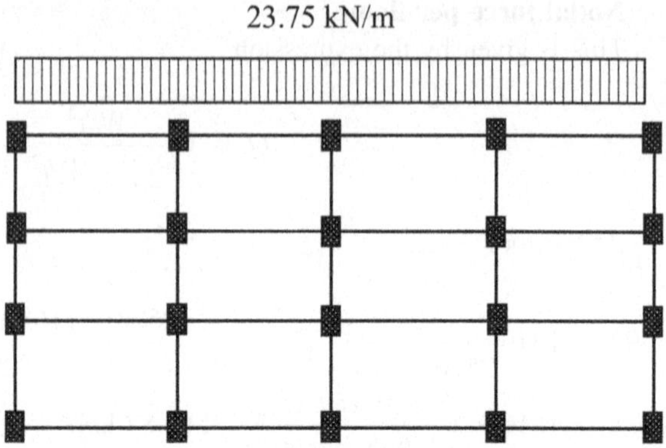

The load is now distributed to each frame, depending on their influence area. The end frames take load

$P_E = 23.75 \times 3.0 = 71.25$ kN, while the internal frames take load PI = $23.75 \times 6 = 142.5$ kN.

In the second case, it is assumed that the whole frame moves as a rigid body in translation and all columns have same displacement, thus considering that all frames have equal displacement, the load per frame can be expressed as

$P_{\mathrm{F}} = \frac{4K_c}{20K_c} \times 570 = 114$ kN on each frame at roof level.

Here P_{F} is the force per frame and K_c is stiffness of column 12 EI_c/H^3.

Authors believe that second approach is more rational for an earthquake. Thus, for a particular frame, the nodal forces can be assessed as shown in Fig. 9.8.

Fig. 9.8 Distribution of load per floor vertically

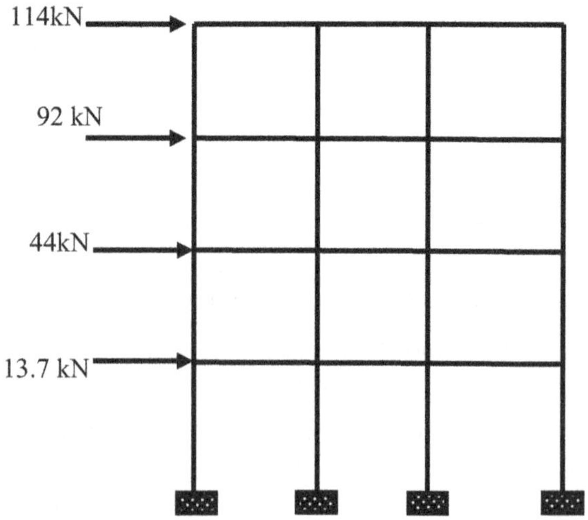

Earthquake force in longitudinal direction:
The time period of building is expressed as $T = 0.09\frac{H}{\sqrt{D}}$
Here H = 16.6 m and D = 24 m
Thus, $T = 0.09\frac{16.6}{\sqrt{24}} = 0.293$ s; for medium soil site S_a/g = 2.5.

$\therefore A_h = \left(\frac{ZI}{2R}\right)\left(\frac{S_a}{g}\right)$, here Z = 0.24 for zone IV, I = 1.0 and R = 3.0 for ordinary moment resistant frame.

$$\rightarrow A_h = \left(\frac{0.24 \times 1.0}{2 \times 3}\right) \times 2.5 = 0.1$$

Thus, the lateral force induced on the building is $V_b = 0.1 \times (11{,}969 + 1188) = 1315.7$ kN.

Proceeding in identical fashion it can be shown that nodal load per frame is as shown below.

Sl No	Floor	Load (kN)
1	1st	17
2	2nd	55
3	3rd	114.5
4	Roof	142.5
	Total	329

9.3 Pseudo-static Analysis of Buildings Constructed with Shear Walls

For multi-storied buildings, especially of commercial type, the structural component adopted to transfer earthquake force is usually shear walls. Beyond a certain height of a building, it is observed that shear walls are much more efficient in transferring load compared to beam column system and are often deployed to resist wind and earthquake loads (Kazimi and Chandra 1980) (Fig. 9.9).

Figure 9.9 shows typical plan of a multi-storied building supported on seven shear walls, S_1 to S_7. Observe that the shear wall S_4 acts as the elevator well and, as the whole floor space is free of columns, gives significant functional advantage.

The lateral loads due to an earthquake in X- or Y-direction, represented by EL_x or EL_y, respectively, are transferred through the floor slab to the shear walls. In this particular load transfer mechanism, it is assumed that the floor is infinitely rigid in its own plane, thus, it transfers the load to shear in such proportion that each has a common displacement δ.

Thus, if k_{xi} and k_{yi} are the stiffness of the ith wall in x- and y-direction, respectively, then

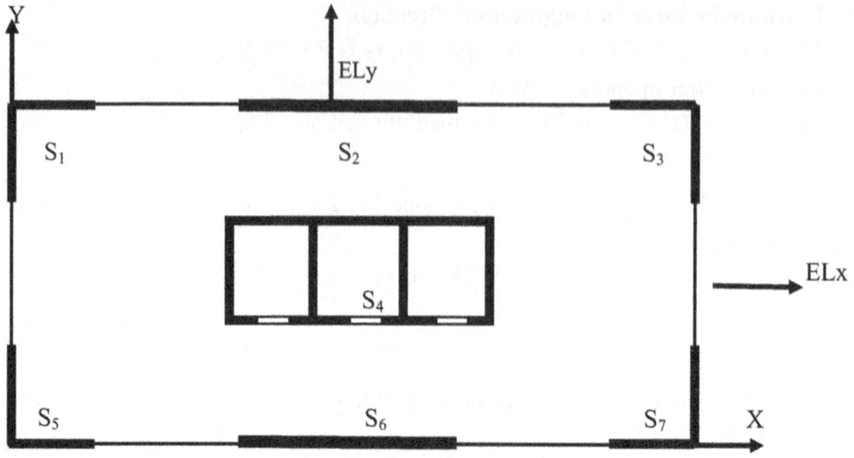

Fig. 9.9 A multi-storied building supported on shear wall

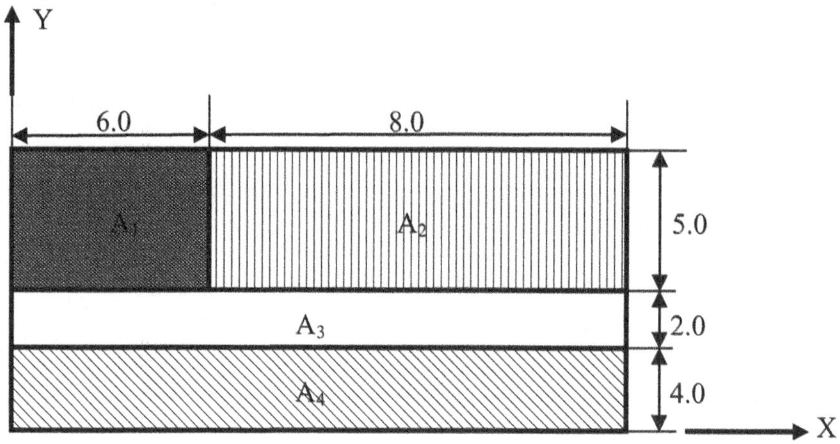

Fig. 9.10 Different loading zones in a building plan

$$\sum_{i=1}^{7} k_{x1}\delta_x = \mathrm{EL}_x \quad \text{and} \quad \sum_{i=1}^{7} k_{y1}\delta_y = \mathrm{EL}_y \tag{9.3.1}$$

Thus, for the *X*-direction, say

$$\delta_x = \frac{\mathrm{EL}_x}{k_{x1} + k_{x2} + k_{x3} + \cdots + k_{x7}} \tag{9.3.2}$$

If now V_{x1} is the shear force in the wall S_1 in the X-direction, then

$$V_{x1} = k_{x1}\delta_x \tag{9.3.3}$$

Now substituting Eq. (9.3.2), we have

$$V_{x1} = \frac{k_{x1}}{k_{x1} + k_{x2} + k_{x3} + \cdots + k_{x7}} \times \mathrm{EL}_x \tag{9.3.4}$$

Equation (9.3.4) can be generalized as

$$V_{xi} = \frac{k_{xi}}{\sum_{i=1}^{n} k_{xi}} \times \mathrm{EL}_x, \quad \text{where} \quad i = 1, 2, \ldots, 0.7 \tag{9.3.5}$$

Similarly, applying a similar argument one can write for the Y-direction

$$V_{yi} = \frac{k_{yi}}{\sum_{i=1}^{n} k_{yi}} \times \mathrm{EL}_y \tag{9.3.6}$$

Considering that the stiffness of the shear wall is a function of its moment of inertia (I) and length (L) and, as the shear walls have same height from floor to floor, Eqs. (9.3.5) and (9.3.6) can also be expressed as

$$V_{xi} = \frac{I_{xi}}{\sum_{i=1}^{n} I_{xi}} \times \mathrm{EL}_x \quad \text{and} \quad V_{yi} = \frac{I_{yi}}{\sum_{i=1}^{n} I_{yi}} \times \mathrm{EL}_y \tag{9.3.7}$$

The above shall only be valid when *center of mass* and *center of stiffness* of the walls match.

9.3.1 What Is Center of Mass (CM) and Stiffness (CS)?

The center of mass (CM) is that point located in the floor slab where all the mass of the slab located in a particular floor, when lumped and multiplied by the acceleration, generates the force EL_x or EL_y.

For a structure with uniform loading, this will obviously be at the geometric center of the floor.

Example 9.2 For an office floor, as shown in Fig. 9.10, the area A_1 which consists of a record room has live load of 10 kN/m^2. Area A_2, which consist of offices, is subjected to a live load of 2.5 kN/m^2. Area A_3, that consists of a corridor, has maximum live load of 5.0 kN/m^2. Area A_4, which consists of conference rooms and utilities, has a maximum live load of 3.0 kN/m^2. Considering the floor slab that is 200 mm thick and the floor finish of 1.5 kN/m^2 computes the center of mass of the floor.

Solution
Computation of dead load (DL):

Thickness of slab = 200 mm

Thus, DL of slab $= 0.2 \times 25 = 5\ \text{kN/m}^2$.
Floor finish = 1.5 kN/m². Thus, total DL of slab = 5 + 1.5 = 6.5 kN/m².

Panel	DL (kN/m²)	LL (kN/m²)	W_i (kN)	x_i	y_i	$W_i x_i$	$W_i y_i$
A_1	6.5	10.0	16.5 × 6 × 5 = 495	3.0	8.5	1485	4207.5
A_2	6.5	2.5	9 × 8 × 5 = 360	10.0	8.5	3600	3060
A_3	6.5	5.0	11.5 × 14 × 2=322	7.0	5.0	2254	1610
A_4	6.5	3.0	9.5 × 14 × 4 = 532	7.0	2.0	3724	1064
			$\sum W = 1709$		Sum →	11,063	9941.5

$$\bar{x} = \frac{11{,}063}{1709} = 6.473\ \text{m} \quad \text{and} \quad \bar{y} = \frac{9941.5}{1709} = 5.817\ \text{m}$$

Thus, the center of mass (CM) is as shown below.

In this case, the acceleration induced due to earthquake shall induce acceleration through the point CM.

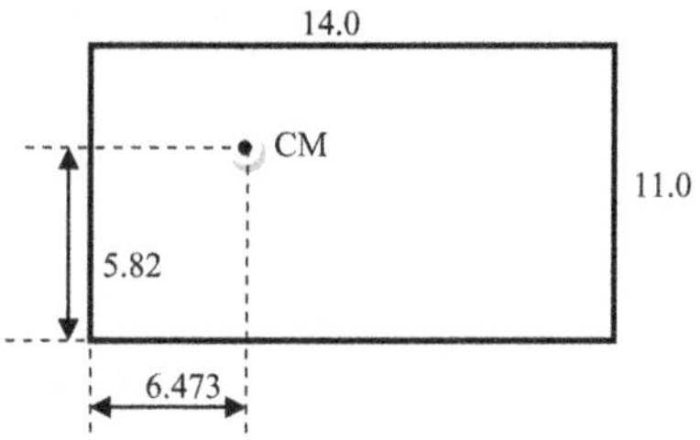

Similar to the center of mass, the center of stiffness (CS) is that point in the building plan through which the resultant resistive force induced by the shear walls or columns passes. Normally, for symmetrically placed walls or columns the center of mass converges with the geometric center of the building plan.

Example 9.3 A typical building is supported on five shear walls S_1 to S_5, as shown in Fig. 9.11. The thickness of shear walls is 300 mm. Locate the center of stiffness in plan view of the building.

Solution
Considering the bottom corner as the origin, the individual moment of inertia of the shear walls and their coordinates from the origin are tabulated hereafter:

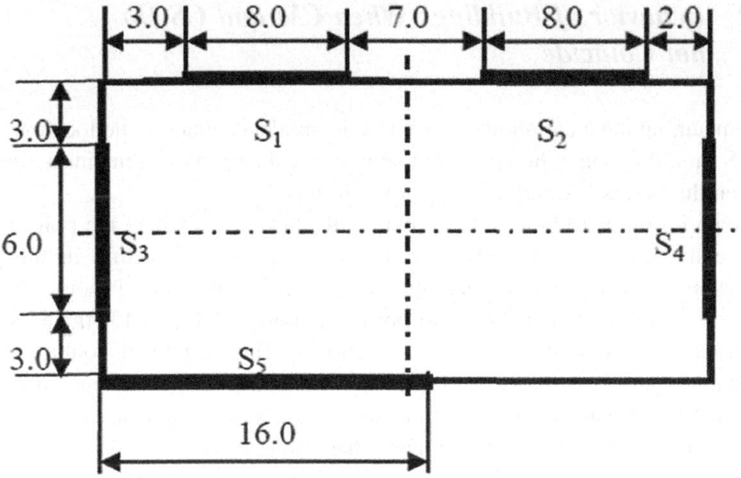

Fig. 9.11 Building plan with shear walls

Wall	I_x (m^4)	I_y (m^4)	Y (m)	X (m)	$Ix.Y$	$I_y.X$
S_1	$(1/12) \times 8^3 \times 0.3 = 12.8$	$(1/12) \times 8 \times (0.3)^3 = 0.018$	7.0	12	89.6	0.216
S_2	12.8	0.018	22	12	281.6	0.216
S_3	$(1/12) \times 0.3^3 \times 6 = 0.0135$	$(1/12) \times 6^3 \times 0.3 = 5.4$	0.0	6	0.0	32.4
S_4	0.0135	5.4	28	6	0.378	32.4
S_5	$(1/12) \times 16^3 \times 0.3 = 102.4$	$(1/12) \times 16 \times (0.3)^3 = 0.036$	8	0	819.2	0.0
	$\sum I_x = 128.027 \text{m}^4$	$\sum I_y = 10.872 \text{m}^4$			1190.778	65.232

Thus, $\bar{X} = \frac{65.232}{10.872} = 6.0$ m and $\bar{Y} = \frac{1190.778}{128.027} = 9.3$ m

Thus, the center of stiffness (CS) is shown in the sketch hereafter (Fig. 9.12).

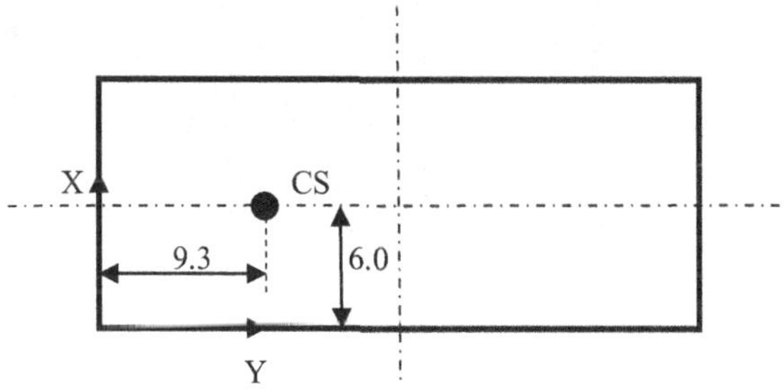

Fig. 9.12 Location of the center of stiffness (CS) in plan

9.3.2 Behavior of Buildings When CM and CS Do not Coincide

While planning the location of shear walls, in an ideal situation, the location of CM and CS should be same; however, in many practical cases there remains a difference between the points CS and CM as shown in Fig. 9.13.

Shown in Fig. 9.13 is a building plan with shear walls where the point CS and CM are not the same. In this case, the seismic force F_y acting in the global Y-direction is resisted by an equal and opposite reaction acting through CS. Now, since CM and CS are not the same point, as shown in Fig. 9.13, there exists an eccentricity in geometry, e_x and e_y, as shown. This results in couple vis-à-vis torsional moment $F_y \times e_x$ which rotates the whole building through an angle θ, say, so that the linear deformation of the ith shear wall is $r_i\theta$. Thus, the displacement of the ith wall is expressed as (Murty and Charleson 2006)

$$\delta_i = r_i\theta \quad (9.3.8)$$

Here r_i = radius vector of the shear wall from the point CS for the ith shear wall as shown in Fig. 9.13, and θ = angular distortion of the building in plan.

Now if the lateral stiffness of the ith shear wall be k_i, then the force on the wall is given by

$$F_i = k_i\delta_i = k_i r_i\theta \quad (9.3.9)$$

Now restoring moment by force on the ith element is expressed as

$$M_i = F_i r_i = k_i r_i^2\theta \quad (9.3.10)$$

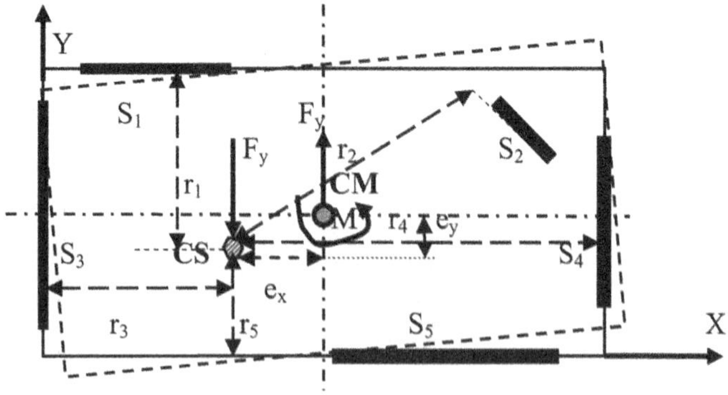

Fig. 9.13 A building plan having different location of CS and CM

Thus, for the total torsional moment M for equilibrium we have

$$M = \theta \sum_{i=1}^{5} k_i r_i^2 \tag{9.3.11}$$

$$\rightarrow \theta = \frac{M}{\sum_{i=1}^{5} k_i r_i^2} \tag{9.3.12}$$

Thus, substituting Eq. (9.3.12) in (9.3.9) we finally have

$$F_i = \frac{k_i r_i}{\sum_{i=1}^{5} k_i r_i^2} M \tag{9.3.13}$$

The shear force, as expressed in Eq. (9.3.13), *will act in addition to the total force F_y that gets distributed in proportion to individual stiffness of the wall.*

9.3.2.1 Estimation of Stiffness of Shear Wall

For cantilever shear wall, as shown in Fig. 9.14, deflection of the wall including shear deformation can be expressed as (Macleod 1990)

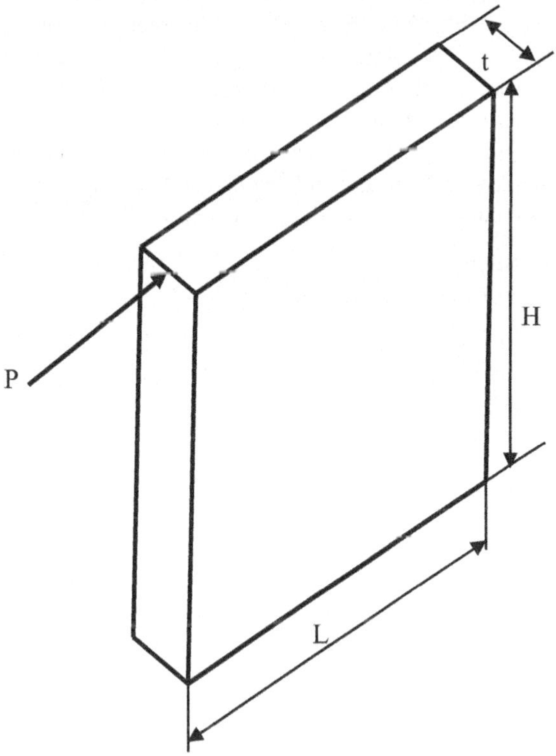

Fig. 9.14 A cantilever shear wall subjected to shear force due to earthquake

$$\Delta = \frac{PH^3}{3EI} + \frac{1.2PH}{AG} \tag{9.3.14}$$

Considering $G = 0.4E$, Eq. (9.3.14) can be expressed as

$$\Delta = \frac{PH^3}{3EI}\left(1 + \frac{9}{\lambda^2}\right) \tag{9.3.15}$$

Considering that stiffness k is the force per unit displacement, Eq. (9.3.15) can be expressed as

$$K_w = \frac{3EI}{H^3}\frac{1}{\left[1 + \left(\frac{3}{\lambda}\right)^2\right]} \tag{9.3.16}$$

In Eq. (9.3.16), E = Young's modulus of wall material, I = moment of inertia of the wall and, with respect to Fig. 9.14, $I = (1/12)tL^3$. $\lambda = H/r$ is the slenderness ratio of the wall. r = radius of gyration $\sqrt{I/A}$, where A = area of cross section $L \times t$.

Example 9.4 Shown in Fig. 9.15 is a G+3-office building, which is supported on RCC shear walls S_1 to S_4, as shown with 100 mm stone fascia. The floor slab thickness is 200 mm; the outer cladding/wall of the building is anodized aluminum frame with glass claddings. The internal partition walls are 125-mm-hollow brick walls. Considering the building is centrally air-conditioned, an air chiller is located at roof level, having dead weight of 500 kN. The building is to be built in Ahmedabad (Gujarat, India). Loads on the building are to be considered as per IS-875, while earthquake load is to be considered as per IS-1893. The thicknesses of shear walls is as tabulated hereafter. Analyze the forces on the shear wall under earthquake force. Grade of concrete used is M25.

Solution
Load calculation for the roof level:

Dead load of slab = $48 \times 16 \times 0.2 \times 25 = 3840$ kN.
1.2 m parapet made of brick 250 mm thick = $2 \times (48 + 16) \times 1.2 \times 0.25 \times 20 = 768$ kN.
Roof finish varying from 125 mm to 75 mm in thickness = $48 \times 16 \times 0.1 \times 20.7 = 1590$ kN.
Load from air chiller = 500 kN.
Live load on roof reduced by 25% due to earthquake load = $48 \times 16 \times 1.0 \times 75 = 576$ kN.
Load from shear wall $S_1 = S_1 = 44 \times 0.25 \times \frac{3.5}{2} \times 25 = 481.25$ kN.
Load from shear wall $S_2 = S_3 = 6 \times 0.35 \times \frac{3.5}{2} \times 25 = 92$ kN.
Load from shear wall $S_4 = 12 \times 0.30 \times \frac{3.5}{2} \times 25 = 158$ kN.

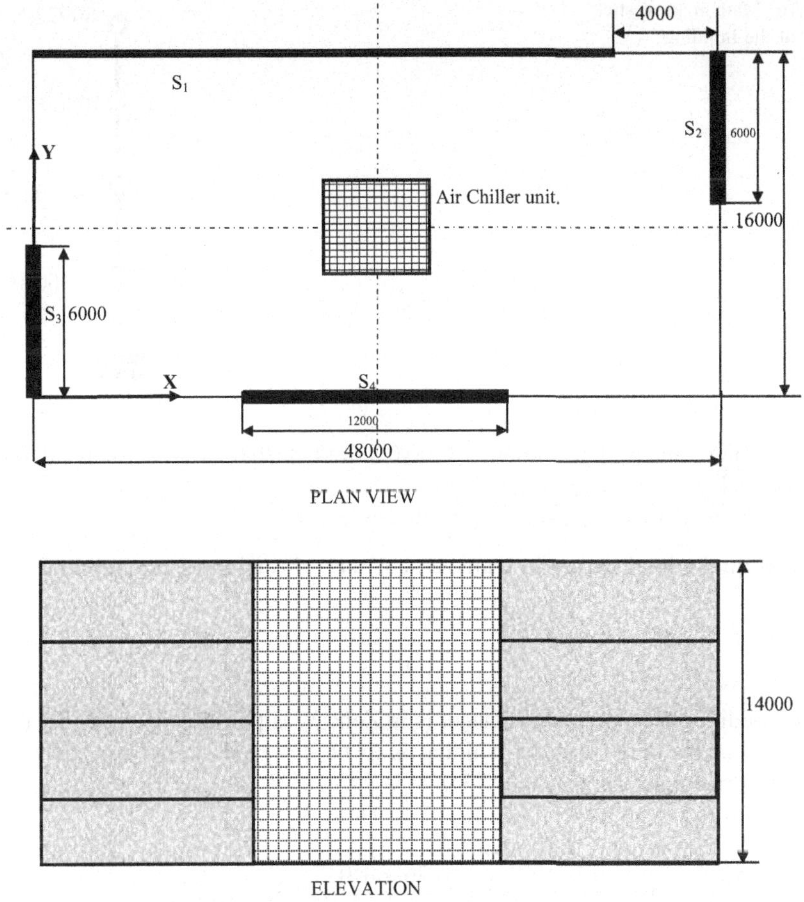

Fig. 9.15 Plan and elevation of G+3 office building

Thus, the total load contribution at roof level = 3840 + 768 + 1590 + 500 + 576 + 481.25 + 2 × 92 + 158 = 8097 kN.

Load calculation for other floors:

Dead load of slab = 3840 kN.
Floor finish = 48 × 16 × 0.1 × 20.7 = 1590 kN.
Load from internal partition wall = 48 × 16 × 1.0 = 768 kN.
Double layered glass mounted in aluminum frame considered as 17.5 kN/m^2 = $(2 \times 10 + 4 + 36) \times 3.5 \times 17.5 = 1575$ kN.
Live load reduced by 25% = 48 × 16 × 2.5 × 0.75 = 1440 kN.
Weight from shear wall = $2 \times [481.25 + 2 \times 92 + 158] = 1647$ kN.

Fig. 9.16 Mathematical stick model of the building

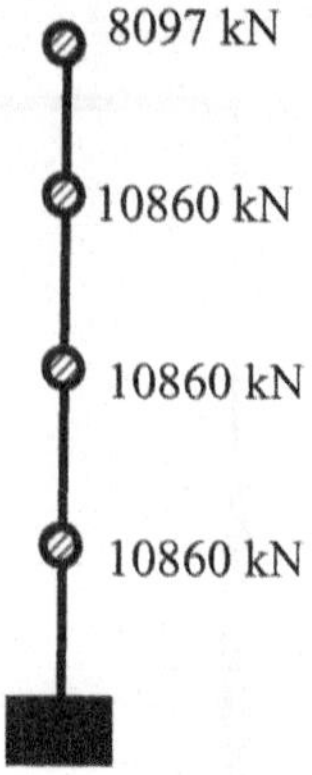

Thus, the load per floor $= 3840 + 1590 + 768 + 1575 + 1440 + 1647 = 10{,}860$ kN.

Thus, the model can be expressed as shown in Fig. 9.16.

For earthquake load acting in transverse (short) direction of the building:

$$T = \frac{0.09H}{\sqrt{D}} = \frac{0.09 \times 14}{\sqrt{16}} = 0.315 \text{ s}.$$

The building, located in Ahmedabad, is in Zone V as per IS code. Consider the importance factor of $I = 1.0$ the building is to be designed as ordinary moment resistance structural component $R = 3.0$.

For $T = 0.315$, S_a/g as per IS code is 2.5. Thus,

$$A_h = \frac{ZI}{2R} \times \left(\frac{S_a}{g}\right) = \frac{0.36 \times 1.0}{2 \times 3} \times 2.5 = 0.15$$

Thus, the base shear $V_b = A_h W = 0.15 \times (8097 + 3 \times 10{,}860) = 6101$ kN.

Distribution of force on each floor:

The distribution of force at each floor is expressed by (Refer Chap. 2)

$$V_i = \frac{W_i h_i^2}{\sum_{i=1}^{4} W_i h_i^2} \times V_b$$

At El 14.0 m

$$V_4 = \frac{8097 \times (14)^2}{(8097 \times 14^2 + 10{,}860 \times 10.5^2 + 10{,}860 \times 7^2 + 10{,}860 \times 3.5^2)} \times 6101$$
$$= 2807 \text{ kN}$$

At El 10.5 m

$$V_3 = \frac{10{,}860 \times (10.5)^2}{(8097 \times 14^2 + 10{,}860 \times 10.5^2 + 10{,}860 \times 7^2 + 10{,}860 \times 3.5^2)} \times 6101$$
$$= 2118 \text{ kN}$$

At El 7.0 m

$$V_2 = \frac{10{,}860 \times (7)^2}{(8097 \times 14^2 + 10{,}860 \times 10.5^2 + 10{,}860 \times 7^2 + 10{,}860 \times 3.5^2)} \times 6101$$
$$= 941 \text{ kN}$$

At El 3.5 m

$$V_1 = \frac{10{,}860 \times (3.5)^2}{(8097 \times 14^2 + 10{,}860 \times 10.5^2 + 10{,}860 \times 7^2 + 10{,}860 \times 3.5^2)} \times 6101$$
$$= 235 \text{ kN}$$

Distribution of force to different shear walls:
For this, the first things that need to be computed are center of mass (CM) and center of stiffness (CS). Considering the load is uniform with respect to building plan it is to be said from inspection that the coordinates of CM will be at (24.0, 8.0) in plan view of the building.

The stiffness of the wall at each level is computed as hereafter

Sl No	Wall	B (m)	t (m)	I_{xx} (m^4)	I_{yy} (m^4)	Area (m^2)	r_{xx} (m)	r_{yy} (m)	$\lambda_{xx}(H_i r_{xx})$	λ_{yy} $(H_i r_{yy})$	K_{wx} (kN/m)	K_{wy} (kN/m)
1	S_1	44	0.25	1774.667	0.057	11	12.70	0.072	0.276	48.50	25,971,365	99,836.63
2	S_2	6	0.35	6.300	0.021	2.1	1.73	0.101	2.021	34.64	3,439,490	37,220.84
3	S_3	6	0.35	6.300	0.021	2.1	1.73	0.101	2.021	34.64	3,439,490	37,220.84
4	S_4	12	0.3	43.200	0.027	3.6	3.46	0.087	1.010	40.41	7,698,248	46,971.5

The center of stiffness is calculated hereafter

Sl No	Wall	K_1 (kN/m)	K_2 (kN/m)	x (m)	y (m)	$K_1.x$	$K_2.y$
1	S_1	25,971,365	99,836.63	22	16	5.71E+08	1,597,386
2	S_2	37,220.84	3,439,490	48	13	1,786,600	44,713,376
3	S_3	37,220.84	3,439,490	0	3	0	10,318,471
4	S_4	7,698,248	46,971.5	24	0	1.85E+08	0
		33,744,055	7,025,789			7.58E+08	56,629,233

$$\widetilde{X} = \frac{7.58 \times 10^8}{33,744,055} = 22.460 \text{ m}$$

$$\widetilde{Y} = \frac{56,629,233}{7,025,789} = 8.06 \text{ m}$$

It is apparent that eccentricity in *Y*-direction is negligible. Thus, the eccentricity in *X*-direction is $e_x = 24 - 22.46 = 1.54$ m and $e_y = 0.0$, as shown in Fig. 9.17.

The force in roof level is shown.

Thus, the total shear at roof level is 2807 kN and torsion is $T = 2807 \times 1.54 = 4323$ kN.m

Thus, the shear force is distributed to the four shear walls as follows at the roof level:

Wall	K_{eff}	Lateral force (kN)
S_1	99,836.63	$\left(\frac{99,836}{7,025,789}\right) \times 2807 = 39.88$
S_2	3,439,490	$\left(\frac{3,439,490}{7,025,789}\right) \times 2807 = 1374$
S_3	3,439,490	$\left(\frac{3,439,490}{7,025,789}\right) \times 2807 = 1374$
S_4	46,971.5	$\left(\frac{46,972}{7,025,789}\right) \times 2807 = 18.76$
	7,025,789	Sum = 2807

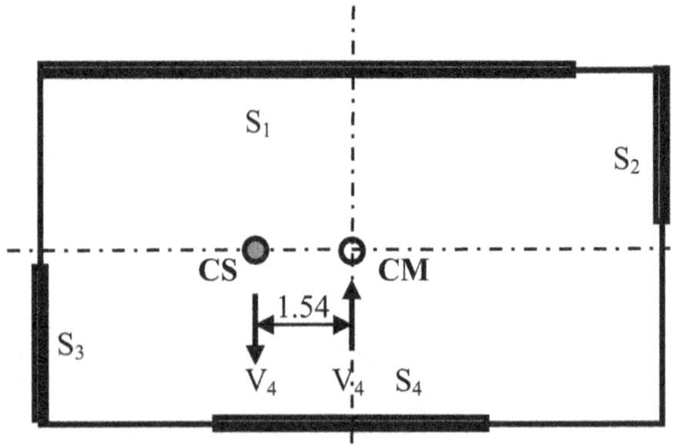

Fig. 9.17 Position of CS and CM and eccentricity in building plan

For torsional force of 4323 kN.m, the shear in four shear walls is computed hereafter

Wall	K_{eff}	r (m)	$k_i r_i$	$k_i r_i^2$	V_i (Eq. 9.3.13)
S_1	99,836.63	8.0	798,693	6,389,544	±540.1
S_2	3,439,490	25.54	87,844,586	2.24E+09	±169.18
S_3	3,439,490	22.46	77,250,955	1.74E+09	±192.38
S_4	46,971.5	8.0	375,772	3,006,176	±540.1

Thus, at El 14.4 m, the forces on the shear walls are as follows (Fig. 9.18):

At El 10.5 m, proceeding in identical manner, we get (Fig. 9.19).
At El 7.0 m, we have (Fig. 9.20).
At El 3.5 m, we have (Fig. 9.21).

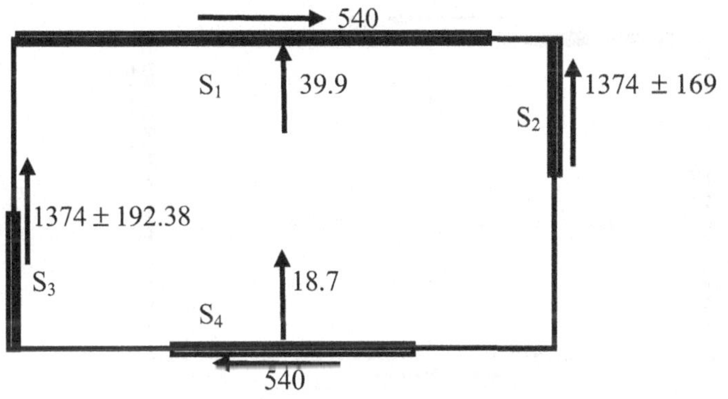

Fig. 9.18 Force on shear walls at EL 14.4 m

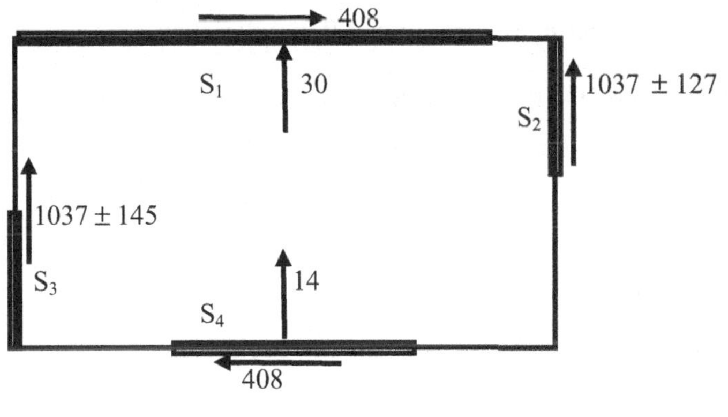

Fig. 9.19 Force on shear walls at EL 10.5 m

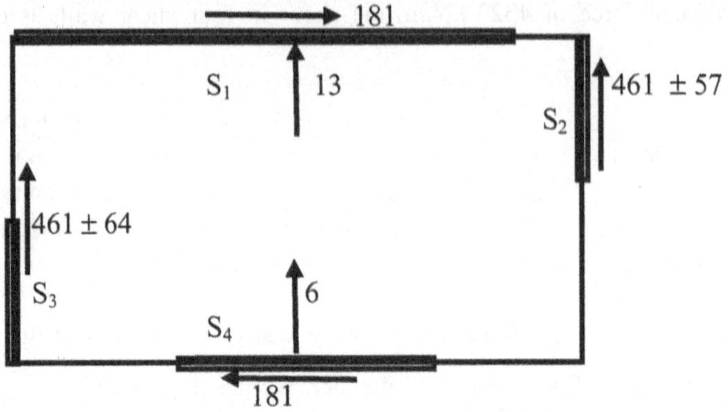

Fig. 9.20 Force on shear walls at EL 7.0 m

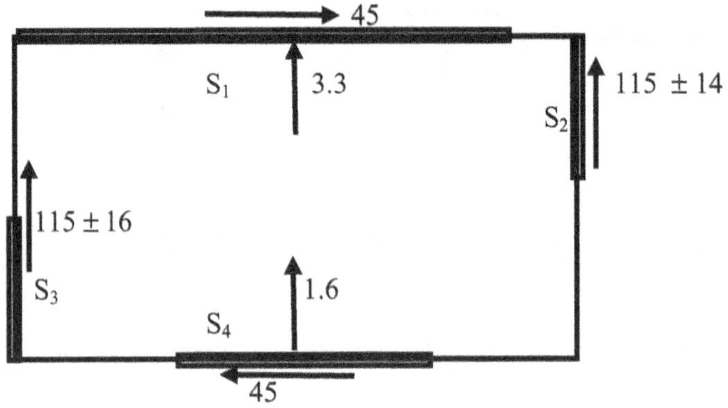

Fig. 9.21 Force on shear walls at EL 3.5 m

Earthquake force acting in longitudinal (long) direction:

$T = \frac{0.09H}{\sqrt{D}} = \frac{0.09 \times 14}{\sqrt{48}} = 0.18$ s, which gives $S_a/g = 2.5$.

Thus, considering all parameters constant for a particular site, $A_h = 0.15$, which gives $V_b = 6101$, as calculated earlier. As the height of the building also remains unaltered, the floor shears remain same as in transverse direction.

Thus, At El 14.0 m, $V_4 = 2807$ kN; at El 10.5 m, $V_3 = 2118$ kN; at El 7.0 m, $V_2 = 941$ kN; and at El 3.5 m, $V_1 = 235$ kN.

However, in this case the orientation of the force in roof level is as shown hereafter, as CS and CM are collinear in Y-direction (Fig. 9.22).

Thus, the distribution of force in roof level (14.4) is as (Fig. 9.23).

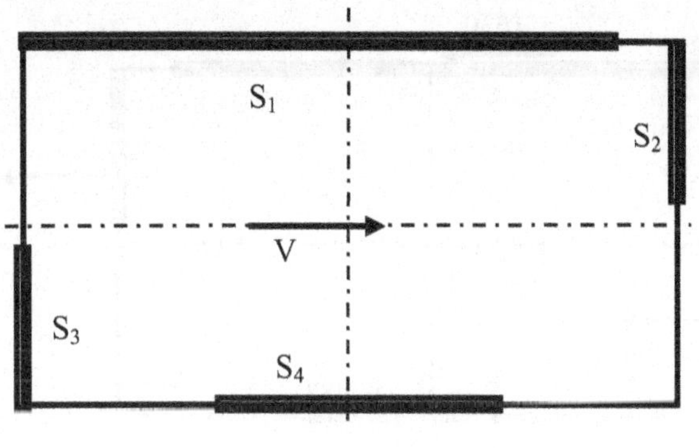

Fig. 9.22 Force in the building in longitudinal direction

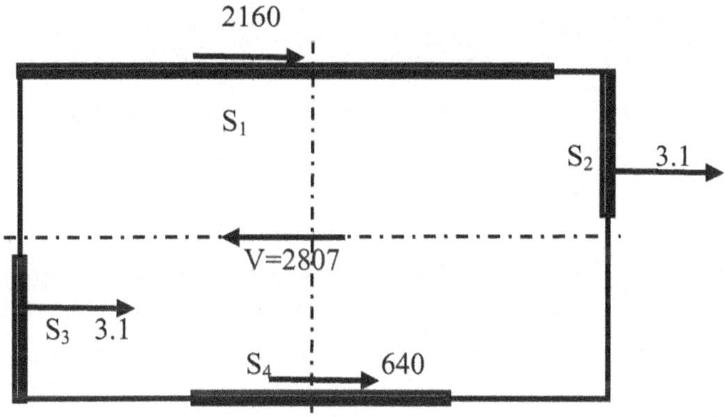

Fig. 9.23 Distribution of force at EL 14.4 m

Sl No	Wall	K	V
1	S_1	25,971,365	$\left(\frac{25{,}971{,}365}{33{,}744{,}055}\right) \times 2807 = 2160$
2	S_2	37,220.84	$\left(\frac{37{,}220.84}{33{,}744{,}055}\right) \times 2807 = 3.1$
3	S_3	37,220.84	$\left(\frac{37{,}220.84}{33{,}744{,}055}\right) \times 2807 = 3.1$
4	S_4	7,698,248	$\left(\frac{7{,}698{,}248}{33{,}744{,}055}\right) \times 2807 = 640$
		33,744,055	

Thus, the distribution of force in roof level (10.5) is as (Fig. 9.24).

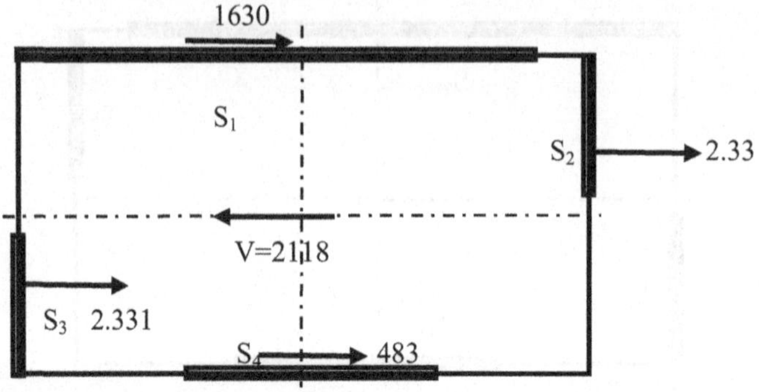

Fig. 9.24 Distribution of force at EL 10.5 m

Sl No	Wall	K	V
1	S_1	25,971,365	$\left(\frac{25{,}971{,}365}{33{,}744{,}055}\right) \times 2118 = 1630$
2	S_2	37,220.84	$\left(\frac{37{,}220.84}{33{,}744{,}055}\right) \times 2118 = 2.33$
3	S_3	37,220.84	$\left(\frac{37{,}220.84}{33{,}744{,}055}\right) \times 2118 = 2.33$
4	S_4	7,698,248	$\left(\frac{7{,}698{,}248}{33{,}744{,}055}\right) \times 2118 = 483$
		33,744,055	

For EL 7.0 m, we have (Fig. 9.25).

For El 3.5 m, we have (Fig. 9.26).

Considering that the force in X-direction is collinear with CM, there is no torsion force in this case.

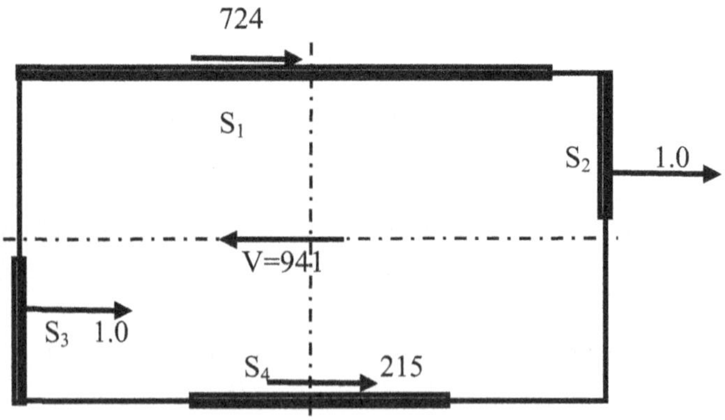

Fig. 9.25 Distribution of force at EL 7.0 m

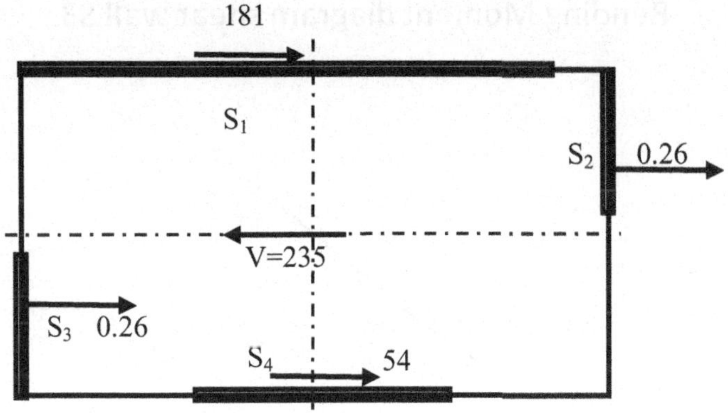

Fig. 9.26 Distribution of force at EL 3.5 m

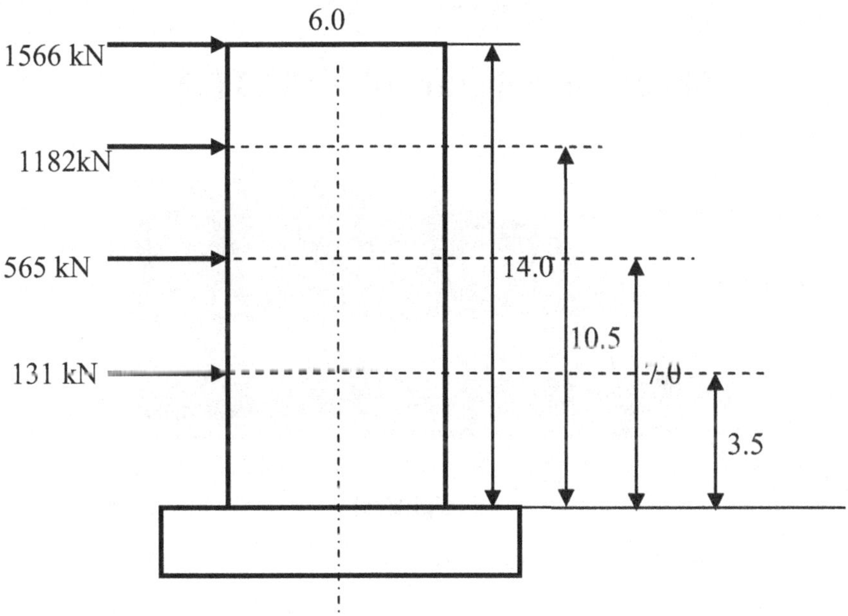

Fig. 9.27 Distribution of force shear wall S_3 in vertical direction

Thus, for an earthquake acting in transverse direction, the maximum design force is on shear wall S_3, as shown hereafter (Fig. 9.27).

The bending moment and shear force diagram of the wall are as shown hereafter (Fig. 9.28).

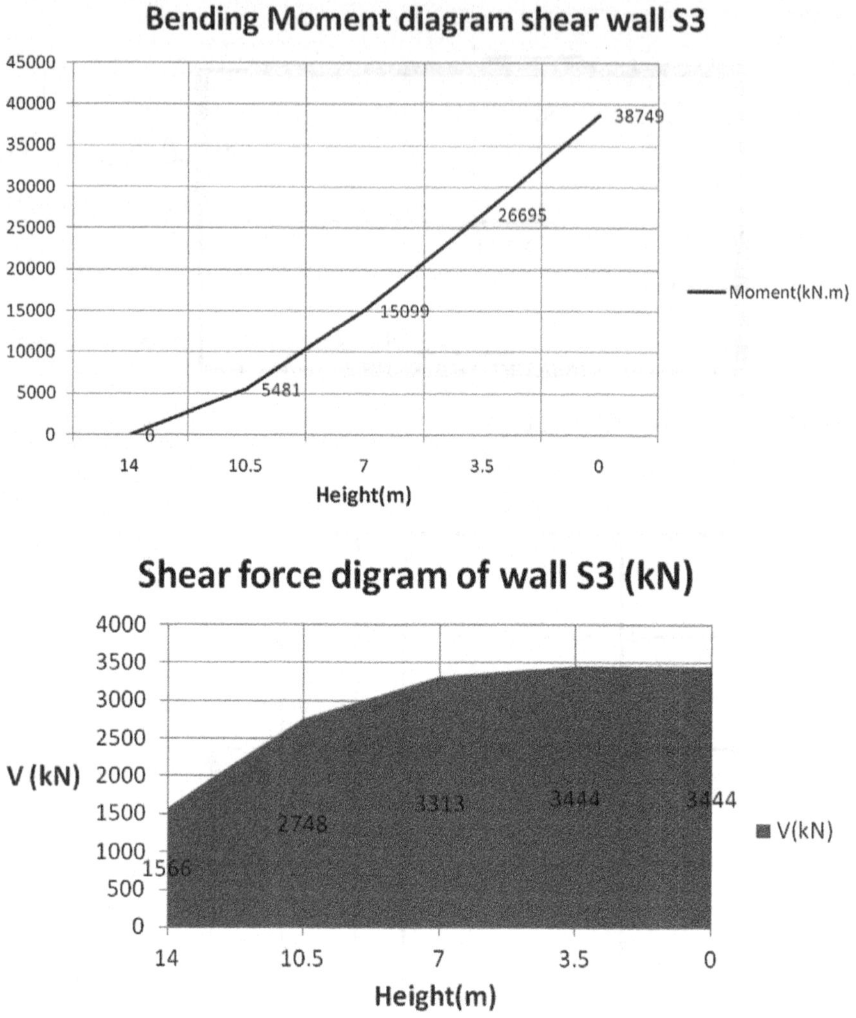

Fig. 9.28 Bending and shear force diagram for wall S_3

9.4 Analysis of Building Frames Based on the Assumed Shape Function

This is a technique in which a multi-degree freedom system is converted into an equivalent system having mass and stiffness of that of a single degree of freedom based on an assumed shape function to find out the time period of a system.

To start with, let us consider a stick model of a system having multi-degree of freedom, as shown in Fig. 9.29.

Fig. 9.29 A stick model having multi-degrees of freedom

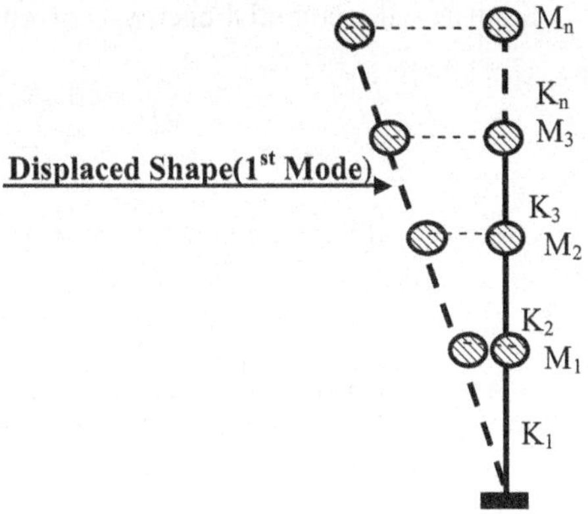

The kinetic energy of the system is given by

$$T(t) = \frac{1}{2}\sum_{i=1}^{n} m_i \left[\frac{\partial y(z,t)}{\partial t}\right]^2 \tag{9.4.1}$$

We consider here $y(z,t) = \phi(z)\xi(t)$ where

$\phi(z)$ admissible shape function which satisfies the boundary condition of system;
$\xi(t)$ generalized coordinate.

$$T(t) = \frac{1}{2}\sum_{i=1}^{n} m_i \left[\sum_{j=1}^{n} \phi_j(z)\dot{\xi}_n(t) \sum_{k=1}^{n} \phi_k(z)\dot{\xi}_k(t)\right] \tag{9.4.2}$$

or

$$T(t) = \frac{1}{2}\sum_{j=1}^{n}\sum_{k=1}^{n} \dot{\xi}_j(t)\dot{\xi}_k(t)\left[m_i \sum_{i=1}^{n} \phi_j(z)\phi_k(z)\right]$$

from which we conclude that the generalized mass of the system is given by

$$M^* = \left[m_i \sum_{i=1}^{n} \phi_j(z)\phi_k(z)\right]$$

Thus, for fundamental mode for $j = k$ we have

$$M^* = \left[\sum_{i=1}^{n} m_i \phi_i^2(z)\right] \tag{9.4.3}$$

Similarly, the potential energy is given by

$$V(t) = \frac{1}{2}\sum_{i=1}^{n} k_i[\Delta y(z,t)]^2 \tag{9.4.4}$$

(Here $\Delta =$ difference in displacement between two adjacent levels)

or

$$V(t) = \frac{1}{2}\sum_{i=1}^{n} k_i\left[\sum_{j=1}^{n} \Delta\phi_j(z)\xi_n(t)\sum_{k=1}^{n} \Delta\phi_k(z)\xi_k(t)\right]$$

or

$$V(t) = \frac{1}{2}\sum_{j=1}^{n}\sum_{k=1}^{n} \xi_j(t)\xi_k(t)\left[k_i\sum_{i=1}^{n} \Delta\phi_j(z)\Delta\phi_k(z)\right] \tag{9.4.5}$$

Thus, for the fundamental mode for $j = k$ we have

$$K^* = \left[\sum_{i=1}^{n} k_i\Delta\phi_i^2(z)\right] \tag{9.4.6}$$

Now knowing $T = 2\pi\sqrt{\frac{m}{K}}$, we have for this generalized case

$$T^* = 2\pi\sqrt{\frac{M^*}{K^*}} \tag{9.4.7}$$

From the above mathematical derivation, it is obvious that if we know what could be the assumed shape function correctly, it is possible to arrive at the fundamental time period of the system to reasonable accuracy.

Based on the aspect ratio (*H*/*D*), Naeim (2004) has proposed the following shape functions, as shown in Table 9.1. These may be considered for buildings modeled as a stick having multi-degrees of freedom.

Here $H =$ height of the building

$D =$ width of building in direction of the earthquake force considered.

Table 9.1 Assumed shape function for building of various aspect ratios

Sl No	*H*/*D*	Shape function
1	$H/D < 1.5$	$\sin\frac{\pi x}{2H}$
2	$1.5 < H/D < 3$	$\frac{x}{H}$
3	$H/D > 3.0$	$1 - \cos\frac{\pi x}{2H}$

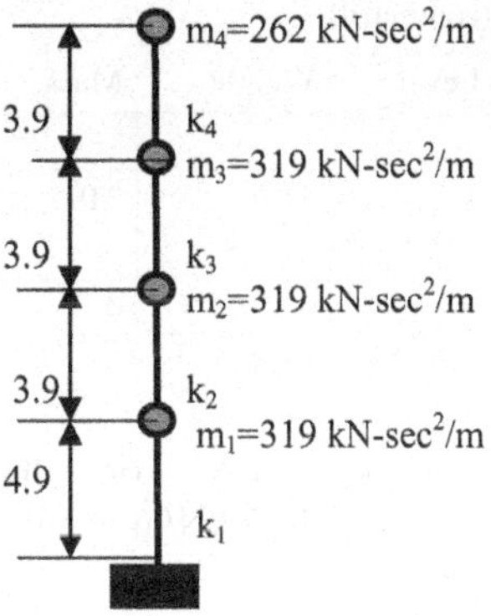

Fig. 9.30 Mathematical model of the building

Example 9.5 Repeat Example 9.1 by the assumed shape function method and compare the base shear based on IS-1893. All other data remain the same (Fig. 9.30).

Column size 300 mm × 300 mm.

$$I_{\text{col}} = \frac{1}{12} \times 0.3 \times (0.3)^3 = 6.75 \times 10^{-4}\text{m}^4.$$

Considering the concrete grade to be M25, $E_c = 2.5 \times 10^7$ kN/m².

Stiffness of one column = $\frac{12E_cI_{\text{col}}}{H^3}$, considering there are 20 columns per floor

$$k_2 = k_3 = k_4 = 20 \times \frac{12E_cI_{\text{col}}}{H^3} = 20 \times \frac{12 \times 2.5 \times 10^7 \times 6.75 \times 10^{-4}}{(3.9)^3}$$
$$= 68275\ \text{kN/m}$$

and

$$k_1 = 20 \times \frac{12E_cI_{\text{col}}}{H^3} = 20 \times \frac{12 \times 2.5 \times 10^7 \times 6.75 \times 10^{-4}}{(4.9)^3} = 34424\ \text{kN/m}$$

In transverse direction, $H/D = 16.6/11 = 1.50$. Let the assumed shape function be $\sin\left(\frac{\pi x}{2H}\right)$

Level	Weight	Mass	Stiffness	ϕ	$\Delta\phi$	$m_i\phi^2$	$k\Delta\phi^2$
4	2570	262		1.0		262	
			68,275		0.077		405
3	3133	319		0.923		272	
			68,275		0.216		3185.4
2	3133	319		0.707		159.45	

(continued)

(continued)

Level	Weight	Mass	Stiffness	ϕ	$\Delta\phi$	$m_i\phi^2$	$k\Delta\phi^2$
			68,275		0.325		7211.5
1	3133	319		0.382		46.55	
			34,424		0.382		5023.3
0	0	0		0		0	
	11,969					740	15,825.2

Thus, based on the above computation, $M^* = 740$ kN s^2/m and $K^* = 15{,}825.2$ kN/m

$$T = 2\pi\sqrt{\frac{M^*}{K^*}} = 2\pi\sqrt{\frac{740}{15{,}825.2}} = 1.36\ \text{s}$$

For medium stiff soil site, $S_a/g = 1.36/T = 1.0$.
Thus

$$A_h = \left(\frac{ZI}{2R}\right)\left(\frac{S_a}{g}\right) = \frac{0.24 \times 1}{6} \times 1 = 0.04$$

Thus, the computed base shear $V_B = 0.04 \times 11969 = 478.76$ kN.

This is much lower than the value given by code formula, which gives $V_B = 1316$ kN. Thus, it is seen that infill panel significantly affects the dynamic response and, for normal buildings, its effect cannot be ignored.

References

Dayaratnam, P., & Jain, S. K. (1991). Behavior of Brick In filled RCC frame under lateral load. *Indian Concrete Journal, 65*(9), 453–457.

Jain, S. K., Saraf, V., & Mehrotra, B. (1995). Experimental Evaluation of fundamental period RC frame buildings with brick infill. *Journal of Structural Engineering India, 22*(2).

Kazimi, S. M. A., & Chandra, R. (1980). *Analysis and design of shear walls*. Kolkata, India: Tor Steel Research Foundation.

Macleod, I. A. (1990). *Analytical model of structural systems*. Chichester, Sussex, UK: Elis Horwood Publications.

Murty, C. V. R., & Charleson, A. W. (2006). *Earthquake design concepts*. Kanpur, India: Indian Institute of Technology Kanpur, NICEEE Publications.

Naeim, F. (2004). *The seismic design handbook*. London, UK: Kluwer Academic Publishers.

Chapter 10
Earthquake Analysis of Tall Chimneys and Stack-Like Structures

10.1 Introduction

Tall chimneys and vertical self-supporting vessels are an important feature of power and petrochemical industry. Damage to them during an earthquake can have a severe consequence both in terms of economy and loss of human life.

While it is expected that a power plant remains functional after an earthquake, as this is essential to fight the aftermath of the disaster, leakage, or damage of vertical vessels in refinery or chemical plants containing flammable or toxic liquid can create significant damage to environment and surrounding life.

Though the analytical procedure herein is discussed in terms of tall reinforced concrete chimneys, it can well be applied for slender, vertical, self-supporting vessels too.

With ever-growing demand for power, engineering industry is churning out power plants of progressively higher capacity globally. In India, fossil fuel power plant capacity has jumped from a single unit having capacity of 210 to 500 MW and more, and a number of plants of such high capacity are being built in different parts of India to cater to its ever increasing demand for power.

To maintain ecological balance, as well as limit the environmental pollution due to fly ash deposition, chimneys emitting spent flue gas are also getting taller every day. In many countries, it is now mandatory that for all fossil fuel power plants, height of chimneys be minimum 220 m for 210 MW power plants and 275 m high for 500 MW units.

Emitting the flue gas at such a great height, though reduces the ground pollution concentration significantly, has posed new challenges to structural engineers to come up with a safe design of these tall and slender chimneys, especially under wind and earthquake, which affect its behavior significantly.

Unlike other tall structures that usually have high redundancy, a chimney is basically a cantilever structure, having one line of defense (the structure itself).

I. Chowdhury and S. P. Dasgupta, *Earthquake Analysis and Design of Industrial Structures and Infra-structures*, GeoPlanet: Earth and Planetary Sciences,
https://doi.org/10.1007/978-3-319-90832-8_10

Thus, during an earthquake, if any portion of it develops a hinge, it would invariably make the system a mechanism with collapse being imminent.

That is why, predicting its dynamic behavior as realistically as possible under wind and earthquake loading is of primary importance. In this chapter, we will limit our discussion to the earthquake behavior.

The normal practice in a design office is to develop a lumped mass model of the smoke and compute its response through a computer; however, considering the fact that these structures have a more-or-less uniform distribution of mass and stiffness, they are quite amenable to classical mathematical treatment and an accurate prediction of their dynamic behavior is quite possible.

One major controversy that still remains unresolved regarding its behavior, *is the level of damping to be considered in the analysis.*

While one school of thought prefers to use standard damping ratio as used for reinforced concrete elements (5–7%), other school of thought is that, due to its huge mass, a major portion of the chimney remains under compression, even under extreme wind or seismic loading, and thus remains un-cracked.

Since the propagation of cracking enhances the damping property of a system and does not occur in major portion of the chimney due to its high compressive stress, a much lower damping ratio, of say 2–3%, should be a more reasonable value to consider during estimation of dynamic forces. Unfortunately, very little field observed instrumented data are available to come to any decisive conclusion on this issue.

Chimneys are usually of two types

- Multi-flue chimneys (used to cater to more than two power units at a time) having a more-or-less uniform cross section or nominal taper, as shown in Fig. 10.3
- Single flues (used to cater one or two units) usually having a tapered profile, where the slope of shell usually changes at 2 or 3 transition points, as shown in Fig. 10.10.

10.2 A Brief History on the Development of Dynamic Response of Chimneys

In earlier days (the 1950–60s), chimneys were mostly made out of masonry bricks, and their height varied from 30 to 40 m. However, with advent of reinforced concrete, up gradation of power plants to higher capacities and tougher legislation in terms of environmental pollution control (the late 1970s), chimney heights jumped from 110 to 220 m and higher, in no time, making engineers realize that performing a pseudo-static analysis based on seismic coefficient method may not be a safe proposition for such tall and slender structures as was the prevalent practice then.

ACI 307 was possibly the first code to come up with the proposition of dynamic analysis of chimneys.

One of the earliest expressions to be used to predict the first two modal time periods of a chimney was that due to Rumman (1963), expressed as

$$T_1 = 5\left(H^2/D_0\right)\sqrt{\rho/E}(t_0/t_b)^{0.22}\left(D_0/D_b\right)^{1.1} \quad (10.2.1)$$

$$T_2 = 0.82\left(H^2/D_0\right)\sqrt{\rho/E}(t_0/t_b)^{0.09}\left(D_0/D_b\right)^{0.78} \quad (10.2.2)$$

Here, D_0 = mean diameter at top, D_b = mean diameter at bottom, t_0 = thickness at top, t_b = thickness at bottom, ρ = mass density of concrete, E = Young's modulus of concrete, H = height of chimney.

Equations (10.2.1) and (10.2.2) are valid for unlined chimneys. For lined chimneys, the above values may be multiplied by a factor $\sqrt{W_1/W}$, where W_1 = weight of shell + weight of lining, and W = weight of unlined chimney (i.e. shell only).

To cater to the peak acceleration and velocity spectra conforming to zone 3, as per Uniform Building Code (UBC), Rumman suggested correction factors J_v and J_m to determine the dynamic moments and shears (Fintel et al. 1986) under earthquake.

Equations (10.2.1) and (10.2.2) had been in practice around the world for almost 20 years, till the late 1980s (era of hand computation) and has also been adopted by ACI 307-08. However, after digital computer made its inroad in design office practice, lumped mass stick model based on finite element analysis (FEM) is usually deployed for dynamic response of such tall chimneys through computers.

The base shear as per ACI 307 can be expressed as

$$V_b = ZW\left(\frac{0.35}{\sqrt{T}}\right) \quad (10.2.3)$$

Here, W = total weight of stack including all other appurtenances connected to it. Z = zone factor where, as per Uniform Building Code for zone 1, Z = 3/16; for zone 2, Z = 3/8; for location 3, Z = 3/4; and Z = 1.0 for zone 5.

As shown in Fig. 10.1, the base shear is distributed along the height of the stack as

$$\mathrm{d}F = \frac{[w(x)\mathrm{d}x]x^2}{\int_0^H [w(x)\mathrm{d}x]x^2} V_b \quad (10.2.3)$$

The shear force at any height h can be expressed as

$$V_h = J_v \int_h^H \mathrm{d}F \quad (10.2.4)$$

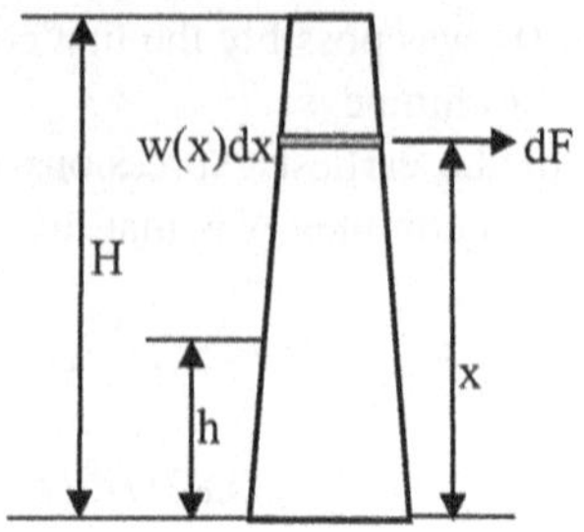

Fig. 10.1 Distribution of shear along chimney height

and

$$M_h = J_m \int_h^H dF(x-h) \tag{10.2.5}$$

The factors J_v and J_m are as defined hereafter.

$$J_v = 8[J_{vt} - J_v(0.5H)]\left[\frac{h}{H} - 0.5\right]^3 + J_v(0.5H) \tag{10.2.6}$$

For $0.5 \le h/H \le 1.0$ and

$$J_v = 1 - 2[1 - J_v(0.5H)]\frac{h}{H} \tag{10.2.7}$$

For $0 \le h/H \le 0.5$ and

When $J_v(0.5H) = 0.43 + \frac{0.5}{T^{1.5}}$ and $J_{vt} = 0.96 + 0.23T$, T is the time period of vibration of the chimney.

$$J_m = 2.78[J_m(0.9H) - J_m(0.3H)]\left[\frac{h}{H} - 0.3\right]^2 + J_m(0.3H) \tag{10.2.8}$$

valid for $0.3 \le \frac{h}{H} \le 1.0$, and

$$J_m = 11.1[J_{mb} - J_m(0.3H)]\left[0.3 - \frac{h}{H}\right]^2 + J_m(0.3H) \tag{10.2.9}$$

valid for $0 \le \frac{h}{H} \le 0.3$.

When $J_m(0.9H) = 1.15 + 0.025T^2$ and

$J_m(0.3H) = 0.3 + 0.004(6 - T)^3$ for $T > 6$ $J_m(0.3H) \to 0.30$

$J_{mb} = 0.40 + \frac{(6-T)^3}{300}$, for $T > 6$ $J_{mb} \rightarrow 0.40$

For a tapered chimney, Rumman (1986) suggested to multiply J_v and J_m by factors as stated hereafter in Table 10.1.

The CICIND model code of tall chimneys has not given any specific formulation regarding time period or any mathematical expression for computation of moment and shear like ACI, the implicit understanding being that the mathematical model and its analysis shall be carried out in computer by finite element method.[1] However, it has recommended that

- While calculating the member stiffness, un-cracked properties be considered.
- The FEM should be so modeled that at least 90% of the chimneys gravity load be accounted for in the modal analysis.
- The Young's modulus of concrete to be considered for dynamic analysis shall be considered as $E_{\text{conc}} = 9500(f_{\text{ck}} + 8)^{0.33}$, where f_{ck} is characteristics strength of concrete in Mpa.

Eurocode 8 has possibly provided most comprehensive recommendations for seismic analysis of chimney towers and masts in Part 6 of the code. The recommendation implicitly assumes that a lumped mass discrete model will be adopted for computer analysis of the chimney.[2]

While developing the model, the code has given some specific recommendations and can be summarized as hereafter.

The mathematical model would be such that:

- It takes into account the rotational and translational springs of the foundations.
- Include sufficient degrees of freedom (and the associated masses) to determine the response of all the important structural component equipments or appendages.
- Take into account relative displacements of supports of equipments or machineries.
- Take into account piping interactions, externally applied structural restraint, etc.
- The selection of lumped masses should be such that it shall be representative of distribution of inertial effect of the seismic force. The masses shall include all permanent parts fitting flues, insulation and any accumulated dust or ash adhering to the surface. Where a coarse distribution of translational mass is used, rotational inertia is to be considered for the corresponding rotational degrees of freedom.
- For estimation of stiffness of the sections, Eurocode advocates estimation of moment of inertia based on cracked section and considering the q factor as 1.0 while carrying out modal analysis. The cracking is a function of strain range to which it is subjected to.

[1]Refer Sects. 10.6 and 10.7 for further details.

[2]Refer Fig. 10.10 for further clarifications.

Table 10.1 Coefficients J_v and J_m for different values of h/H

h/H	0.05	0.15	0.25	0.35	0.45	0.55	0.65	0.75	0.85	0.95
J_v	1.03	0.99	0.89	0.83	0.88	0.97	0.92	0.86	0.94	1.01
h/H	0.0	0.1	0.2	0.3	0.4	0.5	0.6	0.7	0.8	0.9
J_m	0.94	0.91	0.89	0.92	0.9	0.82	0.76	0.78	0.87	0.97

- The code allows for 5% damping in the superstructure and advocates soil–structure interaction if applicable.

Referring to Chap. 2, Sect. 2.3.6, one computes the eigenvalues of the system from the expression $[\mathbf{K}] - [\mathbf{M}]\omega^2 = 0$. Now, knowing that $T_i = 2\pi/\omega_i$, one can compute the time period for j modes (usually 5–6 modes) that contribute to 90% mass participation.

Thus for ith mode, the spectral acceleration $S_{ei}(T_i)$ can be selected from Fig. 2.6. Finally, considering the ground type (refer Table 2.10) the design spectral acceleration is computed from Eqs. (2.1.7)–(2.1.10). The nodal force for the ith mode is thus expressed as

$$\{V_i\} = \kappa_i[\mathbf{M}]\{\phi_i\}\lambda S_{ei}(T_i) \tag{10.2.10}$$

Moment induced at the base is expressed as

$$M_b = \sum_1^n \{V_i\} \times h_n \tag{10.2.11}$$

Here, n = number of mass in the model and h_n are the position vectors of the masses from top of the foundation.

For tall slender towers, Eurocode is the only code (in English language) that advocates the effect of rocking acceleration due to Rayleigh waves, as shown in Fig. 10.2 (Bonev et al. 2010).

In this case, for two mutually orthogonal horizontal axes (X and Y in this case) and the vertical direction Z, Eurocode considers the rocking accelerations as

$$S_x^\theta = 1.7\pi\frac{S_e(T)}{v_s T} \tag{10.2.12}$$

$$S_y^\theta = 1.7\pi\frac{S_e(T)}{v_s T} \tag{10.2.1.3}$$

$$S_z^\theta = 2.0\pi\frac{S_e(T)}{v_s T} \tag{10.2.14}$$

Here, $S_e(T)$ = spectral acceleration as defined previously,

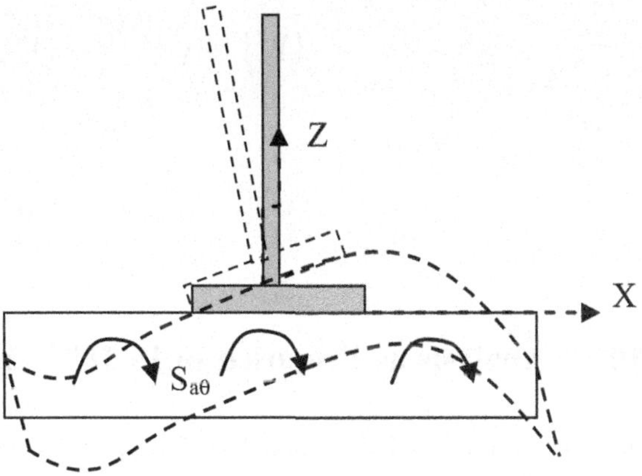

Fig. 10.2 Rocking deformation and acceleration through soil body due to propagation of Rayleigh waves

v_s = low strain (γ_s) average shear wave velocity of soil for the first 30 m, where $\gamma_s \lesssim 10^{-6}$.

Thus, the total mass associated in rocking is expressed as

$$M^{\theta}_{\text{tot}} = \{\xi\}^{\text{T}}[\mathbf{M}]\{\xi\} \tag{10.2.15}$$

Here,

$$\{\xi\}^{\text{T}} = \{z/H\}^{\text{T}} = \{0 \quad \ldots \quad 1\}^{\text{T}} \tag{10.2.16}$$

and

$$\kappa^{\theta}_i = \frac{\{\phi_i\}^{\text{T}}[\mathbf{M}]\{\xi\}}{\{\phi_i\}^{\text{T}}[\mathbf{M}]\{\phi_i\}} \tag{10.2.17}$$

This nodal force V_i is expressed as

$$\{V^{\theta}_i\} = \kappa^{\theta}_i[\mathbf{M}][z_n]\{\phi_i\}S^{\theta}_{ei}(T) \tag{10.2.18}$$

Here,$[z_n]$ =, a diagonal matrix having all off-diagonal terms as zero and the diagonal elements represent the position vectors of masses, 1, 2, 3…*n*, from the top of foundation.

The moment at the base is expressed as

$$M^{\theta}_{bi} = \sum_{1}^{n}\{V^{\theta}_i\}(H - z_n) \tag{10.2.19}$$

The resultant translational and rotational accelerations are finally obtained by SRSS values as

$$\{V_i^R\} = \sqrt{\{V_i^x\}^2 + \{V_i^\theta\}^2} \tag{10.2.20}$$

and

$$\{M_i^R\} = \sqrt{\{M_i^x\}^2 + \{M_i^\theta\}^2} \tag{10.2.21}$$

10.3 Analysis as Proposed in IS 1893 (2005) Part IV

In 2005, Bureau of Indian Standard has upgraded its design procedure for earthquake resistant design of tall reinforced concrete chimneys. This is certainly a significant improvement to the previous code IS 1893 (1984) and is discussed hereafter.

As per IS code, the fundamental time period of such chimneys, as shown in Fig. 10.3, fixed at its base is given by

$$T = C_T \sqrt{\frac{WH}{EAg}} \tag{10.3.1}$$

where W = weight of chimney plus lining and all other accessories; H = height of chimney above the base; E = modulus of elasticity of the structural shell; A = area of cross section of shell at the base; g = acceleration due to gravity; C_T = constant which is a function of the slenderness ratioλ as furnished in Table 10.2. For circular section, $A = 2\pi rt$; r = mean radius of the shell at the base; and t = thickness of the shell at the base.

In Table 10.2, $\lambda = H/r_0$, r_0 = radius of gyration of the structural shell at base section. The code has also suggested an alternate expression of time period as

$$T = 2\pi/\sqrt{g \cdot C_\delta} \tag{10.3.2}$$

Here,

$$C_\delta = \sum_i^N W_i\delta_i \Big/ \sum_i^N W_i\delta_i^2 \tag{10.3.3}$$

In Eq. (10.3.3), W_i = weight lumped at ith location with weights applied simultaneously with the force applied horizontally. δ_i = lateral static deflection under its own lumped weight at ith location (chimney weight to be lumped at ten locations or more). N = number of locations of lumped weight and g = acceleration due to gravity.

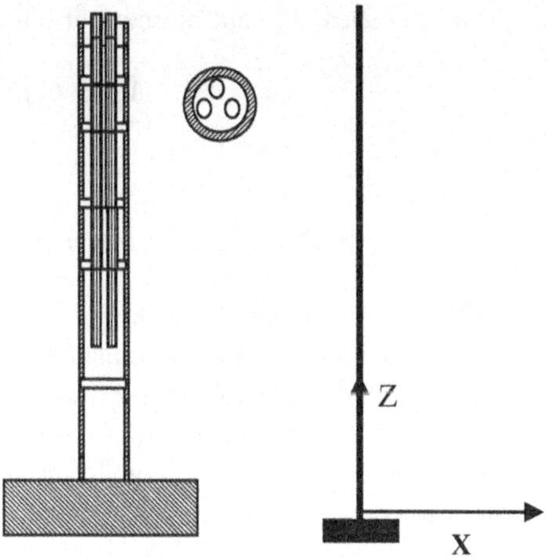

Fig. 10.3 A typical multi-flue RCC stack

Table 10.2 Values of C_T and C_v for computing time period and base shear

$\lambda = H/r_0$	5	10	15	20	25	30	35	40	45	50
C_T	14.4	21.2	29.6	38.4	47.2	56.0	65.0	73.8	82.8	1.8λ
C_v	1.02	1.12	1.19	1.25	1.30	1.35	1.39	1.43	1.47	1.50

Having determined the time period as expressed by Eq. (10.3.1) or (10.3.2), the spectral acceleration can be read off from the chart as furnished in the code (refer Chap. 2 for detail).

The IS code has now recommended a value of damping ratio amounting to 3% for design basis earthquake (DBE) and 5% for the maximum considered earthquake (MCE). As the spectral acceleration furnished in the code is based on 5% damping, S_a/g value has to be multiplied by a factor 1.27 to arrive at the design acceleration spectrum.

Thus,

$$S_{ad}/g = 1.27(S_a/g) \tag{10.3.4}$$

Here, S_a/g = response value read from IS-1893 Part I (2002), and S_{ad}/g is the design response spectrum to be considered for the chimney design.

The code factor β is determined from the expression

$$\beta = ZI/2R \tag{10.3.5}$$

Here, Z = zone factor as elaborated in Chap. 2. For tall concrete chimneys, I = 1.5 and R = 3.0, as per the latest recommendation of code.

In case, MCE is the basis of design, $S_{ad}/g = S_a/g$, as the 5% damping is considered for this case and $\beta \to ZI/R$.

The base shear V_B and moment at base M_B are expressed as

$$V_B = C_v \beta (S_{ad}/g) W \tag{10.3.6}$$

and

$$M_B = \beta (S_{ad}/g) W \bar{h} \tag{10.3.7}$$

In Eqs. (10.3.6) and (10.3.7), C_v is a coefficient which varies as the slenderness ratio of the chimney and is as furnished in Table 10.2, W = total weight of chimney including its lining and all other fixtures at the base, and $\bar{h}$ = height of the center of gravity of the structure, above the base.

Shear force and bending moment at any height z from top of the chimney are expressed as

$$V_z = C_v \beta (S_{ad}/g) W D_v \tag{10.3.8}$$

$$M_z = \beta (S_{ad}/g) W \bar{h} D_m \tag{10.3.9}$$

The coefficients D_v and D_m in Eqs. (10.3.8) and (10.3.9) are functions of the height ratio z/H expressed in Table 10.3 for three types of soil-foundation conditions.

For SPT values less than 50, the code has also recommended to consider dynamic soil–structure interaction (DSSI) by considering the soil spring stiffness (Hall 1967) as

$$K_h = \frac{32 G r_0 (1 - v)}{(7 - 8v)} \tag{10.3.10}$$

$$K_\theta = \frac{8 G r_0^3}{3(1 - v)} \tag{10.3.11}$$

Here, K_x and K_θ are translational and rotation springs of the foundation, respectively. G = dynamic shear modulus of the soil, r_0 = radius of the circular raft, and$v$ = Poisson's ratio of the soil.

For piled foundation, the code recommends only the translational spring K_h^p expressed as

$$K_h^p = \frac{\eta E I_m}{1.2 T_p^3} + \frac{\eta d^2}{2}; T_p = (E I_m/\eta)^{\frac{1}{5}} \tag{10.3.12}$$

Here, K_h^p = lateral stiffness of a single pile, E = Young's modulus of pile material, I_m = moment of inertia of pile section, η = modulus of sub-grade reaction of soil in horizontal direction, d = thickness of pile cap, T_p = characteristic length of pile.

Table 10.3 Factors D_m and D_v for variation of shear and moment along chimney height

Soil-foundation type	D_m^a	D_v^a
Fixed base or raft on hard soil SPT value N $\geq$ 50	$0.4\sqrt{\frac{z}{H}}+0.6\left(\frac{z}{H}\right)^4$	$1.1\sqrt{\frac{z}{H}}+0.75\left(\frac{z}{H}\right)+0.6\left(\frac{z}{H}\right)^4\leq 1.0$
Raft on soil for SPT values <50	$0.6\sqrt{\frac{z}{H}}+0.4\left(\frac{z}{H}\right)^4$	$1.1\sqrt{\frac{z}{H}}-0.75\left(\frac{z}{H}\right)+0.65\left(\frac{z}{H}\right)^4$
Pile foundations	$0.5\sqrt{\frac{z}{H}}+0.5\left(\frac{z}{H}\right)^4$	$0.66\sqrt{\frac{z}{H}}-0.20\left(\frac{z}{H}\right)+0.54\left(\frac{z}{H}\right)^4$

[a]Here, $z = 0$ at tip of the chimney shaft

The maximum tip deflection under service load is restricted to

$$D_{\max} = 0.003 \times H \tag{10.3.13}$$

Here, $D_{\max}$ = maximum deflection at chimney tip, H = height of the chimney from top of raft or pile cap.

10.3.1 Some Comments on Recommendation Made in IS 1893 (Part IV) 2005

Recommendations of IS code as cited above are certainly a significant amelioration to the previous code IS-1893(1984) in terms of seismic analysis of tall chimneys. But it still leaves room for improvement. The recent recommendations yet have some gaps that need to be addressed at the earliest. The shortcomings in the present recommendation can be summarized as hereafter.

- The time period as furnished in the latest code is for fundamental mode only. For design and analysis of tall RCC chimneys, most of the clients and statutory bodies usually insist on calculation of at least three modes (if not more) to ensure that effects of higher mode participation are properly catered to in the analysis. How to find higher time periods? The code is silent on this issue which invariably forces a designer to use the code recommended formula only for a preliminary estimate and resort to an elaborate computer analysis based on FEM during detailed engineering analysis. This, to a certain extent, trivializes the recommendation.

  Detailed study on this issue by Jain et al. (1990) and Chowdhury (2010) shows that unlike a multistory building, considering that the chimney is a far slender structure, higher modes significantly affect its response, especially the shear force whose effect is significant for higher modes. It has also been shown by Jain et al. that for moments and shears obtained based on three-mode analysis and six-mode analysis—*the difference is nominal* and recommends to consider at least three-mode responses for practical design of a chimney.

- In the expression vide Eq. (10.3.1), the basis of arriving at the coefficient C_T has never been made clear either in earlier version [IS-1893(1984, 2002)] or the present version. However, the value is quite reasonable as far as the fundamental mode is concerned. As cited above, IS code does not furnish any values of C_T for higher modes and without these values the expression furnished by Eq. (10.2.1) practically remains unusable as far as design of chimneys of height >220 m is concerned.
- Formula for time period as cited in Eq. (10.3.1) is used indiscriminately for both multi-flue stack of constant diameter and single-flue stack with multi-tapered profile. There is no commentary to explain whether the tapering affects the time period expression or not. Or whether any correction is required for geometrical irregularity of the shaft [like Rumman's expression cited in Eq. (10.2.1)]. This has created significant confusion among chimney designers in Indian industry as to the limitations, if any, on the recommended formula.
- The code now advocates considering DSSI effect for SPT value $N < 50$. It however furnishes no recommendation as to how to arrive at the time period considering the foundation springs. This again forces the user to resort to an elaborate computer analysis based on FEM to arrive at a meaningful result.
- Unique values of D_v and D_m are furnished in the code to cater to DSSI amplification. The recommendation looks a bit simplistic as the amplification factor due to soil–structure interaction will depend on the relative stiffness of the structure and soil/pile. This cannot be a unique value for all types of soil or pile arrangements.
- The code though recommends specific values of soil/pile stiffness; it is silent on the level of damping to be considered for coupled soil–structure interaction. Damping of foundation will play a significant role on the overall dynamic response and has been ignored in the recommendation.
- The pile stiffness expression, furnished as per Eq. (10.3.12), looks like one that is based on lateral stiffness of pile under static load. Moreover, the stiffness furnished is for a single pile with possibly the presumption that, for N piles below a chimney pile cap, the stiffness of a single pile shall be multiplied by N to arrive at the pile group stiffness. Both these assumptions are unrealistic and not correct (For explanation to this, refer Sect. 10.7.1 of this chapter).
- IS 1893 limits the tip deflection of chimney to 0.003H, where H is height of chimney, but no guideline is provided on how to derive it, when, once again, a computer analysis is the only resort to arrive at a correct answer.

10.4 Dynamic Analysis of Tall Chimneys

We present in this section a procedure for dynamic analysis of a multi-flue chimney of uniform or nominally tapered cross section (Chowdhury et al. 2008; Chowdhury 2010), based on continuum approach to arrive at a closed form solution before extending the same to a numerical solution for a multi-tapered cantilever chimney.

Since the shaft of a multi-flue chimney (usually termed as the windshield) is a long and slender structure, we consider it as a cantilever beam fixed at the base. The free vibration equation of such a beam is given by the expression

$$EI\left(\frac{\partial^4 w}{\partial z^4}\right) + \rho A\left(\frac{\partial^2 w}{\partial t^2}\right) = 0 \tag{10.4.1}$$

Here, E = elastic modulus of the beam material; I = moment of inertia of the beam; ρ = mass density of the beam material; A = area of cross section of the beam; and w = displacement of the beam, which is a function of time and geometry:

$$w(z,t) = Y(z) \cdot q(t) \tag{10.4.2}$$

Based on the technique of variable separation, the above partial differential equation can be separated into two linear differential equations, one of which being

$$EI\left(\frac{\mathrm{d}^4 Y}{\mathrm{d}z^4}\right) - \lambda^4 Y = 0 \quad \text{where } \lambda^4 = \frac{\rho A \omega^2}{EI} \tag{10.4.3}$$

The generic solution to this equation is given by

$$Y = C_1 \sin \lambda z + C_2 \cos \lambda z + C_3 \sin\mathrm{h} \lambda z + C_4 \cos\mathrm{h} \lambda z \tag{10.4.4}$$

Imposing the four boundary conditions of a cantilever fixed at the base (for detail refer Chap. 2), we have the shape function solution in the form

$$Y_m = \sin \frac{\mu_m z}{H} - \sin\mathrm{h} \frac{\mu_m z}{H} - \alpha_m\left(\cos \frac{\mu_m z}{H} - \cos\mathrm{h} \frac{\mu_m z}{H}\right) \tag{10.4.5}$$

Here, $\mu_m = 1.875, 4.694, 70.855, \frac{2m-1}{2}\pi$ for m = 1, 2,…m, and

$$\alpha_m = \frac{\sin \mu_m + \sin\mathrm{h} \mu_m}{\cos \mu_m + \cos\mathrm{h} \mu_m} \tag{10.4.6}$$

It has been shown by Chowdhury (2010) that stiffness and mass matrix of the chimney shaft can then be expressed as

$$k_{ij} = EI \int_0^H \frac{\mathrm{d}^2 \phi_i(z)}{\mathrm{d}z^2} \frac{\mathrm{d}^2 \phi_j(z)}{\mathrm{d}z^2} \mathrm{d}z \tag{10.4.7}$$

and

$$m_{ij} = \left[\frac{\gamma A}{g}\int_{0}^{H}\phi_i(z)\phi_j(z)\mathrm{d}z\right] \quad (10.4.8)$$

Considering the shape function as

$$\phi_i = \sin\frac{\mu_i z}{H} - \sin\mathrm{h}\frac{\mu_i z}{H} - \alpha_i\left(\cos\frac{\mu_i z}{H} - \cos\mathrm{h}\frac{\mu_i z}{H}\right)$$

and

$$\phi_j = \sin\frac{\mu_j z}{H} - \sin\mathrm{h}\frac{\mu_j z}{H} - \alpha_j\left(\cos\frac{\mu_j z}{H} - \cos\mathrm{h}\frac{\mu_j z}{H}\right) \quad (10.4.9)$$

the double derivative of the above is given by

$$\phi_i'' = \frac{\mu_i^2}{H^2}\left[-\sin\frac{\mu_i z}{H} - \sin\mathrm{h}\frac{\mu_i z}{H} + \alpha_i\left(\cos\frac{\mu_i z}{H} + \cos\mathrm{h}\frac{\mu_i z}{H}\right)\right]$$

and

$$\phi_j'' = \frac{\mu_j^2}{H^2}\left[-\sin\frac{\mu_j z}{H} - \sin\mathrm{h}\frac{\mu_j z}{H} + \alpha_j\left(\cos\frac{\mu_j z}{H} + \cos\mathrm{h}\frac{\mu_j z}{H}\right)\right] \quad (10.4.10)$$

Before performing integration, we change the above to generalized coordinates by considering

$\xi = \frac{z}{H}$ when $\mathrm{d}\xi = \frac{\mathrm{d}z}{H}$ and as $z \to 0, \xi \to 0$ and as $z \to H, \xi \to 1$; based on the above we can now express the double derivative as

$$\begin{aligned} f''(\xi)_i &= [-\sin\mu_i\xi - \sin\mathrm{h}\mu_i\xi + \alpha_i(\cos\mu_i\xi + \cos\mathrm{h}\mu_i\xi)], \quad \text{and} \\ f''(\xi)_j &= [-\sin\mu_j\xi - \sin\mathrm{h}\mu_j\xi + \alpha_j(\cos\mu_j\xi + \cos\mathrm{h}\mu_j\xi)] \end{aligned} \quad (10.4.11)$$

Thus, the stiffness of the system can now be expressed as

$$k_{ij} = \frac{EI\mu_i^2\mu_j^2}{H^3}\int_0^1 f''(\xi)_i f''(\xi)_j \mathrm{d}\xi \quad (10.4.12)$$

and mass of the system is given by

$$m_{ij} = \frac{\gamma AH}{g}\int_0^1 f(\xi)_i f(\xi)_j \mathrm{d}\xi \quad (10.4.13)$$

where $i = j = 1, 2, 3, \ldots m$.

Thus, for the first three modes, it is sufficient to predict the dynamic response

$$[K]_{ij} = \frac{EI}{H^3}\begin{bmatrix} \mu_1^4 \int_0^1 f''(\xi)_1^2 d\xi & & \text{Symmetric} \\ \mu_2^2\mu_1^2 \int_0^1 f''(\xi)_2 f''(\xi)_1 d\xi & \mu_2^4 \int_0^1 f_2''(\xi)^2 d\xi & \\ \mu_3^2\mu_1^2 \int_0^1 f''(\xi)_3 f''(\xi)_1 d\xi & \mu_3^2\mu_2^2 \int_0^1 f''(\xi)_3 f''(\xi)_2 d\xi & \mu_3^4 \int_0^1 f''(\xi)_3^2 d\xi \end{bmatrix} \quad (10.4.14)$$

and the mass matrix is given by

$$[M]_{ij} = \frac{\gamma AH}{g}\begin{bmatrix} \int_0^1 f(\xi)_1^2 d\xi & \text{Symmetric} & \\ \int_0^1 f(\xi)_2 f(\xi)_1 d\xi & \int_0^1 f_2(\xi)^2 d\xi & \\ \int_0^1 f(\xi)_3 f(\xi)_1 d\xi & \int_0^1 f(\xi)_3 f(\xi)_2 d\xi & \int_0^1 f(\xi)_3^2 d\xi \end{bmatrix} \quad (10.4.15)$$

Solution of the above integrals gives

$$[K]_{3X3} = \frac{EI}{H^3}\begin{bmatrix} 22.936 & -0.002 & 0.006 \\ -0.002 & 468.044 & -0.11 \\ 0.006 & -0.11 & 3812.81 \end{bmatrix} \quad (10.4.16)$$

and the mass matrix is given by

$$[M]_{3X3} = \frac{W}{g}\begin{bmatrix} 1.855 & 0 & 0 \\ 0 & 0.964 & 0 \\ 0 & 0 & 1.002 \end{bmatrix} \quad (10.4.17)$$

Here, W = total weight of the shell + internal slabs + brick linings

Converting the above into standard eigenvalue form of $A\varphi = \lambda\,\varphi$ and applying the generalized Jacobi technique we have

$$[\lambda] = \frac{EIg}{WH^3}\begin{bmatrix} 12.362 & 0 & 0 \\ 0 & 485.519 & 0 \\ 0 & 0 & 3806.5468 \end{bmatrix} \quad (10.4.18)$$

and the scaled eigenvectors are given as

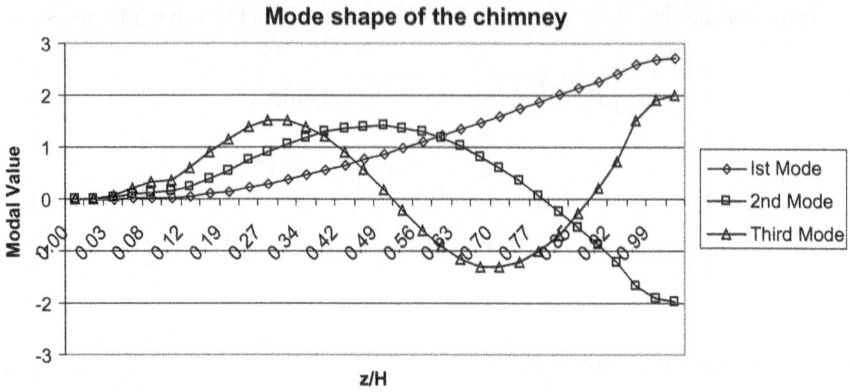

Fig. 10.4 First three modes of the chimney

$$[\varphi] = \begin{bmatrix} -1.0 & 2.278 \times 10^{-6} & 8.528 \times 10^{-7} \\ 4.384 \times 10^{-6} & -1 & -3.437 \times 10^{-5} \\ 1.579 \times 10^{-6} & -3.307 \times 10^{-5} & 1 \end{bmatrix} \begin{bmatrix} f_1(\xi) \\ f_2(\xi) \\ f_3(\xi) \end{bmatrix} \quad (10.4.19)$$

The eigenvector plots for the first three modes are as shown in Fig. 10.4 Since $[\lambda] = \omega^2$ and $T = \frac{2\pi}{\omega}$, we have

$$[T] = \begin{bmatrix} 1.787 & 0 & 0 \\ 0 & 0.285 & 0 \\ 0 & & 0.102 \end{bmatrix} \sqrt{\frac{WH^3}{EIg}} \quad (10.4.20)$$

Thus, for the first three modes we have the time period expressions as shown in Table 10.4.

10.4.1 Transformation to IS Code Format

As per Eq. (10.4.20), in the fundamental mode, the time period is given by

$$T_1 = 1.787\sqrt{\frac{WH^3}{EIg}}, \quad (10.4.21)$$

Now, considering $I = Ar^2$, where A = area of the stack at the base and r = radius of gyration, Eq. (10.4.21) can be expressed in the format of

Table 10.4 Values of time period T_i of chimney for the first three modes

Mode	1	2	3
Time period (s)	$T_1 = 1.787\sqrt{\frac{WH^3}{EIg}}$	$T_2 = 0.285\sqrt{\frac{WH^3}{EIg}}$	$T_2 = 0.102\sqrt{\frac{WH^3}{EIg}}$

Table 10.5 Values C_T for the first three modes by the proposed method

Slenderness ratio(H/r)	C_T (1st mode)	C_T (2nd mode)	C_T (3rd mode)
5	8.935	1.425	0.51
10	17.87	2.85	1.02
15	26.81	4.275	1.53
20	35.74	5.70	2.04
25	44.675	7.125	2.55
30	53.61	8.55	3.06
35	62.54	9.975	3.57
40	71.5	11.4	4.08
45	80.41	12.83	4.59
50	89.35	14.25	5.10

$$T_1 = 1.787\psi\sqrt{\frac{WH}{EAg}} \tag{10.4.22}$$

where, ψ = slenderness ratio of the stack @ H/r.

Considering $C_T - 1.7873\psi$, we have

$$T_1 = C_T\sqrt{\frac{WH}{EAg}} \tag{10.4.23}$$

which is in the same format as presented in the code.

The various values of C_T as proposed by the present method are as listed in Table 10.5.

Values of C_T obtained by the proposed method are compared for fundamental mode of IS code (as furnished in Table 10.5) in Fig. 10.5.

Values are found to be in commendable agreement. The IS code values give time period a little on the higher side. This has also been confirmed by Jain et al. (1990) based on dynamic FEM analysis carried out on ten chimneys.

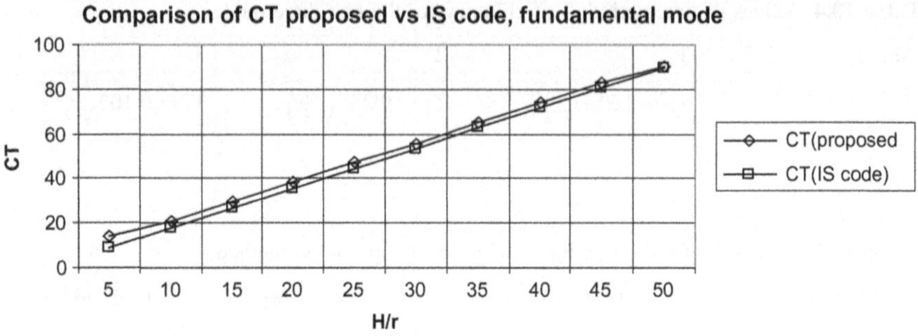

Fig. 10.5 Comparison of coefficient C_T according to IS code versus proposed method

10.4.2 Calculation of Amplitude

In terms of response spectrum analysis, the displacement S_d is given by $S_d = S_a/\omega^2$.

Expressing it in terms of codal formulation, we may write it as

$$S_d = \kappa_i \frac{ZI}{2R} \frac{S_a}{\omega^2} \tag{10.4.24}$$

where Z = zone coefficient; I = importance factor, and R = ductility factor.

κ_i = modal mass participation factor and is given by $\sum_{i=1}^{n} m_i\phi_i / \sum_{i=1}^{n} m_i\phi_i^2$. For an element of length dz, the above can be expressed as

$$\kappa_i = (\gamma A/g)\int_0^H \phi_i \mathrm{d}z / (\gamma A/g)\int_0^H \phi_i^2 \mathrm{d}z = \int_0^1 f_i(\xi)\mathrm{d}\xi / \int_0^1 f_i(\xi)^2 \mathrm{d}\xi \tag{10.4.25}$$

Integration of the mass participation factor within limits 1–0 for the first three modes gives values of κ_i as furnished in Table 10.6

Now considering $\beta = ZI/2R$, an IS code factor, we can write Eq. (10.4.24) as

$$S_d = \kappa_i \beta \frac{S_a}{\omega^2} \tag{10.4.26}$$

Thus, by the theory of modal analysis we have

$$\mathrm{w}(\mathrm{z}, \mathrm{t}) = \kappa_i \beta \frac{S_{ai} T_i^2}{4\pi^2} [\phi_i]^{\mathrm{T}} f_i(\xi) \tag{10.4.27}$$

For the first mode

Table 10.6 Values of modal mass participation factor (κ_i) for the first three modes

Mode	1	2	3
κ_i	0.5692	0.427	0.3081

$$\rightarrow \mathrm{w}(\mathrm{z},\mathrm{t}) = \kappa_1\beta g\frac{S_{a1}T_1^2}{4\pi^2 g}[-f_1(\xi) + 4.384\times 10^{-6}f_2(\xi) + 1.579\times 10^{-6}f_3(\xi)] \tag{10.4.28}$$

Neglecting the influence of the second and third modes, whose influence is negligible, we have

$$w(z,t) = -\left(\frac{\kappa_1\beta g}{4\pi^2}\right)\left(\frac{S_{a1}}{g}\right)T_1^2[f_1(\xi)] \tag{10.4.29}$$

For calculation of moment and shear, we know that $EI\frac{d^2w}{dz^2} = -M_z$, and hence,

$$M_z = -\kappa_i\beta\left(\frac{S_{ai}}{4\pi^2}\right)T_i^2 EI_z\frac{d^2}{dz^2}f(\xi) \tag{10.4.30}$$

Here, I_z = moment of inertia of the chimney cross section at any height z from the top of raft.

Again, considering $V_z = \frac{dM_z}{dz}$, we have

$$V_z = -\kappa_i\beta\left(\frac{S_{ai}}{4\pi^2}\right)T_i^2 EI_z\frac{d^3}{dz^3}f(\xi) \tag{10.4.31}$$

Proceeding in an identical manner for the second and third mode and after some simplifications we finally have

$$w_i(z,t) = (\Delta_i)\beta\left(\frac{S_{ai}}{g}\right)T_i^2 \tag{10.4.32}$$

$$M_{zi} = (C_{FMi})\beta\left(\frac{EI_z}{H^2}\right)\left(\frac{S_{ai}}{g}\right)T_i^2 \tag{10.4.33}$$

and

$$V_z = (C_{FVi})\beta\left(\frac{EI_z}{H^3}\right)\left(\frac{S_{ai}}{g}\right)T_i^2 \tag{10.4.34}$$

Here, Δ_i, C_{FMi}, and C_{FVi} are coefficients along the height of chimney for computation of displacement moment and shear, respectively, in the ith mode

Here, $\Delta_i = \left(\frac{\kappa_i g}{4\pi^2}\right)f(\xi)$

$$C_{FMi} = \left(\frac{\kappa_i g}{4\pi^2}\right)\mu_i^2 f''(\xi)$$

and $C_{FVi} = \left(\frac{\kappa_i g}{4\pi^2}\right)\mu_i^3 f'''(\xi)$

It will be observed that once we know the values of coefficients within parenthesis for i = 1, 2, 3, we can immediately find out the dynamic amplitude, shear, and moments without going through the elaborate process of dynamic analysis.

The coefficients for dynamic amplitude, moment, and shears considering the chimney shell is fixed at the base are as stated in Table 10.7.

An alternate inertial formulation is also possible (Chowdhury 2010), where the displacement and moment shear are expressed in terms of chimney's weight (W) and total height (H)

Based on this, one can express them as

$$w_i(z,t) = \Delta_i \beta \frac{S_{ai} W H^3}{EIg}, \tag{10.4.35}$$

$$M_z = C_{FMi} \beta W H \left(\frac{I_z}{I}\right)\left(\frac{S_{ai}}{g}\right), \tag{10.4.36}$$

and

$$V_z = C_{FVi} \beta W \left(\frac{I_z}{I}\right)\left(\frac{S_{ai}}{g}\right). \tag{10.4.37}$$

Here,

$$\Delta_i = \left(\frac{\kappa_i C_{Ti}^2}{4\pi^2}\right) f(\xi),\ C_{FMi} = \left(\frac{\kappa_i C_{Ti}^2}{4\pi^2}\right)\mu_i^2 f''(\xi), \text{ and } C_{FVi} = \left(\frac{\kappa_i C_{Ti}^2}{4\pi^2}\right)\mu_i^3 f'''(\xi), \tag{10.4.38}$$

Relevant coefficients are as expressed in Table 10.8.

For dynamic analysis, the IS-4998 code for design of reinforced concrete chimney recommends the modulus of elasticity E for dynamic analysis, as shown in Table 10.9. These shall be adopted while performing the computations.

We now elaborate the theory with a suitable numerical solution.

Example 10.1 A multi-flue chimney has a height of 250 m. Its estimated weight, including lining and internal slab, is 185,000 kN. The diameter of the chimney at is 23.0 m, having an average shell thickness of 650 mm. The chimney is situated in a place depicted by zone IV as per IS code resting on stiff soil. Find the deflection, moment, and shear for the first three modes and the maximum design moments and shears. The grade of concrete used is M30. Perform analysis based on dynamic analysis as proposed above and also based on response spectrum method as proposed in IS-1893 (2005), Part IV.

Table 10.7 Design coefficients for amplitude, moment, and shear for fixed base chimney

$\xi = z/H$	Δ_1 (ξ)	Δ_2 (ξ)	Δ_3 (ξ)	C_{FM1} (ξ)	C_{FM2} (ξ)	C_{FM3} (ξ)	C_{FV1} (ξ)	C_{FV2} (ξ)	C_{FV3} (ξ)
0	0.000	0.000	0.000	−1.355	−4.591	−9.455	−1.865	−21.948	−74.211
0.05	0.002	0.005	0.010	−1.262	−3.494	−5.753	−1.864	−21.904	−73.535
0.1	0.006	0.019	0.035	−1.168	−2.405	−2.160	−1.862	−21.621	−69.387
0.15	0.014	0.039	0.065	−1.075	−1.339	1.095	−1.856	−20.919	−59.884
0.2	0.025	0.063	0.093	−0.983	−0.322	3.734	−1.844	−19.684	−44.772
0.25	0.037	0.087	0.111	−0.891	0.620	5.498	−1.825	−17.867	−25.202
0.3	0.053	0.110	0.116	−0.801	1.455	6.216	−1.797	−15.476	−3.370
0.35	0.070	0.129	0.105	−0.712	2.159	5.843	−1.759	−12.574	17.938
0.4	0.089	0.142	0.081	−0.625	2.706	4.479	−1.711	−9.271	35.851
0.45	0.109	0.149	0.045	−0.541	3.082	2.356	−1.650	−5.715	47.912
0.5	0.131	0.149	0.003	−0.460	3.276	−0.187	−1.576	−2.081	52.496
0.55	0.154	0.140	−0.038	−0.383	3.292	−2.759	−1.488	1.438	49.086
0.6	0.178	0.123	−0.073	−0.311	3.138	−4.973	−1.385	4.643	38.382
0.65	0.202	0.098	−0.095	−0.245	2.836	−6.506	−1.268	7.338	22.226
0.7	0.228	0.066	−0.101	−0.185	2.416	−7.150	−1.135	9.342	3.349
0.75	0.253	0.028	−0.089	−0.132	1.916	−6.850	−0.986	10.497	−15.000
0.8	0.280	−0.015	−0.060	−0.087	1.382	−5.715	−0.822	10.670	−29.488
0.85	0.306	−0.061	−0.018	−0.050	0.867	−4.015	−0.641	9.761	−37.127
0.9	0.332	−0.109	0.035	−0.023	0.425	−2.155	−0.444	7.702	−35.600
0.95	0.359	−0.159	0.093	−0.006	0.116	−0.633	−0.230	4.455	−23.440
1	0.385	−0.208	0.153	0.000	0.000	0.000	0.000	0.002	0.018

Solution

The outside diameter of chimney = 23 m;
Shell thickness = 650 mm,
Inside diameter of chimney = 21.7 m.
Young's modulus of concrete = 3.35×10^7 kN/m^2 (refer Table 10.9)
Moment of inertia at the base $= \frac{\pi}{64}(23^4 - 21.7^4) = 2852.15$ m^4
Area of chimney at the base $= \frac{\pi}{4}(23^2 - 21.7^2) = 45.639$ m^2

Radius of gyration $= \sqrt{\frac{I}{A}} = 7.905$ m;

Slenderness ratio $= \frac{250}{7.905} = 31.62$.

Considering Eq. (10.4.22), the time period $= T_1 = 1.787\psi\sqrt{\frac{WH}{EAg}}$

For the **first mode,** we have

$$T_1 = 1.787 \times 31.6\sqrt{\frac{185{,}000 \times 250}{3.35 \times 10^7 \times 45.639 \times 9.81}} = 3.135 \text{ s}$$

This gives $Sa/g = 1.0/T$; for stiff soil $= 0.32 \times 1.27 = 0.405$

Table 10.8 Design coefficients for a fixed base chimney (inertial basis)

$\xi = z/H$	Δ_1 (ξ)	Δ_2 (ξ)	Δ_3 (ξ)	C_{FM1} (ξ)	C_{FM2} (ξ)	C_{FM3} (ξ)	C_{FV1} (ξ)	C_{FV2} (ξ)	C_{FV3} (ξ)
0	0.00000	0.00000	0.00000	0.44100	0.03801	0.01003	−0.60700	−0.18173	−0.07870
0.05	0.00054	0.00004	0.00001	0.41065	0.02893	0.00610	−0.60689	−0.18136	−0.07799
0.1	0.00210	0.00016	0.00004	0.38032	0.01991	0.00229	−0.60612	−0.17902	−0.07359
0.15	0.00462	0.00033	0.00007	0.35006	0.01109	−0.00116	−0.60409	−0.17320	−0.06351
0.2	0.00801	0.00052	0.00010	0.31994	0.00266	−0.00396	−0.60023	−0.16298	−0.04748
0.25	0.01220	0.00072	0.00012	0.29007	−0.00513	−0.00583	−0.59403	−0.14793	−0.02673
0.3	0.01712	0.00091	0.00012	0.26058	−0.01205	−0.00659	−0.58500	−0.12814	−0.00357
0.35	0.02269	0.00107	0.00011	0.23163	−0.01787	−0.00620	−0.57273	−0.10411	0.01902
0.4	0.02883	0.00118	0.00009	0.20337	−0.02241	−0.00475	−0.55684	−0.07677	0.03802
0.45	0.03549	0.00124	0.00005	0.17601	−0.02551	−0.00250	−0.53699	−0.04732	0.05081
0.5	0.04259	0.00123	0.00000	0.14974	−0.02713	0.00020	−0.51289	−0.01723	0.05567
0.55	0.05006	0.00116	−0.00004	0.12479	−0.02725	0.00293	−0.48428	0.01191	0.05206
0.6	0.05784	0.00102	−0.00008	0.10139	−0.02598	0.00527	−0.45096	0.03844	0.04071
0.65	0.06588	0.00081	−0.00010	0.07978	−0.02348	0.00690	−0.41274	0.06076	0.02357
0.7	0.07411	0.00055	−0.00011	0.06020	−0.02000	0.00758	−0.36950	0.07735	0.00355
0.75	0.08250	0.00023	−0.00009	0.04292	−0.01586	0.00726	−0.32111	0.08691	−0.01591
0.8	0.09100	−0.00012	−0.00006	0.02818	−0.01145	0.00606	−0.26751	0.08834	−0.03127
0.85	0.09956	−0.00050	−0.00002	0.01625	−0.00718	0.00426	−0.20863	0.08082	−0.03938
0.9	0.10817	−0.00090	0.00004	0.00740	−0.00352	0.00229	−0.14444	0.06377	−0.03776
0.95	0.11680	−0.00131	0.00010	0.00190	−0.00096	0.00067	−0.07493	0.03689	−0.02486
1	0.12543	−0.00173	0.00016	0.00000	0.00000	0.00000	−0.00009	0.00002	0.00002

Table 10.9 Recommended values of modulus elasticity as per IS-4998

Grade of concrete	*M*25	*M*30	*M*35	*M*40
E (kN/m^2)	3.2×10^7	3.35×10^7	3.5×10^7	3.6×10^7

Note: Here, the Sa/g value is multiplied by a factor 1.27 to cater for 3% damping, as recommended by IS-1893.

Similarly, for the second mode

$T_2 = 0.285 \times 31.6\sqrt{\frac{185{,}000 \times 250}{3.35 \times 10^7 \times 45.639 \times 9.81}} = 0.50$, gives $Sa/g = 2.5$

Thus, the design $S_{ad}/g = 2.5 \times 1.27 = 3.175$

And for the third mode,

$T_2 = 0.102 \times 31.6\sqrt{\frac{185{,}000 \times 250}{3.35 \times 10^7 \times 45.639 \times 9.81}} = 0.178$ s, gives $Sa/g = 2.5$

Thus, the design $S_{ad}/g = 2.5 \times 1.27 = 3.175$

For Zone IV, as per IS-1893

$Z = 0.24$, $I = 1.5$ and $R = 3.0 \rightarrow \beta = \frac{0.24 \times 1.5}{2 \times 3} = 0.06$

Substituting the values of W, H, E, I, Sa/g, and β, and multiplying it by the coefficients as in Table 10.8, we have

ξ	u_1 (mm)	u_2 (mm)	u_3 (mm)	*M*1 (kN m)	*M*2 (kN m)	*M*3 (kN m)	*V*1 (kN)	*V*2 (kN)	*V*3 (kN)
0	0.0000	0.0000	0.0000	495,624	334,914	88,348	2729	6404	2774
0.05	0.3960	0.2520	0.0628	461,516	254,899	53,757	2728	6392	2748
0.1	1.5466	0.9210	0.2136	427,430	175,415	20,186	2725	6309	2593
0.15	3.3963	1.8770	0.3979	393,419	97,685	−10,236	2716	6104	2238
0.2	5.8894	2.9933	0.5662	359,572	23,462	−34,889	2698	5744	1673
0.25	8.9705	4.1486	0.6786	326,005	−45,202	−51,372	2670	5214	942
0.3	12.5848	5.2311	0.7083	292,863	−106,179	−58,083	2630	4516	126
0.35	16.6781	6.1413	0.6444	260,319	−157,476	−54,602	2575	3669	−670
0.4	21.1972	6.7955	0.4926	228,563	−197,419	−41,852	2503	2705	−1340
0.45	26.0903	7.1286	0.2732	197,811	−224,802	−22,018	2414	1668	−1791
0.5	31.3069	7.0959	0.0184	168,292	−239,015	1747	2306	607	−1962
0.55	36.7990	6.6741	−0.2336	140,253	−240,123	25,783	2177	−420	−1835
0.6	42.5207	5.8613	−0.4438	113,953	−228,908	46,471	2027	−1355	−1435
0.65	48.4290	4.6756	−0.5789	89,663	−206,875	60,794	1855	−2141	−831
0.7	54.4843	3.1528	−0.6158	67,660	−176,219	66,813	1661	−2726	−125
0.75	60.6505	1.3427	0.5445	48,232	−139,757	64,005	1444	−3063	561
0.8	66.8960	−0.6957	−0.3697	31,669	−100,840	53,400	1203	−3113	1102
0.85	73.1938	−2.8992	−0.1083	18,267	−63,239	37,521	938	−2848	1388
0.9	79.5219	−5.2068	0.2142	8321	−31,033	20,140	649	−2248	1331
0.95	85.8640	−7.5664	0.5701	2133	−8496	5916	337	−1300	876
1	92.2103	−9.9419	0.9369	0	0	0	0	−1	−1

The design values are obtained by SRSS values of the three modes and are as given hereunder

ξ	$D_{(SRSS)}$ mm	M_{SRSS}	V_{SRSS}
0	0	604,662	7494
0.05	0.40	529,962	7473
0.1	2	462,465	7345
0.15	3	405,494	7046
0.2	6	362,021	6563
0.25	9	333,108	5933
0.3	13	316,886	5227
0.35	17	309,105	4532
0.4	21	304,905	3922
0.45	26	300,250	3437
0.5	31	292,324	3088
0.55	37	279,275	2878
0.6	43	259,891	2829
0.65	48	233,522	2953
0.7	54	200,237	3195
0.75	61	161,106	3432
0.8	67	118,420	3515
0.85	73	75,767	3304
0.9	80	37,920	2691
0.95	86	10,570	1603
1	92	0	1

We show in Figs. 10.6 and 10.7 the modal moments, shear, and SRSS values.

Analysis based on IS 1893(2005) Part IV:

As calculated above, the slenderness ratio $H/r = 31.62$ gives C_T value vide Table 10.2 is 58.916.

Base area $A = 45.639$ m^2, $E = 3.35 \times 10^7$ kN/m^2. $W = 185{,}000$ kN, $H = 250$ m

Thus, $T_1 = C_T\sqrt{\frac{WH}{EAg}} = 58.916 \times \sqrt{\frac{185{,}000\times 250}{3.35\times 10^7\times 45.639\times 9.81}} = 3.27$ s

Thus, the design $S_{ad}/g = (1/3.27) \times 1.27 = 0.388$.

Based on previous calculation, the code factor $\beta = 0.06$ and for $H/r = 31.62$, $C_v = 1.36$

$$V_z = C_v\beta(S_{ad}/g)WD_v$$

$$\rightarrow V_z = C_v\beta(S_{ad}/g)W\left[1.1\sqrt{z/H} + 0.75(z/H) + 0.9(z/H)^4\right]$$

$$V_z = 1.36 \times 0.06 \times 0.388 \times 185{,}000 \times \left[1.1\sqrt{z/H} + 0.75(z/H) + 0.9(z/H)^4\right]$$

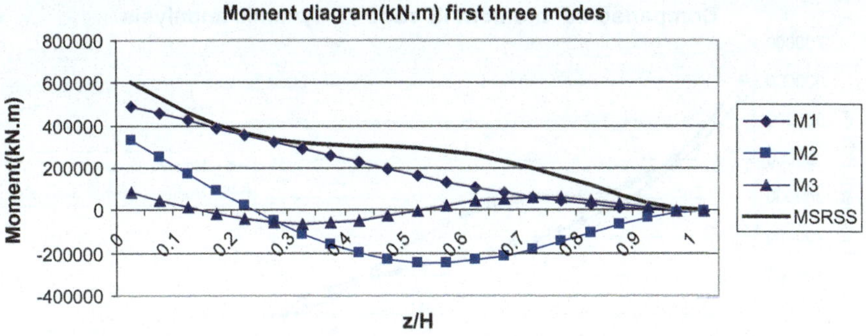

Fig. 10.6 Bending moment diagram of the chimney

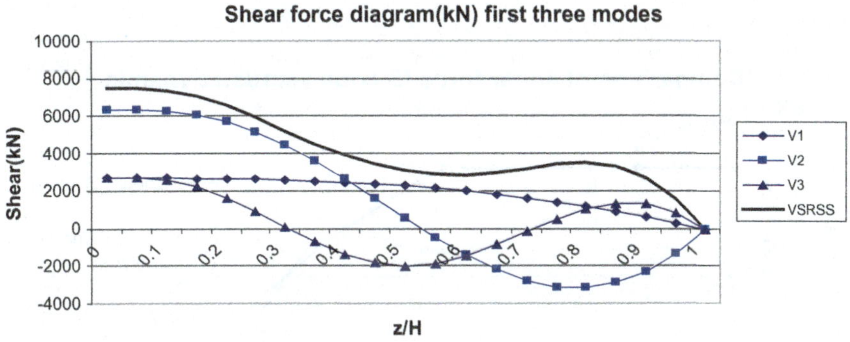

Fig. 10.7 Shear force diagram of the chimney

$$M_z = \beta(S_{ad}/g)W\bar{h}D_m \rightarrow M_z = \beta(S_{ad}/g)W\bar{h}\left[0.4\sqrt{z/H} + 0.6(z/H)^4\right]M_z$$
$$= 0.06 \times 0.388 \times 185{,}000 \times 125 \times \left[0.4\sqrt{z/H} + 0.6(z/H)^4\right]$$

As a comparison, we compare the moment and shear force obtained by dynamic analysis with that proposed by IS-1893(2005), as shown below in Figs. 10.8 and 10.9.

In Fig. 10.9, the D_v value below $z/H < 0.65$ becomes more than 1.0, and thus, as recommended by the code, has been considered as 1.0.

Figures 10.8 and 10.9 clearly show the necessity of considering higher modes, as the moments and shears are found to be higher, almost for the full height of the chimney, compared to the values as derived from IS code recommendations. Figure 10.9 also shows that, as pointed out by, Jain et al. (1990) the higher mode shear contribution is significant compared to fundamental mode response.

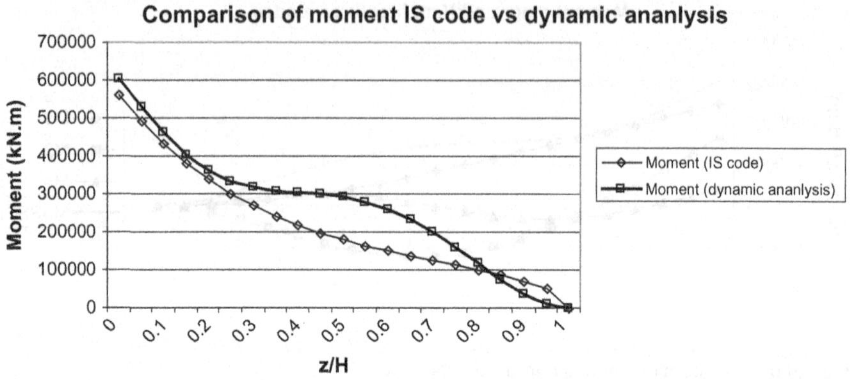

Fig. 10.8 Comparison of moment dynamic analysis versus IS 1893 (2005)

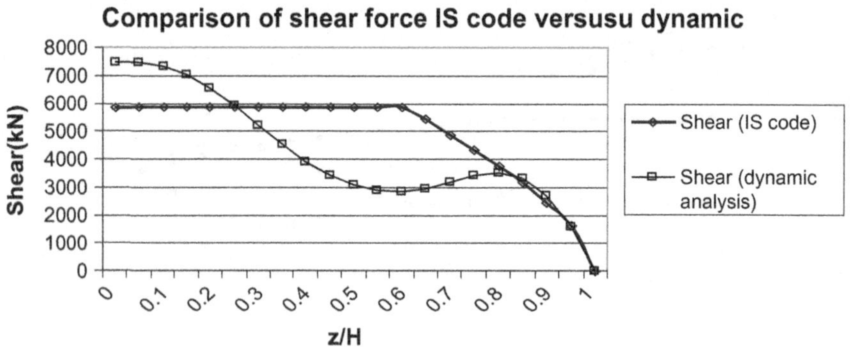

Fig. 10.9 Comparison of shear force dynamic analysis versus IS 1893 (2005)

10.5 Numerical Analysis of Tapered Stacks

Single-flue chimneys, due to thermodynamic reason and to enhance the exit velocity of the flue gas,[3] are usually provided with a tapered section, as shown in Fig. 10.10. Besides the reason cited, the choice is also structurally reasonable for the moment and shear increase from zero at the top to maximum at the base. Based on this, the obvious economic design should be that which has a section minimum at the top and maximum at the base.

As the section has a varying profile (generally linear), the mathematical treatment as shown for a multi-flue chimney with constant EI becomes complex for a

[3]The flue gas needs sufficient exit velocity to reduce the ground level pollution concentration.

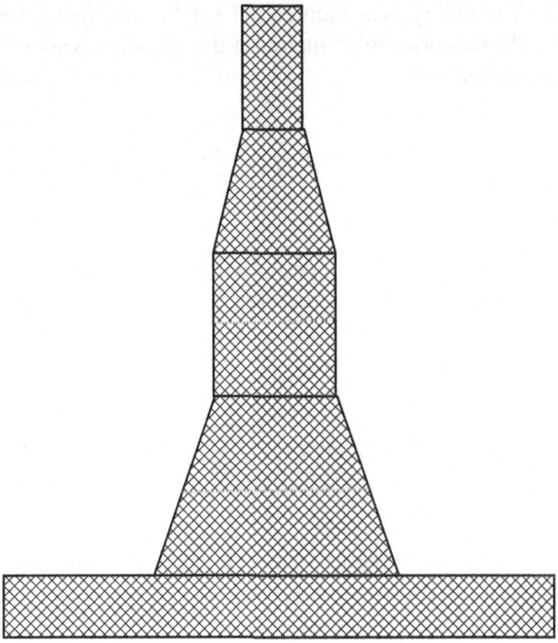

Fig. 10.10 Single-flue stack with varying diameter and shell thickness

closed form solution, except for the fundamental mode. However, this can very easily be solved by applying numerical techniques and arrive at an accurate answer.

In case of tapered chimneys, the numerical solution is often preferred as a design office practice because though in most of the cases, the profile is linear from stress point of view and also to diminish the amplitude at the top, the profile has a number of transition zones (i.e., the slope often changes at two or three positions, thus have varying integral functions with different limits).

Multi-flue chimneys are also constructed of tapered shape nowadays, though slopes are quite low compared to single-flue stacks, with platforms at different levels supporting the internal brick flues. Alternatively, if steel liners are used, they are suspended from one particular level with other platforms used for access and maintenance. Due to this, concentrated loads at different heights from the access platforms and support points combined with tapering of the shell and diameter of the windshield, many engineers feel that analyzing a chimney problem based on a continuum is not correct and a true behavior can only be reflected by a numerical analysis where masses are lumped at different levels considering the varying moment of inertia along the chimney shell. The analysis is usually carried out in computer.

We present here an alternative numerical method catering to the additional masses lumped from platform, liner support, etc. Here, one has to form the stiffness and mass matrix first and then one additional step has to be carried out, i.e., performing the eigenvalue analysis which was already implicit in the previous modal calculation based on continuum.

The theory presented earlier can be modified for numerical analysis as follows:

As the moment of inertia of the section is varying, the stiffness equation can be expressed as

$$k_{ij} = \frac{E\mu_i^2\mu_j^2}{H^4}\int_0^H I_z f''(z)_i f''(z)_j \mathrm{d}z \tag{10.5.1}$$

in which I_z is the moment of inertia considered to be varying at different heights z.

Here,

$$\phi_i'' = \frac{\mu_i^2}{H^2}\left[-\sin\frac{\mu_i z}{H} - \sin\mathrm{h}\frac{\mu_i z}{H} + \alpha_i\left(\cos\frac{\mu_i z}{H} + \cos\mathrm{h}\frac{\mu_i z}{H}\right)\right]$$

$$\rightarrow \phi_i'' = \frac{\mu_i^2}{H^2}F''(z) \text{and } F''(z) = \frac{\mu_i^2}{H^2}\left[-\sin\frac{\mu_i z}{H} - \sin\mathrm{h}\frac{\mu_i z}{H} + \alpha_i\left(\cos\frac{\mu_i z}{H} + \cos\mathrm{h}\frac{\mu_i z}{H}\right)\right] \tag{10.5.2}$$

Thus, the stiffness matrix in natural coordinates can now be written as for numerical computation as

$$[K]_{ij} = \frac{E}{H^3}\begin{bmatrix} \mu_1^4\sum_0^1 I_z f''(\xi)_1^2 & & \text{Symmetrical} \\ \mu_2^2\mu_1^2\sum_0^1 I_z f''(\xi)_2 f''(\xi)_1 & \mu_2^4\sum_0^1 I_z f_2''(\xi)^2 & \\ \mu_3^2\mu_1^2\sum_0^1 I_z f''(\xi)_3 f''(\xi)_1 & \mu_3^2\mu_2^2\sum_0^1 I_z f''(\xi)_3 f''(\xi)_2 & \mu_3^4\sum_0^1 I_z f''(\xi)_3^2 \end{bmatrix} \tag{10.5.3}$$

Similarly, the mass matrix equation can be written as

$$[M]_{ij} = \frac{\gamma_c H}{g}\begin{bmatrix} \sum_0^1 A_z f_1(\xi)^2 + \sum_1^N M f_1(\xi)_{\alpha_g}^2 & & \text{Symmetrical} \\ \sum_0^1 A_z f_2(\xi) f_1(\xi) + \sum_1^N M f_2(\xi)_{\alpha_g} f_1(\xi)_{\alpha_g} & \sum_0^1 A_z f_2(\xi)^2 + \sum_1^N M f_2(\xi)_{\alpha_g}^2 & \\ \sum_0^1 A_z f_3(\xi) f_1(\xi) + \sum_1^N M f_3(\xi)_{\alpha_g} f_1(\xi)_{\alpha_g} & \sum_0^1 A_z f_3(\xi) f_2(\xi) + \sum_1^N M f_3(\xi)_{\alpha_g} f_2(\xi)_{\alpha_g} & \sum_0^1 A_z f_3(\xi)^2 + \sum_1^N M f_3(\xi)_{\alpha_g}^2 \end{bmatrix} \tag{10.5.4}$$

Here, N = number of lumped masses M at various levels and $\alpha_g = H_g/H$, where H_g is the height where the lumped masses are located, γ_c = unit weight of concrete, and A_z = area of shell cross section at height z when $\xi = z/H$.

Each of the above terms of the stiffness and mass matrix is to be obtained by numerical integration within the limits 0–1. For numerical integration, the following values of μ and α are to be adopted.

Mode	1	2	3
μ	1.875	4.694	7.855
α	1.362221	0.981868	1.000776105

After the stiffness and mass matrix are formed, an eigenvalues analysis needs to performed based on the equation

$$[\mathbf{K}] - [\mathbf{M}]\omega^2 = 0 \tag{10.5.5}$$

Once the eigenvalues *vis-à-vis* time periods are known, the S_a/g values can be obtained from the response curve as furnished in the codes.

The dynamic amplitude is thus furnished by the equation

$$S_d = \kappa_i \beta \frac{S_a}{\omega^2} \tag{10.5.6}$$

and the complete solution is given by the expression

$$w_i(z,t) = \kappa_i \beta \frac{S_a}{\omega^2} [\varphi_{ii}] F_i(z). \tag{10.5.7}$$

Here,

$$\kappa_i = \sum_0^1 \gamma_c A_z \xi f(\xi) + Mf(\xi)_{\alpha_g} \Big/ \sum_0^1 \gamma_c A_z \xi f(\xi)^2 + Mf(\xi)^2_{\alpha_g} \tag{10.5.8}$$

Thus,

$$\begin{aligned} w_1(z,t) &= \kappa_1 \beta \frac{S_{a1}}{\omega_1^2} [\phi_{11} F_1(z) + \phi_{12} F_2(z) + \phi_{13} F_3(z)] \\ w_2(z,t) &= \kappa_2 \beta \frac{S_{a2}}{\omega_2^2} [\phi_{21} F_1(z) + \phi_{22} F_2(z) + \phi_{23} F_3(z)] \text{ and} \\ w_3(z,t) &= \kappa_3 \beta \frac{S_{a3}}{\omega_3^2} [\phi_{31} F_1(z) + \phi_{32} F_2(z) + \phi_{33} F_3(z)] \end{aligned} \tag{10.5.9}$$

where $[\phi_{ii}]$ = scaling vector value for the ith mode.

The moment and shear are then obtained from the equation

$EI \frac{d^2 w}{dz^2} = -M_z$ and $EI \frac{d^3 w}{dz^3} = -V_z$, which gives

$$\begin{aligned} M_1(z,t) &= -\kappa_1\beta\frac{S_{a1}EI_z}{\omega_1^2H^2}\left[\phi_{11}\mu_1^2F_1''(z)+\phi_{12}\mu_2^2F_2''(z)+\phi_{13}\mu_3^2F_3''(z)\right] \\ M_2(z,t) &= -\kappa_2\beta\frac{S_{a2}EI_z}{\omega_2^2H^2}\left[\phi_{21}\mu_1^2F_1''(z)+\phi_{22}\mu_2^2F_2''(z)+\phi_{23}\mu_3^2F_3''(z)\right]\text{and} \\ M_3(z,t) &= -\kappa_3\beta\frac{S_{a3}EI_z}{\omega_3^2H^2}\left[\phi_{31}\mu_1^2F_1''(z)+\phi_{32}\mu_2^2F_2''(z)+\phi_{33}\mu_3^2F_3''(z)\right] \end{aligned} \tag{10.5.10}$$

Considering $EI\frac{d^3w}{dz^3} = -V_z$,

$$\begin{aligned} V_1(z,t) &= -\kappa_1\beta\frac{S_{a1}EI_z}{\omega_1^2H^3}\left[\phi_{11}\mu_1^3F_1'''(z)+\phi_{12}\mu_2^3F_2'''(z)+\phi_{13}\mu_3^3F_3'''(z)\right] \\ V_2(z,t) &= -\kappa_2\beta\frac{S_{a2}EI_z}{\omega_2^2H^3}\left[\phi_{21}\mu_1^3F_1'''(z)+\phi_{22}\mu_2^3F_2'''(z)+\phi_{23}\mu_3^3F_3'''(z)\right]\text{ and} \\ V_3(z,t) &= -\kappa_3\beta\frac{S_{a3}EI_z}{\omega_3^2H^3}\left[\phi_{31}\mu_1^3F_1'''(z)+\phi_{32}\mu_2^3F_2'''(z)+\phi_{33}\mu_3^3F_3'''(z)\right] \end{aligned} \tag{10.5.11}$$

where

$$F_i'''(z) = \left[-\cos\frac{\mu_i z}{H} - \cos\text{h}\frac{\mu_i z}{H} + \alpha_i\left(-\sin\frac{\mu_i z}{H} + \sin\text{h}\frac{\mu_i z}{H}\right)\right] \tag{10.5.12}$$

We have solved a complete design example based on numerical analysis for a chimney at the end of this chapter, based on the above principles.

10.6 Computer Analysis of Tall Chimneys

Many engineering consultancy houses have developed in-house computer program for analysis and design of tall chimneys, based on IS-4998 or ACI 307, CICIND, etc. The seismic analysis part, however, can also be done in generic finite element commercially available package like GTSTRUDL, SAP 2000, STAAD PRO.

For dynamic analysis of tall chimneys, *normally a stick model with masses lumped at convenient nodes suffices*. The structural element consists of beam elements with the mass of the shell and brick lining lumped at each end nodes i and j. The computer assembles the stiffness matrix based on the principle of finite element $[K] = \int [B]^T D[B]dz$ and forms the lumped mass matrix $[\mathbf{M}]$ which is diagonal in nature. On formation of these elements, it performs the eigenvalue analysis based on the expression $[K][\varphi] = [M][\varphi]\omega^2 = 0$ and then performs the modal analysis based on response spectrum *method as explained earlier*.

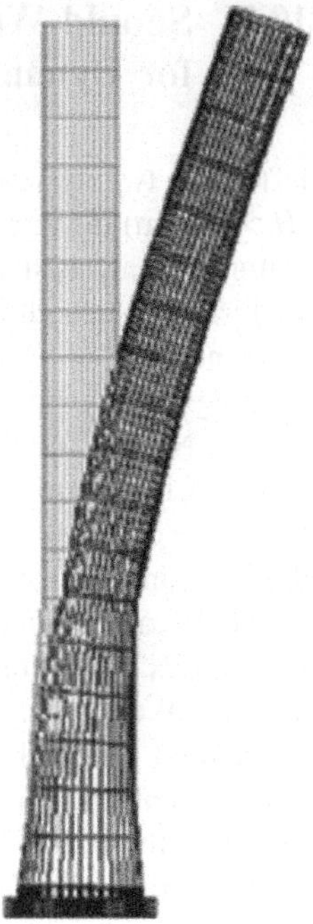

Fig. 10.11 Finite element model of chimney shell considering higher-order iso-parametric shell elements

In some cases, the author has also come across FEM model of chimneys, as shown in Fig. 10.11, where the windshield is modeled by higher-order triangular/quadrilateral shell elements. Though there is nothing mathematically wrong in developing such a detailed model, but, frankly speaking, as a practice, it is simply over killing the issue.

It does not really give any significant advantage over a stick model, either in terms of estimation of time periods, moments, or shears along the shell height in terms of wind or seismic loads. On the contrary, a stick model is much easier and cheaper to model than such detailed FEM model.

Advantages with such FEM model is that the stress concentration around flue duct opening can be better estimated under different load combinations. Moreover, buckling mode frequencies that would be missing in a stick model can also be estimated.

However, the buckling mode hardly governs the behavior of a concrete chimney and rarely is a design criterion. For evaluation of stress around opening, significant mesh refinement is required around it to get a reasonably correct answer and would substantially enhance the computational effort. Stress concentration analysis can well be done locally after the moments, and shear force in the windshield are estimated by a stick model, and the FEM analysis becomes a posterior activity to that by cutting off a part of the shell and doing the stress analysis locally under static loading conditions.

10.7 Should We Consider Soil–Structure Interaction for Dynamic Analysis of Chimneys?

In the last twenty years, a number of chimneys have been built around the country ($H > 220$ m) in various power plants, which has created a significant debate among chimney analysts and designers whether one should consider DSSI in dynamic analysis for such chimneys.

A number of researchers have advocated consideration of DSSI while performing the dynamic analysis of chimneys. Ghose and Batavyal (1985), Luco (1986), Navarro (1992), Sadeghpour and Chowdury (2008), Jaya and Dodagoudar (2009), Chowdhury (2010), Chowdhury et al. (2012) have all suggested to consider DSSI for chimney design, each proposing different techniques to be adopted for dynamic analysis of such tall RCC chimneys.

All researchers have concluded that the fixed base moment and shear of chimneys varies considerably when the effect of foundation compliance is considered in the overall analysis.

IS-1893 (2005) has now provided allowance for DSSI analysis by furnishing translational and rocking spring of soil-foundation system based on Hall (1967); however, the *modus operandi* of the analysis is left open to the discretion of user.

As the time period will get modified due to foundation soil springs, the S_a/g values should correspond to this modified time period and not the fixed base case.

The most popular method that is in vogue in design offices for analysis of such chimneys, including DSSI, is to analyze the same through any finite element software when conceptual mathematical model used for such an analysis is as shown hereafter.

As shown in Fig. 10.12, the chimney is broken into a number of discrete beam elements (minimum 10 as per IS code) with its masses lumped at selected nodes. Soil springs K_h and K_θ are applied at the boundary and then, based on finite element technique, the eigensolution is obtained from the expression

$$[\mathbf{K}] - [\mathbf{M}]\omega^2 = 0 \tag{10.7.1}$$

where [K] = combined stiffness matrix for the soil–structure system, [M] = lumped mass matrix of the superstructure, and ω^2 = eigenvalue of the problem

Application of Eq. (10.7.1) is quite straightforward and is a part and parcel of most of the standard commercially available FEM software packages; it remains an attractive choice to a designer. In spite of its popularity, the mathematical model suffers from three serious lacunae.

1. Though computationally less intense than a continuum finite element analysis, it still requires careful and tedious calculation of nodal mass and geometric beam data for all nodes considered for analysis.

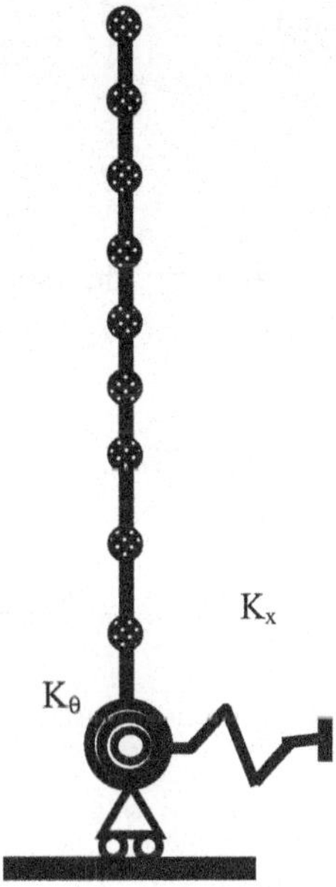

Fig. 10.12 Mathematical model of chimney for DSSI analysis

2. Calls for eigensolution for a significant number of modes through a computer, that can make the solution expensive at times.
3. Requires a guesstimated composite soil-structure damping ratio which is assumed to remain constant for all modes.

It is an established fact that damping values vary with modes and can significantly affect overall dynamic response of a system (Chowdhury and Dasgupta 2003; Giosan 2006; Roussea and van Zijl 2006), especially for the case of DSSI analysis.

To circumvent some of the approximations and difficulties as cited above, the following methodology could be used. It has been shown (Sadeghpour and Chowdhury 2008) that for the first three modes for a tall chimney, the time periods considering the DSSI effect can be expressed as

$$T_1 = 2\pi\sqrt{1.86\frac{W}{g}\left[\frac{H^3}{22.94EI} + \frac{1}{K_h} + \frac{0.5278H^2}{K_\theta}\right]} \tag{10.7.2}$$

$$T_2 = 2\pi\sqrt{0.96\frac{W}{g}\left[\frac{H^3}{468EI} + \frac{1}{K_h} + \frac{0.0438H^2}{K_\theta}\right]} \tag{10.7.3}$$

and

$$T_3 = 2\pi\sqrt{\frac{W}{g}\left[\frac{H^3}{3813EI} + \frac{1}{K_h} + \frac{0.0163H^2}{K_\theta}\right]} \tag{10.7.4}$$

Similarly, the modal damping ratios for the first three modes are expressed as

$$\zeta_{e1} = \frac{\frac{\zeta H^3}{22.94EI} + \frac{\zeta_x}{K_h} + \frac{0.5278\zeta_\theta H^2}{K_\theta}}{\frac{H^3}{22.94EI} + \frac{1}{K_h} + \frac{0.5278H^2}{K_\theta}} \tag{10.7.5}$$

$$\zeta_{e2} = \frac{\frac{\zeta H^3}{468.04EI} + \frac{\zeta_x}{K_h} + \frac{0.0438\zeta_\theta H^2}{K_\theta}}{\frac{H^3}{468.04EI} + \frac{1}{K_h} + \frac{0.0438H^2}{K_\theta}} \tag{10.7.6}$$

$$\zeta_{e3} = \frac{\frac{\zeta H^3}{3813EI} + \frac{\zeta_x}{K_h} + \frac{0.0163\zeta_\theta H^2}{K_\theta}}{\frac{H^3}{3813EI} + \frac{1}{K_h} + \frac{0.0163H^2}{K_\theta}} \tag{10.7.7}$$

Equations (10.7.2)–(10.7.7) are based on analytical solutions and are exact; here K_h and K_θ are Hall's springs as expressed in Eqs. (10.3.10) and (10.3.11), and ζ is the material damping of concrete.

The damping ratios in translation (ζ_x) and rotation are (ζ_θ) are expressed as

$$\zeta_x = \frac{0.288}{\sqrt{B_x}} \tag{10.7.8}$$

where

$$B_x = \frac{(7-8\upsilon)\mathrm{mg}}{32(1-\upsilon)\rho_s r_0^3}. \tag{10.7.9}$$

and

$$\zeta_\theta = \frac{0.15}{(1+\mathrm{B}_\theta)\sqrt{\mathrm{B}_\theta}} \tag{10.7.10}$$

here

$$\mathrm{B}_\theta = \frac{0.375(1-\upsilon)\mathrm{J}_\theta \mathrm{g}}{\rho_s r_\theta^5} \tag{10.7.11}$$

Here, B_x and B_θ are inertial factor contributions to damping ratio, m = mass of raft foundation plus mass coming from super structure, r_0 and r_θ = radii of the circular raft, ρ_s = mass density of soil, $J_\theta = (1/4)m_f r_\theta^2 + m_s\bar{h}^2$, m_f = mass of foundation raft and m_s = mass of superstructure shell including all its accessories, and $\bar{h}$ is the c.g height of the shell.

Applying Eqs. (10.7.5)–(10.7.7), it is observed that modal damping ratio would vary for each mode and, based on time period as derived in Eqs. (10.7.1)–(10.7.4), S_a/g values are to be determined from the code that needs to be scaled up or down depending on whether damping ratio is greater or smaller than 5%.

Based on time period of the coupled soil–structure system, once the acceleration of the system is estimated, displacement of the chimney may be determined as per Eq. (10.4.29); when proceeding in an identical fashion as explained earlier for fixed base case, one obtains for the fundamental mode

$$w_1(z,t) = 1.86\kappa_1\beta\frac{S_{a1}WH^3C_{\eta 1}}{EIg}[f_1(\xi)] \tag{10.7.12}$$

Here, $C_{\eta 1}$ is a factor that takes care of DSSI effect of soil–structure system in the fundamental mode and is expressed as

$$C_{\eta 1} = \frac{1}{22.94} + \frac{EI}{K_h H^3} + \frac{0.53EI}{K_\theta H} \tag{10.7.13}$$

Equation (10.7.12) can be further expressed as

$$w_1(z,t) = 1.06\beta\frac{S_{a1}WH^3C_1(\eta)}{EIg}[f_1(\xi)] \tag{10.7.14}$$

Again, considering $EI_z\frac{d^2w}{dz^2} = -M_z$ and $V_z = \frac{dM_z}{dz}$ we have

$$M_1(z) = -3.73\beta\left(\frac{I_z}{I}\right)\frac{S_{a1}WHC_1(\eta)}{g}[f''(\xi)]$$

$$V_1(z) = -6.98\beta\left(\frac{I_z}{I}\right)\frac{S_{a1}WC_1(\eta)}{g}[f_1'''(\xi)] \tag{10.7.15}$$

Proceeding in identical manner for the second and third modes one may finally have (Chowdhury 2010)

$$w_i(z,t) = (\text{Coeff}_d)\beta C_{\eta i}\frac{S_{ai}WH^3}{EIg}, \tag{10.7.16}$$

$$M_{zi} = (\text{Coeff}_M)\beta WHC_{\eta i}\left(\frac{I_z}{I}\right)\left(\frac{S_{ai}}{g}\right) \tag{10.7.17}$$

$$V_z = (\text{Coeff}_V)\beta WC_{\eta i}\left(\frac{I_z}{I}\right)\left(\frac{S_{ai}}{g}\right). \tag{10.7.18}$$

Here, i = 1, 2, 3… the number of modes.

The DSSI factor $C_{\eta i}$ can be generically expressed for i modes as

$$C_{\eta i} = \frac{1}{A_i} + \frac{EI}{K_h H^3} + \frac{B_i EI}{K_\theta H} \quad (10.7.19)$$

For the first three modes, values of A_i and B_i are furnished in Table 10.10

It is again observed that, based on the proposed method, once coefficients within parenthesis are known, displacement, moment, and shear may be determined very easily without going through an elaborate FEM analysis.

Design coefficients when DSSI is considered in the analysis are furnished in Table 10.11

Table 10.10 Values of A_i and B_i for the first three modes

Mode	1	2	3
A_i	22.94	468	3183
B_i	0.53	0.044	0.016

Table 10.11 Design coefficient for amplitude moment and shear considering DSSI

$\xi = z/H$	Δ_1 (ξ)	Δ_2 (ξ)	Δ_3 (ξ)	M_1 (ξ)	M_2 (ξ)	M_3 (ξ)	V_1 (ξ)	V_2 (ξ)	V_3 (ξ)
0	0.000	0.000	0.000	−10.162	−17.738	−38.290	13.96	84.80	300.50
0.05	0.012	0.020	0.042	−9.463	−13.500	−23.298	13.96	84.63	297.76
0.1	0.048	0.075	0.142	−8.764	−9.291	−8.749	13.94	83.54	280.96
0.15	0.106	0.152	0.264	−8.067	−5.174	4.436	13.89	80.82	242.48
0.2	0.184	0.242	0.375	−7.373	−1.243	15.121	13.80	76.05	181.30
0.25	0.281	0.336	0.450	−6.684	2.394	22.264	13.66	69.03	102.05
0.3	0.394	0.424	0.469	−6.005	5.624	25.173	13.45	59.79	13.64
0.35	0.522	0.497	0.427	−5.338	8.341	23.664	13.17	48.58	−72.64
0.4	0.664	0.550	0.326	−4.686	10.456	18.139	12.81	35.82	−145.17
0.45	0.817	0.577	0.181	−4.056	11.906	9.543	12.35	22.08	−194.01
0.5	0.980	0.575	0.012	−3.451	12.659	−0.757	11.80	8.04	−212.57
0.55	1.152	0.540	−0.155	−2.876	12.718	−11.174	11.14	−5.56	−198.76
0.6	1.332	0.475	−0.294	−2.336	12.124	−20.140	10.37	−17.94	−155.42
0.65	1.517	0.379	−0.384	−1.838	10.957	−26.348	9.49	−28.35	−90.00
0.7	1.706	0.255	−0.408	−1.387	9.333	−28.956	8.50	−36.10	−13.56
0.75	1.899	0.109	−0.361	−0.989	7.402	−27.739	7.39	−40.56	60.74
0.8	2.095	−0.056	−0.245	−0.649	5.341	−23.143	6.15	−41.22	119.41
0.85	2.292	−0.235	−0.072	−0.375	3.349	−16.262	4.80	−37.71	150.34
0.9	2.490	−0.422	0.142	−0.171	1.644	−8.729	3.32	−29.76	144.15
0.95	2.689	−0.613	0.378	−0.044	0.450	−2.564	1.72	−17.21	94.91
1	2.888	−0.805	0.621	0.000	0.000	0.000	0.00	−0.01	−0.07

10.7.1 Soil–Structure Interaction of a Chimney on Pile Foundations

We had shown, vide Eq. (10.3.12), the pile stiffness equation suggested by IS code for considering DSSI effect for a chimney. The formula suggested is for a single pile and to the authors' perception is far too simplistic and the recommendation is certainly not realistic.

Primarily, the stiffness suggested by Eq. (10.3.12) is for a pile under static load. In case of an earthquake, it is the dynamic stiffness of pile that should govern the response. Moreover, to determine the overall stiffness of the pile group, if Eq. (10.3.12) is multiplied by the number N, where N is the number of piles in the group, we will certainly overestimate the stiffness significantly, as it has been observed that the group interaction factor significantly reduces the overall stiffness of a pile group, especially when spacings between the piles are within 2.5–3D, where D is the diameter of pile, (Novak 1974, 1977; Poulos 1979; Gazetas et al. 1991).

The correct method should be to determine the dynamic pile stiffness of a single pile and then determine the overall pile group stiffness as

$$K_{xx}^{g} = \sum_{i=1}^{N} k_{xx} \Big/ \sum_{i=1}^{N} \alpha_x \qquad (10.7.20)$$

Here, N = number of piles in the pile group, α_x = group interaction factor, k_{xx} is lateral dynamic stiffness of a single pile, and K_{xx}^{g} is the lateral stiffness of pile group.

The IS-1893 (2005) part IV provides neither any expression for rocking stiffness of the pile nor any expression for damping of the pile (in translational and rotational modes), both of which have profound influence on dynamic response of such tall structures under DSSI.

Finally, a part of mass of pile itself will also contribute to overall dynamic response, and this has also been ignored in the recommendation.

Thus, as a first step, one needs to determine the dynamic stiffness of pile in lateral and rocking mode. To this end, one can certainly use formulations suggested by Novak (1974) or Gazetas et al. (1991).

However, before application of either of these two methods, one should clearly be aware of their limitations.

Major limitations with both Novak's and Gazetas' formulations are that they ignore the inertial contribution of the piles and that plays a major role in the overall dynamic response.

For a 275-m tall chimney, the number piles in foundation could vary from roughly 300–450 piles (depending on the soil property), whose inertial mass contribution could be significant and cannot be ignored if we wish to arrive at a realistic analysis.

Last but not the least, in international scenario, in many countries where earthquake is active, piles have been found to be driven through soil whose top 6–8 m layer consists of saturated cohesionless soil having SPT value $N < 15$, as such soils are highly susceptible to liquefaction under a major shock.

In such a case, clients often ask for a scenario analysis where the superstructure is analyzed under MCE and simultaneously a part of top portion of soil surrounding the piles is assumed to have undergone liquefaction. In such a case, neither Novak's nor Gazetas formulas are applicable, which also concerns Eq. (10.2.12), as the piles now become partially embedded.

Chowdhury and Dasgupta (2008a, b, 2009) have provided stiffness of piles in lateral and rocking mode for both fully and partially embedded piles, including the inertial effect of piles, which could be used for such DSSI analysis of chimneys on circular pile caps.

In lateral direction, the pile stiffness k_{xx} is expressed as

$$k_{xx} = \frac{E_p I_p}{L^3}\left[\frac{\frac{5X}{4}\left(1 - \mathrm{e}^{-2\beta}\right) - \frac{3Y}{8} - \frac{3\eta}{4}}{(\eta - 1)^3}\right] \tag{10.7.21}$$

Here, $\eta = 1 + 1/\beta, \beta = \sqrt[4]{GS_{x1}L^4/E_pI_p}$, G = dynamic shear modulus of soil, L = length of the pile, E_p = Young's modulus of pile material, I_p = moment of inertia of the pile section. S = Beregdugo's functions and are S_{x1} for translational mode and $S_{\theta 1}$ for rotational mode, respectively.

Values of S_{x1} for v = 0.25 and 0.4 are given in Tables 10.12 and 10.13

The inertial contribution of a pile is expressed as

$$M_{xx} = \frac{\gamma_p A_p L}{4g}\left[\frac{X\left(1 - \mathrm{e}^{-2\beta}\right) + \frac{Y}{2} + \eta}{\beta}\right] \tag{10.7.22}$$

where

γ_{p} = weight density of pile material, A_{p} = area of pile cross section, L = length of the pile, and g = acceleration due to gravity.

The damping of pile in translational mode is expressed as

$$c_{xx} = r_0\sqrt{\rho G}\overline{S}_2 L\left[\frac{X\left(1 - \mathrm{e}^{-2\beta}\right) + \frac{Y}{2} + \eta}{4\beta}\right] \tag{10.7.23}$$

In Eqs. (10.7.20)–(10.7.22), $X = 1 + \eta^2$, and $Y = 1 - \eta^2$, value of Beredugo's function $\overline{S}_{x2}$ is a frequency dependent damping parameter furnished in Table 10.14.

In Table 10.14, $a_0 = \omega r/V_{\mathrm{s}}$, where ω = applied frequency on the system, r = radius of the pile and V_{s} = shear wave velocity of soil, when $V_{\mathrm{s}} = \sqrt{G/\rho}$, where ρ = mass density of soil.

Similarly, for rotational mode

Table 10.12 Suggested value of S_{X1} for Poisson's ratio of soil = 0.25

Poisson's ratio	L/r_0 (slenderness ratio)	S_{x1} (250)	S_{x1} (500)	S_{x1} (1000)	S_{x1} (2500)	S_{x1} (5000)	S_{x1} (10,000)
	25	2.00	1.83	1.66	1.43	1.25	1.07
	40	2.19	2.05	1.90	1.70	1.55	1.39
0.25	60	2.30	2.17	2.05	1.87	1.74	1.60
	80	2.36	2.24	2.12	1.96	1.84	1.71
	100	2.39	2.28	2.17	2.01	1.90	1.78

Note The value in parenthesis after S_{x1} indicates E_p/G_s value of the pile–soil system

Table 10.13 Suggested value of S_{X1} for Poisson's ratio of soil = 0.40

Poisson's ratio	L/r_0 (slenderness ratio)	S_{x1} (250)	S_{x1} (500)	S_{x1} (1000)	S_{x1} (2500)	S_{x1} (5000)	S_{x1} (10,000)
	25	2.27	2.08	1.89	1.63	1.43	1.23
	40	2.48	2.32	2.16	1.94	1.76	1.59
0.40	60	2.60	2.46	2.31	2.12	1.97	1.82
	80	2.66	2.53	2.40	2.22	2.08	1.94
	100	2.70	2.57	2.45	2.28	2.15	2.02

Table 10.14 Beredugo's (1972b) function for S_{x2}

Poisson's ratio	S_{x2}
0.0	$7.334a_0 + \frac{0.8652a_0}{a_0 + 0.00874}$
0.25	$0.83a_0 + \frac{41.59a_0}{a_0 + 3.90}$
0.5	$0.96a_0 + \frac{56.559a_0}{a_0 + 4.68}$

$$k_{\theta\theta} = \frac{E_p I_p}{L}\left[\frac{X(1+\psi)\left(1 - e^{2\beta}\right) + Y\left(\frac{1}{2} + \frac{\psi}{4}\right) - \eta\left(1 - \frac{\psi}{2}\right)}{2(\eta - 1)}\right] \tag{10.7.24}$$

In Eq. (10.7.24), the term $\psi = \left(4G\lambda^2 S_{\theta 1}\right)/\left(\pi E_p \beta^2\right)$ is a dimensionless number and $\lambda = L/r_0$ is the slenderness ratio of the pile. $\beta = \sqrt[4]{GS_{\theta 1}L^4/E_p I_p}$ where $S_{\theta 1}$ is Beredugo's function for an embedded rigid cylinder modified for long piles by Chowdhury and Dasgupta (2009) for v = 0.25 and 0.4; values of these functions are furnished in Tables 10.15 and 10.16.

Parameters like X, Y have same acronym as described earlier for the lateral stiffness of piles.

The inertial contribution is expressed as

Table 10.15 Values of $S_{\theta 1}$ for $\nu = 0.25$

Poisson's ratio	L/r_0 (slenderness ratio)	$S_{\theta 1}$ (250)	$S_{\theta 1}$ (500)	$S_{\theta 1}$ (1000)	$S_{\theta 1}$ (2500)	$S_{\theta 1}$ (5000)	$S_{\theta 1}$ (10,000)
	25	16.968	23.089	30.776	43.412	54.647	66.877
	40	17.358	23.656	31.586	44.678	56.390	69.253
0.25	60	17.567	23.961	32.016	45.333	57.272	70.418
	80	17.674	24.110	32.225	45.648	57.688	70.958
	100	17.736	24.199	32.348	45.833	57.930	71.267

Note The value in parenthesis after $S_{\theta 1}$ depicts the value of E_p/G_s of the pile–soil system

Table 10.16 Values of $S_{\theta 1}$ for $\nu = 0.40$

Poisson's ratio	L/r_0 (slenderness ratio)	$S_{\theta 1}$ (250)	$S_{\theta 1}$ (500)	$S_{\theta 1}$ (1000)	$S_{\theta 1}$ (2500)	$S_{\theta 1}$ (5000)	$S_{\theta 1}$ (10,000)
	25	18.037	24.623	32.937	46.707	59.054	72.614
	40	18.448	25.221	33.794	48.05	60.909	75.145
0.40	60	18.671	25.543	34.249	48.748	61.851	76.393
	80	18.781	25.702	34.471	49.084	62.298	76.974
	100	18.847	25.795	34.603	49.281	62.557	77.307

$$J_{\theta\theta} = \frac{\gamma_p A_p r_0^2 L}{16\beta g}[XF(\lambda) + Y/2 + \eta] \tag{10.7.25}$$

Here,

$$F(\lambda) = 1 - e^{-2\beta} + \frac{\lambda^2}{\beta^2} - 4\lambda^2 e^{-2\beta}\left(2 + \frac{1}{\beta} - \frac{2}{\beta^2}\right) \tag{10.7.26}$$

$$c_{\theta\theta} = r_0^3\sqrt{\rho G}S_{\theta 2}L\left[\frac{X\left(1 - e^{-2\beta}\right) + \frac{Y}{2} + \eta}{4\beta}\right] \tag{10.7.27}$$

and

$$S_{\theta 2} = 0.0144a_0 + 5.262a_0^2 - 4.177a_0^3 + 1.643a_0^4 - 0.2542a_0^5 \tag{10.7.28}$$

Having determined the individual stiffness in lateral and rocking modes, the group stiffness for N piles in the group can be estimated from Eq. (10.7.20). The group damping value should also be computed from similar expression as Eq. (10.7.20).

$$C_{xx}^g = \sum_{i=1}^{N} c_{xx} \Big/ \sum_{i=1}^{N} \alpha_x \quad \text{etc.} \tag{10.7.29}$$

The group interaction factor α can be obtained from the expression as per ACI 351 3R-04.

For a fixed-headed pile (Randolph and Poulos 1982),

$$\alpha_{xf} = 0.6\rho_c\left[E_p/G_c\right]^{1/7}(r/s)\left(1 + \cos^2\beta_p\right) \tag{10.7.30}$$

For a pinned-headed pile,

$$\alpha_{xH} = 0.4\rho_c\left[E_p/G_c\right]^{1/7}(r/s)\left(1 + \cos^2\beta_p\right) \tag{10.7.31}$$

$$\alpha_{\theta H} = \alpha_{xH}^2;\, \alpha_{\theta M} = \alpha_{xH}^3 \tag{10.7.32}$$

This gives for the fixed head pile

$$K_{\theta\theta}^g = \sum_{i=1}^{N} k_{\theta\theta} \Bigg/ \sum_{i=1}^{N} \alpha_{\theta M} \tag{10.7.33}$$

and for the pinned head piles

$$K_{\theta\theta}^g = \sum_{i=1}^{N} k_{\theta\theta} \Bigg/ \sum_{i=1}^{N} \alpha_{\theta H} \tag{10.7.34}$$

where $\rho_c = G_z/G_c$; G_z = shear modulus at depth $L_c/4$; L_c = critical length of pile $\rightarrow 2r\left(E_p/G_c\right)^{2/7}$; G_c = average shear modulus over the critical length L_c (a few iterations may be needed for this to converge); α_{xf} = interaction factor for fixed-headed pile; α_{xH} = interaction factor for pinned-headed pile; $\alpha_{\theta H}$ = rotation due to horizontal force; $\alpha_{\theta M}$ = rotation due to moment; β_p = angle between the direction of loading and the line connecting pile centers (see Fig. 10.13).

When the interaction factor α exceeds 1/3, its value should be replaced by

$$\alpha' = 1 - 2/\sqrt{27\alpha} \tag{10.7.35}$$

The correction is made to avoid $\alpha \rightarrow \infty$, when $s \rightarrow 0$. Same expressions are to be used for damping to determine the group effect. Here, s = center-to-center distance between the reference and other piles in the group. For mass (M_{xx}^g) and mass moment of inertia $(J_{\theta\theta}^g)$ of the pile group, however, the values given in Eqs. (10.7.21) and (10.7.24) are to be multiplied by the factor N—the number of piles in the pile group.

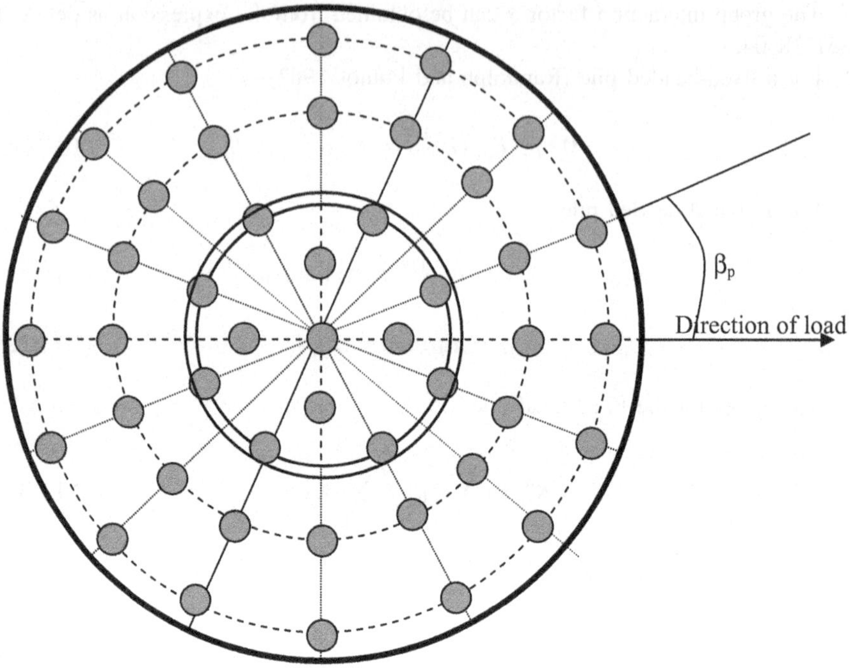

Fig. 10.13 Chimney pile cap with pile layout

10.7.2 Dynamic Stiffness of Partially Embedded Piles

For scenario analysis, in case the piles become partially exposed due to liquefaction of soil at top, so that the piles become partially embedded, as shown in Fig. 10.14, for a length $L_e < L$, Chowdhury and Dasgupta (2008a, b, 2009) have established spring and damping and inertial values of such piles as

$$\beta_e = \sqrt[4]{GS_{x1}L_e^4/E_pI_p}, \eta_e = 1 + 1/\beta_e, X_e = 1 + \eta_e^2, Y_e = 1 - \eta_e^2 \text{and}\, \alpha = L_e/L < 1.0$$

In the above expressions, the subscript "e" stands for embedment

$$k_{xx} = \frac{E_pI_p}{L_e^3}\left[\frac{\begin{Bmatrix} X_e\left(\alpha + \frac{1}{4}\right) + Y_e\left(\frac{1}{8} - \frac{\alpha}{2}\right) + \eta_e\left(\frac{1}{4} - \alpha\right) \\ -X_e e^{-2\beta_e}\left((\alpha/4)e^{-2\beta_e(1-\alpha)} + 1\right) \end{Bmatrix}}{(\eta_e - 1)^3}\right] \quad (10.7.36)$$

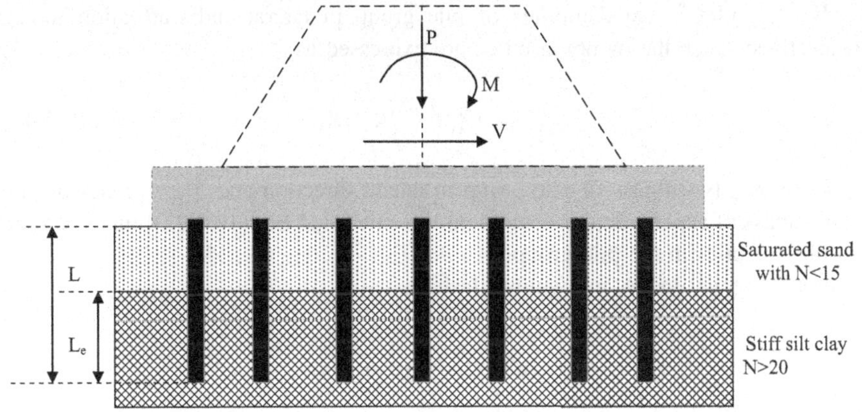

Fig. 10.14 Axial load, moment, and shear due to earthquake loading of a chimney foundation

$$M_{xx} = \frac{\gamma_p A_p L_e}{4g}\left[\frac{X_e\alpha\left(1-e^{-2\beta_e}\right)+\frac{Y_e\alpha}{2}+\eta_e\alpha}{1/(\eta_e-1)}\right] \tag{10.7.37}$$

$$c_{xx} = r_0\sqrt{\rho G}S_{x2}L_e\left[\frac{X_e\left(1-e^{-2\beta_e}\right)+\frac{Y_e}{2}+\eta_e}{4/(\eta_e-1)}\right] \tag{10.7.38}$$

For rocking mode for the partially embedded piles

$$k_{\theta\theta} = \frac{E_p I_p}{L_e}\left[\frac{X_e\left(\alpha+\psi-\alpha e^{-2\beta_e\alpha}-\psi e^{-2\beta_e}\right)+Y_e\left(\frac{\alpha}{2}+\frac{\psi}{4}\right)-\eta_e\left(\alpha-\frac{\psi}{2}\right)}{2(\eta_e-1)}\right] \tag{10.7.39}$$

$$c_{\theta\theta} = r_0^3\sqrt{\rho G}S_{\theta 2}L_e\left[\frac{X_e\left(1-e^{-2\beta_e}\right)+\frac{Y_e}{2}+\eta_e}{4/(\eta_e-1)}\right] \tag{10.7.40}$$

The mass moment of inertia remains same as in Eq. (10.7.24); however, the relevant terms are treated with subscript "e" as appropriate, except λ, which is for the full length of the pile.

Having established the mathematical framework for stiffness and damping of pile, the only question that remains is: what will be the value of applied frequency ω on the system to determine a_0?

This remains an important step in estimation of damping which is frequency dependent.

It should be realized that the moment and shear coming on the foundation, as shown in Fig. 10.14, is dependent on natural frequency of the fixed based structure. Thus, for a conservative estimate of damping, only the first mode frequency (to be computed from time period value furnished in Table 10.4 as $\omega = 2\pi/T$) should be sufficient to estimate the damping of pile.

If C^g_{xx} and $C^g_{\theta\theta}$ are dampings of pile group in lateral and translation mode, respectively, then the damping ratios are expressed as

$$\zeta^g_{xx} = C^g_{xx} \Big/ \sqrt{2K^g_{xx}M^g_{xx}} \qquad (10.7.41)$$

Here, K^g_{xx} is stiffness of pile group in lateral direction and M^g_{xx} = mass of pile cap + mass of superstructure + mass on one pile [vide Eq. (10.7.22)] times N—the number of piles in the pile group.

For rotational mode, in a similar fashion

$$\zeta^g_{\theta\theta x} = C^g_{\theta\theta} \Big/ \sqrt{2K^g_{\theta\theta}J^g_{\theta\theta}}$$

Having obtained the damping ratios of pile group, one can now substitute them in Eqs. (10.7.5) to (10.7.7) and proceed further.

A word of caution in this context in terms of pile damping would perhaps be appropriate at this juncture. Irrespective of the methods used to determine the damping ratio (Novak's, Gazetas', or as above), all the above methods assume that waves are dissipating from the pile in an infinite elastic half space, unhindered.

None of them takes into consideration reflection and diffraction of waves from other piles in a pile group that could significantly attenuate the overall damping factor.

This phenomenon is indeed difficult to model. Thus, to be on safe side, under no circumstance does the author recommends to take a damping ratio of more than 15% in real time design for translational mode and 3–5% in rocking mode—whatever the theoretical values one may arrive at, especially when pile center-to-center distance in a pile group $s \leq 2.5$–3.0 times the diameter of piles, unless there is a strong theoretical reason by which such a decision can be justified.

Example 10.2 A complete analysis for a real 278-m tall multi-flue chimney is shown hereafter by the following methods.

- As per IS 1893 (2005);
- Dynamic analysis as proposed in Sect. 10.4;
- Numerical analysis considering its distributed mass of shell and lumped mass coming from platforms and flue suspension.
- Dynamic analysis considering DSSI as per Sect. 10.7

The chimney is constructed in the area identified as Zone III as per IS-1893 (2005).

Design Calculation for 278-m tall chimney

Input Data:

Geometric Data

Height of chimney = 270.3 m from F.F.L

Height of chimney = 273.825 m from E.G.L (Fig. 10.15)

Fig. 10.15 Construction of the analyzed chimney

Height of the chimney = 278.825 m from the top of raft
Inside diameter at the bottom = 30 m
Inside diameter at the top = 17 m
Shell thickness at the bottom = 800 mm
Shell thickness at the top = 500 mm (varying linearly from (+)79.8 m to (+) 209.5 m and then constant to the top).

Material Data

Concrete grade:
M30 for foundation
M40 for superstructure up to EL (+)50 m
M30 for superstructure from El (+)50 m up to the top
Steel grade: Fe-500 for foundation.

E_{conc} = 3.35 X 10^7 kN/m^2 for superstructure for M30 grade concrete as per IS-4998.

E_{conc} – 3.60 X 10^7 kN/m^2 for superstructure for M40 grade concrete as per IS-4998.

Weighted average design value is thus E_{conc} = 3.4 X 10^7
The density of concrete = 25 kN/m^3
Clear cover to the foundation = 75 mm

Soil–Foundation Data

Bearing capacity of soil = 550 kN/m^2.
Dynamic shear modulus = 590,000 kN/m^2
Diameter of raft foundation = 45.0 m
Self-weight of the raft = 145,350 kN.
Density of soil = 20 kN/m^3.
Poisson's ratio = 0.32

Platform Load Data:

The dead and live loads considered for the platforms at various levels are as shown hereafter:

Inside Platforms

Sl No	Level (m)	Dead load (kN)	Live load (kN)	Remarks
1	25	1600	800	
2	35	600	350	
3	57.5	1050	800	
4	97.5	650	800	
5	137.5	650	650	
6	177.5	650	650	
7	217.5	650	650	
8	261.5	7850	350	The flue is supported from this floor
9	268.8	1950	300	
Total		15,650	5350	

Calculation for the weight of chimney:

Inside diameter at bottom = 30 m

Inside diameter at top = 17 m

Shell thickness at the bottom = 800 mm (constant up to 88.325 m from the top of raft)

Shell thickness at the top = 500 mm (for variation of shell thickness refer GA dwg of chimney)

Thus, the outside diameter at the bottom = 30 + 0.80 × 2 = 31.6 m

The outside diameter at the top = 18 m

Volume of concrete from −8.525 to +79.8 m.

D_{base} = 31.6 m $D_{79.8}$ = 26.61 m for the outside shell

D_{base} = 30 m $D_{79.8}$ = 25.011 m for the inside shell

Volume of a truncated cone is given by

$$V = \frac{\pi H}{3}\left[R^2 + rR + r^2\right]$$

Thus, the outside volume $V = \frac{\pi \times 88.325}{3}\left[15.8^2 + 13.305 \times 15.8 + 13.305^2\right] = 58907.6\ \mathrm{m}^3$

The inside volume $V = \frac{\pi \times 88.325}{3}\left[15^2 + 12.50 \times 15 + 12.50^2\right] = 52624.3\ \mathrm{m}^3$.

The net volume of concrete shell = 58907.6−52624.3 = 6283.3 m^3

The volume of concrete from EL + 79.8 m to +209.5 m.

$D_{79.8}$ = 26.16 m $D_{209.5}$ = 18 m for outside shell

D_{30} = 25.011 m $D_{209.5}$ = 17 m for inside shell

The volume of a truncated cone is given by

$$V = \frac{\pi H}{3}\left[R^2 + rR + r^2\right]$$

Thus, the outside volume $V = \frac{\pi \times 129.7}{3}[13.305^2 + 9 \times 13.305 + 9^2] = 51,309\ \text{m}^3$

The inside volume $V = \frac{\pi \times 129.7}{3}[12.5055^2 + 8.5 \times 12.5055 + 8.5^2] = 45489.008\ \text{m}^3$.

The net volume of concrete shell = 51,309−45,489 = 5820 m^3

The volume of concrete from +209.5 m to 270.3 m.

$$V_{\text{net}} = \pi(R^2 - r^2)H = \pi \times (9^2 - 8.5^2) \times 60.8 = 1671\ \text{m}^3$$

The weight of concrete shell = (6283.3 + 5820 + 1671) × 25 = 344,358 kN

The volume of sand fill from EL −8.525 m to EL +0.00 m.

$D_{-8.525}$ = 30 m and $D_{0.00}$ = 25.49 m (these are internal diameters)

Thus, $V = \frac{\pi H}{3}[R^2 + rR + r^2]$

$$V = \frac{\pi \times 8.525}{3}\left[15^2 + 12.746 \times 15 + 12.746^2\right] = 5165\ \text{m}^3$$

Thus, considering the density of compacted sand as = 22 kN/m^3

The weight of sand fill = 5165 × 22 = 113,648 kN

Thus, for seismic load case we will consider full dead load of shell + dead load of platforms +50% of live load on platforms and the sand fill

or $W_{\text{SI+DL}}$ = 344,358 + 15,650 + 0.5 X 5350 + 113,648 = 476,331 kN

(1) **Calculation as per IS-1893** (2005)

Moment of inertia at the base $I_0 = \pi \frac{31.6^4 - 30^4}{64} = 9185.33\ \text{m}^4$

Area at the base $A_0 = \pi \frac{31.6^2 - 30^2}{4} = 77.408\ \text{m}^2$

Radius of gyration = $\sqrt{I_0/A_0} = 10.89$ m

Slenderness Ratio (λ) = $278.825/10.89 = 25.6$ (Here, the chimney is assumed as a cantilever beam whose fixity is at the top of the raft).

The time period of the chimney for seismic load $T = C_T\sqrt{WH/EAg}$

Here, C_T = 48.25 for λ = 25.6, which gives $T = 48.25 \times \sqrt{\frac{476{,}331 \times 278.825}{3.4 \times 10^7 \times 77.408 \times 9.81}} = 3.46$ s

$Z = 0.16$, $I = 1.75$, $R = 3.0 \Rightarrow \beta = ZI/2R = 0.046$, S_{ad}/g = (1/3.46) × 1.27 = 0.367 considering stiff soil with 3% damping for concrete. For $H/r(l)$ = 25.6 C_v = 1.32, thus

$$V_z = C_v \beta (S_{ad}/g) W D_v$$

$$\rightarrow V_z = C_v \beta (S_{ad}/g) W\left[1.1\sqrt{z/H} + 0.75(z/H) + 0.9(z/H)^4\right]$$

$$V_z = 1.32 \times 0.046 \times 0.37 \times 476,331 \times \left[1.1\sqrt{z/H} + 0.75(z/H) + 0.9(z/H)^4\right]$$

$$M_z = \beta(S_{ad}/g)W\bar{h}D_m$$

$$\rightarrow M_z = \beta(S_{ad}/g)W\bar{h}\left[0.4\sqrt{z/H} + 0.6(z/H)^4\right]$$

$$M_z = 0.046 \times 0.37 \times 476,331 \times 104 \times \left[0.4\sqrt{z/H} + 0.6(z/H)^4\right]$$

Here, the center of gravity $\bar{h} = 104\ m$; the values of moment and shear are as tabulated hereafter (Figs. 10.16 and 10.17).

z/H	Moment (kN m)	Shear (kN)
0	851,251	5857
0.05	747,888	5857
0.1	658,131	5857
0.15	580,541	5857
0.2	513,756	5857
0.25	456,487	5857
0.3	407,514	5857
0.35	365,692	5857
0.4	329,944	5857
0.45	299,259	5857
0.5	272,692	5857
0.55	249,359	5857
0.6	228,427	5857
0.65	209,107	5428
0.7	190,637	4890
0.75	172,245	4340
0.8	153,094	3768
0.85	132,134	3157
0.9	107,727	2477
0.95	76,141	1660
1	0	0

(2) **Analysis based on modal dynamic analysis**

$T = C_T\sqrt{\frac{WH}{EAg}}$; here C_T = 1.787λ, 0.285λ, and 0.102λ, where λ = 25.6.

The time periods for the first three modes are as shown hereafter

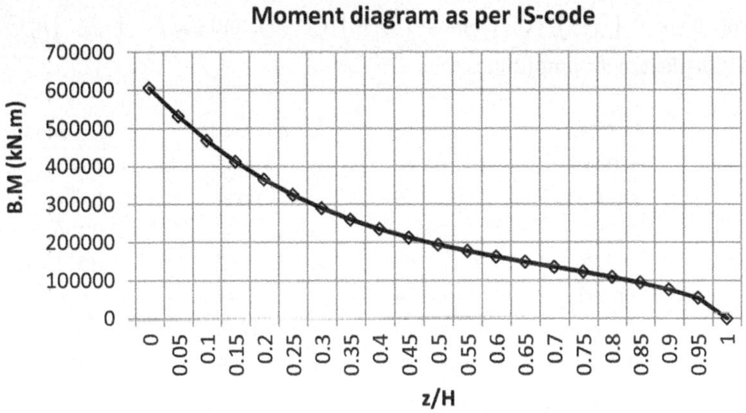

Fig. 10.16 Moment diagram as per IS code

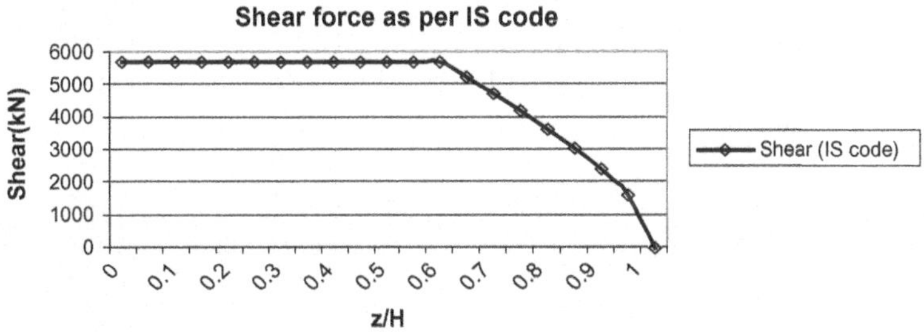

Fig. 10.17 Shear diagram as per IS code

Mode	1	2	3
Time period	3.27	0.53	0.19

Thus, the design spectrum from IS-1893, considering 3% damping for first three modes, is

Mode	1	2	3
S_{ad}/g	0.388	2.39	3.175

Considering Eqs. 10.4.35–10.4.36, we have

$$w_i(z,t) = C_{FDi}\beta \frac{S_{ai}WH^3}{EIg}, M_z = C_{FMi}\beta WH\left(\frac{I_z}{I}\right)\left(\frac{S_{ai}}{g}\right) \text{and } V_z = C_{FVi}\beta W\left(\frac{I_z}{I}\right)\left(\frac{S_{ai}}{g}\right).$$

Here, $C_{FDi} = \left(\frac{\kappa_i C_{Ti}^2}{4\pi^2}\right) f(\xi)$, $C_{FMi} = \left(\frac{\kappa_i C_{Ti}2}{4\pi^2}\right) \mu_i^2 f''(\xi)$, and $C_{FVi} = \left(\frac{\kappa_i C_{Ti}2}{4\pi^2}\right) \mu_i^3 f'''(\xi)$

The displacement amplitudes

z/H	U_1 (mm)	U_2 (mm)	U_3 (mm)	U_{srss} (mm)
0.000	0.0000	0.0000	0.0000	0.000
0.015	0.0302	0.0158	0.0054	0.034
0.031	0.1199	0.0614	0.0208	0.136
0.063	0.5076	0.2496	0.0812	0.571
0.085	0.8941	0.4274	0.1349	1.000
0.088	0.9699	0.4613	0.1448	1.084
0.120	1.7773	0.8076	0.2406	1.967
0.156	2.9436	1.2659	0.3521	3.223
0.192	4.3727	1.7711	0.4541	4.740
0.237	6.5040	2.4258	0.5497	6.963
0.273	8.4649	2.9356	0.5893	8.979
0.309	10.6358	3.4064	0.5899	11.184
0.344	13.0007	3.8161	0.5498	13.560
0.380	15.5440	4.1452	0.4709	16.094
0.416	18.2500	4.3774	0.3585	18.771
0.452	21.1036	4.4997	0.2208	21.579
0.488	24.0900	4.5026	0.0683	24.507
0.524	27.1949	4.3801	−0.0871	27.546
0.560	30.4044	4.1300	−0.2331	30.685
0.595	33.7052	3.7534	−0.3580	33.915
0.631	37.0845	3.2547	−0.4516	37.230
0.667	40.5305	2.6415	−0.5056	40.620
0.703	44.0317	1.9239	−0.5149	44.077
0.739	47.5779	1.1140	−0.4773	47.593
0.775	51.1595	0.2255	−0.3939	51.161
0.811	54.7679	−0.7270	−0.2686	54.773
0.846	58.3958	−1.7286	−0.1080	58.421
0.882	62.0366	−2.7652	0.0798	62.098
0.918	65.6852	−3.8241	0.2861	65.797
0.968	70.7992	−5.3242	0.5907	71.002
0.995	73.4667	−6.1090	0.7518	73.724
1.000	74.0148	−6.2703	0.7849	74.284

The displacement plot is as shown below (Fig. 10.18)

The bending moment values are as shown hereafter

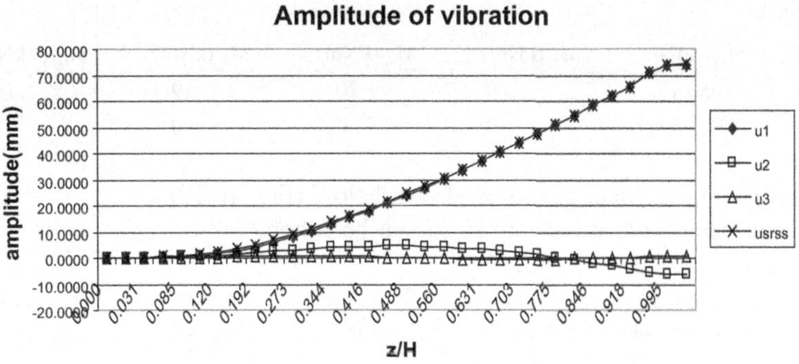

Fig. 10.18 Amplitude of vibration for the first three modes

z/H	I_{xx} (m^4)	M_1 (kNm)	M_2 (kNm)	M_3 (kNm)	M_{srss} (kNm)
0.000	9185.333	1,045,357	555,039	194,506	1,199,446
0.015	8844.339	985,367	495,371	164,812	1,115,125
0.031	8511.97	927,956	439,177	137,016	1,035,737
0.063	7826.766	813,029	329,763	83,657	881,339
0.085	7405.055	744,692	267,018	53,847	792,946
0.088	7335.116	733,541	256,970	49,155	778,802
0.120	6728.67	639,118	174,398	11,852	662,591
0.156	6091.711	544,523	97,160	−19,867	553,480
0.192	5496.286	460,652	34,776	−41,396	463,814
0.237	4808.278	369,688	−24,058	−55,229	374,564
0.273	4301.182	307,083	−57,362	−57,402	317,624
0.309	3831.068	252,709	−79,948	−53,444	270,388
0.344	3346.946	202,843	−91,945	−44,509	227,113
0.380	2895.011	160,205	−95,635	−33,224	189,513
0.416	2489.194	124,906	−93,263	−21,535	157,363
0.452	2126.38	96,000	−86,559	−10,767	129,709
0.488	1803.559	72,615	−77,050	−1814	105,891
0.524	1577.456	56,069	−68,632	5045	88,767
0.560	1543.643	47,877	−66,534	11141	82,723
0.595	1509.982	40,314	−62,771	16218	76,344
0.631	1476.471	33,391	−57,608	19937	69,506
0.667	1443.111	27,117	−51,342	22087	62,122
0.703	1409.9	21,498	−44,297	22605	54,179
0.739	1376.84	16,537	−36,812	21570	45,758
0.775	1343.928	12,236	−29,233	19199	37,052
0.811	1053.17	6903	−17,595	12711	22,777
0.846	1053.17	4618	−12,493	9777	16,522
0.882	1053.17	2761	−7888	6605	10,652
0.918	1053.17	1358	−4079	3616	5618
0.968	1053.17	208	−664	628	937

(continued)

(continued)

z/H	I_{xx} (m^4)	M_1 (kNm)	M_2 (kNm)	M_3 (kNm)	M_{srss} (kNm)
0.995	1053.17	6	−20	19	29
1.000	1053.17	0	0	0	0

The bending moment plot is as shown below (Fig. 10.19)
The shearing force values are as shown hereafter

z/H	$I_{x\,x}$ (m^4)	V_1 (kN)	V_2 (kN)	V_3 (kN)	V_{srss} (kN)
0.000	9185.333	5160	9517	5475	12,132
0.015	8844.339	4969	9163	5271	11,680
0.031	8511.97	4782	8815	5063	11,234
0.063	7826.766	4396	8077	4581	10,273
0.085	7405.055	4157	7602	4234	9643
0.088	7335.116	4117	7521	4172	9535
0.120	6728.67	3771	6796	3580	8557
0.156	6091.711	3404	5981	2854	7450
0.192	5496.286	3057	5169	2099	6361
0.237	4808.278	2652	4177	1185	5088
0.273	4301.182	2350	3422	533	4185
0.309	3831.068	2068	2715	−12	3413
0.344	3346.946	1779	2042	−421	2741
0.380	2895.011	1510	1451	−683	2202
0.416	2489.194	1269	957	−809	1784
0.452	2126.38	1055	559	−824	1451
0.488	1803.559	867	252	−755	1177
0.524	1577.456	730	29	−657	983
0.560	1543.643	683	−152	−590	915
0.595	1509.982	635	−311	−481	855
0.631	1476.471	584	−446	−341	810
0.667	1443.111	532	−553	−184	789
0.703	1409.9	479	−628	−25	790
0.739	1376.84	423	−671	123	803
0.775	1343.928	367	−680	247	811
0.811	1053.17	249	−525	269	641
0.846	1053.17	208	−490	312	617
0.882	1053.17	163	−426	316	555
0.918	1053.17	117	−331	275	446
0.968	1053.17	47	−147	136	206
0.995	1053.17	8	−27	26	38
1.000	1053.17	0	0	0	0

The shear force plot is as shown below (Fig. 10.20)

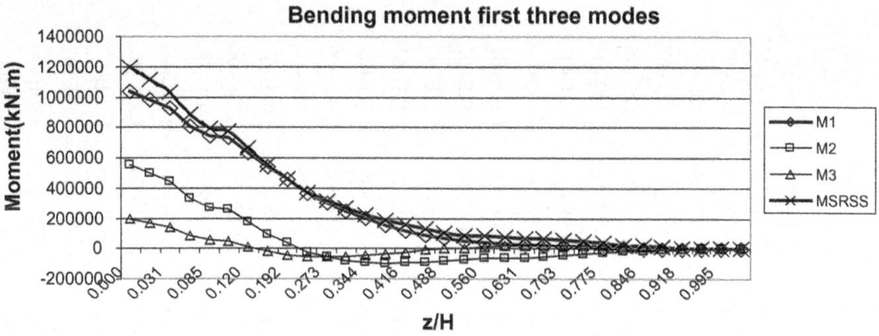

Fig. 10.19 Dynamic moment for the first three modes

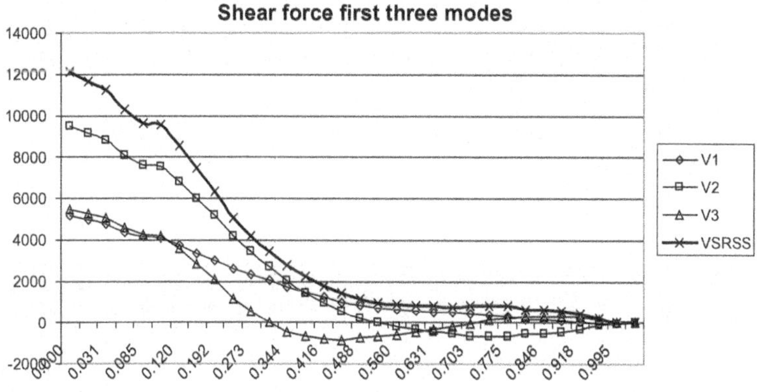

Fig. 10.20 Dynamic shear force for the first three modes

(3) **Numerical analysis of chimneys**

In this case, we take exact variation of the taper, including mass at different platform levels considered as lumped mass, to see how much difference, if any, occurs as compared to the dynamic analytical solution.

As the first step, we determine the stiffness matrix of the chimney as hereafter[4]

[4]Note: For brevity of presentation, the analysis is presented for two places after the decimal point; however, the actual analysis is carried out to the precision up to 15 places after the decimal point.

Node	Z (coord)	I_{xx}	z/H	$f'_1\ (\xi)$	$f'_2\ (\xi)$	$f'_3\ (\xi)$	$I_{xx}f'^2_1$	$I_{xx}f'_1f'_2$	$I_{xx}f'_1f'_3$	$I_{xx}f'^2_2$	$I_{xx}f'_2f'_3$	$I_{xx}f'^2_3$
1	0.00	9185.3	0.00	2.72	1.96	2.00	8.43E+05	3.81E+06	1.09E+07	1.72E+07	4.91E+07	1.40E+08
2	4.26	8844.3	0.02	2.67	1.82	1.76	7.78E+05	3.33E+06	9.01E+06	1.42E+07	3.85E+07	1.04E+08
3	8.53	8512.0	0.03	2.61	1.68	1.52	7.17E+05	2.89E+06	7.33E+06	1.16E+07	2.95E+07	7.50E+07
4	17.68	7826.8	0.06	2.49	1.37	1.01	5.98E+05	2.06E+06	4.27E+06	7.12E+06	1.47E+07	3.04E+07
5	23.58	7405.1	0.08	2.41	1.17	0.69	5.30E+05	1.62E+06	2.66E+06	4.94E+06	8.11E+06	1.33E+07
6	24.58	7335.1	0.09	2.39	1.14	0.63	5.20E+05	1.55E+06	2.41E+06	4.62E+06	7.19E+06	1.12E+07
7	33.53	6728.7	0.12	2.27	0.84	0.17	4.30E+05	9.98E+05	5.53E+05	2.32E+06	1.28E+06	7.10E+05
8	43.53	6091.7	0.16	2.14	0.52	−0.31	3.45E+05	5.23E+05	−8.72E+05	7.95E+05	−1.32E+06	2.20E+06
9	53.53	5496.3	0.19	2.01	0.21	−0.71	2.73E+05	1.76E+05	−1.70E+06	1.13E+05	−1.09E+06	1.06E+07
10	66.03	4808.3	0.24	1.84	−0.16	−1.09	2.01E+05	−1.11E+05	−2.08E+06	6.17E+04	1.15E+06	2.16E+07
11	76.03	4301.2	0.27	1.71	−0.43	−1.26	1.55E+05	−2.47E+05	−2.01E+06	3.92E+05	3.20E+06	2.61E+07
12	86.03	3831.1	0.31	1.58	−0.68	−1.32	1.18E+05	−3.18E+05	−1.73E+06	8.55E+05	4.66E+06	2.54E+07
13	96.03	3346.9	0.34	1.45	−0.89	−1.26	8.71E+04	−3.36E+05	−1.32E+06	1.30E+06	5.11E+06	2.01E+07
14	106.03	2895.0	0.38	1.32	−1.07	−1.08	6.28E+04	−3.19E+05	−9.02E+05	1.62E+06	4.58E+06	1.30E+07
15	116.03	2489.2	0.42	1.20	−1.22	−0.82	4.44E+04	−2.82E+05	−5.30E+05	1.79E+06	3.37E+06	6.34E+06
16	126.03	2126.4	0.45	1.08	−1.32	−0.48	3.07E+04	−2.36E+05	−2.39E+05	1.81E+06	1.83E+06	1.85E+06
17	136.03	1803.6	0.49	0.96	−1.39	−0.10	2.07E+04	−1.87E+05	−3.58E+04	1.69E+06	3.24E+05	6.20E+04
18	146.03	1577.5	0.52	0.85	−1.41	0.30	1.41E+04	−1.47E+05	8.80E+04	1.53E+06	−9.17E+05	5.49E+05
19	156.03	1543.6	0.56	0.74	−1.40	0.68	1.05E+04	−1.24E+05	1.70E+05	1.47E+06	−2.01E+06	2.73E+06
20	166.03	1510.0	0.60	0.64	−1.35	1.02	7.62E+03	−1.01E+05	2.13E+05	1.34E+06	−2.82E+06	5.92E+06
21	176.03	1476.5	0.63	0.54	−1.27	1.28	5.35E+03	−7.85E+04	2.21E+05	1.15E+06	−3.25E+06	9.16E+06
22	186.03	1443.1	0.67	0.45	−1.16	1.45	3.61E+03	−5.81E+04	2.04E+05	9.37E+05	−3.28E+06	1.15E+07
23	196.03	1409.9	0.70	0.37	−1.02	1.52	2.32E+03	−4.07E+04	1.69E+05	7.14E+05	−2.97E+06	1.23E+07
24	206.03	1376.8	0.74	0.29	−0.87	1.48	1.41E+03	−2.66E+04	1.27E+05	5.05E+05	−2.41E+06	1.15E+07
25	216.03	1343.9	0.77	0.22	−0.71	1.35	7.89E+02	−1.60E+04	8.58E+04	3.26E+05	−1.74E+06	9.33E+06

(continued)

(continued)

Node	Z (coord)	I_{xx}	z/H	f'_1 (ξ)	f'_2 (ξ)	f'_3 (ξ)	$I_{xx}f_1^2$	$I_{xx}f'_1f'_2$	$I_{xx}f'_1f'_3$	$I_{xx}f_2^2$	$I_{xx}f'_2f'_3$	$I_{xx}f_{3'}^2$
26	226.03	1053.2	0.81	0.16	−0.54	1.14	3.20E+02	−6.95E+03	4.09E+04	1.51E+05	−8.87E+05	5.22E+06
27	236.03	1053.2	0.85	0.10	−0.39	0.88	1.43E+02	−3.30E+03	2.10E+04	7.60E+04	−4.84E+05	3.09E+06
28	246.03	1053.2	0.88	0.06	−0.24	0.59	5.13E+01	−1.25E+03	8.50E+03	3.03E+04	−2.07E+05	1.41E+06
29	256.03	1053.2	0.92	0.03	−0.13	0.32	1.24E+01	−3.17E+02	2.29E+03	8.10E+03	−5.85E+04	4.22E+05
30	270.03	1053.2	0.97	0.00	−0.02	0.06	2.90E−01	−7.89E+00	6.07E+01	2.15E+02	−1.65E+03	1.27E+04
31	277.33	1053.2	0.99	0.00	0.00	0.00	2.59E−04	−7.20E−03	5.62E−02	2.00E−01	−1.56E+00	1.22E+01
32	278.83	1053.2	1.00	0.00	0.00	0.00	0.00E+00	0.00E+00	0.00E+00	0.00E+00	0.00E+00	0.00E+00
					Sum	→	5799878.5	14306849	27016215	80688917	149235642	575527164

By Eq. 10.5.3, we have

$$[K] = \frac{3.4 \times 10^7}{(278.83)^3}\begin{bmatrix} 5799879 & 14306849 & 27016215 \\ 14306849 & 80688917 & 149235642 \\ 27016215 & 149235642 & 575527164 \end{bmatrix}$$

$$\rightarrow [K] = \begin{bmatrix} 9097094 & 22440254 & 42374862 \\ 22440254 & 126560351 & 234075708 \\ 42374862 & 234075708 & 902712822 \end{bmatrix}$$

Next, we develop the mass matrix as computed hereafter

Node	Z (coord)	A_x	M	z/H	f_1 (ξ)	f_2 (ξ)	f_3 (ξ)	$mf_1^2 + Mf_1^2$	$mf_1f_2 + M f_1f_2$	$mf_1f_3 + M f_1f_3$	$mf_2^2 + Mf_2^2$	$mf_2f_3 + M f_2f_3$	$mf_3^2 + Mf_3^2$
1	0.00	77	0	0.00	0.00	0.00	0.00	0.00	0.00	0.00	0.00	0.00	0.00
2	4.26	76	5792	0.02	0.00	0.00	0.01	0.07	0.15	0.42	0.66	1.85	5.20
3	8.53	75	5792	0.03	0.00	0.02	0.05	1.16	2.28	6.29	9.93	27.42	75.68
4	17.68	73	0	0.06	0.02	0.08	0.2[illegible]	18.20	29.88	79.15	124.98	331.06	876.93
5	23.58	72	0	0.08	0.03	0.13	0.34	55.44	88.47	227.40	359.78	924.78	2377.07
6	24.58	72	0	0.09	0.04	0.14	0.37	65.03	103.27	264.01	417.91	1068.43	2731.55
7	33.53	70	204	0.12	0.07	0.25	0.6[illegible]	213.04	325.26	789.09	1257.55	3050.85	7401.47
8	43.53	68	79	0.16	0.11	0.40	0.90	563.93	811.72	1838.70	2969.97	6727.58	15239.31
9	53.53	65	0	0.19	0.16	0.55	1.16	1200.53	1623.38	3390.41	5594.22	11683.46	24400.79
10	66.03	62	148	0.24	0.24	0.76	1.40	2548.55	3189.74	5887.56	10121.86	18682.70	34484.13
11	76.03	60	0	0.27	0.31	0.92	1.50	4145.60	4799.81	7847.66	14162.21	23155.14	37858.54
12	86.03	58	0	0.31	0.39	1.07	1.50	6296.75	6732.89	9496.81	18346.70	25878.18	36501.40
13	96.03	55	0	0.34	0.48	1.20	1.40	8914.79	8736.07	10251.83	21816.84	25602.21	30044.36
14	106.03	52	107	0.38	0.57	1.30	1.20	12035.50	10763.55	9959.06	24421.05	22595.77	20906.92
15	116.03	48	0	0.42	0.67	1.37	0.91	15543.28	12446.67	8301.64	25400.07	16941.26	11299.43
16	126.03	45	0	0.45	0.78	1.41	0.56	19482.15	13868.27	5542.01	25158.14	10053.65	4017.62
17	136.03	42	0	0.49	0.89	1.41	0.17	23734.19	14810.08	1829.97	23551.11	2910.04	359.57
18	146.03	40	99	0.52	1.00	1.37	−0.22	28662.30	15495.23	−2509.08	21233.71	−3438.29	556.75
19	156.03	39	0	0.56	1.12	1.29	−0.59	34993.51	15869.32	−7295.38	18340.04	−8431.21	3875.96
20	166.03	39	0	0.60	1.24	1.18	−0.91	42134.03	15664.46	−12170.17	14841.20	−11530.55	8958.42
21	176.03	38	0	0.63	1.36	1.02	−1.15	49955.04	14637.23	−16539.92	10929.72	−12350.48	13955.92
22	186.03	37	99	0.67	1.49	0.83	−1.29	58637.26	12833.26	−20006.12	7116.10	−11093.48	17293.93
23	196.03	36	0	0.70	1.62	0.60	−1.31	67466.86	9841.58	−21452.35	3658.56	−7974.81	17383.22
24	206.03	35	0	0.74	1.75	0.35	−1.22	77048.78	6022.73	−21017.59	1199.75	−4186.79	14610.70
25	216.03	35	0	0.77	1.88	0.07	−1.00	87096.64	1281.60	−18233.32	48.06	−683.74	9727.49

(continued)

(continued)

Node	Z (coord)	A_x	M	z/H	f_1 (ξ)	f_2 (ξ)	f_3 (ξ)	$mf_1^2 + Mf_1^2$	$mf_1f_2 + M f_1f_2$	$mf_1f_3 + M f_1f_3$	$mf_2^2 + Mf_2^2$	$mf_2f_3 + M f_2f_3$	$mf_3^2 + Mf_3^2$
26	226.03	27	99	0.81	2.02	−0.23	−0.69	79775.46	−3562.89	−10723.50	402.36	1211.02	3644.88
27	236.03	27	0	0.85	2.15	−0.54	−0.28	90235.00	−8917.53	−4539.83	2245.87	1143.35	582.07
28	246.03	27	0	0.88	2.28	−0.87	0.20	101837.58	−15154.79	3561.73	5747.28	−1350.74	317.46
29	256.03	27	0	0.92	2.42	−1.20	0.73	114168.86	−22190.42	13522.70	10991.43	−6698.11	4081.78
30	270.03	27	818	0.97	2.61	−1.67	1.51	138193.53	−36855.19	33301.09	23580.71	−21306.72	19252.03
31	277.33	27	0	0.99	2.70	−1.91	1.92	142821.30	−39648.95	39741.68	28050.55	−28116.15	28181.91
32	278.83	27	0	1.00	2.72	−1.96	2.00	144960.31	−40999.30	41802.86	29551.18	−30130.36	30720.90
						Sum	→	1352804.7	2647.81	63154.82	351649.5	24697.32	401723.39

By Eq. 10.5.4, the mass matrix can be expressed as

$$[M] = \begin{bmatrix} 1352805 & 2648 & 63155 \\ 2648 & 351650 & 24697 \\ 63155 & 24697 & 401723 \end{bmatrix}$$

The above matrices, $[K]$ and $[M]$, are solved in MATHCAD for eigensolution vis-à-vis time period for the first three modes, vide the expression $[\mathbf{K}] - [\mathbf{M}]\omega^2 = 0$, when we have considering $T = 2\pi/\omega$

$$T = \{\, 3.242 \quad 0.466 \quad 0.129 \,\}\text{s}.$$

The corresponding scaling vectors are given as

$$\phi = \begin{bmatrix} -1 & 0.056 & 0.032 \\ 0.178 & 1 & -0.248 \\ 5 \times 10^{-4} & -0.279 & -1 \end{bmatrix}$$

The modal mass participation is calculated next

Z (coord)	Area	M	z/H	f_1 (ξ)	f_2 (ξ)	f_3 (ξ)	$mi\ fi + M\ fi$	$mi\ fi^2 + M\ fi^2$	$mi\ fi + M\ fi$	$mi\ fi^2 + M\ fi^2$	$mi\ fi + M\ fi$	$mi\ fi^2 + M\ fi^2$
	Az						Mode-1	Mode-1	Mode-2	Mode-2	Mode-3	Mode-3
0	77.41		0.00	0.000	0.000	0.000	0.000	0.000	0.000	0.000	0.000	0.000
4.263	76.44	5792	0.02	0.001	0.005	0.014	66.816	0.074	296.563	1.463	832.896	11.541
8.525	75.47	5792	0.03	0.004	0.019	0.053	262.27	1.158	1143.103	21.992	3155.466	167.577
17.675	73.39		0.06	0.019	0.078	0.207	974.31	18.205	4075.372	318.510	10795.005	2234.777
23.575	72.04		0.08	0.033	0.134	0.344	1684.6	55.438	6850.918	916.871	17609.653	6057.763
24.575	71.81		0.09	0.036	0.144	0.369	1821.6	65.029	7372.009	1065.018	18847.221	6961.131
33.525	69.78	203.8	0.12	0.065	0.253	0.614	3256.7	213.041	12591.390	3184.562	30547.157	18743.208
43.525	67.50	79.00	0.16	0.108	0.396	0.898	5205.0	563.927	19044.616	7549.515	43139.857	38737.535
53.525	65.22		0.19	0.161	0.555	1.158	7459.1	1200.526	25704.603	14256.412	53683.792	62183.449
66.025	62.38	147.8	0.24	0.239	0.760	1.402	10646.0	2548.550	33782.614	25662.671	62355.220	87430.103
76.025	60.10		0.27	0.312	0.919	1.503	13305.77	4145.598	39259.761	36091.257	64189.510	96479.451
86.025	57.83		0.31	0.391	1.067	1.505	16084.97	6296.752	43830.521	46755.096	61823.335	93020.903
96.025	54.79		0.34	0.479	1.195	1.402	18630.16	8914.785	46525.678	55598.477	54598.188	76565.644
106.02	51.60	107.0	0.38	0.572	1.298	1.201	21036.65	12035.50	47729.352	61955.830	44161.966	53040.536
116.02	48.48		0.42	0.672	1.371	0.914	23139.53	15543.27	47221.112	64730.038	31495.388	28795.682
126.02	45.44		0.45	0.777	1.409	0.563	25081.56	19482.15	45499.954	64113.498	18182.617	10238.594
136.02	42.49		0.49	0.887	1.410	0.174	26767.70	23734.18	42566.224	60018.118	5259.600	916.342
146.02	40.12	99.39	0.52	1.001	1.372	−0.222	28634.99	28662.29	39239.643	53822.868	−6353.918	1411.237
156.02	39.32		0.56	1.119	1.293	−0.595	31269.78	34993.51	36138.220	46738.120	−16613.314	9877.573
166.02	38.53		0.60	1.241	1.175	−0.913	33963.34	42134.03	32178.360	37821.603	−25000.289	22829.805
176.02	37.73		0.63	1.365	1.019	−1.152	36598.28	49955.03	27328.207	27853.529	−30880.603	35565.547
186.02	36.94	99.39	0.67	1.492	0.827	−1.290	39306.68	58637.26	21795.730	18029.505	−33977.956	43816.276
196.02	36.15		0.70	1.621	0.602	−1.313	41629.29	67466.85	15475.484	9323.540	−33732.962	44299.756
206.02	35.36		0.74	1.751	0.349	−1.217	43998.18	77048.78	8764.619	3057.472	−30586.007	37234.192

(continued)

(continued)

Z (coord)	Area	M	z/H	f_1 (ξ)	f_2 (ξ)	f_3 (ξ)	$mi\ fi + M\ fi$	$mi\ fi^2 + M\ fi^2$	$mi\ fi + M\ fi$	$mi\ fi^2 + M\ fi^2$	$mi\ fi + M\ fi$	$mi\ fi^2 + M\ fi^2$
	Az						Mode-1	Mode-1	Mode-2	Mode-2	Mode-3	Mode-3
216.02	34.57		0.77	1.883	0.071	−1.005	46254.01	87096.64	1734.495	122.475	−24676.562	24789.735
226.02	27.49	99.39	0.81	2.016	−0.228	−0.685	39574.65	79775.46	−4469.207	1017.410	−13451.295	9216.443
236.02	27.49		0.85	2.149	−0.541	−0.276	41982.45	90234.99	−10573.252	5723.432	−5382.740	1483.359
246.02	27.49		0.88	2.283	−0.866	0.204	44599.94	101837.5	−16914.030	14646.490	3975.189	809.012
256.02	27.49		0.92	2.418	−1.198	0.730	47223.06	114168.8	−23390.684	28010.780	14254.135	10402.102
270.02	27.49	818.0	0.97	2.606	−1.667	1.507	53031.41	138193.5	−33930.595	56572.357	30658.527	46187.439
277.32	27.49		0.99	2.704	−1.913	1.918	52817.37	142821.2	−37366.852	71484.583	37454.243	71819.342
278.82	27.49		1.00	2.724	−1.964	2.002	53211.42	144960.3	−38353.343	75308.814	39105.046	78289.759
					Sum	→	809517.9	1352804.6	441150.58	891772.31	425468.37	1019615.81

Here, $m_i = \gamma_c \times A_z$ and M = lumped mass due to platform and other fixtures at different levels.

The modal mass participation factor for the first three modes is as shown hereafter

Mode	1	2	3
κ_i	0.598	0.494	0.417

Applying Eqs. (10.5.8)–(10.5.10), we have amplitude, moments, and shear for first three modes as hereafter.

Z (coord)	z/H	U_1 (mm)	U_2 (mm)	U_3 (mm)	U_{SRSS}
0	0.00	0.00	0.00	0.00	0.00
4.263	0.02	0.03	0.02	0.00	0.04
8.525	0.03	0.12	0.06	0.01	0.14
17.675	0.06	0.53	0.26	0.05	0.59
23.575	0.08	0.93	0.45	0.09	1.03
24.575	0.09	1.01	0.48	0.09	1.12
33.525	0.12	1.84	0.85	0.15	2.03
43.525	0.16	3.05	1.33	0.23	3.33
53.525	0.19	4.53	1.86	0.29	4.91
66.025	0.24	6.74	2.54	0.35	7.21
76.025	0.27	8.77	3.08	0.38	9.31
86.025	0.31	11.02	3.57	0.38	11.59
96.025	0.34	13.48	4.00	0.35	14.06
106.025	0.38	16.11	4.34	0.30	16.69
116.025	0.42	18.92	4.59	0.23	19.47
126.025	0.45	21.88	4.72	0.14	22.38
136.025	0.49	24.97	4.72	0.04	25.41
146.025	0.52	28.19	4.59	−0.06	28.56
156.025	0.56	31.52	4.33	−0.15	31.81
166.025	0.60	34.94	3.93	−0.23	35.16
176.025	0.63	38.44	3.41	−0.29	38.59
186.025	0.67	42.01	2.77	−0.32	42.11
196.025	0.70	45.64	2.02	−0.33	45.69
206.025	0.74	49.32	1.17	−0.31	49.33
216.025	0.77	53.03	0.24	−0.25	53.03
226.025	0.81	56.77	−0.76	−0.17	56.78
236.025	0.85	60.53	−1.81	−0.07	60.56
246.025	0.88	64.31	−2.90	0.05	64.37
256.025	0.92	68.09	−4.01	0.18	68.21
270.025	0.97	73.39	−5.58	0.38	73.60
277.325	0.99	76.15	−6.40	0.48	76.42
278.825	1.00	76.72	−6.57	0.50	77.00

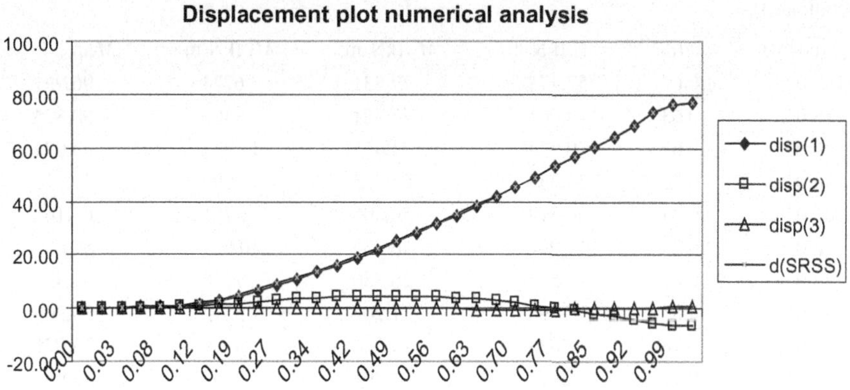

Fig. 10.21 Amplitude of vibration for the first three modes

The displacement plot is shown hereafter (Fig. 10.21)
The bending moment diagrams are plotted hereafter for the first three modes

Z (coord)	*z*/*H*	M_1 (kN m)	M_2 (kN m)	M_3 (kN m)	M_{SRSS}
0	0.00	944,169	−461,443	149,997	1061,548
4.263	0.02	896,991	−421,571	127,953	999,344
8.525	0.03	851,613	−383,838	107,306	940,261
17.675	0.06	759,997	−309,585	67,610	823,413
23.575	0.08	704,938	−266,226	45,367	754,898
24.575	0.09	695,906	−259,205	41,858	743,790
33.525	0.12	618,787	−200,313	13,847	650,549
43.525	0.16	540,144	−142,254	−10,281	558,657
53.525	0.19	468,875	−91,561	−27,102	478,499
66.025	0.24	389,349	−37,737	−38,744	393,087
76.025	0.27	332,800	−1977	−41,640	335,401
86.025	0.31	282,087	27,409	−40,050	286,231
96.025	0.34	233,384	49,736	−34,705	241,135
106.025	0.38	190,087	65,042	−27,431	202,771
116.025	0.42	152,879	74,023	−19,604	170,984
126.025	0.45	121,214	77,404	−12,165	144,334
136.025	0.49	94,573	76,064	−5767	121,504
146.025	0.52	75,300	73,790	−811	105,432
156.025	0.56	66,270	77,129	3265	101,742

(continued)

(continued)

Z (coord)	z/H	M_1 (kN m)	M_2 (kN m)	M_3 (kN m)	M_{SRSS}
166.025	0.60	57,478	77,831	6774	96,991
176.025	0.63	49,001	75,881	9483	90,824
186.025	0.67	40,924	71,415	11,242	83,074
196.025	0.70	33,333	64,715	11,990	73,776
206.025	0.74	26,318	56,205	11,761	63,166
216.025	0.77	19,967	46,433	10,675	51,658
226.025	0.81	11,537	28,950	7170	31,978
236.025	0.85	7896	21,207	5576	23,306
246.025	0.88	4824	13,764	3799	15,071
256.025	0.92	2422	7291	2094	7963
270.025	0.97	380	1221	366	1330
277.325	0.99	11	37	11	41
278.825	1.00	0	0	0	0

The moment plot is shown hereafter (Fig. 10.22)
The shear force is next tabulated for the first three modes

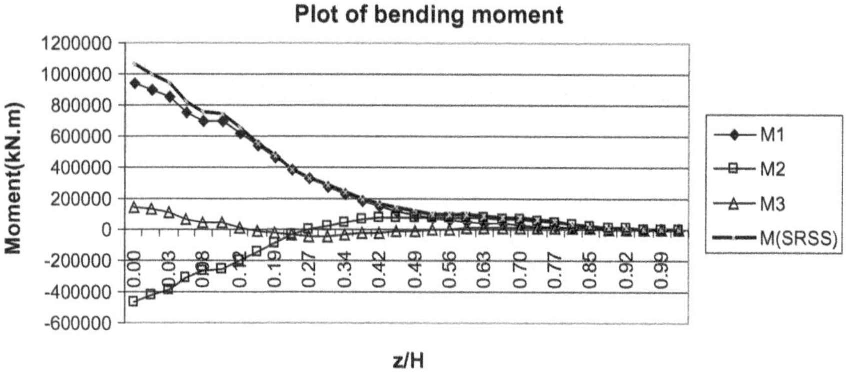

Fig. 10.22 Dynamic moments for the first three modes

Z (co–ord)	z/H	V_1 (kN)	V_2 (kN)	V_3 (kN)	V_{SRSS}
0	0.00	4394	7749	−4280	9883
4.263	0.02	4231	7462	−4120	9516
8.525	0.03	4072	7182	−3958	9156
17.675	0.06	3746	6612	−3590	8404
23.575	0.08	3546	6264	−3328	7930
24.575	0.09	3513	6207	−3282	7851
33.525	0.12	3227	5711	−2842	7148
43.525	0.16	2929	5186	−2309	6387
53.525	0.19	2651	4681	−1757	5659
66.025	0.24	2331	4062	−1087	4808
76.025	0.27	2093	3569	−605	4182
86.025	0.31	1872	3075	−196	3606
96.025	0.34	1640	2546	122	3031
106.025	0.38	1421	2031	339	2501
116.025	0.42	1220	1552	460	2027
126.025	0.45	1038	1119	501	1607
136.025	0.49	873	742	481	1243
146.025	0.52	754	443	436	977
156.025	0.56	724	215	408	858
166.025	0.60	689	−13	351	774
176.025	0.63	651	−231	273	742
186.025	0.67	607	−428	181	764
196.025	0.70	559	−594	84	820
206.025	0.74	506	−720	−8	880
216.025	0.77	449	−798	−88	919
226.025	0.81	311	−660	−119	739
236.025	0.85	264	−650	−151	717
246.025	0.88	212	−589	−160	646
256.025	0.92	154	474	−144	518
270.025	0.97	63	−218	−73	239
277.325	0.99	11	−40	−14	44
278.825	1.00	0	0	0	0

The plot of shear force is shown hereafter (Fig. 10.23).

We compare hereafter the values obtained by analytical method with numerical analysis to check the variation, if any, in the values of displacement, moment, and shear, as plotted hereafter (Fig. 10.24, 10.25 and 10.26).

It is observed that the results are in excellent agreement. It is only at the base level that the numerical method is slightly lower bound to analytical solution. Thus, it can be concluded that masses lumped at different levels (considering that their mass is much less than the total mass of chimney shaft) have marginal effect with respect to analytical solution. So is the variation in geometry in chimney shaft.

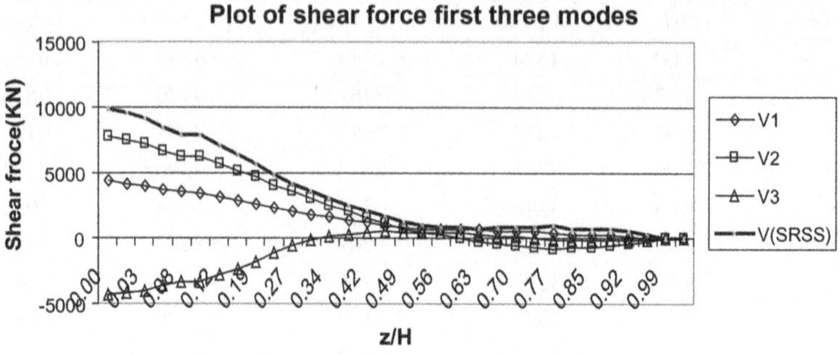

Fig. 10.23 Dynamic shear force for the first three modes

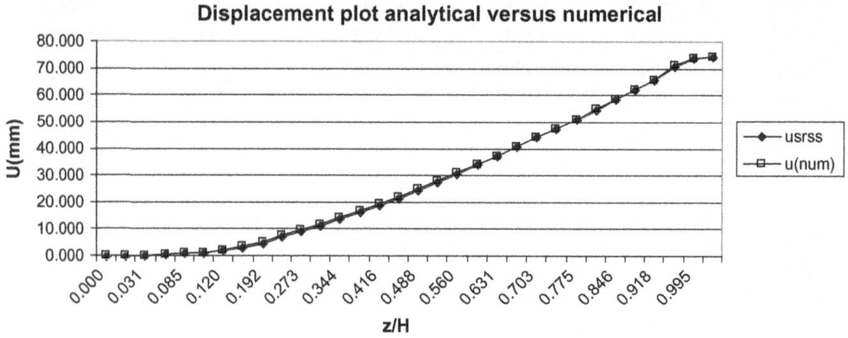

Fig. 10.24 Comparison of amplitude: analytical versus numerical analysis

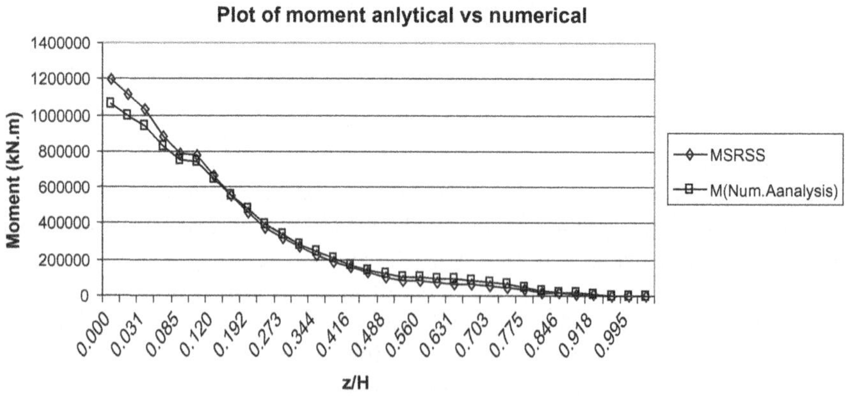

Fig. 10.25 Comparison of moments: analytical versus numerical analysis

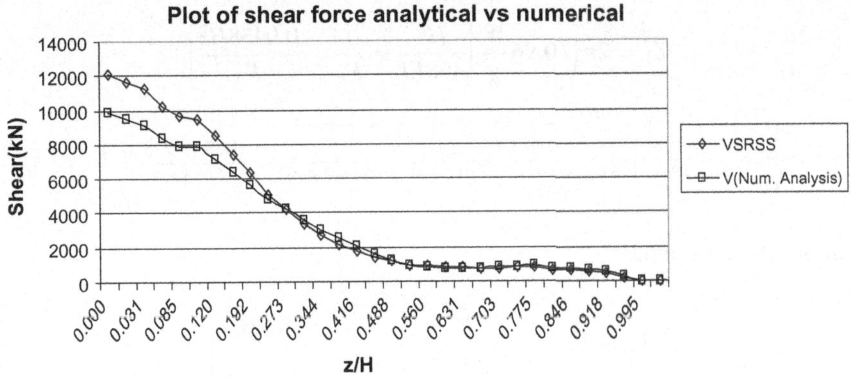

Fig. 10.26 Comparison of shear force: analytical versus numerical analysis

Thus, an exhaustive numerical solution is really not necessary for a chimney, when quick computation for first three modes based on Table 10.7 *or* 10.8 *should be sufficiently accurate for a practical design and would rather be on slightly conservative side.*

(4) **Analysis based on soil–structure interaction**

Diameter of foundation raft = 45 m
Dynamic shear modulus G = 590,000 kN/m^2
Poisson's ratio of soil = 0.32
Weight of raft = 145,350 kN
The soil spring stiffness is expressed as

$$K_h = \frac{32Gr_0(1-v)}{7-8v} = \frac{32\times 590,000\times 22.5\times(1-0.32)}{7-8\times 0.32} = 65,059,459\ \text{kN/m}$$

$$K_\theta = \frac{8Gr_0^3}{3(1-v)} = \frac{8\times 590,000\times(22.5)^3}{3(1-0.32)} = 2.635\times 10^{10}\ \text{kN/m}$$

Thus, for the first mode

$$T_1 = 2\pi\sqrt{1.86\frac{W}{g}\left[\frac{H^3}{22.94EI}+\frac{1}{K_h}+\frac{0.5278H^2}{K_\theta}\right]}$$

$$T_1 = 2\pi\sqrt{1.86\times\frac{476,331}{9.81}\left[\frac{(278.825)^3}{22.94\times 3.4\times 10^7\times 9185.33}+\frac{1}{65,059,459}+\frac{0.5278\times(278.825)^2}{2.635\times 10^{10}}\right]} = 4.05\ \text{s}$$

For the second mode,

$$T_2 = 2\pi\sqrt{0.96\frac{W}{g}\left[\frac{H^3}{468EI} + \frac{1}{K_h} + \frac{0.0438H^2}{K_\theta}\right]}$$

$$T_2 = 2\pi\sqrt{0.96 \times \frac{476,331}{9.81}\left[\frac{(278.825)^3}{468 \times 3.4 \times 10^7 \times 9185.33} + \frac{1}{65,059,459} + \frac{0.0438 \times (278.825)^2}{2.635 \times 10^{10}}\right]} = 0.734\ \text{s}$$

and for the third mode

$$T_3 = 2\pi\sqrt{\frac{W}{g}\left[\frac{H^3}{3813EI} + \frac{1}{K_h} + \frac{0.0163H^2}{K_\theta}\right]}$$

$$T_3 = 2\pi\sqrt{\frac{476,331}{9.81}\left[\frac{(278.825)^3}{3813 \times 3.4 \times 10^7 \times 9185.33} + \frac{1}{65,059,459} + \frac{0.0163 \times (278.825)^2}{2.635 \times 10^{10}}\right]} = 0.395\ \text{s}$$

Damping property B_x is expressed

$$B_x = \frac{(7-8v)mg}{32(1-v)\rho_s r_0^3} = \frac{(7-8\times 0.32)\times(476,331+145,350)}{32\times(1-v)\times(20/9.81)\times(22.5)^3} = 5.46$$

For rocking mode $J_\theta = \frac{m_f r^2}{4} + m_s\bar{h}^2 = \frac{(145,350/9.81)}{4} \times (22.5)^2 + \times (476,331/9.81)(98.86)^2$

This gives $J_\theta = 476{,}424{,}203$ kN s^2.m

$$B_\theta = \frac{0.375(1-v)J_\theta g}{\rho_s r_0^5} = \frac{0.375\times(1-0.32)\times 476,424,203\times 9.81}{(2.04)\times(22.5)^5} = 101.3$$

Thus, $\zeta_x = 0.228/\sqrt{B_x} = 0.123$

$\zeta_\theta = \frac{0.15}{(1+B_\theta)\sqrt{B_\theta}} = 1.456 \times 10^{-4}$(One may note here that tall structures significantly reduce the damping in rocking mode).

The structural damping as recommended by code = 0.03

Thus, the modal damping ratios for the first three modes are expressed as

$$\zeta_{e1} = \frac{\frac{\zeta H^3}{22.94EI} + \frac{\zeta_x}{K_h} + \frac{0.5278\zeta_\theta H^2}{K_\theta}}{\frac{H^3}{22.94EI} + \frac{1}{K_h} + \frac{0.5278H^2}{K_\theta}} = \frac{\frac{0.03\times(278.825)^3}{22.94\times 3.4\times 10^7\times 9185.33} + \frac{0.123}{65{,}059{,}459} + \frac{0.5278\times 1.456\times 10^{-4}\times(278.825)^2}{2.635\times 10^{10}}}{\frac{(278.825)^3}{22.94\times 3.4\times 10^7\times 9185.33} + \frac{1}{65{,}059{,}459} + \frac{0.5278\times(278.825)^2}{2.635\times 10^{10}}}$$

This gives $\zeta_{e1} = 0.02$

Similarly, for the other modes

$$\zeta_{e2} = \frac{\frac{\zeta H^3}{468.04EI} + \frac{\zeta_x}{K_h} + \frac{0.0438\zeta_\theta H^2}{K_\theta}}{\frac{H^3}{468.04EI} + \frac{1}{K_h} + \frac{0.0438H^2}{K_\theta}} \rightarrow 0.022$$

$$\zeta_{e3} = \frac{\frac{\zeta H^3}{3813EI} + \frac{\zeta_x}{K_h} + \frac{0.0163\zeta_\theta H^2}{K_\theta}}{\frac{H^3}{3813EI} + \frac{1}{K_h} + \frac{0.0163H^2}{K_\theta}} \rightarrow 0.030$$

Thus, consider the damping ratio as 2% for the first two modes and 3% for the third mode with amplification to acceleration spectra as 1.4 (for 2% damping) and 1.27 (for 3% damping), respectively.

This gives theS_{ad}/g values and DSSI factor $C\eta$ for the first three modes as mentioned hereafter

Mode	1	2	3
S_{ad}/g	0.345	1.90	3.175
C_η	0.066	4.229×10^{-3}	1.216×10^{-3}

Now, applying Eqs. 10.7.16 through 10.7.18 in conjunction with Table 10.11, we have the SRSS values of amplitude moment and shear along the chimney height as follows

z/H	Amp (mm) SRSS	Moment (kN m) SRSS	Shear (kN) SRSS
0	0	1,890,050	30,292
0.05	0.54	1,378,338	26,362
0.1	2	1,019,882	22,289
0.15	4	766,132	17,181
0.2	7	636,327	12,574
0.25	11	540,939	8253
0.3	15	442,577	5227
0.35	19	362,959	4438
0.4	24	280,400	4569
0.45	29	195,481	4164
0.5	35	147,767	3570
0.55	40	132,969	2997
0.6	47	135,657	2372
0.65	53	136,724	1630
0.7	59	129,674	1186
0.75	66	113,165	1419
0.8	73	76,056	1591
0.85	79	48,214	1667

(continued)

(continued)

z/H	Amp (mm) SRSS	Moment (kN m) SRSS	Shear (kN) SRSS
0.9	86	25,498	1533
0.95	93	7414	989
1	101	0	1

Comparing the values with fixed base analysis it will be observed that the dynamic amplification is moderate and the values have amplified. In this case, the shear wave velocity is 538 m/s, i.e., almost like a rock, yet the amplification for the base moment is about 1.18 times and 2.5 times the fixed base shear.

Some salient data obtained by different analyses are summarized hereafter for readers' benefit.

Time periods obtained by different methods:

Mode	1	2	3
Time period (IS code)	3.46	–	–
ACI-307-08	2.94	0.62	–
Dynamic analysis	3.27	0.53	0.19
Numerical analysis	3.24	0.47	0.13
Considering DSSI	4.05	0.52	0.40
By STAAD (fixed base)	3.37	0.67	0.24

Moments at the base obtained by different methods

Mode	1	2	3	SRSS value
Fixed base moment (IS code)	851,251	–	–	851,251
Fixed base moment (dynamic analysis)	1,045,357	555,039	194,506	1,199,446
Fixed base moment (numerical analysis)	944,169	461,443	149,997	1,061,548
Considering DSSI	1,413,646	870,752	903,152	1,890,050

Shear at the base obtained by different methods

Mode	1	2	3	SRSS value
Fixed base shear (IS code)	5857	–	–	5857
Fixed base shear (dynamic analysis)	5160	9517	5475	12,132
Fixed base shear (numerical analysis)	4394	7749	4280	9883
Considering DSSI	6965	14,930	25,421	30,292

References

Bath, K. J., & Wilson, E. (1980). *Numerical methods in finite element analysis*. New Delhi, India: Prentice Hall.

Bonev, Z., Vaseva, E., Blagov, D., & Mladenov, K. (2010). Seismic design of slender structures including rotational components of the ground acceleration—Eurocode 8 approach. In *Proceedings of 14th European conference in Earthquake Engineering*. Ohrid, Macedonia.

Chowdhury, I. (2010, September). Dynamic analysis of tall multi flue reinforced concrete chimneys under Earthquake force. *Indian Concrete Journal, 84.*

Chowdhury, I., & Dasgupta, S. P. (2003). Computation of Rayleigh damping coefficients for structures with large degrees of freedom. *Electronic Journal of Geotechnical Engineering, Vol-8C* (USA).

Chowdhury, I., & Dasgupta, S. P. (2008). *Dynamics of structures and foundations—A unified approach* (Vol. 1 and 2). Leiden Holland: CRC Press.

Chowdhury, I., & Dasgupta, S. P. (2008, June). Dynamic analysis of piles under lateral loads. *Journal of Indian Geotechnical Society* 249–277.

Chowdhury, I., & Dasgupta, S. P. (2009). Dynamic analysis of piles under rocking motion. *Journal of Indian Geotechnical Society, 39*(4), 360–386.

Chowdhury, I., Singh, J. P., & Tilak, R. (2012). Seismic response of self-supported stack considering foundation compliance. In *15th World Conference in Earthquake Engineering Lisbon Portugal*.

Gazetas, G., Fan, K., Kanyia, A., & Kausel, E. (1991). Dynamic interaction factors for floating pile groups. *Journal of Soil Mechanics and Foundations Division Proceedings ASCE, 117*(10), 1531–1548.

Ghose, D. K., & Batavyal, H. N. (1985). Analysis of structural response to earthquake for 150 m high chimney with soil structure interaction. In *Proceeding National Seminar on Tall Chimneys*. New Delhi: Vigyan Bhavan.

Giosan, I. (2006, December). Building-better free standing structure. *Electrical Lines* (USA).

Hall, J. R. (1967). Coupled rocking and sliding oscillation of rigid circular footing. In *Proceedings International Symposium on Wave Propagation and Dynamic Properties of Earth Material*. Albuquerque New Mexico.

Indian Standard code of practice. (2005). *For design of stack like structures under Earthquake*. New Delhi, India: Bureau of Indian Standards (IS-1893 Part IV).

Jain, S., Singh, B. P., & Gupta, B. P. (1990). IS code provisions for seismic design of tall chimneys. *International Journal of Structures, 10*(2), pp. 103–111.

Jaya, V., Dodagoudar, G. R., Boominathan. A. (2009). Seismic soil structure interaction analysis of ventilation stack structure. *Indian Geotechnical Journal, 39*(1), 116–134.

Luco, J. E. (1986). Soil structure interaction effects on seismic response of tall chimneys. *Journal of Soil Dynamics and Earthquake Engineering, 5*(3), 170–177.

Navarro, C. (1992). Influence of soil flexibility on seismic response of chimneys. *Journal of Soil Dynamics and Earthquake Engineering, 11,* 403–409.

Novak, M. (1974). Dynamic stiffness and damping of piles. *Canadian Geotechnical Journal NRCC, 11,* 574–598.

Novak, M. (1977). Vertical vibration of floating piles. *Journal of Engineering Mechanics Proceedings ASCE, 103*, 153 (EM1).

Poulos, H. (1979). Group factors for pile deflection estimation. *Journal of Geotechnical Engineering Division Proceedings ASCE, 5*, 1489–1509 (GT12).

Roussea, J., & van Zijl, G. P. A. G. (2006). Dynamic evaluation of solar chimney. *Concrete Beton; Journal of Concrete Society of South Africa.*

Sadeghpour, N., & Chowdhury, I. (2008). Dynamic soil-structure interaction analysis of tall multi flue chimney under aerodynamic and seismic force. In *Proceedings 12th International Conference of International Association for Computer Methods and Advancement in Geo-mechanics*. Goa, India.

Chapter 11
Earthquake Analysis of Concrete Gravity Dam

11.1 Introduction

For countries blessed with a number of rivers and lakes, where water management, flood control, and development of hydel power remains a major issue, concrete gravity dams often play a major role in achieving the same.

For countries like Australia, Canada, Egypt, India, Iran, Pakistan, and USA, a number of dams have been built over the years to this end.

Dams of very large magnitude were previously built out of masonry; presently, mass concrete is usually deployed to build such dams. This obviously requires a detailed analysis, as the failure risk for such major dams remains far too catastrophic. However, irrespective of the technology adopted (simplified to most sophisticated finite element analysis), fortunately, no concrete dam in the world has yet undergone a major failure (except developing some few cracks) due to an earthquake, though many of them have faced seismic force as high as 1.0 *g*.

However, in one particular case, major damage occurred to a concrete gravity dam in India in Koyna (near Pune, Maharashtra) on December 11, 1967, where the dam experienced ground acceleration, as high as 0.5 *g*.

Built during the years 1954–1963, the dam is 853 m long and 103 m in height at the point of deepest foundation. It was designed by traditional seismic coefficient method subjected to an acceleration of 0.05 *g*.

But the occurrence of a peak ground acceleration of 0.5 *g* caused significant damage to both upstream and downstream sides of a number of non-overflow monoliths, especially around the elevation where slope of the downward stream changed abruptly. The overflow monoliths were not damaged. Although the dam survived the earthquake without any major collapse or sudden release of water, the cracks appeared to be serious enough for prompting authorities to take retrofitting measures and it was decided to strengthen the non-overflown monoliths with buttresses on downstream side of the dam.

I. Chowdhury and S. P. Dasgupta, *Earthquake Analysis and Design of Industrial Structures and Infra-structures*, GeoPlanet: Earth and Planetary Sciences,
https://doi.org/10.1007/978-3-319-90832-8_11

11.2 Analysis as Per IS-1893

While writing this section, the IS-1893 (2016) code committee is yet to come up with procedural practice for earthquake analysis of such concrete dams. As such, our discussion and benchmarking herein will be as per IS-1893 (1984), and based on which a number of dams have been built in the past.

Shown in Fig. 11.1 is a typical concrete dam with water at the upstream side. During an earthquake, it is usually assumed that the dam vibrates on its own, thus generating a force $W(S_a/g)$, due to its inertia. It is also subjected to a hydrodynamic force due to propagation of waves through the liquid that produces additional force on the dam face over and above the *hydrostatic* force it sustains.

The earthquake analysis for a concrete gravity dam as per IS code is based on certain simplifying assumptions as stated hereunder

- The impounded water the dam supports is *incompressible.*
- The dam is *rigid* and has the same motion throughout its body.
- The dam and the reservoir motion are *uncoupled* and the interaction between them is negligible for all practical purposes.

The code has recommended to adopt *seismic coefficient method* for a dam up to 100 m high, while *response spectrum method* with dynamic analysis (i.e., calculation of time period and taking its effect) for dams that are greater than 100 m.

The seismic coefficient method is quite straightforward and does not require any elaboration as the steps are same for building analysis except of the fact that importance factor to be considered should be as recommended by the code (for dams, $I = 3.0$).

For response spectrum analysis, the IS-1893 code recommends to derive the fundamental time period of a concrete dam as

$$T = C_T \frac{H^2}{B} \sqrt{\frac{\gamma_c}{gE_s}} \tag{11.2.1}$$

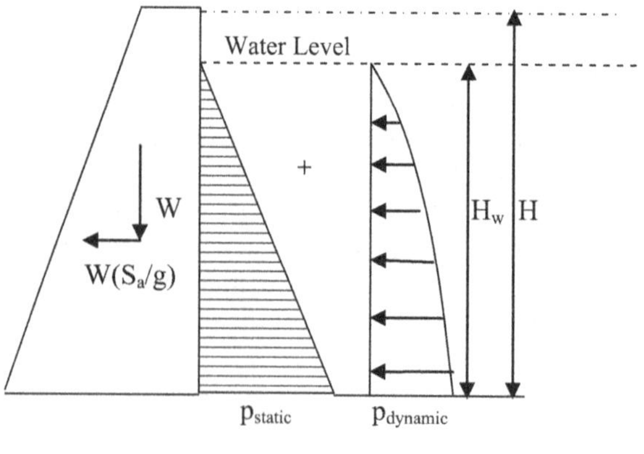

Fig. 11.1 Cantilever dam with a free board

Table 11.1 C_T values for first three modes after Krishna et al. (1994)

Mode	1	2	3
C_T	5.55	2.8	2.25

Here, $C_T = 5.55$ for fundamental mode; B = width of the dam base in m; H = height of the dam in m; γ_c = unit weight of material of dam in kN/m^3; g = acceleration due to gravity, 9.81 m/s^2, and E_s = modulus of elasticity of the material of the dam in kN/m^2.

However, the basis of derivation of this formula has not been elaborated either in the code or in its explanatory manual.[1]

Krishna et al. (1994) have further proposed values of C_T for higher modes expressed as follows:

Looking at the formula, it appears that Eq. (11.2.1) is derived by applying the Rayleigh Ritz method to some assumed shape functions and considering the dam as a cantilever beam having varying cross section. The code also does not give any value of time periods for higher modes (vide Table 11.1) whose effects are considered as non-critical.

The base shear and moment for the dam are given by the expression:

$$V_B = 0.6\alpha_h W; \quad M_B = 0.9\alpha_h W\bar{h} \tag{11.2.2}$$

$$\text{where} \quad \alpha_h = \beta I F_0 \frac{S_a}{g} \tag{11.2.3}$$

Acronyms in Eq. (11.2.3) are as per IS-1893 (1984); β = soil foundation factor, I = importance factor, and F_0 = seismic zone factor.[2]

As per the latest version of the code, IS-1893 (2016), the seismic design coefficient can be taken as

$$\alpha_h = \frac{ZI}{2R}\frac{S_a}{g} \tag{11.2.4}$$

where W = weight of the concrete dam; $\bar{h}$ = height of the center of gravity of the dam, above its base. The parameters Z, I, and R are code factors as per IS-1893 and as explained in Chap. 2.

The value of S_a/g is read from the response spectrum curve furnished in the code and is a function of the time period T of the dam as mentioned in Eq. (11.2.1).

[1]Except for a note that this has been developed based on some research work carried out at the University of Roorkee, India.

[2]For appropriate values, refer to IS-1893 (1984).

Table 11.2 Values of C'_v and C'_M varying with depth z

z/H	0.0	0.2	0.4	0.6	0.8	1.0
C'_v	0.0	0.185	0.400	0.655	0.875	1.0
C'_M	0.0	0.035	0.16	0.365	0.665	1.0

For any horizontal section at a depth z below the top of dam, the shear force V_z and bending moment M_z may be obtained from the expression

$$V_z = C'_v V_B \text{ and } M_z = C'_M M_B \tag{11.2.5}$$

Here, the values of coefficients C'_V and C'_M are as furnished in Table 11.2.

Considering that a concrete dam is a massive structure, its own weight plays a significant part in its stability. It is evident that unlike other structures, the vertical mode of earthquake acceleration plays a significant part in its stability and cannot be ignored.

As per IS code, based on the response spectrum method, the force due to vertical acceleration is considered as 0.75 times the value of α_h at the top of dam and is reducing linearly to zero at the base.

11.2.1 Hydrodynamic Pressure on a Dam Face from the Reservoir (Zangar 1953)

As a wave propagates through the ground, there is an instantaneous hydrodynamic pressure (or suction) exerted on the dam over and above the hydrostatic pressure it sustains.

Based on the assumption that water is incompressible, the hydrodynamic pressure at any depth z below the reservoir is determined by the expression

$$p = C_s \alpha_h \gamma_w h \tag{11.2.6}$$

where p = hydrodynamic pressure of water; C_s = a coefficient which varies with shape and depth; α_h = seismic design coefficient as explained in Eq. (11.2.3) or Eq. (11.2.4); γ_w = unit weight of water, and h = height of water in the reservoir.

The variation of C_s with depth is given by the expression

$$C_s = \frac{C_m}{2}\left[\frac{z}{h}\left(2 - \frac{z}{h}\right) + \sqrt{\frac{z}{h}\left(2 - \frac{z}{h}\right)}\right] \tag{11.2.7}$$

where C_m value varies (almost) linearly from a value of 0.735, when the vertical upstream angle of the dam face varies from 0° (i.e., perfectly vertical), to a value of 0.0, when this angle is 90°.

The approximate values of shear and moment at depth z below the free surface are given by the expression

$$V_z = 0.726\,pz \text{ and } M_z = 0.299\,pz^2. \tag{11.2.8}$$

Here, p = hydrodynamic pressure as given in Eq. (11.2.6).

11.2.2 Some Comments on the IS Code Method

The above method, though has been in practice for quite some time (and a number of dams have been built on this basis), yet it contains some approximations and assumptions that are perhaps not conceptually correct.

Firstly, the value of time period as proposed in Eq. (11.2.1) has been found to underestimate the time period of a dam based on detailed FEM analysis. It actually considers the dam section to be triangular in shape. The top width at crest is ignored in the calculation of the time period, and the expression also ignores the effect of hydrodynamic interaction and the foundation compliance effect.

It has been shown by Chopra (1966) and Chopra et al. (1980a, b, c), Fenves and Chopra (1984, 1986) that hydrodynamic interaction of the impounded reservoir water and foundation compliance significantly affect the time period.

Secondly, while calculating the hydrodynamic pressure, vide Eq. (11.2.6), the IS code uses the value of α_h vis-à-vis S_a/g value based on Eq. (11.2.1).

This is in contradiction to the basic assumption made at the outset that the dam face is rigid and vibrations of the two systems (dam and the fluid in the reservoir) remain uncoupled.

In this case, as per the above-mentioned assumption, the *free field time period* of fluid in the reservoir (assumed tending to infinity) in horizontal direction should govern the value of α_h, rather than the time period of the dam whose stiffness would vary considerably compared to the impounded fluid. The assumptions as made in IS code show variation in the end results, when compared to FEM analysis, and other methods, like that proposed by Okamoto (Japanese practice) or Fenves and Chopra's (1986) simplified method, that are followed in many countries.

11.3 Finite Element Analysis of a Concrete Dam

Presently, with significant development in computational capability of the finite element method (FEM), concrete dams are mostly analyzed by two-dimensional finite element models, where the dam body (the non-overflowing monolith) is discretized into plane stress elements and analyzed under static and dynamic loads, as shown in Fig. 11.2.

The trend possibly started in 1960, when Ray W. Clough (in USA) presented his pioneering paper on finite element analysis (Clough 1960), wherein he solved the

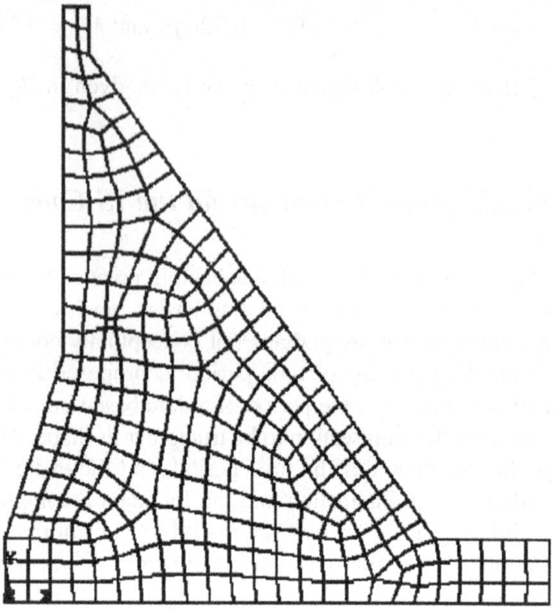

Fig. 11.2 Finite element model of dam (fixed based) with reservoir empty

problem of concrete dam erstwhile solved by Zienkiewicz (1947) using finite difference method and came up with comparable results.

This was followed by his next paper (Clough et al. 1962) on Norfork dam, wherein the nonlinear behavior of cracks under hydrostatic load was again predicted successfully by using the finite element analysis.

University of California, Berkeley in the 1960–80, was pioneering the research in dam engineering and was applying techniques of FEM heavily to predict the behavior of a number of dams across USA, especially under earthquake (Chopra 1966, 1970; Chopra et al. 1980; Chopra and Hall 1980; Fenves and Chopra 1984, 1986).

Finally, the seminal work of Chopra and Chakrabarti (1980), predicting the cause for damage to Koyna dam in India (as described at the outset), firmly established FEM as a standard tool for analysis of such gravity dams in a number of countries around the world. This influence has been so strong that even till date, the application of FEM to concrete dams has continued to be the only sanctimonious tool of analysis, notwithstanding the fact that:

- Lower-order plane strain/stress elements are inherently deficient in predicting flexural behavior of such dams.
- Due to linear polynomial functions chosen for deriving stiffness matrix of either triangular or rectangular elements, parasitic shear locks the element and produces the upper bound stiffness matrix; the more so, since even with the best aspect ratio and mesh refinements, moments and shear may vary from 30 to 43% in reality (Cook et al. 1989).

- So, unless one uses a special non-conforming element and that too further corrected to pass the patch test (Taylor et al. 1976), results could be highly susceptible to error.
- Hydrodynamic pressure that generates additional force on dam's upstream face has been calculated based of frequency domain analysis (Chopra 1966). This hardly eases the problem for professional engineers, as time history response and frequency content of a possible future earthquake at a particular site is always not possible to identify.

To circumvent this, many analysts attempt to consider the fluid also as a 2D finite element to asses the hydrostatic and dynamic pressure on dam's upstream face. But this is not without problem.

- Firstly, the fact that fluid is almost an incompressible medium ($\nu = 0.5$) generates singularity in deriving the stiffness matrix. Even with $\nu = 0.499$, it creates serious difficulty in numerical convergence; thus, unless the software in hand has special fluid elements, the *coupled fluid–structure interaction* becomes a very difficult task to execute.
- An extent in horizontal direction to which the fluid medium needs to be extended to ensure no reflection of waves back to the dam is still not clear; thus, analysts are often forced to extend it to a significant distance to suppress waves from reflecting back which adds to cost and size of the problem, thus making the analysis laborious, if not difficult to interpret.
- Paraxial boundary elements and infinite finite element as applied to fluid elements are always not available in a number of general-purpose finite element software.
- For dams built on a flexible rock, when foundation interaction is important, further modeling of soil as 2D plane strain element and extending it to a significant distance away from the dam/reservoir interface to ensure no back-reflection of waves (often termed as "box effect") makes the overall problem catering to structure, fluid, and soil together prohibitively expensive and time consuming.
- Finally, soil data along the depth to which the soil has to be modeled below a large dam to ensure eradication of this box effect are not always available as geotechnical investigation to such a large depth is not technically or economically viable.

It has been often observed that, in order to cut down the size or complexity of the problem and circumvent the inadequacy of input data (like the lack of SPT values/non-availability of rock core samples at a substantial depth below the dam, site specific response spectra etc.) engineers often make simplifying assumptions (like assuming uncoupled hydrodynamic pressure, fixed base problem, thus ignoring foundation interaction), that makes the whole exercise questionable in spite of detailed FEM analysis, and they arrive at a result that could well be lower bound.

Thus, an efficient tool, that is easy to handle and yet leads to a reasonably sound result within the modal response framework and is less laborious, is still a serious quest in the industry.

Based on research carried out in Berkeley, dynamic finite element analysis is usually carried out for the following four cases depending on the complexity and national importance of a dam.

1. Dynamic analysis of the dam when the reservoir is empty and hydrodynamic pressure remains uncoupled (Fig. 11.2). The two responses are added after a separate analysis. The dam is considered fixed at the base; i.e., the foundation compliance is ignored in this case.
2. Dynamic analysis of the dam with fluid on upstream side assumed incompressible, which is added to its upstream face as a virtual lumped mass (as depicted in Fig. 11.3) and considering the foundation flexibility in the analysis or considering the foundation as fixed at the base.
3. Dynamic analysis of the dam with fluid on upstream side assumed compressible added to its upstream face as fluid elements having both mass and stiffness extended to a substantial distance to invoke the infinite domain properties.
4. Dynamic analysis of cases 2 or 3 considering dam foundation interaction, where the stiffness of foundation properties is again modeled as plane strain finite elements considering the soil–fluid–structure interaction.

It is apparent that of all the models cited above, Fig. 11.4 is the most comprehensive and realistic model, notwithstanding the limitations of FEM as cited above. The model is, however, also computationally expensive and difficult to interpret. Unless a proper care is taken in terms of selection of right elements, adequate refinement of meshes and imposing correct boundary conditions, one can come up with a result that could be erroneous and that an analyst who is not adequately experienced may find quite difficult to diagnose. Finally, preparation of input data,

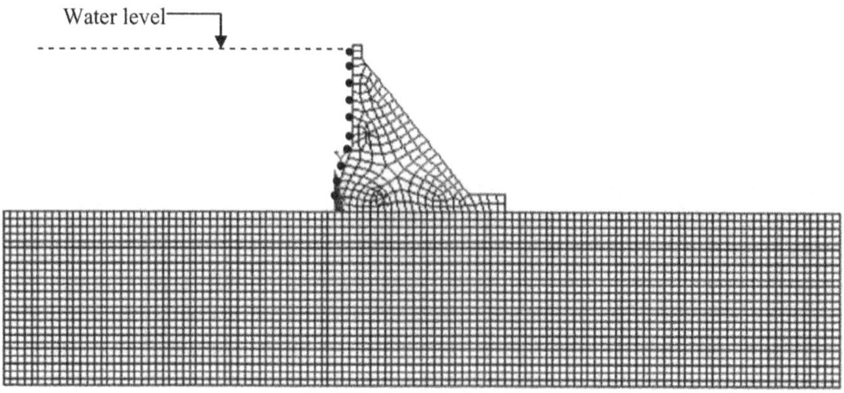

Fig. 11.3 Finite element model of dam with incompressible water added as lumped mass on upstream side and foundation stiffness considered

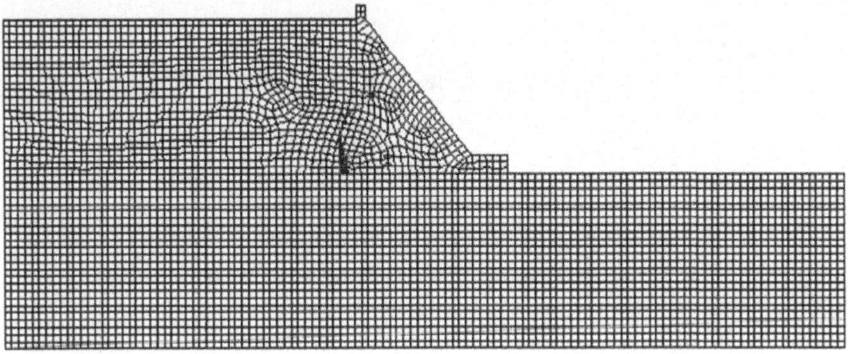

Fig. 11.4 Finite element model of dam, fluid, and foundation for a comprehensive soil–fluid–structure interaction analysis

FEM modeling, and interpretation of the output results would invariably be time consuming.

It is for this, despite FEM's immense popularity quest, is still on within the dam engineering fraternity to find alternative methods that could at least be used as a checking tool to such exhaustive FEM models. A tool, that would be quick to compute and provide comparable result as FEM.

11.4 Chopra and Fenves's Method for Dynamic Analysis of Gravity Dam

Realizing the lacunae as discussed above in terms of design office practice, backed up by exhaustive FEM analysis of a number of dams, Fenves and Chopra (1986) proposed a simplified procedure for computing the dynamic response of dam under earthquake force. The steps for the same may be summarised as hereafter.

Compute the fundamental time period of the dam only (reservoir empty) from the expression

$$T = 0.38H/\sqrt{E_s} \qquad (11.4.1)$$

Here, H = height of the dam in m, and E_s = Young's modulus of the dam material in mega Pascals.

Modify Eq. 11.4.1 to take into consideration the fluid interaction by multiplying it by a factor R_r, when $T_r = R_r \times T$. Values of R_r for different values of H_w/H are furnished in Table 11.3. Here, H_w = height of water in reservoir, and T_r is the elongated time period of the dam considering hydrodynamic interaction with reservoir water.

In Table 11.3, E_s is furnished in *mega Pascals*. The value α is the wave reflection coefficient due to the deposition of bottom silt or alluvium soil on the

Table 11.3 Values of R_r for different values of H_w/H

H_w/H	α	$E_s = 35{,}000$	$E_s = 31{,}500$	$E_s = 28{,}000$	$E_s = 24{,}500$	$E_s = 21{,}000$	$E_s = 17{,}500$	$E_s = 14{,}000$	$E_s = 7000$
		R_r	R_r	R_r	R_r	R_r	R_r	R_r	R_r
1.0	1.0	1.454	1.409	1.370	1.341	1.320	1.301	1.286	1.263
	0.9	1.462	1.416	1.374	1.344	1.319	1.301	1.285	1.263
	0.75	1.456	1.412	1.374	1.341	1.312	1.287	1.284	1.262
	0.50	1.355	1.344	1.333	1.316	1.289	1.283	1.275	1.260
	0.25	1.284	1.285	1.285	1.282	1.264	1.264	1.262	1.256
	0.0	1.261	1.259	1.259	1.256	1.247	1.247	1.247	1.247
0.95	1.0	1.368	1.323	1.289	1.259	1.241	1.224	1.212	1.193
	0.9	1.376	1.330	1.292	1.263	1.240	1.224	1.211	1.193
	0.75	1.366	1.323	1.289	1.259	1.233	1.221	1.210	1.193
	0.50	1.255	1.256	1.247	1.238	1.213	1.209	1.203	1.191
	0.25	1.208	1.208	1.208	1.208	1.194	1.194	1.192	1.187
	0.0	1.192	1.191	1.191	1.188	1.181	1.181	1.181	1.181
0.90	1.0	1.289	1.247	1.214	1.191	1.176	1.164	1.154	1.140
	0.9	1.297	1.253	1.220	1.193	1.176	1.163	1.154	1.140
	0.75	1.284	1.247	1.214	1.193	1.171	1.161	1.152	1.140
	0.50	1.181	1.185	1.179	1.174	1.155	1.152	1.148	1.139
	0.25	1.151	1.152	1.152	1.152	1.141	1.141	1.140	1.136
	0.0	1.139	1.139	1.139	1.136	1.131	1.131	1.131	1.131

reservoir bed. For a newly laid dam, there is no absorption, $\alpha = 1$, where there has been heavy silt deposition over years, and absorption factor α can be considered as zero.

It is quite difficult (if not impossible) to determine the value of α correctly, for the reservoir bottom material may consist of highly variable layers of exposed bedrock, alluvium silt, and other elements, for which appropriate site investigation techniques are yet to be developed. For this, to be on conservative side, it would be justified to select the value of α based on bedrock properties only. Chopra and Hall (1980) have recommended this value to be

$$\alpha = (1 - qc)/(1 + qc) \tag{11.4.2}$$

Here, $q = \rho/(\rho_r c_r)$, c = velocity of sound in water, ρ = density of water, ρ_r = density of bottom rock of reservoir, $c_r = \sqrt{E_r/\rho_r}$, and E_r = Young's modulus of bottom rock material.

Compute the hydrodynamic ratio effect factor R_w, expressed as

$$R_w = R_r \times T \times c/4H \tag{11.4.3}$$

Determine the foundation interaction factor R_f which is a function of ratio E_f/E_s, where E_f is the Young's modulus of the foundation. For $E_f/E_s >= 4.0$, the R_f value is usually considered to be 1.0 (i.e., no interaction). The values of R_f are furnished in Table 11.4 hereafter.

Having determined R_f, compute the modified time period of dam considering the influence of foundation compliance and the effect of impounded water as

$$T^{fr} = R_f \times R_r \times T \tag{11.4.4}$$

Compute the damping ratio of dam, soil, and fluid systems $\tilde{\xi}_1$ as

$$\tilde{\xi}_1 = \frac{1}{R_r} \times \frac{1}{(R_f)^3} \xi_1 + \xi_r + \xi_f \tag{11.4.5}$$

In Eq. 11.4.5, the acronyms are as defined hereafter.

ξ_1 = material damping ratio of dam material usually considered between 0.02 and 0.05.

ξ_r = damping ratio induced by fluid in the reservoir and is a function of α, E_s, and H_w/H and is listed in Table 11.5.

Table 11.4 Foundation interaction factor R_f

E_f/E_s	4.0	3.0	2.0	1.5	1.0	0.5	0.2
R_f	1.054	1.071	1.099	1.129	1.187	1.335	1.678

Table 11.5 Values of ξ_r for different values of H_w/H

H_w/H	α	$E_s = 35{,}000$	$E_s = 31{,}500$	$E_s = 28{,}000$	$E_s = 24{,}500$	$E_s = 21{,}000$	$E_s = 17{,}500$	$E_s = 14{,}000$	$E_s = 7000$
		ξ_r	ξ_r	ξ_r	ξ_r	ξ_r	ξ_r	ξ_r	ξ_r
1.0	1.0	0.0	0.0	0.0	0.0	0.0	0.0	0.0	0.0
	0.9	0.043	0.030	0.021	0.013	0.008	0.005	0.003	0.001
	0.75	0.060	0.051	0.040	0.029	0.021	0.014	0.009	0.004
	0.50	0.067	0.060	0.051	0.042	0.035	0.025	0.018	0.008
	0.25	0.054	0.050	0.045	0.040	0.036	0.030	0.024	0.013
	0.0	0.038	0.036	0.034	0.032	0.030	0.027	0.024	0.017
0.95	1.0	0.0	0.0	0.0	0.0	0.0	0.0	0.0	0.0
	0.9	0.044	0.031	0.020	0.012	0.007	0.005	0.003	0.001
	0.75	0.056	0.049	0.038	0.027	0.019	0.012	0.008	0.003
	0.50	0.060	0.053	0.045	0.036	0.030	0.022	0.015	0.007
	0.25	0.045	0.042	0.038	0.033	0.030	0.025	0.020	0.011
	0.0	0.032	0.030	0.028	0.026	0.025	0.022	0.020	0.014
0.90	1.0	0.0	0.0	0.0	0.0	0.0	0.0	0.0	0.0
	0.9	0.041	0.029	0.017	0.010	0.006	0.004	0.002	0.001
	0.75	0.050	0.042	0.033	0.022	0.015	0.009	0.006	0.002
	0.50	0.050	0.044	0.037	0.029	0.024	0.017	0.012	0.005
	0.25	0.036	0.033	0.030	0.026	0.024	0.020	0.016	0.008
	0.0	0.025	0.023	0.022	0.020	0.019	0.018	0.016	0.011

Table 11.6 Added damping ratio ξ_f for a dam–foundation interaction

E_f/E_s (%)	4.0	3.0	2.0	1.5	1.0	0.5	0.2
$\eta_f = 10$	0.018	0.024	0.035	0.047	0.068	0.121	0.220
$\eta_f = 25$	0.023	0.031	0.046	0.060	0.086	0.151	0.279
$\eta_f = 50$	0.030	0.039	0.057	0.073	0.105	0.186	0.362

ξ_f = damping ratio induced by foundation material as listed in Table 11.6.

In Table 11.6, η_f is the radiation damping ratio property of the dam foundation expressed in percentage.

Having calculated the damping ratio, the hydrodynamic pressure $p(z)$ varying along the depth z of the dam, for the full water height, $H_w/H = 1.0$, for different values of R_w can be determined from the coefficients as expressed in Tables 11.22, 11.23, 11.24, 11.25, 11.26, 11.27—*appendix to this chapter*.

These coefficients are expressed in terms of $gp(z)/\gamma_w H_w$ and need to be multiplied by the term $(H_w/H)^2$, to arrive at the value $p(z)$. For $H_w/H \leq 0.5$, Fenves and Chopra (1986) recommends to consider $p(z) = 0$.

Having computed the hydrodynamic pressure, compute next the generalized mass $\hat{M}_1$ from the equation

$$\hat{M}_1 = (R_r)^2 M_1 \tag{11.4.6}$$

Here, R_r is derived from Table 11.3 and M_1 is expressed as

$$M_1 = (1/g)\int_0^H w(z)\phi(z)^2 dz \tag{11.4.7}$$

In Eq. 11.4.7, $w(z)$ is the weight of the dam per unit height. The fundamental mode shape is as expressed in Table 11.7 and g = acceleration due to gravity 9.81 m/s^2.

Computation of Eq. 11.4.7 may be avoided by considering an approximate value of M_1 = 0.043 W/g, where W is the total weight of dam monolith.

The generalized earthquake force coefficient is determined from the expression

$$\hat{L}_1 = L_1 + \frac{\gamma_w H_w^2}{2g}\left(\frac{H_w}{H}\right)A_p \tag{11.4.8}$$

Table 11.7 Values of the modal shape function $\phi(z)$ for the fundamental mode

z/H	1.0	0.9	0.8	0.7	0.6	0.5	0.4	0.3	0.2	0.1	0.0
$\phi(z)$	1.0	0.735	0.530	0.389	0.284	0.200	0.135	0.084	0.047	0.021	0.0

Table 11.8 **a** Values of A_p for $\alpha = 1$. **b** Values of A_p for $\alpha = 0.9, 0.75, 0.5, 0.25$, and 0

(a)							
R_w	1.0	0.95	0.9	0.85	0.8	0.7	<=0.5
A_p	1.242	0.585	0.431	0.364	0.324	0.279	0.237

(b)						
R_w	1.2	1.0	0.9	0.8	0.7	<=0.5
$\alpha = 0.9$	0.071	0.515	0.417	0.322	0.278	0.237
$\alpha = 0.75$	0.111	0.340	0.361	0.309	0.274	0.236
$\alpha = 0.5$	0.159	0.252	0.274	0.269	0.256	0.231
$\alpha = 0.25$	0.178	0.213	0.224	0.229	0.228	0.222
$\alpha = 0.0$	0.181	0.191	0.195	0.198	0.201	0.206

Here,

$$L_1 = \frac{1}{g}\int_0^H w(z)\phi(z)\mathrm{d}z \tag{11.4.9}$$

Values of A_p for different values of R_w are furnished in Table 11.8a, b. For $H_w/H \le 0.5$, $\hat{L}_1 \cong L_1$. The computation of L_1 can be avoided by considering $L_1 = 0.13$ W/g.

For quick computation, Fenves and Chopra (1986) recommend to consider $\hat{L}_1/\hat{M}_1 = 4.0$ for dams with impounded water and $L_1/M_1 = 3.0$ with an empty reservoir.

Equivalent lateral earthquake force is then expressed as

$$F(z) = \frac{\hat{L}_1}{\hat{M}_1}\frac{S_a\left(T^{fr}, \tilde{\xi}_1\right)}{g}[w(z)\phi(z) + gp(z, T_r)] \tag{11.4.10}$$

The shear force $V(z)$ for any section z from the top can then be expressed as

$$V(z) = \frac{\hat{L}_1}{\hat{M}_1}\frac{S_a\left(T^{fr}, \tilde{\xi}_1\right)}{g}\int_0^H [w(z)\phi(z) + gp(z, T_r)]\mathrm{d}z \tag{11.4.11}$$

In Eq. 11.4.10, $S_a(T^{fr}, \tilde{\xi}_1)/g$ is the spectral acceleration from the response curve as furnished in the code, corresponding to time period T^{fr} and damping ratio $\tilde{\xi}_1$.

11.5 Chowdhury, Singh, and Dasgupta's (CSD) Model for Dynamic Analysis of Dams

Before we delve into the proposed method, as an originator of this model (Chowdhury et al. 2013), it would perhaps not be out of place to discuss what made us take up this research, when FEM has strongly established itself as a standard tool for analysis of such problems.

Major motivation of Clough (1960) to analyze the dam problem by FEM method was to satisfy the irregular geometry of the dam's cross section and identify stresses at any point therein.

Figure 11.5 shows typical first-generation dams solved by FEM based on CST/TCMT triangular elements in the 1960s.[3] It is apparent that due to limited computational capability as available then, meshing was understandably crude and it was observed that though displacements obtained were reasonable, stress outputs were not so good, unless resorted to a very fine mesh refinement.

Computational capability of FEM has certainly improved leaps and bounds from the days of Clough's analysis to where it is today, as shown in Fig. 11.6.

With enhanced graphics and Pentium chips available presently, solving a problem with 35–40 k nodes (or even more) is not a big ordeal.

However, whenever such problems of mega magnitude are solved, an indubitable part of such procedure is cross-checking or vetting the work, where the vetting engineer often faces a daunting task of struggling with 2500 pages of technical report, of which maybe 2200 pages constitute the FEM computer output, and rest is the procedural narrative and graphic output of the problem.

Those who have experienced this arduous task of vetting such voluminous work can testify how difficult a task this can be.

Firstly, it is humanly almost impossible to go through line by line of the output to carry out a number check. Though the graphics can visually testify the nodal parity check, yet every experience analyst is aware that there can be innumerable cracks through which the analysis can sneak through and yet come up with realistic looking numbers (which may or may not be actually the true picture). In most of such cases, the vetting engineer goes through the stress plots and from his past experience tries to assess whether the stress distribution looks realistic or not.

And if he has any doubt on the outcome, the best he can do is to ask for the input file and check the data and rerun the problem himself or do a parallel FEM analysis and check if the results match (more or less).

In practice, very few, and possibly extremely meticulous ones only, would ever opt for a parallel check, and as per authors experience, in 8 out of 10 cases, the vetting engineer would let it go *assuming* the output is correct; however, contractually covering himself by stamping the document with a note that despite his vetting and approving the document, the designer remains completely responsible

[3]CST stands for constant strain triangle, while TCMT is the abbreviation of Turner, Clough, Martin, and Topp.

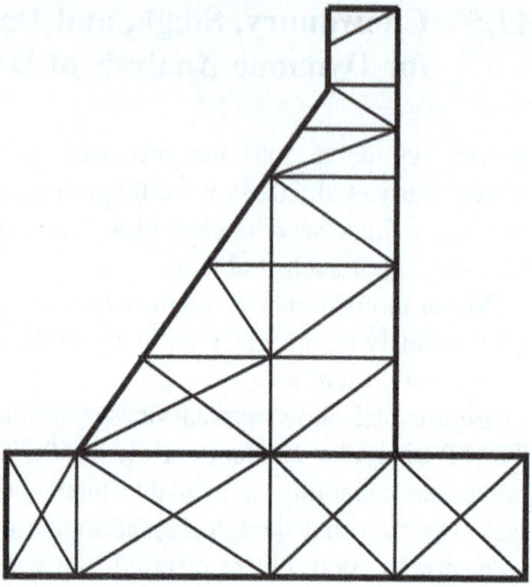

Fig. 11.5 First-generation dam and foundation solved by FEM in the 1960s

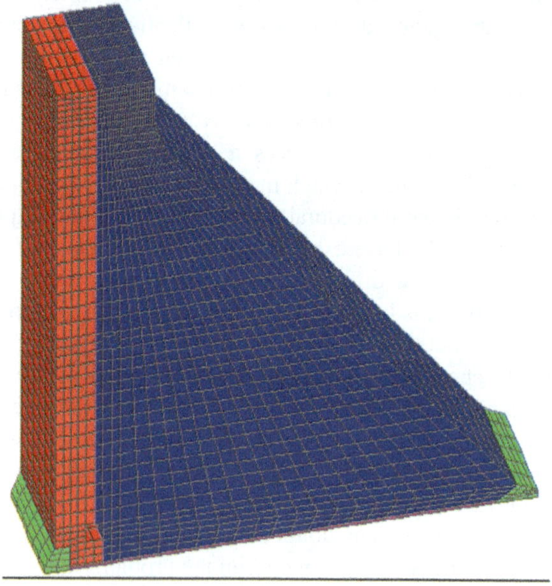

Fig. 11.6 A 3D finite element model of a dam monolith

for the correctness of the design and the vetting engineer remains indemnified against any libel suit, if any, for any failure or damage taking place.

In other words, the whole purpose of cross-checking or value addition to the design boils down to be a charade or a farce in majority of cases.

It is for this, for many years now it has been felt that an alternative tool to FEM should be available which should give a designer or an engineer a firsthand feel of the problem with as much accuracy a detailed FEM result would provide when one can clearly say whether an FEM analysis with 75,000 node, say, is correct or not-without spending too much time while validating the work.

It was in this spirit, the research was carried out to develop an alternative analytical model that can take into cognizance, as realistically as possible, the static and dynamic behavior of a gravity concrete dam vis-à-vis a FEM analysis.

Considering the concrete dam monolith to be usually large in height having limited width in its longitudinal direction and tapering section with very narrow dimension at the top and foundation width less or equal to its height, intuitively it can be stated that the behavior of the dam will be more like a flexural beam element rather than a shear beam or plate loaded in its own plane—though the shear deformation could still play a significant role to affect the final results, as the B/H ratio is usually restricted to 0.7–0.8 (Abel and Popov 1968).

Okamoto (1984) reports that based on instrumented field observation of a number of concrete dams in Japan, it is observed that concrete gravity dams ($H > 100$ m) do behave as a cantilever beam in vertical direction and for such dams the flexural mode dominates. Thus, unless one models a dam cross section to sufficient mesh refinement with lower-order 2D plane strain element, chances of an upper bound solution always exist.

Thus, the fundamental mathematical model we choose to perform the analysis is a beam element—and its validity will be checked against a detailed 2D FEM analysis. For brevity, we start with a dam profile suitable for analytical treatment before converting the same into a numerical model to cater to the irregular geometry of a realistic dam profile.

11.5.1 Static Analysis

Shown in Fig. 11.7 is a dam with upstream water, having its width varying linearly with depth, where B_t is the width at top and B_b the width at base, having height H. For the time being, we ignore any free board of water for simplifying the calculation and will incorporate its effect at a later stage.

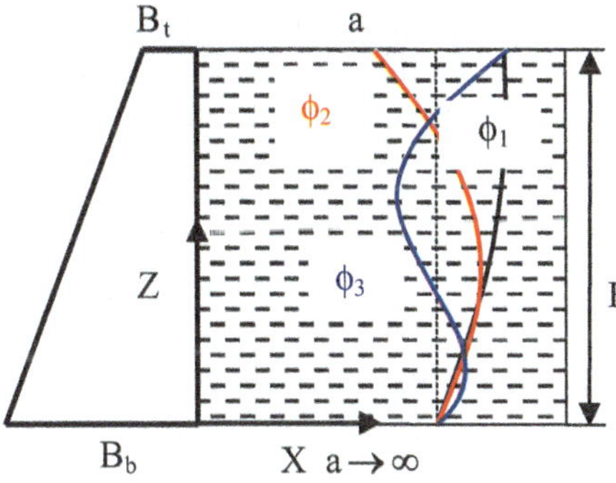

Fig. 11.7 Linearly varying dam with upstream water

Though our principal focus is on seismic loading of gravity dam, however, some discussion regarding its static loading, especially pertaining to its displacement, is essential as it gives a better insight to dynamic behavior that we will take up subsequently.

As the dam profile is idealized as a cantilever beam element having varying moment of inertia

$$\psi_b = (B_t - B_b)/B_b \tag{11.5.1}$$

where ψ_b is a dimensionless term.

The equilibrium equation of the dam profile under static load in flexural mode is expressed as

$$EI_b\left(1+\psi_b\frac{z}{H}\right)^3\frac{d^2u}{dz^2} = \frac{\gamma_w.(H-z)^3}{6} \tag{11.5.2}$$

in which E = Young's modulus of concrete; I_b = moment of inertia of the dam at its base; H = height of the dam; γ_w = weight density of water.

Using $\xi = z/H$, a dimensionless term, the static deflection u in natural coordinates can be expressed as

$$u = \left(\frac{H^5}{6EI_b}\right)\int_0^1\int_0^1 \frac{\gamma_w(1-\xi)^3}{(1+\psi_b\xi)^3}d\xi d\xi \tag{11.5.3}$$

On solving, Eq. 11.5.3 gives

$$u = \left(\frac{\gamma_w H^5}{12EI_b}\right)\left[\frac{-6\psi_b + 6(1+\psi_b)\log_e(1+\psi_b) + \psi_b^3 - 3\psi_b^2}{\psi_b^4}\right] \tag{11.5.4}$$

$$\rightarrow \quad u = \left(\frac{\gamma_w H^5}{12EI_b}\right)J_b \tag{11.5.5}$$

where J_b is a dimensionless term and is expressed by the value within the third parenthesis in Eq. (11.5.4).

In most of the cases, as the H/B_b ratio for a dam varies between 1.25 and 1.42, i.e., <2.0, the shear deformation would affect the final deflection.

Static equilibrium under pure shear deformation for the section varying along its span can be expressed as

$$\eta GA_b\left(1+\psi_b\frac{z}{H}\right)\frac{dv}{dz} = \frac{\gamma_w(H-z)^2}{2} \tag{11.5.6}$$

where η = a shear correction factor, varying from 2/3 to $[10(1+v)/(12+11v)]$ (Shames and Dym 1995) for rectangular sections; v = Poisson's ratio; G = shear modulus of dam material = $E/[2(1 + v)]$; A_b = area of dam at the base, and v = displacement of the shear beam.

The solution of Eq. (11.5.6) is given by

$$\mathrm{v} = \frac{H^3}{2\eta GA_b}\int_0^1 \frac{\gamma_w(1-\xi)^2}{(1+\psi\xi)}d\xi \tag{11.5.7}$$

$$\rightarrow v = \frac{\gamma_w H^3}{4\eta GA_b}\left[\frac{-3\psi_b^2 - 2\psi_b + 2(1+\psi_b^2)\log_e(1+\psi_b) + 4\psi_b\log_e(1+\psi_b)}{\psi_b^3}\right] \tag{11.5.8}$$

$$\rightarrow v = \frac{\gamma_w H^3}{4\eta GA_b}J_s \tag{11.5.9}$$

where J_s is a dimensionless term as expressed by the values within third parenthesis in Eq. (11.5.8).

The total deflection under bending and shear mode combined can be expressed as $\Delta = u + v$, and thus,

$$\Delta = \frac{\gamma_w H^5}{12EI_b}J_b + \frac{\gamma_w H^3}{4\eta GA_b}J_s \tag{11.5.10}$$

Equation (11.5.10) can be further simplified to

$$\Delta = \left(\frac{\gamma_w H^5}{12EI_b}\right)\left[J_b + \left(\frac{6(1+v)}{\eta r_s^2}\right)J_s\right] \tag{11.5.11}$$

in which r_s = slenderness ratio of the dam = H/B_b.

It may be observed in Eq. (11.5.11) that as slenderness ratio of a gravity dam is relatively low, shear deformation plays a significant role in the overall displacement.

Equation (11.5.11) has been checked later against 2D finite element analysis to verify its validity and has been found to be in excellent agreement.

To incorporate the effect of free board, let us assume H_w be the height of water and H the height of dam when $H_w < H$, and hence, let H_W = kH where $k < 1$, as shown in Fig. 11.8.

Using $\psi_{bw} = (B_{tw} - B_b)/B_b$ and $r_{sw} = H_w/B_b$, where B_b, B_t, and B_{tw} are as shown in Fig. 11.8, and the displacement at point P may be exprcssed as

$$\Delta_P = \Delta_o + (H - H_E)\frac{d\Delta_o}{dz} \tag{11.5.12}$$

Fig. 11.8 Cantilever dam with free board

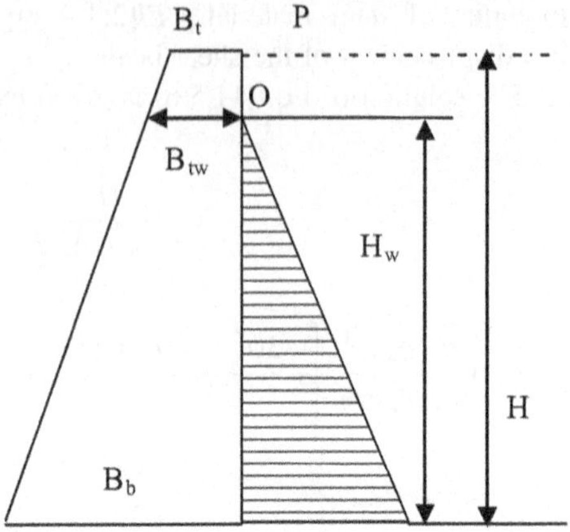

$$\Delta_p = \left(\frac{\gamma_w H_w^5}{12EI_b}\right)\left[J_{bw} + \left(\frac{6(1+v)}{\eta r_{sw}^2}\right)J_{sw}\right]\left(1+\frac{5}{4}\left(\frac{1-k}{k}\right)\right) \tag{11.5.13}$$

Here,

$$J_{bw} = \left[\frac{-6\psi_{bw} + 6(1+\psi_{bw})\log_e(1+\psi_{bw}) + \psi_{bw}^3 - 3\psi_{bw}^2}{\psi_{bw}^4}\right] \tag{11.5.14}$$

and

$$J_{sw} = \left[\frac{-3\psi_{bw}^2 - 2\psi_{bw} + 2(1+\psi_{bw}^2)\log_e(1+\psi_{bw}) + 4\psi_{bw}\log_e(1+\psi_{bw})}{\psi_{bw}^3}\right] \tag{11.5.15}$$

Though we have presented the free board case for clarity and completeness, for practical engineering design considering $H{-}H_{w,}$ a very small term can well be ignored by assuming $H \cong H_w$. Besides simplifying the calculation, the idealization is on a conservative side.

11.5.2 Dynamic Analysis

A dynamic analysis of a dam under earthquake force, as mentioned earlier, is carried out to different levels of complexities, depending on its economic and national importance. Some of the techniques that are in vogue can be summarized as follows:

1. Dynamic analysis of a dam with reservoir empty and hydrodynamic pressure force (assumed to be uncoupled) computed separately. Moments and shears due to this are finally added to the dam moments and shears (reservoir empty case). In this case, in the absence of frequency content data of the earthquake, pressure coefficient as cited by Zangar (1953) is usually adopted.
2. Dynamic analysis of a dam with hydrodynamic pressure on the upstream face is assumed as virtual mass that moves in phase with the dam (Westargaard 1933) —this is usually called coupled analysis, considering incompressible fluid.
3. Coupling the fluid stiffness and mass with that of the dam and performing a combined analysis, which is called coupled fluid–structure interaction (Chopra 1966). However, the solution being in the frequency domain it is essential to know the dominating frequency of the earthquake, which is not always available.
4. Considering foundation stiffness over and above the method 3, as stated above, and performing a comprehensive fluid–soil–structure interaction model (Chopra and Chakrabarti 1980), where a 2D problem having 20,000° and 3D problem with some 75,000° of freedom or more is not uncommon.

11.5.2.1 Flexural/Bending Response of the Dam

We address all these techniques within the framework of the model as proposed herein. For this, we start with a uniform fixed base cantilever beam shown in Fig. 11.9 (ignoring the hydrostatic pressure impinging on it).

Under free vibration condition, the equation of equilibrium is expressed as

$$EI\frac{\partial^4 u}{\partial z^4} = \rho A\frac{\partial^2 u}{\partial t^2} \tag{11.5.16}$$

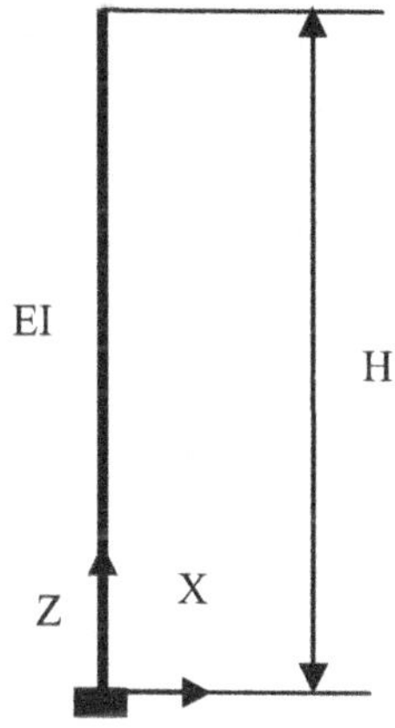

Fig. 11.9 A cantilever beam under free vibration

where ρ = mass density of beam material; A = area of cross section of beam (assumed uniform over its length); I = moment of inertia of the beam element, and E = Young's modulus of the beam material.

For solution of Eq. (11.5.16), we consider

$$u(z,t) = Y(z)q(t) \tag{11.5.17}$$

Based on separation of variable technique, Eq. (11.5.16) can be separated into two linear differential equations, one of which is

$$EI\left(\frac{\mathrm{d}^4 Y}{\mathrm{d}z^4}\right) - \lambda^4 Y = 0 \quad \text{where } \lambda^4 = \frac{\rho A \omega^2}{EI} \tag{11.5.18}$$

The generic solution to this equation is given by

$$Y = C_1 \sin \lambda z + C_2 \cos \lambda z + C_3 \sin h\, \lambda z + C_4 \cos h\, \lambda z \tag{11.5.19}$$

Imposing the four boundary conditions (Fig. 11.9):

$$\begin{aligned} &(1)\ Y = 0 \quad \text{at } z = 0; \quad (2)\ \frac{\mathrm{d}Y}{\mathrm{d}z} = 0 \quad \text{at } z = 0 \\ &(3)\ \frac{\mathrm{d}^3 Y}{\mathrm{d}z^3} = 0 \quad \text{at } z = H; \quad (4)\ \frac{\mathrm{d}^2 Y}{\mathrm{d}z^2} = 0 \quad \text{at } z = H \end{aligned} \tag{11.5.20}$$

we have the shape function as

$$Y_m = \sin\frac{\mu_m z}{H} - \sin h\frac{\mu_m z}{H} - \alpha_m\left(\cos\frac{\mu_m z}{H} - \cos h\frac{\mu_m z}{H}\right) \tag{11.5.21}$$

where m = number of modes 1, 2, 3, ….

$$\begin{aligned} \mu_m &= 1.875,\ 4.694,\ 7.855,\ \frac{2m-1}{2}\pi. \\ \alpha_m &= \frac{\sin\ \mu_m + \sin h\ \mu_m}{\cos\ \mu_m + \cos h\ \mu_m} \end{aligned} \tag{11.5.22}$$

Equation (11.5.21) is actually the eigenvectors of the beam and is usually expressed as

$$\begin{aligned} f_i^{\mathrm{b}}(z) &= \sin\frac{\mu_i z}{H} - \sin h\frac{\mu_i z}{H} - \alpha_i\left(\cos\frac{\mu_i z}{H} - \cos h\frac{\mu_i z}{H}\right) \\ \text{and}\quad f_j^{\mathrm{b}}(z) &= \sin\frac{\mu_j z}{H} - \sin h\frac{\mu_j z}{H} - \alpha_j\left(\cos\frac{\mu_j z}{H} - \cos h\frac{\mu_j z}{H}\right) \end{aligned} \tag{11.5.23}$$

Here, the superscript b stands for bending, and subscripts i and j stand for mode numbers 1, 2, 3....

Equation (11.5.23) will now be used to determine the dynamic stiffness of the dam cross section shown in Fig. 11.7 when

$$I(z) = I_{\mathrm{b}}\left(1 + \psi_b \frac{z}{H}\right)^3 \tag{11.5.24}$$

and

$$A(z) = A_{\mathrm{b}}\left(1 + \psi_b \frac{z}{H}\right) \tag{11.5.25}$$

Now for the dam with variable cross section under a loading q, the equation of equilibrium is expressed as

$$EI_z \frac{\mathrm{d}^4 u}{\mathrm{d}z^4} = q \tag{11.5.26}$$

We assume $u = u_0 f_i^{\mathrm{b}}(z)$ as the displacement profile to Eq. (11.5.26). If it would have been an exact solution, we would have had

$$EI_z \frac{\mathrm{d}^4 u}{\mathrm{d}z^4} - q = 0 \tag{11.5.27}$$

Since it is not, we have a residual error Re expressed as

$$\mathrm{Re} = EI_z \frac{\mathrm{d}^4 u}{\mathrm{d}z^4} - q \tag{11.5.28}$$

To minimize the error based on Galerkin's weighted residual technique we integrate it over the domain, multiplying it by the weighted function $f_j^{\mathrm{b}}(z)$ when

$$\int_0^H \left[EI_z u_0 \frac{\mathrm{d}^4}{\mathrm{d}z^4} f_i^{\mathrm{b}}(z) - q\right] f_j^{\mathrm{b}}(z)\mathrm{d}z = 0 \tag{11.5.29}$$

$$\int_0^H EI_z u_0 \frac{\mathrm{d}^2}{\mathrm{d}z^2}\left(\frac{\mathrm{d}^2 f_i^{\mathrm{b}}(z)}{\mathrm{d}z^2}\right) f_j^{\mathrm{b}}(z)\mathrm{d}z = \int_0^H q f_j^{\mathrm{b}}(z)\mathrm{d}z \tag{11.5.30}$$

On successive integration of Eq. (11.5.30) by parts, one can have

$$\left[EI_z u_0 \frac{\mathrm{d}^2}{\mathrm{d}z^2}\left(\frac{\mathrm{d}f_i^{\mathrm{b}}(z)}{\mathrm{d}z}\right)\right]_0^H - \left[EI_z u_0 \frac{\mathrm{d}^2}{\mathrm{d}z^2}\left(\frac{\mathrm{d}f_i^{\mathrm{b}}(z)}{\mathrm{d}z}\right)\right]_0^H + \int_0^H EI_z u_0 \frac{\mathrm{d}^2}{\mathrm{d}z^2}\left[\frac{\mathrm{d}^2 f_i^{\mathrm{b}}(z)}{\mathrm{d}z^2}\right] f_j^{\mathrm{b}}(z)\mathrm{d}z$$
$$= \int_0^H q f_j^{\mathrm{b}}(z)\mathrm{d}z \tag{11.5.31}$$

The first two terms above cancel out, leaving

$$\int_0^H EI_z u_0 \frac{\mathrm{d}^2}{\mathrm{d}z^2}\left[\frac{\mathrm{d}^2 f_i^{\mathrm{b}}(z)}{\mathrm{d}z^2}\right] f_j^{\mathrm{b}}(z)\mathrm{d}z = \int_0^H q f_j^{\mathrm{b}}(z)\mathrm{d}z \tag{11.5.32}$$

Considering the above in the form $[K]\{\Delta\} = \{P\}$, we conclude that as per Galerkinian's basis, the dynamic stiffness matrix can be expressed as

$$[K]_{ij} = E\int_0^H \frac{\mathrm{d}^2}{\mathrm{d}z^2}\left[I_z \frac{\mathrm{d}^2}{\mathrm{d}z^2} f_i^{\mathrm{b}}(z)\right] f_j^{\mathrm{b}}(z)\mathrm{d}z \tag{11.5.33}$$

For a dam width varying linearly, Eq. (11.5.33) can be expressed as

$$[K]_{ij} = EI_{\mathrm{b}}\int_0^H \frac{\mathrm{d}^2}{\mathrm{d}z^2}\left[\left(1+\psi_b \frac{z}{H}\right)^3 \frac{\mathrm{d}^2}{\mathrm{d}z^2} f_i^{\mathrm{b}}(z)\right] f_j^{\mathrm{b}}(z)\mathrm{d}z \tag{11.5.34}$$

and the mass matrix can be expressed as (Karnovsky and Lebed 2001)

$$[M]_{ij} = \frac{\gamma_c A_{\mathrm{b}}}{g}\int_0^H \left(1+\psi_b \frac{z}{H}\right) f_i^{\mathrm{b}}(z) f_j^{\mathrm{b}}(z)\mathrm{d}z \tag{11.5.35}$$

In natural coordinates, [K] and [M] can be written as

$$[K]_{ij} = \frac{EI_{\mathrm{b}}}{H^3}\int_0^1 \frac{\mathrm{d}^2}{\mathrm{d}\xi^2}\left[(1+\psi_b \xi)^3 \frac{\mathrm{d}^2}{\mathrm{d}\xi^2} f_i^{\mathrm{b}}(\xi)\right] f_j^{\mathrm{b}}(\xi)\mathrm{d}\xi \tag{11.5.36}$$

$$[M]_{ij} = \frac{\gamma_c A_{\mathrm{b}} H}{g}\int_0^1 (1+\psi_b \xi) f_i^{\mathrm{b}}(\xi) f_j^{\mathrm{b}}(\xi)\mathrm{d}\xi \tag{11.5.37}$$

where $f_i^{\rm b}(\xi)$ and $f_j^{\rm b}(\xi)$ are expressed in Eq. (11.5.23), considering $\xi = z/H$.

From Eqs. (11.5.36) and (11.5.37), it is observed that the stiffness and mass matrix are actually dependent on ψ_b which is again a function of the ratio B_t/B_b.

For instance, for $B_t/B_b = 0.1$ (a typical value used in concrete dam), one can have, for the first four modes,

$$[K]_{4\times4} = \frac{EI_b}{H^3}\begin{bmatrix} 13.965 & 24.495 & 23.911 & 17.332 \\ 24.495 & 165.589 & 289.033 & 222.15 \\ 23.911 & 289.033 & 1167 & 1510 \\ 17.332 & 222.15 & 1510 & 4280 \end{bmatrix} \quad (11.5.38)$$

$$[M]_{4\times4} = \frac{\gamma_c A_b H}{g}\begin{bmatrix} 0.509 & 0.185 & -0.025 & 0.011 \\ 0.185 & 0.449 & 0.169 & -0.018 \\ -0.025 & 0.169 & 0.522 & 0.177 \\ 0.011 & -0.018 & 0.177 & 0.535 \end{bmatrix} \quad (11.5.39)$$

Considering the general eigenvalue problem $[K]\{\varphi\} = \lambda[M]\{\varphi\}$ and solving the same by Jacobi's method (Bathe 1980), one can obtain $[\lambda]$, and knowing $\omega = \sqrt{\lambda}$ and $T = 2\pi/\omega$, we finally have

$$[T_f]_{4\times4} = \frac{H^2}{B_b}\sqrt{\frac{\gamma_c}{gE}}\begin{bmatrix} 4.693 & 0 & 0 & 0 \\ 0 & 1.417 & 0 & 0 \\ 0 & 0 & 0.583 & 0 \\ 0 & 0 & 0 & 0.242 \end{bmatrix} \quad (11.5.40)$$

$$[\varphi]_{4\times4} = \begin{bmatrix} -0.984 & -0.24 & 0.12 & 0.071 \\ 0.176 & 0.926 & -0.126 & -0.145 \\ -0.029 & -0.287 & 0.93 & 0.033 \\ 5.099\times10^{-3} & 0.052 & -0.324 & -0.986 \end{bmatrix} \quad (11.5.41)$$

Expanding Eqs. (11.5.38) and (11.5.39) to a 6 × 6 matrix to get a more accurate result, the time period coefficients for the first three modes are tabulated for various B_t/B_b ratios in Table 11.9. Here, the time period is expressed in the form

$$[T_f]_i = C_{{\rm TF}i}\frac{H^2}{B_b}\sqrt{\frac{\gamma_c}{gE}} \quad (11.5.42)$$

Table 11.9 Values $C_{\rm Tf}$ for first three modes

B_t/B_b	Mode-1	Mode-2	Mode-3
0.05	4.431	1.414	0.589
0.1	4.693	1.412	0.583
0.15	4.9	1.397	0.571
0.2	5.069	1.368	0.555
0.25	5.21	1.336	0.539

Table 11.10 Comparison of frequency parameter proposed versus exact solution

B_t/B_b	Proposed method			Gaines and Volterra (1966)		
	Mode-1	Mode-2	Mode-3	Mode-1	Mode-2	Mode-3
0.1	4.632	15.014	33.855	4.631	14.931	32.86
0.2	4.293	15.761	37.178	4.293	15.744	36.92
0.3	4.082	16.630	40.67	4.082	16.626	40.642
0.4	3.934	17.49	44.048	3.934	17.488	44.055

The frequency values obtained herein by the proposed method are compared with exact solution of a linearly varying truncated wedge (Gaines and Volterra 1966) in terms of frequency parameters for various B_t/B_b ratios in Table 11.10.

The results are found to be in excellent agreement.

However, for a concrete dam this will give a lower bound result, and for considering the geometry of the dam, as H/B_b ratio is normally ≤ 2, shear deformation will play a significant role on the final outcome.

11.5.2.2 Shear Beam Response of Dam

The free vibration equation of a shear beam of uniform cross-sectional area A can be expressed as

$$G\frac{\partial^2 u}{\partial z^2} = \rho_c\frac{\partial^2 u}{\partial t^2} \tag{11.5.43}$$

$$\rightarrow \frac{\partial^2 u}{\partial z^2} = \frac{1}{Vs^2}\frac{\partial^2 u}{\partial t^2} \tag{11.5.44}$$

Solving Eq. (11.5.44), it can be shown that the eigenvector of the problem (Chowdhury and Dasgupta 2007) can be expressed as

$$f_i^s = \cos\frac{(2i-1)\pi z}{2H} \tag{11.5.45}$$

Here, the superscript s stands for the term shear and subscript i stands for Modes 1, 2, 3….

Now since in this case the area of dam varies as $A_z = A_t(1 + \psi_t z/H)$, it changes the characteristics of the differential equation. When applying Galerkin's weighted residual technique, it can be shown (Chowdhury and Dasgupta 2012) that the stiffness matrix can be expressed as

$$[K] = -\int_0^H \frac{\mathrm{d}}{\mathrm{d}z}\left[GA_z \frac{\mathrm{d}}{\mathrm{d}z} f_i^{\mathrm{s}}(z)\right] f_j^{\mathrm{s}}(z)\mathrm{d}z \tag{11.5.46}$$

$$\rightarrow [K] = -GA_t \int_0^H \frac{\mathrm{d}}{\mathrm{d}z}\left[\left(1+\psi_t \frac{z}{H}\right) \frac{\mathrm{d}}{\mathrm{d}z} f_i^{\mathrm{s}}(z)\right] f_j^{\mathrm{s}}(z)\mathrm{d}z \tag{11.5.47}$$

where $\psi_t = \frac{B_{\mathrm{b}} - B_{\mathrm{t}}}{B_{\mathrm{t}}}$ a dimensionless term.

In natural coordinates ($\xi = z/H$), this is expressed as

$$[K] = -\frac{GA_t}{H} \int_0^1 \frac{\mathrm{d}}{\mathrm{d}\xi}\left[(1+\psi_t \xi) \frac{\mathrm{d}}{\mathrm{d}\xi} f_i^{\mathrm{s}}(\xi)\right] f_j^{\mathrm{s}}(\xi)\mathrm{d}\xi \tag{11.5.48}$$

and the mass matrix is expressed as

$$[M]_{ij} = \frac{\gamma_c A_t H}{g} \int_0^1 (1+\psi_t \xi) f_i^{\mathrm{s}}(\xi) f_j^{\mathrm{s}}(\xi)\mathrm{d}\xi \tag{11.5.49}$$

where f_i^{s} is as expressed in Eq. (11.5.45) and ψ_t is a dimensionless term as explained above when the origin ($z = 0$) is at the crest of the dam. Expanding the above for a typical value of $B_{\mathrm{t}}/B_{\mathrm{b}} = 0.1$ (i.e., $\psi_t = 9$), we have for the first three modes

$$[K] = \frac{\eta G A_t}{H} \begin{bmatrix} 9.035 & -6.75 & 1.25 \\ -6.75 & 63.32 & -33.75 \\ 1.25 & -33.75 & 171.884 \end{bmatrix} \tag{11.5.50}$$

where η = shear correction factor = 2/3 for rectangular section.

$$[M] = \frac{\gamma_c A_t H}{g} \begin{bmatrix} 1.838 & -0.912 & -0.101 \\ -0.912 & 2.589 & -0.912 \\ -0.101 & -0.912 & 2.714 \end{bmatrix} \tag{11.5.51}$$

On eigensolution, this finally results in the time period for shear deformation as

$$[T_S]_i = \frac{H}{V_s \sqrt{\eta}} \begin{bmatrix} 2.854 & 0 & 0 \\ 0 & 1.219 & 0 \\ 0 & 0 & 0.768 \end{bmatrix} \tag{11.5.52}$$

$$\rightarrow [T_S]_i = [C_S]_i \frac{H}{V_s \sqrt{\eta}} \tag{11.5.53}$$

Table 11.11 Values C_S for first three modes

B_t/B_b	Mode-1	Mode-2	Mode-3
0.05	2.74	1.186	0.753
0.1	2.854	1.22	0.768
0.15	2.958	1.243	0.776
0.2	3.053	1.26	0.782
0.25	3.142	1.273	0.785

where $[C_S]_i$ are the diagonal elements of the matrix in Eq. (11.5.52). Values of C_S, for the first three modes, for different values of B_t/B, are presented in Table 11.11.

The time period of flexure and shear can be now combined as explained hereafter.

If a load vector P is applied on a body, the displacement of the body due to bending as well as shear deformation can be expressed as (Shames and Dym 1995)

$$\delta_t = \delta_b + \delta_s \tag{11.5.54}$$

where δ_t = total displacement of the system; δ_b = displacement due to bending; and δ_s = displacement due to shear.

Equation (11.5.54) can be expressed as

$$\frac{P}{K_e} = \frac{P}{K_b} + \frac{P}{K_s} \tag{11.5.55}$$

where P = total load; K_b and K_s are stiffness due to bending and shear deformation, respectively.

Equation (11.5.55) can be expressed as

$$m/K_e = m/K_b + m/K_s \tag{11.5.56}$$

$$\rightarrow 1/\omega_e^2 = 1/\omega_b^2 + 1/\omega_s^2 \tag{11.5.57}$$

where ω_e = combined frequency of the system due to bending and shear; ω_b = frequency due to bending only; ω_s = frequency due to shear only.

Equation (11.5.57) is precisely the classical Dunkerley's equation (Dunkerley 1894) by which combined frequency of multi-body systems can be derived by treating each system in isolation and is often used in dynamic soil–structure interaction (DSSI) problems (Chowdhury and Singh 2010; Chowdhury and Dasgupta 2011, 2012).

Equation (11.5.57) finally yields

$$T_{ei}^2 = T_{bi}^2 + T_{si}^2 \tag{11.5.58}$$

Substituting Eqs. (11.5.42) and (11.5.53) in Eq. (11.5.58), one can have

$$T_{ei} = C_{\mathrm{TF}i}\frac{H^2}{B_\mathrm{b}}\sqrt{\frac{\gamma_c}{gE}}\left[1+\left(\frac{C_{Si}}{C_{\mathrm{TF}i}}\right)^2\frac{2(1+\nu)}{\eta(H/B_\mathrm{b})^2}\right]^{0.5} \tag{11.5.59}$$

Considering $r_\mathrm{s} = H/B_\mathrm{b}$, the slenderness ratio of the dam, the time period is

$$T_{ei} = C_{\mathrm{TF}i}\frac{H^2}{B_\mathrm{b}}\sqrt{\frac{\gamma_c}{gE}\left[1+\left(\frac{C_{Si}}{C_{\mathrm{TF}i}}\right)^2\frac{2(1+\nu)}{\eta r_\mathrm{s}^2}\right]} \tag{11.5.60}$$

Equation (11.5.60) is of particular interest as it explains the effect of shear deformation that goes on to elongate the time period. It will be observed from Eq. (11.5.60) that it is strongly dependent on the slenderness ratio r_s. As B_b reduces, the value of r_s increases rapidly (in terms of its square), and for large values of r_s makes the second term within the parenthesis insignificant when the flexural mode dominates the response.

For varying moment of inertia that is not linear with depth, the stiffness and mass matrix for first three modes can be numerically computed.

For flexural mode:

$$[K_b]_{ij} = \frac{E}{H^3}\left[\sum_{\xi=1}^{0} I(\xi)\left(\frac{\mathrm{d}^2 f_i^\mathrm{b}(\xi)}{\mathrm{d}\xi^2}\right)\left(\frac{\mathrm{d}^2 f_j^\mathrm{b}(\xi)}{\mathrm{d}\xi^2}\right)\right] \tag{11.5.61}$$

$$[M_b]_{ij} = \frac{\gamma_c H}{g}\left[\sum_{\xi=1}^{0} A(\xi) f_i^\mathrm{b}(\xi) f_j^\mathrm{b}(\xi)\right] \tag{11.5.62}$$

Here, the shape function $f_i^\mathrm{b}(\xi)$ is as expressed in Eq. (11.5.23).

For shear deformation mode:

$$[K_s]_{ij} = \frac{G}{H}\left[\sum_{\xi=1}^{0} A(\xi)\frac{\mathrm{d}}{\mathrm{d}\xi}f_i^\mathrm{s}(\xi)\frac{\mathrm{d}}{\mathrm{d}\xi}f_j^\mathrm{s}(\xi)\right] \tag{11.5.63}$$

$$[M_b]_{ij} = \frac{\gamma_c H}{g}\left[\sum_{\xi=1}^{0} A(\xi) f_i^\mathrm{s}(\xi) f_j^\mathrm{s}(\xi)\right] \tag{11.5.64}$$

The shape function $f_i^\mathrm{s}(\xi)$ is expressed in Eq. (11.5.45).

Applying a 41-point numerical integration between 0 and 1 based on trapezoidal rule, each of the terms of the stiffness and mass matrix can be computed. Of course, in this case it is not possible to derive a straightforward formula like Eq. (11.5.60) and a proper eigenvalue analysis has to be carried out to determine the time period. However, for a matrix of the order 4×4, it can be computed through the standard utility package like MATLAB or Mathcad.

The final time period of the dam can be obtained by using Eq. (11.5.58).

The time period expression obtained by the proposed method has been compared for the celebrated Pine Flat Dam on which sufficient research data is available, especially from University of California, Berkeley.

Gogoi and Athani (2010) in India have carried out a detailed finite element analysis of the same dam in ANSYS 10 (2010) using solid quadrilateral PLANE 42 elements 300 in numbers having 341 nodes, as shown in Fig. 11.10. The outcome is as presented hereafter in Table 11.12.

It is observed that the time periods by the proposed method are in excellent agreement with the established research and FEM analysis using ANSYS.

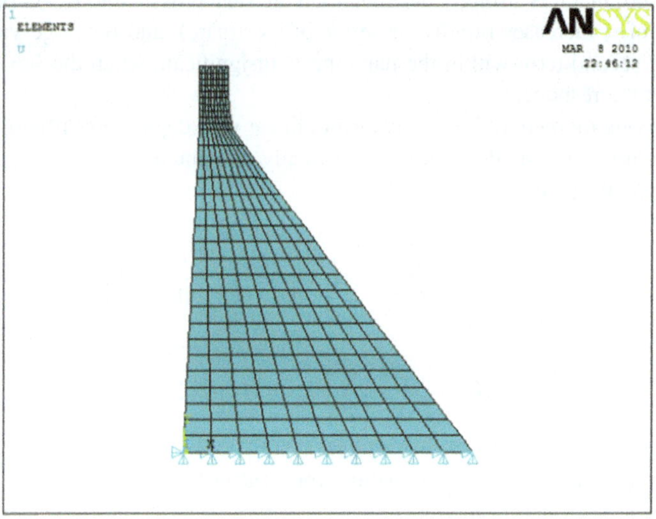

Fig. 11.10 Finite element model of Pine Flat Dam after Gogoi et al. (Reproduced with kind permission of the author)

Table 11.12 Comparison of fundamental time period proposed versus established data

Method	Fundamental time period (s)	Remarks
Fenves and Chopra (1986)	0.311	Simplified procedure
Fenves and Chopra (1986)	0.317	Refined procedure
Gogoi and Athani (2010)	0.313	FEM analysis (Fig. 11.10)
Okamoto (1984)	0.305	Japanese practice
IS-1893-1984	0.289	Recommended for H > 100 m
Proposed method	0.312	Vide Eq. (11.5.60)
Proposed numerical method	0.294	Three-mode analysis
Proposed numerical method	0.305	Four-mode analysis

As per modal response analysis,

$$S_d = S_a/\omega^2 \tag{11.5.65}$$

where S_d = modal amplitude; S_a = spectral acceleration corresponding to time period T.

Thus as per norms prescribed in the code, the amplitude of vibration for the fundamental mode is expressed as

$$u_{di} == \kappa_i \beta \frac{S_a}{4\pi^2} C_{Ti}^2 \frac{H^4}{B_b^2} \frac{\gamma_c}{gE} \left[1 + \left(\frac{C_{Si}}{C_{TFi}}\right)^2 \frac{2(1+\nu)}{\eta r_s^2}\right] [\phi_i]^T \{f_i^b(\xi)\} \tag{11.5.66}$$

Considering $M_z = -EI\left(d^2 u_{di}/dz^2\right)$, we have for the ith mode,

$$M_{zi} = -\kappa_i \beta \gamma_c (1 + \psi_b \xi)^3 \left(\frac{S_{ai}}{g}\right) \left(\frac{C_{Ti}^2}{48\pi^2}\right) B_b H^2 \left[1 + \left(\frac{C_{Si}}{C_{TFi}}\right)^2 \frac{2(1+\nu)}{\eta r_s^2}\right] [\phi_i]^T \left\{\frac{d^2}{d\xi^2} f_i^b(\xi)\right\} \tag{11.5.67}$$

Similarly,

$$V_{zi} = -\kappa_i \beta \gamma_c (1 + \psi_b \xi)^3 \left(\frac{S_{ai}}{g}\right) \left(\frac{C_{Ti}^2}{48\pi^2}\right) B_b H \left[1 + \left(\frac{C_{Si}}{C_{TFi}}\right)^2 \frac{2(1+\nu)}{\eta r_s^2}\right] [\phi_i]^T \left\{\frac{d^3 f_i^b(\xi)}{d\xi^3}\right\} \tag{11.5.68}$$

where $[\phi_i]^T$ is the transpose of the scaling matrix of the eigenvectors, as shown in Eq. (11.5.41), and κ_i = modal mass participation factor and for the ith mode is expressed as

$$\kappa_i = \int_0^1 (1 + \psi\xi) f_i^b(\xi) d\xi \bigg/ \int_0^1 (1 + \psi\xi) f_i^b(\xi)^2 d\xi \tag{11.5.69}$$

where β = code factor = $ZI/2R$, Z = zone factor, I = importance factor = 3.0, and R = response reduction factor = 1.5.

The modal participation factor κ is found to be dependent on the factor ψ_b vis-à-vis B_t/B_b ratio.

Values of κ for different B_t/B_b ratio values for the fundamental mode are furnished below in Table 11.13.

For varying moment of inertia that is not linear with depth,

Table 11.13 Modal mass participation factor for fundamental mode for B_t/B_b ratio

Bt/Bb	0.05	0.1	0.15	0.2	0.25
κ	0.762	0.726	0.699	0.679	0.62

$$S_{di} = \frac{S_a}{4\pi^2} T_{ei}^2 \tag{11.5.70}$$

Thus,

$$u_{di} == \kappa_i \beta \frac{S_{ai}}{4\pi^2} T_{ei}^2 [\phi_i]^{\mathrm{T}} \{f_i^{\mathrm{b}}(\xi)\} \tag{11.5.71}$$

$$M_{zi} = -\kappa_i \beta \frac{EI(\xi)}{H^2} \left(\frac{S_{ai} T_{ei}^2}{4\pi^2}\right) [\phi_i]^{\mathrm{T}} \left\{\frac{\mathrm{d}^2 f_i^{\mathrm{b}}(\xi)}{\mathrm{d}\xi^2}\right\} \tag{11.5.72}$$

$$V_{zi} = -\kappa_i \beta \frac{EI(\xi)}{H^3} \left(\frac{S_{ai} T_{ei}^2}{4\pi^2}\right) [\phi_i]^{\mathrm{T}} \left\{\frac{\mathrm{d}^3 f_i^{\mathrm{b}}(\xi)}{\mathrm{d}\xi^3}\right\} \tag{11.5.73}$$

where

$$\kappa_i = \sum_{\xi=0}^{1} B(\xi) f_i^{\mathrm{b}}(\xi) \Big/ \sum_{\xi=0}^{1} B(\xi) f_i^{\mathrm{b}}(\xi)^2 \tag{11.5.74}$$

11.5.2.3 Hydrodynamic Pressure on Dam Upstream

When seismic waves travel through the reservoir water, it is apparent that the reservoir medium will get excited and start vibrating. The general character of such mode shapes of vibrations is as shown in Fig. 11.7 by the curves ϕ_1, ϕ_2, ϕ_3, etc.

In two dimensions (Fig. 11.7), the propagation of waves through the water medium can be expressed as

$$\frac{\partial^2 u(x,z,t)}{\partial x^2} + \frac{\partial^2 u(x,z,t)}{\partial z^2} = \frac{1}{c^2} \frac{\partial^2 u(x,z,t)}{\partial t^2} \tag{11.5.75}$$

where c = sound wave velocity through water expressed as $\sqrt{B_m/\rho_w}$ = 1439 m/s; B_m = bulk modulus of water = 2.11 x 10^6 kN/m^2, and ρ_w = mass density of water; $u(x, z, t)$ = the displacement function and can be considered as $u = H(x) \cdot Q(z) \cdot P(t)$ [H, P, Q are the three independent functions of x, z, and t, respectively].

It can be shown that Eq. (11.5.75) can be broken down into three ordinary differential equations of second order, given by

$$\frac{d^2P}{dt^2} + \lambda^2 P = 0 \tag{11.5.76}$$

where $\lambda = ic$ where i is a constant

$$\frac{d^2H(x)}{dx^2} + k^2 H(x) = 0, \tag{11.5.77}$$

where k is another constant

$$\frac{d^2Q(z)}{dz^2} + p^2 Q(z) = 0 \tag{11.5.78}$$

where p, i, and k are related through $p^2 = i^2 - k^2$.

The solutions of Eqs. (11.5.77) and (11.5.78) are given by

$$H(x) = A \cos kx + B \sin kx \tag{11.5.79}$$

$$Q(z) = C \cos pz + D \sin pz \tag{11.5.80}$$

Imposing the boundary conditions

At $x = 0$, $u = 0 \rightarrow H(x) = 0$, which implies $A = 0$.

At $x = a$ (where a may be very large), $u = 0 \rightarrow H(a) = 0$ which implies $H(a) = B \sin ka = 0$

$$\rightarrow k = m\pi/a \quad \text{and} \quad \text{hence} \tag{11.5.81}$$

$$H_m(x) = \sin \frac{m\pi x}{a}. \tag{11.5.82}$$

At the free surface, i.e., where $z = 0$ (the crest of the dam in this case), the boundary conditions are

At $z = 0$, shear strain, $\frac{\partial u}{\partial z} = 0$ or $\frac{dQ(z)}{dz} = 0$ which implies $D = 0$.

At $z = H$, displacement, $u = 0$, i.e., $Q(H) = 0$.

This implies that

$$p = [(2n - 1)\pi]/2H \tag{11.5.83}$$

and hence

$$Q(z) = \cos \frac{(2n - 1)\pi z}{2H} \tag{11.5.84}$$

The eigenvectors of the problem can thus be established as

$$f^{\mathrm{f}}(x,z) = H(x)Q(z) = \sin\frac{m\pi x}{a}\cos\frac{(2n-1)\pi z}{2H} \tag{11.5.85}$$

where superscript f stands for free field, and m, n are mode numbers = 1, 2, 3…

Again, from the description of Eqs. (11.5.76) and (11.5.78)

$$\lambda = c\sqrt{p^2 + k^2} \tag{11.5.86}$$

Substituting the value of p and k from Eqs. (11.5.83) and (11.5.81), one can have

$$\lambda = c\pi\sqrt{\frac{m^2}{a^2} + \frac{(2n-1)^2}{4H^2}}. \tag{11.5.87}$$

For the fundamental mode, considering that m, n = 1 and $\lim a \to \infty$, the value of λ reduces to

$$\lambda = \omega = \pi c\sqrt{0 + \frac{1}{4H^2}} \text{and} \quad \omega = \frac{\pi c}{2H} \tag{11.5.88}$$

The period T can be derived from Eq. (11.5.88) as $T = 4H/[(2n-1)c]$ which is basically the free field time period in one dimension for the reservoir.

For $\lim a \to \infty$, the first term of eigenfunction (in x-direction) can be dropped in Eq. (11.5.85) which gives the eigenfunction in one dimension as

$$f_n^{\mathrm{f}}(z) = \cos\frac{(2n-1)\pi z}{2H} \tag{11.5.89}$$

Considering modal analysis, the amplitude of displacement can be expressed as:

$$S_d = \frac{\widehat{S}_a}{\omega^2} \tag{11.5.90}$$

Here, $\widehat{S}_a$= spectral acceleration corresponding to free field time period of the reservoir.

Based on code definition, the amplitude of vibration may be expressed as

$$u_n = \kappa_n\beta\frac{\widehat{S}_{an}}{\omega^2}\cos(2n-1)\frac{\pi z}{2H} \tag{11.5.91}$$

where κ_n is the modal mass participation factor and is expressed as

Table 11.14 Values of modal mass participation factor

Mode	1	2	3	4
κ	$8/(\pi + 2)$	$-8/(3\pi - 2)$	$8/(5\pi + 2)$	$-8/(7\pi - 2)$

$$\kappa_n = \sum_{i=1}^{n} m_n f_n^{\mathrm{f}} \Big/ \sum_{i=1}^{n} m_n f_n^{\mathrm{f}2} \tag{11.5.92}$$

For the first four modes, this is expressed as given in Table 11.14.

Here, β is a code factor and is expressed as $\beta = ZI/2R$ where Z = zone factor, I = importance factor, and R = response reduction factor as per IS-1893 (2016).[4]

The strain within the water body is expressed as

$\varepsilon_{xx} = 0$ and $\varepsilon_{zz} = \frac{\partial u}{\partial z}$, which gives

$$\varepsilon_{zzn} = -\kappa_n \frac{(2n-1)\pi}{2H} \beta \frac{\widehat{S}_{an} T_n^2}{4\pi^2} \sin (2n-1)\frac{\pi z}{2H} \tag{11.5.93}$$

Equation (11.5.93), on simplification for the first four modes, finally gives

$$\varepsilon_{zz1} = -\frac{16}{\pi(\pi+2)} \beta\gamma_s \left(\frac{\widehat{S}_{a1}}{g}\right) \frac{H}{B_m} \sin\frac{\pi z}{2H} \tag{11.5.94}$$

$$\varepsilon_{zz2} = \frac{16}{3\pi(3\pi-2)} \beta\gamma_s \left(\frac{\widehat{S}_{a2}}{g}\right) \frac{H}{B_m} \sin\frac{3\pi z}{2H} \tag{11.5.95}$$

$$\varepsilon_{zz3} = -\frac{16}{5\pi(5\pi+2)} \beta\gamma_s \left(\frac{\widehat{S}_{a2}}{g}\right) \frac{H}{B_m} \sin\frac{5\pi z}{2H} \tag{11.5.96}$$

$$\varepsilon_{zz4} = \frac{16}{7\pi(7\pi-2)} \beta\gamma_s \left(\frac{\widehat{S}_{a4}}{g}\right) \frac{H}{B_m} \sin\frac{7\pi z}{2H} \tag{11.5.97}$$

Considering $p = B_m \frac{\partial u}{\partial z}$ = the dynamic pressure on the dam's face, for the first four modes with 15% damping for fluid (Newmark 1971) (multiplying by a scaling factor of 3/4 as per IS-1893-2002), we finally have

[4]Refer to Chap. 2 for further details.

$$p_1 = -\frac{12}{\pi(\pi+2)}\beta\left(\frac{\hat{S}_{a1}}{g}\right)\gamma_w H \sin\frac{\pi z}{2H} \tag{11.5.98}$$

$$p_2 = \frac{12}{3\pi(3\pi-2)}\beta\left(\frac{\hat{S}_{a2}}{g}\right)\gamma_w H \sin\frac{3\pi z}{2H} \tag{11.5.99}$$

$$p_3 = -\frac{12}{5\pi(5\pi+2)}\beta\left(\frac{\hat{S}_{a3}}{g}\right)\gamma_w H \sin\frac{5\pi z}{2H} \tag{11.5.100}$$

$$p_4 = \frac{12}{7\pi(7\pi-2)}\beta\left(\frac{\hat{S}_{a4}}{g}\right)\gamma_w H \sin\frac{7\pi z}{2H} \tag{11.5.101}$$

Equations (11.5.98)–(11.5.101) can also be generically expressed as

$$p = \text{coeff} \times \beta \times \left(\frac{\hat{S}_a}{g}\right) \times \gamma_w \times H \tag{11.5.102}$$

Here, H can also be considered as H_w (free head of water in the reservoir), and in case, it is considered that the height of water in the reservoir and the height of the dam (H) are different.

The pressure coefficients are plotted, as shown in Fig. 11.11.

Based on Fig. 11.11, one may observe that

- The higher-order modes have significantly reduced response.
- The SRSS values of pressure for the first three modes vis-à-vis fundamental mode—there is a negligible difference.

Thus, if one works with fundamental mode only, for practical engineering problems, it is more than adequate.

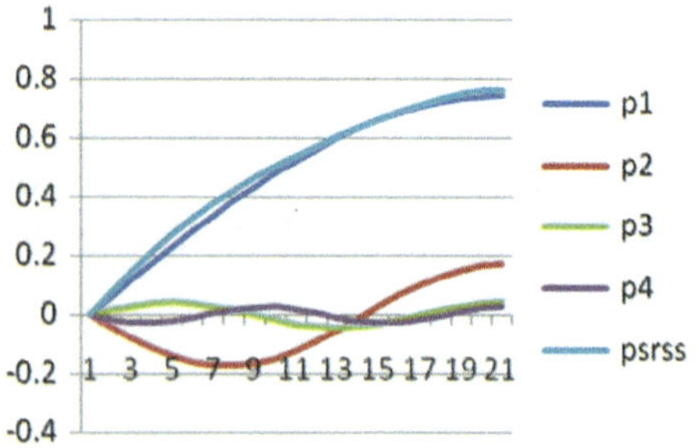

Fig. 11.11 Pressure coefficient of hydrodynamic pressure for the first four modes including SRSS values

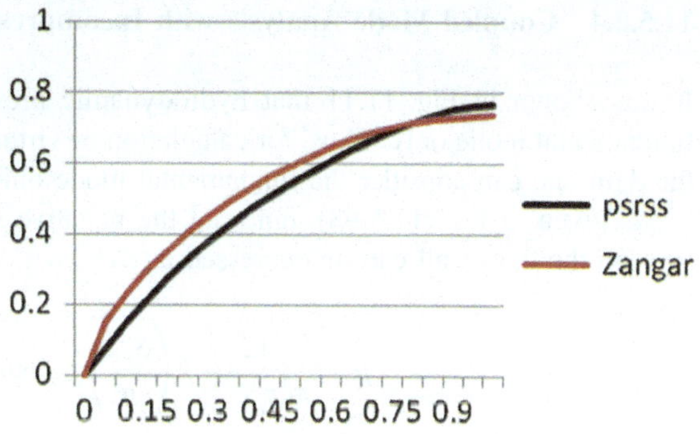

Fig. 11.12 Hydrodynamic pressure coefficients proposed versus Zangar's (1953) solution

The SRSS values of pressure coefficients as derived herein have been compared with Zangar's (1953) in Fig. 11.12; Zangar developed coefficients of hydrodynamic pressure based on experimentation and analog model.

It is observed that the values are in quite close agreement.

Considering $V_z = \int p dz$, for the fundamental mode we have

$$V_z = \frac{24}{\pi^2(\pi+2)}\beta\left(\frac{\hat{S}_{a1}}{g}\right)\gamma_w H^2 \cos\left(\frac{\pi z}{2H}\right) \tag{11.5.103}$$

Similarly for the fundamental mode,

$$M_z = \frac{48}{\pi^3(\pi+2)}\beta\left(\frac{\hat{S}_{a1}}{g}\right)\gamma_w H^3 \left[1 - \sin\left(\frac{\pi z}{2H}\right)\right] \tag{11.5.104}$$

Equations (11.5.103) and (11.5.104), when added to SRSS values of Eqs. (11.5.68) and (11.5.67), give the complete dynamic response of the dam under seismic force. One of the major advantages with this model is that the hydrodynamic pressure is obtained within the modal response framework, without pre-knowing the frequency content of the earthquake force.

However, one primary lacunae in this model is that the fluid and the dam are treated separately (uncoupled), and finally, their independent values are added together. When a body immersed in liquid starts vibrating, a part of the fluid gets locked with the body and starts moving in the same phase. This coupling was first pointed out by Westargaard (1933). Based on this, many researchers prefer to assume the pressure as derived by Zangar (1953) to be the added virtual mass that gets locked to the dam upstream and moves in the same phase. This method is usually considered as coupled mode with incompressible fluid, as only the fluid inertial effect is considered in the analysis.

11.5.2.4 Coupled Mode Analysis with Incompressible Fluid

It was shown in Fig. 11.11 that hydrodynamic pressure is most critical for the fundamental mode only. Thus, for calculation of virtual mass moving in phase with the dam one can consider the fundamental mode only for the fluid.

As given in Eq. (11.5.98), ignoring the negative sign, the hydrodynamic pressure on the dam wall can be expressed as

$$p = \frac{12}{\pi(\pi+2)}\beta\left(\frac{\widehat{S}_{a1}}{g}\right)\gamma_w H \sin\frac{\pi z}{2H} \tag{11.5.105}$$

Converting the axis of origin to the base (to be consistent with dynamic analysis model) as chosen in Fig. 11.9, it gives

$$p = \frac{12}{\pi(\pi+2)}\beta\left(\frac{\widehat{S}_{a1}}{g}\right)\gamma_w H \cos\frac{\pi z}{2H} \tag{11.5.106}$$

Thus, for an area dz × 1 the force due to this pressure can be written as

$$P = \frac{12}{\pi(\pi+2)}\beta\left(\frac{\widehat{S}_{a1}}{g}\right)\gamma_w H\left(\cos\frac{\pi z}{2H}\right)\mathrm{d}z \tag{11.5.107}$$

The net fluid mass acting on this area is given by

$$m(z) = \frac{12}{\pi(\pi+2)}\beta\left(\frac{\widehat{S}_{a1}}{g}\right)\frac{\gamma_w}{g}H\left(\cos\frac{\pi z}{2H}\right)\mathrm{d}z \tag{11.5.108}$$

Thus, the coupled mass of dam and fluid in the bending mode can be expressed as

$$\begin{aligned}[M_b]_{ij} &= \frac{\gamma_c A_\mathrm{b}}{g}\int_0^H \left(1+\psi_b\frac{z}{H}\right) f_i^\mathrm{b}(z) f_j^\mathrm{b}(z)\mathrm{d}z \\ &+ \frac{12}{\pi(\pi+2)}\beta\left(\frac{\widehat{S}_{a1}}{g}\right)\frac{\gamma_w}{g}\int_0^H f_i^\mathrm{b}(z) f_j^\mathrm{b}(z)\left(\cos\frac{\pi z}{2H}\right)\mathrm{d}z\end{aligned} \tag{11.5.109}$$

and in natural coordinates this can be expressed as

$$[M_b]_{ij} = \frac{\gamma_c A_b H}{g} \left[\int_0^1 (1 + \psi_b \xi) f_i^b(\xi) f_j^b(\xi) d\xi + \Re_{bf} \int_0^1 \left[\cos\left(\frac{\pi\xi}{2}\right) \right] f_i^b(\xi) f_i^b(\xi) d\xi \right] \tag{11.5.110}$$

where $\Re_{bf}$ = a dimensionless fluid coupling factor in bending mode and may be expressed as

$$\Re_{bf} = \frac{12}{\pi(\pi + 2)} \beta \left(\frac{S_{a1}}{g}\right) \left(\frac{\gamma_w}{\gamma_c}\right) \left(\frac{H}{B_b}\right) \tag{11.5.111}$$

Similarly, for the shear mode, the mass matrix is expressed as

$$[M_S]_{ij} = \frac{\gamma_c A_t H}{g} \begin{bmatrix} \int_0^1 (1 + \psi_t \xi) \cos(2i-1) \frac{\pi\xi}{2} \cos(2j-1) \frac{\pi\xi}{2} d\xi + \\ R_{sf} \int_0^1 \sin\left(\frac{\pi\xi}{2}\right) \cos(2i-1) \frac{\pi\xi}{2} \cos(2j-1) \frac{\pi\xi}{2} d\xi \end{bmatrix} \tag{11.5.112}$$

where $\Re_{Sf}$ = a dimensionless fluid coupling factor in shear deformation mode expressed as

$$\Re_{St} - \frac{12}{\pi(\pi + 2)} \beta \left(\frac{S_{a1}}{g}\right) \left(\frac{\gamma_w}{\gamma_c}\right) \left(\frac{H}{B_t}\right) \tag{11.5.113}$$

The stiffness matrix for bending and shear modes, $[K_b]$ and $[K_s]$, remains the same as Eqs. (11.5.36) and (11.5.48), based on which, from the expression $[K]\{\phi\} = \lambda[M]\{\phi\}$, we determine the eigenvalues for the respective bending and shear modes. Now, considering $\omega = \sqrt{\lambda}$ and $T = 2\pi/\omega$, we find out the time periods T_b and T_s in bending and shear modes, respectively. Based on Dunkerley's equation as cited in Eq. (11.5.58), we finally find out the effective time period T_{ei} of the dam when fluid is coupled to the dam and moving in same phase. Determining the spectral acceleration from the code corresponding to this time period, the displacement, moment, and shear force for the fundamental mode are obtained from the equations

$$u_{di} == \kappa_i \beta \frac{S_{ai}}{4\pi^2} T_{ei}^2 [\phi_i]^T \{f_i^b(\xi)\} \tag{11.5.114}$$

$$M_{zi} = -\kappa_i \beta \frac{EI_b}{H^2} (1 + \psi_b \xi)^3 \left(\frac{S_{ai} T_{ei}^2}{4\pi^2} \right) [\phi_i]^T \left\{ \frac{d^2 f_i^b(\xi)}{d\xi^2} \right\} \tag{11.5.115}$$

$$V_{zi} = -\kappa_i \beta \frac{EI_b}{H^3} (1 + \psi_b \xi)^3 \left(\frac{S_{ai} T_{ei}^2}{4\pi^2} \right) [\phi_i]^T \left\{ \frac{d^3 f_i^b(\xi)}{d\xi^3} \right\} \tag{11.5.116}$$

Here,

$$\kappa_i = \frac{\int_0^1 (1 + \psi_b \xi) f_i^b(\xi) d\xi + \Re_{bf} \int_0^1 \cos\left(\frac{\pi\xi}{2}\right) f_i^b(\xi) d\xi}{\int_0^1 (1 + \psi_b \xi) f_i^{b2}(\xi) d\xi + \Re_{bf} \int_0^1 \cos\left(\frac{\pi\xi}{2}\right) f_i^{b2}(\xi) d\xi} \tag{11.5.117}$$

11.5.2.5 Coupled Mode Analysis with Compressible Fluid

In the previous section, we have considered the inertial property of the fluid where its stiffness has not come into play. The analysis is now carried out where compressible property of the fluid (bulk modulus B_m) is considered in the analysis.

Shown in Fig. 11.13 is an elemental body of length dx, dy, and dz in space.

For the element under load, the strain energy equation in three dimensions is given by the expression

$$V = \frac{\lambda e^2}{2} + G\left(\varepsilon_x^2 + \varepsilon_y^2 + \varepsilon_z^2\right) + \frac{G}{2}\left(\gamma_{xy}^2 + \gamma_{yz}^2 + \gamma_{xz}^2\right) \tag{11.5.118}$$

If the element is a fluid body, then $G = 0$, and one has

$V = \frac{B_m e^2}{2}$ where B_m = bulk modulus of the water and $e = e_x + e_y + e_z$.

It has been shown by Chowdhury and Dasgupta (2008) that the stiffness and mass of fluid for the fundamental mode can be expressed as

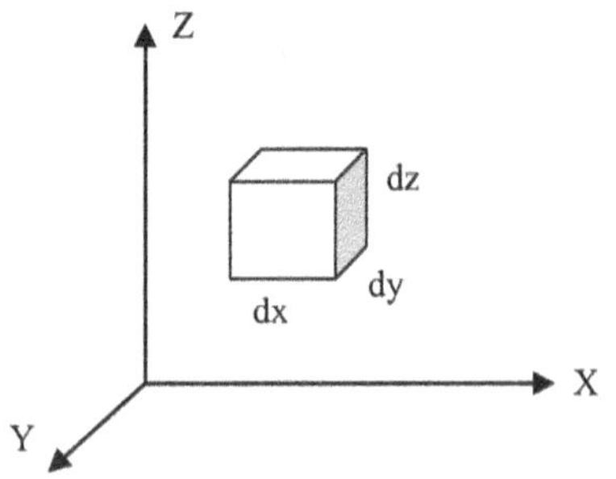

Fig. 11.13 Elastic elemental body in space

$$K_{11} = \frac{\pi^2 B_m}{8H_w} \text{ and } M_{11} = \frac{\gamma_w H_w}{2g} \qquad (11.5.119)$$

It may be observed that the unit of stiffness derived here is kN/m^3, which means that the expression gives *stiffness per unit area*. Thus, to determine the total stiffness of the water in contact with dam's face one has to multiply this by the contact area which in this case is ($H_w \times 1$). This gives $K_{11} = \pi^2 B_m/8$ kN/m per meter width.

Similarly, the effective mass is given by $M_{11} = \gamma_w H^2/2g$ kN s^2/m per meter width.

It may be also be noted that considering $T = 2\pi\sqrt{M/K}$ and substituting the above values, one can arrive at the same expression as $T = 4H_w/c$ which is the free field time period of the reservoir.

Ignoring the small difference between H and H_w and assuming $H \cong H_w$, one has the combined stiffness of the dam and fluid under bending mode as

$$[K_b]_{ij} = EI_b \int_0^H \frac{d^2}{dz^2}\left[\left(1+\psi_b \frac{z}{H}\right)^3 \frac{d^2 f_i^b(z)}{dz^2}\right] f_j^b(z)dz + \frac{\pi^2 B_m}{8H} \int_0^H f_i^b(z) f_j^b(z)dz \qquad (11.5.120)$$

In natural coordinates, the stiffness can be expressed as

$$[K_b]_{ij} = \frac{EI_b}{H^3} \int_0^1 \frac{d^2}{d\xi^2}\left[(1+\psi_b \xi)^3 \frac{d^2 f_i^b(\xi)}{d\xi^2}\right] f_j^b(\xi)dz + \frac{\pi^2 B_m}{8} \int_0^1 f_i^b(\xi) f_j^b(\xi)d\xi \qquad (11.5.121)$$

The damping matrix of fluid for bending mode can be expressed as

$$[C_{bf}]_{ij} = \left[\frac{\gamma_w}{g}\right] cH \int_0^1 f_i^b(\xi) f_j^b(\xi)d\xi \qquad (11.5.122)$$

where c = sound velocity of water; $f_i^b(z)$ is expressed in Eq. (11.5.23).

The mass matrix of the dam and fluid in bending mode can be expressed as

$$[M]_{ij} = \frac{\gamma_c A_b H}{g} \int_0^1 (1+\psi_b \xi) f_i^b(\xi) f_j^b(\xi)d\xi + \frac{\gamma_w H^2}{2g} \int_0^1 f_i^b(\xi) f_j^b(\xi)d\xi \qquad (11.5.123)$$

In the shear deformation mode, the stiffness matrix of the dam is expressed as

$$\rightarrow [K_s] = -GA_t \int_0^H \frac{\mathrm{d}}{\mathrm{d}z}\left[\left(1+\psi_t \frac{z}{H}\right)\frac{\mathrm{d}}{\mathrm{d}z} f_i^{\mathrm{s}}(z)\right] f_j^{\mathrm{s}}(z)\mathrm{d}z + \frac{\pi^2 B_{\mathrm{m}}}{8H}\int_0^H f_i^{\mathrm{s}}(z) f_j^{\mathrm{s}}(z)\mathrm{d}z \tag{11.5.124}$$

$$\rightarrow [K_s] = \frac{-GA_t}{H}\int_0^1 \frac{\mathrm{d}}{\mathrm{d}\xi}\left[(1+\psi_t\xi)\frac{\mathrm{d}}{\mathrm{d}\xi} f_i^{\mathrm{s}}(\xi)\right] f_j^{\mathrm{s}}(\xi)\mathrm{d}\xi + \frac{\pi^2 B_{\mathrm{m}}}{8}\int_0^1 f_i^{\mathrm{s}}(\xi) f_j^{\mathrm{s}}(\xi)\mathrm{d}\xi \tag{11.5.125}$$

The damping matrix of fluid for shear mode can be expressed

$$[C_{sf}]_{ij} = \left[\frac{\gamma_w}{g}\right] cH \int_0^1 f_i^{\mathrm{s}}(\xi) f_j^{\mathrm{s}}(\xi)\mathrm{d}\xi \tag{11.5.126}$$

The mass matrix of the dam and fluid in shear mode is expressed as

$$[M]_{ij} = \frac{\gamma_c A_t H}{g}\int_0^1 (1+\psi_t\xi) f_i^{\mathrm{s}}(\xi) f_j^{\mathrm{s}}(\xi)\mathrm{d}\xi + \frac{\gamma_w H^2}{2g}\int_0^1 f_i^{\mathrm{s}}(\xi) f_j^{\mathrm{s}}(\xi)\mathrm{d}\xi \tag{11.5.127}$$

where $f_i^{\mathrm{s}}(z)$ is expressed in Eq. (11.5.45).

Now, from the expression $[K]\{\phi\} = \lambda[M]\{\phi\}$, one can determine the eigenvalues for the respective bending and shear modes. Again, considering $\omega = \sqrt{\lambda}$ and $T = 2\pi/\omega$, we find out T_{b} and T_{s}, the time periods in bending and shear modes, respectively. Based on Dunkerley's equation as cited in Eq. (11.5.58), we finally find out the effective time period T_{fsi} (coupled fluid–structure interaction) of the dam. Determining the spectral acceleration corresponding to this time period, the displacement, moment, and shear force are obtained from the equations

$$u_{di} == \kappa_i \beta \frac{S_{ai}}{4\pi^2} T_{\mathrm{fsi}}^2 [\phi_i]^{\mathrm{T}} f_i^{\mathrm{b}}(\xi) \tag{11.5.128}$$

$$M_{zi} = -\kappa_i \beta \frac{EI_{\mathrm{b}}}{H^2}(1+\psi_b\xi)^3 \left(\frac{S_{ai} T_{\mathrm{fsi}}^2}{4\pi^2}\right)[\phi_i]^{\mathrm{T}} \left\{\frac{\mathrm{d}^2 f_i^{\mathrm{b}}(\xi)}{\mathrm{d}\xi^2}\right\} \tag{11.5.129}$$

$$V_{zi} = -\kappa_i \beta \frac{EI_{\mathrm{b}}}{H^3}(1+\psi_b\xi)^3 \left(\frac{S_{ai} T_{\mathrm{fsi}}^2}{4\pi^2}\right)[\phi_i]^{\mathrm{T}} \left\{\frac{\mathrm{d}^3 f_i^{\mathrm{s}}(\xi)}{\mathrm{d}\xi^3}\right\} \tag{11.5.130}$$

where

$$\kappa_i = \frac{\int_0^1 (1+\psi_b\xi) f_i^{\mathrm{b}}(\xi)\mathrm{d}\xi + \aleph_f \int_0^1 f_i^{\mathrm{b}}(\xi)\mathrm{d}\xi}{\int_0^1 (1+\psi_b\xi) f_i^{\mathrm{b}2}(\xi)\mathrm{d}\xi + \aleph_f \int_0^1 f_i^{\mathrm{b}2}(\xi)\mathrm{d}\xi} \tag{11.5.131}$$

Here, $\aleph_f$ is a dimensionless fluid coupling term expressed as

$$\aleph_f = \frac{1}{2}\left(\frac{\gamma_w}{\gamma_c}\right)\left(\frac{H}{B_{\mathrm{b}}}\right) \cong \frac{r_{\mathrm{s}}}{5} \tag{11.5.132}$$

11.5.2.6 Soil–Fluid–Structure Interaction

This is by far the most comprehensive procedure for the dynamic analysis of dam and is usually carried out when Young's modulus of the dam's material $\mathrm{E} \geq E_f$, where E_f is the elastic modulus of the foundation material.

When $E = E_f$, the soil is discretized into plane strain finite element usually extending up to a depth equal to height H of the dam. In case $E_f < E$ by 0.25–0.5, the depth is usually extended to 4–5 times the height of the dam.

This is to ensure that waves are not reflected from the soil boundary and if the software in hand has provisions, infinite finite element, or paraxial boundaries, these are applied at the soil boundaries to ensure no waves are reflected back to the system.

However, the basic point to be remembered is that, in this case, what we are interested to know is: how does the soil stiffness affect the structural behavior of the dam?

For this, the deformation in soil at the dam–soil interface is actually the most important criterion, for one would not be really interested to know the deformation within the soil, until and unless, some special geotechnical reason makes it an essential study.

Thus, the model by which we propose to idealize the soil is by linear spring element where nonlinearity, if any, due to large strain, can be catered to by scaling down the dynamic shear modulus G as proposed by Seed et al. (1984). If the dam in the Y-direction is very long compared to its width (B_{b}) and height (H), we consider spring values as proposed for strip foundation by Wolf (1985), expressed as

$K_x = 0.94\frac{\pi G_s}{2-v_s}$ in translational motion; and $K_\theta = 0.94\frac{\pi G_s b^2}{2(1-v_s)}$ in rocking mode.

In case of dam, the rocking stiffness can be expressed as

$$K_\theta = 0.1175 \frac{\pi G_s B_b^2}{(1 - \nu_s)} \tag{11.5.133}$$

where G_s = dynamic shear modulus of the foundation; ν_s = Poisson's ratio of foundation material

The dynamic soil–structure interaction (DSSI) effect is considered as follows:

If T_{fsi} is the time period of the dam in ith mode due to fluid–structure interaction (considering the dam base fixed, Fig. 11.14), as explained in earlier section, then as per modal analysis the amplitude of vibration is expressed as

$$S_d = \frac{S_{a\text{fsi}}}{4\pi^2} T_{\text{fsi}}^2,$$

Here, $S_{a\text{fsi}}$ is the spectral acceleration corresponding to the time period T_{fsi}. Then, as per code norm, the displacement can be expressed as

$$u_{\text{fsi}} = \kappa_i \beta \frac{S_{a\text{fsi}}}{4\pi^2} T_{\text{fsi}}^2 [\phi_i]^{\mathrm{T}} f_i^{\mathrm{b}}(\xi) \tag{11.5.134}$$

This gives $M_z = -EI \frac{\mathrm{d}^2 u_{\text{fsi}}}{\mathrm{d}z^2}$

$$\to M_{zi} = -\kappa_i \beta \frac{S_{a\text{fsi}}}{4\pi^2} \left(\frac{EI_{\mathrm{b}}(1 + \psi_b \xi)^3}{H^2} \right) T_{\text{fsi}}^2 [\phi_i]^{\mathrm{T}} \left\{ \frac{\mathrm{d}^2}{\mathrm{d}\xi^2} f_i^{\mathrm{b}}(\xi) \right\} \tag{11.5.135}$$

and

$$V_z = -\kappa_i \beta \frac{S_{a\text{fsi}}}{4\pi^2} \left(\frac{EI_{\mathrm{b}}(1 + \psi_b \xi)^3}{H^3} \right) T_{\text{fsi}}^2 [\phi_i]^{\mathrm{T}} \left\{ \frac{\mathrm{d}^3}{\mathrm{d}\xi^3} f_i^{\mathrm{b}}(\xi) \right\} \tag{11.5.136}$$

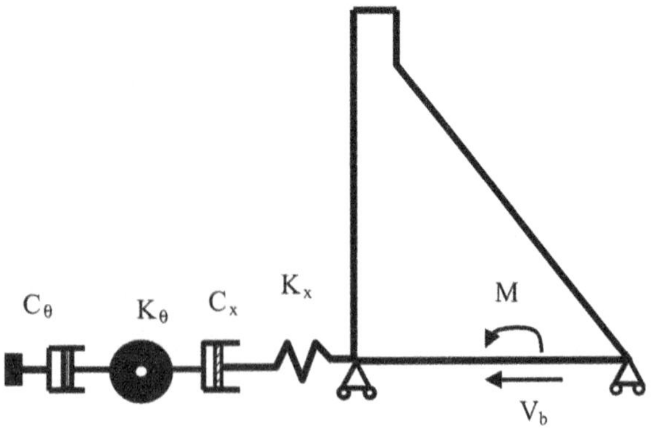

Fig. 11.14 Gravity dam with foundation springs

Thus, the base moment and shear at the bottom of dam, at $\xi = 0$, can be expressed as

$$M_{bi} = -\kappa_i \beta \frac{S_{\text{afsi}}}{4\pi^2} \left(\frac{EI_{\text{b}}}{H^2}\right) T_{\text{fsi}}^2 [\phi_i]^{\text{T}} \left\{ \frac{\mathrm{d}^2 f_i^{\text{b}}(0)}{\mathrm{d}\xi^2} \right\} \tag{11.5.137}$$

$$\text{and, } V_{bi} = -\kappa_i \beta \frac{S_{\text{afsi}}}{4\pi^2} \left(\frac{EI_{\text{b}}}{H^3}\right) T_{\text{fsi}}^2 [\phi_i]^{\text{T}} \left\{ \frac{\mathrm{d}^3 f_i^{\text{b}}(0)}{\mathrm{d}\xi^3} \right\} \tag{11.5.138}$$

Considering the foundation as a massless elastic half-space represented by a spring and dashpot, the equilibrium equation in lateral direction at the base can be expressed as

$$c_x \dot{x} + K_x x = V_{bi} \tag{11.5.139}$$

Equation (11.5.139) can be expressed as

$$\rightarrow \frac{\mathrm{d}x}{\mathrm{d}t} + Px = Q \tag{11.5.140}$$

Here, $Q = V_{bi}/c_x$ and $P = K_x/c_x$.

Equation (11.5.140) is a linear differential equation in x whose solution is expressed as

$$x = \mathrm{e}^{-\int P\mathrm{d}t} \left[\int Q \mathrm{e}^{\int P\mathrm{d}t} \mathrm{d}t + C \right] \tag{11.5.141}$$

Here, C is integration constant.

Integrating Eq. (11.5.141) gives

$$x = \mathrm{e}^{\frac{-K_x t}{c_x}} \left[\frac{V_{bi}}{K_x} \mathrm{e}^{\frac{K_x t}{c_x}} + C \right] \tag{11.5.142}$$

To determine the integration constant C, we impose the boundary condition at $t = 0$ $\ddot{x} = -S_{ax}$

$$\rightarrow C = -\frac{S_{ax} K_x^2}{c_x^2} = -\frac{V_{bi} c_x^2}{M K_x^2} \tag{11.5.143}$$

Thus, the displacement x in lateral direction at any time t is given by

$$x = \frac{V_{bi}}{K_x} \left[1 - \frac{c_x^2}{M K_x} \mathrm{e}^{\frac{-K_x t}{c_x}} \right] \tag{11.5.144}$$

Considering $c_x = \zeta_x C_c = \zeta_x 2\sqrt{K_x M}$, we finally have

$$u_x = \frac{V_{bi}}{K_x}\left[1 - 4\zeta_x^2 e^{\frac{-\omega_x t}{2\zeta_x}}\right] \tag{11.5.145}$$

where u_x = amplitude of displacement in the x-direction, ζ_x = damping ratio of foundation in translation, and ω_x = frequency of foundation in translational mode.

Similarly, in rocking mode it can be shown that

$$\theta = \frac{M_{bi}}{K_\theta}\left[1 - 4\zeta_\theta^2 e^{\frac{-\omega_\theta t}{2\zeta_\theta}}\right] \tag{11.5.146}$$

Total deflection at the crest of the dam is expressed as $U_t = u_{\text{fsi}} + u_x + H\theta$,

$$U_t = u_{\text{fsi}} + \frac{V_{bi}}{K_x}\left[1 - 4\zeta_x^2 e^{\frac{-\omega_x t}{2\zeta_x}}\right] + H\frac{M_{bi}}{K_\theta}\left[1 - 4\zeta_\theta^2 e^{\frac{-\omega_\theta t}{2\zeta_\theta}}\right] \tag{11.5.147}$$

where H = dam height.

On computation, it will be observed that the second term within the parenthesis dies down very fast to zero (see Fig. 11.15), and under a steady state, we are finally left with

$$U_t = u_{\text{fsi}} + \frac{V_{bi}}{K_x} + H\frac{M_{bi}}{K_\theta} \tag{11.5.148}$$

Equation (11.5.148) can be simplified to

$$U_t = u_{fsi}\left[1 + \frac{V_{bi}}{K_x u_{fsi}} + H\frac{M_{bi}}{K_\theta u_{fsi}}\right] \tag{11.5.149}$$

The term within the parenthesis in Eq. (11.5.149) is a dimensionless term and is an amplification factor due to coupling of foundation stiffness that enhances the fixed base response.

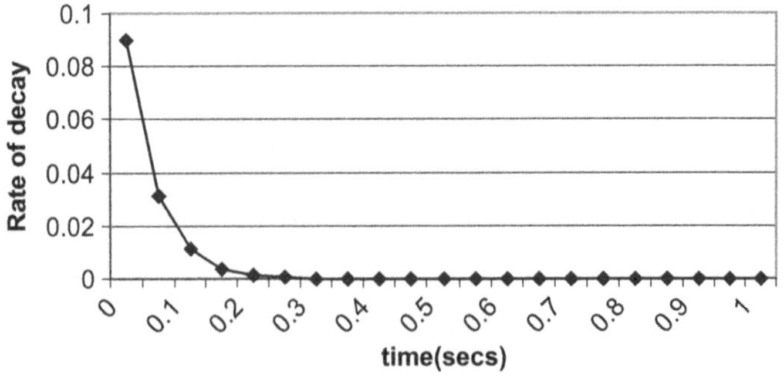

Fig. 11.15 Rate of decay of damping term with time

Thus, the SRSS amplification factor (AF) of the dam due to foundation compliance can be expressed as

$$\mathrm{AF} = \left[1 + \left(\frac{V_{\mathrm{SRSS}}}{K_x u_{\mathrm{srss}}}\right) + H\left(\frac{M_{\mathrm{SRSS}}}{K_\theta u_{\mathrm{srss}}}\right)\right] \tag{11.5.150}$$

The effect of foundation on displacement, moment, and shear can be expressed as

$$u_{\mathrm{fsi}} = \kappa_i \beta[\mathrm{AF}]\frac{S_{\mathrm{afsi}}}{4\pi^2}T_{\mathrm{fsi}}^2[\phi_i]^{\mathrm{T}} f_i^{\mathrm{b}}(\xi) \tag{11.5.151}$$

$$M_{zi} = -\kappa_i \beta[\mathrm{AF}]\frac{S_{\mathrm{afsi}}}{4\pi^2}\left(\frac{EI_{\mathrm{b}}(1+\psi\xi)^3}{H^2}\right)T_{\mathrm{fsi}}^2[\phi_i]^{\mathrm{T}}\left\{\frac{\mathrm{d}^2 f_i^{\mathrm{b}}(\xi)}{\mathrm{d}\xi^2}\right\} \tag{11.5.152}$$

$$V_{zi} = -\kappa_i \beta[\mathrm{AF}]\frac{S_{\mathrm{afsi}}}{4\pi^2}\left(\frac{EI_{\mathrm{b}}(1+\psi\xi)^3}{H^3}\right)T_{\mathrm{fsi}}^2[\phi_i]^{\mathrm{T}}\left\{\frac{\mathrm{d}^3 f_i^{\mathrm{b}}(\xi)}{\mathrm{d}\xi^3}\right\} \tag{11.5.153}$$

It is may be observed that the foundation effect enhances the dynamic response. The scaling of amplification varies with height H. With increase in the dam height, the rocking mode starts playing a significant role. For a massless foundation, the coupling is static in nature. This particular phenomenon of static coupling has also been confirmed by Wolf (1985), Hall and Kissenpfennig (1976), and Kramer (Kramer 2002), who stated that under seismic load, interaction with soil is basically static in nature, where dimensionless frequency number $a = \omega r/V_s$ plays an insignificant role.

The time period in this case gets elongated by a factor which is the square root of peak amplification. This may be expressed as

$$T_{\mathrm{el}} = T_{\mathrm{fsi}}\sqrt{\left[1 + \left(\frac{V_{\mathrm{SRSS}}}{K_x u_{\mathrm{srss}}}\right) + H\left(\frac{M_{\mathrm{SRSS}}}{K_\theta u_{\mathrm{srss}}}\right)\right]} \tag{11.5.154}$$

11.5.2.7 Validation of the Theory

To further consolidate the issue, a real-life dam (constructed in India) has been studied in detail based on finite element and the proposed method and salient design outputs are compared as hereafter. The basic design data of the dam is: height of dam = 76.5 m; width at base = 69.75 m, top width = 13.5 m; Poisson's ratio = 0.15; Young's modulus of concrete = 2.5×10^7 kN/m^2, unit weight of concrete = 25 kN/m^3, unit weight of water = 10 kN/m^3; sound velocity in water = 1439 m/s; bulk modulus of water = 2.11×10^6 kN/m^2. As per IS-1893, the dam is situated in Zone-V.

The foundation properties are as follows:

$$E_{\text{soil}} = 8 \times 10^6 \text{ kN/m}^2; n_{\text{soil}} = 0.15 \ \gamma_{\text{soil}} = 27.8 \text{ kN/m}^3.$$

A detailed FEM analysis is carried out in ANSYS 10 (2010) and compared with the proposed method for different cases.

In the finite element method, the dam is discretized by a four-node iso-parametric element (Plane 42) as shown in Fig. 11.16. The number of elements used is 11,513, having the number of active nodes as 11,770, which gives 23,540 degrees of freedom.

Case-1 Static Deflection of Dam Under Hydrostatic Load

Referring to Fig. 11.16, the maximum static tip deflections, as obtained by ANSYS and the proposed method, are as shown in Table 11.15.

Case-2 Dynamic Analysis of Dam Only (Reservoir Empty Case)

Time period of the dam only (No water in reservoir) by FEM analysis vis-à-vis the proposed method is as shown in Table 11.16.

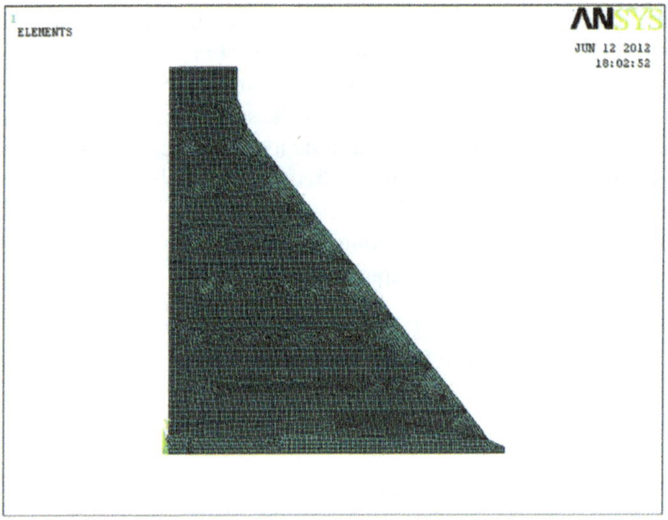

Fig. 11.16 ANSYS finite element model of the dam

Table 11.15 Comparison of deflections

Analyzed by	Tip deflection (mm)	Remarks
ANSYS	5.003	
By Eq. (11.5.11)	5.277	$\eta = 2/3$
By Eq. (11.5.11)	4.872	$\eta = \frac{10(1+\nu)}{12+11\nu}$

Table 11.16 Comparison of time periods

Time period	Mode 1	Mode 2	Mode 3	Remarks
By ANSYS	0.191	0.08	0.044	
By Eq. (11.5.60)	0.193	0.068	0.038	$\eta = 2/3$
By Eq. (11.5.60)	0.183	0.063	0.035	$\eta = \frac{10(1+\nu)}{12+11\nu}$
Chopra et al. (1986)	0.184	–	–	
Okamoto (1984)	0.157	–	–	
IS-1893 (1984)	0.148	–	–	

The mode shapes of the dam by FEM analysis are shown in Fig. 11.17.

The bending moment and shear force diagram for the dam under combined inertial and hydrodynamic pressure are as furnished hereafter. In this case, the moment and shear due to dam inertia and due to hydrodynamic pressure are considered separately. The moment due to dam only is shown by thin firm line and that due to hydrodynamic pressure is shown by dotted line, while there combined effect is shown by firm thick line in Fig. 11.18.

Case-3 Dynamic Analysis with Coupled Incompressible Fluid

Time periods of the dam with virtual mass added to the dam face (i.e., ignoring fluid stiffness) analyzed by FEM vis-à-vis the proposed method are as shown in Table 11.17.

The bending moment and shear force diagram for the dam under combined inertial and hydrodynamic pressure are as furnished in Fig. 11.19. In this case, the fluid mass is added to the dam as a virtual mass moving in same phase.

Case-4 Dynamic Analysis with Coupled Compressible Fluid

Time period of the dam with coupled fluid stiffness and mass added to the dam face analyzed by FEM vis-à-vis the proposed method is as shown in Fig. 11.20.

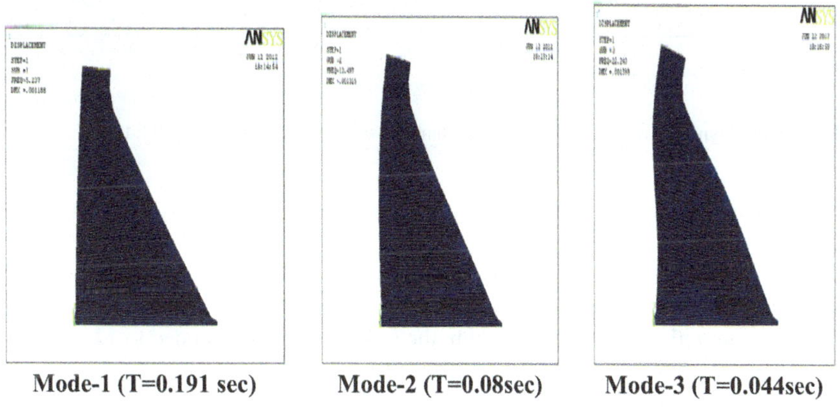

Fig. 11.17 Modal deformation of dam only, first three modes

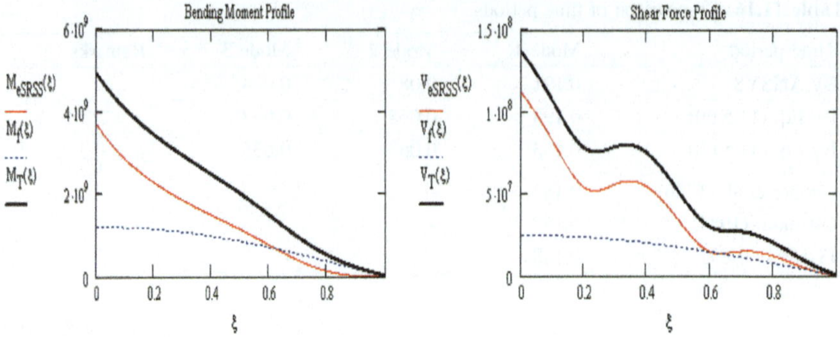

Fig. 11.18 Dynamic moment and shear on dam decoupled case

Table 11.17 Comparison of time period

Time period	Mode 1	Mode 2	Mode 3	Remarks
By ANSYS	0.227	0.063	0.043	
Proposed method	0.213	0.075	0.042	$\eta = 2/3$
Proposed method	0.201	0.069	0.038	$\eta = \frac{10(1+\nu)}{12+11\nu}$

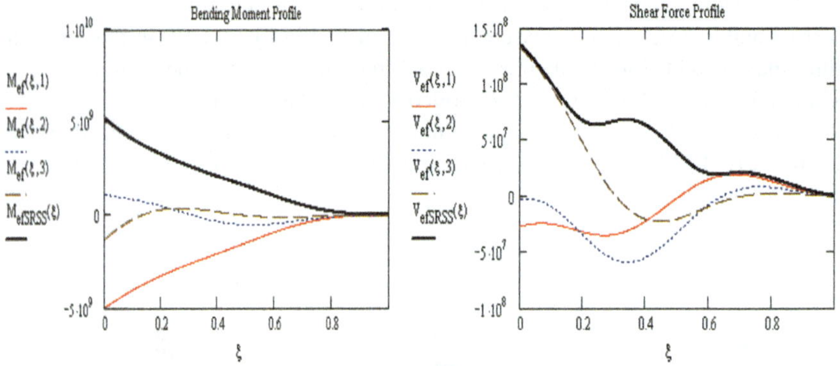

Fig. 11.19 Dynamic moment and shear on dam considering incompressible fluid

The mode shapes for the first three modes, including time period for this case, are as shown in Fig. 11.21.

In this case, the fluid is modeled by the fluid 72 element as available in ANSYS with master–slave node option to cater to fluid–structure interaction. To ensure that no waves are reflected back to the dam, the fluid domain is extended to 12 times the height of the dam as suggested by Gogoi and Athani (2010).

Comparative time periods are as shown in Table 11.18.

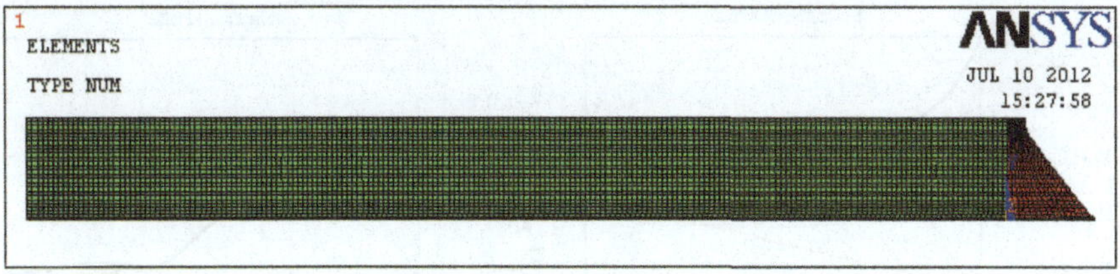

Fig. 11.20 Finite element model of dam, with compressible fluid–structure interaction

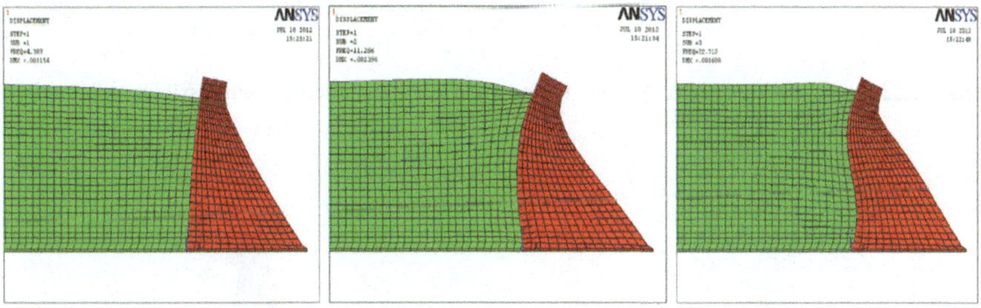

Fig. 11.21 First three modes of the dam, with compressible fluid

Table 11.18 Comparison of time periods

Time period	Mode 1	Mode 2	Mode 3	Remarks
By ANSYS	0.228	0.089	0.044	
Proposed method	0.216	0.080	0.044	$\eta = 2/3$
Proposed method	0.205	0.074	0.039	$\eta = \frac{10(1+\nu)}{12+11\nu}$

The bending moment and shear force diagram for the dam under combined inertial and hydrodynamic pressure are as furnished in Fig. 11.22. In this case, both the fluid mass and stiffness are added to the dam and act together.

Case-5 Dynamic Analysis with Coupled Incompressible Fluid including Foundation Interaction

The FEM model and response with ANSYS are shown in Fig. 11.23. The bending moment and shear force diagram for the dam in this case are as shown Fig. 11.24.

Comparative time periods are as shown in Table 11.19.

It is to be noted that the moment and shear, $M_{\mathrm{emT}}(\xi)$ and $V_{\mathrm{emT}}(\xi)$, are the moment and shear on the dam at any height ξ due to the effect of foundation interaction.

Case-6 Dynamic Analysis with Coupled Compressible Fluid including Foundation Interaction

The periods for different modes are shown in Table 11.20. The bending moment and shear force diagram for the dam in this case are as shown in Fig. 11.25.

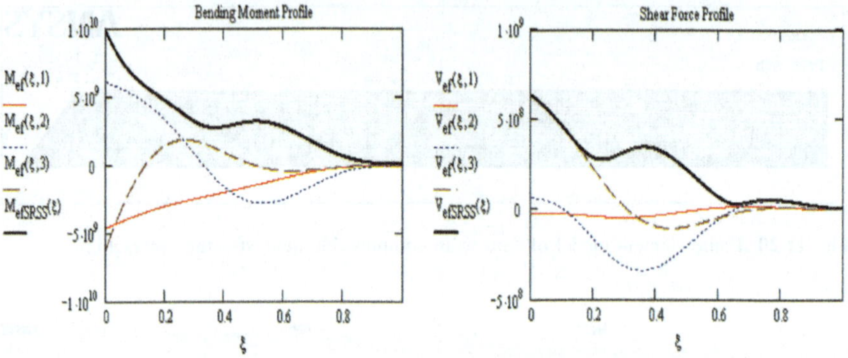

Fig. 11.22 Bending moment and shear for coupled fluid considered compressible

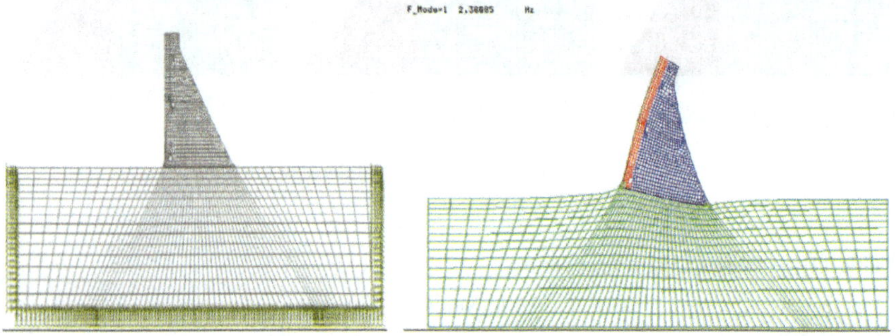

Fig. 11.23 Finite element model in ANSYS of a dam with incompressible fluid added as virtual mass with foundation interaction

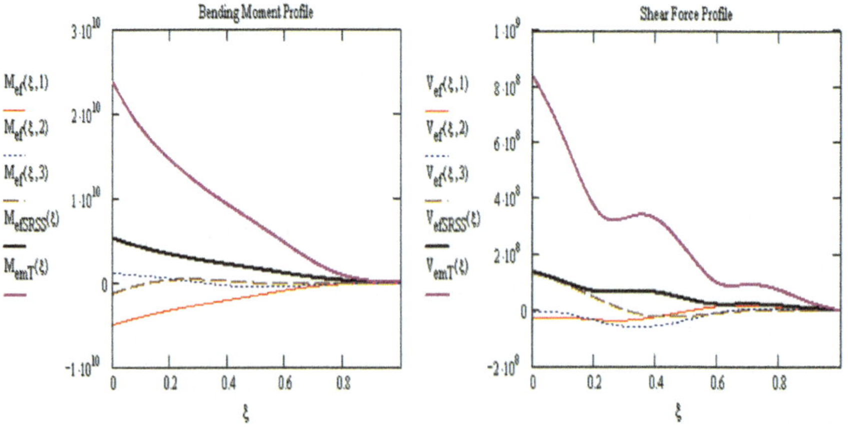

Fig. 11.24 Dynamic moment and shear on dam considering incompressible fluid with DSSI effect

Table 11.19 Comparison of time periods

Time period	Mode 1	Mode 2	Mode 3	Remarks
By ANSYS	0.423	0.159	0.085	
Proposed method by Eq. (11.5.154)[a]	0.446	0.157	0.087	$\eta = 2/3$
Proposed method by Eq. (11.5.154)[a]	0.421	0.144	0.080	$\eta = \frac{10(1+\nu)}{12+11\nu}$

[a]Elongation factor for time period is 2.09 in this case

Table 11.20 Comparison of time period

Time period	Mode 1	Mode 2	Mode 3	Remarks
Proposed method by Eq. (11.5.154)	0.682	0.253	0.137	$\eta = 2/3$
Proposed method by Eq. (11.5.154)	0.647	0.233	0.123	$\eta = \frac{10(1+\nu)}{12+11\nu}$

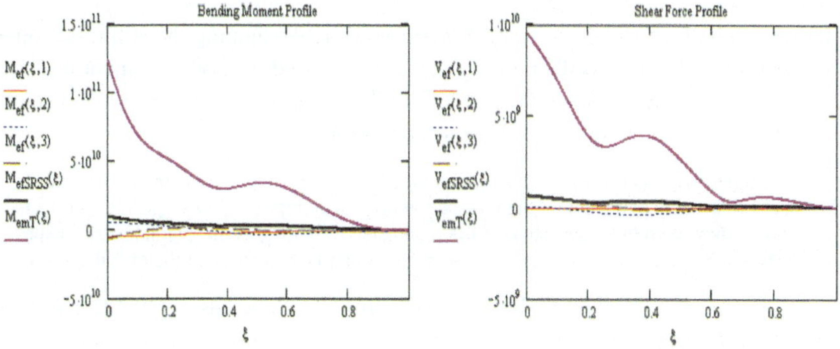

Fig. 11.25 Dynamic moment and shear on dam considering compressible fluid with DSSI effect

The salient design outputs with and without dynamic soil–structure interaction for the different cases are as furnished in Table 11.21.

It is thus seen that for almost all cases, the key dynamic parameter, the time period (T), computed by the proposed method is well comparable to 2D FEM analysis. It would, however, be difficult to say with authenticity as to which gives more accurate results, for both (FEM and the proposed method) have some idealization or approximations. While FEM is essentially a numerical method where some inherent numerical error would invariably be present, especially with the selective reduced integration that is often undertaken to make the stiffness of plane stress elements converge lower bound. The dam section in this case (being actually a 2D plane stress element) with irregular geometry has been idealized in the proposed method by a beam element with variable cross section. However, with this idealization, the error induced, if any, is minimized by Galerkin's weighted residual technique and compounded by trigonometric shape functions which are in reality infinite series (unlike FEM where polynomial functions of limited order are chosen to define a plane stress

Table 11.21 Comparison of amplitude, moment, and shear force

Method of analysis	Amplitude (mm)	Shear (kN) SRSS	Moment (kN m) SRSS
Conventional decoupled analysis without DSSI	16	1.36×10^{05}	4.92×10^{06}
Conventional decoupled analysis with DSSI	242	9.60×10^{06}	1.25×10^{08}
Coupled analysis, incompressible fluid without DSSI	23	1.36×10^{05}	5.20×10^{06}
Coupled analysis, incompressible fluid with DSSI	102	8.40×10^{05}	2.38×10^{07}
Coupled analysis, compressible fluid without DSSI	24	6.37×10^{05}	9.84×10^{06}
Coupled analysis, compressible fluid with DSSI	93	9.31×10^{05}	2.23×10^{07}

element), which generates no truncation error while developing the stiffness matrix on integration. It would suffice to say that the proposed method is comparable to a 2D FEM analysis at the advantage of cutting down significant computation cost and come up with a comparable result for time period *T*.

If the time period (*T*) of the system is computed correctly, amplitude moment and shear, which are actually a function of the time period only, will implicitly converge to the correct result as they are only a derivative of the parameter *T*, as long as the trigonometric shape functions chosen for the system satisfy the natural boundary condtion of the global system.

A few transition points or slight variation in dam slope has negligible effect on the overall dynamic response of the dam as long as in general the top and bottom widths remain linearly variable. For large dams (especially ones having $H > 100$ m) resting on a flexible rock having $E_f < E$, the DSSI analysis with compressible fluid coupled to the dam is the most realistic analysis.

The presence of openings due to galleries or inspection tunnel (refer Fig. 11.23) has negligible effect on the overall dynamic response of the dam, as the gallery dimensions are usually negligible compared to the overall dimensions. However, stress concentration around the gallery can be computed by taking effective moment of inertia at the gallery level ($I_e = I_z - A_g r^2$) and then compute the stresses. Here, A_g = gallery area (in plan), and r is the radius of gyration with respect to the centroid of the dam cross section in plan.

Example 11.1 A concrete dam of height of 100 m, base width of 70 m, has a top width of 7 m. The maximum height of water it contains is 97.5 m. The dam is built on site of hard rock and is in zone IV as per IS-1893. The grade of concrete used is M25. Consider unit weight of water as 10 kN/m^3 and that of concrete as 25 kN/m^3. Bulk modulus of water (B_m) = 2.11×10^6 kN/m^2. The shear wave velocity of foundation rock = 1100 m/s and $\nu_f = 0.30$. Unit weight of rock = 22 kN/m^3. Perform the earthquake analysis based on

- IS-1893 1984.
- IS-1893 2002/2016.
- Chopra and Fenves's method.
- Dynamic Analysis by CSD model.

Solution

(1) **Analysis based on IS-1893 (1984)**

As per IS-1893, the time period of the dam is expressed as

$$T = C_T \frac{H^2}{B}\sqrt{\frac{\gamma_c}{gE_s}} = 5.55 \times \frac{(100)^2}{70}\sqrt{\frac{25}{2.85 \times 10^7 \times 9.81}} = 0.237 \text{ s}$$

As per code $\alpha_h = \beta I F_0 \left(\frac{S_a}{g}\right)$

Here, for zone IV, $\beta = 1.0$, $I = 3.0$, $F_0 = 0.25$; this gives $\alpha_h = 1.0 \times 3.0 \times 0.25(S_a/g) = 0.75(S_a/g)$.

For $T = 0.237$ s, from the response spectrum curve furnished in code, $S_a/\text{g} = 0.2 \rightarrow \alpha_h = 0.15$.

Weight (W) of dam per meter width is expressed as

$$W = (1/2) \times (70 + 7) \times 100 \times 25 = 96,250 \text{ kN}$$

$$\bar{h} = \frac{7\times100\times50 + (1/2)\times63\times100\times(100/3)}{7\times100 + (1/2)\times63\times100} = 36.36 \text{ m from bottom of dam}$$

$$V_B = 0.6\alpha_h W = 0.6 \times 0.15 \times 96,250 = 8663 \text{ kN}$$

$$M_B = 0.9\alpha_h W\bar{h} = 0.9 \times 0.15 \times 96,250 \times 36.36 = 472,453 \text{ kN m}$$

Multiplying by the coefficients furnished in Table 11.2, we have

z/H	0	0.1	0.2	0.3	0.4	0.5	0.6	0.7	0.8	0.9	1
V_z	0	801	1603	2534	3465	4570	5674	6627	7580	8122	8663
M_z	0	8268	16,536	46,064	75,592	124,019	172,445	243,313	314,181	393,317	472,453

The above gives the shear and moment along the dam height for inertial response due to earthquake. Here, $z = 0$ at the top of dam.

(1a) **Computation of hydrodynamic pressure**

Considering velocity of sound $c = 1440$ m/s in water,

$$T = \frac{4H}{c} = 4 \times \frac{97.5}{1440} = 0.27 \text{ s}$$

As per IS-1893 $S_a/g = 0.20$

$$\therefore \alpha_h = 1.0 \times 3.0 \times 0.25 \times 0.20 = 0.15$$

Now, based on Eqs. (11.2.6)–(11.2.8), the pressure and corresponding moment and shear are as tabled hereafter.

z/H	0	0.1	0.2	0.3	0.4	0.5	0.6	0.7	0.8	0.9	1
C_s	0	0.230	0.353	0.450	0.529	0.594	0.646	0.685	0.713	0.729	0.735
p_h	0	35	53	67	79	89	97	103	107	109	110
V_z	0	250	768	1470	2305	3234	4218	5222	6211	7150	8004
M_z	0	1032	6329	18,159	37,975	66,590	104,226	150,539	204,624	265,010	329,648

The above gives the shear and moment along height of the dam due to hydrodynamic pressure.

The total moment and shear due to the dam inertia and hydrodynamic pressure are now combined as shown below.

z/H	0	0.1	0.2	0.3	0.4	0.5	0.6	0.7	0.8	0.9	1
V_z	0	1052	2371	4004	5770	7803	9892	11849	13791	15271	16667
M_z	0	9300	22865	64223	113568	190609	276671	393852	518805	658327	802101

(2) **Analysis based on IS-1893 (2002)**

In this case, the time period remains same as in the previous case; thus, $T = 0.237$ s.

As $T > 0.1$ s, hence, $S_a/g = 2.5$ for 5% damping.

$\alpha_h = \frac{ZI}{2R}$; for zone IV, $Z = 0.24$, $I = 3.0$, and $R = 1.5$ (assuming un-reinforced concrete section)

Hence, $\alpha_h = \frac{0.24 \times 3}{2 \times 1.5} \times 2.5 = 0.60$[$\alpha_h$ is four times the value obtained by IS-1893 (1984)]

$$V_B = 0.6\alpha_h W = 0.6 \times 0.6 \times 96,250 = 34,650 \text{ kN}$$

$$M_B = 0.9\alpha_h W\bar{h} = 0.9 \times 0.6 \times 96,250 \times 36.36 = 1,889,811 \text{ kN m}$$

Multiplying by the coefficients furnished in Table 11.2, we have

z/H	0	0.1	0.2	0.3	0.4	0.5	0.6	0.7	0.8	0.9	1
V_z	0	3222	6410	10,152	13,860	18,295	22,696	26,507	30,319	32,502	34,650
M_z	0	34,017	66,143	185,201	302,370	497,020	689,781	973,253	1,256,724	1,574,213	1,889,811

(2a) **Computation of hydrodynamic pressure**

Considering velocity of sound $c = 1440$ m/s in water,

$$T = 4\frac{H}{c} = 4 \times \frac{97.5}{1440} = 0.27\ \text{s}$$

$$\therefore \alpha_h = \frac{0.24 \times 3}{2 \times 1.5} \times 2.5 = 0.60$$

Now, based on Eqs. (11.2.6)–(11.2.8), the pressure and the corresponding moment and shear are as tabled hereafter.

z/H	0	0.1	0.2	0.3	0.4	0.5	0.6	0.7	0.8	0.9	1
C_s	0.000	0.230	0.353	0.450	0.529	0.594	0.646	0.685	0.713	0.729	0.735
p_h	0	138	212	270	318	356	387	411	428	438	441
V_z	0	1002	3074	5879	9221	12,935	16,871	20,887	24,842	28,599	32,017
M_z	0	4126	25,317	72,636	151,902	266,359	416,902	602,154	818,495	1,060,041	1,318,590

Total moment (kN m) and shear (kN) due to the dam inertia and hydrodynamic pressure are combined as shown below.

z/H	0	0.1	0.2	0.3	0.4	0.5	0.6	0.7	0.8	0.9	1
V_z	0	4224	9484	16,031	23,081	31,230	39,567	47,394	55,161	61,101	66,667
M_z	0	38,143	91,460	257,837	454,272	763,379	1,106,683	1,575,407	2,075,219	2,634,254	3,208,401

It will be observed that by the latest IS-1893 (2002), the moments and shears have significantly increased (4 times) compared to the 1984 version.

(3) **Fenves and Chopra's** (1986) **Method**

We apply Chopra and Fenves's method, in this case considering the response spectrum curves as furnished in IS-1893 (2016).

Case 1: For reservoir empty case

Time period as per Chopra and Fenves is

$T = 0.38H/\sqrt{E_s}$; here, E_s is in mPa.

$$T = 0.38 \times 100 \Big/ \sqrt{28,500} = 0.225\text{s}.$$

$\alpha = \frac{1-qc}{1+qc}$, where $q = \rho/\rho_r c_r = \frac{10}{25 \times 1100} = 3.64 \times 10^{-4}$

$\rightarrow \alpha = \frac{1-qc}{1+qc} = \frac{1-3.64\times10^{-4}\times1100}{1+3.64\times10^{-4}\times1100} = 0.428 \cong 0.5$ say.

$\rightarrow R_r = 1.30$ Vide Table 11.3.

Thus, $T_r = 1.3 \times 0.225 = 0.2925$ s—the elongated time period due to fluid–structure interaction.

Hydrodynamic ratio effect factor R_w is expressed as

$R_w = \frac{T \times c}{4H} = \frac{0.225 \times 1440}{4 \times 100} = 0.81$(In this case, we have ignored the free board of water level.)

Weight of the dam per unit height $w(z) = 9625$ kN/m.

Here, $T = 0.225$ s, when the reservoir is empty when $S_a/g = 2.5$

The lateral force due to earthquake is given by

$$F(z) = (L_1/M_1) \times \beta \times (S_a/g) \times w(z) \times \phi(z)$$

As recommended by Chopra and Fenves, $L_1/M_1 = 3.0$ for an empty reservoir and β = code factor (ZI/2R), calculated previously as 0.24.

Thus, the shear value along height of the dam is expressed as

$$F(z) = 3.0 \times 0.24 \times 2.5 \times 9625 \times \phi(z)$$

Given the value of $\phi(z)$ as furnished in Table 11.7, the lateral and shear force due to earthquake are expressed as

z/H	0	0.1	0.2	0.3	0.4	0.5	0.6	0.7	0.8	0.9	1
F(z)	0	364	814	1455	2339	3465	4920	6739	9182	12,734	17,325
V(z)	0	364	1178	2633	4972	8437	13,357	20,096	29,278	42,012	59,337

Case 2: With soil–fluid–structure interaction

The foundation shear modulus is expressed as $G_f = \rho_f \times V_s^2 = \frac{22}{9.81} \times (1100)^2 = 2,713,557$ kN/m^2.

Young's modulus $E_f = G_f \times 2 \times (1 + v_f) = 2,713,557 \times 2 \times 1.3 = 7,055,250$ kN/m^2.

$\therefore \frac{E_f}{E_s} = \frac{7,055,250}{2.85 \times 10^7} = 0.25$, which gives foundation elongation factor $R_f = 1.62$

Thus, the time period T_{fr} of the dam considering soil–fluid–structure interaction is expressed as

$$T_{fr} = R_r \times R_f \times T = 1.3 \times 1.62 \times 0.225 = 0.4738 \text{ s.}$$

The damping ratio is expressed as

$$\tilde{\xi}_1 = \frac{1}{R_r} \times \frac{1}{(R_f)^3} \xi_1 + \xi_r + \xi_f = \frac{1}{1.3} \times \frac{1}{(1.62)^3} \times 0.05 + 0.051 + 0.20 = 0.26$$

Thus, for 26% damping scaling factor for S_a/g is 0.58 and for $T_{fr} = 0.4738$ s

$$S_a/g = (1/T_{fr}) \times 0.58 = 1.224 \times 0.24 = 0.293.$$

$$F(z) = \frac{\hat{L}_1}{\hat{M}_1} \frac{S_a\left(T^{fr}, \tilde{\xi}_1\right)}{g} [w(z)\phi(z) + gp(z, T_r)]$$

As per Chopra and Fenves, $\hat{L}_1/\hat{M}_1 = 4.0$, and the value of $gp(z)$ is furnished below from tables furnished in the appendix for $R_w = 0.8$ and $\alpha = 0.5$

z/H	0	0.1	0.2	0.3	0.4	0.5	0.6	0.7	0.8	0.9	1
$gp(z)$	0	1180	1430	1600	1570	1550	1450	1400	1300	1250	1190

This finally gives $F(z)$ and $V(z)$ for this case in kN as

z/H	0	0.1	0.2	0.3	0.4	0.5	0.6	0.7	0.8	0.9	1
$F(z)$	0	1622	2209	2826	3367	4078	4909	6037	7512	9769	12691
$V(z)$	0	1622	3831	6657	10,024	14,102	19,011	25,048	32,560	42,329	55,020

The moment at any section can now be obtained as $M_i(z) = \sum F_i h_i$ where $h_i = H - z_i$.

(4) **Analysis based on CSD model.**

(4a) **Uncoupled analysis**

As a first illustration, we solve the problem for the reservoir empty case.

Based on Eq. (11.5.60) and substituting the values of C_{TFi} and C_{Si} from Tables 11.9 and 11.10, and considering $\eta = 2/3$, we have time periods for the first three modes as

$T_1 = 0.267$ s, $T_2 = 0.093$ s, $T_3 = 0.047$ s.

Considering 5% damping, the S_a/g values as per IS-1893 (2002) are given as

Mode	1	2	3
Time period	0.267	0.093	0.04
S_a/g (Spectral acceleration)	2.5	2.395	1.705
Design acceleration (g)[a]	0.6	0.57	0.41

[a]Design acceleration is obtained by multiplying the spectral acceleration by the code factor $\alpha_h = 0.24$ calculated earlier

The scaling vectors ϕ for the first three modes (solved in Mathcad) are expressed as

$$\phi = \begin{bmatrix} 1 & 0.256 & -0.13 \\ -0.169 & -1 & 0.173 \\ 0.021 & 0.23 & -1 \end{bmatrix}$$

Here, $\kappa_i = \int_0^1 (1+\psi\xi) f_i^{\mathrm{b}}(\xi)\mathrm{d}\xi \Big/ \int_0^1 (1+\psi\xi) f_i^{\mathrm{b}}(\xi)^2 \mathrm{d}\xi$

Here, $f_i^{\mathrm{b}}(\xi) = \sin\,\mu_i\xi - \sin h\,\mu_i\xi - \alpha_i(\cos\,\mu i\xi - \cos h\,\mu_i\xi)$, where μ_i for $i = 1$, 2, 3 are 1.875,4.6941, and 7.855, respectively, and $\alpha_i = \frac{\sin\,\mu_i + \sin h\,\mu_i}{\cos\,\mu_i + \cos h\,\mu_i}$.

This gives on integration $\kappa_i = 0.726, 0.771$ and 0.432, respectively, for the first three modes.

The values of moments and shears are expressed as

$$M_{zi} = -\kappa_i\beta\frac{EI_{\mathrm{b}}}{H^2}(1+\psi_b\xi)^3\left(\frac{S_{ai}T_{ei}^2}{4\pi^2}\right)[\phi_i]^{\mathrm{T}}\left\{\frac{\mathrm{d}^2 f_i^{\mathrm{b}}(\xi)}{\mathrm{d}\xi^2}\right\}$$

$$V_{zi} = -\kappa_i\beta\frac{EI_{\mathrm{b}}}{H^3}(1+\psi_b\xi)^3\left(\frac{S_{ai}T_{ei}^2}{4\pi^2}\right)[\phi_i]^{\mathrm{T}}\left\{\frac{\mathrm{d}^3 f_i^{\mathrm{b}}(\xi)}{\mathrm{d}\xi^3}\right\}$$

Here, $E = 2.85\times10^7$ kN/m^2; $I = 2.858\times10^4$ m^4; $\psi_b = -0.9$, $H = 100$ m, $\beta = 0.24$,

$$f_i^{\prime\prime\mathrm{b}}(\xi) = \mu_i^2(-\sin\,\mu_i - \sin h\,\mu_i - \alpha_i(-\cos\,\mu_i - \cos h\,\mu_i))$$

$$f_i^{\prime\prime\prime\mathrm{b}}(\xi) = \mu_i^3(-\cos\,\mu_i - \cos h\,\mu_i - \alpha_i(-\sin\,\mu_i - \sin h\,\mu_i))$$

The SRSS values of moment and shear force for the first three modes, including the added hydrodynamic force and moment considered uncoupled in this case, based on Eqs. (11.5.103) and (11.5.104), are as shown hereafter (Fig. 11.26).

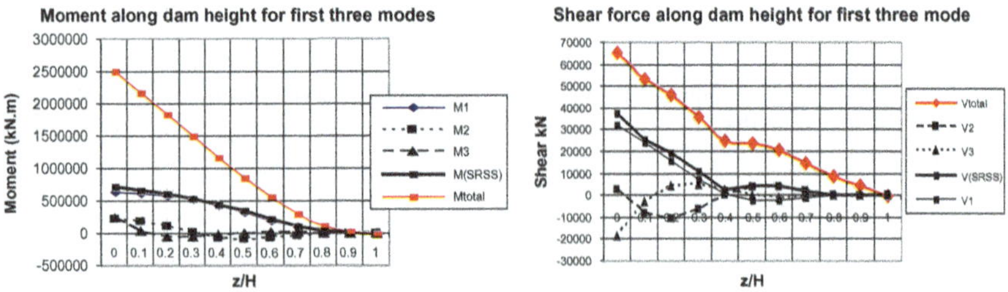

Fig. 11.26 Moment and shear along the dam height; first three modes uncoupled case

z/H	Moment (first mode)	Moment (second mode)	Moment (third mode)	$M_{(SRSS)}$	Moment due to hydrodynamic effect	M_{total}
0	647,785	215,427	232,323	721,116	1,806,530	2,527,646
0.1	636,984	179,454	38,136	662,878	1,523,927	2,186,804
0.2	591,089	112,888	−52,490	604,057	1,248,282	1,852,338
0.3	528,522	16,911	−59,790	532,162	986,383	1,518,544
0.4	442,417	−65,980	−27,447	448,152	744,678	1,192,830
0.5	329,319	−97,872	2976	343,568	529,120	872,688
0.6	206,837	−79,346	13,754	221,960	345,017	566,977
0.7	103,901	−41,056	9893	112,155	196,900	309,055
0.8	39,060	−12,439	3399	41,134	88,418	129,551
0.9	9918	−1540	409	10,045	22,241	32,287
1	1277	−43	2	1278	0	1278
z/H	Shear (first mode)	Shear (second mode)	Shear (third mode	$V_{(SRSS)}$	Shear due to hydrodynamic effect	V_{total}
0	3.22E+04	2.37E+03	−1.85E+04	37,227	22,412	59,639
0.1	2.40E+04	−8.33E+03	−2.98E+03	25,552	20,019	45,571
0.2	1.57E+04	−1.08E+04	4.26E+03	19,500	17,390	36,890
0.3	7.17E+03	−6.45E+03	4.88E+03	10,810	14,621	25,431
0.4	5.06E+02	−1.74E+02	2.34E+03	2402	11,810	14,212
0.5	−2.58E+03	3.55E+03	−6.94E+01	4388	9054	13,442
0.6	−2.42E+03	3.62E+03	−9.38E+02	4449	6455	10,904
0.7	−9.35E+02	1.79E+03	−6.53E+02	2125	4115	6240
0.8	1.25E+02	2.65E+02	−1.65E+02	337	2152	2489
0.9	2.95E+02	−1.91E+02	3.28E+01	353	701	1054
1	8.57E+01	−7.44E+01	1.99E+01	115	0	115

(4b) **Response considering the fluid–structure interaction**

Based on the methodology explained in Sect. 11.5, considering fluid compressibility, the elongated time period and corresponding spectral acceleration as per IS-1893 (2002) are as shown hereafter.

Mode	1	2	3
Time period	0.302	0.106	0.056
S_a/g (spectral acceleration)	2.5	2.5	1.84
Design acceleration (g)	0.6	0.6	0.44

The scaling vector ϕ_{bf} in this case is expressed as

$$\phi_{bf} = \begin{bmatrix} 1 & 0.314 & -0.179 \\ -0.162 & -1 & 0.24 \\ 0.019 & 0.226 & -1 \end{bmatrix}$$

Computing Eq. (11.5.131) by Mathcad, we have $\kappa_i = 0.835$, 0.609, and 0.368 for the first three modes.

Based on the above, substituting in Eqs. (11.5.128), (11.5.129), and (11.5.130) gives

z/H	Amplitude (first mode)	Amplitude (second mode)	Amplitude (third mode)	$u_{(SRSS)}$
0	0	0	0	0
0.1	0.0004446	0.0001002	−5.426E−05	0.000459
0.2	0.0016931	0.0003257	−0.0001438	0.00173
0.3	0.0036179	0.0005692	−0.0001799	0.003667
0.4	0.0060938	0.0007394	−0.0001251	0.00614
0.5	0.0090001	0.0007721	−4.663E−06	0.009033
0.6	0.0122238	0.0006377	0.0001127	0.012241
0.7	0.0156631	0.0003431	0.0001564	0.015668
0.8	0.0192312	−7.57E−05	9.391E−05	0.019232
0.9	0.0228609	−0.000567	−5.441E−05	0.022868
1	0.0265085	−0.001082	−0.000238	0.026532
z/H	Moment (first mode)	Moment (second mode)	Moment (third mode)	$M_{(SRSS)}$
0	1,090,971	246,310	284,131	1,153,957
0.1	1,077,679	199,549	45,670	1,096,949
0.2	998,701	121,460	−65,190	1,008,169
0.3	885,723	12,602	−73,334	888,843
0.4	733,048	−79,942	−32,865	738,127
0.5	540,156	−114,271	4762	552,131
0.6	337,105	−91,746	17,768	349,819
0.7	169,104	−47,360	12,605	176,062
0.8	63,863	−14,406	4318	65,610
0.9	16,385	−1830	523	16,495
1	2126	−60	4	2127
z/H	Shear (first mode)	Shear (second mode)	Shear (third mode)	$V_{(SRSS)}$
0	5.48E+04	1.84E+03	−2.27E+04	59,313
0.1	3.93E+04	−9.79E+03	−3.57E+03	40,697

(continued)

(continued)

z/H	Amplitude (first mode)	Amplitude (second mode)	Amplitude (third mode)	$u_{(SRSS)}$
0.2	2.50E+04	−1.23E+04	5.35E+03	28,340
0.3	1.12E+04	−7.25E+03	6.07E+03	14,631
0.4	7.32E+02	−1.31E+02	2.88E+03	2975
0.5	−3.96E+03	4.06E+03	−1.19E+02	5678
0.6	−3.66E+03	4.11E+03	−1.19E+03	5626
0.7	−1.37E+03	2.03E+03	−8.20E+02	2580
0.8	2.40E+02	2.95E+02	−2.06E+02	433
0.9	4.73E+02	−2.19E+02	4.22E+01	523
1	1.35E+02	−8.48E+01	2.53E+01	162

(4c) **Response considering dynamic soil–structure interaction (DSSI)**
Shear wave velocity of foundation = 1100 m/s.
Density of foundation = 22 kN/m^3.
Thus, shear modulus $G_f = (22/9.81) \times 1100^2 = 2,713,557\ \text{kN/m}^2$.
Poisson's ratio = 0.3

$$K_x = 0.94\frac{\pi G_f}{2 - \nu_f} = 0.94 \times \frac{\pi \times 2,713,557}{2 - 0.3} = 4,713,763\ \text{kN/m}$$

$$K_\theta = 0.1175\frac{\pi G_f B_b^2}{1 - \nu_f} = 0.1175 \times \frac{\pi \times 2,713,557 \times 70^2}{1 - 0.3} = 7.0117 \times 10^9\ \text{kN/m}$$

The amplification factor due to soil–fluid–structure interaction is expressed as

$$\text{AF} = \left[1 + \left(\frac{V_{\text{SRSS}}}{K_x u_{\text{srss}}}\right) + H\left(\frac{M_{\text{SRSS}}}{K_\theta u_{\text{srss}}}\right)\right]$$

$$\text{AF} = \left[1 + \left(\frac{59,313}{4,713,763 \times 0.026532}\right) + 100 \times \left(\frac{1,153,957}{7.0117 \times 10^9 \times 0.026532}\right)\right]$$
$$= 2.094$$

Thus, the time period is elongated to

$$T_1 = 0.302 \times \sqrt{2.094} = 0.437\ \text{s}, T_2 = 0.106 \times \sqrt{2.094} = 0.153\ \text{s},$$
$$T_3 = 0.056 \times \sqrt{2.094} = 0.081\ \text{s}$$

The amplified amplitude, moment, and shear due to DSSI are as shown hereafter

z/H	Amplitude (first mode)	Amplitude (second mode)	Amplitude (third mode)	$u_{(SRSS)}$ (m)
0	0	0	0	0
0.1	0.000931	0.0002098	−0.0001136	0.000961
0.2	0.0035453	0.000682	−0.0003011	0.003623
0.3	0.0075758	0.0011919	−0.0003767	0.007678
0.4	0.0127603	0.0015483	−0.000262	0.012857
0.5	0.0188462	0.0016167	−9.765E−06	0.018915
0.6	0.0255967	0.0013354	0.000236	0.025633
0.7	0.0327985	0.0007183	0.0003275	0.032808
0.8	0.0402702	−0.000159	0.0001966	0.040271
0.9	0.0478707	−0.001186	−0.0001139	0.047886
1	0.0555089	−0.002265	−0.0004983	0.055557
z/H	Moment (first mode)	Moment (second mode)	Moment (third mode)	$M_{(SRSS)}$ (kN m)
0	2,284,494	515,772	594,969	2,416,386
0.1	2,256,659	417,856	95,634	2,297,011
0.2	2,091,280	254,337	−136,508	2,111,107
0.3	1,854,704	26,388	−153,561	1,861,238
0.4	1,535,003	−167,399	−68,820	1,545,637
0.5	1,131,087	−239,284	9972	1,156,163
0.6	705,899	−192,117	37,206	732,521
0.7	354,103	−99,171	26,396	368,674
0.8	133,730	−30,166	9041	137,388
0.9	34,311	−3832	1096	34,542
1	4452	−126	9	4454
z/H	Shear (first mode)	Shear (second mode)	Shear (third mode)	$V_{(SRSS)}$ (kN)
0	1.15E+05	3.85E+03	−4.76E+04	124,201
0.1	8.24E+04	−2.05E+04	−7.49E+03	85,218
0.2	5.23E+04	−2.57E+04	1.12E+04	59,344
0.3	2.34E+04	−1.52E+04	1.27E+04	30,638
0.4	1.53E+03	−2.75E+02	6.03E+03	6229
0.5	−8.30E+03	8.51E+03	−2.50E+02	11,889
0.6	−7.66E+03	8.60E+03	−2.49E+03	11,781
0.7	−2.86E+03	4.25E+03	−1.72E+03	5402
0.8	5.03E+02	6.19E+02	−4.31E+02	906
0.9	9.91E+02	−4.58E+02	8.85E+01	1096
1	2.83E+02	−1.78E+02	5.29E+01	338

It is observed that though the uncoupled analysis gives comparable moment at the base; yet over the whole height of the dam the DSSI effect significantly amplifies the bending moment magnitude. Similarly, for a height of 25 m from the base, the shear force gets significantly amplified. Considering fluid as incompressible with 15% damping gives a much lower value than the uncoupled analysis.

In the present IS code, based on the acceleration spectra furnished, considering importance factor of $I = 3$ gives a far too conservative result. The IS-1893 (2002) part IV (industrial and stack-like structure) gives the category of structures based on its economic importance. Of which, highest is category one type of structures. Adopting the same concept for dam and considering it also as a category one type of structure importance factor I can be considered as = 2.0, instead of 3.0 as considered presently. This looks more rational and would possibly give more sensible result.

Appendix

See Tables (11.22, 11.23, 11.24, 11.25, and 11.26, 11.27).

Table 11.22 Standard values of the hydrodynamic pressure function $p(z)$ for a full reservoir $H_w/H = 1.0$ and $\alpha = 1.0$, after Fenves and Chopra (1986)

$\hat{y} = \frac{y}{H}$	Value of $\frac{gp(\hat{y})}{w(H)}$												
	$R_w \le 0.5$	$R_{w=}0.7$	$R_w = 0.8$	$R_w = 0.85$	$R_w = 0.90$	$R_w = 0.92$	$R_w = 0.93$	$R_w = 0.94$	$R_w = 0.95$	$R_w = 0.96$	$R_w = 0.97$	$R_w = 0.98$	$R_w = 0.99$
1.0	0	0	0	0	0	0	0	0	0	0	0	0	0
0.95	0.070	0.073	0.076	0.079	0.083	0.086	0.088	0.090	0.092	0.096	0.102	0.111	0.133
0.90	0.112	0.118	0.124	0.129	0.138	0.143	0.147	0.151	0.157	0.164	0.176	0.195	0.238
0.85	0.127	0.135	0.144	0.152	0.164	0.172	0.178	0.184	0.193	0.204	0.221	0.249	0.313
0.80	0.133	0.144	0.155	0.165	0.182	0.193	0.200	0.208	0.220	0.235	0.257	0.295	0.379
0.75	0.141	0.154	0.168	0.180	0.201	0.214	0.223	0.234	0.248	0.267	0.294	0.340	0.445
0.70	0.145	0.161	0.178	0.192	0.216	0.232	0.242	0.255	0.272	0.294	0.327	0.382	0.506
0.65	0.143	0.161	0.180	0.197	0.224	0.242	0.254	0.269	0.288	0.313	0.351	0.414	0.558
0.60	0.139	0.159	0.180	0.199	0.230	0.250	0.264	0.280	0.301	0.330	0.373	0.444	0.605
0.55	0.137	0.159	0.183	0.203	0.237	0.260	0.274	0.293	0.316	0.348	0.395	0.473	0.651
0.50	0.135	0.159	0.184	0.206	0.244	0.269	0.284	0.304	0.329	0.364	0.415	0.500	0.694
0.45	0.130	0.155	0.182	0.206	0.246	0.272	0.289	0.310	0.338	0.375	0.430	0.522	0.730
0.40	0.124	0.151	0.179	0.204	0.247	0.275	0.293	0.315	0.345	0.384	0.442	0.540	0.762
0.35	0.121	0.149	0.179	0.205	0.250	0.279	0.298	0.322	0.353	0.395	0.456	0.559	0.793
0.30	0.118	0.147	0.178	0.206	0.252	0.283	0.303	0.328	0.360	0.403	0.467	0.575	0.820
0.25	0.113	0.143	0.175	0.204	0.252	0.284	0.304	0.330	0.363	0.408	0.475	0.587	0.840
0.20	0.109	0.139	0.172	0.202	0.252	0.284	0.305	0.332	0.366	0.412	0.481	0.596	0.856
0.15	0.107	0.138	0.172	0.202	0.252	0.286	0.307	0.334	0.369	0.417	0.487	0.604	0.871
0.10	0.106	0.137	0.172	0.202	0.252	0.286	0.309	0.337	0.372	0.420	0.491	0.611	0.881
0.05	0.103	0.135	0.169	0.200	0.252	0.286	0.308	0.336	0.372	0.420	0.492	0.613	0.886
0.0	0.010	0.133	0.168	0.198	0.251	0.285	0.307	0.335	0.371	0.420	0.492	0.613	0.886

Table 11.23 Standard values of the hydrodynamic pressure function $p(z)$ for a full reservoir $H_w/H = 1.0$ and $\alpha = 0.9$, after Fenves and Chopra (1986)

$\hat{y} = \frac{y}{H}$	$\frac{gp(\hat{y})}{w(H)}$								
	$R_w \leq 0.5$	$R_w = 0.7$	$R_w = 0.8$	$R_w = 0.9$	$R_w = 0.95$	$R_w = 1.0$	$R_w = 1.05$	$R_w = 1.1$	$R_w = 1.2$
1.0	0.0	0.0	0.0	0.0	0.0	0.0	0.0	0.0	0.0
0.95	0.070	0.073	0.076	0.082	0.088	0.089	0.069	0.064	0.062
0.90	0.112	0.118	0.124	0.136	0.149	0.149	0.110	0.100	0.095
0.85	0.127	0.135	0.144	0.162	0.181	0.181	0.123	0.108	0.101
0.80	0.133	0.144	0.155	0.179	0.204	0.205	0.127	0.107	0.098
0.75	0.141	0.154	0.168	0.197	0.228	0.229	0.133	0.108	0.097
0.70	0.145	0.161	0.177	0.212	0.249	0.249	0.135	0.105	0.092
0.65	0.143	0.161	0.179	0.219	0.261	0.262	0.130	0.096	0.081
0.60	0.139	0.159	0.179	0.234	0.271	0.272	0.124	0.085	0.067
0.55	0.137	0.159	0.182	0.231	0.283	0.283	0.119	0.076	0.057
0.50	0.135	0.159	0.183	0.236	0.293	0.292	0.114	0.067	0.046
0.45	0.130	0.155	0.181	0.238	0.299	0.298	0.106	0.055	0.032
0.40	0.124	0.150	0.178	0.238	0.303	0.302	0.097	0.044	0.019
0.35	0.121	0.148	0.177	0.241	0.309	0.307	0.091	0.035	0.009
0.30	0.118	0.146	0.177	0.243	0.313	0.311	0.086	0.027	0.000
0.25	0.113	0.142	0.174	0.242	0.315	0.312	0.078	0.017	0.000
0.20	0.109	0.139	0.171	0.241	0.316	0.312	0.071	0.008	0.000
0.15	0.107	0.137	0.170	0.242	0.318	0.313	0.067	0.003	0.000
0.10	0.106	0.136	0.170	0.242	0.320	0.313	0.064	0.000	0.000
0.05	0.103	0.134	0.167	0.241	0.318	0.311	0.059	0.000	0.000
0.0	0.101	0.133	0.166	0.239	0.317	0.309	0.056	0.000	0.000

Table 11.24 Standard values of the hydrodynamic pressure function $p(z)$ for a full reservoir $H_w/H = 1.0$ and $\alpha = 0.75$, after Fenves and Chopra (1986)

$\hat{y} = \frac{y}{H}$	$\frac{gp(\hat{y})}{w(H)}$								
	$R_w \leq 0.5$	$R_w = 0.7$	$R_w = 0.8$	$R_w = 0.9$	$R_w = 0.95$	$R_w = 1.0$	$R_w = 1.05$	$R_w = 1.1$	$R_w = 1.2$
1.0	0.0	0.0	0.0	0.0	0.0	0.0	0.0	0.0	0.0
0.95	0.070	0.073	0.075	0.079	0.80	0.078	0.073	0.068	0.065
0.90	0.112	0.118	0.122	0.129	0.132	0.128	0.118	0.101	0.101
0.85	0.127	0.133	0.140	0.151	0.154	0.150	0.134	0.121	0.110
0.80	0.133	0.143	0.152	0.166	0.171	0.163	0.142	0.125	0.110
0.75	0.140	0.153	0.164	0.181	0.187	0.177	0.151	0.130	0.110
0.70	0.145	0.159	0.173	0.193	0.200	0.188	0.157	0.131	0.108
0.65	0.143	0.159	0.174	0.197	0.205	0.191	0.155	0.126	0.099
0.60	0.139	0.157	0.174	0.199	0.208	0.192	0.151	0.118	0.088
0.55	0.137	0.157	0.175	0.203	0.213	0.195	0.150	0.113	0.079
0.50	0.135	0.156	0.176	0.206	0.216	0.196	0.147	0.107	0.070
0.45	0.129	0.152	0.173	0.205	0.216	0.194	0.140	0.097	0.058
0.40	0.123	0.147	0.170	0.203	0.214	0.191	0.134	0.088	0.045
0.35	0.120	0.145	0.169	0.204	0.215	0.190	0.129	0.080	0.036
0.30	0.117	0.143	0.168	0.204	0.215	0.188	0.125	0.074	0.027
0.25	0.112	0.139	0.164	0.201	0.212	0.184	0.118	0.065	0.016
0.20	0.108	0.135	0.161	0.199	0.209	0.180	0.111	0.056	0.007
0.15	0.106	0.134	0.159	0.198	0.208	0.177	0.107	0.051	0.001
0.10	0.104	0.133	0.158	0.197	0.207	0.175	0.103	0.046	0.000
0.05	0.102	0.130	0.156	0.194	0.204	0.171	0.098	0.040	0.000
0.0	0.100	0.128	0.154	0.192	0.201	0.167	0.093	0.036	0.000

Table 11.25 Standard values of the hydrodynamic pressure function $p(z)$ for a full reservoir $H_w/H = 1.0$ and $\alpha = 0.5$, after Fenves and Chopra (1986)

$\hat{y} = \frac{y}{H}$	$\frac{gp(\hat{y})}{w(H)}$								
	$R_w \leq 0.5$	$R_w = 0.7$	$R_w = 0.8$	$R_w = 0.9$	$R_w = 0.95$	$R_w = 1.0$	$R_w = 1.05$	$R_w = 1.1$	$R_w = 1.2$
1.0	0.0	0.0	0.0	0.0	0.0	0.0	0.0	0.0	0.0
0.95	0.071	0.072	0.073	0.074	0.074	0.073	0.072	0.070	0.068
0.90	0.112	0.116	0.118	0.119	0.119	0.118	0.116	0.113	0.108
0.85	0.125	0.132	0.135	0.136	0.135	0.134	0.130	0.127	0.120
0.80	0.132	0.130	0.143	0.146	0.145	0.143	0.138	0.133	0.123
0.75	0.139	0.148	0.153	0.156	0.155	0.152	0.146	0.139	0.127
0.70	0.144	0.154	0.160	0.163	0.162	0.158	0.151	0.143	0.128
0.65	0.141	0.150	0.159	0.163	0.161	0.156	0.148	0.138	0.122
0.60	0.137	0.149	0.157	0.162	0.160	0.153	0.143	0.132	0.113
0.55	0.135	0.148	0.156	0.161	0.158	0.151	0.141	0.128	0.107
0.50	0.133	0.147	0.155	0.159	0.156	0.148	0.137	0.123	0.099
0.45	0.127	0.142	0.150	0.154	0.151	0.142	0.129	0.115	0.088
0.40	0.121	0.136	0.145	0.149	0.145	0.136	0.122	0.106	0.077
0.35	0.117	0.133	0.143	0.146	0.142	0.131	0.116	0.099	0.069
0.30	0.114	0.131	0.140	0.143	0.137	0.126	0.110	0.092	0.060
0.25	0.109	0.126	0.135	0.137	0.131	0.119	0.102	0.083	0.050
0.20	0.104	0.121	0.130	0.132	0.125	0.112	0.094	0.074	0.040
0.15	0.102	0.119	0.127	0.128	0.121	0.108	0.089	0.068	0.033
0.10	0.100	9,117	0.125	0.125	0.118	0.104	0.083	0.062	0.026
0.05	0.098	0.114	0.121	0.121	0.113	0.098	0.077	0.055	0.018
0.0	0.096	0.111	0.119	0.117	0.108	0.093	0.072	0.049	0.012

Table 11.26 Standard values of the hydrodynamic pressure function $p(z)$ for a full reservoir $H_w/H = 1.0$ and $\alpha = 0.25$, after Fenves and Chopra (1986)

$\hat{y} = \frac{y}{H}$	$\frac{gp(\hat{y})}{w(H)}$								
	$R_w \leq 0.5$	$R_w = 0.7$	$R_w = 0.8$	$R_w = 0.9$	$R_w = 0.95$	$R_w = 1.0$	$R_w = 1.05$	$R_w = 1.1$	$R_w = 1.2$
1.0	0.0	0.0	0.0	0.0	0.0	0.0	0.0	0.0	0.0
0.95	0.069	0.070	0.071	0.071	0.071	0.071	0.070	0.070	0.070
0.90	0.111	0.113	0.114	0.114	0.114	0.114	0.113	0.113	0.111
0.85	0.124	0.127	0.128	0.129	0.129	0.128	0.127	0.127	0.125
0.80	0.130	0.133	0.134	0.135	0.135	0.134	0.133	0.132	0.129
0.75	0.137	0.141	0.142	0.143	0.142	0.141	0.140	0.138	0.135
0.70	0.141	0.145	0.147	0.147	0.146	0.145	0.143	0.141	0.137
0.65	0.137	0.142	0.144	0.144	0.143	0.142	0.140	0.137	0.131
0.60	0.133	0.138	0.140	0.139	0.138	0.136	0.134	0.131	0.124
0.55	0.131	0.136	0.137	0.136	0.135	0.133	0.130	0.126	0.118
0.50	0.128	0.133	0.134	0.133	0.131	0.128	0.125	0.121	0.112
0.45	0.121	0.126	0.127	0.126	0.124	0.120	0.116	0.112	0.101
0.40	0.115	0.120	0.120	0.118	0.115	0.112	0.107	0.102	0.091
0.35	0.111	0.116	0.116	0.113	0.110	0.106	0.100	0.095	0.082
0.30	0.107	0.111	0.111	0.107	0.104	0.099	0.093	0.087	0.074
0.25	0.101	0.105	0.104	0.100	0.096	0.090	0.084	0.077	0.063
0.20	0.096	0.099	0.098	0.093	0.088	0.082	0.076	0.068	0.052
0.15	0.094	0.096	0.094	0.088	0.083	0.076	0.069	0.061	0.044
0.10	0.092	0.096	0.090	0.083	0.078	0.071	0.063	0.054	0.037
0.05	0.088	0.088	0.085	0.077	0.071	0.064	0.055	0.046	0.028
0.0	0.086	0.085	0.081	0.072	0.065	0.057	0.048	0.039	0.020

Table 11.27 Standard values of the hydrodynamic pressure function $p(z)$ for a full reservoir $H_w/H = 1.0$ and $\alpha = 0.0$, after Fenves and Chopra (1986)

$\hat{y} = \frac{y}{H}$	$\frac{gp(\hat{y})}{w(H)}$								
	$R_w \leq 0.5$	$R_w = 0.7$	$R_w = 0.8$	$R_w = 0.9$	$R_w = 0.95$	$R_w = 1.0$	$R_w = 1.05$	$R_w = 1.1$	$R_w = 1.2$
1.0	0.0	0.0	0.0	0.0	0.0	0.0	0.0	0.0	0.0
0.95	0.069	0.069	0.069	0.069	0.069	0.069	0.070	0.070	0.070
0.90	0.109	0.110	0.110	0.111	0.111	0.111	0.112	0.112	0.112
0.85	0.122	0.123	0.124	0.125	0.125	0.125	0.126	0.126	0.126
0.80	0.127	0.128	0.128	0.129	0.129	0.129	0.130	0.130	0.130
0.75	0.133	0.134	0.134	0.135	0.135	0.135	0.136	0.136	0.136
0.70	0.135	0.136	0.137	0.138	0.138	0.138	0.139	0.139	0.139
0.65	0.132	0.133	0.133	0.133	0.133	0.133	0.134	0.134	0.134
0.60	0.127	0.127	0.127	0.127	0.127	0.127	0.127	0.127	0.127
0.55	0.123	0.123	0.123	0.123	0.123	0.123	0.122	0.122	0.121
0.50	0.120	0.119	0.118	0.118	0.118	0.117	0.116	0.116	0.115
0.45	0.113	0.111	0.110	0.109	0.109	0.108	0.107	0.106	0.105
0.40	0.105	0.103	0.102	0.100	0.099	0.098	0.097	0.096	0.094
0.35	0.101	0.098	0.096	0.094	0.092	0.091	0.090	0.088	0.085
0.30	0.096	0.092	0.090	0.087	0.085	0.084	0.082	0.080	0.076
0.25	0.090	0.085	0.082	0.078	0.076	0.074	0.072	0.069	0.065
0.20	0.084	0.078	0.074	0.070	0.067	0.065	0.062	0.059	0.053
0.15	0.080	0.073	0.068	0.064	0.061	0.058	0.055	0.051	0.045
0.10	0.077	0.069	0.064	0.058	0.054	0.051	0.048	0.044	0.036
0.05	0.073	0.063	0.057	0.050	0.046	0.043	0.039	0.035	0.026
0.0	0.070	0.058	0.052	0.044	0.040	0.036	0.031	0.027	0.017

References

Abel, J., & Popov, E. P. (1968). Static and dynamic finite element analysis of sandwhich structures. In *Proceedings, 2nd Conference in matrix analysis of structures*. Ohio USA: Wright Patterson Air force base.

ANSYS 10. (2010). *User operation manual for ANSYS multiphysics software.*

Bathe, K. J. (1980). *Finite element procedures in enginering*. USA: Prentice Hall Publication.

Chopra, A. K. (1966). *Hydrodynamic pressure on dams during earthquakes* (EERC Research Report # 66-2A). Berkeley, USA: University of California.

Chopra, A. K. (1970). *Earthquake response of concrete gravity dams* (EERC Research Report # 70-1). Berkeley, USA: University of California.

Chopra, A. K., & Chakrabarti, P. (1980a). Earthquake analysis of gravity dams including dam-water- foundation rock interaction. *Journal of Earthquake engineering and structural dynamics, 9*, 363–383.

Chopra, A. K., & Hall, J. F. (1980b). *Dynamic response of embankment, concrete gravity and arch dams including hydrodynamic effect* (EERC Research report # 80/39). Berkeley USA: University of California.

Chopra, A. K., Chakrabarti, P., & Gupta, S. (1980c). *Earthquake response of concrete dams including hydrodynamic and foundation interaction effect* (EERC Research Report # 80/01). Berkeley USA: University of California.

Chowdhury, I., & Dasgupta, S. P. (2007). Dynamic pressure on rigid unyielding walls under earthquake force. *Indian Geotechnical Journal, 37*, 2.

Chowdhury, I., & Dasgupta, S. P. (2008). *Dynamics of structures and foundation a unified approach* (Vol. 1 and 2). Leiden, Holland: Taylor and Francis publication.

Chowdhury, I., & Dasgupta, S. P. (2011). Dynamic response of cantliver retaining walls with coupled soil structure interaction. In *Proceedings, 13th International Conference in Computer Methods and Advancement in Geomecahnics*, Melbourne, Australia.

Chowdhury, I., & Dasgupta, S. P. (2012). Seismic response of piles under earthquake force. *Indian Geotechnical Journal, 42*(2), 57–74.

Chowdhury, I., & Singh, J. P. (2010) Do DSSI attenuate dynamic response of Building? In *14th International Symposium on Earthquake Engineering*, IIT (Roorkee), India.

Chowdhury, I., Singh, J. P., & Dasgupta, S. P. (2013, April). Static and dynamic response of concrete gravity dam. *International Journal of Dam Engineering, XXIII*(3), 129–182.

Clough, R. W. (1960). The finite element method in plane stress analysis. *Proceedings 2nd ASCE Conference in Electronic Computation*, Pittsburg, USA (1964).

Clough, R. W. (1962) *The stress distribution of Norfork Dam* (Institute of Engineering research final report to the corps of engineer).

Cook, R. D., Malkus, D. S., & Plehsa, M. E. (1989). *Concepts and application in finite element analysis* (3rd ed.). USA: Wiley.

Dunkerley, S. (1894). On whirling and vibration of shafts. *Philosophical transactions Royal Society of London.*

Fenves, G., & Chopra, A. K. (1984). *Earthquake analysis and response of concrete gravity dams* (EERC Research report # 84/10). Berkeley, USA: University of California.

Fenves, G., & Chopra, A. K. (1986). *Simplified analysis for earthquake resistant design of concrete gravity dams* (EERC Research report # 85/10). Berkeley, USA: University of California.

Gogoi, I., & Athani, S. (2010). Effect of modeling far and interface boundary on seismic performance of dam. In *14th International Symposium on Earthquake Engineering*, IIT (Roorkee), India.

Gaines, J. H., & Volterra, E. (1966). Transverse vibrations of cantilever bars of variable cross section. *The, Journal of Acoustical Society of America, 39*(4), 674–679.

Hall, J. R., & Kissenpfennig, J. F. (1976). Special topics on Soil-structure interaction. *Nuclear Engineering design, 38.*

IS-1893. (2016). *Earthquake resistant design of structures*. New Delhi, India: Bureau of Indian Standards.

Karnovsky, I., & Lebed, O. (2001). *Formulas for structural dynamics*. New York, NY: McGrawHill Publication.

Kramer, S. (2002). *Dynamic stiffnes of soil in liquefiable ground (Technical report # WA-RD-514.1)*. University of Washington, USA.

Krishna, J., Chandrashekharan, A. R., & Chandra, B. (1994). *Elements of Earthquake Engineering*. New Delhi, India: South Asian Publishers.

Okamoto, S. (1984). *Introduction to earthquake engineering*. University Press of Tokyo.

Shames, I. H., & Dym, C. L. (1995). *Energy and finite element methods in structural mechanics*. New Delhi, India: NewAge International publication.

Seed, H. B., Wong, R. T., Idriss, I. M., & Tokimatsu, K. (1984). Modulii and damping factor for dynamic analysis of cohesionless soil. *Journal of Geotechnical engineering, ASCE, 112*(11).

Taylor, R. L., Bersford, P. J., & Wilson, E. L. (1976). A nonconforming element for stress analysis. *International Journal for Numerical Methods in Engineeering, 10*.

Westergaard, H. M. (1933). Water pressure on dams during earthquakes. *Transactions ASCE, 98* (1850).

Wolf, J. P. (1985). *Dynamic soil structure inetraction*. New Jersey, USA: Prentice Hall Publication.

Zangar, C. N. (1953) Hydrodynamic pressure on dam due to horizontal earthquake. In *Proceedings, Society of Experimental Stress Analysis, 10*.

Zienkiewicz, O. C. (1947). Stress distribution in gravity dams. In *Proceedings Institiution of Civl Engineers* (Vol. 27, pp. 244–271), U.K.

Chapter 12
Earthquake Response of Bridge Piers and Their Foundations

12.1 Introduction

Bridge piers and their foundation remain the primary supporting structures to support the bridge decks against all possible loads, including earthquakes. Figure 12.1 shows a typical multi-span plate girder-type bridge supported on piers across a river.

Figure 12.2 shows a schematic sketch of a two-way road bridge section supported on bridge pier.

It is apparent from Figs. 12.1 and 12.2 that during a strong ground motion, the load from the superstructure, including any traffic plus the flange load of the pier, acts as a lumped mass on the top of the pier column. The vibration induced in it generates dynamic moments and shear in the pier column, which the pier should be able to sustain without any distress.

12.2 Earthquake Analysis as Recommended in Codes

Three codes are briefly reviewed in this section:

1. AASHTO (2015) Standard Specification for seismic design of Highways Bridges. American Authority of State Highway and Transport Officials;
2. CALTRANS (2004) Bridge Design Specification. California Department of Transportation Sacramento California;
3. IRC:06 (2010) Standard Specifications and Code of Practice for Road Bridges, Loads and Specification, Indian Road Congress.

Present recommendation in IRC:06 (2010) for design of bridges under seismic force is based on a study by Murthy and Jain (1997) that made a significant improvement to the IRC codes compared to its previous version. The study and

I. Chowdhury and S. P. Dasgupta, *Earthquake Analysis and Design of Industrial Structures and Infra-structures*, GeoPlanet: Earth and Planetary Sciences,
https://doi.org/10.1007/978-3-319-90832-8_12

Fig. 12.1 A multi-span plate girder bridge supported on piers

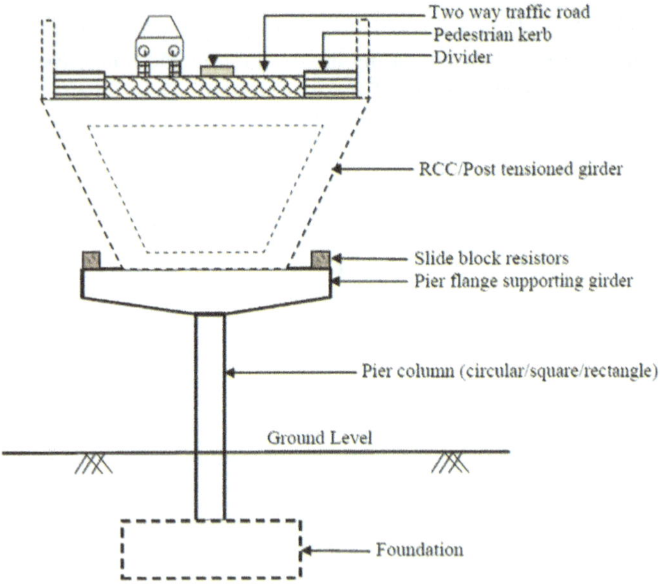

Fig. 12.2 Schematic sectional view of a two-way road bridge

subsequent recommendation are in line with the procedure as furnished in AASHTO (2015) and CALTRANS (2004) and was carried out to bring it in par with international practice.

In this particular study, though a number of recommendations were made on different aspects of bridge design under seismic loading, like adapting multiple response reduction factor (R) for different parts of the bridge, development of plastic hinges, damping properties, in this chapter, the overview is restricted to seismic response of the bridge pier part and its foundation only.

As per IRC:06 (2010), time period of the pier in fundamental mode is computed from the expression,

$$T = 2\sqrt{\frac{D}{1000F}} \tag{12.2.1}$$

In Eq. (12.2.1), D = dead load reaction from superstructure in kN; F = horizontal force in kN, to be applied at the center of mass of the superstructure for 1 mm horizontal deflection of the bridge along the considered direction of horizontal force.

The above expression is similar to what has been proposed in AASHTO (2015) and CALTRANS (2004) that recommend the expression,

$$T = 0.32\sqrt{\frac{D}{F}} \tag{12.2.2}$$

Here D and F are same as defined in Eq. (12.2.1) except that values are in FPS units.

The basis of these expressions is actually $T = 2\pi\sqrt{m/k}$, where m is the lumped mass considered at the top of pier and k is the stiffness of pier in fundamental mode.

Figure 12.3 shows typical bridge piers deployed in practice to support the superstructure of a bridge/flyover. Section 1-1 could be circular, square, or even rectangular in shape.

Load from the superstructure, including the flared part at the top (the pier cap), is considered as a lumped mass over a column of height H, and based on the expression $T = 2\pi\sqrt{m/k}$, Eqs. (12.2.1) and (12.2.2) have been derived.

Once the time period is computed, the design acceleration is computed from the spectral acceleration curve of the respective codes. For example, as per IRC:06 (2010), the spectral acceleration curve is as shown in Fig. 12.4.

The maximum elastic shear force V_e is computed as

$$V_e = \left(\frac{Z}{2}\right) I \left(\frac{S_a}{g}\right) W \tag{12.2.3}$$

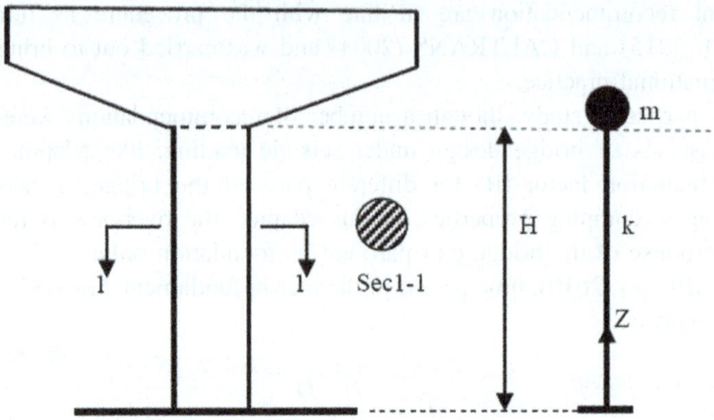

Fig. 12.3 Fixed bridge pier and its perceived mathematical model

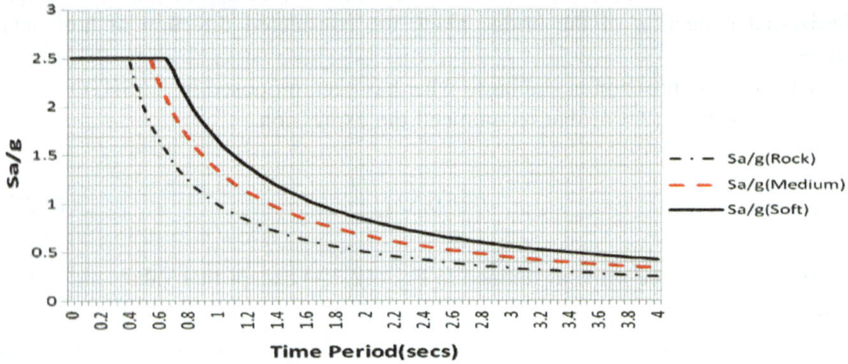

Fig. 12.4 Response spectrum curve as per IRC:06 (2010) for 5% material damping

Here

Z zone factor as explained in Chap. 2, Table 2.4;

I importance factor as depicted in Table 12.1;

S_a/g is the spectral acceleration as shown in Fig. 12.4, which is a function of the time period T;

W weight on top of pier usually considered as lumped mass (refer Fig. 12.3) constituting the dead load of the bridge deck and its appurtenances coming on the pier, including appropriate percentage of live/traffic load.

Having computed the maximum elastic shear, the design shear force is computed as

Table 12.1 Importance factor I as per IRC:06 (2010)

Seismic class	Illustrative example	Importance factor
Normal bridges	All bridges except those mentioned in other classes	1.0
Important bridges	(a) River bridges and flyovers inside cities (b) Bridges on national and state highways (c) Bridge crossing railway lines	1.2
Large critical bridges in all seismic zones	(a) Long bridges more than 1 km in length over perennial rivers and creeks (b) Bridges for which alternative routes are not available	1.5

Table 12.2 Response reduction factor R

Bridge component	R with ductile detailing	R without ductile detailing
Superstructure	2.0	N.A.
Sub-structure		
(1) Masonry/ PCC piers, abutments	–	1.0
(2) RCC short plate piers, where plastic hinge cannot develop in direction of length and RCC abutments	3.0	2.5
(3) RCC long piers, where hinges can develop	4.0	3.3
(4) Column	4.0	3.3
(5) Beam of RCC potal frames supporting bearings	1.0	1.0
Bearings	2.0	2.0
Connectors and stoppers (reaction blocks) Those restraining dislodgement or drifting away of bridge elements	When connectors and stoppers are designed to withstand seismic forces primarily $R = 1.0$ When connectors and stoppers are designed as additional safety measures in the event of failure of bearings, the R value to be considered for appropriate *sub-structure* as given above	

$$V_D = V_e/R \tag{12.2.4}$$

Here V_D = design shear force; R = response reduction factor as explained in Chap. 2. However, the bridges, unlike buildings which are designed for a unique R value, are as suggested in Table 2.8, and designed for different values of R for different components. The values of R to be considered are shown in Table 12.2.

Fig. 12.5 A bridge pier-supporting superstructure

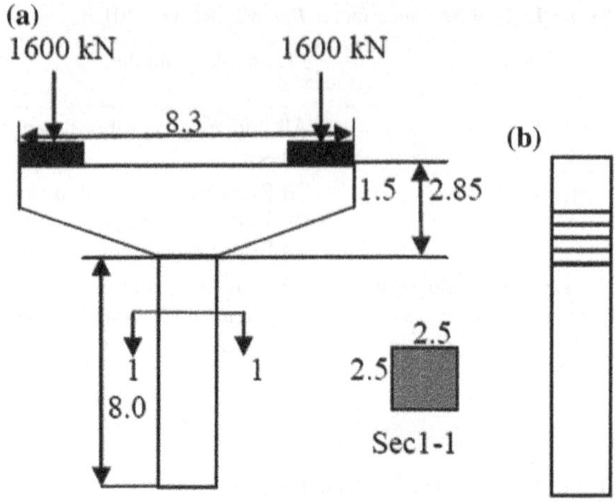

Example 12.1 A bridge pier having dimensions as shown in Fig. 12.5 supports 3200 kN of load from superstructure. Considering the flyover located in Guwahati (Assam) marked zone V as per IS-1893, compute the shear force and moment in the pier and size the foundation. Given the pier is constructed with M35 grade concrete and supported on soil having bearing capacity of 250 kN/m^2.

Solution:

As a first step, we compute the dead weight of the flanged part,

$$W_{\mathrm{F}} = \left[8.3 \times 1.5 \times 2.5 + \frac{1}{2}(8.3+2.5) \times (2.85 - 1.5) \times 2.5\right] \times 25$$
$$= 1233.75\ \mathrm{kN}.$$

Thus, the total lumped weight at top of pier = $1600 \times 2 + 1234 = 4434\,\mathrm{kN}$.

Young's modulus of concrete $= 5000\sqrt{f_{ck}} = \frac{5000\times\sqrt{35}}{1000} \times 10^6 = 2.95 \times 10^7\ \mathrm{kN/m^2}$.

Moment of inertia $= \frac{1}{12} \times 2.5 \times 2.5^3 = 3.255\ \mathrm{m}^4$.

For cantilever beam fixed at the base, the deflection Δ is given by,

$$\Delta = \frac{FL^3}{3EI}, \quad \text{for } \Delta = 1.0\ \mathrm{mm}$$

$$F = \frac{3EI\Delta}{L^3} = \frac{3 \times 2.95 \times 10^7 \times 3.255 \times 1}{1000 \times (8)^3} = 562.6\ \mathrm{kN}$$

Considering $T = 2\sqrt{\frac{D}{1000F}}$, we have,

$$T = 2\sqrt{\frac{4434}{1000 \times 562.6}} = 0.178 \text{ s.}$$

Spectral acceleration, as per Fig. 12.4, $S_a/g = 2.5$.

Thus, the maximum elastic shear $V_e = \left(\frac{Z}{2}\right) I \left(\frac{S_a}{g}\right) W$, here for zone V, as per IS-1893, $Z = 0.36$ and $I = 1.0$; thus,

$$V_e = \left(\frac{0.36}{2}\right) \times 1.0 \times 2.5 \times 4434 = 1995.3 \text{ kN.}$$

Considering that the pier column is designed for ductility, as per Table 12.2, $R = 4.0$.

Thus, the design shear force acting on the column is $V = \frac{1995.3}{4} = 499$ kN.

This maximum moment on top of foundation = 499 × 8 = 3992 kN.m

Sizing of foundation

Load from superstructure = 4434 kN.
Load of column pier = 2.5 × 2.5 × 8 × 25 = 1250 kN.
Assume foundation slab as 6 m × 7 m × 1.2 m.
Thus, the weight of foundation = 1260 kN.
Thus, the total load on soil = 4434 + 1250 + 1260 = 6944 kN.
Thus, the pressure on soil,

$$\sigma = \frac{P}{L \times B} \pm \frac{6M}{BL^2}$$

$$\sigma = \frac{6944}{7 \times 6} \pm \frac{6 \times 3992}{6 \times 49} = 165.3 \pm 81.46$$

$$= 246.76 \text{ kN/m}^2 \text{ or } 83.84 \text{ kN/m}^2. < 250 \text{ kN/m}^2. \text{ OK}$$

12.2.1 Some Comments on the Procedure as Proposed in Codes

The formulation proposed in Sect. 12.2, though adopted universally (i.e. IRC and other codes), has two significant lacunae in it.

1. It ignores the self-weight of the pier which can have significant contribution to the time period due to its inertial contribution.
2. It ignores shear deformation of the pier, whose effect could be significant for flyovers with variable height, especially when pier heights get progressively shorter near the ends.

We thus look into a derivation below where these two effects are incorporated in the calculation (Chowdhury et al. 2017).

12.3 A Proposed Dynamic Analysis Model for Bride Piers

For a bridge pier as shown in Section 1-1, the mathematical model can be expressed as shown in Fig. 12.3.

Considering the pier fixed at the base, i.e., at the top of foundation, it may be assumed as a cantilever beam whose equation of free vibration under flexure can be expressed as (Hurty and Rubenstein 1967),

$$EI\frac{\partial^4 u}{\partial z^4} = \rho A\frac{\partial^2 u}{\partial t^2} \tag{12.3.1}$$

Here ρ = mass density of beam material, A = cross-sectional area, E = Young's modulus of beam, I = moment of inertia of the beam, u = displacement in lateral direction.

Equation (12.3.1) has already been solved in Chaps. 2 and 10, wherein we have shown the solution to be,

$$Y_m = \sin\frac{\mu_m z}{H} - \sinh\frac{\mu_m z}{H} - \alpha_m\left(\cos\frac{\mu_m z}{H} - \cosh\frac{\mu_m z}{H}\right) \tag{12.3.2}$$

here m = number of modes 1, 2, 3,…

$$\mu_m = 1.875,\ 4.694,\ 7.855,\ \frac{2m-1}{2}\pi \quad \text{for } m \geq 4 \text{ and}$$

$$\alpha_m = \frac{\sin\mu_m + \sinh\mu_m}{\cos\mu_m + \cosh\mu_m} \tag{12.3.3}$$

Equation (12.3.2) represents eigenvectors of the beam and is expressed as,

$$f_i^{\mathrm{b}}(z) = \sin\frac{\mu_i z}{H} - \sinh\frac{\mu_i z}{H} - \alpha_i\left(\cos\frac{\mu_i z}{H} - \cosh\frac{\mu_i z}{H}\right) \tag{12.3.4}$$

Here superscript "b" stands for bending and subscript i stand for mode numbers 1, 2, 3….

Based on Rayleigh's technique, without going into the details of mathematical derivation (considering it has already been derived in Chap. 2), the stiffness matrix of the pier can be expressed as,

$$[K] = EI \int_{0}^{H} f_i^{b''}(z) f_j^{b''}(z) dz \tag{12.3.5}$$

Considering that the flanged pier cap plus dead and live load from top deck act as a mass (W_d/g) lumped at the tip of pier, the mass matrix can be expressed as,

$$[M] = \frac{\gamma_c A}{g} \int_{0}^{H} f_i^{b}(z) f_j^{b}(z) dz + \frac{W_d}{g} f_i^{b}(H) f_j^{b}(H) \tag{12.3.6}$$

Here E = Young's modulus of bridge pier material, A = area of cross section of pier, I = moment of inertia of pier cross section, γ_c = weight density of pier material, H = height of pier, W_d = weight coming from the top deck on the pier, g = acceleration due to gravity, i = mode number 1, 2, 3,...

Considering $\xi = z/H$, we convert Eqs. (12.3.5) and (12.3.6) from global to natural coordinates when we have,

$$[\mathbf{K}] = \frac{EI}{H^3} \int_{0}^{1} f_i^{b''}(\xi) f_j^{b''}(\xi) d\xi \tag{12.3.7}$$

$$[\mathbf{M}] = \frac{\gamma_c AH}{g} \int_{0}^{1} f_i^{b}(\xi) f_j^{b}(\xi) d\xi + \frac{W_d}{g} f_i^{b}(1) f_j^{b}(1) \tag{12.3.8}$$

$$\rightarrow [\mathbf{M}] = \frac{\gamma_c AH}{g} \left[\int_{0}^{1} f_i^{b}(\xi) f_j^{b}(\xi) d\xi + \Omega f_i^{b}(1) f_j^{b}(1) \right] \tag{12.3.9}$$

Here Ω is the ratio of the mass lumped at its tip (W_d/g) to that of mass of the pier ($\gamma_c AH/g$) itself.

Considering the general eigenvalue problem $[\mathbf{K}]\{\phi\} = \lambda[\mathbf{M}]\{\phi\}$ and solving the same by Jacobi's method (Bathe 1980), one can obtain $[\lambda]$, and knowing $\omega = \sqrt{\lambda}$ and $T = 2\pi/\omega$, we finally have,

$$T_i^{b} = \frac{2\pi}{\sqrt{\lambda_i^{b}}} H^2 \sqrt{\frac{\gamma_c A}{EIg}} \tag{12.3.10}$$

$$\rightarrow T_i^{b} = C_{Ti}^{b} H^2 \sqrt{\frac{\gamma_c A}{EIg}} \tag{12.3.11}$$

Table 12.3 Values of factor C_T^b for the first three modes**

Mode	$\Omega = 1$	$\Omega = 2$	$\Omega = 4$	$\Omega = 6$	$\Omega = 8$	$\Omega = 10$
1	4.033	5.422	7.461	9.053	10.403	11.598
2	0.384	0.394	0.399	0.401	0.402	0.402
3	0.121	0.122	0.122	0.122	0.122	0.122

**intermediate values can be linearly interpolated without any significant errors

Values of C_{Ti}^b for different values of Ω are furnished in Table 12.3 for the first three modes.

Equation (12.3.11) will give an accurate value of time period of the bridge pier which is tall and slender, when flexural deformation predominates, but would underestimate the response, when pier height is short and rigid, where shear deformation would predominate, or for intermediate height when both flexure and shear deformation play almost an equal role in its dynamic behavior.

12.3.1 *Shear Deformation Behavior of the Pier*

For a cantilever shear beam (which is usually short and wide) fixed at the base of uniform area A, equation of free vibration is expressed as (refer Chap. 2 for details).

$$G\frac{\partial^2 u}{\partial z^2} = \rho_c \frac{\partial^2 u}{\partial t^2} \tag{12.3.1.1}$$

$$\rightarrow \frac{\partial^2 u}{\partial z^2} = \frac{1}{V_s^2}\frac{\partial^2 u}{\partial t^2} \tag{12.3.1.2}$$

Here G = dynamic shear modulus of the bridge pier material, ρ_c = mass density of bridge pier material, V_s = shear wave velocity of the bridge pier material @ $G = \rho_c . V_s^2$.

Solving Eq. (12.3.1.2), the eigenvector of the problem can be expressed as (Chowdhury and Dasgupta 2007),

$$f_i^s = \cos\frac{(2i-1)\pi z}{2H} \tag{12.3.1.3}$$

Here superscript s stands for the term shear and subscript i stands for mode 1, 2, 3,…

Based on Rayleigh's technique, it can be shown that the stiffness matrix of the system for such shear beams can be expressed as,

$$[\mathbf{K}] = \eta GA \int_0^H f_i^{s'}(z) f_j^{s'}(z) \mathrm{d}z \tag{12.3.1.4}$$

Equation (12.3.1.4) in natural coordinates $\xi = z/H$ can be expressed as,

$$[\mathbf{K}] = \frac{\eta GA}{H} \int_0^1 f_i^{s'}(\xi) f_j^{s'}(\xi) \mathrm{d}\xi \tag{12.3.1.5}$$

Mass matrix in natural coordinates can be expressed as,

$$[\mathbf{M}] = \frac{\gamma_c AH}{g} \left[\int_0^1 f_i^{s}(\xi) f_j^{s}(\xi) \mathrm{d}\xi + \Omega f_i^{s}(0) f_j^{s}(0) \right] \tag{12.3.1.6}$$

In Eq. (12.3.1.5), η is the shear correction factor for the pier cross section and is dependent on shape of the pier cross section. Values of η, for different types of pier cross sections, can be obtained from various standard references and also furnished in Chap. 7 for *rectangular cross sections*.

Considering the general eigenvalue problem $[\mathbf{K}]\{\phi\} = \lambda[\mathbf{M}]\{\phi\}$ and solving the same by Jacobi's method, one can obtain $[\lambda]$, and knowing $\omega = \sqrt{\lambda}$ and $T = 2\pi/\omega$, we finally have,

$$T_i^s = \frac{2\pi}{\sqrt{\lambda_i^s}} \sqrt{\frac{\gamma_c H^2}{\eta G g}} \tag{12.3.1.7}$$

$$\rightarrow T_i^s = C_{Ti}^s \frac{H}{V_s \sqrt{\eta}} \tag{12.3.1.8}$$

Here V_s = shear wave velocity of concrete and values of C_{Ti}^s for different values of Ω are as furnished in Table 12.4.

Time period of flexure and shear deformation can now be combined, as explained hereafter.

Table 12.4 Value of factor C_T^s for the first three modes**

Mode	$\Omega = 1$	$\Omega = 2$	$\Omega = 4$	$\Omega = 6$	$\Omega = 8$	$\Omega = 10$
1	7.171	9.383	12.712	15.337	17.574	19.558
2	1.731	1.789	1.823	1.836	1.842	1.846
3	0.909	0.916	0.92	0.922	0.922	0.923

**intermediate values can be linearly interpolated without any significant errors

If a load vector P is applied on a body, displacement of the body due to bending as well as shear deformation can be expressed as (Shames and Dym 1995),

$$\delta_t = \delta_b + \delta_s \tag{12.3.1.9}$$

where δ_t = total displacement of the system; δ_b = displacement due to bending; and δ_s = displacement due to shear.

Equation (12.3.1.9) can be expressed as,

$$\frac{P}{K_e} = \frac{P}{K_b} + \frac{P}{K_s} \tag{12.3.1.10}$$

where P = total load; K_b and K_s are stiffness due to bending and shear deformation, respectively.

Equation (12.3.1.10) can be expressed as,

$$m/K_e = m/K_b + m/K_s \tag{12.3.1.11}$$

$$\rightarrow 1/\omega_e^2 = 1/\omega_b^2 + 1/\omega_s^2 \tag{12.3.1.12}$$

where ω_e = combined frequency of the system due to bending and shear; ω_b = frequency due to bending only; ω_s = frequency due to shear only.

Equation (12.3.1.12) is actually the classical Dunkerley equation (Dunkerley 1894) by which combined frequency of multi-body systems can be derived by treating each system in isolation; it is often used in dynamic soil–structure interaction (DSSI) problems.

Equation (12.3.1.12) finally yields,

$$T_{ei}^2 = T_{bi}^2 + T_{si}^2 \tag{12.3.1.13}$$

Substituting Eqs. (12.3.11) and (12.3.1.8) in Eq. (12.3.1.13), one can have the combined effective time period T_{ei} as

$$T_{ei} = C_{Ti}^{b} \sqrt{\frac{W_p H^3}{EIg}\left[1 + \left(\frac{C_{Ti}^{s}}{C_{Ti}^{b}}\right)^2 \frac{2(1+v)}{\eta r_s^2}\right]} \tag{12.3.1.14}$$

Here W_p = weight of the pier expressed as $\gamma_c.A.H$; r_s = slenderness ratio of pier (H/R); R = radius of gyration of pier cross section ($I = AR^2$); v = Poisson's ratio; and η = shear correction factor.

Equation (12.3.1.14) which is the general equation of time period considering combined effect of flexure and shear is of particular interest, as it explains the effect of shear deformation that goes on to elongate the time period of the pier. It will be observed from Eq. (12.3.1.14) that it is strongly dependent on slenderness ratio r_s. As R reduces, the component r_s increases rapidly (in terms of its square) and, for large values of r_s, makes the second term within the parenthesis insignificant when flexural mode dominates the overall response.

As per modal response analysis,

$$S_d = \frac{S_a}{\omega^2} \tag{12.3.1.15}$$

Here S_d = modal amplitude, S_a = spectral acceleration corresponding to time period T, and ω = natural frequency of the system when $\omega = 2\pi/T$.

Thus, as per modal analysis, the amplitude of vibration is expressed as,

$$S_{di} = \frac{S_{ai}}{4\pi^2} T_{ei}^2 \tag{12.3.1.16}$$

$$u_{si} = \kappa_i \beta \frac{S_{ai}}{4\pi^2} T_{ei}^2 \left[\phi_i^b\right]^T \left\{f_i^b(\xi)\right\} \tag{12.3.1.17}$$

$$M_{zi} = -\kappa_i \beta \frac{EI}{H^2} \left(\frac{S_{ai} T_{ei}^2}{4\pi^2}\right) \left[\phi_i^b\right]^T \left\{\frac{d^2 f_i^b(\xi)}{d\xi^2}\right\} \tag{12.3.1.18}$$

$$V_{zi} = -\kappa_i \beta \frac{EI}{H^3} \left(\frac{S_{ai} T_{ei}^2}{4\pi^2}\right) \left[\phi_i^b\right]^T \left\{\frac{d^3 f_i^b(\xi)}{d\xi^3}\right\} \tag{12.3.1.19}$$

where

$$\kappa_i = \frac{\left[\int_0^1 \left\{\phi_i^b\right\}^T \left\{f_i^b(\xi)\right\} d\xi + \Omega \left\{\phi_i^b\right\}^T \left\{f_i^b(1)\right\}\right]}{\left[\int_0^1 \left[\left\{\phi_i^b\right\}^T \left\{f_i^b(\xi)\right\}\right]^2 d\xi + \Omega \left[\left\{\phi_i^b\right\}^T f_i^b(1)\right]^2\right]} \tag{12.3.1.20}$$

Here the values of κ_i for the first three modes for different values of Ω are shown in Table 12.5.

Here κ_i is the modal mass participation factor, and $\beta = ZI/2$, where Z = zone factor and I = importance factor. Thus, for different modes, amplitude, shear, and moment over height of the pier can be obtained from Eqs. (12.3.1.17) to (12.3.1.19), and SRSS values for n number of modes can be obtained from the equation.

Table 12.5 Values of factor κ_i for the first three modes**

Mode	$\Omega = 1$	$\Omega = 2$	$\Omega = 4$	$\Omega = 6$	$\Omega = 8$	$\Omega = 10$
1	−0.421	−0.402	−0.391	−0.387	−0.385	−0.383
2	0.192	0.243	0.273	0.284	0.289	0.293
3	0.369	0.386	0.397	0.401	0.403	0.404

**intermediate values can be linearly interpolated without any significant errors

$$\aleph(t) = \sqrt{\sum_{i=1}^{n} \aleph_i^2(t)} \qquad (12.3.1.21)$$

where $\aleph(t) = u_s(t)$, $V_b(t)$ or $M_b(t)$, as the case may be, are the SRSS values of amplitude, shear, or moment, respectively.

Example 12.2 Repeat the problem in Example 12.1 considering it as an end pier of a flyover of height 4.0 m, having all other conditions remaining same. Compare the results based on the procedure as furnished in the code and considering the shear deformation and self-weight of the pier (Fig. 12.6).

Solution:
Analysis as per IRC:06

$$W_F = \left[8.3 \times 1.5 \times 2.5 + \frac{1}{2}(8.3 + 2.5) \times (2.85 - 1.5) \times 2.5\right] \times 25$$
$$= 1233.75 \text{ kN}.$$

Here $\gamma = 25$ kN/m^3.

Thus, the total lumped weight at the top of pier = $1600 \times 2 + 1234 = 4434$ kN.

Young's modulus of concrete $(E) = 5000\sqrt{f_{ck}} = \frac{5000 \times \sqrt{35}}{1000} \times 10^6 = 2.95 \times 10^7$ kN/m^2.

Moment of inertia of column $(I) = \frac{1}{12} \times 2.5 \times 2.5^3 = 3.255$ m^4.

Area of column $(A) = 6.25$

For cantilever beam fixed at the base, the deflection Δ is given by,

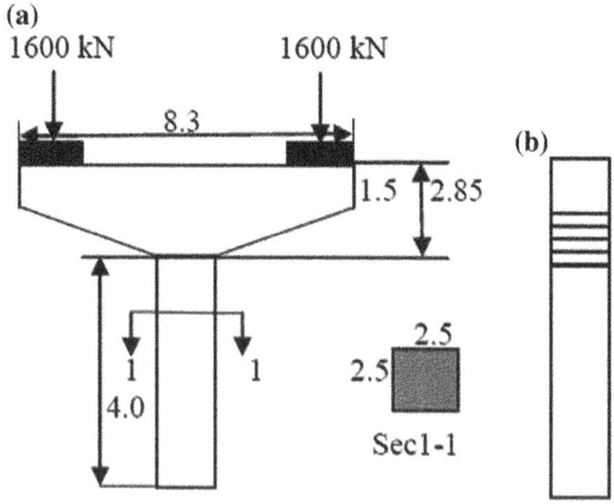

Fig. 12.6 Bridge pier of a height of 4.0 m

$$\Delta = \frac{FH^3}{3EI}, \quad \text{for } \Delta = 1.0 \text{ mm}$$

$$F = \frac{3EI\Delta}{H^3} = \frac{3 \times 2.95 \times 10^7 \times 3.255 \times 1}{1000 \times (4)^3} = 4501 \text{ kN}$$

Considering $T = 2\sqrt{\frac{D}{1000F}}$, we have

$$T = 2\sqrt{\frac{4434}{1000 \times 4501}} = 0.062 \text{ s.}$$

Spectral acceleration, as per Fig. 12.4, $S_a/g = 2.5$.

Thus, the maximum elastic shear $V_e = \left(\frac{Z}{2}\right) I \left(\frac{S_a}{g}\right) W$, here for zone V, as per IS-1893, $Z = 0.36$ and $I = 1.0$, thus,

$$V_e = \left(\frac{0.36}{2}\right) \times 1.0 \times 2.5 \times 4434 = 1995.3 \text{ kN.}$$

Considering that the pier column is designed for ductility, as per Table 12.2, $R = 4.0$

Thus, the design shear force acting on the column is $V = \frac{1995.3}{4} = 499$ kN.

This maximum moment on top of foundation = 499 × 4 = 1996 kN.m

Dynamic analysis considering shear deformation and self-weight of the pier:

Self-weight of pier = 4.0 × 2.5 × 2.5 × 25 = 625 kN.

Total load lumped at top of pier = 4434 kN (refer calculation above).

Thus $\Omega = \frac{4434}{625} = 7.09$

- **For the bending mode:**

Considering $[\mathbf{K}] = \frac{EI}{H^3} \int_0^1 f_i^{b''}(\xi) f_j^{b''}(\xi) d\xi$, we have,

$$[\mathbf{K}] = \frac{EI}{H^3} \begin{bmatrix} 22.936 & -3.894 \times 10^{-3} & 5.842 \times 10^{-3} \\ -3.894 \times 10^{-3} & 468.074 & -0.063 \\ 5.842 \times 10^{-3} & -0.063 & 3813 \end{bmatrix} \quad \text{and}$$

$$[\mathbf{M}] = \frac{\gamma_c AH}{g} \left[\int_0^1 f_i^s(\xi) f_j^s(\xi) d\xi + \Omega f_i^s(0) f_j^s(0) \right]$$

$$[\mathbf{M}] = \frac{\gamma_c AH}{g} \begin{bmatrix} 54.473 & -37.93 & 38.669 \\ -37.93 & 28.305 & -27.875 \\ 38.669 & -27.875 & 29.419 \end{bmatrix}$$

Considering $g = 9.81$ m/s^2 and $[\mathbf{K}] - [\mathbf{M}]\omega^2 = 0$, where $T = 2\pi/\omega$, we have

$\left\{T^{\mathrm{b}}_{i=1,2,3}\right\}^T = \left\{0.064 \quad 2.615\times10^{-3} \quad 7.967\times10^{-4}\right\}$ s, and the corresponding eigenvectors are

$$[\boldsymbol{\varphi}] = \begin{bmatrix} -1 & -0.745 & -0.326 \\ 0.034 & -1 & 0.551 \\ -4.273\times10^{-3} & 0.066 & 1 \end{bmatrix}$$

Considering the shear mode:

$$[\mathbf{K}] = \frac{\eta GA}{H}\int_0^1 f_i^{\mathrm{s}'}(\xi) f_j^{\mathrm{s}'}(\xi)\,\mathrm{d}\xi$$

$$\rightarrow [\mathbf{K}] = \frac{\eta GA}{H}\begin{bmatrix} 1.234 & 7.55\times10^{-15} & -2.554\times10^{-14} \\ 7.55\times10^{-15} & 11.103 & 7.672\times10^{-14} \\ -2.554\times10^{-14} & 7.672\times10^{-14} & 30.843 \end{bmatrix}$$

The corresponding mass is

$$[\mathbf{M}] = \frac{\gamma_c AH}{g}\left[\int_0^1 f_i^{\mathrm{s}}(\xi) f_j^{\mathrm{s}}(\xi)\,\mathrm{d}\xi + \Omega f_i^{\mathrm{s}}(0) f_j^{\mathrm{s}}(0)\right]$$

$$[\mathbf{M}] = \frac{\gamma AH}{g}\begin{bmatrix} 7.59 & 7.09 & 7.09 \\ 7.09 & 7.59 & 7.09 \\ 7.09 & 7.09 & 7.59 \end{bmatrix}$$

Performing the eigensolution, the time period in shear mode is expressed as

$$\left\{T^{\mathrm{s}}_{i=1,2,3}\right\}^T = \left\{0.038 \quad 4.189\times10^{-3} \quad 2.099\times10^{-3}\right\}\ \mathrm{s}$$

The effective time period considering flexure and shear mode is thus,

$T_{ei}^2 = T_{bi}^2 + T_{si}^2$ which gives,

$$\{T_{ei}\}^T = \left\{0.074 \quad 4.938\times10^{-3} \quad 2.245\times10^{-3}\right\}\ \mathrm{s}.$$

The S_a/g value for all three modes is thus 2.5 as per IS code.

The modal mass participation is expressed as
$$\kappa_i = \frac{\left[\int_0^1 \{\phi_i^b\}^T \{f_i^b(\xi)\} d\xi + \Omega \{\phi_i^b\}^T \{f_i^b(1)\}\right]}{\left[\int_0^1 \left[\{\phi_i^b\}^T \{f_i^b(\xi)\}\right]^2 d\xi + \Omega \left[\{\phi_i^b\}^T f_i^b(1)\right]^2\right]}$$

This gives for the first mode,

$$\kappa_1 = \frac{\left[\int_0^1 \left[-f_1^b(\xi) + 0.034 f_2^b(\xi) + 4.273 \times 10^{-3} f_3^b\right] d\xi + 7.09\left[-f_1^b(1) + 0.034 f_2^b(1) + 4.273 \times 10^{-3} f_3^b(1)\right]\right]}{\int_0^1 \left[-f_1^b(\xi) + 0.034 f_2^b(\xi) + 4.273 \times 10^{-3} f_3^b(\xi)\right]^2 d\xi + 7.09\left[-f_1^b(1) + 0.034 f_2^b(1) + 4.273 \times 10^{-3} f_3^b(1)\right]^2}$$

$\kappa_1 = -0.364$; here the function $f_i^b(\xi)$ is as explained in Eq. (12.3.4) to be considered for first mode, taking $\mu = 1.875$.

For the second mode,

$$\kappa_2 = \frac{\left[\int_0^1 \left[-0.745 f_1^b(\xi) - f_2^b(\xi) + 0.066 f_3^b\right] d\xi + 7.09\left[-0.745 f_1^b(1) - f_2^b(1) + 0.066 f_3^b(1)\right]\right]}{\int_0^1 \left[-0.745 f_1^b(\xi) - f_2^b(\xi) + 0.066 f_3^b(\xi)\right]^2 d\xi + 7.09\left[-0.745 f_1^b(1) - 1 f_2^b(1) + 0.066 f_3^b(1)\right]^2}$$

$\kappa_2 = -0.356$ when $\mu = 4.694$.

Proceeding in identical fashion for the third mode, we have $\kappa_3 = 0.242$.

Now considering $u_{si} = \kappa_i \beta \frac{S_{ai}}{4\pi^2} T_{ei}^2 \left[\phi_i^b\right]^T \{f_i^b(\xi)\}$, substituting the appropriate values for the first three modes, the values are shown in Table 12.6.

The bending moment along the depth of the pier is expressed as,

$$M_{zi} = -\kappa_i \beta \frac{EI}{H^2} \left(\frac{S_{ai} T_{ei}^2}{4\pi^2}\right) \left[\phi_i^b\right]^T \left\{\frac{d^2 f_i^b(\xi)}{d\xi^2}\right\}$$

Thus for the first mode, we have,

$$M_{z1} = -0.364 \times \left(\frac{0.36 \times 1}{2 \times 4}\right)\left(\frac{2.95 \times 10^7 \times 3.255}{4^2}\right)\left(\frac{24.525 \times 0.074^2}{4\pi^2}\right)$$
$$\begin{bmatrix} -f_1''(\xi) \times (1.875)^2 + 0.034 f_2''(\xi) \times (4.694)^2 \\ -4.273 \times 10^{-3} f_3''(\xi) \times (7.855)^2 \end{bmatrix} \text{kN/m}$$

For the second mode,

Table 12.6 Amplitude, moment, and shear in pier for first three modes

z/H	u_1 (m)	u_2 (m)	u_3 (m)	M_1 (kN.m)	M_2 (kN.m)	M_3 (kN.m)	V_1 (kN)	V_2 (kN)	V_3 (kN)
0	0.00E+00	0.00E+00	0.00E+00	−2887.49	−61.54	−29.51	−860.40	−55.60	−55.20
0.05	5.89E−07	1.20E−08	5.38E−09	−2715.52	−50.41	−18.49	−858.21	−55.66	−54.73
0.1	2.31E−06	4.51E−08	1.85E−08	−2544.95	−39.25	−7.78	−845.07	−55.98	−51.89
0.15	5.09E−06	9.45E−08	3.48E−08	−2378.58	−28.00	2.01	−815.86	−56.54	−45.35
0.2	8.86E−06	1.56E−07	5.04E−08	−2219.63	−16.64	10.10	−771.43	−57.03	−34.93
0.25	1.36E−05	2.24E−07	6.19E−08	−2070.65	−5.23	15.77	−717.55	−56.90	−21.37
0.3	1.91E−05	2.94E−07	6.69E−08	−1932.64	6.03	18.54	−663.37	−55.49	−6.15
0.35	2.55E−05	3.61E−07	6.43E−08	−1804.61	16.83	18.25	−619.43	−52.15	8.85
0.4	3.26E−05	4.22E−07	5.41E−08	−1683.52	26.73	15.14	−595.49	−46.35	21.69
0.45	4.04E−05	4.72E−07	3.78E−08	−1564.61	35.19	9.83	−598.45	−37.81	30.69
0.5	4.89E−05	5.07E−07	1.74E−08	−1442.18	41.68	3.20	−630.66	−26.64	34.71
0.55	5.80E−05	5.24E−07	−4.34E−09	−1310.60	45.70	−3.70	−688.89	−13.33	33.36
0.6	6.76E−05	5.23E−07	−2.46E−08	−1165.48	46.92	−9.81	−764.16	1.23	27.04
0.65	7.77E−05	5.03E−07	−4.07E−08	−1004.74	45.21	−14.26	−842.34	15.77	16.95
0.7	8.82E−05	4.63E−07	−5.11E−08	−829.53	40.71	−16.46	−905.58	28.83	4.85
0.75	9.91E−05	4.07E−07	−5.46E−08	−644.81	33.88	−16.21	−934.24	38.83	−7.12
0.8	1.10E−04	3.37E−07	−5.15E−08	−459.43	25.48	−13.76	−909.13	44.29	−16.77
0.85	1.22E−04	2.56E−07	−4.27E−08	−285.87	16.56	−9.79	−813.51	43.92	−22.14
0.9	1.33E−04	1.68E−07	−2.98E−08	−139.58	8.37	−5.30	−634.70	36.75	−21.69
0.95	1.44E−04	7.65E−08	−1.47E−08	−38.08	2.35	−1.57	−364.75	22.20	−14.45
1	1.56E−04	−1.60E−08	1.09E−09	0.00	0.00	0.00	−0.12	0.00	0.01

$$M_{z2} = -0.356 \times \left(\frac{0.36 \times 1}{2 \times 4}\right)\left(\frac{2.95 \times 10^7 \times 3.255}{4^2}\right)\left(\frac{24.525 \times (4.94 \times 10^{-3})^2}{4\pi^2}\right)$$
$$\left[\begin{array}{l}-0.745 f_1''(\xi) \times (1.875)^2 - f_2''(\xi) \times (4.694)^2 \\ +0.066 f_3''(\xi) \times (7.855)^2\end{array}\right]$$

For the third mode,

$$M_{z3} = 0.242 \times \left(\frac{0.36 \times 1}{2 \times 4}\right)\left(\frac{2.95 \times 10^7 \times 3.255}{4^2}\right)\left(\frac{24.525 \times (2.25 \times 10^{-3})^2}{4\pi^2}\right)$$
$$\left[\begin{array}{l}-0.326 f_1''(\xi) \times (1.875)^2 + 0.551 f_2''(\xi) \times (4.694)^2 \\ + f_3''(\xi) \times (7.855)^2\end{array}\right]$$

For shear force, considering $V_{zi} = -\kappa_i \beta \frac{EI}{H^3}\left(\frac{S_{ai} T_{ei}^2}{4\pi^2}\right)\left[\phi_i^b\right]^T \left\{\frac{d^3 f_i^b(\xi)}{d\xi^3}\right\}$, it can be expanded like as shown above for the moment expression when shear forces for the first three modes can be computed. The final values of displacement, moment, and shear are tabulated as hereafter.

Moment values computed by dynamic analysis are graphically compared in Fig. 12.2.2 with the method as proposed in the code in Fig. 12.7.

It is observed from Fig. 12.7 that the method as proposed in code significantly underrates the moment vis-à-vis the proposed dynamic analysis which goes on to show that self-load of the pier and the shear deformation can significantly affect the results.

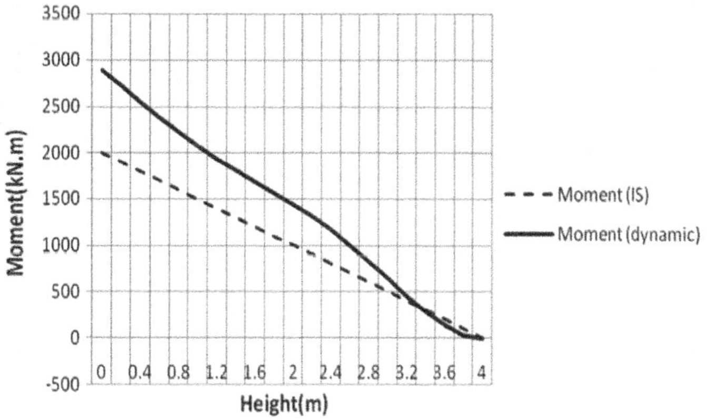

Fig. 12.7 Comparison of moment along the pier height code versus the proposed method

12.4 Effect of Foundation on Response of Pier and Superstructure

Shown in Fig. 12.8 are the types of foundations usually deployed for the bridge piers. While isolated and pile foundations (Fig. 12.8a, b) are usually used in road bridges, pile foundations and caisson/well foundations (Fig. 12.8b, c) are normally constructed for piers built on waterways/rivers.

12.4.1 Piers Resting on a Shallow Foundation

Shown in Fig. 12.9 is a bridge pier mounted on a foundation of mass M_f and moment of inertia J_θ. The soil medium supporting the foundation is represented by springs and dashpots. Let the body in ith mode induce a moment and shear at foundation base expressed as M_{bi} and V_i, respectively, which is a time-dependent loading. Then equation of equilibrium can be expressed as,

$$\left(M_f + m_i\right)\ddot{u}_{xi} + C_x\dot{u}_{xi} + K_x u_{xi} = V_i \tag{12.4.1.1}$$

$$\left(J_\theta + m_i h_i^2\right)\ddot{\theta}_i + C_\theta\dot{\theta}_i + K_\theta\theta_i = M_{bi} \tag{12.4.1.2}$$

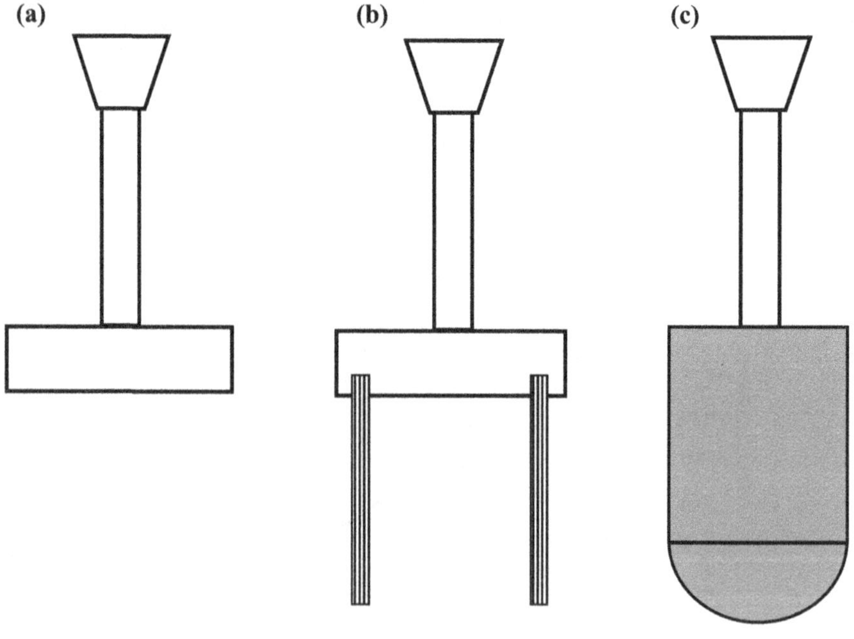

Fig. 12.8 Different foundations deployed for bridge piers in the field

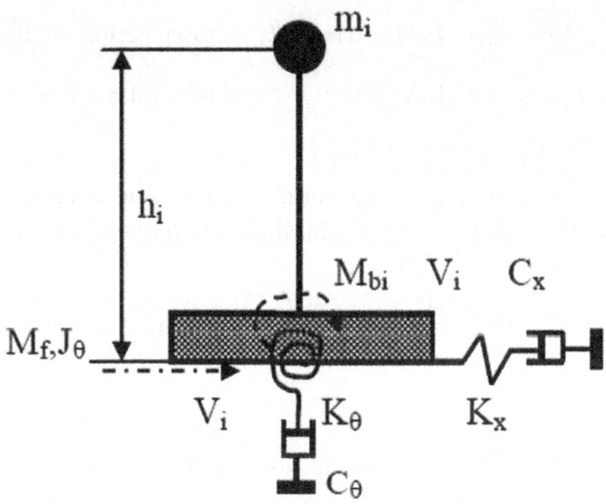

Fig. 12.9 A bridge pier mounted on a foundation with soil spring and dashpot

In Eq. (12.4.1.1) and (12.4.1.2), $m_i = \frac{\gamma_c AH}{g}\left[\int_0^1 f_i^{\mathrm{b}}(\xi)^2 \mathrm{d}\xi + \Omega f_i^{\mathrm{b}}(0)^2\right]$ is the modal mass contribution from superstructure in ith mode. Soil stiffness and damping values K_x, K_θ, C_x, C_θ are as furnished in Table 7.3 in Chap. 7.

Solution to Eq. (12.4.1.1) constitutes of a complimentary function and a particular integral, and this is expressed as,

$$u_{xi} = \mathrm{e}^{-\zeta_x \omega_x t}\left(A \cos \omega_x \sqrt{1-\zeta_x^2 t} + B \sin \omega_x \sqrt{1-\zeta_x^2 t}\right) + V_i/K_x \qquad (12.4.1.3)$$

Now considering the boundary conditions (i) at $t = 0$, $x = 0$ and (ii) at $t = 0$, $\mathrm{d}x/\mathrm{d}t = S_\mathrm{d}\omega_x$, finally gives,

$$u_{xi} = \frac{V_i}{K_x}\left(1 + \mathrm{e}^{\zeta_x \omega_x t}\left\{\sqrt{\frac{1-\zeta_x}{1+\zeta_x}} \sin \omega_x \sqrt{1-\zeta_x^2 t} - \cos \omega_x \sqrt{1-\zeta_x^2 t}\right\}\right)$$
(12.4.1.4)

Now one should remember that here V_i is actually a *frequency-dependent term and is a function of dominant modal frequency of the superstructure*. Based on this, Eq. (12.4.1.4) can be expressed as,

$$u_{xi} = \frac{V_i}{K_x\sqrt{\left(1-r_x^2\right)^2 + \left(2\zeta_x r_x\right)^2}}\left(1 + \mathrm{e}^{-\zeta_x \omega_x t}\left\{\sqrt{\frac{1-\zeta_x}{1+\zeta_x}} \sin \omega_x \sqrt{1-\zeta_x^2 t} - \cos \omega_x \sqrt{1-\zeta_x^2 t}\right\}\right)$$
(12.4.1.5)

Here $r_x = T_x/T_i$, where T_x = time period of foundation in translation, $T_x = 2\pi\sqrt{(m_f + m_i)/K_x}$ and T_i = fixed base time period of the structure in ith mode, and $\zeta_x = C_x/2\sqrt{K_x(m_f + m_i)}$.

In earthquake engineering, considering everything that is computed in time domain, Eq. (12.4.1.5) can thus be expressed as,

$$u_{xi} = \frac{V_i}{K_x\sqrt{(1-r_x^2)^2 + (2\zeta_x r_x)^2}}\left(1 + e^{-\zeta_x\frac{2\pi t}{T_x}}\left\{\sqrt{\frac{1-\zeta_x}{1+\zeta_x}}\sin 2\pi\sqrt{1-\zeta_x^2}\frac{t}{T_x} - \cos 2\pi\sqrt{1-\zeta_x^2}\frac{t}{T_x}\right\}\right) \tag{12.4.1.6}$$

Proceeding in identical fashion, it can be shown that,

$$\theta_i = \frac{M_{bi}}{K_\theta\sqrt{(1-r_\theta^2)^2 + (2\zeta_x r_\theta)^2}}\left(1 + e^{-\zeta_\theta\frac{2\pi t}{T_\theta}}\left\{\sqrt{\frac{1-\zeta_\theta}{1+\zeta_\theta}}\sin 2\pi\sqrt{1-\zeta_\theta^2}\frac{t}{T_\theta} - \cos 2\pi\sqrt{1-\zeta_\theta^2}\frac{t}{T_\theta}\right\}\right) \tag{12.4.1.7}$$

Here $r_\theta = T_\theta/T_i$ and $T_\theta = 2\pi\sqrt{[J_\theta + (m_i)\bar{h}_i^2]/K_\theta}$ and $\zeta_\theta = C_\theta/2\sqrt{K_\theta(J_\theta + (m_i)\bar{h}_i^2)}$

Thus, as explained in Chap. 7 Eq. (7.9.9)

$$u_{Ti} = u_{si}\left[\begin{array}{l} 1 + \frac{V_i}{\hat{K}_x u_{si}}\left(1 + e^{-\zeta_x\frac{2\pi t}{T_x}}\left\{\sqrt{\frac{1-\zeta_x}{1+\zeta_x}}\sin 2\pi\sqrt{1-\zeta_x^2}\frac{t}{T_x} - \cos 2\pi\sqrt{1-\zeta_x^2}\frac{t}{T_x}\right\}\right) \\ + \frac{M_{bi}\bar{h}_i}{\hat{K}_\theta u_{si}}\left(1 + e^{-\zeta_\theta\frac{2\pi t}{T_\theta}}\left\{\sqrt{\frac{1-\zeta_\theta}{1+\zeta_\theta}}\sin 2\pi\sqrt{1-\zeta_\theta^2}\frac{t}{T_\theta} - \cos 2\pi\sqrt{1-\zeta_\theta^2}\frac{t}{T_\theta}\right\}\right) \end{array}\right] \tag{12.4.1.8}$$

Here $\hat{K}_x = K_x\sqrt{(1-r_x^2)^2 + (2\zeta_x r_x)^2}$ and $\hat{K}_\theta = K_\theta\sqrt{(1-r_\theta^2)^2 + (2\zeta_\theta r_\theta)^2}$, the dynamic stiffness of soil which is frequency-dependent, and u_{si} = amplitude of the fixed based structure at centroidal height $\bar{h}_i = M_{bi}/V_i$, Thus, the dynamic amplification AF due to DSSI is expressed as,

$$\mathrm{AF}_i = \left[\begin{array}{l} 1 + \frac{V_i}{\hat{K}_x u_{si}}\left(1 + e^{-\zeta_x\frac{2\pi t}{T_x}}\left\{\sqrt{\frac{1-\zeta_x}{1+\zeta_x}}\sin 2\pi\sqrt{1-\zeta_x^2}\frac{t}{T_x} - \cos 2\pi\sqrt{1-\zeta_x^2}\frac{t}{T_x}\right\}\right) \\ + \frac{M_{bi}^2}{V_i\hat{K}_\theta u_{si}}\left(1 + e^{-\zeta_\theta\frac{2\pi t}{T_\theta}}\left\{\sqrt{\frac{1-\zeta_\theta}{1+\zeta_\theta}}\sin 2\pi\sqrt{1-\zeta_\theta^2}\frac{t}{T_\theta} - \cos 2\pi\sqrt{1-\zeta_\theta^2}\frac{t}{T_\theta}\right\}\right) \end{array}\right] \tag{12.4.1.9}$$

Variation of static stiffness K_x and K_θ with respect to frequency and damping ratio are as furnished in Fig. 12.10.

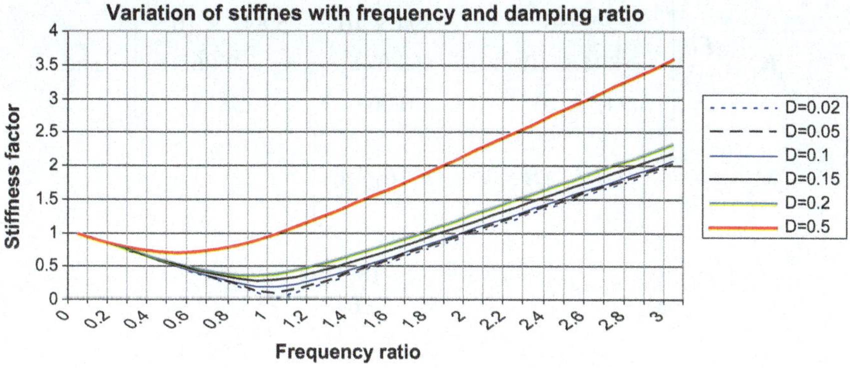

Fig. 12.10 Variation of stiffness factor with frequency and damping ratio

As cited by Gazetas (1983) and Wolf (1985), it is observed in Fig. 12.10 that static stiffness is dependent on the frequency ratio of the structure and foundation as well as the damping ratio of soil. For a frequency ratio from zero to roughly 2.0, the static soil spring stiffness value is ≤ 1.0. However, beyond frequency ratio of 2.0, the stiffness value increases almost linearly. Thus, it may be concluded that for higher modes, the amplification will not be as profound as the first few modes which remain most significant in terms of DSSI.

The following issues are to be observed at this point:

- The amplification factor is dependent on time and varies with it but can never be less than 1.0, irrespective of if there is stiffness degradation in shear modulus G of soil due to large strain vis-à-vis increase in damping due to radiation.
- Each mode has a unique amplification pattern with respect to time and may not necessarily occur at the same time; i.e., the phase lags in different modes are distinctly visible.

Example 12.3 Repeat the problem as cited in Example 12.1 based on multi-mode dynamic analysis considering fixed base and dynamic soil–structure interaction. Consider bearing capacity of soil as 250 kN/m^2 and shear wave velocity of soil as 450 m/s having Poisson's ratio = 0.33 and weigh density of soil = 20.42 kN/m^3.

Solution:

Self-weight of pier = 8.0 × 2.5 × 2.5 × 25 = 1250 kN

Total load lumped at top of pier = 4434 kN (Refer calculation in Example 12.1)

Thus $\Omega = \frac{4434}{1250} = 3.54$

- **For bending mode**:

Considering $[\mathbf{K}] = \frac{EI}{H^3}\int_0^1 f_i^{b''}(\xi) f_j^{b''}(\xi)\mathrm{d}\xi$ we have,

$$[\mathbf{K}] = \frac{EI}{H^3}\begin{bmatrix} 22.936 & -3.894\times10^{-3} & 5.842\times10^{-3} \\ -3.894\times10^{-3} & 468.074 & -0.063 \\ 5.842\times10^{-3} & -0.063 & 3813 \end{bmatrix}$$ and

$$[\mathbf{M}] = \frac{\gamma_c AH}{g}\left[\int_0^1 f_i^s(\xi)f_j^s(\xi)\mathrm{d}\xi + \Omega f_i^s(0)f_j^s(0)\right]$$

$$[\mathbf{M}] = \frac{\gamma_c AH}{g}\begin{bmatrix} 28.179 & -18.975 & 19.346 \\ -18.975 & 14.642 & -13.945 \\ 19.356 & -13.945 & 15.219 \end{bmatrix}$$

Considering g = 9.81 m/s^2 and $[\mathbf{K}] - [\mathbf{M}]\omega^2 = 0$, where $T = 2\pi/\omega$, we have,

$\left\{T^{\mathrm{b}}_{i=1,2,3}\right\}^T = \{0.184 \quad 0.01 \quad 3.18\times10^{-3}\}$ s and the corresponding eigen-vectors are,

$$[\varphi] = \begin{bmatrix} -1 & -0.721 & -0.321 \\ 0.033 & -1 & 0.543 \\ -4.136\times10^{-3} & 0.065 & 1 \end{bmatrix}$$

Considering shear mode:

$$[\mathbf{K}] = \frac{\eta GA}{H}\int_0^1 f_i^{s'}(\xi)f_j^{s'}(\xi)\mathrm{d}\xi$$

$$\rightarrow [\mathbf{K}] = \frac{\eta GA}{H}\begin{bmatrix} 1.234 & 7.55\times10^{-15} & -2.554\times10^{-14} \\ 7.55\times10^{-15} & 11.103 & 7.672\times10^{-14} \\ -2.554\times10^{-14} & 7.672\times10^{-14} & 30.843 \end{bmatrix}$$

The corresponding mass,

$$[\mathbf{M}] = \frac{\gamma_c AH}{g}\left[\int_0^1 f_i^s(\xi)f_j^s(\xi)\mathrm{d}\xi + \Omega f_i^s(0)f_j^s(0)\right]$$

$$[\mathbf{M}] = \frac{\gamma AH}{g}\begin{bmatrix} 4.047 & 3.547 & 3.547 \\ 3.547 & 4.047 & 3.547 \\ 3.547 & 3.547 & 4.047 \end{bmatrix}$$

Performing the eigensolution, the time period in shear mode is expressed as,

$$\left\{T^{\mathrm{s}}_{i=1,2,3}\right\}^T = \left\{0.055 \quad 8.282\times10^{-3} \quad 4.188\times10^{-3}\right\}\mathrm{s}$$

The effective time period considering flexure and shear mode is thus,

$T_{ei}^2 = T_{bi}^2 + T_{si}^2$ which gives,

$$\{T_{ei}\}^T = \left\{0.192 \quad 0.013 \quad 5.259\times10^{-3}\right\}\mathrm{s}.$$

The S_a/g values for all three modes are thus 2.5 as per IS code.

The modal mass participation is expressed as

$$\kappa_i = \frac{\left[\int_0^1 \{\phi_i^{\mathrm{b}}\}^T\{f_i^{\mathrm{b}}(\xi)\}\mathrm{d}\xi + \Omega\{\phi_i^{\mathrm{b}}\}^T\{f_i^{\mathrm{b}}(1)\}\right]}{\left[\int_0^1 \left[\{\phi_i^{\mathrm{b}}\}^T\{f_i^{\mathrm{b}}(\xi)\}\right]^2\mathrm{d}\xi + \Omega\left[\{\phi_i^{\mathrm{b}}\}^T f_i^{\mathrm{b}}(1)\right]^2\right]}.$$

This gives for first three modes (like those computed in Example 12.2)

$\kappa_{i=1.2.3} = -0.371,\ -0.357$, and 0.242, respectively.

Now considering $u_{si} = \kappa_i\beta\frac{S_{ai}}{4\pi^2}T_{ei}^2\left[\phi_i^{\mathrm{b}}\right]^T\{f_i^{\mathrm{b}}(\xi)\}$, substituting the appropriate values for the first three modes, the values are tabulated below.

The bending moment along the pier height is expressed as,

$$M_{zi} = -\kappa_i\beta\frac{EI}{H^2}\left(\frac{S_{ai}T_{ei}^2}{4\pi^2}\right)\left[\phi_i^{\mathrm{b}}\right]^T\left\{\frac{\mathrm{d}^2 f_i^{\mathrm{b}}(\xi)}{\mathrm{d}\xi^2}\right\}$$

Thus, for the first mode, we have,

$$M_{z1} = -0.371\times\left(\frac{0.36\times1}{2\times4}\right)\left(\frac{2.95\times10^7\times3.255}{8^2}\right)\left(\frac{24.525\times0.192^2}{4\pi^2}\right)$$
$$\begin{bmatrix}-f_1''(\xi)\times(1.875)^2 + 0.033f_2''(\xi)\times(4.694)^2\\ -4.136\times10^{-3}f_3''(\xi)\times(7.855)^2\end{bmatrix}$$

For the second mode,

$$M_{z2} = -0.357\times\left(\frac{0.36\times1}{2\times4}\right)\left(\frac{2.95\times10^7\times3.255}{8^2}\right)\left(\frac{24.525\times(0.013)^2}{4\pi^2}\right)$$
$$\begin{bmatrix}-0.7215f_1''(\xi)\times(1.875)^2 - f_2''(\xi)\times(4.694)^2\\ +0.065f_3''(\xi)\times(7.855)^2\end{bmatrix}$$

For the third mode,

$$M_{z3} = 0.242 \times \left(\frac{0.36 \times 1}{2 \times 4}\right)\left(\frac{2.95 \times 10^7 \times 3.255}{8^2}\right)\left(\frac{24.525 \times \left(5.259 \times 10^{-3}\right)^2}{4\pi^2}\right)$$

$$\begin{bmatrix} -0.321 f_1''(\xi) \times (1.875)^2 + 0.543 f_2''(\xi) \times (4.694)^2 \\ + f_3''(\xi) \times (7.855)^2 \end{bmatrix}$$

For the shear force, considering $V_{zi} = -\kappa_i \beta \frac{EI}{H^3}\left(\frac{S_{ai}T_{ei}^2}{4\pi^2}\right)\left[\phi_i^{b}\right]^T \left\{\frac{d^3 f_i^{b}(\xi)}{d\xi^3}\right\}$, it can be expanded like shown for the moment expression as elaborated above, and forces for the first three modes can be computed. The final values of displacement, moment, and shear are tabulated as hereafter.

z/H	u_1 (m)	u_2 (m)	u_3 (m)	M_1 (kN.m)	M_2 (kN.m)	M_3 (kN.m)	V_1 (kN)	V_2 (kN)	V_3 (kN)
0	0.00E+00	0.00E+00	0.00E+00	-4968.23	-106.65	-40.42	-743.32	-48.51	-37.82
0.05	4.06E-06	8.35E-08	2.95E-08	-4671.09	-87.24	-25.32	-741.50	-48.56	-37.50
0.1	1.59E-05	3.12E-07	1.01E-07	-4376.27	-67.78	-10.64	-730.56	-48.82	-35.55
0.15	3.50E-05	6.54E-07	1.91E-07	-4088.44	-48.16	2.78	-706.25	-49.27	-31.07
0.2	6.10E-05	1.08E-06	2.76E-07	-3812.98	-28.37	13.86	-669.23	-49.63	-23.92
0.25	9.33E-05	1.55E-06	3.39E-07	-3554.12	-8.53	21.62	-624.29	-49.44	-14.62
0.3	1.32E-04	2.03E-06	3.66E-07	-3313.60	11.03	25.40	-578.98	-48.13	-4.18
0.35	1.75E-04	2.49E-06	3.51E-07	-3089.82	29.75	24.99	-542.05	-45.14	6.11
0.4	2.24E-04	2.91E-06	2.96E-07	-2877.76	46.86	20.72	-521.57	-40.03	14.91
0.45	2.78E-04	3.25E-06	2.06E-07	-2669.60	61.46	13.42	-523.24	-32.56	21.07
0.5	3.36E-04	3.48E-06	9.42E-08	-2455.94	72.61	4.32	-548.99	-22.84	23.82
0.55	3.98E-04	3.60E-06	-2.48E-08	-2227.53	79.48	-5.15	-596.12	-11.29	22.88
0.6	4.64E-04	3.58E-06	-1.35E-07	-1977.17	81.50	-13.53	-657.15	1.30	18.54
0.65	5.33E-04	3.43E-06	-2.24E-07	-1701.51	78.45	-19.63	-720.33	13.86	11.61
0.7	6.05E-04	3.15E-06	-2.80E-07	-1402.59	70.59	-22.64	-770.80	25.12	3.30
0.75	6.80E-04	2.75E-06	-2.99E-07	-1088.74	58.71	-22.29	-792.27	33.73	-4.91
0.8	7.56E-04	2.26E-06	-2.81E-07	-774.78	44.13	-18.92	-768.77	38.40	-11.54
0.85	8.34E-04	1.69E-06	-2.31E-07	-481.59	28.66	-13.45	-686.40	38.04	-15.22
0.9	9.12E-04	1.07E-06	-1.60E-07	-234.93	14.48	-7.28	-534.62	31.81	-14.90
0.95	9.91E-04	4.31E-07	-7.63E-08	-64.04	4.06	-2.15	-306.84	19.20	-9.93
1	1.07E-03	-2.18E-07	1.15E-08	0.00	0.00	0.00	-0.11	0.00	0.01

Comparison of the moment based on recommendation of code vis-à-vis modal damping analysis is plotted hereafter in Fig. 12.11.

It is again observed that the procedure recommended in the code significantly underrates the response.

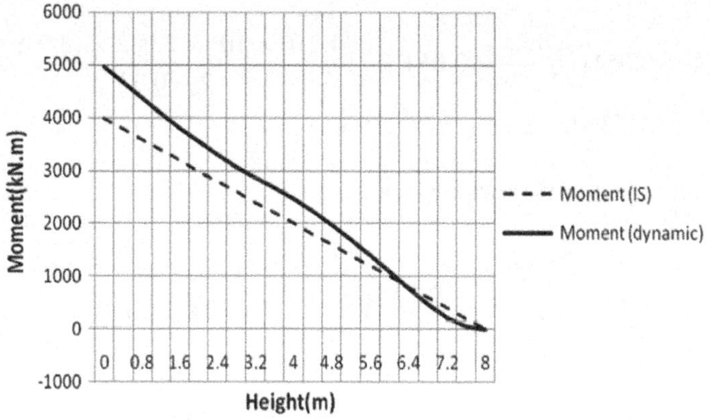

Fig. 12.11 Comparison of bending moment

Design of foundation:
Load from superstructure = 4434 kN.
Load of column pier = 2.5 × 2.5 × 8 × 25 = 1250 kN.
Assume foundation slab as 7 m × 7 m × 1.5 m.
Thus, the weight of foundation = 1834 kN.
Thus, the total load on soil = 4434 + 1250 + 1834 = 7518 kN.
Thus, the pressure on soil,

$$\sigma = \frac{P}{L \times B} \pm \frac{6M}{BL^2}$$
$$\sigma = \frac{7518}{7 \times 7} \pm \frac{6 \times 4970}{7 \times 49} = 153.4 \pm 86.93$$
$$= 240.33\ \text{kN/m}^2 \text{ or } 66.47\ \text{kN/m}^2 < 250\ \text{kN/m}^2\ \text{OK.}$$

Equivalent radius of a circular footing $R_f = \sqrt{\frac{49}{\pi}} = 3.949$ m.

Dynamic shear modulus of soil $G = \rho \times V_s^2 = \left(\frac{20.42}{9.81}\right) \times 450^2 = 4.216 \times 10^5$ kN/m^2.

Stiffness of soil in translation $K_x = \frac{8GR_f}{2-\nu} = \frac{8 \times 4.216 \times 10^5 \times 3.949}{2-0.33} = 7.976 \times 10^6$ kN/m.

Stiffness of soil in rocking $K_\theta = \frac{8GR_f^3}{3(1-\nu)} = \frac{8 \times 4.216 \times 10^5 \times (3.949)^3}{3 \times (1-0.33)} = 1.033 \times 10^8$ kN m.

Damping of soil in translational mode,

$$C_x = 3.054\sqrt{\frac{G\rho R_f^4}{2-v}} = 3.054 \times \sqrt{\frac{4.216 \times 10^5 \times 2.082 \times (3.949)^4}{2-0.33}}$$
$$= 34530 \text{ kN-s/m}$$

Damping of soil in the rocking mode,

$$C_\theta = 1.33\sqrt{\frac{G\rho R_f^8}{2-v}} = 3.054 \times \sqrt{\frac{4.216 \times 10^5 \times 2.082 \times (3.949)^8}{2-0.33}}$$
$$= 2.345 \times 10^5 \text{ kN-s/rad}$$

Mode	1	2	3
M_{bi} (kN-m)	4968.23	106.65	40.42
V_{bi}(kn)	743.32	48.51	37.82
S_{ai}(m/s^2)	(0.36/8) × 2.5 × 9.81 = 1.103	1.103	1.103
Mi (kN-s^2/m)	673.5	43.955	34.368
hi (M_{bi}/V_{bi})m	6.683	2.198	1.068
ζ_x	0.208	0.402	0.410
ζ_θ	0.065	0.375	0.416

Thus, the natural frequency of the foundation in,

Translation $\omega_x = \sqrt{\frac{K_x}{M_f + m_i}}$ rad/s

Here

$$M_f = \pi \times R_f^2 \times D_f \times \left(\frac{\gamma}{g}\right) = \pi \times (3.949)^2 \times 1.5 \times \left(\frac{25}{9.81}\right) = 187.278 \text{ kN-s}^2/\text{m}.$$

$$\omega_{x1} = \sqrt{\frac{7.976 \times 10^6}{187.278 + 673.5}} = 96.281 \text{ rad/s for mode 1}$$

$$\omega_{x2} = \sqrt{\frac{7.976 \times 10^6}{187.278 + 43.955}} = 185.875 \text{ rad/s for mode 2}$$

and $\omega_{x3} = \sqrt{\frac{7.976 \times 10^6}{187.278 + 34.368}} = 189.86$ rad/s for mode 3

Mass moment of inertia $J_\theta = \frac{1}{4} M_f R_f^2 = \frac{1}{4} \times 187.278 \times (3.949)^2 = 730.131$ kN-s^2-m.

In rocking, $\omega_\theta = \sqrt{\frac{K_\theta}{J_\theta + m_i h_i^2}} = \sqrt{\frac{1.033 \times 10^8}{730.131 + 673.35 \times (6.683)^2}} = 57.896$ rad/s for mode 1

Similarly, $\omega_\theta = 331.04$ and 366.42 rad/s, for mode 2 and mode 3, respectively.

Damping ratio in translation $\zeta_x = \frac{C_x}{2\sqrt{K_x(M_f + m_i)}}$ and damping ratio in rocking $\zeta_\theta = \frac{C_\theta}{2\sqrt{K_\theta\left(J_\theta + m_i h_i^2\right)}}$; the values are tabulated in the above table for the three modes.

Substituting it in the equation,

$$\mathrm{AF}_i = \left[\begin{array}{l} 1+\dfrac{V_i}{\hat{K}_x u_{si}}\left(1+e^{-\zeta_x\frac{2\pi}{T_x}t}\left\{\sqrt{\dfrac{1-\zeta_x}{1+\zeta_x}}\sin 2\pi\sqrt{1-\zeta_x^2}\dfrac{t}{T_x}-\cos 2\pi\sqrt{1-\zeta_x^2}\dfrac{t}{T_x}\right\}\right) \\ +\dfrac{M_{bi}^2}{V_i\hat{K}_\theta u_{si}}\left(1+e^{-\zeta_\theta\frac{2\pi}{T_\theta}t}\left\{\sqrt{\dfrac{1-\zeta_\theta}{1+\zeta_\theta}}\sin 2\pi\sqrt{1-\zeta_\theta^2}\dfrac{t}{T_\theta}-\cos 2\pi\sqrt{1-\zeta_\theta^2}\dfrac{t}{T_\theta}\right\}\right) \end{array}\right]$$

we finally have,

Taking time steps of t = 0.001 up to 0.5 s we have for the first three modes amplification factors as shown in Figs. 12.12, 12.13, and 12.14.

It is observed that the peak amplification (marked bold) has a time lag, and for each mode, they do not act at same time.

The peak amplifications are tabulated hereafter.

Time step	AF_1	AF_2	AF_3	V_{b1}	V_{b2}	V_{b3}	V_{srss}	M_{b1}	M_{b2}	M_{b3}	M_{srss}
0.037	**2.073**	11.81	14.66	-1541	-573	-555	1735	10,298	-1259	592	10,392
0.008	1.318	**16.166**	19.044	-980	-784	-720	1447	6549	-1724	769	6816
0.011	1.445	15.371	**19.799**	-1075	-746	-749	1507	7183	-1639	800	7411

Fig. 12.12 Amplification factor for the first mode

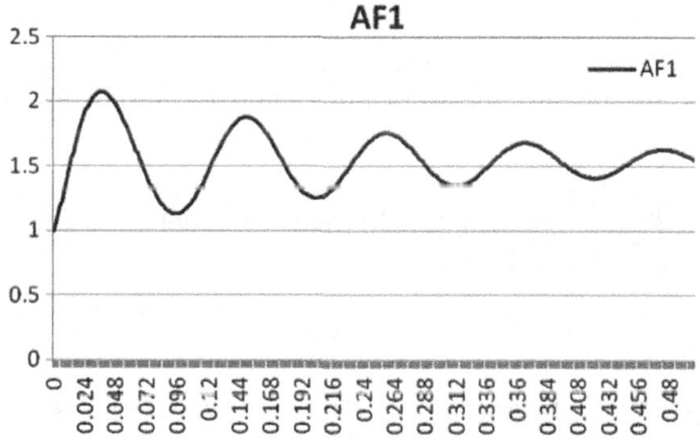

Fig. 12.13 Amplification factor for the second mode

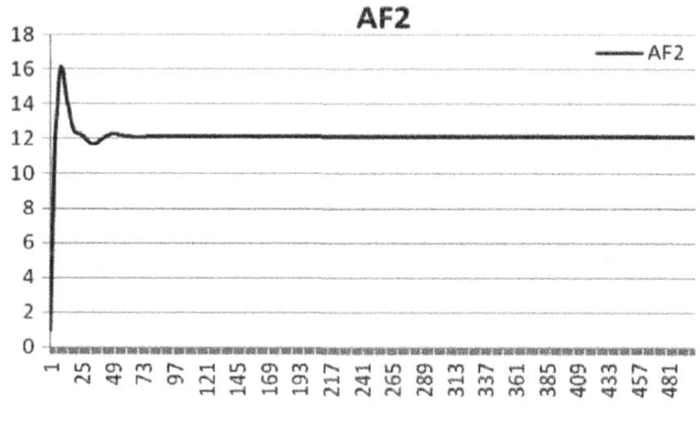

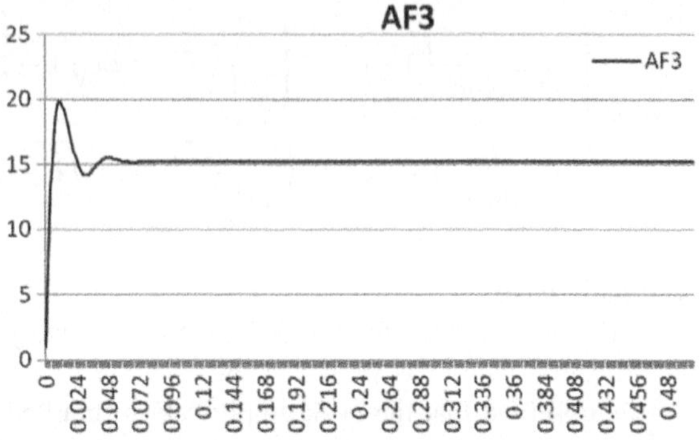

Fig. 12.14 Amplification factor for the third mode

The corresponding fixed base moment shall be multiplied by the above amplification factor to arrive at the design moment considering the effect of DSSI.

It will be observed that despite the fact that the second and third modes have high amplification values, they do not affect the final outcome due to low modal mass participation; the fundamental mode basically governs the design.

12.4.2 Piers Resting on Pile Foundations

In this case, the mechanics remains same as elaborated in Sect. 12.4.1 for shallow foundation except the fact that the pile spring stiffness and damping values as cited in Sect. 10.7.1 of Chap. 10 are to be used.

Once the pile stiffness and damping values are computed, identical steps as followed in Example 12.3 may be adopted to arrive at the dynamic soil amplification due to DSSI.

12.4.3 Bridge Pier Resting on Caisson or Well Foundation

Investigation on the soil–structure interaction effect between well and pier has not drawn much attention, though considerable research has been carried out on the behavior of rigid caissons/cylindrical foundation embedded in soil under earthquake force.

Thakkar et al. (2002) studied the inertial effect of well on overall response of the pier–well system. They used a lumped mass stick model for this study considering both the pier and the foundation.

Gerloymos and Gazetas (2006) used a Winkler-type model for dynamic soil–structure interaction of rigid caisson foundation under seismic loading for linear analysis.

In this work, the focus was more on estimating a realistic value of the soil springs that can be used to estimate the foundation's dynamic response. The validation of soil springs was carried out by comparing the results against three-dimensional finite element analyses.

Mandal and Jain (2008) made a study on pier and well interaction in nonlinear domain to assess the load capacity of pier.

Chowdhury et al. (2012) used a finite beam on elastic foundation without considering the shear deformation effect to determine the effect of well and soil on overall dynamic response of the pier. The results showed that considering pier–well and soil interaction, there was an amplification in the moment and shear in pier cross sections. It demonstrated that the fixed based assumption of the pier as recommended in codes is certainly not realistic.

Varun et al. (2009) also developed a simplified model for lateral response of large diameter caisson foundation. Using a macroscopic model of the well foundation, they developed spring stiffness for both rigid and semi-rigid foundation, validating the same against detailed finite element analysis.

Other than this, stiffness and damping expressions for shallow embedded cylindrical foundations have also been developed by Novak and Beredugo (1972), Kausel and Roesset (1975), Wolf (1985), Dominguez (1978), Gazetas (1983), Gazetas and Dobry (1986), Mita and Luco (1989) and others that have extensive application in industry. However, most of these researches are focused on developing effective spring and damping stiffness of the embedded footing, and the response is sought for an external dynamic load which is harmonic in nature like that in a machine foundation.

Tsigginos et al. (2008) were possibly the first to do a systematic DSSI analysis for a pier–well foundation system under seismic load based on numerical analysis, though the problem was first attempted long back analytically by Tajimi (1969) in Japan.

Shown in Fig. 12.15 is a typical bridge pier supported on well foundation. Two mathematical models are presented herein (Chowdhury et al. 2017).

- A simplified model;
- A generalized solution.

12.4.3.1 Simplified Model

In this case, mathematical model perceived is as shown in Fig. 12.16. The well foundation is assumed as a rigid cylinder partially embedded in ground. Under what condition a well foundation behaves as a rigid block or otherwise, there are no guidelines furnished in any of the codes, though in practice in most cases spring stiffness is derived based on foundations—*to be infinitely stiff.*

Gerloymos et al. (2008) have pointed out that for caissons when $D_f/D \leq 4.0$ the foundation behaves as rigid foundation, for $D_f/D = 3.0 - 8.0$ the well foundation behavior is semi-rigid while for $D_f/D \geq 8.0$ the foundation is flexible and behaves

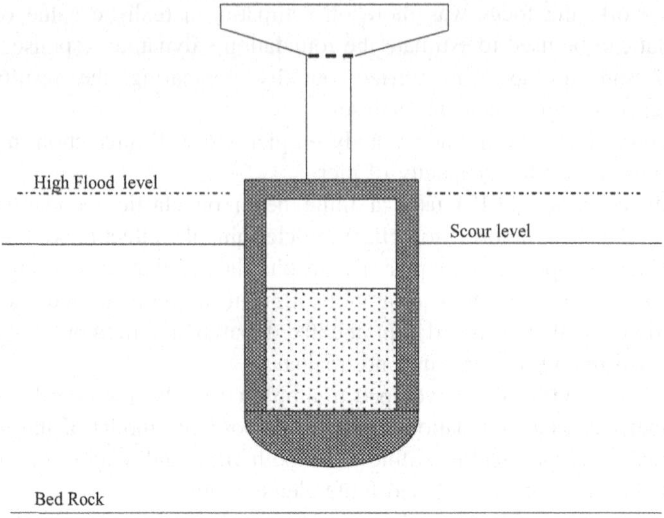

Fig. 12.15 Bridge pier supported on caisson foundation

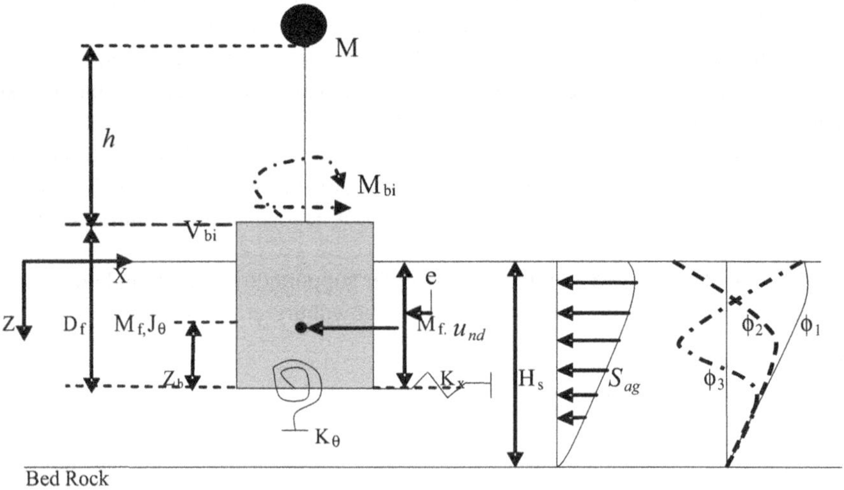

Fig. 12.16 Mathematical model adapted for simplified model

mostly like a long pile. Here D_f = depth of foundation (refer Fig. 12.16) and D = diameter of the well foundation. For most of the well foundations constructed around the world, the aspect ratio (D_f/D) is somewhere between 2 and 8.

Let M_{bi} and V_{bi} be the base shear and moment on top of the foundation in ith mode from the superstructure pier, considering that its base is fixed.

Equation of equilibrium along with well foundation can then be expressed as,

$$\begin{bmatrix} M_f + m_i & (M_f Z_b + m_i \bar{h}_i) \\ (M_f Z_b + m_i \bar{h}_i) & J_\theta + M_f Z_b^2 + m_i \bar{h}_i^2 \end{bmatrix} \begin{Bmatrix} \ddot{u}_{xi} \\ \ddot{\theta}_i \end{Bmatrix} + \begin{bmatrix} C_x & 0 \\ 0 & C_\theta \end{bmatrix} \begin{Bmatrix} \dot{u}_{xi} \\ \dot{\theta}_i \end{Bmatrix} + \begin{bmatrix} K_x & 0 \\ 0 & K_\theta \end{bmatrix} \begin{Bmatrix} u_{xi} \\ \theta_i \end{Bmatrix} = \begin{Bmatrix} V_{bi} \\ M_{bi} \end{Bmatrix} \tag{12.4.3.1.1}$$

In Eq. (12.4.3.1.1), M_f = mass of foundation, J_θ = mass moment of inertia of the foundation, Z_b is c.g. height of foundation only from center of stiffness, $\bar{h}_i$ = c.g. height of effective mass and can be expressed as $|M_{bi}/V_{bi}| + D_f$, and m_i is modal mass contribution from pier plus superstructure in ith mode and is expressed as,

$$m_i = \frac{|V_{bi}|}{\left(\frac{ZI}{2R}\right) \times S_{ai}} \tag{12.4.3.1.2}$$

Here Z, R, and I are code factors and S_{ai} is spectral acceleration of the fixed based pier in ith mode.

Equation (12.4.3.1.1) can be expressed in compact matrix notation as,

$$[\mathbf{M}]\{\ddot{\mathbf{U}}\} + [\mathbf{C}]\{\dot{\mathbf{U}}\} + [\mathbf{K}]\{\mathbf{U}\} = \{\mathbf{P}\} \tag{12.4.3.1.3}$$

For un-damped free vibration, considering the right-hand side as zero, we have,

$$\begin{aligned} &([\mathbf{K}] - [\mathbf{M}]\omega_i^2)[\varphi] = 0 \\ &i = 1, 2, \ldots \end{aligned} \tag{12.4.3.1.4}$$

Here ω_i are natural frequencies of the system while $[\varphi]$ is a 2 × 2 matrix representing the eigenvectors of the system.

Now consider $\{U_i\} = \{\phi\}_i q_i(t)$, where $q_i(t)$ is time-dependent amplitude in generalized coordinates in ith mode.

Equation (12.4.3.1.3) can thus be expressed as,

$$[\mathbf{M}]\{\phi_i\}\ddot{q}_i(t) + [\mathbf{C}]\{\phi_i\}\dot{q}_i(t) + [\mathbf{K}]\{\phi_i\}q_i(t) = \{\mathbf{P}\} \tag{12.4.3.1.5}$$

Pre-multiplying Eq. (12.4.3.1.5) by $\{\phi_i\}^T$, we have,

$$\{\phi_i\}^T[\mathbf{M}]\{\phi_i\}\ddot{q}_i(t) + \{\phi_i\}^T[\mathbf{C}]\{\phi\}\dot{q}_i(t) + \{\phi_i\}^T[\mathbf{K}]\{\phi_i\}q_i(t) = \{\phi_i\}^T\{\mathbf{P}\} \tag{12.4.3.1.6}$$

Based on the theory of vibration, Eq. (12.4.3.1.6) can be expressed as,

$$\ddot{q}_i(t) + 2\zeta_i\omega_i\dot{q}_i(t) + \omega_i^2 q_i(t) = p_i(t) \tag{12.4.3.1.7}$$

Equation (12.4.3.1.7) gives two uncoupled equations,

$$\ddot{q}_x(t) + 2\zeta_x\omega_x\dot{q}_x(t) + \omega_x^2 q_x(t) = p_x(t) \tag{12.4.3.1.8}$$

$$\ddot{q}_\theta(t) + 2\zeta_\theta\omega_\theta\dot{q}_\theta(t) + \omega_\theta^2 q_\theta(t) = p_\theta(t) \tag{12.4.3.1.9}$$

Solution of Eq. (12.4.3.1.8) will give

$$q_x(t) = \mathrm{e}^{-\zeta_x\omega_x t}\left(A\,\cos\omega_x\sqrt{1-\zeta_x^2} + B\,\sin\omega_x\sqrt{1-\zeta_x^2 t}\right) + p_x(t)/\omega_x^2 \tag{12.4.3.1.10}$$

Imposing the boundary conditions (i) at t = 0, $q_x(t)$ = 0, and (ii) at t = 0 $\mathrm{d}q_x(t)/\mathrm{d}t = S_\mathrm{d}\omega_x$, finally gives,

$$q_x(t) = \frac{p_x(t)}{\omega_x^2}\left(1 + \mathrm{e}^{-\zeta_x\omega_x t}\left\{\sqrt{\frac{1-\zeta_x}{1+\zeta_x}}\sin\omega_x\sqrt{1-\zeta_x^2}t - \cos\omega_x\sqrt{1-\zeta_x^2 t}\right\}\right) \tag{12.4.3.1.11}$$

Now imposing the frequency-dependent soil springs on the dominant modal frequency of the superstructure, we have,

$$q_x(t) = \frac{p_x(t)}{\omega_x^2\sqrt{\left(1-r_x^2\right)^2 + \left(2\zeta_x r_x\right)^2}}\left(1 + \mathrm{e}^{-\zeta_x\omega_x t}\left\{\sqrt{\frac{1-\zeta_x}{1+\zeta_x}}\sin\omega_x\sqrt{1-\zeta_x^2}t - \cos\omega_x\sqrt{1-\zeta_x^2 t}\right\}\right) \tag{12.4.3.1.12}$$

Now considering the denominator outside the bracket as $\tilde{k}_x(\omega_i)$—the frequency-dependent stiffness of the system in uncoupled mode—Eq. (12.4.3.1.12) can be expressed in a more condensed form as,

$$q_x(t) = \frac{p_x(t)}{\tilde{k}_x(\omega_i)}\left(1 + \mathrm{e}^{-\zeta_x\omega_x t}\left\{\sqrt{\frac{1-\zeta_x}{1+\zeta_x}}\sin\omega_x\sqrt{1-\zeta_x^2}t - \cos\omega_x\sqrt{1-\zeta_x^2 t}\right\}\right) \tag{12.4.3.1.13}$$

Proceeding in identical fashion, one can derive,

$$q_\theta(t) = \frac{p_\theta(t)}{\tilde{k}_\theta(\omega_i)}\left(1 + \mathrm{e}^{-\zeta_\theta\omega_\theta t}\left\{\sqrt{\frac{1-\zeta_\theta}{1+\zeta_\theta}}\sin\omega_\theta\sqrt{1-\zeta_\theta^2}t - \cos\omega_\theta\sqrt{1-\zeta_\theta^2 t}\right\}\right) \tag{12.4.3.1.14}$$

Now, let eigenvector $[\varphi]$ for the translation and rocking mode be expressed as,

$$[\varphi] = \begin{bmatrix} \phi_{xx} & \phi_{\theta x} \\ \phi_{x\theta} & \phi_{\theta\theta} \end{bmatrix} \tag{12.4.3.1.15}$$

Then for translation and rocking mode, the foundation amplitude is expressed as,

$$[\mathbf{U_x}]^i = \begin{Bmatrix} \phi_{xx} q_x(t) \\ \phi_{x\theta} q_x(t) \end{Bmatrix} \quad \text{and} \quad [\mathbf{U_\theta}]^i = \begin{Bmatrix} \phi_{\theta x} q_\theta(t) \\ \phi_{\theta\theta} q_\theta(t) \end{Bmatrix} \tag{12.4.3.1.16}$$

As explained in Chap. 7 Sect. 7.9.2, it has been shown that dynamic amplification due to DSSI can be finally expressed as,

$$\text{AF} = 1 + \frac{u_x}{u_s} + \bar{h}\frac{\theta}{u_s} \tag{12.4.3.1.17}$$

In Eq. (12.4.3.1.17), u_s = fixed based amplitude of the pier at its c.g. height $\bar{h}_i$, u_x is the translational displacement of the foundation, and θ = rotational amplitude of the foundation.

Thus for the present problem,

$$\text{AF}_x^i = 1 + \frac{\phi_{xx} q_x(t)}{u_{si}} + \bar{h}_i \frac{\phi_{x\theta} q_x(t)}{u_{si}} \tag{12.4.3.1.18}$$

Equation (12.4.3.1.18) can be expressed in an expanded format as,

$$\begin{aligned} \text{AF}_x^i = 1 &+ \frac{\phi_{xx}}{u_{si}} \left[\frac{p_x(t)}{\bar{k}_x(\omega_i)} \left(1 + \mathrm{e}^{-\zeta_x \omega_x t} \left\{ \sqrt{\frac{1-\zeta_x}{1+\zeta_x}} \sin \omega_x \sqrt{1-\zeta_x^2} t - \cos \omega_x \sqrt{1-\zeta_x^2} t \right\} \right) \right] \\ &+ \left(\frac{M_{bi}}{V_{bi}} \right) \frac{\phi_{x\theta}}{u_{si}} \left[\frac{p_x(t)}{\bar{k}_x(\omega_i)} \left(1 + \mathrm{e}^{-\zeta_x \omega_x t} \left\{ \sqrt{\frac{1-\zeta_x}{1+\zeta_x}} \sin \omega_x \sqrt{1-\zeta_x^2} t - \cos \omega_x \sqrt{1-\zeta_x^2} t \right\} \right) \right] \end{aligned} \tag{12.4.3.1.19}$$

In Eq. (12.4.3.1.19), the subscript x to the term AF depicts translational mode of the foundation while superscript i depicts ith mode of the superstructure.

For the rocking mode, proceeding in identical fashion, we derive,

$$\begin{aligned} \text{AF}_\theta^i = 1 &+ \frac{\phi_{\theta x}}{u_{si}} \left[\frac{p_\theta(t)}{\bar{k}_\theta(\omega_i)} \left(1 + \mathrm{e}^{-\zeta_\theta \omega_\theta t} \left\{ \sqrt{\frac{1-\zeta_\theta}{1+\zeta_\theta}} \sin \omega_\theta \sqrt{1-\zeta_\theta^2} t - \cos \omega_\theta \sqrt{1-\zeta_\theta^2} t \right\} \right) \right] \\ &+ \left(\frac{M_{bi}}{V_{bi}} \right) \frac{\phi_{\theta\theta}}{u_{si}} \left[\frac{p_\theta(t)}{\bar{k}_\theta(\omega_i)} \left(1 + \mathrm{e}^{-\zeta_\theta \omega_\theta t} \left\{ \sqrt{\frac{1-\zeta_\theta}{1+\zeta_\theta}} \sin \omega_\theta \sqrt{1-\zeta_\theta^2} t - \cos \omega_\theta \sqrt{1-\zeta_\theta^2} t \right\} \right) \right] \end{aligned} \tag{12.4.3.1.20}$$

The resultant amplification factor can thus be expressed as,

$$\mathrm{AF}_r = 1 + \left(\mathrm{AF}_x^i - 1\right) + \left(\mathrm{AF}_\theta^i - 1\right) \tag{12.4.3.1.21}$$

The logic behind Eq. (12.4.3.1.21) is that, say, if the foundation is completely fixed (i.e., resting on rock) then both AF_x and AF_θ will have value 1.0. But SRSSR value would be 1.732 which is not logical as it should remain 1.0 in this case (i.e., no amplification). Thus, for any value of AF_x and $\mathrm{AF}_\theta > 1.0$, we deduct 1.0 and then do ABSSUM to arrive at a realistic result. Direct SRSS will overestimate the response.

It is observed in this case that the amplification factor due to DSSI is affected by the scaling vectors that give a coupled response.

Equations (12.3.1.17) to (12.3.1.19) multiplied by this amplification factor will give the augmented value of amplitude moment and shear due to soil–structure interaction.

In the proposed procedure, considering that the basic concept of modal analysis is adapted, it is apparent that the analysis is linear when we assume perfect bondage between the soil and the well where no slippage occurs and the strain is deemed low.

For a major earthquake, however, the strain can be large (especially near the surface) and would result in stiffness degradation and augmentation of damping of soil. It is possible to fit this phenomenon within the above framework by a special free field analysis.[1] The reduced G value and the enhanced damping ratio can then be considered to compute K_x, K_θ, C_x, C_θ and use them in Eq. (12.4.3.1.1).

As cited in footnote, refer to Sect. 8.7.1 of Chap. 8 for more details on the procedure of determining the reduced G and augmented damping ratio.

12.4.3.2 Effect of Kinematical Interaction

When earthquake waves propagate through soil, the soil starts vibrating in different modes $\phi_1, \phi_2 \ldots \phi_n$, as shown in Fig. 12.16. This is called the free field vibration of soil and shows how the soil is expected to vibrate in the absence of any structure supported on it. However, when a structure is supported on a foundation, especially an embedded foundation like a pile group or a caisson, due to the presence of foundation, the wave patterns get altered, thus modifying the soil deformation and inducing additional strain in the soil. This results in scattering of seismic waves and the seismic response to which the foundation is subjected to differ from that of the free field motion.

For a long flexible foundation like a pile, the pile follows the deformation characteristics of the ground, but when the foundation is rigid/semi-rigid, like that

[1]Refer Chap. 8 for details of this method.

of caisson as in present case, the free field motion and the foundation displacement can differ significantly.

The analysis is done in two steps here.

As a first step, the free field response of the site is computed without the presence of caisson. This free field response generates shear and moment on the well foundation which is then imposed on the caisson spring and dampers to finally compute the response of the pier.

For vertically propagating shear waves, the equation of equilibrium for the site, as shown in Fig. 12.16, can be expressed as,

$$\frac{\partial^2 u(x,z,t)}{\partial x^2} + \frac{\partial^2 u(x,z,t)}{\partial z^2} = \frac{1}{V_\mathrm{s}^2}\frac{\partial^2 u(x,z,t)}{\partial t^2} \quad (12.4.3.2.1)$$

For the soil mass having depth H_s over the bedrock and extending to infinity in horizontal direction, it can be shown (refer Chap. 3 for details), that the free field time period of the medium can be expressed as,

$$T_n = \frac{4H_\mathrm{s}}{(2n-1)V_\mathrm{s}} \quad (12.4.3.2.2)$$

In Eqs. (12.4.3.2.1) and (12.4.3.2.2), V_s is the shear wave velocity of the soil medium.

It has been shown in Chap. 8 vide Eq. (8.6.1.6) that the amplitude of vibration of the site can be expressed as,

$$u_n = \kappa_n \beta \frac{S_{an} T_n^2}{4\pi^2} \cos(2n-1)\frac{\pi z}{2H_\mathrm{s}} \quad (12.4.3.2.3)$$

Here u_n = amplitude in nth mode, S_{an} = spectral acceleration for the free field time period T_n in nth mode.

$\beta =$ code factor $ZI/2$ and $\kappa_n =$ modal mass participation factor and can be expressed as $8/(\pi + 2)$, $-8(3\pi - 2)$, and $8/(5\pi + 2)$ for the first three modes of vibration for G constant with depth.

Based on the definition of modal analysis, the acceleration generated within the ground, varying with depth, can thus be expressed as,

$$\ddot{u}_n = \kappa_n \beta S_{an} \cos(2n-1)\frac{\pi z}{2H_\mathrm{s}} \quad (12.4.3.2.3)$$

Thus, at $z = 0$, $\ddot{u}_n = \kappa_n \beta S_{an}$ and at a depth e below scour level, as shown in Fig. 12.16,

$$\ddot{u}_{ne} = \kappa_n \beta S_{an} \cos(2n-1)\frac{\pi e}{2H_s} \tag{12.4.3.2.4}$$

Thus, the acceleration over the embedded depth e can be *averaged* out as,

$$\ddot{u}_{n_d} = \frac{1}{2}\kappa_n \beta S_{an}\left(1 + \cos\frac{(2n-1)\pi e}{2H_s}\right) \tag{12.4.3.2.5}$$

Thus, under free field motion, the waves generated within the soil would try to move the foundation and superstructure with an acceleration $\ddot{u}_{n_d}$, as shown in Fig. 12.16 inducing a lateral force.

$$V_i = M_T \times \ddot{u}_{n_d} \tag{12.4.3.2.6}$$

and a moment about center of stiffness as,

$$M_i = M_T \times \ddot{u}_{n_d} \times Z_c \tag{12.4.3.2.7}$$

Here Z_c = distance of the common center of gravity of the foundation and superstructure measured from the center of stiffness and M_T = total mass from superstructure and the foundation.

Equation of equilibrium with respect to Fig. 12.16 can now be expressed as,

$$\begin{bmatrix} M_f + M & M_f Z_b + M(\bar{h} + D_f) & M \\ M_f Z_b + M(\bar{h} + D_f) & J_\theta + M_f Z_b^2 + M(\bar{h} + D_f)^2 & M(\bar{h} + D_f) \\ M & M(\bar{h} + D_f) & M \end{bmatrix} \begin{Bmatrix} \ddot{u}_x \\ \ddot{\theta} \\ \ddot{u}_s \end{Bmatrix} + \begin{bmatrix} C_x & 0 & 0 \\ 0 & C_\theta & 0 \\ 0 & 0 & C_s \end{bmatrix} \begin{Bmatrix} \dot{u}_x \\ \dot{\theta} \\ \dot{u}_s \end{Bmatrix} + \begin{bmatrix} K_x & 0 & 0 \\ 0 & K_\theta & 0 \\ 0 & 0 & K_s \end{bmatrix} \begin{Bmatrix} u_x \\ \theta \\ u_s \end{Bmatrix} = \begin{Bmatrix} V_i \\ M_i \\ 0 \end{Bmatrix} \tag{12.4.3.2.8}$$

In Eq. (12.4.3.2.8), only the first mode of the superstructure is considered as it can be shown that it is the fundamental mode that has major contribution. The structural stiffness K_s can be considered simply as $3EI/H^3$ and M is the mass from top deck plus effective self *wt* of pier.

Solving Eq. (12.4.3.2.8) one can determine u_s, the structural deformation due to kinematical interaction; thus, the additional force the pier experiences is,

$$V_{KI} = K_s . u_s \tag{12.4.3.2.9}$$

Moment due to kinematical interaction at any depth z along the pier height is expressed as,

$$M_{\mathrm{KI}} = V_{\mathrm{KI}} \times z \tag{12.4.3.2.10}$$

These forces now are to be added to inertial interaction forces to determine the total force.

One may note that kinematical interaction was predominantly developed while estimating the dynamic response of piles (Waas and Haartman 1984; Wolf and Von Arx 1982; Novak 1991).

However, in all these cases, the mass of pile is ignored. Thus, in such a case, the load vector is computed by approaching the problem considering the foundation stiffness and multiplying the same by the free field displacement of the site.

For a well foundation, the mass is significant and cannot be neglected, and it becomes computationally much simpler to derive the load vectors from the basic exprcssion:

Force (P) = $m \times a$, and moment (M) = $P \times z$. It can be proved that the load vector approached from stiffness or from inertia gives the same result as at any point of time for a dynamic analysis, i.e., $P = K \times \delta = m \times a$.

The derivation of free field response was shown for G constant with depth. But in most cases, G is found to vary with depth; when considering the characteristics of the differential equation changes, it affects the time period as well as the acceleration.

These values have been computed earlier by Chowdhury et al. (2015a, b) and are tabulated below for ready reference.

Considering $T = C_T.(H_s/V_s)$, values of C_T for different types of soil are furnished in Table 12.7.

The acceleration for different types of soil is as expressed below.

1. For $G = G_0(z/H)$

$$\ddot{u}_1 - \kappa_n \beta S_{\mathrm{a1}} \left[0.971.\cos\frac{\pi z}{2H} - 0.217.\cos\frac{3\pi z}{2H} - 0.086.\cos\frac{5\pi z}{2H} \right] \tag{12.4.3.2.11}$$

Table 12.7 Values of coefficient C_T for various soil profiles

Mode →	1	2	3
$G = G_0(z/H)$	5.19	2.133	1.228
$G = G_0(z/H)^{0.5}$	4.486	1.668	1.005
$G = G_0(z/H)^2$	7.826	3.301	1.66
$G = G_0(1 + z/H)$	3.094	1.095	0.66
$G = G_0(1 + z/H)^2$	2.421	0.906	0.545

2. For $G = G_0\sqrt{z/H}$

$$\ddot{u}_1 = \kappa_n \beta S_{a1}\left[-1.0\cos\frac{\pi z}{2H} - 0.079\cos\frac{3\pi z}{2H} - 0.013\cos\frac{5\pi z}{2H}\right] \quad (12.4.3.2.12)$$

3. For $G = G_0(z/H)^2$

$$\ddot{u}_1 = \kappa_n \beta S_{a1}\left[1.0\cos\frac{\pi z}{2H} + 0.427\cos\frac{3\pi z}{2H} + 0.177\cos\frac{5\pi z}{2H}\right] \quad (12.4.3.2.13)$$

4. For $G = G_0(1 + z/H)$

$$\ddot{u}_1 = \kappa_n \beta S_{a1}\left[1.0\cos\frac{\pi z}{2H} + 0.051\cos\frac{3\pi z}{2H} + 0.00161\cos\frac{5\pi z}{2H}\right] \quad (12.4.3.2.14)$$

5. For $G = G_0(1 + z/H)^2$

$$\ddot{u}_1 = \kappa_n \beta S_{a1}\left[1.0\cos\frac{\pi z}{2H} + 0.109\cos\frac{3\pi z}{2H} + 0.00955\cos\frac{5\pi z}{2H}\right] \quad (12.4.3.2.15)$$

The modal mass participation factors κ_n for the first three modes are as furnished in Table 12.8.

12.4.4 Generalized Model for Pier and Well Foundation

Primary assumption made in the previous model was that the well foundation is infinitely rigid. For a solid concrete block, this would certainly be true. But in reality, the well is actually a hollow shell, whose weight is increased by filling it up

Table 12.8 Values of κ_n for first three modes

Mode →	κ_1	κ_2	κ_3
$G = G_0(z/H)$	−1.534	1.27	0.529
$G = G_0(z/H)^{0.5}$	−1.586	−1.159	0.531
$G = G_0(z/H)^2$	1.480	0.864	0.465
$G = G_0(1 + z/H)$	1.571	1.126	0.513
$G = G_0(1 + z/H)^2$	1.581	1.143	0.536

with available river sand. Thus, depending on its geometry and shell thickness it may behave as a rigid, flexible, or even a semi-rigid foundation, i.e., as explained in Sect. 12.4.3.1, the semi-rigid model is valid when $3 \leq D_f/D \leq 8$.

Considering there are no guidelines existing to assess the rigidity of the well foundation, a generalized formulation is certainly necessary that can take care of this issue automatically during computation. The endeavor in this section is to provide such a formulation.

A generalized model for the pier and semi-rigid well foundation is shown in Fig. 12.17.

In this case, the well is considered as a deformable body, with two base springs representing translation (K_x) and rotation (K_θ) of soil below well plug that are attached to the base.

To start with, the well is considered as a finite beam on elastic foundation that also has considerable shear rigidity.

The general solution for such rigid beams on elastic foundation can be represented by (Karnovsky and Lebed 2001),

$$u(z) = C_0V_0(pz) + C_1V_1(pz) + C_2V_2(pz) + C_3V_3(pz) \tag{12.4.4.1}$$

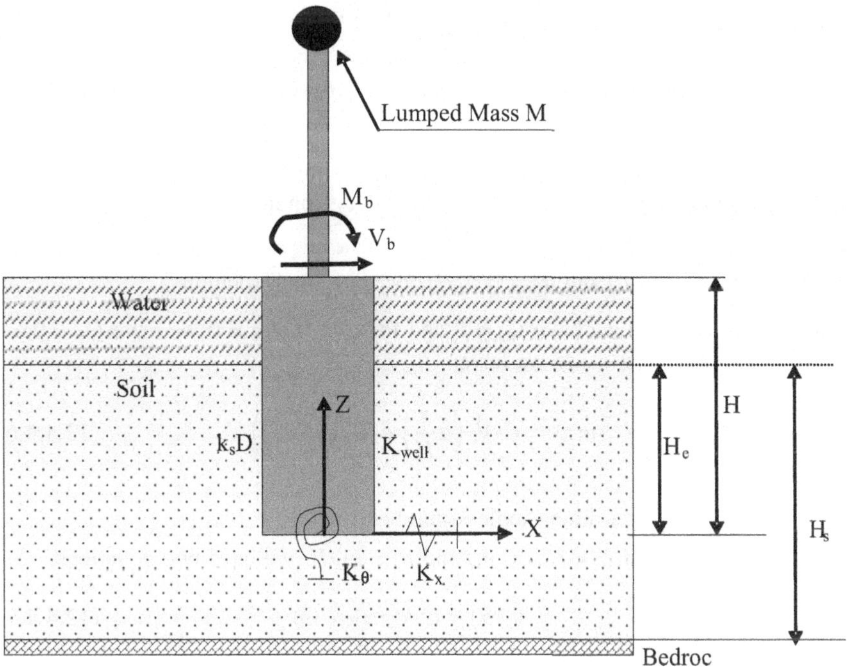

Fig. 12.17 A generalized model for the pier and well semi-rigid foundation

Here

$u(z)$ displacement of body in horizontal direction and varies with depth z;

C_0, C_1, C_2, C_3 integration constants.

$p = 4\sqrt{\frac{k_s D_{eff}}{4EI}}$ an interaction parameter and is function of relative stiffness of the well and surrounding soil.

k_s dynamic sub-grade modulus of soil;
E Young's modulus of well material;
I moment of inertia of well cross section;
D_{eff} effective width of well expressed as $0.5\pi D_w$;
D_w diameter of well;
$V_0(pz)$, $V_1(pz)$, $V_2(pz)$, $V_3(pz)$ Puzrevsky functions and are expressed as,

$$V_0(pz) = \cosh(pz)\cos(pz) \tag{12.4.4.2}$$

$$V_1(pz) = \frac{1}{\sqrt{2}}[\cosh(pz)\sin(pz) + \sinh(pz)\cos(pz)] \tag{12.4.4.3}$$

$$V_2(pz) = \sinh(pz)\sin(pz) \tag{12.4.4.4}$$

$$V_3(pz) = \frac{1}{\sqrt{2}}[\cosh(pz)\sin(pz) - \sinh(pz)\cos(pz)] \tag{12.4.4.5}$$

When the pier induces dynamic moment due to earthquake, the well will deform as semi-rigid cantilever beam when the boundary conditions are,

- At $z = 0$, $u(z) = 0$ and $u'(z) = 0$.
- At $z = H$, $u(z) = u_H$ and $u'(z) = u_H/H$ for very small rotations.

For $u = 0$ at $z = 0$ we have $C_0 = 0$ and for $u' = 0$ at $z = 0$ we have $C_1 = 0$. Now considering $\alpha = pH$ we have,

$$u(z) = C_2 V_2\left(\alpha\frac{z}{H}\right) + C_3 V_3\left(\alpha\frac{z}{H}\right) \tag{12.4.4.6}$$

where $\alpha = 4\sqrt{\frac{k_s D_{eff} H^4}{4EI}}$ a dimensionless number.

Now considering at $z = H$ $u = u_H$ and $u' = u_H/H$, we have two equations.

$$C_2 V_2(\alpha) + C_3 V_3(\alpha) = u_H \tag{12.4.4.7}$$

and

$$C_2\frac{\alpha\sqrt{2}}{H}V_1(\alpha)+C_3\frac{\alpha\sqrt{2}}{H}V_2(\alpha)=\frac{u_H}{H} \tag{12.4.4.8}$$

Equation (12.4.4.7) and (12.4.4.8) can be expressed in matrix form as,

$$\begin{bmatrix} V_2(\alpha) & V_3(\alpha) \\ \frac{\alpha\sqrt{2}}{H}V_1(\alpha) & \frac{\alpha\sqrt{2}}{H}V_2(\alpha) \end{bmatrix}\begin{Bmatrix} C_2 \\ C_3 \end{Bmatrix}=\begin{Bmatrix} u_H \\ \frac{u_H}{H} \end{Bmatrix} \tag{12.4.4.9}$$

Solution of Eq. (12.4.4.9) finally gives,

$$C_2=\left[\frac{\alpha\sqrt{2}}{H}V_2(\alpha)-\frac{V_3(\alpha)}{H}\right]\frac{u_H}{\Delta} \tag{12.4.4.10}$$

and

$$C_3=\left[\frac{-\alpha\sqrt{2}}{H}V_1(\alpha)+\frac{V_2(\alpha)}{H}\right]\frac{u_H}{\Delta} \tag{12.4.4.11}$$

Here,

$$\Delta=\frac{\alpha\sqrt{2}}{H}\left[V_2^2(\alpha)-V_1(\alpha)V_3(\alpha)\right] \tag{12.4.4.12}$$

Now substituting the value of C_2 and C_3 in Eq. (81), we finally have,

$$\begin{aligned}\frac{u(z)}{u_H}=&\left[\frac{\alpha\sqrt{2}V_2(\alpha)-V_3(\alpha)}{H\Delta}\right]V_2\left(\frac{\alpha z}{H}\right)\\&+\left[\frac{V_2(\alpha)-\alpha\sqrt{2}V_1(\alpha)}{H\Delta}\right]V_3\left(\frac{\alpha z}{H}\right)\end{aligned} \tag{12.4.4.13}$$

Equation (12.4.4.13) is a dimensionless term and is the shape function of well deformation and can be expressed as,

$$\begin{aligned}\phi(z)=&\left[\frac{\alpha\sqrt{2}V_2(\alpha)-V_3(\alpha)}{\alpha\sqrt{2}(V_2^2(\alpha)-V_1(\alpha)V_3(\alpha))}\right]V_2\left(\frac{\alpha z}{H}\right)\\&+\left[\frac{V_2(\alpha)-\alpha\sqrt{2}V_1(\alpha)}{\alpha\sqrt{2}(V_2^2(\alpha)-V_1(\alpha)V_3(\alpha))}\right]V_3\left(\frac{\alpha z}{H}\right)\end{aligned} \tag{12.4.4.14}$$

Considering $\xi=z/H$, Eq. (12.4.4.14) can be expressed in natural coordinates as,

$$\begin{aligned}\phi(\xi) &= \left[\frac{\alpha\sqrt{2}V_2(\alpha) - V_3(\alpha)}{\alpha\sqrt{2}\left(V_2^2(\alpha) - V_1(\alpha)V_3(\alpha)\right)}\right]V_2(\alpha\xi) \\ &+ \left[\frac{V_2(\alpha) - \alpha\sqrt{2}V_1(\alpha)}{\alpha\sqrt{2}\left(V_2^2(\alpha) - V_1(\alpha)V_3(\alpha)\right)}\right]V_3(\alpha\xi) \\ \rightarrow \phi(\xi) &= AV_2(\alpha\xi) + BV_3(\alpha\xi)\end{aligned} \tag{12.4.4.15}$$

Equation (12.4.4.15) is the shape function when the well is fully embedded; however, due to scour as it is embedded up to a depth H_e, we have $\alpha_e = 4\sqrt{\frac{k_s D_{eff} H_e^4}{4EI}}$ when

$$\begin{aligned}\phi_e(\xi) &= \left[\frac{\alpha_e\sqrt{2}V_2(\alpha_e) - V_3(\alpha_e)}{\alpha_e\sqrt{2}\left(V_2^2(\alpha_e) - V_1(\alpha_e)V_3(\alpha_e)\right)}\right]V_2(\alpha_e\xi) \\ &+ \left[\frac{V_2(\alpha_e) - \alpha_e\sqrt{2}V_1(\alpha_e)}{\alpha_e\sqrt{2}\left(V_2^2(\alpha_e) - V_1(\alpha_e)V_3(\alpha_e)\right)}\right]V_3(\alpha_e\xi) \\ \rightarrow \phi_e(\xi) &= A_e V_2(\alpha_e\xi) + B_e V_3(\alpha_e\xi)\end{aligned} \tag{12.4.4.16}$$

The stiffness of well in fundamental mode can thus be expressed as,

$$K_w = EI\int_0^H \phi_e''(z)^2 dz + k_s D_{eff}\int_0^{H_e} \phi_e^2(z) dz \tag{12.4.4.17}$$

Equation (12.4.4.17) in natural coordinates can be expressed as

$$K_w = \frac{4EI\alpha_e^4}{H^3}\int_0^1 \phi_e''(\xi)^2 d\xi + k_s D_{eff} H_e \int_0^{\eta_e} \phi_e^2(\xi) d\xi \tag{12.4.4.18}$$

$$\rightarrow K_w = \frac{4EI\alpha_e^4}{H^3}\left[\int_0^1 [A_e V_0(\alpha_e\xi) + B_e V_1(\alpha_e\xi)]^2 d\xi + k_s D_{eff} H_e \int_0^{\eta_e} [A_e V_2(\alpha_e\xi) + B_e V_3(\alpha_e\xi)]^2 d\xi\right] \tag{12.4.4.19}$$

Here η_e is the ratio of embedded depth to full depth of well H_e/H, where $0 \leq \eta_e \leq 1.0$.

Equation (12.4.4.19) can be written in a simplified compact form as,

$$K_w = \frac{4EI\alpha_e^4}{H^3}[I_1 + \eta_e I_2] \rightarrow k_s D_{eff} H_e \eta_e^3 [I_1 + \eta_e I_2] \tag{12.4.4.20}$$

Equation (12.4.4.20) can be expressed as

$$K_w^b = \frac{4EI}{H^3}\chi^b \qquad (12.4.4.21)$$

where $\chi^b = \alpha^4(I_1 + \eta_e I_2)$ is a stiffness coefficient and the superscript b stands for the term bending.

Here,

$I_1 = \int_0^1 [A_e V_0(\alpha_e \xi) + B_e V_1(\alpha_e \xi)]^2 d\xi$ and $I_2 = \int_0^{\eta_e} [A_e V_2(\alpha_e \xi) + B_e V_3(\alpha_e \xi)]^2 d\xi$ are integration constants that can be computed numerically.

12.4.4.1 Stiffness Due to Shear Deformation

Considering the large diameter of well, the well shaft can have significant stiffness in shear deformation. Equation of equilibrium in this case can be expressed as

$$\eta GA \frac{d^2 u}{dz^2} + k_s D_{\text{eff}} u = 0 \qquad (12.4.4.1.1)$$

Here G = shear modulus of well, A = area of cross section of well, k_s = dynamic sub-grade modulus.

Equation (95) is a standard second-order linear differential equation whose solution can be expressed as (Meirovitch 2001),

$$u(z) = C \cos qz + D \sin qz \qquad (12.4.4.1.2)$$

Here $q = \sqrt{k_s D_{\text{eff}}/GA}$ a standard parameter. Now considering $\beta = qH$, Eq. (12.4.4.1.2) can be expressed as

$$u(z) = C \cos \frac{\beta z}{H} + D \sin \frac{\beta z}{H} \qquad (12.4.4.1.3)$$

Applying the boundary condition at z = 0, u = 0, we have C = 0. Again considering at $z = H$ $u = u_H$, we have $D = u_H/\sin\beta$. From which we have

$$\frac{u(z)}{u_H} = \frac{1}{\sin\beta} \sin\left(\frac{\beta z}{H}\right) \qquad (12.4.4.1.4)$$

Converting Eq. (12.4.4.1.4.) to natural coordinates and also taking the embedment up to depth H_e we have

$$\phi_e^{\rm s}(\xi) = \frac{1}{\sin\beta_{\rm e}}\sin\left(\frac{\beta_{\rm e} z}{H}\right) \tag{12.4.4.1.5}$$

Here $\beta_{\rm e} = \sqrt{k_{\rm s} D_{\rm eff} H_{\rm e}^2 / GA}$, where e stands for embedment and superscript s stands for shear.

The shear stiffness of the well can be expressed as,

$$\begin{aligned} K_{\rm w}^{\rm s} &= \eta GA \int_0^H (\phi_e'^{\rm s})^2(z){\rm d}z + k_{\rm s} D_{\rm eff} \int_0^{H_{\rm e}} (\phi_e^{\rm s})^2(z){\rm d}z \\ \rightarrow K_{\rm w}^{\rm s} &= \frac{\eta GA}{H} \int_0^1 (\phi_e'^{\rm s}(\xi))^2 {\rm d}\xi + k_{\rm s} D_{\rm eff} H \int_0^{\eta_e} (\phi_e'^{\rm s}(\xi))^2 {\rm d}\xi \end{aligned} \tag{12.4.4.1.6}$$

Considering that the shape functions are simple, in this case, the equation can be analytically solved directly, which finally gives

$$K_{\rm w}^{\rm s} = \frac{\eta GA \beta_{\rm e}^2}{2H\sin^2\beta_{\rm e}}\left[1 + \frac{\sin 2\beta_{\rm e}}{2\beta_{\rm e}} + \frac{1}{\eta_{\rm e}^2}\left(\eta_e - \frac{\sin 2\beta_{\rm e}\eta_e}{2\beta_{\rm e}}\right)\right] \tag{12.4.4.1.7}$$

Equation (12.4.4.1.7) can be expressed in a compact form as,

$$K_{\rm w}^{\rm s} = \frac{\eta GA}{2H}\chi^{\rm s} \tag{12.4.4.1.8}$$

Here,

$$\chi^{\rm s} = \frac{\beta_{\rm e}^2}{\sin^2\beta_{\rm e}}\left[1 + \frac{\sin 2\beta_{\rm e}}{2\beta_{\rm e}} + \frac{1}{\eta_{\rm e}^2}\left(\eta_e - \frac{\sin 2\beta_{\rm e}\eta_e}{2\beta_{\rm e}}\right)\right] \tag{12.4.4.1.9}$$

Now considering Eqs. (12.3.1.10) to (12.3.1.11), the effective stiffness of the well can be expressed as

$$\frac{1}{K_{\rm w}^{\rm e}} = \frac{1}{K_{\rm w}^{\rm b}} + \frac{1}{K_{\rm w}^{\rm s}} \tag{12.4.4.1.10}$$

Substituting Eqs. (12.4.4.21) and (12.4.4.1.8) in (12.4.4.10) will finally give,

$$K_{\rm w}^{\rm e} = \frac{4EI}{H^3}\chi^{\rm b}\left[\frac{1}{1 + \frac{16(1+\nu)}{\eta\lambda^2}\left(\frac{\chi^{\rm b}}{\chi^{\rm s}}\right)}\right] \tag{12.4.4.1.11}$$

In Eq. (12.4.4.1.11), $\lambda = H/r$ is the slenderness ratio of the well shaft, ν = Poisson's ratio, and η = shear correction factor for hollow circular shell. It will be observed that when the well is wide and short, the shear deformation significantly affects the stiffness as λ is small.

The only value that needs to be computed now is the dynamic sub-grade modulus of soil k_s, as derived hereafter.

12.4.4.2 Determination of Dynamic Sub-grade Modulus K_s

For a rigid cylinder embedded in ground, the lateral stiffness is expressed as,

$$K_x = \frac{8Gr_0}{2-\nu}\left[1 + \frac{e}{r_0}\right] \tag{12.4.4.2.1}$$

where the first term within the bracket depicts the stiffness from bearing and the second term represents the stiffness from the foundation wall.

Thus, ignoring the first term and assuming e = dz, an infinitely small segment, we can express Eq. (12.4.4.2.1) as,

$$\mathrm{d}K_x = \frac{8G\mathrm{d}z}{2-\nu} \tag{12.4.4.2.2}$$

Thus, the second term of Eq. (12.4.4.19) that represents soil stiffness can be expressed as (Chowdhury et al. 2016),

$$K_{\mathrm{soil}} = \frac{8GH_e}{2-\nu}\int_0^{\eta_e}[A_e V_2(\alpha_e\xi) + B_e V_3(\alpha_e\xi)]^2 \mathrm{d}\xi \tag{12.4.4.2.3}$$

Equating this to the second term of Eq. (12.4.4.19), one finally obtains,

$$k_s = \frac{16G}{\pi D(2-\nu)} \tag{12.4.4.2.4}$$

Here G = dynamic shear modulus of soil, ν = Poisson's ratio of soil, D = diameter of well.

The mass matrix of the well can be expressed as,

$$M_w = \frac{\gamma_{\mathrm{eff}}\mathrm{AH}}{g}\int_0^1 \left[\phi_e^b(z)\right]^2 \mathrm{d}z \rightarrow \frac{\gamma_{\mathrm{eff}}\mathrm{AH}}{g} I_3. \tag{12.4.4.2.5}$$

Here γ_{eff} = effective density that represents the total weight of well and river sand inside it, including all other fixtures connected to it.

The damping matrix is expressed as,

$$C_\mathrm{w} = \frac{\gamma_\mathrm{s} D_\mathrm{eff} H_\mathrm{e} V_\mathrm{s}}{g} \int_0^{\eta_e} \left[\phi_e^\mathrm{b}(z)\right]^2 \mathrm{d}z \rightarrow \frac{\gamma_\mathrm{s} D_\mathrm{eff} H_\mathrm{e} V_\mathrm{s}}{g} I_3. \quad (12.3.2.4.6)$$

The equilibrium equation under superstructure moment and shear with respect to the center of stiffness can now be expressed as,

$$\begin{bmatrix} M_\mathrm{w} + M_\mathrm{s} & M_\mathrm{w} Z_\mathrm{b} + M_\mathrm{s}(\bar{h} + H) & M_\mathrm{w} \\ M_\mathrm{w} Z_\mathrm{b} + M_\mathrm{s}(\bar{h} + H) & J_\theta + M_\mathrm{w} Z_\mathrm{b}^2 + M_\mathrm{s}(\bar{h} + H)^2 & M_\mathrm{w} Z_\mathrm{b} \\ M_\mathrm{w} & M_\mathrm{w} Z_\mathrm{b} & M_\mathrm{w} \end{bmatrix} \begin{Bmatrix} \ddot{x} \\ \ddot{\theta} \\ \ddot{u}_w \end{Bmatrix} + \begin{bmatrix} C_x & 0 & 0 \\ 0 & C_\theta & 0 \\ 0 & 0 & C_w \end{bmatrix} \begin{Bmatrix} \dot{x} \\ \dot{\theta} \\ \dot{u}_w \end{Bmatrix} + \begin{bmatrix} K_x & 0 & 0 \\ 0 & K_\theta & 0 \\ 0 & 0 & K_w^e \end{bmatrix} \begin{Bmatrix} x \\ \theta \\ u_w \end{Bmatrix} = \begin{Bmatrix} V_\mathrm{b} \\ V_\mathrm{b} H + M_\mathrm{b} \\ V_\mathrm{b} \end{Bmatrix} \quad (12.4.4.2.7)$$

In Eqs. (12.4.4.2.5) to (12.4.4.2.7), M_s = mass of superstructure or pier, Z_b is c.g. height of base and is expressed as,

$$Z_\mathrm{b} = \frac{\int_0^1 \xi \phi_e^b(\xi) \mathrm{d}\xi}{\int_0^1 \phi_b^e(\xi)} H \quad (12.4.4.2.8)$$

$\bar{h} = M_\mathrm{b}/V_\mathrm{b}$, the c.g height of pier. J_θ = moment of inertia of the well foundation about its c.g. $K_x = \frac{8Gr_0}{2-v}$, $K_\theta = \frac{8Gr_\theta^3}{3(1-v)}$, $C_x = 3.054\sqrt{\frac{G\rho r_0^4}{2-v}}$, $C_\theta = 1.33\sqrt{\frac{G\rho r_\theta^8}{2-v}}$, r_0 and r_θ are radii of well foundation, G = dynamic shear modulus of soil, and ρ = mass density of soil.

Since we are using the SRSS value of the superstructure moment and shear, the dominant frequency of the applied load is obviously the fundamental mode. Thus, computing the displacement by standard modal analysis technique cited earlier, the total displacement at c.g. height of well is given by,

$$u_\mathrm{w}(\mathrm{tot}) = u_\mathrm{w}(t) + x(t) + Z_\mathrm{b}\theta(t) \quad (12.4.4.2.9)$$

Thus, the amplification factor for the moment and shear in well is expressed as,

$$\mathrm{AF}_w(t) = 1 + \frac{x(t)}{u_\mathrm{w}(t)} + \frac{Z_\mathrm{b}\theta(t)}{u_\mathrm{w}(t)} \quad (12.4.4.2.10)$$

Thus, while designing the well section, the bending moment and shear forces in well foundation has to be multiplied by the peak value of AF_w.

Moment and shear along the well shaft are expressed as,

$$M_z = \frac{2EI\alpha_e^2}{H^2}\left[C_2 V_0\left(\frac{\alpha_e z}{H}\right) + C_3 V_1\left(\frac{\alpha_e z}{H}\right)\right] \tag{12.4.4.2.11}$$

$$V_z = \frac{2\sqrt{2}EI\alpha_e^3}{H^3}\left[-C_2 V_3\left(\frac{\alpha_e z}{H}\right) + C_3 V_0\left(\frac{\alpha_e z}{H}\right)\right] \tag{12.4.4.2.12}$$

where,

$$C_2 = \frac{\left[\frac{M_b H^2}{2EI\alpha_e^2} V_0(\alpha_e) - \frac{V_b H^3}{2\sqrt{2}EI\alpha_e^3} V_1(\alpha_e)\right]}{V_0^2(\alpha_e) + V_1(\alpha^e) V_3(\alpha_e)} \quad \text{and}$$

$$C_3 = \frac{\left[\frac{M_b H^2}{2EI\alpha_e^2} V_3(\alpha_e) + \frac{V_b H^3}{2\sqrt{2}EI\alpha_e^3} V_0(\alpha_e)\right]}{V_0^2(\alpha_e) + V_1(\alpha^e) V_3(\alpha_e)}$$

12.4.5 Kinematical Interaction of Well and Pier

When the well foundation is considered as a rigid block, we have, as per Eq. (12.4.3.2.8), force induced at the center of stiffness due to the motion of soil. Thus, in this case, the mathematical model conceived is as shown in Fig. 12.18. The equilibrium equation in this case is expressed as,

$$\begin{bmatrix} M_w + M_s & M_w.Z_b + M_s(\bar{h}+H) & M_w & M_s \\ M_s(\bar{h}+H) & J_\theta + M_w Z_b^2 + M_s(\bar{h}+H)^2 & M_w Z_b & M_s(\bar{h}+H) \\ M_w & M_w Z_b & M_w & 0 \\ M_s & M_s(\bar{h}+H) & 0 & M_s \end{bmatrix} \begin{Bmatrix} \ddot{x} \\ \ddot{\theta} \\ \ddot{u}_w \\ \ddot{u}_s \end{Bmatrix} + \begin{bmatrix} C_x & 0 & 0 & 0 \\ 0 & C_\theta & 0 & 0 \\ 0 & 0 & C_w + C_s & -C_s \\ 0 & 0 & -C_s & C_s \end{bmatrix} \begin{Bmatrix} x \\ \dot{\theta} \\ \dot{u}_w \\ \dot{u}_s \end{Bmatrix} + \begin{bmatrix} K_x & 0 & 0 & 0 \\ 0 & K_\theta & 0 & 0 \\ 0 & 0 & K_w^e + k_p & -k_p \\ 0 & 0 & -k_p & k_p \end{bmatrix} \begin{Bmatrix} x \\ \theta \\ u_w \\ u_s \end{Bmatrix} = \begin{Bmatrix} V_b \\ M_b \\ 0 \\ 0 \end{Bmatrix} \tag{12.4.5.1}$$

Solving Eq. (117), one can find out u_w and u_s from which the nodal forces can be computed from the expression,

$$\begin{Bmatrix} V_w \\ V_p \end{Bmatrix} = \begin{bmatrix} K_w^e + k_p & -k_p \\ -k_p & k_p \end{bmatrix} \begin{Bmatrix} u_w \\ u_s \end{Bmatrix} \tag{12.4.5.2}$$

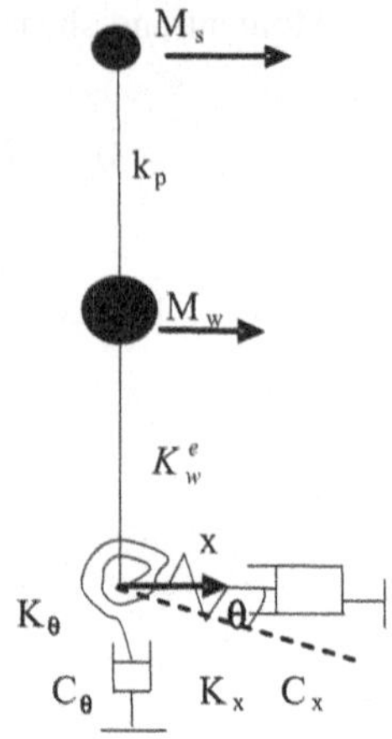

Fig. 12.18 Mathematical model for kinematical interaction of semi-rigid well foundation

12.5 A Case Study

To elaborate the procedure, a bridge pier in Assam (India)/Bangladesh border, supported on a circular well foundation is taken up as a case study. The dimensions of the well and the pier are as shown in Fig. 12.19.

Shown in Fig. 12.19 is basic dimension of the well foundation and the pier. The well is partially filled up with saturated compacted river sand up to the scour line level. The dead plus relevant live load coming from the top deck is 3200 kN. The grade of concrete used is M25. Bottom plug is M20. The soil consists of loose fine river sand of SPT value of 8 at scour level with relative density increasing progressively with depth, having N value ≥ 50 at 25 m below the scour line. Average density of sand is 19 kN/m^3. Shear wave velocity of soil was observed to be 145 m/s at scour level and 550 m/s at base of well plug giving an average shear wave velocity of 347.5 m/s over the well shaft. Design codes adopted IRC:06 (2010), IRC:45 (1992), and IRC:78 (2000).

As per IS-1893(2002), the site is Zone V having $Z = 0.36$ and importance factor = 1.5. The response reduction factor considered is $R = 3.0$.

The computed weight of well foundation (including sand inside) = 11,381 kN. Center of gravity of well only = 10.58 m above the bottom of well. Center of gravity of the full system = 16.33 m above the bottom of well.

The fixed base time periods of the pier, computed by different methods, are as shown in Table 12.9.

The comparative bending moment diagram considering proposed dynamic analysis and method as proposed in code is as elaborated hereafter in Fig. 12.20.

Figure 12.20 clearly shows that the method proposed by code would all underestimate the moment in pier.

Code recommendation underestimates the moment by as much as 40% at the base of pier compared to the proposed method. Considering the first mode dynamic analysis only, it underestimates the moment by about 14%; this possibly

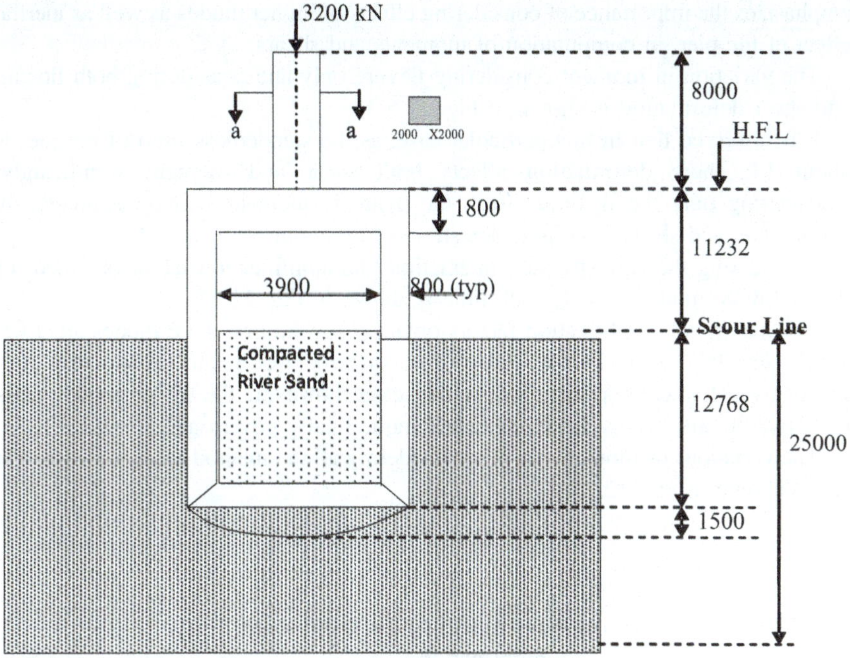

Fig. 12.19 Bridge pier supported on caisson foundation

Table 12.9 Fixed base time period of the pier

Mode	1	2	3
IRC:06 (2010)	0.256	–	–
Proposed method considering flexure only [Vide Eq. (12.3.11)]	0.264	0.014	0.0043
Proposed method considering combined flexure and shear [Vide Eq. (12.3.1.14)]	0.271	0.017	0.0061

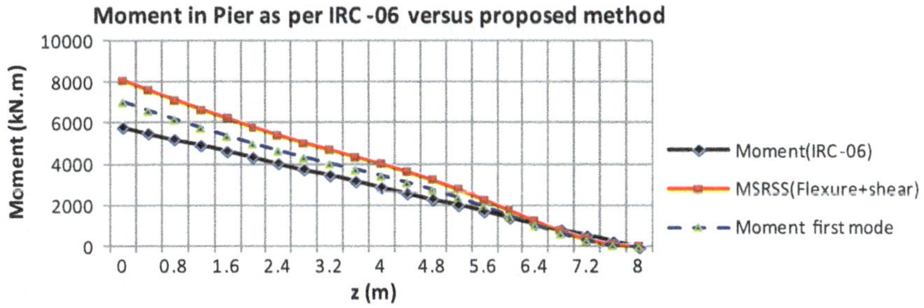

Fig. 12.20 Bending moment diagram of the fixed based pier IRC:06 versus the proposed method

emphasizes the importance of considering effects of higher modes as well as inertial effect of the pier, in computation of moments and shears.

The variation in moment considering flexure only and considering both flexure and shear deformation is shown in Fig. 12.5.3.

It is observed that in this particular case, as the slenderness ratio of the pier is about 14, shear deformation affects both shear and moment significantly. Considering only the flexural effect, the dynamic moment is underestimated by about 17%, and shear force is underestimated by about 6% (Fig. 12.21).

Considering the soil–structure interaction, the amplification factor obtained for the first three modes varying with time is shown in Fig. 12.22.

The maximum amplification factors obtained for the first three modes are 1.05, 1.567, and 1.52 respectively. The predominant rocking in the second and third mode gives higher amplification; in this case, however, the lower modal mass participation still makes the fundamental mode finally governing.

The variations of moments in pier considering all the methods considered herein are as shown in Fig. 12.23.

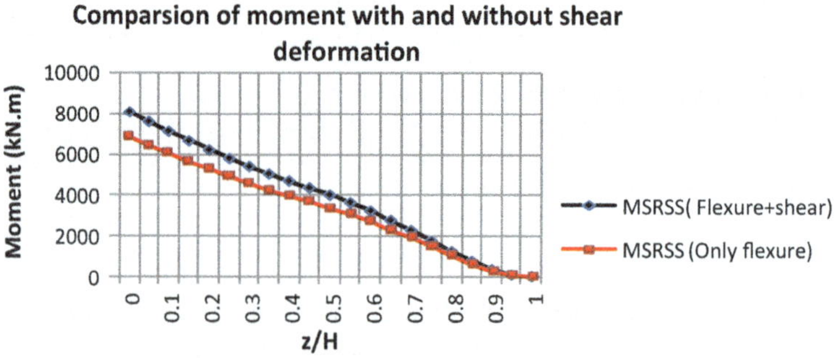

Fig. 12.21 Comparison of bending moment with and without the shear deformation effect

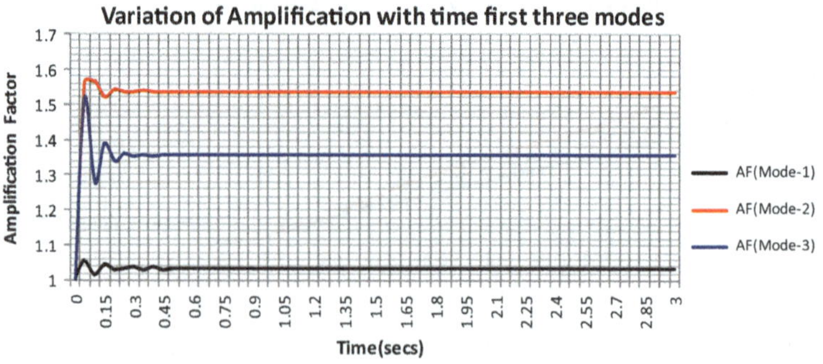

Fig. 12.22 Variation of dynamic amplification factor with time; first three modes

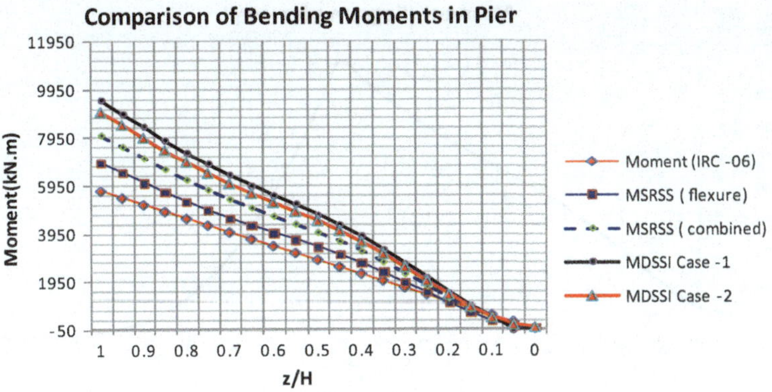

Fig. 12.23 Variation of bending moment considering DSSI and fixed based analysis

It is evident from Fig. 12.23 that DSSI augments the fixed based response. In this case, the amplification on SRSS value is almost two times compared to code values. In Fig. 12.23, MDSSI Case-1 is when the well foundation is considered as rigid and case-2 depicts the situation when well foundation is considered flexible. From Fig. 12.23, it is apparent that MDSSI (Case-1) > MDSSI (Case-2) > MSRSS the fixed based response. Flexibility in the foundation reduces the moment and shear force to a certain extent; however, the system still experiences more force than a fixed based case which goes on to highlight the importance in considering DSSI effect for pier–well combination.

The variation of shear force in a pier is as shown Fig. 12.24.

In Fig. 12.24, it is again observed that shear force values are amplified due to DSSI effect.

Considering the well as a semi-rigid body, the complete moment and shear over the length of pier and well are shown in Figs. 12.25 and 12.26.

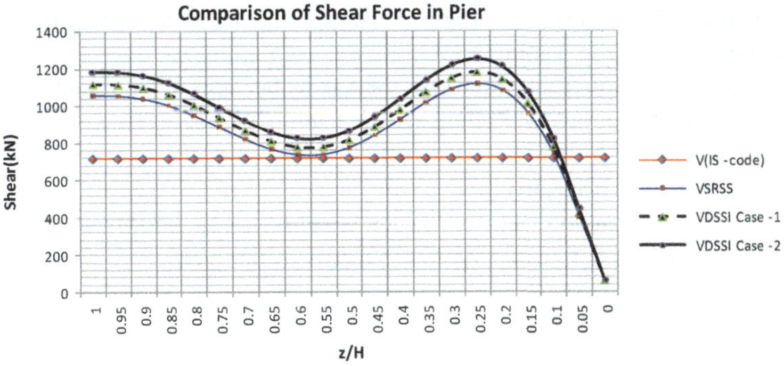

Fig. 12.24 Variation of force on the pier considering fixed base and DSSI

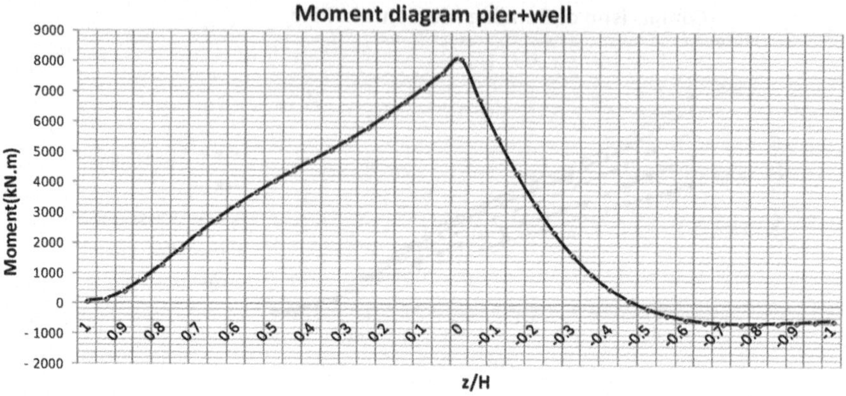

Fig. 12.25 Moment diagram of the complete pier and well in tandem

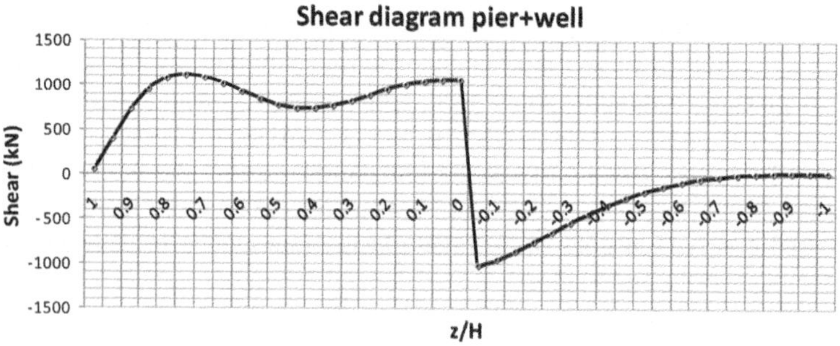

Fig. 12.26 Shear diagram of the complete pier and well in tandem

In Figs. (12.5.7) and (12.5.8), $z/H > 0$ represents the pier part and $z/H < 0$ represents the well foundation part.

The kinematical interaction gives a very interesting insight to the problem. The tip deflections of the pier, considering the well behaving as a rigid and semi-rigid foundation, are given in Table 12.10.

The results shown in Table 12.10 are logical, as because in the foundation being rigid the tip deflection is lower compared to a semi-rigid case. However, the force induced in the pier is 2400 kN vis-à-vis 492 kN only when the well is considered as rigid and semi-rigid, respectively.

In the latter case, as the well foundation is subjected to deformation, it absorbs significant energy, thus transferring lower forces to the pier.

Table 12.10 Deflection at the pier tip

Foundation type	Tip deflection (mm)
Rigid	12
Semi-rigid	26

Thus, from the above one can infer that structures supported on deep heavy foundations can be subjected to significant force due to kinematic interaction and thus foundations in such cases should preferably be conceived in a manner so that its mass is reduced as much as feasible.

References

AASHTO. (2015). *Standard specification for seismic design of highways bridges*. Washington USA: American Authority of State Highway and Transport Officials.

Bathe, K. J. (1980). *Finite element analysis in engineering procedures*. New Delhi: Prentice Hall of India.

CALTRANS. (2004). *Bridge design specification*. Sacramento, CA: California Department of Transportation.

Chowdhury, I., & Dasgupta, S. P. (2007). Dynamic earth pressure on rigid unyielding walls under earthquake forces. *Indian Geotechnical Journal, 37*(2), 81–93.

Chowdhury, I., Singh, J. P., & Tilak, R. (2012). Seismic response of well foundation with dynamic soil structure interaction. In *Proceedings 15th World Conference in Earthquake Engineering*. Lisbon, Portugal.

Chowdhury, I., Tarafdar, R., & Ghosh, A. (2015a). An analytical solution to kinematic and inertial interaction of building with deep basements. In: *Proceedings, 23rd International Conference in Structural Mechanics in Reactor Technology*, Manchester, UK.

Chowdhury, I., Tarafdar, R., Ghosh, A., & Dasgupta, S. P. (2015b). Stiffness degradation and damping augmentation of soil under earthquake loading. *Electronic Journal of Geotechnical Engineering, 20*(9). Oklahama, USA.

Chowdhury, I., Tarafdar, R., Ghosh, A., & Dasgupta, S. P. (2016). Dynamic response of cylindrical structures with coupled soil structure interaction. *Bulletin of Earthquake Engineering Europe*.

Chowdhury, I., Tarafdar, A., Ghosh, A., & Dasgupta, S. P. (2017). Dynamic soil structure interaction of bride piers supported on well foundation. *Journal of Soil Dynamics and Earthquake engineering*, May (Online).

Dominguez, J. (1978). *Dynamic stiffness of rectangular foundations* (Research Report) 78–20. USA: MIT.

Dunkerley, S. (1894). On the whirling and vibration of shafts. *Philosophical Transactions of the Royal Society, London, UK, 185*, 279–360.

Gazetas, G. (1983). Analysis of machine foundation vibrations; state of the art. *Journal of Soil Dynamics and Earthquake Engineering, 2*, 2–42.

Gazetas, G., & Dobry, R. (1986). Dynamic response of arbitrarily shaped foundations. *Journal of Geotechnical Engineering ASCE, 112*, #GT2.

Gazetas, G., & Gerolymos, N. (2006). Winkler model for lateral response of rigid caisson foundations in linear soil. *Journal of Soil Dynamics and Earthquake Engineering, 26*, 347–361.

Hurty, W., & Rubenstein, M. F. (1967). *Structural dynamics*. New Delhi, India: Prentice Hall India.

IRC:06. (2010). *Standard specifications and code of practice for road bridges. Section II loads and stresses*. New Delhi, India: Indian Roads Congress.

IRC:45. (1992). *Recommendations for estimating the resistance of soil below the maximum scour level in the design of well foundations for bridges*. New Delhi, India: Indian Roads Congress.

IRC:78. (2000). *Standard specifications and code of practice for road bridges, foundations and substructures*. New Delhi, India: Indian Roads Congress.

IS-1893. (2002). *Code of practice for earthquake resistant design of buildings*. New Delhi India: Bureau of Indian Standards.

Japanese Regulatory Authority (JRA). part V Design Specifications of Highway bridges for seismic design in Japan.

Karnovsky, I., & Lebed, O. (2001). *Formulas for structural dynamics*. NY, USA: McGraw-Hill.

Kausel, E., & Rosset, J. M. (1975). Dynamic stiffness of circular foundation. *Journal of Engineering Mechanics division ASCE, 98*(SM12), 770–785.

Mandal, G., & Jain, S. K. (2008). Effect of non-linearity in pier and well foundation on seismic response of bridges. In *14th World conference on earthquake engineering*. Beijing, China.

Meirovitch, L. (2001). *Elements of vibration analysis*. New Delhi, India: Allied Publishers.

Mita, A., & Luco, J. E. (1989). Dynamic response of square foundations embedded in elastic half space. *Journal of Soil Dynamics and Earthquake Engineering, 8*(2), 54–67.

Murty, C. V. R. & Jain, S. K. (1997). A study on seismic design codes for highway bridges (*Research report # 3*). Department of Civil Engineering, Indian Institute of Technology Kanpur, India.

Novak, M. (1991). Pile under dynamic loads. In *Proceedings 2nd International Conference in Recent advances in Geotechnical Earthquake Engineering St Louis USA Paper #SOA14*.

Novak, M., Beredugo, Y. O., & Novak, M. (1972). Coupled horizontal and rocking vibration of embedded footings. *Canadian Geotechnical Journal, 9*(4).

Shames, I. H., & Dym, C. L. (1995). *Energy and finite element method in structural mechanics*. New Delhi, India: New Age Publishers.

Tajimi, H. (1969). Dynamic analysis of a structure embedded in an elastic stratum. In: *Proceeding of IVth World Conference in Earthquake Engineering*, Chile.

Thakkar, S. K., Dubey, R. N. & Singh, J. P. (2002). Effect of inertia of embedded portion of well foundation on seismic response of bridge structure. In *12th Symposium on Earthquake Engineering*, I.I.T. Roorkee, India, pp. 1055–1061.

Tsigginos, C., Gerolymos, N., Assimaki, D., & Gazetas, G. (2008). Seismic response of bridge pier on rigid caisson foundation in soil stratum. *Journal of Earthquake Engineering and Engineering Vibration, 7*, 33–44.

Varun, Assimaki, D., & Gazetas, G. (2009). A simplified model for lateral response of large diameter Caisson foundations - linear elastic formulation. *Journal of Soil Dynamics and Earthquake Engineering, 29*, 268–291.

Waas, G., & Hartmann, H. G. (1984). Seismic analysis of pile foundations including pile soil pile interaction. In *Proceeding VIIIth World Conference in Earthquake Engineering*, San Francisco, USA.

Wolf, J. P. (1985). *Dynamic soil structure interaction in time domain*. NY, USA: Prentice Hall.

Wolf, J. P., & Von Arx, G. A. (1982). Horizontally travelling waves in a group of piles taking pile-soil pile interaction into account. *Journal of Earthquake Engineering and Structural Dynamics, 16*(2), 235–237.

Chapter 13
Dynamic Response of Bunkers and Circular Silos Under Earthquake Force

13.1 Introduction

Bunkers and silos often form an important storage facility in fossil fuel power plants, steel plants, oil and gas, cement industry for preserving raw products.

It is also extensively used in agricultural farms for stockpiling grains.

In industrial plants, they are commonly used to store coal (for fossil fuel plants), coke (in steel pants), and sulfur pellets extracted from oil and sour gas in oil and gas industry, etc.

Bunkers are normally feasible when storage requirement is large, and industrial bunkers thus deployed are often of large capacity.

Failure of such bunkers under earthquake not only could create a significant environmental hazard but can also have a major effect on the production output of the plant.

For earthquake analysis of such bunkers, the bunker and its content are usually considered as a lumped mass connected by rigid links to its supporting frame and force on the frame is then considered for design under earthquake.

No methodology exists till date to estimate the dynamic force/pressure that gets induced on bunker wall when earthquake force excites the content itself.

This local vibration could be significant for large bunkers, and in earthquake-prone zone can become a potential source of damage to its wall.

Global analysis of bunkers based on FEM is a difficult proposition, for it is inordinately complex (if not impossible) to assess the in situ elastic property and density of stored materials that are at various level of compaction, and this often keeps changing with space and time, as bunkers are frequently emptied and filled.

It is not that engineers and analysts are unaware of this additional pressure on bunker wall due to seismic force; however, the complexity of the model compels them to consider only a simplified solution (affecting the supporting frame only) and ignore the dynamic pressure on bunker walls. The lacunae in design process are attempted to be overcome by a conservative design of the wall.

I. Chowdhury and S. P. Dasgupta, *Earthquake Analysis and Design of Industrial Structures and Infra-structures*, GeoPlanet: Earth and Planetary Sciences, https://doi.org/10.1007/978-3-319-90832-8_13

We present here three mathematical models based on which the dynamic pressure, moments and shears can be estimated on bunker walls (both shallow and deep) under earthquake force. The derivations are finally reduced to design coefficients which are amenable to practical design office applications.

The derivation though has been based on IS-1893 (2016) "The Indian Code of practice for Earthquake-Resistant Design of Structures," can very well be adapted for other codes like Euro and UBC with some minor modification.

13.2 A Proposed Model to Estimate Seismic Effect on Bunker Walls (Chowdhury 2009)

A number of methods exist for estimation of static pressure induced on the wall of a bunker-like (Gray and Manning 1973),

- Rankine's method;
- Airy's method;
- Janssens's method.

Model proposed herein uses Airy's method of wedge failure as the basis for evaluation of dynamic pressure on the wall of a bunker under earthquake.

To start with, Airy's method for evaluation of static pressure is discussed briefly hereafter as it forms the basis of calculation of dynamic pressure and forces subsequently.

As shown in Fig. 13.1, static pressure induced on the bunker wall is due to formation of wedge ABC within the bunker content.

Based on static equilibrium condition of the wedge ABC, it can be shown (Gray and Manning 1973) that the horizontal force on the wall may be expressed as,

$$H = W\frac{\tan\theta - \mu}{(1-\mu\mu') + (\mu+\mu')\tan\theta} \tag{13.2.1}$$

For maximum value of H setting the expression $\frac{dH}{d\theta} = 0$, we have,

$$\tan\theta = \mu + \sqrt{\mu\frac{1+\mu^2}{\mu+\mu'}} \tag{13.2.2}$$

Here

- W weight of the wedge ABC;
- H horizontal force on the bunker wall;
- θ angle subtended by the wedge as shown in Fig. 13.1;
- μ coefficient of friction of the bunker material;
- μ' coefficient of friction between the wall and the bunker material.

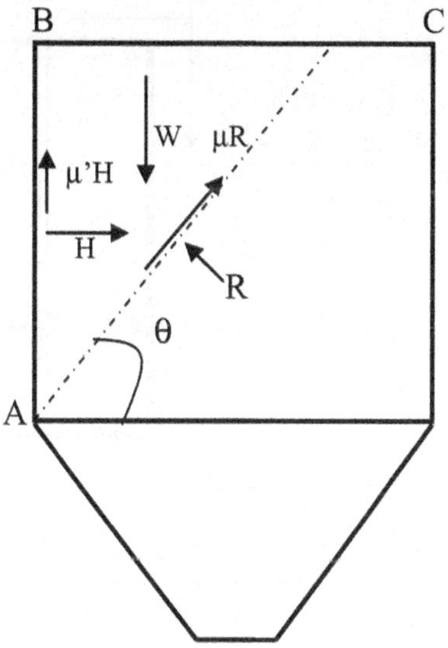

Fig. 13.1 Wedge formation in a shallow bunker by Airy's theory

Based on the above expression, it may be argued that the material within the bunker having already formed the wedge is under a state of limiting failure, having no stiffness of its own. Thus, when excited under an earthquake force, this wedge gets locked to the wall and only contributes its inertial effect to vibration. The wall will contribute to stiffness of the system including its own mass which also gets excited in the process. However, since walls of such bunkers are usually quite thin, mass contribution of these is quite low (compared to the wedge ABC) and can be neglected for all practical purposes. It is assumed that the wall contributes to stiffness of the system only.

Under static load, pressure on wall is determined by considering the slab as a plate having four sides fixed, for in many cases the top of bunker is monolithically cast with a RCC roof with access holes to cover its content.

For dynamic case, thus, the problem constitutes of vibration of a plate supporting the wedge ABC having its four sides fixed.

Shown in Fig. 13.2 is a bunker wall of height H and width b. The failed mass wedge as per Airy's theory is shown by the triangular wedge that contributes its inertial effect on the wall where γ is weight density of the material stored in the bunker, having unit kN/m^3.

Displacement function of the wall can thus be represented by,

$$w(z,y,t) = \sum_{i=1}^{n} \phi_i(z,y)q_i(t) \tag{13.2.3}$$

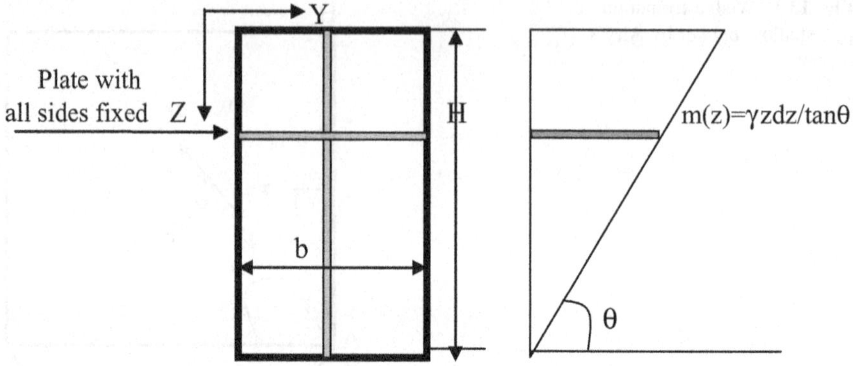

Fig. 13.2 Bunker wall (four edges fixed) with wedge mass content under earthquake

The kinetic energy of the system (Meirovitch 1975, 1993) is given by,

$$T = \frac{1}{2}\int_0^H \int_0^b m(z,y)\left[\frac{\partial w(z,y,t)}{\partial t}\right]^2 \tag{13.2.4}$$

Substituting Eq. (13.2.3) in (13.2.4), we have,

$$\begin{aligned} T &= \frac{1}{2}\int_0^H \int_0^b m(z,y)\left[\sum_{i=1}^{n} \phi_i(z,y)\dot{q}_i(t)\right]\left[\sum_{j=1}^{n} \phi_j(z,y)\dot{q}_j(t)\right] \\ &= \frac{1}{2}\sum_{i=1}^{n}\sum_{j=1}^{n} \dot{q}_i(t)\dot{q}_j(t)\left[\int_0^H \int_0^b m(z,y)\phi_i(z,y)\phi_j(z,y)\right] \end{aligned}$$

From the above, we deduce that the mass coefficient is given by,

$$m_{ij} = \left[\int_0^H \int_0^b m(z,y)\phi_i(z,y)\phi_j(z,y)\right] \mathrm{dy.dz} \quad \text{for } i,j = 1,2,3\ldots n \tag{13.2.5}$$

Strain energy equation of the plate (Timoshenko and Kreiger 1985) is given by,

$$V = \frac{1}{2}\int_0^H \int_0^b D(z,y)\left[\left\{\frac{\partial^2 w}{\partial z^2} + \frac{\partial^2 w}{\partial y^2}\right\} - 2(1-\nu)\left\{\frac{\partial^2 w}{\partial z^2}\frac{\partial^2 w}{\partial y^2} - \left(\frac{\partial^2 w}{\partial z \partial y}\right)^2\right\}\right] \tag{13.2.6}$$

Here w is expressed as per Eq. (13.2.3).

It can be proved, based on Lagrange's equation (Hurty and Rubenstein 1967), that on differentiation of Eq. (13.2.6) with respect to $q_j(t)$ the stiffness coefficient of an isotropic plate is given by,

$$K_{ij} = \int_0^H \int_0^b D \left[\begin{array}{c} \left(\frac{\partial^2 \phi_i}{\partial z^2} + \frac{\partial^2 \phi_j}{\partial y^2} \right) \left(\frac{\partial^2 \phi_i}{\partial y^2} + \frac{\partial^2 \phi_j}{\partial z^2} \right) \\ -(1-v) \left(\frac{\partial^2 \phi_i}{\partial z^2} \frac{\partial^2 \phi_j}{\partial y^2} \right) - 2 \frac{\partial^2 \phi_i}{\partial z \partial y} \frac{\partial^2 \phi_j}{\partial z \partial y} \end{array} \right] dy.dz \tag{13.2.7}$$

$$\text{Here } D = \frac{Et^3}{12(1-v^2)} \text{ is the flexural stiffness of thin plate,} \tag{13.2.8}$$

E = Young's modulus of concrete, t = thickness of the bunker wall, v = Poisson's ratio of concrete, ϕ_i, ϕ_j = mode shapes at different modes whose values have to be chosen for the appropriate boundary condition and shall be a function of both z and y.

To select the appropriate mode shape let us imagine two thin strips of element as shown in Fig. 13.2, in vertical and horizontal directions.

Let us now try to visualize how they will deflect under a dynamic loading.

It is obvious that both the strips in orthogonal directions will behave as a beam element with its both ends fixed.

Based on the above assumption, the shape function for the first mode can be considered as (Chowdhury and Dasgupta 2004),

$\phi(z, y) = F(z).F(y)$, for the first mode analysis.

$$F(z) = \sin\frac{3\pi z}{2H} - \sinh\frac{3\pi z}{H} - \alpha_z \left(\cos\frac{3\pi z}{2H} - \cosh\frac{3\pi z}{2H} \right) \tag{13.2.9}$$

$$F(y) = \sin\frac{3\pi y}{2b} - \sinh\frac{3\pi y}{2b} - \alpha_y \left(\cos\frac{3\pi y}{2b} - \cosh\frac{3\pi y}{2b} \right) \tag{13.2.10}$$

$$\alpha_z = \alpha_y = \frac{\sin\frac{3\pi}{2} - \sinh\frac{3\pi}{2}}{\cos\frac{3\pi}{2} - \cosh\frac{3\pi}{2}} \tag{13.2.11}$$

13.2.1 Calculation of Mass Coefficient

For first mode analysis, as $\phi_i(z, y) = \phi_j(z, y)$, we have,

$$m = \int_0^H \int_0^b \frac{\gamma z}{g \tan\theta} \phi^2 \mathrm{dy.dz}$$
$$= \int_0^H \int_0^b \frac{\gamma z}{g \tan\theta} F(z)^2 F(y)^2 \mathrm{dy.dz}, \tag{13.2.1.1}$$

where $F(z)$ and $F(y)$ are as defined in Eqs. (13.2.9) and (13.2.10).

For ease of analysis, we convert the integration basis from local to natural coordinates when,

$z = H\xi$ and $y = b\eta$, which implies $\mathrm{d}z = H\mathrm{d}\xi$ and $\mathrm{d}y = b\mathrm{d}\eta$.
As $z \to 0,\ \xi \to 0;\ z \to H,\ \xi \to 1$.
Similarly as $y \to 0,\ \eta \to 0;\ y \to b,\ \eta \to 1$.
Thus,

$$F(\xi) = \sin\frac{3\pi\xi}{2} - \sinh\frac{3\pi\xi}{2} - \alpha_z\left(\cos\frac{3\pi\xi}{2} - \cosh\frac{3\pi\xi)}{2}\right) \tag{13.2.1.2}$$

and

$$F(\eta) = \sin\frac{3\pi\eta}{2} - \sinh\frac{3\pi\eta}{2} - \alpha_y\left(\cos\frac{3\pi\eta}{2} - \cosh\frac{3\pi\eta}{2}\right) \tag{13.2.1.3}$$

Substituting Eqs. (13.2.1.2) and (13.2.1.3) in Eq. (13.2.1.1), we have,

$$m = \int_0^1 \int_0^1 \frac{\gamma H^2 b}{g \tan\theta} \xi F(\xi)^2 F(\eta)^2 \mathrm{d}\eta \mathrm{d}\xi \tag{13.2.1.4}$$

The above can be further simplified to,

$$m = \frac{\gamma H^2 b}{g \tan\theta} C_1, \quad \text{where} \tag{13.2.1.5}$$

$$C_1 = \int_0^1 \int_0^1 \xi F(\xi)^2 F(\eta)^2 \mathrm{d}\eta \mathrm{d}\xi \tag{13.2.1.6}$$

13.2.2 Derivation of Stiffness Coefficient

For derivation of stiffness coefficient for the first mode, as $\phi_i(z,y) = \phi_j(z,y)$, we have from Eq. (13.2.7).

$$K = \int_0^H \int_0^b D\left[\left(\frac{\partial^2\phi}{\partial z^2} + \frac{\partial^2\phi}{\partial y^2}\right)^2 - 2(1-v)\left\{\frac{\partial^2\phi}{\partial z^2}\frac{\partial^2\phi}{\partial y^2} - \left(\frac{\partial^2\phi}{\partial z\partial y}\right)^2\right\}\right] \mathrm{dy.dz} \tag{13.2.2.1}$$

As $\phi(z,y) = F(z).F(y)$, we have

$$\frac{\partial^2\phi}{\partial z^2} = \ddot{F}(z)F(y)$$

$$\frac{\partial^2\phi}{\partial y^2} = \ddot{F}(y)F(z) \quad \text{and}$$

$$\frac{\partial^2\phi}{\partial y\partial z} = \dot{F}(y)\dot{F}(z)$$

Substituting the above in Eq. (13.2.2.1) and on expansion, we have,

$$K = \int_0^H \int_0^b D\left[\begin{array}{l}\{F(y)\ddot{F}(z)\}^2 + \{F(z)\ddot{F}(y)\}^2 \\ +2v\{F(y)F(z)\ddot{F}(y)\ddot{F}(z)\} + 2(1-v)\{\dot{F}(y)\dot{F}(z)\}^2\end{array}\right] \tag{13.2.2.2}$$

where $F(z)$ and $F(y)$ are as defined in Eqs. (13.2.1.2) and (13.2.1.3).

Equation (13.2.2.2), after some simplification and transformation to natural coordinates, can be expressed as,

$$K = DHb\left[\begin{array}{l}\dfrac{81\pi^4}{16H^4}\displaystyle\int_0^1\int_0^1 \{f(\eta)f''(\xi)\}^2 \mathrm{d}\xi \mathrm{d}\eta + \dfrac{81\pi^4}{16b^4}\displaystyle\int_0^1\int_0^1 \{f(\xi)f''(\eta)\}^2 \mathrm{d}\xi \mathrm{d}\eta \\ + \dfrac{81v\pi^4}{8H^2b^2}\displaystyle\int_0^1\int_0^1 f(\eta)f(\xi)f''(\xi)f''(\eta)\mathrm{d}\xi \mathrm{d}\eta + \dfrac{81\pi^4(1-v)}{8H^2b^2}\displaystyle\int_0^1\int_0^1 \{f'(\xi)f'(\eta)\}^2 \mathrm{d}\xi \mathrm{d}\eta\end{array}\right] \tag{13.2.2.3}$$

Here relationships between $F(\xi) \Leftrightarrow f(\xi)$, $F(\eta) \Leftrightarrow f(\eta)$ and their first and second derivatives are given as,

$F(\xi) = f(\xi)$, $\dot{F}(\xi) = \frac{3\pi}{2H}f'(\xi)$, $\ddot{F}(\xi) = \frac{9\pi^2}{4H^2}f''(\xi)$ and,

$F(\eta) = f(\eta)$, $\dot{F}(\eta) = \frac{3\pi}{2b}f'(\eta)$ and $\ddot{F}(\eta) = \frac{9\pi^2}{4b^2}f''(\eta)$.
Considering,

$$C_2 = \int_0^1\int_0^1 [f(\eta)f''(\xi)]^2 d\eta d\xi \tag{13.2.2.4}$$

$$C_3 = \int_0^1\int_0^1 [f(\xi)f''(\eta)]^2 d\eta d\xi \tag{13.2.2.5}$$

$$C_4 = \int_0^1\int_0^1 [f(\eta)f(\xi)f''(\eta)f''(\xi)] d\eta d\xi \quad \text{and} \tag{13.2.2.6}$$

$$C_5 = \int_0^1\int_0^1 [f'(\eta)f'(\xi)]^2 d\eta d\xi \tag{13.2.2.7}$$

this gives

$$K = DHb\left[\begin{array}{r} \dfrac{81\pi^4}{16H^4}C_2 + \dfrac{81\pi^4}{16b^4}C_3 \\ + \dfrac{81\pi^4 v}{8b^2H^2}C_4 + \dfrac{81(1-v)\pi^4}{8H^2b^2}C_5 \end{array}\right] \tag{13.2.2.8}$$

13.2.3 Derivation of Fundamental Time Period

Considering the time period as $T = 2\pi\sqrt{\frac{m}{K}}$, substituting Eqs. (13.2.1.5) and (13.2.2.8) gives,

$$T = 2\pi\sqrt{\frac{\gamma H^2 bC_1}{DgHb\tan\theta\left[\frac{81\pi^4}{16H^4}C_2 + \frac{81\pi^4}{16b^4}C_3 + \frac{81\pi^4 v}{8H^2b^2}C_4 + \frac{81(1-v)\pi^4}{8H^2b^2}C_5\right]}} \tag{13.2.3.1}$$

Considering the value of D as given in Eq. (13.2.8) and $r = H/b$, the aspect ratio of the plate, Eq. (13.2.3.1) on some manipulation and simplification may be expressed as,

Table 13.1 Values of integration constants C_1 to C_5

Coefficients	Integral Function	Value
C_1	$\int_0^1\int_0^1 \xi F(\xi)^2 F(\eta)^2 d\eta d\xi$	0.9911026
C_2	$\int_0^1\int_0^1 [f(\eta)f''(\xi)]^2 d\eta d\xi$	0.977953
C_3	$\int_0^1\int_0^1 [f(\xi)f''(\eta)]^2 d\eta d\xi$	0.977953
C_4	$\int_0^1\int_0^1 [f(\eta)f(\xi)f''(\eta)f''(\xi)] d\eta d\xi$	0.977953
C_5	$\int_0^1\int_0^1 [f'(\eta)f'(\xi)]^2 d\eta d\xi$	0.409343

$$T = 0.693\sqrt{\frac{\gamma H^5(1-v^2)}{Et^3 g \tan\theta(X_1 r^4 + X_2 r^2 + X_3 v r^2 + X_4)}} \tag{13.2.3.2}$$

where

$$\begin{aligned} X_1 &= C_3/2C_1, \\ X_2 &= C_5/C_1, \\ X_3 &= (C_4 - C_5)/C_1, \\ X_4 &= C_2/2C_1 \end{aligned} \tag{13.2.3.3}$$

It will be observed that Eq. (13.2.3.2), which gives the fundamental time period of the plate element, is strongly affected by aspect ratio (r) and Poisson's ratio (v).

It should also be noted that this is not the fundamental time period of the overall system where the vibration of bunker frame in translation mode would guide. This is the local vibration of the bunker wall and will occur at a higher mode when wall itself gets excited under seismic force.

The integration constants C_1 to C_5 are solved numerically based on Simpson's method and are furnished in Table 13.1 for reference.

Similarly, the constants X_1, X_2, X_3, X_4, vide Eq. (13.2.3.3), are given by,
$X_1 = 0.49336$, $X_2 = 0.41302$, $X_3 = 0.57371$, $X_4 = 0.49336$

13.2.4 Calculation of Dynamic Amplitude of the Bunker Wall

As shown at the outset, the displacement is defined by,

$$w(z,y,t) = \sum_{i=1}^{n} \phi_i(z,y)q_i(t) \tag{13.2.4.1}$$

For structural systems under earthquake, the dynamic amplitude can be calculated from the expression as depicted in Chap. 2 as,

$$w(z,y) = \kappa_i \phi(z_i y_i) S_\mathrm{d} \tag{13.2.4.2}$$

where,

$$S_\mathrm{d} = \frac{S_\mathrm{a}}{\omega^2} \tag{13.2.4.3}$$

Here

S_a acceleration spectrum and is a function of time period; and
ω^2 square of the natural frequency.

Modal mass participation factor κ_i is expressed as,

$$\kappa_i = \frac{\sum_{i=1}^{n} m_i \phi_i}{\sum_{i=1}^{n} m_i \phi_i^2} = \frac{\int_0^1 \int_0^1 \xi f(\xi)^3 f(\eta)^3 \mathrm{d}\xi \mathrm{d}\eta}{\int_0^1 \int_0^1 \xi f(\xi)^4 f(\eta)^4 \mathrm{d}\xi \mathrm{d}\eta} \tag{13.2.4.4}$$

The above, on numerical integration, gives a unique value of $\kappa_i = 0.438$.
Based on codal provision, the displacement amplitude can now be expressed as,

$$S_\mathrm{d} = \frac{\beta \kappa_i S_a}{\omega^2} \tag{13.2.4.5}$$

where,

$\beta = \frac{ZI}{2R}$ the code factor.
Here

Z zone factor depending on the earthquake intensity;
I importance factor; and
R ductility factor.

It is to be noted that here β mentioned above is as per IS code and needs to be suitably modified for other codes like UBC, IBC, Euro. All other steps are generic and independent of any specific code.

Considering $\omega^2 = \frac{4\pi^2}{T^2}$, substituting Eq. (13.2.3.2), we have,

$$\omega^2 = \frac{81\pi^4 Et^3 g \tan\theta (X_1 r^4 + X_2 r^2 + \upsilon X_3 r^2 + X_4)}{96(1 - v^2)\gamma H^5} \quad (13.2.4.6)$$

$$\therefore S_d = \frac{96(1 - v^2)\beta\kappa_i S_a \gamma H^5}{81\pi^4 Et^3 g \tan\theta (X_1 r^4 + X_2 r^2 + vX_3 r^2 + X_4)} \quad (13.2.4.7)$$

Substituting Eq. (13.2.4.7) in (13.2.4.2), we have

$$\therefore \quad w(z,y) = \frac{96(1 - v^2)\kappa_i \beta S_a \gamma H^5}{81\pi^4 Et^3 g \tan\theta (X_1 r^4 + X_2 r^2 + vX_3 r^2 + X_4)} F(z)F(y) \quad (13.2.4.8)$$

Here $F(z)$ and $F(y)$ are as defined in Eqs. (13.2.9) and (13.2.10).

Considering $f(r) = X_1 r^4 + X_2 r^2 + vX_3 r^2 + X_4$, a function of aspect ratio, expression (13.2.4.8) can be modified and written as,

$$w(z,y) = \frac{96(1 - v^2)\kappa_i \beta S_a \gamma H^5}{81\pi^4 Et^3 g \tan\theta f(r)} F(z)F(y) \quad (13.2.4.9)$$

Expression (13.2.4.9) can now be expressed as,

$$w(z,y) = \frac{96(1 - v^2)\kappa_i \beta S_a}{81\pi^4 Et^3 g \tan\theta} \left[\frac{f(\xi)f(\eta)}{f(r)}\right] \gamma H^5 \quad (13.2.4.10)$$

It is thus observed that dynamic amplitude $w(z, y)$ is a function of flexural stiffness of the plate (Et^3), $f(r)$ is the aspect ratio function of the plate, and $F(z)$ and $F(y)$ are the shape functions.

The variation of $f(r)$ with respect to aspect ratio is as shown hereafter (Fig. 13.3).

Based on Eq. (13.2.4.10) shown below is displacement envelope of the plate with typical aspect ratio of $H/b = 2$ and Poisson's ratio of 0.25 (Fig. 13.4).

13.2.5 Calculation of Dynamic Moments and Shear

The bending moment of a thin plate is given by the expression,

$$M_z = D\left[\frac{\partial^2 w}{\partial z^2} + v\frac{\partial^2 w}{\partial y^2}\right] \quad (13.2.5.1)$$

$$M_y = D\left[\frac{\partial^2 w}{\partial y^2} + v\frac{\partial^2 w}{\partial z^2}\right] \quad (13.2.5.2)$$

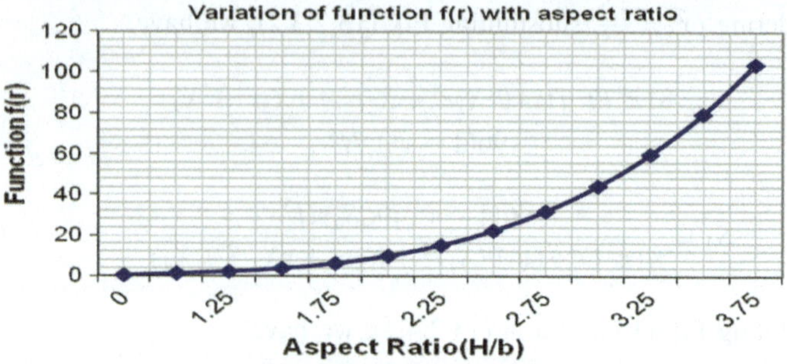

Fig. 13.3 Variation of function $f(r)$ with aspect ratio of the wall

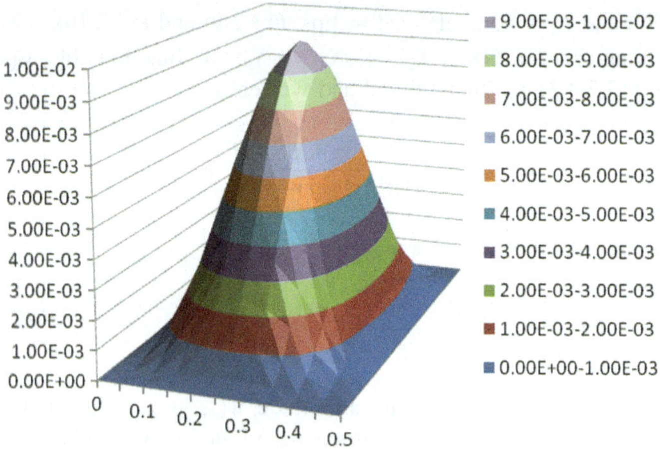

Fig. 13.4 Amplitude function of wall for aspect ratio $H/b = 2$

$$Q_z = D\frac{\partial}{\partial z}\left[\frac{\partial^2 w}{\partial z^2} + \frac{\partial^2 w}{\partial y^2}\right] \tag{13.2.5.3}$$

$$Q_y = D\frac{\partial}{\partial y}\left[\frac{\partial^2 w}{\partial z^2} + \frac{\partial^2 w}{\partial y^2}\right] \tag{13.2.5.4}$$

where $w(z, y)$ is as per Eq. (13.2.4.10).

Substituting Eq. (13.2.4.10) in (13.2.5.1), we have,

$$M_z = \frac{2\kappa_i \beta S_a}{9\pi^2 g \tan \theta f(r)}\left[f''(\xi)f(\eta) + vr^2 f(\xi)f''(\eta)\right]\gamma H^3 \tag{13.2.5.5}$$

From the above expression, we observe that moment varies as the cube of the height and is a function of the basic shape functions and their derivatives, Poisson's

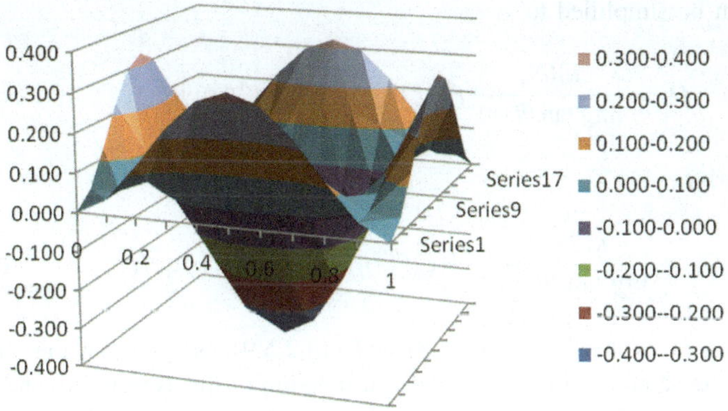

Fig. 13.5 Variation of dynamic moment in vertical (*Z*) direction for *H*/*b* = 2

ratio *v*, and aspect ratio *r*. A typical envelope for aspect ratio *H*/*b* = 2 and Poisson's ratio = 0.25 is shown in Fig. 13.5.

Similarly, for *y*-direction we have,

$$M_y = \frac{2\kappa_i \beta S_a}{9\pi^2 g \tan \theta f(r)} \left[vf''(\xi)f(\eta) + r^2 f(\xi)f''(\eta) \right] \gamma H^3 \quad (13.2.5.6)$$

We show below a typical envelope of moment (M_y) for aspect ratio *r*(*H*/*b*) = 2 and Poisson's ratio = 0.25 (Fig. 13.6).

Similarly, for the shear force in *Z*- and *Y*-direction we have,

$$Q_z = \frac{96(1-v^2)\kappa_i \beta S_a Et^3}{81\pi^4 Et^3 g \tan \theta 12(1-v^2) f(r)} \left[\frac{27\pi^3}{8H^3} f'''(\xi)f(\eta) - \frac{27\pi^3}{8b^2 H} f'(\xi)f''(\eta) \right| \gamma H^5 \quad (13.2.5.7)$$

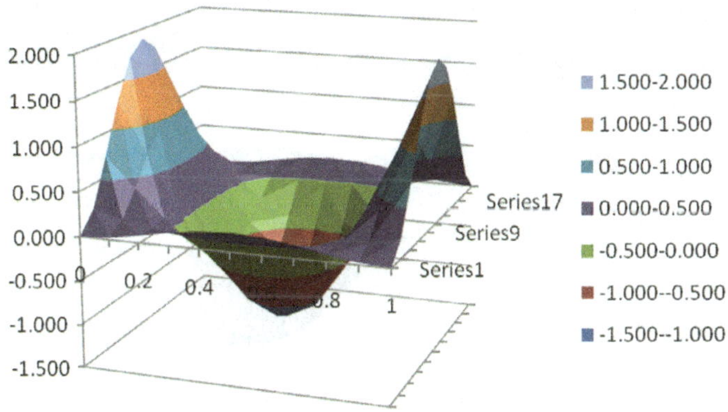

Fig. 13.6 Variation in dynamic moment in horizontal (*Y*) direction for *H*/*b* = 2

which can be simplified to,

$$Q_z = \frac{\kappa_i \beta S_a}{3\pi g \tan\theta f(r)} [f'''(\xi) f(\eta) - f'(\xi) f''(\eta)] \gamma H^2 \tag{13.2.5.8}$$

Similarly,

$$Q_y = \frac{\kappa_i \beta S_a}{3\pi g \tan\theta f(r)} \left[r^3 f(\xi) f'''(\eta) - r^2 f'(\eta) f''(\xi)\right] \gamma H^2 \tag{13.2.5.9}$$

Looking at expressions (13.2.5.8) and (13.2.5.9), we see that the shear force varies as the square of the height and is a function of the aspect ratio and the basic shape function and their derivatives.

We show below a typical variation of the dynamic shear force in *Z*- and *Y*-direction for a typical case of aspect ratio of $H/b = 2$ (Figs. 13.7 and 13.8).

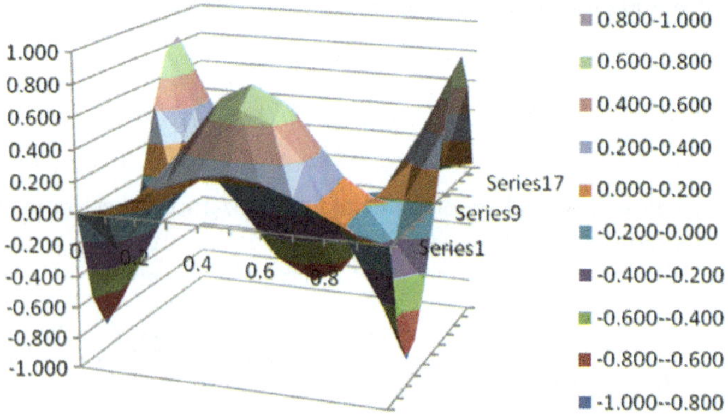

Fig. 13.7 Variation in dynamic shear in vertical (*Z*) direction for $H/b = 2$

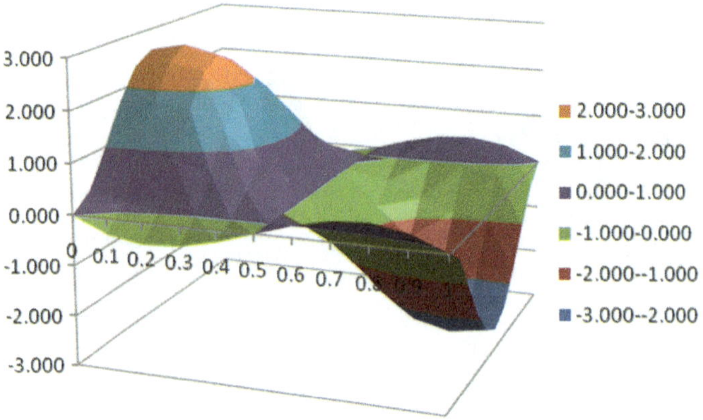

Fig. 13.8 Variation in dynamic shear in horizontal (*Y*) direction for $H/b = 2$

Dynamic pressure on wall

Sometimes it is necessary to do a local FEM analysis of the wall when one wants to transfer the edge loads to the top slab or otherwise. In such cases, it is important to know the dynamic pressure which, when multiplied by the influence area surrounding a particular node, gives an equivalent nodal force at each of the nodes for which a FEM analysis may be carried out.

The dynamic pressure induced on the wall may be calculated as,

$$p = D\left[\frac{\partial^4 w}{\partial z^4} + 2\frac{\partial^4 w}{\partial z^2 \partial y^2} + \frac{\partial^4 w}{\partial y^4}\right] \quad (13.2.5.10)$$

Substitution of the shape functions and their derivatives finally gives,

$$p = \frac{\kappa_i \beta S_a}{24 g \tan\theta f(r)}\left[f''''(\xi)f(\eta) + 2r^2 f''(\xi)f''(\eta) + r^4 f(\xi)f''''(\eta)\right]\gamma H \quad (13.2.5.11)$$

A typical dynamic pressure profile of the wall of aspect ratio $H/b = 2$ under earthquake force is shown hereafter (Fig. 13.9).

13.2.6 *Design Coefficients for Dynamic Moments and Shear*

Referring to Eq. (13.2.5.5), the dynamic moment coefficient can be expressed as,

$$M_z = \frac{2\kappa_i \beta S_a}{9\pi^2 g \tan\theta f(r)}\left[f''(\xi)f(\eta) + \nu r^2 f(\xi)f''(\eta)\right]\gamma H^3 \quad (13.2.6.1)$$

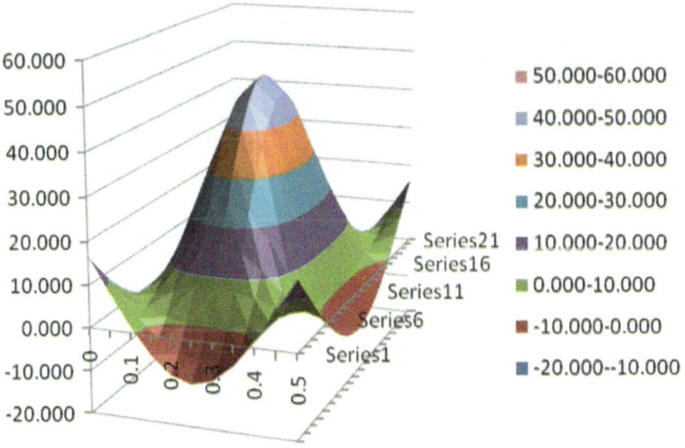

Fig. 13.9 Variation of dynamic pressure on bunker wall under earthquake force

This can be simplified to

$$\text{or} \quad M_z = \text{Coeff}_{mz} \frac{2\kappa_i \beta S_a}{9\pi^2 g \tan\theta} \gamma H^3 \tag{13.2.6.2}$$

where

$$\text{Coeff}_{mz} = \left[\frac{f''(\xi)f(\eta) + vr^2 f(\xi)f''(\eta)}{f(r)} \right] \tag{13.2.6.3}$$

Typical coefficients for aspect ratio from 0.5 to 2.0 and Poisson's ratio = 0.25 are given in Table 13.2.

Similarly, for horizontal y-direction,

$$M_y = \text{Coeff}_{my} \frac{2\kappa_i \beta S_a}{9\pi^2 g \tan\theta} \gamma H^3 \tag{13.2.6.4}$$

$$\text{Coeff}_{my} = \left[\frac{vf''(\xi)f(\eta) + r^2 f(\xi)f''(\eta)}{f(r)} \right] \tag{13.2.6.5}$$

Typical coefficients for aspect ratio 0.5 to 2.0 and Poisson's ratio = 0.25 are given in Table 13.3.

For the dynamic shear, the expression is,

$$Q_z = \frac{\kappa_i \beta S_a}{3\pi g \tan\theta f(r)} [f'''(\xi)f(\eta) - f'(\xi)f''(\eta)] \gamma H^2, \tag{13.2.6.6}$$

$$\rightarrow Q_z = \text{Coeff}_{Qz} \frac{\kappa_i \beta S_a}{3\pi g \tan\theta} \gamma H^2 \tag{13.2.6.7}$$

where

$$\text{Coeff}_{Qz} = \left[\frac{f'''(\xi)f(\eta) - f'(\xi)f''(\eta)}{f(r)} \right] \tag{13.2.6.8}$$

Typical coefficients for aspect ratio 0.5 to 2.0 and Poisson's ratio = 0.25 are given in Table 13.4.

For horizontal direction,

$$Q_y = \frac{\kappa_i \beta S_a}{3\pi g \tan\theta f(r)} \left[r^3 f(\xi)f'''(\eta) - r^2 f'(\eta)f''(\xi) \right] \gamma H^2 \tag{13.2.6.9}$$

Table 13.2 Design moment coefficients in vertical direction for bunker wall

$\eta \rightarrow$ $\xi \downarrow$	0.0	0.1	0.2	0.3	0.4	0.5	0.6	0.7	0.8	0.9	1.0
***H/b* = 0.5**											
0.0	0.000	0.587	1.923	3.408	4.534	4.960	4.564	3.459	1.980	0.627	0.000
0.1	0.037	0.336	1.040	1.827	2.424	2.651	2.440	1.854	1.070	0.357	0.036
0.2	0.120	0.124	0.206	0.311	0.395	0.427	0.397	0.315	0.210	0.126	0.118
0.3	0.213	−0.043	−0.495	−0.972	−1.327	−1.461	−1.337	−0.988	−0.513	−0.057	0.209
0.4	0.283	−0.151	−0.967	−1.840	−2.494	−2.740	−2.511	−1.870	−1.001	−0.176	0.278
0.5	0.310	−0.190	−1.138	−2.155	−2.916	−3.203	−2.936	−2.189	−1.177	−0.219	0.305
0.6	0.285	−0.154	−0.979	−1.863	−2.524	−2.773	−2.541	−1.893	−1.013	−0.180	0.280
0.7	0.216	−0.048	−0.517	−1.013	−1.383	−1.522	−1.392	−1.030	−0.536	−0.063	0.212
0.8	0.124	0.117	0.177	0.258	0.323	0.348	0.325	0.261	0.180	0.118	0.122
0.9	0.039	0.327	1.007	1.767	2.345	2.564	2.361	1.794	1.036	0.347	0.038
1.0	0.000	0.576	1.890	3.348	4.455	4.874	4.484	3.399	1.945	0.616	0.000
***H/b* = 1.0**											
0.0	0.000	0.278	0.911	1.615	2.149	2.351	2.163	1.639	0.938	0.297	0.000
0.1	0.070	0.187	0.498	0.852	1.122	1.224	1.129	0.864	0.511	0.196	0.068
0.2	0.228	0.151	0.115	0.102	0.099	0.098	0.099	0.101	0.114	0.149	0.224
0.3	0.404	0.143	−0.204	−0.542	−0.786	−0.876	−0.792	−0.553	−0.217	0.131	0.397
0.4	0.537	0.146	−0.418	−0.980	−1.390	−1.543	−1.401	−0.999	−0.440	0.127	0.528
0.5	0.588	0.148	−0.495	−1.139	−1.610	−1.786	−1.623	−1.161	−0.520	0.126	0.577
0.6	0.541	0.146	−0.423	−0.992	−1.406	−1.561	−1.417	−1.011	−0.445	0.126	0.531
0.7	0.410	0.143	−0.214	−0.563	−0.814	−0.908	−0.821	−0.574	−0.228	0.131	0.403
0.8	0.235	0.150	0.101	0.075	0.062	0.058	0.062	0.074	0.100	0.148	0.230
0.9	0.074	0.185	0.483	0.823	1.083	1.181	1.089	0.834	0.496	0.194	0.073
1.0	0.000	0.273	0.896	1.587	2.111	2.309	2.125	1.610	0.922	0.292	0.000

(continued)

Table 13.2 (continued)

$\eta \rightarrow$ $\xi \downarrow$	0.0	0.1	0.2	0.3	0.4	0.5	0.6	0.7	0.8	0.9	1.0
***H/b* = 1.5**											
0.0	0.000	0.099	0.325	0.577	0.767	0.839	0.772	0.585	0.335	0.106	0.000
0.1	0.056	0.084	0.181	0.296	0.384	0.418	0.387	0.300	0.185	0.086	0.055
0.2	0.183	0.109	0.051	0.009	−0.017	−0.027	−0.018	0.008	0.049	0.106	0.180
0.3	0.324	0.148	−0.055	−0.242	−0.374	−0.422	−0.377	−0.248	−0.062	0.141	0.319
0.4	0.432	0.181	−0.125	−0.414	−0.620	−0.697	−0.626	−0.424	−0.137	0.170	0.424
0.5	0.472	0.194	−0.150	−0.477	−0.711	−0.797	−0.717	−0.488	−0.163	0.182	0.464
0.6	0.434	0.182	−0.127	−0.419	−0.627	−0.704	−0.632	−0.428	−0.138	0.171	0.427
0.7	0.329	0.150	−0.058	−0.250	−0.385	−0.435	−0.389	−0.256	−0.066	0.142	0.323
0.8	0.188	0.110	0.047	−0.001	−0.032	−0.043	−0.033	−0.003	0.045	0.107	0.185
0.9	0.060	0.084	0.176	0.285	0.369	0.402	0.372	0.289	0.180	0.086	0.059
1.0	0.000	0.098	0.320	0.567	0.754	0.825	0.759	0.575	0.329	0.104	0.000
***H/b* = 2.0**											
0.0	0.000	0.039	0.127	0.225	0.300	0.328	0.302	0.229	0.131	0.041	0.000
0.1	0.039	0.042	0.072	0.111	0.141	0.153	0.142	0.112	0.074	0.043	0.038
0.2	0.127	0.072	0.026	−0.011	−0.036	−0.044	−0.036	−0.013	0.024	0.070	0.125
0.3	0.225	0.111	−0.011	−0.121	−0.197	−0.225	−0.199	−0.124	−0.016	0.106	0.221
0.4	0.300	0.141	−0.036	−0.197	−0.310	−0.352	−0.313	−0.202	−0.042	0.135	0.294
0.5	0.328	0.153	−0.044	−0.225	−0.352	−0.398	−0.355	−0.231	−0.052	0.146	0.322
0.6	0.302	0.142	−0.036	−0.199	−0.313	−0.355	−0.316	−0.204	−0.043	0.136	0.296
0.7	0.229	0.112	−0.013	−0.124	−0.202	−0.231	−0.204	−0.128	−0.017	0.108	0.225
0.8	0.131	0.074	0.024	−0.016	−0.042	−0.052	−0.043	−0.017	0.022	0.072	0.128
0.9	0.041	0.043	0.070	0.106	0.135	0.146	0.136	0.108	0.072	0.043	0.041
1.0	0.000	0.038	0.125	0.221	0.294	0.322	0.296	0.225	0.128	0.041	0.000

Table 13.3 Design moment coefficients in horizontal direction for bunker wall

$\eta \rightarrow$ $\xi \downarrow$	0.0	0.1	0.2	0.3	0.4	0.5	0.6	0.7	0.8	0.9	1.0
***H/b* = 0.5**											
0.0	0.000	0.147	0.481	0.852	1.133	1.240	1.141	0.865	0.495	0.157	0.000
0.1	0.147	0.158	0.274	0.420	0.535	0.579	0.538	0.425	0.279	0.161	0.144
0.2	0.481	0.274	0.097	−0.043	−0.135	−0.167	−0.137	−0.048	0.091	0.267	0.472
0.3	0.852	0.420	−0.043	−0.457	−0.745	−0.851	−0.753	−0.471	−0.060	0.402	0.837
0.4	1.133	0.535	−0.135	−0.745	−1.174	−1.331	−1.185	−0.765	−0.159	0.510	1.114
0.5	1.240	0.579	−0.167	−0.851	−1.331	−1.507	−1.343	−0.873	−0.195	0.552	1.218
0.6	1.141	0.538	−0.137	−0.753	−1.185	−1.343	−1.196	−0.773	−0.162	0.513	1.121
0.7	0.865	0.425	−0.048	−0.471	−0.765	−0.873	−0.773	−0.485	−0.065	0.407	0.850
0.8	0.495	0.279	0.091	−0.060	−0.159	−0.195	−0.162	−0.065	0.084	0.272	0.486
0.9	0.157	0.161	0.267	0.402	0.510	0.552	0.513	0.407	0.272	0.163	0.154
1.0	0.000	0.144	0.472	0.837	1.114	1.218	1.121	0.850	0.486	0.154	0.000
***H/b* = 1.0**											
0.0	0.000	0.070	0.228	0.404	0.537	0.588	0.541	0.410	0.235	0.074	0.000
0.1	0.278	0.187	0.151	0.143	0.146	0.148	0.146	0.143	0.150	0.185	0.273
0.2	0.911	0.498	0.115	−0.204	−0.418	−0.495	−0.423	−0.214	0.101	0.483	0.896
0.3	1.615	0.852	0.102	−0.542	−0.980	−1.139	−0.992	−0.563	0.075	0.823	1.587
0.4	2.149	1.122	0.099	−0.786	−1.390	−1.610	−1.406	−0.814	0.062	1.083	2.111
0.5	2.351	1.224	0.098	−0.876	−1.543	−1.786	−1.561	−0.908	0.058	1.181	2.309
0.6	2.163	1.129	0.099	−0.792	−1.401	−1.623	−1.417	−0.821	0.062	1.089	2.125
0.7	1.639	0.864	0.101	−0.553	−0.999	−1.161	−1.011	−0.574	0.074	0.834	1.610
0.8	0.938	0.511	0.114	−0.217	−0.440	−0.520	−0.445	−0.228	0.100	0.496	0.922
0.9	0.297	0.196	0.149	0.131	0.127	0.126	0.126	0.131	0.148	0.194	0.292
1.0	0.000	0.068	0.224	0.397	0.528	0.577	0.531	0.403	0.230	0.073	0.000

(continued)

Table 13.3 (continued)

$\eta \rightarrow$ $\xi \downarrow$	0.0	0.1	0.2	0.3	0.4	0.5	0.6	0.7	0.8	0.9	1.0
***H/b* = 1.5**											
0.0	0.000	0.025	0.081	0.144	0.192	0.210	0.193	0.146	0.084	0.027	0.000
0.1	0.223	0.134	0.066	0.018	−0.012	−0.023	−0.013	0.016	0.064	0.131	0.219
0.2	0.732	0.397	0.082	−0.182	−0.360	−0.424	−0.364	−0.190	0.071	0.385	0.719
0.3	1.297	0.692	0.109	−0.387	−0.723	−0.845	−0.732	−0.403	0.088	0.670	1.275
0.4	1.726	0.917	0.132	−0.538	−0.993	−1.158	−1.005	−0.560	0.104	0.887	1.696
0.5	1.888	1.003	0.141	−0.595	−1.094	−1.275	−1.107	−0.618	0.110	0.969	1.855
0.6	1.737	0.923	0.133	−0.542	−1.000	−1.166	−1.012	−0.564	0.104	0.893	1.707
0.7	1.317	0.703	0.110	−0.394	−0.735	−0.859	−0.744	−0.410	0.089	0.680	1.294
0.8	0.754	0.408	0.083	−0.190	−0.374	−0.440	−0.378	−0.199	0.071	0.395	0.740
0.9	0.239	0.141	0.066	0.011	−0.023	−0.036	−0.024	0.009	0.064	0.138	0.234
1.0	0.000	0.024	0.080	0.142	0.188	0.206	0.190	0.144	0.082	0.026	0.000
***H/b* = 2.0**											
0.0	0.000	0.010	0.032	0.056	0.075	0.082	0.075	0.057	0.033	0.010	0.000
0.1	0.155	0.089	0.033	−0.011	−0.040	−0.050	−0.041	−0.013	0.031	0.086	0.152
0.2	0.508	0.275	0.054	−0.131	−0.256	−0.301	−0.259	−0.137	0.047	0.266	0.499
0.3	0.901	0.483	0.082	−0.257	−0.486	−0.569	−0.492	−0.268	0.068	0.467	0.885
0.4	1.198	0.641	0.104	−0.351	−0.659	−0.771	−0.667	−0.365	0.085	0.620	1.177
0.5	1.311	0.701	0.113	−0.386	−0.724	−0.847	−0.733	−0.402	0.092	0.678	1.288
0.6	1.206	0.645	0.105	−0.353	−0.664	−0.776	−0.672	−0.368	0.086	0.624	1.185
0.7	0.914	0.490	0.083	−0.261	−0.494	−0.579	−0.500	−0.272	0.069	0.474	0.898
0.8	0.523	0.283	0.055	−0.136	−0.264	−0.311	−0.268	−0.142	0.047	0.274	0.514
0.9	0.166	0.094	0.033	−0.015	−0.047	−0.058	−0.047	−0.017	0.031	0.092	0.163
1.0	0.000	0.010	0.031	0.055	0.074	0.081	0.074	0.056	0.032	0.010	0.000

Table 13.4 Design shear coefficients in vertical direction for bunker wall

$\eta \rightarrow$ / $\xi \downarrow$	0.0	0.1	0.2	0.3	0.4	0.5	0.6	0.7	0.8	0.9	1.0
***H/b* = 0.5**											
0.0	0.000	0.576	1.890	3.348	4.455	4.874	4.484	3.398	1.945	0.616	0.000
0.1	−0.566	0.262	1.803	3.448	4.678	5.141	4.710	3.504	1.866	0.311	−0.556
0.2	−0.799	0.083	1.603	3.197	4.381	4.826	4.413	3.251	1.664	0.132	−0.785
0.3	−0.731	0.001	1.221	2.491	3.431	3.784	3.456	2.534	1.270	0.040	−0.718
0.4	−0.433	−0.016	0.668	1.378	1.902	2.099	1.916	1.402	0.696	0.006	−0.425
0.5	−0.008	0.000	0.013	0.026	0.036	0.040	0.036	0.027	0.013	0.000	−0.008
0.6	0.419	0.016	−0.645	−1.330	−1.837	−2.027	−1.850	−1.353	−0.672	−0.005	0.411
0.7	0.724	0.001	−1.203	−2.455	−3.383	−3.731	−3.407	−2.498	−1.252	−0.038	0.711
0.8	0.801	−0.078	−1.592	−3.179	−4.357	−4.800	−4.389	−3.232	−1.653	−0.127	0.787
0.9	0.581	−0.253	−1.798	−3.445	−4.677	−5.141	−4.710	−3.502	−1.861	−0.302	0.570
1.0	0.028	−0.561	−1.887	−3.355	−4.468	−4.890	−4.498	−3.406	−1.942	−0.601	0.027
***H/b* = 1.0**											
0.0	0.000	0.273	0.896	1.587	2.111	2.309	2.125	1.610	0.922	0.292	0.000
0.1	−1.073	−0.309	0.773	1.850	2.633	2.925	2.654	1.886	0.816	−0.272	−1.054
0.2	−1.514	−0.572	0.645	1.820	2.664	2.977	2.686	1.859	0.692	−0.530	−1.487
0.3	−1.386	−0.560	0.474	1.459	2.164	2.425	2.182	1.492	0.513	−0.523	−1.362
0.4	−0.821	−0.339	0.255	0.818	1.220	1.369	1.231	0.837	0.277	−0.318	−0.806
0.5	−0.016	−0.007	0.005	0.016	0.023	0.026	0.023	0.016	0.005	−0.006	−0.016
0.6	0.794	0.328	−0.246	−0.790	−1.179	−1.322	−1.189	−0.808	−0.267	0.308	0.780
0.7	1.372	0.555	−0.467	−1.440	−2.136	−2.393	−2.154	−1.472	−0.505	0.519	1.348
0.8	1.519	0.577	−0.639	−1.812	−2.655	−2.967	−2.677	−1.851	−0.686	0.534	1.492
0.9	1.101	0.325	−0.769	−1.854	−2.644	−2.938	−2.664	−1.891	−0.812	0.288	1.081
1.0	0.052	−0.245	−0.890	−1.600	−2.138	−2.341	−2.152	−1.625	−0.917	−0.264	0.051

(continued)

Table 13.4 (continued)

$\eta \rightarrow$ $\xi \downarrow$	0.0	0.1	0.2	0.3	0.4	0.5	0.6	0.7	0.8	0.9	1.0
***H/b* = 1.5**											
0.0	0.000	0.098	0.320	0.567	0.754	0.825	0.759	0.575	0.329	0.104	0.000
0.1	−0.862	−0.368	0.228	0.789	1.188	1.335	1.198	0.808	0.250	−0.347	−0.847
0.2	−1.216	−0.568	0.162	0.831	1.301	1.474	1.313	0.853	0.189	−0.542	−1.195
0.3	−1.113	−0.533	0.107	0.687	1.093	1.242	1.103	0.706	0.130	−0.509	−1.094
0.4	−0.659	−0.319	0.054	0.390	0.625	0.711	0.631	0.401	0.068	−0.305	−0.648
0.5	−0.013	−0.006	0.001	0.007	0.012	0.014	0.012	0.008	0.001	−0.006	−0.012
0.6	0.638	0.308	−0.052	−0.377	−0.604	−0.687	−0.610	−0.388	−0.065	0.295	0.626
0.7	1.102	0.528	−0.105	−0.678	−1.079	−1.227	−1.090	−0.697	−0.128	0.505	1.083
0.8	1.220	0.571	−0.160	−0.829	−1.299	−1.471	−1.311	−0.851	−0.187	0.544	1.199
0.9	0.884	0.381	−0.225	−0.794	−1.198	−1.348	−1.209	−0.813	−0.248	0.359	0.869
1.0	0.042	−0.075	−0.315	−0.578	−0.775	−0.850	−0.781	−0.587	−0.325	−0.082	0.041
***H/b* = 2.0**											
0.0	0.000	0.038	0.125	0.221	0.294	0.322	0.296	0.225	0.128	0.041	0.000
0.1	−0.598	−0.285	0.063	0.378	0.599	0.681	0.605	0.389	0.075	−0.272	−0.588
0.2	−0.844	−0.421	0.026	0.424	0.699	0.800	0.706	0.437	0.042	−0.404	−0.829
0.3	−0.773	−0.390	0.008	0.359	0.602	0.690	0.608	0.370	0.022	−0.375	−0.759
0.4	−0.458	−0.232	0.001	0.206	0.348	0.400	0.352	0.213	0.009	−0.224	−0.450
0.5	−0.009	−0.004	0.000	0.004	0.007	0.008	0.007	0.004	0.000	−0.004	−0.009
0.6	0.443	0.225	−0.001	−0.199	−0.336	−0.386	−0.340	−0.206	−0.009	0.216	0.435
0.7	0.765	0.386	−0.007	−0.355	−0.595	−0.682	−0.601	−0.366	−0.021	0.372	0.752
0.8	0.847	0.423	−0.025	−0.423	−0.699	−0.800	−0.706	−0.436	−0.041	0.406	0.832
0.9	0.614	0.293	−0.061	−0.382	−0.607	−0.689	−0.613	−0.393	−0.074	0.280	0.603
1.0	0.029	−0.022	−0.122	−0.229	−0.309	−0.340	−0.311	−0.233	−0.126	−0.025	0.029

which can be simplified to,

$$Q_y = \text{Coeff}_{Qy} \frac{\kappa_i \beta S_a}{3\pi g \tan\theta} \gamma H^2 \tag{13.2.6.10}$$

where

$$\text{Coeff}_{Qy} = \left[\frac{r^3 f(\xi) f'''(\eta) - r^2 f'(\xi) f''(\eta)}{f(r)} \right] \tag{13.2.6.11}$$

Typical coefficients for aspect ratio 0.5 to 2.0 and Poisson's ratio = 0.25 are given in Table 13.5.

The above type of design coefficients can be developed for various aspect ratios of the wall and can be effectively used for rapid evaluation of design moments and shears of such bunkers subjected to dynamic pressure under earthquake force.

13.2.7 Dynamic Response of Bunkers with Walls that Is Deep

As shown in Fig. 13.10, there could be cases when the bunker wall is so deep that the failure wedge intersects the opposite wall.

As shown in Fig. 13.10, the angle θ is given by (Grey and Manning 1973).

$$\tan\theta = \sqrt{\frac{2H}{b}\frac{1+\mu^2}{\mu+\mu'} + \frac{1+\mu^2}{\mu+\mu'}\frac{1-\mu\mu'}{\mu+\mu'}} - \frac{1-\mu\mu'}{\mu+\mu'} \tag{13.2.7.1}$$

In this case, the wedge constitutes of a triangle and a rectangle whose inertial mass would contribute to the dynamic response.

To bring it within the framework of formulation, as done above, the area of failed wedge is converted to an equivalent triangle of same height (H) but different base dimension b' and angle θ' so that the net mass content remains equal.

The area of the failed wedge is given by,

$$\text{Area} = b\left[H - \frac{b}{2}\tan\theta\right] \tag{13.2.7.2}$$

Considering an equivalent triangle of height h and base b', we have,

$$\frac{1}{2}b'h = b\left[H - \frac{b}{2}\tan\theta\right] \Rightarrow b' = 2b\left[1 - \frac{b}{2H}\tan\theta\right] \tag{13.2.7.3}$$

Table 13.5 Design shear coefficients in horizontal direction for bunker wall

$\eta \rightarrow$ / $\xi \downarrow$	0.0	0.1	0.2	0.3	0.4	0.5	0.6	0.7	0.8	0.9	1.0
***H/b* = 0.5**											
0.0	0.000	−1.132	−1.597	−1.462	−0.866	−0.017	0.837	1.447	1.603	1.161	0.055
0.1	0.072	−0.539	−0.797	−0.739	−0.440	−0.008	0.425	0.731	0.800	0.555	−0.042
0.2	0.236	0.118	0.049	0.014	0.002	0.000	−0.001	−0.014	−0.048	−0.115	−0.231
0.3	0.419	0.716	0.802	0.679	0.390	0.007	−0.377	−0.671	−0.801	−0.723	−0.433
0.4	0.557	1.134	1.323	1.138	0.658	0.013	−0.636	−1.125	−1.322	−1.148	−0.585
0.5	0.609	1.288	1.513	1.306	0.756	0.015	−0.731	−1.291	−1.513	−1.304	−0.643
0.6	0.561	1.145	1.336	1.150	0.665	0.013	−0.643	−1.137	−1.336	−1.159	−0.589
0.7	0.425	0.735	0.826	0.701	0.403	0.008	−0.389	−0.693	−0.825	−0.743	−0.440
0.8	0.243	0.143	0.080	0.042	0.018	0.000	−0.017	−0.041	−0.078	−0.140	−0.238
0.9	0.077	−0.515	−0.765	−0.710	−0.423	−0.008	0.409	0.703	0.768	0.530	−0.048
1.0	0.000	−1.112	−1.569	−1.437	−0.851	−0.016	0.823	1.422	1.575	1.141	0.054
***H/b* = 1.0**											
0.0	0.000	−1.073	−1.514	−1.386	−0.821	−0.016	0.794	1.372	1.519	1.101	0.052
0.1	0.273	−0.309	−0.572	−0.560	−0.339	−0.007	0.328	0.555	0.577	0.325	−0.245
0.2	0.896	0.773	0.645	0.474	0.255	0.005	−0.246	−0.467	−0.639	−0.769	−0.890
0.3	1.587	1.850	1.820	1.459	0.818	0.016	−0.790	−1.440	−1.812	−1.854	−1.600
0.4	2.111	2.633	2.664	2.164	1.220	0.023	−1.179	−2.136	−2.655	−2.644	−2.138
0.5	2.309	2.925	2.977	2.425	1.369	0.026	−1.322	−2.393	−2.967	−2.938	−2.341
0.6	2.125	2.654	2.686	2.182	1.231	0.023	−1.189	−2.154	−2.677	−2.664	−2.152
0.7	1.610	1.886	1.859	1.492	0.837	0.016	−0.808	−1.472	−1.851	−1.891	−1.625
0.8	0.922	0.816	0.692	0.513	0.277	0.005	−0.267	−0.505	−0.686	−0.812	−0.917
0.9	0.292	−0.272	−0.530	−0.523	−0.318	−0.006	0.308	0.519	0.534	0.288	−0.264
1.0	0.000	−1.054	−1.487	−1.362	−0.806	−0.016	0.780	1.348	1.492	1.081	0.051

(continued)

Table 13.5 (continued)

$\eta \rightarrow$ $\xi \downarrow$	0.0	0.1	0.2	0.3	0.4	0.5	0.6	0.7	0.8	0.9	1.0
***H/b* = 1.5**											
0.0	0.000	−0.574	−0.811	−0.742	−0.440	−0.008	0.425	0.735	0.813	0.589	0.028
0.1	0.329	0.014	−0.144	−0.174	−0.113	−0.002	0.109	0.174	0.147	−0.006	−0.314
0.2	1.079	1.004	0.879	0.665	0.362	0.007	−0.350	−0.655	−0.873	−1.001	−1.076
0.3	1.912	2.036	1.921	1.509	0.838	0.016	−0.809	−1.488	−1.911	−2.037	−1.919
0.4	2.544	2.801	2.686	2.127	1.186	0.023	−1.145	−2.098	−2.672	−2.804	−2.558
0.5	2.783	3.088	2.972	2.358	1.316	0.025	−1.270	−2.326	−2.958	−3.092	−2.800
0.6	2.561	2.821	2.706	2.144	1.195	0.023	−1.154	−2.114	−2.693	−2.825	−2.575
0.7	1.941	2.071	1.956	1.538	0.854	0.016	−0.825	−1.517	−1.946	−2.072	−1.948
0.8	1.111	1.044	0.920	0.698	0.381	0.007	−0.368	−0.687	−0.914	−1.041	−1.108
0.9	0.352	0.046	−0.110	−0.146	−0.097	−0.002	0.094	0.146	0.113	−0.038	−0.337
1.0	0.000	−0.564	−0.797	−0.729	−0.432	−0.008	0.418	0.722	0.799	0.579	0.027
***H/b* = 2.0**											
0.0	0.000	−0.299	−0.422	−0.386	−0.229	−0.004	0.221	0.382	0.424	0.307	0.015
0.1	0.305	0.139	0.044	0.001	−0.009	0.000	0.008	0.000	−0.041	−0.134	−0.297
0.2	0.999	0.953	0.847	0.646	0.353	0.007	−0.341	−0.636	−0.841	−0.950	−0.997
0.3	1.770	1.822	1.690	1.317	0.728	0.014	−0.703	−1.298	−1.680	−1.821	−1.773
0.4	2.355	2.472	2.316	1.814	1.006	0.019	−0.971	−1.788	−2.303	−2.472	−2.362
0.5	2.576	2.717	2.551	2.000	1.110	0.021	−1.071	−1.972	−2.537	−2.717	−2.584
0.6	2.370	2.490	2.332	1.827	1.013	0.019	−0.978	−1.801	−2.320	−2.489	−2.377
0.7	1.796	1.852	1.718	1.339	0.741	0.014	−0.715	−1.320	−1.709	−1.851	−1.800
0.8	1.028	0.986	0.880	0.671	0.368	0.007	−0.355	−0.662	−0.874	−0.984	−1.027
0.9	0.325	0.164	0.070	0.021	0.003	0.000	−0.003	−0.020	−0.067	−0.160	−0.318
1.0	0.000	−0.294	−0.415	−0.380	−0.225	−0.004	0.217	0.376	0.416	0.302	0.014

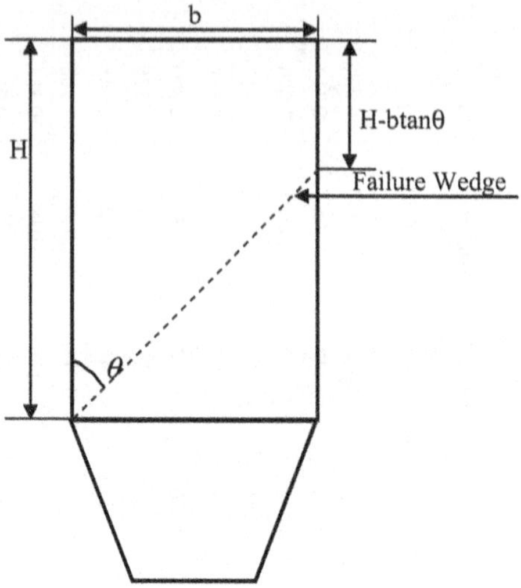

Fig. 13.10 Failure wedge profile for deep bunker

Considering $\frac{H}{b'} = \tan\,\theta'$, we have,

$$\tan\,\theta' = \frac{H}{2b\left[1 - \frac{b}{2H}\tan\,\theta\right]} \tag{13.2.7.4}$$

Thus, the mass coefficient for the equivalent triangular wedge is given by,

$$m = \int_0^1\int_0^1 \frac{\gamma H^2 b}{g\,\tan\,\theta'}\xi F(\xi)^2 F(\eta)^2 \mathrm{d}\eta\mathrm{d}\xi \tag{13.2.7.5}$$

where $\xi, F(\xi)$, and $F(\eta)$ are as explained earlier.

Since the wall stiffness remains unchanged, the stiffness value as furnished in Eq. (13.2.2.8) and rest of the steps would remain the same.

The above procedure gives a comprehensive analytic solution for dynamic forces on the bunker walls having all sides fixed subjected to seismic force where the moments, shears, and amplitude, etc., are dependent on the time period, geometry of the wall, its boundary condition as well as the material and engineering property of the bunker content.

It also provides finally design coefficients which can be used for quick computation of the design dynamic moments and shears which need to be added to the static moment to arrive at design moments and shears for which the bunker wall need to be locally checked under earthquake force.

13.2.8 Some Comments on the Above Method

Method as cited in Sect. 13.2 effectively computes local vibration of the bunker slab that induces additional dynamic pressure on the wall over and above its static pressure. In this case, however, the effect of the supporting frame is ignored. Thus, the method can effectively be used in those cases only, when the bunker is located near or on ground or the supporting frame/skirt is of short height compared to the bunker height, as shown in Fig. 13.11a, b.

When bunkers are supported at a significant height above ground (e.g., bunker bays in powerhouse) on multistoried frames, additional interaction with frame takes place which further amplifies the dynamic pressure and shall be dealt subsequently.

Example 13.1 A bunker made out of concrete wall has internal dimension 4 m 4 m with wall thickness of 300 mm. It carries coal of density 10.5 kN/m^3. The bunker is located in Gujarat (India) having zonal attribute as *V*. Considering grade of concrete as M25, determine the dynamic moments and shear in the bunker wall. Friction angle of coal = 0.510. Consider friction between wall and the fill material as also 0.510 (Fig. 13.12).

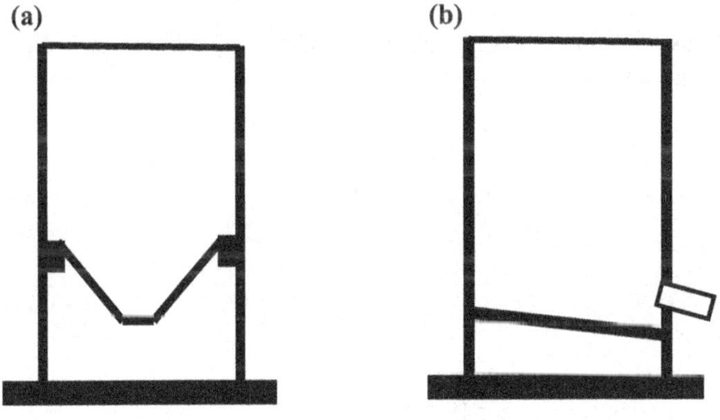

Fig. 13.11 Bunkers supported at low height above ground level

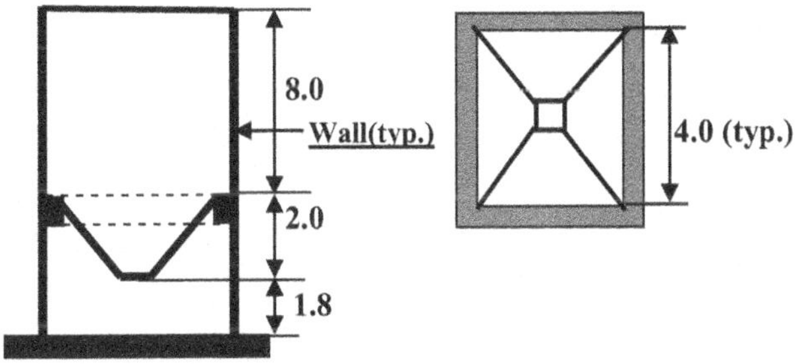

Fig. 13.12 Geometric dimension of the bunker

Solution

Height of wall = 8.0 m, width of wall = 4.0 m, thickness = 0.3 m

$E_{\text{conc}} = 5000\sqrt{f_{ck}}$ N/mm$^2 \rightarrow 2.5 \times 10^7$ kN/m^2, $g = 9.81$ m/s^2, $\gamma_c = 10.5$ kN/m^3, $v = 0.25$, $r = H/b = 2.0$.

In this case, the failure angle cuts the opposite wall as the wall is deep; thus,

$$\tan\theta = \sqrt{\frac{2H}{b}\frac{1+\mu^2}{\mu+\mu'} + \frac{1+\mu^2}{\mu+\mu'}\frac{1-\mu\mu'}{\mu+\mu'}} - \frac{1-\mu\mu'}{\mu+\mu'}$$

$$\rightarrow \tan\theta = \sqrt{\left[\frac{2\times 8}{4}\right]\left[\frac{1+0.51^2}{2\times 0.51}\right] + \left[\frac{1+0.51^2}{2\times 0.51}\right]\left[\frac{1-0.51^2}{2\times 0.51}\right]} - \frac{1-0.51^2}{2\times 0.51} = 1.691$$

Thus,

$$\tan\theta' = \frac{H}{2b\left[1-\frac{b}{2H}\tan\theta\right]} = \frac{8}{2\times 4\left[1-\frac{4}{16}\times 1.691\right]} = 1.732$$

Computation of time period

Here, as stated earlier, $X_1 = 0.49336$, $X_2 = 0.41302$, $X_3 = 0.57371$, $X_4 = 0.49336$, and

$$T = 0.693\sqrt{\frac{\gamma H^5(1-v^2)}{Et^3 g \tan\theta'(X_1 r^4 + X_2 r^2 + X_3 v r^2 + X_4)}}$$

$$T = 0.693\sqrt{\frac{10.5\times(8)^5(1-0.25^2)}{2.5\times 10^7\times(0.3)^3\times 9.81\times 1.732(0.49336(2)^4+0.41302\times(2)^2+0.57371\times 0.25(2)^2+0.49336)}}$$
$$= 0.036\,\text{s}$$

Spectral acceleration as per IS 1893 (2002) $S_a = (1+15T)g = 15.059$ m/s^2.

For zone V as per IS code $Z = 0.36$, $I = 1.5$ and $R = 3.0$ (ordinary moment resistant wall).

Thus, the design spectra $S_{ad} = \left(\frac{ZI}{2R}\right)S_a = \left[\frac{0.36\times 1.5}{2\times 3}\right]\times 15.059 = 1.355$ m/s^2

Thus, as per Eq. (13.2.6.2), we have,

$$M_z = \text{Coeff}_{mz}\frac{2\kappa_i\beta S_a}{9\pi^2 g\tan\theta'}\gamma H^3$$

$$\rightarrow M_z = \text{Coeff}_{mz}\left[\frac{2\times 0.438\times 1.355}{9\pi^2\times 9.81\times 1.732}10.5\times(8)^3\right] = \text{Coeff}_{mz}\times 4.229\ \text{kN.m/m}$$

Multiplying by the appropriate coefficient as furnished in Table 13.2, we have moment in vertical direction (M_z) as,

	$B = 0.0$	$B = 0.4$	$B = 0.8$	$B = 1.2$	$B = 1.6$	$B = 2.0$	$B = 2.4$	$B = 2.8$	$B = 3.2$	$B = 3.6$	$B = 4.0$
$H = 0.0$[a]	0.000	0.163	0.534	0.945	1.258	1.376	1.266	0.960	0.549	0.174	0.000
$H = 0.8$	0.163	0.175	0.304	0.466	0.594	0.643	0.597	0.472	0.310	0.179	0.160
$H = 1.6$	0.534	0.304	0.108	−0.048	−0.149	−0.186	−0.152	−0.053	0.101	0.296	0.524
$H = 2.4$	0.945	0.466	−0.048	−0.508	−0.827	−0.944	−0.835	−0.523	−0.067	0.447	0.929
$H = 3.2$	1.258	0.594	−0.149	−0.827	−1.302	−1.477	−1.315	−0.849	−0.177	0.566	1.236
$H = 4.0$	1.376	0.643	−0.186	−0.944	−1.477	−1.673	−1.491	−0.969	−0.216	0.612	1.352
$H = 4.8$	1.266	0.597	−0.152	−0.835	−1.315	−1.491	−1.327	−0.858	−0.180	0.569	1.244
$H = 5.6$	0.960	0.472	−0.053	−0.523	−0.849	−0.969	−0.858	−0.538	−0.072	0.452	0.943
$H = 6.4$	0.549	0.310	0.101	−0.067	−0.177	−0.216	−0.180	−0.072	0.094	0.301	0.540
$H = 7.2$	0.174	0.179	0.296	0.447	0.566	0.612	0.569	0.452	0.301	0.181	0.171
$H = 8.0$	0.000	0.160	0.524	0.929	1.236	1.352	1.244	0.943	0.540	0.171	0.000

[a]here H is considered as 0.0 at top of bunker wall

Moment in horizontal direction is expressed as $M_y = \text{Coeff}_{my} \frac{2\kappa_i \beta S_a}{9\pi^2 g \tan\theta'} \gamma H^3$ and this gives

	$B = 0.0$	$B = 0.4$	$B = 0.8$	$B = 1.2$	$B = 1.6$	$B = 2.0$	$B = 2.4$	$B = 2.8$	$B = 3.2$	$B = 3.6$	$B = 4.0$
$H = 0.0$	0.000	0.104	0.342	0.605	0.805	0.881	0.811	0.614	0.352	0.111	0.000
$H = 0.8$	0.938	0.562	0.279	0.074	−0.052	−0.095	−0.055	0.068	0.270	0.549	0.921
$H = 1.6$	3.075	1.667	0.344	−0.764	−1.510	−1.780	−1.530	−0.800	0.298	1.615	3.021
$H = 2.4$	5.448	2.908	0.457	−1.625	−3.036	−3.548	−3.073	−1.692	0.371	2.812	5.352
$H = 3.2$	7.248	3.852	0.554	−2.259	−4.169	−4.861	−4.218	−2.349	0.437	3.724	7.121
$H = 4.0$	7.929	4.210	0.592	−2.497	−4.594	−5.355	−4.648	−2.596	0.463	4.069	7.791
$H = 4.8$	7.296	3.877	0.556	−2.276	−4.199	−4.896	−4.248	−2.367	0.438	3.748	7.168
$H = 5.6$	5.529	2.950	0.462	−1.654	−3.088	−3.608	−3.125	−1.722	0.373	2.854	5.433
$H = 6.4$	3.164	1.714	0.348	−0.798	−1.569	−1.848	−1.589	−0.834	0.300	1.660	3.109
$H = 7.2$	1.002	0.594	0.279	0.047	−0.098	−0.149	−0.102	0.040	0.269	0.581	0.984
$H = 8.0$	0.000	0.102	0.336	0.595	0.791	0.866	0.796	0.604	0.345	0.109	0.000

Shear in vertical direction (kN) is expressed as Eq. (13.2.6.8).

	$B = 0.0$	$B = 0.4$	$B = 0.8$	$B = 1.2$	$B = 1.6$	$B = 2.0$	$B = 2.4$	$B = 2.8$	$B = 3.2$	$B = 3.6$	$B = 4.0$
$H = 0.0$	0.000	0.094	0.309	0.547	0.728	0.797	0.733	0.556	0.318	0.101	0.000
$H = 0.8$	−1.480	−0.705	0.155	0.936	1.483	1.684	1.497	0.962	0.187	−0.673	−1.454
$H = 1.6$	−2.089	−1.042	0.065	1.048	1.730	1.979	1.748	1.080	0.105	−1.000	−2.052
$H = 2.4$	−1.912	−0.966	0.019	0.888	1.489	1.708	1.504	0.917	0.055	−0.929	−1.879
$H = 3.2$	−1.133	−0.575	0.002	0.510	0.861	0.989	0.870	0.527	0.023	−0.553	−1.113
$H = 4.0$	−0.022	−0.011	0.000	0.010	0.017	0.019	0.017	0.010	0.000	−0.011	−0.021
$H = 4.8$	1.095	0.556	−0.002	−0.493	−0.832	−0.955	−0.840	−0.509	−0.022	0.535	1.076
$H = 5.6$	1.892	0.956	−0.018	−0.878	−1.471	−1.688	−1.487	−0.906	−0.053	0.919	1.859
$H = 6.4$	2.096	1.046	−0.062	−1.047	−1.729	−1.979	−1.747	−1.079	−0.102	1.004	2.059
$H = 7.2$	1.519	0.726	−0.150	−0.945	−1.501	−1.706	−1.516	−0.971	−0.183	0.693	1.492
$H = 8.0$	0.072	−0.055	−0.302	−0.567	−0.765	−0.840	−0.771	−0.576	−0.312	−0.063	0.071

Shear in horizontal direction (kN) is expressed as Eq. (13.2.6.10).

	B = 0.0	B = 0.4	B = 0.8	B = 1.2	B = 1.6	B = 2.0	B = 2.4	B = 2.8	B = 3.2	B = 3.6	B = 4.0
H = 0.0	0.000	−0.740	−1.044	−0.956	−0.566	−0.011	0.548	0.946	1.048	0.759	0.036
H = 0.8	0.754	0.343	0.109	0.001	−0.021	0.000	0.021	0.001	−0.102	−0.331	−0.734
H = 1.6	2.471	2.358	2.096	1.597	0.874	0.017	−0.843	−1.574	−2.082	−2.351	−2.467
H = 2.4	4.378	4.508	4.180	3.257	1.802	0.034	−1.739	−3.211	−4.157	−4.505	−4.387
H = 3.2	5.825	6.117	5.729	4.487	2.488	0.047	−2.402	−4.424	−5.698	−6.116	−5.843
H = 4.0	6.373	6.723	6.312	4.948	2.745	0.052	−2.650	−4.879	−6.278	−6.723	−6.394
H = 4.8	5.864	6.159	5.770	4.519	2.506	0.048	−2.419	−4.456	−5.739	−6.159	−5.882
H = 5.6	4.444	4.582	4.251	3.313	1.833	0.035	−1.770	−3.267	−4.227	−4.579	−4.454
H = 6.4	2.543	2.440	2.176	1.661	0.910	0.017	−0.878	−1.637	−2.162	−2.434	−2.540
H = 7.2	0.805	0.406	0.172	0.053	0.008	0.000	−0.007	−0.050	−0.166	−0.395	−0.786
H = 8.0	0.000	−0.727	−1.026	−0.939	−0.556	−0.011	0.538	0.930	1.029	0.746	0.035

13.3 Analysis of Bunker with Supporting Frame Interaction and Adapting Rankine's Earth Pressure (Chowdhury and Singh 2012)

For very large bunkers, like that in cement or a power plant, the bunkers are usually located at a height enabling trucks to get below it to clear its content. For this, bunkers are often supported on frames to give clear access to vehicles and also for periodic internal maintenance of the roof slab that constitutes of precast slab panels or checkered steel plates allowing easy access inside the bunker. In this case, the conventional analysis is to consider the whole bunker as a lumped mass and transfer the force to the frame, ignoring any deformation within the bunker body itself.

A typical industrial bunker supported on frame is as shown in Fig. 13.13

In this proposed method, the analysis is done in two parts: (i) Analysis of the bunker frame and (ii) finally the dynamic response of the bunker wall—that are elaborated hereafter.

The mathematical model as perceived for dynamic analysis of the bunker frame is shown in Fig. 13.14.

The stiffness matrix $[K_F]$ and lumped mass matrix $[M_F]$ of the frame are given as,

$$[K_F] = \begin{bmatrix} K_1 + K_2 & -K_2 \\ -K_2 & K_2 \end{bmatrix} \quad \text{and} \quad [M_F] = \begin{bmatrix} M_1 & 0 \\ 0 & M_2 \end{bmatrix} \tag{13.3.1}$$

where $K_1 = n\frac{12EI_1}{L_1^3}$ and $K_2 = n\frac{12EI_2}{L_2^3}$ are flexural stiffness of frame below and above the tie-level.

n number of columns in the frame;
M_2 mass of container and fill material;
M_1 effective mass of column and beam at tie-level.

Fig. 13.13 Frame supported industrial bunker under construction. By kind courtesy of Petrofac International Limited

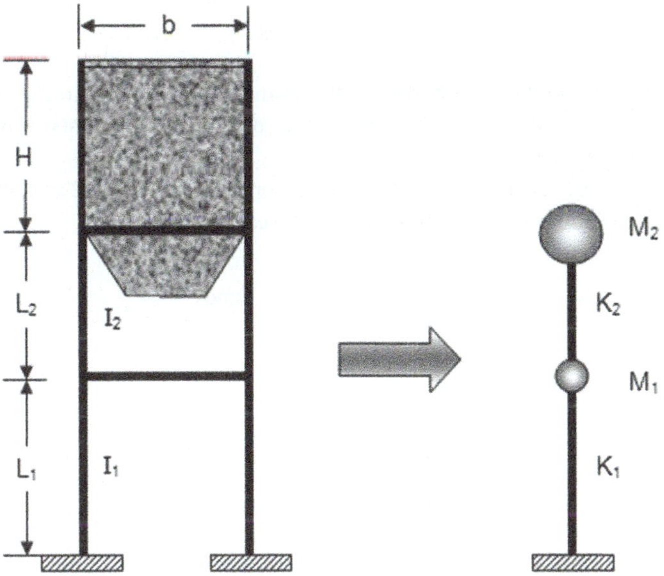

Fig. 13.14 Mathematical model of frame of a rectangular bunker

The eigenvalue of the problem is given by,

$$[\lambda] = \begin{bmatrix} K_1 + K_2 - M_1\lambda & -K_2 \\ -K_2 & K_2 - M_2\lambda \end{bmatrix} \tag{13.3.2}$$

Let λ_1 and λ_2 be the eigenvalues and let the corresponding eigenvectors be expressed as

$$[\phi] = \begin{bmatrix} \phi_{11} & \phi_{12} \\ \phi_{21} & \phi_{22} \end{bmatrix} \tag{13.3.3}$$

Normalized eigenvectors for the first and second mode of vibration are given by,

$$\begin{bmatrix} \phi_{11}^n \\ \phi_{21}^n \end{bmatrix} = \frac{1}{\sqrt{M_1\phi_{11}^2 + M_2\phi_{21}^2}} \begin{bmatrix} \phi_{11} \\ \phi_{21} \end{bmatrix} \quad \text{and}$$
$$\begin{bmatrix} \phi_{12}^n \\ \phi_{22}^n \end{bmatrix} = \frac{1}{\sqrt{M_1\phi_{12}^2 + M_2\phi_{22}^2}} \begin{bmatrix} \phi_{12} \\ \phi_{22} \end{bmatrix} \tag{13.3.4}$$

Modal participation factors are expressed as,

$$\kappa_1 = \frac{M_1\phi_{11}^n + M_2\phi_{21}^n}{M_1\left(\phi_{11}^n\right)^2 + M_2\left(\phi_{21}^n\right)^2} \quad \text{and}$$
$$\kappa_2 = \frac{M_1\phi_{12}^n + M_2\phi_{22}^n}{M_1\left(\phi_{12}^n\right)^2 + M_2\left(\phi_{22}^n\right)^2} \tag{13.3.5}$$

Let T_1 and T_2 be the time period of the frame corresponding to eigenvalues λ_1 and λ_2. Let S_{a1}/g and S_{a2}/g be the spectral acceleration coefficients from code for T_1 and T_2, respectively.

Design spectral acceleration coefficients at top of frame (node 2 of lumped mass model) for two modes of vibration are expressed as,

$$\frac{D_{a1}}{g} = \kappa_1\beta\frac{S_{a1}}{g}\phi_{21}^n \quad \text{and}$$
$$\frac{D_{a2}}{g} = \kappa_2\beta\frac{S_{a1}}{g}\phi_{22}^n \tag{13.3.6}$$

where $\beta = ZI/2R$ is a code factor for seismic zone, importance and response reduction factor.

The above accelerations will work on the container as shown in Fig. 13.15.

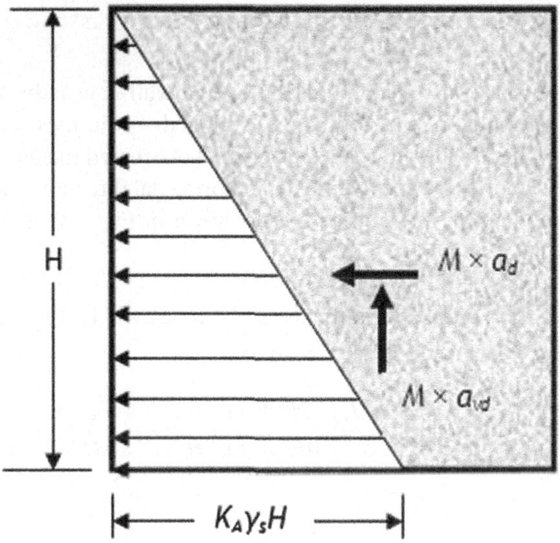

Fig. 13.15 Variation of pressure on vertical wall of the bunker

During earthquake, pressure at any depth z from top is expressed as,

$$p_h = K_A \gamma_s z + \gamma_s z \left(\frac{D_{ai}}{g} \right) + K_A \gamma_s z \left(\frac{2}{3} \frac{D_{ai}}{g} \right) \quad (13.3.7)$$

The last expression in Eq. (13.3.7) represents the effect of vertical acceleration on horizontal pressure on bunker wall,

or

$$p_h = K_A \gamma_s z \left[1 + \frac{1}{K_A} \frac{D_{ai}}{g} + \frac{2}{3} \frac{D_{ai}}{g} \right] \quad (13.3.8)$$

or

$$p_h = K_A \gamma_s z \Lambda_i \quad (13.3.9)$$

where

$$\Lambda_i = 1 + \frac{1}{K_A} \frac{D_{ai}}{g} + \frac{2}{3} \frac{D_{ai}}{g} \quad (13.3.10)$$

Here Λ is a dimensionless amplification factor for two modes of vibration (i = 1, 2) contributing to enhance the static wall pressure and shall always be ≥ 1.0.

13.3.1 Analysis of Bunker Wall

For analysis of the bunker wall, the wall is usually considered as three sides fixed and one side (the top edge) as hinged—as in most of the cases the roof is covered by removable precast slabs or steel checkered plates or gratings. Now considering a strip of the wall in vertical and horizontal direction, the boundary conditions and the loading on these strips are as shown in Fig. 13.16.

Shape Function of Plate

For propped cantilever strip as shown in Fig. 13.16a, displacement at depth z from the top is expressed as,

$$\delta_z = \frac{K_A \gamma_s H^5 A_i}{30EI}\left[\frac{1}{4}\left(\frac{z}{H}\right)^5 - \frac{1}{2}\left(\frac{z}{H}\right)^3 + \frac{1}{4}\left(\frac{z}{H}\right)\right] \tag{13.3.1.1}$$

or

$$\delta_z = \delta_{\text{static}} \times f(\xi) \tag{13.3.1.2}$$

where $\xi = z/H$ is a non-dimensional term that varies between 0 and 1 and $f(\xi)$ is the generic shape function of propped cantilever beam given as,

$$f(\xi) = \frac{\xi^5}{4} - \frac{\xi^3}{2} + \frac{\xi}{4} \tag{13.3.1.3}$$

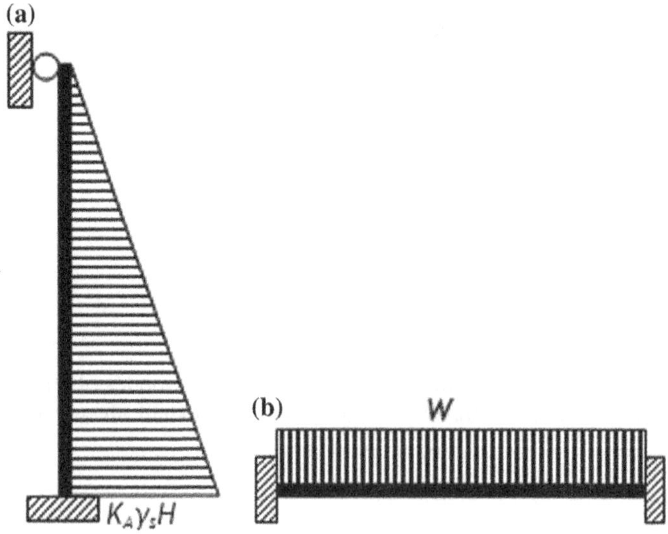

Fig. 13.16 **a** Propped cantilever and **b** fixed supported strips

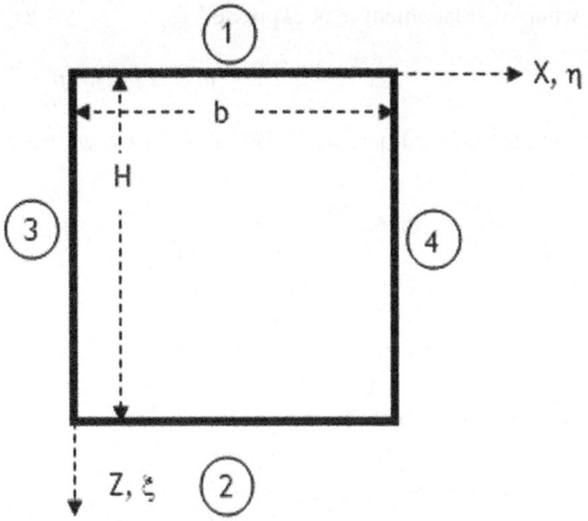

Fig. 13.17 Coordinate system of bunker wall

Similarly, for a strip which is fixed at both ends in horizontal direction, the generic shape function of displacement can be expressed as,

$$f(\eta) = 16\eta^4 - 32\eta^3 + 16\eta^2 \tag{13.3.1.4}$$

where $\eta = x/b$ is a non-dimensional term that varies between 0 and 1 (Fig. 13.17).

The shape functions derived in Eqs. (13.3.1.3) and (13.3.1.4) will generically satisfy plate equilibrium equation having boundary conditions of three sides fixed and one side propped but will have residual error (R_e) as they are not derived from exact analysis of the fourth-order partial differential equation of an isotropic plate expressed as (Timoshenko and Kreiger 1985)

$$\frac{\partial^4 u}{\partial x^4} + 2\frac{\partial^4 u}{\partial x^2 \partial z^2} + \frac{\partial^4 u}{\partial z^4} = \frac{q}{D} \tag{13.3.1.5}$$

where u is the displacement of the plate under a pressure load q. D is flexural stiffness of the plate expressed as $D = Et^3/12(1 - v^2)$, and v the Poisson ratio of bunker wall material.

In natural coordinates, as expressed in Eqs. (13.3.1.3) and (13.3.1.4), Eq. (13.3.1.5) can be expressed as,

$$\frac{1}{b^4}\frac{\partial^4 u}{\partial \eta^4} + \frac{2}{b^2 H^2}\frac{\partial^4 u}{\partial \eta^2 \partial \xi^2} + \frac{1}{H^4}\frac{\partial^4 u}{\partial \xi^4} = \frac{q}{D} \tag{13.3.1.6}$$

where displacement u is expressed as,

$$u = \Delta.f(\xi).f(\eta) \tag{13.3.1.7}$$

Substituting Eq. (13.3.1.7) in (13.3.1.6), we have,

$$\frac{\Delta}{b^4}\frac{\partial^4}{\partial \eta^4}f(\xi).f(\eta) + \frac{2\Delta}{b^2H^2}\frac{\partial^4}{\partial \eta^2 \partial \xi^2}f(\xi).f(\eta) + \frac{\Delta}{H^4}\frac{\partial^4}{\partial \xi^4}f(\xi).f(\eta) - \frac{q}{D} = 0 \tag{13.3.1.8}$$

The shape functions $f(\xi)$ and $f(\eta)$ will generically satisfy Eq. (13.3.1.8) as they conform to boundary condition of the given plate; however, they will have residual error R_e which may be expressed as,

$$R_e = \frac{\Delta}{b^4}\frac{\partial^4}{\partial \eta^4}f(\xi).f(\eta) + \frac{2\Delta}{b^2H^2}\frac{\partial^4}{\partial \eta^2 \partial \xi^2}f(\xi).f(\eta) + \frac{\Delta}{H^4}\frac{\partial^4}{\partial \xi^4}f(\xi).f(\eta) - \frac{q}{D} \tag{13.3.1.9}$$

The residual error R_e is now minimized by Galerkin's (Chowdhury and Dasgupta 2008) weighted residual method based on which,

$$\int_0^1\int_0^1 [R_e].f(\xi).f(\eta).\mathrm{d}\xi.\mathrm{d}\eta = 0 \tag{13.3.1.10}$$

Equation (13.3.1.10) can thus be expressed as,

$$\begin{aligned}&\tfrac{\Delta}{b^4}\int_0^1\int_0^1 f(\xi)^2.f'''(\eta).f(\eta).\mathrm{d}\xi.\mathrm{d}\eta + \tfrac{2\Delta}{b^2H^2}\int_0^1\int_0^1 f''(\xi)f''(\eta)f(\xi)f(\eta)\mathrm{d}\xi.\mathrm{d}\eta \\ &+ \tfrac{\Delta}{H^4}\int_0^1\int_0^1 f'''(\xi)f(\xi)f(\eta)^2\mathrm{d}\xi\mathrm{d}\eta - \tfrac{1}{D}\int_0^1\int_0^1 qf(\xi)f(\eta)\mathrm{d}\xi.\mathrm{d}\eta = 0\end{aligned} \tag{13.3.1.11}$$

Considering $q = K_A.\gamma_s.z.\Lambda_i$, Eq. (13.3.1.11) can be expressed in a simplified form as,

$$\Delta\left[X_1 + \frac{2H^2}{b^2}X_2 + \frac{H^4}{b^4}X_3\right] = \frac{K_A.\gamma_s.\Lambda_i.H^5}{D}X_4 \tag{13.3.1.12}$$

Considering $r = H/b$, the aspect ratio of the plate, Eq. (13.3.1.12) can be expressed as,

$$\Delta = \frac{K_A.\gamma_s.\Lambda_i.H^5}{D(X_1 + 2r^2X_2 + r^4X_3)}X_4 \tag{13.3.1.13}$$

Table 13.6 Values of integral functions X_1 to X_4

Integral functions	Values
$X_1 = \int_0^1\int_0^1 f(\xi)^2 \cdot \left[\frac{\partial^4}{\partial\eta^4} f(\eta)\right] \cdot f(\eta).\mathrm{d}\xi \cdot \mathrm{d}\eta$	0.473
$X_2 = \int_0^1\int_0^1 \left[\frac{\partial^2}{\partial\xi^2} f(\xi)\right]\left[\frac{\partial^2}{\partial\eta^2} f(\eta)\right] f(\xi) \cdot f(\eta) \cdot \mathrm{d}\xi \cdot \mathrm{d}\eta$	0.124
$X_3 = \int_0^1\int_0^1 \left[\frac{\partial^4}{\partial\xi^4} f(\xi)\right] \cdot f(\xi) \cdot f(\eta)^2 \cdot \mathrm{d}\xi \cdot \mathrm{d}\eta$	0.232
$X_4 = \int_0^1\int_0^1 \xi \cdot f(\eta) \cdot f(\xi) \cdot \mathrm{d}\xi \cdot d\eta$	0.010

where X_1, X_2, X_3, and X_4 are integral functions that can be solved numerically or explicitly and are as expressed in Table 13.6.

The maximum static displacement enhanced by the vibration of the frame for two modes of vibration of bunker wall are calculated using Eq. (13.3.1.13) as,

$$\Delta_1 = \frac{K_A.\gamma_s.\Lambda_1.H^5}{D(X_1 + 2r^2X_2 + r^4X_3)} X_4 \tag{13.3.1.14}$$

$$\Delta_2 = \frac{K_A.\gamma_s.\Lambda_2.H^5}{D(X_1 + 2r^2X_2 + r^4X_3)} X_4 \tag{13.3.1.15}$$

The displacement profile u of container wall for the first two modes of vibration can thus be expressed as,

$$u_1(\xi,\eta) = \Delta_1.f(\xi).f(\eta) \quad \text{and}$$
$$u_2(\xi,\eta) = \Delta_2.f(\xi).f(\eta) \tag{13.3.1.16}$$

Here $f(\xi)$ and $f(\eta)$ are as expressed in Eqs. (13.3.1.3) and (13.3.1.4), respectively. The maximum deflection will occur at $\xi = 0.447$ and $\eta = 0.5$, which gives

$$u_1(0.447, 0.5) = \Delta_1.f(0.447).f(0.5) \quad \text{and}$$
$$u_2(0.447, 0.5) = \Delta_2.f(0.447).f(0.5) \tag{13.3.1.17}$$

The time period of the wall for first two modes of vibration can be expressed as,

$$T_{w1} = 2\pi\sqrt{\frac{u_1(0.447, 0.5)}{g}} \quad \text{and}$$
$$T_{w2} = 2\pi\sqrt{\frac{u_2(0.447, 0.5)}{g}} \tag{13.3.1.18}$$

Let S_{aw1} and S_{aw2} be the spectral acceleration coefficients from code for T_{w1} and T_{w2}, respectively.

For modal analysis, maximum amplitudes (S_{wd}) in terms of the code for the first two modes of vibration are expressed as, (Clough and Penzien 1984)

$$S_{dw1} = \kappa_{w1}.\beta.\frac{S_{aw1}.g}{\omega_1^2} \quad \text{and}$$
$$S_{dw2} = \kappa_{w2}.\beta.\frac{S_{aw2}.g}{\omega_2^2} \tag{13.3.1.19}$$

where $\omega_1 = 2\pi/T_{w1}$ and $\omega_2 = 2\pi/T_{w2}$ are natural frequencies of the wall; β = code factor given earlier; κ_{w1} and κ_{w2} = mode participation factors for the first two modes of wall given in Eq. (13.3.1.20).

$$\kappa_{w1} = \kappa_{w2} = \frac{\int_0^1 \int_0^1 \xi.f(\xi).f(\eta).d\xi.d\eta}{\int_0^1 \int_0^1 \xi.(f(\xi).f(\eta))^2.d\xi.d\eta} \tag{13.3.1.20}$$

Equation (13.3.1.20) on computation gives a value of κ_{w1} = 24.

Thus, based on Eq. (13.3.1.9), the combined spectral displacement of the wall can be calculated using SRSS method as,

$$S_{dw} = \sqrt{(S_{dw1} + \Delta_1)^2 + (S_{dw2} + \Delta_2)^2} \tag{13.3.1.21}$$

Finally, the dynamic SRSS displacement of wall can be expressed as,

$$u(\xi, \eta) = S_{dw}.f(\xi).f(\eta) \tag{13.3.1.22}$$

Equation (13.3.1.22) gives a complete dynamic displacement profile of the wall between limits 0 and 1 in both *X*- and *Z*-directions for the first two modes.

Dynamic Bending Moment and Shear Force

Bending moment and shear force induced in the wall can be expressed as (Timoshenko and Kreiger 1985),

$$M_x = -D\left[\frac{\partial^2 u}{\partial x^2} + \nu\frac{\partial^2 u}{\partial z^2}\right] \tag{13.3.1.23}$$

$$M_z = -D\left[\nu\frac{\partial^2 u}{\partial x^2} + \frac{\partial^2 u}{\partial z^2}\right] \tag{13.3.1.24}$$

$$Q_x = -D\left[\frac{\partial^3 u}{\partial x^3} + \frac{\partial^3 u}{\partial x.\partial z^2}\right] \tag{13.3.1.25}$$

$$Q_z = -D\left[\frac{\partial^3 u}{\partial x^2 \partial z} + \frac{\partial^3 u}{\partial z^3}\right] \tag{13.3.1.26}$$

Transferring the above Eq. (13.3.1.23) to (13.3.1.26) in natural coordinates and substituting in Eq. (13.3.1.22), dynamic moments and shears can be calculated using the following equations.

$$M_x(\xi,\eta) = -\frac{D.S_{\mathrm{dw}}}{H^2}\left[r^2.f(\xi).f''(\eta) + v.f''(\xi).f(\eta)\right] \tag{13.3.1.27}$$

$$M_z(\xi,\eta) = -\frac{D.S_{\mathrm{dw}}}{H^2}\left[v.r^2.f(\xi).f''(\eta) + f''(\xi).f(\eta)\right] \tag{13.3.1.28}$$

$$Q_x(\xi,\eta) = -\frac{D.S_{\mathrm{dw}}}{H^3}\left[r^3.f(\xi).f'''(\eta) + r.f''(\xi).f(\eta)\right] \tag{13.3.1.29}$$

$$Q_z(\xi,\eta) = -\frac{D.S_{\mathrm{dw}}}{H^3}\left[r^2.f'(\xi).f''(\eta) + f'''(\xi).f(\eta)\right] \tag{13.3.1.30}$$

The successive derivatives of $f(\xi)$,$f(\eta)$ are as shown hereafter.

$f(\xi)$	$\frac{\xi^5}{4} - \frac{\xi^3}{2} + \frac{\xi}{4}$	$f(\eta)$	$16\eta^4 - 32\eta^3 + 16\eta^2$
$f'(\xi)$	$\frac{5\xi^4}{4} - \frac{3\xi^2}{2} + \frac{1}{4}$	$f'(\eta)$	$64\eta^3 - 96\eta^2 + 32\eta$
$f''(\xi)$	$5\xi^3 - 3\xi$	$f''(\eta)$	$192\eta^2 - 192\eta + 32$
$f'''(\xi)$	$15\xi^2 - 3$	$f'''(\eta)$	$384\eta - 192$

Example 13.2 A square bunker is 4.57 m internally in plan and 3.66 m deep from the top of the hopper slopes. The pyramidal hopper is 2.29 m deep internally. Perform the dynamic analysis under earthquake given the following parameters (Fig. 13.18):

Data:

Size of columns = 305 mm × 305 mm
Number of columns = 4
Grade of concrete of walls, beam, and columns = M15

Fill Materials:

Angle of internal friction = 35°
Weight density of fill material = 7.85 kN/m^3
Thickness of concrete wall = 127 mm

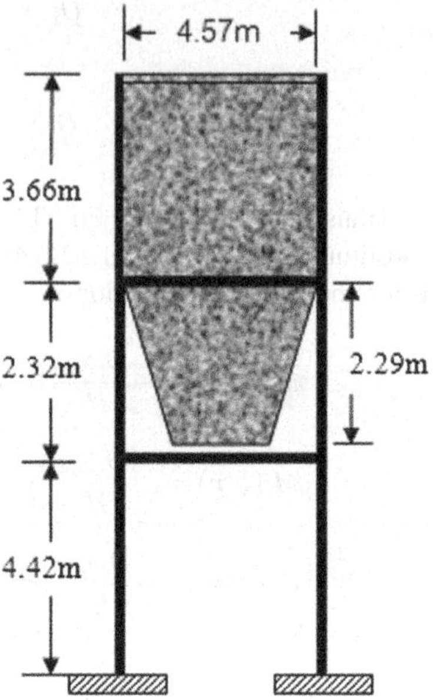

Fig. 13.18 Bunker with dimensions

Stiffness of frame:

stiffness of all columns below tie-level $K_1 = 7.76 \times 10^3$ kN/m;
stiffness of all columns above tie-level $K_2 = 5.34 \times 10^4$ kN/m.

Lumped masses:

$$M_1 = 6808\,\text{kg} \quad \text{and} \quad M_2 = 114000\,\text{kg}$$

Hence, mass and stiffness matrix of the frame are given as,

$$M = \begin{bmatrix} 6.808 \times 10^3 & 0 \\ 0 & 1.14 \times 10^5 \end{bmatrix} \text{kg}$$

$$K = \begin{bmatrix} 6.117 \times 10^4 & -5.34 \times 10^4 \\ -5.34 \times 10^4 & 5.34 \times 10^4 \end{bmatrix} \frac{\text{kN}}{\text{m}}$$

The eigensolution of the above mass and stiffness matrix is

$$T = \begin{Bmatrix} 0.833 \\ 0.065 \end{Bmatrix} \text{s}$$

$$\phi = \begin{bmatrix} -0.660 & 0.999 \\ -0.751 & -0.052 \end{bmatrix}$$

Normalized eigenvectors are

$$\phi_N = \begin{bmatrix} -2.544 \times 10^{-3} & 0.012 \\ -2.896 \times 10^{-3} & -6.217 \times 10^{-4} \end{bmatrix}$$

Mode participation factors for two modes of vibrations are given as,

$$\kappa_1 = -347.436 \quad \text{and} \quad \kappa_2 = -249.443$$

Beta factor from IS code,

$$\beta = \frac{ZI}{2R} = 0.06$$

Spectral acceleration coefficients corresponding to two time periods from IS code for 5% damping ratio.

$$\frac{S_{a1}}{g} = 2.004 \quad \text{and}$$
$$\frac{S_{a2}}{g} = 1.972$$

Design acceleration coefficients at the top of frame (node 2) corresponding to two modes of vibration.

$$\frac{D_{a1}}{g} = \kappa_1 \times \beta \times \frac{S_{a1}}{g} \times \phi_{N_{21}} = 0.121 \quad \text{and}$$
$$\frac{D_{a2}}{g} = \kappa_2 \times \beta \times \frac{S_{a2}}{g} \times \phi_{N_{22}} = 0.018$$

Amplification factors for two modes of vibration.

$$\Lambda_1 = 1.527 \quad \text{and} \quad \Lambda_2 = 1.08$$

Amplitude of vibration for first mode and second mode of vibrations,

$$\Delta_1 = \frac{K_A.\gamma_s.\Lambda_1.H^5}{D(X_1 + 2r^2X_2 + r^4X_3)} X_4 = 5.522 \times 10^{-4}\ \text{m}$$

$$\Delta_2 = \frac{K_A.\gamma_s.\Lambda_2.H^5}{D(X_1 + 2r^2X_2 + r^4X_3)} X_4 = 3.905 \times 10\ ^{4}\text{m}$$

Deflection profile of vertical walls in two modes of vibrations,

$$u_1(\xi,\eta) = 5.522 \times 10^{-4} f(\xi).f(\eta) \quad \text{and}$$
$$u_2(\xi,\eta) = 3.905 \times 10^{-4} f(\xi).f(\eta)$$

Time periods of walls in two modes of vibration,

$$T_1 = 2.\pi.\sqrt{\frac{u_1(0.447, 0.5)}{g}} = 0.047\text{s} \quad \text{and}$$

$$T_2 = 2.\pi.\sqrt{\frac{u_2(0.447, 0.5)}{g}} = 0.040\text{s}$$

Spectral accelerations coefficients corresponding to two time periods from IS code for 5% damping ratio,

$$\frac{S_{aw1}}{g} = 1.707 \quad \text{and} \quad \frac{S_{aw2}}{g} = 1.595$$

Spectral displacements of walls corresponding to two modes of vibration,

$$S_{dw1} = \kappa_1 \times \beta \times \left(\frac{S_{a1}}{g}\right) \times \frac{g}{\omega_1^2} = 1.358 \times 10^{-3} \quad \text{and}$$
$$S_{dw2} = \kappa_2 \times \beta \times \left(\frac{S_{a2}}{g}\right) \times \frac{g}{\omega_2^2} = 8.968 \times 10^{-4}$$

Hence, spectral displacement profile over the wall is,

$$S_{dw}(\xi,\eta) = \sqrt{(S_{dw1} + \Delta_1)^2 + (S_{dw2} + \Delta_2)^2} f(\xi).f(\eta)$$
$$= 0.011.f(\xi).f(\eta)$$

Maximum SRSS deflection is $S_{dw}(0.447, 0.5) = 0.794$ mm.
Maximum moments:

$M_x(0.5, 0) = -4.61$ kN-m/m, $M_x(0.5, 0.5) = 2.72$ kN-m/m and $M_x(0.5, 1) = -4.61$ kN-m/m.
$M_y(0, 0.5) = 0.00$ kN-m/m, $M_y(0.5,0.5) = 3.14$ kN-m/m and $M_y(1, 0.5) = -6.39$ kN-m/m.

Maximum shear forces:

$Q_x(0.5, 0) = 6.05$ kN/m, $Q_x(0.5, 0.5) = 0.00$ kN/m and $Q_x(0.5, 1) = -6.05$ kN/m.
$Q_y(0, 0.5) = 4.86$ kN/m, $Q_y(0.5,0.5) = -1.07$ kN/m and $Q_y(1, 0.5) = -10.47$ kN/m.

Bending moments and shear force diagram on the wall in horizontal and vertical direction, vide Eqs. (13.3.1.27) to (13.3.1.30), are shown below (Fig. 13.19).

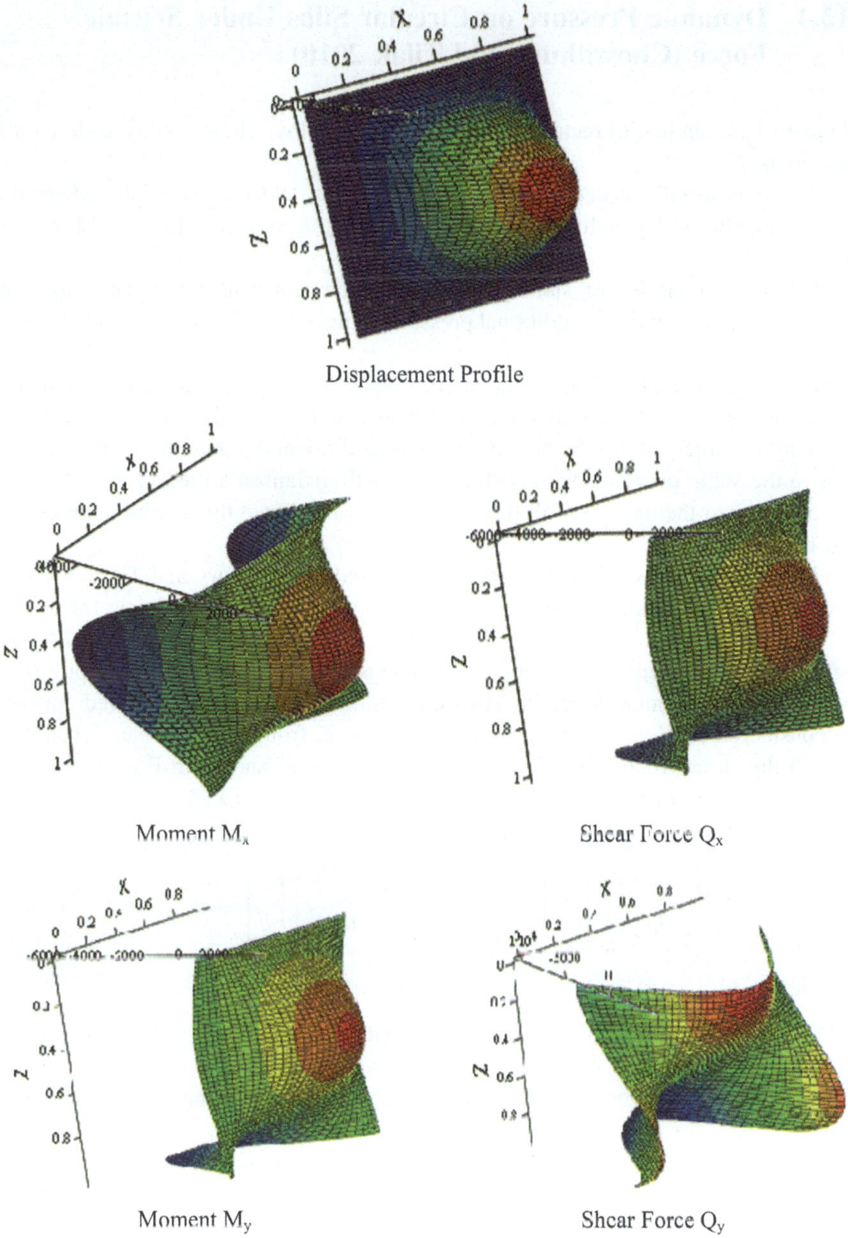

Fig. 13.19 Dynamic amplitude, moments, and shear of the bunker wall

13.4 Dynamic Pressure on Circular Silos Under Seismic Force (Chowdhury and Tilak 2010)

In many cases in lieu of rectangular bunkers, one deploys circular silos to store bulk materials.

A silo is usually a deep circular container used to store materials, where the failure profile will usually cut the opposite wall. A typical silo is as shown in Fig. 13.20.

In conventional design, static pressure on the circular wall is estimated based on Janssen's theory[1] and any additional pressure on the silo wall due to seismic force is ignored.

It is a well-established fact that when seismic waves due to a major earthquake hit a site, the whole system (i.e., the frame and the container together with its content) is subjected to vibration and induces additional dynamic pressure over and above the static pressure that is estimated based on Jansen's theory.

No rational theory exists till date in practice to estimate this additional pressure on the wall.

In the following section, a method is proposed (Chowdhury and Tilak 2010) to estimate this additional dynamic pressure on a silo wall due to seismic force.

As a first step, the circular silo can be modeled as shown in the Fig. 13.21. The silo being supported on columns and bracings have been replaced by springs in vertical and horizontal direction. The mass of the silo including the stored material is considered as M acting at the center of gravity Z_c from the ring beam level from which the superstructure is suspended on the frame, as shown in Fig. 13.21.

Based on the mathematical model, as proposed in Fig. 13.21, the free vibration equation for the above system can be expressed as

$$\begin{bmatrix} M & 0 \\ 0 & J_\theta \end{bmatrix} \begin{bmatrix} \ddot{x} \\ \ddot{\theta} \end{bmatrix} + \begin{bmatrix} K_x & -K_x.Z_c \\ -K_x Z_c & K_x Z_c^2 + K_z R^2 \end{bmatrix} \begin{bmatrix} x \\ \theta \end{bmatrix} = 0 \tag{13.4.1}$$

Here

M mass of the silo including its content;

J_θ mass moment of inertia of the silo and its content;

K_x, K_z lateral and vertical stiffness of the supporting frame (magnitude elaborated later for different structural configuration);

Z_c center of gravity of the silo including its content above the collar or ring beam;

R radius of the silo.

[1]Refer Design of Bunkers and Silos by Gray and Manning; Cement and Concrete Association, UK.

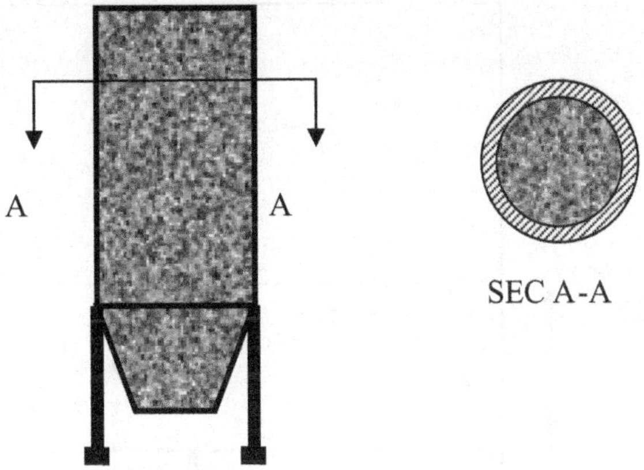

Fig. 13.20 Typical circular silo and its sectional plan

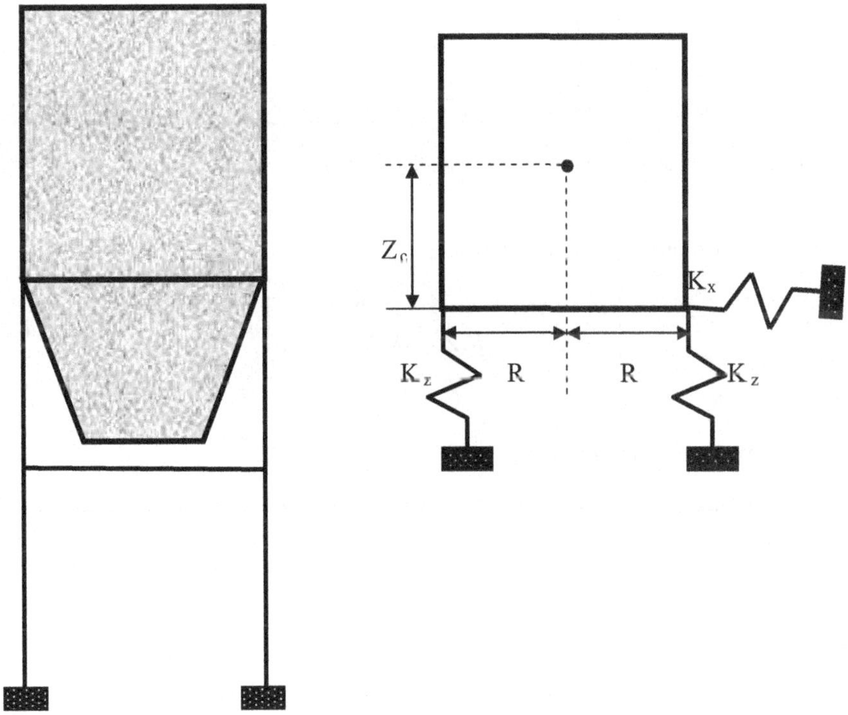

Fig. 13.21 Structural arrangement of circular silo and its equivalent mathematical model

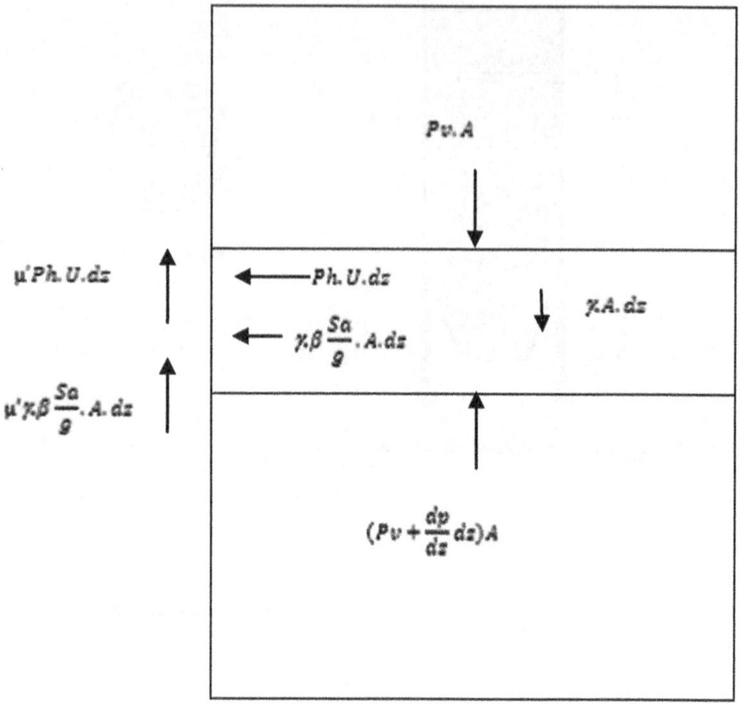

Fig. 13.22 Free body diagram of pressure equilibrium for an element of strip dz

The eigensolution of the problem is given by

$$\begin{bmatrix} K_x - M\Lambda & -K_x.Z_c \\ -K_x Z_c & K_x Z_c^2 + K_z R^2 - J_\theta \Lambda \end{bmatrix} = 0 \tag{13.4.2}$$

Here

Λ eigenvalue of the problem.

Considering $\Lambda = \omega^2$ and $T = 2\pi/\omega$, we can find out the first two fundamental time periods of the system.

Considering time periods for the given two modes, T_1 and T_2, we can find out the value of S_{a1}/g and S_{a2}/g for the first two modes for 5% damping, say, for a reinforced concrete structural system and 2–3% damping for a steel structure system.

Having estimated the acceleration the silo is subjected to based on its first two fundamental time periods, the same can be fitted into Jansen's theory based on Fig. 13.22.

We take here a strip of the material stored in silo of depth dz and apply the applicable forces as shown in the figure below. The difference here is the additional force exerted on the wall due to the seismic force generated by the stored material, which is also included.

When these forces are applied on the wall, using coefficient of friction μ', we have the free body diagram as shown in Fig. 13.22.

For equilibrium, summation of forces in vertical direction leads to the expression (Mallick and Rangaswamy 1980),

$$-P_v.A - \gamma.A.\mathrm{d}z + \mu'\gamma\beta\frac{S_{al}}{g}A\mathrm{d}z + \mu' P_h U\mathrm{d}z + \left(P_v + \frac{\mathrm{d}P_v}{\mathrm{d}z}\mathrm{d}z\right)A = 0 \quad (13.4.3)$$

$$\Rightarrow -\gamma.\mathrm{d}z + \mu'\gamma\beta\frac{S_{ai}}{g}\mathrm{d}z + \mu' P_h\frac{U}{A}.\mathrm{d}z + \left(\frac{\mathrm{d}P_v}{\mathrm{d}z}\mathrm{d}z\right) = 0 \quad (13.4.4)$$

Considering $R_h = A/U$ and $\lambda = \frac{P_h}{P_v}$ and substituting in Eq. (13.4.4), we have,

$$\mathrm{d}P_v = \left\{\gamma\left(1 - \mu'\beta\frac{S_{ai}}{g}\right) - \frac{\mu'\lambda}{R_h}P_v\right\}\mathrm{d}z \quad (13.4.5)$$

Rearranging and integrating over the entire depth of fill with appropriate boundary condition, we finally get the total static plus dynamic pressure due to seismic force as,

$$P_v = \frac{\gamma R_h\left(1 - \mu'\beta\frac{S_{ai}}{g}\right)}{-\mu'\lambda}\left(-1 + \mathrm{e}^{-\frac{\mu'\lambda}{R_h}z}\right) \quad (13.4.6)$$

Horizontal intensity of pressure on the silo wall is then finally given by,

$$P_{he} = \frac{\gamma R_h\left(1 \pm \mu'\beta\frac{S_{ai}}{g}\right)}{\mu'}\left(1 - \mathrm{e}^{-\frac{\mu'\lambda}{R_h}z}\right) \quad (13.4.7)$$

The sign components within the brackets will fluctuate depending upon the direction of the seismic force. Therefore, to arrive at the maximum possible value of P_{he}, we must take positive sign.

$$\therefore P_{he} = \frac{\gamma R_h\left(1 + \mu'\beta\frac{S_{ai}}{g}\right)}{\mu'}\left(1 - \mathrm{e}^{-\frac{\mu'\lambda}{R_h}z}\right) \quad (13.4.8)$$

Considering the vertical seismic component also, along with the previously defined horizontal component of seismic force on the wall of the silo, we can modify Eq. (13.4.8) to,

$$P_{\text{hev}} = \frac{\gamma R_h}{\mu'}\left(1 + \mu'\beta\frac{S_{\text{ai}}}{g} + \frac{2}{3}\beta\frac{S_{\text{ai}}}{g}\right)\left(1 - e^{-\frac{\mu'\lambda}{R_h}z}\right) \tag{13.4.9}$$

In Eq. (13.4.9), the third term in the first parenthesis represents the vertical component of seismic acceleration as per Clause 6.4.5 of IS-1893 (2002).

In the above derivation, the acronyms used are as defined hereafter,

A	area of cross section of the silo;
g	acceleration due to gravity;
I	importance factor as per IS code;
P_v	vertical pressure on silo wall;
P_{he}	horizontal pressure on silo wall;
P_{hev}	horizontal pressure on silo wall considering vertical acceleration also;
R	ductility factor as per IS code;
R_h	hydraulic mean depth of the section @A/U;
S_{ai}	horizontal acceleration due to seismic force in the ith mode where i = 1,2;
U	perimeter of the section;
Z	zone factor as per IS code;
z	any depth from top of silo;
$\beta = \frac{ZI}{2R}$	a code factor;
γ	weight density of the fill material;
μ'	coefficient of friction between wall and the fill material.

$$\lambda = \frac{P_h}{P_v} = \frac{1 - \sin\phi}{1 + \sin\phi}$$

φ angle of repose of the fill material.

The final SRSS design pressure is given by

$$P_{hd} = \sqrt{\sum_{i=1}^{2} P_{hi}^2} \tag{13.4.10}$$

13.4.1 A Case Study

A silo of circular cross section is being analyzed here for the pressure distribution along the wall of the silo, using Janssen's Formulation and is being compared with the increase or decrease in pressure on the walls on introduction of seismic forces.

Seismic forces have been further split into two cases, i.e.,

(i) With horizontal acceleration only and;
(ii) With both the horizontal and vertical accelerations.

The geometry of the silo considered has been taken as defined below:

diameter of silo	10 m;
height of silo	10 m;
number of columns	8;
number of panels	2;
number of bracings	16@8 each panel;
column height	8 m@4 m each panel;
column cross section	0.5 m × 0.5 m;
beam length	3.83 m;
beam cross section	0.5 m × 0.5 m;
bracing length	4.57 m;
bracing cross section	0.5 m × 0.5 m.

In order to study the behavior of different structural arrangements, two types of structural arrangements have been studied as given below:

Case 1: Staging configuration with circumferential beams (infinite stiffness).
Case 2: Staging configuration with circumferential beams and diagonal braces.

Case 1: For staging type as shown in Fig. 13.23a, stiffness of the frame is given by

The structural stiffness K_z and K_x as used in Eq. (13.4.1) are adopted from Dutta et al. (2003) for above-defined configuration or cases and have been used here.

$$K_x = \frac{12E_c.I_c.N_c}{h_p^3 N_p} \tag{13.4.1.1}$$

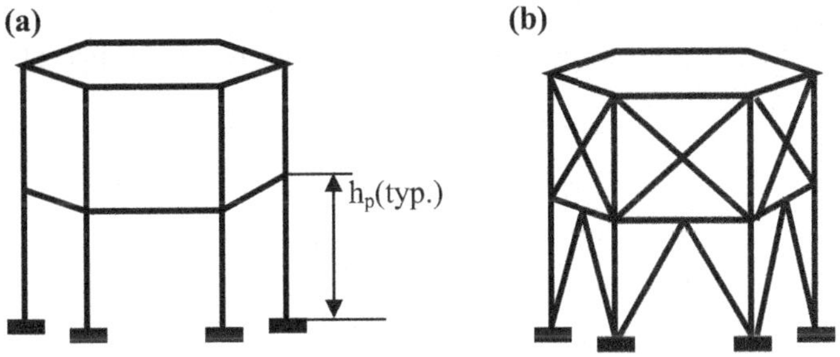

Fig. 13.23 Different structural configuration used as the staging frame

Here

E_c Young's modulus of the column material;
I_c moment of inertia of the column cross section;
N_c number of columns;
N_p number of panels;
h_p height of panel.

Axial stiffness of columns may be represented as below,

$$K_z = \frac{E_c.I_c.N_c}{h_p N_p} \tag{13.4.1.2}$$

Case 2: For staging type as shown in Fig. 13.23b**, stiffness of the frame is given by**

$$K_x = \frac{12E_cI_c}{h_p^3}\left(\frac{1}{N_p + 2(N_p - 1)K_r}\right) + \frac{\left(\frac{N_cA_0E_0}{2N_pL_0}\right)\cos^2\theta_v}{1 + C_1K_{\text{arc}} + C_2K_{\text{arb}}} \tag{13.4.1.3}$$

Here

A_0 area of cross section of diagonal braces;
E_0 modulus of elasticity of diagonal braces;
L_0 length of diagonal braces;
L_b length of panel beams;
θ_v angle of inclination of diagonal brace with horizontal direction.

$$K_r = \frac{E_cI_c/h_p}{E_bI_b/L_b}$$

K_{arc} relative axial stiffness of diagonal braces, with respect to that of the column $@\frac{A_0E_0h_p}{A_cE_cL_0}$;
K_{arb} relative axial stiffness of diagonal braces, with respect to that of the beams $@\frac{A_0E_0L_b}{A_bE_bL_0}$.

$$C_1 = \left[\frac{4\left(N_p^2 - 1\right)}{3}\sin^2\left(\frac{\pi}{N_c}\right) + 1\right]\sin^2\theta_v$$

$$C_2 = \frac{N_p - 1}{N_p}\cos^2\theta_v$$

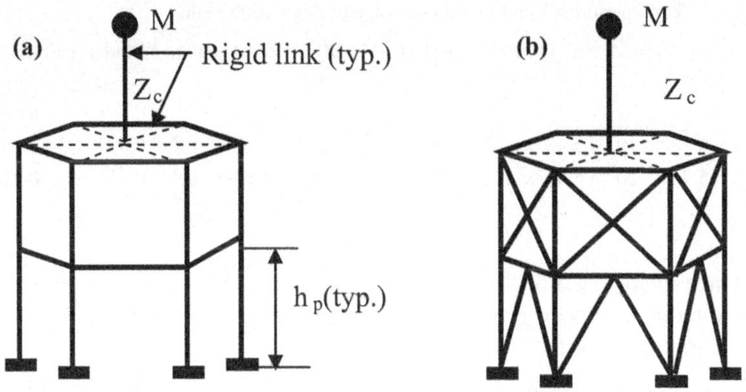

Fig. 13.24 Mathematical model to determine K_z and K_x of the staging frame

Axial stiffness of columns may be represented as below

$$K_z = \frac{E_c.I_c.N_c}{h_p N_p} \tag{13.3.1.4}$$

Alternatively, the vertical and lateral stiffness of the staging frame can be determined as mentioned hereafter shown in Fig. 13.24 is the three-dimensional mathematical model of the staging frame that can be easily modeled in any commercially available structural software like STAAD, SAP 2000, GTSTRUDL.

The mass of the silo including its content can be lumped at the node M at a height Z_c from the ring beam and is connected to the staging frame by massless rigid link of high stiffness ($K \approx 10^{10}$). We apply a unit load successively in the vertical and horizontal direction and find out the displacement at the ring beam level as δ_z and δ_x, say. Once the displacement is known, the corresponding stiffness can be obtained from the expression $P = K.\delta$, where P is the applied unit load.

Material Properties:

modulus of elasticity of columns, beams, and bracings: 21.7 GPa;
coefficient of friction of fill material: 0.7;
density of fill material: 14.1 kN/m^3;
horizontal to vertical pressure ratio (λ): 0.27.

Seismic Parameters:

Here IS 1893: 2002 is being used to calculate the seismic force, using the following parameters,

seismic zone factor for Zone V (Z): 0.36;
type of spectra: III (soft soil);

Table 13.7 Time periods of first two modes of vibrations of the silo

	K_x (N/m)	K_z (N/m)	m (kg)	Z_c (m)	J_θ (kg-m^2)	Modal periods	
						1st mode (s)	2nd mode (s)
Case 1	8.48×10^7	5.43×10^9	1.38×10^6	2.85	1.62×10^8	0.804	0.217
Case 2	5.74×10^8	5.43×10^9	1.38×10^6	2.85	1.62×10^8	0.318	0.210

importance factor (I): 1.5;
response reduction factor (R): 3 (ordinary moment resisting frame);
modal damping ratio: 5%.

Calculated time periods of silo for two types of staging are listed in Table 13.7.

13.4.2 Results and Discussion

Pressure profile along the depth has been calculated using the proposed method and compared with Janssen's pressure profile. Pressure profiles in Case 1 and Case 2 have been compared with Janssen's pressure profile in Figs. 13.25, 13.26, and 13.27, respectively (Fig. 13.28).

Perusing the above data, we can come to some very important and interesting conclusions mentioned hereafter:

1. Ignoring the seismic effect, we significantly under design the silo wall design pressure.

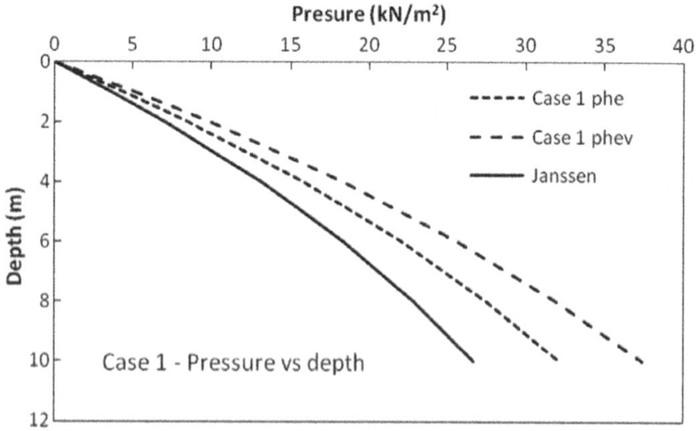

Fig. 13.25 Comparison of effective pressure profile for Case 1, considering seismic analysis without vertical seismic force (ph_e), with vertical seismic force (ph_{ev}) and Janssen's formula

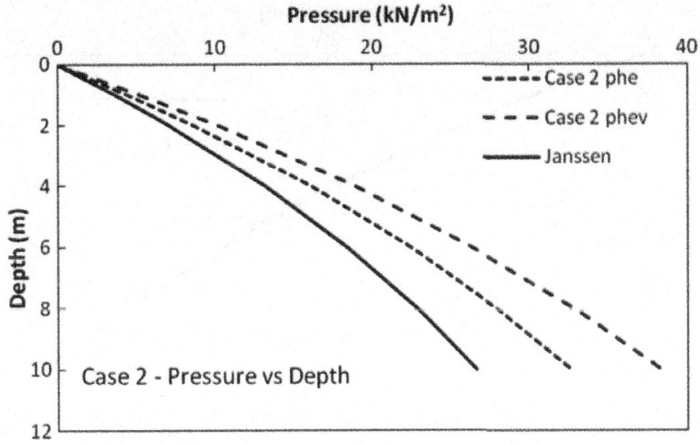

Fig. 13.26 Comparison of effective pressure profile for Case 2, considering seismic analysis without vertical seismic force (ph_e), with vertical seismic force (ph_{ev}) and Janssen's formula

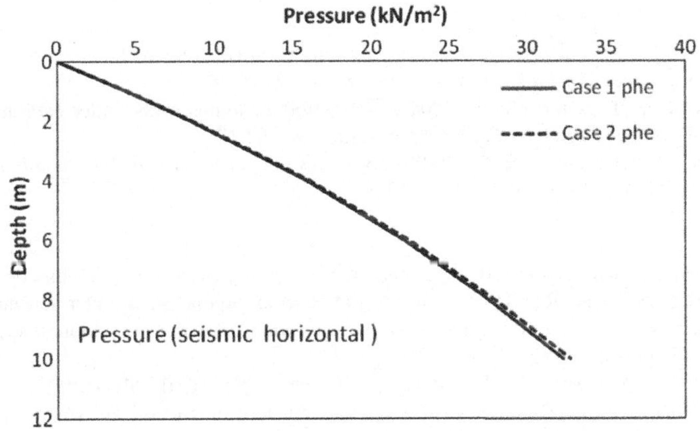

Fig. 13.27 Comparison of effective pressure profile for all cases, considering horizontal component of seismic force (ph_e)

2. The dynamic pressure can be as high as 37–40% over Jansen's static pressure depending on the staging configuration and type of foundation on which it is resting like soft, medium stiff soil, or rock.
3. The vertical component of earthquake that is usually ignored in conventional structural design significantly enhances the lateral dynamic pressure on the silo wall and should not be ignored, especially when the silo is of large capacity.
4. The mathematical model proposed herein, while sound in logic, is easy to apply within a design office framework and does not need an elaborate FEM analysis and can very well be adopted in a spreadsheet or a Mathcad shell.

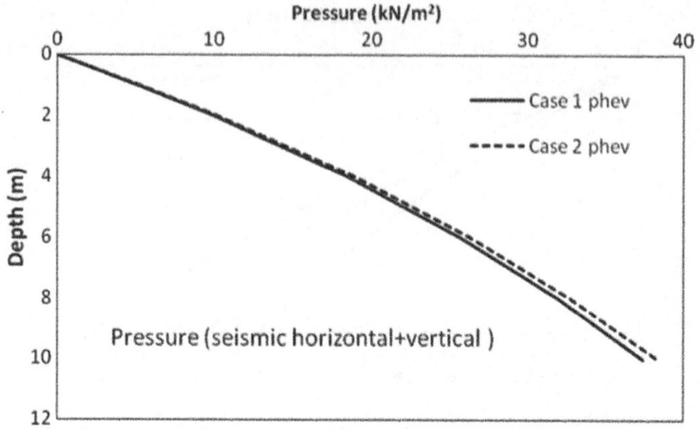

Fig. 13.28 Comparison of effective pressure profile for all cases, considering horizontal and vertical components of seismic force (ph_{ev})

References

Chowdhury, I. (2009). Dynamic response of reinforced concrete rectangular bunkers under earthquake force. *The Indian Concrete Journal, 83*(2), 7–18.

Chowdhury, I., & Dasgupta, S. P. (2004). Counterfort retaining walls under earthquake force. *Electronic Journal of Geotechnical Engineering, 9A*, USA.

Chowdhury, I., & Dasgupta, S. P. (2008). *Dynamics and Structures and Foundation-a unified approach*. Leiden, Holland: Taylor and Francis.

Chowdhury, I., & Singh, J. P. (2012). Dynamic response of rectangular bunker walls considering earthquake force. In *Proceedings International Symposium Engineering under Uncertainty and Safety Management*. India: Indian Institute of Engineering Science and Technology.

Chowdhury, I., & Tilak, R. (2010). Dynamic pressure on circular silos under seismic force. In *Proceedings 14th International Symposium on Earthquake Engineering*. India: Indian Institute of Technology Roorke.

Clough, R. W., & Penzien, J. (1984). *Structural dynamics*. New York: McGraw-Hill Publication.

Dutta, S., Mandal, A., & Dutta, S. C. (2003). Soil structure interaction in dynamic behavior of elevated tanks with alternate frame staging configurations. *Journal of Sound and Vibration*, 1–29.

Gray, W. S., & Manning, G. P. (1973). *Reinforced concrete water tower bunkers and silos*. London: Concrete Publications limited.

Hurty, W., & Rubenstein, M. (1967). *Dynamics of structures*. USA: Prentice Hall Publication.

IS-1893. (2002). *India standard code of practice for earthquake resistant design of structures*. Bureau of Indian Standards: New Delhi.

IS-1893. (2016). *Indian standard code of practice for design of earthquake resistant design of structures and foundations*.

Mallick, S. K., & Rangaswami, K. S. (1980). *Design of reinforced concrete structures*. New Delhi, India: Khanna Publishers.

Meirovitch, L. (1975). *Elements of vibration analysis*. New York, USA: Allied Publishers Inc.

Meirovitch, L. (1993). *Analytical methods in vibration*. UK: Macmillan Publication.

Timoshenko, S., & Kreiger, S. W. (1985). *Theory of plates and shells*. New York: McGraw-Hill Publication.

Chapter 14
Analysis of Water Tanks Under Earthquake Force

14.1 Introduction

Water has always remained an essential natural commodity for survival of a human habitat. One observes that, from early dawn of history, human civilization has always thrived near water bodies.

Mesopotamian civilization flourished besides rivers Euphrates (presently called Al Furat) and Tigris (Al Dijlah). In Indus valley civilization, cities like Mohenjo-Daro and Harappa thrived by river Indus, while the great Pharaohs ruled Egyptian empire from the banks of river Nile.

As human dwellings spread, storage of water became a necessity for daily use and consumption. Initially, holes dug in ground were used to store water, called *reservoirs*, like one shown in Fig. 14.1.

It was however soon realized that loss of water due to evaporation and ground percolation often does not make it an economic proposition—especially in areas where rainfall is sparse.

With advent of modern science, it also became apparent that water exposed to sunlight promotes development of algae and planktons in water, generating biological activities that often make the water unfit for human consumption and use with course of time.

Thus, storing water in a manner so that it can be long preserved without degeneration of its quality became necessary and brought in the trend of development of vessels to store it, which are called water tanks in modern phraseology.

In modern era, tanks are mostly built out of reinforced concrete, pre-stressed concrete, steel, and even high-density plastics (HDPE). They either rest on ground or are constructed underground and are sometimes built at a height above ground (to take advantage of the natural gravity force). A typical modern-day concrete water tank is as shown in Fig. 14.2.

The mechanics for design of these tanks under its own weight including hydrostatic pressure is an established technology; however, uncertainties still

I. Chowdhury and S. P. Dasgupta, *Earthquake Analysis and Design of Industrial Structures and Infra-structures*, GeoPlanet: Earth and Planetary Sciences,
https://doi.org/10.1007/978-3-319-90832-8_14

Fig. 14.1 Reservoirs built in city of Harappa for public use during 2600–1900 BC

Fig. 14.2 A modern-day overhead water tank built out of concrete

prevail on how the tanks behave under earthquake force, for often tanks are found to undergo damages under earthquake, despite being an essential service to a community, especially under a post-earthquake relief scenario.

In the present chapter, we look into the technology in vogue and also on some recent developments on earthquake resistant design of liquid retaining structures.

14.2 Rigid Tanks Resting on Ground

Tanks resting on ground are usually rectangular or circular in plan. Oily water separators, raw water/sewage water tanks, etc., are usually built of concrete having rectangular shapes, while oil storage tanks, chemical tanks, etc., in industrial plants are usually circular in plan and made of steel.

Ever since Jacobsen (1949), Graham and Rodriguez (1952) published solution for a fixed-based liquid retaining tank subjected to seismic force, the topic has been investigated by a number of researchers like Housner (1957, 1963), Edwards (1969), Veletsos and Yang (1977), Haroun and Housner (1981), Veletsos (1984), Tang (1986), to name some of the pioneering few.

Of these, Housner and Veletsos's work remains most popular and has been the backbone of Eurocode-8 Part IV, IS-1893 Part 2 (2016) and ACI 350.3-01 (2001) codes.

Shown in Fig. 14.3 is a tank, either circular (of diameter *D*) or rectangular (having width *L*), resting on ground and subjected to ground acceleration $\ddot{u}_g$ due to an earthquake.

It is obvious that the liquid will start vibrating having deformed surface, as shown by dotted line in Fig. 14.3a.

It can be shown (Housner 1963) that the liquid during vibration breaks up into two parts as shown in Fig. 14.3b.

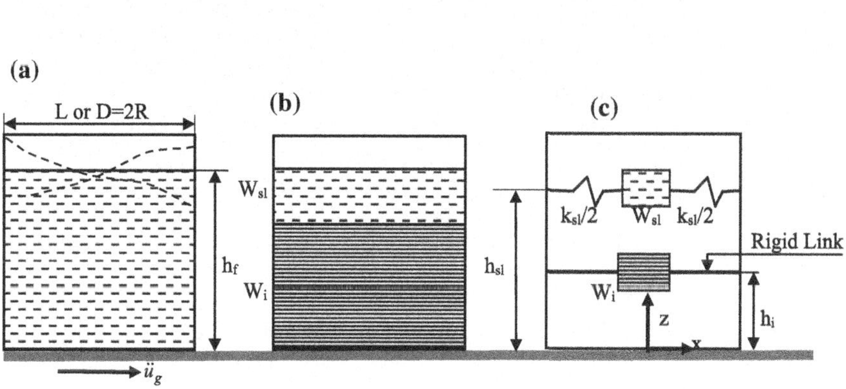

Fig. 14.3 Water tanks resting on ground subjected to ground acceleration

The bottom part, called the impulsive weight W_i, moves in the same phase as the tank itself as a rigid body, while the top part, called sloshing weight W_{sl}, undergoes free vibration which is a function of the fluid stiffness and mass. The total weight of fluid in tank $W = W_i + W_{sl}$.

Thus, the mathematical model for analysis can be expressed as shown in Fig. 14.3c, where the sloshing weight W_{sl} is connected to the tank wall through the fluid spring k_{sl} at a height h_{sl} from the bottom of tank. The impulsive weight W_i is connected by rigid links to the tank wall at a height h_i from the bottom of the tank.

The parameters as defined above are further elaborated in Table 14.1.

For fixed-based rigid tank, one assumes the tank to be subjected to peak ground acceleration (PGA) based on which the base shear is expressed as,

$$V_b = (m_i + m_w + m_r)S_e(T_i) + m_{sl}S_e(T_{sl}) \tag{14.2.1}$$

Moment at the base of vertical wall is expressed as,

$$M_b = (m_i h_i + m_w h_w + m_r h_r)S_e(T_i) + m_{sl}h_{sl}S_e(T_{sl}) \tag{14.2.2}$$

Table 14.1 Design parameters for dynamic analysis of tanks with fluid (after Housner)

Item	Relationship (rectangular tank)	Relationship (circular tank)
Impulsive weight	$W_i = W\frac{\tanh(0.866L/h_f)}{0.866L/h_f}$	$W_i = W\frac{\tanh(\sqrt{3}R/h_f)}{\sqrt{3}R/h_f}$
Application point of impulsive force	$h_i = 0.375h_f$ for $h_f/L \le 0.75$ and $h_i = \left[0.5 - \frac{0.09375}{h_f/L}\right]h_f$ for $h_f/L > 0.75$—Excluding base pressure (EBP)	$h_i = 0.375h_f$ for $h_f/D \le 0.75$ and $h_i = \left[0.5 - \frac{0.1875}{h_f/R}\right]h_f$ for $h_f/D > 0.75$—Excluding base pressure (EBP)
	$h'_i = h_f\left[\frac{0.866L/h_f}{2\tanh(0.866L/h_f)} - \frac{1}{8}\right]$ for $h_f/L \le 1.33$ and $h'_i = 0.45h_f$ for $h_f/L > 1.33$—Including base pressure (IBP)	$h'_i = h_f\left[\frac{\sqrt{3}R/h_f}{2\tan(\sqrt{3}R/h_f)} - \frac{1}{8}\right]$ for $h_f/D \le 1.33$ and $h'_i = 0.45h_f$ for $h_f/D > 1.33$—Including base pressure (IBP)
Sloshing weight	$W_{sl} = 0.264W\frac{L}{h_f}\tanh\left(3.16\frac{h_f}{L}\right)$	$W_{sl} = 0.46W\frac{R}{h_f}\tanh\left(1.84\frac{h_f}{R}\right)$
Application point of sloshing force	$h_{sl} = h_f\left[1 - \frac{\cosh(3.16h_f/L)-1}{3.16\frac{h_f}{L}\sinh\left(3.16\frac{h_f}{L}\right)}\right]$— Excluding base pressure (EBP)	$h_{sl} = h_f\left[1 - \frac{\cosh(1.84h_f/R)-1}{1.84\frac{h_f}{R}\sinh\left(1.84\frac{h_f}{R}\right)}\right]$— Excluding base pressure (EBP)
	$h'_{sl} = h_f\left[1 - \frac{\cosh(3.16h_f/L)-2.01}{3.16\frac{h_f}{L}\sinh\left(3.16\frac{h_f}{L}\right)}\right]$— Including base pressure (IBP)	$h'_{sl} = h_f\left[1 - \frac{\cosh(1.84h_f/R)-2.01}{1.84\frac{h_f}{R}\sinh\left(1.84\frac{h_f}{R}\right)}\right]$— Including base pressure (IBP)
Stiffness	$k_{sl} = \frac{0.833W}{h_f}\tanh^2\left(3.16\frac{h_f}{L}\right)$	$k_{sl} = 0.836\frac{W}{h_f}\tanh^2\left(1.84\frac{h_f}{R}\right)$
Damping	$C_{sl} = \left(\frac{0.5}{100}\right)2\sqrt{k_{sl}m_{sl}}$	Same as rectangular case

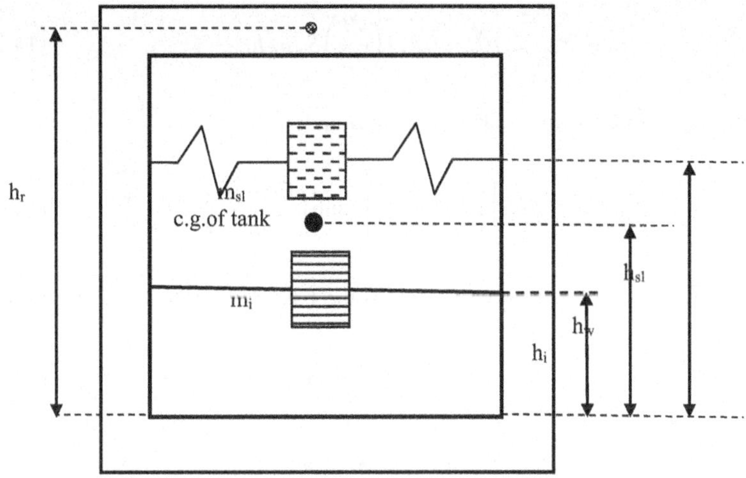

Fig. 14.4 Rectangular tank with lumped mass and geometric position

Moment at the foundation level is expressed as,

$$M_{\mathrm{bf}} = \left(m_{\mathrm{i}}h'_{\mathrm{i}} + m_{\mathrm{w}}h'_{\mathrm{w}} + m_{\mathrm{r}}h'_{\mathrm{r}}\right)S_{\mathrm{e}}(T_{\mathrm{i}}) + m_{\mathrm{sl}}h'_{\mathrm{sl}}S_{\mathrm{e}}(T_{\mathrm{sl}}) \tag{14.2.3}$$

In Eqs. (14.2.1), (14.2.2), (14.2.3), and Fig. 14.4:

$m_{\mathrm{i}}, m_{\mathrm{sl}}$ impulsive and sloshing mass of water, respectively;
m_{w} mass of wall of the tank;
m_{r} contributory mass from roof on tank wall;
$h_{\mathrm{i}}, h_{\mathrm{sl}}, h_{\mathrm{w}}, h_{\mathrm{r}}$ height of impulsive and sloshing mass excluding base pressure (EBP), centroidal height of wall at top of base raft, and height of roof from top of base raft, respectively;
$h'_{\mathrm{i}}, h'_{\mathrm{sl}}, h'_{\mathrm{w}}, h'_{\mathrm{r}}$ height of impulsive and sloshing mass including base pressure (IBP), centroidal height of wall to bottom of base raft, and height of roof to bottom of base raft, respectively;
$S_{\mathrm{e}}(T_{\mathrm{sl}})$ spectral acceleration due to time period of sloshing mass m_{sl}. Value of T_{sl} is furnished by the expression $T_{\mathrm{sl}} = 2\pi\sqrt{m_{\mathrm{sl}}/k_{\mathrm{sl}}}$;
$S_{\mathrm{e}}(T_{\mathrm{i}})$ spectral acceleration due to time period of impulsive mass m_{i}. For rigid tank, T_{i} is equal to zero.

14.2.1 Hydrodynamic Pressure on Tank Wall and Base

For rectangular tank (refer Fig. 14.3), the impulsive pressure $(p_{\mathrm{i}}^{\mathrm{w}})$ on wall in lateral direction is expressed as (Veletsos 1984),

$$p_{\mathrm{i}}^{\mathrm{w}} = \psi_{\mathrm{i}}(z)\left(\frac{ZI}{2R}\right)S_{\mathrm{ai}}\rho_{\mathrm{w}}h_{\mathrm{f}} \tag{14.2.1.1}$$

Here, ρ_{w} = mass density of water/liquid and,

$$\psi_{\mathrm{i}}(z) = 0.866\left[1 - \left(\frac{z}{h_{\mathrm{f}}}\right)^2\right]\tanh\left(\frac{0.866L}{h_{\mathrm{f}}}\right) \tag{14.2.1.2}$$

Impulsive pressure acting vertically on the base slab ($p_{\mathrm{i}}^{\mathrm{bs}}$) is expressed as,

$$p_{\mathrm{i}}^{\mathrm{bs}} = \psi_{\mathrm{i}}(x)\left(\frac{ZI}{2R}\right)S_{\mathrm{ai}}\rho_{\mathrm{w}}h_{\mathrm{f}} \tag{14.2.1.3}$$

Here,

$$\psi_{\mathrm{i}}(x) = \frac{\sinh\left(\frac{1.732x}{h_f}\right)}{\cosh\left(\frac{0.866L}{h_f}\right)} \tag{14.2.1.4}$$

As shown in Fig. 14.3c, $x = 0$ at center of the tank in plan with extreme ends as $(\pm)L/2$.

The sloshing pressure acting on the vertical wall $\left(p_{\mathrm{sl}}^{\mathrm{w}}\right)$ in horizontal direction is expressed as,

$$p_{\mathrm{sl}}^{\mathrm{w}} = \psi_{\mathrm{sl}}(z)\left(\frac{ZI}{2R}\right)S_{\mathrm{asl}}\rho_{\mathrm{w}}L \tag{14.2.1.5}$$

Here,

$$\psi_{\mathrm{sl}}(z) = 0.4165\frac{\cosh\left(\frac{3.162z}{L}\right)}{\cosh\left(\frac{3.162h_{\mathrm{f}}}{L}\right)} \tag{14.2.1.6}$$

The sloshing pressure on the base slab is expressed as,

$$p_{\mathrm{sl}}^{\mathrm{bs}} = \psi_{\mathrm{sl}}(x)\left(\frac{ZI}{2R}\right)S_{\mathrm{asl}}\rho_{\mathrm{w}}L \tag{14.2.1.7}$$

Here,

$$\psi_{\mathrm{sl}}^{\mathrm{b}}(x) = 1.25\left[\frac{x}{L} - \frac{4}{3}\left(\frac{x}{L}\right)^3\right]\frac{1}{\cosh\left(\frac{3.162h_{\mathrm{f}}}{L}\right)} \tag{14.2.1.8}$$

The impulsive pressure on wall due to its own inertia is given by,

$$p_{\mathrm{w}}^{\mathrm{w}} = \left(\frac{ZI}{2R}\right) S_{\mathrm{ai}} t_{\mathrm{w}} \rho_{\mathrm{m}} \tag{14.2.1.9}$$

Here, t_{w} = thickness of the tank wall and ρ_{m} = mass density of the wall.

The hydrodynamic pressure on tank wall due to vertical acceleration is considered as,

$$p_{\mathrm{v}}^{\mathrm{w}} = \frac{2}{3}\left(\frac{ZI}{2R}\right) S_{\mathrm{ai}} \rho_{\mathrm{w}} h_{\mathrm{f}} \left(1 - \frac{z}{h_{\mathrm{f}}}\right) \tag{14.2.1.10}$$

The resultant pressure on the wall is expressed as SRSS values of all above as,

$$p_{\mathrm{r}}^{\mathrm{w}} = \sqrt{\left(p_{\mathrm{i}}^{\mathrm{w}} + p_{\mathrm{w}}^{\mathrm{w}}\right)^2 + \left(p_{\mathrm{sl}}^{\mathrm{w}}\right)^2 + \left(p_{\mathrm{v}}^{\mathrm{w}}\right)^2} \tag{14.2.1.11}$$

The resultant pressure on the base slab is given by,

$$p_{\mathrm{r}}^{\mathrm{bs}} = \sqrt{\left(p_{\mathrm{i}}^{\mathrm{bs}}\right)^2 + \left(p_{\mathrm{sl}}^{\mathrm{bs}}\right)^2} \tag{14.2.1.12}$$

For circular tanks as shown in Fig. 14.5.

$$p_{\mathrm{i}}^{\mathrm{w}} = \psi_{\mathrm{i}}(z)\left(\frac{ZI}{2R}\right) S_{\mathrm{ai}} \rho_{\mathrm{w}} h_{\mathrm{f}} \cos\phi \tag{14.2.1.13}$$

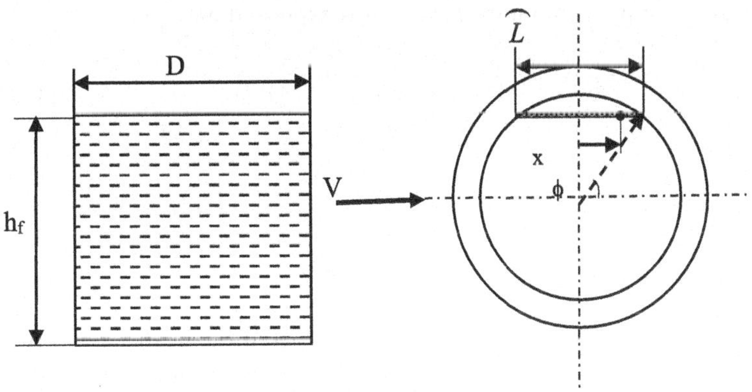

Fig. 14.5 Circular tank with its geometric definition

Here, ρ_w = mass density of water/liquid and

$$\psi_i(z) = 0.866\left[1 - \left(\frac{z}{h_f}\right)^2\right] \tanh\left(\frac{0.866D}{h_f}\right) \tag{14.2.1.14}$$

Impulsive pressure acting vertically on the base slab is expressed as,

$$p_i^{bs} = 0.866\psi_i(x)\left(\frac{ZI}{2R}\right) S_{ai}\rho_w h_f \tag{14.2.1.15}$$

Here,

$$\psi_i(x) = \frac{\sinh\left(\frac{1.732x}{h_f}\right)}{\cosh\left(\frac{0.866L}{h_f}\right)} \tag{14.2.1.16}$$

The sloshing pressure acting on the vertical wall in horizontal direction is expressed as,

$$p_{sl}^{w} = \psi_{sl}(z)\left(\frac{ZI}{2R}\right) S_{asl}\rho_w D\left(1 - \frac{1}{3}\cos^2\phi\right)\cos\phi \tag{14.2.1.17}$$

Here,

$$\psi_{sl}(z) = 0.5625\frac{\cosh\left(\frac{3.674z}{D}\right)}{\cosh\left(\frac{3.674h_f}{D}\right)} \tag{14.2.1.18}$$

The sloshing pressure on the base slab is expressed as,

$$p_{sl}^{bs} = \psi_{sl}(x)\left(\frac{ZI}{2R}\right) S_{asl}\rho_w D \tag{14.2.1.19}$$

Here,

$$\psi_{sl}^{b}(x) = 1.25\left[\frac{x}{D} - \frac{4}{3}\left(\frac{x}{D}\right)^3\right]\frac{1}{\cosh\left(\frac{3.674h_f}{D}\right)} \tag{14.2.1.20}$$

The impulsive pressure on wall due to its own inertia is as given in Eq. (14.2.1.9), and the hydrodynamic pressure on tank wall due to vertical acceleration is given by Eq. (14.2.1.10).

The resultant pressure on the wall and base slab is expressed as SRSS values given by Eqs. (14.2.1.11) and (14.2.1.12), respectively.

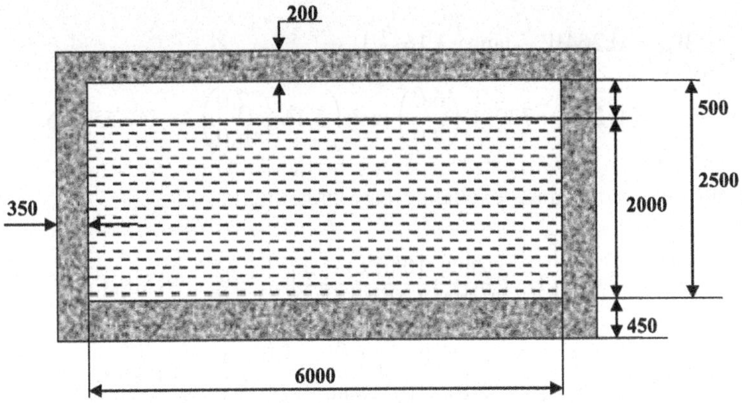

Fig. 14.6 Rectangular tank resting on ground—geometric dimensions

Example 14.1 A reinforced concrete firewater tank (square in plan) of size 6.0 m × 6.0 m in plan, of 2.5 m height, with monolithically cast roof slab is resting on ground, as shown in Fig. 14.6. The tank is situated in a site attributed as zone IV as per IS 1893. Compute the moment in the wall and the foundation including the dynamic pressures on the wall. Consider $\gamma_c = 25$ kN/m^3 and that of water as 10 kN/m^3. Compute moment and shears as per Euorcode8 and IS code. Assume the tank behaving as a rigid box.

Solution

Weight of water in the tank = 6.0 × 6.0 × 2.0 × 10.0 = 720 kN

Weight of tank wall = (6.7 × 6.7 − 6.0 × 6.0) × 2.5 × 25 = 555.625 kN

Weight of roof slab = 6.7 × 6.7 × 0.2 × 25 = 224.45 kN

Weight of base slab = 6.7 × 6.7 × 0.45 × 25 = 505.013 kN

Thus, weight of empty tank = 555.625 + 224.450 + 505.013 = 1285 kN

Impulsive weight of water,

$$W_i = W\frac{\tanh(0.866L/h_f)}{0.866L/h_f} = 720\left[\frac{\tanh\left(\frac{0.866\times 6}{2.0}\right)}{\frac{0.866\times 6}{2.0}}\right]$$

$$= 274.083 \text{ kN}.$$

Height of impulsive mass ($h_f/L = 0.333 < 0.75$),

$h_i = 0.375h_f = 0.375 \times 2 = 0.75$ m, excluding base pressure.

Again $$h_i' = h_f\left[\frac{0.866L/h_f}{2\tanh(0.866L/h_f)} - \frac{1}{8}\right] = 2.0\left[\frac{\frac{0.866\times 6.0}{2.0}}{2\tanh\left(\frac{0.866\times 6}{2}\right)} - 0.125\right] = 2.377 \text{ m}$$

including base pressure.

Weight of sloshing mass.

$$\begin{aligned} W_{\mathrm{sl}} &= 0.264W\frac{L}{h_{\mathrm{f}}}\tanh\left(3.16\frac{h_{\mathrm{f}}}{L}\right) \\ &= 0.264 \times 720 \times \left(\frac{6.0}{2.0}\right)\tanh\left(3.16 \times \frac{2.0}{6.0}\right) = 446.554\ \mathrm{kN} \end{aligned}$$

Height of sloshing mass excluding base pressure,

$$\begin{aligned} h_{\mathrm{sl}} &= h_{\mathrm{f}}\left[1 - \frac{\cosh(3.16h_{\mathrm{f}}/L) - 1}{3.16\frac{h_{\mathrm{f}}}{L}\sinh\left(3.16\frac{h_{\mathrm{f}}}{L}\right)}\right] \\ &= 2.0\left[1 - \frac{\cosh\left(\frac{3.16\times2.0}{6.0}\right) - 1.0}{3.16 \times \left(\frac{2.0}{6.0}\right)\sinh\left(\frac{3.16\times2.0}{6.0}\right)}\right] = 1.083\ \mathrm{m}. \end{aligned}$$

Height of sloshing mass including base pressure,

$$\begin{aligned} h_{\mathrm{sl}} &= h_{\mathrm{f}}\left[1 - \frac{\cosh(3.16h_{\mathrm{f}}/L) - 2.01}{3.16\frac{h_{\mathrm{f}}}{L}\sinh\left(3.16\frac{h_{\mathrm{f}}}{L}\right)}\right] \\ &= 2.0\left[1 - \frac{\cosh\left(\frac{3.16\times2.0}{6.0}\right) - 2.01}{3.16 \times \left(\frac{2.0}{6.0}\right)\sinh\left(\frac{3.16\times2.0}{6.0}\right)}\right] = 2.606\ \mathrm{m}. \end{aligned}$$

Spring stiffness of sloshing fluid,

$$\begin{aligned} k_{\mathrm{sl}} &= \frac{0.833W}{h_{\mathrm{f}}}\tanh^2\left(3.16\frac{h_{\mathrm{f}}}{L}\right) \\ &= \frac{0.833 \times 720}{2}\tanh^2\left(\frac{3.16 \times 2}{6}\right) = 183.9\ \mathrm{kN/m}. \end{aligned}$$

Time period due to sloshing mass,

$$T_{\mathrm{sl}} = 2\pi\sqrt{\frac{M_{\mathrm{sl}}}{k_{\mathrm{sl}}}}$$

Here, $M_{\mathrm{sl}} = \frac{W_{\mathrm{sl}}}{g} = \frac{446.554}{9.81} = 45.52\ \mathrm{kN\text{-}s^2/m}$.

Thus, $T_{\mathrm{sl}} = 2\pi\sqrt{\frac{45.52}{183.9}} = 3.126\ \mathrm{s}$.

Considering the tank is resting on soft soil,

$$\frac{S_{\mathrm{asl}}}{g} = \left(\frac{ZI}{2R}\right) \times \left(\frac{1.67}{T_{\mathrm{sl}}}\right)$$

Here, $Z = 0.24$, as per IS-1893. $I = 1.5$ for RCC tank $R = 2.0$.

Thus,

$$S_{asl} = \left(\frac{ZI}{2R}\right) \times \left(\frac{1.67}{T_{sl}}\right) \times g$$
$$= \left(\frac{0.24 \times 1.5}{2 \times 2}\right) \times \left(\frac{1.67}{3.126}\right) \times 9.81 = 0.472 \text{ m/s}^2.$$

The above is valid for 5% damping. Considering that for water we take 0.5% damping, the design spectral acceleration above is modified by a scaling factor, which gives

$$S^{d}_{asl} = \sqrt{\frac{10}{5+\zeta}} \times S_{asl}$$
$$= \sqrt{\frac{10}{5+0.5}} \times 0.472 = 0.636 \text{ m/s}^2.$$

For impulsive mass, considering the tank as rigid, the tank will move with the ground at peak ground acceleration which is expressed as,

$$\frac{S_{ai}}{g} = \left(\frac{ZI}{2R}\right)(1+15T)$$

where at $T = 0$, as the tank is considered to have infinite stiffness,

$$S_{ai} = \left(\frac{ZI}{2R}\right)g = \left(\frac{0.24 \times 1.5}{2 \times 2}\right) \times 9.81 = 0.883 \text{ m/s}^2$$

Moment at base of vertical wall is expressed as,

$$M_{bw} = (m_i h_i + m_w h_w + m_r h_r) S_e(T_i) + m_{sl} h_{sl} S_e(T_{sl})$$

$$M_{bw} = \left(\frac{274.083}{9.81} \times 0.75 + \frac{555.625}{9.81} \times 1.25 + \frac{224.45}{9.81} \times 2.6\right) \times 0.883$$
$$+ \frac{446.554}{9.81} \times 1.083 \times 0.636 = 164.898 \text{ kN.m}$$

Moment at top of base slab,

$$M_{bf} = (m_i h'_i + m_w h'_w + m_r h'_r) S_e(T_i) + m_{sl} h'_{sl} S_e(T_{sl})$$

$$M_{bf} = \left(\frac{274.083}{9.81} \times 2.377 + \frac{555.625}{9.81} \times 1.25 + \frac{224.45}{9.81} \times 2.6\right) \times 0.883$$
$$+ \frac{446.554}{9.81} \times 2.606 \times 0.636 = 250.67 \text{ kN.m}$$

The base shear is expressed as,

$$V_b = (m_i + m_w + m_r)S_e(T_i) + m_{sl}S_e(T_{sl})$$

$$V_b = \left(\frac{274.083}{9.81} + \frac{555.625}{9.81} + \frac{224.45}{9.81}\right) \times 0.883 + \frac{446.554}{9.81} \times 0.636 = 123.83 \text{ kN}.$$

One should realize here that the whole tank is considered as a rigid block in the above case.

Thus, for V_b = 123.83 kN inducing a moment of M_{bw} = 164.89 kN.m.

$\bar{h} = \frac{164.89}{123.83} = 1.33$ m above base of wall.

Thus, the shear acting on each wall = 123.83/2 = 61.9 kN.

Thus, the moment acting at base of each wall = 61.9 × 1.33 = 82.34 kN/m

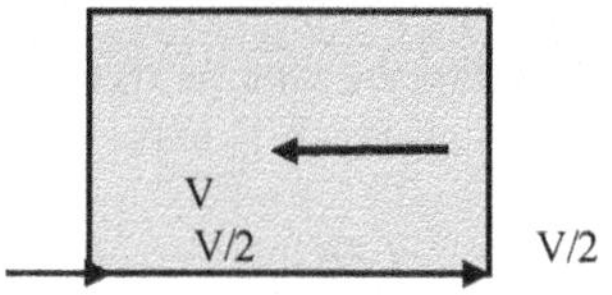

The above expressions are followed by Eurocode 8 that essentially adopts the absolute sum technique. IS-1893 follows the SRSS method, when as per IS-1893.

- **Moment in vertical wall**

$$M_{bw} = \sqrt{[(m_ih_i + m_wh_w + m_rh_r)S_e(T_i)]^2 + [m_{sl}h_{sl}S_e(T_{sl})]^2}$$

$$M_{bw} = \sqrt{(133.54)^2 + (31.354)^2} = 137.17 \text{ kN.m}$$

Thus, the moment per wall = 68.6 kN.m

- **Moment at top of foundation**

$$M_{bf} = \sqrt{(175.224)^2 + (75.44)^2} = 190.77 \text{ kN.m}$$

- **Base shear**

$$V_b = \sqrt{(94.88)^2 + (28.95)^2} = 99.19 \text{ kN}.$$

Thus, shear at base of each wall = 49.6 kN.

The sloshing wave height is expressed as,

$$\Delta = \left(\frac{ZI}{2}\right) \times \frac{S_{\text{asl}}^{\text{d}}}{g} \times \frac{L}{2}$$
$$= \left(\frac{0.24 \times 1.5}{2}\right) \times \left(\frac{1.67}{3.126}\right) \times \sqrt{\frac{10}{5+0.5}} \times \frac{6}{2} = 0.388 \text{ m}$$

Thus, the expected height of sloshing is 388 mm.

Computation of dynamic pressure on wall (kN/m^2)

Based on the formulation as shown in Sect. 14.2.1, the dynamic pressures on wall are tabulated hereafter.

z/h_f	$\psi_i(z)$	$p_i(z)$	$\psi_{sl}(z)$	$p_{sl}(z)$	p_w	p_{vi}	p_{srss}
0	0.85646	1.526244	0.258885	0.996877	0.787589	1.193966	2.788036
0.05	0.854319	1.522429	0.259245	0.998262	0.787589	1.134267	2.760303
0.1	0.847895	1.510982	0.260324	1.002419	0.787589	1.074569	2.728181
0.15	0.83719	1.491904	0.262127	1.009362	0.787589	1.014871	2.691628
0.2	0.822201	1.465195	0.264658	1.019108	0.787589	0.955173	2.650655
0.25	0.802931	1.430854	0.267925	1.031685	0.787589	0.895474	2.605329
0.3	0.779378	1.388882	0.271935	1.047129	0.787589	0.835776	2.555783
0.35	0.751544	1.339279	0.276701	1.065481	0.787589	0.776078	2.502222
0.4	0.719426	1.282045	0.282236	1.086793	0.787589	0.716379	2.444934
0.45	0.683027	1.21718	0.288554	1.111124	0.787589	0.656681	2.384308
0.5	0.642345	1.144683	0.295675	1.138541	0.787589	0.596983	2.320849
0.55	0.597381	1.064555	0.303616	1.169122	0.787589	0.537285	2.255207
0.6	0.548134	0.976796	0.312401	1.20295	0.787589	0.477586	2.188203
0.65	0.494606	0.881406	0.322054	1.24012	0.787589	0.417888	2.120866
0.7	0.436795	0.778385	0.332602	1.280735	0.787589	0.35819	2.054472
0.75	0.374701	0.667732	0.344073	1.324907	0.787589	0.298491	1.990587
0.8	0.308326	0.549448	0.3565	1.372761	0.787589	0.238793	1.931104
0.85	0.237668	0.423533	0.369918	1.424427	0.787589	0.179095	1.878266
0.9	0.162727	0.289986	0.384363	1.480051	0.787589	0.119397	1.834659
0.95	0.083505	0.148809	0.399876	1.539786	0.787589	0.059698	1.803149
1	−3.8E−16	−6.8E−16	0.4165	1.603798	0.787589	−2.7E−16	1.786747

The pressure diagrams are plotted in Fig. 14.7.

14.3 Flexible Circular Tanks Resting on Ground

Figure 14.8 shows typical flexible circular oil tanks resting on ground, a very common occurrence in oil and gas industry.

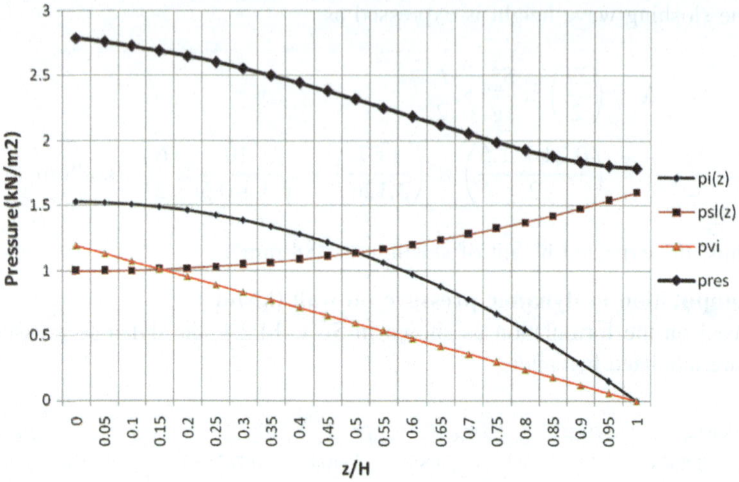

Fig. 14.7 Dynamic pressure plot of fluid on tank wall

Fig. 14.8 Flexible circular oil tanks resting on ground

For flexible circular tanks resting on ground with wall firmly fixed to base slab, the time period due to impulsive mass is expressed as,

$$T_{\mathrm{i}} = C_{\mathrm{i}} \frac{h_{\mathrm{f}}/\sqrt{\rho}}{\sqrt{t_{\mathrm{w}}/D} \times \sqrt{E_{\mathrm{t}}}} \tag{14.3.1}$$

Here,

$$C_{\mathrm{i}} = \frac{1}{\sqrt{\frac{h_{\mathrm{f}}}{D}}\left[0.46 - 0.3\left(\frac{h_{\mathrm{f}}}{D}\right) + 0.067\left(\frac{h_{\mathrm{f}}}{D}\right)^2\right]} \tag{14.3.2}$$

where h_{f} = depth of fluid in tank, D = diameter of tank, t_{w} = thickness of tank wall, E_{t} = Young's modulus of the tank material, and ρ = mass density of liquid (Fig. 14.9).

Time period of the sloshing mass is expressed as,

$$T_{\mathrm{sl}} = C_{\mathrm{sl}}\sqrt{D/g} \tag{14.3.3}$$

where

$$C_{\mathrm{sl}} = \frac{2\pi}{\sqrt{3.68\tanh\left(3.68\frac{h_{\mathrm{f}}}{D}\right)}} \tag{14.3.4}$$

Having computed the time period, the seismic coefficient is expressed as,

$$A_{\mathrm{h}} = \frac{ZI}{2R}\left(\frac{S_{\mathrm{a}}}{g}\right) \tag{14.3.5}$$

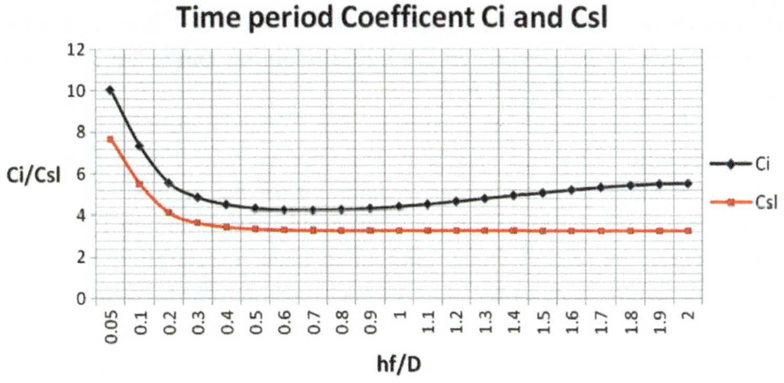

Fig. 14.9 Graphical plot of time period coefficients for impulsive and sloshing force

Table 14.2 Importance factor I for water tanks

Type of liquid storage tank	I
Tanks storing highly inflammable or hazardous liquid that can harm human life due to spillage	1.75
Tanks storing drinking water, non-volatile material, low inflammable petrochemicals, emergency services like firefighting, post-earthquake relief	1.5
Tanks with no hazard to human life or environment with insignificant consequence to economy and environment	1.0

Table 14.3 Response reduction factor R for tanks resting on ground

Ground supported tank	R
1.0. Masonry Tank	
1.1. Masonry wall reinforced with horizontal bands	1.3[b]
1.2. Masonry wall reinforced with horizontal bars with vertical bars at corners jambs and openings	1.5[b]
2.0. RCC/pre-stressed tanks	
2.1. Fixed or pinned-based tanks	2.0
2.2. Anchored flexible-based tank	2.5
2.3. Un-anchored restrained or unrestrained tanks	1.5
3.0 Steel tanks	
Un-anchored base	2.0
Anchored base	2.5
4.0 Underground tanks	4.0[a]

[a]for partially embedded tanks, value of R can be linearly interpolated between tanks resting on ground and fully embedded
[b]not permissible for construction in zone IV and V

The acronyms in Eq. (14.3.5) have already been defined in Chap. 2, Sect. 2.2.1; however, liquid retaining tanks have specific values of I and R as mentioned below.

Importance factors I for water tanks are as listed in Table 14.2.

The zone factors remain same as furnished in Table 2.4, Chap. 2. Values of R, the ductility factor to be considered for liquid retaining tanks, are as furnished in Table 14.3.

Based on the time periods as derived in Eqs. (14.3.1) and (14.3.3), the design spectral acceleration S_{ai}/g and S_{asl}/g can be computed from Fig. 2.5 of Chap. 2.

Code recommends to consider 5–7% damping for the impulsive force for concrete tanks and 2–3% damping for steel tanks and 0.5% damping for the sloshing force for which the S_{asl}/g value computed need to multiplied by a factor $\sqrt{10/(5+D)}$ while computing the design spectral acceleration. Here, D is damping ratio of the sloshing fluid.

14.3.1 Computation of Base Shear

The base shear at bottom of wall due to impulsive force is expressed as,

$$V_{bi} = A_{hi}(m_i + m_w + m_r + m_b)g \tag{14.3.1.1}$$

where m_b = mass of base slab,
and for sloshing force,

$$V_{bsl} = A_{hsl} m_{sl} g \tag{14.3.1.2}$$

The resultant base shear is finally expressed as,

$$V_{bR} = \sqrt{V_{bi}^2 + V_{bsl}^2} \tag{14.3.1.3}$$

14.3.2 Computation of Moment at the Base of Wall

Moment at the base of wall due to impulsive force is expressed as,

$$M_{bi} = A_{hi}(m_i h_i + m_w h_w + m_r h_r + m_b h_b)g \tag{14.3.2.1}$$

Here, m_b and h_b are mass and centeroidal height of base slab above the soil line and below the foundation.

The moment due to sloshing force is expressed as,

$$M_{bsl} = A_{hsl} m_{sl} h_{sl} \tag{14.3.2.2}$$

where symbols m_i, m_w, m_r, m_{sl}, h_i, h_w, h_r, h_{sl}, etc., have been defined previously in Sect. 14.2.

The resultant moment at the base of wall is defined as,

$$M_{bR} = \sqrt{M_{bi}^2 + M_{bsl}^2} \tag{14.3.2.3}$$

For computation of moment at the base of foundation due to impulsive force,

$$M_{bif} = A_{hi}\left(m_i h_i' + m_w h_w' + m_r h_r + m_b h_b\right)g \tag{14.3.2.4}$$

$$M_{bslf} = A_{hsl} m_{sl} h_{sl}' g \tag{14.3.2.5}$$

The acronyms in Eqs. (14.3.2.4) and (14.3.2.5) have already been defined in Sect. 14.2.

The resultant force on soil/foundation is expressed as,

$$M_{\mathrm{bRf}} = \sqrt{M_{\mathrm{bif}}^2 + M_{\mathrm{bslf}}^2} \quad (14.3.2.6)$$

14.4 Flexible Rectangular Tanks Resting on Ground

For flexible rectangular tanks resting on ground (Fig. 14.10), the time period due to impulsive force is given by,

$$T_{\mathrm{i}} = 2\pi\sqrt{\delta/g} \quad (14.4.1)$$

Here, δ = deflection of wall at vertical center at a height $\bar{h}$ when loaded by a uniformly distributed pressure q, where

$$\bar{h} = \frac{\frac{m_{\mathrm{i}}}{2} \times h_{\mathrm{i}} + \hat{m}_{\mathrm{w}} \times \frac{h_{\mathrm{w}}}{2}}{\frac{m_{\mathrm{i}}}{2} + \hat{m}_{\mathrm{w}}} \quad (14.4.2)$$

and

$$q = \frac{\left(\frac{m_{\mathrm{i}}}{2} + \hat{m}_{\mathrm{w}}\right)g}{B \times h_{\mathrm{w}}} \quad (14.4.3)$$

Fig. 14.10 Rectangular concrete tank resting on ground

Here,

m_i impulsive mass as defined in Table 14.1;
$\hat{m}_w$ mass of one tank wall perpendicular to the direction of the earthquake force;
B width of tank wall;
h_w height of tank wall.

Based on the above, an equivalent concentrated force P is assumed to work at a height $\bar{h}$, where $P = qh_f$ and the deflection δ is expressed as,

$$\delta = \frac{Ph^3}{3EI}$$

where E = Young's modulus of the tank wall and $I = 1 \times t_w^3/12$, where t_w = thickness of tank wall.

The sloshing time period can be expressed as,

$$T_{sl} = 2\pi\sqrt{m_{sl}/k_{sl}} \quad (14.4.4)$$

Amplitude of liquid sloshing Δ may be expressed as,

$$\Delta = \left(\frac{ZI}{2}\right)\left(\frac{S_{asl}}{g}\right)\sqrt{\frac{10}{5+D}} \times \frac{L}{2} \quad (14.4.5)$$

Here, m_{sl} and k_{sl} are as furnished in Table 14.1.

Once the time periods are computed, the rest of steps remain same as elaborated in Sects. 14.3.1 and 14.3.2 for circular tanks.

14.5 Some Comments on the Method as Proposed for Flexible Tanks

The time period as furnished by Eq. (14.3.1) is strictly valid for walls monolithically cast with base or welded as a moment connection with the base plate. They are not valid when the base is hinged or the wall is supported on elastomeric bearing. Thus, one has to be careful in applying this formula in these cases as they may give a wrong estimate of the period of vibration.

Similarly, the time period expression as furnished in Eq. (14.4.1) is again strictly valid for rectangular tanks having walls cast monolithically with base slab (can transfer moment) and open at top. It is not valid when,

- The base is hinged or supported on bearing pads.
- The roof is covered, preventing the wall to behave as a free cantilever.

- When the aspect ratio of the wall, L/h_w or B/h_w, is ≤ 2.0. In such cases, the wall behaves as a two-way slab and time period computed by Eq. (14.4.1) will give erroneous result as the wall in this case is considered as a i meter wide beam spanning vertically.

Example 14.2 A reinforced concrete firewater tank of size 6.0 m × 6.0 m in plan, 2.5 m high, without any roof covering is resting on ground, as shown in Fig. 14.11. The tank is situated in a site attributed as zone IV as per IS 1893. Compute the moment in the wall and the foundation including the dynamic pressures on the wall. Consider $\gamma_c = 25$ kN/m^3 and that of water as 10 kN/m^3. Compute moment and shears as per Euorcode8 and IS code. Assume the tank behaving as a flexible tank. Grade of concrete used is M25.

Solution

Weight of water in the tank = 6.0 × 6.0 × 2.0 × 10.0 = 720 kN

Weight of one tank wall = 6.0 × 2.5 × 0.35 × 25 = 131.25 kN

Impulsive weight of water,

$$W_i = W\frac{\tanh(0.866L/h_f)}{0.866L/h_f}$$

$$= 720\left[\frac{\tanh\left(\frac{0.866\times 6}{2.0}\right)}{\frac{0.866\times 6}{2.0}}\right] = 274.083 \text{ kN.}$$

Height of impulsive mass ($h_f/L = 0.333 < 0.75$).

$h_i = 0.375h_f = 0.375 \times 2 = 0.75$ m, excluding base pressure.

Again, $$h_i' = h_f\left[\frac{0.866L/h_f}{2\tanh(0.866L/h_f)} - \frac{1}{8}\right] = 2.0\left[\frac{\frac{0.866\times 6.0}{2.0}}{2\tanh\left(\frac{0.866\times 6}{2}\right)} - 0.125\right] = 2.377 \text{ m}$$

including base pressure.

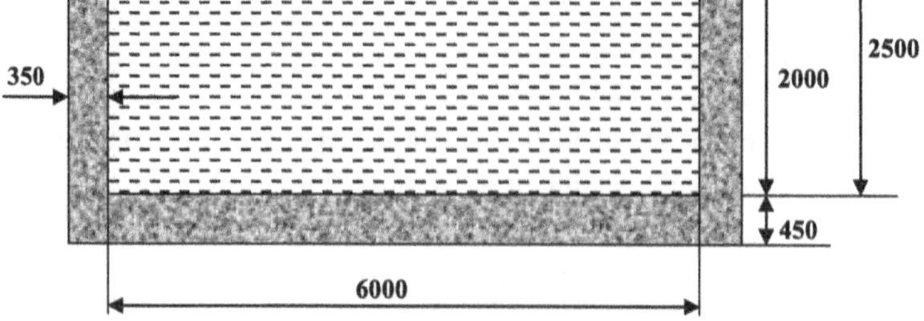

Fig. 14.11 Rectangular tank resting on ground—geometric dimension

Weight of sloshing mass,

$$W_{\rm sl} = 0.264W\frac{L}{h_{\rm f}}\tanh\left(3.16\frac{h_{\rm f}}{L}\right)$$
$$= 0.264 \times 720 \times \left(\frac{6.0}{2.0}\right)\tanh\left(3.16 \times \frac{2.0}{6.0}\right) = 446.554\ \text{kN}$$

Height of sloshing mass excluding base pressure,

$$h_{\rm sl} = h_{\rm f}\left(1 - \frac{\cosh(3.16h_{\rm f}/L) - 1}{3.16\frac{h_{\rm f}}{L}\sinh\left(3.16\frac{h_{\rm f}}{L}\right)}\right)$$
$$= 2.0\left[1 - \frac{\cosh\left(\frac{3.16\times 2.0}{6.0}\right) - 1.0}{3.16 \times \left(\frac{2.0}{6.0}\right)\sinh\left(\frac{3.16\times 2.0}{6.0}\right)}\right] = 1.083\ \text{m}.$$

Height of sloshing mass including base pressure,

$$h_{\rm sl} = h_{\rm f}\left(1 - \frac{\cosh(3.16h_{\rm f}/L) - 2.01}{3.16\frac{h_{\rm f}}{L}\sinh\left(3.16\frac{h_{\rm f}}{L}\right)}\right)$$
$$= 2.0\left[1 - \frac{\cosh\left(\frac{3.16\times 2.0}{6.0}\right) - 2.01}{3.16 \times \left(\frac{2.0}{6.0}\right)\sinh\left(\frac{3.16\times 2.0}{6.0}\right)}\right] = 2.606\ \text{m}.$$

Spring stiffness of sloshing fluid,

$$k_{\rm sl} = \frac{0.833W}{h_{\rm f}}\tanh^2\left(3.16\frac{h_{\rm f}}{L}\right)$$
$$= \frac{0.833 \times 720}{2}\tanh^2\left(\frac{3.16 \times 2}{6}\right) = 183.9\ \text{kN/m}.$$

Time period due to sloshing mass,

$$T_{\rm sl} = 2\pi\sqrt{\frac{M_{\rm sl}}{k_{\rm sl}}}$$

Here, $M_{\rm sl} = \frac{W_{\rm sl}}{g} = \frac{446.554}{9.81} = 45.52\ \text{kN-s}^2/\text{m}$.

Thus, $T_{\rm sl} = 2\pi\sqrt{\frac{45.52}{183.9}} = 3.126\ \text{s}$

Here, $Z = 0.24$, as per IS-1893. $I = 1.5$ for RCC tank $R = 2.0$.

Thus,

$$S_{\mathrm{asl}} = \left(\frac{ZI}{2R}\right) \times \left(\frac{1.67}{T_{\mathrm{sl}}}\right)$$
$$= \left(\frac{0.24 \times 1.5}{2 \times 2}\right) \times \left(\frac{1.67}{3.126}\right) \times 9.81 = 0.472 \text{ m/s}^2.$$

The above is valid for 5% damping. Considering that for water, we take 0.5% damping, the design spectral acceleration above is modified by a scaling factor which gives

$$S^{\mathrm{d}}_{\mathrm{asl}} = \sqrt{\frac{10}{5+\zeta}} \times S_{\mathrm{asl}}$$
$$= \sqrt{\frac{10}{5+0.5}} \times 0.472 = 0.636 \text{ m/s}^2.$$

Time period of impulsive mass for the wall,

$$\bar{h} = \frac{\frac{m_{\mathrm{i}}}{2} \times h_{\mathrm{i}} + \hat{m}_{\mathrm{w}} \times \frac{h_{\mathrm{w}}}{2}}{\frac{m_{\mathrm{i}}}{2} + \hat{m}_{\mathrm{w}}}$$
$$= \frac{\frac{27.939}{2} \times 0.75 + 13.38 \times 1.25}{\frac{27.939}{2} + 13.38} = \frac{27.202}{27.349} = 0.994 \text{ m}$$

Equivalent uniformly distributed load q can be calculated from the expression,

$$q = \frac{\left(\frac{m_{\mathrm{i}}}{2} + \hat{m}_{\mathrm{w}}\right) g}{B \times h_{\mathrm{w}}}$$
$$= \frac{\left(\frac{27.939}{2} + 13.38\right) \times 9.81}{6 \times 2.5} = 17.88 \text{ kN/m}^2.$$

Equivalent concentrated load P acting on the wall at $\bar{h} = 0.994$ m is,
$P = q \times h_{\mathrm{f}} = 17.88 \times 2.0 = 35.77$ kN per meter width.
Young modulus $E = 5000\sqrt{f_{\mathrm{ck}}}$ N/mm². $\rightarrow E = 2.5 \times 10^7$ kN/m².

$$I = \frac{1}{12} \times 1.0 \times (0.35)^3 = 3.573 \times 10^{-3} \text{ m}^4.$$

Thus, the deflection is given by,

$$\delta = \frac{P\bar{h}^3}{3EI} = \frac{35.77 \times (0.994)^3}{3 \times 2.5 \times 10^7 \times 3.573 \times 10^{-3}} = 1.31 \times 10^{-4} \text{ m}$$

Now considering $T_i = 2\pi\sqrt{\delta/g}$, we have,

$$T = 2\pi\sqrt{\frac{1.31 \times 10^{-4}}{9.81}} = 0.022 \text{ s}$$

$$\frac{S_{ai}}{g} = \left(\frac{ZI}{2R}\right)(1 + 15T)$$

where at $T = 0.022$ s.

$$S_{ai} = \left(\frac{ZI}{2R}\right) \times 1.33g = \left(\frac{0.24 \times 1.5}{2 \times 2}\right) \times 1.33 \times 9.81 = 1.174 \text{ m/s}^2$$

Moment at the base of vertical wall based on absolute sum is expressed as,

$$M_{bw} = (m_i h_i + m_w h_w + m_r h_r) S_e(T_i) + m_{sl} h_{sl} S_e(T_{sl})$$

$$M_{bw} = \left(\frac{274.083}{2 \times 9.81} \times 0.75 + \frac{131.25}{9.81} \times 1.25\right) \times 1.174 + \frac{446.554}{9.81} \times 1.083 \times 0.636 = 63.28 \text{ kN m}$$

[Since the tank is open to sky, $m_r = 0.0$],

Moment at top of base slab,

$$M_{bf} = (m_i h_i' + m_w h_w') S_e(T_i) + m_{sl} h_{sl}' S_e(T_{sl})$$

$$M_{bf} = \left(\frac{274.083}{2 \times 9.81} \times 2.377 + \frac{131.25}{9.81} \times 1.25\right) \times 1.174 + \frac{446.554}{9.81} \times 2.606 \times 0.636 = 132.1 \text{ kN m}$$

The base shear at base of wall is expressed as,

$$V_b = (m_i + m_w) S_e(T_i) + m_{sl} S_e(T_{sl})$$

$$V_b = \left(\frac{274.083}{2 \times 9.81} + \frac{131.25}{9.81}\right) \times 1.174 + \frac{446.554}{9.81} \times 0.636 = 61.05 \text{ kN.}$$

Based on SRSS method, the moments and shears are as shown hereafter,

- **Moment in vertical wall**

$$M_{bw} = \sqrt{[(m_i h_i + m_w h_w) S_e(T_i)]^2 + [m_{sl} h_{sl} S_e(T_{sl})]^2}$$

$$M_{bw} = \sqrt{(31.926)^2 + (31.354)^2} = 44.74 \text{ kN m}$$

- **Moment at top of foundation**

$$M_{bf} = \sqrt{(56.65)^2 + (75.44)^2} = 94.34 \text{ kN m}$$

- **Base shear**

$$V_b = \sqrt{(32.1)^2 + (28.95)^2} = 43.23 \text{ kN.}$$

The sloshing wave height is expressed as,

$$\Delta = \left(\frac{ZI}{2}\right) \times \left(\frac{S_{asl}}{g}\right) \times \frac{L}{2}$$

$$= \left(\frac{0.24 \times 1.5}{2}\right) \times \left(\frac{1.67}{3.126}\right) \times \sqrt{\frac{10}{5.5}} \times \frac{6}{2} = 0.388 \text{ m}$$

Thus, the expected height of sloshing is 388 mm.

Computation of dynamic pressure on wall (kN/m²)

Based on the formulation as shown in Sect. 14.2.1, the dynamic pressure on wall are tabulated hereafter,

z/h_f	$\psi_i(z)$	$p_i(z)$	$\psi_{sl}(z)$	$p_{sl}(z)$	p_w	p_{vi}	p_{srss}
0	0.85646	2.029231	0.258885	0.996877	1.047146	1.587447	3.602478
0.05	0.854319	2.024158	0.259245	0.998262	1.047146	1.508075	3.564228
0.1	0.847895	2.008939	0.260324	1.002419	1.047146	1.428702	3.51933
0.15	0.83719	1.983573	0.262127	1.009362	1.047146	1.34933	3.467673
0.2	0.822201	1.948062	0.264658	1.019108	1.047146	1.269958	3.409199
0.25	0.802931	1.902404	0.267925	1.031685	1.047146	1.190585	3.343907
0.3	0.779378	1.8466	0.271935	1.047129	1.047146	1.111213	3.271855
0.35	0.751544	1.78065	0.276701	1.065481	1.047146	1.031841	3.193176
0.4	0.719426	1.704554	0.282236	1.086793	1.047146	0.952468	3.108081
0.45	0.683027	1.618312	0.288554	1.111124	1.047146	0.873096	3.016878
0.5	0.642345	1.521923	0.295675	1.138541	1.047146	0.793724	2.919998
0.55	0.597381	1.415388	0.303616	1.169122	1.047146	0.714351	2.818017
0.6	0.548134	1.298708	0.312401	1.20295	1.047146	0.634979	2.7117

(continued)

(continued)

z/h_f	$\psi_i(z)$	$p_i(z)$	$\psi_{sl}(z)$	$p_{sl}(z)$	p_w	p_{vi}	p_{srss}
0.65	0.494606	1.171881	0.322054	1.24012	1.047146	0.555606	2.602052
0.7	0.436795	1.034908	0.332602	1.280735	1.047146	0.476234	2.490387
0.75	0.374701	0.887788	0.344073	1.324907	1.047146	0.396862	2.378413
0.8	0.308326	0.730523	0.3565	1.372761	1.047146	0.317489	2.268342
0.85	0.237668	0.563112	0.369918	1.424427	1.047146	0.238117	2.163012
0.9	0.162727	0.385554	0.384363	1.480051	1.047146	0.158745	2.066005
0.95	0.083505	0.19785	0.399876	1.539786	1.047146	0.079372	1.98173
1	−3.8E−16	−9E−16	0.4165	1.603798	1.047146	−3.5E−16	1.915381

14.6 Elevated Water Tanks

Shown in Fig. 14.12 is a typical elevated water tank deployed in urban area for water distribution. Here, the tanks are constructed at an elevated height to take advantage of gravity pressure that considerably economize the distribution cost.

Considering the fact that the elevated water tanks are essentially inverted pendulum-type structures, they remain quite susceptible to earthquake forces.

Figure 14.13 shows one such case where an elevated water tank supported on circular shaft collapsed during an earthquake in India.

The mathematical model usually adopted is as shown in Fig. 14.14.

Two types of mathematical model are possible: One is a simplified model as proposed in code considering a model having single degree of freedom where the sloshing and the impulsive mass effect are considered separately, and the other is considering two mass lumped model, as shown in Fig. 14.14.

14.6.1 Simplified Method as Proposed in the Code

Time period for impulsive force is expressed by,

$$T_i = 2\pi\sqrt{\frac{m_i + m_t + m_s/3}{K_s}} \tag{14.6.1.1}$$

Here,

m_i mass due to impulsive force of the liquid;
m_t mass of the tank and all-inclusive appurtenances;
m_s mass of the staging, assume 1/3rd of it is weight contribute to the vibration;
K_s stiffness of the staging frame.

The time period for sloshing mass can be derived from Eq. (14.4.4).

Fig. 14.12 Typical elevated water tanks deployed for drinking water distribution in urban area

Fig. 14.13 Collapse of an elevated water tank under earthquake in India

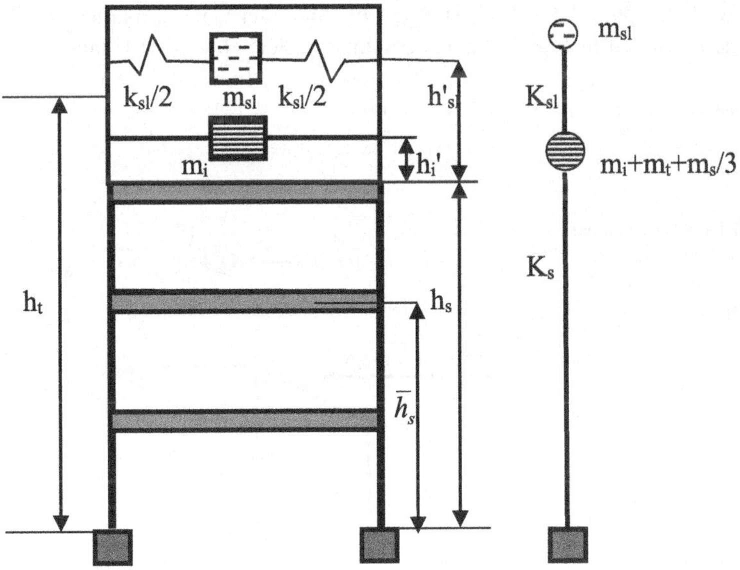

Fig. 14.14 Mathematical model for earthquake analysis including base pressure

Knowing the two time periods, spectral accelerations S_{ai} and S_{asl} can now be derived from Fig. 2.5 of Chap. 2.

The impulsive force V_{bi} acting at the top of foundation level can be expressed as,

$$V_{bi} = \left(m_i + m_t + \frac{m_s}{3}\right) S_{ai} \tag{14.6.1.2}$$

The impulsive force is expressed as,

$$V_{bsl} = (m_{sl}) S_{asl} \tag{14.6.1.3}$$

The resultant shear is expressed as per Eq. (14.3.1.3).

Moment at the top of foundation due to impulsive force is expressed as (refer Fig. 14.13)

$$M_{bif} = \left[m_i \times \left(h_i' + h_s\right) + m_t(h_t + h_s) + m_s \bar{h}_s\right] \tag{14.6.1.4}$$

Here, h_t = center of gravity of empty tank and $\bar{h}_s$ = center of gravity of the staging frame. The notation prime on h represents the including base pressure case.

The moment at the top of foundation due to sloshing force is expressed as,

$$M_{bslf} = V_{bsl} \times \left(h_{sl}' + h_s\right) \tag{14.6.1.5}$$

How to compute the staging stiffness by computer using a mathematical model like one as shown in Fig. 14.15 is explained in Sect. 13.4.1 of Chap. 13.

For direct formulation of the stiffness of the frame, one can also use the expressions as cited in Eqs. (13.4.11) through (13.4.13).

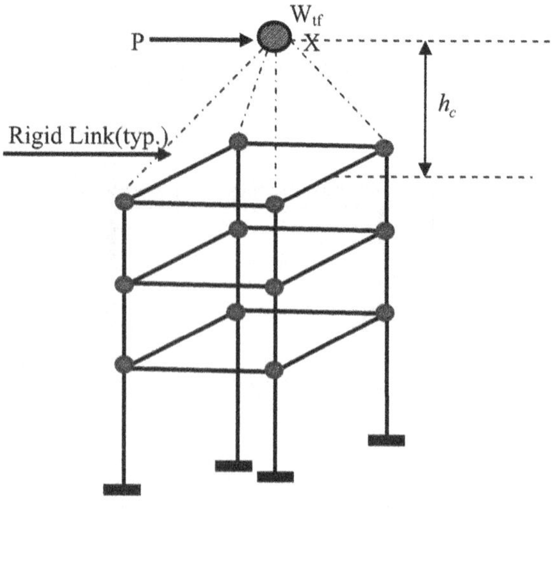

Fig. 14.15 Mathematical model for derivation of staging stiffness through computer analysis

14.6.1.1 Two Mass Lumped Model Having Coupled Mode

An alternate two-mass model of the water tank can also be considered, vide Fig. 14.13 as,

$$\begin{bmatrix} m_i + m_t + m_s/3 & 0 \\ 0 & m_{sl} \end{bmatrix} \begin{Bmatrix} \ddot{x}_i \\ \ddot{x}_{sl} \end{Bmatrix} + \begin{bmatrix} K_s + K_{sl} & -K_{sl} \\ -K_{sl} & K_{sl} \end{bmatrix} \begin{Bmatrix} x_i \\ x_{sl} \end{Bmatrix} = 0 \qquad (14.6.1.6)$$

Equation (14.6.1.6) can be expressed in compact form as,

$$[\mathbf{M}]\{\ddot{X}\} + [\mathbf{K}]\{X\} = 0 \qquad (14.6.1.7)$$

Now considering $[\mathbf{K}] - [\mathbf{M}]\omega^2 = 0$, one can find out the natural frequency vis-à-vis time period for the sloshing and structural mode. Now referring to Fig. 2.5 (Chap. 2), one can find out the spectral accelerations S_{ai} for impulsive mode and S_{asl} for the sloshing mode of the tank, considering damping ratio somewhere between 5 and 7% for RCC tank and 2–3% for steel tank, and 0.5% for sloshing fluid.

The nodal force can be expressed as,

$$\{f_i\} = [\mathbf{M}]\{\phi_i\}\left(\frac{ZI}{2R}\right)\left(\frac{L_i}{M_i}\right)[\bar{S}_{ai}] \qquad (14.6.1.8)$$

Here, $L_i = [\phi_i]^T[\mathbf{M}]\{\mathbf{I}\}$, $M_i = [\phi_i]^T[\mathbf{M}][\phi_i]$ and $\bar{S}_{ai}$ = the design response spectrum as obtained from Fig. 2.5. $\{\mathbf{I}\}$ = an identity matrix.

The base shear V_b is calculated as,

$$V_b = \left(\frac{ZI}{2R}\right)\left(\frac{L_n^2}{M_n}\right)\bar{S}_{an} \qquad (14.6.1.9)$$

Example 14.3 An overhead water tank, having dimensions 7.0 m × 4.5 m × 6.5 m (high), is supported on a staging frame, as shown in Fig. 14.16. The beams supporting the tank have sectional area of 600 mm × 800 mm. The beams constituting the staging frames have sectional area of 400 mm × 600 mm at all levels. The columns have typical dimensions of 600 mm × 600 mm. The tanks walls are 450 mm thick while the base slab is 575 mm thick. The roof slab is 175 mm thick. Maximum level of water allowed in tank is 6.0 m. Grade of concrete used for the tank is M30, while that for the frame is M25.

The tank is located in Delhi (India) marked as zone IV as per IS 1893.

Consider density of concrete = 25 kN/m^3 and that of water is 10 kN/m^3. Perform a seismic analysis of the tank based on the method as proposed in code and also by the two mass mathematical model.

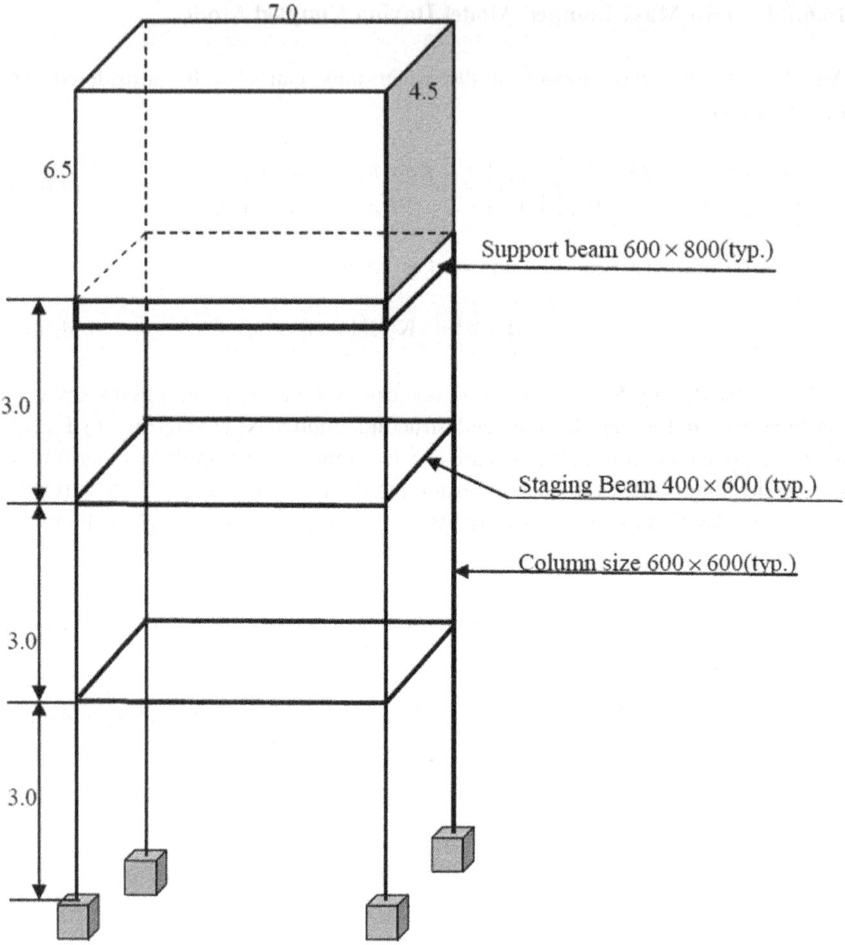

Fig. 14.16 Elevated water tank and its geometric dimensions

Solution

Weight of water in tank = $7.0 \times 4.5 \times 6.0 \times 10 = 1890$ kN

Computation of tank weight:

Weight of tank wall = $(7.9 \times 5.4 - 7.0 \times 4.5) \times 6.5 \times 25 = 1813.5$ kN

Weight of roof slab = $7.9 \times 5.4 \times 0.175 \times 25 = 186.64$ kN

Weight of base slab = $7.9 \times 5.4 \times 0.575 \times 25 = 613.24$ kN

Total weight of empty tank $W_t = 1813.5 + 186.64 + 613.24 = 2613.40$ kN

Weight of staging:

Weight of columns = $0.6 \times 0.6 \times 9.0 \times 4 \times 25 = 324$ kN

Weight of supporting beam = $2 \times (7.45 + 4.95) \times 0.6 \times 0.8 \times 25 = 297.6$ kN

Weight of staging beams = 2 × 2 × (7.45 + 4.95) × 0.4 × 0.6 × 25 = 297.6 kN.

Total weight of staging frame W_{st} = 324 + 297.6 + 297.6 = 919.2 kN.

Computation of impulsive weight of water:

$$W_i = W\frac{\tanh(0.866L/h_f)}{0.866L/h_f}$$
$$= 1890\left[\frac{\tanh(0.866 \times 7/6)}{0.866 \times 7/6}\right] = 1432.746 \text{ kN}$$

Computation of sloshing weight of water:

$$W_{sl} = 0.264W\frac{L}{h_f}\tanh\left(3.16\frac{h_f}{L}\right)$$
$$= 0.264 \times 1890 \times \left(\frac{7}{6}\right) \times \tanh\left(\frac{3.16 \times 6}{7}\right) = 576.9738 \text{ kN}$$

Height of impulsive mass (excluding base pressure):

$$h_i = \left[0.5 - \frac{0.09375}{h_f/L}\right]h_f$$
$$= \left[0.5 - \frac{0.09375}{6/7}\right] \times 6.0 = 2.3438 \text{ m}$$

Height of sloshing mass (excluding base pressure):

$$h_{sl} = h_f\left[1 - \frac{\cosh(3.16h_f/L) - 1}{3.16\frac{h_f}{L}\sinh\left(3.16\frac{h_f}{L}\right)}\right]$$
$$= 6.0 \times \left[1 - \frac{\cosh(3.16 \times 6/7) - 1.0}{3.16 \times \left(\frac{6}{7}\right) \times \sinh\left(\frac{3.16 \times 6}{7}\right)}\right] = 4.0615 \text{ m}$$

Height of impulsive mass (including base pressure):

$$h'_i = h_f\left[\frac{0.866L/h_f}{2\tanh(0.866L/h_f)} - \frac{1}{8}\right]$$
$$= 6.0 \times \left[\frac{0.866 \times 7/6}{2 \times \tanh(0.866 \times 7/6)} - \frac{1}{8}\right] = 2.932 \text{ m}$$

Height of sloshing mass (including base pressure):

$$h'_{\rm sl} = h_{\rm f}\left[1 - \frac{\cosh(3.16h_{\rm f}/L) - 2.01}{3.16\frac{h_{\rm f}}{L}\sinh(3.16\frac{h_{\rm f}}{L})}\right]$$

$$= 6.0 \times \left[1 - \frac{\cosh(3.16 \times 6/7) - 2.01}{3.16 \times (\frac{6}{7}) \times \sinh(\frac{3.16\times 6}{7})}\right] = 4.361\ {\rm m}$$

Stiffness of sloshing fluid,

$$k_{\rm sl} = \frac{0.833W}{h_{\rm f}}\tanh^2\left(3.16\frac{h_{\rm f}}{L}\right)$$

$$= \frac{0.833 \times 1890}{6}\tanh^2\left(\frac{3.16 \times 6}{7}\right) = 257.776\ {\rm kN/m}$$

Referring to Eq. (13.4.11) in Chap. 13, the lateral stiffness of frame is expressed as,

$$K_{\rm x} = \frac{12E_{\rm c}.I_{\rm c}.N_{\rm c}}{h_{\rm p}^3 N_{\rm p}}$$

Here,

$$E_{\rm c} = 5000 \times \sqrt{25} \times 1000 = 2.5 \times 10^7\ {\rm kN/m^2},$$

$$I = \frac{1}{12} \times 0.6 \times (0.6)^3 = 0.0108\ {\rm m^4}, N_{\rm c} = 4, N_{\rm p} = 2, h_{\rm p} = 3.0.$$

Substituting the values, we have,

$$K_{\rm x} = \frac{12 \times 2.5 \times 10^7 \times 0.0108 \times 4}{(3.0)^3 \times 2.0} = 240{,}000\ {\rm kN/m}.$$

Effective mass vibrating with frame,

$$m_{\rm e} = \left(\frac{W_{\rm i} + W_{\rm t} + W_{\rm st}/3}{g}\right)$$

$$= \left(\frac{1432.746 + 2613.4 + 919.2/3}{9.81}\right) = 443.6508\ {\rm kN\text{-}s^2/m}$$

Time period of vibration under impulsive force,

$$T_i = 2\pi\sqrt{\frac{m_e}{K_x}} = 2 \times \pi \times \sqrt{\frac{443.6508}{240,000}} = 0.270\ \text{s}.$$

Design spectral acceleration from Fig. 2.4 of Chap. 2: $S_a/g = 2.5$.

For zone IV, $Z = 0.24$, $I = 1.5$, $R = 2.5$ (assumed average of 2.0 for the tank and 3.0 for the frame considered as ordinary moment resistant frame).

Thus, the design spectrum for 5% damping is,

$$\frac{S_{ai}}{g} = \left(\frac{ZI}{2R}\right) \times 2.5 = 0.18$$
$$\rightarrow S_{ai} = 0.18 \times 9.81 = 1.7658\ \text{m/s}^2.$$

Time period of vibration under sloshing force,

$$T_i = 2\pi\sqrt{\frac{m_{sl}}{K_{sl}}} = 2 \times \pi \times \sqrt{\frac{58.815}{257.776}} = 3.00\ \text{s}.$$

Thus, the design spectrum for 5% damping and $R = 2.0$ is,

$$\frac{S_{asl}}{g} = \left(\frac{ZI}{2R}\right) \times \left(\frac{1.67}{3}\right) = 0.05$$
$$\rightarrow S_{asl} = 0.05 \times 9.81 = 0.4912\ \text{m/s}^2.$$

Considering 0.5% damping for fluid,

$$S_{asl} = 0.4912 \times \sqrt{\frac{10}{5 + 0.5}} = 0.662\ \text{m/s}^2.$$

Impulsive shear force acting at top of foundation,

$$V_{bi} = \left(m_i + m_t + \frac{m_s}{3}\right) S_{ai}$$
$$= 443.6508 \times 1.7658 = 783.4\ \text{kN}$$

Sloshing shear force acting at top of foundation,

$$V_{bsl} = (m_{sl}) S_{asl} = 58.815 \times 0.662 = 38.93\ \text{kN}$$

Resultant shear at top of foundation,

$$V_R = \sqrt{V_{bi}^2 + V_{bsl}^2}$$
$$= \sqrt{(783.4)^2 + (38.93)^2} = 784.36 \text{ kN.}$$

Computation of moment at top of foundation,

Lever arm distance of impulsive mass to the top of foundation = $h_i' + h_{st} = 2.932 + 9 = 11.932$ m

The c.g. distance of empty tank,

$$h_t = \frac{1813.5 \times (3.25 + 0.575) + 186.64 \times (6.5875 + 0.575) + 613.24 \times 0.2875}{1813.5 + 186.64 + 613.24}$$
$$h_t = \frac{6936.6375 + 1336.809 + 176.3065}{1813.5 + 186.64 + 613.24} = 3.233 \text{ m}$$

Above bottom of base slab of tank.

Thus, lever arm distance from top of foundation = 3.233 + 9 = 12.233 m

For the staging, c.g distance is computed as,

$$\bar{h}_{st} = \frac{324 \times 4.5 + 297.6 \times 9 + (297.6/2) \times 6 + (297.6/2) \times 3}{324 + 297.6 + 297.6}$$
$$= \frac{5473.8}{919.2} = 5.955 \text{ m}$$

Above the top of foundation,

Lever arm for sloshing mass = 4.361 + 9 = 13.361 m above foundation.

Impulsive moment at the top of foundation,

$$M_{fi} = [192.6606 \times 11.932 + 266.4 + 12.233 + 93.7 \times 5.955] \times 1.7658 = 10,799 \text{ kN.m}$$

Sloshing moment at the top of foundation,

$$M_{fsl} = 58.815 \times 0.662 \times 13.361 = 520.22 \text{ kN.m}$$

Thus, resultant moment at the top of foundation,

$$M_R = \sqrt{M_{fi}^2 + M_{fsl}^2} = \sqrt{(10799)^2 + (520)^2} = 10,812 \text{ kN.m}$$

Sloshing amplitude is expressed as,

$$\Delta = \left(\frac{ZI}{2}\right)\left(\frac{S_{asl}}{g}\right)\sqrt{\frac{10}{5+D}}\left(\frac{L}{2}\right)$$

$$\Delta = \left(\frac{0.24 \times 1.5}{2}\right) \times \left(\frac{1.67}{3.0}\right) \times \sqrt{\frac{10}{5.5}} \times \left(\frac{7.0}{2}\right) \times 1000 = 473 \text{ mm}$$

Considering a two mass lumped model we have (refer the sketch)

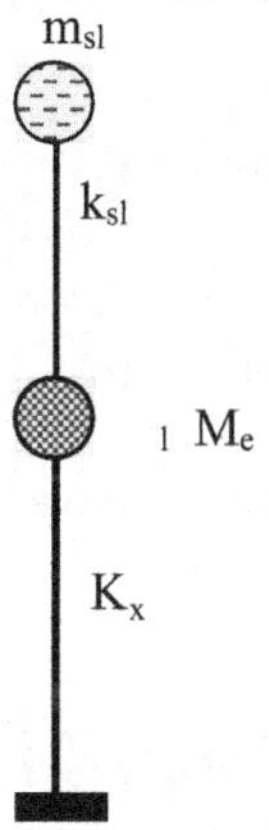

$$[\mathbf{K}] = \begin{bmatrix} K_x + K_{sl} & -K_{sl} \\ -K_{sl} & K_{sl} \end{bmatrix}$$

$$[\mathbf{K}] = \begin{bmatrix} 240,258 & -257.776 \\ -257.776 & 257.776 \end{bmatrix}$$

The mass matrix is represented by,

$$[\mathbf{M}] = \begin{bmatrix} 443.651 & 0 \\ 0 & 58.815 \end{bmatrix}$$

For eigenvalue solution, we have,

$$\begin{bmatrix} 240,258 - 443.651\lambda & -257.776 \\ -257.776 & 257.776 - 58.815\lambda \end{bmatrix} = 0$$

Solution of the above gives,

$\lambda = \begin{Bmatrix} 541.552 \\ 4.378 \end{Bmatrix}$, considering natural frequency $\omega = \sqrt{\lambda}$ and time period $T = 2\pi/\omega$ we have

$$T = \begin{Bmatrix} 0.27 \\ 3.003 \end{Bmatrix} \text{s}$$

The eigenvectors of the problem are given by,

$$[\varphi] = \begin{bmatrix} 1 & 1.082 \times 10^{-3} \\ -8.159 \times 10^{-3} & 1 \end{bmatrix}$$

For sloshing mode,

$T_{sl} = 3.003$ s and $\{\phi_{sl}\} = \begin{Bmatrix} 1.082 \times 10^{-3} \\ 1.0 \end{Bmatrix}$

For $T = 3.0$ s, as calculated previously above for 0.5% damping, $S_{asl} = 0.662$ m/s^2.

Considering $\{I\} = \begin{Bmatrix} 1 \\ 1 \end{Bmatrix}$,

$$L_{sl} = \{\phi_{sl}\}^T[\mathbf{M}]\{I\} \rightarrow L_{sl} = 59.295$$
$$M_{sl} = \{\phi_{sl}\}^T[\mathbf{M}]\{\phi_{sl}\} \rightarrow L_{sl} = 58.816$$

$$\{f_{sl}\} = [\mathbf{M}]\{\varphi_{sl}\}\left(\frac{L_i}{M_i}\right)[\bar{S}_{asl}]$$

$$\{f_{sl}\} = \begin{bmatrix} 443.651 & 0 \\ 0 & 58.815 \end{bmatrix}\begin{Bmatrix} 1.082 \times 10^{-3} \\ 1.0 \end{Bmatrix}\left(\frac{59.295}{58.816}\right) \times 0.662$$
$$= \begin{Bmatrix} 0.32 \\ 39.253 \end{Bmatrix} \text{kN}$$

Base shear $V_{bsl} = 0.32 + 39.253 = 39.573$ kN.

For impulsive mode

$T_{sl} = 0.27$ s and $\{\phi_i\} = \begin{Bmatrix} 1 \\ -8.159 \times 10^{-3} \end{Bmatrix}$

For $T = 0.27$ s, as calculated previously above for 5% damping, $S_{asl} = 1.7658$ m/s^2.

Considering $\{I\} = \begin{Bmatrix} 1 \\ 1 \end{Bmatrix}$,

$$L_i = \{\phi_i\}^T[\mathbf{M}]\{I\} \rightarrow L_i = 443.171$$
$$M_i = \{\phi_i\}^T[\mathbf{M}]\{\phi_i\} \rightarrow L_{sl} = 443.655$$

$$\{f_{\text{sl}}\} = [\mathbf{M}]\{\phi_{\text{sl}}\}\left(\frac{L_{\text{i}}}{M_{\text{i}}}\right)[\bar{S}_{\text{asl}}]$$

$$\{f_{\text{sl}}\} = \begin{bmatrix} 443.651 & 0 \\ 0 & 58.815 \end{bmatrix}\begin{Bmatrix} 1 \\ -8.159 \times 10^{-3} \end{Bmatrix}\left(\frac{443.171}{443.655}\right) \times 1.7658$$
$$= \begin{Bmatrix} 782.544 \\ -0.846 \end{Bmatrix} \text{kN}$$

Base shear $V_{\text{bsl}} = 782.544 - 0.846 = 781.698$ kN.
Thus, the resultant base shear is expressed as,

$$V_{\text{R}} = \sqrt{V_{\text{bi}}^2 + V_{\text{bsl}}^2} = \sqrt{(781.698)^2 + (39.573)^2} = 782.7 \text{ kN}$$

It will be observed that difference in results with respect to previous computation (V_{b} = 784.36 kN) is marginal. Computation of moment at base of foundation remains same as illustrated previously. Dynamic pressure on wall as per Sect. 14.2.1 is shown hereafter. The values are presented in kN/m^2.

z/h	$\psi_i(z)$	$p_i(z)$	$\psi_{sl}(z)$	$p_{sl}(z)$	p_w	p_{vi}	p_{res}
0	0.663269	7.164119	0.055165	0.260587	2.025229	7.23682	11.69973
0.05	0.661611	7.146209	0.055673	0.262983	2.025229	6.874979	11.46515
0.1	0.656637	7.092478	0.057204	0.270217	2.025229	6.513138	11.20833
0.15	0.648346	7.002927	0.059787	0.28242	2.025229	6.151297	10.92821
0.2	0.636739	6.877555	0.06347	0.299817	2.025229	5.789456	10.6239
0.25	0.621815	6.716362	0.06832	0.322729	2.025229	5.427615	10.29459
0.3	0.603575	6.519349	0.074427	0.351576	2.025229	5.065774	9.939592
0.35	0.582019	6.286515	0.081903	0.38689	2.025229	4.703933	9.558335
0.4	0.557146	6.01786	0.090885	0.42932	2.025229	4.342092	9.150375
0.45	0.528957	5.713385	0.101539	0.479645	2.025229	3.980251	8.715424
0.5	0.497452	5.37309	0.11406	0.538793	2.025229	3.61841	8.253382
0.55	0.46263	4.996973	0.128679	0.60785	2.025229	3.256569	7.764409
0.6	0.424492	4.585036	0.145665	0.688087	2.025229	2.894728	7.249036
0.65	0.383038	4.137279	0.16533	0.780979	2.025229	2.532887	6.708349
0.7	0.338267	3.653701	0.188036	0.888235	2.025229	2.171046	6.144319
0.75	0.29018	3.134302	0.2142	1.011828	2.025229	1.809205	5.560376
0.8	0.238777	2.579083	0.244304	1.15403	2.025229	1.447364	4.962493
0.85	0.184057	1.988043	0.278901	1.317458	2.025229	1.085523	4.36124
0.9	0.126021	1.361183	0.318627	1.505117	2.025229	0.723682	3.775828

(continued)

(continued)

z/h	$\psi_i(z)$	$p_i(z)$	$\psi_{sl}(z)$	$p_{sl}(z)$	p_w	p_{vi}	p_{res}
0.95	0.064669	0.698502	0.364214	1.720458	2.025229	0.361841	3.241854
1	−2.9E−16	−3.2E−15	0.4165	1.967442	2.025229	−1.6E−15	2.823541

The horizontal pressure on the floor slab is presented in tabular form hereafter and graphically in Figs. 14.17 and 14.18.

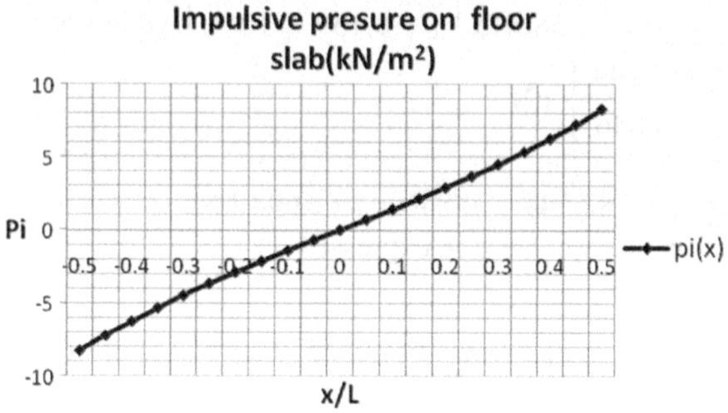

Fig. 14.17 Impulsive pressure on base slab of the tank

Fig. 14.18 Sloshing pressure on base of the water tank

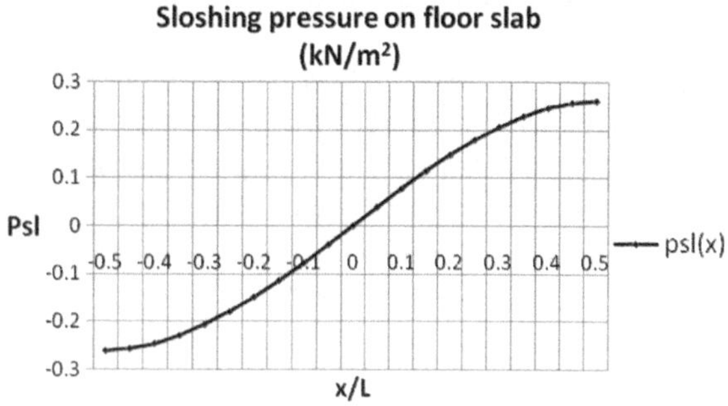

x/L	$\psi_i(x)$	$p_i(x)$	$\psi_{s1}(x)$	$p_{s1}(x)$
−0.5	−0.7659	−8.27266	−0.05519	−0.26069
−0.45	−0.66861	−7.22177	−0.05439	−0.25691
−0.4	−0.57814	−6.24467	−0.0521	−0.24609
−0.35	−0.49359	−5.33137	−0.04848	−0.22902
−0.3	−0.41408	−4.47253	−0.04371	−0.20647
−0.25	−0.33879	−3.65939	−0.03794	−0.17923
−0.2	−0.26697	−2.88363	−0.03135	−0.14807
−0.15	−0.19788	−2.13733	−0.02409	−0.11379
−0.1	−0.13081	−1.41287	−0.01634	−0.07716
−0.05	−0.06507	−0.70285	−0.00825	−0.03897
0	0	0	0	0
0.05	0.065071	0.702846	0.008251	0.038973
0.1	0.130807	1.412872	0.016335	0.077165
0.15	0.197879	2.137333	0.024089	0.113792
0.2	0.266973	2.88363	0.031346	0.148073
0.25	0.338794	3.659387	0.037941	0.179225
0.3	0.414076	4.47253	0.043708	0.206467
0.35	0.493589	5.331367	0.048482	0.229017
0.4	0.578145	6.244671	0.052097	0.246092
0.45	0.668607	7.221773	0.054387	0.256911
0.5	0.7659	8.272655	0.055187	0.260691

14.7 Effect of Foundation Stiffness on Tanks Resting on Ground

In Sects. 14.2 to 14.4, we have shown seismic analysis of rectangular and circular tanks that are resting on ground, where the effect of foundation stiffness is neglected.

As discussed in previous chapters (especially Chap. 7), it has been mentioned that it is a simplification to the reality and would possibly be valid only when the tank is resting on soil, which is rock or hard as rock having shear wave velocity $V_s > 600$ m/s. For normal soil, the tank will interact with the foundation stiffness to give a coupled response.

Effect of foundation compliance on seismic response of such a structure has been investigated by a number of researchers and engineers, like Haroun and Hafiz (1986), Veletsos and Yang (1990), Livaoglu (2008), Ghaemmaghami (2010), Jeong (2011), Chowdhury and Tarafdar (2015), Chowdhury et al. (2017).

Despite all these researches, little has permeated in daily design office practice, where simplified and approximate fixed-based model to analyze such tanks has been dominating the practice.

On the other hand, a comprehensive three-dimensional solution considering fluid, the tank and soil, based on finite element method (FEM) is possible (mostly practiced in nuclear industry while analyzing structures like spent fuel chamber.).

However, for other industries, such exhaustive analyses are often not in trend or feasible as neither cost nor schedule permits such intricate and detailed analysis to be carried out for such structural systems.

Thus, a workable yet realistic mathematical model that can effectively predict behavior of such soil structure system is still under active quest in both industry and academics. In subsequent section, we propose two mathematical models that can be used to overcome the constraints as mentioned above.

14.7.1 Practices as in Trend

As a first step, it would perhaps be enlightening to delve a little on practice—*as in vogue* in design offices around the world in terms considering foundation compliance into design methodology and also the various assumptions made in the mathematical model as elaborated previously.

Eurocode 8 Part IV (Design of structures for earthquake resistance for silos, tanks, and pipelines) has certainly dealt this topic in significant detail, especially for circular tanks. But the recommendations are somewhat sparse in clarity—in terms of *rectangular tanks.*

For estimation of impulsive and sloshing mass, it uses the method as proposed by Veletsos and Tang (1990) (for this, charts are available in the code) and for fixed-based rigid tank assumes the body to be subjected to peak ground acceleration (PGA) based on which the base shear and moments are derived.

For a rigid water tank, dynamic soil–structure interaction (DSSI) is considered adapting a simplified approach in line with Veletsos and Meek (1974), based on equivalent time period and damping where the effect of sloshing mass is ignored in the formulation. Though code acknowledges that the effect of DSSI on sloshing mass is negligible, yet, for a dynamically coupled system as that like water tank with foundation spring, it is mathematically an oversimplification to ignore the sloshing mass *coupled effect.*

Secondly, there exists an ambiguity in the formulation in terms of effect of foundation. While the effect of foundation mass (m_0) is considered in translational mode, the effect of mass moment of inertia of the foundation raft (J_θ) is missing in the formulation. Depending on aspect ratio of the tank in plan, J_θ can have significant effect on overall response and should not be ignored.

Finally, depending on thickness of the raft and height of liquid in tank, translational and rocking mode due to soil structure interaction gets coupled. This could well amplify the response, especially for sloshing and the rocking mode and this has been ignored in the time period formulation.

Thus, based on these arguments it may be concluded that the formulation presented in Eurocode 8 is at best an approximation, and has definite room for improvement.

For flexible tank, Eurocode 8 expresses the effective time period as,

$$\hat{T}_{\mathrm{i}} = T_{\mathrm{f}}\sqrt{1 + \frac{k_{\mathrm{f}}}{k_{\mathrm{x}}\alpha_{\mathrm{x}}}\left(1 + \frac{k_{\mathrm{x}}h_{\mathrm{f}}^{2}}{k_{\theta}\alpha_{\theta}}\right)} \tag{14.7.1.1}$$

In Eq. (14.7.1.1),

T_{f} time period of flexible tank wall considering its base as fixed;
h_{f} height of impulsive mass in the tank;
k_{f} stiffness of fixed-based flexible tank wall;
k_{x} lateral stiffness of foundation;
k_{θ} rotational stiffness of foundation;
$\alpha_{\mathrm{x}}, \alpha_{\theta}$ frequency dependent factors converting static stiffness to dynamic stiffness.

Fixed base time period T_{f} of the flexible wall is expressed as,

$$T_{\mathrm{f}} = 2\pi\sqrt{d_{\mathrm{f}}/g} \tag{14.7.1.2}$$

Here, d_{f} is the deflection of the wall along vertical center line, and at a height of the impulsive mass, when the wall is loaded by a load uniform in direction of ground motion and is of magnitude $(m_{\mathrm{i}} + m_{\mathrm{w}})g/2BH$. Here, H is the height of wall and B its total width. Stiffness k_{f} is expressed as,

$$k_{\mathrm{f}} = 4\pi^{2}m_{\mathrm{f}}^{2}/T_{\mathrm{f}}^{2} \tag{14.7.1.3}$$

In Eq. (14.7.1.2), deflection is computed assuming the wall as essentially a one-way cantilever slab spanning vertically. This can be only true for large shallow tanks (when ratio of width/height ≥ 2.0).

But for most of such tanks, it is more a rule than an exception that walls of such flexible tanks behave as a two-way slab. How to effectively cater to this plate behavior, code is silent on this issue, notwithstanding the fact that in this case again the sloshing mass effect due to inertial coupling with wall and foundation is not catered for.

Thus, the procedure as recommended in Eurocode 8 Part IV for seismic analysis of flexible tank can only be attributed as an approximation to its actual behavior.

ACI 350.3-01 American code of practice for a liquid retaining structure follows a similar procedure as suggested by Eurocode 8, Part IV, except the fact that the code does not have any recommendation for DSSI effect.

Earthquake analysis is carried out considering the tank as a fixed-based system.

For tanks considered as perfectly rigid, the code assumes the tank to move along with the ground having peak ground acceleration (PGA) and expressions for bending moment and shear force induced in the wall the same as considered in Eurocode 8.

For a flexible tank, the time period of the tank including impulsive mass is expressed based on the assumption that the tank acts as a cantilever beam and uses Housner's (1963) model to compute the sloshing and impulsive mass.

Ignoring the DSSI effect, it also makes two, somewhat simplified approximations.

- It considers sloshing and impulsive mass effect as uncoupled.
- Stiffness of wall is derived on the basis of a beam. It neither caters to two-way slab effect nor does it cater to other boundary conditions, like wall fixed at top which is as much a reality as a tank *open to sky*.

Thus, it is seen that ACI 351.3-01 is also at best a simplified approximation that can well digress from reality and can certainly be improved upon.

Water tanks in India are designed as per recommendations of IS-1893 Part 2 (2016) and a design guideline developed by Indian Institute of Technology (Kanpur) available at http://www.iitk.ac.in/nicee/IITK-GSDMA/EQ08.pdf.

IS-1893 Part 2 (2016) is the same as what has been illustrated in Sects. 14.2 to 14.4

The design guideline IITK-GSDMA.EQ08, is though a definite improvement over the recommendation furnished in IS 1893 1984 version used previously (and the present 2016 code developed), yet ignores the effect of compliant foundation considering it as fixed base and also treats the wall as a cantilever beam/one way slab spanning vertically only that may not again be a reality in many cases.

In Sect. 14.7.2, we present a procedure (Chowdhury and Tarafdar 2015; Chowdhury et al. 2017) by which a number of limitations as cited above may be circumvented.

14.7.2 Dynamic Analysis of a Water Tank Constructed on Ground, Considering DSSI

Two mathematical models are proposed herein,

- One for rigid tank;
- The other for flexible tank.

That can well overcome a good number of the limitations as discussed in the previous section.

14.7.2.1 Dynamic Analysis of Rigid Tank Considering DSSI

Shown in Fig. 14.19a is a rectangular tank resting on ground of width L, which is assumed to be rigid and resting on an elastic half-space. A schematic representation of the rigid mass (m_t), sloshing mass and stiffness (m_{sl}, k_{sl}), soil springs (k_x, k_θ), and damping contributions (C_x and C_θ) for the DSSI model is shown in Fig. 14.19b.

The tank is subjected to ground acceleration $\ddot{u}_g$ at its base, where point S is the center of stiffness where soil springs are connected. The system is assumed to have three degrees of freedom.

- Translation of sloshing mass u_{sl},
- Translation of tank plus impulsive fluid mass and foundation u_x, and
- Rotation of foundation θ.

Now if T and U are kinetic and potential energy of the system, respectively, then as per Lagrange's equation (Meirovitch 1985).

$$d(T+U) = \sum_{n=1}^{3}\left[\frac{d}{dt}\left(\frac{\partial T}{\partial \dot{q}_n}\right) - \frac{\partial T}{\partial \dot{q}_n} + \frac{\partial U}{\partial q_n}\right] dq_n = 0 \tag{14.7.2.1}$$

Kinetic energy of the system is expressed as,

$$T = \frac{1}{2}m_t(\dot{u}_x + Z_c\dot{\theta})^2 + \frac{1}{2}J_0\dot{\theta}^2 + \frac{1}{2}m_{sl}(\dot{u}_{sl} + \dot{u}_x + h_{sl}\dot{\theta})^2 \tag{14.7.2.2}$$

Potential energy of the system is expressed as,

$$U = \frac{1}{2}k_{sl}u_{sl}^2 + \frac{1}{2}k_x u_x^2 + \frac{1}{2}k_\theta \theta^2 \tag{14.7.2.3}$$

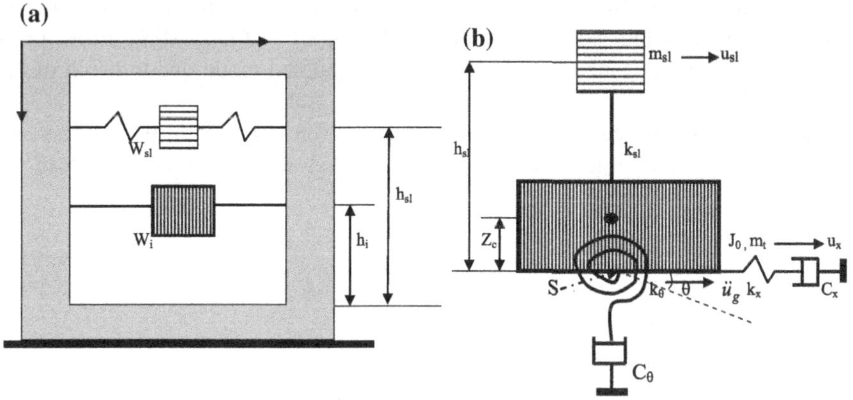

Fig. 14.19 Housner's model adding the foundation compliance and damping

where

m_t mass of empty tank, including roof slab with any live load and equipment load on it if any, plus the impulsive mass of fluid;
m_{sl} mass of sloshing fluid;
h_{sl} height above center of stiffness point S, where mass of sloshing fluid is considered;
Z_c centroid of rigid mass from the center of stiffness point S, and is expressed as,

$$Z_c = \frac{W_i \times h_i + W_t \times h_t}{W_i + W_t} \tag{14.7.2.4}$$

W_i impulsive weight of fluid connected rigidly with tank moving in the same phase;
W_t empty weight of tank including roof slab if any and equipments on the roof;
h_i height above point S where impulsive fluid mass is located;
h_t height above point S where centroid of the empty tank is located;
J_θ mass moment of inertia of the empty tank.

Now applying Eqs. (14.7.2.1) and (14.7.2.2), the free un-damped equation of motion can be expressed as,

$$\begin{bmatrix} m_t + m_{sl} & m_t Z_c + m_{sl} h_{sl} & m_{sl} \\ m_t Z_c + m_{sl} h_{sl} & m_t Z_c^2 + m_{sl} h_{sl}^2 + J_\theta & m_{sl} h_{sl} \\ m_{sl} & m_{sl} h_{sl} & m_{sl} \end{bmatrix} \begin{Bmatrix} \ddot{u}_x \\ \ddot{\theta} \\ \ddot{u}_{sl} \end{Bmatrix} + \begin{bmatrix} k_x & 0 & 0 \\ 0 & k_\theta & 0 \\ 0 & 0 & k_{sl} \end{bmatrix} \begin{Bmatrix} u_x \\ \theta \\ u_{sl} \end{Bmatrix} = 0 \tag{14.7.2.5}$$

From Eq. (14.7.2.5), it is noted that mass matrix is coupled but symmetrical, while the stiffness matrix has remained diagonal. This shows that the equation of equilibrium is a dynamically coupled system.

One may also note that, in this case, unlike formulation as proposed by Eurocode 8 Part IV, it takes into cognizance the coupled effect of the sloshing mass, mass moment of inertia of foundation (J_θ), as well as coupled mode due to depth of the bottom raft Z_c.

Now, considering the fact that the resistance given to a system by stiffness and damping is of the same nature, the free damped motion of equation can be expressed as

$$\begin{bmatrix} m_t + m_{sl} & m_t Z_c + m_{sl} h_{sl} & m_{sl} \\ m_t Z_c + m_{sl} h_{sl} & m_t Z_c^2 + m_{sl} h_{sl}^2 + J_\theta & m_{sl} h_{sl} \\ m_{sl} & m_{sl} h_{sl} & m_{sl} \end{bmatrix} \begin{Bmatrix} \ddot{u}_x \\ \ddot{\theta} \\ \ddot{u}_{sl} \end{Bmatrix} + \begin{bmatrix} C_x & 0 & 0 \\ 0 & C_\theta & 0 \\ 0 & 0 & C_{sl} \end{bmatrix} \begin{Bmatrix} \dot{u}_x \\ \dot{\theta} \\ \dot{u}_{sl} \end{Bmatrix} + \begin{bmatrix} k_x & 0 & 0 \\ 0 & k_\theta & 0 \\ 0 & 0 & k_{sl} \end{bmatrix} \begin{Bmatrix} u_x \\ \theta \\ u_{sl} \end{Bmatrix} = 0 \quad (14.7.2.6)$$

While stiffness and damping of sloshing fluid is furnished in Table 14.1, stiffness and damping of soil can be estimated from the expressions furnished in Table 7.3 of Chap. 7.

For modal analysis, eigenvalue expression for Eq. (14.7.2.6) is expressed as,

$$[\mathbf{K}] - [\mathbf{M}]\omega^2 = 0 \quad (14.7.2.7)$$

Here, [K] and [M] are stiffness and mass matrix, respectively.

The squares of natural frequencies (eigenvalues λ) are thus obtained as

$$[\boldsymbol{\lambda}] = \begin{bmatrix} \omega_x^2 & 0 & 0 \\ 0 & \omega_\theta^2 & 0 \\ 0 & 0 & \omega_{sl}^2 \end{bmatrix} \quad (14.7.2.8)$$

and

$$T_i = \frac{2\pi}{\omega_i} \quad (14.7.2.9)$$

ω_i circular frequency of vibration in *i*th mode;
T_i time period of vibration in *i*th mode.

Mode shapes associated with the different modes of vibration can be obtained from the eigenvectors [φ] of Eq. (14.7.2.6). Here, [φ] is a matrix of order 3 × 3.

Having determined the time period and eigenvectors for each of the three modes, spectral acceleration factor S_a/g can be obtained from the curve as furnished in the code (refer Fig. 2.4 in Chap. 2), and further analysis can be carried out as outlined below.

The zone factor (*Z*), importance factor (*I*), and response reduction factor (*R*) are first to be estimated from provisions of respective code under design consideration. As per IS code, for example,

$$\beta = \frac{Z.I}{2.R} \quad (14.7.2.10)$$

The *R* factor which is also called the response reduction factor, as depicted in Table 14.3, is the same as that in ACI 350.3-01, while it is represented by *q*, the behavior factor, as per Eurocode 8. Appropriate value of the reduction factor has to be taken for computing the spectral acceleration as per the respective code.

Amplitude of vibration for *i*th mode can then be expressed as (Clough 1985).

$$\{u_i\} = [\phi]_i \left[\frac{L_i}{M_i}\right] \beta S_{\text{ai}} \frac{T_i^2}{4\pi^2} \tag{14.7.2.11}$$

where

$$L_i = [\phi]_i^{\text{T}}[\mathbf{M}]\{I\} \tag{14.7.2.12}$$

Here, I is an identity matrix,

$$M_i = [\phi]_i^{\text{T}}[\mathbf{M}][\phi]_i \tag{14.7.2.13}$$

Modal damping ratio per mode is expressed as,

$$\xi_i = \frac{[\phi]_i^{\text{T}}[C][\phi]_i}{2M_i\omega_i} \tag{14.7.2.14}$$

For the estimated modal damping and time period of motion, spectral acceleration coefficient (S_{ai}/g) is determined from response spectrum as furnished in code and corrected for damping as per the respective code. For example, the expression $\sqrt{10/(5+\zeta)}$ as cited in Eurocode 8, where ζ the modal damping ratio, has been used herein.

Modal forces can now be estimated using the expression,

$$\{V_i\} = \begin{bmatrix} k_{\text{x}} & 0 & 0 \\ 0 & k_\theta & 0 \\ 0 & 0 & k_{\text{sl}} \end{bmatrix} \begin{Bmatrix} u_{i\text{x}} \\ \theta_i \\ u_{i\text{sl}} \end{Bmatrix} \tag{14.7.2.15}$$

As per Eq. (14.7.2.15)

$$V_{i\text{sl}} = k_{\text{sl}}.u_{i\text{sl}} \tag{14.7.2.16}$$

Thus, the resultant value of V_{sl} is expressed as,

$$V_{\text{sl}} = \sqrt{\sum_{i=1}^{3} V_{i\text{sl}}^2} \tag{14.7.2.17}$$

Here, V_{sl} is the resultant shear induced on tank due to sloshing force for the three modes.

For impulsive force,

$$V_{ix} = k_x . u_{ix} \tag{14.7.2.18}$$

Thus, the resultant value of V_x is expressed as,

$$V_x = \sqrt{\sum_{i=1}^{3} V_{ix}^2} \tag{14.7.2.19}$$

Here, the resultant V_x is the shear induced at base of tank wall due to impulsive force and tank weight for first three modes.

In this case, h_{sl} and h_i are to be considered as IBP case. Moment at base of wall due to sloshing is computed as,

$$M_{sl} = V_{sl} \times h_{sl} \tag{14.7.2.20}$$

and the impulsive moment is expressed as,

$$M_\theta = \sqrt{\sum_{i=1}^{3} (k_\theta \times \theta_i)^2} \tag{14.7.2.21}$$

Thus, the resultant shear and moment at tank wall base of foundation can be expressed as,

$$V_R = \sqrt{V_{sl}^2 + V_x^2} \tag{14.7.2.22}$$

$$M_R = \sqrt{M_{sl}^2 + M_\theta^2} \tag{14.7.2.23}$$

Equations (14.7.2.1) to (14.7.2.23) are based on closed-form analytical solutions and the results are exact.

The sloshing amplitude of fluid with respect to Fig. 14.19 is estimated by equating the kinetic and potential energy of sloshing mass and can be expressed as,

$$\delta_{vi} = v_i^2 / 2g \tag{14.7.2.24}$$

$$\rightarrow \delta_{vi} = 2\pi^2 u_{sli}^2 / gT_i^2 \tag{14.7.2.25}$$

Sloshing amplitude δ_{vi} shall be lower than the freeboard, especially if the liquid inside tank is deemed hazardous.

14.7.3 Dynamic Analysis of Flexible Tank Behaving as a One-Way Slab

For flexible tanks, it is perceived that the tank wall vibrates locally rather than the whole tank vibrating as a rigid body. Analysis in this case becomes trifle complicated as the stiffness of wall comes into picture and this can be widely varying. For instance, for the tank as shown in Fig. 14.20, it is apparent that the long wall has aspect ratio (L/H) > 2.0, when the wall will behave as a one-way slab spanning vertically as cantilever, while the short wall having an aspect ratio less than two will behave as a two-way slab with three sides fixed and one (top) edge free.

When the wall is considered as a one-way slab behaving as a beam, mathematical model perceived considering it as a fixed-based problem, or considering soil–structure interaction are as shown in Fig. 14.21.

For fixed-based case, equation of equilibrium under free vibration can be expressed as,

$$\begin{bmatrix} M & 0 \\ 0 & m_{sl} \end{bmatrix} \begin{Bmatrix} \ddot{u}_i \\ \ddot{u}_{sl} \end{Bmatrix} + \begin{bmatrix} k_w + k_{sl} & -k_{sl} \\ -k_{sl} & k_{sl} \end{bmatrix} \begin{Bmatrix} u_i \\ u_{sl} \end{Bmatrix} = 0 \tag{14.7.3.1}$$

It may be observed that in Eq. (14.7.3.1) the additional stiffness of wall k_w has come into play.

Referring to Eq. (2.6.2.8) of Chap. 2, stiffness of such beam type walls for the fundamental mode can be expressed as,

$$k_w = EI \int_0^H \phi_i''(z)\phi_j''(z)dz \tag{14.7.3.2}$$

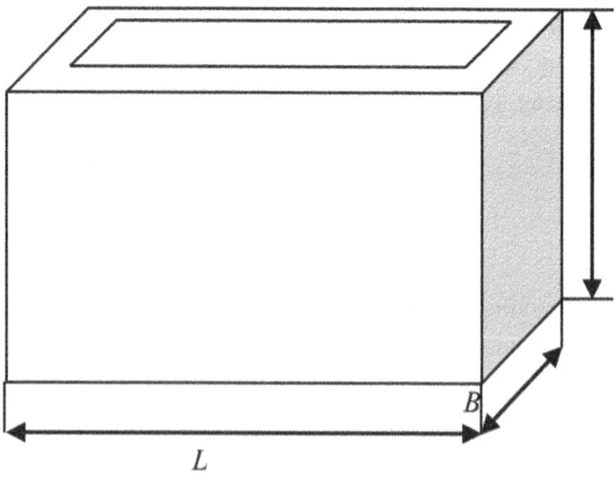

Fig. 14.20 A water tank resting on ground open at top

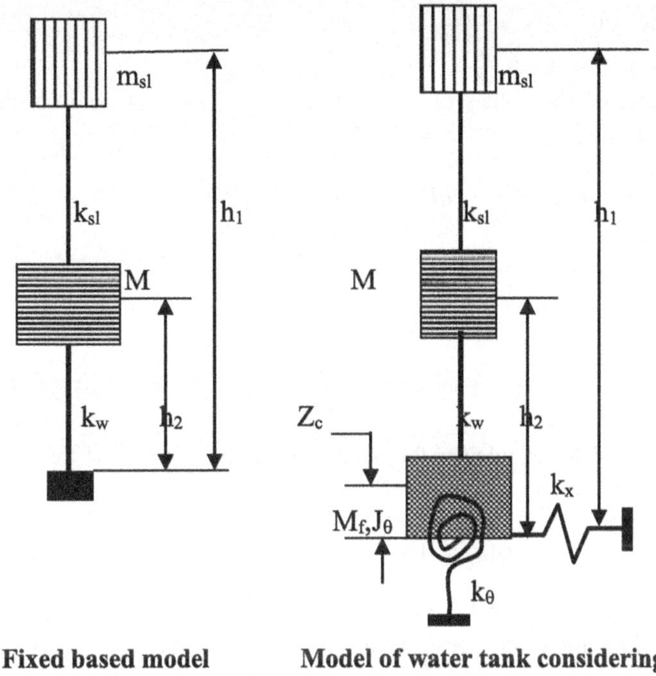

Fig. 14.21 Mathematical model as perceived for a flexible water tank, fixed based, and DSSI case

In Eq. (14.7.3.2), for fundamental mode $i = j = 1$, E = Young's modulus of wall material I = moment of inertia of the wall @ $Bt^3/12$, where t is the thickness of wall, and the shape function $\phi(z)$ is expressed as hereafter.

For walls fixed at base free at top considering it as beam subjected to uniform load q, equation of equilibrium can be expressed as,

$$EI\frac{d^4u}{dz^4} = -q \tag{14.7.3.3}$$

On successive integration of Eq. (14.7.3.3), we finally have

$$EIu = \int\left[\int\left\{\int\left(\int -q dz\right)dz\right\}dz\right]dz \tag{14.7.3.4}$$

$$\rightarrow EIu = -\frac{qz^4}{24} + \frac{C_1z^3}{6} + \frac{C_2z^2}{2} + C_3z + C_4$$

For a cantilever beam, considering the boundary conditions
At $z = 0, u''' = 0 \rightarrow C_1 = 0$;
At $z = 0, u'' = 0 \rightarrow C_2 = 0$

At $z = H, u' = 0 \rightarrow C_3 = \frac{-qH^3}{6}$
At $z = H, u = 0 \rightarrow C_4 = \frac{qH^4}{8}$
Back substituting the values of constants into Eq. (14.7.3.4) and considering $\eta = \frac{z}{H}$, we have

$$u_z = \frac{qH^4}{8EI}\left[1 - \frac{4}{3}\eta + \frac{1}{3}\eta^4\right] \tag{14.7.3.5}$$

This gives $u_z = u_0\phi(z)$ when $\phi(z) = 1 - \frac{4z}{3H} + \frac{z^4}{3H^4}$ and $u_0 = \frac{qH^4}{8EI}$.

$$\Rightarrow \phi(\eta) = 1 - \frac{4}{3}\eta + \frac{\eta^4}{3} \tag{14.7.3.6}$$

Proceeding in identical manner for walls fixed at both base and top it can be shown that,

$$\phi(\eta) = 16\eta^4 - 32\eta^3 + 16\eta^2 \tag{14.7.3.7}$$

Substituting these values in Eq. (14.7.3.2), stiffness of wall can be expressed as, $k_w = 16EI/5H^3$ for cantilever wall, and $k_w = 1024EI/5H^3$ for walls fixed at both ends.

The sloshing stiffness k_{sl} is obtained from Table 14.1.

Sloshing mass is expressed as,

$$m_{sl} = \frac{W_{sl}}{g}\phi[\eta_{sl}]^2 \tag{14.7.3.8}$$

In Eq. (14.7.3.8), $\eta_{sl} = \frac{H - h_{sl}}{H}$

Mass of wall M_w is expressed as,

$$M_w = \frac{\gamma_c AH}{g}\int_0^1 \phi(\eta)^2 d\xi = 0.257\frac{\gamma_c AH}{g} \tag{14.7.3.9}$$

And impulsive mass,

$$M_i = \frac{W_i}{g}\phi(\eta_i)^2 \tag{14.7.3.10}$$

Here, again $\eta = z/H$ and $\eta_i = (H - h_i)/H$, where h_i is a height at which impulsive mass acts from bottom of base raft and can be obtained from values furnished in Table 14.1.

Here, γ_c = weight density of tank wall, A = area of tank cross section $2B \times t$, t = thickness of tank wall, H = height of tank wall, and B = width of the wall. Here $M = M_w + M_i$

Having derived the stiffness and mass matrix [**K**] and [**M**] as elaborated above one can now follow the steps as furnished in Eqs. (14.7.2.7) through (14.7.2.25) to compute the shear and moment induced in tank wall.

In this case, as recommended in Eurocode 8, 0.5% damping may be considered for sloshing fluid and 5–7% damping to be considered for concrete wall.

For DSSI effect, considering Eqs. (14.7.2.2) to (14.7.2.4) in conjunction with Fig. 14.21, one can write the equation of equilibrium as,

$$\begin{bmatrix} M_f + M + m_{sl} & M_f Z_c + M h_2 + m_{sl} h_1 & M & m_{sl} \\ M_f Z_c + M h_2 + m_{sl} h_1 & J_\theta + M_f Z_c^2 + M h_2^2 + m_{sl} h_1^2 & M h_2 & m_{sl} h_1 \\ M & M h_2 & M & 0 \\ m_{sl} & m_{sl} h_1 & 0 & m_{sl} \end{bmatrix} \begin{Bmatrix} \ddot{u}_f \\ \ddot{\theta} \\ \ddot{u}_i \\ \ddot{u}_{sl} \end{Bmatrix} + \begin{bmatrix} C_x & 0 & 0 & 0 \\ 0 & C_\theta & 0 & 0 \\ 0 & 0 & C_w + C_{sl} & -C_{sl} \\ 0 & 0 & -C_{sl} & C_{sl} \end{bmatrix} \begin{Bmatrix} \dot{u}_f \\ \dot{\theta} \\ \dot{u}_i \\ \dot{u}_{sl} \end{Bmatrix} + \begin{bmatrix} K_x & 0 & 0 & 0 \\ 0 & K_\theta & 0 & 0 \\ 0 & 0 & k_w + k_{sl} & -k_{sl} \\ 0 & 0 & -k_{sl} & k_{sl} \end{bmatrix} \begin{Bmatrix} u_f \\ \theta \\ u_i \\ u_{sl} \end{Bmatrix} = 0 \tag{14.7.3.11}$$

In Eq. (14.7.3.11), the nomenclatures are as defined hereafter.

- M_f J_θ mass and mass moment of inertia of base slab of foundation, respectively;
- M mass of impulsive water plus mass of the tank wall;
- m_{sl} sloshing mass of fluid (W_{sl}/g) is as furnished in Table 14.1;
- Z_c center of gravity of foundation from center of stiffness where the foundation springs are connected;
- $h_1 = h_{sl}$ the sloshing height of fluid as furnished in Table 14.1. Considering that the foundation is also taken into cognizance, the IBP case to be considered in this case is $h_2 = \frac{W_w \times h_w + W_i \times h_i(\mathrm{IBP})}{W_w + W_i}$ effective center of gravity of the tank wall and impulsive mass;
- h_w center of gravity of the wall from center of stiffness S;
- W_w weight of wall per meter width;
- W_i impulsive weight of water;
- K_x, K_θ stiffness of foundation as furnished in Table 7.3 (Chap. 7);
- k_w, k_{sl} stiffness of wall and sloshing fluid as furnished in the above and Table (14.1), respectively;
- C_x, C_θ damping values of soil as furnished in Table 7.3;
- C_w, C_{sl} damping values of the wall and sloshing fluid, respectively, that can be expressed as $0.1\sqrt{k_w m_w}$ and $0.01\sqrt{k_{sl} m_{sl}}$, i.e., considering 5 and 0.5% damping, respectively.

Again applying Eqs. (14.7.2.7) through (14.7.2.11), one can find out the displacement. Bending moments and shears in tank wall and foundation can finally be computed from the expression $M = EIu\,\phi''(z) \rightarrow 4(EI/H)u\eta^2$ for cantilever wall and $(32EI/H^2)u(6\eta^2 - 6\eta + 1)$ for fixed end walls. Shear force is expressed as $V = EIu\,\phi''(z) \rightarrow (8EI/H^3)\eta$ for cantilever wall and $[192EI/H^3](2\eta - 1)$ for wall fixed at the top.

14.7.4 Dynamic Analysis of Flexible Tank Behaving as a Two-Way Slab

In most of the tanks constructed in real world (e.g., API oily water separators), the walls behave as a two-way slab as the aspect ratio is usually <2.0. In this case, the tank wall behaves as a plate with different boundary conditions like,

- Three sides fixed, one side free, i.e., tank open to sky.
- All sides fixed, i.e., roof slab monolithically cast with tank wall.

This certainly makes computation rather complicated, especially under earthquake load. *No solution exists for the same till date as per any recommendations as furnished in the code, as the state of the art is still to assume the wall as a one-way beam for analysis.*

One of the main reasons for this could be that, as the fluid mass is perceived as a lumped mass it would act as a concentrated load and concentrated load generates singularity in thin plate analysis. An attempt is made herein to provide solution to this problem.

Eurocode 8 converts the lumped mass into an equivalent uniformly distributed load to determine the deflection of the flexible wall vis-à-vis time periods, vide Eq. (14.4.4). A similar approach is used herein for the two-way slab. Shown in Fig. 14.22 is a two slab panel ($L'/H < 2.0$) with three sides fixed and one side free subjected to a *uniformly distributed pressure*(udp) q, where q is the equivalent pressure due to the impulsive force and self-weight of wall, say.

Now if we take a small segment of the wall of width dx say, we can then assume this strip as a cantilever beam which is fixed at base and free at top. Assuming it to be loaded by uniformly distributed pressure q for the width dx, we can assume the displacement function same as the beam derived earlier, i.e.,

$$u_z = \frac{qH^4}{8EI}\left[1 - \frac{4}{3}\eta + \frac{1}{3}\eta^4\right] \tag{14.7.4.1}$$

Similarly, in horizontal direction if we take a strip dz as shown in Fig. 14.22 and proceed in identical fashion for a beam fixed at both ends, we have

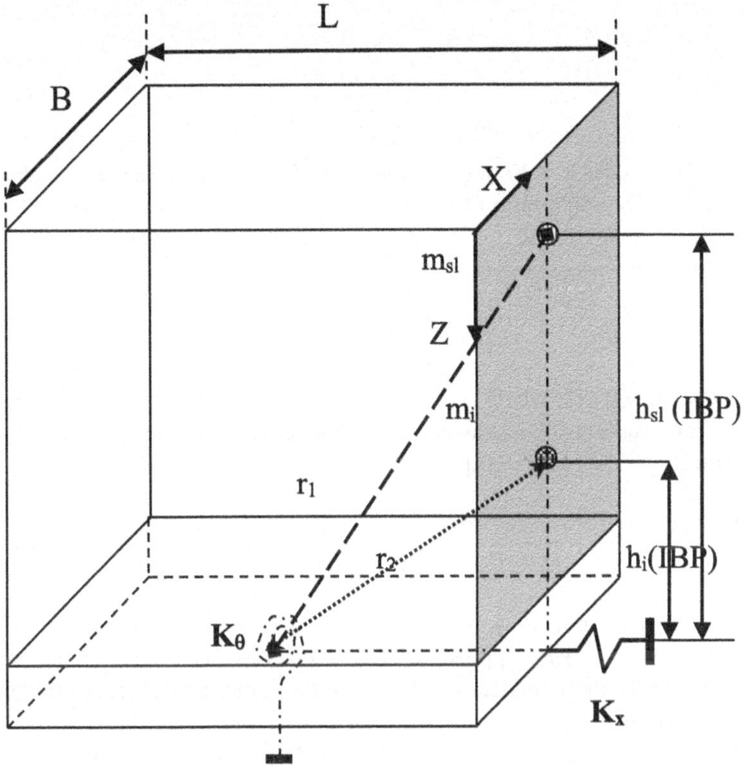

Fig. 14.22 Tank with wall behaving as a two-way slab; three sides fixed one side free

$$u_x = \frac{q(L)^4}{384EI}\left[16\xi^4 - 32\xi^3 + 16\xi^2\right] \tag{14.7.4.2}$$

In Eq. (14.7.4.2), $\xi = x/L$.

In Eqs. (14.7.4.1) and (14.7.4.2), the terms within the parenthesis are shape functions of the beam of width dz and dx, respectively, and will generically satisfy the boundary conditions of a plate also having three sides fixed and one side free, as depicted in Fig. 14.22.

Under a distributed pressure/load "q", the standard plate equation can be written (Timoshenko and Krieger 1985) as;

$$\frac{\partial^4 u}{\partial x^4} + 2\frac{\partial^4 u}{\partial x^2 \partial z^2} + \frac{\partial^4 u}{\partial z^4} = \frac{q}{D} \tag{14.7.4.3}$$

where

$$D = \frac{Et^3}{12(1 - \nu^2)} \tag{14.7.4.4}$$

In Eq. (14.7.4.4), E = Young's modulus of the plate material, t = thickness of plate, and ν = Poisson's ratio of the plate.

Considering $x = \xi L$ and $z = \eta H$, Eq. (14.7.4.3) can be written as;

$$\frac{1}{(L)^4}\frac{\partial^4 u}{\partial \xi^4} + \frac{2}{(L)^2 H^2}\frac{\partial^4 u}{\partial \xi^2 \partial \eta^2} + \frac{1}{H^4}\frac{\partial^4 u}{\partial \eta^4} = \frac{q}{D} \tag{14.7.4.5}$$

Let $u = \Delta f(\xi) f(\eta)$ where $f(\xi) f(\eta)$ are shape functions determined from Eqs. (14.7.4.1) and (14.7.4.2) as per given boundary conditions.

Substituting value of u in Eq. (14.7.4.5), we have

$$\frac{\Delta}{(L)^4}\frac{\partial^4 f(\xi) f(\eta)}{\partial \xi^4} + \frac{2\Delta}{(L)^2 H^2}\frac{\partial^4 f(\xi) f(\eta)}{\partial \xi^2 \partial \eta^2} + \frac{\Delta}{H^4}\frac{\partial^4 f(\xi) f(\eta)}{\partial \eta^4} - \frac{q}{D} = 0 \tag{14.7.4.6}$$

Now the shape functions, $f(\xi)$ and $f(\eta)$, have been determined from beam equations and have been applied to plate equations. The shape functions will generically satisfy the boundary conditions of the same. However, considering the fact that functions are approximate will not exactly satisfy Eq. (14.7.4.6) (i.e., the right-hand side of Eq. (14.7.4.6) will have a finite value $\neq 0$), giving rise to a residual error (R_e).

The residual error can be expressed as,

$$R_e = \frac{\Delta}{(L)^4} f^{iv}(\xi) f(\eta) + \frac{2\Delta}{(L)^2 H^2} f''(\xi) f''(\eta) + \frac{\Delta}{H^4} f(\xi) f^{iv}(\eta) - \frac{q}{D} \tag{14.7.4.7}$$

To arrive at a correct solution, the residual error is minimized by Galerkin's technique when $R_e \to 0$ yields,

$$\int_0^1 \int_0^1 \left[\frac{\Delta}{(L)^4} f^{iv}(\xi) f(\eta) + \frac{2\Delta}{(L)^2 H^2} f''(\xi) f''(\eta) + \frac{\Delta}{H^4} f(\xi) f^{iv}(\eta) - \frac{q}{D} \right] f(\xi) f(\eta) \mathrm{d}\xi \mathrm{d}\eta = 0 \tag{14.7.4.8}$$

Expanding the above expression, we have

$$\frac{\Delta}{(L)^4}\int_0^1\int_0^1 f^{iv}(\xi)f^2(\eta)f(\xi)\mathrm{d}\xi\mathrm{d}\eta+\frac{2\Delta}{(L)^2H^2}\int_0^1\int_0^1 f''(\xi)f''(\eta)f(\xi)f(\eta)\mathrm{d}\xi\mathrm{d}\eta$$
$$+\frac{\Delta}{H^4}\int_0^1\int_0^1 f^2(\xi)f(\eta)f^{iv}(\eta)\mathrm{d}\xi\mathrm{d}\eta-\frac{q}{D}\int_0^1\int_0^1 f(\xi)f(\eta)\mathrm{d}\xi\mathrm{d}\eta=0 \tag{14.7.4.9}$$

Equation (14.7.4.9) can be written in more compact form as,

$$\Delta\left[\frac{X_1}{(L)^4}+\frac{2X_2}{(L)^2H^2}+\frac{X_3}{H^4}\right]=\frac{qX_4}{D} \tag{14.7.4.10}$$

Considering $\lambda = H/L$ as the aspect ratio, Eq. (14.7.4.10) can be written as,

$$\frac{\Delta}{H^4}\left[\lambda^4X_1+2\lambda^2X_2+X_3\right]=\frac{qX_4}{D} \tag{14.7.4.11}$$

Thus,

$$\Delta=\frac{qH^4}{D}\left[\frac{X_4}{\lambda^4X_1+2\lambda^2X_2+X_3}\right]=u_{\max} \tag{14.7.4.12}$$

Thus, we have

$$\begin{aligned} u&=\Delta f(\xi)f(\eta)\\ &=\frac{qH^4}{D}\left[\frac{X_4}{\lambda^4X_1+2\lambda^2X_2+X_3}\right]f(\xi)f(\eta) \end{aligned} \tag{14.7.4.13}$$

In Eq. (14.7.4.13), X_1 to X_4 are integral functions expressed as,

$$X_1=\int_0^1\int_0^1 f^{iv}(\xi)f^2(\eta)f(\xi)\mathrm{d}\xi\mathrm{d}\eta \tag{14.7.4.14}$$

$$X_2=\int_0^1\int_0^1 f''(\xi)f''(\eta)f(\xi)f(\eta)\mathrm{d}\xi\mathrm{d}\eta \tag{14.7.4.15}$$

$$X_3 = \int_0^1 \int_0^1 f^2(\xi) f(\eta) f^{iv}(\eta) \mathrm{d}\xi \mathrm{d}\eta \tag{14.7.4.16}$$

$$X_4 = \int_0^1 \int_0^1 f(\xi) f(\eta) \mathrm{d}\xi \mathrm{d}\eta \tag{14.7.4.17}$$

From the theory of vibration, it is known that

$$T = 2\pi \sqrt{\frac{u_{max}}{g}} \tag{14.7.4.18}$$

Now $u = u_{max}$ at $\eta = 0$ and $\xi = 0.5$, when both $f(\eta) = f(\xi) = 1.0$, thus,

$$T = 2\pi \sqrt{\frac{qH^4}{Dg} \left[\frac{X_4}{\lambda^4 X_1 + 2\lambda^2 X_2 + X_3} \right]} \tag{14.7.4.19}$$

The contributing mass matrix of the plate can be expressed as (Hurty and Rubenstein 1967),

$$m = \frac{q}{g} \int_0^L \int_0^H [f(x) \cdot f(z)]^2 \mathrm{d}x \cdot \mathrm{d}z \tag{14.7.4.20}$$

In natural co-ordinates, Eq. (14.7.4.20) can be expressed as,

$$m = \frac{qHL}{g} \int_0^1 \int_0^1 [f(\xi) \cdot f(\eta)]^2 \mathrm{d}\xi \cdot \mathrm{d}\eta \tag{14.7.4.21}$$

Contributing mass of impulsive and sloshing mass can be expressed as,

$$\begin{aligned} m_i &= \frac{W_i}{g} [f(\eta_i) f(0.5)]^2 \quad \text{and} \\ m_{sl} &= \frac{W_{sl}}{g} [f(\eta_{sl}) f(0.5)]^2. \end{aligned} \tag{14.7.4.22}$$

Here, η_i and η_{sl} are dimensionless parameters in terms of height H already explained earlier.

Based on Eq. (14.7.2.7), one can say that

$$k_w = m\omega^2 \rightarrow \frac{4\pi^2 m}{T^2} \tag{14.7.4.23}$$

Substituting the values of D, T, and m from Eqs. (14.7.4.4), (14.7.4.18), and (14.7.4.21), respectively, we finally have

$$k_w = \frac{Et^3 L}{12(1-\nu^2)H^3}\left(\frac{X_5}{X_4}\right)\left[\lambda^4 X_1 + 2\lambda^2 X_2 + X_3\right] \tag{14.7.4.24}$$

In Eq. (14.7.4.24),

$$X_5 = \int_0^1\int_0^1 [f(\xi)\cdot f(\eta)]^2 \mathrm{d}\xi \cdot \mathrm{d}\eta \tag{14.7.4.25}$$

It is observed above that stiffness of the wall, k_w, is independent of q and is strongly dependent on the aspect ratio (H/L) of the plate, while coefficients X_1 to X_5 are unique properties and are dependent on the boundary conditions of the plate. Values of X_1 to X_5 for various conditions are furnished in Table 14.5.

Now substituting the value of k_w in Eq. (14.7.3.1) as a fixed-based case one can find out the response $\{\mathrm{U}\} = \{u_f, \theta, u_i, u_{sl}\}$ of the flexible rectangular tank. For effect of DSSI, the equation of equilibrium is expressed as,

$$\begin{bmatrix} M_f + M + m_{sl} & M_f Z_c + M r_2 + m_{sl} r_1 & M & m_{sl} \\ M_f Z_c + M r_2 + m_{sl} r_1 & J_\theta + M_f Z_c^2 + M r_2^2 + m_{sl} r_1^2 & M h_2 & m_{sl} r_1 \\ M & M r_2 & M & 0 \\ m_{sl} & m_{sl} r_1 & 0 & m_{sl} \end{bmatrix} \begin{Bmatrix} \ddot{u}_f \\ \ddot{\theta} \\ \ddot{u}_i \\ \ddot{u}_{sl} \end{Bmatrix} + \begin{bmatrix} C_x & 0 & 0 & 0 \\ 0 & C_\theta & 0 & 0 \\ 0 & 0 & C_w + C_{sl} & -C_{sl} \\ 0 & 0 & -C_{sl} & C_{sl} \end{bmatrix} \begin{Bmatrix} \dot{u}_f \\ \dot{\theta} \\ \dot{u}_i \\ \dot{u}_{sl} \end{Bmatrix} + \begin{bmatrix} K_x & 0 & 0 & 0 \\ 0 & K_\theta & 0 & 0 \\ 0 & 0 & k_w + k_{sl} & -k_{sl} \\ 0 & 0 & -k_{sl} & k_{sl} \end{bmatrix} \begin{Bmatrix} u_f \\ \theta \\ u_i \\ u_{sl} \end{Bmatrix} = 0 \tag{14.7.4.26}$$

Here r_1 and r_2, as shown in Fig. 14.22, are the radius vectors of the sloshing and impulsive mass expressed as $\sqrt{h_1^2 + (L/2)^2}$ and $\sqrt{h_2^2 + (L/2)^2}$, respectively. For determination of moment and shear in the wall, we proceed as hereafter.

Vide Timoshenko and Kreiger (1985), it is known that

$$M_x = -D\left[\frac{\partial^2 u}{\partial x^2} + \nu \frac{\partial^2 u}{\partial z^2}\right] \tag{14.7.4.27}$$

$$M_z = -D\left[\nu\frac{\partial^2 u}{\partial x^2} + \frac{\partial^2 u}{\partial z^2}\right] \tag{14.7.4.28}$$

$$Q_x = -D\frac{\partial}{\partial x}\left[\frac{\partial^2 u}{\partial x^2} + \frac{\partial^2 u}{\partial z^2}\right] \tag{14.7.4.29}$$

$$Q_z = -D\frac{\partial}{\partial z}\left[\frac{\partial^2 u}{\partial x^2} + \frac{\partial^2 u}{\partial z^2}\right] \tag{14.7.4.30}$$

Considering $u = \Delta f(x)f(z)$, where $\Delta = \sqrt{\sum_1^4 u_i^2}$ is the SRSS value of the four modes, as obtained from solution of Eq. (14.7.4.26) and subsequent computation as per Eq. (14.7.4.13), the maximum amplitude can be computed at point $\eta = 0$, $\xi = 0.5$; from Eqs. (14.7.4.27) to (14.7.4.30) can thus be expressed as,

$$M_x = -\frac{Du_i}{H^2}\left[\lambda^2 f''(\xi)f(\eta) + \nu f(\xi)f''(\eta)\right] \tag{14.7.4.31}$$

$$M_z = -\frac{Du_i}{H^2}\left[\lambda^2 \nu f''(\xi)f(\eta) + f(\xi)f''(\eta)\right] \tag{14.7.4.32}$$

$$Q_x = -\frac{Du_i}{H^3}\left[\lambda^3 f'''(\xi)f(\eta) + \lambda f'(\xi)f''(\eta)\right] \tag{14.7.4.33}$$

$$Q_z = -\frac{Du_i}{H^3}\left[\lambda^2 f''(\xi)f'(\eta) + f(\xi)f'''(\eta)\right] \tag{14.7.4.34}$$

The units of moments and shears are kN.m and kN, per meter width, respectively

14.7.5 Estimation of Response Due to Sloshing Force and Vertical Amplitude

The primary difficulty in estimating force on the wall due to sloshing effect is that u_{sl} as obtained from Eq. (14.7.4.26) is the amplitude of the fluid and not of the tank. So the question remains as to how does this influence the displacement of the tank, especially when the wall behavior is like a two-way slab?

The problem may be approached as explained hereafter.

For sloshing mode, the SRSS value of the nodal force, V_{sl}, acting at a height h_{sl} above the base raft can be expressed as,

$$V_{sl} = k_{sl}\sqrt{\sum_1^4 u_{sl}^2} \tag{14.7.5.1}$$

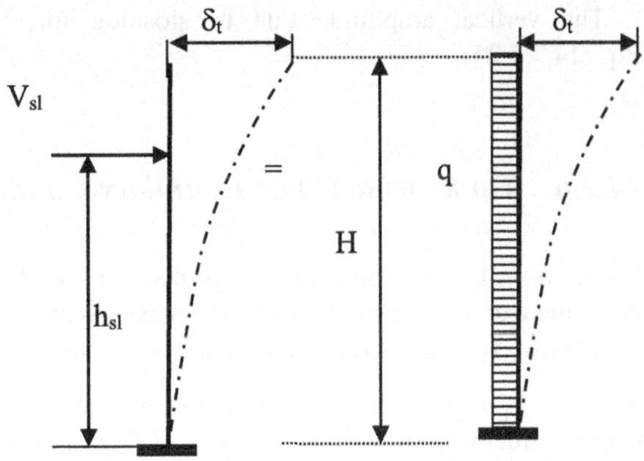

Fig. 14.23 Section of tank wall subjected to nodal force V_{sl} due to sloshing mass

Shown in Fig. 14.23 is section of the tank wall subjected to nodal force V_{sl} due to sloshing mass acting at a height h_{sl} from the base.

The displacement at free end can be expressed as,

$$\delta_T = \frac{V_{sl} h_{sl}^3}{3EI}\left[1 + \frac{3}{2}\left(\frac{H - h_{sl}}{h_{sl}}\right)\right] \quad (14.7.5.2)$$

For a uniformly distributed loading, q displacement at the free end can be expressed as,

$$\delta_T = \frac{qH^4}{8EI} \quad (14.7.5.3)$$

For the equivalent sloshing pressure, the displacement must be compatible. Thus, equating Eqs. (14.7.5.2) and (14.7.5.3), we have pressure per unit width as,

$$q_{sl} = \frac{8 V_{sl} h_{sl}^3}{3H^4 L'}\left[1 + \frac{3}{2}\left(\frac{H - h_{sl}}{h_{sl}}\right)\right] \quad (14.7.5.4)$$

Once the sloshing pressure q_{sl} is determined, same can be substituted in Eq. (14.7.4.13) to determine displacement u of the wall. Back substituting the value of u in Eqs. (14.7.4.27) through (14.7.4.30) the moments and shears in the wall can be computed.

Resultant shear and moment in the wall due to impulsive and sloshing force can finally be determined from the equation,

$$Q_R = \sqrt{Q_i^2 + Q_{sl}^2} \quad (14.7.5.5)$$

$$M_R = \sqrt{M_i^2 + M_{sl}^2} \quad (14.7.5.6)$$

The vertical amplitude due to sloshing force may be computed as per Eq. (14.7.2.25).

14.7.6 Tanks with Other Boundary Conditions

For tanks with other boundary conditions, steps as elaborated earlier remain exactly the same except for the fact that the shape functions $f(\xi), f(\eta)$, integral functions from X_1 to X_5, and q_{sl} are different and are as listed in Tables 14.4, 14.5 and 14.6.

Example 14.4 A square reinforced concrete water tank resting on ground having overall dimension of 8.0 × 8.0 × 6.2 m (high), as shown in Fig. 14.24, has been analyzed by the proposed method considering DSSI for soil having shear wave velocities of 200, 450, 750, and 1500 m/s.

Results are also shown for the case when base of the tank is assumed to be completely fixed. The analysis has been carried for two cases when the moment at base of side walls is estimated for both rigid and flexible tank, further classified into wall behaving as one-way and two-way slab. The tank is assumed to be in zone IV as per IS-1893.

Considering it is zone IV, as per IS-1893 Part 1 (2016), Z = 0.24, the importance factor I for the tank is considered as 1.25. For the response reduction factor *R*, the IS code does not have any specific guidelines for liquid retaining tanks. Thus, as a

Table 14.4 Shape functions for different boundary conditions

Boundary conditions	Three sides fixed one side free	All sides fixed
$f(\eta)$	$1-\frac{4}{3}\eta+\frac{1}{3}\eta^4$	$16\eta^4-32\eta^3+16\eta^2$
$f(\xi)$	$16\xi^4-32\xi^3+16\xi^2$	$16\xi^4-32\xi^3+16\xi^2$

Table 14.5 Values of X_1 to X_5 for different boundary conditions

Boundary conditions	X_1	X_2	X_3	X_4	X_5
Three sides fixed one side free	52.591	−0.929	1.3	0.213	0.104
All sides fixed	83.22	23.77	83.22	0.284	0.165

Table 14.6 Equivalent sloshing pressure q_{sl} for different boundary conditions

Boundary conditions	Equivalent sloshing pressure (q_{sl})	Remarks
Three sides fixed one side free	$q_{sl}=\frac{8V_{sl}h_{sl}^3}{3H^4L'}\left[1+\frac{3}{2}\left(\frac{H-h_{sl}}{h_{sl}}\right)\right]$	Displacement compatible at height *H*
All sides fixed	$q_{sl}=\frac{8V_{sl}h_{sl}^3(H-h_{sl})^3}{H^7L'}$	Displacement compatible at height h_{sl}

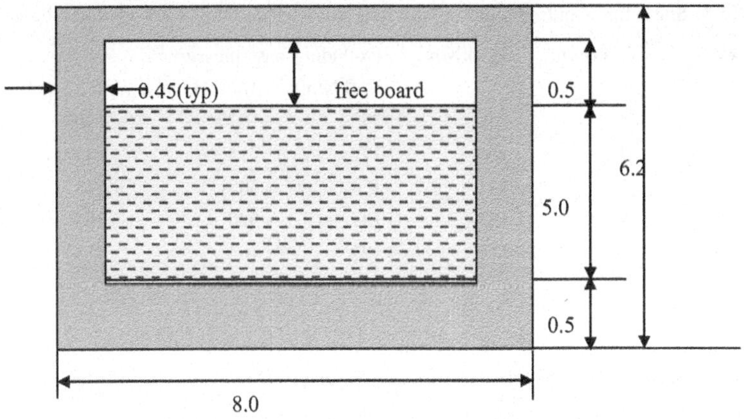

Fig. 14.24 Geometric dimension of water tank resting on ground

conservative estimate, $R = 2.0$ has been considered. This is in line with recommendations as furnished in Eurocode 8.

It is assumed herein that over the range of shear wave velocity considered the foundation size does not change, as the selection of tank size is more functional rather than soil bearing capacity guiding its foundation size.

Case-1 Tank considered as a rigid body resting on ground

Time periods of the tank for first three modes for EBP and IBP case are as shown in Table 14.7

It will be observed from Table 14.7 that the first mode, which is the sloshing mode of fluid, remains unchanged with variation of foundation stiffness both for EBP and IBP case. From this, we conclude that *sloshing mode of fluid remains unaffected by dynamic soil–structure interaction effect.*

However, for translational and rocking mode of tank along with the impulsive mass W_i, there is a progressive increase in time period as the foundation stiffness diminishes.

The sloshing and impulsive forces at base of wall are as shown Table 14.8.

Table 14.7 Time period for different shear wave velocities for rigid tank

Shear wave velocity(m/s)	Time period (EBP)			Time period (IBP)		
	Mode 1	Mode 2	Mode 3	Mode 1	Mode 2	Mode 3
200	3.043	0.119	0.062	3.043	0.111	0.066
450	3.043	0.053	0.028	3.043	0.049	0.023
750	3.043	0.032	0.017	3.043	0.03	0.018
1500	3.043	0.016	8.252×10^{-3}	3.043	0.015	8.834×10^{-3}
Base fixed	3.043	0.0	–	3.043	0.0	–

Table 14.8 Shear force and moment at wall base considering DSSI vis-à-vis fixed base

Shear wave velocity (m/s)	V_{sl} (kN/m)	V_x (kN/m)	Excluding base pressure		
			V_r (kN/m)	M_θ (kN.m/m)	M_{wall} (kN.m/m)
200	61	248	255	1077	1100
450	61	166	177	715	749
750	61	141	154	603	643
1500	61	123	137	520	566
Fixed base	61	111	127	464	515

Table 14.9 Shear force and moment in foundation due to DSSI vis-à-vis fixed base

Shear wave velocity (m/s)	V_{sl} (kN/m)	V_x (kN/m)	Including base pressure		
			V_r (kN/m)	M_w (kN.m/m)	M_{fdn} (kN.m/m)
200	66	219	228	845	888
450	66	151	165	590	651
750	66	131	147	513	582
1500	66	116	133	458	534
Fixed base	66	106	125	422	503

The sloshing and impulsive forces at foundation level are as shown Table 14.9.

Perusing the data furnished in Tables 14.8 and 14.9, it is seen that as the foundation stiffness decreases there is a significant increase in both moment and shear in wall as well as foundation. The DSSI amplification is as high as 2.0 compared to fixed base case and soil with shear wave velocity of 200 m/s. Maximum vertical sloshing height for the water is 373 mm and is invariant with soil stiffness.

Case-2 Tank considered as a flexible tank resting on ground, free at top, acting as a two-way slab

Variation of time period considering DSSI and fixed based are as shown in Table 14.10.

Table 14.10 Time period of flexible tank free at top; two-way behavior

Shear wave velocity(m/s)	Time period (IBP)			
	Mode 1	Mode 2	Mode 3	Mode 4
200	1.289	0.056	0.036	0.025
450	1.288	0.034	0.023	0.013
750	1.288	0.03327	0.01394	0.00765
1500	1.288	0.03295	0.00701	0.00384
Base fixed	1.288	0.03286	0	0

Table 14.11 Amplitude of flexible tank free at top; two-way behavior

Shear wave velocity (m/s)	U_{sl} (mm)	U_i (mm)	U_f (mm)	θ (rad)	δ_v (mm)
200	372	0.61	0.024	0.000113	168
450	303	0.536	0.004	0.000017	111
750	223	0.422	0.001	0.000005	60.3
1500	223	0.417	0	0.000001	60.3
Base fixed	32.4	0.06	0	0	1.3

Table 14.12 Shear force and moment in wall and foundation for tank free at top–two way

Shear wave velocity (m/s)	V_{sl} (wall) kN	V_i (wall) kN	V_r (wall) kN	M_{fdn} (kN.m)	V_{fdn} (kN)	M_x (kN. m/m)	M_y (kN. m/m)
200	136.414	26.727	139	733.783	200.159	11.752	47.008
450	111.037	51.06	122.21	657.113	156.224	10.312	41.248
750	81.638	51.164	96.35	520.113	118.266	8.111	32.444
1500	81.635	48.881	95.11	504.332	111.749	8.017	32.068
Base fixed	11.861	6.849	13.69	–	–	1.155	4.62

Deflection of the tank and foundation for various soils are as shown in Table 14.11 and forces in wall as per Table 14.12.

The deflection diagram of the wall is a shown Fig. 14.25 as per Eq. (14.7.4.13).

Shear force and moment plots in x- and z-direction are as shown in Fig. 14.26 and 14.27 as per Eqs. (14.7.4.31) to (14.7.4.34).

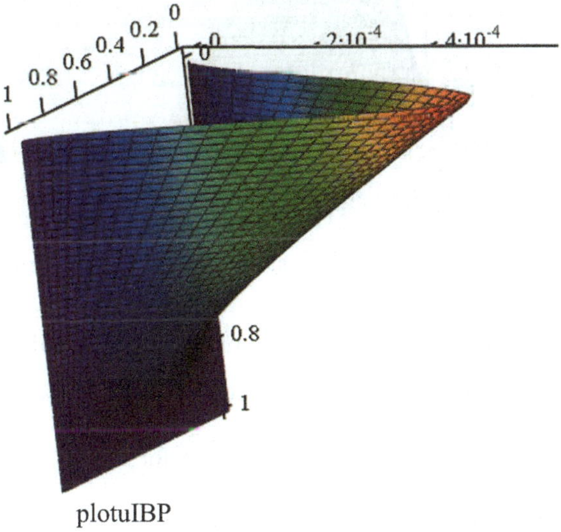

Fig. 14.25 Typical deflection profile of cantilever flexible tank

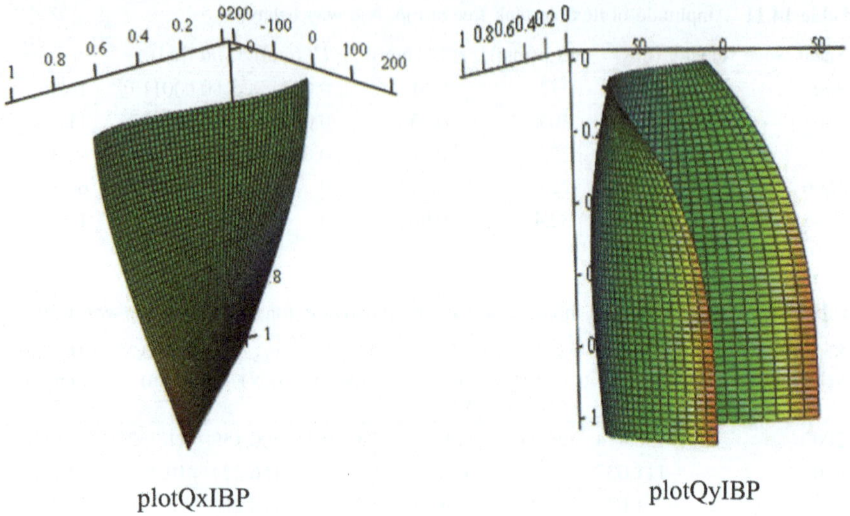

Fig. 14.26 Shear force plot of cantilever flexible wall in horizontal and vertical direction

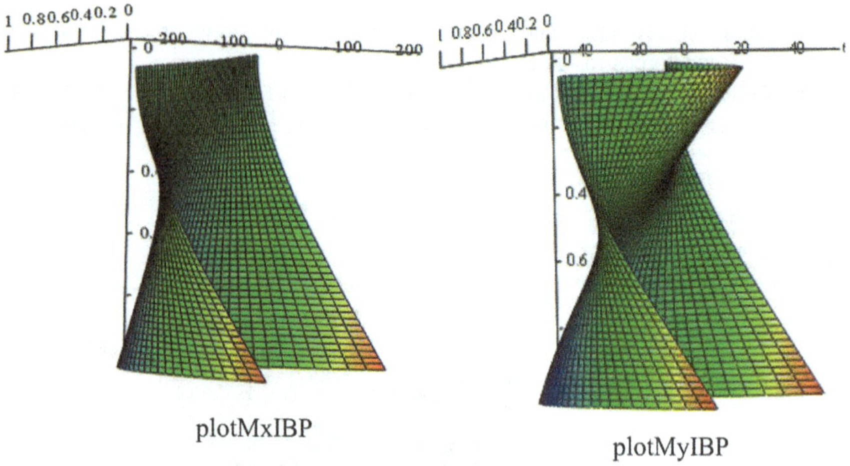

Fig. 14.27 Moment plot of cantilever flexible wall in horizontal and vertical direction

Case-3 Tank considered as a flexible tank resting on ground fixed at top

Variation of time period considering DSSI and fixed base are as shown in Table 14.13.

Deflection of tank and foundation for various soils are as shown in Tables 14.14 and 14.15.

Deflection diagram of the wall is a shown Fig. 14.28.

Table 14.13 Time period of flexible tank fixed at top

Shear wave velocity (m/s)	Time period (IBP)			
	Mode 1	Mode 2	Mode 3	Mode 4
200	2.88	0.057	0.012	0.033
450	2.879	0.026	0.017	0.011
750	2.8792	0.018	0.0126	0.0073
1500	2.8792	0.0158	0.0069	0.0038
Base fixed	2.8792	0.01543	0	0

Table 14.14 Amplitude of flexible tank fixed at top

Shear wave velocity (m/s)	U_{sl} (mm)	U_i (mm)	U_f (mm)	θ (rad)	δ_v (mm)
200	831	0.151	0.05	0.000197	168
450	677	0.124	0.008	0.000032	111
750	498	0.094	0.002	0.000009	60.2
1500	498	0.098	0.001	0.000002	60.2
Base fixed	72.331	0.014	0	0	1.3

Table 14.15 Shear force and moment in wall and foundation for tank fixed at top

Shear wave velocity (m/s)	V_{sl} (wall) kN	V_i (wall) kN	V_r (wall) kN	M_{fdn} (kN.m)	V_{fdn} (Kn)	M_x (kN. m/m)	M_y (kN. m/m)
200	304.89	28.098	306.20	1536.93	347.54	39.975	54.289
450	248.301	33.104	250.49	1246.81	286.67	32.056	43.534
750	182.575	53.808	199.08	970.96	233.32	24.299	33
1500	182.575	79.384	199.08	995.19	213.76	25.328	34.397
Base Fixed	26.528	11.343	28.85	0	0	3.628	4.927

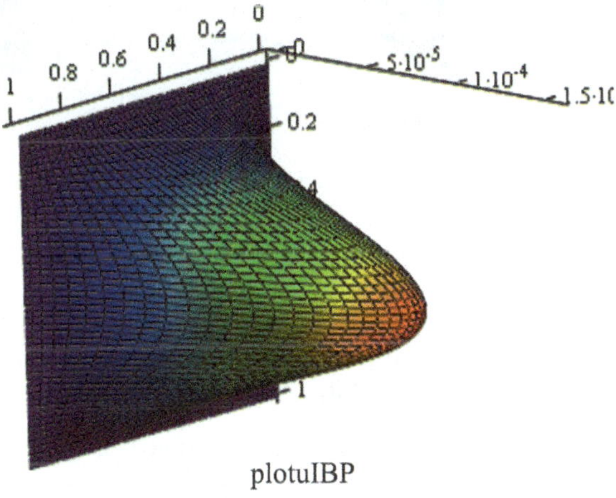

Fig. 14.28 Typical deflection profile of flexible tank fixed at base and top

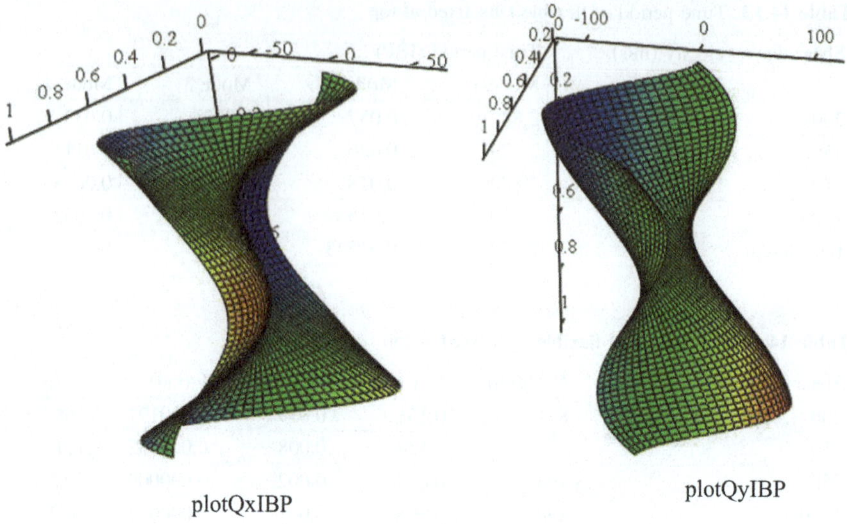

Fig. 14.29 Shear force plot of cantilever flexible wall in horizontal and vertical direction

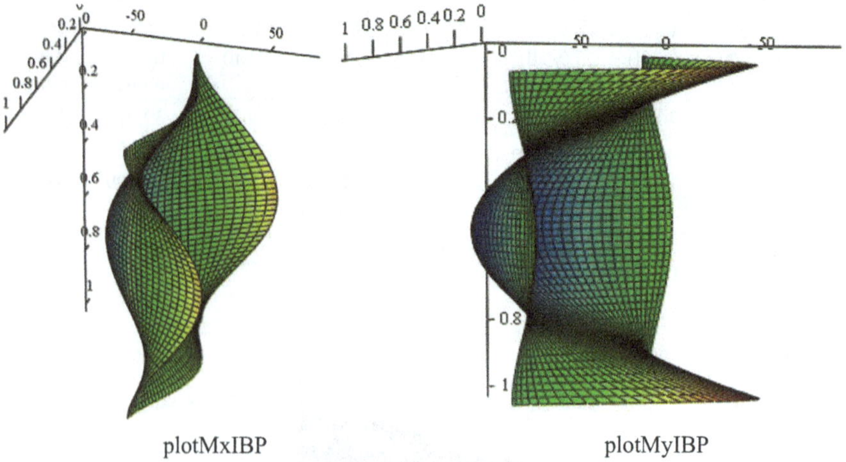

Fig. 14.30 Moment plot of flexible tanks fixed at top and bottom

Shear force and moment plot in *x*- and *z*-direction are as shown in Fig. 14.29 and 14.30.

Case-4 Tank considered as a flexible tank resting on ground free at top considered as one-way slab

Variation of time period considering DSSI and cantilever one-way slab are as shown in Tables 14.16, 14.17, and 14.18.

Table 14.16 Time period of flexible tank free at top

Shear wave velocity (m/s)	Time period (IBP)			
	Mode 1	Mode 2	Mode 3	Mode 4
200	1.292	0.111	0.051	0.028
450	1.292	0.107	0.023	0.013
750	1.292	0.10674	0.01402	0.00768
1500	1.292	0.10654	0.00702	0.00384
Base fixed	1.2917	0.10647	–	–

Table 14.17 Amplitude of flexible tank free at top acting as one-way slab

Shear wave velocity (m/s)	U_{sl} (mm)	U_i (mm)	U_f (mm)	θ (rad)	δ_v (mm)
200	376	4.557	0.042	0.000172	170
450	306	4.295	0.007	0.000028	112
750	225	4.012	0.002	0.000009	61
1500	225	3.993	0.001	0.000002	61
Base fixed	32.678	0.571	–	–	1.3

Table 14.18 Shear force and moment in wall and foundation for tank free at top

Shear wave velocity (m/s)	V_{sl} (wall) kN	V_i (wall) kN	V_r (wall) kN	M_{fdn} (kN.m)	V_{fdn} (kN)	M_y (kN. m/m)
200	136.97	199.479	242	1300.888	303.128	239.583
450	111.466	199.183	228	1165.83	246.135	224.42
750	81.955	197.137	213	1082.783	224.601	208.186
1500	81.951	196.041	212	1071.673	219.867	207.223
Base fixed	11.906	27.93	30.4	–	–	29.643

Variations of deflection, shear and moment with height of cantilever one-way wall and soils with different shear wave velocity are shown in Fig. 14.31, 14.32, and 14.33, respectively.

Case-5 Tank considered as a flexible tank resting on ground fixed at top considered as one-way slab

Variations of time period considering DSSI and cantilever one-way slab are as shown in Tables 14.19, 14.20, and 14.21.

Variation of deflection, shear and moment with height for fixed one-way wall and soils with different shear wave velocity are shown in Figs. 14.34, 14.35, and 14.36, respectively.

Perusing the results for different cases, as mentioned above, some very interesting points emerge.

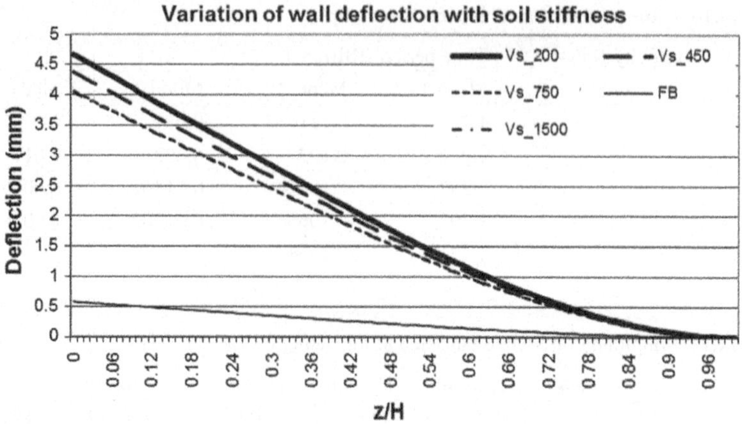

Fig. 14.31 Variation of deflection with height for cantilever one-way slab

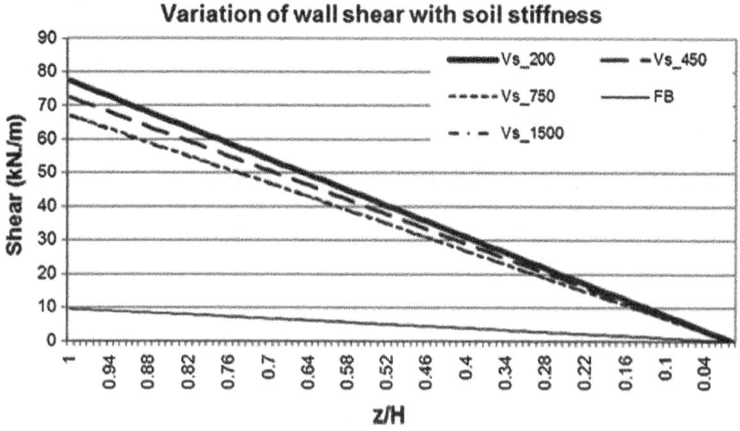

Fig. 14.32 Variation of shear force with height for cantilever one-way slab

For a rigid tank, sloshing amplitude and time period is independent of the soil stiffness. However, contrary to popular perception, forces, and moment on tank wall increases as the soil gets softer.

For a flexible tank, the sloshing amplitude increases with decrease in soil stiffness; however, the amplitude of liquid excitation is lower than that of a rigid tank. Boundary condition of the wall or its aspect ratio has practically no effect on this.

Moments and shears in the wall for a two-way slab panel is lower than that of the same slab considered as a one-way beam. Thus, a two-way slab designed based on the present recommendation of code would be much more conservative than reality. The trend looks logical as because in two-way slab panel loads are distributed in both directions, while for a one-way slab panel the loads are distributed only in

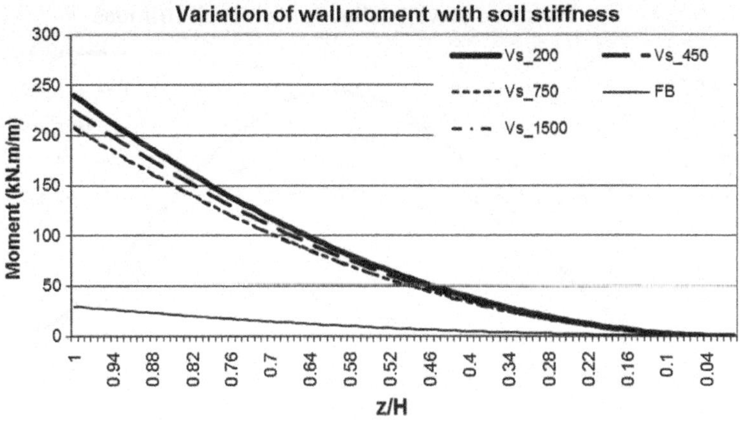

Fig. 14.33 Variation of moment with height for cantilever one-way slab

Table 14.19 Time period of flexible tank fixed at top

Shear wave velocity (m/s)	Time period (IBP)			
	Mode 1	Mode 2	Mode 3	Mode 4
200	2.88	0.062	0.036	0.012
450	2.88	0.03	0.018	0.0099
750	2.879	0.0214	0.0126	0.0071
1500	2.879	0.0184	0.0069	0.0038
Base fixed	2.879	0.0176	–	–

Table 14.20 Amplitude of flexible tank fixed at top acting as one-way slab

Shear wave velocity (m/s)	U_{sl} (mm)	U_i (mm)	U_f (mm)	θ (radian)	δ_v (mm)
200	831	0.093	0.052	0.000231	167
450	677	0.077	0.009	0.000038	111
750	498	0.076	0.003	0.000012	60
1500	498	0.074	0.001	0.000003	60
Base fixed	72.3	0.011	–	–	1.26

Table 14.21 Shear force and moment in wall and foundation for tank fixed at top acting as one-way slab

Shear wave velocity (m/s)	V_{sl} (wall) kN	V_i (wall) kN	V_r (wall) kN	M_{fdn} (kN.m)	V_{fdn} (kN)	M_y (kN. m/m)
200	304.932	75.761	314	1615.498	408.031	202.25
450	248.316	82.674	262	1341.196	340.231	164.742
750	182.584	178.612	255	1183.702	298.498	122.049
1500	182.583	173.66	252	1276.523	278.608	122.819
Base fixed	26.529	24.949	36.36	–	–	17.817

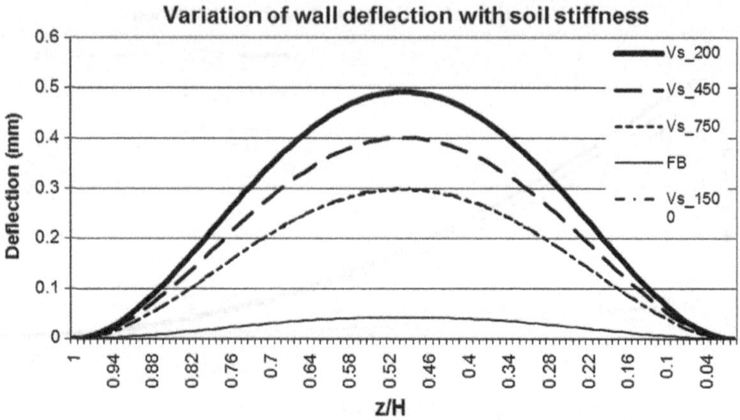

Fig. 14.34 Variation of deflection with height for fixed ended one-way slab

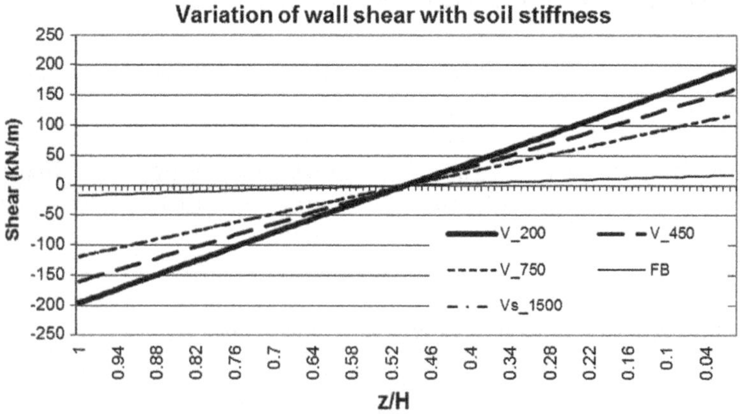

Fig. 14.35 Variation of shear force with height for fixed ended one-way slab

vertical direction. Thus, a two-way slab panel designed as per code would remain over-designed in vertical direction and under-designed in the horizontal direction. The magnitude of shear and moment is observed to be increasing with DSSI effect vis-à-vis tanks resting on rock-like soil, i.e., fixed based.

The most significant points to be noted herein are the results when the system is considered fixed based (i.e., foundation mass is ignored) and when the same is considered while the tank is resting on bedrock (i.e., $V_s \geq 600$ m/s). It is observed that though the time period remains practically same, yet the magnitude of force is completely different when the foundation mass is considered. This is because even when the time period of the foundation limit $T_f \rightarrow 0$, the foundation is subjected to peak ground acceleration (PGA) or zero period acceleration and this significantly influences the response of the wall. On the contrary, for a classical fixed-based

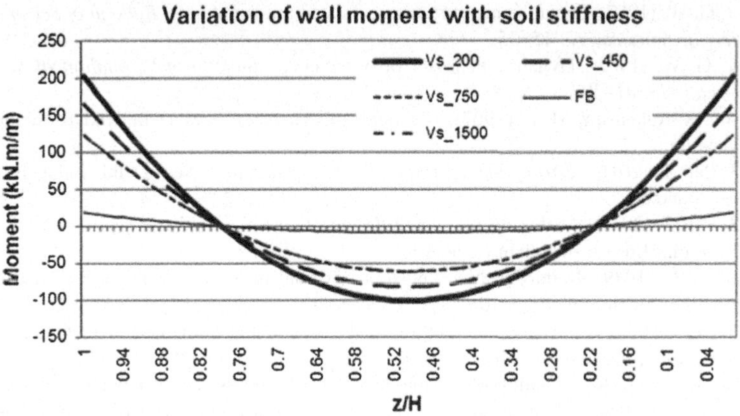

Fig. 14.36 Variation of moment with height for fixed ended one-way slab

analysis as the foundation mass is ignored this important effect does not come into play and seriously underrates the response. Thus, it may be concluded that for structures mounted on a heavy foundation, the foundation inertia significantly affects the overall response, irrespective of *whether the soil is soft or hard and cannot be ignored.*

References

ACI 350.3-01. (2001). *Seismic design of liquid containing concrete structures*. Washington, USA: American Code committee #350.

Chowdhury, I., & Tarafdar, R. (2015, January). Dynamic soil structure interaction analysis of rigid reinforced concrete water tank resting on ground. *Indian Concrete Journal, 89.*

Chowdhury, I., Tarafdar, R., Ghosh, A., & Dasgupta, S. P. (2017). Seismic response of rectangular liquid retaining structures resting on ground considering coupled soil-structure interaction. *Journal of Bulletin of Earthquake Engineering Europe, 15*(9), 3695–3726.

Clough, R. W. (1985). *Dynamics of structures*. New York, USA: McGraw-Hill Publications.

Edwards N. W. (1969). *A procedure for dynamic analysis of thin walled liquid storage cylindrical tanks subjected to lateral ground motion* (Ph.D. thesis). University of Michigan, Ann Arbor, USA.

Eurocode 8, Part IV. (2006). *Design of structures for earthquake resistance for silos, tanks and Pipe Lines*. Brussels, Belgium.

Ghaemmaghami, A. R. (2010). *Dynamic time history response of concrete rectangular liquid storage tanks* (Ph.D. Thesis). University of Reyerson, Toronto, Canada.

Graham, E. W., & Rodriguez, A. M. (1952). Characteristics of fuel motion that affects airplane dynamics. *Journal of Applied Mechanics, 19*(38), 1–8.

Haroun, M. A., & Housner, G. W. (1981). Seismic design of liquid storage tanks. *Journal of Technical Council ASCE New York, 107*(1), 191–207.

Haroun, M. A., & Hafiz Abdel, E. A. (1986). A simplified analysis of rigid based liquid storage tanks under vertical excitation considering soil structure interaction. *International Journal of Soil Dynamics and Earthquake Engineering, 5*(4), 217–225.

Housner, G. W. (1957). Dynamic pressure on accelerated fluid container. *Bulletin of Seismological Society of America, 47,* 15–35.

Housner, G. W. (1963). Dynamic behavior of water tanks. *Bulletin of Seismological Society of America, 53,* 381–387.

Hurty, W., & Rubenstin, M. F. (1967). *Dynamics of structures*. New Delhi, India: Prentice Hall Publication.

IS-1893. Part 1. (2016). *Earthquake resistant design of buildings*. New Delhi, India: Bureau of Indian Standards.

IS-1893. Part 2. (2016). *Earthquake resistant design of structures (Liquid retaining tanks)*. New Delhi, India: Bureau of Indian Standards.

Jacobsen, L. S. (1949). Impulsive hydrodynamics of fluid inside a cylindrical tank and of fluid surrounding a cylindrical pier. *Bulletin of Seismological Society of America, 39,* 189–203.

Jeong, K. L. (2011). Hydro-elastic vibration analysis of liquid contained in rectangular tanks. *Structural Engineering and Mechanics an International Journal, 40,* 665–688.

Livaoglu, R. (2008). Investigation of seismic behavior of fluid–rectangular tanks–soil/foundation systems in frequency domain. *Journal of Soil Dynamics and Earthquake Engineering, 28,* 132–146.

Meirovitch, L. (1985). *Elements of vibration analysis*. New Delhi, India: Allied Publishers.

Tang, Y. (1986). *Studies of dynamic response of liquid storage tanks* (Ph.D. Thesis). Rice University, Houston, USA.

Timoshenko, S., & Krieger, W. (1985). *Theory of plates and shells*. New York, USA: McGraw-Hill Publications.

Veletsos A. S. (1984). Seismic response and design of liquid storage tanks. In *Guidelines for the seismic design for Oil and gas pipeline system*, pp. 255–370, ASCE, New York.

Veletsos, A. S., & Meek, J. W. (1974). Dynamic behavior of building foundation systems. *Journal of Earthquake Engineering and Structural Dynamics, 3,* 121–138.

Veletsos, A. S., & Tang, Y. (1990). Soil structure interaction effects of laterally excited liquid storage tanks. *Journal of Earthquake Engineering and Structural Dynamics, 19*(4), 473–496.

Veletsos, A. S., & Yang, J. Y. (1977). Earthquake response of liquid storage tanks. In *Proceedings of the Second Engineering Mechanics Specialty Conference ASCE*, pp. 1–24, Raleigh.

Index

A
Absolute sum (ABSSUM), 109
Acceleration
absolute, 231, 232, 241
ground, 3, 6, 8, 21, 25, 34–38, 43, 45, 50, 54, 56, 58, 59, 64, 83, 87–91, 171, 211, 230, 231, 235, 236, 238, 260, 265, 266, 301, 303, 305–309, 390, 417, 424, 543, 789, 975, 983, 1012, 1014, 1015, 1042
integral random response, 44, 238, 367
peak ground, 6, 8, 25, 34, 36, 37, 43, 50, 54, 56, 58, 59, 64, 236, 266, 303, 305–308, 417, 424, 789, 983, 1012, 1014, 1042
pseudo, 45, 237
spectral, 45, 47, 56, 85, 105, 114, 235, 236, 266, 314, 337, 363, 426, 507, 551, 556, 557, 571, 666, 671, 722, 725, 802, 819, 822, 827, 830, 832, 847, 849, 865, 866, 869, 875, 877, 895, 899, 946, 950, 956, 959, 960, 977, 983, 988, 994, 1000, 1001, 1005, 1017, 1018
Amplification
dynamic, 520, 558, 786, 914
Gibson type soil, 254, 360, 435, 500, 668
seed and Idriss, 269, 270, 378, 643, 654, 657, 666, 831
shake method, 242
soil overlying bedrock, 210, 539
soil structure interaction, 306, 377, 493, 531, 533, 539, 547, 548, 558, 563, 571, 600, 628, 629, 728, 885, 892, 914, 1034
ANSYS, 13, 307, 308, 818, 836–841
Asthenosphere, 15, 16
Attenuation, 35, 36, 548, 556, 596

B
Bearing capacity of shallow foundation
bearing capacity factors (N_c,N_q,N_γ), 294, 405, 416, 417, 420, 424, 432, 433
Meyerhoff's Method, 446, 448
Terzaghi's Method, 407, 408, 414, 415, 422, 424, 425, 432, 433
under seismic load
Chowdhury and Dasgupta, 249, 266, 311, 356, 388, 399, 425, 426, 498, 505, 555, 556, 569, 670, 749, 754, 755, 758, 814, 828, 870, 893, 923
dynamic bearing capacity factors, 416, 417, 424, 433
Richard Elms and Budhu, 416
Block Vibration Test
IS 5429, 645
forced vibration, 153, 646, 648
free vibration, 74, 649
Bridge piers
AASHTO, 863, 865
CALTRANS, 863, 865
dynamic analysis
flexural deformation, 872
shear deformation, 369, 806, 816, 872–875, 877, 881, 893, 907, 909, 914
IRC-06, 865–867, 876, 913
pile foundation, 499, 505, 514, 515, 882, 892

I. Chowdhury and S. P. Dasgupta, *Earthquake Analysis and Design of Industrial Structures and Infra-structures*, GeoPlanet: Earth and Planetary Sciences, https://doi.org/10.1007/978-3-319-90832-8

Bridge piers (*cont.*)
shallow foundation, 292, 294, 295, 405, 407, 408, 438, 586, 65, 642, 882, 892
well foundation
generalized model, 902, 903
kinematical interaction, 507, 605–607, 611, 617, 625, 626, 628, 898, 900, 901, 911, 912, 916
Puzrevsky functions, 904
rigid model, 903
simplified method, 367, 793, 997
Building
assumed shape function, 255, 712, 714, 715
center of mass, 698, 705, 910
center of stiffness, 608, 697, 698, 699, 705
code method, 793
pseudo static, 535, 690, 691, 695
shear walls, 63, 68, 688, 689, 695, 697, 698, 699, 700, 702, 705, 708
Bunkers
airy's method, 920
integral functions, 955
moment coefficients for design, 935, 937
on supporting frame, 919, 945, 948, 949, 951, 953, 955, 957, 959
plate theory, 350
Rankine's Method, 920
shape functions for walls, 1032
shear coefficients for design, 939, 942
time period, 919, 927

C
Chimneys/Stacks
ACI 307, 719, 746, 786
design coefficients, 738
dynamic analysis, 555, 718, 719, 721, 728, 742, 746, 748, 760
IS 1893 Part IV (2005), 69, 725
number of modes, 110, 379, 391, 749, 751, 810, 870, 875
numerical analysis, 728, 742, 744, 746, 760, 769
pile foundations, 727, 728
soil structure interaction, 555, 661, 727, 728, 748, 754
Code
CICIND, 721, 746
eurocode 8, 48, 56, 57, 63, 64, 70, 71, 235, 721, 984, 1012–1014, 1018, 1023, 1024, 1033
ICOLD, 13
IS-1893(2016), 230, 266
UBC 97, 48, 64, 65, 67–70, 72, 74
Complete quadratic combination(CQC), 109, 110
Cone penetration value(CPT), 278
Concrete gravity dam
Chopra and Fenves, 797, 843, 845–847
coupled mode, 102, 585, 600, 611, 621, 825, 826, 828, 896, 1001, 1016
dynamic analysis, 68, 110, 304, 719, 721, 736, 761, 790, 816, 957
finite element (FEM), 322, 345, 356, 415, 462, 477, 533, 719, 793, 1012
flexural deformation, 872
hydrodynamic pressure, 399, 792, 793, 795, 801, 809, 820, 824, 825, 826, 837, 839, 843, 844, 854, 977, 979, 980
IS code method, 793
shear deformation, 166, 369, 701, 806, 807, 814–817, 827, 830, 869, 872–875, 877, 881, 893, 907, 909, 914
soil fluid structure interaction, 796, 797, 831, 846, 851
static analysis, 304, 718, 805
Continental drift, 19
Continuum model of beam
different boundary conditions, 1024, 1032
flexure, 166
shear, 153, 748
Cyclic
plate load test, 644, 650, 651
Resistance Ratio (CRR), 267, 277
Stress Ratio (CSR), 264, 277

D
Damping
critical, 80, 81, 230, 329, 375, 574
increase in, 235, 885
material, 55, 78, 245, 328, 330, 424, 493, 505, 552, 606, 617, 631, 643, 647, 648, 661, 662, 667, 668, 673, 750, 799, 866
geometric, 661
Design Basis Earthquake (DBE), 53, 725
Duhamel Integral, 84, 85
Dynamic Soil Structure Interaction (DSSI)
amplification factor, 210, 558, 597, 835, 851, 892, 910, 914
equivalent time period, 1012
kinematical interaction, 507, 605–607, 611, 617, 625, 626, 628, 898, 900, 901, 911, 912, 916
Lagrangian formulation, 543
Veletsos and Meek, 536, 538, 545, 547, 1012

E
Earth dam
Chowdhury & Dasgupta's method, 249, 266, 311, 332, 356, 377, 388, 399, 425, 426, 498, 505, 531, 555, 556, 569, 586, 666, 670, 728, 749, 754, 755, 758, 803, 814, 828, 870, 893, 923, 954
dam with clay core, 402
factor of safety (seismic), 270, 271, 278, 279, 284, 293, 294, 296, 390, 395, 399, 400, 402, 407, 415, 424, 434–438, 441, 452
Gazetas's Method, 374
Makdisi & Seed's method, 375–378, 382, 389, 391, 401
Mononobe's method, 330, 375
stability analysis, 393
Earthquake
fault, 66
intensity, 928
EMS, 21, 22
JMA, 21, 22
MCS, 21, 22
MM, 21, 26, 651
MSK, 21, 22
magnitude
moment (M_W), 28
richter (M_L), 26, 27, 28
surface wave (MS), 27, 35, 222, 224
stability analysis, 393
Eigen values, 92, 93, 97, 100–102, 113, 121–126, 129, 131, 134, 135, 137, 138, 140, 146, 148, 251, 253, 257, 357, 360, 381, 382, 387, 500, 506, 598, 619, 627, 668, 731, 743, 745, 746, 748, 813, 817, 827, 830, 871, 873, 1007, 1017
Eigen value techniques
Jacobi, 139, 140, 142, 145, 146, 871, 873
generalized, 146–148, 731
standard, 139, 142, 873
matrix deflation, 125, 126, 135, 146
Newton Raphson, 98, 123, 124, 129, 131
Stodola Vianello, 134, 135
vector transformation, 138
visual inspection, 121, 122
Eigen vectors, 93, 95, 97, 99–102, 113, 127–129, 131–135, 137, 138, 153, 157, 159, 161, 162, 164, 165, 168, 169, 250, 251, 255, 257, 258, 357, 360, 361, 382, 383, 426, 435, 557, 582, 590, 732, 870, 897, 950, 959, 1008, 1017
Elastic rebound, 15, 17, 18, 651
Epicenter, 9, 20, 27, 35–37, 265, 297, 539
Epicentric distance, 171, 276

F
Fine content, 267–269, 271, 272, 279, 280, 283, 297
Focus, 2, 7, 9, 13, 20, 36, 48, 78, 450, 534, 805, 893
Frequency
damped, 81, 86, 103, 231, 239, 242, 244, 246, 375, 610
natural, 76, 81, 85, 86, 122, 133, 157, 161, 165, 226, 230, 231, 233, 239, 242, 244, 247, 259, 267, 314, 329, 337, 364–376, 494, 505, 510–512, 542, 549, 599, 611, 646, 648, 759, 875, 890, 928, 1001, 1008

G
Galerkin's weighted residual method, 336
Giant Atlas, 14, 15
Gibson type soil, 254, 360, 435, 500, 668
GTSTRUDL, 11, 746, 969

H
Hydrodynamic pressure, 399, 792, 793, 795, 801, 809, 820, 824–826, 837, 839, 843, 844, 854, 977, 979, 980
Hypocenter, 2

I
Inertial interaction, 507, 518, 605, 606, 607, 612, 626, 628, 901

K
Kinematical interaction, 507, 605–607, 611, 617, 626, 628, 898, 900, 901, 911, 912, 916

L
Lagrange's Equation, 597, 923, 1015
Liquefaction
bearing capacity reduction, 296
estimation from SPT, 444, 654
estimation by CPT, 278, 284, 285, 288, 289
general shear failure, 294, 405, 407, 418, 434
ground subsidence, 296, 297
settlement, 263, 287, 288, 291, 292, 516
Lithosphere, 15–19

M
Mass, 38, 43, 53, 69, 74, 75, 82, 83, 87, 91, 100–102, 104, 106–108, 111, 112, 115, 116, 118, 119, 122, 126, 135, 138, 139, 142, 146, 148, 153, 156, 161, 162, 164–166, 169, 170, 171, 175, 177, 182,

187, 198, 210, 230, 247, 252–254, 256, 257, 303, 304, 310, 312, 314, 319, 321, 324, 331, 333, 338, 344, 348, 357, 358–360, 367, 369, 378, 380, 381, 386–391, 416, 435, 494, 498, 500, 504, 506, 510, 511, 516, 523, 534, 536, 542, 543, 545, 550, 552–555, 557, 559, 561, 564, 566, 568–570, 586, 588, 590, 592, 601, 613, 617, 618, 625, 646–649, 661, 665, 668, 674, 675, 678, 680, 697, 698, 705, 712, 713, 715, 716, 718, 719, 721, 729–731, 734, 735, 744, 745, 748, 750, 753, 754, 760, 769

Mass participation, 107, 108, 115, 116, 118, 119, 165, 171, 254, 314, 319, 321, 324, 331, 338, 344, 348, 358, 388, 389, 435, 506, 553, 564, 625, 665, 722, 734, 735, 775, 778, 819, 820, 822, 823, 875, 879, 887, 892, 899, 902, 928

Maximum considered earthquake (MCE), 50, 54, 236, 246, 725

Mesosphere, 15, 16

Modal analysis, 11, 55, 153, 172, 254, 314, 318, 320, 322, 324, 337, 358, 426, 506, 512, 550, 551, 552, 558, 559, 571, 607, 612, 613, 617, 628, 721, 734, 746, 822, 832, 875, 898, 899, 910, 956, 1017

Modal mass, 107, 115, 116, 118, 119, 165, 171, 254, 314, 319, 321, 324, 331, 338, 344, 348, 358, 388, 389, 435, 506, 553, 564, 665, 734, 735, 775, 778, 819, 820, 822, 823, 875, 879, 883, 887, 899, 902, 914, 928

Mode

Molten magma, 15, 16, 17

Mononobe-Okabe (M-O), 13, 301, 303, 305, 48, 55, 93–95, 99, 100, 102, 105–107, 109, 110, 113–120, 133–135, 137, 146, 157, 158, 162, 163, 165, 169, 171, 246, 248–250, 252, 254, 257–259, 266, 267, 315, 356, 359, 361, 372, 374–377, 382, 383, 389, 401, 402, 424, 426, 429, 431, 432, 434, 435, 437, 500, 502, 505, 507, 509, 511, 517–519, 542, 551–553, 556, 557, 560–563, 565–568, 574, 575, 577–587, 591–597, 599, 600, 601, 602–605, 607, 611, 614–616, 620–628, 631, 633–636, 661, 666, 668–670, 675, 679, 713, 727, 728, 732–735, 737, 739, 741, 743, 745, 747, 751–754, 759, 760, 765–778, 783–786, 791, 792, 801, 805–807, 811, 813, 814, 816–820, 822, 824–831, 834, 835, 837–839, 841, 847–850–852, 865, 870–873, 875, 877–879, 881, 882, 883

N

Namazu, 14, 16

O

Ostadan's method 366, 367

Over-damped, 80

P

Pangaea, 18

Pile foundations
- vertical load
  - clayey soil, 445, 508, 518, 523, 663
  - granular soil, 420
  - SPT value 444, 445, 447, 518, 654, 664
- lateral load
  - Brinch Hansen, 450, 452, 455
  - Broms, 450, 455–457, 460–462
  - Chowdhury & Dasgupta, 249, 266, 311, 356, 388, 399, 425, 426, 505, 555, 556, 816
  - Finite Difference (FDM), 86, 173, 174, 472, 477, 491, 536
  - Finite Element (FEM), 322, 345, 356, 415, 462, 477, 483, 533, 719, 1012
  - IS-2911, 462–464
  - JRA recommendations, 514–516
  - kinematical interaction, 613, 901, 916
  - liquefied soil, 509, 515
  - Matlock & Rees, 472, 473, 477
  - Prakash & Chandrashekhar, 494, 495, 498
  - stiffness, 43, 472, 483, 486, 487, 500–502, 545, 547
  - STAAD model, 488
  - sub-grade modulus, 455, 466, 469, 472, 476, 477, 480

Plate tectonics, 15, 18

R

Response
- multi degree, 104, 505, 555, 557
- single degree, 43, 58, 83, 91, 101, 104, 543, 599

Response reduction factor (R), 53, 314, 865, 867, 988, 1017

Response spectrum
- design, 87, 229
- normalized design, 46

pseudo spectral, 45, 235
tripartite plot, 237, 239
Retaining wall
cantilever type
cohesion less backfill, 254, 311, 312, 315, 319, 324, 333, 508
c-ϕ soil, 308, 321, 327
c-ϕ soil with overburden q, 330
c-ϕ soil partially submerged, 325, 326
inclined slope, 304, 308, 316, 317
intact clay, 321, 344
special case with tension at top, 310, 319, 327
counterfort type
cohesion less backfill, 311
c-ϕ soil, 343
c-ϕ soil with overburden q, 352
c-ϕ soil partially submerged, 325, 349
inclined slope, 317, 341
intact clay 321, 344
special case with tension at top, 310, 327
validation of plate theory, 350
gravity type (M-O), 301, 303, 304, 311
Seed & Whitman, 308
Steedman & Zeng, 309

S
SASSI, 356, 361, 366
STAAD, 488, 746, 786, 969
Seismograph, 1, 23, 26–28, 38, 211
Seismic cross hole, 644, 649, 683
Shear Modulus, 166, 169, 215, 222, 244, 247, 251, 258, 260, 261, 363, 370, 374, 375, 377, 378, 284, 385, 387, 435, 518, 522, 547, 576, 581, 592, 601, 629, 641, 643, 647–656, 659, 660, 666, 667, 673, 675–677, 726, 754, 757, 761, 783, 831, 832, 846, 851, 872, 885, 889, 907, 909, 910
Shindo, 10, 42, 209
Silo
dynamic pressure, 919, 962, 965, 971
Jansen's theory, 962, 964
staging stiffness, 1000
Soil
dynamic earth pressure, 356, 361, 362, 500
layered, 246, 443, 450, 480, 642, 674, 676
mechanical analogue, 628, 629
strain due to earthquake, 378, 629, 643, 663, 667
Square Root of Sum Square (SRSS), 109, 119–121, 359, 360, 376, 401, 402, 563, 567, 597, 605, 617, 625, 723, 740, 741, 766–769, 778–783, 785, 786, 824, 825, 835, 842, 848–852, 875, 876, 891, 898, 910, 913–915, 956, 960, 966, 979, 980, 984, 985, 995–997, 1030
Standard penetration value (N), 263, 267
Stiffness
fluid, 809, 837, 976
plate, 335, 351
soil, 49, 545, 612, 831, 883, 909, 1034
staging frame, 969, 997, 1000
structure, 49, 69, 122, 552, 728, 929, 953
Stiffness degradation, 246, 378, 631, 664, 667, 885, 898
Substructures, 356

T
Transient impulse, 82, 210
Time history analysis
central difference, 174, 187, 198
Newmark-β, 197, 199
Wilson-θ, 184, 197, 198, 206
Time period, 26, 36, 37, 43–45, 48–50, 53, 55, 58, 59, 61, 63, 67, 68, 71, 72, 74, 76, 82, 85, 93, 95, 97, 168, 184, 226, 233, 235–237, 242, 246, 250, 251, 257, 259, 266, 267, 304, 306, 311, 313, 315, 320, 327, 337, 343, 346, 363, 364, 372, 382–384, 387, 391, 398, 400, 401, 417, 425, 426

U
Under-damped, 79, 81
Underground structure, 24, 356, 357

V
Vibration
free, 74, 75, 77–79, 91, 95, 96, 126, 138, 140, 153, 156, 157, 166, 167, 570, 574, 580, 602, 649, 729, 809, 814, 870, 872, 895, 962, 976, 1020
forced, 83, 153, 646, 648
transient, 43, 45, 83, 261, 291, 508

W
Water tank
design paramcters, 976
flexible tanks
circular, 985, 987
rectangular, 990, 991, 1012
soil structure interaction, 1013, 1014, 1020, 1021, 1029, 1034
overhead tanks, 974, 1001
resting on ground
circular tanks, 979, 985, 987

rectangular tanks, 990, 991
rigid tank
soil structure interaction, 1012, 1033
Waves
compressive (P), 214
in polar coordinate, 227
Rayleigh, 211, 217, 221–223, 256, 533, 722, 723
shear (S), 214
love, 211, 224, 226
Wronskian, 106

Z
Zangar's method, 825
Zone factor (Z), 64, 1017

The manufacturer's authorised representative in the EU is Springer Nature Customer Service Centre GmbH, Europaplatz 3, 69115 Heidelberg, Germany. If you have any concerns regarding our products, please contact ProductSafety@springernature.com

Printed and bound by CPI Group (UK) Ltd, Croydon, CR0 4YY
30/09/2026
02225663-0002